Applied Science

Applied Science

Science and Medicine

Editor

Donald R. Franceschetti, Ph.D.

The University of Memphis

SALEM PRESS

A Division of EBSCO Information Services, Inc.

Ipswich, Massachusetts

GREY HOUSE PUBLISHING

Library of Congress Cataloging-in-Publication Data

Publisher's Cataloging-In-Publication Data
(Prepared by The Donohue Group, Inc.)

Applied science. Science and medicine / editor, Donald R. Franceschetti.
  -- [1st ed.].
          p. : ill. ; cm.

  Comprises articles on science and medicine extracted from the five-volume reference set Applied Science. Salem Press, 2012.
  Includes bibliographical references and index.
  ISBN: 978-1-61925-240-0

  1. Science. 2. Medicine. I. Franceschetti, Donald R., 1947- II. Title: Science and medicine

TA45 .A671 2013
600.

CONTENTS

Publisher's Note. vii
Editor's Introduction. ix
Contributors. xvii
Common Units of Measure. xxi

Acoustics. 1
Agricultural Science. 10
Agroforestry. 16
Agronomy. 21
Anesthesiology. 27
Animal Breeding and Husbandry. 32
Archaeology. 38
Areology. 45
Atmospheric Sciences. 52
Audiology and Hearing Aids. 61

Barometry. 68
Bioengineering. 75
Bioinformatics. 83
Biomathematics. 90
Biomechanics. 96
Biophysics. 102
Biosynthetics. 108

Cardiology. 114
Cell and Tissue Engineering. 120
Ceramics. 127
Clinical Engineering. 133
Cloning. 138
Computed Tomography. 145
Criminology. 151

Demography and Demographics. 158
Dentistry. 164
Dermatology and Dermatopathology. 170
Diffraction Analysis. 177
DNA Analysis. 184
DNA Sequencing. 191

Economic Geology. 197
Electrochemistry. 203
Electron Microscopy. 211
Electron Spectroscopy Analysis. 215
Electronic Commerce. 220
Electronic Materials Production. 226
Electronics and Electronic Engineering. 232

Endocrinology. 238
Environmental Chemistry. 245
Environmental Microbiology. 252
Epidemiology. 259
Ergonomics. 266

Gastroenterology. 272
Gemology and Chrysology. 279
Genetically Modified Organisms. 285
Genetic Engineering. 291
Genomics. 299
Geriatrics and Gerontology. 305

Hematology. 313
Histology. 319
Human Genetic Engineering. 325
Hydroelectric Power Plants. 334
Hydrology and Hydrogeology. 343
Hypnosis. 346

Immunology and Vaccination. 351
International System of Units. 357

Kinesiology. 363

Landscape Ecology. 369
Land-Use Management. 375

Magnetic Resonance Imaging. 382
Measurement and Units. 388
Metabolic Engineering. 395
Meteorology. 401
Microscopy. 408
Military Sciences and Combat Engineering. . . . 416
Mineralogy. 424
Mirrors and Lenses. 430

Nephrology. 436
Neural Engineering. 442
Neurology. 449
Nursing. 457
Nutrition and Dietetics. 463

Obstetrics and Gynecology. 469
Occupational Health. 475
Oceanography. 482

Ophthalmology 488
Optics 493
Optometry 499
Orthopedics 505
Osteopathy 511
Otorhinolaryngology 516

Paleontology 523
Parasitology 530
Pathology 535
Pediatric Medicine and Surgery 541
Penology 548
Pharmacology 553
Photonics 560
Planetology and Astrogeology 565
Propulsion Technologies 572
Prosthetics 578
Psychiatry 584
Pulmonary Medicine 591

Radio Astronomy 597
Radiology and Medical Imaging 603
Rehabilitation Engineering 610
Reproductive Science and Engineering 616
Rheumatology 622

Soil Science 627
Somnology 633
Space Science 638

Space Stations 645
Spectroscopy 651
Speech Therapy and Phoniatrics 658
Stem Cell Research and Technology 664
Superconductivity and Superconducting Devices . 670
Surface and Interface Science 677
Surgery 682

Telescopy 688
Toxicology 696
Tribology 701

Ultrasonic Imaging 706
Urology 711

Veterinary Science 718
Virology 724

Xenotransplantation 730

Biographical Dictionary of Scientists 738
Glossary 755
Timeline 765
General Bibliography 815
Subject Index 838

PUBLISHER'S NOTE

Applied Science: Science and Medicine is a comprehensive volume that examines the many ways in which science and medicine affect daily life. Drawn from the science and medicine articles in Salem's five-volume *Applied Science* reference set, this single volume is designed to help those specifically planning for a career in science or medicine understand the interconnectedness of the different and varied branches of these fields. Toward that end, essays look beyond basic principles to examine a wide range of topics, including industrial and business applications, historical and social contexts, and the impact a particular field of science or technology will have on future jobs and careers. A career-oriented feature details core jobs within the field, with a focus on fundamental and recommended coursework.

Applied Science: Science and Medicine is specifically designed for a high-school audience and is edited to align with secondary or high-school curriculum standards. The content is readily accessible, as well, to patrons of both academic and university libraries. Librarians and general readers alike will also turn to this reference work for both basic information and current developments, from stem cell engineering to renewable energy production, presented in accessible language with copious reference aids. Pedagogical tools and elements, including a bibliographical directory of scientists, a timeline of major scientific milestones, and a glossary of key terms and concepts, round out this unique resource.

SCOPE OF COVERAGE

Comprising 116 lengthy and alphabetically arranged essays on a broad range of science and medicine subjects, this excellent reference work addresses fields as long-established as archaeology, nursing, and hypnosis, and as cutting edge as cloning, landscape ecology and space stations. You will find entries in areas as diverse as genetics and soil science, as well as a multitude of entries that fall under the emerging fields of allied health, genetic engineering, and electronics. *Applied Science: Science and Medicine* also includes charts and diagrams, as well as "Fascinating Facts" related to each applied science field.

ESSAY LENGTH AND FORMAT

Essays in the encyclopedic set range in length from 3,000 to 4,000 words. All the entries begin with ready-reference top matter, including an indication of the relevant discipline or field of study and a summary statement in which the writer explains why the topic is important to the study of the applied sciences. A selection of key terms and concepts within that major field or technology is presented, with basic principles examined. Essays then place the subject field in historical or technological perspective and examine its development and implication. Also discussed are applications and applicable products and the field's impact on industry. Cross-references direct readers to related topics, and further reading suggestions accompany all articles.

SPECIAL FEATURES

Several features continue to distinguish this reference series from other works on applied science. The front matter includes a table detailing common units of measure, with equivalent units of measure provided for the user's convenience. Additional features are provided to assist in the retrieval of information. An index illustrates the breadth of the reference work's coverage, directing the reader to appearances of important names, terms, and topics throughout the text.

The back matter includes several appendixes and indexes, including a biographical dictionary of scientists, a compendium of the people most influential in shaping the discoveries and technologies of applied science. A time line provides a chronology of all major scientific events across significant fields and technologies, from agriculture to computers to engineering to medical to space sciences. Additionally, a complete glossary covers all technical and other specialized terms used throughout the set, while a general bibliography offers a comprehensive list of works on applied science for students seeking out more information on a particular field or subject. A general index ends the volume.

ACKNOWLEDGMENTS

Many hands went into the creation of this work. Special mention must be made of editor Donald R.

Franceschetti, who played a principal role in shaping this volume. Thanks are also due to the many academicians and professionals who worked to communicate their expert understanding of the applied sciences to the general reader; a list of these individuals and their affiliations appears at the beginning of the volume. The contributions of all are gratefully acknowledged.

EDITOR'S INTRODUCTION: SCIENCE AND MEDICINE

This volume presents the science and medicine articles from the five-volume reference set *Applied Science*, for the benefit of students considering careers in science or medicine, their teachers, and their counselors. Other volumes cover technology, and engineering and mathematics.

Something should be said at the outset about the relationship between science and technology. Terence Kealey states in his book *The Economic Laws of Scientific Research* (1996) that technology is the activity of manipulating nature, while science is the activity of learning about nature. For much of history, the two realms were relatively separate, technologies being developed by artisans, craftsmen, and farmers and the sciences by natural philosophers. Leonard Mlodinow, in *Feynman's Rainbow* (2003) makes a distinction between the Greek way of approaching science and the Babylonian way. The Greeks, the great mathematicians of the ancient world greatly admired pure thought. The Babylonians, technologists at heart, didn't care much about fine theoretical points but made important practical discoveries. The two realms began coming together at the time of the Industrial Revolution. In England the process was tied to the emergence of the Royal Institution, founded in 1799, which made it possible for the general public to learn about technical matters in their spare time and without the training in classical languages and the social standing expected at the universities of Oxford and Cambridge. At about the same time, these august institutions were joined by the "red brick" institutions, which emphasized the training of students for industrial leadership. In the United States, the venerable Ivy League schools, including Harvard, Yale, and Princeton, were joined by the newly established state universities. Some of these were designated as land-grant colleges under the Morrill Act (1862) and were set up to advance agriculture and the mechanical arts. The Morrill Act provided that each state in the Union designate a parcel of land, income from the sale of which would be used to support public colleges and provide for agricultural experimental stations where new ideas in agriculture could be tested.

Educational practice is sometimes slow to recognize changing patterns in society. Colleges in the United States still award bachelor's, master's, and doctoral degrees that have their roots in the Middle Ages. In fact the Bachelor of Arts, Master of Arts, and Doctor of Philosophy degrees are still awarded by the majority of American universities, although the coursework required and the major subjects offered differ greatly from those of a century ago. The bachelor's degree is still awarded to individuals who have completed a four-year course of study, though the emphasis may be on computer science or psycholinguistics instead of the liberal arts. Many schools now award the Bachelor of Science degree to graduates with majors in the sciences (although some, particularly in the Ivy League, award the traditional B.A. degree to all graduates). Additional study is required for the master's degree, and a period of intensive research and the publication of a dissertation is required for the Doctor of Philosophy degree.

Despite the antiquity of the bachelor's, master's, and doctoral designations, what the degrees actually meant was, for a time, quite flexible in early-twentieth-century America. Eventually a measure of quality control was achieved, with colleges being chartered by state governments and subject to review by regional accrediting agencies and, in some fields, by additional specialized agencies.

In the sciences the Doctor of Philosophy degree (Ph.D.) became the standard expected of individuals seeking academic positions or leadership roles in government or industrial laboratories. Prior to World War II, many scientists had to complete their doctoral studies at a European university. Now, as a result of the growth of higher education, doctoral programs can be found on many university campuses. Doctoral education in the United States began at Johns Hopkins University and typically follows the German model, in which the student works under the close supervision of his or her dissertation adviser on a single research question. Doctoral programs generally include some form of comprehensive examinations testing broad knowledge of one's field of interest and often reading knowledge of a foreign language. A Ph.D. in science can lead to many career options. The late Dr. Sally Ride, for instance, the first American woman to go into space, received her Ph.D. in physics from Stanford University.

Students with a strong interest in one of the sciences can also find many satisfactory job placements below the Ph.D. level. Careers in science teaching are open to students with a bachelor's or master's degree. There is currently a critical need for science teachers with majors in the sciences. Students with bachelor's or master's degrees can work as technicians in industrial or governmental laboratories. There are also positions in the armed forces, both civilian and in uniform, and with defense contractors. Students with science degrees often move rapidly into management positions.

Regarding medical fields, the Flexner report of 1910, done with the support of the Carnegie Foundation, did a great deal to standardize medical education. Prior to the work of Abraham Flexner and his colleagues, fly–by-night medical schools existed side by side with reputable institutions, and the patient in need of a doctor could not be sure that the impressive-looking documents on the wall of a medical office guaranteed competence. The Flexner report fixed the Doctor of Medicine degree, awarded after four years at an accredited medical college, generally requiring an accredited bachelor's degree for admission. Remarkably, after the report became public, about half of the medical schools in the United States shut down. Similar standards were imposed for dentists, pharmacists, and veterinarians. Each profession has its own organization, the American Medical Association being the most prominent.

Students interested in the healing professions might also consider becoming a registered nurse, nurse practitioner or physician's assistant. The course of study for these occupations generally takes fewer years than the M.D. degree and emphasizes practical skills a bit more than theory.

TO THE STUDENT

Your decision to enter a science based or medical field is one that, with persistence, will lead to a comfortable income and, more importantly, a variety of interesting challenges in your career. While in junior high and high school, be sure to practice writing and organizing presentations. If possible do not interrupt your study of mathematics, as mathematics courses build on each other. Resist the temptation to take an easy semester. Start or join a science or health careers club at your school.

Talk with the physicians and dentists you visit and any professional scientists who might be in local industry or at nearby colleges. Visit nearby sites that employ scientists.

You are likely to be affected by two recent trends. One is the recent growth in on-line education. Although the widespread availability of computers and the Internet make it possible for students almost anywhere to study almost anything, there is still much to be said for the pre-professional socialization that goes on in college. For science majors, needing to keep to a schedule of assignments, work collaboratively with other students, develop presentation skills, and find information by research are all established aspects of college life. The second trend is the need for continuing professional education. Once you have your professional degree, you will need to attend appropriate professional society meetings. In the health-related professions, this is often a formal continuing-education requirement to retain licensure.

A BRIEF SURVEY OF THE SCIENCES AND MEDICINE

A clear distinction between science and technology is not possible. Nor is it really possible to separate the sciences and medicine. The International Museum of Surgical Science in Chicago, for example, has a hall devoted to statues of individuals who have had the greatest impact on medical practice today. Three of the six—Louis Pasteur, Marie Curie, and Konrad Roentgen—were academic scientists without medical degrees. Today every large hospital has a Department of Nuclear Medicine; most offer magnetic resonance imaging and positron emission tomography, technologies that have moved from the nuclear physics lab into medical practice in about 20 years' time.

At the close of World War II, there was considerable debate as to the proper relationship between science and government in peacetime. Some conservatives wanted to impose strong military control over scientific research. In 1945 Vannevar Bush, who had been director of the wartime Office of Scientific Research and Development, published the report *Science: The Endless Frontier,* which painted a very rosy picture of the gains to be derived by public investment in basic scientific research. Among other things, Bush emphasized the role that government could play in supporting research in universities, research to fill in the middle ground between purely academic research and research directed toward immediate objectives. Further, universities had a natural role to play in training the next generation of scientists, engineers, and physicians. Bush's arguments led to establishment of the National Science Foundation and to expanded research funding within and by the National Institutes of Health in the United States. This trend was accelerated by the Soviet Union's launch of Sputnik 1, the first Earth satellite, in 1957, an event that shook the American public's faith in the inevitable superiority of American technology. In the twenty-first century, science and technology are supported by many sources, public and private, and the pace of development is perhaps even greater than at any time in the past.

The Preliterate World

According to archaeologists and anthropologists, the development of written language occurred relatively recently in human history, perhaps about 3000 B.C.E. Many of the basic components of technology date to those preliterate times, when humans struggled to secure the basic necessities—food, clothing, and shelter—against a background of growing population and changing climate. When food collection was limited to hunting and gathering, knowledge of the seasons and animal behavior was important for survival. The development of primitive stone tools and weapons greatly facilitated both hunting and obtaining meat from animal carcasses, as well as the preservation of the hides for clothing and shelter. Sometime in the middle Stone Age, humans obtained control over fire, making it possible to soften food by cooking and to separate some metals from their ores. Control over fire also made it possible to harden earthenware pottery and keep predatory animals away at night.

Even without language a primitive human had to be something of a naturalist and anatomist. It is certain that the medicinal properties of certain plants were discovered, and in the process of extracting meat from killed animals, a basic knowledge of anatomy was gained. There is also evidence that primitive humans performed trepanations, opening a hole in a person's skull in order to cure headaches. Such drastic procedures were not necessarily fatal. Skulls with healed trepanations have been found in ancient burial sites.

With gradually improved living conditions, human fertility and longevity both increased, as did competition for necessities of life. Spoken language, music, magical thinking, and myth developed as a means of coordinating activity. Warfare, along with more peaceful approaches, was adopted as a means of settling disputes, while society was reorganized to ensure access to the necessities of life, including protection from military attack.

The Ancient World

With the invention of written language, it became possible to enlarge and coordinate human activity on an unprecedented scale. Several new areas of technology and engineering arose. Cities were established so that skilled workers could be freed from direct involvement in food production. Logistics and management became functions of the scribal class, the members of which could read and write. Libraries were built and manuscripts collected. The beginnings of mathematics may be seen in building, surveying, and wealth tabulation. Engineers built roads so that a ruler could oversee his enlarged domain and troops could move rapidly to where they were needed. Taxes were imposed on the public to support the central government, and accounting methods were introduced. Aqueducts were needed to bring fresh water to the cities.

By 500 BCE a number of cultures were advanced enough to have a body of medical knowledge and recognize the need for physicians. The first of these conditions occurred in ancient Egypt, where medical manuscripts can be dated to 3000 BCE and surgery was practiced by 2750 BCE. The Egyptians' anatomical knowledge undoubtedly grew as a result of belief in an afterlife for which the deceased had to be prepared by mummification. The ancient Babylonians also left extensive medical texts, as did the ancient Hindus and the Chinese.

Hippocrates of Cos (450–370) BCE is often considered to be the "father" of modern Western medicine. Hippocrates is known as the author of a code of medical practice and was widely considered to be the author of the so-called "Oath of Hippocrates," versions of which are still administered by many medical schools. Hippocrates was also among the first to establish a theoretical basis for health. Building on the four-element theory of Empedocles (air, earth, fire, and water), he considered health to be a balance of four "humors": blood, phlegm, black bile, and yellow bile. Illness arose when one of the humors was present in excess. This belief led to the practice of bloodletting, one that continued until the 1850s.

The Greek physician Galen (169–217 CE) was a major influence on modern surgery. By the time of Galen, strictures had arisen against the dissection of the human body, so Galen's writings are based on his dissection of Barbary apes. In medieval universities Galen, like Aristotle, would become one of the ancients not to be questioned. By the Renaissance, however, the attitude toward dissecting humans had changed, and medical schools again allowed dissections. The pendulum may have swung a bit too far. In Renaissance Florence public dissections were conducted at Carnival time (the period leading up to penetential Ash Wednesday), with the public attending in costume, as if at a modern Halloween party.

The Scientific Revolution

The Renaissance and the Protestant Reformation marked something of a rebirth of scientific thinking. This "scientific revolution" would not have been possible without Gutenberg's printing press and the technology of printing with movable type. With wealthy patrons, natural philosophers felt secure in challenging the authority of Aristotle and Galen. Galileo published arguments in favor of the Copernican solar system. In the *Novum Organum* (1620; "New Instrument"), Sir Francis Bacon formalized the inductive method, by which generalizations could be made from observations, which then could be tested by further observation or experiment. In England in 1660, with the nominal support of the British Crown, the Royal Society was formed to serve as a forum for the exchange of scientific ideas and the support and publication of research results. The need for larger-scale studies brought craftsmen into the sciences, culminating in the recognition of the professional scientist. Earlier Bacon had proposed that the government undertake the support of scientific investigation for the common good. Bacon himself tried his hand at frozen-food technology. While on a coach trip, he conceived the idea that low temperatures could preserve meat. He stopped the coach, purchased a chicken from a farmer's wife, and stuffed it with snow. Unfortunately he contracted pneumonia while doing this experiment and died forthwith.

The Industrial Revolution followed on the heels of the Scientific Revolution in England. Key to the Industrial Revolution was the technology of the steam engine, the first portable source of motive power that was not dependent on human or animal muscle. The modern form of the steam engine owes much to James Watt, a self-taught technologist. The steam engine powered factories, ships, and, later, locomotives. In the case of the steam engine, technological advance preceded the development of the pertinent science—thermodynamics and the present-day understanding of heat as a random molecular form of energy.

It is not possible, of course, to do justice to the full scale of applied science and technology in this short space. In the remainder of this introduction, consideration will be given to only a few representative fields, highlighting the evolution of each area and its interconnectedness with fundamental and applied science as a whole.

Atomic Theory and the Nucleus

If a single idea permeates modern science, it is the atomic theory, the idea that all matter is composed of atoms. While we may owe this idea to the ancient Greek philosopher Democritus, it did not make much headway until the French savant Antoine Lavoisier published his *Elementary Treatise on Chemistry* in 1789 and the English schoolteacher John Dalton suggested, in the mid-nineteenth century, that pure chemical substances were composed of atoms in a fixed ratio. From this realization it was just a short time until the great Russian chemist Dmitri Mendeleev had arranged the atoms in the now-familiar periodic table.

That atoms were made of component parts was well established by 1900. The existence of the atomic nucleus, however, was not suspected for another decade. Once it was known that most of the mass of the atom is concentrated in the tiny atomic nucleus, held together by forces many hundreds of times stronger than that between protons, nuclear physics came into its own. While nuclear physics made possible the atomic bomb that destroyed two Japanese cities in 1945, ending World War II in the Pacific, the ability to create isotopes in the laboratory was quickly applied to the treatment of cancer, and later medical imaging, saving a great many lives.

The Beginnings of Chemotherapy

In 1792 the Scottish inventor William Murdock discovered a way to produce illuminating gas by the destructive distillation of coal, producing a cleaner and more dependable source of light than previously was available and bringing about the gaslight era. The production of illuminating gas, however, left behind a nasty residue called coal tar. A search was launched to find an application for this major industrial waste. An early use, the waterproofing of cloth, was discovered by the Scottish chemist Charles Macintosh, resulting in the raincoat that now carries his name. In 1856 English chemist William Henry Perkin discovered the first of the coal-tar dyes, mauve. The color mauve, a deep lavender-lilac purple, had previously been obtained from plant sources and had become something of a fashion fad in Paris by 1857. The demand for mauve outstripped the supply of vegetable sources. The discovery of several other dyes followed.

The possibility of dyeing living tissue was rapidly seized on and applied to the tissues of the human body and the microorganisms that afflict it. German bacteriologist Paul Ehrlich proposed that the selective adsorption of dyes could serve as the basis for a chemically based therapy to kill infectious disease-bearing organisms. Although Ehrlich's chemical theory is laughable today, it led to the first effective chemotherapy: the use of Salvarsan, an arsenic-based synthetic drug developed by Ehrlich and his team, as a treatment for syphilis.

Optical Technology, the Microscope, and Microbiology

The use of lenses as an aid to vision may date to China in 500 B.C.E. Marco Polo, in his journeys more than 1,700 years later, reported seeing many Chinese wearing eyeglasses. In 1665 English physicist (and curator of experiments for the Royal Society) Robert Hooke published his book *Micrographia* ("Tiny Handwriting"), which included many illustrations of living tissue. Antonie van Leeuwenhoek was influenced by Hooke, and he reported many observations of microbial life to the Royal Society. The simple microscopes of Hooke and van Leeuwenhoek suffered from many forms of aberration or distortion. Subsequent investigators introduced combinations of lenses to reduce the aberrations, and good compound microscopes became available for the study of microscopic life around 1830.

While van Leeuwenhoek had reported the existence of microorganisms, the notion that they might be responsible for disease or agricultural problems met considerable resistance. Louis Pasteur, an accomplished physical chemist, became best known as the father of microbiology. Pasteur was drawn into applied research by problems arising in the fermentation industry. In 1857 he announced that fermentation was the result of microbial action. He also showed that the souring of milk resulted from microorganisms, leading to the development of "pasteurization"—heating to a certain temperature for a specific amount of time—as a technique for preserving milk. As a sequel to his work on fermentation, Pasteur brought into question the commonly held idea that living organisms could generate spontaneously. Through carefully designed experiments, he demonstrated that broth could be maintained sterile indefinitely, even when exposed to the air, provided that bacteria-carrying dust was excluded.

Pasteur's further research included investigating the diseases that plagued the French silk industry. He also developed a means of vaccinating sheep against infection by *Bacillus anthracis* and a vaccine to protect chickens against cholera. Pasteur's most impressive achievement may have been the development of a treatment effective against the rabies virus for people bitten by rabid dogs or wolves.

Pasteur's scientific achievements illustrate the close interplay of fundamental and applied advances that occurs in many scientific fields. Political scientist Donald Stokes has termed this arena of application-driven scientific research "Pasteur's Quadrant," to distinguish it from purely curiosity-driven research (as in modern particle physics); advance by trial and error (for example, Edison's early work on the electric light); and the simple cataloging of properties and behaviors (as in classical botany and zoology). The study of applied science is a detailed examination of Pasteur's Quadrant.

Electromagnetic Technology

The history of electromagnetic devices provides an excellent example of the complex interplay of fundamental and applied science. The phenomena of static electricity and natural magnetism were described by Thales of Miletus in ancient times, but they remained curiosities through much of history. The magnetic compass was developed by Chinese explorers in about 1100 B.C.E., and the nature of Earth's magnetic field was explored by William Gilbert (physician to Queen Elizabeth I) around 1600. By the late eighteenth century, a number of devices for producing and storing static electricity were being used in popular demonstrations, and the lightning rod, invented by Benjamin Franklin, greatly reduced the damage due to lightning strikes on tall buildings. In 1800 Italian physicist Alessandro Volta developed the first electrical battery. Equipped with a source of continuous electric current, scientists made electrical and electromagnetic discoveries, practical and fundamental, at a breakneck pace.

The voltaic pile, or battery, was employed by British scientist Sir Humphry Davy to isolate a number of chemical elements for the first time. In 1820 Danish physicist Hans Christian Ørsted discovered that any current-carrying wire is surrounded by an electric field. In 1831 English physicist Michael Faraday discovered that a changing magnetic field would induce an electric current in a loop of wire, thus paving the way for the electric generator and the transformer. In Albany, New York, schoolteacher Joseph Henry set his students the challenge of building the strongest possible electromagnet. Henry would move on to become professor of natural philosophy at Princeton University, where he invented a primitive telegraph.

The basic laws of electromagnetism were summarized in 1865 by Scottish physicist James Clerk Maxwell in a set of four differential equations that yielded a number of practical results almost immediately. These equations described the behavior of electric and magnetic fields in different media, including in empty space. In a vacuum it was possible to find wavelike solutions that appeared to move in time at the speed of light, which was immediately realized to be a form of electromagnetic radiation. Further, it turned out that visible light covered only a small frequency range. Applied scientists soon discovered how to transmit messages by radio waves, electromagnetic waves of much lower frequency. Higher frequency waves included Roentgen's X-rays and the gamma rays now used in radiation therapy.

Heredity, Evolution, DNA, and the Genetic Code

The theory of evolution by natural selection was put forth by Charles Darwin and Alfred Russell Wallace in 1859. Although initially unpopular with religious leaders, because it seemed to require millions of years of Earth history instead of the few thousands posited in the Bible, it gained in acceptance over time. The theory, in a nutshell, proposes that populations of organisms produced offspring with some variation in their ability to survive in their environment, and, over time, only the organisms fittest for their environment survive. The theory did not explain how the variations occur or how they are transmitted from parent to offspring. Unbeknownst to Darwin and Wallace, the transmission of traits from one generation to the next was being studied in pea plants by an Austrian monk, Gregor Mendel. Mendel defined dominant and recessive genes and derived the classical laws of inheritance by observation, over generations, of his plants.

The mechanism of variation was identified in part by Thomas Hunt Morgan, who studied the genetics of fruit flies extensively. Morgan saw under the microscope that the nucleus of each of his fruit-fly cells contained chromosomes (so-called because they took up colored dyes easily), and each chromosome involved a number of colored bands that were duplicated when the cell divided into offspring cells. The chromosomes contained long molecules of deoxyribose nucleic acid (DNA). The molecules were arranged as a double helix, and they differed in composition from species to species and gene to gene. It was then realized that the sequence of base pairs in a particular DNA molecule was actually a set of instructions for assembling amino acids into protein molecules, in which the proteins were characteristic of each species.

The double-helical structure for DNA was proposed by J. D. Watson and F. H. C. Crick in 1955 and published in a four-page paper in the journal *Nature*, to almost universal acclaim. This was followed by the development of techniques for assembling a DNA strand to order, and making millions of copies in a very short time. The discovery of these techniques of microbiology allowed for the introduction of DNA evidence into court and to the determination of which genes were responsible for various genetic diseases. It is impossible to say where this technology will lead over the careers of those now in medical school. In principle one drop of blood, or one cheek cell, can give a complete genetic profile of an individual.

The Computer

One of the most clearly useful modern artifacts the digital electronic computer, as it has come to be known, has a lineage that includes the most abstract of mathematics, the automated loom, the vacuum tubes of the early twentieth century, and the modern sciences of semiconductor physics and photochemistry. Although computing devices such as the abacus and slide rule themselves have long histories, the programmable digital computer has advanced computational power by many orders of magnitude. The basic logic of the computer and the computer program, however, arose from a mathematical logician's attempt to answer a problem arising in the foundations of mathematics.

From the time of the ancient Greeks to the end of the nineteenth century, mathematicians had assumed that their subject was essentially a study of the real world, the part amenable to purely deductive reasoning. This included the structure of space and the basic rules of counting, which lead to the rules of arithmetic and algebra. With the discovery of non-Euclidean geometries and the paradoxes of set theory, mathematicians felt the need for a closer study of the foundations of mathematics, to make sure that the objects that might exist only in their minds could be studied and talked about without risking inconsistency.

David Hilbert, a professor of mathematics at the University of Göttingen, was the recognized leader of German mathematics. At a mathematics conference in 1928, Hilbert identified three questions about the foundations of mathematics that he hoped would be resolved in short order. The third of these was the so-called decidability problem: Was there was a foolproof procedure to determine whether a mathematical statement was true or false? Essentially, if one had the statement in symbolic form, was there a procedure for manipulating the symbols in such a way that one could determine whether the statement was true in a finite number of steps?

British mathematician Alan Turing presented an analysis of the problem by showing that any sort of mathematical symbol manipulation was in essence a computation and thus a manipulation of symbols not unlike the addition or multiplication one learns in elementary school. Any such symbolic manipulation could be emulated by an abstract machine that worked with a finite set of symbols that would store a simple set of instructions and process a one-dimensional array of symbols, replacing it with a second array of symbols. Turing showed that there was no

solution in general to Hilbert's decision problem, but in the process he also showed how to construct a machine (now called a Turing machine) that could execute any possible calculation. The machine would operate on a string of symbols recorded on a tape and would output the result of the same calculation on the same tape. Further, Turing showed the existence of machines that could read instructions given in symbolic form and then perform any desired computation on a one-dimensional array of numbers that followed. The universal Turing machine was a programmable digital computer. The instructions could be read from a one-dimensional tape, a magnetically stored memory, or a card punched with holes, as was used for mechanized weaving of fabric.

The earliest electronic computers were developed at the time of World War II and involved numerous vacuum tubes. Since vacuum tubes are based on thermionic emission—the so-called Edison effect—they produced immense amounts of heat and involved the possibility that the heating element in one of the tubes might well burn out during the computation. In fact, it was standard procedure to run a program, one that required proper function of all the vacuum tubes, both before and after the program of interest. If the results of the first and last computations did not vary, one could assume that no tubes had burned out in the meantime.

World War II ended in 1945. In addition to the critical role of computing machines in the design of the first atomic bombs, computational science had played an important role in predicting the behavior of targets. The capabilities of computing machines would grow rapidly following the invention of the transistor by John Bardeen, Walter Brattain, and William Shockley in 1947. In this case fundamental science led to tremendous advances in applied science.

The 1960s saw the production of integrated circuits – many transistors and other circuit elements on a single silicon wafer, or chip. Currently hundreds of thousands of circuit elements are available on a single chip, and anyone who buys a laptop computer will command more computational power than any government could control in 1950. Further, the "lab on a chip" is becoming a reality, able to diagnose and, in time, prepare medicine for most medical conditions.

Donald R. Franceschetti

FURTHER READING

Bell Telephone Laboratories. *A History of Engineering and Science in the Bell System: Electronics Technology, 1925–1975*. Ed. by M. D. Fagen. 7 vols. New York: Bell Laboratories, 1975–1985. Provides detailed information on the development of the transistor and the integrated circuit.

Bodanis, David. *Electric Universe: How Electricity Switched On the Modern World*. New York: Three Rivers Press, 2005. Popular exposition of the applications of electronics and electromagnetism from the time of Joseph Henry to the microprocessor age.

Burke, James. *Connections*. Boston: Little Brown, 1978; reprint, New York: Simon & Schuster, 2007. Describes linkages among inventions throughout history.

Cobb, Cathy, and Harold Goldwhite. *Creations of Fire: Chemistry's Lively History from Alchemy to the Atomic Age*. New York: Plenum Press, 1995; reprint, Cambridge, Mass.: Perseus, 2001. History of pure and applied chemistry from the beginning through the late twentieth century.

Garfield, Simon. *Mauve: How One Man Invented a Color That Changed the World*. New York: W. W. Norton, 2000. Focuses on how the single and partly accidental discovery of coal tar dyes led to several new areas of chemical industry

Kealey, Terence. *The Economic Laws of Scientific Research*. New York: St. Martin's Press, 1996. Makes the case that government funding of scientific research is relatively inefficient and emphasizes the role of private investment and hobbyist scientists.

Mlodinow, Leonard, *Feynman's Rainbow: A Search for Beauty in Physics and in Life*. New York: Warner Books, 2003; reprint, New York: Vintage Books, 2011.

Schlager, Neil, ed. *Science and Its Times: Understanding the Social Significance of Scientific Discovery*. 8 vols. Detroit: Gale Group, 2000–2001. Massive reference work on the impact of scientific and technological developments from the earliest times to the present.

Stokes, Donald E. *Pasteur's Quadrant: Basic Science and Technological Innovation*. Washington, D.C.: Brookings Institution Press, 1997. Presents an extended argument that many fundamental scientific discoveries originate in application-driven research and that the distinction between pure and applied science is not, of itself, very useful.

CONTRIBUTORS

Richard Adler
University of Michigan-Dearborn

Mihaela Avramut
Verlan Medical Communications

Dana K. Bagwell
Memory Health and Fitness Institute

Craig Belanger
Journal of Advancing Technology

Raymond D. Benge
Tarrant County College

Michael A. Buratovich
Spring Arbor University

Byron D. Cannon
University of Utah

Richard P. Capriccioso
University of Phoenix

Christine M. Carroll
American Medical Writers Association

Robert L. Cullers
Kansas State University

Jeremy Dugosh
American Board of Internal Medicine

Elvira R. Eivazova
Vanderbilt University School of Medicine

David Elliott
Northern Arizona University

Renée Euchner
American Medical Writers Association

Jennifer L. Gibson
Marietta, Georgia

James S. Godde
Monmouth College, Illinois

Dalton R. Gossett
College of Arts and Sciences at Louisiana State University, Shreveport

Gina Hagler
Washington, D.C.

Howard V. Hendrix
California State University, Fresno

Robert M. Hordon
Rutgers University

Carly L. Huth
Publication Services, Inc.

April D. Ingram
Kelowna, British Columbia

Micah L. Issitt
Philadelphia, Pennsylvania

Domingo M. Jariel, Jr.
Louisiana State University, Eunice

Cheryl Pokalo Jones
Townsend, Delaware

Marylane Wade Koch
Loewenberg School of Nursing, University of Memphis

Narayanan M. Komerath
Georgia Institute of Technology

Jeanne L. Kuhler
Benedictine University

Lisa LaGoo
Medtronic

Dawn A. Laney
Atlanta, Georgia

Jeffrey Larson
Tioga Medical Center, Tioga, North Dakota

M. Lee
Independent Scholar

Arthur J. Lurigio
Loyola University Chicago

Marianne M. Madsen
University of Utah

Mary E. Markland
Argosy University

Sergei A. Markov
Austin Peay State University

Jordan M. Marshall
Indiana University-Purdue University

Amber M. Mathiesen
University of Utah

Laurence W. Mazzeno
Alvernia University

Roman Meinhold
Assumption University, Bangkok, Thailand

Mary R. Muslow
Louisiana Tech University

Holly Nyple
San Jose, California

David Olle
Eastshire Communications

Ayman Oweida
University of Western Ontario

Robert J. Paradowski
*Rochester Institute of
Technology*

Ellen E. Anderson Penno
Western Laser Eye Associates

George R. Plitnik
Frostburg State University

Cynthia F. Racer
*American Medical Writers
Association*

Diane C. Rein
University at Buffalo

Richard M. J. Renneboog
Independent Scholar

Joseph Di Rienzi
*College of Notre Dame of
Maryland*

Carol A. Rolf
Rivier College

Julia A. Rosenthal
Chicago, Illinois

Sibani Sengupta
*American Medical Writers
Association*

Linda R. Shoaf
American Dietetic Association

Paul P. Sipiera
Harper College, Palatine, Illinois

Bethany Thivierge
Technicality Resources

Christine Watts
University of Sydney

Shawncey Jay Webb
Taylor University

Judith Weinblatt
New York, New York

Jessica C.Y. Wongt
Environment Canada

Robin L. Wulffson
*Faculty, American College of
Obstetrics and Gynecology*

Susan M. Zneimer
U.S. Labs, Irvine, California

COMMON UNITS OF MEASURE

Common prefixes for metric units—which may apply in more cases than shown below—include *giga-* (1 billion times the unit), *mega-* (one million times), *kilo-* (1,000 times), *hecto-* (100 times), *deka-* (10 times), *deci-* (0.1 times, or one tenth), *centi-* (0.01, or one hundredth), *milli-* (0.001, or one thousandth), and *micro-* (0.0001, or one millionth).

Unit	Quantity	Symbol	Equivalents
Acre	Area	ac	43,560 square feet 4,840 square yards 0.405 hectare
Ampere	Electric current	A *or* amp	1.00016502722949 international ampere 0.1 biot *or* abampere
Angstrom	Length	Å	0.1 nanometer 0.0000001 millimeter 0.000000004 inch
Astronomical unit	Length	AU	92,955,807 miles 149,597,871 kilometers (mean Earth-Sun distance)
Barn	Area	b	10^{-28} meters squared (approx. cross-sectional area of 1 uranium nucleus)
Barrel (dry, for most produce)	Volume/capacity	bbl	7,056 cubic inches; 105 dry quarts; 3.281 bushels, struck measure
Barrel (liquid)	Volume/capacity	bbl	31 to 42 gallons
British thermal unit	Energy	Btu	1055.05585262 joule
Bushel (U.S., heaped)	Volume/capacity	bsh *or* bu	2,747.715 cubic inches 1.278 bushels, struck measure
Bushel (U.S., struck measure)	Volume/capacity	bsh *or* bu	2,150.42 cubic inches 35.238 liters
Candela	Luminous intensity	cd	1.09 hefner candle
Celsius	Temperature	C	1° centigrade
Centigram	Mass/weight	cg	0.15 grain
Centimeter	Length	cm	0.3937 inch
Centimeter, cubic	Volume/capacity	cm^3	0.061 cubic inch
Centimeter, square	Area	cm^2	0.155 square inch
Coulomb	Electric charge	C	1 ampere second
Cup	Volume/capacity	C	250 milliliters 8 fluid ounces 0.5 liquid pint

Unit	Quantity	Symbol	Equivalents
Deciliter	Volume/capacity	dl	0.21 pint
Decimeter	Length	dm	3.937 inches
Decimeter, cubic	Volume/capacity	dm³	61.024 cubic inches
Decimeter, square	Area	dm²	15.5 square inches
Dekaliter	Volume/capacity	dal	2.642 gallons 1.135 pecks
Dekameter	Length	dam	32.808 feet
Dram	Mass/weight	dr *or* dr avdp	0.0625 ounce 27.344 grains 1.772 grams
Electron volt	Energy	eV	$1.5185847232839 \times 10^{-22}$ Btus $1.6021917 \times 10^{-19}$ joules
Fermi	Length	fm	1 femtometer 1.0×10^{-15} meters
Foot	Length	ft *or* '	12 inches 0.3048 meter 30.48 centimeters
Foot, square	Area	ft²	929.030 square centimeters
Foot, cubic	Volume/capacity	ft³	0.028 cubic meter 0.0370 cubic yard 1,728 cubic inches
Gallon (British Imperial)	Volume/capacity	gal	277.42 cubic inches 1.201 U.S. gallons 4.546 liters 160 British fluid ounces
Gallon (U.S.)	Volume/capacity	gal	231 cubic inches 3.785 liters 0.833 British gallon 128 U.S. fluid ounces
Giga-electron volt	Energy	GeV	$1.6021917 \times 10^{-10}$ joule
Gigahertz	Frequency	GHz	—
Gill	Volume/capacity	gi	7.219 cubic inches 4 fluid ounces 0.118 liter
Grain	Mass/weight	gr	0.037 dram 0.002083 ounce 0.0648 gram
Gram	Mass/weight	g	15.432 grains 0.035 avoirdupois ounce

Unit	Quantity	Symbol	Equivalents
Hectare	Area	ha	2.471 acres
Hectoliter	Volume/capacity	hl	26.418 gallons 2.838 bushels
Hertz	Frequency	Hz	$1.08782775707767 \times 10^{-10}$ cesium atom frequency
Hour	Time	h	60 minutes 3,600 seconds
Inch	Length	in *or* "	2.54 centimeters
Inch, cubic	Volume/capacity	in³	0.554 fluid ounce 4.433 fluid drams 16.387 cubic centimeters
Inch, square	Area	in²	6.4516 square centimeters
Joule	Energy	J	$6.2414503832469 \times 10^{18}$ electron volt
Joule per kelvin	Heat capacity	J/K	$7.24311216248908 \times 10^{22}$ Boltzmann constant
Joule per second	Power	J/s	1 watt
Kelvin	Temperature	K	-272.15° Celsius
Kilo-electron volt	Energy	keV	$1.5185847232839 \times 10^{-19}$ joule
Kilogram	Mass/weight	kg	2.205 pounds
Kilogram per cubic meter	Mass/weight density	kg/m³	$5.78036672001339 \times 10^{-4}$ ounces per cubic inch
Kilohertz	Frequency	kHz	—
Kiloliter	Volume/capacity	kl	—
Kilometer	Length	km	0.621 mile
Kilometer, square	Area	km²	0.386 square mile 247.105 acres
Light-year (distance traveled by light in one Earth year)	Length/distance	lt-yr	5,878,499,814,275.88 miles 9.46×10^{12} kilometers
Liter	Volume/capacity	L	1.057 liquid quarts 0.908 dry quart 61.024 cubic inches
Mega-electron volt	Energy	MeV	—
Megahertz	Frequency	MHz	—
Meter	Length	m	39.37 inches
Meter, cubic	Volume/capacity	m³	1.308 cubic yards

Unit	Quantity	Symbol	Equivalents
Meter per second	Velocity	m/s	2.24 miles per hour 3.60 kilometers per hour
Meter per second per second	Acceleration	m/s²	12,960.00 kilometers per hour per hour 8,052.97 miles per hour per hour
Meter, square	Area	m²	1.196 square yards 10.764 square feet
Metric. *See* unit name			
Microgram	Mass/weight	mcg *or* μg	0.000001 gram
Microliter	Volume/capacity	μl	0.00027 fluid ounce
Micrometer	Length	μm	0.001 millimeter 0.00003937 inch
Mile (nautical international)	Length	mi	1.852 kilometers 1.151 statute miles 0.999 U.S. nautical miles
Mile (statute or land)	Length	mi	5,280 feet 1.609 kilometers
Mile, square	Area	mi²	258.999 hectares
Milligram	Mass/weight	mg	0.015 grain
Milliliter	Volume/capacity	ml	0.271 fluid dram 16.231 minims 0.061 cubic inch
Millimeter	Length	mm	0.03937 inch
Millimeter, square	Area	mm²	0.002 square inch
Minute	Time	m	60 seconds
Mole	Amount of substance	mol	6.02×10^{23} atoms or molecules of a given substance
Nanometer	Length	nm	1,000,000 fermis 10 angstroms 0.001 micrometer 0.00000003937 inch
Newton	Force	N	0.224808943099711 pound force 0.101971621297793 kilogram force 100,000 dynes
Newton meter	Torque	N·m	0.7375621 foot-pound
Ounce (avoirdupois)	Mass/weight	oz	28.350 grams 437.5 grains 0.911 troy or apothecaries' ounce

Unit	Quantity	Symbol	Equivalents
Ounce (troy)	Mass/weight	oz	31.103 grams 480 grains 1.097 avoirdupois ounces
Ounce (U.S., fluid or liquid)	Mass/weight	oz	1.805 cubic inch 29.574 milliliters 1.041 British fluid ounces
Parsec	Length	pc	30,856,775,876,793 kilometers 19,173,511,615,163 miles
Peck	Volume/capacity	pk	8.810 liters
Pint (dry)	Volume/capacity	pt	33.600 cubic inches 0.551 liter
Pint (liquid)	Volume/capacity	pt	28.875 cubic inches 0.473 liter
Pound (avoirdupois)	Mass/weight	lb	7,000 grains 1.215 troy or apothecaries' pounds 453.59237 grams
Pound (troy)	Mass/weight	lb	5,760 grains 0.823 avoirdupois pound 373.242 grams
Quart (British)	Volume/capacity	qt	69.354 cubic inches 1.032 U.S. dry quarts 1.201 U.S. liquid quarts
Quart (U.S., dry)	Volume/capacity	qt	67.201 cubic inches 1.101 liters 0.969 British quart
Quart (U.S., liquid)	Volume/capacity	qt	57.75 cubic inches 0.946 liter 0.833 British quart
Rod	Length	rd	5.029 meters 5.50 yards
Rod, square	Area	rd^2	25.293 square meters 30.25 square yards 0.00625 acre
Second	Time	s or sec	$1/60$ minute $1/3600$ hour
Tablespoon	Volume/capacity	T or tb	3 teaspoons 4 fluid drams
Teaspoon	Volume/capacity	t or tsp	0.33 tablespoon 1.33 fluid drams

Unit	Quantity	Symbol	Equivalents
Ton (gross or long)	Mass/weight	t	2,240 pounds 1.12 net tons 1.016 metric tons
Ton (metric)	Mass/weight	t	1,000 kilograms 2,204.62 pounds 0.984 gross ton 1.102 net tons
Ton (net or short)	Mass/weight	t	2,000 pounds 0.893 gross ton 0.907 metric ton
Volt	Electric potential	V	1 joule per coulomb
Watt	Power	W	1 joule per second 0.001 kilowatt $2.84345136093995 \times 10^{-4}$ ton of refrigeration
Yard	Length	yd	0.9144 meter
Yard, cubic	Volume/capacity	yd^3	0.765 cubic meter
Yard, square	Area	yd^2	0.836 square meter

Applied Science

Science and Medicine

ACOUSTICS

FIELDS OF STUDY

Electrical, chemical, and mechanical engineering; architecture; music; speech; psychology; physiology; medicine; atmospheric physics; geology; oceanography.

SUMMARY

Acoustics is the science dealing with the production, transmission, and effects of vibration in material media. If the medium is air and the vibration frequency is between 18 and 18,000 hertz (Hz), the vibration is termed "sound." Sound is used in a broader context to describe sounds in solids and underwater and structure-borne sounds. Because mechanical vibrations, whether natural or human induced, have accompanied humans through the long course of human evolution, acoustics is the most interdisciplinary science. For humans, hearing is a very important sense, and the ability to vocalize greatly facilitates communication and social interaction. Sound can have profound psychological effects; music may soothe or relax a troubled mind, and noise can induce anxiety and hypertension.

KEY TERMS AND CONCEPTS

- **Cochlea:** Inner ear, which converts pressure waves of sound into electric impulses that are transmitted to the brain via the auditory nerves.
- **Decibel (dB):** Unit of sound intensity used to quantify the loudness of a vibration.
- **Destructive Interference:** Interference that occurs when two waves having the same amplitude in opposite directions come together and cancel each other.
- **Doppler Effect:** Apparent change in frequency of a wave because of the relative motions of the source and an observer. Wavelengths of approaching objects are shortened, and those of receding objects are lengthened.
- **Hertz (Hz):** Unit of frequency; the number of vibrations per second of an oscillation.
- **Infrasound:** Air vibration below 20 hertz; perceived as vibration.
- **Physical Acoustics:** Theoretical area concerned with the fundamental physics of wave propagation and the use of acoustics to probe the physical properties of matter.
- **Resonance:** Large amplitude of vibration that occurs when an oscillator is driven at its natural frequency.
- **Sound:** Vibrations in air having frequencies between 20 and 20,000 hertz and intensities between 0 and 135 decibels and therefore perceptible to humans.
- **Sound Spectrum:** Representation of a sound in terms of the amount of vibration at a each individual frequency. Usually presented as a graph of amplitude (plotted vertically) versus frequency (plotted horizontally).
- **Spectrogram:** Graph used in speech research that plots frequency (vertical axis) versus the time of the utterance (horizontal axis). The amplitude of each frequency component is represented by its darkness.
- **Transducer:** Device that transmutes one form of energy into another. Acoustic examples include microphones and loudspeakers.
- **Ultrasound:** Frequencies above 20,000 hertz used by bats for navigation and by humans for industrial applications and nonradiative ultrasonic imaging.

DEFINITION AND BASIC PRINCIPLES

The words "acoustics," and "phonics" evolved from ancient Greek roots for hearing and speaking, respectively. Thus, acoustics began with human communication, making it one of the oldest if not the most basic of sciences. Because acoustics is ubiquitous in human endeavors, it is the broadest and most interdisciplinary of sciences; its most profound contributions have occurred when it is commingled with an independent field. The interdisciplinary nature of acoustics has often consigned it to a subsidiary role as an minor subdivision of mechanics, hydrodynamics, or electrical engineering. Certainly, the various technical aspects of acoustics could be parceled out to larger and better established divisions of science, but then acoustics would lose its unique strengths and its source of dynamic creativity. The main difference between acoustics and more self-sufficient branches of science is that acoustics depends on physical laws developed in and borrowed from other fields. Therefore, the primary task of acoustics is to take these divergent principles and integrate them into a coherent whole in order to understand, measure, and control vibration phenomena.

The Acoustical Society of America subdivides acoustics into fifteen main areas, the most important of which are ultrasonics, which examines high-frequency waves not audible to humans; psychological acoustics, which studies how sound is perceived in the brain; physiological acoustics, which looks at human and animal hearing mechanisms; speech acoustics, which focuses on the human vocal apparatus and oral communication; musical acoustics, which involves the physics of musical instruments; underwater sound, which examines the production and propagation of sound in liquids; and noise, which concentrates on the control and suppression of unwanted sound. Two other important areas of applied acoustics are architectural acoustics (the acoustical design of concert halls and sound reinforcement systems) and audio engineering (recording and reproducing sound).

BACKGROUND AND HISTORY

Acoustics arguably originated with human communication and music. The caves in which the prehistoric Cro-Magnons displayed their most elaborate paintings have resonances easily excited by the human voice, and stalactites emit musical tones when struck or rubbed with a stick. Paleolithic societies constructed flutes of bird bone, used animal horns to produce drones, and employed rattles and scrapers to provide rhythm.

In the sixth century b.c.e., Pythagoras was the first to correlate musical sounds and mathematics by relating consonant musical intervals to simple ratios of integers. In the fourth century b.c.e., Aristotle deduced that the medium that carries a sound must be compressed by the sounding body, and the third century b.c.e. philosopher Chrysippus correctly depicted the propagation of sound waves with an expanding spherical pattern. In the first century b.c.e., the Roman architect and engineer Marcus Vitruvius Pollio explained the acoustical characteristics of Greek theaters, but when the Roman civilization declined in the fourth century, scientific inquiry in the West ceased for the next millennium.

In the seventeenth century, modern experimental acoustics originated when the Italian mathematician Galileo explained resonance as well as musical consonance and dissonance, and theoretical acoustics got its start with Sir Isaac Newton's derivation of an expression for the velocity of sound. Although this yielded a value considerably lower than the experimental result, a more rigorous derivation by Pierre-Simon Laplace in 1816 obtained an equation yielding values in complete agreement with experimental results.

During the eighteenth century, many famous mathematicians studied vibration. In 1700, French mathematician Joseph Sauveur observed that strings vibrate in sections consisting of stationary nodes located between aggressively vibrating antinodes and that these vibrations have integer multiple frequencies, or harmonics, of the lowest frequency. He also noted that a vibrating string could simultaneously produce the sounds of several harmonics. In 1755, Daniel Bernoulli proved that this resultant vibration was the independent algebraic sum of the various harmonics. In 1750, Jean le Rond d'Alembert used calculus to obtain the wave equation for a vibrating string. By the end of the eighteenth century, the basic experimental results and theoretical underpinnings of acoustics were extant and in reasonable agreement, but it was not until the following century that theory and a concomitant advance of technology led to the evolution of the major divisions of acoustics.

Although mathematical theory is central to all acoustics, the two major divisions, physical and applied acoustics, evolved from the central theoretical core. In the late nineteenth century, Hermann von Helmholtz and Lord Rayleigh (John William Strutt), two polymaths, developed the theoretical aspects. Helmholtz's contributions to acoustics were primarily in explaining the physiological aspects of the ear. Rayleigh, a well-educated wealthy English baron, synthesized virtually all previous knowledge of acoustics and also formulated an appreciable corpus of experiment and theory.

Experiments by Georg Simon Ohm indicated that all musical tones arise from simple harmonic vibrations of definite frequency, with the constituent components determining the sound quality. This gave birth to the field of musical acoustics. Helmholtz's studies of instruments and Rayleigh's work contributed to the nascent area of musical acoustics. Helmholtz's knowledge of ear physiology shaped the field that was to become physiological acoustics.

Underwater acoustics commenced with theories developed by the nineteenth-century mathematician Siméon-Denis Poisson, but further development had to await the invention of underwater transducers in the next century.

Two important nineteenth-century inventions, the telephone (patented 1876) and the mechanical phonograph (invented 1877), commingled and evolved into twentieth-century audio acoustics when united with electronics. Some products in which sound production and reception are combined are microphones, loudspeakers, radios, talking motion pictures, high-fidelity stereo systems, and public sound-reinforcement systems. Improved instrumentation for the study of speech and hearing has stimulated the areas of physiological and psychology acoustics, and ultrasonic devices are routinely used for medical diagnosis and therapy, as well as for burglar alarms and rodent repellants. Underwater transducers are employed to detect and measure moving objects in the water, while audio engineering technology has transformed music performance as well as sound reproduction. Virtually no area of human activity has remained unaffected by continually evolving technology based on acoustics.

HOW IT WORKS

Ultrasonics. Dog whistles, which can be heard by dogs but not by humans, can generate ultrasonic frequencies of about 25 kilohertz (kHz). Two types of transducers, magnetostrictive and piezoelectric, are used to generate higher frequencies and greater power. Magnetostrictive devices convert magnetic energy into ultrasound by subjecting ferric material (iron or nickel) to a strong oscillating magnetic field. The field causes the material to alternately expand and contract, thus creating sound waves of the same frequency as that of the field. The resulting sound waves have frequencies between 20 Hz and 50 kHz and several thousand watts of power. Such transducers operate at the mechanical resonance frequency where the energy transfer is most efficient.

Piezoelectric transducers convert electric energy into ultrasound by applying an oscillating electric field to a piezoelectric crystal (such as quartz). These transducers, which work in liquids or air, can generate frequencies in the megahertz region with considerable power. In addition to natural crystals, ceramic piezoelectric materials, which can be fabricated into any desired shape, have been developed.

Physiological and Psychological Acoustics. Physiological acoustics studies auditory responses of the ear and its associated neural pathways, and psychological acoustics is the subjective perception of sounds through human auditory physiology. Mechanical, electrical, optical, radiological, or biochemical techniques are used to study neural responses to various aural stimuli. Because these techniques are typically invasive, experiments are performed on animals with auditory systems that are similar to the human system. In contrast, psychological acoustic studies are noninvasive and typically use human subjects.

A primary objective of psychological acoustics is to define the psychological correlates to the physical parameters of sound waves. Sound waves in air may be characterized by three physical parameters: frequency, intensity, and their spectrum. When a sound wave impinges on the ear, the pressure variations in the air are transformed by the middle ear to mechanical vibrations in the inner ear. The cochlea then decomposes the sound into its constituent frequencies and transforms these into neural action potentials, which travel to the brain where the sound is evidenced. Frequency is perceived as pitch, the intensity level as loudness, and the spectrum determines the timbre, or tone quality, of a note.

Another psychoacoustic effect is masking. When a person listens to a noisy version of recorded music, the noise virtually disappears if the music is being enjoyed. This ability of the brain to selectively listen has had important applications in digitally recorded music. When the sounds are digitally compressed, such as in MP3 (MPEG-1 audio layer 3) systems, the brain compensates for the loss of information; thus one experiences higher fidelity sound than the stored content would imply. Also, the brain creates

Fascinating Facts About Acoustics

- Scientists have created an acoustic refrigerator, which uses a standing sound wave in a resonator to provide the motive power for operation. Oscillating gas particles increase the local temperature, causing heat to be transferred to the container walls, where it is expelled to the environment, cooling the interior.
- A cochlear implant, an electronic device surgically implanted in the inner ear, provides some hearing ability to those with damaged cochlea or those with congenital deafness. Because the implants use only about two dozen electrodes to replace 16,000 hair cells, speech sounds, although intelligible, have a robotic quality.
- MP3 files contain audio that is digitally encoded using an algorithm that compresses the data by a factor of about eleven but yields a reasonably faithful reproduction. The quality of sound reproduced depends on the data sampling rate, the quality of the encoder, and the complexity of the signal.
- Sound cannot travel through a vacuum, but it can travel four times faster through water than through air.
- The cocktail party effect refers to a person's ability to direct attention to one conversation at a time despite the many conversations taking place in the room.
- Continued exposure to noise over 85 decibels will gradually cause hearing loss. The noise level on a quiet residential street is 40 decibels, a vacuum cleaner 60-85, a leafblower 110, an ambulance siren 120, a rifle 163, and a rocket launching from its pad 180.

information when the incoming signal is masked or nonexistent, producing a psychoacoustic phantom effect. This phantom effect is particularly prevalent when heightened perceptions are imperative, as when danger is lurking.

Psychoacoustic studies have determined that the frequency range of hearing is from 20 to about 20,000 Hz for young people, and the upper limit progressively decreases with age. The rate at which hearing acuity declines depends on several factors, not the least of which is lifetime exposure to loud sounds, which progressively deteriorate the hair cells of the

cochlea. Moderate hearing loss can be compensated for by a hearing aid; severe loss requires a cochlear implant.

Speech Acoustics. Also known as acoustic phonetics, speech acoustics deals with speech production and recognition. The scientific study of speech began with Thomas Alva Edison's phonograph, which allowed a speech signal to be recorded and stored for later analysis. Replaying the same short speech segment several times using consecutive filters passing through a limited range of frequencies creates a spectrogram, which visualizes the spectral properties of vowels and consonants. During the first half of the twentieth century, Bell Telephone Laboratories invested considerable time and resources to the systematic understanding of all aspects of speech, including vocal tract resonances, voice quality, and prosodic features of speech. For the first time, electric circuit theory was applied to speech acoustics, and analogue electric circuits were used to investigate synthetic speech.

Musical Acoustics. A conjunction of music, craftsmanship, auditory science, and vibration physics, musical acoustics analyzes musical instruments to better understand how the instruments are crafted, the physical principles of their tone production, and why each instrument has a unique timbre. Musical instruments are studied by analyzing their tones and then creating computer models to synthesize these sounds. When the sounds can be recreated with minimal software complications, a synthesizer featuring realistic orchestral tones may be constructed. The second method of study is to assemble an instrument or modify an existing instrument to perform nondestructive (or on occasion destructive) testing so that the effects of various modifications may be gauged.

Underwater Sound. Also know as hydroacoustics, this field uses frequencies between 10 Hz and 1 megahertz (MHz). Although the origin of hydroacoustics can be traced back to Rayleigh, the deployment of submarines in World War I provided the impetus for the rapid development of underwater listening devices (hydrophones) and sonar (sound navigation ranging), the acoustic equivalent of radar. Pulses of sound are emitted and the echoes are processed to extract information about submerged objects. When the speed of underwater sound is known, the reflection time for a pulse determines the distance to an object. If the object is moving, its speed of approach or recession is deduced from the frequency shift of the reflection, or the Doppler effect. Returning pulses have a higher frequency when the object approaches and lower frequency when it moves away.

Noise. Physically, noise may be defined as an intermittent or random oscillation with multiple frequency components, but psychologically, noise is any unwanted sound. Noise can adversely affect human health and well-being by inducing stress, interfering with sleep, increasing heart rate, raising blood pressure, modifying hormone secretion, and even inducing depression. The physical effects of noise are no less severe. The vibrations in irregular road surfaces caused by large rapid vehicles can cause adjacent buildings to vibrate to an extent that is intolerable to the buildings' inhabitants, even without structural damage. Machinery noise in industry is a serious problem because continuous exposure to loud sounds will induce hearing loss. In apartment buildings, noise transmitted through walls is always problematic; the goal is to obtain adequate sound insulation using lightweight construction materials.

Traffic noise, both external and internal, is ubiquitous in modern life. The first line of defense is to reduce noise at its source by improving engine enclosures, mufflers, and tires. The next method, used primarily when interstate highways are adjacent to residential areas, is to block the noise by the construction of concrete barriers or the planting of sound-absorbing vegetation. Internal automobile noise has been greatly abated by designing more aerodynamically efficient vehicles to reduce air turbulence, using better sound isolation materials, and improving vibration isolation.

Aircraft noise, particularly in the vicinity of airports, is a serious problem exacerbated by the fact that as modern airplanes have become more powerful, the noise they generate has risen concomitantly. The noise radiated by jet engines is reduced by two structural modifications. Acoustic linings are placed around the moving parts to absorb the high frequencies caused by jet whine and turbulence, but this modification is limited by size and weight constraints. The second modification is to reduce the number of rotor blades and stator vanes, but this is somewhat inhibited by the desired power output. Special noise problems occur when aircraft travel at supersonic speeds (faster than the speed of sound), as this propagates a large pressure wave toward the

ground that is experienced as an explosion. The unexpected sonic boom startles people, breaks windows, and damages houses. Sonic booms have been known to destroy rock structures in national parks. Because of these concerns, commercial aircraft are prohibited from flying at supersonic speeds over land areas.

Construction equipment (such as earthmoving machines) creates high noise levels both internally and externally. When the cabs of these machines are not closed, the only feasible manner of protecting operators' hearing is by using ear plugs. By carefully designing an enclosed cabin, structural vibration can be reduced and sound leaks made less significant, thus quieting the operator's environment. Although manufacturers are attempting to reduce the external noise, it is a daunting task because the rubber tractor treads occasionally used to replace metal are not as durable.

APPLICATIONS AND PRODUCTS

Ultrasonics. High-intensity ultrasonic applications include ultrasonic cleaning, mixing, welding, drilling, and various chemical processes. Ultrasonic cleaners use waves in the 150 to 400 kHz range on items (such as jewelry, watches, lenses, and surgical instruments) placed in an appropriate solution. Ultrasonic cleaners have proven to be particularly effective in cleaning surgical devices because they loosen contaminants by aggressive agitation irrespective of an instrument's size or shape, and disassembly is not required. Ultrasonic waves are effective in cleaning most metals and alloys, as well as wood, plastic, rubber, and cloth. Ultrasonic waves are used to emulsify two nonmiscible liquids, such as oil and water, by forming the liquids into finely dispersed particles that then remain in homogeneous suspension. Many paints, cosmetics, and foods are emulsions formed by this process.

Although aluminum cannot be soldered by conventional means, two surfaces subjected to intense ultrasonic vibration will bond—without the application of heat—in a strong and precise weld. Ultrasonic drilling is effective where conventional drilling is problematic, for instance, drilling square holes in glass. The drill bit, a transducer having the required shape and size, is used with an abrasive slurry that chips away the material when the suspended powder oscillates. Some of the chemical applications of ultrasonics are in the atomization of liquids, in electroplating, and as a catalyst in chemical reactions.

Low-intensity ultrasonic waves are used for nondestructive probing to locate flaws in materials for which complete reliability is mandatory, such as those used in spacecraft components and nuclear reactor vessels. When an ultrasonic transducer emits a pulse of energy into the test object, flaws reflect the wave and are detected. Because objects subjected to stress emit ultrasonic waves, these signals may be used to interpret the condition of the material as it is increasingly stressed. Another application is ultrasonic emission testing, which records the ultrasound emitted by porous rock when natural gas is pumped into cavities formed by the rock to determine the maximum pressure these natural holding tanks can withstand.

Low-intensity ultrasonics is used for medical diagnostics in two different applications. First, ultrasonic waves penetrate body tissues but are reflected by moving internal organs, such as the heart. The frequency of waves reflected from a moving structure is Doppler-shifted, thus causing beats with the original wave, which can be heard. This procedure is particularly useful for performing fetal examinations on a pregnant woman; because sound waves are not electromagnetic, they will not harm the fetus. The second application is to create a sonogram image of the body's interior. A complete cross-sectional image may be produced by superimposing the images scanned by successive ultrasonic waves passing through different regions. This procedure, unlike an X ray, displays all the tissues in the cross section and also avoids any danger posed by the radiation involved in X-ray imaging.

Physiological and Psychological Acoustics. Because the ear is a nonlinear system, it produces beat tones that are the sum and difference of two frequencies. For example, if two sinusoidal frequencies of 100 and 150 Hz simultaneously arrive at the ear, the brain will, in addition to these two tones, create tones of 250 and 50 Hz (sum and difference, respectively). Thus, although a small speaker cannot reproduce the fundamental frequencies of bass tones, the difference between the harmonics of that pitch will re-create the missing fundamental in the listener's brain.

Another psychoacoustic effect is masking. When a person listens to a noisy version of recorded music, the noise virtually disappears if the individual is enjoying the music. This ability of the brain to selectively listen

has had important applications in digitally recorded music. When sounds are digitally compressed, as in MP3 systems, the brain compensates for the loss of information, thus creating a higher fidelity sound than that conveyed by the stored content alone.

As twentieth-century technology evolved, environmental noise increased concomitantly; lifetime exposure to loud sounds, commercial and recreational, has created an epidemic of hearing loss, most noticeable in the elderly because the effects are cumulative. Wearing a hearing aid, fitted adjacent to or inside the ear canal, is an effectual means of counteracting this handicap. The device consists of one or several microphones, which create electric signals that are amplified and transduced into sound waves redirected back into the ear. More sophisticated hearing aids incorporate an integrated circuit to control volume, either manually or automatically, or to switch to volume contours designed for various listening environments, such conversations on the telephone or where excessive background noise is present.

Speech Acoustics. With the advent of the computer age, speech synthesis moved to digital processing, either by bandwidth compression of stored speech or by using a speech synthesizer. The synthesizer reads a text and then produces the appropriate phonemes on demand from their basic acoustic parameters, such as the vibration frequency of the vocal cords and the frequencies and amplitudes of the vowel formants. This method of generating speech is considerably more efficient in terms of data storage than archiving a dictionary of prerecorded phrases.

Another important, and probably the most difficult, area of speech acoustics is the machine recognition of spoken language. When machine recognition programs are sufficiently advanced, the computer will be able to listen to a sentence in any reasonable dialect and produce a printed text of the utterance. Two basic recognition strategies exist, one dealing with words spoken in isolation and the other with continuous speech. In both cases, it is desirable to teach the computer to recognize the speech of different people through a training program. Because recognition of continuous speech is considerably more difficult than the identification of isolated words, very sophisticated pattern-matching models must be employed. One example of a machine recognition system is a word-driven dictation system that uses sophisticated software to process input speech.

This system is somewhat adaptable to different voices and is able to recognize 30,000 words at a rate of 30 words per minute. The ideal machine recognition system would translate a spoken input language into another language in real time with correct grammar. Although some progress is being made, such a device has remained in the realm of speculative fantasy.

Musical Acoustics. The importance of musical acoustics to manufacturers of quality instruments is apparent. During the last decades of the twentieth century, fundamental research led, for example, to vastly improved French horns, organ pipes, orchestral strings, and the creation of an entirely new family of violins.

Underwater Sound. Applications for underwater acoustics include devices for underwater communication by acoustic means, remote control devices, underwater navigation and positioning systems, acoustic thermometers to measure ocean temperature, and echo sounders to locate schools of fish or other biota. Low-frequency devices can be used to explore the seabed for seismic research.

Although primitive measuring devices were developed in the 1920's, it was during the 1930's that sonar systems began incorporating piezoelectric transducers to increase their accuracy. These improved systems and their increasingly more sophisticated progeny became essential for the submarine warfare of World War II. After the war, theoretical advances in underwater acoustics coupled with computer technology have raised sonar systems to ever more sophisticated levels.

Noise. One system for abating unwanted sound is active noise control. The first successful application of active noise control was noise-canceling headphones, which reduce unwanted sound by using microphones placed in proximity to the ear to record the incoming noise. Electronic circuitry then generates a signal, exactly opposite to the incoming sound, which is reproduced in the earphones, thus canceling the noise by destructive interference. This system enables listeners to enjoy music without having to use excessive volume levels to mask outside noise and allows people to sleep in noisy vehicles such as airplanes. Because active noise suppression is more effective with low frequencies, most commercial systems rely on soundproofing the earphone to attenuate high frequencies. To effectively cancel high frequencies, the microphone and emitter would have to be situated adjacent to the user's eardrum, but this is not technically feasible. Active noise control is also being considered as a means of controlling low-frequency airport noise, but because of its complexity and expense, this is not yet commercially feasible.

IMPACT ON INDUSTRY

Acoustics is the focus of research at numerous governmental agencies and academic institutions, as well as some private industries. Acoustics also plays an important role in many industries, often as part of product design (hearing aids and musical instruments) or as an element in a service (noise control consulting).

Government Research. Acoustics is studied in many government laboratories in the United States, including the U.S. Naval Research Laboratory (NRL), the Air Force Research Laboratory (AFRL), the Los Alamos National Laboratory, and the Lawrence Livermore National Laboratory. Research at the NRL and the AFRL is primarily in the applied acoustics area, and Los Alamos and Lawrence Livermore are oriented toward physical acoustics. The NRL emphasizes fundamental multidisciplinary research focused on creating and applying new materials and technologies to maritime applications. In particular, the applied acoustics division, using ongoing basic scientific research, develops improved signal processing systems for detecting and tracking underwater targets. The AFRL is heavily invested in research on auditory localization (spatial hearing), virtual auditory display technologies, and speech intelligibility in noisy environments. The effects of high-intensity noise on humans, as well as methods of attenuation, constitute a significant area of investigation at this facility. Another important area of research is the problem of providing intelligible voice communication in extremely noisy situations, such as those encountered by military or emergency personnel using low data rate narrowband radios, which compromise signal quality.

Academic Research. Research in acoustics is conducted at many colleges and universities in the United States, usually through physics or engineering departments, but, in the case of physiological and psychological acoustics, in groups that draw from multiple departments, including psychology, neurology, and linguistics. The Speech Research

Laboratory at Indiana University investigates speech perception and processing through a broad interdisciplinary research program. The Speech Research Lab, a collaboration between the University of Delaware and the A. I. duPont Hospital for Children, creates speech synthesizers for the vocally impaired. A human speaker records a data bank of words and phrases that can be concatenated on demand to produce natural-sounding speech.

Academic research in acoustics is also being conducted in laboratories in Europe and other parts of the world. The Laboratoire d'Acoustique at the Université de Maine in Le Mans, France, specializes in research in vibration in materials, transducers, and musical instruments. The Andreyev Acoustics Institute of the Russian Acoustical Society brings together researchers from Russian universities, agencies, and businesses to conduct fundamental and applied research in ocean acoustics, ultrasonics, signal processing, noise and vibration, electroacoustics, and bioacoustics. The Speech and Acoustics Laboratory at the Nara Institute of Science and Technology in Nara, Japan, studies diverse aspects of human-machine communication through speech-oriented multimodal interaction. The Acoustics Research Centre, part of the National Institute of Creative Arts and Industries in New Zealand, is concerned with the impact of noise on humans. A section of this group, Acoustic Testing Service, provides commercial testing of building materials for their noise attenuation properties.

Industry and Business. Many businesses (such as the manufacturers of hearing aids, ultrasound medical devices, and musical instruments) use acoustics in their products or services and therefore employ experts in acoustics. Businesses also are involved in many aspects of acoustic research, particularly controlling noise and facilitating communication. Raytheon BBN technologies (Cambridge, Massachusetts) has developed low data rate Noise Robust Vocoders (electronic speech synthesizers) that generate comprehensible speech at data rates considerably below other state-of-the-art devices. Acoustic Research Laboratories in Sydney, Australia, designs and manufactures specialized equipment for measuring environmental noise and vibration, in addition to providing contract research and development services.

CAREERS AND COURSE WORK

Career opportunities occur in academia (teaching and research), industry, and national laboratories. Academic positions dedicated to acoustics are few, as are the numbers of qualified applicants. Most graduates of acoustics programs find employment in research-based industries in which acoustical aspects of products are important, and others work for government laboratories.

Although the subfields of acoustics are integrated into multiple disciplines, most aspects of acoustics can be learned by obtaining a broad background in a scientific or technological field, such as physics, engineering, meteorology, geology, or oceanography. Physics probably provides the best training for almost any area of acoustics. An electrical engineering major is useful for signal processing and synthetic speech research, and a mechanical engineering background is requisite for comprehending vibration. Training in biology is expedient for physiological acoustic research, and psychology course work provides essential background for psychological acoustics. Architects often employ acoustical consultants to advise on the proper acoustical design of concert halls, auditoriums, or conference rooms. Acoustical consultants also assist with noise reduction problems and help design soundproofing structures for rooms. Although background in architecture is not a prerequisite for becoming this type of acoustical consultant, engineering or physics is.

Acoustics is not a university major; therefore, specialized knowledge is best acquired at the graduate level. Many electrical engineering departments have at least one undergraduate course in acoustics, but most physics departments do not. Nevertheless, a firm foundation in classical mechanics (through physics programs) or a mechanical engineering vibration course will provide, along with numerous courses in mathematics, sufficient underpinning for successful graduate study in acoustics.

SOCIAL CONTEXT AND FUTURE PROSPECTS

Acoustics affects virtually every aspect of modern life; its contributions to societal needs are incalculable. Ultrasonic waves clean objects, are routinely employed to probe matter, and are used in medical diagnosis. Cochlear implants restore people's ability to hear, and active noise control helps provide quieter listening environments. New concert halls are routinely designed with excellent acoustical properties, and vastly improved or entirely new musical instruments have made their debut. Infrasound from earthquakes is used to study the composition of Earth's mantle, and sonar is essential to locate submarines and aquatic life. Sound waves are used to explore the effects of structural vibrations. Automatic speech recognition devices and hearing aid technology are constantly improving.

Many societal problems related to acoustics remain to be tackled. The technological advances that made modern life possible have also resulted in more people with hearing loss. Environmental noise is ubiquitous and increasing despite efforts to design quieter machinery and pains taken to contain unwanted sound or to isolate it from people. Also, although medical technology has been able to help many hearing- and speech-impaired people, other individuals still lack appropriate treatments. For example, although voice generators exist, there is considerable room for improvement.

George R. Plitnik, B.A., B.S., M.A., Ph.D.

FURTHER READING

Bass, Henry E., and William J. Cavanaugh, eds. *ASA at Seventy-five*. Melville, N.Y.: Acoustical Society of America, 2004. An overview of the history, progress, and future possibilities for each of the fifteen major subdivisions of acoustics as defined by the Acoustical Society of America.

Beyer, Robert T. *Sounds of Our Times: Two Hundred Years of Acoustics*. New York: Springer-Verlag, 1999. A history of the development of all areas of acoustics. Organized into chapters covering twenty-five to fifty years. Virtually all subfields of acoustics are covered.

Crocker, Malcolm J., ed. *The Encyclopedia of Acoustics.* 4 vols. New York: Wiley, 1997. A comprehensive work detailing virtually all aspects of acoustics.

Everest, F. Alton, and Ken C. Pohlmann. *Master Handbook of Acoustics*. 5th ed., New York: McGraw-Hill, 2009. A revision of a classic reference work designed for those who desire accurate information on a level accessible to the layperson with limited technical ability.

Rossing, Thomas, and Neville Fletcher. *Principles of Vibration and Sound*. 2d ed. New York: Springer-Verlag, 2004. A basic introduction to the physics of sound and vibration.

Rumsey, Francis, and Tim McCormick. *Sound and Recording: An Introduction*. 5th ed. Boston: Elsevier/Focal Press, 2004. Presents basic information on the principles of sound, sound perception, and audio technology and systems.

Strong, William J., and George R. Plitnik. *Music, Speech, Audio*. 3d ed. Provo, Utah: Brigham Young University Academic Publishing, 2007. A comprehensive text, written for the layperson, which covers vibration, the ear and hearing, noise, architectural acoustics, speech, musical instruments, and sound recording and reproduction.

Swift, Gregory. "Thermoacoustic Engines and Refrigerators." *Physics Today* (July, 1995): 22-28. Explains how sound waves may be used to create more efficient refrigerators with no moving parts.

WEB SITES

Acoustical Society of America
http://asa.aip.org

Institute of Noise Control Engineering
http://www.inceusa.org

International Commission for Acoustics
http://www.icacommission.org

National Council of Acoustical Consultants
http://www.ncac.com

See also: Speech Therapy and Phoniatrics; Ultrasonic Imaging.

AGRICULTURAL SCIENCE

FIELDS OF STUDY

Animal breeding and husbandry; plant breeding and propagation; agroforestry; agronomy; horticulture; soil science.

SUMMARY

Agriculture, the practice of producing foods, fibers, and other useful products from domesticated plants and animals, has developed to its present state through the application of scientific discoveries that have been made throughout history. Agricultural science (the study of agriculture) gathers and analyzes information from basic research conducted at federal, state, and private facilities and applies it to agricultural settings. Agriculture encompasses a wide spectrum of applications; therefore, agricultural science is extremely diverse, covering almost every aspect of plant and animal science as well as the humans involved in the process.

KEY TERMS AND CONCEPTS

- **Agribusiness:** Corporation that controls the entire agricultural process from production to marketing.
- **Agricultural Chemical:** Chemical such as a fertilizer or pesticide that is used to enhance agricultural production.
- **Agricultural Scientist:** Person who studies some aspect of agriculture to understand and improve it.
- **Agriculturist:** Person such as a farmer or rancher who is directly involved in agricultural production.
- **Animal Husbandry:** Practice of breeding or caring for farm animals.
- **Biotechnology:** Manipulation of the genetic material of plants and animals in order to change the expressed characteristics.
- **Monoculture:** Practice of devoting large acreages to the production of only one crop.
- **Sustainable Agriculture:** Agriculture that is diversified, ecologically sound, economically viable, socially just, and culturally appropriate.

DEFINITION AND BASIC PRINCIPLES

Agricultural science is the study of agriculture (also called farming), the practice of producing foods, fibers, and other useful products from domesticated plants and animals. The production of food is one of the oldest professions in the history of humankind, and even in the twenty-first century, agriculture remains the most common occupation worldwide. The term "agriculture" covers a wide spectrum of practices, beginning with sustenance agriculture, in which a farmer produces a variety of crops sufficient to feed the farmer and his or her family as well as a small excess, which can be sold for cash or bartered for other goods or services. At the other end of the scale, commercialized industrial agriculture involves many acres of land and large numbers of livestock; a considerable input of resources such as fuel, pesticides, and fertilizers; and a high level of mechanization to maximize profit from the enterprise.

Agricultural science uses information gained from basic research and applies it to agricultural settings to increase the yield and quality of agricultural products. The overall goal is to produce sufficient food and fibers to feed and clothe the population and increase the profit margin for those involved in the production process. Agricultural science covers almost every aspect of plant and animal science, and the application of scientifically sound agricultural principles can enhance production regardless of the level of agriculture being practiced.

BACKGROUND AND HISTORY

The beginnings of agriculture predate recorded history. No one knows when the first crop was cultivated, but beginning around 8500 b.c.e., humans began the gradual transition from a hunter-gatherer to an agricultural society. By 4000 b.c.e., agriculture had been firmly established in Asia, India, Mesopotamia, Egypt, Mexico, Central America, and South America. The transition to agriculture occurred because humans discovered that seeds from certain wild grasses could be collected and planted on land that could be controlled, and that the resulting crop could be harvested for food. This process represents the earliest stage of agricultural science. Through continued observation and selection of preferred plant and animal characteristics, agriculture continued to develop, and by the start of the Bronze Age (around 3000 b.c.e. in the Middle East), humankind had become fully dependent on domestic crops.

Agriculture developed slowly, and even in the nineteenth century, most agricultural enterprises practiced sustenance agriculture. The Industrial Revolution changed the agriculture industry with the invention and production of agricultural machinery such as the cotton gin (by Eli Whitney in 1793), mechanical reaper (by Cyrus Hall McCormick in 1833) and steel plow (by John Deere in 1837). Mechanization led to the development of commercial agriculture and an increase in interest in all areas of agricultural science.

In the twentieth century, the combined action of many agricultural scientists in the Green Revolution led to the development of high-yielding varieties of numerous crops. As a result of numerous other developments in agricultural science, a modern agricultural unit requires relatively few employees, is highly mechanized, devotes large amounts of land to the production of only one crop, and is highly reliant on agricultural chemicals.

HOW IT WORKS

Research. Agricultural science can be divided into two broad categories, research and extension. Agricultural scientists who engage in research are involved in designing experiments, gathering and analyzing data, drawing conclusions, and publishing their results. In the United States, formal training in agricultural science can be traced back to 1862, with the establishment of the Department of Agriculture and the passage of the Morrill Act, which provided the means for the establishment of land-grant colleges in each state. One of the major roles of the land-grant colleges was to provide training in agriculture. The Hatch Act of 1887 provided the means to fund the establishment of agricultural research stations in each state. In most states, the agricultural research station is connected with a land-grant college or university, and the major function of the research station is to conduct agricultural research.

The research being conducted at land-grant colleges and universities and state and federal agricultural research stations is as varied as agriculture itself. Some of the disciplines involved in the study

of agriculture include agronomy and horticulture, weed science, forestry, animal and poultry science, dairy and food science, agricultural engineering, soil chemistry and physics, rural sociology, agricultural economics, and plant and animal physiology, pathology, and genetics.

Extension. Basic research usually is conducted at land-grant institutions, and applied research typically takes place at agricultural research stations. Basic research is designed to study some fundamental principle and generally leads to a better understanding of how nature works. Applied research usually involves taking information discovered in basic research and using it in real-world situations. In agricultural science, this means applying the knowledge in an agricultural setting. When applied research shows that a particular technique or development is an improvement over the existing practice (for example, resulting in higher yield or quality), the information is passed on to those who can actually use it. This is most often done through pamphlets and other materials published by the state's Cooperative Extension Service. The Cooperative Extension Service was established by the Smith-Lever Act of 1914. In most states, the Cooperative Extension Service is a component of the agricultural research station, and its primary function is to transfer technology from those who discover it (agricultural scientists) to those who produce agricultural commodities. In many areas, agricultural extension agents will visit individually with the producers to ensure that the technology is transferred.

APPLICATIONS AND PRODUCTS

Food and Fiber Production. In a sense, humans have been applying some aspects of agricultural science for more than 10,000 years. Even though they were not aware of it, people were acting in accordance with basic scientific principles every time they selected seeds for planting and animals for breeding in hopes of reproducing a desired trait. In the modern world, essentially no food, fiber, or other agricultural product has not been affected by the application of agricultural science. Many of these applications resulted in higher yields. For example, in the later part of the nineteenth century, soil scientists learned that soil erosion could be dramatically reduced by planting windbreaks and plowing perpendicular to the slope rather than in the direction of the slope.

This helped conserve topsoil and increased yields. Another practice that has increased yields has been the use of fertilizers. In the nineteenth century, plant scientists discovered which nutrients are required by plants, and agriculturalists used this knowledge to return fertility to depleted soil and increase yields.

The use of agricultural machinery led to tremendous increases in crop yields. To use their machinery more efficiently, agriculturalists turned to monoculture. Although this practice raised yields, it also made crops more susceptible to damage by pests. Scientists have developed many different agricultural chemicals to combat these pests and maintain high yields.

Animal husbandry has also benefitted from the efforts of agricultural scientists. In the poultry industry, scientists discovered that by controlling the temperature and lighting in the poultry house, the number of consecutive days on which hens lay eggs can be increased from around fifty to more than three hundred. This has resulted in tremendous yields in egg production.

Discoveries in genetics have also contributed significantly to increases in agricultural yields. Plant breeders have been able to use knowledge of genetics to produce numerous varieties of high-yielding crops, and animal breeders have produced breeds that grow much larger than the parental stock. In combination with genetics, biotechnology is being used to enhance the quality of many agricultural products. For example, genetic engineering has produced a variety of tomatoes that maintains its flavor during storage, thereby increasing its shelf life. In the future, the quality of many other agricultural products is likely to be improved through genetic engineering.

Agricultural Products. The products of agricultural science are the foods, fibers, and other commodities that are produced by agricultural endeavors around the world and that affect the lives of everyone on a daily basis. Agricultural products are generally divided into those that come from animals and those that are derived from plants. Each of these products exists in its present state as a result of the application of knowledge gained from agricultural science.

The major animal-derived products can be divided into edible and nonedible red meat products, milk and milk products, poultry and egg products, and wool and mohair. Edible red meat products primarily come from cattle, swine, sheep, goats, and animals such as horses and Asian or African buffalo. The major nonedible red meat products include rendered fat, which is used to make soap and formula animal feeds; bone meal, which is used in fertilizer and animal feeds; manures used as fertilizers; and hides and skins, which are tanned and used to make leather products. Milk and milk products (also known as dairy products) are produced primarily by dairy cattle and include whole milk, evaporated and condensed milk, cultured milk products, cream products, butter, cheese, and ice cream. Poultry (chickens, turkeys, ducks, geese, pigeons, and guinea hens) and egg products are nutritious, relatively inexpensive, and used by humans throughout the world. The hair covering the skin of some farm animals (wool and mohair) is also considered an agricultural product.

Plant-derived agricultural products are very diverse. Timber products include those materials derived from the trees of renewable forests. Grain crops such as corn, rice, and wheat are grasses that produce edible seed. Cotton, flax, and hemp are the principal fiber plants grown in the United States, although less important crops such as ramie, jute, and sisal are also grown. Fruit crops refer to the fruit from a variety of perennial plants that are harvested for their refreshing flavors and nourishment. Nut crops such as pecan, walnut, and almond refer to those woody plants that produce seed with firm shells and an inner kernel. Vegetable crops are extremely diverse and range from starchy calorie sources (such as potatoes) to foods that supply mainly vitamins and minerals (such as lettuce). The world's three most popular nonalcoholic beverages are derived from coffee, tea, and cocoa plants. Spice crops are plants grown for their strong aroma or flavor, and drug crops are plants that have a medicinal property. Ornamental crops are grown for aesthetic purposes and are divided into florist crops (flower and foliage plants) and landscape crops (nursery plants). Forage crops include clover, alfalfa, and other small grain grasses that are grown to feed livestock and a variety of straw crops for haymaking. Specialty crops, grown for their high cash value, include sugarcane, tobacco, artichokes, and rubber.

IMPACT ON INDUSTRY

Throughout the world, agricultural scientists can be found working for governments, colleges and universities, international organizations, and industry.

Government Regulation and Research. Most developed countries have an agency responsible for overseeing the nation's food and fiber production. Often this agency is a federally funded agricultural agency similar to the United States Department of Agriculture (USDA) that is responsible for conducting agricultural research and ensuring that the latest developments are made available to the agricultural industry. Agencies such as the USDA generally have research facilities located in various parts of the country. These facilities may stand alone or be located on a university campus.

In most developed countries, each state or province will have a state agency that oversees regional agricultural research stations and extension specialists. The type of agriculture practiced in a given area is influenced by a number of factors, including water availability, soil, and climate; therefore, these facilities are usually in different geographical areas, and the agricultural scientists who are employed at these facilities generally focus their research on the agricultural commodities produced in that area. In the United States, for example, a research station in the dairy region of Wisconsin may conduct research related to the dairy industry; in the corn belt of Iowa, the focus may be on corn research; and cotton research may receive the major research emphasis in the high plains of Texas.

Academic Research. Throughout the world, a significant amount of agricultural research takes place at colleges and universities. Much of this research is conducted in institutions such as the land-grant universities, which were partially founded on the premise that training in agricultural science would be provided; however, research projects related to agriculture can also be found at other educational institutions. Scientists at these institutions often conduct basic research, but discoveries made at this level often lead to practical developments in agricultural science. Regardless of the institution, research is expensive and in most cases is supported by grant funds primarily provided by government agencies and to a lesser extent by private industry.

International Organizations. In underdeveloped countries, agricultural development through the application of scientific agricultural practices is often accomplished through the actions of international organizations. One such organization, the International Fund for Agricultural Development, was established in 1977 as a specialized agency of the United Nations. Its primary function is to provide financial support for agricultural development projects, particularly those involving food production in developing countries. Although the agency may fund only agricultural development, other groups such as the Consultative Group on International Agricultural Research may actually conduct agricultural research. This group's mission is to use scientific research and research-related activities in the fields of agriculture, forestry, fisheries, and the environment to create sustainable food sources and reduce poverty in developing countries. Through the activities of organizations such as these, agricultural science is being applied to help reduce poverty and hunger and improve human health and nutrition throughout the world.

Industry. Agricultural science plays a major role in the financial health of many businesses, ranging from small feed, seed, and fertilizer retail outlets to major corporations. At the same time, industry has contributed to the advancement of agricultural science. Sometimes this contribution is in the form of grants to finance agricultural research by federal or state scientists or university faculty, in the hope that the research will be useful in developing an agricultural product. However, many companies employ agricultural scientists to conduct research specifically for the company.

Agricultural chemicals and machinery are two of the areas in which industry has played a major role. Because of the low availability and high cost of labor, mechanization is a must; therefore, agricultural engineers are constantly developing new cost- and time-saving machinery. Two prominent companies producing agricultural equipment are Case IH and Deere & Company.

Modern agricultural is highly reliant on agricultural chemicals such as fertilizers and pesticides, and large-scale agriculture would be impossible without the use of these chemicals; therefore, agricultural chemical companies are constantly conducting research on new, environmentally safe, and effective chemicals. Leaders in the production of agricultural chemicals include Monsanto and Bayer CropScience.

CAREERS AND COURSE WORK

Careers in agricultural science are as varied as the field of agriculture itself. A large number of agricultural scientists work in basic or applied research and development for federal or state governmental agencies, colleges and universities, or private industry. Some people trained in agricultural science advance to administrative positions, where they manage research and development programs or marketing or production operations in companies that produce food products or agricultural chemicals, supplies, and machinery. Other agricultural scientists work as consultants to agricultural enterprises. The actual nature of the work being performed by agricultural scientists can be broadly divided into food science and technology, plant science, soil science, and animal science. Of the roughly 31,000 people employed in agricultural science in the United States in 2008, about 43 percent were employed as food scientists or technologists, 45 percent were working as plant and soil scientists, and 12 percent were animal scientists. These numbers do not include those college and university faculty members who were trained in one of the areas of agricultural science. As of the 2010's, employment opportunities in agricultural science were expected to be higher than the average for all occupations.

The course work required for majors in agricultural science varies considerably. The specific curriculum depends on the area of agricultural science in which the student is interested, but a strong math and science background is generally needed. Most entry-level positions in the farming or food-processing industry require a bachelor's degree; however, to acquire a research position in most universities, state or federal agencies, or private industry, a master's or doctoral degree in agricultural science, in an engineering specialty, or in a related science such as biology, chemistry, or physics is usually mandatory.

SOCIAL CONTEXT AND FUTURE PROSPECTS

Agriculture has been and will continue to be extremely important to the development of human culture. Modern civilization would not have developed without agriculture, and agriculture has been successful primarily because of myriad discoveries made in various disciplines associated with agricultural science. The continued advancement of civilization depends on the ability to produce sufficient food and fiber to feed and clothe the world's populations; therefore, agricultural science will continue to be of paramount importance in the future as

---

Fascinating Facts About Agricultural Science

- In 1810, 90 percent of the people in the United States lived on farms. In 1860, only about 60 percent of American people were involved in agriculture, and by 1972, only 4.6 percent of the U.S. population was involved in farming. In 2010, less than 2 percent of Americans were actively engaged in agricultural production.

- In 1800, a single American farmer produced enough food to feed five people; in 2010, each farmer produced enough food to sustain ninety-seven people.

- In the United States, more than $9 billion per year is awarded in grants to fund research in agricultural science.

- As a result of research in agricultural science, modern American farmers can produce 76 percent more crops than their parents did on the same amount of land.

- American supermarkets offer 6,000 to 8,000 agricultural products, more than half of which were not available in 1980.

- More than 40 percent of the world's laborers are employed in some aspect of food production. This means that more than 2 billion people depend on agricultural science for their livelihood.

---

scientists try to determine the best methods to produce enough food to feed an ever-growing number of people.

Biotechnology is likely to have a tremendous impact on agricultural science. Agricultural scientists have begun to use biotechnology in an attempt to improve a wide variety of characteristics in various plants and animals. Some genetically modified plants are already commercially available. As genetically modified foods are created, scientists must deal with the social issues and scientific questions that they create.

Although agricultural science has increased the yields of agricultural products, some of these gains have come at a cost to the environment. Agricultural scientists are studying ways to practice sustainable agriculture and to raise crop yields and quality without causing damage to the environment. Environmental concerns have also sparked interest in the production of biofuels, fuels manufactured from agricultural derivatives. This raises the question of how much agricultural land should be devoted to food production and how much to growing crops as energy sources.

Dalton R. Gossett, Ph.D.

FURTHER READING

Brown, Robert C. *Biorenewable Resources: Engineering New Products from Agriculture.* Hoboken, N.J.: Wiley-Blackwell, 2003. Describes the interface between agricultural science and process engineering.

Conkin, Paul K. *A Revolution Down on the Farm: The Transformation of American Agriculture Since 1929.* Lexington: University Press of Kentucky, 2009. Discusses changes in agriculture during the twentieth century and presents a concise introduction to agriculture in the United States.

Gardner, Bruce L. *American Agriculture in the Twentieth Century: How It Flourished and What It Cost.* Cambridge, Mass.: Harvard University Press, 2006. Describes how both mechanical and biotechnical inventions and innovations have contributed to increases in productivity in American agriculture.

Hurt, R. Douglas. *American Agriculture: A Brief History.* Rev. ed. West Lafayette, Ind.: Purdue University Press, 2002. An introductory work that describes the beginning of agriculture in the United States.

Janick, Jules. *Horticulture Science.* 4th ed. San Francisco: W. H. Freeman, 1986. Describes vegetable crop production and contains sections on the major horticultural crops.

Metcalfe, Darrel S., and D. M. Elkins. *Crop Production: Principles and Practices.* 4th ed. New York: Macmillan, 1980. One of the most valuable sources available on the practical aspects of crop production.

Rasmussen, R. Kent, ed. *Agriculture in History.* Pasadena, Calif.: Salem Press, 2010. A collection of essays on events in agricultural history, arranged in chronological order to demonstrate trends.

Smith, Bruce D. *Emergence of Agriculture.* San Francisco: W. H. Freeman, 1999. Examines the development of agriculture as it occurred in the Middle East, Europe, China, Africa, and the Americas.

Taylor Robert E., and Thomas G. Field. *Scientific Farm Animal Production.* 9th ed. New York: Prentice Hall, 2007. An excellent overview of animal production and animal products.

WEB SITES

Consultative Group on International Agricultural Research
http://www.cgiar.org

International Fund for Agricultural Development
http://www.ifad.org

U.S. Department of Agriculture
http://www.usda.gov

See also: Agroforestry; Animal Breeding and Husbandry; Bioengineering; Genetic Engineering; Soil Science.

AGROFORESTRY

FIELDS OF STUDY

Forestry; silviculture; agriculture; botany; soil science; horticulture; animal husbandry; hydrology; sustainability.

SUMMARY

Agroforestry is an interdisciplinary scientific field focused on combining and applying forestry and agricultural principles and techniques to create a sustainable land-use system. Agroforestry provides a wide range of ecological benefits, including improvements to water quality and wildlife habitat and reductions in soil erosion. By combining trees and shrubs with crops and livestock, agroforestry systems attempt to optimize the benefits of short-term crop and livestock rotations and long-term forest rotations. Blending of production practices, which has occurred for centuries, allows a landowner to reap the economic benefits of annual crops while waiting for forest products. In addition to economic benefits, agroforestry results in ecological benefits that may not occur in a traditional agricultural system, including the formation of windbreaks, wildlife corridors, and riparian buffers.

KEY TERMS AND CONCEPTS

- **Agriculture:** Practice of establishing, growing, managing, and harvesting crops and livestock.
- **Agronomy:** Breeding and raising of crops.
- **Carrying Capacity:** Number of individuals of a plant or animal species that a given piece of land can support.
- **Competition:** In plants, the interaction between two individuals in the attempt to use the same resource, resulting in negative impacts on both.
- **Husbandry:** Breeding and raising of livestock.
- **Monocultures:** Single crops grown on large farms, typical of industrialized agriculture since the early twentieth century.
- **Mycorrhiza:** Mutualistic relationship between a fungus and a plant (tree or crop) wherein the plant gains access to limited soil nutrients from the fungus and the fungus gains energy from the plant.
- **Silviculture:** Practice of establishing, growing, managing, and harvesting trees in a forest.
- **Taungya:** Burmese word for an agroforestry technique of growing crops alongside trees during tree-plantation establishment.

DEFINITION AND BASIC PRINCIPLES

In many situations, land is managed using a single-approach (or monoculture) system, such as a farm devoted solely to raising crops or livestock or a forest that is cultivated for pulp or lumber production. These separate approaches can be managed successfully to produce a sustainable product. However, through the application of key forestry and agricultural techniques, agroforestry practitioners attempt to develop a sustainable land-use practice producing both short-term and long-term benefits. Although many combinations of forest- and crop-livestock-management techniques exist, agroforestry typically takes the forms categorized as tree-crop, tree-animal, shelterbelt, riparian forest, and natural-specialty crop systems.

Agroforestry provides the framework for a systems approach to land management, allowing for understanding how the physical structure, ecological influences, and economic outputs of trees and forests are interconnected with those same characteristics in crops and livestock. In many climatic regions, from tropical to semiarid zones, constant cropping can place a drain on soil nutrients eventually creating infertile land. By incorporating fallow periods in the crop rotation, farmers can replenish these soil nutrients. However, during the fallow period no crops are produced and, depending on local policies and food-production demand, the fallow period may be excessively shortened and no longer serve its purpose. The goal of agroforestry is to incorporate deeper-rooted perennials (trees and shrubs) in order to maintain nutrient cycling and soil fertility so agriculture can continue for longer periods of time.

BACKGROUND AND HISTORY

Agroforestry as a technique for land management has been used for centuries, but it was not until the mid-1970's that the term was coined and defined. Early adoptions of what would later be defined as agroforestry probably incorporated the production of crops, animals, and building materials. Although there was a subsequent shift to farming and forestry with a focus on monocultures, agroforestry continued to exist. In later decades, agroforestry has become an important approach to land management in developing regions of the world. Focusing on sustainable land-management practices that allow for increases in economic and ecological diversity has facilitated survival and development of regions where heavy forestry or heavy agriculture could have destroyed resources. Because of the large economic and ecological impact on local communities, government and nongovernmental organizations have expanded research since the 1990's to refine agroforestry practices and disseminate information to the public.

HOW IT WORKS

Practitioners cannot assume that all forms of agroforestry will succeed in all environments. The surrounding ecological and economic environment, previous land-use practices, and goals of the manager, owner, and community all play into the success of any one of the numerous and overlapping agroforestry strategies. Although the techniques used may be structurally different, the outputs are the same: Establish an ecologically and economically sustainable system that provides short- and long-term products in the forms of agricultural crops and livestock, as well as tree and non-tree forestry products. The five common agroforestry systems (tree-crop, tree-animal, shelterbelt, riparian forests, and natural-specialty crops) are broad, and there is overlap between each one. The application of a specific system is typically related to the ecological and economic environment surrounding an individual owner and piece of land, which includes the selection of particular crop and tree species. Because of the broad nature of these systems, the individual goals defined by the practitioner may separate or merge the agroforestry practice being conducted, resulting in several agroforestry systems existing on a single piece of land.

Adding Trees to Agriculture. Two important decisions a practitioner must make regarding the incorporation of trees into an agricultural operation concern are spacing and timing of tree growth. First, trees can be grown in zones or scattered across the land. Zones can be in lines, plots, blocks, or any other systematic arrangement on the land. Spacing

between trees is a powerful management tool, in both zoned and scattered arrangements, because it determines the level and intensity of competition and facilitates management. If the trees are growing too close together, growth may be reduced as a result of competition for moisture and nutrients, as well as issues related to crop planting, fertilizing, and harvesting by machinery.

Second, timing can influence competition between trees and crops, as well as labor costs for planting and harvesting. If trees and crops are grown sequentially, direct competition between the two groups of plants can be greatly reduced. However, there may be increases to required labor and associated costs by staggering the land-management techniques. If trees and crops are grown simultaneously, then direct competition between the two groups of plants may be increased, but labor costs may be reduced because of the combined management.

Adding Crops and Livestock to Silviculture. As with adding trees to establish agriculture, the same issues of spacing and timing exist when adding crops to a forest. With application of agroforestry techniques to land under more forest-focused management, the overall goals shift from long-term forest management to include short-term crop rotations. One different strategy of adding crops to silviculture is called *taungya*, which is a Burmese term for growing crops during the first few years of tree-plantation establishment. The goal of taungya is to gain short-term crop benefits during the early stages of tree establishment. After one to three years, the crops are no longer sown and the focus is turned to the management of the trees. Whether cropping ends after a few years or is continued indefinitely, there is a need to space trees effectively so that the future plantation can successfully reach the long-term goals defined at the outset.

APPLICATIONS AND PRODUCTS

Tree-Crop Systems. Planting trees in rows, or any spacing configuration that allows for sowing of crops between the trees, is known as intercropping, alleycropping, and agrosilviculture. Often, the spacing is arranged to allow modern agricultural machinery use and limit the need for physical labor. Two main approaches exist in tree-crop systems, which include planting trees on cropped land and planting crops on forested land. The former typically results in planting straight rows of trees while the latter may result in establishing tree plantations or utilizing natural forests, which may add difficulty in properly spacing trees. In either approach, there is a need for practitioners to make management decisions related to the pattern of trees on the land (zoned or scattered), spacing (between zones, between individual trees), and diversity of tree species. Crops may be temporarily sown during the establishment phase of a forest, which can have an added complication of multiple owners between trees and crops. Tree-crop systems are common in both temperate and tropical zones and have been used for fruit, nut, olive, and grape production for centuries in Europe and North and South America.

Tree-Animal Systems. Silvopasture is the management of livestock, their forage, and trees. If a pasture has trees, but those trees are not part of the active land management, then it is not agroforestry. The application of tree-animal agroforestry may occur as an agriculture-dominated, integrated, or forestry-dominated approach. In an agriculture-dominated approach, the focus is placed on maximizing the livestock and food-product outputs. This approach may occur when trees are planted in a pasture to provide shelter for livestock or when livestock are allowed to graze in an orchard. With an integrated approach, the focus is balanced between the management of livestock outputs and tree-product outputs. In temperate areas, such as the southeastern United States, this approach is used for the management of pine plantations with grazing for cattle. Forestry-dominated approaches place management emphasis on tree outputs and utilizing livestock to maximize those tree products. In tropical areas where tree growth can be rapid, introduction of livestock may be beneficial to the trees. However, in temperate areas, tree growth may be slower. If livestock are introduced too early, there may be grazing and damage to the trees, reducing survival and success of the tree-animal system.

Shelterbelt Systems. When planted upwind from croplands, windbreaks provide linear shelter that reduces soil erosion. These windbreaks may include mixtures of trees and shrubs. A single row of trees does not provide a solid wall of protection, so it is necessary for multiple rows to be planted in order to create a stratified, continuous canopy to block wind. To reduce excess runoff and soil erosion related to water flow, contour hedgerows are planted along contours of moderate to steep slopes and can slow water movement. Contour hedgerows can be a less costly and less time-consuming strategy for reducing water erosion than the construction of terraces along a slope. Depending on strategies and goals for the trees, shelterbelts can function merely as filter strips, reducing the volume of soil erosion, or they can function as alley crops, providing mulch or tree crops mixed in with the annual row crops.

Riparian Forest Systems. Riparian forests are stands of trees adjacent to a linear, flowing body of water (a river or stream). The soils of these forests are either continuously saturated or repeatedly inundated with water. Hydrology and soil types play very important roles in determining which trees can survive and be productive. Riparian forest systems can benefit more than just the immediate community of landowners, as compared with tree-crop or tree-animal systems. Forested zones adjacent to rivers and streams can maintain water quality by filtering agricultural chemicals in runoff through infiltration and immobilization by plants and soil microbes; stabilize banks and reduce erosion by creating a soil-retaining structure with roots and coarse woody debris, thereby improving water infiltration, which reduces the quantity of runoff and the erosive energy; and reduce average and fluctuations of water temperature, maintaining fish and other aquatic-organism populations. Riparian forest systems benefit both the land adjacent to the river and adjacent land downstream.

Natural-Specialty Crop Systems. Forest farming takes advantage of the shade produced by a forest canopy to produce special, high-value non-timber forest products. In temperate agroforestry, these products include fruits, nuts, and mushrooms as food crops; ginseng, catnip, and echinacea as herbal medicines; and ferns and flowers for decorative and ornamental uses. Often, these specialty crops are added as products to forests where high-value timber is being produced in order to diversify the economic outputs.

IMPACT ON INDUSTRY

Incorporation of agroforestry techniques into an agricultural or silvicultural market can have significant impacts on the market value of products, as well as the economic development of a village. With external support from both governmental and nongovernmental organizations, such growth and development can improve standards of living; increase and improve infrastructure related to product storage, transportation, and market stability; and improve food quality and quantity. Although the agricultural industry being affected by agroforestry initially is localized to a village or community, expansion of agroforestry techniques into a region, state, or nation, can impact the industry at those scales, greatly improving the economic and ecological aspects.

Government and University Research. Research conducted by government agencies, universities, and nongovernmental organizations is essential to the improvement of agroforestry strategies. These studies provide a basis for understanding competition levels, improvements to growth and production, and ecological and economic benefits. Since not every agroforestry practice is effective in all environments, it is important to understand what components of a particular practice will work and what components fail within an environment. Modifying strategies may mean the difference between marginal or no gains in crops or tree products, and complete success of a farm incorporating agroforestry. Questions related to the operational side of agroforestry (species selection and combinations, spacing, timing), as well as the social side (market fluctuations, lifestyle improvements, economic development), are addressed by research supported by different organizations, both governmental and nongovernmental.

Industry and Business Sectors. The commercial aspect of agroforestry is most typically related to local and regional economic growth and development, with the exception of some extremely high-value specialty crops. The addition of a crop to a tree system, or vice versa, can be quite lucrative for local farmers. However, a lack of knowledge or inadequate infrastructure may pose obstacles to getting those products to a suitable market. To overcome barriers related to market knowledge, timing of sales, production processing, or product transport, local farmers may need to form cooperative associations. In some cases, this effort requires external support in the form of micro loans to individual farms from banks or government agencies in order to meet production needs while products are warehoused until markets improve.

CAREERS AND COURSE WORK

Because of the combination of techniques and principles used in agroforestry, practitioners benefit from a strong educational background in both agriculture and silviculture. Specialized degree programs in agroforestry, both undergraduate and graduate, are more likely found at universities in tropical regions where agroforestry is more commonly practiced. Courses in plant and animal biology, soil and water resources, crop science, and forestry are necessary to understand the application of agroforestry techniques. In addition to the actual production side of agroforestry, a social aspect related to the communities surrounds the practice, so courses in social and economic issues related to natural resources and rural areas are beneficial.

Careers in agroforestry will most likely not have the job title of "agroforester." Typically, a professional in the field will either be a landowner actually practicing agroforestry or an outside consultant assisting landowners in the application of techniques. The latter may be an individual working for a government agency as a local, state, or regional forester or extension specialist, or for a nongovernmental organization with a mission to assist local landowners. Within the United States Department of Agriculture there is the National Agroforestry Center and in Canada there is the Agroforestry Development Centre within the country's Agriculture and Agri-Food department. Outside governmental employment, opportunities do exist with organizations such as the World Agroforestry Centre, which is headquartered in Nairobi, Kenya. Because of the scale and scope of such an organization, as well as with national-level agencies, professionals usually work as part of interdisciplinary groups. Individuals with specialized education and experience in tree, crop, animal, or social sciences can be involved with agroforestry research and applications, adding to the growth and development of the field.

SOCIAL CONTEXT AND FUTURE PROSPECTS

Implementation of agroforestry techniques has improved the monetary, social, and ecological economies of local villages and regions in many tropical and temperate areas. However, in order to garner those benefits at the national level there is a need to scale up the techniques and practices used in agroforestry. To expand agroforestry and its benefits, national governmental agencies, as well as multinational nongovernment organizations, must take an active role in the dissemination of information and support to farmers. This has been demonstrated in several countries, including Cameroon, Indonesia, and Uganda, where increasing involvement of farmers in agroforestry programs has significantly improved the quality of products and the return on farmer investments. In each case, cooperative efforts by governmental agencies, as well as international relief organizations and other nongovernmental organizations, has expanded the application of agroforestry techniques to magnify the geographic scope of monetary, social, and ecological benefits.

Jordan M. Marshall, Ph.D

FURTHER READING

Gordon, Andrew M., and Steven M. Newman, eds. *Temperate Agroforestry Systems.* Wallingford, England: CAB International, 1997. Includes chapters on the application of agroforestry approaches in several countries in North and South America, Europe, and Asia.

Huxley, Peter A. *Tropical Agroforestry.* Hoboken, N.J.: Wiley-Blackwell, 1999. Discusses general agroforestry topics, strategies, and concerns with strong focus on tree-specific aspects.

Jose, Shibu. "Agroforestry for Ecosystem Services and Environmental Benefits: An Overview." *Agroforestry Systems* 76, no. 1 (May, 2009): 1-10. Reviews the environmental benefits of agroforestry in terms of carbon sequestration, biodiversity conservation, soil enrichment, and air and water quality, which are often more difficult to quantify compared to the economic benefits of agroforestry.

Smith, David M., et al. *The Practice of Silviculture: Applied Forest Ecology.* 9th ed. New York: John Wiley & Sons, 1997. Plantation silviculture as well as broader forestry theories are covered in easy-to-understand language.

WEB SITES

Association for Temperate Agroforestry
Agroforestry: An Integrated Land-Use Management System for Production and Farmland Conservation
http://www.aftaweb.org/resources1.php?page=36

National Association of University Forest Resources Programs
http://www.naufrp.org

Society of American Foresters
http://www.safnet.org

U.S. Forest Service
http://www.fs.fed.us

World Agroforestry Centre
http://www.worldagroforestrycentre.org

See also: Agricultural Science; Agronomy; Animal Breeding and Husbandry; Land-Use Management; Soil Science.

AGRONOMY

FIELDS OF STUDY

Biology; chemistry; earth science; biotechnology; mineralogy; ecology; field crop production; soil management; horticulture; meteorology; climatology; entomology; plant physiology; plant genetics; turf science.

SUMMARY

Agronomy is the interdisciplinary field in which plant and soil sciences are applied to the production of crops. Agronomists develop ways in which crop yields can be increased and their quality improved. Some agronomists specialize in soil management and land use, which seeks to protect existing farmland and reclaim land for future use in growing crops. Other specialties cover areas such as weed and pest management, meteorology, and the impact of climate change on crop production. The growing importance of biofuels such as ethanol has increased interest in agronomy as a scientific and professional field.

KEY TERMS AND CONCEPTS

- **Biomass:** Plant or animal matter, particularly when used as an energy source.
- **Crop:** Plant product grown for use as food, animal feed, fiber, or fuel.
- **Forage:** Crop category that includes grasses and is used primarily for feeding animals.
- **Herbicide:** Product that kills or controls weeds and other plants that reduce crop yields.
- **Input:** Product added to soil to increase or improve crop yield, such as fertilizer.
- **Irrigation:** Watering of fields to supplement rainfall.
- **Rotation:** System under which the types of crops grown in a field are changed from one season to the next to improve yields.
- **Yield:** Amount of a crop produced within a defined geographic area, such as a field, during one growing season.

DEFINITION AND BASIC PRINCIPLES

Agronomy is the study of plants grown as crops for food, animal feed, and nonfood uses such as energy. In the United States, these crops include wheat, corn, soybeans, grasses, cotton, and a wide variety of fruits and vegetables. Leading crops in other countries vary widely, depending on the nature of the local soil, geography, and growing season.

Plant science is a major component of agronomy. Many agronomists look for ways to grow stronger, hardier plants with higher yields. New types of plants are bred by agronomists to contain specific improvements, such as increased nutrient levels or resistance to pests, over earlier breeds. An area of strong interest is the development of plant types that require fewer inputs such as fertilizers and insecticides to perform well.

The field of agronomy also covers the many factors in the environment that play a role in whether a crop succeeds or fails. The chemical makeup and water balance within a crop's soil are leading factors. Weather and climate patterns, both within a single season and over many years, affect the quantity and quality of crop yields. Technology and economics influence demand for certain types of crops, which in turn pushes market prices up and down. Agronomists help producers respond to these factors.

BACKGROUND AND HISTORY

Agronomy is nearly as old as human civilization. According to archaeological findings, people have been growing plants for food for more than 10,000 years, starting in the western Asian regions of what was Mesopotamia and the Levant.

Many historians believe that plant cultivation, the earliest form of farming, led to a major change in human culture. The growing season required people to live in one place for long periods of time. Permanent settlements near fields most likely evolved into some of the first villages. These settlements were often near water sources such as rivers, which were needed to irrigate field crops. Some of the first developments in agronomy involved the design and building of water-delivery systems.

The industrial revolution brought widespread change to the field of agronomy. Steam-powered farming equipment replaced draft animals such as horses and mules. Plant scientists developed and standardized new breeds of field crops, which increased yields. By the mid-twentieth century, nearly all corn grown in the United States was from hybrid stock.

The use of inputs such as fertilizers and pesticides also increased, but in some cases caused significant environmental harm. Since the 1990's, agronomists have focused more closely on ways to improve crops without damaging local ecosystems.

HOW IT WORKS

Field crops require the right plant type and breed, healthy soil, adequate water and nutrients, appropriate growing temperatures and rainfall, and the control of disease and pests in order to succeed.

Plant Breeding and Genetics. When choosing a type of field crop to plant, farmers and growers consider factors such as the hardiness of certain breeds and their expected yields at the end of the harvest season. Buyers of agricultural products such as food-manufacturing companies look for products that are high in quality and contain specific nutritional or chemical properties. Agronomists who specialize in plant breeding and genetics support the needs of both farms and buyers.

Multiple methods are used to create hybrid plants. Some hybrid strains are created by planting one breed next to another and allowing the two breeds to cross-pollinate. Plant scientists also use in vitro techniques, in which plant tissues are combined in a laboratory setting to create strains that would not occur in nature. One technique that has received significant media attention is genetic modification. Genetically modified plants contain genes introduced directly from other sources that create changes in the plant much more quickly than could be generated through traditional breeding.

Soil Health. To support a crop with the highest possible yields, the soil in which the seeds are planted must be in good condition and must match the needs of the particular plant breed. The health of soil can be measured on the basis of its physical properties, its chemical makeup, and the biological material it contains. These qualities are tracked by soil surveys that are conducted and published regularly. Many farmers switch the types of crops they grow in a particular field every few years in order to keep certain soil nutrients from being depleted. This practice is known as crop rotation.

Hydration and Irrigation. Field crops require vast amounts of water. When sources such as rainfall do not provide enough water for healthy plant growth, hydration must be supplemented by irrigation systems. These systems are often based on networks of pipes or hoses connected to sprinklers or drip mechanisms that can supply water to an entire field.

Weather and Climate. Even when a hardy breed of plant is grown in healthy soil and receives enough water, an entire crop can be damaged or destroyed by unexpected weather patterns. Many farmers and growers protect themselves against weather-related risks by purchasing crop insurance, which covers losses in situations such as storms or early freezes. While climate has less variance for individual farmers on a season-by-season basis, changes in climate over the long term can affect the types of plants that grow successfully in a given area. In some cases, climate change can increase or decrease the amount of land suited for growing crops.

Pathology and Pest Control. Like any living organism, field crops are susceptible to natural threats such as disease, predators, and competition from other plants. Agronomists specializing in plant pathology look for ways to fight disease through direct treatment as well as the breeding of new, hardier strains for future crops. Pests such as insects are controlled through the application of inputs such as insecticides. Pest control also influences plant genetics, as in the case of cotton bred to contain a natural compound toxic to boll weevils. Weeds are managed through a combination of herbicides, adjustments to soil properties such as adding or removing water, and the development of plant breeds resistant to weeds.

APPLICATIONS AND PRODUCTS

Agronomic crops can be broken down into categories in a number of different ways, such as plant type or climate in which the crop is most likely to be found. One of the most common ways in which field crops are grouped is by the end use of the raw material. Nearly all crops in the world can be considered a form of food, animal feed, fiber, energy, or tools for environmental preservation. Some major crops such as corn can be classified in multiple ways, as corn is used for feeding both people and animals and is also refined into ethanol.

Food. Food represents one of the most diverse

categories of agronomic crops in the United States and worldwide. When people think of field crops and food, the types of plants that come to mind first are grains such as corn, wheat, and rice. When measured by acres of land planted, grains make up the largest share of agronomic crops grown throughout the world. This category also includes fruits, tree nuts, vegetables, plants grown to be refined into sugar and sweeteners (such as beets and sugarcane), and plants from which oil is made (such as soybeans). While tobacco is not considered a food product, tobacco crops are often included in this category because cigarettes and other items made from tobacco leaves can be consumed only once.

The demand for food crops worldwide grows only as fast as the global population. Individual types of food crops may face sharp increases and drops in demand, however. These changes are influenced by factors such as weather patterns and crop failures, prices set by the commodities markets (which, in turn, make the prices of consumer items rise and fall), and the changing tastes of food buyers. The spike in popular interest in low-carbohydrate diets in the late 1990's had an impact on crops used to make products such as flours and sugars. Similarly, populations in developing countries where incomes are rising often change their food-buying habits and choose items more prominent in North American and European diets than in local ones. This pattern can push up prices for crops such as corn and lower demand for locally grown fruits and vegetables. Interestingly, this trend also works in reverse. Consumers in affluent economies such as the United States have become more interested in buying produce from crops grown locally in an effort to reduce the overall impact on the environment. These kinds of changes can lead to rapid shifts in demand for individual types of agronomic crops.

Animal Feed. As with food, the category of agronomic crops grown for animal feed is dominated by grains. In the United States, the leading feed grains tracked by the U.S. Department of Agriculture (USDA) are corn, sorghum, barley, and oats. Corn makes up the largest share of this category by volume. The USDA estimates that the 2010-2011 growing season will yield a total of nearly 12.5 billion bushels of corn, much of which will be used for animal feed. The combined yield for sorghum, barley, and oats is estimated at less than 1 billion bushels.

A second type of animal-feed crop is hay. Much of the hay grown and harvested in the United States comes from alfalfa plants or a mixture of grass types. Demand for hay is influenced in part by the weather. Farmers feed more hay to their livestock—particularly cattle—in drier conditions. Hay is often grouped by the USDA with a type of crop known as silage. Silage is not a unique plant, but rather the plant stalks and leaves left after the harvesting and processing of grains such as corn and sorghum. Hay and silage together belong to a category of crops known as forage. While forage was once defined as plant matter eaten by livestock grazing in fields, it has been expanded to include plants that are cut, dried, and brought to the animals.

Fiber. Plants grown for nonfood use often fall into the category of fiber. Fiber crops are processed for use in making cloth, rope, paper and packaging, and composite materials such as insulation for homes. In the United States, most of the yearly agronomic fiber crop is made up of cotton plants. The United States is the third-largest grower of cotton in the world. Other plants in this category are jute, sisal, and flax. Jute and sisal are frequently used to make rope, burlap, and rugs. Flax is refined into linen and used in a wide variety of applications ranging from fine clothing and home-decorating products to high-grade papers. Flax fiber is also used in making rope and burlap.

Energy and Environmental Preservation. Most of the agronomic crops in these categories also appear in one of the three categories above. Of the crops grown in the United States as sources of bioenergy, corn tops the list as a source of ethanol. Ethanol sources in other countries include crops such as sugarcane and grasses. Vegetable-based oils made from soybeans are blended into diesel fuel to make a composite known as biodiesel. The refining processes for turning crops into bioenergy sources are not yet as cost-effective as traditional sources of fuel such as petroleum and coal, but this situation is likely to change as technologies improve.

Environmental preservation from the standpoint of agronomic crops is a broad and developing category. It includes the strategic planting and rotating of crops to return depleted nutrients to the soil. It also includes the growth of plants near fields to minimize soil runoff and to protect endangered areas such as wetlands.

IMPACT ON INDUSTRY

Government agencies, academic institutions, and the private sector all play an important role in the field of agronomy. Funding from government sources provides much of the financial support needed for technological development. Agencies also lead many of the research initiatives, which are supplemented by the work of scholars at colleges and universities. Private corporations help to spread innovation from one country to another and devote a portion of their profits to research.

Government Agencies. Nearly every country in the world has a national-level government department or agency devoted to agriculture. The departments frequently oversee agencies at the state or local level. The missions of these departments and agencies are to manage each country's agricultural policies and to ensure that public funding is spent on projects that improve the performance of the farming sector. Information on the country's agricultural practices and yields is gathered, published, and used to support policy decisions. The USDA is one of the largest agencies of this type in the world, employing a significant number of agronomists. The USDA also holds regulatory powers over farms and agricultural businesses by setting standards and ensuring that approved practices are followed.

Academic Institutions. A significant amount of the research and development conducted in agronomy takes place at colleges and universities. Academic institutions offer advanced programs of study in subfields of agricultural science such as crop science, environmental management and land use, and soil science. In the United States, state universities established as land-grant institutions are required to fund departments in agricultural science. Many of the members of professional associations such as the American Society of Agronomy and the European Society for Agronomy are faculty members at academic institutions.

Corporations and the Private Sector. Because agriculture is a part of every country's economy, the business of agronomy is one of the most global in scope of any industry. Some of the largest private-sector corporations in the world, such as Cargill, Monsanto and ADM, are focused on agricultural science. These firms develop new seed hybrids to meet goals such as higher yields per growing season and greater resistance to pests such as insects and weeds.

Private-sector firms also design and manufacture soil inputs and manage the processing and shipping of raw materials from crops.

Because many of the largest firms in this industry have operations throughout the world, they are well positioned to spread new technological developments quickly from one region to another. However, these firms also receive negative attention from the media and from consumers because of concerns about issues such as the environmental impact of new technologies. The development of genetically modified plants has been a topic of heightened interest in modern times, as the plants do not occur in nature and their long-term environmental effects are not fully documented. Other advances have been less controversial. These include the creation of new crop breeds that require fewer inputs or offer a higher concentration of nutrients such as vitamins, minerals, and proteins.

CAREERS AND COURSE WORK

A career in agronomy requires a solid background in agricultural science. Most agronomists hold bachelor's degrees, while many specialists—particularly those in research or teaching positions—have master's degrees or doctorates. All state colleges and universities established as land-grant institutions offer programs in agronomy or agricultural science as part of their educational mission. These programs allow students to specialize in fields such as plant genetics and breeding, soil science, meteorology and climatology, and agronomic finance and business management.

Students majoring in agronomy take courses in a wide variety of areas. Common fields are mathematics (particularly calculus and geometry), physics, and mechanics. Depending on the field a student pursues, advanced course work in biology, botany and plant science, and organic chemistry may be needed. Many courses are highly focused in scope, such as plant pathology or the physical properties of soil.

Demand for professionals with agronomy degrees is rising, according to the U.S. Department of Labor. Job growth in the field of agricultural and food science is projected to be 16 percent between 2008 and 2018. The need for reliable, efficient, environmentally sound sources of plant-based food is a major factor. The increasing use of biomass as an energy source is also contributing to the need for

Fascinating Facts About Agronomy

- The Weed Science Society of America tracks nearly 3,500 types of weeds in its database. Many have colorful names such as kangaroo thorn and sneezewort yarrow.

- Some historians say that people in the first farming villages appeared not only to tend field crops but also to brew beer and other alcoholic beverages.

- More food is going straight from the farm to the fork. From 1997 to 2007, the amount of food sold by farmers directly to consumers more than doubled, helped by farmers' markets and community-supported agriculture (CSA) programs.

- One of the leading sources of sweeteners in U.S. food products is corn. One bushel of corn provides enough syrup to sweeten more than 400 cans of soda.

- A combine, one of the most common types of farm equipment in the United States, can harvest enough wheat in nine seconds to make seventy loaves of bread.

- Soybeans are used in many nonfood products. Soy-based wax is used to make crayons, which have brighter colors and are easier to use than those made from petroleum-based wax.

- The top three growers of fiber crops in the world are China, India, and the United States. Cotton makes up the largest share in all three countries. In the United States, more than half of all cotton goes into clothing production, which is led by jeans.

agronomists with an up-to-date knowledge of science and technology.

Many agronomists work for companies that serve farms. These companies manufacture inputs such as fertilizers, create new breeds of field crops, and process raw materials such as grains and fibers. Other agronomists work for government agencies, primarily within the USDA, or teach and conduct research at colleges and universities. An estimated 12 percent of agricultural scientists are independent consultants.

SOCIAL CONTEXT AND FUTURE PROSPECTS

One of the turning points in public consciousness about agronomy was the release of Rachel Carson's book *Silent Spring* in 1962. Carson's book linked a number of ecological problems, particularly the deaths of wild plants and birds, to the widespread use of pesticides such as DDT. Many of these pesticides were used on field crops. The book led to the United States ban on DDT in 1972 and increased public awareness of the potential environmental harm in certain farming practices.

Consumer interest in the quality of food sources has been on the rise since the 1990's. While there has been demand for products such as organic foods for much of the twentieth century, the category has grown most quickly at the beginning of the twenty-first century. Some consumers participate in community-supported agriculture (CSA) programs in which fruits, vegetables, dairy, and other items are delivered directly from local farmers. Gourmet and chain restaurants are more likely to advertise their use of locally grown and environmentally sound food ingredients. This demand extends to nonfood items ranging from organic cotton fibers in clothing and linens to plant-based, biodegradable home products. It has also influenced the growth of plant-based, non-petroleum energy sources such as ethanol.

Agronomists are well positioned to benefit professionally from these trends. Upcoming issues of interest for agronomists are likely to include the impact of changing weather and climate patterns and the ways in which crop yields can be increased without causing environmental harm.

Julia A. Rosenthal, B.A.

FURTHER READING

Carson, Rachel. *Silent Spring.* 1962. Reprint. New York: Houghton Mifflin, 2002. A landmark examination of the impact of pesticides in agriculture, particularly DDT, and on the environment that sharply increased consumer awareness when it was first published—and continues to do so.

Fageria, Nand Kumar, Virupax C. Baligar, and Charles Allan Jones. 3d ed. *Growth and Mineral Nutrition of Field Crops.* Boca Raton, Fla.: CRC Press, 2011. Covers the biology of crops and the factors that affect soil quality.

Kingsbury, Noel. *Hybrid: The History and Science of Plant Breeding.* Chicago: University of Chicago Press, 2009. An engaging overview of the history of plants and their cultivation for human use throughout the world.

Miller, Fred P. "After 10,000 Years of Agriculture,

Whither Agronomy?" *Agronomy Journal* 100, No. 1 (2007): 22-34. Available at https://www.agronomy.org/files/about-agronomy/future-of-agronomy.pdf.

Reed, Matthew. *Rebels for the Soil: The Rise of the Global Organic Food and Farming Movement.* London: Earthscan, 2010. An extensively researched history of organic farming and its political implications.

Vandermeer, John H. *The Ecology of Agroecosystems.* Sudbury, Mass.: Jones and Bartlett, 2011. A reference on the relationship between agronomy and environmental issues, supported by case studies on historical crises.

WEB SITES

Agricultural Council of America
http://www.agday.org/index.php

American Society of Agronomy
http://www.agronomy.org

Crop Science Society of America
http://www.crops.org

Soil Science Society of America
https://www.soils.org

United States Department of Agriculture
http://www.usda.gov

Weed Science Society of America
http://www.wssa.net

See also: Agricultural Science; Meteorology; Soil Science.

ANESTHESIOLOGY

FIELDS OF STUDY

Pain management; pharmacology; cardiac and pulmonary resuscitation; veterinary anesthesiology; sedation; pathophysiology; airway management; advanced life support; pediatric anesthesiology; geriatric anesthesiology; intensive care; end-of-life care; surgical anesthesiology; local anesthesiology; general anesthesiology.

SUMMARY

Anesthesiology is a specialty of medical science concerned with the management and control of acute or chronic pain as well as the care of patients before, during, and after surgical procedures. Anesthesiologists are physicians who have specialized training in the administration of drugs or other treatments that can induce the various forms of anesthesia. The word "anesthesiology" is of Greek origin, *an* meaning "without" and *aisth'sis* meaning "sensation." Careful assessment, monitoring, and pharmaceutical treatments allow patients to undergo medical procedures without pain or distress. Anesthesia can induce temporary pain relief, amnesia, loss of responsiveness, and loss of muscle reflexes in localized areas or in the entire body.

KEY TERMS AND CONCEPTS

- **Anesthesia:** Partial or complete loss of sensation, usually brought about by injection or inhalation.
- **Anesthesiologist:** Physician specializing in anesthesiology.
- **Anesthetist:** Person who administers anesthetic. May be a physician or specially trained nurse.
- **Emergence:** Transition from the sleep (anesthetized) state to full consciousness.
- **General Anesthesia:** Temporary anesthesia that works on the brain, affecting the entire body by creating a full loss of consciousness and responsiveness.
- **Induction:** Period from the initial introduction of an anesthetic drug, by injection or inhalation, until the optimum level of anesthesia is reached.
- **Infiltration Anesthesia:** Local anesthesia injected directly into the tissues, such as into the gums during dental procedures.
- **Local Anesthesia:** Temporary pharmacological inhibition of nerve impulses to a specific body part, typically used to treat small lacerations or perform minor surgery.
- **Nerve Block:** Regional anesthetic injected directly into a nerve (intraneural) or adjacent to the nerve (paraneural).
- **Post Anesthesia Care Unit (PACU):** Area where patients go to recover from the immediate effects of anesthesia and surgery.
- **Twilight:** State of light anesthesia.

DEFINITION AND BASIC PRINCIPLES

Anesthesiology is a specialized division of medicine that uses drugs or other agents to cause insensibility to pain, reduced sensation, amnesia, or loss of reflexes. This branch of science involves a critical balance of biology, chemistry, physiology, and pharmacology to produce safe and effective pain management for patients. Anesthesiology has evolved beyond the operating room, affecting many other specialized areas of patient care and expanding its role in health care delivery. Anesthesiologists provide consultation and medical support in intensive care units, emergency departments, and pain management clinics, as well as for diagnostic and cardiac procedures. The goal of anesthesiology is to decrease the amount of pain and emotional distress to a patient while effectively monitoring his or her vital signs and safety. Research in this field is focused on developing more efficient administration methods, providing continuous monitoring of vital sign information, decreasing induction and emergence time, and reducing harmful side effects. Advances in anesthesiology have provided tremendous opportunities to surgeons, improving their ability to safely and effectively treat their patients.

BACKGROUND AND HISTORY

Although the term "anesthesia" was not used until the mid-1800's, many herbal and alcoholic remedies had been used for thousands of years to dull sensation and relieve pain. Records from 1500 b.c.e. describe the use of inhaled opium preparations, and later Indian and Chinese texts encourage the use of inhaled cannabis paired with wine before medical procedures. The German physician Valerius Coruds described the synthesis of ether in 1540, and British-born American scientist Joseph Priestly discovered nitrous oxide (laughing gas) in 1772.

The impact of these discoveries and their application to medicine would not be realized until the 1800's, as ether and nitrous oxide were most commonly used for entertainment in shows known as the "ether frolics." Exactly who should be credited with the discovery of anesthesiology remains controversial. In 1844, American dentist Horace Wells attended a laughing gas show and noted that one of the performers had hurt himself during the show but did not feel any pain until the effects of the gas had worn off. Wells tested the effects of nitrous oxide by inhaling it and having one of his teeth removed. In 1846, William Thomas Green Morton, an apprentice of Wells, publicly demonstrated a painless tumor removal while his patient inhaled ether. Crawford Williamson Long argued that he had performed the same procedure using ether in 1842, but his report was not published until 1849. Physicians once used combinations of salt, ice, and ether to numb small areas, but in 1884, Austrian Karl Koller used the first cocaine-derived local anesthetic in ophthalmic surgery.

Since the scientific basics of anesthesiology were discovered in the mid-1800's, improvements in the methods of administering anesthesia and the discovery of better drugs have led to increased patient safety, improved pain control, and applications beyond the operating room.

How It Works

Millions of patients every year, both human and animal, undergo surgical or other procedures that involve anesthesiology. The type of anesthesia and its method of administration are carefully selected based on the procedure and the patient's general state of health. The most common types of anesthesia are general, regional, and local.

General anesthesia is often described a being asleep, but it is very different and much more complicated than sleep. It is a carefully controlled state of unconsciousness, amnesia, analgesia, and paralysis that requires constant monitoring and adjustment. Surgeons require that their patients do not move during an operation, so control over voluntary and involuntary reflexes is crucial. Anesthesiologists may use a combination of three to fifteen different drugs during general anesthesia, depending on the case. These drugs, including isoflurane and desflurlane, are extremely potent, allowing the patient to feel no pain during an operation, remember nothing about the procedure, and recover safely afterward. Typically, in the first phase of general anesthesia (induction), the patient is given an initial intravenous injection that causes a drowsy or unconscious state. Oxygen is continuously provided, first through a mask fitted over the nose and mouth. When the patient reaches an unconscious state, a breathing tube is inserted through the mouth into the windpipe, and a ventilator is attached. To keep the patient in a painless, unconscious state, a carefully selected combination of narcotics, muscle relaxants, and anesthetic gases are administered. This phase is called maintenance. During the state of general anesthesia, the patient's bodily functions are also carefully monitored and controlled. When the procedure is complete, the anesthetic gases are stopped, and a combination of different drugs is given to reverse the effects of the induction drugs. This phase is called emergence. As the patient regains consciousness and is able to breathe on his or her own, the breathing tube is also removed.

Regional anesthesia, also called a nerve block, provides anesthesia to only the part of the body involved in a procedure rather than the entire body. The anesthesiologist injects local anesthetics through a needle, close to the nerves of the involved part of the body. Usually, to reduce discomfort to the patient, the skin and tissues that the needle goes through are first numbed with local anesthetic. These drugs temporarily stop the nerves from working so that no pain, sensation, or movement occurs. The most common type of regional anesthesia is spinal anesthesia, such as an epidural, which can anesthetize the abdomen and legs and is often used during childbirth.

Local anesthesia refers to injecting anesthetic into the skin to temporarily numb a small area so that a minor procedures can be done painlessly. This type of anesthetic is normally used for stitching small lacerations or for dental procedures. Drugs commonly used as local anesthetics include lidocaine and prilocaine.

APPLICATIONS AND PRODUCTS

Anesthesiology encompasses the entire range of patients, from premature infants to the elderly, as well as individuals with complicated medical challenges. Advances in applications, drugs, and equipment have expanded the role of practitioners beyond the operating room; however, the ultimate goal of safely preventing the patient from feeling pain and distress remains the same. Anesthesiology has evolved into a number of subspecialties to address the needs of patients and health care systems. Subspecialties include critical care, pain management, and pediatric, cardiovascular, ambulatory, bariatric, geriatric, neurologic, and obstetric anesthesiology. Anesthesiology research is creating growth and development in delivery systems, vital sign and blood gas monitoring systems, respiratory accessories, pulmonary-function testing products, ventilators, and pharmaceutical agents.

Integrated anesthesia systems are at the forefront of anesthesiology research and market development. These systems combine anesthesia delivery and patient monitoring. Monitoring technologies are critical to patient safety and include medical devices with a wide array of sensors that enable earlier detection and treatment of potentially life-threatening conditions.

Critical care anesthesiology involves the care and monitoring of patients who have been admitted to the intensive care unit because of critical illness as a result of serious injury or before or after complicated major surgery. In the past, many seriously ill patients developed lung problems and died from acute respiratory failure. However, modern critical care units use mechanical ventilation to assist patients to breathe while they recover from serious injuries or major surgery. Anesthesiologists are experts in breathing assistance and resuscitation methods, and they are the attending physician in more than 30 percent of intensive care units.

Obstetric or maternity anesthesiology is a common subspecialty. Anesthesiologists are frequently consulted to provide pain relief to women in labor, anesthesia for obstetric procedures such as a cesarean section, and resuscitation to newborn infants. Safe and reliable pain relief during childbirth can be provided using epidural analgesia, a type of regional anesthesia. Small amounts of local anesthetic are injected through a small plastic tube or catheter into the woman's back, near the nerves that supply the parts of the body involved in childbirth. Almost 20 percent of births are by cesarean section, which requires anesthesia while maintaining the safety and comfort of mother and baby.

Acute pain management involves the treatment of pain following surgery or trauma, or for patients with chronic pain from terminal cancer. Anesthesiologists are able to treat most pain and provide relief to these patients. Regional anesthesia, as in epidural analgesia, was originally developed for pain during childbirth but has since been extended to a vast range of surgical procedures. A small amount of anesthesia given to patients through an epidural catheter after leg, abdominal, or chest surgery allows them to remain comfortable without the need for large amounts of narcotics, such as morphine. Advances in anesthesiology also provide other treatments for acute pain management using combinations of nonnarcotic drugs and nerve blocks to minimize the amount of narcotic medication needed and keep patients comfortable. Also, special pumps have been developed to allow patients to control the amount of narcotic medication they receive, depending on their level of pain.

IMPACT ON INDUSTRY

Research and medical science continue to produce new and innovative surgical procedures, instrumentation, and drug treatments. To perform these procedures, patients often require some sort of sedation, pain management, or localized sensation block; therefore, anesthesiology research has had to progress to accommodate these surgical and diagnostic advances. Increased regulations from the American Society of Anesthesiologists and health care facilities necessitate the continuous evaluation and documentation of patient's oxygenation, ventilation, circulation, and temperature, which has increased the demand for efficient and effective monitoring equipment. Biotechnology and pharmaceutical companies are competing to develop drugs and devices for anesthesiology applications that will improve patient care and safety. The future of the anesthesiology industry includes automated and feedback-controlled anesthesia workstations that are in the development and approval stages. These systems promise to allow anesthesiology professionals to focus more on the status of the patient than on the functioning of the equipment and the numerous monitoring screens. These

Fascinating Facts About Anesthesiology

- The stages of delivering general anesthesia have been compared to piloting an airplane: Taking off (induction), keeping the airplane in the air (maintenance), and landing smoothly (emergence).
- Researchers from the University of Louisville, Kentucky, found that people born with red hair require about 20 percernt more anesthesia for sedation.
- In 1853, Queen Victoria took chloroform to provide some pain relief during the birth of her seventh child.
- A monument to ether, probably the world's only monument to a drug, stands in a prominent place in Boston's Public Garden. This 40-foot-tall tribute commemorates the first use of ether as an anesthetic under the Etherdome at Massachusetts General Hospital on October 16, 1846.
- In China, acupuncture is often added to or used in place of Western anesthesia.
- It is estimated that 40 million anesthetics are administered each year in the United States.
- Bed bugs inject their saliva, containing anesthetics and anticoagulants, into their victims, making their bites initially painless.
- Patients used to decline the use of anesthesia during medical procedures for religious reasons. They thought the pain was God's will.
- Anesthesiology was the first medical specialty to specifically focus on patient safety.
- Originally ether and nitrous oxide (laughing gas) were not given to people for anesthesia or pain relief but were taken by performers to entertain an audience. These shows were called "ether frolics."

in high demand as more 40 million procedures involving anesthesia are performed in the United States each year. Careers in the field require extensive training but are highly transportable as the methodology is applicable in medical settings all over the world. Anesthesiology research is improving the technology and available methods, expanding the roles of anesthesiology personnel into new areas of specialty care.

The most recognized profession in anesthesiology is an anesthesiologist, a specialized medical doctor. To become an anesthesiologist, a student must complete the necessary undergraduate degree requirements to gain admission to medical school. Upon completion of medical school, the student completes an internship and a three-year anesthesiology residency. Some students complete an additional year of fellowship to specialize further. During the residency training, anesthesiologists will work toward their certification from the American Society of Anesthesiologists or the American Board of Anesthesiology by passing the required board examination. Dental and veterinary anesthesiologists are doctors of their specific professions who have specialized in the delivery of anesthetic to their patients.

Anesthetists are individuals who can administer anesthetics. They may or may not be physicians. Many anesthetists in the United States are certified registered nurse anesthetists (CRNAs) who work with other medical professionals. Educational requirements for a career as a certified registered nurse anesthetist begin with a bachelor's degree and at least one year of acute care nursing experience. This is followed by completion of a nurse anesthesia program (twenty-four to thirty-six months) leading to a master's degree, after which the student must pass the mandatory certification exam.

Anesthesiologist assistants provide anesthesia care under the direction of an anesthesiologist. These professionals have specialized master's degree training and require licensing, certification, or physician delegation, depending on where they work.

Anesthesia technicians are biomedical personnel who assist anesthesiologists, nurse anesthetists, and anesthesiologist assistants in the operating room with monitoring equipment, supplies, and patient care procedures.

stations are considerably more complex, and therefore, thorough training is essential. Devices to improve the economics of anesthesiology by reducing medical gas leakage and drug flow are also a focus of industry development.

CAREERS AND COURSE WORK

Professionals in the field of anesthesiology are

SOCIAL CONTEXT AND FUTURE PROSPECTS

Many of the advances in medical and veterinary technology, diagnosis, and treatment could not occur without parallel advancements in anesthesiology. Careers in anesthesiology have evolved beyond the operating room as professionals in this field manage a wider range of patients and provide consultation in many departments. Most anesthesiologists specialize their practice to areas such as critical care, pain management, or pediatric, geriatric, ambulatory, or cardiovascular anesthesia. Anesthesiologists are also consulted to address many societal and often controversial issues in bioethics such as addiction to pain medication and patient's right to die.

Research in anesthesiology focuses on both scientific and practical areas. Some key areas include anesthetic safety, medical quality assurance, ambulatory anesthesia, automated delivery systems, and monitoring equipment. Veterinary anesthesiology is a field of particular interest to researchers as it is extremely challenging to apply effective anesthesia techniques to the wide range of species and sizes of animals encountered by veterinarians. The great majority of anesthesiology researchers are also practicing anesthesiologists, which means that finding time and funding to conduct research in addition to their medical duties can be quite a challenge.

April D. Ingram, B.Sc.

FURTHER READING

Maltby, Roger. *Notable Names in Anaesthesia.* New York: Oxford University Press, 2002. A collection of biographies of scientists who have affected and influenced anesthesiology.

Snow, Stephanie. *Blessed Days of Anaesthesia: How Anaesthetics Changed the World.* New York: Oxford University Press, 2008. Provides an interesting account of anesthesia history and how it influenced society and beliefs about enduring pain.

Stoelting, Robert K., and Ronald D. Miller. *Basics of Anesthesia.* Philadelphia: Churchill Livingstone/Elsevier, 2007. An introductory text that can help familiarize readers with the language and concepts of anesthesiology by building on basic science knowledge.

Sweeny, Frank. *The Anesthesia Fact Book: Everything You Need to Know Before Surgery.* Cambridge, Mass.: Perseus, 2003. A very experienced anesthesiologist from California writes in a readily understandable way about anesthesia. He discusses what happens before and after surgery, explains general surgery, and describes several other types of anesthesia.

WEB SITES

American Society of Anesthesiologists
http://www.asahq.org

International Anesthesia Research Society
http://www.iars.org

MedlinePlus
Anesthesia
http://www.nlm.nih.gov/medlineplus/anesthesia.html

See also: Pediatric Medicine and Surgery; Pharmacology; Surgery.

ANIMAL BREEDING AND HUSBANDRY

FIELDS OF STUDY

Animal science; genetics; statistics; genomics; biotechnology; animal nutrition.

SUMMARY

Animal husbandry is the production and care of animals. Animal husbandry is usually called animal science in universities, since academic studies involve research and the application of scientific principles. Animal breeding is often considered part of husbandry and is the application of genetic principles in the development of breeds and lines of animals for human purposes. Animal breeding principles are also used in captive breeding programs to propagate endangered wildlife species. The development of a leaner line of pigs and a strain of chickens that produces more eggs are examples of animal breeding.

KEY TERMS AND CONCEPTS

- **Biotechnology:** Application of biological techniques to practical uses.
- **Breed:** Population of animals within a species that have similar identifying characteristics.
- **Complimentarity:** Improvement in performance of offspring from parents with different but complimentary breeding values.
- **Correlated Characters:** Traits that change together, either in the same or opposite directions.
- **Genotype:** Genetic makeup of an animal.
- **Heritability:** Measure of the relationship between phenotype and breeding value.
- **Hybrid:** Offspring of breeding different species, lines, or breeds.
- **Inbreeding:** Mating of closely related animals.
- **Line:** Group of related individuals within a breed.
- **Mating System:** Set of rules for mating selected males with selected females.
- **Outbreeding:** Mating of unrelated animals, such as animals in different breeds or lines.
- **Phenotype:** Observed physical appearance or performance of a trait.
- **Polygenic Trait:** Trait affected by many genes, with no gene having a dominating influence.
- **Population:** Group of intermating individuals within a herd, breed, or species.
- **Trait:** Any observable or measurable characteristic of an animal.

DEFINITION AND BASIC PRINCIPLES

Animal husbandry is concerned with all aspects of the management, care, and breeding of farm animals. The goal of animal husbandry is to provide the best conditions (given economic constraints) to maximize productivity in terms of body weight, wool, milk, or eggs. The animals must remain in good health to attain this productivity and to reproduce. Animal husbandry involves the choice of proper feeds, housing, and suitable animals.

Animal breeding begins with a measurement of desirable traits (phenotype) that relate to improved animal production. The breeding value of an animal, however, is the degree to which its underlying genotype can be transmitted to its offspring. Modern methods of breeder selection combine traditional measurements of quantitative traits with the new technology of genome analysis, which aids in determining the breeder's genotype. The rate of genetic change (in animal populations) is directly related to accuracy of selection, selection intensity, and genetic variation in the population and is inversely related to generation interval. There are two primary types of breeding programs: development of breeds or lines that can be used as breeders (seedstock) and development of crossbreeds for production. Crossbreeds demonstrate improved productivity because of hybrid vigor and complimentary traits exhibited by their parents.

BACKGROUND AND HISTORY

Animal husbandry began with the domestication of animals for human purposes from around 10,000 to 5000 b.c.e. Sheep were the first to be domesticated, followed by cattle, horses, pigs, goats, and finally chickens and turkeys. A relatively small number of species have been domesticated because they must possess several suitable characteristics that allow them to adapt to interaction with humans. Their diet must be simple to provide (the early domesticated animals depended on grazing and foraging for their food). They must be able to breed in captivity and must grow and reproduce over a relatively short time interval. They must have a calm, predictable behavior and a cooperative type of social structure.

Animal breeding started in the Roman Empire or perhaps earlier. Early breeders recognized desirable traits in animals that they wanted to propagate, so they selected those animals for mating. The characteristics of domesticated animals began to vary greatly from those of their wild cousins and became totally dependent on their human captors. Systematic selective breeding methods began with the English sheep farmer Robert Bakewell in the late 1700's. Bakewell sought to increase the growth rate of sheep so that they could be slaughtered at an earlier age, to increase the proportion of muscle, and to improve feed efficiency. The application of genetics in animal breeding began in the twentieth century. Jay Lush, a professor at Iowa State University, is considered to have been a pioneer in the application of genetic techniques in animal breeding. His *Animal Breeding Plans* (1937) advocated breeding based on quantitative measures and genetics in addition rather than just on the animal's appearance.

How It Works

Husbandry. Farm animal production is an economic venture, undertaken to produce food (meat, milk, and eggs) or other animal products, such as wool, hides, hair, and pelts. Through the process of animal husbandry, growers seek to create conditions that maximize production of animal products at the lowest cost. With advancing technology and improvements in breeds, animal production has evolved from extensive systems to increasingly intensive systems.

Intensive systems put more demands on good husbandry practices, because the animals are often under more stress and depend more on humans for their well-being. Extensive systems involve keeping animals on pastures or in small pens with minimal housing. Intensive systems are most advanced in the case of poultry. Broilers (meat animals) are kept in total confinement indoors, while laying hens are kept completely in cages. Swine are also commonly kept in confinement, usually on slat or grid floors made of metal. Confinement operations require closer attention to the requirements of ventilation, sanitation, and animal interactions. Beef cattle are still grown on range or pasture, but it is more common to finish

them in large feedlots concentrating thousands of animals. Dairy cattle usually have pastures for grazing but are practically always milked by machine in parlors. Sheep are still largely grazed on range or pasture for most of their growing cycle.

Traits and Breeding Value. The selection of animals in the early days of animal breeding depended on physical or quantitative traits exhibited by the animal without any understanding of the underlying genetic principles. These observed or measured traits are known as the phenotype of the animal. The animal's phenotype is a result of the interaction of its genotype (genetic makeup) with the environment. The goal of animal breeding is to produce animals in a herd, flock, line, strain, or breed that possess superior phenotypes that can be passed on to future generations. The degree to which observed phenotypes can be transmitted to offspring is known as heritability and is a measure of the breeding value of the animal. The selection of desired traits depends on the species of animal and the intended purposes for raising them. The selection also depends on the management practices adopted by the farmer and the relationships between farm inputs and the value of the animals. Examples of traits include calving interval for beef cattle, milk yield for dairy cattle, litter size for swine, first-year egg numbers for hens, and breast weight for meat chickens. The performance of traits can depend on the environment. A high-producing Holstein cow may not produce as well in the tropics because it is not heat tolerant.

Rate of Genetic Change. Progress in a breeding program is related to the rate of genetic change in a population. There are several factors that affect this rate of change: accuracy of selection, selection intensity, genetic variation, and generation interval. The accuracy of selection relates true breeding values to their prediction for a trait under selection. Selection intensity refers to the proportion of individuals in a population that are selected. Populations selected more intensely will be genetically better than the average, leading to a faster rate of genetic change. Populations exhibiting greater genetic variation among individuals have the potential for more rapid genetic change. Finally, species having a short generation interval will have a faster rate of genetic change.

Multiple Trait Selection. Breeders seldom select just one trait for improvement, since a combination of traits is important for the economic success of the enterprise. In fact, selection for one trait usually affects the response to traits not selected for because of the phenomenon of correlated response. The major cause of correlated response is pleiotrophy, the situation in which one gene influences more than one trait.

Breeders practice multiple trait selection by three primary means. Tandem selection involves selecting for one trait, then another. Independent culling levels set minimum standards for traits undergoing selection, and animals are rejected that do not meet all the standards. Finally, the method of economic selection indexes assigns weighted values to the various traits.

APPLICATIONS AND PRODUCTS

Seedstock. A term commonly applied to breeding stock is "seedstock." The purpose of breeding stock is to provide genes to the next generation rather than to be producers of meat, milk, wool, or eggs. Traditionally, seedstock have been purebreds, but the number of nonpurebred stock is increasing. Seedstock animals are obtained by programs of inbreeding. These programs result in an increase in homozygous or similar genotypes. As a result, seedstock have a greater tendency to pass on performance characteristics to their offspring, an ability known as prepotency. One risk of inbreeding is the expression of deleterious genes resulting in reduced performance, known as inbreeding depression. Outcrossing or linebreeding is a milder form of inbreeding and involves mating animals from different lines or strains within the same breed. This process still maintains a degree of relationship to highly regarded ancestors but is less intense than breeding first-degree relatives. Outcrossing allows for the return of vigor that can be lost by inbreeding, while still maintaining the genetic gains obtained by inbreeding.

Crossbred Animals. Mating animals from different species is known as crossbreeding. The resultant offspring are known as hybrids. In modern animal husbandry, even hybrids are commonly used in crossbreeding systems. Crossbred animals are used for production and are designed to take advantage of hybrid vigor and breed complimentarity. Hybrid vigor is the increased performance of hybrid offspring over either purebred parent, especially in traits such as fertility and survivability. A classic example of complimentarity is the crossing of specialized male and female lines of broiler chickens. Individuals from male lines are heavily muscled and fast growing but not great egg producers, while individuals from the female line are outstanding egg producers.

Artificial Insemination. Artificial insemination is a reproductive technology that has been used for a long time. Semen is collected from males and is used to breed females. Because semen can be frozen, it can be used to eventually sire thousands of offspring. This expanded use of superior males can markedly increase the rate of genetic change. Estrus synchronization facilitates artificial insemination by ensuring that a group of females come into estrus at the same time.

Embryo Transfer. Embryo transfer involves collecting embryos from donor females and transferring them to recipient females. Although the motive for embryo transfer is to propagate valuable genes from females, the number of progeny is much fewer, and the procedure is more difficult and costly than artificial insemination.

A variation of embryo transfer is the emerging technology of in vitro fertilization. This technology involves collecting eggs from donor females, which are then matured, fertilized, and cultured in the laboratory. The embryos can then be transferred to recipient females or frozen for later use. The procedure is very expensive and time-consuming. However, it has the potential to aid genetic selection and crossbreeding programs. The genotype of the embryo could be determined before pregnancy. Knowing the genotype could be particularly important for dairy cattle, which frequently have fertility problems.

Cloning. Cloning is the production of genetically identical animals. Cloning allows the breeder to predict the characteristics of offspring, to increase uniformity of breeding stock, and to preserve and extend superior genetics. The preferred method of cloning, somatic cell nuclear transfer, involves removing the nuclei from multiple unfertilized eggs, followed by the transfer of somatic cells from the animal to be cloned. If the process is successful, the resulting embryo is placed in the uterus of a surrogate mother for development.

Genetic Marker Technology. Genetic marker technology was made possible with the development of reasonably inexpensive and efficient genomic analysis of farm animals. The term commonly used in the

Fascinating Facts About Animal Breeding and Husbandry

- Genomic estimated breeding values are expected to revolutionize dairy breeding programs. The calculation combines traditional parent average evaluation programs with the new genetic marker discoveries.

- Using estimated breeding values, a farmer can select for cows that calve easily or have calves with lower birth weight.

- The Animal Improvement Programs Laboratory of the U.S. Department of Agriculture conducts research into the genetic evaluation of dairy cattle and goats, directed at improving yield traits and nonyield traits that affect health and profitability.

- Since 1950, milk production per cow has risen from 5,313 to 16,400 pounds; age to market weight of broiler chickens has decreased from 12 weeks to 7.3 weeks; eggs per hen per year has risen from 174 to 254 eggs.

- Holstein dairy cows originated in Europe and were imported into the United States in the 1800's. Early breeding was for cows that would make the best use of grass.

- Captive breeding has saved many wild animal species from extinction. These species include wolves, the bison, the Peregrine falcon, the California condor, and the whooping crane.

- Most poultry and swine grown for meat are crossbred, while purebred stock are used as breeders.

genetic marker field is quantitative trait loci(QTI). Animal breeders select for traits of economic importance that are largely quantitative traits. These traits are usually controlled by a large number of genes, even thousands of genes. Each gene can contribute a small portion to the total genetic variation of the trait. Since the location (locus) and identity of these genes on the DNA molecule is frequently unknown, the use of genetic markers has become important. Genetic markers are associated with quantitative genes and can be identified in the laboratory.

Single Nucleotide Polymorphisms. Another term associated with genetic markers is single nucleotide polymorphisms (SNPs). Nucleotides are the building blocks of DNA, and polymorphism means "many forms." Nucleotides are made of one of four different

bases. Genes are made of many nucleotides. The exchange of one base for another in a nucleotide is an SNP, which can change the expression of a gene. Instead of analyzing the entire genome of an animal, the dense SNP array test measures around 50,000 SNPs, which is then related to the genetic merit of the animal. With traditional breeding programs, each offspring is assumed to have inherited an average sample of genes from his or her sire (father) and dam (mother). Full siblings have equal parent average (PA) but are expected to share only half of their genes as copies of the same genes in their parents. Considerable improvements in breeding value have been demonstrated by the use of a genomic predicted transmitting ability (gPTA) calculation, which combines genomic data with the traditional parent average data.

IMPACT ON INDUSTRY

In 1966, the World Congress on Genetics Applied to Livestock Production was established to provide an avenue for researchers to present their research findings. The areas of research reflect the concerns or problems of the respective countries. Most of the papers presented at the congress every four years have been on breeding for meat or milk production, estimating genetic parameters, and designing sustainable breeding programs. In spite of widespread dissemination of research findings worldwide, there is a large gap in biotechnology applications between developing and developed countries. Research partnerships between developed and developing countries are much fewer than those between developed countries. The only exception is artificial insemination, which is not really a new technology. The more complex technologies such as embryo transfer and genetic markers are adopted less frequently.

Government and University Research. Animal breeding research at the federal level takes place within the Agricultural Research Service (ARS), a branch of the U.S. Department of Agriculture (USDA). Research takes place at one hundred stations located throughout the country as well as in some foreign countries. The service welcomes collaboration with businesses, state and local governments, and universities. Many of its accomplishments are a result of these joint efforts. ARS researchers adopted quantitative measures for evaluating breeding stock early on, leading to calculations that predict the average performance of offspring, such as "expected progeny difference" for beef cattle and "predicted transmitting ability" for dairy cattle. They have been deeply involved in genome sequencing of farm animals. By the use of the Illumina Bovine SNP 50 Bead Chip, the genotypes of more than 40,000 animals have been determined. In 2007, the USDA Animal Genomics Strategic Planning Task Force, consisting of members of the ARS and the Cooperative Research, Education, and Extension Service, as well as university collaborators, developed the "Blueprint for USDA Efforts in Agricultural Animal Genomics." The blueprint identifies three major areas of focus: outreach, discovery, and infrastructure.

The land-grant college system inaugurated a close relationship between the federal government and the states. The Morrill Act (1862) and Hatch Act (1887) provided federal funds for the establishment of state colleges of agriculture and for associated agricultural research stations. The Cooperative Extension Service was also established to disseminate research information to producers and consumers of agricultural products. Practically all animal science departments have faculty that have had assignments in foreign countries, with the largest number being in Latin America.

Industry Research and Applications. The new technology of SNP markers is revolutionizing the selection of animal breeders. This technology was developed in a partnership among the companies Illumina and Merial, the Agricultural Research Service, the National Association of Animal Breeders, and researchers at several universities and institutes. The researchers found that the breeding value of an animal could be determined by association of these genetic markers with production traits. The genetic evaluation of breeding stock has two advantages—speed and lower costs—over the traditional parent average method. Instead of waiting for proof of breeding value through progeny performance, the animals can be selected very early in their lifetimes. Selection by genetic markers is particularly useful for identifying males with low heritability traits such as fertility and longevity.

The nature of animal breeding companies is changing, and many do not maintain their own breeding stock. They also provide services such as semen collection or embryo transfer, and maintain and sell the semen and embryos. The private sector is playing an increasingly important role in livestock genetic improvement. Specialized breeding firms supply virtually all commercial poultry breeding stock as well as increasing amounts of genetic material for swine, beef, and dairy cattle. Private investment in livestock breeding is affected by demand from producers, market structure, intellectual property protection, new technologies, and market globalization.

CAREERS AND COURSE WORK

The field of animal husbandry is called animal science in colleges and universities to reflect scientific study and applications in the field. Students specifically interested in animal husbandry should concentrate on course work and experience related to production and management. Animal science also encompasses areas such as agribusiness, government, and research and teaching. Job titles in animal husbandry can include livestock or dairy herdsperson, stable manager, veterinary technician, feed mill supervisor, or farm manager. Previously, on-farm experience was enough to work in animal husbandry, but it has become a more complex field. A two-year associate's degree should be considered minimal for the field, while a four-year bachelor's degree would be beneficial for managerial positions. Course work can include animal production, biology, chemistry, animal growth and development, physiology, animal nutrition, biotechnology, farm management, and economics.

A career in animal breeding and genetics requires a doctoral degree, whether employment is in academia or industry. The careers can include such specialties as quantitative or molecular genetics, bioinformatics, immunogenetics, and functional genomics. The prerequisites for graduate studies typically include undergraduate course work in animal science. Graduate courses can include animal breeding, statistics, endocrinology, genome analysis, population and quantitative genetics, animal breeding strategies, statistical methods, and physiology and metabolism. Specific course work varies depending on the school.

SOCIAL CONTEXT AND FUTURE PROSPECTS

Some controversy has arisen because state agricultural experiment stations (SAES) are increasingly entering into collaboration with and receiving funding from private firms. Because the agricultural experiment stations are public institutions, some people feel that they may be compromising their independence and objectivity. However, these stations are primarily involved in basic research, and the private firms conduct the necessary practical research leading to commercialization of new products.

The modern factory farming methods, with poultry, swine, and other farm animals kept in confinement and in crowded conditions, has been condemned by animal rights activists. They claim that because the animals are often not able to perform their natural and instinctive behaviors, they are suffering. Animal science departments have been aware of these criticisms and have developed a new field of farm animal welfare. Faculty positions in the emerging field have been filled, and students are being introduced to the concepts. Animal welfare is being studied academically in a manner that is validated and measured objectively and, therefore, is reliable. The discipline considers the relationship of farm animal welfare to the animals' environment in three areas: how the animals feel, their fitness and health, and their natural behaviors.

One issue addressed by the Food and Agriculture Organization of the United Nations is animal diversity. As globalization of agriculture encourages breeding for high-input, high-output animals, some breeds of livestock are becoming extinct. This leads to the existence of fewer breeds, which means less flexibility when confronted with an emerging disease or changed environmental conditions. Another problem is that genetic breeding takes place mostly in advanced countries, under the conditions of intensive agriculture and the local environment. These conditions and farming methods are not the same as in some lesser developed countries, and animals that produce well in one country may not do as well in another because of environmental factors.

The potential benefits and profits from improvements in animal breeding and husbandry are great, and the field is likely to remain active. Genetic engineering is likely to be part of animal breeding, despite social concerns, although these areas of concern may affect how genetic engineering is used.

David Olle, B.S., M.S.

FURTHER READING

Bourdon, Richard. *Understanding Animal Breeding.* 2d ed. Upper Saddle River, N.J.: Prentice Hall, 2000. An excellent basic text on animal breeding that presents concepts with a minimum of mathematics.

Herren, Ray V. *The Science of Animal Agriculture.* 3d ed. Clifton, N.Y.: Thomson Delmar Learning, 2007. Covers all aspects of the sciences involved in animal agriculture, including breeding.

Taylor, Robert, and Thomas Field. *Scientific Farm Animal Production.* 9th ed. Upper Saddle River, N.J.: Pearson Prentice Hall, 2009. Provides an excellent introduction to all aspects of animal husbandry and production of all major farm animals.

Turner, Jacky. *Animal Breeding, Welfare, and Society.* Washington, D.C.: Earthscan, 2010. Examines how the trend toward human intervention in animal breeding is affecting animal behavior, health, and well-being.

WEB SITES

American Livestock Breeds Conservancy
http://www.albc-usa.org

Food and Agriculture Association of the United Nations
http://www.fao.org

Sustainable Animal Production
http://www.agriculture.de/acms1/conf6/index.htm

U.S. Department of Agriculture
Agricultural Research Service
http://www.ars.usda.gov/main/main.htm

World Congress on Genetics Applied to Livestock Production
http://www.wcgalp2010.org

See also: Agricultural Science; Genetically Modified Organisms; Genetic Engineering; Genomics; Veterinary Science.

ARCHAEOLOGY

FIELDS OF STUDY

History; geography; geology; ecology; anthropology; sociology; economics; chemistry; mineralogy; sedimentology; geophysics; geographic information systems; evolutionary science; geomorphology; paleontology; ethnography; environmental studies; zooarchaeology (or archaeozoology); archaeobotany (or paleoethnobotany); paleobotany; palynology; paleoclimatology; archaeoastronomy; geoarchaeology; Egyptology; Phoeniciology; classical archaeology; Assyriology; historical archaeology; prehistoric archaeology; industrial archaeology; experimental archaeology; ethnoarchaeology; archaeometry; mathematics; statistics.

SUMMARY

Archaeology is the field of applied science concerned with the techniques and practice of collecting, preserving, and analyzing the artifacts and physical remains left behind by human civilizations. The central purpose of archaeology is to study the lives and cultural practices of past societies, both ancient and modern. Archaeology has a deep intangible impact on human experience, as a means of understanding and maintaining the multifaceted cultural heritage produced by the human species. It also has many practical applications; for example, archaeological studies of rural agriculture can help farmers increase the yield of their crops, and urban archaeologists can advise municipal administrators on issues such as transit patterns and garbage disposal.

KEY TERMS AND CONCEPTS

- **Absolute Dating:** Dating an artifact using a time scale (years b.c.e. and c.e.); also known as chronometric dating.
- **Arbitrary Level:** Predetermined depth to which all digging at any given level of an archaeological excavation will descend.
- **Artifact:** Physical object that was made or altered by human activity; one that is not naturally occurring.
- **Ecofact:** Artifact consisting of organic or environmental remains; often something used by humans (for example, food remains).
- **Feature:** Artifacts and ecofacts that together form an identifiable entity (such as a burial ground).
- **Judgmental Sampling:** Process of excavating and analyzing only artifacts from certain areas of an archaeological site that have been selected based on the judgment of an archaeologist.
- **Lithics:** Study of stone artifacts, such as tools or weapons.
- **Matrix:** Physical surroundings or materials in which an artifact is found, such as soil.
- **Probabilistic Sampling:** Process of excavating and analyzing only artifacts from certain areas of an archaeological site that have been selected, at least partially, by chance.
- **Radiocarbon Dating:** Method of absolute dating based on the radioactive decay of carbon in organic materials.
- **Relative Dating:** Process of determining how old an artifact or feature is in relation to some other object or event.
- **Stratigraphy:** Study of stratifications, or layers of deposits that have built up sequentially over time.
- **Typology:** Practice of classifying artifacts into groups with similar characteristics; the study of such groups through time.
- **Use-Wear Analysis:** Technique of determining how an artifact was used by examining it microscopically, looking for marks indicating wear or damage.

DEFINITION AND BASIC PRINCIPLES

Archaeology is the study of past human cultures, both historic and prehistoric, through the systematic excavation, inspection, and interpretation of material remains such as tools, toys, clothing, bones, buildings, and other artifacts. In the United States, archaeology is considered a subfield of anthropology, which is the science concerned with the origin, evolution, behavior, beliefs, culture, and physical features of humankind.

The traditional image of archaeology held by many people (and reinforced by television programs and motion pictures featuring adventurous archaeologists such as Indiana Jones) is that it is a field largely concerned with examining ancient artifacts drawn from the extremely distant past. This is certainly true of many subfields of archaeology, such as Egyptology, Assyriology (study of ancient Assyria), classical archaeology, and prehistoric archaeology. However, archaeological tools and techniques can be and frequently are applied to the analysis of human cultures across all segments of time, from the very beginning of the species to much later epochs. For example, industrial archaeologists study the development and use of industrial methods, most of which did not truly come into being on a large scale until the eighteenth century, while urban archaeologists study the patterns of life revealed by the material past of metropolises—such as New York or Paris—that still exist and that have a long history of human habitation.

Archaeology as practiced in the modern world is characterized by several key principles. First, the field is not simply descriptive but rather explanatory. Archaeologists are concerned not just with the question of what happened in the past but also how and why it happened. To that end, they attempt to interpret the artifacts and features they uncover for clues about the belief systems, behaviors, traditions, and social, political, and economic lives of the cultures at hand. For example, archaeological excavations of the artifacts buried in the tombs of ancient Egyptian pharaohs have revealed a great deal of specific knowledge about what Egyptians believed to be true about the afterlife. Second, it is multiscalar, meaning that the examination of a particular culture takes place simultaneously on many scales, each of which is intertwined with the others. For instance, an archaeologist may combine the small-scale analysis of individual pots, vases, and other clay artifacts—which give insight into the specific production processes used by the artisans who created them—with large-scale evidence about how those objects were handled on trade routes and marketplace procedures—which give insight into the overall historical trajectory of an entire civilization. By the same token, no matter which geographic region an archaeologist is studying, he or she is likely to examine collected data in the light of wider global trends.

BACKGROUND AND HISTORY

People have always been fascinated by the cultures that came before them. Even in ancient Babylon, for example, relics such as ruined temples were objects

Fascinating Facts About Archaeology

- Some archaeologists spend their days working in scuba-diving suits. The field of underwater archaeology focuses on physical remains of human culture found in the sea or other bodies of water, such as those created by shipwrecks or airplane crashes.

- Archaeologists make use of any scientific tool that can help them uncover information about artifacts—including sophisticated medical imaging. In 2009, computed tomography (CT) scans of 3,500-year-old mummies showed that many of them suffered from heart disease, something that was believed to be a modern affliction.

- Some archaeologists use incredibly tiny artifacts to construct huge ideas. Palynologists study preserved pollen and spores—which can last for thousands of years—to come up with ideas about what the environment was like in ancient times.

- Many elaborate archaeological hoaxes have been perpetrated over the course of the field's history. For more than two decades, for instance, the amateur Japanese archaeologist Shinichi Fujimura buried and then unearthed, to great fanfare, hundreds of objects that he claimed were genuine artifacts from Japanese prehistory.

- Food archaeologists may try to recreate the diet of a culture long past, using the physical evidence that remains from plant and animal meals, as well as written recipes that survive. In 2009, archaeologists managed to analyze and reproduce an ancient Chinese wine using chemical analyses of the wine residue found in pottery jars.

- It may not be the most flattering nickname, but the private-sector archaeologists who work for CRM firms in the United States bear it with pride. What do they call themselves? Shovelbums.

- Archaeologists do not always have to dig deep to unearth artifacts. Aerial archaeology involves conducting bird's-eye view reconnaissance missions using cameras mounted on kites, remote-controlled parachutes, hot-air balloons, model airplanes, and helicopters.

of interest. During the European Renaissance, an early form of archaeology arose that was mainly focused on classical antiquities and the investigation of large prehistoric sites and monuments, such as

Stonehenge. The eighteenth century witnessed the large prehistoric sites and monuments, such as Stonehenge. The eighteenth century witnessed the first formal archaeological excavations, one of which was conducted by future president Thomas Jefferson on an Indian burial mound located on his Virginia estate. In the nineteenth century, the site of the ancient city Pompeii, destroyed by a sudden volcanic eruption, became the focus of intensive archaeological investigations. Innovative techniques, such as the use of plaster of Paris to create molds of bones and other artifacts, were developed by early practitioners such as Italian scholar Giuseppe Fiorelli.

During the nineteenth century, archaeology began to develop the formality of a scientific discipline. The simultaneous maturation of the field of geology, with its theory of the stratification of rocks, helped establish similar ideas in archaeology. In addition, archaeologists began to work with the principle of uniformitarianism for the first time; this was the idea that past and presents societies had more commonalities with each other than differences. Finally, the nineteenth century was the age in which naturalist Charles Darwin's theory of evolution, combined with the accumulating weight of physical evidence, led archaeologists to search for material evidence of human activity in the very distant past, far beyond the age of the Earth as defined by biblical scholars.

Archaeology underwent a transformation during the twentieth century, when many useful field techniques were developed that helped transform the field into a true science. Detailed records were kept at every excavation, including the location and description of each artifact. Sites were also divided into grids and searched systematically, and drawings and models were made of every dig. Advances were also made in absolute dating; particularly significant was the invention of radiocarbon dating by chemist Willard Libby in the United States. Partly as a result of these techniques, archaeology became far more quantitative, or driven by data and statistical analysis, than it had previously been. Hypotheses and theories were carefully tested before being accepted.

HOW IT WORKS

Archaeological Surveys. The first step in conducting any archaeological investigation is to identify and take stock of the location in which the study will be held. This is known as a survey. The choice of location can be determined in several ways. Documentary sources, such as old maps or other written materials, can sometimes be used to accurately discover the location of an archaeological site. For example, clues in the writings of the ancient Greek poet Homer helped scientists find the ruined city of Troy (in northwest Turkey). In other cases, archaeologists attempt to survey sites on which new developments such as roads or buildings are planned to unearth important remains before they are destroyed. This is known as salvage archaeology. On occasion, developers have integrated remains found by salvage archaeologists into their construction plans, as in the case of the Aztec temple dug up in Mexico City and used as part of the subway station that was built on that site. Other archaeological sites of interest are discovered through ground reconnaissance, in which either a judgmental or probabilistic sampling method is used to search for physical evidence, or aerial reconnaissance, which can sometimes turn up traces of human activity that cannot be seen from the ground.

Once a site has been identified as being of archaeological interest and an excavation begins, a deeper survey is conducted of the selected area. The same kinds of sampling methods used in ground reconnaissance are available to archaeologists at this point. Judgmental sampling is based solely on an archaeologist's judgment about which locations within the site should be searched for evidence. For example, since humans tend to settle near sources of water, judgmental sampling might concentrate the excavation along a river bed that runs through the site. Probabilistic sampling, in contrast, is subject to chance and variation, and may help archaeologists turn up unexpected discoveries or make more accurate predictions about areas not sampled. Probabilistic sampling can take the form of simple random, stratified random, or systematic sampling. Simple random sampling is where any given location within a site has an equal chance of being sampled. Stratified random sampling is where the site is divided into naturally occurring regions, such as forest versus cultivated land, and the number of searches conducted in each region is based on its size in relation to the entire site. Systematic sampling is where a site is divided into equal parts and then sampled at consistent intervals (for example, a search is made every 100 meters).

Types of Archaeological Evidence. The evidence used by archaeologists to form a picture of an ancient or past society can be classified into four major categories: material, environmental, documentary, and oral evidence. Of the four, material evidence is perhaps the most fundamental. It can consist of either organic or inorganic remains that have been used by or constructed by people. Material evidence includes buildings, tools, pottery, toys, textiles, baskets, and food remains. Environmental evidence can take many forms, including soil samples, minerals, pollen, spores, animal bones, shells, and fossils.

Because the Earth has undergone many significant geological and climate-related changes over the course of its history, environmental evidence can help reveal what the world looked like during the time of a particular group's existence in an area. Environmental evidence can also offer clues about people's relationships with the landscape around them. For example, if the materials in a lower layer of a dig consisted mostly of wood, charcoal, and pollen, while those in a higher layer of a dig consisted mostly of cattle bones, one hypothesis might be that a region that had once been wooded had at some point been cleared by the inhabitants of the area and used for the rearing of animals.

Documentary evidence, or written records, can be extremely useful to archaeologists whenever it is available. Depending on the epoch, or span of history, from which these documents arise, they may consist of text inscribed on stone slabs, clay tablets, papyrus, or other types of materials. They may record laws, serve as proof of legal contracts (such as marriages or business agreements), list births and deaths in a city, or inventory commodities such as grain or shells. Finally, oral and ethnographic evidence—interviews and oral histories with modern-day inhabitants of a particular area—can help supplement the physical remains collected in an excavation. Certain practices such as the techniques of rural architecture or traditions of marriage, for instance, may have been preserved without much variation over time. In such cases, living members of a culture that has been around for many generations may be able to provide key information about the habits of their ancestors.

Dating Archaeological Evidence. Archaeologists have a plethora of techniques for placing objects in time. Relative dating methods identify artifacts as being older or younger than others, thus placing them into a rough inferred sequence. One of the most straightforward techniques of relative dating is to note the depth at which a particular object was found at a site; that is, in which strata, or deposit layer, it was contained. This method, known as stratigraphy, is an effective way to gauge the relative age of objects because layers build up on top of other layers over time. Other means of relative dating include pollen analysis, ice-core sampling, and seriation. Seriation uses the association of artifacts with known dates of use, along with knowledge about how the frequency of use of those artifacts changed over time, as markers for other items found in close proximity.

Absolute dating methods are far more accurate than relative dating techniques (which can be confused by the effects of natural disasters or animal activity on the organization of strata). They also produce more specific results. The absolute dating technique that is used more than any other by archaeologists is radiocarbon dating, which is capable of assigning an accurate age to artifacts of biological origin (including bone, charcoal, leather, shell, hair, textiles, paper, and glues) that are 50,000 years old or younger. Radiocarbon dating relies on the fact that the element carbon has a particular isotope—carbon 14—which is radioactive. Radiocarbon dating is also known as carbon-14 dating. This radioactive carbon combines with oxygen in the atmosphere to form carbon dioxide, which is absorbed by plants as they photosynthesize. Because plants form the base of every food chain, all living things on Earth contain some amount of radiocarbon within their bodies. Because scientists know the half-life of radiocarbon (the time it takes for half the amount in a given sample to decay), they are able to measure the radiocarbon in any organic material and calculate how long it has been around. Besides radiocarbon dating, other absolute dating techniques include dendochronology, which relies on the annual growth rings present in the trunks of long-lived trees, and thermoluminescent dating, in which trace amounts of radioactive atoms in rock, soil, and clay can be heated to produce light, the intensity of which varies depending on the age of the object. This method is primarily used to date pottery and other human-created artifacts.

Interpreting Evidence. Three basic concepts dominate the approach an archaeologist takes when he or she approaches an artifact, a feature, or a site: context, classification, and chronology. Context, or where an object is found and what other items surround it, can reveal more of a complete story than any single artifact. For example, a decorated container found within a tomb could serve one purpose; the same container found in the kitchen of a dwelling could have been intended for an entirely different one. Classification (also known as typology) also helps place unearthed artifacts within a particular frame of reference. For example, determining that all the items found at a particular location belong to the same type—such as cooking utensils, hunting supplies, or bathing vessels—could indicate how a location was used or who used it. Finally, chronology is essential to understanding the relationship between the elements of any archaeological investigation. For example, unless they have been disturbed, objects located at lower levels, or strata, of a particular excavation site will be older than those located at higher levels. This can help archaeologists trace the development of a particular tool over time or to see and identify important changes in the cultural practices of a given society.

APPLICATIONS AND PRODUCTS

Cultural Resource Management. The major practical application of archaeological tools and techniques can be found in the area of cultural resource management (CRM). In a sense, CRM is a form of institutionally legislated salvage archaeology. It involves the use of archaeological skills to identify, preserve, and maintain important features of historic and prehistoric culture to benefit the public interest. In most countries, including the United States and Canada, any planned development of any significance, including the building of new gas or oil pipelines, residences, highways, golf courses, or any other construction, requires the developer—whether private or public—to conduct a CRM survey to comply with legal requirements. The goal of CRM is to ensure that important pieces of a nation's past are conserved. In the United States, the major pieces of legislation mandating CRM include the National Historic Preservation Act, the National Environmental Policy Act, the Archaeological Resources Protection Act, and the Native American Graves Protection and Repatriation Act. CRM studies are generally performed by trained archaeologists serving as consultants and involve reconnaissance and sampling intended to scan the area for the presence of significant archaeological sites,

features, or artifacts. If any are found, the developers are responsible for safely excavating and preserving them. If this is impossible or the site is identified as especially historically significant, the project may be terminated.

Waste Management. The field of garbage archaeology, or garbology, was pioneered by American urban archaeologist William Rathje in the 1970's. By applying the tools of archaeology to the waste dumped in landfills, Rathje and other scientists have been able to provide many practical insights that are useful to waste management specialists and city, state, and federal administrators concerned with reducing waste and conserving energy. For instance, they showed that even supposedly biodegradable artifacts, such as paper, wood, and even food, are preserved for far longer when packed tightly together in a landfill than they otherwise would be. This is because the dense conditions reduce the amount of oxygen available for decomposition to take place. One of the implications of this discovery is that simply switching from nonbiodegradable to biodegradable materials in the production process—banning plastic disposable cups in favor of paper ones, for example—will not be enough to reduce the amount of effectively permanent waste that is generated by consumers and filling landfills to the point of overflowing.

Agriculture. In some parts of the world, environmental archaeologists have discovered evidence that areas of land that are marked by barren soil in fact used to be rich and fertile. For example, the Negev Desert in Israel, with its high temperatures and meager rainfall, used to be the site of an ancient urban society that cultivated crops such as wheat and grapes. Archaeologists also established that the climate in the area had not changed over the 2,000-year period in between. Using aerial reconnaissance techniques, scientists were able to show that the ancient farmers had employed a water-delivery system composed of terraces and cisterns to collect and redirect rainwater from the infrequent flash floods that occur in the area. This insight has proved to be of immense importance to modern-day agricultural scientists and farmers in Israel and has played a role in the country's successful efforts to "green" the Negev. Similarly, local farmers in rural Peru have begun using prehistoric field technologies unearthed by environmental archaeologists—primarily elevated fields that improve drainage and help protect plants from chilly nights—to dramatically increase their crop yields.

Crime Investigation. Forensic archaeologists apply the methodologies of archaeology to the investigation of crimes, working closely with coroners and police officers to collect and analyze physical evidence from the scene. They investigate fragments of bone, teeth, soil, fabric, jewelry, or other artifacts, while more subtle clues such as disturbed soil or markings from tools or weapons may also provide important leads. Just as in other forms of archaeology, the forensic archaeologist first establishes the boundaries of the site to be surveyed, divides it into grids, and performs a thorough excavation of the material and environmental evidence found in the area. Stratigraphy is an important element of forensic archaeology, particularly so when a body has been buried before being found; the deeper into the soil digging proceeds, the older the remains, personal artifacts, or other evidence that is found will be.

IMPACT ON INDUSTRY

Public Archaeological Research. Governments across the world invest in archaeological studies designed to identify, excavate, and preserve the cultural and geological history of their nations. This kind of archaeological research is especially vital in countries with an extremely long history of human habitation, such as Israel, Egypt, China, and Greece. In the United States, federal entities such as the Forest Service, the National Park Service, the Bureau of Land Management, and the U.S. Army Corps of Engineers all engage in regular archaeological field studies. For example, in 2003, the National Park Service completed a survey of Yellowstone National Park that included fieldwork on how American Indians used land before the arrival of Europeans. In addition, each state has an office dedicated to historical preservation that employs archaeologists to perform similar research within state lines. In the United States, the National Council on the Humanities and the National Science Foundation are the two major government agencies that provide funding to support archaeological research. Important public research also takes place under the auspices of museums. Among the most active are the Smithsonian Institution, the Penn Museum, and the Field Museum of Chicago.

Industry and Business. Through the mechanism of cultural resource management (CRM), archaeological research has a major impact on both private

and public sector development. Any municipality or state wishing to build new infrastructure, as well as any company wishing to engage in construction, whether building new homes or offices or installing telecommunications cables, must hire a consulting archaeologist or CRM firm to ensure that full compliance with legislation regarding the preservation of cultural resources. Unlike academic research, this type of archaeological research is not only robust but also growing. It is estimated to have an industry worth of about $1 billion per year and to employ a total of about 14,000 people—the majority of them trained archaeologists—within the United States alone. CRM research is especially active in the western and southwestern parts of the country, largely as a result of interest in developing these areas to exploit their natural oil and gas reserves.

Major Corporations. Hundreds of small to midsized CRM firms operate within the United States alone. Because knowledge of local geography, geology, and history are essential, in general the market tends to be fragmented, with small firms each controlling only a tiny portion of the total available work. Some CRM companies, however, have managed to grow into nationwide firms serving multiple regions, including Statistical Research Inc. (SRI), and engineering and construction firms such as the Louis Berger Group and Parsons offer CRM services.

CAREERS AND COURSE WORK

The typical archaeology career path begins with an undergraduate degree in anthropology, with an emphasis on course work covering the fields of biological, linguistic, and cultural anthropology. In addition, the aspiring archaeologist should be sure to take a set of courses in related sciences—such as geology, chemistry, ecology, environmental studies, evolutionary science, geophysics, anatomy, and paleontology. Mathematics and statistics are equally important subjects, since much archaeological analysis requires a keen understanding of these areas. In the humanities, history courses—preferably focused on a particular geographic area or epoch in time—are essential, as is an advanced knowledge of at least one foreign language. In addition, business management and technical writing skills will assist any archaeologist who chooses to go into the field of CRM. Finally, if at all possible, students should attempt to pursue archaeological internships, fieldwork placements, or other forms of practical work experience during their undergraduate careers. Such experiences not only are a means of gaining hands-on knowledge of the field techniques used in archaeology but also provide students with professional contacts.

A bachelor's degree will serve as sufficient minimum qualification for many professional archaeological positions in the private sector, particularly in consulting firms hired by developers to perform CRM surveys—in such settings, practical work experience is more important. However, additional graduate study at the master's or doctoral level is necessary to obtain a role as a crew supervisor and enter higher pay brackets. For archaeologists who wish to become faculty in academic institutions, a doctoral degree in archaeology and evidence of original research is required. Other organizations for which archaeologists work include museums, city and state governments, and the federal government. Typical roles for archaeologists include conducting field investigations, performing analyses of found artifacts, curating museum archaeology collections, teaching courses, and publishing research papers.

SOCIAL CONTEXT AND FUTURE PROSPECTS

Archaeology is one of the most important and influential lenses through which scientists, historians, and other scholars view pieces of the past that otherwise would be lost. Its findings often reveal stunning insights into the nature of human civilization and the growth of human culture. For example, archaeologists piecing together skeletons found in Africa have been able to uncover the point at which the proto-human species Neanderthal died out and *Homo sapiens* (anatomically identical to modern humans) replaced it, while the ancient cave paintings discovered in Lascaux, an area in the southwest of France, demonstrate that human beings were creating art as long ago as 15,000 b.c.e. In other words, archaeology is a science dedicated to telling the story of humanity itself.

Although archaeology is at its core a science concerned with the investigation of material objects, the future of archaeology may be surprisingly metaphysical: A small but growing movement exists that is determined to treat the World Wide Web as a treasure trove of archaeological artifacts. Internet archaeology, as this nascent subfield of archaeology is called, seeks to archive and interpret Web sites, Web pages, and graphics that were created during the early years of the Internet and are no longer being updated by their owners.

M. Lee, B.A., M.A

FURTHER READING

Fagan, Brian. *Before California: An Archaeologist Looks at Our Earliest Inhabitants.* Lanham, Md.: Rowman & Littlefield, 2004. Examines the indigenous cultures that existed in California before the arrival of Europeans. Organized chronologically, contains complete notes and references.

Gamble, Clive. *Archaeology: The Basics.* 2d ed. New York: Routledge, 2008. A succinct, accessible guide for the beginning archaeology student that includes figures, boxes, and a comprehensive index.

Kelly, Robert L., and David Hurst Thomas. *Archaeology.* 5th ed. Belmont, Calif.: Wadsworth, 2010. A comprehensive introductory textbook for the undergraduate level. Each chapter includes full-color photographs, a running glossary, a summary, additional reading, and themed sidebars.

Murray, Tim. *Milestones in Archaeology: A Chronological Encyclopedia.* Santa Barbara, Calif.: ABC-CLIO, 2007. Covers five centuries of archaeological discovery through two hundred entries. Includes maps, photographs, drawings, and bibliographies.

Renfrew, Colin, and Paul G. Bahn, eds. *Archaeology: The Key Concepts.* New York: Routledge, 2005. Presents the central ideas in archaeology, including concepts of time, experimental archaeology, and the archaeology of gender.

WEB SITES

Archaeological Institute of America
http://www.archaeological.org

National Parks Service
Federal Archaeology Program
http://www.nps.gov/archeology/sites/fedarch.htm

Smithsonian Tropical Research Institute
Archaeology and Anthropology
http://www.stri.org/english/research/
programs/archeology_anthropology/index.php

See also: Mineralogy; Paleontology.

AREOLOGY

FIELDS OF STUDY

Atmospheric chemistry; astronomy; biology; climatology; engineering; cosmochemistry; global physiography; astrobiology; cartography; biogeophysics; chemistry; robotics; computer science; geochemistry; petrology; geodesy; mineralogy; geomorphology; geophysics; glaciology; hydrology; meteorology; planetary engineering; space exploration; space physics; soil science; volcanology.

SUMMARY

Areology, from the words *areo* (Ares, the Greek god of war) and *logy* (theory), is the interdisciplinary study of Mars. Most of the earth science disciplines can be applied to areology. As an interdisciplinary endeavor, areology also includes the study of the technologies for Mars exploration, both by robotic and manned craft, and the history of human speculation concerning the prospects for life on Mars, including the scientific principles, expectations, and designs for human colonization, and the engineering of the Martian planetary surface to support human life.

KEY TERMS AND CONCEPTS

- **Absolute Age:** Age of a geological unit measured in years.
- **Bombardment:** Repeated collision of a planet with asteroids, usually over geologic time scales.
- **Chaotic Terrain:** Low region within heavily cratered uplands that appears to consist of irregular, blocky, fractured landscape.
- **Chasma:** Canyonlike feature on Mars, from the Latin for "large canyon or gorge"; the plural is *chasmata*.
- **Crustal Dichotomy:** Pronounced hemispheric contrast in physical characteristics of a planet's crust.
- **Ejecta:** Material blasted loose during the formation of an impact crater and deposited around that crater.
- **Flyby:** Mission procedure in which a spacecraft on its way to another destination examines a planet as it flies past that planet.

- **Gravity Map:** A map that shows variations in gravitational attraction across a planetary surface that results from variations found in the internal density of the planet.
- **Mascon:** Acronym for "mass concentration," which describes a zone of anomalously high density within Mars.
- **Mons:** Term used in names of mountainous features on Mars, from the Latin for "mountain"; the plural is *montes*.
- **Planitia:** Term used to indicate Martian regions composed of plains, from the Latin for "plains"; the plural is *planitiae*.
- **Province:** Region of similar terrain or a grouping of geological units with similar or related origins.
- **Relative Age:** Age of a feature or geological unit in relation to other features or geological units.
- **Rover:** Self-propelled, robotically operated vehicle used for exploring the surface of a body distant in space.
- **Terraforming:** Transformation of an alien landscape into one more suitable for human beings.
- **Vallis:** Used in naming valleylike features on the surface of Mars, from the Latin for "valley"; the plural is *valles*.

DEFINITION AND BASIC PRINCIPLES

Areology is sometimes narrowly defined as the study of the geology of Mars, but it more properly involves not only most of the other earth sciences (from meteorology to hydrology to mineralogy) but also space physics, cosmochemistry, and astrobiology. Given that it deals with largely speculative prospects for life on Mars—both indigenous and imported, in the past, present, or future—areology must also take into account both the history of science and the literature of science fiction.

Although the term "areology" was in fact popularized by science fiction author Kim Stanley Robinson in his Mars trilogy (*Red Mars, Green Mars, Blue Mars*), the debate in the scientific community has largely swung between the poles of "wet Mars" (Mars once had water) and "white Mars" (Mars never had water). American astronomer Percival Lowell, who claimed to see through his telescope visions of supposedly water-filled canals on Mars, established one pole of the debate: Mars was a dynamic planet warm and wet enough to support life at the present time.

From its zenith in Lowell's work of the 1890's, this vision of Mars declined to its nadir after the Mariner 4 flyby in 1965. Mariner 4 showed a cratered, dusty ball clad in only the most diaphanous of atmospheres—one whose white polar regions were declared to be most likely covered in dry ice (carbon dioxide rather than water). As the data from the Viking landers of the 1970's proved inconclusive and controversial, the vision of dry, white Mars dominated discussion of the planet for decades.

After the failures of several probes, the successes of a growing armada of orbiters, landers, and rovers began to suggest in the 1990's and 2000's that, cold as it might be, Mars was not as dry and white as many in the planetology community had long contended. The notion that water ice was an important component on the Martian surface made a comeback, along with physical and chemical evidence of a potentially watery past.

The successes and failures of these unmanned spacecraft, along with the discoveries made possible by their successes, have set the parameters for the continuing discussion of the efficacy, expense, and likelihood of manned missions to Mars and eventual human colonization of the planet.

BACKGROUND AND HISTORY

Scientific interest in Mars goes back to the seventeenth century and the work of Galileo Galilei, Johannes Kepler, and Giovanni Domenico Cassini—the last of whom, in 1666, observed the Martian polar caps and calculated the length of the Martian day. The apparent Earth-like nature of Mars led French author Bernard le Bouvier de Fontenelle in 1688 and British astronomer William Herschel in 1784 to speculate on the nature of life on Mars.

Despite this, in the scientific community before the end of the nineteenth century Mars was generally not seen as the best candidate for a second life-supporting world in the solar system. Venus—significantly closer to Earth in terms of size, mass, gravity, distance from the Sun, and actual travel time—at first seemed the more likely choice, and was still argued to be such until the advent of radar and radio telescopy, which pierced the thick Venusian atmosphere. Probes then confirmed the planet's merciless heat.

In literary history, too, the case was similar. Lucian

of Samosata wrote of a trip to the Moon in his *True History* as early as the second century, and in the eighteenth century both Jonathan Swift (in *Gulliver's Travels*) and Voltaire (in *Micromégas*) hypothesized the existence of two as-yet-undiscovered Martian moons. It was not until 1877, after Italian astronomer Giovanni Schiaparelli claimed to see on Mars a network of straight lines he called *canali* (canals), that writers began to examine Mars as the solar system's other main abode of life. This shift began with Percy Greg's *Across the Zodiac* in 1880 and continued most prominently with H. G. Wells's *War of the Worlds* in 1898.

Since Wells, the scientific understanding of Mars has been reflected in—and shaped by—the writings of Aleksandr Bogdanov, Edgar Rice Burroughs, J. H. Rosny, Stanley G. Weinbaum, Ray Bradbury, Leigh Brackett, Robert Heinlein, Isaac Asimov, Philip K. Dick, Frederik Pohl, Kim Stanley Robinson, and many more. Nowhere is the relationship between scientific speculation and speculative fiction clearer than the future Mars projects and programs put forward by space scientists from Wernher von Braun to Robert Zubrin.

HOW IT WORKS

Telescopy. Although areology is a relatively new term, the roots of a general discipline of Mars studies stretch back nearly four centuries. For the first three and a half centuries, however, these studies were exclusively telescopic. Mars was an object viewed through an eyepiece from Earth. The power of telescopes and the levels of resolution they offered grew steadily over time, and telescopic studies remain very important in areological research, but recognition of the inherent limitations of such studies led the push to move scientific instrumentation closer to Mars via flyby, then linger in orbit to gather more detailed data. Eventually this led to landing scientific payloads on the planet's surface, then to making those payloads capable of self-propulsion across that surface.

Flyby. Mars 1, also known as Sputnik 23, was launched on November 1, 1962, and was intended to fly past Mars at a distance of about 11,000 kilometers or 7,000 miles. It was the first Soviet Mars probe and carried a package of scientific instrumentation that included television photographic equipment, a magnetometer probe, a spectral reflectometer, a spectrograph, a micrometeoroid impact instrument, and radiation sensors. Data from this instrumentation package were to be broadcast back to Earth via radio and television transmitters. Although Mars 1 lost contact with Earth before accomplishing its flyby, the configuration of its scientific instrumentation package (for collecting data) and transmission capabilities (for returning that collected data to Earth) became the standard for all Mars flyby missions.

The American craft Mariner 4, launched on November 28, 1964, completed the first successful flyby of Mars. Mariner 4's television pictures of the Martian surface were the first images of another planet sent back from deep space and changed the way the scientific community viewed the possibility of life on Mars. Mariner 6 and Mariner 7, in 1969, were similarly successful flyby missions, making closer approaches and providing more photographic and other data to that already compiled by Mariner 4.

Orbiter. In 1971, Mariner 9 was launched and, once inserted into orbit around Mars, became the first spacecraft to orbit another planet. It was followed soon after by two successful Russian orbiters, Mars 2 and Mars 3.

In orbiting the planet, Mariner 9 photographed 100 percent of the Martian surface and was able to wait out a prolonged dust storm that obscured much of the planet's surface—something a flyby mission could not have done. Mariner 9's successful data collection laid the groundwork not only for the later Viking orbiter/lander missions but also for successful later-generation orbiters with more advanced instrument packages, including the Mars Global Surveyor in 1996, and Mars Odyssey, Mars Express, and Mars Reconnaissance Orbiter during the first decade of the twenty-first century.

Lander. The Soviet Mars 3, whose orbiter was successful, also had a partially successful lander component in that its descent module, which contained both a lander and a rover, was able to utilize aerodynamic braking, parachutes, and retro-rockets to make a soft landing. Unfortunately, twenty seconds after touching down the lander stopped transmitting, and it was unable to deploy its rover component.

Considerably more successful were the American Viking 1 and Viking 2 craft, whose orbiters achieved orbit and whose landers, again through a combination of aerodynamic braking, parachutes, and retro-rockets, landed softly and stayed in operation for years, completing scientific objectives that included

not only photographic imaging at the planet's surface but also soil analysis and biological-assay experiments for evidence of organic compounds and, potentially, the presence of life.

Later successful U.S. landers included the Mars Pathfinder lander/rover mission (which utilized air bags rather than retro-rockets during the last phase of its landing) and the Phoenix, which studied the geologic history of water on Mars, its involvement in Martian climate change, and the planet's past or future habitability.

Rover. Although Russian Mars 2 and Mars 3 descent modules brought rovers with them as early as 1971, no rover was successfully deployed on Mars until the U.S. Pathfinder mission of 1997 deployed its Sojourner rover. Able to travel about a half kilometer, or one-third of a mile from the lander, the Sojourner rover returned 550 photographs to Earth and the data from chemical analyses of sixteen locations on the Martian surface.

Mars Exploration Rovers (MER) Spirit and Opportunity landed on opposite sides of Mars in 2004. Both vehicles were intended to engage in geologic, hydrologic, and biologic assessment activities: to examine rocks and soils for evidence of past water activity, as well as assess whether the environments that prevailed when water was present were conducive to life.

The Spirit and Opportunity rovers have been tremendously successful, their missions lasting more than twenty times the planned duration. Opportunity is still operational and holds the record for longest Mars surface mission. The two rovers have covered far more terrain and have provided far more data than any previous mission.

APPLICATIONS AND PRODUCTS

The National Aeronautics and Space Administration (NASA) lists more than 2,000 applications and products on its spin-off database. These spin-offs from space research contribute to national security, the economy, productivity, and lifestyle not only in the United States but also throughout the world. These spin-offs are so numerous and ubiquitous that people are scarcely aware of them and too often take them for granted. Below is a sampling of those specifically related to Mars research, many of which were developed in response to areological studies of Martian surface conditions.

Fascinating Facts About Areology

- The launch patch for Mars Exploration Rover Spirit features Marvin the Martian from the Looney Tunes cartoons.
- The launch patch for Mars Exploration Rover Opportunity features Duck Dodgers, an avatar of Daffy Duck.
- A dog killed in Nakhla, Egypt, in 1911 is reported to be the only known casualty of a Martian meteorite.
- In 1984, a meteorite of Martian origin (ALH84001) was discovered in Antarctica and contained what looked like fossil bacteria. The evidence remains inconclusive and controversial.
- A Viking orbiter's photograph of a low knoll on the Cydonia plateau caused scientists on the imaging team to joke that the image looked like a human face, but the Cydonia landform, in fact, depicts what looks like a human face.
- Writers Jonathan Swift and Voltaire both independently "predicted" that Mars had two moons, considerably more than a century before they were discovered.
- The seemingly high percentage of Mars mission failures has been attributed to a Mars curse (alternatively called the "Galactic Ghoul" and the "Mars Triangle") but is most likely caused by more prosaic circumstances, such as the use of complex advanced technologies in unprecedentedly severe environments.

Sensors. NASA research into detecting biological traces on Mars has resulted in biosensor technology monitoring water quality. Sensors incorporating carbon nanotubes tipped with single strands of nucleic acid from waterborne pathogens can detect minute amounts of disease-causing bacteria, viruses, and parasites and be used to alert organizations to potential biological hazards in water used for agriculture, food and beverages, showers, and at beaches and lakes.

NASA's Jet Propulsion Laboratory (JPL) developed a bacterial spore-detection system for Mars-bound spacecraft that can also recognize anthrax and other harmful, spore-forming bacteria on Earth and alert people of the impending danger.

JPL also developed a laser diode-based gas analyzer

as part of the 1999 Mars Polar Lander mission to explore the possibility of life-giving elements on Mars. It has since been used on aircraft and on balloons to study weather and climate, global warming, emissions from aircraft, and numerous other areas where chemical-gas analysis is needed.

Computing and Imaging. NASA Advanced Supercomputing (NAS) division, which includes the Columbia supercomputer, is responsible for a wide range of products, from the development of computational fluid dynamics (CFD) computer codes to novel immersive visualization technologies used to pilot the Spirit and Opportunity rovers. Wide-screen panoramic photography technologies developed for the Mars rovers' Pancam robotic platform is in production as a GigaPan platform for automating the creation of highly detailed digital panoramas in consumer cameras.

Materials. Multilayer textiles developed for the air bags used in the Mars Pathfinder and Exploration Rovers are being used in Warwick Mills' puncture- and impact-resistant TurtleSkin product line of metal flex armor (MFA) vests, which are comparable to rigid steel plates but far more comfortable.

The thin, shiny insulation material used extensively in the Mars rover missions—a strong lightweight plastic, vacuum-metallized film that minimizes weight impact on vehicle payload while also protecting spacecraft, equipment, and personnel from the extreme temperature fluctuations of space—is found in applications ranging from reflective thermal blankets to party balloons.

IMPACT ON INDUSTRY

The annual budget for the entire American space program is about $19 billion, or 0.8 percent of the $2.4 trillion budget. Collectively, however, secondary applications (spin-offs) represent a substantial return on the national investment in aerospace research: $7 from come back from spin-offs for every $1 spent on research. This surplus is generated by taxes from increased jobs in aerospace as well as all the other fields that produce spin-off goods.

Although it is difficult to sort out the actual worldwide spending on research and development relating specifically to Mars exploration and to separate out space-related research from other aerospace and defense spending, the NASA budget for Mars exploration in the first part of the twenty-first century has generally averaged around $500 million per year. Given multiplier effects and the share of Mars-related research in many aerospace-industrialized nations worldwide, including Japan, France, United Kingdom, Canada, Belgium, Russia, and China, the total value of global Mars-related research and development is estimated at roughly $13 billion.

Government and University Research. In Mars exploration-related research, NASA has a robust international relationship with agencies like the European Space Agency and the Japanese Aerospace Exploration Agency and with governmental and university scientific researchers from the United Kingdom, France, Italy, Australia, Belgium, Canada, Japan, Sweden, and Switzerland.

In Mars exploration-related research within the United States government, NASA maintains close ties with many Defense Department units, including the Naval Research Lab but particularly Defense Advanced Research Projects Agency (DARPA) and the U.S. Army, whose work involving tracked vehicles and robotics have paralleled JPL's work with rovers.

In Mars exploration-related research within NASA itself, JPL is the most important of NASA's dozen nationwide centers. JPL, which was established by the California Institute of Technology, has formed strategic relationships with ten schools that have major commitments to space exploration: Arizona State University; Carnegie Mellon University; Dartmouth College; Massachusetts Institute of Technology; Princeton University; Stanford University; University of Arizona; University of California, Los Angeles; University of Michigan; and University of Southern California. Through such relationships, JPL and its university collaborators facilitate joint access to particular capabilities in science, technology, and engineering and encourage better understanding of the state of research in the broader scientific community. Such collaborations also support students in space exploration topics, including graduate research on topics of interest to JPL/NASA, student participation in JPL summer programs, and input regarding courses of strong interest to NASA. Such collaborations also cultivate JPL's future workforce and ensure a pipeline to meet future technical challenges.

Industry and Business. Mars exploration-related research is most closely connected to aerospace and robotics. A single NASA/JPL program in development as of this writing, Distributed Spacecraft

Technology Program for Precision Format Flying, involves companies as varied as Guidance Dynamics Corporation, DI-TEC International, Ball Aerospace, Applied Physics Technologies, Tera Semicon Corporation, and Pacific Wave Industries. The true impact, however, is less direct and found largely through the role secondary applications or spin-offs play in the wide variety of industries that make use of sensing, computing, imaging, and advanced materials.

CAREERS AND COURSE WORK

Courses in astronomy, biology, chemistry, computer science, engineering, geology, and mathematics are foundational for students wishing to pursue careers in areology.

Master's and doctoral degrees are often the necessary minimum qualification for more advanced academic, governmental, or industrial careers in Mars exploration-related science. More specialized courses may include astrobiology, biochemistry, geophysics, climatology, hydrology, geodesy, and robotics, as well as a number of specializations within engineering, particularly mechanical, electrical, human factors, or systems.

Although areology is geological at its root, it is also the general study of a world other than that known to humans and at this point in its development is strongly interdisciplinary, so background in a diversity of fields, including the history of science and the study of literature concerning Mars, can also prove very helpful.

SOCIAL CONTEXT AND FUTURE PROSPECTS

For areology, the twentieth century was shaped by two important movements. One was the transition from an understanding of Mars based on telescopy to one characterized by spacecraft with scientific instrument payloads flying by, orbiting, landing, and discharging mobile quasi-autonomous vehicles onto the surface to "follow the water" and look for evidence of life. The other was the movement from an understanding of Mars based primarily in fictional speculation to one increasingly based in science.

The question of past or present life on Mars, however, remains in the realm of speculation. The great debates in this century for areology will begin with whether remotely controlled or increasingly autonomous robotic vehicles can conclusively decide the question of past or present life or whether that question can be conclusively decided only through expensive, potentially dangerous (and perhaps infeasible) manned missions to Mars. That in itself, however, presents a problem: If there is no life on Mars, should the planet be preserved in its pristine state? Conversely, if there is life on Mars, should people risk causing the extinction of that life through contamination from Earth—or humanity's own extinction, through contamination from something on Mars?

These sound more and more like the speculations of science fiction, and matters become only more speculative as people contemplate the efficacy and feasibility of expensive, dangerous, and longer-term effects of colonization and terraforming of Mars by humans.

In trying to find Mars analogues on Earth, scientists are learning more about the limits to life on the world. By setting up microbial observatories in environments that may be in at least one way or another like certain environments on Mars, people have broadened their understanding of the diversity of life on Earth, ultimately serving to make Mars and Earth look more like each other at their extremes than previously assumed.

Howard V. Hendrix, B.A., M.A., Ph.D.

FURTHER READING

Brandenburg, John E., and Monica Rix Paxson. *Dead Mars, Dying Earth.* Freedom, Calif.: Crossing Press, 1999. Controversial text arguing that Mars was much warmer and wetter until roughly a half billion years ago, when catastrophic climate change ended its ability to support life, and how understanding the death of Mars may save Earth.

Chapman, Mary G., and Laszlo P. Keszthelyi. *Preservation of Random Megascale Events on Mars and Earth: Influence on Geologic History.* Boulder, Colo.: Geological Society of America, 2009. Illustrated, multicontributor volume of essays by professional geologists regarding the preservation of large-scale geologic events on Earth and Mars.

Harland, David M. *Water and the Search for Life on Mars.* Chichester, England: Praxis, 2005. Richly illustrated examination of the "follow the water" approach to Mars exploration and the implications of the possibility of life existing (or having existed) on the planet.

Kargel, Jeffrey S. *Mars: A Warmer Wetter Planet.* Chichester, England: Praxis, 2004. A well-researched,
thoroughly illustrated, and extensive examination of the wet-Mars hypothesis and what it means to human expectations and realities concerning the Red Planet.

Morton, Oliver. *Mapping Mars: Science, Imagination, and the Birth of a World.* New York: Picador, 2002. Cartography meets philosophy in this illustrated text that explores the natural history and topography of Mars.

Tokano, Tetsuya, ed. *Water on Mars and Life.* New York: Springer-Verlag, 2005. This collection of essays by professional scientists details the role of water in the planetary evolution of early Mars, water reservoirs on Mars, and the possible astrobiological importance of terrestrial analogues of putative aqueous environments on Mars.

Turner, Martin J. L. *Expedition Mars.* Chichester, England: Praxis, 2004. A thoroughly illustrated history of space exploration (both manned and robotic) tending toward a future Mars landing and the challenges inherent in undertaking an expedition to the planet.

Zubrin, Robert. *Mars on Earth: The Adventures of Space Pioneers in the High Arctic.* New York: Jeremy P. Tarcher/Penguin, 2003. An account of the Flashline Mars Arctic Research Station, a Mars analogue habitat on Devon Island in the Canadian Arctic, which was inhabited by volunteers during a simulation of human habitation on Mars that began in 2001.

WEB SITES

Association of Mars Explorers
http://marsexplorers.org

Jet Propulsion Laboratory
http://www.jpl.nasa.gov

The Mars Society
http://www.marssociety.org

National Aeronautics and Space Administration Spinoffs
http://www.sti.nasa.gov/tto

Institute of Electrical and Electronics Engineers
http://www.ieee.org

Massachusetts Institute of Technology
Computer Science and Artificial Intelligence
Laboratory
http://www.csail.mit.edu

Society for the Study of Artificial Intelligence and
Simulation of Behaviour
http://www.aisb.org.uk

ATMOSPHERIC SCIENCES

FIELDS OF STUDY

Climatology; meteorology; climate change; climate modeling; hydroclimatology; hydrometeorology; physical geography; chemistry, physics.

SUMMARY

Atmospheric sciences includes the fields of physics and chemistry and the study of the composition and dynamics of the layers of air that constitute the atmosphere. Related topics include climatic processes, circulation patterns, chemical and particulate deposition, greenhouse gases, oceanic temperatures, interaction between the atmosphere and the ocean, the ozone layer, precipitation patterns and amounts, climate change, air pollution, aerosol composition, atmospheric chemistry, modeling of pollutants both indoors and outdoors, and anthropogenic alteration of land surfaces that in turn affect conditions within the ever-changing atmosphere.

KEY TERMS AND CONCEPTS

- **Atmosphere:** Gaseous layer that surrounds the Earth and is held in place by gravity.
- **Atmospheric Pressure:** Gravitational force caused by the weight of overlying layers of air.
- **Atmospheric Water:** Water in the atmosphere; it can be in a gaseous, liquid, or frozen form.
- **Coriolis Effect:** Rightward (Northern Hemisphere) and leftward (Southern Hemisphere) deflection of air caused by the Earth's rotation.
- **Energy Balance:** Return to space of the Sun's energy that was received by the Earth.
- **Front:** Boundary between airmasses that differ in temperature, moisture, and pressure.
- **Greenhouse Effect:** Process by which longwave radiation is trapped in the atmosphere and then returned to the Earth's surface by counterradiation.
- **Occluded Front:** Front produced when a cold front overtakes a warm front and forces the air upward.
- **Troposphere:** Lowest level of the Earth's atmosphere that contains water vapor that can condense into clouds.

- **Urban Heat Island:** Urbanized area that experiences higher temperatures than surrounding rural lands.

DEFINITION AND BASIC PRINCIPLES

Atmospheric sciences is the study of various aspects of the nature of the atmosphere, including its origin, layered structure, density, and temperature variation with height; natural variations and alterations associated with anthropogenic impacts; and how it is similar to or different from other atmospheres within the solar system. The present-day atmosphere is in all likelihood quite dissimilar from the original atmosphere. The form and composition of the present-day atmosphere is believed to have developed about 400 million years ago in the late Devonian period of the Paleozoic era, when plant life developed on land. This vegetative cover allowed plants to take in carbon dioxide and release oxygen as part of the photosynthesis process.

The atmosphere consists of a mixture of gases that remain in place because of the gravitational attraction of the Earth. Although the atmosphere extends about 6,000 miles above the Earth's surface, the vast proportion of its gases (97 percent) are located in the lower 19 miles. The bulk of the atmosphere consists of nitrogen (78 percent) and oxygen (21 percent). The last 1 percent of the atmosphere contains all the remaining gases, including an inert gas (argon), which accounts for 0.93 percent of the 1 percent, and carbon dioxide (CO_2), which makes up a little less than 0.04 percent. Carbon dioxide has the ability to absorb longwave radiation leaving the Earth and shortwave radiation from the Sun; therefore, any increase in carbon dioxide in the atmosphere has profound implications for global warming.

BACKGROUND AND HISTORY

Evangelista Torricelli, an Italian physicist, mathematician, and secretary to Galileo, invented the barometer, which measures barometric pressure, in 1643. The first attempt to explain the circulation of the global atmosphere was made in 1686 by Edmond Halley, an English astronomer and mathematician. In 1735, George Hadley, an English optician, described a pattern of air circulation that became known as a Hadley cell. In 1835, Gustave-Gaspard Coriolis, a French engineer and mathematician, analyzed the movement of air on a rotating Earth, a pattern that became known as the Coriolis effect. In 1856, William Ferrel, an American meteorologist, developed a model of hemispheric circulation of the atmosphere that became known as a Ferrel cell. Christophorus Buys Ballot, a Dutch meteorologist, explained the relationship between the distribution of pressure, wind speed, and direction in 1860.

Manned hot-air balloon flights beginning in the mid-nineteenth century facilitated high-level observations of the atmosphere. For example, in 1862, English meteorologist James Glaisher and English pilot Henry Coxwell reached 29,000 feet, at which point Glaisher became unconscious and Coxwell was partially paralyzed so that he had to move the control valve with his teeth. In 1902, Léon Teisserenc de Bort of France was able to determine that air temperatures begin to level out at 39,000 feet and actually increase at higher elevations. In the twentieth century, additional information about the upper atmosphere became available through radio waves, rocket flights, and satellites.

How It Works

A knowledge of the basic structure and dynamics of the atmosphere are a necessary foundation for understanding applications and practical uses based on atmospheric science.

Layers of the Atmosphere. The heterosphere and the homosphere form the two major subdivisions of the Earth's atmosphere. The uppermost subdivision (heterosphere) extends from about 50 miles above the Earth's surface to the outer limits of the atmosphere at about 6,000 miles. Nitrogen and oxygen, the heavier elements, are found in the lower layers of the heterosphere, and lighter elements such as hydrogen and helium are found at the uppermost layers of the atmosphere. The homosphere (or lowest layer) contains gases that are more uniformly mixed, although their density decreases with height. Some exceptions to this statement occur with the existence of an ozone layer at an altitude of 12 to 31 miles and with variations in concentrations of carbon dioxide, water vapor, and air pollutants closer to the Earth's surface.

The atmosphere can be divided into several zones based on decreasing or increasing temperatures as elevation increases. The lowest zone is the troposphere, where temperatures decrease from sea level up to an altitude of 10 miles in equatorial and tropical regions and up to an altitude of 4 miles at the poles. This lowermost zone holds substantial amounts of water vapor, aerosols that are very small, and light particles that originate from volcanic eruptions, desert surfaces, soot from forest and brush fires, and industrial emissions. Clouds, storms, and weather systems occur in the troposphere.

The tropopause marks the boundary between the troposphere and the next higher layer, the stratosphere, which reaches an altitude of 30 miles above the Earth's surface. Circulation in this layer occurs with strong winds that move from west to east. There is limited circulation between the troposphere and the stratosphere. However, manned balloons, certain types of aircraft (Concorde and the U-2), volcanic eruptions, and nuclear bomb tests are able to break through the tropopause and enter the stratosphere.

The gases in the stratosphere are generally uniformly mixed, with the major exception of the ozone layer, which is found at an altitude range of 12 to 31 miles above the Earth. This layer is extremely important because it shields life on Earth from the intense and harmful ultraviolet radiation from the Sun. The ozone layer has been diminishing because of the release of chlorofluorocarbons (CFCs), organic compounds containing chlorine, fluorine, and carbon, used as propellants in aerosol sprays and in synthetic chemical compounds used for refrigeration purposes. In 1978, the use of CFCs in aerosol sprays was banned in the United States, but they are still used in some refrigeration systems. Other countries continue to use CFCs, which eventually get into the ozone layer and result in ozone holes of considerable size. In 1987, members of the international community took steps to reduce CFC production through the Montreal protocol, and by 2003, the rate of ozone depletion had began to slow down. Although the manufacture and use of CFCs can be controlled, natural events that are detrimental to the ozone layer cannot be prevented. For example, the 1991 eruption of Mount Pinatubo in the Philippines reduced the ozone layer in the midlatitudes nearly 9 percent.

Temperatures decrease with elevation at the stratopause, where the mesosphere layer begins at about 30 miles and continues to an altitude of about 50 miles. The mesopause at about 50 miles marks the beginning of the thermosphere, where the density of the air is very low and holds minimal amounts of heat. However, even though the atmospheric density is minimal at altitudes above 155 miles, there is enough atmosphere to have a drag effect on spaceships.

Atmospheric Pressure. The gas molecules in the atmosphere exert a pressure due to gravity that amounts to about 15 pounds per square inch on all surfaces at sea level. As the distance from the Earth gets larger, in contrast to the various increases and decreases in atmospheric temperature, atmospheric pressure decreases at an exponential rate. For example, air pressure at sea level varies from about 28.35 to 31.01 inches of mercury, averaging 29.92 inches. The pressure at the top of Mount Everest at 20,029 feet can get as low as 8.86 inches. This means that each inhalation of air at this altitude is about one-third of the pressure at sea level, producing severe shortness of breath.

Earth's Global Energy Balance. The Earth's elliptical orbit about the Sun ranges from 91.5 million miles at perihelion (closest point to the Sun) on January 3 to 94.5 million miles at aphelion (furthest from the Sun) on July 4, averaging 93 million miles. The Earth intercepts only a tiny fraction of the total energy output of the Sun. Upon reaching the Earth, part of the incoming radiation is reflected back into space, and part is absorbed by the atmosphere, land, or oceans. Over time, the incoming shortwave solar radiation is balanced by a return to outer space of longwave radiation.

Earth-Moon Differences. Scientists believe that the moon's surface has a large number of craters formed by the impact of meteorites. In contrast, there are relatively few meteorite craters on the Earth, even though, based simply on its size, the Earth is likely to have been hit by as many or even more meteorites than the Moon. This notable difference is attributed to the Earth's atmosphere, which burns up incoming meteorites, particularly small ones (the Moon does not have an atmosphere). Larger meteorites can pass through the Earth's atmosphere, but their impact craters may have been filled in or washed away over millions of years. Only the more recent ones, such as Meteor Crater in northern Arizona, with a diameter of 4,000 feet and a depth of 600 feet, remain easily recognizable.

Air Masses. Different types of air masses within the troposphere, the lowest layer of the atmosphere, can be delineated on the basis of their similarity in temperature, moisture, and to a certain extent, air pressure. These air masses develop over continental and maritime locations that strongly determine their physical characteristics. For example, an air mass starting in the cold, dry interior portion of a continent develops thermal, moisture, and pressure differences that can be substantially different from an air mass that develops over water. Atmospheric dynamics also allow air masses to modify their characteristics as they move from land to water and vice versa.

Air mass and weather front terminology were developed in Norway during World War I. Norwegian meteorologists were unable to get weather reports from the Atlantic theater of operations; consequently, they developed a dense network of weather stations that led to impressive advances in atmospheric modeling that are still being used.

The Radiation Budget. The incoming solar energy that reaches the Earth is primarily in the shortwave (or visible) portion of the electromagnetic spectrum. The Earth's energy balance is attained by about one-third of this incoming energy being reflected back to space and the other two-thirds leaving the Earth as outgoing longwave radiation. This balance between incoming and outgoing energy is known as the Earth's radiation budget. The decades-long, ongoing National Aeronautics and Space Administration (NASA) program known as Clouds and the Earth's Radiant Energy System (CERES) is designed to measure how much shortwave and longwave radiation leaves the Earth from the top of the atmosphere.

Clouds play a very important role in the global radiation balance. For one thing, they constantly change over time and in type. Some clouds, such as high cirrus clouds found near the top of the troposphere at 40,000 feet, can have a substantial impact on atmospheric warming. Accordingly, the value of CERES is based on its ability to observe if human or natural changes in the atmosphere can be measured even if they are smaller than large-scale energy variations.

Greenhouse Effect. Selected gases in the lower parts of the atmosphere trap heat and then radiate some of that heat back to Earth. If there was no natural greenhouse effect, the Earth's overall average temperature would be close to 0 degrees Fahrenheit, rather than the existing 57 degrees Fahrenheit.

The burning of coal, oil, and gas makes carbon

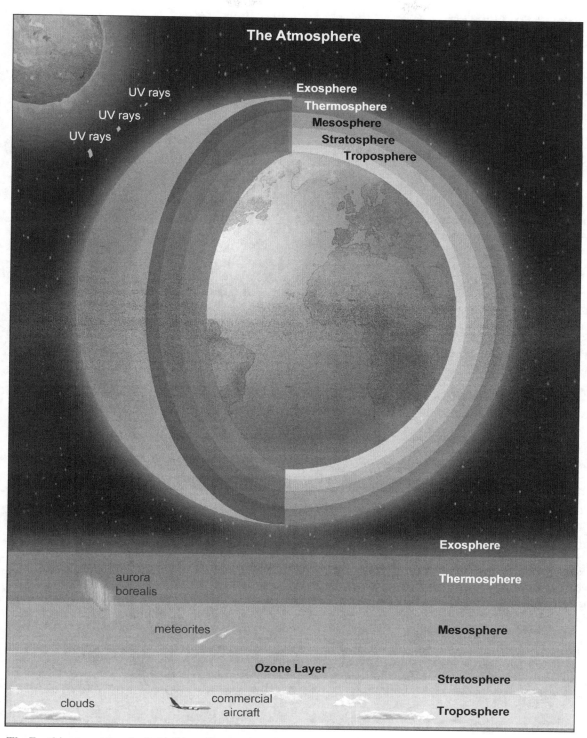

The Earth's atmosphere is divided into five layers.

dioxide the major green house gas, accounting for nearly half of the total amount of heat-producing gases in the atmosphere. Before the Industrial Revolution in Great Britain in the mid-eighteenth century, the estimated level of carbon dioxide in the atmosphere was about 280 parts per million by volume (ppmv). Estimates for the natural range of carbon dioxide for the past 650,000 years range from 180 to 300 ppmv. All these values are less than the 391 ppmv recorded in January, 2011. Carbon dioxide levels have been increasing since 2000 at a rate of 1.9 ppmv each year. The radiative effect of carbon dioxide accounts for about one-half of all the factors that affect global warming. Estimates of carbon dioxide levels at the end of the twenty-first century range from 490 to 1,260 ppmv.

The second most important greenhouse gas is methane (CH_4), which accounts for about 14 percent of all global warming factors. The origin of this gas is attributed to the natural decay of organic matter in wetlands, but anthropogenic activity—rice paddies, manure from farm animals, the decay of bacteria in sewage and landfills, and biomass burning (both natural and human induced)—results in a doubling of the amount of this gas over what would be produced solely by wetland decay.

Chlorofluorocarbons (CFCs) absorb longwave energy (warming effect), but they also have the ability to destroy stratospheric ozone (cooling effect). The warming radiative effect is three times greater than the cooling effect. CFCs account for about 10 percent of all global warming factors. Tropospheric ozone from air pollution and nitrous oxide (N_2O) from motor vehicle exhaust and bacterial emissions from nitrogen fertilizers account for about 10 percent and 5 percent, respectively, of all global warming factors.

Several kinds of human actions lead to a cooling of the Earth's climate. For example, the burning of fossil fuels results in the release of tropospheric aerosols, which acts to scatter incoming solar radiation back into space, thereby lowering the amount of solar energy that can reach the Earth's surface. These aerosols also lead to the development of low and bright clouds that are quite effective in reflecting solar radiation back into space.

APPLICATIONS AND PRODUCTS

Atmospheric science is applied in many ways. It is used to help people better understand their global and interplanetary environment and to make it possible for them to live safely and comfortably within that environment. By using the principles of this field, researchers, engineers, and space scientists have developed a vast number of applications. Among the most important are those used to track and predict weather cycles and climate.

Remote Sensing Techniques. Oceans cover about 71 percent of the Earth's surface, which means that large portions of the world do not have weather stations or places where precipitation can be measured with standard rain gauges. To provide more information about precipitation in the equatorial and tropical parts of the world, NASA and the Japan Aerospace Exploration Agency initiated the Tropical Rainfall Monitoring Mission (TRMM) in 1997. The orbit of the TRMM satellite monitors the Earth between 35 degrees north and 35 degrees south latitude. The goal of the study is to obtain information about the extent of precipitation, along with its intensity and length of occurrence. The major instruments on the satellite are radar to detect rainfall, a passive microwave imager that can acquire data about precipitation intensity and the extent of water vapor, and a scanner that can examine objects in the visible and infrared portions of the electromagnetic spectrum. The goal of data collection is to obtain the necessary climatological information about atmospheric circulation in this portion of the Earth to develop better mathematical models for determining large-scale energy movement and precipitation.

Geostationary Satellites. Geostationary operational environmental satellites (GOES) enable researchers to view images of the planet from what appears to be a fixed position above the Earth. The satellites are actually circling the globe at a speed that is in step with the Earth's rotation. This means that a satellite at an altitude of 22,200 miles will make one complete revolution in the same twenty-four hours and direction that the Earth is turning above the equator. At this height, the satellite is in a position to view nearly one-half of the planet at any time. On-board instruments can be activated to look for special weather conditions such as hurricanes, flash floods, and tornadoes. On-board instruments are also used to make precipitation estimates during storm events.

Doppler Radar. Doppler radar was first used in

England in 1953 to pick up the movement of small storms. The basic principle guiding this type of radar is that back-scattered radiation frequency detected at a certain location changes over time as the target, such as a storm, moves. A transmitter is used to send short but powerful microwave pulses. When a foreign object (or target) is intercepted, some of the outgoing energy is returned to the transmitter, where a receiver can pick up the signal. An image (or echo) from the target can then be enlarged and shown on a screen. The target's distance is revealed by the time that elapses between transmission and return. The radar screen cannot only indicate where the precipitation is taking place but also reveal the intensity of the rain by the amount of the echo's brightness. In short, Doppler radar has become a very useful device for determining the location of a storm and the intensity of its precipitation and for obtaining good estimates of the total amount of precipitation.

Responses to Climate Change. Since the 1970's, many scientists have pointed out the possibility that human activity is having more than a short-term impact on the atmosphere and therefore on weather and climate. Although much debate continues on the full impact of human activities and greenhouse gas emissions, the atmospheric sciences have led to conferences, United Nations conventions, and agreements among nations on ways that human beings can alter their behavior to halt or at least mitigate the possibility of global climate change. The impact of these agreements, still in their infancy, remains unknown—as does the overall effect of human activity on weather and climate (the models for which are highly complex). However, the insights contributed by the atmospheric sciences to the overall debate on whether climate change is primarily anthropogenic (human caused)—and whether global warming is actually taking place—have caused many nations and individuals to modify their attitudes toward human relationships with the global environment, resulting in national and intergovernmental changes in policies concerning carbon emissions, as well as personal decisions ranging from the consumption of "green" building materials to the purchase of vehicles fueled by noncarbon sources of energy.

IMPACT ON INDUSTRY

Global Perspective. The World Meteorological Organization, headquartered in Geneva, Switzerland,

was established to encourage weather station networks to acquire many types of atmospheric data. Accordingly, in 2007, its members decided to expand the Global Observing System (GOS) and other related observing systems, including the Global Ocean Observing System (GOOS), Global Terrestrial Observing System (GTOS), and the Global Climate Observing System (GCOS). Data are being collected from some 10,000 manned and automatic surface weather stations, 1,000 upper-air stations, more than 7,000 ships, 100 moored and 1,000 floating buoys that can drift with the currents, several hundred radars, and more than 3,000 commercial airplanes that can acquire key data on aspects of the atmosphere, land, and ocean surfaces on a daily basis.

Government Research. About 180 countries maintain meteorological departments. Although many of these departments are small, the larger countries tend to have well-established governmental organizations. The major U.S. agency involved in the atmospheric sciences is the National Oceanic and Atmospheric Administration (NOAA). The agency's National Climatic Data Center (NCDC) in Asheville, North Carolina, has meteorological records going back to 1880 for both the world and the United States. These records provide invaluable historical information. For example, sea ice in the Arctic Ocean typically reaches its maximum extent in March. The coverage at the end of March, 2010, was 5.8 million square miles, which marked the seventeenth consecutive March with below-average areal extent. The National Climatic Data Center issues monthly temperature and precipitation summaries for all of the states in addition to many specialized climate data publications. It also publishes monthly mean climatic data for temperature, precipitation, barometric pressure, sunshine, and vapor pressure (that portion of atmospheric pressure that is attributed to water vapor at the time of measurement) on a global scale for about 2,000 surface sites. In addition, monthly mean upper air temperatures, dew point depressions, and wind velocities are collected and published for about 500 locations scattered around the world.

University Research. Forty-eight states (all but Rhode Island and Tennessee) have either a state climatologist or someone with comparable responsibility. Most of the state climatologists are connected with state universities, in particular, the land grant institutions. The number of cooperative weather stations established since the late nineteenth century to take daily readings of temperature and precipitation in each of the forty-eight states has varied over time. These cooperative weather stations include public and private water supply facilities, colleges and universities, airports, and interested citizens.

Industry and Business Sectors. The number, size, and capability of private consulting firms has increased since the latter part of the twentieth century. Perhaps among the earliest entrants into this market were frost-warning service providers for citrus and egetable growers in Arizona, Florida, and California. These private companies expanded as better forecasting and warning techniques were developed. One example of a private enterprise using atmospheric sciences is AccuWeather.com, which has seven global and fourteen regional forecast models for the United States and North America and prepares a daily weather report for *The New York Times*.

Careers and Course Work

The study of the physical characteristics of the atmosphere falls within the purview of atmospheric scientists. Those interested in a career in this technical area should recognize that there are several categories of specialization. The major group of specialists are operational meteorologists, who are responsible for weather forecasts. They have to carefully study the temperature, humidity, wind speed, and barometric pressure from a number of weather stations to make daily and long-range forecasts. They use data from weather satellites, radar, special sensors, and observation stations in other locations to make forecasts.

In contrast to meteorologists, who focus on short-term weather forecasts, the study of changes in weather over longer periods of time such as months, years, and in some cases centuries, is handled by climatologists. Other atmospheric scientists concentrate on research. For example, physical meteorologists are concerned with various aspects of the atmosphere, such as its chemical and physical properties, energy transfer, severe storm mechanics, and the spread of air pollutants over urbanized areas. The growing interest in air pollution and water shortages has led to another group of research scientists known as environmental meteorologists.

Given the importance of weather forecasting on a daily basis, operational meteorologists who work in weather stations may work on evenings, weekends,

---

Fascinating Facts About Atmospheric Sciences

- In April, 1934, a surface wind gust of 231 miles per hour was recorded at Mount Washington Observatory in New England.
- The coldest temperature in the world, −127 degrees Fahrenheit, was recorded at the Russian weather station at Vostok in Antarctica in 1958.
- The world's lowest barometric pressure (25.69 inches of mercury) was recorded on October 12, 1979, during Typhoon Tip in the western Pacific Ocean, and the highest (32.06 inches) was recorded in Tosontsengel, Mongolia, on December 19, 2001.
- One study found that cloud formation and rain occurred more frequently on weekdays than weekends, because of higher levels of air pollutants, at least in humid regions with cities.
- Acid rain, a mixture containing higher than normal amounts of nitric and sulfuric acid, occurs when emissions of sulfuric acid and nitrogen dioxide (released from power plants and decaying vegetation) react with oxygen, water, and other chemicals, forming acidic compounds.
- Mean sea levels for the world increased 0.07 inches per year in the twentieth century, an amount that is much larger than the average rate of increase for the last several thousand years.
- The specific chemicals in particulate matter, an air pollutant, depend on the source and the geographic location. They include inorganic compounds such as sulfate, nitrate, and ammonia; organic compounds formed by the incomplete burning of wood, gasoline, and diesel fuels; and secondary organic aerosols, new compounds formed when pollutants combine.
- Winds that reach the center of a low-pressure area rise, causing condensation, cloud formation, and storms. Dry winds are drawn down the center of a high-pressure area, resulting in fair weather.

---

and holidays. Research scientists who are not engaged in weather forecasts may work regular hours.

In 2009, the American Meteorological Society es In 2009, the American Meteorological Society estimated that there are about one hundred undergraduate and graduate atmospheric science programs in the United States that offer courses is such departments as physics, earth science, environmental science, geography, and geophysics. Entry-level positions usually require a bachelor's degree with at least one hundred undergraduate and graduate atmospheric science programs in the United States that offer courses is such departments as physics, earth science, environmental science, geography, and geophysics. Entry-level positions usually require a bachelor's degree with at least twenty-four semester hours in courses covering atmospheric science and meteorology. The acquisition of a master's degree enhances the chances of employment and usually means a higher salary and more opportunities for advancement. A doctorate is required only for those who want a research position at a university.

In 2008, excluding research positions in colleges and universities, about 9,400 atmospheric scientists were working, and about one-third were employed by the federal government. Above-average employment growth is projected until 2018, representing a 15 percent increase from 2008. The median annual average salary for atmospheric scientists in May, 2008, was $81,290. The middle 50 percent had earnings between $55,140 and $101,340.

SOCIAL CONTEXT AND FUTURE PROSPECTS

Climate change may be caused by both natural internal/external processes in the Earth-Sun system or by human-induced changes in land use and the atmosphere. Article 1 of the United Nations Framework Convention on Climate Change (entered into force March, 1994) states that the term "climate change" should refer to anthropogenic changes that affect the composition of the atmosphere rather than natural causes, which should be referred to "climate variability." An example of natural climate variability is the global cooling of about 0.5 degrees Fahrenheit in 1992-1993 that was caused by the 1991 eruption of Mount Pinatubo in the Philippines. The 15 million to 20 million tons of sulfuric acid aerosols ejected into the stratosphere reflected incoming radiation from the sun, thereby creating a cooling effect. Many suggest that the above-normal temperatures experienced in the first decade of the twenty-first century provide evidence of climate change caused by human activity. Based on a variety of techniques that allow scientists to estimate the temperature in previous centuries, the year 2005 was the warmest in the

last thousand years. A 2009 article published by the American Geophysical Union suggests that human intervention in Earth systems has reached a point where the Holocene epoch of the past 12,000 years is becoming a new Anthropocene epoch in which human systems have become primary Earth systems rather than simply influencing natural systems.

Numerous observations strongly suggest a continuing warming trend. Snow and ice have retreated from areas such as Mount Kilimanjaro in Tanzania, which at 19,340 feet is the highest mountain in Africa. Glaciated areas in Switzerland also provide evidence of this warming trend. The Special Report on Emission Scenarios issued in 2001 by the Intergovernmental Panel on Climate Change (IPCC) examined the broad spectrum of possible concentrations of greenhouse gases by considering the growth of population and industry along with the efficiency of energy use. The IPCC computer climate models were used to estimate future trends. For example, a global temperature increase of 35.2 to 39.2 degrees Fahrenheit by the year 2100 is a IPCC standard estimate.

Robert M. Hordon, B.A., M.A., Ph.D.

FURTHER READING

Christopherson, Robert W. *Geosystems: An Introduction to Physical Geography.* 8th ed. Upper Saddle River, N.J.: Pearson Prentice Hall, 2012. Covers many topics in atmospheric sciences. Color illustrations.

Coley, David A. *Energy and Climate Change: Creating a Sustainable Future.* Hoboken, N.J.: John Wiley & Sons, 2008. A detailed review of energy topics and their relationship to climate change and energy technologies.

Ellis, Erle C., and Peter K. Haff. "Earth Science in the Anthropocene: New Epoch, New Paradigm, New Responsibilities." *EOS, Transactions, American Geophysical Union* 90, no. 49 (2009): 473. Makes the point that human systems are no longer simply influencing the natural world but are becoming part of it.

Gautier, Catherine. *Oil, Water, and Climate: An Introduction.* New York: Cambridge University Press, 2008. A good discussion of the impact of fossil fuel burning, particularly on climate change.

Lutgens, Frederick K., and Edward J. Tarbuck. *The Atmosphere: An Introduction to Meteorology.* 11th ed. Upper Saddle River, N.J.: Prentice Hall, 2010.

A very good textbook written with considerable clarity.

Strahler, Alan. *Introducing Physical Geography.* 5th ed. Ho-boken, N.J.: John Wiley, & Sons 2011. Covers weather and climate; contains superlative illustrations, clear maps, and lucid discussions of the material.

Wolfson, Richard. *Energy, Environment, and Climate.* New York: W. W. Norton, 2008. Provides an extensive discussion of the relationship between energy and climate change.

WEB SITES

American Geophysical Union
http://www.agu.org

American Meteorological Society
http://www.ametsoc.org

Intergovernmental Panel on Climate Change
http://www.ipcc.ch

International Association of Meteorology and Atmospheric Sciences
http://www.iamas.org

National Oceanic and Atmospheric Administration
National Climatic Data Center
http://www.Ncdc.noaa.gov

National Weather Association
http://www.nwas.org

U.S. Environmental Protection Agency
Air Science
http://www.epa.gov/airscience

U.S. Geological Survey
Climate and Land Use Change
http://www.usgs.gov/climate_landuse

World Meteorological Organization
Climate
http://www.wmo.int/pages/themes/climate/index_en.php

See also: Barometry; Measurement and Units; Meteorology.

AUDIOLOGY AND HEARING AIDS

FIELDS OF STUDY

Biology; chemistry; physics; mathematics; anatomy; physiology; genetics; pharmacology; neurology; head and neck anatomy; acoustics.

SUMMARY

Audiology is the study of hearing, balance, and related ear disorders. Hearing disorders may be the result of congenital abnormalities, trauma, infections, exposure to loud noise, some medications, and aging. Some of these disorders may be corrected by hearing aids and cochlear implants. Hearing aids amplify sounds so that the damaged ears can discern them. Some fit over the ear with the receiver behind the ear, and some fit partially or completely within the ear canal. Cochlear implants directly stimulate the auditory nerve, and the brain interprets the stimulation as sound.

KEY TERMS AND CONCEPTS

- **Analogue Signal Processing:** Process in which sound is amplified without additional changes.
- **Audiologist:** Licensed professional who assesses hearing loss and oversees treatment.
- **Auditory Nerve:** Nerve that carries stimuli from the cochlea to the brain.
- **Cochlea:** Coiled cavity within the ear that contains nerve endings necessary for hearing.
- **Deafness:** Full or partial inability to detect or interpret sounds.
- **Digital Signal Processing:** Process in which sound is received and mathematically altered to produce clearer, sharper sound.
- **Hearing:** Ability to detect and process sounds.
- **Ototoxicity:** Damage to the hair cells of the inner ear, resulting in hearing loss.
- **Sound:** Waves of pressure that a vibrating object emits through air or water.

DEFINITION AND BASIC PRINCIPLES

Audiology is the study of hearing, balance, and related ear disorders. Audiologists are licensed professionals who assess hearing loss and related sensory input and neural conduction problems and oversee treatment of patients.

Hearing is the ability to receive, sense, and decipher sounds. To hear, the ear must direct the sound waves inside, sense the sound vibrations, and translate the sensations into neurological impulses that the brain can recognize.

The outer ear funnels sound into the ear canal. It also helps the brain determine the direction from which the sound is coming. When the sound waves reach the ear canal, they vibrate the eardrum. These vibrations are amplified by the eardrum's movement against three tiny bones behind the eardrum. The third bone rests against the cochlea; when it transmits the sound, it creates waves in the fluid of the cochlea.

The cochlea is a coiled, fluid-filled organ that contains 30,000 hairs of different lengths that resonate at different frequencies. Vibrations of these hairs trigger complex electrical patterns that are transmitted along the auditory nerve to the brain, where they are interpreted.

BACKGROUND AND HISTORY

The first hearing aids, popularized in the sixteenth century, were large ear trumpets. In the nineteenth century, small trumpets or ear cups were placed in acoustic headbands that could be concealed in hats and hairstyles. Small ear trumpets were also built into parasols, fans, and walking sticks.

Electrical hearing devices emerged at the beginning of the twentieth century. These devices, which had an external power source, could provide greater amplification than mechanical devices. The batteries were large and difficult to carry; the carrying cases were often disguised as purses or camera cases. Zenith introduced smaller hearing aids with vacuum tubes and batteries in the 1940's.

In 1954, the first hearing aid with a transistor was introduced. This led to hearing aids that were made to fit behind the ear. Components became smaller and more complex, leading to the marketing of devices that could fit partially into the ear canal in the mid-1960's and ones that could fit completely into the ear canal in the 1980's.

HOW IT WORKS

Audiologists are concerned with three kinds of hearing loss: conductive hearing loss, in which sound waves are not properly transmitted to the inner ear; sensorineural hearing loss, in which the cochlea or the auditory nerve is damaged; and mixed hearing loss, which is a combination of these.

Conductive Hearing Loss. Otosclerosis, inefficient movement of the three bones in the middle ear, results in hearing loss from poor conduction. This disease is treatable with surgery to replace the malformed, misaligned bones with prosthetic pieces to restore conductance of sound waves to the cochlea.

Meniere's disease is thought to result from an abnormal accumulation of fluid in the inner ear in response to allergies, blocked drainage, trauma, or infection. Its symptoms include vertigo with nausea and vomiting and hearing loss. The first line of treatment is motion sickness and antinausea medications. If the vertigo persists, treatment with a Meniett pulse generator may result in improvement. This device safely applies pulses of low pressure to the middle ear to improve fluid exchange.

Hearing loss may result from physical trauma, such as a fracture of the temporal bone that lies just behind the ear or a puncture of the eardrum. These injuries typically heal on their own, and in most cases, the hearing loss is temporary.

A gradual buildup of earwax may block sound from entering the ear. Earwax should not be removed with a cotton swab or other object inserted in the ear canal; that may result in infection or further impaction. Earwax should be softened with a few drops of baby oil or mineral oil placed in the ear canal twice a day for a week and then removed with warm water squirted gently from a small bulb syringe. Once the wax is removed, the ear should be dried with rubbing alcohol. In stubborn cases, a physician or audiologist may have to perform the removal.

Foreign bodies in the ear canal, most commonly toys or insects, may block sound. If they can be seen clearly, they may be carefully removed with tweezers. If they cannot be seen clearly or moved, they may be floated out. For toys, the ear canal is flooded with warm water squirted gently from a small bulb syringe. For insects, the ear canal should first be filled with mineral oil to kill the bug. If the bug still cannot be seen clearly, the ear canal may then be flooded with warm water squirted gently from a small bulb syringe.

Once the object is removed, the ear canal should be dried with rubbing alcohol.

Sensorineural Hearing Loss. Some medications have adverse effects on the auditory system and may cause hearing loss. These medications include large doses of aspirin, certain antibiotics, and some chemotherapy agents. Doses of antioxidants such as vitamin C, vitamin E, and ginkgo biloba may ameliorate these ototoxic effects.

Exposure to harmful levels of noise, either over long periods or in a single acute event, may result in hearing loss. If the hair cells in the cochlea are destroyed, the hearing loss is permanent. Hearing aids or cochlear implants may be necessary to compensate.

An acoustic neuroma is a noncancerous tumor that grows on the auditory nerve. It is generally surgically removed, although patients who are unable to undergo surgery because of age or illness may undergo stereotactic radiation therapy instead.

Presbycusis is the progressive loss of hearing with age as a result of the gradual degeneration of the cochlea. It may be first noticed as an inability to hear high-pitched sounds. Hearing aids are an appropriate remedy.

Mixed Hearing Loss. Infections of the inner ear by the viruses that cause mumps, measles, and chickenpox and infections of the auditory nerve by the viruses that cause mumps and rubella may cause permanent hearing loss. Because fluid builds up and viruses do not respond to antibiotics, viral ear infections may require surgical treatment. In a surgical procedure called myringotomy, a small hole is created in the eardrum to allow the drainage of fluid. A small tube may be inserted to keep the hole open long enough for drainage to finish.

Some children are born with hearing loss as the result of congenital abnormalities. Screening to determine the nature and severity of the hearing loss is difficult. Children are not eligible for cochlear implants before the age of twelve months. Young children often do well with behind-the-ear hearing aids, which are durable and can grow with them.

APPLICATIONS AND PRODUCTS

Audiological products consist mainly of hearing aids and cochlear implants.

Hearing Aids. Although hearing aids take many forms, basically, all of them consist of a microphone that collects sound, an amplifier that magnifies the sound, and a speaker that sends it into the ear canal. They require daily placement and removal, and must be removed for showering, swimming, and battery replacement. Most people do not wear them when sleeping.

Hearing aids that sit behind the ear consist of a plastic case, a tube, and an earmold. The case contains components that collect and amplify the sound, which is then sent to the earmold through the tube. The earmold is custom-made to fit comfortably. This type of hearing aid is durable and well suited for children, although their earmolds must be replaced as they grow.

Hearing aids that sit partially within the ear canal are self-contained and custom-molded to the ear canal, so they are not recommended for growing children. Because of the short acoustic tube that channels the amplified sound, they are prone to feedback, which makes them less than ideal for people with profound hearing loss. Newer models offer feedback cancellation features.

Hearing aids that sit completely within the ear canal are self-contained and custom-molded. However, thin electrical wires replace the acoustic tube, so this type of hearing aid is free of feedback and sound distortion. They are not well suited for elderly people because their minimal size limits their volume capabilities.

The first hearing aids were analogue, a process that amplified sound without changing its properties. Although amplification was adjustable, the sound was not sorted, so background noise was also amplified. New hearing aids are digital, a process in which sound waves are converted into binary data that can be cleaned up to deliver clearer, sharper sounds.

The choice of hearing aid depends on the type and severity of hearing loss. Hearing aids will not restore natural hearing. However, they increase a person's awareness of sounds and the direction from which they are coming and heighten discernment of words.

Cochlear Implants. The U.S. Food and Drug Administration approved cochlear implants in the mid-1980's. A cochlear implant differs from a hearing aid in its structure and its function. It consists of a microphone that collects sound, a speech processor that sorts the incoming sound, a transmitter that relays the digital data, a receiver/stimulator that converts the sound into electrical impulses, and an electrode

array that sends the impulses from the stimulator to the auditory nerve. Hearing aids compensate for damage within the ear; cochlear implants avoid the damage within the ear altogether and directly stimulate the auditory nerve.

Whereas hearing aids are completely removable, cochlear implants are not. The internal receiver/stimulator and electrode array are surgically implanted. Three to six weeks after implantation, when the surgical incision has healed, the external portions are fitted. The transmitter and receiver are held together through the skin by magnets. Thus, the external portions may be removed for sleep, showering, and swimming.

IMPACT ON INDUSTRY

Audiology is practiced in many clinics and hospitals throughout the United States, and research in the field is conducted at governmental agencies and facilities associated with universities and medical institutions. Hearing aids, cochlear implants, and other devices designed to assist people with hearing loss are the focus of many medical equipment manufacturers.

University Research. A Massachusetts Institute of Technology research team, led by electrical engineering professor Dennis M. Freeman, is studying how the ear sorts sounds in an effort to create the next generation of hearing aids. This team is studying how the tectorial membrane of the cochlea carries sound waves that move from side to side, while the basilar membrane of the cochlea carries sound waves that move up and down. Together, these two membranes detect individual sounds. This team has genetically engineered mice that lack a crucial gene and protein of the tectorial membrane to study the compromised hearing.

Second-year audiology students at the University of Western Australia in Perth are conducting research projects in the development of new hearing assessment tests for children of all ages. These tests include electrophysiological studies for infants, efficient screening for children beginning school, and audiovisual games for grade school students.

Government Research. Researchers funded by the National Institute on Deafness and Other Communication Disorders (NIDCD) are looking at ways that hearing aids can selectively enhance sounds to improve the comprehension of speech. A related research team is studying ears on animals to fine-tune directional microphones that may make conversations more easily understood. Another related research team is learning to use computer-aided design programs to create and fabricate better fitting and performing hearing aids.

The U.S. Department of Defense oversees the Military Audiology Association, an organization that strives to provide hearing health care, education, and research to American fighting forces. Its Committee for Audiology Research is investigating alternative

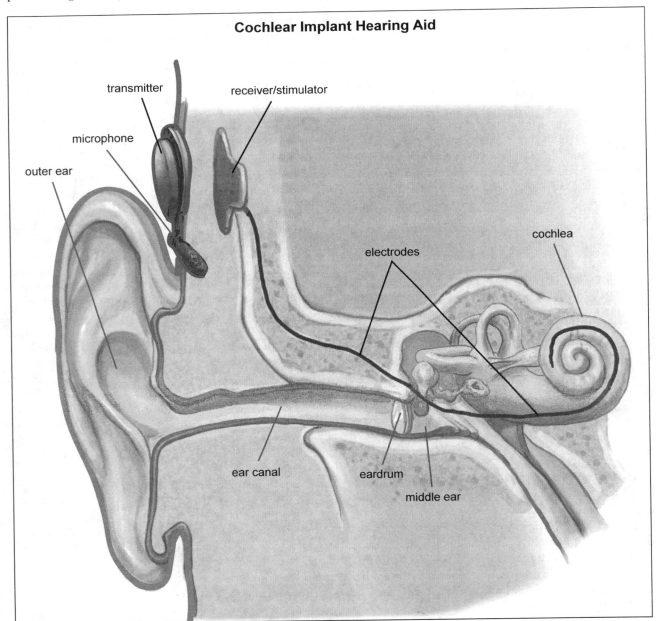

Cochlear Implant Hearing Aid

A cochlear implant differs from a hearing aid, delivering a utilitarian representation of sound.

helmet designs to protect hearing without blocking auditory cues and also developing a bone conduction microphone for insertion into helmets to facilitate hands-free radio operation.

Medical Clinics. In addition to providing standard audiological care, medical clinics put into practice the knowledge obtained by research. William H. Lippy, founder of the Lippy Group for Ear, Nose, and Throat in Warren, Ohio, is one of the few surgeons in the world who has perfected the stapedectomy, an operation in which the stapes (the bone closest to the cochlea) is replaced with a prosthesis to restore hearing. He created the artificial stapes and many surgical instruments for this procedure. He is sharing his knowledge through peer-reviewed articles, book chapters, and an online video library.

Researchers at the Mayo Clinic are working to improve hearing aid use in younger persons, who are less likely to wear amplification devices than older people. They are developing remote programming of digital hearing aids over the Internet, disposable hearing aids (because children break or lose things easily), and devices that wearers can adjust themselves.

Assistive Device Manufacturers. Companies are continuing to design, manufacture, and market assistive devices for individuals with hearing loss. These devices can be divided into two categories: listening devices and alerting devices. Listening devices include amplified telephones and cell phones, wireless headsets for listening to television, and FM or infrared receivers for use in theaters and churches. Alerting devices signal users through flashing lights, vibration, or increased sound. They work in combination with smoke and carbon monoxide detectors, baby monitors, alarm clocks, and doorbells.

Some companies offer patch cords that connect the speech processors of cochlear implants to assistive listening devices for improved effect. Such patch cords make it easier to use cell phones and enjoy music on MP3 players. They are available with and without volume controls.

CAREERS AND COURSE WORK

Audiologists are licensed professionals but not medical doctors. Physicians who specialize in ears, nose, and throat disorders are called otorhinolaryngologists. Audiologists must have a minimum of a master's degree in audiology; a doctoral (Au.D.) degree

is becoming increasingly desirable in the workplace and is required for licensure in eighteen states. State licensure is required to practice in all fifty states. To obtain a license, an applicant must graduate from an accredited audiology program, accumulate 300 to 375 hours of supervised clinical experience, and pass a national licensing exam. To get into a graduate program in audiology, applicants must have had undergraduate courses in biology, chemistry, physics, anatomy, physiology, math, psychology, and communication. Forty-one states have continuing education requirements to renew a license. An audiologist must pass a separate exam to mold and place hearing aids. The American Board of Audiology and the American Speech-Language-Hearing Association (AHSA) offer professional certification programs for licensed audiologists.

Audiologists can go into private practice as a sole proprietor or as an associate or partner in a larger practice. They may also work in hospitals, outpatient clinics, and rehabilitation centers. Some are employed in state and local health departments and school districts. Some teach at universities and conduct academic research, and others work for medical device manufacturers. Audiologists may specialize in working with specific age groups, conducting hearing protection programs, or developing therapy programs for patients who are newly deaf or newly hearing.

SOCIAL CONTEXT AND FUTURE PROSPECTS

In 2008, there were 12,800 audiologists working in the United States. About 64 percent were employed in health care settings and 14 percent in educational settings. The number of audiologists was expected to grow by 25 percent until 2018. The projected increased need can be traced to several factors.

As the population of older people continues to grow, so will the incidence of hearing loss from aging. In addition, the market demand for hearing aids is expected to increase as devices become less noticeable and existing wearers switch from analogue to digital models. At the same time, advances in medical treatment are increasing the survival rates of premature infants, trauma patients, and stroke patients, populations who may experience hearing loss.

Hearing aid manufacturers are on their fourth generation of products, and digital devices are becoming smaller and providing increasingly better

Fascinating Facts About Audiology and Hearing Aids

- More than 31.5 million Americans have some degree of hearing loss.
- Three out of every 1,000 babies in the United States are born with hearing loss.
- Only one in five people who would benefit from a hearing aid actually wears one.
- Approximately 188,000 people around the world have cochlear implants. In the United States, 41,500 adults and 25,500 children have had the device implanted.
- Noise exposure is the most common cause of hearing loss. One study has shown that people who eat substantial quantities of salt are more susceptible to hearing damage from noise.
- Addiction to Vicodin (hydrocodone with acetaminophen) can result in complete deafness.
- After its discovery in 1944, streptomycin was used to treat tuberculosis with great success; however, many patients experienced irreversible cochlear dysfunction as a result of the drug's ototoxicity.
- Medications that can cause hearing loss in adults are aspirin in large quantities, antibiotics such as streptomycin and neomycin, and chemotherapy agents such as cisplatin and carboplatin. Taking antioxidants such as vitamin C, vitamin E, zinc, and ginkgo biloba helps combat ototoxicity.
- The first nationally broadcast television program to show open captioning was *The French Chef with Julia Child*, which appeared on station WGBH from Boston on August 5, 1972.
- The National Captioning Institute broadcast the first closed-captioned television series on March 16, 1980. Real-time closed captioning was developed in 1982.
- The Television Decoder Circuitry Act of 1990 mandated that by mid-1993 all television sets 13 inches or larger must have caption-decoding technology.

specialized structures of the inner ear. They hope to discover the biological mechanisms behind hearing loss in order to interrupt them or compensate for them on the molecular level.

Bethany Thivierge, B.S., M.P.H.

FURTHER READING

Dalebout, Susan. *The Praeger Guide to Hearing and Hearing Loss: Assessment, Treatment, and Prevention.* Westport, Conn.: Praeger, 2009. Guides those who are experiencing hearing loss through the process of assessment and describes possible treatments and assistive devices.

DeBonis, David A., and Constance L. Donohue. *Survey of Audiology: Fundamentals for Audiologists and Health Professionals.* 2d ed. Boston: Pearson/ Allyn and Bacon, 2008. Provides excellent coverage of audiology, focusing on assessment and covering pediatric audiology.

Gelfand, Stanley A. *Essentials of Audiology.* 3d ed. New York: Thieme Medical Publishers, 2009. A comprehensive introductory textbook with abundant figures and study questions at the end of each chapter.

Kramer, Steven J. *Audiology: Science to Practice.* San Diego, Calif.: Plural, 2008. Examines the basics of audiology and how they relate to practice. Contains a chapter on hearing aids by H. Gustav Mueller and Earl E. Johnson and a chapter on the history of audiology by James Jerger.

Lass, Norman J., and Charles M. Woodford. *Hearing Science Fundamentals.* Philadelphia: Mosby, 2007. Covers the anatomy, physiology, and physics of hearing. Contains figures, learning objectives, and chapter questions.

Moore, Brian C. J. *Cochlear Hearing Loss: Physiological, Psychological and Technical Issues.* 2d ed. Hoboken, N.J.: John Wiley & Sons, 2007. Comprehensive coverage of issues associated with cochlear hearing loss, including advances in pitch and speech perception.

Roeser, Ross J., Holly Hosford-Dunn, and Michael Valente, eds. *Audiology.* 2d ed. 3 vols. New York: Thieme, 2008. Volumes in this set cover diagnosis, treatment, and practice management. Focus is on clinical practice.

sound processing. Neurosurgical techniques also are improving. Public health programs are promoting hearing protection. Excessive noise is the most common cause of hearing loss, and one-third of noise-related hearing loss is preventable with proper protective equipment and practices. Researchers are continuing to study the genes and proteins related to

WEB SITES

American Academy of Audiology
http://www.audiology.org

American Speech-Language-Hearing Association
http://www.asha.org

Audiological Resource Association
http://www.audresources.org

Educational Audiology Assocation
http://www.edaud.org

Military Audiology Association

http://www.militaryaudiology.org

National Institute on Deafness and Other Communication Disorders
http://www.nidcd.nih.gov

See also: Geriatrics and Gerontology; Occupational Health; Otorhinolaryngology; Pediatric Medicine and Surgery; Prosthetics; Rehabilitation Engineering; Speech Therapy and Phoniatrics.

B

BAROMETRY

FIELDS OF STUDY

Atmospheric science; physics; chemistry; fluid mechanics; electromagnetics; signal processing, meteorology;

SUMMARY

The science and engineering of pressure measurement in gases take its practitioners far beyond its original realm of weather prediction. The sensors used in barometry range from those for the near vacuum of space and the small amplitudes of soft music to those for ocean depths and the shock waves of nuclear-fusion explosions.

KEY TERMS AND CONCEPTS

- **Aneroid Barometer:** Instrument that measures atmospheric pressure without using liquid.
- **Bar:** Unit of pressure equal to 100,000 newtons per square meter.
- **Pascal:** Unit of pressure equal to 1 newton per square meter.
- **Piezoelectric Material:** Any material that generates a voltage when the pressure acting on it changes.
- **Pounds per square inch absolute (PSIA):** Unit of pressure in pounds per square in absolute.
- **Pounds Per Square Inch Gauge (PSIG):** Unit of pressure in pounds per square inch relative to some reference pressure.
- **Pressure-sensitive paint:** Liquid mixture that, when painted and dried on a surface and illuminated with ultraviolet light, emits radiation, usually infrared, the intensity of which changes with the pressure of air acting on the surface.
- **Torr:** Unit of pressure equal to one part in 760 of a standard atmosphere, also equal to the pressure at the bottom of a column of mercury 1 millimeter high with vacuum above it.
- **Torricelli Barometer:** Instrument to measure absolute atmospheric pressure, consisting of a graduated tube closed at the top end, containing a vacuum at the top above a column of mercury, and standing in a reservoir of mercury.
- **U-Tube Manometer:** Instrument consisting of a graduated tube containing liquid and shaped like the letter U with each arm connected to a source of pressure, so that the difference in liquid levels between the two tubes indicates the differential pressure.

DEFINITION AND BASIC PRINCIPLES

Barometry is the science of measuring the pressure of the atmosphere. Derived from the Greek words for "heavy" or "weight" (*baros*) and "measure" (*metron*), it refers generally to the measurement of gas pressure. In gases, pressure is fundamentally ascribed to the momentum flowing across a given surface per unit time, per unit area of the surface. Pressure is expressed in units of force per unit area. Although pressure is a scalar quantity, the direction of the force due to pressure exerted on a surface is taken to be perpendicular and directed onto the surface. Therefore, methods to measure pressure often measure the force acting per unit area of a sensor or the effects of that force. Pressure is expressed in newtons per square meter (pascals), in pounds per square foot (psf), or in pounds per square inch (psi). The pressure of the atmosphere at standard sea level at a temperature of 288.15 Kelvin (K) is 101,325 pascals, or 14.7 psi. This is called 1 atmosphere. Mercury and water barometers have become such familiar devices that pressure is also expressed in inches of water, inches of mercury, or in torrs (1 torr equals about 133.3 pascals).

The initial weather-forecasting barometer, the Torricelli barometer, measured the height of a liquid column that the pressure of air would support, with a vacuum at the closed top end of a vertical tube. This barometer is an absolute pressure instrument. Atmospheric pressure is obtained as the product of the height, the density of the barometric liquid, and the acceleration because of gravity at the Earth's surface. The aneroid barometer uses a partially evacuated box the spring-loaded sides of which expand or contract depending on the atmospheric pressure, driving a clocklike mechanism to show the pressure on a circular dial. This portable instrument was convenient to carry on mountaineering, ballooning, and mining expeditions to measure altitude by the change in atmospheric pressure. A barograph is an aneroid barometer mechanism adapted to plot a graph of the variation of pressure with time, using a stylus moving on a continuous roll of paper. The rate of change of pressure helps weather forecasters to predict the strength of approaching storms.

The term "manometer" derives from the Greek word *manos*, meaning "sparse," and denotes an instrument used to measure the pressure relative to a known pressure. A U-tube manometer measures the pressure difference from a reference pressure by the difference between the height of a liquid column in the leg of the U-tube that is connected to a known pressure source and the height of the liquid in the other leg, exposed to the pressure of interest. Manometers of various types have been used extensively in aerospace engineering experimental-test facilities such as wind tunnels. The pitot-static tubes used to measure flow velocity in wind tunnels were initially connected to water or mercury manometers. Later, electronic equivalents were developed. Inclined tube manometers were used to increase the sensitivity of the instrument in measuring small pressure differences amounting to fractions of an inch of water.

BACKGROUND AND HISTORY

In 1643, Italian physicist and mathematician Evangelista Torricelli proved that atmospheric pressure would support the weight of a thirty-five-foot water column leaving a vacuum above that in a closed tube and that this height would change with the weather. Later Torricelli barometers used liquid mercury to reduce the size of the column and make such instruments more practical.

The technology of pressure measurement has evolved gradually since then, with the aneroid barometer demonstrating the reliability of deflecting a diaphragm. This led to electrical means of measuring the amount of deflection. The most obvious

method was to place strain gauges on the diaphragm and directly measure the strain. Later electrical methods used the change in capacitance caused by the changing gap between two charged plates. Such sensors dominated the market until the 1990's at the low end of the measurement range. Piezoresistive materials expanded the ability of miniaturized strain gauge sensors to measure high pressures changing at high frequency. Microelectromechanical system (MEMS) technology enabled miniaturized solid-state sensors to challenge the market dominance of the diaphragm sensors. In the early twenty-first century, pressure-sensitive paints allowed increasingly sensitive and faster-responding measurements of varying pressure with very fine spatial resolution.

HOW IT WORKS

Barometry measures a broad variety of pressures using an equally broad variety of measurement techniques, including liquid column methods, elastic element methods, and electrical sensors. Electrical sensors include resistance strain gauges, capacitances, piezoresistive instruments, and piezoelectric devices. The technologies range from those developed by French mathematician Blaise Pascal, Greek mathematician Archimedes, and Torricelli to early twenty-first century MEMS sensors and those used to conduct nanoscale materials science.

Pressures can be measured in environments from the near vacuum of space to more than 1,400 megapascals (MPa) and from steady state to frequencies greater than 100,000 cycles per second. Sensors that measure with respect to zero pressure or absolute vacuum are called absolute pressure sensors, whereas those that measure with respect to some other reference pressure are called gauge pressure sensors. Vented gauge sensors have the reference side open to the atmosphere so that the pressure reading is with respect to atmospheric pressure. Sealed gauges report pressure with respect to a constant reference pressure.

Where rapid changes in pressure must be measured, errors due to the variation of sensitivity with the rate of change must be considered. A good sensor is one whose frequency response is constant over the entire range of frequency of fluctuations that might occur. Condenser microphones with electromechanical diaphragms have long been used to measure acoustic pressure in demanding applications such as

Fascinating Facts About Barometry

- Italian physicist and mathematician Evangelista Torricelli built a tall water barometer protruding through the roof of his house in 1643 to display his invention. His neighbors accused him of practicing sorcery.
- The sonic boom generated on the ground by an aircraft flying at supersonic speeds produces a pressure change shaped like the letter N: a sharp increase, a more gradual decrease and a sharp recovery at the end.
- The mean atmospheric pressure at the surface of Mars is roughly 600 pascals, compared with 101,300 pascals at Earth's surface. This is roughly equal to the atmospheric pressure at 34,500 meters above Earth.
- The atmospheric static pressure inside a hurricane may go down to only 87 percent of the normal atmosphere. Pressure in the core of a tornado is believed to be similar to this; however, the changes occur within a few seconds as opposed to hours in the case of a hurricane.
- The pressure at the bottom of the Mariana Trench—at 11,034 meters, the lowest surveyed point of the Pacific Ocean—is roughly 111 megapascals, or 1,099 times the pressure at the surface.
- Solar radiation at Earth's orbit around the Sun exerts a pressure of roughly 4.56 micropascals.
- The threshold of human hearing is a pressure fluctuation of roughly twenty micropascals, while the threshold of pain is a pressure change of 100 pascals.
- The pressure inside the core of the Sun is calculated to be around 250 billion bars, while that occurring during the explosion of an American W80 nuclear-fusion weapon is roughly 64 billion bars.

music recording, with flat frequency response from 0.1 cycles per second (hertz) to well over 20,000 hertz, covering the range of human hearing. Pressure-sensitive paint in certain special formulations has been shown to achieve excellent frequency response to more than 1,600 hertz but only when the fluctuation amplitude is quite large, near the upper limit of human tolerance. Using digital signal processing, inexpensive sensors can be corrected to produce signals with frequency response quality approaching that of much more expensive sensors.

In the 1970's, devices based on the aneroid barometer principle were developed, in which the deflection of a diaphragm caused changes in electrical capacitance that then were indicated as voltage changes in a circuit. In the 1980's, piezoelectric materials were developed, enabling electrical voltages to be created from changes in pressure. Micro devices based on these largely replaced the more expensive but accurate diaphragm-based electromechanical sensors. Digital signal processing enabled engineers using the new small, inexpensive devices to recover most of the accuracy possessed by the more expensive devices.

APPLICATIONS AND PRODUCTS

Barometry has ubiquitous applications, measured by a broad variety of sensors. It is key to weather prediction and measuring the altitude of aircraft as well as to measuring blood pressure to monitor health. Pressure-sensitive paints enable measurement of surface pressure as it changes in space and time. The accuracy of measuring and controlling gas pressure is fundamental to manufacturing processes.

Weather Forecasting. Scientists learned to relate the rate of change of atmospheric pressure to the possibility of strong winds, usually bringing rain or snow. For example, if the pressure drops by more than three millibar per hour, winds of up to fifty kilometers per hour are likely to follow. Powerful storms may be preceded by drops of more than twenty-four millibar in twenty-four hours. If the pressure starts rising, clear calm weather is expected. However, these rules change with regional conditions. For instance in the Great Lakes region of the United States, rising pressure may indicate an Arctic cold front moving in, causing heavy snowfall. In other regions, a sharply dropping pressure indicates a cold front moving in, followed by a quick rise in pressure as the colder weather is established. As a warm front approaches, the pressure may level out and rise slowly after the front passes. Modern forecasters construct detailed maps showing contours of pressure from sensors distributed over the countryside and use these to predict weather patterns. Aircraft pilots use such maps to identify safe routes and areas to avoid. Using Doppler radar wind measurements, infrared temperature maps and cloud images from satellites, and computational fluid dynamics, modern weather bureaus are able to issue warnings about severe weather several hours in advance for smaller local weather fronts and storms and several days ahead for major storms moving across continents or oceans. However, the number of weather-monitoring pressure sensors available to forecasters is quite inadequate to issue accurate predictions for minor weather changes, particularly when predicting rain or snow.

Electrical Gauges. Gauges operating on the electrical changes induced by deflection of a diaphragm are used in industrial process monitoring and control where computer interfacing is required. Unsteady pressure transducers come in many ranges of amplitude and frequency. Piezoresistive sensors are integrated into an electrical-resistance bridge and constructed as miniature self-contained, button-like sensors. These are suitable for high amplitudes and frequencies, such as those encountered in shock waves and explosions, and transonic or supersonic wind-tunnel tests. Condenser microphones are used in acoustic measurements. As computerized data-acquisition systems became common, but pressure sensors remained expensive, pressure switches enabled dozens of pressure-sensing lines connected through the switches to each sensor to be measured one at a time. This required a long time to collect data from all the sensors, spending enough time at each to capture all the fluctuations and construct stable averages, making it unsuitable for rapidly changing conditions. Inexpensive, miniaturized, and highly sensitive solid-state piezoelectric sensors and fast, multichannel analogue-digital converters have made it possible to connect each pressure port to an individual sensor, vastly reducing the time between individual measurements at each sensor.

Aircraft Testing. Water and mercury manometers were used extensively in aerospace test facilities such as wind tunnels, where banks of manometers indicated the distribution of pressure around the surfaces of models from pressure-sensing holes in the models. Pressure switches connecting numerous pressure-sensing ports to a single sensor became common in the 1970's. In the 1990's, inexpensive sensors based on microelectromechanical systems technology enabled numerous independent sensing channels to be monitored simultaneously.

Sphygmomanometers for Blood Pressure. Other than weather forecasting, the major common application of pressure measurement is in measuring blood pressure. The device used is called a sphygmomanometer. The high and low points of pressure reached in the heartbeat cycle are noted on a mercury manometer tube synchronized with the heartbeat sounds detected through a stethoscope.

Bourdon Tubes for Household Barometry. The Bourdon tube is a pressure-measuring device in which a coiled tube stretches and uncoils depending on the difference between pressures inside and outside the tube and drives a levered mechanism connected to an indicator dial. Diaphragm-type pressure gauges and Bourdon-tube gauges are still used in the vast majority of household and urban plumbing. These instruments are highly reliable and robust, but they operate over fairly narrow ranges of pressure.

Pressure-Sensitive Paints to Map Pressure Over Surfaces. So-called pressure-sensitive paints (PSPs) offer an indirect technique to measure pressure variations over an entire surface, using the fact that the amount of oxygen felt at a surface is proportional to the density and thus to the pressure if the temperature does not change. These paints are luminescent dyes dispersed in an oxygen-permeable binder. When illuminated at certain ultraviolet wavelengths, the dye molecules absorb the light and move up into higher energy levels. The molecules then release energy in the infrared wavelengths as they relax to equilibrium. If the molecule collides with an oxygen molecule, the energy gets transferred without emission of radiation. Therefore, the emission from a surface becomes less intense if the number of oxygen molecules being encountered increases. This occurs when the pressure of air increases. The observed intensity from a painted surface is inversely proportional to the pressure of oxygen-containing air. Light-intensity values at individual picture element (pixel) are converted to numbers, compared with values at some known reference pressure, and presented graphically as colors. Typically, an accurate pressure sensor using either piezoelectric or other technology is used for reference. As of 2011, pressure-sensitive paints had reached the sensitivity required to quantify pressure distributions over passenger automobiles at moderate highway speeds, given expert signal processing and averaging a large number of images.

Smart Pressure Transmitters for Automatic Control Systems. Wireless pressure sensors are used in remote applications such as weather sensing. Modern automobiles incorporate tire pressure transmitters.

Manifold pressure sensors send instantaneous readings of the pressure inside automobile engine manifolds so that a control computer can calculate the best rate of fuel flow to achieve the most efficient combustion. Smart pressure transmitters incorporating capacitance-type diaphragm pressure sensors and microprocessors can be configured to adjust their settings remotely, perform automatic temperature compensation of data, and transmit pressure data and self-diagnosis data in digital streams.

Nuclear Explosion Sensors. Piezoresistive transducers have been developed to report the extremely high overpressure, as high as sixty-nine megapascals, of an air blast of a nuclear weapon, with the microsecond rise time required to measure the blast wave accurately. One design uses a silicon disk with integral diffused strain-sensitive regions and thermal barriers. Another design uses the principle of Fabry-Perot interferometry, in which laser light reflecting in a cavity changes intensity depending on the shape of the cavity when the diaphragm bounding the cavity flexes because of pressure changes. This sensor has the response speed and ruggedness required to operate in a hostile environment, where there may be very large electromagnetic pulses. In such environments a capacitance-based sensor or piezoelectric sensor may not survive.

Extreme Applications of Barometry. The basic origins of pressure can be used to explain the pressure due to radiation as the momentum flux of photons. At Earth's orbit around the Sun, the solar intensity of 1.38 kilowatts per square meter causes a radiation pressure of roughly 4.56 micropascals. Solar sails have been proposed for long-duration missions in space, driven by this pressure. Close to the center of the Earth, the pressure reaches 3.2 to 3.4 million bars. Inside the Sun, pressure as high as 250 billion bars is expected, while the explosion of a nuclear-fusion weapon may produce a quarter of that. Metallic solid hydrogen is projected to form at pressures of 250,000 to 500,000 bars.

IMPACT ON INDUSTRY

Government and University Research. Barometers enabled rapid development of scientific weather forecasting. Weather forecasting, in turn, has had a tremendous effect on emergency preparedness. Research sponsored by the defense research offices provides fertile opportunities and challenges for new pressure-measurement techniques. Any experiment that uses fluids, either flowing or in containers, requires monitoring and often rapid measurement of pressure. The development of pressure-sensitive paint technology is a frontier in research in the early twenty-first century. Both the sensitivity and the frequency response of such paints need substantial improvement before they can be routinely used in laboratory measurements and transitioned to industrial measurements.

Industry and Business. Quantitative knowledge of the detailed surface pressure distribution on wind tunnel models of flight vehicles enables engineers to develop modifications to improve the performance of the vehicle and reduce fuel consumption. The ability to monitor pressure is critical in the nuclear and petroleum industries as well. Submarine and oil-rig crews depend on pressure measurements for their lives. Oil exploration involves several steps in which the pressure must be tracked with extreme care, especially when there is a danger of gas rising through drilling tubes from subterranean reservoirs. Pressure buildup in steam or other gas circuits is critical in the nuclear industry and in most of the chemical industry wherever leaks of gas into the atmosphere must be strictly controlled.

CAREERS AND COURSE WORK

Because barometry is so important to so many industries and so many branches of scientific research, most students who are planning on a career in engineering, other technological jobs, and the sciences must understand it.

Modern pressure-measurement technology integrates ideas from many branches of science and engineering derived from physics and chemistry. The pressure-measurement industry includes experts in weather forecasting, plumbing, atmospheric sciences, aerospace wind-tunnel experimentation, automobile-engine development, the chemical industry, chemists developing paint formulations, electrical and electronics engineers developing microelectromechanical sensors, software engineers developing smart sensor logic, and the medical community interested in using barometry to monitor patients' health and vital signs.

Pressure measurement therefore comes up as a subject in courses offered in schools of mechanical, chemical, civil, and aerospace engineering. The numerous other related issues come up in specialized courses in materials science, electronics, atmospheric sciences, and computer science.

SOCIAL CONTEXT AND FUTURE PROSPECTS

Instrumentation for measuring pressure, normal and shear stresses, and flow rate from numerous sensors are becoming integrated into computerized measurement systems. In many applications, such sensors are mass-produced using facilities similar to those for making chips for computers. Very few ideas exist for directly measuring pressure, as it changes rapidly at a point in a flowing fluid, without intrusive probes of some kind. Such nonintrusive measurements, if they become possible, could help us to understand the nature of turbulence and assist in a major breakthrough in fluid dynamics.

Measurement of pressure is difficult to make inside flame environments, where density and temperature fluctuate rapidly. Better methods of measuring pressure in biological systems, such as inside blood vessels, would have major benefits in diagnosing heart disease and improving health.

As of 2011, pressure-measurement systems are still too expensive to allow sufficient numbers to be deployed to report pressure with enough spatial and time resolution to permit development of a real-time three-dimensional representation. Research in this area will doubtless improve the resolution and response and hopefully bring down the cost. With more pressure sensors distributed over the world, weather prediction will become more accurate and reliable.

Narayanan M. Komerath, Ph.D.

FURTHER READING

American Society of Mechanical Engineers. *Pressure Measurement.* New York: American Society of Mechanical Engineers, 2010. Authoritative document with guidance on determining pressure values, according to the American Society of Mechanical Engineers performance test codes. Discusses how to choose and use the best methods, instrumentation, and corrections, as well as the allowable uncertainty.

Avallone, Eugene A., Theodore Baumeister III, and Ali M. Sadegh. *Marks' Standard Handbook for Mechanical Engineers.* 11th ed. New York: McGraw-Hill, 2006. Best reference for solving mechanical engineering problems. Discusses pressure sensors and measurement techniques and their applications in various parts of mechanical engineering.

Benedict, Robert P. *Fundamentals of Temperature, Pressure, and Flow Measurements.* 3d ed. New York: John Wiley & Sons, 1984. Suited for practicing engineers in the process control industry.

Burch, David. *The Barometer Handbook: A Modern Look at Barometers and Applications of Barometric Pressure.* Seattle: Starpath Publications, 2009. Written to assist the practicing weather forecaster, with chapters on weather forecasting on land and sea. Contains an excellent history of the field as well as methods for instrument calibration and maintenance.

Gillum, Donald R. *Industrial Pressure, Level, and Density Measurement.* 2d ed. Research Triangle Park, N.C.: International Society for Automation, 2009. Teaching and learning resource on the issues and methods of pressure measurement, especially related to industrial control systems. Contains assessment questions at the end of each section.

Green, Don W., and Robert H. Perry. *Perry's Chemical Engineers' Handbook.* 8th ed. New York: McGraw-Hill, 2008. Still considered the best source for bringing together knowledge from various parts of the field of chemical engineering, where the student can find the different applications of pressure measurement and many other things. Contains a succinct, illustrated explanation of pressure-measurement techniques.

Ryans, J. L. "Pressure Measurement." In *Kirk-Othmer Encyclopedia of Chemical Technology.* 5th ed. Hoboken, N.J.: John Wiley & Sons, 2000. Describes mechanical and electronic sensors and instrumentation for pressure measurements from 1,380 megapascals to near vacuum.

Taylor, George Frederic. *Elementary Meteorology.* New York: Prentice Hall, 1954. This classic remains an excellent resource for the basic methods of weather prediction, including the methods for measuring temperature, pressure, and humidity, as well as the methods for using these measurements in predicting the weather.

WEB SITES

American Meteorological Society
http://www.ametsoc.org

American Society of Mechanical Engineers
http://www.asme.org

National Oceanic and Atmospheric Administration
Office of Oceanic and Atmospheric Research
http://www.oar.noaa.gov

See also: Atmospheric Sciences; Chemical Engineering; Meteorology.

BIOENGINEERING

FIELDS OF STUDY

Cell biology; molecular biology; biochemistry; physiology; ecology; microbiology; pharmacology; genetics; medicine; immunology; neurobiology; biotechnology; biomechanics; bioinformatics; physics; mechanical engineering; electrical engineering; materials science; buildings science; architecture; chemical engineering; genetic engineering; thermodynamics; robotics; mathematics; computer science; biomedical engineering; tissue engineering; bioinstrumentation; bionics; agricultural engineering; human factors engineering; environmental health engineering; biodefense; nanotechnology; nanoengineering.

SUMMARY

Bioengineering is the field in which techniques drawn from engineering are used to tackle biological problems. For example, bioengineers may use mechanics principles—knowledge about how to design and construct mechanical objects using the most ideal materials—to create drug delivery systems. They may work on developing efficient ways to irrigate and drain land for growing crops, or they may be involved in building artificial environments that can support life even in the harsh climate of outer space. A highly interdisciplinary, collaborative field that synthesizes expertise from multiple research areas, bioengineering has had a significant impact on many fields of study, including the health sciences, technology, and agriculture.

KEY TERMS AND CONCEPTS

- **Biocompatible Material:** Material used to replace or repair tissues in the body, or to perform a biological function in a living organism.
- **Bioinformatics:** Application of data processing, retrieval, and storage techniques to biological research, especially in genomics.
- **Biomechanics:** Application of mechanical principles to questions of motor control in biological systems.
- **Bioreactor:** Tool or device that generates chemical reactions to create a product.
- **Bioremediation:** Use of bacteria and other microorganisms to solve environmental problems, such as neutralizing hazardous waste.
- **Geoengineering:** Use of engineering techniques to modify environmental or geological processes, such as the weather, on a global scale.
- **Prosthetic Device:** Artificial part or implant designed to replace the function of a lost or damaged part of the body.
- **Regenerative Medicine:** Therapies that aim to restore the function of tissues that have been damaged or lost through injury or disease by using tissue or cells grown in laboratories or compounds created in laboratories.
- **Systems Biology:** Theoretical branch of bioengineering that creates models of complex biological processes or systems, using them to predict future behavior.
- **Transgenic Organism:** Plant or animal containing genetic information taken from another species.

DEFINITION AND BASIC PRINCIPLES

Bioengineering is an interdisciplinary field of applied science that deals with the application of engineering methods, techniques, design approaches, and fundamental knowledge to solve practical problems in the life sciences, including biology, geology, environmental studies, and agriculture. In many contexts, the term bioengineering is used to refer solely to biomedical engineering. This is the application of engineering principles to medicine, such as in the development of artificial limbs or organs. However, the field of bioengineering has many applications beyond the field of health care. For example, genetically modified crops that are resistant to pests, suits that protect astronauts from the ultra-low pressures in space, and brain-computer interfaces that may allow soldiers to exercise remote control over military vehicles all fall under the wide umbrella of bioengineering.

Each of the subdisciplines within bioengineering relies on different sets of basic engineering principles, but a few fundamental approaches can be said to apply broadly across the entire field. From an engineering perspective, three basic steps are involved in solving any problem: an analysis of how the system in question works, an attempt to synthesize the information gathered from this analysis and generate potential solutions, and finally an attempt to design and test a useful product. Bioengineers apply this three-stage problem-solving process to problems in the life sciences. What is somewhat novel about this approach is that it is a holistic one. In other words, it treats biological entities as systems—sets of parts that work together and form an integrated whole—rather than looking at individual parts in isolation. For example, to develop an artificial heart, bioengineers need to consider not just the structure of the heart itself on a cellular or tissue level but also the complex dynamics of the organ's interactions with the rest of the body through the circulatory system and the immune system. They must build a device whose parts can mimic the functionality of a healthy heart and whose materials can be easily integrated into the body without triggering a harmful immune response.

BACKGROUND AND HISTORY

Principles of chemical and mechanical engineering have been applied to problems in specific biological systems for centuries. For example, bioengineering applications include the fermentation of alcoholic beverages, the use of artificial limbs (which are documented as far back as 500 b.c.e.), and the building of heating and cooling systems that regulate human environments.

Bioengineering did not emerge as a formal scientific discipline, however, until the middle of the twentieth century. During this period, more and more scientists began to be interested in applying new technologies from electronic and mechanical engineering to the life sciences. As the United States, Japan, and Europe began to enter a period of economic recovery and growth following World War II, governments increased funding for bioengineering efforts. The cardiac pacemaker and the defibrillator, both developed during this postwar period, were two of the earliest and most significant inventions to come out of the quickly developing field. In 1966, the Engineers Joint Council Committee on Engineering Interaction with Biology and Medicine first used the term "bioengineering." At about the same time, academic institutions began to form specialized departments and programs of study to train professionals in the application of engineering principles to

biological problems. In the twenty-first century, rapid technological advances continue to produce growth in the field of bioengineering.

HOW IT WORKS

Because bioengineering is such a large and diverse field, it would be impossible to enumerate all the processes involved in creating the totality of its applications. The following are a few of the most significant examples of the types of technological tools used in bioengineering.

Materials Science. One of the most important areas of bioengineering is the intersection of materials science and biology. Scientists working in this field are charged with developing materials that, although synthetic, are able to successfully interact with living tissues or other natural biological systems without impeding them. (For example, it is vital that biocompatible materials not allow blood platelets to adhere to them and form clots, which can be fatal.) Depending on the specific application in question, other properties, such as tensile strength, resistance to wear, and permeability to water, gases, and small biological molecules, are also important. To manipulate these properties to achieve a desired end, engineers must carefully control both the chemical structure and the molecular organization of the materials. For this reason, biocompatible materials are generally made out of some kind of synthetic polymer—substances with simple and extremely regular molecular structures that repeat again and again. In addition, additives may be incorporated into the materials, such as inorganic fillers that allow for greater mechanical flexibility or stabilizers and antioxidants that keep the material from becoming degraded over time.

Biochemical Engineering. Since living cells are essentially chemical systems, the tools of chemical engineering are especially applicable to biology. Biochemical engineers study and manipulate the behavior of living cells. Their basic tool for doing this is a fermenter, a large reactor within which chemical processes can be carried out under carefully controlled conditions. For example, the modern production of virtually all antibiotics, such as penicillin and tetracycline, takes place inside a fermenter. A central vessel, sealed tight to prevent contamination and surrounded by jackets filled with coolants to control its temperature, contains propellers that stir around the nutrients, culture ingredients, and catalysts that are associated with the reaction at hand.

Genetic engineering is a subfield of biochemical engineering that is growing increasingly significant. Scientists alter the genetic information in one cell by inserting into it a gene from another organism. To do this, a vector such as a virus or a plasmid (a small strand of DNA) is placed into the cell nucleus and combines with the existing genes to form a new genetic code. The technology that enables scientists to alter the genetic information of an organism is called gene splicing. The new genetic information created by this process is known as recombinant DNA. Genetic engineering can be divided into two types: somatic and germ line. Somatic genetic engineering is a process by which gene splicing is carried out within specific organs or tissues of a fully formed organism; germ-line genetic engineering is a process by which gene splicing is carried out within sex cells or embryos, causing the recombinant DNA to exist in every cell of the organism as it grows.

Electrical Engineering. Electrical engineering technologies are an essential part of the bioengineering tool kit. In many cases, what is required is for the bioengineer to find some way to convert sensory data into electric signals, and then to produce these electric signals in such a way as to enable them to have a physiological effect on a living organism.

The cochlear implant is an example of one such development. The cochlea is the part of the brain that interprets sounds, and a cochlear implant is designed for people who are profoundly deaf. A cochlear implant uses electronic devices that capture sounds and relay them to the cochlea. The implant has four parts: a microphone, a tiny computer processor, a radio transmitter, and a receiver, which surgeons implant in the user's skull. The microphone picks up nearby sounds, such as human speech or music emerging from a pair of stereo speakers. Then the processor converts the sounds into digital

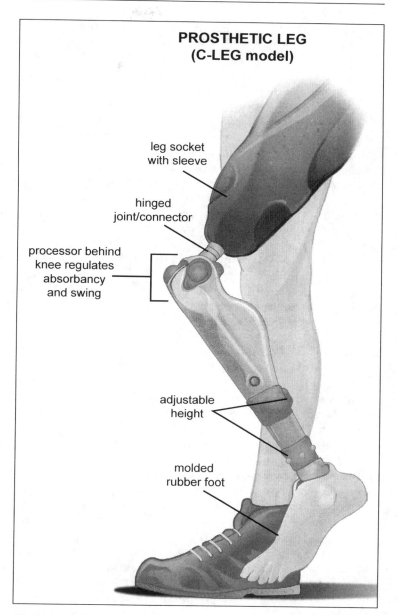

PROSTHETIC LEG (C-LEG model)

leg socket with sleeve

hinged joint/connector

processor behind knee regulates absorbancy and swing

adjustable height

molded rubber foot

Bioengineers have made significant advances in the field of prosthetics; the C-Leg prosthesis was introduced in 1997.

information that can be sent through a wire to the radio transmitter. The software used by the processor separates sounds into different channels, each representing a range of frequencies. In turn, the radio transmitter translates the digital information into radio signals, which it relays through the skull to the receiver. The receiver then turns the radio signals into electric impulses, which directly stimulate the nerve endings in the cochlea. It is these electric signals that the brain is able to interpret as sounds, allowing even profoundly deaf people to hear.

Another example of how electric signals can be used to direct biological systems can be found in brain-computer interfaces (BCIs). BCIs are direct channels of communication between a computer and the neurons in the human brain. They work because activity in the brain, such as that produced by thoughts or sensory processing, can be detected by bioinstruments designed to record electrophysiological signals. These signals can then be transmitted to a computer and used to generate commands. For example, BCIs allow stroke victims who have lost the use of a limb to regain mobility; a patient's thoughts about movement are transmitted to an external machine, which in turn transmits electric signals that precisely control the movements of a cradle holding his or her paralyzed arm.

APPLICATIONS AND PRODUCTS

Biomedical Applications. Biomedical engineering is a vast subdiscipline of bioengineering, which itself encompasses multiple fields of interest. The many clinical areas in which applications are being developed by biomedical engineers include medical imaging, cell and tissue engineering, bioinstrumentation, the development of biocompatible materials and devices, biomechanics, and the emerging field of bionanotechnology.

Medical imaging applications collect data about patients' bodies and turn that data into useful images that physicians can interpret for diagnostic purposes. For example, ultrasound scans, which map the reflection and reduction in force of sounds as they bounce off an object, are used to monitor the development of fetuses in the wombs of pregnant women. Magnetic resonance imaging (MRI), which measures the response of body tissues to high-frequency radio waves, is often used to detect structural abnormalities in the brain or other body parts.

Cell and tissue engineering is the attempt to exploit the natural characteristics of living cells to regenerate lost or damaged tissue. For example, bioengineers are working on creating viable replacement heart cells for people who have suffered cardiac arrests, as well as trying to discover ways to regenerate brain cells lost by patients with neurodegenerative

disorders such as Alzheimers disease. Genetic engineering is a closely related area of biomedicine in which DNA from a foreign organism is introduced into a cell so as to create a new genetic code with desired characteristics.

Bioinstrumentation is the application of electrical engineering principles to develop machines that can sense and respond to biological or physiological signals, such as portable devices for diabetics that measure and report the level of glucose in their blood. Other common examples of bioinstrumentation include electroencephalogram (EEG) machines that continuously monitor brain waves in real time, and electrocardiograph (ECG) machines that perform the same task with heartbeats.

Many biomedical engineers work on developing materials and devices that are biocompatible, meaning that they can replace or come into direct contact with living tissues, perform a biological function, and refrain from triggering an immune system response. Pacemakers, small artificial devices that are implanted within the body and used to stimulate heart muscles to produce steady, reliable contractions, are a good example of a biocompatible device that has emerged from the collaboration of engineers and clinicians.

Biomechanics is the study of how the muscles and skeletal structure of living organisms are affected by and exert mechanical forces. Biomechanics applications include the development of orthotics (braces or supports), such as spinal, leg, and foot braces for patients with disabling disorders such as cerebral palsy, multiple sclerosis, or stroke. Prostheses (artificial limbs) also fall under the field of biomechanics; the sockets, joints, brakes, and pneumatic or hydraulic controls of an artificial leg, for example, are manufactured and then combined in a modular fashion, in much the same way as are the parts of an automobile in a factory.

Bionanotechnology. Nanotechnology is a fairly young field of applied science concerned with the manipulation of objects at the nanoscale (about 1-100 nanometers, or about one-thousandth the width of a strand of human hair) to produce machinery. Bionanotechnological applications within medicine include microscopic biosensors installed on small chips; these can be specialized to recognize and flag specific proteins or antibodies, helping physicians conduct extremely fast and inexpensive diagnostic tests. Bioengineers are also developing microelectrodes on a nanoscale; these arrays of tiny electrodes can be implanted into the brain and used to stimulate specific nerve cells to treat movement disorders and other diseases.

Military Applications. Bioengineering applications are making themselves felt as a powerful presence on the front lines of the military. For example, bioengineering students at the University of Virginia designed lighter, more flexible, and stronger bulletproof body armor using specially created ceramic tiles that are inserted into protective vests. The armor is able to withstand multiple impacts and distributes shock more evenly across the wearer's body, preventing damaging compression to the chest. Others working in the field are creating sophisticated biosensors that soldiers can use to detect the presence of potential pathogens or biological weapons that have been released into the air.

One of the most significant contributions of bioengineering to the military is in the development of treatments for severe traumas sustained during warfare. For example, stem cell research may one day enable military physicians to regenerate functional tissues such as nerves, bone, cartilage, skin, and muscle—an invaluable tool for helping those who have lost limbs or other body parts as a result of explosives. The United States military was responsible for much of the early research done in creating safe, effective artificial blood substitutes that could be easily stored and relied on to be free of contamination on the battlefield.

Agriculture. Agricultural engineering involves the application of both engineering technologies and knowledge from animal and plant biology to problems in agriculture, such as soil and water conservation, food processing, and animal husbandry. For example, agricultural engineers can help farmers maximize crop yields from a defined area of land. This technique, known as precision farming, involves analyzing the properties of the soil (factors such as drainage, electrical conductivity, pH [acidity] level, and levels of chemicals such as nitrogen) and carefully calibrating the type and amount of seeds, insecticides, and fertilizers to be used.

Farm machinery and implements represent another area of agriculture in which engineering principles have made a big impact. Tractors, harvesters, combines, and grain-processing equipment, for example, have to be designed with mechanical and electrical principles in mind and also must take into account the characteristics of the land, the needs of the human operators, and the demands of working with particular agricultural products. For example, many crops require specialized equipment to be successfully mechanically harvested. Thus a pea harvester may have several components—one that lifts the vines and cuts them from the plant, one that strips pea pods from the stalk, and one that threshes the pods, causing them to open and release the peas inside them. Another example of an agricultural engineering application is the development of automatic milking machines that attach to the udders of a cow and enable dairy farmers to dispense with the arduous task of milking each animal by hand.

The management of soil and water is also an important priority for bioengineers working in agricultural settings. They may design structures to control the flow of water, such as dams or reservoirs. They may develop water-treatment systems to purify wastewater coming out of industrial agricultural production centers. Alternatively, they may use soil walls or cover crops to reduce the amount of pesticides and nutrients that run off from the soil, as well as the amount of erosion that takes place as a result of watering or rainfall.

Environmental and Ecological Applications. Environmental and ecological engineers study the impact of human activity on the environment, as well as the ways in which humans respond to different features of their environments. They use engineering principles to clean, control, and improve the quality of natural spaces, and find ways to make human interactions with environmental resources more sustainable. For example, the reduction and remediation of pollution is an important area of concern. Therefore, an environmental engineer may study the pathways and rates at which volatile organic compounds (such as those found in many paints, adhesives, tiles, wall coverings, and furniture) react with other gases in the air, causing smog and other forms of air pollution. They may design and build sound walls in residential areas to cut down on the amount of noise pollution caused by airplanes taking off and landing or cars racing up and down highways.

The life-support systems designed by bioengineers to enable astronauts to survive in the harsh conditions o outer space are also a form of environmental

Fascinating Facts About Bioengineering

- Bioengineering has enabled scientists to grow replacement human skin, tracheas, bladders, cartilage, and other tissues and organs in the laboratory.
- Materials scientists and clinical researchers are working together to develop contact lenses that can deliver precise doses of drugs directly into the eye.
- By genetically engineering crops that are naturally resistant to insects, bioengineers have helped reduce the need to use harmful pesticides in industrial farming.
- Bacteria whose genetic information has been carefully reengineered may eventually provide an endless supply of crude oil, helping meet the world's energy needs without engaging in damaging drilling.
- In 2009, an MIT bioengineer invented a new way to pressurize space suits that does not use gas, making them far sleeker and less bulky than conventional astronaut gear.
- One military application of bioengineering is a robotic system that seeks out and identifies tiny pieces of shrapnel lodged within tissue, then guides a needle to those precise spots so that the shrapnel can be removed.
- Bionic men and women are not just the stuff of television and motion-picture fantasy. In fact, anyone who has an artificial body part, such as a prosthetic leg, a pacemaker, or an implanted hearing aid, can be considered bionic.
- Some bioengineers are working on developing artificial noses that can detect and diagnose disease by smell–literally sniffing out infections and cancer, for example.
- One day, it may be possible to "print out" artificial organs using a three-dimensional printer. Layer by layer, cells would be deposited onto a glass slide, building up specialized tissues that could be used to replace damaged kidneys, livers, and other organs.
- Each year, more women choose to enter the field of biomedical engineering than any other specialty within engineering.

engineering. For example, temperatures around a space shuttle can vary wildly, depending on which side of the vehicle is facing the Sun at any given moment. A complex system of heating, insulation, and ventilation helps regulate the temperature inside the cabin. Because space is a vacuum, the shuttle itself must be filled with pressurized gas. In addition, levels of oxygen, carbon dioxide, and nitrogen within the cabin must be controlled so that they resemble the atmosphere on Earth. Oxygen is stored on board in tanks, and additional supplies of the essential gas are produced from electrolyzed water; in turn, carbon dioxide is channeled out of the shuttle through vents.

Geoengineering. Geoengineering is an emerging subfield of bioengineering that is still largely theoretical. It would involve the large-scale modification of environmental processes in an attempt to counteract the effects of human activity leading to climate change. One proposed geoengineering project involves depositing a fine dust of iron particles into the ocean in an attempt to increase the rate at which algae grows in the water. Since algae absorbs carbon dioxide as it photosynthesizes, essentially trapping and containing it, this would be a means of reducing the amount of this greenhouse gas in the atmosphere. Other geoengineering proposals include the suggestion that it might be possible to spray sulfur dust into the high atmosphere to reflect some of the Sun's light and heat back into space, or to spray drops of seawater high up into the air so that the salt particles they contain would be absorbed into the clouds, making them thicker and more able to reflect sunlight.

IMPACT ON INDUSTRY

Bioengineering is a global industry that boasts consistent revenues. With an aging public, biomedical engineering looks poised for significant growth. The global focus on reversing climate change, ensuring an adequate supply of food and clean water, and improving and maintaining health means that bioengineering is likely to be an ever-expanding field. Those nations that embrace its possibilities will find themselves reaping the rewards in a better quality of life for their citizens.

Government and University Research. The United States is generally considered to be the world leader in bioengineering research, especially within the field of biomedical engineering. However, significant strides are being made in many European countries, including France and Germany, as well as by many growing economies in Asia, such as China, Singapore, and Taiwan. In the United States, the main governmental organization funding studies in this area is the National Institute of Biomedical Imaging and Bioengineering, a branch of the National Institutes of Health. Among the United States universities whose faculty and students are recognized as conducting the most leading-edge research in bioengineering are The Johns Hopkins School of Medicine, the Massachusetts Institute of Technology (MIT), and the University of California, San Diego—all of which were ranked at the top of the 2009 *U.S. News and World Report's* list of the best biomedical/bioengineering schools in the country.

Major Corporations. Major biomedical engineering corporations include Medtronic, Abbot, Merck, and Glaxo-Smith Kline, all international producers of products such as pharmaceuticals and medical devices. Medtronic, for example, manufactures items such as defibrillators, pacemakers, and heart valves, while Merck produces drugs to treat cancer, heart disease, diabetes, and infections. Within the field of agricultural biotechnology, Monsanto and DuPont are industry leaders. Both corporations produce, patent, and market seeds for transgenic crops. The plants grown from these seeds possess traits attractive to industrial farmers, such as resistance to pesticides, longer ripening times, and higher yield.

Industry and Business. Bioengineering is considered a strong growth industry with a great deal of potential for expansion in the twenty-first century. The production of biomedical devices and biocompatible materials alone, for example, is a market worth an estimated $170 billion per year. In the United States, Department of Labor statistics indicate that in 2006, engineers working in the biomedical, agricultural, health and safety, and environmental engineering fields—all of which fall under the bioengineering umbrella—held a total of nearly 100,000 jobs nationwide. In the following years, the department estimated that each of these sectors would add jobs at a rate that either keeps pace with or far exceeds the national average for all occupations.

CAREERS AND COURSE WORK

Although bioengineering is a field that exists at the intersection between biology and engineering, the most common path for professionals in the field is to first become trained as engineers and later apply their technical knowledge to problems in the life sciences. (A less common path is to pursue a medical degree and become a clinical researcher.) At the high school level, it is important to cover a broad range of mathematical topics, including geometry, calculus, trigonometry, and algebra. Biology, chemistry, and physics should also be among an aspiring bioengineer's course work. At the college level, a student should pursue a bachelor of science in engineering. At many institutions, it is possible to further concentrate in a subfield of engineering: Appropriate subfields include biomedical engineering, electrical engineering, mechanical engineering, and chemical engineering. Students should continue to take electives in biology, geology, and other life sciences wherever possible. In addition, English and humanities courses, especially writing classes, can provide the aspiring bioengineer with strong communication skills—important for working collaboratively with colleagues from many different disciplines.

Many, though not all, choose to pursue graduate-level degrees in biomedical engineering, agricultural engineering, environmental engineering, or another subfields of bioengineering. Others go through master's of business administration programs and combine this training with their engineering background to become entrepreneurs in the bioengineering industry. Additional academic training beyond the undergraduate level is required for careers in academia and higher-level positions in private research and development laboratories, but entry-level technical jobs in bioengineering may require only a bachelor's degree. Internships (such as at biomedical companies) or evidence of experience conducting original research will be helpful in obtaining one's first job.

A variety of career options exist for bioengineers; many work as researchers in academic settings, private industry, government institutions, or research hospitals. Some are faculty members, and some are administrators, managers, supervisors, or marketing consultants for these same organizations. Others are engaged in designing, developing, and conducting safety and performance testing for bioengineering instruments and devices.

SOCIAL CONTEXT AND FUTURE PROSPECTS

Bioengineering is a field with the capacity to exert

a powerful impact on many aspects of social life. Perhaps most profound are the transformations it has made in health care and medicine. By treating the body as a complex system—looking at it almost as if it were a machine—bioengineers and physicians working together have enabled countless patients to overcome what once might have seemed to be insurmountable damage. After all, if the body is a machine, its parts might be reengineered or replaced entirely with new ones—as when the damaged cilia of individuals with hearing impairments are replaced with electro-mechanical devices. Some aspects of bioengineering, however, have drawn concern from observers who worry that there may be no limit to the scientific ability to interfere with biological processes. Transgenic foods are one area in which a contentious debate has sprung up. Some are convinced that the ecological and health ramifications of growing and ingesting crops that contain genetic information from more than one species have not yet been fully explored. Stem cell research is another area of controversy; some critics are uncomfortable with the fact that human embryonic stem cells are being obtained from aborted fetuses or fertilized eggs that are left over from assisted reproductive technology procedures.

One aspect of bioengineering that has been the subject of both fear and hope in the twenty-first century is the question of whether it might be possible to stop or even reverse the harmful effects of climate change by carefully and deliberately interfering with certain geological processes. Some believe that geoengineering could help the international community avoid the devastating effects of global warming predicted by scientists, such as widespread flooding, droughts, and crop failure. Others, however, warn that any attempt to interfere with complex environmental systems on a global scale could have wildly unpredictable results. Geoengineering is especially controversial because such projects could potentially be carried out unilaterally by countries acting without international agreement and yet have repercussions that could be felt all across the world.

M. Lee, B.A., M.A.

FURTHER READING

Artmann, Gerhard M., and Shu Chien, eds. *Bioengineering in Cell and Tissue Research.* New York: Springer, 2008. Examines bioengineering's role in cell research. Heavily illustrated with diagrams and figures; includes a comprehensive index and references after each section.

Enderle, John D., Susan M. Blanchard, and Joseph D. Bronzino, eds. *Introduction to Biomedical Engineering.* 2d ed. Boston: Elsevier Academic Press, 2005. A broad introductory textbook designed for undergraduates. Each chapter contains an outline, objectives, exercises, and suggested reading.

Huffman, Wallace E., and Robert E. Evenson. *Science for Agriculture: A Long-Term Perspective.* 2d ed. Ames, Iowa: Blackwell, 2006. A history of agricultural engineering research within the United States. Includes a glossary and list of relevant acronyms.

Madhavan, Guruprasad, Barbara Oakley, and Luis G. Kun, eds. *Career Development in Bioengineering and Biotechnology.* New York: Springer, 2008. An extensive guide to careers in bioengineering, biotechnology, and related fields, written by active practitioners. Covers both traditional and alternative job opportunities.

Nemerow, Nelson Leonard, et al., eds. *Environmental Engineering.* 3 vols. 6th ed. Hoboken, N.J.: John Wiley & Sons, 2009. Discusses topics such as food protection, soil management, waste management, water supply, and disease control. Each section includes references and a bibliography.

WEB SITES

Biomedical Engineering Society
http://www.bmes.org

National Institutes of Health
National Institute of Biomedical Imaging and Bioengineering
http://www.nibib.nih.gov

Society for Biological Engineering
http://www.aiche.org/sbe

See also: Agricultural Science; Audiology and Hearing Aids; Biomathematics; Cell and Tissue Engineering; Genetically Modified Organisms; Genetic Engineering; Human Genetic Engineering; Military Sciences and Combat Engineering; Rehabilitation Engineering.

BIOINFORMATICS

FIELDS OF STUDY

Molecular biology; genetics; molecular genetics; phylogenetics; cell biology; physics; biochemistry; biophysics; biostatistics; computational biology; computer science; evolutionary biology; structural biology; systems biology; mathematics.

SUMMARY

Bioinformatics is simultaneously a new type of research practice and a rapidly emerging new discipline. It has shifted the practice of scientific research from traditional laboratory bench research to computer-based data analysis and experimentation with massive datasets available on the Internet. To support this research, bioinformaticians develop the software required for data analyses, design the biodatabases, organize and manage the data within them, create the online computing environment, and develop the highly specialized mathematical algorithms and statistical packages to search, retrieve, and analyze biodata by bioresearchers. Bioinformatics has accelerated the pace and understanding of biological systems and molecules exponentially, with its greatest promise indicated in medical and environmental applications.

KEY TERMS AND CONCEPTS

- **Genome-Wide Association Studies:** Scanning of a human genome biocomputationally to associate genetic variations to specific diseases, drug reactions, and other relevant issues.
- **Genomics:** Study of the structure, function, and interaction of the complete set of genetic material found in an organism.
- **Multiple Sequence Alignment:** Algorithm that aligns three or more sequences to one another according to their best sequence similarities.
- **Pharmacogenomics:** Identification and study of the genes involved in an organism's response to drugs.
- **Phylogenetics:** Bioinformatically, the evolutionary relationships between biosequences to trace sequence ancestries of genes or organisms.
- **Protein Modeling:** Manipulation of three-dimensional structures of biosequence molecules computationally and through visualization software to correlate structure with function.
- **Proteomics:** Study and comparison of the entire complement of proteins (proteome) under a given condition: a cell's proteome, comparison of a normal liver proteome before and after drug treatment, comparison of proteomes from normal tissues with cancerous tissues.
- **Single Nucleotide Polymorphisms (SNPs):** Normal variations found in individual genomes at a single nucleotide position; rare variations are often disease-causing mutations.

DEFINITION AND BASIC PRINCIPLES

Bioinformatics is simultaneously a relatively new type of research practice and a rapidly emerging discipline. As research, bioinformatics is defined as the manipulation and the varied analyses performed by laboratory-based researchers on massive biological datasets residing in thousands of Internet-based databases, each with a distinct set of data and a specific purpose. Originating from molecular biology, bioinformatics has rapidly spread to cell biology, chemistry, statistics, computer sciences, physics, biomedical engineering, psychology, and even anthropology.

As a discipline, bioinformatics draws on those professionals with advanced skills from the computer sciences, information sciences, and mathematics disciplines to bear on biological problems posed by laboratory-based bioresearchers. Collectively, these specialists are referred to as "bioinformaticians" to distinguish them from the laboratory-based science researchers carrying out experiments with the products bioinformaticians have created.

Researchers often define bioinformatics from within the perspective of their specific discipline or individual research efforts. Some biologists view bioinformatics as only involving DNA or protein sequencing. Chemists and physicists tend to view bioinformatics as involving protein molecular structures. Computer scientists describe bioinformatics from a programming or information infrastructure perspective. Pharmacologists often define bioinformatics from the viewpoint of drug-protein interactions. All of these variations share the concept of applying computational analyses to biological processes.

A definition that encompasses bioinformatics both as a profession and a research practice and also takes into account the multitude of disciplines involved is still very much a work in progress as bioinformatics continues to evolve. A unified definition views bioinformatics as the convergence of the biological sciences and computer technologies and the integration of statistics and probability mathematics to understand biological processes of molecules on a very large scale. In turn, collecting, cataloging, classification, storage, organization, management, and retrieval of these massive biodatasets requires information theory and practice (informatics) from the information sciences disciplines to make them available for problem solving.

BACKGROUND AND HISTORY

Bioinformatics originates from within the fields of genetics and molecular biology. The computational, mathematical, and biodatabase origins of bioinformatics arose not from within the biological, computer, or mathematical sciences but rather from two individuals who had a fascination with the computer technologies being introduced in the 1960's: Robert Ledley, a dentist turned theoretical physicist, and Margaret Dayhoff, a quantum chemist. Ledley, the inventor of the whole-body computerized tomography machine, founded the National Biomedical Research Foundation (NBRF) in 1960 to research and discover possible uses of computers in biomedical research. He recruited Dayhoff to apply her knowledge and skills at data entry and processing toward protein sequencing, which, at that time was taking more than a year to sequence a single protein by traditional laboratory methods.

Using computational analyses, Dayhoff discovered sequence patterns that identified similar proteins and predicted possible functions. She created a series of mathematical scoring matrices and defined a set of mathematical expressions that accurately reflected these similarities across evolutionary distances. In so doing, she created the first bioinformatics algorithms. Her sequence similarity matrices and rules still provide the basis for contemporary sequence similarity searching algorithms, most notably the suite of Basic Local Alignment Sequence Tools (BLAST) created by the National Center for Biotechnology Information (NCBI) in 1997.

Fascinating Facts About Bioinformatics

- During the 2009 H1N1 swine flu epidemics, more than 24,000 individual virus genomes were sequenced immediately from infected patients worldwide. These data played a major role in the autumn 2009 vaccine development.
- GenBank has become a major sequence resource containing 150 million sequence records in 2009 and giving rise to hundreds of secondary, specialized databases worldwide. More than 500 million records in thirty-plus secondary databases exist at the National Center for Biotechnology Information alone. Worldwide, the number of bioinformatics records based on GenBank is in the billions.
- The genomes between different humans are 99.9 percent similar. The 0.1 percent difference is due to single DNA nucleotide variations at very specific points within the human genome, making each person different from all others physically, behaviorally, and physiologically.
- Before the Human Genome Project, scientists thought that the human genome contained up to 100,000 genes because of the large number of proteins that are known to exist in humans. Scientists now know that there are fewer than 30,000 genes in humans, with each gene estimated to give rise to 3 to 8 different proteins. There are exceptions. The *dscam* gene is involved in the development of neural circuits. In humans, this gene codes for more than 16,000 variations of the dscam protein, and in the fruit fly, 38,016 isoforms of the dscam protein have been proven to exist.
- Human DNA is 98 percent similar in sequence to chimpanzees. However, the genetic difference between any two chimpanzees is four to five times greater than the difference between any two humans.
- Of the greater than 3 billion nucleotides in the human genome, less than 3 percent actually codes for protein molecules. The functions of the vast majority of the human genome and what it does or does not do remain unknown.

In 1963, Dayhoff began compiling protein sequences into a series of books titled *Atlas of Protein*

Structure and Function. By 1978, the *Atlas of Protein Structure and Function* had grown too large to make comparisons and perform analyses. Her second major contribution was to create a database infrastructure to convert the atlas to the first online biological database accessible to researchers who could use it to sort, manipulate, and align multiple protein sequences. The database created, the Protein Information Resource (PIR), has become the major Internet UniProt protein bioinformatics resource at the European Bioinformatics Institute (EBI).

Although the National Institutes of Health (NIH) founded the DNA bioinformatics database, GenBank, in 1982 to specifically accelerate nucleic acid sequence experimentation, progress in DNA sequencing and gene cloning technologies lagged behind protein sequencing. In 1985, then Chancellor of the University of California, Santa Cruz and molecular biologist Robert Sinsheimer convened a workshop of prominent scientists and made what was considered a radical and controversial proposal. He proposed to sequence the entire human genome and then use computational analyses to discover unknown genes and their functions and interrelationships. Thus, the Human Genome Project was initiated in 1990 by the National Institutes of Health. It soon became obvious that existing computational power and hardware were insufficient to process or hold the data being generated. Major engineering innovations were needed to process larger sample numbers, faster. The computer sciences and engineering disciplines responded. Within a few years, specialized robotics, miniaturization of samples, faster computers and processors, larger data storage capacity, and new kinds of software engineering tools were in use, greatly accelerating DNA sequencing.

How It Works

How bioinformatics is practiced depends on whether the research is conducted on small data as typified by individual research laboratories or on a mega-scale. Small-scale data handling is often called low throughput, while large-scale is always called high throughput.

Low-Throughput Bioinformatics. In a simple sequencing scenario, researchers working to identify a protein or a gene perform "wet research" experiments ultimately yielding DNA or protein candidates. The candidates are sequenced. The researcher

accesses the appropriate bioinformatics databases over the Internet and searches for similar sequences using sequence similarity algorithms, analyzing the results to provide clues to the function and identity of the candidate sequences. Once the researcher has clues to possible functions, additional bioinformatics databases are searched to aid in the development of the next experiment to be performed. In the process, many different bioinformatics databases and tools are used. Learning what databases and tools exist and are best is part of the process of learning bioinformatics research. There are times when the tool may not exist or existing databases are not sufficient. The bench researcher may ask the local bioinformatician to help design a more specific programming tool. If this becomes a critical problem for this area of research in general, bioinformaticians develop new tools and/or databases. These are published in the peer-reviewed literature and tried out by the scientific community. Those that work eventually become established as bioinformatics resources.

Sequences recovered by laboratory researchers with federally funded grants must be uploaded to a sequence repository, along with any information discovered. In the United States, this is NCBI GenBank. Data uploaded in the United States, Europe, and Asia are shared among the countries daily, permitting rapid access to the biodata generated worldwide. NCBI curators then work on the uploaded sequences to integrate and incorporate them into any of the thirty-plus databases at the National Center for Biotechnology Information. When new types of data are being uploaded as research progresses into new areas, the center or its European and Asian counterparts design new kinds of databases and algorithms or fund others to do so.

High-Throughput Bioinformatics. This research typically involves massive generation of data, such as large-scale genome sequencing efforts, or the simultaneous analyses of very large datasets. An example of the latter would be clinical data arising from the identification of proteins unique to a specific cancer isolated from many patients. In these scenarios, millions of sequences need to be processed daily. This kind of bioinformatics requires robotic bioinstrumentation and different algorithms to process. It typically is carried out by supercomputing facilities supported by bioinformaticians with experience in parallel computing, networking, grid computing, advanced algorithms, statistical programming skills, and advanced database modeling and design. Any sequence data recovered from research supported from federal funds must be uploaded to the National Center for Biotechnology Information. In this case, since the functions of the sequences are unknown, the center computationally processes these data to different databases than GenBank, making them available for others to search and identify the function of the sequences.

APPLICATIONS AND PRODUCTS

Biological Databases. Biosequence databases are at the very foundation of bioinformatics research and discovery. The National Center for Biotechnology Information, the European Bioinformatics Institute, and the DNA Database of Japan are the major biosequence spaces; each has an extensive suite of hyperlinked protein, genomes, nucleotide, genes, gene expressions, disease, and chromosome databases. Scientific organizations, government agencies, and research institutes have collaborated to create other databases.

The major protein databases are UniProt of EBI and the three-dimensional structural protein resource at Protein DataBank. Online Mendelian Inheritance of Man (OMIM) and Animal (OMIA) correlate mutations and their inheritance patterns with disease phenotypes. PharmGKB is a major pharmacogenetics database that monitors human genetic variations to specific patient drug reactions and their symptoms. Biological pathway databases, including BRENDA, Reactome, and KEGG, enable researchers to locate proteins that interact with each other and determine how protein sequence alterations could give rise to abnormal biological processes.

Genomes of many different organisms have been sequenced, each representing biodata that detail a model biological system or disease process. Finally, there are thousands of smaller "boutique" biodatabases for specific diseases, the different functional or structural components of genes or proteins, and similar topics.

Algorithms. Although there are many mechanisms to search biodatabases, the most critical and extensively used is sequence similarity searching. Needleman-Wunsch, Smith-Waterman, FASTA, and BLAST represent the major similarity algorithms. They differ in algorithmic mechanism and

computational speed, with Needleman-Wunsch being the most accurate but also the most computationally intense. At the time of its publication in 1970, it took days to return results. BLAST is the least accurate of the set but computationally the fastest, taking only minutes to return results. BLAST supported laboratory bench research in real time and is the major sequence similarity algorithm in use. However, as personal computers have advanced to faster processors, the Needleman-Wunsch and Smith-Waterman algorithms have been reengineered and made available at the National Center for Biotechnology Information and the European Bioinformatics Institute.

Molecular Visualization and Modeling. A combination of software engineering and sequence algorithm, three-dimensional molecular viewers enable researchers to manipulate and computationally model proteins. They are particularly important in drug design and analyses of mutant proteins involved in disease, as researchers can introduce changes *in silico* and view how they alter drug interaction or structure or compare a mutant protein directly with a normal protein superimposed in three dimensions. The two most important molecular viewers are Cn3D at NCBI and RasMol for the other protein and nucleic acid databases.

Biodiversity. Microbes (bacteria, fungi, protozoans, and viruses) represent half of the Earth's biomass. It is estimated that there are at least 10 million bacterial species alone, with only a few thousand described. Since the 1990's, it has become clear that most microbes live in mixed communities with other microbes, with any given species present in a small number, none of which can be cultured in a laboratory.

Metagenomics is that part of bioinformatics that determines and then studies what organismal communities are present in various environmental samples such as soil or oceans. It can also detect the organisms present in animal organs or tissues such as the digestive tract or skin. The present-day state of bioinformatics technology and data acquisition and storage permits the identification of microorganismal communities only. This includes mixed communities containing bacteria, fungi, and viruses. The samples are collected and all the organisms present in the sample are recovered. Without an attempt to isolate, culture, or identify any of the organisms present, all the DNA from all the organisms is extracted in mass, sequenced, and reassembled into the original genomes, thereby identifying what organisms are present.

Initiated in 2008, the Human Microbiome Project aims to identify all the microorganisms present in five areas of the human body—the digestive tract, the mouth, the skin, the nose, and the vagina—from samples taken from healthy human volunteers. Once the healthy human microbiome has been characterized, the human microbiome will be studied in different disease, nutritional, or treatment states. The aim is to use the human microbiome to identify particular diseases and to study the effectiveness of probiotics, pharmaceutical drug treatments, and other therapies. In 2010, 900 microbial genomes had been identified as components of the human microbiome, of which 178 have been fully sequenced. The data indicate that the human microbiome is massive and at least one hundred times larger than the human genome itself. It contains nearly twice the microbial diversity already identified in public domain databases.

Understanding the oceanic microbiome and how it responds to climate and human impact is an important step to oceanic conservation. In addition, adding to the catalog of known proteins enhances the ability to discover new proteins that could be reengineered or repurposed for medicinal or bioremediation uses. In a metagenomics approach similar to that taken by the Human Microbiome Project, oceanic samples have been collected, all the microorganisms recovered, DNA extracted, sequenced, and genomically reassembled. This Global Ocean Sampling expedition has identified at least 400 new microbial species and 6 million predicted proteins, doubling the total number of proteins previously identified.

Bioinformatics' contribution to biodiversity is not limited to the present day. Museums worldwide contain unique specimens (both plant and animal) that can be sequenced, genomically cataloged, and characterized. Ancient DNA, the DNA recovered from fossil organisms trapped in underground ancient lake beds or water droplets trapped in various geological samples, is also available for genomic analyses, adding to the publicly available Neanderthal and *Mastodon* genomes.

Personal Genomes. In less than seven years, the cost of sequencing a human genome dropped from almost $3 billion for the Human Genome Project to less than $30,000 in 2010 because of rapid advances in computational and engineering technologies in bioinformatics. By the end of 2011, it is estimated that more 30,000 different human genomes will have been sequenced by various genomic centers and institutes worldwide. As costs continue to drop, sequencing a human genome will be within the reach of individual research laboratories in several years and affordable by many private citizens in possibly five more years. Several private companies are already advertising (at a cost ranging from $400 to $1,500) to scan people's genome for common DNA sequence variations that are associated with specific diseases or conditions such as diabetes or high cholesterol or are known to reduce or enhance the metabolism of pharmaceutical drugs. Some of these companies offer services that trace an individual's ancestry through his or her inheritance of specific DNA patterns now known to be specific to particular ethnicities or to have originated in distinct geographical areas around the world. This new branch of genomics, in which individual human genomes are sequenced and analyzed, is called personal genomics and carries with it evolving ethical issues that are themselves undergoing rapid debate and analyses.

At the academic research level, large consortiums are being formed to analyze vast numbers of individual human genomes to first catalog and then study all the known genetic differences among both individuals and different kinds of populations. The 1000 Genomes project is a consortium of more than seventy-five universities, institutes, and companies worldwide. Regardless of its name, it aims to sequence the genomes of 2,300 individuals with ancestry from Europe, east Asia, south Asia, West Africa, and the Americas. Each genome will be independently analyzed as well as compared to genomes within the same populations. Early studies indicate that each person may carry 250 to 300 mutations in genes known to cause disease, as well as 50 to 100 sequence variations known to be implicated in inherited disorders. Not all genetic variations give rise to disease. The 1000 Genomes catalog has already identified several candidate genetic differences in two genes inherited within one family group that may be responsible for this family having very low cholesterol levels. The hope is that the study of these genes can lead to new cholesterol-lowering strategies.

IMPACT ON INDUSTRY

Bioinformatics is creating new industries and services. Industrial applications are very much in their infancy and in the research and development phase for the most part.

Government and University Research. The National Institutes of Health is a major source of research funding for basic science, biomedical, and clinical research. The National Science Foundation funds more in the environmental and educational sectors. Both are active policy setters and enforcers of data sharing and integration, cyberinfrastructure, supercomputing, and bioethical issues. Because of a federal mandate, the National Institutes of Health focuses on issues related to health and disease, which is the reason that the majority of bioinformatics research in the United States is related to medical and clinical research. The Department of Energy, through the Office of Biological and Environmental Research (BER), funds research on selected organisms, as well as on environmental genomics and proteomics projects related to bioremediation. BER also studies the ethical, legal, and social ramifications of genome projects through the Ethical, Legal, and Social Issues Program. The Department of Energy's Advanced Scientific Computing Research office funds computer science, networking, and mathematics research.

Industry and Business Sectors. The Howard Hughes Medical Institute (HHMI) is a nonprofit independent research institution that both funds and performs bioinformatics research. Its funding program entails appointing scientists as Hughes Investigators, providing them with long-term funding at their home institutions with the freedom to explore research projects as they choose. It is influential in recommending policies and standards for undergraduate science and medical education, including incorporating bioinformatics into the curriculum.

Many independent research institutions (including the Broad Institute, the J. Craig Venter Institute, and the Sanger Institute) carry out bioinformatics research. Several are large biodata producers, performing only high-throughput genomic and metagenomic computationally intense bioinformatics research.

In the for-profit sector, the biotechnology and pharmaceutical companies are significantly invested in protein engineering and modeling for pharmaceutical drug and diagnostic kit development. Bioproduct companies produce the enzymes, reagents, and kits needed to support the laboratory-based molecular biology research related to bioinformatics. Bioinstrumentation companies research, develop, and provide the highly specialized automated genomics and proteomics sequence analyzers and processors needed by both for-profit and nonprofit research efforts. They are an important source of innovation in sequencing methodologies that is largely responsible for continuing to advance bioinformatics into practical applications including metagenomics and personalized medicine initiatives.

CAREERS AND COURSE WORK

Optimum bioinformatics practice needs bioinformaticians trained or experienced in biocomputational research and development. As bioinformatics continues to evolve and expand, the employment outlook is excellent. Jobs are available as algorithmic developers, programmers, data analysts/integrators, software engineers, or database designers in private industry, academic environments, government agencies, and nonprofit research institutions. Positions exist at all levels, from entry programmers to senior-level scientists to research directors.

Formal undergraduate and graduate academic bioinformatics or computational biology educational programs are just becoming available. Curricula try to provide a broad foundational core of understanding of the mathematics, computer, molecular, cellular, and genetics sciences. Course work in PERL programming languages, data structures, database design, algebra, probability and statistics, calculus, and discrete mathematics are designed to fill the existing gap in the marketplace. Laboratory courses provide exposure to the basic data and tools used, including DNA and protein sequence similarity and alignment, protein structure, phylogenetic analyses, and finding and cloning genes. Bioethics, justice courses, or a senior research project/thesis are not uncommon requirements. Graduate level curricula add courses in bioinstrumentation, protein engineering, population genetics, molecular diagnostics and prediction, the emerging field of genetic association studies, statistical computing packages such as R, and topically specialized seminars.

SOCIAL CONTEXT AND FUTURE PROSPECTS

Metagenomics, personalized medicine, and future bioinformatics initiatives yet to be discovered will rapidly affect nearly all individuals. It is not surprising that with these major advances in bioinformatics technologies comes a caution by many for careful introspection and debate of their implications. Interest in bioethics is on the rise and has been added to the curricula of not only bioinformatics educational programs but those of many other disciplines as well.

Diane C. Rein, Ph.D., M.L.S.

FURTHER READING

Baxevanis, Andreas D., and B. F. Francis Ouellette, eds. *Bioinformatics: A Practical Guide to the Analysis of Genes and Proteins.* 3d ed. Hoboken, N.J.: John Wiley & Sons, 2005. Covers bioinformatics from the database and searching perspective. Contains chapters on various biological databases and their search interfaces.

Gu, Jenny, and Phillip E. Bourne, eds. *Structural Bioinformatics.* 2d ed. Hoboken, N.J.: John Wiley & Sons, 2009. Combination textbook and manual covering all aspects of protein bioinformatics, including the major protein databases, visualization, mass spectrometry, and protein modeling.

Lesk, Arthur M. *Introduction to Bioinformatics.* 3d ed. New York: Oxford University Press, 2008. Comprehensive overview of genomes, proteomics, protein structure, databases, phylogenetics, programming languages, and more.

Zvelebil, Marketa, and Jeremy Baum. *Understanding Bioinformatics.* New York: Garland Science, 2008. Intermediate text with detailed descriptions on sequence alignments, phylogenetics, genomics, proteomics, and protein structure and modeling.

WEB SITES

American Medical Informatics Association
http://www.amia.org

Bioinformatics Organization
http://www.bioinformatics.org

European Bioinformatics Institute
http://www.ebi.ac.uk

National Center for Biotechnology Information
http://www.ncbi.nlm.nih.gov

See also: Bioengineering; Biomathematics; DNA Analysis; DNA Sequencing; Genomics; Human Genetic Engineering.

BIOMATHEMATICS

FIELDS OF STUDY

Algebra; geometry; calculus; probability; statistics; cellular biology; genetics; differential equations; molecular biology; oncology; immunology; epidemiology; prokaryotic biology; eukaryotic biology

SUMMARY

Biomathematics is a field that applies mathematical techniques to analyze and model biological phenomena. Often a collaborative effort, mathematicians and biologists work together using mathematical tools such as algorithms and differential equations in order to understand and illustrate a specific biological function. Biomathematics is used in a wide variety of applications from medicine to agriculture. As new technologies lead to a rise in the amount of biological data available, biomathematics will become a discipline that is increasingly in demand to help analyze and effectively utilize the data.

KEY TERMS AND CONCEPTS

- **Algorithm:** Use of symbols and a set of rules for solving problems.
- **Biology:** Study of living things.
- **Cell:** Unit that is the basis of an organism and encompasses genetic material, as well as other molecules, and is defined by a cell membrane.
- **Deoxyribonucleic Acid (DNA):** Nucleic acid that forms the molecular basis for heredity.
- **Differential Equation:** Mathematic expression that uses variables to express changes over time. Differential equations can be linear or nonlinear.
- **Genetics:** Study of an organism's traits, including how they are passed down through generations.
- **Matrix:** Mathematical structure for arranging numbers or symbols that have particular mathematical rules for use.
- **Oncology:** Field of science that studies the cause and treatment of cancer.

DEFINITION AND BASIC PRINCIPLES

Biomathematics is a discipline that quantifies biological occurrences using mathematical tools.

Biomathematics is related to and may be a part of other disciplines including bioinformatics, biophysics, bioengineering, and computational biology, as these disciplines include the use of mathematical tools in the study of biology.

Biologists have used different ways to explain biological functions, often employing words or pictures. Biomathematics allows biologists to illustrate these functions using techniques such as algorithms and differential equations. Biological phenomena vary in both scale and complexity, encompassing everything from molecules to ecosystems. Therefore, the creation of a model requires the scientist to make some assumptions in order to simplify the process. Biomathematical models vary in length and complexity and several different models may be tested.

The use of biomathematics is not limited to modeling a biological function and includes other techniques, such as structuring and analyzing data. Scientists may use biomathematics to organize data or analyze data sets, and statistics are often considered an integral tool.

BACKGROUND AND HISTORY

As early as the 1600's, mathematics was used to explain biological phenomena, although the mathematical tools used date back even farther. In 1628, British physician William Harvey used mathematics to prove that blood circulates in the body. His model changed the belief at that time that there were two kinds of blood. In the mid-1800's, Gregor Mendel, an Augustinian monk, used mathematics to analyze the data he obtained from his experiments with pea plants. His experiments would become the basis for genetics. In the early 1900's, British mathematician R. A. Fisher applied statistical methods to population biology, providing a better framework for studying the field. In 1947, theoretical physicist Nicolas Rashevsky argued that mathematical tools should be applied to biological processes and created a group dedicated to mathematical biology. Despite the fact that some dismiss Rashevsky's work as being too theoretical, many view him as one of the founders of mathematical biology. In the 1950's, the Hodgkin-Huxley equations were developed to describe a cellular function known as ion channels. These equations are still used. In the 1980's, the Smith-Waterman algorithm was created to aid scientists in comparing DNA sequences. While the algorithm was not particularly efficient, it paved the way for the BLAST (Basic Local Alignment Search Tool) software, a program that has allowed scientists to compare DNA sequences since 1990. Despite the fact that mathematical tools have been applied to some biological problems during the second half of the twentieth century, the practice has not been all-inclusive. In the twenty-first century, there has been a renewed interest in biology becoming more quantitative, due in part to an increase in new data.

HOW IT WORKS

Basic Mathematical Tools. Biomathematicians may use mathematical tools at different points during the investigation of a biological function. Mathematical tools may be used to organize data, analyze data, or even to generate data. Algorithms, which use symbols and procedures for solving problems, are employed in biomathematics in several ways. They may be used to analyze data, as in sequence analysis. Sequence analysis uses specifically developed algorithms to detect similarity in pieces of DNA. Specifically developed algorithms are also used to predict the structure of different biological molecules, such as proteins. Algorithms have led to the development of more useful biological instruments such as specific types of microscopy. Statistics are another common way of analyzing biological data. Statistics may be used to analyze data, and this data may help create an equation to describe a theory: Statistics was used to analyze the movement of single cells. The data taken from the analysis was then used to create partial equations describing cell movement.

Differential equations, which use variables to express changes over time, are another common technique in biomathematics. There are two kinds of differential equations: linear and nonlinear. Nonlinear equations are commonly used in biomathematics. Differential equations, along with other tools, have been used to model the functions of intercellular processes. Differential equations are utilized in several of the important systems used in biomathematics for modeling, including mean field approaches. Other modeling systems include: patch models, reaction-diffusion equations, stochastic models, and interacting particle systems. Each modeling system provides a different approach based on different assumptions. Computers have helped in this area by providing an easier way to apply and solve complex equations. Computer modeling of dynamic systems, such as the motion of proteins, is also a work in progress.

New methods and technology have increased the amount of data being obtained from biological experimentation. The data gained through experimentation and analysis may be structured in different ways. Mathematics may be used to determine the structure. For example, phylogenetic trees (treelike graphs that illustrate how pieces of data relate to one another) use different mathematical tools, including matrices, to determine their structure. Phlyogenic trees also provide a model for how a particular piece of data evolved. Another way to organize data is a site graph, or hidden Markov model, which uses probability to illustrate relationships between the data.

Modeling a Biological Function. The scientist may be at different starting points when considering a mathematical model. He or she may be starting with data already analyzed or organized by a mathematical technique or already described by a visual depiction or written theory. However, there are several considerations that scientists must take into account when creating a mathematical model. As biology covers a wide range of matter, from molecules to ecosystems, when creating a model the scale of phenomena must be considered. The time scale and complexity must also be considered, as many biological systems are dynamic or interact with their environment. The scientist must make assumptions about the biological phenomena in order to reduce the parameters used in the model. The scientist may then define important variables and the relationships between them. Often, more than one model may be created and tested.

APPLICATIONS AND PRODUCTS

The field of biomathematics is applicable to every area of biology. For example, biomathematics has been used to study population growth, evolution, and genetic variation and inheritance. Mathematical models have also been created for communities, modeling competition or predators, often using differential equations. Whether the scale is large or small, biomathematics allows scientists a greater understanding of biological phenomena.

Molecules and Cells. Biomathematics has been

applied to various biological molecules, including DNA, ribonucleic acid (RNA), and proteins. Biomathematics may be used to help predict the structure of these molecules or help determine how certain molecules are related to one another. Scientists have used biomathematics to model how bacteria can obtain new, important traits by transferring genetic material between different strains. This information is important because bacteria may, through sharing genetic material, acquire a trait such as a resistance to an antibiotic. To model the sharing of a trait, scientists have combined two of the ways to structure data: the phylogenetic tree and the site graph. The phylogenetic tree illustrates how the types of bacteria are related to one another. The site graph illustrates how pieces of genetic material interact. Then, scientists use a particular algorithm to determine the parameters of the model. By using such tools, scientists can predict which areas of genetic material are most likely to transfer between the bacterium.

Biomathematics has been used in cellular biology to model various cellular functions, including cellular division. The models can then be used to help scientists organize information and gain a deeper understanding about cellular functions. Cellular movement is one example of an application of biomathematics to cellular biology. Cellular movements can be seen as a set of steps. The scientists first considered certain cellular steps or functions, including how a cell senses a signal and how this signal is used within the cell to start movement. Scientists also considered the environment surrounding the cell, how the signal was provided, and the processes that occurred within the cell to read the signal and start movement. The scientists were then able to build a mathematical function that takes these steps into account. Depending on the particular question, the scientists may chose to focus on any of these steps. Therefore, more than one model may be used.

Organisms and Agriculture. Biomathematics has been used to create mathematical models for different functions of organisms. One popular area has been organism movement, where models have been created for bacteria movement and insect flight. A more complete understanding of organisms through mathematical models supports new technologies in agriculture. Biomathematics may also be used to help protect harvests. For example, biomathematics has been used to model a type of algae bloom known as

Fascinating Facts About Biomathematics

- Scientists are using biomathematics to create a virtual patient. First, biomathematics is used to model the human body. Using the virtual patient, scientists can test cancer-prevention drugs. The result is a quicker and cheaper way to develop drugs.
- Biomathematics has applications in nature. By using biomathematics, the pigment patterns of a leopard's spots or the patterns of seashells can be modeled.
- Scientists and mathematicians have found that by using biomathematics and computers they can simulate kidney functions. This kidney simulation helps doctors understand kidney disease better and can help them provide more effective treatments.
- Biomathematics is being used to model the workings of a heart in order to improve artificial heart valves. With biomathematics models, designs can be tested more quickly and efficiently.
- Mathematical models of biofilms systems, which are layers of usually nonresistant microorganisms such as bacteria that attach to a surface, are critical to the medical and technical industries. Biofilm systems can cause infections in humans and corrosion and deterioration in technical systems. Biomathematics can be used to model biofilms systems to help understand and prevent their formation.
- Biomathematics was used to sort and analyze data from the Human Genome Project. Completed in 2003, the thirteen-year-long project identified the entire human genome.

brown tide. In the late 1980's, brown tide appeared in the waters near Long Island, New York, badly affecting the shellfish population by blocking sunlight and depleting oxygen. Four years later, the algae blooms receded. Both mathematicians and scientists collaborated in order to create a model of the brown tide in order to understand why it bloomed and whether it will bloom in the future. To create a model, the collaborators used differential equations. They focused on the population density, which included factors such as temperature and nutrients. The collaborators had to consider many variables and re-move the ones they did not consider important. For instance, they hypothesized that a period of drought followed by rain may have affected growth. They also considered fertilizers and pesticides that were used in the area. A better understanding of the brown tide may help protect the shellfish harvest in future years.

Medical Uses. Biomathematical models have been developed to illustrate various functions within the human body, including the heart, kidneys, and car-diac and neural tissue. Biomathematics is useful in modeling cancer, enabling scientists to learn more about the type of cancer, thereby allowing them to study the efficacy of different types of treatment. One project has focused on modeling colon cancer on a genetic and molecular level. Not only did scientists gain information about the genetic mutations that are present during colon cancer, but they also devel-oped a model that predicted when tumor cells would be sensitive to radiation, which is the most common way to treat colon cancer. Studies such as this can be built on in future experimentations, the results of which may someday be used by doctors to create more effective cancer treatments.

Biomathematics has also been used to organize and analyze data from experiments dealing with drug efficacy and gene expression in cancer cells. Using matrices, statistics, and algorithms, scientists have been able to understand if a particular drug is more likely to work based on the patient's cancer cell's gene expression. Biomathematics has also been inte-gral in epidemiology, the field that studies diseases within a population. Biomathematics may be used to model various aspects of a disease such as human im-munodeficiency virus (HIV), allowing for more com-prehensive planning and treatment.

IMPACT ON INDUSTRY

Government and University Research. Biomath-ematics is often developed through a collaborative effort between biologists and mathematicians. In the United States, the National Science Foundation has a mathematical biology program that provides grants for research to develop mathematical applications related to biology. In addition, many U.S. universi-ties offer biomathematics programs. Some universi-ties have biomathematics departments, such as Ohio State University's Mathematical Biosciences Institute, which focuses on creating mathematical applications

to help solve biological problems. This institute provides research and education opportunities, in-cluding workshops and public lectures.

Each biomathematics department or program may emphasize a different aspect of biomathematics: Some focus on medical applications, others focus on the need to quantify biological phenomena using mathematical tools. UCLA's department of biomath-ematics conducts research in areas such as statistical genetics, evolutionary biology, molecular imaging, and neuroscience.

Biomathematics is also being developed interna-tionally. Many universities in the United Kingdom have biomathematics programs. The University of Oxford has a Centre for Mathematical Biology. There are also independent research institutions and orga-nizations that focus on biomathematics. The Institute for Medical BioMathematics, in Israel, works on de-veloping analytical approaches to treating cancer and infectious diseases. The Society for Mathematical Biology, in Boulder, Colorado, has provided an inter-national forum for biomathematics for more than twenty years. The International Biometric Society in Washington, D.C., also addresses the application of mathematical tools to biological phenomena. The European Society for Mathematical and Theoretical Biology, founded in 1991, promotes biomathe-matics, and the Society for Industrial and Applied Mathematics in Philadelphia has an activity group on the life sciences that provides a platform for re-searchers working in the area of biomathematics.

Industry. Biomathematics may lead to develop-ments in medical treatments or technology, which may then be marketed. The pharmaceutical in-dustry is a good example of this development. Biomathematics is often used to create models of diseases that can lead to a deeper understanding of the disease and new medicines. In addition, biomath-ematics provides tools that may be used throughout the drug-creation process and can be used to predict how well a drug will work or how safe a drug will be for a group of patients with a particular genetic makeup. The engineering of microorganisms has also bene-fited from biomathematics. Biomathematics is being used to create models to understand the fundamen-tals of microorganisms better. The end result of this understanding may be the changed metabolism or structure of an organism to produce more milk or a sweeter wine. Finally, some companies are targeting software to aid with biological modeling, and others provide consulting in the field of biomathematics. BioMath Solutions in Austin, Texas, is a company that provides analytical software in the area of molecular biology.

CAREERS AND COURSE WORK

Degree programs in biomathematics are gaining popularity in universities. Some schools have biomathematics departments, and others have biomathematics programs within the mathematics or biology departments. Undergraduate course work for a B.S. in biomathematics encompasses classes in mathematics, biology, and computer science, including: calculus, chemistry, genetics, physics, software development, probability, statistics, organic chemistry, epidemiology, population biology, molecular biology, and physiology. A student may also choose to receive a B.S. in biology or mathematics. In addition, students may seek additional opportunities outside their program. Ohio State's Mathematical Biosciences Institute offers summer programs for undergraduate and graduate students.

In the field of biomathematics, a doctorate is required for many careers. Doctoral programs in biomathematics include course work in statistics, biology, probability, differential equations, linear algebra, cellular modeling, genetics modeling, computer programming, pharmacology, and clinical research methods. In addition, doctoral candidates often will perform biomathematics research with support from departmental faculty. As with the undergraduate degrees, a student may also choose to pursue a doctorate in biology or mathematics.

With the influx of biological data from new technologies and tools, a degree in biomathematics is imperative. Those who receive a Ph.D. may choose to enter a postdoctoral program, such as the one at Ohio State's Mathematical Biosciences Institute, which offers postdoctoral fellowships as well as mentorship and research opportunities. Other career paths include research in medicine, biology, and mathematics with universities and private research institutions; work with software development and computer modeling; teaching; or collaborating with other professionals and consulting in an industry such as pharmaceuticals or bioengineering.

SOCIAL CONTEXT AND FUTURE PROSPECTS

While mathematical tools have been applied to biology for some time, many scientists believe there is still a need for increased quantitative analysis of biology. Some call for more emphasis on mathematics in high school and undergraduate biology classes. They believe that this will advance biomathematics. As more universities develop biomathematics departments and degrees, more mathematics classes will be added to the curriculum. A concern has been raised in the biomathematics field about the assumptions used to create simplified mathematical models. More complex and accurate models will likely be developed.

Important future applications for biomathematics will be in the bioengineering and medical industries. The development of mathematical models for complex biological phenomena will aid scientists in a deeper understanding that can lead to more effective treatments in such areas such as tumor therapy. As new tools and methods continue to develop, biomathematics will be a field that expands to sort and analyze the large influx of data.

Carly L. Huth, B.S., J.D.

FURTHER READING

Hochberg, Robert, and Kathleen Gabric. "A Provably Necessary Symbiosis." *The American Biology Teacher* 72, No. 5 (2010): 296-300. This article describes some mathematics that can be taught in biology classrooms.

Misra, J. C., ed. *Biomathematics: Modelling and Simulation.* Hackensack, N.J.: World Scientific, 2006. This book provides an in-depth guide to several modern applications of biomathematics and includes many helpful illustrations.

Schnell, Santiago, Ramon Grima, and Philip Maini. "Multiscale Modeling in Biology." *American Scientist* 95 (March-April, 2007): 134-142. This article gives an overview of how biological models are created and provides several modern examples of biomathical applications.

WEB SITES

International Biometric Society
http://www.tibs.org

National Science Foundation
http://www.nsf.gov

Ohio State University Mathematical Biosciences Institute
http://mbi.osu.edu

Society for Industrial and Applied Mathematics
http://www.siam.org

Society for Mathematical Biology
http://www.smb.org

See also: Bioengineering; Bioinformatics;
Biophysics.

BIOMECHANICS

FIELDS OF STUDY

Kinesiology; physiology; kinetics; kinematics; sports medicine; technique/performance analysis; injury rehabilitation; modeling; orthopedics; prosthetics; bioengineering; bioinstrumentation; computational biomechanics; cellular/molecular biomechanics; veterinary (equine) biomechanics; forensic biomechanics; ergonomics.

SUMMARY

Biomechanics is the study of the application of mechanical forces to a living organism. It investigates the effects of the relationship between the body and forces applied either from outside or within. In humans, biomechanists study the movements made by the body, how they are performed, and whether the forces produced by the muscles are optimal for the intended result or purpose. Biomechanics integrates the study of anatomy and physiology with physics, mathematics, and engineering principles. It may be considered a subdiscipline of kinesiology as well as a scientific branch of sports medicine.

KEY TERMS AND CONCEPTS

- **Angular Motion:** Motion involving rotation around a central line or point known as the axis of rotation.
- **Dynamics:** Branch of mechanics that studies systems in motion, subject to acceleration or deceleration.
- **Kinematics:** Study of movement of segments of a body without regard for the forces causing the movement.
- **Kinesiology:** Study of human movement.
- **Kinetics:** Study of forces associated with motion.
- **Lever:** Rigid bars (in the body, bones) that move around an axis of rotation (joint) and have the ability to magnify or alter the direction of a force.
- **Linear Motion:** Motion involving all the parts of a body or system moving in the same direction, at the same speed, following a straight (rectilinear) or curved (curvilinear) line.
- **Mechanics:** Branch of physics analyzing the resulting actions of forces on particles or systems.
- **Qualitative Movement:** Description of the quality of movement without the use of numbers.
- **Quantitative Movement:** Description or analysis of movement using numbers or measurement.
- **Sports Medicine:** Branch of medicine studying the clinical and scientific characteristics of exercise and sport, as well as any resulting injuries.
- **Statics:** Branch of mechanics that studies systems that are in a constant state of motion or constant state of rest.
- **Torque:** Turning effect of a force applied in a direction not in line with the center of rotation of a nonmoving axis (eccentric).

DEFINITION AND BASIC PRINCIPLES

Biomechanics is a science that closely examines the forces acting on a living system, such as a body, and the effects that are produced by these forces. External forces can be quantified using sophisticated measuring tools and devices. Internal forces can be measured using implanted devices or from model calculations. Forces on a body can result in movement or biological changes to the anatomical tissue. Biomechanical research quantifies the movement of different body parts and the factors that may influence the movement, such as equipment, body alignment, or weight distribution. Research also studies the biological effects of the forces that may affect growth and development or lead to injury. Two distinct branches of mechanics are statics and dynamics. Statics studies systems that are in a constant state of motion or constant state of rest, and dynamics studies systems that are in motion, subject to acceleration or deceleration. A moving body may be described using kinematics or kinetics. Kinematics studies and describes the motion of a body with respect to a specific pattern and speed, which translate into coordination of a display. Kinetics studies the forces associated with a motion, those causing it and resulting from it. Biomechanics combines kinetics and kinematics as they apply to the theory of mechanics and physiology to study the structure and function of living organisms.

BACKGROUND AND HISTORY

Biomechanics has a long history even though the actual term and field of study concerned with mechanical analysis of living organisms was not internationally accepted and recognized until the early 1970's. Definitions provided by early biomechanics specialists James G. Hay in 1971 and Herbert Hatze in 1974 are still accepted. Hatze stated, "Biomechanics is the science which studies structures and functions of biological systems using the knowledge and methods of mechanics."

Highlights throughout history have provided insight into the development of this scientific discipline. The ancient Greek philosopher Aristotle was the first to introduce the term "mechanics," writing about the movement of living beings around 322 b.c.e. He developed a theory of running techniques and suggested that people could run faster by swinging their arms. In the 1500's, Leonardo da Vinci proposed that the human body is subject to the law of mechanics, and he contributed significantly to the development of anatomy as a modern science. Italian scientist Giovanni Alfonso Borelli, a student of Galileo, is often considered the father of biomechanics. In the mid-1600's, he developed mathematical models to describe anatomy and human movement mechanically. In the late 1600's, English physician and mathematician Sir Isaac Newton formulated mechanical principles and Newtonian laws of motion (inertia, acceleration, and reaction) that became the foundation of biomechanics.

British physiologist A. V. Hill, the 1923 winner of the Nobel Prize in Physiology or Medicine, conducted research to formulate mechanical and structural theories for muscle action. In the 1930's, American anatomy professor Herbert Elftman was able to quantify the internal forces in muscles and joints and developed the force plate to quantify ground reaction. A significant breakthrough in the understanding of muscle action was made by British physiologist Andrew F. Huxley in 1953, when he described his filament theory to explain muscle shortening. Russian physiologist Nicolas Bernstein published a paper in 1967 describing theories for motor coordination and control following his work studying locomotion patterns of children and adults in the Soviet Union.

HOW IT WORKS

The study of human movement is multifaceted, and biomechanics applies mechanical principles to the study of the structure and function of living things. Biomechanics is considered a relatively new field of applied science, and the research being done is of considerable interest to many other disciplines, including zoology, orthopedics, dentistry, physical education, forensics, cardiology, and a host of other medical specialties. Biomechanical analysis for each particular application is very specific; however, the basic principles are the same.

Newton's Laws of Motion. The development of scientific models reduces all things to their basic level to provide an understanding of how things work. This also allows scientists to predict how things will behave in response to forces and stimuli and ultimately to influence this behavior.

Newton's laws describe the conservation of energy and the state of equilibrium. Equilibrium results when the sum of forces is zero and no change occurs, and conservation of energy explains that energy cannot be created or destroyed, only converted from one form to another. Motion occurs in two ways, linear motion in a particular direction or rotational movement around an axis. Biomechanics explores and quantifies the movement and production of force used or required to produce a desired objective.

Seven Principles. Seven basic principles of biomechanics serve as the building blocks for analysis. These can be applied or modified to describe the reaction of forces to any living organism.

1. The lower the center of mass, the larger the base of support; the closer the center of mass to the base of support and the greater the mass, the more stability increases.
2. The production of maximum force requires the use of all possible joint movements that contribute to the task's objective.
3. The production of maximum velocity requires the use of joints in order, from largest to smallest.
4. The greater the applied impulse, the greater increase in velocity.
5. Movement usually occurs in the direction opposite that of the applied force.
6. Angular motion is produced by the application of force acting at some distance from an axis, that is, by torque.
7. Angular momentum is constant when an athlete

or object is free in the air.

Static and dynamic forces play key roles in the complex biochemical and biophysical processes that underlie cell function. The mechanical behavior of individual cells is of interest for many different biologic processes. Single-cell mechanics, including growth, cell division, active motion, and contractile mechanisms, can be quite dynamic and provide insight into mechanisms of stress and damage of structures. Cell mechanics can be involved in processes that lie at the root of many diseases and may provide opportunities as focal points for therapeutic interventions.

APPLICATIONS AND PRODUCTS

Biomechanics studies and quantifies the movement of all living things, from the cellular level to body systems and entire bodies, human and animal. There are many scientific and health disciplines, as well as industries that have applications developed from this knowledge. Research is ongoing in many areas to effectively develop treatment options for clinicians and better products and applications for industry.

Dentistry. Biomechanical principles are relevant in orthodontic and dental science to provide solutions to restore dental health, resolve jaw pain, and manage cosmetic and orthodontic issues. The design of dental implants must incorporate an analysis of load bearing and stress transfer while maintaining the integrity of surrounding tissue and comfortable function for the patient. This work has lead to the development of new materials in dental practices such as reinforced composites rather than metal frameworks.

Forensics. The field of forensic biomechanical analysis has been used to determine mechanisms of injury after traumatic events such as explosions in military situations. This understanding of how parts of the body behave in these events can be used to develop mitigation strategies that will reduce injuries. Accident and injury reconstruction using biomechanics is an emerging field with industrial and legal applications.

Biomechanical Modeling. Biomechanical modeling is a tremendous research field, and it has potential uses across many health care applications. Modeling has resulted in recommendations for prosthetic design and modifications of existing devices.

Deformable breast models have demonstrated capabilities for breast cancer diagnosis and treatment. Tremendous growth is occurring in many medical fields that are exploring the biomechanical relationships between organs and supporting structures. These models can assist with planning surgical and treatment interventions and reconstruction and determining optimal loading and boundary constraints during clinical procedures.

Materials. Materials used for medical and surgical procedures in humans and animals are being evaluated and some are being changed as biomechanical science is demonstrating that different materials, procedures, and techniques may be better for reducing complications and improving long-term patient health. Evaluation of the physical relationship between the body and foreign implements can quantify the stresses and forces on the body, allowing for more accurate prediction of patient outcomes and determination of which treatments should be redesigned.

Predictability. Medical professionals are particularly interested in the predictive value that biomechanical profiling can provide for their patients. An example is the unpredictability of expansion and rupture of an abdominal aortic aneurysm. Major progress has been made in determining aortic wall stress using finite element analysis. Improvements in biomechanical computational methodology and advances in imaging and processing technology have provided increased predictive ability for this life-threatening event.

As the need for accurate and efficient evaluation grows, so does the research and development of effective biomechanical tools. Capturing real-time, real-world data, such as with gait analysis and range of motion features, provides immediate opportunities for applications. This real-time data can quantify an injury and over time provide information about the extent that the injury has improved. High-tech devices can translate real-world situations and two-dimensional images into a three-dimensional framework for analysis. Devices, imaging, and modeling tools and software are making tremendous strides and becoming the heart of a highly competitive industry aimed at simplifying the process of analysis and making it less invasive.

IMPACT ON INDUSTRY

Many companies have discovered the benefit of

Fascinating Facts About Biomechanics

- Italian artist and scientist Leonardo da Vinci, born in the fifteenth century, called the foot "a masterpiece of engineering and a work of art."
- To take a step requires about two hundred muscles.
- When walking, one's feet bear a force of one and one-half times one's body weight; when running, the force on one's feet increases to three to four times one's body weight.
- During a one-mile walk, a person's feet strike the ground an average of 1,800 times.
- Just to overcome air resistance, a person running a 5,000-meter race uses 9 percent of his or her total energy expenditure, while a sprinter in a 100-meter race uses 20 percent.
- Pound for pound, bone is six times stronger than steel.
- The muscles that power the fingers are strong enough in some people to allow them to climb vertical surfaces by supporting their entire weight on a few fingertips.
- The muscles that bend the finger joints are located in the palm and in the mid-forearm and are connected to the finger bones by tendons, which pull on and move the fingers like the strings of a marionette.
- It is physically impossible for a person to lick his or her elbow.

biomechanics in various facets of their operations, including the development of products and of workplace procedures and practices. Most products that are made to assist or interact with people or any living being have probably been designed with the input of a biomechanical professional. Corporations want to protect their investment and profits by ensuring that their products will effectively meet the needs of the consumer and comply with strict safety standards. Biomechanics personnel work with product development engineers and designers to create new products and improve existing ones. Athletic equipment has been redesigned to produce better results with the same exertion of force. Two major sports products that have received international attention and led to world-record-breaking performances have been clap

skates (used in speed skating) and the LZR Racer swimsuit, designed to reduce drag on swimmers.

Sporting equipment for athletes and the general public is constantly being redesigned to enhance performance and reduce the chance of injury. Sport-specific footwear is designed to maximize comfort and make it easier for athletes to perform the movements necessary for their sport. Equipment such as bicycles and golf clubs are designed using lighter and stronger materials, using optimal angles and maximizing strength to provide athletes with the best experience possible. Sports equipment goes a step further by analyzing the individual athlete and adjusting a piece of equipment specifically to his or her body. A small change to an angle or lever can produce dramatic results for an individual. This customization goes beyond sporting equipment to rehabilitation implements, wheelchairs, and prosthetic devices.

Biomechanics has a profound influence on many industries that are outside of sports. Most products used or handled on a daily basis have undergone biomechanical evaluation. Medicine bottle tops have been redesigned for easy opening by those with arthritic hands, and products from kitchen gadgets to automobiles have all been altered to improve comfort, safety, and effectiveness and to reduce the need for physical exertion.

Corporations are facing stricter regulations regarding the workplace environment. Safety and injury prevention are key to keeping productivity optimal. In a process called ergonomic assessment, equipment and workstations are biomechanically assessed, and adjustments are made that will limit acute or chronic injuries. Employees also receive instruction on the proper procedures to follow, such as lifting techniques. Industry procedures need to be in place and diligently followed to protect both employees and the company.

Accident litigation is becoming more common, and biomechanical science is playing a large role in accident re-creation and law-enforcement training. Investigations at accident or crime scenes can reveal more evidence, with greater accuracy than ever before, leaving less room for speculation when reconstructing the event.

CAREERS AND COURSE WORK

Careers in biomechanics can be dynamic and can take many paths. Graduates with accredited degrees may pursue careers in laboratories in universities or in private corporations researching and developing ways of improving and maximizing human performance. Beyond research, careers in biomechanics can involve working in a medical capacity in sport medicine and rehabilitation. Biomechanics experts may also seek careers in coaching, athlete development, and education.

Consulting and legal practices are increasingly seeking individuals with biomechanics expertise who are able to analyze injuries and reconstruct accidents involving vehicles, consumer products, and the environment.

Biomechanical engineers commonly work in industry, developing new products and prototypes and evaluating their performance. Positions normally require a biomechanics degree in addition to mechanical or biomedical engineering degrees.

Private corporations are employing individuals with biomechanical knowledge to perform employee fitness evaluations and to provide analyses of work environments and positions. Using these assessments, the biomechanics experts advise employers of any ergonomic changes or job modifications that will reduce the risk of workplace injury.

Individuals with a biomechanics background may chose to work in rehabilitation and prosthetic design. This is very challenging work, devising and modifying existing implements to maximize people's abilities and mobility. Most prosthetic devices are customized to meet the needs of the patient and to maximize the recipient's abilities. This is an ongoing process because over time the body and needs of a patient may change. This is particularly challenging in pediatrics, where adjustments become necessary as a child grows and develops.

SOCIAL CONTEXT AND FUTURE PROSPECTS

Biomechanics has gone from a narrow focus on athletic performance to become a broad-based science, driving multibillion dollar industries to satisfy the needs of consumers who have become more knowledgeable about the relationship between science, health, and athletic performance. Funding for biomechanical research is increasingly available from national health promotion and injury prevention programs, governing bodies for sport, and business and industry. National athletic programs want to ensure that their athletes have the most advanced training methods, performance analysis methods, and equipment to maximize their athletes' performance at global competitions.

Much of the existing and developing technology is focused on increasingly automated and digitized systems to monitor and analyze movement and force. The physiological aspect of movement can be examined at a microscopic level, and instrumented athletic implements such as paddles or bicycle cranks allow real-time data to be collected during an event or performance. Force platforms are being reconfigured as starting blocks and diving platforms to measure reaction forces. These techniques for biomechanical performance analysis have led to revolutionary technique changes in many sports programs and rehabilitation methods.

Advances in biomechanical engineering have led to the development of innovations in equipment, playing surfaces, footwear, and clothing, allowing people to reduce injury and perform beyond previous expectations and records.

Computer modeling and virtual simulation training can provide athletes with realistic training opportunities, while their performance is analyzed and measured for improvement and injury prevention.

April D. Ingram, B.Sc.

FURTHER READING

Hamill, Joseph, and Kathleen Knutzen. *Biomechanical Basis of Human Movement.* 3d ed. Philadelphia: Lippincott, Williams & Wilkins, 2009. This introductory text integrates basic anatomy, physics, and physiology as it relates to human movement. It also includes real-life examples and clinically relevant material.

Hatze, H. "The Meaning of the Term 'Biomechanics.'" *Journal of Biomechanics* 7, no. 2 (March, 1974): 89-90. Contains Hatze's definition of biomechanics, then an emerging field.

Hay, James G. *The Biomechanics of Sports Techniques.* 4th ed. Englewood Cliffs, N.J.: Prentice Hall, 1993. A seminal work by an early biomechanics expert, first published in 1973.

Kerr, Andrew. *Introductory Biomechanics.* London: Elsevier, 2010. Provides a clear, basic understanding of major biomechanical principles in a workbook style interactive text.

Peterson, Donald, and Joseph Bronzino. *Biomechanics:*

Principles and Applications. Boca Raton, Fla.: CRC Press, 2008. A broad collection of twenty articles on various aspects of research in biomechanics.

Watkins, James. *Introduction to Biomechanics of Sport and Exercise.* London: Elsevier, 2007. An introduction to the fundamental concepts of biomechanics that develops knowledge from the basics. Many applied examples, illustrations, and solutions are included.

WEB SITES

American Society of Biomechanics
http://www.asbweb.org

European Society of Biomechanics
http://www.esbiomech.org

International Society of Biomechanics
http://www.isbweb.org

World Commission of Science and Sports
http://www.wcss.org.uk

See also: Bioengineering; Biophysics; Dentistry; Ergonomics; Kinesiology; Orthopedics; Prosthetics; Rehabilitation Engineering.

BIOPHYSICS

FIELDS OF STUDY

Physics; physical sciences; chemistry; mathematics; biology; molecular biology; chemical biology; engineering; biochemistry; classical genetics; molecular genetics; cell biology.

SUMMARY

Biophysics is the branch of science that uses the principles of physics to study biological concepts. It examines how life systems function, especially at the cellular and molecular level. It plays an important role in understanding the structure and function of proteins and membranes and in developing new pharmaceuticals. Biophysics is the foundation for molecular biology, a field that combines physics, biology, and chemistry.

KEY TERMS AND CONCEPTS

- **Circular Dichroism (CD):** Differential absorption of left- and right-handed circularly polarized light.
- **Electromagnetic Waves:** Waves that can transmit their energy through a vacuum.
- **Molecular Genetics:** Branch of genetics that analyzes the structure and function of genes at the molecular level.
- **Polarized Light:** Light waves that vibrate in a single plane.
- **Quantum Mechanics:** Physical analysis at the level of atoms or subatomic fundamental particles.
- **Thermodynamics:** Branch of physics that studies energy conversions.
- **Vector:** Quantity that has both magnitude and direction.

DEFINITION AND BASIC PRINCIPLES

The word "biophysics" means the physics of life. Biophysics studies the functioning of life systems, especially at the cellular and molecular level, using the principles of physics. It is known that atoms make up molecules, molecules make up cells, and cells in turn make up tissues and organs that are part of an organism, or a living machine. Biophysicists use this knowledge to understand how the living machine works.

In photosynthesis, for instance, the absorption of sunlight by green plants initiates a process that culminates with synthesis of high-energy sugars such as glucose. To fully understand this process, one needs to look at how it begins—light absorption by the photosystems. Photosystems are groups of energy-absorbing pigments such as chlorophyll and carotenoids that are located on the thylakoid membranes inside the chloroplast, the photosynthetic organelle in the plant cell. Biophysical studies have shown that once a chlorophyll molecule captures solar energy, it gets excited and transfers the energy to a neighboring unexcited chlorophyll molecule. The process repeats itself, and thus, packets of energy jump from one chlorophyll molecule to the next. The energy eventually reaches the reaction center, where it begins a chain of high-energy electron-transfer reactions that lead to the storage of the light energy in the form of adenosine triphosphate (ATP) and nicotinamide adenine dinucleotide phosphate (NADPH). In the second half of photosynthesis, ATP and NADPH provide the energy to make glucose from carbon dioxide.

Biophysics is often confused with medical physics. Medical physics is the science devoted to studying the relationship between human health and radiation exposure. For example, a medical physicist often works closely with a radiation oncologist to set up radiotherapy treatment plans for cancer patients.

BACKGROUND AND HISTORY

In comparison with other branches of biology and physics, biophysics is relatively new and, therefore, still evolving. Even though the use of physical concepts and instrumentation to explain the workings of life systems had begun as early as the 1840's, biophysics did not emerge as an independent field until the 1920's. Some of the earliest studies in biophysics were conducted in the 1840's by a group known as the Berlin school of physiologists. Among its members were pioneers such as Hermann von Helmholtz, Ernst Heinrich Weber, Carl F. W. Ludwig, and Johannes Peter Müller. This group used well-known physical methods to investigate physiological issues, such as the mechanics of muscular contraction and the electrical changes in a nerve cell during impulse transmission. The first biophysics textbook was written in 1856 by Adolf Fick, a student of Ludwig. Although these early biophysicists made significant advances, subsequent research focused on other areas.

In the 1920's, the first biophysical institutes were established in Germany and the first textbook with the word "biophysics" in its title was published. However, through the 1940's, biophysics research was primarily aimed at understanding the biophysical impact of ionizing radiation. In 1944, Austrian physicist Erwin Schrödinger published *What Is Life ? The Physical Aspect of the Living Cell,* based on a series of lectures that addressed biology from the viewpoint of a classical physicist. This cross-disciplinary work motivated several physicists to become interested in biology and thus laid the foundation for the field of molecular biology. From 1950 to 1970, the field of biophysics experienced rapid growth, tremendously accelerated by the discovery in 1953 of the double helix structure of DNA by James D. Watson and Francis Crick. Both Watson and Crick have stated that they were inspired by Schrödinger's work.

HOW IT WORKS

Biophysicists study life at all levels, from atoms and molecules to cells, organisms, and environments. They attempt to describe complex living systems with the simple laws of physics. Often, biophysicists work at the molecular level to understand cells and their processes.

The work of Gregor Mendel in the late nineteenth century laid the foundation for genetics, the science of heredity. His studies, rediscovered in the twentieth century, led to the understanding that the inheritance of certain traits is governed by genes and that the alleles of the genes are separated during gamete formation. Experiments in the 1940's revealed that genes are made of DNA, but the mechanisms by which genes function remained a mystery. Watson and Crick's discovery of the double helix structure of DNA in 1953 revealed how genes could be translated into proteins.

Biophysicists use a number of physical tools and techniques to understand how cellular processes work, especially at the molecular level. Some of the important tools are electron microscopy, nuclear magnetic resonance (NMR) spectroscopy, circular dichroism (CD) spectroscopy, and X-ray

crystallography. For example, the discovery of Watson and Crick's double helix model was possible in part because of the X-ray images of DNA that were taken by Rosalind Franklin and Maurice H. F. Wilkins. Franklin and Wilkins, both biophysicists, made DNA crystals and then used X-ray crystallography to analyze the structure of DNA. The array of black dots arranged in an X-shaped pattern on the X-ray photograph of wet DNA suggested to Franklin that DNA was helical.

Electron Microscopy. Electron microscopes use beams of electrons to study objects in detail. Electron microscopy can be used to analyze an object's surface texture (topography) and constituent elements and compounds (composition), as well as the shape and size (morphology) and atomic arrangements (crystallographic details) of those elements and compounds. Electron microscopes were invented to overcome the limitations posed by light microscopes, which have maximum magnifications of 500x or 1000x and a maximum resolution of 0.2 millimeter. To see and study subcellular structures and processes required magnification capabilities of greater than 10,000x. The first electron microscope was a transmission electron microscope (TEM) built by Max Knoll and Ernst Ruska in 1931. The invention of the scanning electron microscope (SEM) was somewhat delayed (the first was built in 1937 by Manfred von Ardenne) because the field had to figure out how to make the electron beam scan the sample.

NMR Spectroscopy. Nuclear magnetic resonance (NMR) spectroscopy is an extremely useful tool for the biophysicist to study the molecular structure of organic compounds. The underlying principle of NMR spectroscopy is identical to that of magnetic resonance imaging (MRI), a common tool in medical diagnostics. The nuclei of several elements, including the isotopes carbon-12 and oxygen-16, have a characteristic spin when placed in an external magnetic field. NMR focuses on studying the transitions between these spin states. In comparison with mass spectroscopy, NMR requires a larger amount of sample, but it does not destroy the sample.

CD Spectroscopy. Circular dichroism (CD) spectroscopy measures differences in how left-handed and right-handed polarized light is absorbed. These differences are caused by structural asymmetry. CD spectroscopy can determine the secondary and tertiary structure of proteins as well as their thermal

Fascinating Facts About Biophysics

- The concept of biophysics was first developed by ancient Greeks and Romans who were trying to analyze the basis of consciousness and perception.
- Human sight begins when the protein rhodopsin absorbs a unit of light called the quanta. This energy absorption triggers an enzymatic cascade that culminates in an amplified electric signal to the brain and enables vision.
- Crystallin, the lens protein, is made only in the lens of the human eye, and melanin, the skin pigment, is made only in skin cells, or melanocytes, even though all the cells in the human body have the genes to make crystallin and melanin.
- A technique called footprinting allows scientists to determine exactly where a protein binds on DNA and how much of the protein actually binds.
- Genes in human cells are selectively turned on and off by proteins called regulators. A defect in this regulatory mechanism can cause diseases such as cancer.

stability. It is usually used to study proteins in solution.

X-Ray Crystallography. X rays are electromagnetic waves with wavelengths ranging from 0.02 to 100 angstroms (Å). Even before X rays were discovered in 1895 by Wilhelm Conrad Röntgen, scientists knew that atoms in crystals were arranged in definite patterns and that a study of the angles therein could provide clues to the crystal structure. As is true of all forms of radiation, the wavelength of X rays is inversely proportional to its energy. Because the wavelength of X rays is smaller than that of visible light, X rays are powerful enough to penetrate most matter. As X rays travel through an object, they are diffracted by the atomic arrangements inside and thus provide a guideline for the electron densities inside the object. Analysis of this electron density data offers a glimpse into the internal structure of the crystal. As of 2010, about 90 percent of the structures in the Worldwide Protein Data Bank had been elucidated through X-ray crystallography.

APPLICATIONS AND PRODUCTS

Biophysical tools and techniques have become extremely useful in many areas and fields. They have furthered research in protein crystallography, synthetic biology, and nanobiology, and allowed scientists to discover new pharmaceuticals and to study biomolecular structures and interactions and membrane structure and transport. Biophysics and its related fields, molecular biology and genetics, are rapidly developing and are at the center of biomedical research.

Biomolecular Structures. Because structure dictates function in the world of biomolecules, understanding the structure of the biomolecule (with tools such as X-ray crystallography and NMR and CD spectroscopy), whether it is a protein or a nucleic acid, is critical to understanding its individual function in the cell. Proteins function as catalysts and bind to and regulate other downstream biomolecules. Their functional basis lies in their tertiary structure, or their three-dimensional form, and this function cannot be predicted from the gene sequence. The sequence of nucleotides in a gene can be used only to predict the primary structure, which is the amino acid sequence in the polypeptide. Once the structure-function relationship has been analyzed, the next step is to make mutants or knock out the gene via techniques such as ribonucleic acid interference (RNAi) and confirm loss of function. Subsequently, a literature search is performed to see if there are any known genetic disorders that are caused by a defect in the gene being studied. If so, the structure-function relationship can be examined for a possible cure.

Membrane Structure and Transport. In 1972, biologist S. J. Singer and his student Garth Nicolson conceived the fluid mosaic model of the plasma membrane. According to this model, the plasma membrane is a fluid lipid bilayer largely made up of phospholipids arranged in an amphipathic pattern, with the hydrophobic lipid tails buried inside and the hydrophilic phosphate groups on the exterior. The bilayer is interspersed with proteins, which help in cross-membrane transport. Because membranes control the import and export of materials into the cell, understanding membrane structure is key to coming up with ways to block transport of potentially harmful pathogens across the membrane.

Electron microscopes were used in the early days of membrane biology, but fluorescence and confocal microscopes have come to be used more frequently. The development of organelle-specific vital stains has rejuvenated interest in evanescent field (EF) microscopy because it permits the study of even the smallest of vesicles and the tracking of the movements of individual protein molecules.

Synthetic Biology. In the 2000's, the term "systems biology" became part of the field of life science, followed by the term "synthetic biology." To many people, these terms appear to refer to the same thing, but they do not, even though they are indeed closely related. While systems biology focuses on using a quantitative approach to study existing biological systems, synthetic biology concentrates on applying engineering principles to biology and constructing novel systems heretofore unseen in nature. Clearly, synthetic biology benefits immensely from research in systems and molecular biology. In essence, synthetic biology could be described as an engineering discipline that uses known, tested functional components (parts) such as genes, proteins, and various regulatory circuits in conjunction with modeling software to design new functional biological systems, such as bacteria that make ethanol from water, carbon dioxide, and light. The biggest challenge to synthetic biologists is the complexity of lifeforms, especially higher eukaryotes such as humans, and the possible existence of unknown processes that can affect the synthetic biological systems.

Drug Discovery. In the pharmaceutical world, the initial task is to identify the aberrant protein, the one responsible for generating the symptoms in any disease or disorder. Once that is done, a series of biophysical tools are used to ensure that the target is the correct one. First, the identity of the protein is confirmed using techniques such as N-terminal sequencing and tandem mass spectroscopy (MS-MS). Second, the protein sample is tested for purity (which typically should be more than 95 percent) using methods such as denaturing sodium dodecyl sulfate polyacrylamide gel electrophoresis (SDS-PAGE). Third, the concentration of the protein sample is determined by chromogenic assays such as the Bradford or Lowry assay. The fourth and probably the most important test is that of protein functionality. This is typically carried out by either checking the ligand binding capacity of the protein (with biacore ligand binding assays) or by testing the ability of the protein to carry out its biological function. All

these thermodynamic parameters need to be tested to develop a putative drug, one that could somehow correct or restrain the ramifications of the protein's malfunction.

Nanobiology. With the aid of biophysical tools and techniques, the field of biology has moved from organismic biology to molecular biology to nanobiology. To get a feel for the size of a nanometer, picture a strand of hair, then visualize a width that is 100,000 times thinner. Typically nanoparticles are about the size of either a protein molecule or a short DNA segment. Nanomedicine, or the application of technology that relies on nanoparticles to medicine, has become a popular area for research. In particular, the search for appropriate vectors to deliver drugs into the cells is an endless pursuit, especially in emerging therapeutic approaches such as RNA interference. Because lipid and polymer-based nanoparticles are extremely small, they are easily taken up by cells instead of being cleared by the body.

IMPACT ON INDUSTRY

As one would expect, biophysics research worldwide has progressed faster in countries that traditionally have had a large base of physicists. The Max Planck Institute of Biophysics, one of the earliest pioneers in this field, was established in 1921 in Frankfurt, Germany, as the Institut für Physikalische Grundlagen der Medizin. The aim of the first director, Friedrich Dessauer, was to look for ways to apply the knowledge of radiation physics to medicine. He was followed by Boris Rajewsky, who coined the term "biophysics." In 1937, Rajewsky established the Kaiser Wilhelm Institute for Biophysics, which incorporated the institute led by Dessauer. The Max Planck Institute of Biophysics has become one of the world's foremost biophysics research institutes, with scientists and students analyzing a wide array of topics in biophysics such as membrane biology, molecular neurogenetics, and structural biology. In addition to Germany, countries active in biophysics research include the United States, Japan, France, Great Britain, Russia, China, India, and Sweden.

The International Union of Pure and Applied Biophysics (IUPAB) was created to provide a platform for biophysicists worldwide to exchange ideas and set up collaborations. The IUPAB in turn is a part of the International Council for Science (ICSU). The primary goal of IUPAB is to encourage and support students and researchers so that the field continues to grow and flourish. By 2010, the national biophysics societies of about fifty countries had become affiliated with the IUPAB. In the United States, the Biophysical Society was created in 1957 to facilitate propagation of biophysics concepts and ideas.

Government and University Research. To further broaden its mission, the Biophysical Society includes members from universities as well those in government research agencies such as the National Institutes for Health (NIH) and National Institute of Standards and Technology (NIST), many of whom are also part of the American Institute for Advancement of Sciences (AAAS) and the National Science Foundation (NSF). These members provide useful feedback to federal agencies such as the National Science and Technology Council, National Science Board, and the White House's Office of Science and Technology Policy, which are responsible for formulating national policies and initiatives.

Industry and Business. Most countries at the forefront of pharmaceutical breakthroughs have industries that are heavily invested in biophysical and biochemical research. The pharmaceutical industry has been spending billions of dollars to find treatments and cures for diseases and disorders that affect millions of people, including stroke, arthritis, cancer, heart disease, and neurological disorders. Biophysics is at the forefront of the field of drug discovery because it provides the tools for conducting research in proteomics and genomics and allows the scientific community to identify opportunities for drug design. The next step after drug design is to plan the method of drug delivery, and biophysical research can help provide suitable vectors, including nanovectors. The pharmaceutical industry in the United States—companies such as Novartis, Eli Lilly, Bristol-Myers Squibb, and Pfizer—and the National Institutes of Health have a combined budget of about $60 billion per year. However, neither the industry nor the government support basic research at the interface of life sciences with physics and mathematics. Without this support, biophysics is unlikely to produce new tools to revolutionize or accelerate the ten-to-twelve-year drug development cycle. If this impediment can be overcome, the number of new drugs added to the market every year is likely to grow.

Careers and Course Work

Few universities offer undergraduate majors in biophysics, but several universities offer graduate programs in biophysics. Students interested in pursuing a career in biophysics can major in molecular biology, physics, mathematics, or chemistry and supplement that with courses outside their major but relevant to biophysics. For example, a mathematics major would take supplementary courses in biology and vice versa. An ideal undergraduate curriculum for the future biophysicist would include courses in biology (genetics, cell biology, molecular biology), physics (thermodynamics, radiation physics, quantum mechanics), chemistry (organic, inorganic, and analytical), and mathematics (calculus, differential equations, computer programming, and statistics). The student should also have hands-on research experience, preferably as a research intern in a laboratory. To become an independent biophysicist, a graduate degree is required, usually a doctorate in biophysics, although some combine that with a medical degree. While in graduate school, the student should determine an area of research to pursue and engage in postdoctoral research for several years. Typically, this is the last step before one becomes an independent biophysicist running his or her own laboratory.

Social Context and Future Prospects

The discovery of the structure of DNA set off a revolution in molecular biology that has continued into the twenty-first century. In addition, modern scientific equipment has made study at the molecular level possible and productive. Many biophysicists, especially those who have also had course work in genetics and biochemistry, are working in molecular biology, which promises to be an active and exciting area for the foreseeable future.

Organisms are believed to be complex machines made of many simpler machines, such as proteins and nucleic acids. To understand why an organism behaves or reacts a certain way, one must determine how proteins and nucleic acids function. Biophysicists examine the structure of proteins and nucleic acids, seeking a correlation between structure and function. Once proper function is understood, scientists can prevent or treat diseases or disorders that result from malfunctions. This understanding of how proteins function enables scientists to develop pharmaceuticals and to find better means of delivering drugs to patients, and someday, this knowledge may allow scientists to design drugs specifically for a patient, thus avoiding many side effects. In addition, the scientific equipment developed by biophysicists in their research has been adapted for use in medical imaging for diagnosis and treatment. This transformation of laboratory equipment to medical equipment is likely to continue.

Biophysics applications have played and will continue to play a large role in medicine and health care, but future biophysicists may be environmental scientists. Biophysics is providing ways to improve the environment. For example, scientists are modifying microorganisms so that they produce electricity and biofuels that may lessen the need for fossil fuels. They are also using microorganisms to clean polluted water. As biophysics research continues, its applications are likely to cover an even broader range.

Sibani Sengupta, B.S., M.S., Ph.D.

FURTHER READING

Bischof, Marco. "Some Remarks on the History of Biophysics and Its Future." In *Current Development of Biophysics*, edited by Changlin Zhang, Fritz Albert Popp, and Marco Bischof. Hangzhou, China: Hangzhou University Press, 1996. This paper delivered at a 1995 symposium on biophysics in Neuss, Germany, examines how the field of biophysics got its start and predicts future developments.

Claycomb, James R., and Jonathan Quoc P. Tran. *Introductory Biophysics: Perspectives on the Living State.* Sudbury, Mass.: Jones and Bartlett, 2011. This textbook considers life in relation to the universe. Contains a compact disc that allows computer simulation of biophysical phenomena. Relates biophysics to many other fields and subjects, including fractal geometry, chaos systems, biomagnetism, bioenergetics, and nerve conduction.

Glaser, Roland. *Biophysics.* 5th ed. New York: Springer, 2005. Contains numerous chapters on the molecular structure, kinetics, energetics, and dynamics of biological systems. Also looks at the physical environment, with chapters on the biophysics of hearing and on the biological effects of electromagnetic fields.

Goldfarb, Daniel. *Biophysics Demystified.* Maidenhead, England: McGraw-Hill, 2010. Examines anatomical, cellular, and subcellular biophysics as well as tools and techniques used in the field. Designed as a self-teaching tool, this work contains ample examples, illustrations, and quizzes.

Herman, Irving P. *Physics of the Human Body.* New York: Springer, 2007. Analyzes how physical concepts apply to human body functions.

Kaneko, K. *Life: An Introduction to Complex Systems Biology.* New York: Springer, 2006. Provides an introduction to the field of systems biology, focusing on complex systems.

WEB SITES

Biophysical Society
http://www.biophysics.org

International Union of Pure and Applied Biophysics
http://iupab.org

Worldwide Protein Data Bank
http://www.wwpdb.org

See also: Biosynthetics; DNA Analysis; DNA Sequencing; Genetic Engineering; Genomics; Human Genetic Engineering; Magnetic Resonance Imaging; Pharmacology; Radiology and Medical Imaging.

BIOSYNTHETICS

FIELDS OF STUDY

Organic chemistry; biochemistry; bio-organic chemistry; bioinorganic chemistry; medicinal chemistry; pharmaceutical chemistry; pharmacology; analytical chemistry; nanotechnology; biomedical engineering; genetic engineering; genetics; synthetic biology; biology; molecular biology.

SUMMARY

Biosynthesis is the process of using small, simple molecules to make larger, more complex molecules, either inside the body or in the laboratory. Numerous applications for drug development and medicine include the synthesis of proteins, hormones, dietary supplements, blood products, and surgical dressings for wounds. Additional techniques to facilitate the diagnosis and treatment of disease include protein biomarkers for immune assays, the development of proteomics to analyze changes in proteins in response to a drug, the development of polyclonal and monoclonal antibodies, immunizations, and various drug delivery systems.

KEY TERMS AND CONCEPTS

- **Amino Acid:** Building block of proteins.
- **Antibody:** Glycoprotein that binds to and immobilizes a substance that the cell recognizes as foreign.
- **Antigen:** Substance that triggers an immune response.
- **Binding Assay:** Experimental method for selecting one molecule out of a number of possibilities by specific binding.
- **DNA (Deoxyribonucleic Acid):** Molecule that contains the genetic code.
- **Enzyme:** Biological catalyst, usually a globular protein.
- **Gene:** Individual unit of inheritance that consists of a sequence of DNA.
- **Hormone:** Substance produced by endocrine glands and delivered by the bloodstream to target cells, producing a desired effect.
- **Hydrophilic:** Property of tending to dissolve in water.
- **Insulin:** Hormone released from the pancreas.
- **Monoclonal Antibody:** Antibody produced from the progeny of a single cell and specific for a single antigen.
- **Peptide:** Molecule formed by linking two to several dozen amino acids.
- **Protein:** Macromolecule formed by polymerization of amino acids.

DEFINITION AND BASIC PRINCIPLES

In general, the term "biosynthetic" refers to any type of material produced via a biosynthetic process. A biosynthetic process uses enzymes and energetic molecules to transform small molecules into larger molecules within the cells of organisms. The two types of metabolites produced from cellular biosynthetic pathways include the primary metabolites of fatty acids and DNA needed by cells and the secondary metabolites of pheromones, antibiotics, and vitamins that assist the entire organism. Additional small molecules, such as adenosine triphosphate (ATP), provide the energetic driving force for the biosynthetic pathways, and other small molecules, including enzymes, further facilitate the reactions in these pathways. Thus, there have been many possibilities for numerous types of scientists, including chemists, biochemists, biologists, and geneticists, to create innovations.

The term "biosynthetic" differs from the term "chemosynthetic," because chemosynthetic indicates the production of materials that cannot take place within a living organism. Scientists generally begin the process of developing a new medical application or dietary supplement by first isolating and characterizing the DNA of the proteins or other small molecules directly involved in the biological process. They then try to duplicate this naturally occurring biological process to produce massive quantities of the desired material, and ultimately they combine these naturally occurring processes with chemicals that can mimic the process during laboratory manufacturing processes.

BACKGROUND AND HISTORY

The biochemical pharmacologist Hermann Karl Felix "Hugh" Blaschko was a trailblazer whose discoveries in the 1930's initiated the field of biosynthetics. His work elucidated the biosynthetic pathway for adrenaline, which is often called the fight-and-flight hormone, and encompassed the study of the enzymes important for regulation of this hormone. This work led the way toward the development of syntheses using amino acids for therapeutic applications.

Another key development in biosynthetics was the discovery of the role of the amino acid L-arginine in the synthesis of creatine, an important biomolecule, by G. L. Foster, Rudolf Schoenheimer, and D. Rittenberg in 1939. Since that time, L-arginine has also been shown to be a precursor to nitrous oxide and nitric oxide, as well as a component of the urea cycle, which is important for ammonia regulation and thus influences the operation of the kidneys and other organs. Nitric oxide is important in the regulation of blood flow to muscles. These discoveries involving L-arginine have led to dietary supplements useful to bodybuilders who wish to enhance their weight-lifting performance.

Throughout the 1940's, 1950's, and 1960's, progress was made toward understanding the genetic composition of organisms, enzymes, and biosynthetic pathways. Researchers made contributions to understanding pyrimidine, galactosidase, *Escherichia coli*, and chlorophyll. Practical biosynthetic applications that were made possible by these fundamental discoveries began to manifest themselves throughout the 1970's, 1980's, and 1990's, with the development of surgical dressings, therapeutic hormones, and plant supplements for increased nutritional value.

HOW IT WORKS

General Process. Often the isolation and characterization of a specific gene responsible for producing an important enzyme or other small molecule is the first step in a lengthy process toward synthesis of a product that undergoes lengthy clinical trials before the final, approved product is ready for manufacture. Once the gene has been characterized, its DNA is further characterized to facilitate the process of peptide synthesis (the process of producing long peptides is known as protein biosynthesis).

The process of peptide synthesis involves the general concepts of antigenicity, hydrophilicity, and surface probability, as well as flexibility indexes. The process involves an analysis of the peptide's characteristics, the use of software and databases to determine hydrophilicity (affinity for water), study of the antigenicity (capacity to stimulate the production of antibodies) to assist with antibody production, the study of surface probability (which determines the likelihood of inducing the formation of antibodies), the determination of the protein sequence, phosphorylation (process that activates or deactivates many protein enzymes), and then selection of two to three peptides, followed by comparison of their homology (similarity of structure).

In a general process called screening, the efficacy of an antibiotic is first tested using bacterial cultures, followed by injection of the antibiotic into laboratory animals, such as rats, rabbits, or guinea pigs; then clinical trials are conducted according to protocols established by the Food and Drug Administration (FDA). Combinatorial chemistry, a faster screening method, is often used instead. FDA-approved products are then manufactured on a larger scale.

Antibody Production. The application of a binding assay is used for isolation of the purified protein that is to be the source of an antigen. This antigen is then used as a conjugate to a carrier protein, such as kehole limpet hemocyanin (KLH), to produce a target peptide with a length of thirteen to twenty amino acids to stimulate the immune system. A carrier protein is a membrane protein that can bind to a substance to facilitate the substance's passive transport into a cell. Injection into a laboratory animal occurs next, and then the animals undergo a series of four to six immunizations separated by about twenty days. Enzyme-linked immunosorbent assay (ELISA) is used to detect antibodies. ELISA is based on the antibody-antigen binding interaction and often uses color to visually indicate the concentration of antibodies. Purification of antibodies obtained from the antiserum for specific antigen binding completes the antibody production process.

Antigen Preparation. This process is facilitated through bioinformatics analysis to choose the appropriate two to three peptides based on the protein sequence provided by a customer. KLH conjugation used for immunization, and bovine serum albumin (BSA) conjugation is carried out for screening. After immunization protocols and specific antibodies have been selected during fusion and screening, a cell can be cryopreserved.

Combinatorial Chemistry. In combinatorial chemistry synthesis, a high-throughput screening method, the starting small molecule is attached to a type of polymeric resin, followed by different permutations of reagents, to produce large libraries containing hundreds of unique products that can be rapidly screened for enzymatic activity, specific antigen-binding, or protein-protein interactions. Often the process is controlled by a computer and completed through the application of robotics. A customer can specify antigen details, and a pharmaceutical company can design a protocol involving the general phases of preparation of antigen, immunization, fusion and screening of assays, and finally selection, purification, and production of antibodies.

APPLICATIONS AND PRODUCTS

Biosensors. Biosensors are microelectronic devices that use antibodies, enzymes, or other biological molecules to interact with an optical device or electrode to record data electronically. These devices can be operated by home health care providers to transmit data obtained from blood or urine samples, for example, to a clinical laboratory some distance away.

Therapeutic Proteins. Plasmids are used to transfer human genes that provide the code for proteins important for growth hormones, blood clotting, and insulin production to bacterial cells.

Disposable Micropumps for Drug Delivery. Disposable micropumps manufactured by Acuros in Germany are capable of delivering a preset amount of liquid hormones, proteins, antibodies, or other medications. An osmotic microactuator, based on osmotic pressure, is used to regulate the amount of drug delivered, and there are no moving parts or power supply components.

High-Throughput Screening. High-throughput screening can assay more than twenty thousand potentially useful drugs per week by using multiwell plates, standard binding assay methodologies, and robotics.

Protein Biomarker Assays. NextGen Sciences has developed a mass spectrometry method for protein biomarker assays that does not depend on antibodies but instead uses surrogate proteins to facilitate development of assays. The mass spectrometer measures the amount of surrogate peptides and applies statistical evaluation to assess each biomarker. This first stage requires that a protein be confirmed; then only these selected proteins are used for the second stage

of validation of these protein biomarkers. The mass spectrometry data are used along with carbon-13 or nitrogen-15 isotopically labeled standards to calculate protein concentrations. Reporting the protein biomarkers in terms of concentration is important to allow batches containing hundreds of samples to be analyzed and validated. This technique uses proteomics (the quantitative analysis of proteins based on a physiological response) to allow for much faster development of assays than immunoassays. A wide range of at least 500 plasma proteins and 3,000 tissue proteins can be analyzed at once.

Gene Expression Databases. Gene Logic's BioExpress System is a comprehensive genome-wide gene expression database. The BioExpress System allows cells from a patient to be collected and analyzed to develop a useful biomarker profile for comparison with a database sample to indicate a therapeutic target. This process is made possible by the use of high-throughput gene expression profiling of the mononuclear cell fractions present in a blood sample. The software is capable of mining a database that has access to more than 18,000 samples containing biomarkers for the expression of the gene associated with ovarian cancer. This system is also capable of developing biomarker profiles to help diagnosis autoimmune diseases. Autoimmune diseases include rheumatoid arthritis, Crohn's disease, multiple sclerosis, systemic lupus erythematosus, and psoriasis, which affect about 20 million people in the United States.

Biosynthetic Temporary Skin Substitute. A biosynthetic skin substitute is a useful treatment for partial-thickness wounds, including skin tears, burns, and abrasions. After applying a gel to the surface of the wound, a semipermeable membrane of biosynthetic skin is used to cover the wound for protection from infection. Before the development of biosynthetic skin grafts, a physician had to choose between an allograft, which uses cadaver skin, and a xenograft, which uses tissue from another species. Biosynthetic dressings have also been developed. The dressing called Hydrofiber contains ionic silver and has been shown to prevent the spread of bacteria.

Needle-Free Drug Delivery Systems. The three types of needle-free drug delivery systems are liquid, powder, and depot injections. Each of these types uses some form of mechanical compression to create enough pressure to force the medication into the skin. Although these needle-free delivery systems cost more initially and require more technical expertise because of their complexity, they also have many advantages. In addition to eliminating pain from needle injections and reducing physician visits, these needle-free delivery systems decrease the frequency of incorrect doses. They are being used to deliver anesthetics, chemotherapy injections, vaccines, and hormones.

Nanoparticles. DNA nanotechnology uses discoveries involving nanoparticles and nanomaterials to manipulate DNA's molecular recognition abilities to build tiny medical robots that mimic bond parts or function within cells.

IMPACT ON INDUSTRY

The United States has been the world leader in biosynthetic innovation, followed closely by India. Many other countries, including Germany, France, the United Kingdom, the Czech Republic, Switzerland, Belgium, Denmark, Australia, Canada, Italy, Spain, Japan, and China, also make significant contributions.

Government and University Research. Because the fundamental starting point for the identification and synthesis of a biosynthetic target involves the isolation and characterization of a gene, the Human Genome Project has led to an explosion of developments in this field. The U.S. Department of Energy along with the National Institutes of Health financially supported the Human Genome Project, which was formally proposed in 1985 and officially initiated in 1990. The goal of this project was to identify all the genes in human DNA and organize this information into databases. Among the most significant contributors to the project were the U.S. Department of Energy Joint Genome Institute in Walnut Creek, California; the Baylor College of Medicine Human Genome Sequencing Center in Houston, Texas; the Washington University School of Medicine Genome Sequencing Center in St. Louis, Missouri; and the Whitehead Institute/MIT Center for Genome Research in Cambridge, Massachusetts. This project quickly became international in scope with the establishment of the International Human Genome Sequencing Consortium. Participating research institutes were in the United States—New York, Tennessee, California, Oklahoma, Texas, and Washington—and the United Kingdom, China, France, Germany, and Japan. In addition to identifying the genes in the human genome, these government and university research centers led the way in developing better methods for analyzing and manipulating genes, important for the initial stages of any biosynthetic project. Although the Human Genome Project was completed in 2003, the data obtained continue to be analyzed and used as the starting point for numerous innovations involving biosynthetic targets.

Industry and Major Corporations. Although companies based in the United States, such as Amgen, Genzyme, Gilead Sciences, Eli Lilly, Johnson & Johnson, Abbott Laboratories, Wyeth, Bristol-Myers Squibb, and Pfizer, have traditionally led the way in biosynthetic developments, they have been expanding worldwide, contributing to the rapid growth taking place in other countries. For example, Johnson & Johnson has subsidiaries in hundreds of countries and is responsible for a wide range of products, including surgical dressings, bandages, contact lenses, medications, and various hygiene products. Abbott Laboratories has subsidiaries in more than one hundred different countries and generated more than $29 billion in revenue in 2008. Abbott Laboratories has been a leader in the development of medical devices and medical diagnostic systems, including the first blood screening test for the human immunodeficiency virus (HIV) in 1985. In the 1950's, Wyeth became the first company to apply biosynthetic principles to manufacture the vaccine for polio, a freeze-dried smallpox vaccine. It has evolved to produce more than $3 billion in annual revenue, primarily from the sale of oral contraceptives and over-the-counter medications. Eli Lilly had revenues of more than $20 billion in 2008 and was the first company to apply biosynthetic principles to manufacture penicillin. It has expanded its range of manufactured products to include several successful psychiatric drugs, as well as pacemakers for the heart and systems to monitor intravenous fluid infusions. Pfizer became famous for its application of biosynthetic principles to develop fermentation technology, and it later manufactured penicillin, several other antibiotics, several anti-inflammatory medications, and kinase inhibitors to block the metabolic pathways of several cancers.

CAREERS AND COURSE WORK

A bachelor of science degree is adequate training

for an entry-level position in the biosynthetics field, but an advanced degree, such as a master's of science or a doctorate, is required to have the opportunity to lead research project teams in research and development, whether in academia, industry, or government. Because the field of biosynthetics involves several disciplines, college courses in chemistry, biology, genetics, microbiology, biochemistry, biomedical engineering, molecular biology, or biochemical engineering are the most helpful. Degrees in any of these disciplines would be appropriate preparation for entry into the biosynthetic field.

Many researchers with a doctorate in one of the appropriate fields work in academia and teach in addition to pursuing research. A significantly larger number of employment opportunities exist in the pharmaceutical industry and require either a bachelor's of science or master's of science degree. These positions are in research and development and various areas of manufacturing, including quality control, quality assurance, and process development. There are also opportunities for technicians who do not have a bachelor's degree. Technicians primarily record and analyze data while monitoring experiments and are often responsible for the maintenance of laboratory equipment.

According to the U.S. Bureau of Labor Statistics, employment in pharmaceutical and medical product manufacturing is expected to be one of the fastest growing manufacturing areas, with a growth rate of about 6 percent expected through 2018. The majority of these jobs within the United States are projected to be found in New Jersey, New York, Pennsylvania, Indiana, California, North Carolina, and Illinois.

SOCIAL CONTEXT AND FUTURE PROSPECTS

The Human Genome Project has facilitated the mapping of genes, which has been instrumental to the development of vaccines to treat influenza, cervical cancer, and malaria, as well as the creation of new diagnostic tools for analysis. As a result, the pharmaceutical industry in the United States has become a multibillion-dollar industry. The generation of biosynthetic products has enhanced the lives of thousands of people through the development of treatments for many types of cancer, pneumonia, cardiovascular diseases, diabetes, tuberculosis, neurological disorders, strokes, blood disorders, and many other diseases.

Fascinating Facts About Biosynthetics

- The generation of biosynthetic products has led to the development of successful treatments for many types of cancer, pneumonia, cardiovascular diseases, diabetes, tuberculosis, neurological disorders, strokes, blood disorders, and many other diseases and conditions.
- Biosynthetic corneas were used to restore vision in people with keratoconus, a condition that causes corneal scarring. These biosynthetic corneas replaced rejection-prone, scarce cadaver corneas.
- The J. Craig Venter Institute synthesized the first self-replicating synthetic bacteria cell in 2010. Synthesis of such cells may aid researchers and help develop new drugs.
- Synthetic genomics has made it possible to design and assemble chromosomes and genes and gene pathways, which may be used in creating green biofuels, pharmaceuticals, and vaccines.
- Scientists at the University of Sheffield are mapping the metabolism of the *Nostic* bacterium, which fixes nitrogen and releases hydrogen, which could be used as fuel. Once they understand the metabolic process thoroughly, they hope to be able to genetically engineer an organism that can produce hydrogen more efficiently.
- Scientists have identified biosynthetic gene clusters for many aminoglycoside antibiotics, including streptomycin, kanamycin, butirosin, neomycin and gentamicin. A full understanding of how these antibiotics work may enable scientists to get around the problem of antibiotic-resistant bacteria.
- Mass-produced biosynthetic bovine growth hormone, which when injected into dairy cows raises milk production, has been used in many developing countries. However, its use is controversial as questions have arisen regarding its effects on the health of the cows and the people who drink the milk.

Combinatorial chemistry has allowed for rapid screening of potentially successful medications that may enhance and extend the lives of many people.

Normally, only one out of every 5,000 to 10,000 compounds screened makes it through the multiyear process of clinical trials to become an FDA-approved drug. However, the desire to recoup the money spent during the years of research required to bring a drug to market has caused some pharmaceutical companies to launch a product as early as possible, which has resulted in serious litigation because some drugs proved to have harmful side effects. The application of biosynthetic growth hormones for nonmedical applications, such as bodybuilding, has also caused ethical and medical controversy. However, as the global population continues to grow and the percentage of elderly persons increases, the need for the products of biosynthetic research will continue to grow.

Jeanne L. Kuhler, B.S., M.S., Ph.D.

FURTHER READING

Arya, Dev. *Aminoglycoside Antibiotics: From Chemical Biology to Drug Discovery.* New York: Wiley-Interscience, 2007. Describes the design and synthesis of antibiotics and the process of antibiotic resistance.

Dewick, Paul. *Medicinal Natural Products: A Biosynthetic Approach.* New York: John Wiley & Sons, 2009. Comprehensive textbook describing biosynthetic methods and processes, including new techniques in genetic engineering and isolation of genes.

Lazo, John, and Peter Wipf. "Combinatorial Chemistry and Contemporary Pharmacology." *The Journal of Pharmacology and Experimental Therapeutics* 293, no. 3 (February, 2000): 705-709. Describes the process of combinatorial chemistry. Includes experimental strategies and flow charts describing the screening of compounds.

Pettit, George. *Biosynthetic Products for Cancer Chemotherapy.* Vol. 5 London: Elsevier Science, 1985. A discussion of the fundamental processes involved with screening for antitumor agents.

Savageau, Michael. *Biochemical Systems Analysis: A Study of Function and Design in Molecular Biology.* New York: CreateSpace, 2010. Detailed textbook describing the immune system and gene regulation.

Spentzos, Dimitri. "Gene Expression Signature with Independent Prognostic Significance in Epithelial Ovarian Cancer." *Journal of Clinical Oncology* 22, no. 23 (December, 2004): 4648-4658. The research article describes the diagnosis of ovarian cancer and the use of biomarkers for detection.

Stanforth, Stephen. *Natural Product Chemistry at a Glance.* New York: Wiley-Blackwell, 2006. An introductory textbook that describes much of the organic chemistry involved in biosynthesis.

WEB SITES

American Chemical Society
http://acs.org

Society for Industrial Microbiology
http://www.simhq.org

See also: Bioengineering; DNA Sequencing; Genetically Modified Organisms; Genetic Engineering; Pharmacology; Xenotransplantation.

C

CARDIOLOGY

FIELDS OF STUDY

Biology; chemistry; anatomy; physiology; biochemistry; neurology; pharmacology; pathology.

SUMMARY

Cardiology is the study of the heart. By understanding the heart's normal functional state, applied science can address its abnormalities, designing stents and bypasses for blocked coronary vessels, replacements for faulty valves, and pacemakers to compensate for electrical abnormalities. In addition, cardiopulmonary bypass machines allow the patient to undergo open-heart surgery, and artificial hearts briefly extend the life of a patient waiting for a transplant. Automated external defibrillators for applying an electrical shock to reset a heart are becoming increasingly available in public places such as airplanes and shopping malls.

KEY TERMS AND CONCEPTS

- **Angiogram:** X ray of blood vessels using radioactive dye.
- **Angioplasty:** Surgical procedure that unblocks blood vessels.
- **Aorta:** Largest artery that carries oxygenated blood to the body.
- **Arrhythmia:** Irregular heartbeat.
- **Atrium:** One of two upper chambers of the heart.
- **Bradycardia:** Lower than normal heart rate.
- **Catheter:** Thin tube that fits within blood vessels and other body tunnels.
- **Defibrillator:** Machine that delivers an electrical shock to the heart.
- **Echocardiogram:** Image of the heart using ultrasound.
- **Electrocardiogram (EKG):** Recording over time of the heart's electrical activity.
- **Pacemaker:** Device implanted in the chest to regulate the heartbeat.
- **Stent:** Implanted device that keeps arteries open for smooth blood flow.
- **Stethoscope:** Medical device for listening to sounds inside the body such as blood flow.
- **Tachycardia:** Higher than normal heart rate.
- **Vena Cava:** Largest vein that delivers deoxygenated blood from the body to the heart.
- **Ventricle:** One of two lower chambers of the heart.

DEFINITION AND BASIC PRINCIPLES

Cardiology is the branch of medicine that deals with the heart and the cardiovascular system. The heart is a muscular organ in the middle of the chest and consists of four chambers with valves that send blood to the lungs for oxygenation and then send the oxygenated blood throughout the body. It is essentially a pump; the muscle contracts in response to electrical signals that arise from the sinoatrial node, a patch of specialized tissue located at the top of the right atrium. Without oxygenated blood circulating to the body's tissues, a person would die.

Blood that is poor in oxygen enters the right atrium from the vena cava. The right atrium collects this oxygen-poor blood and sends it through the tricuspid valve into the right ventricle. The right ventricle sends this oxygen-poor blood through the pulmonary valve and pulmonary artery to the lungs for oxygenation. Oxygen-rich blood returns to the heart and enters the left atrium through the pulmonary vein. The left atrium collects this oxygen-rich blood and sends it though the mitral valve into the left ventricle. The left ventricle sends this oxygen-rich blood through the aortic valve to the rest of the body. A wall of muscle, called the septum, separates the left and right halves of the heart.

BACKGROUND AND HISTORY

A great deal of applied technology related to the heart began to evolve in the twentieth century. In 1903, Dutch physiologist Willem Einthoven invented the electrocardiograph; for this, he was awarded

The Human Heart

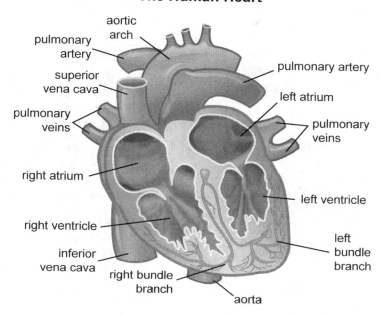

the Nobel Prize in Physiology or Medicine in 1924. Werner Forssmann performed the first human cardiac catheterization in Eberswalde, Germany, in 1929. Twelve years later, in New York, André F. Cournand and Dickinson W. Richards first used cardiac catheter techniques to measure cardiac output for diagnostic purposes. In 1956, these three men shared the Nobel Prize in Physiology or Medicine for their work.

In 1952, at the University of Minnesota, American surgeons F. John Lewis and C. Walton Lillehei successfully performed the first open-heart surgery on a human patient. After reducing the body temperature of a five-year-old girl, they repaired a hole in her heart in a ten-minute operation. In 1967, the first whole-heart transplant was performed by South African surgeon Christiaan Barnard. The patient, Louis Washkansky, died eighteen days later of pneumonia.

In 1982, American surgeon William C. DeVries implanted the first permanent artificial heart, the Jarvik-7, invented by American physician Robert K. Jarvik, into Seattle dentist Barney Clark. The dentist lived for another 112 days.

How It Works

Blockage of Coronary Vessels. Blockage in the blood vessels leading to the heart may result from blood clots or fatty deposits called plaques. Blockages may be found on an electrocardiogram, echocardiogram, or angiogram. Small blockages may be managed by lifestyle changes such as increased exercise, smoking cessation, and a diet low in cholesterol. Medications are also available that lower cholesterol levels, lower blood pressure, or thin the blood, minimizing blood clot formation. Applied science has contributed to the development of surgical procedures: An artery can be kept open for adequate blood flow by the insertion of a stent, blood vessels can be expanded by balloon angioplasty, and oxygenated blood can bypass a blocked coronary artery and reach the heart through an artery from another part of the body that is used to connect the blocked artery and the aorta.

Faulty Heart Valves. A heart valve may function improperly as the result of a birth defect, an infection, or age-related changes. It may also become deformed as the result of damage and scar tissue. Properly functioning valves permit only the one-way flow of blood. One malfunction, called regurgitation, occurs when the valve does not close tightly, allowing blood to reenter the chamber from which it came. Another malfunction, called stenosis, occurs when the valve becomes thick and stiff so it does not open completely and insufficient blood passes through with each heartbeat. The third malfunction, called atresia, occurs when the valve is fused shut, not allowing blood to pass. Untreated valve malfunctions may progress to cause sudden cardiac arrest and death. Less severe valve malfunctions may be managed by lifestyle changes and medications to relieve the symptoms. More severe valve malfunctions may be surgically corrected by repair or replacement. Replacement tissue valves may come from cows, pigs, or human cadavers. They last ten to fifteen years in older, less-active patients but wear out and must be replaced sooner in younger, more active patients. Applied science has developed mechanical replacement valves that are intended to last beyond a patient's lifetime, so that only one surgery is needed. However, the risk of blood clot formation is higher, so the patient must remain on anticoagulant drugs.

Electrical Irregularities. Electrical irregularities of the heart may result in a heartbeat that is too fast (tachycardia), too slow (bradycardia), or erratic (arrhythmia). Such dysfunction may be detected using a stethoscope, feeling peripheral pulses, or generating an electrocardiogram. Treatment depends on the stability of the patient's condition. In some cases, physical maneuvers may be employed to regulate occasional palpitations. In other cases, medications that prevent arrhythmia may be prescribed; these must be taken in conjunction with anticoagulant drugs to reduce the risk of blood clot formation. In cases of chronic bradycardia, an electrical pacemaker may be implanted to deliver a shock to the heart when the heart rate falls too low.

APPLICATIONS AND PRODUCTS

Stethoscopes and Electrocardiograms. Electrical irregularities such as an erratic heartbeat may be detected through a stethoscope, felt in a peripheral pulse, or seen as an abnormal tracing on electrocardiography. The electrocardiography invented by Dutch physiologist Einthoven in 1903 was a large tablelike contraption. The patient put his or her hands and feet in buckets of saltwater to facilitate electrical conduction. Four decades later, the machine had become smaller and received input from wire leads on metal disks attached to the patient's wrists and ankles. Modern machines are compact and easily transported on a wheeled cart from one exam room to another. The wire leads clip to disposable,

---

Fascinating Facts About Cardiology

- The average adult heart weighs 7 to 15 ounces and is slightly smaller than two clenched fists.
- Each day, the heart beats 100,000 times and pumps 2,000 gallons of blood. Most adults have a total blood volume of 10 pints, which is 1.25 gallons, in their body.
- A healthy heart beats with enough pressure to shoot blood 30 feet.
- In 1949, a crude prototype of an artificial heart was built by two Yale doctors from an erector set, small cheap toys, and mismatched household items. It kept a dog alive for more than an hour.
- A modern version of the Jarvik-7 total artificial heart has been implanted in more than eight hundred people since 1982 but each device was removed when a donor heart became available.
- In the 1980's, the external pneumatic power sources that drove artificial hearts were large and based on milking machines.
- The prevalence of heart disease increased so dramatically between 1940 and 1967 that the World Health Organization called it the world's most serious epidemic.
- Surgeons got the idea to lower the body temperatures of patients to slow their metabolism and heart rate during surgery from observing hibernating groundhogs.

---

self-adhesive disks for easy placement and removal.

Pacemakers. Heartbeats that flutter instead of beating strongly and regularly may be corrected by electrical stimulation from a cardiac pacemaker. In 1950, Canadian John A. Hopps invented the first cardiac pacemaker; it was external because it was simply too heavy to be implanted into the chest. His background in electrical engineering led him to be called the father of biomedical engineering. When transistors replaced vacuum tubes, the pacemaker became less cumbersome, and in 1958, Colombian scientist Jorge Reynolds Pombo designed the first internal cardiac pacemaker.

Defibrillators. A heart that has just stopped beating or that is pumping with no apparent rhythm may be restarted with an electrical charge delivered by a defibrillator. The shock disrupts the chaotic heart action and allows the sinoatrial node to resume its regulatory function. Automated external defibrillators are becoming increasingly available for emergency situations in public places. When activated, these machines deliver audible instructions so that untrained bystanders may use them effectively.

Stents. A coronary stent is a wire-mesh tube that is inserted into an artery to improve blood flow. It is placed as part of an angioplasty to remove a blood clot or plaque deposit and remains permanently. Charles Dotter invented the first coronary stent in 1969 and implanted it in a dog. Stents were implanted in humans in Europe as early as 1986. The U.S. Food and Drug Administration (FDA) approved the Palmaz-Schatz stent, a balloon-expandable coronary stent, for human use in 1994.

Angiography, Angioplasty, and Catheters. A cardiac catheter is a long, small-diameter tube that is threaded to the heart through the femoral artery. When contrast dye is injected through the tube, the coronary arteries may be visualized on X rays, indicating the location of any blockage; this process is called coronary angiography. These catheters and specific techniques for using them were developed by American radiologist Melvin Judkins in the 1960's.

One such catheter, a balloon-tipped catheter, is used to open a blocked artery. When the catheter with a balloon fitted on its tip reaches the site of a blockage, the balloon is inflated to enlarge the interior of the artery by flattening plaque deposits against the vessel wall. In 1977, German cardiologist Andreas Gruentzig performed the first balloon angioplasty in a human in Zurich, Switzerland.

When angioplasty cannot sufficiently open a closed artery, coronary artery bypass surgery may be performed. In this procedure, an artery from another part of the body is grafted to the heart to reroute blood flow around the blockage. This procedure requires open-heart surgery, for which the patient must be put on a heart-lung machine (cardiopulmonary bypass machine) to remain alive while the heart is not beating. The patient's blood is pumped from the body through the vena cava into the machine, where it is filtered, cooled, diluted with a specific solution to lower its viscosity, and oxygenated. The blood is then returned to the body through the ascending aorta. The patient is also given an anticoagulant to prevent the formation of blood clots.

Heart Valves. The cardiopulmonary bypass machine, which allows blood flow to bypass the heart and lungs, is also used in other surgical procedures, such as valve repair and replacement, repairs of septal defects and congenital heart defects, and heart transplantation. Surgeons repair or replace heart valves in 99,000 operations per year in the United States. The valves most commonly affected are those on the left side of the heart, namely the mitral and aortic valves, because they are exposed to higher blood pressure than those on the right side of the heart to pump oxygenated blood into the body.

When any one of the four heart valves requires replacement, the transplanted valve may be from a cow, pig, or deceased human, or it may be mechanical. Biological valves of animal tissue last only about ten years and are better suited for use in older, less-active patients. Mechanical valves are typically fashioned from plastic, carbon, or metal. Although they last longer than tissue valves and seldom require replacement, blood clots may form on their surface, so the recipient must take an anticoagulant for the rest of his or her life.

Artificial Hearts. An artificial heart is a machine that substitutes for a heart in which both halves no longer function properly. It is generally used as a temporary measure until a healthy human heart becomes available for transplantation. However, the goal of ongoing development is to create a lightweight, durable, functional machine that does not need to be replaced and will not be rejected by the body. Various models of artificial hearts are available; some have tubes for pumping pressure and wires for electrical charging that attach to equipment outside of the body; however, later models are designed to be completely enclosed within the chest. Novel biomaterials lessen the risk of foreign-body rejection and complete enclosure reduces the risk of infection.

Impact on Industry

Private Industry. Private industry has created a market known as worldwide cardiac rhythm management, which covers the research and development, manufacturing, and marketing of cardiac-related medical devices such as internal and external pacemakers, defibrillators, and monitoring leads. Regulatory issues involve countries that trade parts and processes, including the United States, Japan, China, Korea, India, Australia, and France. Companies are also developing technology such as wireless heart monitors so that physicians may receive a patient's cardiac data over the Internet and stents coated with drugs that discourage scarring and collapsing of arteries that have been unblocked with angioplasty. Because the demand for transplantable organs consistently exceeds the available supply, companies are continuing to develop artificial heart valves and artificial hearts from biomaterials that are compatible with magnetic resonance imaging as well as the body's immune system. Biometric studies are being conducted to expand product lines to accommodate the smaller sizes of women, children, and infants.

The countries in the European Union and the European Free Trade Area have an average of fifty-eight cardiologists per million inhabitants. In the United States, there are seventy cardiologists per million inhabitants. A density below fifty is considered insufficient and a density greater than eighty is excessive. However, shortages are predicted in the future. Cardiologists and cardiology researchers are needed in various settings.

National Agencies. Heart disease is the number-one killer in the United States. The American Heart Association is committed to research and education aimed at reducing deaths from heart disease. Funding is available for research in such areas as cardiovascular aging, pediatric cardiomyopathy, and Friedreich's ataxia (a genetic cause of heart disease).

The National Institutes of Health oversees the National Heart Lung and Blood Institute, which in June, 1991, launched the National Heart Attack Alert Program. The goals of this program are to reduce the severity of heart attacks through early detection and treatment and to improve the quality of life for patients and their families.

The Food and Drug Administration, the agency that oversees pharmaceuticals, alerts consumers to heart-related risks associated with newly developed medications such as antidiabetes drugs, attention deficit hyperactivity disorder drugs, and diet pills. The FDA oversees the manufacturing of medical devices and takes action when devices such as stents or pacemakers are suspected of acting in a faulty manner.

Research Hospitals. At the Covenant Heart and Vascular Institute, a member of the Covenant Health System in West Texas, research and clinical trials are conducted to develop therapeutics, procedures, and medical devices for the treatment of heart disease. It is an accredited Cycle III Chest Pain Center with Percutaneous Coronary Intervention, one of only

fourteen in the state. This certificate is for excellence in providing acute cardiac medicine.

The Heart Institute of Childrens Hospital Los Angeles is part of a fifteen-member consortium of pediatric cardiac institutes called the Pediatric Heart Network. Its research focus is the optimal survival of babies born with hypoplastic left heart syndrome, in which the left ventricle is missing or underdeveloped; without corrective surgery, the condition is fatal.

The Jim Moran Heart and Vascular Research Institute at Holy Cross Hospital in Fort Lauderdale, Florida, conducts cardiac research in collaboration with private corporations, university medical centers, and other research institutes.

University Medical Centers. The Dorothy M. Davis Heart & Lung Research Institute at Ohio State University Medical Center engages in research into the basic science of heart disease, including such processes as natural cell death and inflammation. Researchers are also seeking gene-based information that will allow therapies to be customized.

The Penn Cardiovascular Institute at the University of Pennsylvania provides education, research, and patient care. The researchers there have been awarded grant money to study stem cell therapies.

Nonprofit Organizations. The Florida Heart Research Institute is a nonprofit organization dedicated to stopping heart disease through research, education, and prevention. It facilitates collaborative research nationally and offers free community services locally.

The Heart Disease Research Institute is an international nonprofit organization based in Arizona that offers information on the prevention and treatment of heart disease to the general public.

The American College of Cardiology is a national medical society that makes transparent its members' relationships with drug and device manufacturers, especially regarding research funding, publications, and continuing medical education for practitioners.

CAREERS AND COURSE WORK

According to the U.S. Bureau of Labor Statistics, cardiologists will see a 27 percent growth in their field by 2014. To become a cardiologist, a person must complete an undergraduate college degree, majoring in a life science such as biology, chemistry, or biochemistry. That graduate must then complete a medical degree, studying detailed anatomy, physiology, pathology, and pharmacology. That physician must then complete additional training in cardiology, through residencies and fellowships, and become board certified as a cardiologist. Cardiologists may further specialize in areas such as cardiac diagnosis, pediatric cardiology, and electrophysiology. Cardiologists commonly have a private practice with hospital privileges.

Cardiac researchers must also complete an undergraduate college degree, majoring in a life science. That graduate may follow the same path as a cardiologist but most choose instead to pursue additional education in master's and doctoral programs. They then do postdoctoral work and become published authors before they apply for and receive funding for their own research projects. Most cardiac research laboratories are associated with university medical schools, although some are associated with government agencies or private companies such as pharmaceutical and medical device manufacturers.

Jobs are available for cardiac technicians who set up electrocardiograms, echocardiograms, and other diagnostic procedures in a cardiologist's office or a hospital cardiology department. Educational opportunities vary; some schools offer specific vocational programs with job placement, although some physicians prefer to train a college graduate who majored in a life science.

SOCIAL CONTEXT AND FUTURE PROSPECTS

Researchers are continuing their efforts to develop an artificial heart that will sustain patients for longer periods while they wait for a transplant, with the ultimate goal of implanting a mechanical heart that will not need replacement. The AbioCor artificial heart, manufactured by Abiomed of Danvers, Massachusetts, was the first self-contained implantable device; previous versions of an artificial heart required patients to remain in bed connected to machines with tubes and electrodes. The AbioCor device is powered by an external battery pack, allowing the patient to be ambulatory. In human trials, this device was implanted in patients whose life expectancy was thought to be less than 30 days. The goal was to extend their lives by an additional 30 days. This device received FDA approval on September 5, 2006. So far, fifteen patients have received an AbioCor artificial heart; the longest a recipient lived was 512 days.

A modern prototype of a fully implantable

artificial heart contains cutting-edge electronic sensors, synthetic microporous skins, and other novel biomaterials.

Bethany Thivierge, B.S., M.P.H.

FURTHER READING

Holler, Teresa. *Cardiology Essentials*. Sudbury, Mass.: Jones and Bartlett Learning, 2007. Presents cardiology with a practical clinical orientation for medical personnel working in a cardiology office.

Mueller, Richard L., and Timothy A. Sanborn. "The History of Interventional Cardiology: Cardiac Catheterization, Angioplasty, and Related Interventions." *American Heart Journal* 129, no. 1 (January, 1995): 146-172. Contains plenty of names, dates, and details of interest in cardiology history.

Murphy, Joseph G. *Mayo Clinic Cardiology: Concise Textbook*. 3d ed. London: Informa Healthcare Communications, 2006. This easy-to-read textbook features information on all aspects of cardiology.

Topol, Eric J., ed. *Textbook of Cardiovascular Medicine*. 3d ed. Philadelphia: Lippincott Williams & Wilkins, 2006. A complete, well-organized, user-friendly reference book, complete with audio and visual aids.

WEB SITES

Alliance of Cardiovascular Professionals
http://www.acp-online.org

American College of Cardiology
http://www.cardiosource.org

American Heart Association
http://www.heart.org/HEARTORG

Heart Disease Research Institute
http://heart-research.org

National Institutes of Health
National Heart Lung and Blood Institute
http://www.nhlbi.nih.gov

See also: Bioengineering; Cell and Tissue Engineering; Geriatrics and Gerontology; Pediatric Medicine and Surgery.

CELL AND TISSUE ENGINEERING

FIELDS OF STUDY

Cell biology; cardiology; cardiac surgery; biochemistry; organic chemistry; developmental biology; physiology; transplant surgery; stem cell research; biomaterial science; drug and gene delivery; neuroscience; bioinformatics; molecular engineering; orthopedic surgery; mechanical engineering; physical therapy; biophysical tools; medical ethics; public health.

SUMMARY

Cell and tissue engineering are fields dedicated to discovering the mechanisms that underlie cellular function and organization to develop biological or hybrid biological and nonbiological substitutes to restore or improve cellular tissues. The most immediate goal of cell and tissue engineering is to allow physicians to replace damaged or failing tissues within the body. The field was first recognized as a distinct branch of bioengineering in the 1980's and has since grown to attract participation from numerous medical and biological disciplines.

Engineered cellular materials may be used to grow new tissue within a patient's heart or to replace damaged bone, cartilage, or other tissues. In addition, research into the mechanisms affecting cellular organization and development may aid in the treatment of congenital and developmental disorders. Cell and tissue engineering has developed in conjunction with stem cell research and is therefore subject to debate over the ethics of stem cell research.

KEY TERMS AND CONCEPTS

- **Bioartificial Device:** Substance, tissue, or organ that combines biological and synthetic components.
- **Bioengineering:** Medical or biological application of engineering principles, including the process of engineering biological tissues and components from raw materials.
- **Biomaterial:** Cellular or synthetic material that can be introduced into living tissues; often part of a medical device.
- **Cardiology:** Branch of medicine concerned with the disorders, diseases, and function of the heart and associated systems.
- **Cell Therapy:** Introduction of cells or tissues to treat disease or other physiological disorders.
- **Differentiation:** Processes by which cells change morphology and function to fill a specific role within an organism or tissue.
- **Drug And Gene Delivery:** Field of study dedicated to the methods involved in introducing medical chemicals and genes to an organism.
- **Extracellular Matrix:** Substance surrounding cellular tissues in which connecting tissues are fixed.
- **Growth Factor:** Substance that stimulates growth of a cell, tissue, or other part of an organism.
- **Heterologous:** Derived from an organism of a different species.
- **In Vitro:** Outside a living body or organism, in an artificial environment.
- **In Vivo:** Within a living body or organism.
- **Orthopedics:** Branch of medicine concerned with disorders, diseases, and injuries to the skeleton and associated tissues.
- **Regenerative Medicine:** Branch of medicine concerned with applying techniques and tools from a variety of disciplines to restore or repair damaged tissues, cells, and organs by stimulating the biological healing and regeneration processes.
- **Rejection:** Immune response in which the host's immune system attempts to defend against cells or tissues introduced from a foreign organism.
- **Stem Cell:** Undifferentiated cell type that gives rise to specialized cells within the body; most are derived from embryonic tissues.
- **Transplant:** Transfer of an organ, tissue, or other cellular material from one individual to another.

DEFINITION AND BASIC PRINCIPLES

Cell and tissue engineering is a branch of bioengineering concerned with two basic goals: studying and understanding the processes that control and contribute to cell and tissue organization and developing substitutes to replace or improve existing tissues in an organism. Substitute tissues can be composed either of biological materials or of a blend of biological and nonbiological materials.

The basic goal of cell and tissue engineering is to create more effective treatments for tissue degeneration and damage resulting from congenital disorders, disease, and injury. Engineers may, for instance, introduce foreign tissues that have been modified to stimulate healing within the patient's own tissues, or they may implant synthetic structures that help control and stimulate cellular development. Another goal in cell and tissue engineering is to create tissues that are resistant to rejection from the host organism's immune system. Rejection is one of the primary difficulties in organ transplant and limb replacement surgery.

One of the basic principles of cell and tissue engineering is to use and enhance an organism's innate regenerative capacity. Engineers therefore examine the ways that tissues grow and change during development. Using cutting-edge development in genomics and gene therapy, engineers are working to develop ways to stimulate a patient's immune system and enhance healing.

Cell and tissue engineering have a wide variety of potential applications. In addition to creating new therapies, engineering principles can be used to create new methods for delivering drugs and engineered cells to target locations within a patient. The potential applications of cell and tissue engineering depend on the capability to create cultures of cells and tissues to use for experimentation and transplantation. Research on cell growth is a major facet of the bioengineering field.

BACKGROUND AND HISTORY

Cell and tissue engineering emerged from a field of study known as regenerative medicine, a branch concerned with developing and using methods to enhance the regenerative properties of tissues involved in the healing process. Ultimately, cell and tissue engineering became most closely associated with transplant medicine and surgery.

Medical historians have found documents from as early as 1825 recording the successful transplantation of skin. The first complete organ transplants occurred in the 1950's, and the first heart transplant was completed successfully in 1964.

The science of cell and tissue engineering arose from attempts to combat the problems that affect transplantation, including scarcity of organs and frequent issues involving rejection by the host's immune system. In the 1970's and 1980's, scientists began working on ways to build artificial or semi-artificial substitutes for organ transplants. Most early work in tissue engineering involved the search for a suitable artificial substitute for skin grafts.

By the mid-1980's, physicians were using semi-synthetic compounds to anchor and guide transplanted tissues. The first symposium for tissue engineering was held in 1988, by which time the field had adherents around the world. The rapid advance of research into the human genome and genetic medicine in the mid-1990's had a considerable effect on bioengineering. In the twenty-first century, cell and tissue engineers work closely with genetic engineers in an effort to create new and better tissue substitutes.

HOW IT WORKS

Broadly speaking, cell and tissue engineering involves creating cell cultures and tissues that are introduced to an organism to repair damaged or degenerated tissues. There are a wide variety of techniques and specific applications for cell and tissue engineering, ranging from cellular manipulation at the chemical or genetic level to the creation of artificial organs for transplant.

Most cell and tissue engineering methods share several common procedures. First, scientists must produce cells or tissues. Next, engineers must tell the cells what to do. This can be done in a variety of ways, from physically manipulating cellular development and tissue formation to altering the genes of cells in such a way as to direct their function. Finally, engineered tissues and cells must be integrated into the body of the host organism under controlled conditions to limit the potential for rejection. Cell and tissue engineering can be divided into two main categories, in vitro engineering and in vivo engineering.

In Vitro Engineering. In vitro engineering is the development of cell cultures and tissues outside of the body in a controlled laboratory environment. This method has several advantages. Producing tissues in a laboratory has the potential for growing large amounts of tissue and eventually entire organs. This could help solve a major issue with transplant surgery: the scarcity of viable organs for transplantation. Scientists can more precisely control the growing environment and can therefore exert greater control over developing cells and tissues. In vitro engineering allows engineers to modify and

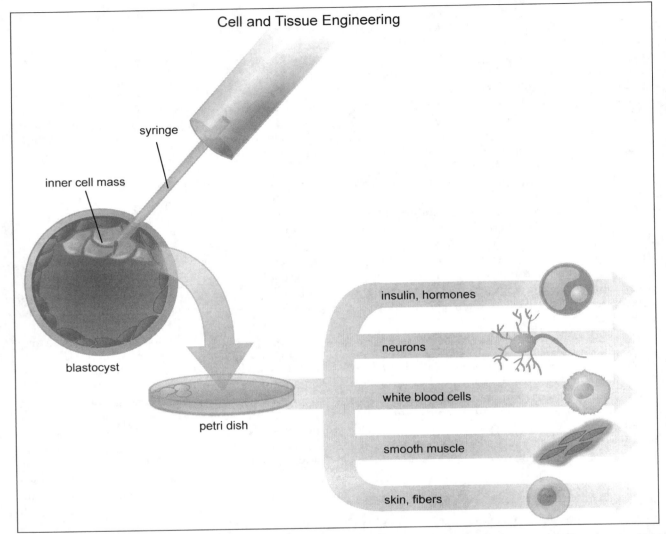

One of the most immediate goals of cell and tissue engineering is to allow physicians to replace damaged or failing tissues within the body.

and adjust cellular properties without the need for surgery or invasive techniques.

In vitro engineering is commonly used in the creation of skin tissues, cartilage, and some bone replacement tissues. Although in vitro techniques have certain advantages, they have serious drawbacks, including a higher rejection rate for cells and tissues created in vitro. In addition, there are physiological advantages to engineering within the host organism's body, including the presence of accessible cellular nutrients.

In Vivo Engineering. In vivo engineering is the family of techniques that involves creating engineered cellular cultures or tissues within the host's body. It involves the use of chemicals to alter cellular function and the use of synthetic materials that interact with the host's body to stimulate or direct cellular growth.

In vivo procedures typically involve introducing only minor changes to the host's internal environment, and therefore, these tissues are more likely to be resistant to rejection. In addition, working in vivo allows engineers to take full advantage of the host's existing cellular networks and the physiological environment of the body. The body provides the essential nutrients, exchange of materials, and disposal of waste, helping create healthy tissues.

The primary disadvantages of the in vivo approach are that engineers have less direct control over the development of the cells and tissues and cannot make exact changes to the microenvironment during development. In addition, in vivo engineering does not allow for the production of mass quantities of cells and is therefore not an avenue toward addressing the shortage of available tissues and organs for transplant.

APPLICATIONS AND PRODUCTS

Hundreds of bioengineers are working around the world, and they have created a wide variety of applications using cell and tissue engineering research. Among the most promising applications are cell matrices and bioartificial organ assistance devices.

Cell Matrices. In an effort to improve the success of tissue transplants, bioengineers have developed a method for using artificial matrices, also called "scaffolds," to control and direct the growth of new tissues. Using cutting-edge microengineering techniques and materials, engineers create three-dimensional structures that are implanted into an organism and thereafter serve as a "guide" for developing tissues.

The scaffold acts like an extracellular matrix that anchors growing cells. New cells anchor to the artificial matrix rather than to the organism's own extracellular material, allowing engineers to exert control over the eventual size, shape, and function of the new tissue. In addition, scaffolds can aid in the diffusion of resources within the growing tissue and can help engineers direct the placement of functional cells, as the scaffold can be installed directly at the site of an injury.

Matrices may be constructed from a variety of materials, including entirely synthetic combinations of polymers and other structures that are created from derivatives of the extracellular matrix. Many researchers have been designing scaffolds that dissolve as the tissues form and are then absorbed into the organism. These biodegradable scaffolds allow engineers to avoid further surgical procedures to remove implanted material.

Cellular scaffolds represent a middle ground between in vivo and in vitro engineering. Engineers can create a scaffold in a laboratory environment and can allow tissue to anchor and grow around the matrix before implantation, or they can place a scaffold in their target area within the organism and allow the organism's own cells to populate the matrix.

Scaffolds have been used successfully in cardiac repair, especially in conjunction with stem cells. A scaffold seeded with stem cells may be implanted directly into a heart valve, roughly at the site where a cardiac infarction has occurred. The scaffold then directs the growing cells toward the injured area and facilitates regeneration of damaged tissue.

Artificial matrices have also been successful in treating disorders that affect the kidney, bone, and cartilage. Researchers are hopeful that cellular scaffolds could eventually allow the creation of entire organs by coaxing cells to develop around a scaffold designed as an organ template.

Bioartificial Organs. One of the major areas of research in tissue engineering is the creation of machines that assist organs damaged by disease or injury. Made from a combination of synthetic and organic materials, these machines are sometimes called bioartificial devices.

One of the most promising organ assistance devices is the bioartificial liver (BAL), which has been developed to help patients suffering from congenital liver disease, acute liver failure, and other metabolic disorders affecting the liver. The BAL consists of cells incorporated into a bioreactor, which is a small machine that provides an environment conducive to biological processes. Cells growing within the BAL receive optimal nutrients and are exposed to hormones and growth factors to stimulate development. The bioreactor is also designed to facilitate the delivery of any chemicals produced by the developing tissues to surrounding areas.

The BAL performs some of the functions usually performed by the liver: It processes blood, removes impurities, produces proteins, and aids in the synthesis of digestive enzymes. The BAL is not intended to permanently replace the liver but rather to supplement liver function or to allow a patient to survive until a liver transplant can be arranged. The bioartificial liver enables patients to forgo dialysis treatments, and some researchers hope to develop BAL devices that may function as a permanent replacement for patients in need of dialysis.

Researchers are working on bioartificial kidney devices that would aid patients with diabetes and other disorders leading to kidney failure. Again, the bioartificial kidney devices are bioreactors, using stem cells and kidney cells to perform some of the purification and detoxification functions of the kidney. Researchers are also developing bioartificial devices to treat disorders of the pancreas and the heart and to help patients suffering from nervous system or circulatory disorders. Taken as a whole, the development of organ assistance devices may be a step toward the development of bioartificial devices that can function to fully replace a patient's malfunctioning organ.

IMPACT ON INDUSTRY

Bioengineering has become one of the fastest growing fields in medical and biological research. As of 2008, the global market was estimated to have an approximate value of $1.5 billion, according to a study by Life Science Intelligence.

Analysts have said that the bioengineering market has not reached its full potential, as research organizations are only beginning to use the potential funding and resources available for cell and tissue engineering applications. According to Life Science Intelligence, the global market for tissue and cell engineering may reach $118 billion by 2013.

In 2008, the United States controlled more than 90 percent of the bioengineered products market. Europe and Japan were the next leading providers, and both the European and the Japanese markets were expected to continue to grow. There are more than two hundred companies in the United States working to provide research, equipment, and products in the industry.

Government Funding and Research. In the United States, a variety of public funding organizations, including the National Institutes of Health, the National Science Foundation, and the National Institute of Standards and Technology support cell and tissue research.

Within the National Institutes of Health, the government's largest research and funding organization, there are six divisions that provide funding for projects dealing with bioengineering, including the National Institute of Biomedical Imaging and Bioengineering. The National Institute of Standards and Technology has two divisions that provide

bioengineering grants and the National Science Foundation's Division of Bioengineering and Environmental Systems also provides grants and assistance for tissue and cell engineers.

The America COMPETES Act, passed by Congress in August of 2007, provides funding for emerging research to promote American dominance in developing scientific fields. Funding from the federal government was used to create the Technology Innovation Program in the National Institute of Standards and Technology, which in turn provides funding for bioengineering programs in addition to a variety of other research initiatives. The Technology Innovation Program received more than $65 million of the institute's 2009 budget of $813 million.

Regulation and Legal Issues. Each country maintains its own laws and systems for regulating biotechnology. As much of the technology involved in cell and tissue engineering is newly developed or emerging, regulatory agencies and government officials are still debating the best way to monitor and regulate biotechnology.

The European health ministers completed a set of regulatory guidelines in 2007 designed to increase patient access to emerging treatments from biotechnology and to provide a framework for determining the health risks of biomedical research and applications. The legislation was accompanied by increased European funding for researchers investigating the benefits of bioengineering in the medical sciences.

In the United States, biotechnology is one of the most heavily regulated fields, and most emerging technology falls under the auspices of the Food and Drug Administration. While some researchers believe that United States regulations are too stringent and unnecessarily delay distribution of new technologies and therapies, the United States continues to be a world leader in the field.

Biotech Corporations. Biotechnology is experiencing rapid growth because in addition to the research being conducted in universities and government-sponsored programs, for-profit companies are investing in the field. As of 2009, more than two hundred companies in the United States were engaging in biotechnology research. A number of profit-driven bioengineering firms centered in Europe and Japan also have begun competing in the global market.

Major American companies involved in tissue and cell engineering include medical industry leaders, such as Pfizer, Johnson & Johnson, and Novartis, and dozens of smaller corporations. Although government funding for bioengineering has increased in the twenty-first century, private and corporate investment, largely from companies with other medical investments including pharmaceuticals and medical equipment manufacturing, accounts for more than 90 percent of available funding in the field.

CAREERS AND COURSE WORK

Students interested in cell and tissue engineering might start at the undergraduate level, working toward a degree in biology or biochemistry, with a focus on cellular biology. Students might also enter the bioengineering field with a background in engineering, though students will still need a significant background in biology and medical science.

After achieving an undergraduate education, students can progress in the field by pursuing graduate studies in cell biology, bioengineering, or related fields. Many professionals working in other disciplines, such as orthopedic medicine, dermatology, and cardiac surgery may also become involved with cell and tissue engineering during their careers. Graduate institutions are increasingly trying to introduce programs that focus on cell and tissue engineering. The University of Pittsburgh and Duke University are among the universities offering specializations in tissue and cell engineering for qualified graduate candidates.

Professionals seeking work in the cell and tissue engineering field can seek employment with non-profit research institutions, such as those in many universities. Positions in universities are generally funded by a combination of public and private funding. Additionally, those interested in bioengineering careers can find employment within a large number of corporations. The more than two hundred companies in the United States involved in biotechnology employ chemists, mechanical engineers, physicians, and individuals trained specifically in bioengineering.

SOCIAL CONTEXT AND FUTURE PROSPECTS

Bioengineering is intended to improve daily life, both for those suffering from injury and illness and for the population at large. Cell and tissue engineers are focusing on ways to replace damaged tissues, providing, for instance, new skin where skin has been

---

Fascinating Facts About Cell and Tissue Engineering

- In 2008, scientists from a number of European countries reported on the first procedure to install a bioengineered trachea, constructed from synthetic materials and cultures produced from stem cells, in a woman with a failing respiratory system.

- In August of 2009, a team of Italian scientists announced they were working on an innovative method to replace damaged bone with substitute bone made from wood.

- The Russ Prize, given by the National Academy of Engineering since 1999, is considered the Nobel Prize for bioengineering. Past winners have come from both the engineering and the medical fields.

- Scientists at the Fraunhofer Institute in Germany are working on creating artificial human organs that can be used to replace animal subjects in clinical experiments, allowing scientists to achieve more accurate results and to avoid costly and controversial animal trials.

- The large number of soldiers who lost limbs while serving in Iraq has prompted the U.S. Department of Defense to invest in a University of Michigan program aimed at creating prosthetic hands that can transmit touch sensations to a patient's brain.

- In 2009, scientists in Germany revealed a plan to institute an automated process to produce synthetic skin. Automation would be a first necessary step toward producing sufficient quantities of skin to meet all needs.

---

destroyed, and technology to supplement the function of essential organs. One of the ultimate goals of the industry is to create artificial organs that can fully and permanently replace damaged organs. Bioengineers are confident that in the future it will be possible to provide patients with a variety of organs including a heart, liver, or pancreas.

Although most cell and tissue engineers focus on combating physical illness and injury, bioengineering also has the potential to produce technology that will allow humans to improve their functional abilities. At some point, combinations of synthetic computer technology and biological components could be used to improve human visual capacity or to endow humans with more precise access to memory.

Humans are not the only targets for bioengineers, as other organisms may also be altered to improve their basic physiological functions. Take, for instance, a 2008 project from the Australian Center for Plant Functional Genomics in which researchers are attempting to bioengineer plants that can withstand higher levels of salt in the soil, a breakthrough that could turn into a major benefit for agriculture. Salt-resistant strains of important agricultural crops could grow where agriculture was previously impossible because of the soil's alkalinity.

As a distinct discipline, bioengineering is relatively new and scientists have only begun to investigate the potential applications and discoveries possible with further research. As the field has begun to expand, so too have opportunities for scientists, engineers, and physicians interested in exploring the future of medicine and science. The bioengineering field has already created billions in revenue and is still in a state of rapid growth. Universities, hospitals, and biomedical corporations are likely to increase their investment in these emerging technologies and techniques, creating a strong and growing industry for many years to come.

Micah L. Issitt, B.S.

FURTHER READING

Chien, Shu, Peter C. Y. Chen, and Y. C. Fung, eds. *An Introductory Text to Bioengineering.* Hackensack, N.J.: World Scientific Publishing, 2008. While definitely written with advanced science students in mind, this text is one of the most basic and yet comprehensive texts available as an introduction to all types of bioengineering.

De Gray, Aubrey, and Michael Rae. *Ending Aging: The Rejuvenation Breakthroughs That Could Reverse Aging in Our Lifetime.* New York: St. Martin's Griffin, 2008. An investigation of research programs in bioengineering, nutrition, and other fields of medicine that are aimed at prolonging life. Provides interesting coverage of organ transplantation and cellular manipulation.

Mataigne, Fen. *Medicine by Design: The Practice and Promise of Biomedical Engineering.* Baltimore: The Johns Hopkins University Press, 2006. An introduction to and investigation of bioengineering

and the potential future of the field. Provides discussions of issues such as bioreactors and organ replacements.

Rose, Nickolas. *The Politics of Life Itself: Biomedicine, Power, and Subjectivity in the Twenty-first Century.* Princeton, N.J.: Princeton University Press, 2006. An introduction to the moral, ethical, and political issues that surround medical engineering, genetic manipulation, and bioengineering. Addresses several prominent fields in cell and tissue engineering.

Valentinuzzi, Max. *Understanding the Human Machine: A Primer for Bioengineering.* Hackensack, N.J.: World Scientific Publishing, 2004. An accessible reference designed to give students much of the biological knowledge needed to pursue studies in bioengineering. Also provides useful information about the nature, goals, and development of the bioengineering field.

Zenios, Stefanos, Josh Makower, and Paul Yock, eds. *Biodesign: The Process of Innovating Medical Technologies.* New York: Cambridge University Press, 2010. Covers the biomedical industry, with a particular focus on the process of creating and marketing medical technology. Also provides information about the future of biotechnology and bioengineered products.

WEB SITES

Johns Hopkins University School of Medicine
Institute for Cell Engineering
http://www.hopkins-ice.org

Tissue Engineering International and Regenerative Medical Society
http://www.termis.org

See also: Bioengineering; Biosynthetics; Genetic Engineering; Human Genetic Engineering; Prosthetics; Stem Cell Research and Technology.

CERAMICS

FIELDS OF STUDY

Calculus; chemistry; glass engineering; materials engineering; physics; statistics; thermodynamics.

SUMMARY

Ceramics is a specialty field of materials engineering that includes traditional and advanced ceramics, which are inorganic, nonmetallic solids typically created at high temperatures. Ceramics form components of various products used in multiple industries, and new applications are constantly being developed. Examples of these components are rotors in jet engines, containers for storing nuclear and chemical waste, and telescope lenses.

KEY TERMS AND CONCEPTS

- **Ceramic:** Inorganic, nonmetallic solid processed or used at high temperatures; made by combining metallic and nonmetallic elements.
- **Coke:** Solid, carbon-rich material derived from the destructive distillation of low-ash, low-sulfur bituminous coal.
- **Glazing:** Process of applying a layer or coating of a glassy substance to a ceramic object before firing it, thus fusing the coating to the object.
- **Kiln:** Thermally insulated chamber, or oven, in which a controlled temperature is produced and used to fire clay and other raw materials to form ceramics.

DEFINITION AND BASIC PRINCIPLES

Ceramic engineering is the science and technology of creating objects from inorganic, nonmetallic materials. A specialty field of materials engineering, ceramic engineering involves the research and development of products such as space shuttle tiles and rocket nozzles, building materials, as well as ball bearings, glass, spark plugs, and fiber optics.

Ceramics can be crystalline in nature; however, in the broader definition of ceramics (which includes glass, enamel, glass ceramics, cement, and optical fibers) they can also be noncrystalline. The most distinguishing feature of ceramics is their ability to resist extremely high temperatures. This makes ceramics very useful for tasks where materials such as metals and polymers alone are unsuitable. For example, ceramics are used in the manufacture of disk brakes for high-performance cars (such as race cars) and for heavy vehicles (such as trucks, trains, and aircraft). These brakes are lighter and more durable and can withstand greater heat and speed than the conventional metal-disk brakes. Ceramics can also be used to increase the efficiency of turbine engines used to operate helicopters. These aircraft have a limited travel range and cannot carry a great deal of weight because of the stress these activities place on engines made of metallic alloys. However, turbine engines using ceramic parts and thermal barrier coatings are currently in development—they already show superior performance when compared with existing engines. The ceramic engine parts are from 30 to 50 percent lighter than their metallic counterparts, and the thermal coatings increase the engine operating temperatures to 1,650 degrees Celsius (C). These qualities will enable future helicopters to travel farther and carry more weight.

BACKGROUND AND HISTORY

One of the oldest industries on Earth, ceramics date back to prehistoric times. The earliest known examples of ceramics, animal and clay figures that were fired in kilns, date back to 24,000 b.c.e. These ceramics were earthenware that had no glaze. Glazing was discovered by accident and the earliest known glazed items date back to 5000 b.c.e. Chinese potters studied glazing and first developed a consistent glazing technique. Glass was first produced around 1500 b.c.e. The development of synthetic materials with better resistance to very high temperatures in the 1500's enabled the creation of glass, cement, and ceramics on an industrial scale. The ceramics industry has grown in leaps and bounds since then.

Many notable innovators have contributed to the growth of advanced ceramics. In 1709, Abraham Darby, a British brass worker and key player in the Industrial Revolution, first developed a smelting process for producing pig iron using coke instead of wood as fuel. Coke is now widely used in the production of carbide ceramics. In 1888, Austrian chemist

Karl Bayer first separated alumina from bauxite ore. This method, known as the Bayer process, is still used to purify alumina. In 1893, Edward Goodrich Acheson, an American chemist, electronically fused carbon and clay to create carborundum, also known as synthetic silicon carbide, a highly effective abrasive.

Other innovators include brothers Pierre and Jacques Curie, French physicists who discovered piezoelectricity around 1880; French chemist Henri Moissan, who combined silicon carbide with tungsten carbide around the same time as Acheson; and German mathematician Karl Schröter, who in 1923 developed a liquid-phase sintering method to bond cobalt with the tungsten-carbide particles created by Moissan.

The need for high-performance materials during World War II helped accelerate ceramic science and engineering technologies. Development continued throughout the 1960's and 1970's, when new types of ceramics were created to facilitate advances in atomic energy, electronics, and space travel. This growth continues as new uses for ceramics are researched and developed.

How It Works

There are two main types of ceramics: traditional and advanced. Traditional ceramics are so called because the methods for producing them have been in existence for many years. The familiar methods of creating these ceramics—digging clay, molding the clay by hand, or using a potter's wheel, firing, and then decorating the object—have been around for centuries, and have only been improved and mechanized to meet increasing demand. Advanced ceramics, which cover the more recent developments in the field, focus on products that make full use of specific properties of ceramic or glass materials. For example, ferrites, a type of advanced ceramics, are very good conductors of electricity and are typically used in electrical transformers and superconductors. Zirconia, another type of advanced ceramics, is strong, tough, very resistant to wear and tear, and does not cause an adverse reaction when introduced to biological tissue. This makes it ideal for use in creating joint replacements in humans. It works particularly well in hip replacements, but it is also useful for knee, shoulder, and finger-joint replacements. The unique qualities of these, and other advanced ceramics, and the research into the variety of ways they can be applied, are what differentiate them from traditional ceramics.

There are seven basic steps to creating traditional ceramics. They are described in detail below.

Raw Materials. In this first stage, the raw materials are chosen for creating a ceramic product. The type of ceramic product to be created determines the type of raw materials required. Traditional ceramics use natural raw materials, such as clay, sand, quartz, and flint. Advanced ceramics require the use of chemically synthesized powders.

Beneficiation. Here, the raw materials are treated chemically or physically to make them easier to process.

Batching and Mixing. In this step, the parts of the ceramic product are weighed and combined to create a more chemically and physically uniform material to use in forming, the next step.

Forming. The mixed material is consolidated and molded to create a cohesive body of the determined shape and size. Forming produces a "green" part, which is soft, pliable and, if left at this stage, will lose its shape over time.

Green Machining. This step eliminates rough surfaces, smooths seams, and modifies the size and shape of the green part to prepare for sintering.

Drying. Here, the water or other binding agent is removed from the formed material. Drying is a carefully controlled process that should be done as quickly as possible. After drying, the product will be smaller than the green part. It is also very brittle and must be handled carefully.

Firing or Sintering. The dried parts now undergo a controlled heating process. The ceramic becomes denser during firing, as the spaces between the individual particles of the ceramic are reduced as they heat. It is during this stage that the ceramic product acquires its heat-resistant properties.

Assembly. This step occurs only when ceramic parts need to be combined with other parts to form a complete product. It does not apply to all ceramic products.

This is not a comprehensive list. More steps may be required depending on the type of ceramic product being made. For advanced ceramics production, this list of steps will either vary or expand. For example, an advanced ceramic product may need to have forming or additives processes completed in addition to the standard forming processes. It may also require a post-sintering process such as machining or annealing.

APPLICATIONS AND PRODUCTS

Traditional Ceramics. Applications and products include whiteware, glass, structural clay items, cement, refractories, and abrasives. Whiteware, so named because of its white or off-white color, includes dinnerware (plates, mugs, and bowls), sanitary ware (bathroom sinks and toilets), floor and wall tiles, dental implants, and decorative ceramics (vases, figurines, and planters). Glass products include containers (bottles and jars), pressed and blown glass (wineglasses and crystal), flat glass (windows and mirrors), and glass fibers (home insulation). Structural clay products include bricks, sewer pipes, flooring, and wall and roofing tiles. Cement is used in the construction of concrete roads, buildings, dams, bridges, and sidewalks. Refractories are materials that retain their strength at high temperatures. They are used to line furnaces, kilns, incinerators, crucibles, and reactors. Abrasives include natural materials such as diamonds and garnets, and synthetic materials such as fused alumina and silicon carbide, which are used for precision cutting as well.

Advanced Ceramics. Advanced ceramics focus on specific chemical, biomedical, mechanical, or optical uses of ceramic or glass materials.

Advanced ceramics fully came into being within the last few decades (beginning in the 1960's), and research and development is ongoing. The field has produced a wide range of applications and products. In aerospace, ceramics are used in space shuttle tiles, aircraft instrumentation and control systems, missile nose cones, rocket nozzles, and thermal insulation. Automotive applications include spark plugs, brakes, clutches, filters, heaters, fuel pump rollers, and emission control devices. Biomedical uses for ceramics include replacement joints and teeth, artificial bones and heart valves, hearing aids, pacemakers, dental veneers, and orthodontics. Electronic devices that use ceramics include insulators, magnets, cathodes, antennae, capacitors, integrated circuit packages, and superconductors. Ceramics are used in the chemical and petrochemical industry for ceramic catalysts, catalyst supports, rotary seals, thermocouple protection tubes, and pumping tubes. Laser and fiber-optics applications for ceramics include glass optical fibers (used for very fast data transmission), laser

materials, laser and fiber amplifiers, lenses, and switches. Environmental uses of ceramics include solar cells, nuclear fuel storage, solid oxide fuel cells, hot gas filters (used in coal plants), and gas turbine components. Ceramic coatings include self-cleaning coatings for building materials, coatings for engine components, cutting tools, industrial wear parts, optical materials (such as lenses), and antireflection coatings.

Other products in the advanced ceramics segment include water-purification devices, particulate or gas filters, and glass-ceramic stove tops.

IMPACT ON INDUSTRY

Ceramic engineering has come a long way from its humble beginnings in the 1800's. Various government agencies, universities, businesses, and the general public have contributed to, or benefited from, the innovations made in advanced ceramics. The North American market for advanced structural ceramics was worth $ 2.7 million in 2007 and is predicted to grow to $ 3.7 million by 2012. This represented a compound annual growth of 6 percent. In 2010, the global advanced ceramics market was predicted to grow to $ 56.4 million by 2015. These predictions are evidence that the field of advanced ceramics continues to grow even in the face of economic difficulties.

Japan is the largest regional market for advanced ceramics, and the market is expected to grow as it is driven by rapid developments in the information technology and electronics industries. The United States remains the largest market for military applications of advanced ceramics. The United Kingdom along with India, China, and Germany are making significant progress in research and development.

Government and University Research. Fraunhofer-Gesellschaft, a German research organization that is partially funded by the German government, has been making great strides in developing new advanced ceramic products. Recently, the company developed pump components that use a diamond-ceramic composite it invented. Scientists at the company also created digital projector lenses using flat arrays of glass microlenses as well as a credit card-size platform that uses magnetic nanoparticles to detect sepsis. Each of these products was a result of the joint efforts of the various institutes that fall under Fraunhofer's substantial umbrella.

The United States Department of Energy is a major investor in advanced ceramics research and development. The agency recently awarded funding for three U.S.-China Clean Energy Research Centers (CERCs). These CERCs, which will focus on clean vehicle and clean coal technologies, will receive $50 million in funding (governmental and nongovernmental) over a period of five years.

A joint Chinese research team from Tsinghua University and Peking University has developed a lightweight, durable floating sponge for use in cleaning up oil spills. The sponge is made of randomly oriented carbon nanotubes and attracts only oil. It expands to hold nearly 200 times its weight and 800 times the volume of the oil. It automatically moves toward higher concentrations of the oil and can be squeezed clean and reused dozens of times.

Industry and Business. Nearly all the major industries and business sectors rely to some degree on advanced ceramics. Corning, Alcoa, Boeing, and Motorola are just a few of the major corporations that use advanced ceramics in their products. DuPont recently opened a photovoltaic applications lab at its Chestnut Run facility in Wilmington, Delaware. The lab will support materials development for the fast-growing photovoltaic energy market. PolyPlus in Berkeley, California, is developing lightweight, high-energy, single-use batteries that can use the surrounding air as a cathode. The company is currently developing these batteries for the government, but it expects them to be on the market within a few years. The drive toward efficient, inexpensive, clean technology ensures continued advances in research and development in the field.

CAREERS AND COURSE WORK

Ceramic scientists and engineers work in a variety of industries, as their skills are applicable in various contexts. These fields include aerospace, medicine, mining, refining, electronics, nuclear technology, telecommunications, transportation, and construction. A bachelor's degree is required for entrance into the above-mentioned fields. A master's degree in ceramic engineering qualifies the holder for managerial, consulting, sales, research, development, and administrative positions.

Pursuing a career in this field requires one have an aptitude for the sciences. Most colleges require that high school course work include four years of English,

---

Fascinating Facts About Ceramics

- Advanced ceramics are the basis of the lightest, most durable body armor used by soldiers for small- to medium-caliber protection.
- Snowboards have become stronger and tougher, thanks to special composite materials that combine innovative glass laminates and carbon fiber materials.
- Prior to the 1700's, potters were criticized for digging holes in the roads to obtain more clay. The name of this offense is still used: pothole.
- Ceramics tend to be hard but brittle. To reduce the incidence of cracks, they are often applied as coatings on other materials that are resistant to cracking.
- Clay bricks are the only building product that will not burn, melt, dent, peel, warp, rot, rust, or succumb to termites. Buildings built with bricks are also more resistant to hurricanes.
- Radioactive glass microspheres are being used to treat patients with liver cancer. The microspheres, which are about one-third the diameter of a human hair, are made radioactive in a nuclear reactor. They are inserted into the artery supplying blood to the tumor using a catheter, and the radiation destroys the tumor and causes only minimal damage to the healthy tissue.

---

four years of math (at least one of which should be an advanced math course), at least three years of science (one of which should be a laboratory science), and at least two years of a foreign language.

Typical course work for a bachelor's degree in ceramics engineering includes calculus, physics, chemistry, statistics, materials engineering, and glass engineering. It also includes biology, mechanics, English, English, four years of math (at least one of which should be an advanced math course), at least three years of science (one of which should be a laboratory science), and at least two years of a foreign language.

Typical course work for a bachelor's degree in ceramics engineering includes calculus, physics, chemistry, statistics, materials engineering, and glass engineering. It also includes biology, mechanics, English, four years of math (at least one of which should be an advanced math course), at least three years of science (one of which should be a laboratory science), and at least two years of a foreign language.

Typical course work for a bachelor's degree in ceramics engineering includes calculus, physics, chemistry, statistics, materials engineering, and glass engineering. It also includes biology, mechanics, English composition, process design, and ceramic processing. There are fewer than ten universities in the United States that offer bachelor's degrees in ceramic engineering.

These include the Inamori School of Engineering at Alfred University and Missouri University of Science and Technology. Other universities, such as Iowa State University and Ohio State University, offer a bachelor's degree in materials engineering with a specialization in ceramics engineering. Additionally, some schools offer a combined-degree option: Undergraduate students can combine undergraduate and graduate course work to earn a bachelor's and a master's or doctorate in materials engineering with a concentration in ceramics simultaneously.

However, many American universities offer a bachelor's degree in materials engineering with a specialization in ceramics engineering.

An alternative to acquiring a bachelor's degree in ceramics engineering is to acquire a degree in a related field and then pursue a master's degree in ceramics engineering. Examples of related fields include biomedical engineering, chemical engineering, materials engineering, chemistry, physics, and mechanical engineering.

About ten universities in the United States offer master's degrees in ceramic engineering. As a result, admission into these programs is extremely competitive. Graduate students focus primarily on research and development, though they are required to take classes. Doctoral candidates also focus on research and development and after they have been awarded their degree, they can choose to reach or continue working in research

SOCIAL CONTEXT AND FUTURE PROSPECTS

As mentioned in preceding sections, the advanced ceramics segment of the field, which is the primary focus of ceramic engineering, has plenty of room for growth. In 2008, there were about 24,350 materials engineers (including ceramic engineers) employed nationally; and a small number of ceramic engineers taught in colleges. The number of job openings is expected to exceed the number of engineers available.

Many of the industries in which ceramic engineers work—stone, clay, and glass products; primary metals; fabricated metal products; and transportation equipment industries—are expected to experience little employment growth through the year 2018. However, employment opportunities are expected to grow in service industries (research and testing, engineering and architectural). This is primarily because more firms, and by extension, more ceramic engineers, will be hired to develop improved materials for industrial customers.

Ezinne Amaonwu, LL.B., M.A.P.W

FURTHER READING

Barsoum, M. W. *Fundamentals of Ceramics.* London: Institute of Physics, 2003. This text provides a detailed yet easy to understand overview of ceramics engineering. It is a good introductory and reference text, especially for readers with experience in other fields of science such as physics and chemistry, as it approaches ceramics engineering from these viewpoints.

Callister, William D., Jr., and David G. Rethwisch. *Materials Science and Engineering: An Introduction.* 8th ed. Hoboken, N.J.: John Wiley & Sons, 2010. This text provides an overview of materials engineering, with useful content on ceramics.

King, Alan G. *Ceramic Technology and Processing.* Norwich, N.Y.: Noyes, 2002. Published posthumously by the author's son, this text provides a technical, detailed description of every step in the advanced ceramics-production process. It focuses on implementing the production process in a laboratory, describes common problems associated with each step, and offers solutions for these problems.

Kingery, W. D., H. K. Bowen, and D. R. Uhlmann. *Introduction to Ceramics.* 2d ed. New York: John Wiley & Sons, 1976. A well-written guide to the basics of ceramics engineering, with an exhaustive index and questions to test the reader's retention.

Rahaman, M. N. *Ceramic Processing and Sintering.* 2d ed. New York: Marcel Dekker, 2003. Provides a clear description of all the steps in ceramics processing.

WEB SITES

American Ceramic Society
http://ceramics.org

American Society for Testing and Materials
http://www.astm.org

ASM International
http://www.asminternational.org

CLINICAL ENGINEERING

FIELDS OF STUDY

Biomedical engineering; medicine; computer engineering; electrical engineering; mechanical engineering; systems engineering; biomechanics; mathematics; physiology; measurement; quality control; strategic planning; systems analysis.

SUMMARY

Clinical engineering is one of many subfields of biomedical engineering. In this subspecialty, practitioners support, monitor, and advance patient care by applying engineering and managerial skills to health care technology and delivery. A clinical engineer can function as a bridge between modern medicine and modern technology, interpreting and implementing sophisticated technology and complicated equipment to improve and enhance patient care and delivery of that care. Possibly the most important characteristic of a clinical engineer is the ability to identify and solve problems, translating solutions into usable clinical care to improve patient outcomes.

KEY TERMS AND CONCEPTS

- **Biomechanics:** Study of the mechanics of a living body and how the forces exerted on it, including those provided as part of clinical care, affect it and its function.
- **Biomedical Engineering:** Application of engineering techniques and principles to the understanding of biological systems, the development of therapeutic technologies and devices, and solutions to medical problems experienced by patients.
- **Clinical Outcome:** End outcome of patient's treatment; includes any problems experienced by delivery of health care.
- **Health Care Technology:** Equipment or device that interacts with the human body to deliver clinical care or diagnose a condition.
- **Human Factors:** Human-machine interface and the psychological, social, physical, biological, and safety factors involved in a system that interfaces with a human body.
- **Strategic Planning:** Process whereby an organization defines the direction it plans to take and allocates time, money, and people to accomplishing this goal.
- **Systems Analysis:** Method of analyzing a sequence of activities or operations to determine which are necessary and how they can best and most effectively be accomplished.

DEFINITION AND BASIC PRINCIPLES

Clinical engineering is a field involving many aspects of biomedical engineering. It is a combination of technology and medicine that focuses on the practical side of implementing sophisticated medical technology into clinical care of patients. A clinical engineer's job involves monitoring the interaction of humans and medical equipment to ensure that the machines or equipment are providing service to a patient in the most comfortable or efficient way.

As part of their job responsibilities, clinical engineers may train and supervise other equipment technicians who handle biomedical equipment in a hospital-type setting. In a manufacturing setting, a clinical engineer may advise manufacturers of medical devices on how to improve the design of such equipment to function better in a clinical situation. The field of clinical engineering tends toward redesign rather than new design, because it focuses on improving patient care.

Clinical engineers can be confused with biomedical equipment technicians (BMETs). The basic difference between these two jobs is that BMETs are usually responsible for service and repair of any medical equipment when it fails to function properly. Clinical engineers may supervise and, possibly, even perform these functions depending on the size of the hospital in which the clinical engineer is employed. Typically, however, the clinical engineer is more involved in developing equipment or suggesting changes to existing equipment to improve patient care and delivery of health care services.

BACKGROUND AND HISTORY

Medicine and engineering have been partners for many years, nearly from the beginning of both fields. In the modern age, this relationship possibly started in the early eighteenth century with English physiologist Stephen Hales and his work, which led to the invention of a ventilator and the discovery of blood pressure. The first meeting focusing on the collaboration between medicine and engineering is thought to have been held in 1948 by the Alliance for Engineering in Medicine and Biology. The term "clinical engineer" was first used in 1969 in a paper published by cardiologist Cesar Caceres, who is generally credited with coining the term.

The first formal accreditation process for clinical engineers was started in the early 1970's by the Association for the Advancement of Medical Instrumentation (AAMI). This body formed the International Certification Commission for Clinical Engineering and Biomedical Technology (ICC) to provide an avenue for clinical engineers to be formally certified. Another body, the American Board of Clinical Engineering, started a similar program, but this version of the program was based in academic institutions offering graduate degrees in clinical engineering. This program was dissolved in 1979, and those who had been certified under its program were absorbed into the ICC. However, in the early 1980's, this body had certified only 350 clinical engineers, and in the late 1990's it suspended its program.

In 2002, the American College of Clinical Engineering started a clinical engineering certification program sponsored by its Healthcare Technology Foundation. This body awarded certification to several individuals who had certified previously under other programs and has continued awarding certifications since then.

The field of clinical engineering as a subfield of bioengineering has had a relatively rocky history. The delineation between the fields has always been hazy and continues to be so. Many of those who function as clinical engineers have a background in biomedical engineering or one of its other subdisciplines.

The Healthcare Technology Foundation offers an Excellence in Clinical Engineering Leadership Award (ExCEL), which identifies clinical engineering professionals who demonstrate leadership in best practices in the management and advancement of health care technology. AAMI also has a Clinical Engineering Management Committee, which follows the career development of clinical engineers and has opportunities for career development and recognition. The *Journal of Clinical Engineering* is an industry publication that provides articles about career development and innovations in the field. There are also many journals covering the biomedical engineering field that include items of interest to clinical engineers.

HOW IT WORKS

Clinical engineers focus on the interaction of medical technology and the human body. They may perform systems analysis to understand how this interaction takes place and suggest or implement changes to make sure the equipment is providing the necessary service to a particular patient. For example, different types of equipment have certain sizes of tubing that enter a patient's body through a variety of methods. A clinical engineer may examine how this tubing enters the body of a particular patient and make suggestions that may include ways to incorporate larger or smaller tubing to ensure that the device is functioning correctly and providing the patient with the best service. If the clinical engineer sees that a different type of delivery, such as different size tubing or any other equipment modification, makes a functional difference to many patients, he or she may be involved in decisions to modify equipment or suggest a different type of equipment or another manufacturer who would help provide consistently better outcomes for patients.

A clinical engineer may also be involved in implants or prostheses and ensuring that these devices are working correctly and providing the patient with the best possible outcome. He or she may make suggestions as to device modifications that may help the patient. Engineers may also be involved with manufacturers who make these types of equipment to help incorporate changes.

APPLICATIONS AND PRODUCTS

Clinical engineers are necessary for the safe operation of the diagnostic and therapeutic equipment that is necessary for the operation of a health care facility to deliver top-notch care to its patients.

Evaluate. Clinical engineers evaluate equipment before it is purchased to ensure it meets standards of current patient care and safety. They may make recommendations on the equipment a health care system should purchase to provide patients with cutting-edge care. They may manage or coordinate service contracts or purchase negotiations and participate in strategic planning or systems analysis to assure that the recommended equipment is the best possible solution at the best possible price.

Install and Test. When new equipment is purchased, clinical engineers may be involved in the installation or in supervising the technicians who actually install the equipment. They make sure that the equipment meets the requirements of any regulatory agency and provide or supervise inspection, installation, and preventive or corrective maintenance.

Repair and Maintain. Clinical engineering departments oversee or coordinate the maintenance and continued functionality of technological medical equipment. This extends beyond the mere functionality of the equipment itself and into the realm of warranting that the equipment is functioning at the best level for patient care and comfort and suggesting any changes that would lead to improvement.

Improve. Clinical engineers are on the front lines of patient care and are charged with ensuring that technological advances improve that care. As such, they may make adjustments to equipment or even suggest changes to device makers to make sure that patient's needs are met. If incidents occur, clinical engineers investigate the cause and correct and improve the situation. They provide continuous quality assurance and control.

IMPACT ON INDUSTRY

The field of clinical engineering has had a wide impact on the delivery of health care. As health care delivery becomes more and more technologically advanced, the impact of this field will only increase. The United States and Europe have possibly the best and most comprehensive programs for those interested in biomedical engineering in general and clinical engineering specifically. The Food and Drug Administration, World Health Organization, Department of Veterans Affairs, and U.S. military organizations all have interest in and programs for biomedical and clinical engineers. The Indian Health Service, the Federal Health Program for American Indians and Alaska Natives, established a clinical engineering program in 1973 to service all Indian Health Service facilities. All these types of facilities have recognized the importance of clinical engineering and have jobs available for clinical engineers.

In addition to government agencies, many universities have clinical engineering departments that feed into health care centers. Some of these include Duke University, University of Kentucky, University of Connecticut, University of Arkansas, California State University at Long Beach, Wayne State University, and University of Toronto.

Worldwide, India continues to offer a number of clinical engineering programs at a university level. Australia also has recently developed programs to enhance the education of clinical engineers.

CAREERS AND COURSE WORK

A clinical engineer generally completes at least a bachelor's degree in a field of engineering then completes specific training related to clinical engineering. As this field is still evolving, there are not many programs that specifically relate to clinical engineering, so often clinical engineers have a background in biomedical engineering with further training in human factors engineering or a similar field.

After course work is completed, clinical engineers often complete an internship in a teaching hospital, which gives him or her a practical background in hospital functions. He or she may also become certified by the American College of Clinical Engineering, which involves written and oral tests to demonstrate specific knowledge and a portfolio review to determine applicable experience.

A clinical engineer may then begin a career in nearly any aspect of the medical field, including academia, design, or research, but usually this type of engineer puts his or her skills to work in a practical clinical setting, where he or she assesses, manages, and solves the problems that occur when complicated, technologically advanced medical equipment and the human body interact.

In a hospital setting, a clinical engineer may be the technological manager of medical equipment and related systems. This type of job requires knowledge of budget and finance, service agreements, data-processing systems, planning, assessing the effectiveness and efficacy of new equipment, quality control, incident investigation, training, and hospital operations in general. An executive-level job for a clinical engineer might be a chief technology officer position.

Another possible career for a clinical engineer involves research and development of new medical equipment. He or she may work in clinical trials during the development of new equipment to help evaluate and improve products or may even work in a medical-equipment company to develop and design equipment.

A clinical engineer may be asked to consult in a variety of situations, including acting as an expert witness or serve on governmental commissions overseeing the rapidly changing world of medical technology.

SOCIAL CONTEXT AND FUTURE PROSPECTS

When the term "clinical engineering" was first coined in the late 1960's, expectations were high as to the number of clinical engineers that would be needed in the upcoming years in the medical field. However, not nearly as many clinical engineering positions were created as expected. This lack of positions was partly due to the confusion as to what clinical engineering actually is and how it differs from biomedical engineering in general and biomedical equipment technicians specifically. These fields have been muddled and interchanged since their beginnings.

As health care technology becomes more and more complex and complicated, a clinical engineer can act as a translator between information systems, medical equipment, and the patient. Clinical engineers bring creativity, curiosity, and analytical, communication, and problem-solving skills to the medical field. They may serve on health care system-wide committees that relate to process and performance improvement, quality control and inspection, new technology purchases, and patient safety.

As evidence-based medicine becomes implemented into health care services, the convergence of information technology and medical technology becomes more and more important to control health care costs and improve clinical outcomes for patients. Clinical engineers are perfectly positioned to bridge the gap between developing and implementing emerging technologies in the intricate web of interconnected machinery and equipment.

The American College of Clinical Engineering is a major professional organization that promotes clinical engineering as a career of the future. It has established a code of ethics, advanced workshops, and certification programs to pursue recognition of the field. This organization has committees and task forces to educate and improve the field of clinical engineering as it relates to different pressing issues such as quality control and medical errors. This committee and its

members work internationally to improve knowledge of and education in this field.

Marianne M. Madsen, M.S.

FURTHER READING

Carr, Joseph J., and John M. Brown. *Introduction to Biomedical Equipment Technology*. 4th ed. Upper Saddle River, N.J.: Prentice Hall, 2001. Discusses quality improvement and continuous quality assurance for those who use, monitor, or modify medical equipment.

Chan, Anthony Y. K. *Biomedical Device Technology: Principles and Design*. Springfield, Ill.: Charles C. Thomas, 2008. Discusses basic medical equipment and the principles behind their use and operation.

David, Yadin, et al. *Clinical Engineering*. Boca Raton, Fla.: CRC Press, 2003. Essays about the discipline of clinical engineering from leading practitioners in the field; includes graphics, tables, figures, formulas, definitions of terms, and comprehensive index.

Dyro, Joseph F. *Clinical Engineering Handbook*. Burlington, Mass.: Academic Press, 2004. Comprehensive guide to clinical engineering, including history and worldwide practice in the field, ethics, and future trends.

Enderle, John, Susan Blanchard, and Joseph Bronzino. *Introduction to Biomedical Engineering*. 2d ed. Burlington, Mass.: Academic Press, 2005. A reference book for the field of biomedical engineering with practical examples.

McGill, Jennifer. "Clinical Engineering." In *Career Development in Bioengineering and Biotechnology*, edited by Guruprasad Madhavan, Barbara Oakley, and Luis Kun. New York: Springer Science + Business, 2008. Overview of clinical engineering as a profession including background, required course work, possible careers, and first-person stories about being a clinical engineer.

Montaigne, Fen. *Medicine by Design: The Practice and Promise of Biomedical Engineering*. Baltimore: The Johns Hopkins University Press, 2006. Nontechnical introduction to the field in which each chapter focuses on a different subspecialty of the field; includes stories and case studies.

Saltzman, W. Mark. *Biomedical Engineering: Bridging Medicine and Technology*. New York: Cambridge University Press, 2009. Basic overview of the field of biomedical engineering including discussion of range of fields available; includes further reading, suggested Web sites, and appendixes.

WEB SITES

American College of Clinical Engineering
http://www.accenet.org

American Society for Healthcare Engineering
http://www.ashe.org

Biomedical Engineering Society
http://www.bmes.org/aws/BMES/pt/sp/home_
page

Healthcare Technology Foundation
http://thehtf.org

See also: Biomechanics

CLONING

FIELDS OF STUDY

Biology; genetics; biotechnology; animal husbandry; aquaculture; veterinary medicine; developmental biology; embryology; biochemistry; theriogenology; agriculture; horticulture; botany; cell biology; conservation biology; viticulture; enology; medicine; pharmacology; microbiology; pomology; toxicology; zoology.

SUMMARY

Cloning is any type of biological reproduction that produces offspring that are genetically identical to their parents. Cloning occurs naturally, since many organisms routinely reproduce through natural cloning processes. Artificial cloning technologies include molecular cloning, which reproduces large quantities of discrete segments of DNA; reproductive cloning, which uses assisted reproductive technologies to produce animals that share the same desirable genetic characteristics as another living or previously existing organism; and therapeutic cloning, which uses the same techniques as reproductive cloning but instead derives useful cell lines from cloned embryos.

KEY TERMS AND CONCEPTS

- **Clone:** Organism whose genetic information is identical to the donor organism from which it was created, or a macromolecule that is an exact replicate of another macromolecule.
- **Embryonic Stem Cell:** Stem cell made from the inner cell mass of very young mammalian embryos.
- **Enucleation:** Microsurgical technique that removes nuclei from cells.
- **Genome:** Sum total of the DNA stored in the cells of an organism.
- **Parthenogenesis:** Biological process whereby an egg initiates embryonic development without having first undergone fertilization.
- **Pharming:** Use of genetic engineering to express cloned genes that encode useful pharmaceutical products in host animals or plants.
- **Pluripotent:** Ability of a cell to differentiate into any fetal or adult cell type.
- **Restriction Endonuclease:** Special enzyme that cuts DNA at a specific sequence motif.
- **Somatic Cell Nuclear Transfer:** Implantation of nuclei from somatic (body) cells into an egg to make a cloned embryo.
- **Transgenic Organism:** Biological entity that has had a foreign gene inserted into its genome. Commercially available transgenic organisms are often called genetically modified organisms (GMOs).

DEFINITION AND BASIC PRINCIPLES

Cloning is a means of producing biological organisms, cells, or DNA molecules that are genetically identical to their progenitors. There are natural forms of cloning and three main types of artificial cloning: molecular, reproductive, and therapeutic cloning.

Natural mechanisms of cloning occur in organisms such as bacteria that simply split or fragment into identical copies of themselves. In other organisms, reproductive cells, or gametes, undergo a process called parthenogenesis, in which they initiate development without the benefit of fertilization. Cloning is uncommon in mammals, but rarely, early mammalian embryos undergo a form of cloning called twinning, in which the embryo splits into two embryos, which develop into genetically identical twins.

Molecular cloning, also known as recombinant DNA technology or DNA cloning, involves the transfer of an isolated fragment of DNA from an organism of interest to a host cell that replicates it. Such isolated DNA fragments are known as cloned DNA or genes.

Reproductive cloning uses assisted reproductive technologies to generate animals with the same nuclear genome as another animal. The particular procedure used during reproductive cloning is called somatic cell nuclear transfer (SCNT). Cloned embryos are gestated in the womb of a surrogate mother until they come to term. Cloned organisms are not genetically modified organisms but are simply produced through a type of assisted reproduction.

Therapeutic cloning uses the same procedures as reproductive cloning; however, instead of transferring the cloned embryo into the womb of a surrogate mother, the embryo is further manipulated in the laboratory to make cell cultures of embryonic cells for basic or clinical research.

BACKGROUND AND HISTORY

Sea urchins were the first animal cloned in the laboratory. In 1894, Hans Dreisch isolated sea urchin embryo cells and watched them develop into small, separate larvae. In 1902, Hans Spemann used the same procedure, embryo splitting, to isolate cells from salamander embryos, which also developed into identical adult salamanders. In 1903, U.S. Department of Agriculture employee Herbert Webber coined the word "clon" for asexually produced cells or organisms, which later evolved into "clone." This term comes from the Greek *klon*, which means "trunk" or "branch." Horticulturists have used this term for more than a century, since an entire new plant can grow from a cutting, resulting in a plant that is genetically identical to the plant from which the cutting was taken.

In 1928, Spemann cloned salamanders by transferring the nucleus, the subcellular compartment that houses the chromosomes, from one salamander embryo into the egg of another. Since Spemann's seminal experiments, scientists have adapted nuclear transfer technology to clone other organisms. In 1952, frogs were cloned, and in 1963, the Chinese embryologist Tong Dizhou cloned a carp to produce the first cloned fish. During the 1980's and 1990's, sheep, cows, and mice were cloned. However, all these animals were cloned by using nuclei from embryos. In 1996, Ian Wilmut and his team at the Roslin Institute in Edinburgh, Scotland, cloned a sheep from an adult cell, demonstrating that adult cells could serve as the source of genetic material for animal clones. This technological feat was followed by the cloning of goats, mules, gaurs (an endangered species), horses, pigs, mouflons (a wild sheep), mice, rats, dogs, cats, water buffalos, camels, rabbits, deer, wolves, and African wildcats, and even embryos from nonhuman primates and humans.

HOW IT WORKS

Molecular Cloning. To clone a gene, the DNA of the model organism is selectively fragmented by enzymes called restriction endonucleases (REs) and inserted into another piece of DNA called a cloning vector. Cloning vectors are either small circles of DNA called plasmids, bacterial viruses, or bacterial or yeast artificial chromosomes. They ferry the DNA fragments from the genome of the model organism into a host cell (either a bacterium or yeast). This population of host cells collectively carries the entire genome of the model organism in small fragments, and is called a gene library.

To isolate a gene from a gene library requires a probe, which is a fragment of DNA or RNA of any length that has a sequence that is complementary to the sequence of the gene that is to be isolated. Probes can be made synthetically or can come from the genes of closely related organisms. By screening the gene library with the probe, the gene of interest is cloned, which simply means to isolate it from all the other sequences found in the genome of the model organism.

Alternatively, scientists can synthesize small strands of DNA called primers, whose sequences are complementary to different locations in the gene. These primers can be used to specifically amplify the gene from the library by means of a polymerase chain reaction (PCR). A polymerase chain reaction makes large quantities of the gene of interest from a very small amount of starting material, and the amplified DNA can also be cloned into a cloning vector or analyzed directly.

Reproductive Cloning. To clone an animal, mature eggs are isolated from females of the animal species that is to be cloned. The egg is enucleated by piercing it with a microscopically narrow (0.0002-inch-wide) glass tube that is used to vacuum out the egg nucleus. The enucleated egg is fused with a cell from the body of the animal to be cloned and activated with either chemicals or an electric current. This procedure is called somatic cell nuclear transplantation (SCNT).

After activation, the egg divides and grows like a newly formed embryo. However, if the animal is a mammal, the embryo can survive only for a limited period of time before it must implant into the inner layer of the mother's womb. Therefore, a surrogate female from the same species of the animal to be cloned, or a closely related species, is made pseudopregnant by feeding her hormones, and the embryo is released into her receptive womb, where it implants. Barring any technical or biological mishap, the cloned embryo will develop, and the process will result in a live birth.

Therapeutic Cloning. To make embryonic cell

cultures, cloned embryos are made by means of somatic cell nuclear transplantation. They are then either disassembled in the laboratory and used to establish embryonic cell cultures or gestated in a surrogate mother to the fetal stage, at which time the fetus is aborted, and cells from the fetus are used to establish fetal cell cultures.

By culturing specific cells from cloned embryos, scientists can make embryonic stem cell (ESC) cultures. During mammalian development, two distinct cell populations form after the first few days of embryonic development. The trophoblast, or the flattened, outer layer of cells, will eventually form the placenta and its associated structures. The inner cell mass (ICM) is the round, inner clump of cells that develop to form the embryo proper and a few structures associated with the placenta. If ICM cells are isolated and cultured on feeder cells, a layer of nondividing skin cells that secrete a cocktail of growth-promoting chemicals, the ICM cells will grow and spread over the surface of the culture dish. Such a culture is an embryonic stem cell culture, and these cells are pluripotent, which means that they can differentiate into any cell type in the adult body.

APPLICATIONS AND PRODUCTS

Molecular Cloning. Organisms that express cloned genes make many useful pharmaceuticals such as human insulin, growth hormone, clotting factors, fertility drugs, and vaccines. Cloned genes are also used to genetically screen individuals for genetic diseases. Pharmacologists even use cloned genes for pharmacogenetics, which screens patients for the presence of gene variants that can profoundly affect the efficacy and toxicity of particular drugs. This allows clinicians to tailor treatment to the exact genetic makeup of the patient to maximize treatment efficacy and minimize side effects. Such a strategy is called personalized medicine. Cloned genes are also used in gene therapy, which delivers cloned genes into the bodies of patients who suffer from genetic diseases in an attempt to cure them. Patients with cancer and inherited deficiencies of the immune system, blindness, and blood-based defects have been treated with gene therapy protocols.

In agriculture, the introduction of cloned genes into plants that are used as food crops has generated transgenic crops. These crops display several advantageous traits: reduced dependence on agrochemical applications (for example, Bt-corn and herbicide-resistant crops), increased nutritional value (for example, Golden Rice), increased resistance to environmental stresses, and reduced spoilage (for example, the Flavr Savr tomato).

Reproductive Cloning. When farmers identify food animals with desirable traits, they typically breed those animals as much as possible to improve the genetic quality of their herds and flocks. However, such prize animals inevitably die. Propagating these animals by reproductive cloning and mating them to as many animals as possible preserves the exceptional genetic content of a prize animal and allows it to produce far more offspring. This significantly raises the genetic quality of the flock or herd, and commercial dissemination of such cloned animals to other farmers raises the overall genetic quality of food animals. Reproductive cloning also eliminates the need for artificial insemination, which is often expensive and inconvenient.

Cloning effectively maintains high-quality animal stocks. Reproductive cloning of only the healthiest and most productive animals increases their numbers and improves the gene pool (sum total of genetic diversity) and overall health of food animals. This results in safer and healthier food and reduces the use of growth hormones, antibiotics, and other chemicals in the raising of animals.

In the field of conservation biology, the numbers of endangered species are often increased by captive breeding programs. However, not all endangered species can effectively breed in captivity. Reproductive cloning can aid in the preservation of those organisms that do not reproduce in captivity. Cloning can also resurrect genetic material from dead animals and potentially expand the gene pool of endangered species. In 2001, scientists at the University of Teramo, Italy, cloned the European mouflon, an endangered sheep, from cells sampled from a dead animal. When combined with other reproductive technologies, cloning can help save endangered species.

Cloned animals also serve as excellent research models. Because each cloned animal is genetically identical, experiments on cloned animals are devoid of differences caused by heterogeneous genetic backgrounds. Genetic manipulation of cloned animals allows researchers to modify genes of interest and more completely analyze their contribution to development and disease. Modifying particular genes of cloned animals also generates model systems for particular genetic diseases. Cloned, transgenic mice and cloned knockout mice, which have had a specific gene inactivated, are examples of the vast usefulness of such model systems.

Of enormous interest is modifying the genomes of cloned animals so that they can produce clinically and pharmaceutically significant products. By genetically modifying pigs, it is possible to make cloned pigs that contain organs that are fit for transplantation into humans (xenotransplantation). Also, producing antibodies, clotting factors, or even vaccines in the blood or milk of farm animals provides a means to mass-produce potentially expensive pharmaceutical agents at a fraction of the normal cost. This process is called pharming.

Therapeutic Cloning. Therapeutic cloning has tremendous potential for numerous clinical applications. Embryonic stem cells (ESCs) made from therapeutic cloning procedures are pluripotent. Therefore, injured, diseased, or failing tissues or organs could potentially be replaced by tissues or organs manufactured from embryonic stem cells in the laboratory or fetal cells from cloned fetuses. Furthermore, embryonic stem cells made from cloned embryos, or any tissues or organs fashioned from these cells, would not be regarded by the patient's body as foreign. Experiments in laboratory animals have shown that such scenarios are possible. Therapeutic cloning, coupled with embryonic stem cells technology, could christen a new era of regenerative medicine.

Embryonic stem cells from cloned embryos have toxicological applications. Toxicologists typically use laboratory animals or cultured cells to gauge the biological effects of natural or industrially produced molecules on human beings. Unfortunately, laboratory animals show limited utility as a model for human toxicology, and cultured cells do not represent the response of an organ or tissue to foreign molecules. Furthermore, neither of these model systems can assess the individual responses people will have to such molecules, because the genetic variation between individual humans causes differential responses to drugs, toxins, or environmental pollutants. However, cultured embryonic stem cells from cloned embryos can test the biological effects of drugs or environmental pollutants on cells made

---

Fascinating Facts About Cloning

- Scientists at Advanced Cell Technology used fetal heart muscle cells from cloned cow fetuses to reverse the effects of heart attacks in adult cows.
- The first cloned cat, CC (CopyCat), made at Texas A & M University in 2001, has a completely different personality than the donor cat. Even though CC is genetically identical to her donor, she is shy and timid whereas the donor cat is outgoing and playful.
- In 2008, BioArts International held an essay contest that invited people to argue why their dog should be cloned. The winner was Trakr, a German Shepherd police dog, who discovered the last survivor of the September 11, 2001, terrorist attacks on the World Trade Center in New York City.
- By cloning vaccines into plants, scientists have made edible vaccines against digestive diseases such as cholera, the Norwalk virus, some food poisonings, and enterotoxigenic *Escherichia coli*. These vaccines are not injected but rather eaten.
- Ingo Potrykus and Peter Beyer invented Golden Rice in the 1990's. This genetically engineered strain of rice produces beta-carotene, a precursor for vitamin A biosynthesis, which is not found in normal rice in appreciable quantities. Children who live in countries where rice is the main food staple are at higher risk for vitamin A deficiency, and Golden Rice was developed to help prevent this deficiency. Subsequent development has increased the nutritional value of Golden Rice even further. Even though the makers of Golden Rice want to give it to farmers completely free of charge, opposition to genetically modified organisms has prevented it from ever being cultivated for food.

---

from a specific person. In addition, because these cells can be differentiated into various tissues and even organs, they can be used to evaluate the individual and tissue-specific responses people might have to particular drugs or pollutants.

IMPACT ON INDUSTRY

Biotechnology companies that use cloning technology in the United States, Europe, Canada, and Australia reported a combined net profit of $3.7

billion in 2009. The United States has been the leader in cloning research, but there are many high-quality laboratories that study cloning technology in the United Kingdom, continental Europe, South Korea, Australia, China, Canada, Iran, and Israel.

Governmental Regulatory Agencies. The U.S. governmental agencies that regulate cloning are the Food and Drug Administration (FDA), Environmental Protection Agency (EPA), and the Department of Agriculture (USDA). The FDA regulates any foods made by genetically modified organisms. This agency concerns itself with only the safety of foods and not the manner in which they are made. The EPA has regulatory authority over all pest-resistant plants to ensure that genetically engineered crops do not adversely affect the environment. Field testing of genetically modified organisms is overseen by a division of the USDA, the Animal and Plant Health Inspection Service.

Government and University Research. The largest funder of cloning research is the National Institutes of Health (NIH). Other governmental funding agencies include the National Science Foundation and the USDA. The NIH not only funds the research of other laboratories but also houses many of its own laboratories, some of which use investigate cloning technologies.

Most of the cloning on university campuses is basic research. Many universities house cloning research centers on their campuses. In other cases, the cloning centers are extensions of state universities. The Roslin Institute, for example, where Dolly the cloned sheep was made, is an extension of the University of Edinburgh. Some universities have even formed partnerships with biotechnology companies that allow the company to work on university property in exchange for funds and increased collaboration between the company and the university.

Industry and Business. Biotechnology companies from all over the world participate in cloning research. Many of these companies have even formed associations. Ausbiotech represents Australian biotechnology companies, the European Federation of Biotechnology represents institutions from European and non-European countries, and BIOTECanada represents more than 250 Canadian biotechnology companies. These trade associations represent the interests of biotechnology to governing bodies.

Pharmaceuticals made by transgenic organisms that express cloned genes constitute the largest proportion of products developed and manufactured by biotechnology companies. These products include diabetes treatments, vaccines, cytokines (special proteins that signal to white blood cells), and other medicines. The demand for new medicines drives research and development in this area, and the understanding of the entire human genome is ushering in many previously unknown medical treatment strategies.

Agricultural biotechnology companies focus largely on developing new crops with improved characteristics. Some of their work is focused on making crops that can grow in underdeveloped countries. For example, Monsanto has begun field trials of a genetically engineered cassava plant that is virus resistant, less poisonous, and much more nutritious than its native counterpart. Some 800 million people globally rely on cassava as their main food staple, but viral infections, poor processing that tends to generate poisonous cyanides, and a lack of nutritional content tend to limit the food potential of cassava.

A few animal cloning industries market techniques for cloning pets. Genetic Savings and Clone is one such company. A related company, Viagen, which is part of Exeter Life Sciences, offers commercial cloning services for farm animals.

Human cloning industries are working toward therapeutic cloning strategies. Advanced Cell Technologies (ACT) works on human cloning. Based in Worchester, Massachusetts, ACT seeks to produce patient-specific stem cells from cloned embryos and cloned fetuses that can cure degenerative diseases without the risk of rejection by the immune system.

CAREERS AND COURSE WORK

Students who wish to pursue a career in cloning should possess a foundational grasp of biology and chemistry. Of cardinal importance is advanced course work in cell, molecular, and developmental biology. Many entry-level jobs exist in academic and industrial laboratories for those with a bachelor's degree in biology, biochemistry, or chemistry. Such jobs are usually laboratory technician positions, and good laboratory skills are required for such work. For those who wish to work as a leader of a research group, a Ph.D. in either cell or developmental biology, a D.V.M., or M.D. degree is required.

Cloning work requires highly skilled technicians

who are very dexterous and can look through microscopes for a long period of time while performing extremely fine manipulations. Such techniques often require many weeks of practice to perfect, and a patient, forbearing personality greatly helps people in the cloning field. Because cloning experiments are also very labor intensive, a collaborative mind-set is also very helpful.

Many scientists are involved in cloning work in industry, and several biotechnology companies have divisions that investigate cloning technology. Because such companies normally attempt to clone organisms for profit, cloning divisions of biotechnology companies usually examine improving the efficiency and cost-effectiveness of cloning procedures to standardize the manufacture of clones.

A large percentage of cloning scientists work in academic laboratories, where they devote their time to more basic research questions. Scientists in academic institutions must split their time between teaching and research. Other scientists who work for government laboratories such as the NIH can work on cloning, safety testing of cloned products, and other aspects of cloning without teaching responsibilities.

SOCIAL CONTEXT AND FUTURE PROSPECTS

Despite the reservations of some people, cloning is a part of everyday life. Many of the foods Americans consume contain some genetically engineered products. Physicians prescribe medicines, give vaccines, and apply other biological products made by genetically engineered microorganisms on a quotidian basis. Given the inroads molecular cloning has already made into people's lives, it is unlikely that people would suffer any revulsion from eating meat from cloned cattle or sheep or having their lives saved by the transplantation of an organ that came from a cloned pig. People would also probably not protest seeing cloned versions of endangered species at their local zoos.

Nevertheless, many people have raised concerns over cloning technologies. First, conservation biologists have suggested that cloning endangered species does not address the habitat destruction and environmental degradation that pushed these species to near extinction in the first place. Second, cloning only makes one species and does not re-create an ecosystem. For example, cloning cannot recapitulate a coral reef or an old growth forest. Thus, it is the

wrong solution for the problem.

Genetically modified organisms have become the focal point of concern for several environmental activism groups. Such groups oppose GMOs because they believe that the cloned genes inserted into them can spread to other species and cause severe environmental disruption and that genetically engineered foods have not been sufficiently tested and are potentially dangerous to human health.

The most contentious aspect of cloning technologies is human genetic engineering and reproductive cloning. Transhumanists are some of the most energetic proponents of human cloning and genetic enhancement. As a movement, Transhumanism regards infirmity, disease, aging, and death as undesirable and unnecessary and views science and technology as the means to defeat human limitations. Transhumanists' main argument for human cloning is that reproductive freedoms extend to everyone, and therefore, every human being has an inherent right to clone himself or herself.

Opponents of human cloning object to the manufacturing of human beings. Cloned children are made to be identical to someone else and therefore will always live in the shadow of the original person and never be completely the person they choose to be. These unreasonable expectations can psychologically damage them and violate their human dignity and individuality. Cloning would also alter the concept of human nature and therefore undermine the very foundation of liberal democracy.

In the future, the argument over cloning will not dissipate, but cloning research will certainly advance and provide more and more examples of the utility of this remarkable technology.

Michael A. Buratovich, B.S., M.A., Ph.D.

FURTHER READING

Alexander, Brian. *Rapture: A Raucous Tour of Cloning, Transhumanism, and the New Era of Immortality.* New York: Basic Books, 2004. A reporter examines the fringe groups that support human cloning and genetic enhancement and finds people who want to defeat the effect of entropy and live forever.

Fukuyama, Francis. *Our Posthuman Future: Consequences of the Biotechnology Revolution.* New York: Picador, 2003. A historian's admonition of the consequences of the biotechnology revolution and its potential to abolish human rights and erode the foundations of liberal democracy.

Mitchell, C. Ben, et al. *Biotechnology and the Human Good.* Washington D.C.: Georgetown University Press, 2007. A distinctly Christian assessment of the application of biotechnology to humans that remains optimistic but cautious and concerned.

Shanks, Pete. *Human Genetic Engineering: A Guide for Activists, Skeptics, and the Very Perplexed.* New York: Nation Books, 2005. A helpful explication of the science behind cloning, coupled with stern warnings against it, by a noted social activist.

Silver, Lee. *Challenging Nature: The Clash Between Biotechnology and Spirituality.* New York: Harper Perennial, 2006. A Princeton stem cell scientist explains the science behind biotechnology and stem cells. He offers some rather harsh critiques of more conservative thinkers who do not agree with his optimistic views of genetic enhancement and embryonic stem cells.

_____. *Remaking Eden: How Genetic Engineering and Cloning Will Transform the American Family.* New York: Harper Perennial, 2007. A very readable introduction to the science of cloning and genetic engineering by a noted mammalian embryologist, who believes that humans should be cloned and that people should welcome the profound changes that it will invoke within human societies.

Wilmut, Ian, Keith Campbell, and Colin Trudge. *The Second Creation: Dolly and the Age of Biological Control.* New York: Farrar, Straus and Giroux, 2000. The two researchers who made Dolly team up with a noted British science writer to give a personal but rigorous explanation and thoughtful examination of cloning. Contains a helpful glossary of terms.

WEB SITES

Human Cloning Foundation
http://www.humancloning.org

MedlinePlus
Cloning
http://www.nlm.nih.gov/medlineplus/cloning.html

National Human Genome Research Institute
Cloning
http://www.genome.gov/25020028

See also: Agricultural Science; Animal Breeding and Husbandry; Bioengineering; DNA Sequencing; Genetically Modified Organisms; Genetic Engineering; Reproductive Science and Engineering; Stem Cell Research and Technology.

COMPUTED TOMOGRAPHY

FIELDS OF STUDY

Physics; physiology; medical physics; radiology; mathematics; computer science; electrical engineering; biomedical engineering; electronics; health care; medicine.

SUMMARY

Computed tomography (CT) is an imaging modality that relies on the use of X rays and computer algorithms to provide high-quality image data. CT scanners are an integral part of medical health care in the developed world. Physicians rely on CT scanners to acquire important anatomical information on patients. CT images provide a three-dimensional view of the body that is both qualitative and quantitative in nature. CT scans are often used in the diagnosis of cancer, stroke, bone disorders, lung disease, atherosclerosis, heart problems, inflammation, and a range of other diseases and physical ailments, such as a herniated disc and digestive problems.

KEY TERMS AND CONCEPTS

- **Attenuation Coefficient:** Physical quantity that characterizes the response of a specific material to incoming radiation and the associated reduction in intensity.
- **Back Projection:** Process of reversing the acquired projection measurements from the detector to reconstruct an image set.
- **Ionization:** Process through which an atom loses one or more of its electrons or acquires an electron. The resulting atom is ionized while the radiation that caused the ionization is termed ionizing radiation.
- **Nondestructive Evaluation (NDE):** Process of testing the integrity of a material according to specified standards; usually done using a form of X-ray imaging. In contrast, destructive testing is the process of placing the test material under extensive forces (hence destructive) to evaluate the integrity and strength of the material.
- **Photon:** Elementary particle that carries electromagnetic radiation from all wavelengths, including those of visible light and X rays. Photons travel at the speed of light and interact with matter in different ways depending on their characteristic frequency.
- **Picture Archiving And Computer System (PACS):** Computer-based system of managing radiological information, including patient-specific data and images. PACS is used in all imaging modalities to facilitate the transfer, retrieval, and presentation of images.
- **Tomography:** Method of X-ray photography in which a single section or plane is imaged while eliminating other sections or planes. When this process is done through a computer, it is known as computed tomography.
- **X Ray:** Form of electromagnetic radiation that is characterized by its ionizing potential and short wavelength, ranging from 0.01 to 10 nanometers.

DEFINITION AND BASIC PRINCIPLES

Computed tomography (CT; also known as computer-aided tomography, or CAT) is an imaging modality that uses X rays and computational and mathematical processes to generate detailed image data of a scanned subject. X rays are generated through an evacuated tube containing a cathode, an anode, and a target material. The high voltage traveling through the tube accelerates electrons from the cathode toward the anode. This is very similar to a light bulb, with the addition of a target that generates the X rays and directs them perpendicularly to the tube. As electrons interact with the target material, small packets of energy, called photons, are produced. The photons have energies ranging from 50 to 120 kilovolts, which is characteristic of X-ray photons. The interaction of X-ray photons with a person's body produces a planar image with varying contrast depending on the density of the tissue being imaged. Bone has a relatively high density and is more readily absorbed by X rays, resulting in a bright image on the X-ray film. Less dense tissues, such as lungs, do not absorb X rays as readily, and as a result, the image produced is only slightly exposed and therefore dark.

In computed tomography, X rays are directed toward the subject in a rotational manner, which generates orthogonal, two-dimensional images. The X-ray tube and the X-ray detector are placed on a rotating gantry, which allows the X rays to be detected at every possible gantry angle. The resulting two-dimensional images are processed through computer algorithms, and a three-dimensional image of the subject is constructed. Because computed tomography relies on the use of ionizing radiation, it has an associated risk of inducing cancer in the patient. The radiation dose obtained from CT procedures varies considerably depending on the size of the patient and the type of imaging performed.

BACKGROUND AND HISTORY

X rays were first discovered by Wilhelm Conrad Röntgen in November, 1895, during his experiments with cathode-ray tubes. He noticed that when an electrical discharge passed through these tubes, a certain kind of light was emitted that could pass through solid objects. Over the course of his experimentations with this light, Röntgen began to refer to it as an "X ray," as x is the mathematical term for an unknown. By the early 1900's, medical use of X rays was widespread in society. They were also used to entertain people by providing them with photographs of their bodies or other items. In 1901, Röntgen was awarded the first Nobel Prize in Physics for his discovery of X rays. During World War II, X rays were frequently used on injured soldiers to locate bullets or bone fractures. The use of X rays in medicine increased drastically by the mid-twentieth century.

The development of computer technology in the 1970's made it possible to invent computed tomography. In 1972, British engineer Godfrey Hounsfield and South African physicist Allan Cormack, who was working at Tufts University in Massachusetts, independently developed computed tomography. Both scientists were awarded the 1979 Nobel Prize in Physiology or Medicine for their discovery.

HOW IT WORKS

The first generation of CT scanners was built by Electric and Musical Industries (EMI) in 1971. The CT scanner consisted of a narrow X-ray beam (pencil beam) and a single detector. The X-ray tube moved linearly across the patient and subsequently rotated to acquire data at the next gantry angle. The process of data acquisition was lengthy, taking several minutes for a single CT slice. By the third generation of CT scanners, numerous detectors were placed on an arc across from the X-ray source. The X-ray beam in these scanners is wide (fan beam) and is covered by the entire area of detectors. At any single X-ray emission, the entire subject is in the field of view of the detectors, and therefore linear movement of the X-ray source is eliminated. The X-ray tube and detectors remain stationary, while the entire apparatus rotates about the patient, resulting in a drastic reduction in scan times. Most medical CT scanners in the world are of the third-generation type.

Mode of Operation. The process of CT image acquisition begins with the X-ray beam. X-ray photons are generated when high-energy electrons bombard a target material, such as tungsten, that is placed in the X-ray tube. At the atomic level, electrons interact with the atoms of the target material through two processes to generate X rays. One mode of interaction is when the incoming electron knocks another electron from its orbital in the target atom. Another electron from within the atom fills the vacancy, and as a result, an X-ray photon is emitted. Another mode of interaction occurs when an incoming electron interacts with the nucleus of the target atom. The electron is scattered by the strong electric field in the nucleus of the target atom, and as a result, an X-ray photon is emitted. Both modes of interaction are very inefficient, resulting in a considerable amount of energy that is dissipated as heat. Cooling methods need to be considered in the design of X-ray machines to prevent overheating in the X-ray tube. The resulting X-ray beam has a continuous energy spectrum, ranging from low-energy photons to the highest energy photons, which corresponds with the X-ray tube potential. However, since low-energy photons increase the dose to the body and do not contribute to image quality, they are filtered from the X-ray beam.

Once a useful filtered X-ray beam is generated, the beam is directed toward the subject, while an image receptor (film or detector) is placed in the beam direction past the subject to collect the X-ray signal and provide an image of the subject. As X rays interact with the body's tissues, the X-ray beam is attenuated by different degrees, depending on the density of the material. High-density materials, such as bone, attenuate the beam drastically and result in a brighter X-ray image. Low-density materials, such as lungs, cause minimal attenuation of the X-ray beam and appear dark on the X-ray image because most of the X rays strike the detector.

Image Acquisition. In CT, X-ray images of the subject are taken from many angles and reconstructed into a three-dimensional image that provides an excellent view of the scanned subject. At each angle, X-ray detectors measure the X-ray beam intensities, which are characteristic of the attenuation coefficients of the material through which the X-ray beam passes. Generating an image from the acquired detector measurements involves determining the attenuation coefficients of each pixel within the image matrix and using mathematical algorithms to reconstruct the raw image data into cross-section CT image data.

APPLICATIONS AND PRODUCTS

The power of computed tomography to provide detailed visual and quantitative information on the object being scanned has made it useful in many fields and suitable for numerous applications. Aside from disease diagnosis, CT is also commonly used as a real-time guide for surgeons to accurately locate their target within the human body. It is also used in the manufacturing industry for nondestructive evaluation of manufactured products.

Disease Diagnosis. The most common use for CT is in radiological clinics, where it is used as an initial procedure to evaluate specific patients' complaints or symptoms, thereby providing information for a diagnosis, and to assess surgical or treatment options. Radiologists, medical professionals specialized in reading and analyzing patient CT data, look for foreign bodies such as stones, cancers, and fluid-filled cavities that are revealed by the images. Radiologists can also analyze CT images for size and volume of body organs and detect abnormalities that suggest diseases and conditions involving changes in tissue density or size, such as pancreatitis, bowel disease, aneurysms, blood clots, infections, tuberculosis, narrowing of blood vessels, damaged organs, and osteoporosis.

In addition to disease diagnosis, CT has been used by private radiological clinics to provide full-body scans to symptom-free people who desire to obtain a CT image of their bodies to ascertain their health and to detect any conditions or abnormalities that might indicate a developing problem. However, the use of CT imaging for screenings in the absence of symptoms is controversial because the X-ray radiation used in CT has an associated risk of cancer. The dose of radiation deposited by a CT scan is between fifty and two hundred times the dose deposited by a conventional X-ray image. Although the association between CT imaging and cancer induction is not well established, its casual use remains an area of considerable debate.

Guided Biopsy. A biopsy is a time-consuming, sometimes inaccurate, and invasive procedure. Traditionally, doctors obtained a biopsy (sample) of the tissue of interest by inserting a needle into a patient at the approximate location of the target tissue. However, real-time CT imaging allows doctors to observe the location of the biopsy needle within the patient. Therefore, doctors can obtain a more accurate tissue biopsy in a relatively short time and without using invasive procedures.

CT Microscopy. The resolution of clinical CT scanners is limited by practical scan times for the patients and the size of X-ray detectors used. In the case of small animals, higher resolutions can be obtained by use of smaller detectors and longer scan time. The field of microCT, or CT microscopy, has rapidly developed in the early twenty-first century as a way of studying disease pathology in animal models of human disease. Numerous disorders can be modeled in small animals, such as rats and mice, to obtain a better understanding of the disease biology or assess the efficacy of emerging treatments or drugs. Traditionally, animal studies would involve killing the animal at a specific time point and processing the tissue for viewing under the microscope. However, the development of CT microscopy has allowed scientists and researchers to investigate disease pathology and treatment efficacy at very high resolutions while the animal is alive, reaching one-fifth of the resolution of a light microscope.

Nondestructive Evaluation. Computed tomography has gained wide use in numerous manufacturing industries for nondestructive evaluation of composite materials. Nondestructive evaluation is used to inspect specimens and ensure the integrity of manufactured products, either through sampling or through continuous evaluation of each product. CT requirements for nondestructive testing differ from those for medical imaging. For medical imaging, scan times have to be short, exposure to radiation has to be minimal, and patient comfort throughout the procedure must be taken into consideration. For nondestructive evaluation, patient comfort and exposure to radiation are no longer important issues. However, keeping scan times short is advantageous, especially for large-scale industries. Furthermore, the X-ray energy for scanning industrial samples can vary significantly from the energy used for patient imaging, since patient composition is primarily water and industrial samples can have a wide range of compositions and associated densities. Engineers have custom-designed CT scanners for specific materials, including plastics, metals, wood, fibers, glass, soil, concrete, rocks, and various composites. The capability of CT to provide excellent qualitative image data and accurate quantitative data on the density of the specimen has made it a powerful tool for non-destructive evaluation. CT is used in the aerospace industry to ensure the integrity of various airplane components and in the automotive industry to evaluate the structure of wheels and tires. In addition to industrial applications, CT is commonly used in research centers to further their imaging and analytical power.

IMPACT ON INDUSTRY

CT has revolutionized many industries by providing detailed information on specific physical parameters of materials along with high-quality three-dimensional images. CT has been successfully incorporated into the workflow of many companies—including aerospace and aviation, plastic, casting, electronic, medical device, pharmaceutical, and dentistry industries—and research laboratories, museums, and archaeological centers. The continued development of CT scanners will allow for improved spatial resolution, reduced scan times, and customized designs.

Academic Research. Computed tomography has found many scientific applications in academic and research centers across North America and Western Europe. Many engineering departments have dedicated research centers that incorporate CT for research into nondestructive testing.

Medical research into the application of CT for robotic surgery and biopsy is becoming increasingly popular. Furthermore, the manufacturing of compact CT scanners used for animal imaging has allowed the incorporation of CT into basic science laboratories for investigating disease mechanisms and progression and for evaluating new drugs. Using CT microscopy, researchers can track the progression of disease and the effects of treatment in animals while they are alive, rather than killing the animals to obtain the necessary tissues for analysis with conventional light microscopy.

Major Corporations. Some of the major corporations that have been involved in the manufacture and development of CT scanners are GE Healthcare, Hitachi Medical Corporation, Philips Medical Systems, Siemens Medical Systems, Shimadzu Medical Systems, and Toshiba Medical Systems.

CAREERS AND COURSE WORK

CT is an imaging modality that exists primarily in the health care sector and is available in every major hospital and in most small hospitals in the developed world. CT is also widely distributed in hospitals in the developing world. Experts in the operation and maintenance of CT scanners who live in developed nations are often recruited to work in less-developed nations.

Those who wish to pursue careers in CT can take many paths. A degree in physics or engineering with a specialization in biomedical or electrical engineering can provide a solid grounding in the electronics and hardware of CT. Engineers can find work in hospitals or in industries that design and manufacture CT scanners. Degrees in mathematics and computer science provide the necessary background for working with image reconstruction algorithms and advancing software-related operation of CT scanners. Computer programmers can find work in the research and development sector of CT industries.

A graduate degree in medical physics provides theoretical and practical experience in medical imaging, from data acquisition to image reconstruction, interpretation, and troubleshooting. Medical physicists often work in imaging facilities and hospitals, where they usually supervise the personnel operating CT scanners. A degree in medicine with specialization in radiology provides theoretical and practical experience in understanding human anatomy, pathology, and physiology; in interpreting CT images; and in diagnosing specific disease conditions. Technical colleges can provide education in the operation of CT scanners. Graduates with a technical degree in CT imaging often work as technologists in hospitals, where they are responsible for patient scheduling and CT operation. CT is invaluable in the medical sector and career   prospects   have   been   good.

Fascinating Facts About Computed Tomography

- Computed tomography (CT) is one of the milestone discoveries of the twentieth century that are based on X-ray imaging.
- CT provides three-dimensional and sectional views of the human body. CT images of a patient's body can be correlated to a cross section of the human body in real life. The ability of CT to see inside the body has revolutionized modern medicine.
- CT can produce images of tissues in small animals, such as mice and rats, that approximate the resolution of a microscope. CT microscopy has become an integral tool for understanding disease pathology and evaluation of treatment efficacy in animal models of disease.
- The automotive industry uses CT scanners to evaluate the integrity of parts, such as rims and tires, without destroying them.
- CT produces not only high-quality images but also quantitative data on the density of the scanned material.
- In the United States, more than 6,000 CT scanners are being used to create more than 62 million CT scans per year.

SOCIAL CONTEXT AND FUTURE PROSPECTS

The number of CT scanners and scans being performed have risen considerably since the 1980's. By the early twenty-first century, there were more than 6,000 scanners in the United States, and more than 62 million CT scans were being performed every year. The rapid and wide acceptance of CT scanners in health care institutions has sparked controversy in the media and among health care practitioners regarding the radiation doses being delivered through the scans. Risk of cancer induction rises with increased exposure to radiation, and this risk has to be carefully weighed against the benefits of a CT scan. The issue of cancer induction is more alarming when CT procedures are performed on young children or infants. Studies have recommended that CT scanning of children should not be performed using the same protocols as used for adults because the children are generally more sensitive to radiation than adults. More strict federal and state regulations are being instituted to better control the use of CT in health care.

Ayman Oweida, B.Sc., M.Sc.

FURTHER READING

Aichinger, H., et al. *Radiation Exposure and Image Quality in X-ray Diagnostic Radiology: Physical Principles and Clinical Applications.* New York: Springer, 2003. Provides the reader with a strong foundation in radiation physics, especially in relation to X-ray medical imaging. It also describes many of the concepts of X-ray production, as well as its interaction with matter.

Bossi, R. H., K. D. Friddell, and A. R. Lowrey. "Computed Tomography." In *Non-destructive Testing of Fibre-reinforced Plastics Composites,* edited by John Summerscales. New York: Elsevier Science, 1990. Covers various applications of CT in relation to composite material. Provides a basic review of various CT concepts important to nondestructive testing of materials, as well as various illustrations and data on CT measurements in fiber-reinforced plastics.

Brenner, David, and Eric Hall. "Computed Tomography: An Increasing Source of Radiation Exposure." *New England Journal of Medicine* 357, no. 22 (2007): 2277-2284. An analytical overview of the use of CT and the risk of cancer induction from radiation exposure. Illustrations and statistical data on various CT procedures.

Hsieh, Jiang. *Computed Tomography: Principles, Design, Artifacts, and Recent Advances.* 2d ed. Bellingham, Wash.: International Society for Optical Engineering, 2009. A comprehensive overview of the fundamentals of CT in medical imaging. Reviews concepts in physics and mathematics that are necessary for a detailed understanding of CT operation and development and image analysis.

Otani, Jun, and Yuzo Obara, eds. *X-ray CT for Geomaterials: Soils, Concrete, Rocks.* Rotterdam, the Netherlands: A. A. Balkema, 2004. This collection of lectures and papers presented at a workshop on X-ray CT for geomaterials in 2003 is a broad and technical overview of the numerous applications of CT in imaging composite materials. Abundant diagrams and illustrations of the various applications.

WEB SITES
MedlinePlus
CT Scan
http://www.nlm.nih.gov/medlineplus/ency/
article/003330.htm

U.S. Food and Drug Administration
What Is Computed Tomography?
http://www.fda.gov/Radiation-EmittingProducts/
RadiationEmittingProductsandProcedures/Medi-
calImaging/MedicalX-Rays/ucm115318.htm

See also: Magnetic Resonance Imaging; Radiology and Medical Imaging; Ultrasonic Imaging.

CRIMINOLOGY

FIELDS OF STUDY

Criminal justice; law enforcement; law; psychology; philosophy; sociology; anthropology; political science; social work; forensic science; psychiatry.

SUMMARY

Criminology is the study of crime causation and control. Criminologists attempt to elucidate the characteristics and motivation of criminals—from the most mundane and innocuous (scofflaws and petty thieves) to the most obscure and heinous (mass murderers and serial killers)—and how they differ from noncriminals. Why do people commit crimes? Are people born criminals, or do their experiences dictate whether they will break the law? Can crime be controlled or eliminated? Why are some communities safer than others? What causes changes in crime rates? Criminologists search for answers to these and related questions.

Understanding who commits crimes and why can directly affect the passage of laws and the operations and practices of the criminal justice system, which comprises the agencies authorized to respond to criminal acts (law enforcement, courts, and corrections). Criminological theories can provide the basis for the creation of new and more effective programs and interventions designed to help lower crime rates, thereby making communities safer.

KEY TERMS AND CONCEPTS

- **Biological Theory:** Explanations of crime that focus on genetically based, inherited, or physical characteristics.
- **Crime:** Act that constitutes the violation of a law or criminal statute.
- **Crime Typology:** Classification of crimes according to different types of offenses and criminal motivations as well as offender and victim characteristics.
- **Criminal Career:** Frequency, type, and duration of criminal activities committed by an individual offender over a period of time.
- **Criminality:** Strong psychological propensity or tendency to commit crime.

- **Desistance:** Cessation of criminal activity after a period of offending.
- **Deterrence:** Attempt to prevent crime by making punishments match the severity of the crime committed.
- **Positivism:** Application of empirical scientific methods to study crime and criminals.
- **Psychological Theory:** Explanations of crime that focus on early-childhood experiences and personality traits.
- **Sentence:** Punishment that follows a conviction for a crime, such as a fine, community supervision, incarceration, or the death penalty.
- **Social Theory:** Explanation of crime that focuses on environmental processes or structures and the relationships between and among social groups.
- **Theory:** Set of testable, interrelated propositions intended to describe, explain, and predict an event or activity.

DEFINITION AND BASIC PRINCIPLES

Criminology draws from writings in the areas of law and crime control as well as from research in the social and behavioral sciences, such as sociology (particularly the subspecialty of the sociology of social-anthropology deviance), medicine (particularly the subspecialty of psychiatry), and psychology (particularly the subspecialty of clinical psychology).

BACKGROUND AND HISTORY

Edwin H. Sutherland, who came to be known as the "dean of American criminology," wrote the first modern-day textbook on criminology in 1924. *Criminology* paved the way for twentieth century criminological academic pursuits, and his contributions to the discipline are still relevant. Sutherland recognized that crime was a complicated phenomenon affected by political, social, economic, and geographic variables, and he rejected the notion that criminals were simpleminded. In two of his classic textbooks, *The Professional Thief* and *White Collar Crime*, Sutherland presented in-depth ethnographic analyses of a professional thief and white-collar criminals, demonstrating the humanity and complexity of criminals.

HOW IT WORKS

Criminologists are theoreticians and researchers. In studying the distribution and causes of crime, criminologists conduct research and other quantitative studies, analyzing large public surveys and data sets and asking offenders to report their histories, experiences, and decisions to pursue criminal activities. Criminologists also use qualitative research, such as ethnographic studies, to explore firsthand how people move into and out of criminal lifestyles and develop their skills and expertise in criminal specialties.

Criminologists develop theories using a deductive or an inductive approach. The deductive method of theory development starts with general propositions that are used to generate hypotheses, that is, testable questions. These questions are examined in carefully conducted studies, and the data are tested to see if they support the researcher's theory. The inductive method of theory development starts with data used to generate propositions, which are the building blocks of theory. As the theory becomes increasingly elaborate, more data are collected in a process of theory refinement. As both of these approaches suggest, the key elements of a good criminological theory are testability and empirical support.

Criminological theories are primarily derived from schools of thought that provide basic frameworks for formulating theories and their propositions as well as testable hypotheses. The schools also provide a general perspective and level of analysis for studying crime. Theories of crime are built on the concepts, variables, methodologies, and traditions of a particular discipline. The major disciplines in criminology include sociology, psychology, and biology. Sociology-based theories examine social structures and processes as well as the relationship between social groups. One sociological theory, strain theory, suggests that people without access to legitimate means to achieve success turn to crime. These people are feeling a certain amount of strain (economic, social, emotional, mental) in their lives, and they turn to crime to alleviate it. If members of society are unable to attain their goals because of lack of resources (money, education) those members may engage in illegitimate actions to achieve their goals. Some will retreat from mainstream society to join deviant subcultures, such as gangs and communities of drug users.

Psychology-based theories examine individual differences in characteristics and traits. Michael Gottfredson and Travis Hirschi, authors of *A General Theory of Crime*, posit that people commit crimes because of poorly developed or low self-control, which explains all delinquent and criminal behaviors and is the single, most important cause of crime at the individual level. People who are in control of themselves are able to consider the consequences of their behavior and are careful and deliberate in their decision-making. Such individuals self-monitor effectively and conform in socially desirable ways. In contrast, people with low self-control are present-oriented, impulsive, reckless, and lack empathy. They prefer physical rather than mental activities and engage in antisocial behaviors to meet their selfish needs. Self-control is learned during childhood and once learned it is difficult to change.

Biology-based theories examine the genetic contribution to crime, known as heritability, by employing several different methods of research. For example, twin studies compare identical and fraternal twins raised in the same household to determine the degree to which they are similar regarding crime—that is, their concordance rates. If both twins in identical pairs, who share all the same genes, are more likely to engage in crime than those in fraternal pairs, who share half of their genes, the evidence suggests that crime has a genetic component.

Another type of biology-based investigation strategy is the adoption study. In this type of research, investigators compare the criminality of parents and children who were separated at birth. The children in these investigations were reared by adoptive parents and had no contact with their biological parents. If the criminal involvement of the adoptive children was more like that of their biological parents than their adoptive parents, the data suggest that crime is affected more by nature (biology) than by nurture (environment).

Criminological theories are published in books. Even more important, the work of criminologists is published in journals, which are repositories for the body of knowledge in the field. Publication in a journal lends a degree of prestige and respectability to a study because it has been scrutinized by the author's peers prior to publication. Journals that publish the research of criminologists include *Criminology, Justice Quarterly, Crime and Delinquency, Criminal Justice and Behavior, Journal of Research in Crime and Delinquency,*

Journal of Criminal Law and Criminology, International Journal of Offender Therapy and Comparative Criminology, and *Journal of Quantitative Criminology.*

APPLICATIONS AND PRODUCTS

Several schools of thought are responsible for most of the major theories of crime. These include the Classical School, the Positivist School, and the Chicago (Ecological) School.

Classical School. This school of thought came about during the age of Enlightenment and was fueled by major reforms in penology and law in which imprisonment replaced corporal punishment as the predominant sanction. Cesare Beccaria (who argued against the death penalty in *On Crimes and Punishments,* published in 1764), Jeremy Bentham (English philosopher and the inventor of the panopticon, an early, revolutionary prison design), and other Classical School theorists argued that all people—including criminals—act on the basis of free will. Classical criminologists advocated for a humane penology.

In the Classical framework, people are considered fundamentally rationale beings who maximize pleasure and minimize pain as well as weigh the costs and benefits of each action in a process of rational choice. Thus, people decide to commit crimes using a straightforward, cost-benefit analysis. Beccaria wrote that threat of sufficient punishment can deter people from committing crimes, and that the most effective punishments are swift and consistent, and the severity is commensurate with the seriousness of the crime.

In addition to being against the death penalty, Beccaria also stood against torture and the inhumane treatment of prisoners—punishments he considered to be nonrational (ineffective) deterrents. Bentham similarly considered punishment for crimes only as useful as they served as deterrents for future offenses.

Positivist School. Formed in the late nineteenth century, the Positivist School maintained that criminal behavior stemmed from factors beyond a person's control, both internal (biological and psychological) and external (sociological). Positivists believed that crime and criminals could be best understood through the application of scientific techniques, and that biological, personal, and environmental factors determine criminal behaviors in a cause-and-effect relationship. Biological positivism was first proposed by the Italian physician Cesare Lombroso. Lombroso described criminals as "atavistic throwbacks," who acted on their primitive urges because they had failed to evolve fully. Their brains were underdeveloped, rendering them incapable of comporting their behaviors to the rules and regulations of society. They could also be differentiated from noncriminals by their physical characteristics. For example, murderers had glassy eyes, aquiline noses, and thin lips. Enrico Ferri, a student of Lombroso, believed that social as well as biological factors played a role in criminality. He argued that criminals should not be held responsible for their crimes when the factors causing their criminality were patently deterministic.

As an example of psychological positivism, Hans Eysenck, a British psychologist and author of *Crime and Personality* (1964), contended that certain psychological traits were at the root of what drove one to crime. His model of personality also contained the dimension of psychoticism, which consists of traits similar to those found in profiles of people with psychopathy, a set of behaviors and characteristics (the lack of empathy, conscience, and impulse control) that predispose people to commit serious crimes. Eysenck's model also acknowledged the influence of early parental socialization on childhood and adult tendencies to engage in criminal behaviors. His approach bridged the gaps among biological (William H. Sheldon), environmental (B.F. Skinner), and social learning-based (Albert Bandura) explanations of criminal behavior.

One of the tenets of sociological positivism is that societal factors (poverty, membership in subcultures, low levels of education) create a predisposition to crime. Belgian mathematician Adolphe Quetelet was one of the first to explore the relationship between crime and societal factors. He reported that poverty and low educational levels were important components in crime. British statistician Rawson W. Rawson linked population density and crime rates with statistics. He theorized that crowded cities create an environment conducive to crime and violence. French sociologist Émile Durkheim viewed crime as an inevitable consequence of the uneven distribution of wealth among the social classes.

Chicago (Ecological) School. Between 1915 and the early 1940's, sociological research in the United States was dominated by various academic disciplines at the University of Chicago—most notably, political s

Fascinating Facts About Criminology

- Early criminologists meticulously measured the facial and bodily features of criminals and non-criminals to identify distinguishable differences between the two groups. Based on so-called somatotyping criteria, men were classified along certain physical dimensions (endomorph, mesomorph, and ectomorph) that purportedly made them more or less susceptible to criminal behavior.

- Spurious research in the 1960's suggested that violent male criminals had an extra Y chromosome, which made them hypermasculine and contributed to their criminality.

- Crime rates have been declining steadily since the early 1990's, especially in big cities. However, the public remains highly fearful of crime.

- The incarceration rate in the United States is the highest in the world among industrialized countries. The explosion in the prison population in the past thirty years is not due to an increase in the number of crimes committed but rather to changes in laws and crime-control policies.

- The United States has 5 percent of the world's population but nearly 25 percent of the number of people incarcerated worldwide.

- No theory of crime has ever fully explained criminal behavior. Each theory has limitations, and no criminologist has ever formulated a fully integrated theory of criminal behavior. The causes of crime are simply too multifarious and complex to elucidate within a single theoretical framework.

- The relationship between the economy and crime is complicated and confounding. Crime has gone up during periods of economic prosperity, down during periods of economic woe, and vice versa.

- Called the "immigrant paradox," data suggest that immigrants (both legal and illegal) are less likely to commit crimes than are native-born Americans.

their talents and applied their intellects to examine the harsh sequelae of urbanism, particularly those problems generated by inner-city living. The Chicago School brought to its research on urbanism many innovative, trenchant, and eclectic methods of social scientific analyses. Members of the Chicago School used a wide array of methodologies in their research, which was conducted in the field (streets, housing developments, opium dens, brothels, alleys, and parks) rather than in the sterility of a library, laboratory, or faculty office.

The Chicago School is exemplified in the work of urban sociologists at the University of Chicago, notably Robert E. Park and Ernest Burgess. In their 1925 book, *The City*, Park and Burgess identified five zones that often form as cities develop. The business district is in the center of the Park-Burgess model, and the zones that appear concentrically include the factory zone, the zone of transition (which was often crime prone), followed by the working-class, residential, and commuter zones. While researching juvenile delinquency in the 1940's, Chicago School sociologists Henry McKay and Clifford R. Shaw discovered that most troubled and troublesome adolescents were concentrated in the zone of transition.

Chicago School sociologists adopted a social-ecology approach in their study of cities, postulating that urban neighborhoods with high levels of poverty often experience a breakdown in social structure and institutions (families and schools). In the ecological model, such areas are hotbeds of social pathology, including disorganization, disorder, and decline—all of which render social institutions unable to control behavior, resulting in a downward spiral of neighborhood decay and creating an environment ripe for criminal and other deviant behaviors. Such neighborhoods tend to experience high rates of population turnover, which does not allow informal social structures to develop adequately and, in turn, makes it difficult to maintain social order in the community.

IMPACT ON INDUSTRY

Criminology has greatly affected crime-control policies and practices. Theories of crime have implications for punishments and crime-prevention techniques. The adoption of utilitarian-based (Classical) approaches to punishment led to the creation of determinate sentencing structures in which sanctions—in particular, lengths of incarceration—are

science and sociology. To journalists, social reformers, and sociologists, the everyday struggles associated with living in Chicago became a microcosm of the human condition and an encapsulation of human suffering. In this atmosphere of urban despair and blight, many creative scholars combined pre-established and based on the seriousness of the offense. The punishment fits the crime in terms of severity, while characteristics of the offender are given little weight in the sentencing decision.

Another application of the Classical School of criminology stems from rational-choice theory, which argues that criminals, like everyone else, weigh costs and benefits when deciding whether to commit a crime. An economic calculus is their primary decision-making modality. Rational-choice theories also suggest that increasing the likelihood of being caught through added surveillance (cameras), the visible presence of police officers or security guards, more voluminous street lighting, and other environmental measures is effective in reducing crime.

Based on the Positivist School of thought, social-disorganization theory suggests strategies to reduce crime by strengthening communities. The Chicago Area Project (CAP) exemplifies this strategy and has achieved legendary status in the annals of criminology. Founded in 1934 by Chicago School sociologist Clifford R. Shaw, CAP has a long history of community building in which low-income residents assume responsibility for addressing critical neighborhood problems, such as delinquency, gang violence, substance abuse, and unemployment. Skeptical of psychological explanations of delinquency and programs aimed solely at reforming individuals, Shaw created CAP as a new form of grassroots community organization. Its goal was to prevent delinquency by encouraging local residents' active participation in community self-renewal. CAP rests on a powerful network of organizations and special projects that promote positive youth development and prevent juvenile delinquency through community building.

CAREERS AND COURSE WORK

Educational programs in criminology focus largely on crime and deviant behavior and include courses in criminal law and procedures, psychology, sociology, research methodology, and statistics. Criminology students also study the components and operations of the criminal justice system. Academic programs in criminology differ from those in criminal justice, which focus more on the criminal justice system itself and often provide training and job placement for particular careers in the field.

Criminologists are mostly doctoral-level academicians and policy analysts. The individuals who work in the criminal justice system are most properly described as criminal justice professionals or practitioners. These include a wide variety of personnel: judges, state's attorneys, public defenders, police officers, probation officers, parole agents, correctional officers, and victims' advocates. Criminologists can also be confused with criminalists, who specialize in the collection and analyses of the physical evidence deposited at a crime scene (also known as ballistic, fingerprint, shoe-print experts; crime laboratory technicians; and crime scene investigators and photographers).

Primarily involved in theory construction, research, teaching, writing, and policy analysis, criminologists contribute a great deal of expertise to the study of policing, police administration and policies, juvenile justice and delinquency, corrections, correctional administration and policies, drug addiction and enforcement, criminal subcultures, typologies of criminals, and victimology. In addition, they examine the various biological, sociological, and psychological factors related to criminal trajectories, which are pathways into and out of criminal behavior. Some criminologists also engage directly in community initiatives as well as in evaluation and policy projects with local, state, and federal criminal justice agencies.

Criminologists conduct their own research while teaching courses in psychology, legal studies, criminal justice, criminology, sociology, and pre-law at two- and four-year colleges and universities. Others work for state and federal justice agencies as policy advisers or researchers. These agencies include the Bureau of Justice Statistics, the Bureau of Justice Assistance, the National Institute of Justice, the National Institute of Corrections, the Office of National Drug Control Policy, the National Criminal Justice Reference Service, the Office of Juvenile Justice and Delinquency Prevention, and the Federal Bureau of Investigation. Criminologists can also be found in the private sector, where they provide consulting services on various issues such as crime statistics, juvenile and adult correctional programming, crime prevention and security protocols, legal reform, and justice initiatives, or they can work for large think tanks such as RAND or Abt Associates, pursuing policy-oriented research and evaluations.

Becoming a criminologist requires a minimum of a master's degree in criminology. However, criminologists who work in university settings typically possess doctoral degrees and postdoctoral training. In addition, criminologists can receive their doctoral degrees in other disciplines, such as sociology, psychology, political science, or public policy, and specialize in crime and criminals during the course of their advanced studies.

SOCIAL CONTEXT AND FUTURE PROSPECTS

The precipitous increases in crime that accompanied the creation of the urban ghetto and the alienation of the immigrant populations during the turbulent 1960's, as well as the crime-accelerating effects of unstable drug markets in the 1990's, provided further impetus for criminological theory and research and the continual search for solutions to Americans' crime problems. Despite the admonitions of European criminologist Hermanus Bianchi, who warned criminologists to ply their trade away from politics, modern criminologists are interested in informing political agendas and public policies relating to crime control and justice issues. Indeed, many practitioners increasingly seek to incorporate into their everyday activities evidence-based practices grounded in scientific knowledge and solid theorizing. Nonetheless, as historian Lawrence Friedman notes, enduring cultural taboos have become obstacles to the implementation of lasting and effective crime-control efforts; these include widespread resistance to adopting strict gun-control laws, legalizing/decriminalizing drugs, and increasing taxes to pay for social and rehabilitative programs.

Arthur J. Lurigio, Ph.D.

FURTHER READING

Beccaria, Cesare. *On Crimes and Punishments.* Translated by David Young. Indianapolis, Ind.: Hackett, 1986. This is the definitive, seminal work that launched the Classical School of criminology and remains relevant to the field despite its origins nearly 250 years ago.

Gaines, Larry K., and Roger Leroy Miller. *Criminal Justice in Action: The Core.* Belmont, Calif.: Wadsworth, 2010. An engaging textbook that covers the criminal justice process from arrest to incarceration and describes its three major components of operation: law enforcement, courts, and corrections. Contains text boxes with "fast facts," landmark criminal cases, and information about careers in criminal justice. Many illustrations and photographs enliven the text.

Gottfredson, Michael R., and Travis Hirschi. *A General Theory of Crime.* Stanford, Calif.: Stanford University Press, 1990. The authors explore the real reasons people commit crimes and bring together classic and modern theories on criminology in an effort to provide answers to the questions criminologists have been asking for centuries.

Hayward, Keith, Shadd Maruna, and Jayne Mooney, eds. *Fifty Key Thinkers in Criminology.* Abingdon, England: Routledge, 2010. This comprehensive collection includes essays on the earliest proponents of the science (Cesare Beccaria and Jeremy Bentham) as well as contemporary practitioners (Frances Heidensohn and Travis Hirschi).

Jacoby, Joseph, ed. *Classics of Criminology.* 3d ed. Long Grove, Ill.: Waveland Press, 2004. An anthology of the most influential papers published in the field from the 1700's through the 1990's, exploring numerous criminological theories and featuring the most prominent theoreticians in the discipline.

Schmalleger, Frank. *Criminology Today.* 6th ed. Upper Saddle River, N.J.: Prentice Hall, 2011. A highly readable tome that contains a wealth of colorful illustrations and boxed text that provides interesting case studies and contemporary takes on criminology

Shoemaker, Donald J. *Theories of Delinquency: An Examination of Explanations of Delinquent Behavior.* 6th ed. New York: Oxford University Press, 2010. This well-researched book presents the best known sociological theories of crime, which are each described in a separate chapter, as well as chapters that summarize psychological and biological theories of crime and issues in the field, such as female delinquency and radical criminology.

Sutherland, Edwin H., and Donald R. Cressey. *Criminology.* Philadelphia: Lippincott, 1978. This is an updated edition of the first recognized textbook in the field, which defined the discipline and articulated basic concepts and methodologies for the study of crime and criminals.

WEB SITES

American Society of Criminology
http://www.asc41.com

American Sociological Association
http://www.asanet.org

Federal Bureau of Investigation
http://www.fbi.gov

U.S. Department of Justice
http://www.justice.gov

See also: Penology; Psychiatry.

D

DEMOGRAPHY AND DEMOGRAPHICS

FIELDS OF STUDY

Sociology; political science; social psychology; statistics; probability; data collection; data modeling; economics; anthropology; biology; population dynamics; urban studies; public policy.

SUMMARY

Demography and demographics both involve the study of human populations. Demography refers to the discipline of measuring and analyzing factors about groups of people, such as group size, composition, density, and growth. These factors may be studied at a single point in time or followed over years. Demographics commonly refers to the factors themselves but can also mean the study of populations as a discipline. Many demographers work in professional fields related to marketing. Others seek to understand populations to predict political activity such as voting patterns, public health trends, or the expansion and decline of cities.

KEY TERMS AND CONCEPTS

- **Census:** Gathering of data on all members of a population.
- **Cohort:** Population that experienced the same type of life-shaping event, such as birth or marriage, within a defined time period.
- **Migration:** Movement of a population from one place to another to establish new homes.
- **Morbidity:** Rate at which disease and injuries affect a population.
- **Mortality:** Rate at which deaths occur within a population.
- **Natality:** Rate at which births occur within a population.
- **Population:** Group of people studied by demographers that share a characteristic, such as living in a region.
- **Survey:** Gathering of data on selected, but not all, members of a population.

DEFINITION AND BASIC PRINCIPLES

Demography is the study of human populations and the ways in which they grow and change over time. Although a population might be a group of people as small as a single rural village, most demographic studies look at groups living in a metropolitan area, a region, or even a country or continent. Demographers are also likely to use data categories such as age, family status, ethnicity, household size, education, and income level in answering questions about population change.

Three primary factors govern changes in population size. Fertility is the rate at which new members of the population are born. Migration shows the rate at which members move in and out of the population during their lifetimes. Mortality is the rate at which population members die. Demography is the social science in which these factors, along with many others, are analyzed to better understand why a population is growing or declining.

Demographics is a term that refers to the data gathered by demographic analyses. Government agencies are some of the leading users of demographic data as they create and implement public policy. Companies also rely on demographic data as they design products and services for growing markets.

BACKGROUND AND HISTORY

Philosophers and scholars have discussed ideas about population growth for thousands of years. In ancient Greece, Plato and Aristotle advocated the concept that civilizations should strive to reach certain population levels determined by the maximum quality of life that could be achieved under such numbers. Both Plato and Aristotle examined questions about migration, population and the environment, and fertility control. Similar issues were explored by writers in China and India.

Demography was further developed in Europe in the sixteenth through eighteenth centuries under a school of thought known as mercantilism. Under mercantilism, population growth was seen as a necessary force behind the increase of a country's trading power and wealth. In 1798, this philosophy was challenged by British professor Thomas Malthus, who anonymously published a pamphlet in which he argued that human biology would, by nature, create a future society that would be overpopulated and struggle with heightened problems such as poverty and famine.

Many nineteenth century demographers such as William Farr and Louis-Adolphe Bertillon made advancements in the field through their work in public health. By the late 1800's, statistical analysis had become a key component of demographic study. Actuarial science, led by insurance providers to better understand mortality and risk, also contributed to the discipline.

HOW IT WORKS

Demographic studies seek to answer questions about human populations. To conduct a study, demographers must define their questions and design a survey or census to collect the necessary data. Once the data are gathered, demographers apply statistical tools to measure and analyze the information. The results allow demographers to determine whether their question has been answered or whether further research is needed.

Question Definition. Every demographic study begins with a question about a group of people that can be answered by data analysis. A well-designed study needs a question that is thoroughly defined. For example, a city's board of education might ask the question, "How many students are expected to attend public school in the city each year over the next ten years?" Demographers working for the board would choose the items to be measured in the study and define each item in detail, such as the number of days in a year that a student would need to be present at school in order to be counted.

Study Design and Data Collection. Once the question is defined, demographers decide what sources of data to use and how to collect the information. Many demographic studies rely on census data, which involves gathering information on each member of a population. Census data are thorough but difficult and expensive to compile. Many demographers instead rely on surveys, which collect information from people chosen as a representative sample of the population. The survey data are analyzed statistically to infer conclusions about the population as a whole. In the example of the school board, the demographers might gather data through a census or survey of households with at least one child below the age of nine.

Measurement and Analysis. Demographers use tools that store the information gathered in surveys or censuses and assist with statistical analysis. The demographers in this example might transfer their survey data into a database or other application that would allow them to create multiyear growth forecasts. Additional information might come from population growth projections for the city or region. The demographers would use this information to compensate for areas not covered by the survey, such as households where there are not yet any children or households living outside the city limits at the time of the survey's mailing. Once these sources of data are combined and analyzed, demographers document their findings and make recommendations about additional research, if needed.

The types of data gathered by demographers often fall into categories. Fertility measures the rate at which a population grows because of the number of babies being born to population members. Migration is the rate at which people join or leave a population by moving to new homes inside or outside its borders. Immigration refers to people moving into an area, while emigration describes people leaving the area. Some studies follow patterns known as domestic migration, in which people move from one part of an area to another. Mortality measures a population's rate of death. Aside from these categories, demographers seek to understand factors such as the distribution of age and sex across a population. These factors are particularly important when using population data to build forecasts.

APPLICATIONS AND PRODUCTS

Demographic data are used in a wide range of contexts. Some of the most common tools for demographic research are population census data, social surveys, and commercial and marketing surveys.

Population Census Data. Countries throughout the world track the number of people living within their borders. Most national governments have a department or agency that uses a tool such as a census to gather information. Census data are used by other government agencies to track population growth, decline, and migration. Aggregate data from censuses are often published by governments for use by businesses and the public. These data are released in the form of reports and electronic databases. In many countries, a full population census is conducted roughly once every five to fifteen years. Countries such as France gather population data on a yearly basis. Other countries such as Germany do not conduct a census of all residents but use statistical sampling to estimate information about their populations.

Social Surveys. Social surveys are conducted on selected members of a population to gather data that will be used to understand the population as a whole. Although they present a higher rate of error than a census, they cost less and can be carried out more quickly. Many large-scale social surveys focus on specific issues such as education, housing, employment, or health. Others are focused by geography. The American Community Survey from the Census Bureau is one of the best known social surveys in the United States. The United Nations conducts a large number of social surveys that are focused by both topic and region, such as the economic and social survey of Asia and the Pacific.

Commercial and Marketing Surveys. Demographic information about groups of people is useful to companies as well as government agencies. Commercial and marketing surveys gather some of the same types of data as social surveys. Companies focusing on a new target market or expanding their hold on an existing group of customers are likely to track factors such as average age, ethnic background, family structure, education, and income levels. When analyzed, these factors are used to guide the company's decisions about which products and services to offer and how to communicate them to the customers most likely to buy them. Market data publisher Nielsen's Pop-Facts reports are an example of demographic data available commercially.

Areas of Focus. Professionals working within the field of demography often hold jobs that require them to focus by subject area. Common areas in which demographers are employed are public health and epidemiology, immigration and emigration

---

Fascinating Facts About Demography and Demographics

- The terms "demography" and "demographics" come from the Greek words for "people" and "writing."
- Greek philosophers Plato and Aristotle were in favor of giving rewards such as medals to parents for increasing a city's population.
- Mercantilism, a popular school of thought in sixteenth- through eighteenth-century Europe, linked population growth to economic strength. It supported the idea that more people—and therefore more labor—enabled a country to arm for war more quickly than its enemies could.
- Characters in Aldous Huxley's novel *Brave New World* (1932) wear contraceptive Malthusian belts, named after Malthus, as a form of population control.
- The U.S. POPClock and World POPClock are maintained by the U.S. Census Bureau on the agency's Web site. The clocks estimate the U.S. and global population and are updated once a minute.
- Aside from counting people, the U.S. Census Bureau studies many things about the way they live, work, and talk. The bureau found that the number of U.S. residents who speak a language other than English at home rose 140 percent from 1980 to 2007. The population as a whole grew 34 percent.

---

policy, and urban and environmental planning.

Public Health and Epidemiology. Demographers working in the field of public health and epidemiology seek to answer questions such as which public health problems pose the greatest risk to a population, based on its age structure and sex ratio, and what factors will define a population's public health needs in ten to twenty years. They conduct surveys to gather data, which they may combine with other sources, such as statistics from hospitals, to assess which health issues are most critical for a selected group of people.

One of the largest subfields of demography in public health concerns fertility and family planning. In many of the world's most developed economies, the overall birth rate is high enough to support stable growth without posing unusual challenges to a country's public resources such as health care. However, developing economies, such as those in many African and Asian countries, have been experiencing rapid growth in their populations because of high fertility levels. Demographers studying these trends are likely to be involved with health issues concerning mothers and children. They may also be employed by government agencies as advisers on the development of public policies shaping long-term population growth, such as economic incentives for having more or fewer offspring.

Immigration and Emigration Policy. Demographers advising on immigration and emigration policy examine issues such as how much of a country's population growth is caused by people moving into the country and why a country's most educated residents are leaving for other countries. Although questions of migration are important in a wide range of contexts, public policy is developed by government agencies and specialized research firms, where demographers focusing on the issues are most likely to work. Migration policy has become a topic of great public interest in the twenty-first century in areas such as the European Union, where residents may live and work not only in the country of which they are a citizen but also in member countries. Regions such as the Caribbean and the countries that were formerly Soviet republics or satellite states are studying emigration to better understand how to retain their most skilled citizens, who have moved in large numbers to countries with stronger economies in a trend known as brain drain.

Urban and Environmental Planning. Population change has a profound effect on the environment, especially where housing and employment are concerned. As people live and work in an area, they influence its natural resources and the types of infrastructure needed to sustain further growth, such as houses, roads, schools, stores, and offices. Demographers who work in the field of urban and environmental planning hold a variety of jobs. They may work for city planning agencies or provide research for national offices such as the U.S. Department of Housing and Urban Development. They also may specialize in specific categories of growth, such as families with young children, to advise boards and departments of education in planning school systems. Many demographers choose to work in private industry for companies such as homebuilders, where they support the design of new types of houses or make forecasts about where and when population growth is likely to be highest.

IMPACT ON INDUSTRY

Demography is, by nature, a multidisciplinary field. Few areas of government and industry are not affected by population change in some way. Changes in age ranges, family types, sex ratios, and migration patterns have an impact on public services of all kinds as well as the types of products and services that can be sold to the members of a population.

The Population Association of America, the leading professional organization for demographic specialists in the United States, provides a forum in which demographers can share their expertise. Although the association does not release statistics on the types of jobs held by its members or the sectors in which they work, the organization has active connections in a number of areas. These relationships provide insight into the influence held by demographers in many professional spheres. They include the Ad Hoc Group for Medical Research, a group that provided $35 billion in federal funding for the National Institutes of Health in fiscal year 2011; the Census Project, a team of public- and private-sector leaders in demography that were active in lobbying for $14 billion to support the work of the 2010 U.S. Census; the Coalition for Health Funding, made up of fifty national professional associations in the health care field that seek funding for the U.S. Public Health Service and its many agencies; and the Council of Professional Associations on Federal Statistics, a group of associations, businesses, research institutions, and individual members who foster discussion and collaboration on uses of the statistics released by agencies of the federal government.

Demographic data, while useful, can be difficult and expensive to compile. Information must be mined from existing resources such as government reports, or it must be collected directly from population members through censuses and surveys. Because the same demographic data can be valuable in multiple applications, an industry of demographic research firms has evolved to serve the needs of companies. These research firms gather information and sell it to clients in the form of raw data, published studies, or customized consulting services.

Some providers of demographic data are large and

multinational in scope. Leading commercial firms of this type include Nielsen, Information Resources, Kantar, and Ipsos. These firms concentrate on data about not only consumers themselves but also their behavior, specifically in purchasing goods and services. Nielsen and several of its competitors also monitor the ways in which consumers use media such as television and the Internet. Firms such as Nielsen invite consumers, known as members, to participate in surveys. Information about buying habits and media use is gathered through a variety of channels. These channels range from the voluntary answering of questions posed to consumers to the automated collection of data through monitoring devices. In 2009, Nielsen, the largest consumer demographics research firm worldwide, reported $4.8 billion in revenues.

Other demographic research providers are smaller and concentrate on specific populations. These populations might be defined by age, such as those studied by Teenage Research Unlimited or Childwise, or by ethnic background, such as Multicultural Marketing Resources. Specialist firms provide data to companies that allow products and services to be tailored to the needs of a more focused market.

CAREERS AND COURSE WORK

Demographers have a wide range of career options in government, industry, and nonprofit settings. Because there is no association that certifies college and university programs in demography, few numbers are available when it comes to students or full-time employees in the field. However, the professional demands of studying population change make a bachelor's degree or graduate degree an asset when looking for a job.

Students interested in demography—also known as population studies at many institutions—are likely to approach it as a topic within a major in sociology, anthropology, political science, economics, geography, or biology. Course work varies widely, depending on the student's focus and department. Most students interested in demography take classes on topics such as population theory and human development. Multiple courses on research methods, study design, and statistical analysis are required. Issue-specific classes cover topics such as poverty, reproductive policy and health, and the environmental impact of human populations. Students who earn

graduate degrees often take managerial positions or work as professors or researchers at academic institutions.

Government agencies and nonprofit organizations employ many demographers. The U.S. Census Bureau is one of the best-known agencies, but population studies play a key role in aspects of government ranging from health care policy to land use. In industry, jobs in marketing and new product development often call for experience in gathering and interpreting demographic data. This experience can come through college-level course work as well as internships with consumer products manufacturers.

SOCIAL CONTEXT AND FUTURE PROSPECTS

The demand for professionals with a background in demography and population studies is expected to increase significantly in the 2010's. Job candidates with bachelor's or graduate degrees and years of focused experience in demographic analysis are likely to benefit most. There will also be many opportunities for new college graduates with course work in demography.

The U.S. Bureau of Labor Statistics reports that sociologists and political scientists, a group to which demographers belong, will see job growth of 21 percent from 2008 to 2018—a rate considered much faster than average. Sociologists earned a median annual income of $68,570 in 2008, and political scientists reported a median income of $104,130. Many of the opportunities open to demographic specialists are not called "demographer" but instead carry titles such as "policy analyst" or "research analyst."

Government agencies will continue to hire demographers as elected officials look for more efficient ways in which to use tax funds. Demographers can help evaluate whether a particular program is reaching its target audience most effectively. Consumer products manufacturers and marketing firms will hire more demographic specialists as new product development and advertising focus on smaller customer niches. Research centers and management consulting firms are likely to be interested in demographers for their experience working with complex data on population change.

Julia A. Rosenthal, B.A., M.S.

FURTHER READING
Buchholz, Todd G., and Martin Feldstein. "Malthus:

Prophet of Doom and Population Boom." Chapter 3 in *New Ideas from Dead Economists: An Introduction to Modern Economic Thought*. New York: Penguin, 2007. A brief overview of the life of Thomas Malthus and his landmark contributions to demographics.

Magnus, George. *The Age of Aging: How Demographics Are Changing the Global Economy and Our World*. New York: John Wiley & Sons, 2008. A discussion of the effects of decreasing birth rates and increasing average ages worldwide.

Malthus, Thomas. *An Essay on the Principle of Population*. New York: Oxford University Press, 2008. A scholarly edition of Malthus's most famous work.

Poston, Dudley L., Jr., and Leon F. Bouvier. *Population and Society: An Introduction to Demography*. New York: Cambridge University Press, 2010. A college-level textbook that covers demography as a science, written for a nonspecialist audience.

Siegel, Jacob S., and David A. Swanson, eds. *The Methods and Materials of Demography*. 2d ed. San Diego, Calif.: Elsevier, 2004. Provides the tools and techniques used by demographers.

Yaukey, David, Douglas L. Anderton, and Jennifer Hickes Lundquist. *Demography: The Study of Human Population*. 3d ed. Long Grove, Ill.: Waveland Press, 2007. Examines the causes of changes in the human population and the effects of these changes. Features an international focus and long-term projections.

WEB SITES
The Census Project
http://www.thecensusproject.org

Council of Professional Associations on Federal Statistics
http://www.copafs.org

Population Association of America
http://www.populationassociation.org

Population Reference Bureau
http://www.prb.org

U.S. Bureau of the Census
http://www.census.gov

See also: Land-Use Management; Reproductive Science and Engineering.

DENTISTRY

FIELDS OF STUDY

Biology; health science; physiology; anatomy; pharmacology; biochemistry; chemistry; mathematics; microbiology; physics.

SUMMARY

Dentistry involves the diagnosis, treatment, and prevention of disorders and diseases of the teeth, mouth, jaw, and face. Dentistry includes instruction on proper dental care, removal of tooth decay, teeth straightening, cavity filling, and corrective and reconstructive work on teeth and gums. Dentistry is recognized as an important component of overall health. Practitioners of dentistry are called dentists. Dental hygienists, technicians, and assistants aid dentists in the provision of dental care.

KEY TERMS AND CONCEPTS

- **Appliance:** Removable restorative or corrective dental or orthodontic device.
- **Bite:** Contact of the upper and lower teeth; also known as occlusion.
- **Caries:** Tooth decay; also known as cavities.
- **Cleaning:** Removal of plaque and tartar from the teeth, generally above the gum line.
- **Enamel:** Hard ceramic that covers and protects the exposed part of the tooth.
- **Gingiva:** Soft, pink tissue surrounding the base of the teeth; also known as gums.
- **Permanent Teeth:** The thirty-two teeth that appear after the loss of the primary teeth, beginning around the age of six years; also known as adult teeth.
- **Plaque:** Sticky film of food particles, bacteria, and saliva that forms on the teeth and can eventually turn into tartar.
- **Primary Teeth:** The first set of teeth that appear between the ages of six months and one year that help children learn to speak and chew; also known as baby teeth or deciduous teeth.
- **Pulp:** Soft, inner part of the tooth that contains nerves and blood vessels.
- **Root:** Part of the tooth that is embedded in the gums.
- **Tartar:** Hard deposit that adheres to teeth and attracts plaque; also known as dental calculus.

DEFINITION AND BASIC PRINCIPLES

Dentistry is a branch of medicine that focuses on diseases and disorders of the teeth, mouth, face, and oral cavity. Dentistry includes examining the teeth, gums, mouth, head, and neck to evaluate dental health. The examination may include a variety of dental instruments, imaging techniques, and other diagnostic equipment. Dentistry involves diagnosing oral or dental diseases or disorders and formulating treatment plans. Dentistry is instrumental in teaching patients about the importance of maintaining oral health and instructing patients on proper oral hygiene techniques.

Although dentistry is an independent health care field, it is not entirely detached from other health care services and collaboration between dentistry and other health care providers ensures positive outcomes for patients. Dentists often see patients more often than physicians and may be the first to diagnose systemic diseases, including inflammatory conditions, autoimmune diseases, and cardiovascular risk factors. Dentists also work with pharmacists to prescribe the best antibiotics or anesthetics for dental patients, as well as to understand how certain medications affect dental care and oral health.

Dentistry not only prevents and treats serious oral health disorders but also provides cosmetic services to enhance facial features and correct signs of aging. Dentistry strives to promote oral health as a part of overall health and applies principles of basic medicine, pharmacology, and psychology to dental care.

BACKGROUND AND HISTORY

Before the seventeenth century, dental care was crude, unrefined, and most often provided by physicians. Through the eighteenth and nineteenth centuries, dentistry emerged as its own medical discipline, and most dentists trained through apprenticeships.

Pierre Fauchard is credited as the founder of modern dentistry. In 1728, the French surgeon published *Le Chirurgien Dentiste: Ou, Traité des dents* (*The Surgeon Dentist: Or, Treatise on the Teeth*, 1946), which summarized all available knowledge of dental anatomy, diseases of the teeth, and the construction of dentures.

In 1840, Horace Hayden and Chapin Harris established the world's first dental school, the Baltimore College of Dental Surgery, in Baltimore, Maryland. In 1867, Harvard University became the first university to establish a university-affiliated dental school.

Several scientific milestones transformed dentistry in the nineteenth century. In 1844, Horace Wells administered nitrous oxide to a patient before a tooth extraction, becoming the first dentist to use anesthesia. In 1890, dentist Willoughby Dayton Miller connected microbes to the decay process, extending the germ theory to dental disease. In 1898, William Hunter introduced the term "oral sepsis" to the profession of dentistry and called attention to the contaminated practices and instruments used by dentists. In 1918, radiology was added to dental school curricula, and by the 1930's, most dentists in the United States were using X rays as part of routine dental diagnostics.

Advances in science and technology, including the sequencing of the human genome and the arrival of the digital age, have revolutionized dentistry, rendering it nearly unrecognizable when compared with nineteenth-century dentistry and improving the diagnostic and treatment capabilities within the field.

How It Works

Dental Tools. Many common dental tools are available for home use as part of a daily oral care routine. The most basic of dental tools is the toothbrush. Toothbrushes come in a variety of sizes, shapes, and stiffness. Patient age and oral condition determine the best toothbrush for each individual. Toothbrushes usually consist of a plastic handle with nylon bristles that remove food, bacteria, and plaque that can lead to tartar and dental caries. Toothpaste is usually added to a toothbrush to aid in cleaning the teeth and freshening the mouth. Toothpaste is available in a variety of flavors and compositions and may contain polishing or bleaching agents. Dental floss is another basic tool used to remove food and debris from between the teeth. Floss is available in waxed and unwaxed formulations and in a variety of widths and thicknesses. Mouthwash is a rinse that prevents gum disease. Mouthwash is available in many flavors, but all types reduce the number of germs in the mouth that cause gingivitis.

More sophisticated dental tools are used by dentists during dental examinations and procedures. A routine dental cleaning removes stains on the teeth, as well as tartar that brushing and flossing cannot remove. Polishing the teeth aids the dentist in visualizing the teeth and makes it more difficult for plaque to accumulate on the surface of the teeth. Mirrors, scrapers, scalers, and probes are essential in-office dental tools.

Dental Therapy and Devices. Countless therapies and devices are available to diagnose, prevent, and treat disorders and diseases of the teeth and mouth. Extraction, previously the mainstay of dentistry, involves simply removing the affected tooth. Fillings are used to replace a portion of a tooth that is missing or decayed. Fillings are often made of gold or silver but may also be made of composite resins or amalgam depending on the size and location of the filling. A dental implant is the extension or replacement of a tooth or its root by inserting a post made of metal or other material into the bone to support a new artificial tooth. Crownwork involves covering a damaged tooth with porcelain or other alloy to restore the tooth's original size and shape. A denture is a removable prosthetic appliance that replaces missing teeth. Dentures may replace all or just some teeth. In contrast, a bridge is a tooth-replacement device that cannot be removed. A bridge is made of one or more artificial crowns that are cemented to adjacent teeth.

Orthodontic appliances are necessary to correct and prevent irregularities in the alignment of the teeth, face, and jaw. Braces are among the most common orthodontic appliances, along with headgear and retainers. Conventional braces have metal brackets that are attached to the outer surfaces of the teeth. Wires are attached to the brackets, and manipulation of the wire allows movement and rotation of the teeth into the desired position. The braces may be attached to headgear to help move teeth or secure them into position. Retainers are often worn after braces are removed to maintain the new position of the teeth. Retainers may be permanent or removable. Removable retainers consist of a wire attached to a resin base that is worn at all times (except during meals) to hold the teeth in place for up to several years after braces are removed. A permanent retainer is a metal wire attached to the tongue side of the lower teeth that can maintain the desired position of severely crowded or rotated teeth.

APPLICATIONS AND PRODUCTS

Most dentists practice dentistry as general practitioners. In addition, the American Dental Association recognizes nine specialties within the field of dentistry, each of which requires additional education or training beyond dental school.

General Dentistry. A general practitioner of dentistry deals with the overall maintenance of patients' teeth, gum, and mouth health. Ideally, general dentistry is preventive in nature, focusing on the maintenance of oral health and hygiene to avoid the occurrence of disorders and diseases of the mouth. Dentists who practice general dentistry encourage regular checkups to ensure proper functioning of the mouth and teeth. A general dentist will provide individualized treatment plans that include dental examinations, tooth cleanings, and X rays or other diagnostic tests to prevent or treat disorders of the mouth as early as possible. General dentists also repair and restore injuries of the teeth and mouth that result from decay, disease, or trauma. All dentists are able to prescribe medicines and treatments to diagnose, prevent, or treat diseases of the mouth and teeth.

Orthodontics. The largest specialty within dentistry is orthodontics. Orthodontics focuses on straightening teeth and correcting misalignment of the bite, usually using braces and retainers. Misalignment of the teeth or bite can cause eating or speaking disorders, making orthodontics an important part of overall health. Also, orthodontics may be aesthetic in nature, focusing on improving the structure and appearance of the teeth, mouth, and face to improve a patient's self-esteem. Most orthodontic patients are children because corrective procedures of the teeth are most effective when started early. However, an increasing number of adults are seeking orthodontic care, owing to the development of new methods and techniques in orthodontics that allow minimal discomfort and improved healing.

Oral and Maxillofacial Surgery. Commonly referred to as oral surgery, oral and maxillofacial surgery is the application of surgical techniques to the diagnosis and treatment of disorders of the teeth, mouth, face, and jaw. An oral surgeon may remove damaged or decayed teeth under intravenous sedation or general anesthesia; place dental implants to replace missing or damaged teeth; repair facial trauma, including injuries to soft tissues, nerves, and bones; evaluate and treat head and neck cancers; alleviate facial pain; perform cosmetic surgery of the face; perform corrective and reconstructive surgery of the face and jaw; and correct sleep apnea.

Pedodontics. Also known as pediatric dentistry, pedodontics focuses on dental care and oral hygiene of children and adolescents. Pediatric dentists apply the principles of dentistry to the growth and development of young patients, oral disease prevention, and child psychology. Some pediatric dentists also specialize in the treatment of patients with developmental or physical disabilities. Pediatric dentists emphasize proper oral hygiene, beginning with baby teeth, because healthy teeth allow for proper chewing and correct speech. Pediatric dentists also stress the importance of proper nutrition for its role in oral health, as well as overall growth and development. Early dental care facilitates lifelong oral health.

Periodontics. The field of dentistry called periodontics studies the bone and connective tissues that surround the teeth. Periodontics also involves the placement of dental implants. Periodontists prevent, diagnose, and treat periodontal disorders and infections, including gingivitis and periodontitis. Most periodontal diseases are inflammatory in nature, as are some cardiovascular diseases, and a connection between these two disease states has prompted physicians and periodontists to work together to treat patients at risk for either condition.

Prosthodontics. Also known as prosthetic dentistry, prosthodontics is the specialized field of dentistry that focuses on restoring and replacing teeth with dental implants, bridges, dentures, and crowns. Although general dentists can perform simple restoration or replacement of teeth, prostodontists handle severe or extreme cases of tooth loss because of trauma, disease, congenital defects, and age.

Endodontics. The field of dentistry that studies abnormal tooth pulp and focuses on the prevention, diagnosis, and treatment of diseases of the tooth pulp is called endodontics. Endodontic treatment is also known as root canal therapy. Endodontic therapy may also include surgery necessary to save a diseased tooth. Endodontists are often able to treat the diseased or damaged inside of a tooth instead of extracting it completely.

Oral and Maxillofacial Pathology. In oral and maxillofacial pathology, the principles of dentistry are applied to investigating the causes and effects of diseases of the mouth, head, and neck. Oral pathologists are trained to diagnose and treat such diseases, as well as to expose the connection between oral disease and systemic disease.

Oral and Maxillofacial Radiology. The use of advanced imaging techniques to diagnose and treat disorders of the mouth, teeth, head, and neck is known as oral and maxillofacial radiology. An oral and maxillofacial radiologist is a dentist who uses radiographic images to diagnose disease and guide treatment plans. Radiologists may use X rays, computed tomography (CT) scans, magnetic resonance imaging (MRI), ultrasound, and positron emission tomography (PET) to visualize the oral cavity or maxillofacial regions. Specialized sialography images the salivary glands. Intra-oral radiographs are used routinely by general dentists as part of regular dental checkups.

Dental Public Health. The field of dental public health is involved in the epidemiology of dental diseases and applies the principles of dentistry to populations rather than individuals. Dental public health specialists have been involved in promoting fluoridation of drinking water and examining the links between commercial mouthwash and cancer. Dental health specialists assess the oral health needs of communities, develop programs to teach and promote oral health, and implement policies and regulations to address oral health issues.

IMPACT ON INDUSTRY

Globally, demand for dental care and consumer oral health care products, including toothbrushes, toothpaste, dental floss, and mouth rinse, is growing at an annual rate of at least 2 percent, with stronger growth in Latin America and Europe. In 2004, sales of oral care products in the United States—the world's strongest oral health care market—reached almost $3 billion. Globally, the dentistry industry reaches revenues of nearly $20 billion annually, dominated by markets in the United States, Europe, and Japan. The industry is highly segmented, with primarily small office practices and clinics and no suppliers of dental technology or products attaining a majority of market share.

Government Initiatives. In the United States, the government is increasingly involved in the provision and regulation of health care and medical services, including dentistry, in order to maintain the safety and health of patients, consumers, and workers within the industry.

The Occupational Safety and Health Administration (OSHA), a division of the United States Department of Labor, maintains workplace safety standards for the dentistry industry. Primarily, dental professionals may be exposed to toxic or harmful chemicals, materials, infectious substances, or medications as part of their professional duties. OSHA establishes guidelines to recognize and prevent situations that place workers in the dentistry industry at risk for exposure to harmful substances or pathogens. OSHA also provides safety training for all health care workers.

The United States Department of Health and Human Services established the Health Insurance Portability and Accountability Act (HIPAA) in 1996, which, in part, protects the privacy and security of private health information. Dental care and history is considered private health information and is, therefore, protected by HIPAA regulations. Dental practices must adhere to strict guidelines to maintain patient confidentiality and improve patient safety.

The National Institute of Dental and Craniofacial Research (NIDCR), part of the National Institutes of Health, is the government-sponsored research arm within the field of dentistry. The NIDCR conducts research covering the entire spectrum of dental-related diseases and disorders and encourages interdisciplinary approaches to oral health. The NIDCR is a major contributor to the government's Healthy People 2010 program, along with the Centers for Disease Control and Prevention, the Indian Health Service, and the Health Resources and Services Administration. Overall, the goal of the oral health arm of the program is to prevent oral and craniofacial diseases and injuries by promoting oral hygiene and health maintenance and improving access to oral health care for underserved populations.

Consumer Initiatives. Consumer demand for oral health care products is increasing as people recognize the importance of oral health in overall health and well-being. There are several major corporations that contribute to the worldwide market for dental products, but no single company dominates the landscape. The largest players include Colgate, Procter & Gamble, and Johnson and Johnson. Many consumer

Fascinating Facts About Dentistry

- In the mid-1850's, a textbook on dental surgery taught that asbestos could be placed under the filling of a sensitive tooth because asbestos is unable to conduct heat or electricity.

- The earliest evidence of dental caries dates back 100 million years to the Cretaceous period; dinosaur and fish fossils from this period show signs of dental decay.

- Toothpicks were in use at least 3,000 years ago and were made of wood, metal, thorns, or porcupine quills; ornate metal toothpicks were a sign of wealth in ancient Egypt.

- Saint Appollonia, a Christian martyr whose teeth were removed by her Roman captors, is the patron saint of dental pain sufferers, and her intercession is thought to bring healing to all oral pain and afflictions.

- The structure and anatomy of the teeth and mouth are unique to each individual, and dental records and examinations are used to identify victims of accidents, terrorism, or natural disasters. Paul Revere was the first dentist to suggest using bridgework to identify remains—namely, a Revolutionary War general—and he became a pioneer of forensic dentistry in the United States.

- More than 90 percent of systemic diseases exhibit oral manifestations; oral signs and symptoms may come before, after, or at the same time as signs and symptoms elsewhere in the body but are often the first signs of systemic illness.

oral health product manufacturers are involved in sponsoring dentistry research and education. In addition to educating dentistry scientists and professionals, these large corporations sponsor education programs for school-age children and the community at large to promote awareness of oral health and teach proper oral hygiene. Industry-supported programs also bring dental care to underserved patients.

Consumers demand safe and effective oral health products and rely on the Seal of Acceptance of the American Dental Association (ADA) to choose goods. All major manufacturers strive to receive the ADA seal, which denotes extensive clinical and laboratory research to ensure product safety and effectiveness. More than three hundred manufacturers of dental care products volunteer their products for scrutiny by the ADA. Although the Food and Drug Administration establishes and enforces safety and effectiveness guidelines for the manufacture and use of health care products, the ADA surpasses these guidelines to offer reassurance in product choice for consumers.

CAREERS AND COURSE WORK

All fifty states in the United States, plus the District of Columbia, require a license to practice dentistry. To obtain a license, applicants must graduate from an accredited school of dentistry and pass written and practical exams. There are more than fifty dental schools in the United States, offering either the doctor of dental surgery (D.D.S.) or the doctor of dental medicine (D.M.D.) degree, which are equivalent degrees.

To apply to dental school, students should be proficient in basic sciences, including biology, chemistry, physics, health science, and mathematics. A minimum of two years of college education is necessary to apply to dental school, but most applicants have completed an undergraduate degree in a science discipline. The Dental Admission Test is also required for applicants to dental school.

Dental school is a four-year program consisting of two years of didactic learning and two years of clinical training. Education in anatomy, microbiology, biochemistry, physiology, and pharmacology is essential to training in dentistry. Clinical practice experience takes place under the supervision of licensed dentists.

Dentists must possess superb diagnostic skills, supreme visual memory, and excellent manual dexterity. Most dentists work in private practice, either alone or with partners. Therefore, dentists also need business management talents, self-discipline, and communication skills.

Dental hygienists, assistants, and laboratory technicians are related professions within the field of dentistry. They each work closely with a dentist to perform the technical duties associated with oral care and teach patients about proper hygiene and good nutrition. The educational and licensing requirements for these dental careers vary by state, although formal education is encouraged and favorable in a competitive job market.

SOCIAL CONTEXT AND FUTURE PROSPECTS

The connection between oral health and overall health has led to an increase in oral home care as well as in professional dentistry services. Patients seek dental care for routine maintenance of oral health and cosmetic procedures to improve the appearance of the face and mouth. In the future, dentistry will increasingly play a fundamental role in people's overall health and wellness. From preventing childhood tooth decay and age-related tooth loss to improving self-esteem through a brighter, straighter smile, dentistry has evolved from a fearful, painful process of tooth extraction to a respected field of medicine that is associated with comfortable care and daily hygiene.

Dentistry of the future will emphasize less painful therapy and disease prevention. It will seek to identify at-risk groups and to provide services to underserved populations to improve dental public health, which will have lasting benefits in education and overall disease morbidity and mortality. Emerging research is focused on mouthwashes that prevent the buildup of plaque on teeth, vaccines that prevent decay and dental caries, and long-lasting pellets that deliver a continuous dose of fluoride to the teeth. Braces may soon be replaced or aided by small, battery-operated paddles that deliver an undetectable electric current to the gums to rearrange bone and tissue structures of the mouth. Lasers will replace existing surgical techniques, allowing for painfree treatment of dental disease. Further, new enzymes and plastics are emerging as options for tooth restoration and dental diagnostics.

Dentistry will continue to be a collaborative and interdisciplinary practice that meets the growing and changing needs of dental health.

Jennifer L. Gibson, B.S., D.P.

FURTHER READING

Kendall, Bonnie. *Opportunities in Dental Care Careers.* Rev. ed. New York: McGraw-Hill, 2006. A review of the educational requirements and professional expectations for all specialties of dentistry and dental-related careers.

Picard, Alyssa. *Making the American Mouth: Dentists and Public Health in the Twentieth Century.* New Brunswick, N.J.: Rutgers University Press, 2009. Presents a history of dentistry as well as essays on issues such as dental hygiene, dental economics, and the American diet.

Pyle, Marsha, et al. "The Case for Change in Dental Education." *Journal of Dental Education* 70, no. 9 (September, 2006): 921-924. The American Dental Education Association's Commission on Change and Innovation in Dental Education examines the need for change in dental education. It takes into account the financial expense of a dental education and the professional responsibilities of meeting all individual and public health needs.

Rossomando, Edward F., and Mathew Moura. "The Role of Science and Technology in Shaping the Dental Curriculum." *Journal of Dental Education* 72, no. 1 (January, 2008): 19-25. Offers a history of the changing dental school curricula in the United States and offers perspectives for the future of dentistry education.

Wynbrandt, James. *The Excruciating History of Dentistry: Toothsome Tales and Oral Oddities from Babylon to Braces.* New York: St. Martin's Press, 1998. An entertaining history of the development of the dental profession, offering humorous anecdotes and macabre tales of the profession.

WEB SITES

American Dental Association
http://www.ada.org

American Dental Education Association
http://www.adea.org

See also: Anesthesiology; Cardiology; Pediatric Medicine and Surgery; Radiology and Medical Imaging; Surgery.

DERMATOLOGY AND DERMATOPATHOLOGY

FIELDS OF STUDY

Medicine; pathology; surgery; surgical pathology; biology; histology; chemistry; physics; immunodermatology; pediatric dermatology; cosmetic dermatology; surgical dermatology; veterinary dermatology; Mohs micrographic surgery.

SUMMARY

Dermatologists diagnose and treat medical conditions of the skin, including acne, rosacea, psoriasis, warts, hair loss, and various forms of skin cancer. Dermatopathologists analyze the mechanisms of skin diseases and perform microscopic diagnoses based on the tissue samples submitted by dermatologists. Skin disorders have a high prevalence and can affect patients of all ages, from neonates to elderly people. Because of the great variety and dynamic nature of the lesions, specialties focusing on skin are among the most complex in medicine.

KEY TERMS AND CONCEPTS

- **Botulinum Toxin:** Neurotoxin produced by the bacterium *Clostridium botulinum;* commonly known as Botox, its trade name.
- **Epidermis:** Upper (outer) skin layer.
- **Flow Cytometry:** Technique for separating and counting cells or chromosomes by suspending them in fluid and passing them by a focused light.
- **Immunohistochemistry:** Antibody-based method of detecting a specific protein in a tissue sample.
- **Keratinocyte:** Common epidermal cell that synthesizes keratin and changes while moving upward from basal to superficial layers.
- **Macule:** Flat, colored skin area that measures less than 10 millimeters in diameter.
- **Melanocyte:** Epidermal cell that produces the skin pigment melanin.
- **Papule:** Solid, raised spot on the skin that measures less than 10 millimeters in diameter.
- **Plaque:** Broad, raised area of skin.
- **Pustule:** Small skin swelling filled with pus.
- **Retinoids:** Class of compounds chemically related to vitamin A.

DEFINITION AND BASIC PRINCIPLES

Dermatology is the branch of medicine dedicated to the diagnosis, treatment, and prevention of diseases and conditions of the skin, the hair, the nails, and mucous membranes. A subspecialty of pathology and dermatology, dermatopathology focuses on studying the mechanisms of skin diseases and on the microscopic examination of cutaneous tissue.

Dermatologists assess the appearance and distribution of any abnormalities in the skin, identifying primary and secondary lesions. These lesions can manifest in numerous forms, including macules, papules, plaques, nodules, pustules, vesicles, wheals (hives), scales, fissures, and scars. The patient may complain of itchiness (pruritus), pain, or hair loss, or may be uncomfortable with the appearance of a skin area. If a diagnosis is not readily apparent, the dermatologist performs a skin biopsy. A dermatopathologist examines the tissue under a microscope and renders a pathological diagnosis.

The skin is the largest and most visible organ of the human body, with essential functions in storage, absorption, thermoregulation, vitamin D synthesis, and protection against pathogens. It is readily accessible to the examiner; however, the potential abnormalities are numerous and the differential diagnoses extensive, rendering dermatology one of the most complex medical disciplines. Although the field has been morphologically oriented for centuries, advances in molecular medicine and genetics have opened new opportunities for understanding the pathogenesis of skin diseases and for improved diagnosis strategies. An evolving interrelationship with other disciplines such as plastic surgery and endocrinology has been expanding the frontiers of this medical specialty.

BACKGROUND AND HISTORY

People have been concerned with the health and appearance of their skin throughout history. Egyptian physicians used arsenic applications to treat skin cancer and sandpaper to smooth scars. Queen Cleopatra was known for her cosmetic knowledge. Geoffrey Chaucer's *The Canterbury Tales* (1387-1400) and William Shakespeare's plays contain numerous references to unsightly skin afflictions, such as boils, carbuncles, and scabs. Not surprisingly, their appearance is frequently a metaphor for character flaws.

Some of the first skin treatments were undoubtedly borrowed from the plant world, making use of leaves, flowers, and roots. The juice of the aloe vera, for example, is an ancient and effective remedy that continues to be used for some skin conditions. For centuries, physicians treated a wide range of afflictions, from rashes to wounds, using oils, powders, and salves they mixed themselves. Sunlight was used by European physicians in the eighteenth and nineteenth centuries to treat psoriasis and eczema.

Starting in the nineteenth century, a true revolution in biology galvanized the progress of skin sciences. The terms "dermatology" and "dermatosis" were introduced. In the late 1800's, dermatologists began using a variety of chemicals to smooth facial wrinkles and scars. Cryosurgery and electrosurgery came into use. Soon after the development of the laser in the 1950's, dermatologists used it to treat skin conditions. The surge of innovations has continued, making dermatology an exciting and rapidly evolving specialty.

How It Works

Skin diseases and conditions affect patients of all ages and ethnicities. Physicians may specialize in a specific age group, such as children, or a category of conditions. Some dermatologists focus on cosmetic disorders of the skin and may be certified to perform procedures such as injections of botulinum toxin, chemical peels, and laser therapy. Others concentrate on skin cancers or immunological conditions. Regardless of the focus of a dermatologist's practice, the day-to-day work can be divided into three main areas: diagnosis, treatment, and management.

Diagnosis. Dermatologists obtain the patient's medical history and assess his or her status. They examine the affected skin and adjacent areas to determine the nature and extent of the lesions. A frequently used method is dermoscopy (or epiluminescent microscopy), which employs a quality magnifying lens and a powerful lighting system to allow a close examination of the skin's structure. It is useful in evaluating pigmented skin lesions and can facilitate the diagnosis of melanoma.

Some skin conditions are more readily diagnosable than others. Acne and psoriasis, for example, often do not necessitate further tests. The lesions,

however, may be of an ambiguous nature or potentially malignant. In these cases, the physician takes a tissue sample (for example, a biopsy or nail clippings) and submits it, usually with a differential diagnosis, to a laboratory. There, the sample undergoes a dermatopathological evaluation.

Dermatopathologists interpret tissue samples on specially prepared slides using light, fluorescent, and sometimes electron microscopy. They first determine how the specimen was obtained (for example, a punch or shave biopsy), then establish if the condition appears to be infectious, inflammatory, degenerative, or neoplastic (benign or malignant). Often, consultation with other dermatopathologists and the attending dermatologist or primary care physician is necessary. Additional sections of the specimen may be required before a diagnosis can be rendered and the report sent to the clinician. The work needs to be extremely thorough; no part of the microscopy slide can be left unexamined. Ancillary methods used by dermatopathologists include immunohistochemistry and flow cytometry.

Additional tests that may be undertaken in the dermatologist's office include a potassium hydroxide examination for fungi, bacterial stains, fungal and bacterial cultures, skin scrapings for scabies, patch tests (for contact allergies), and blood tests.

Treatment. Once the diagnosis has been made, treatment options are considered and discussed at length with the patient or caregiver. Dermatopathologists often play an active role in this process. Treatment may involve medications to be administered externally or internally, injections, or surgical procedures. Punch biopsy, shave biopsy, electrodesiccation and curettage, blunt dissection, and simple excision and suture closure are the basic techniques that dermatologists master. They are also familiar with more sophisticated techniques, such as Mohs micrographic surgery and, if appropriate, may refer patients to physicians who perform these techniques.

Management. Skin conditions can be lifelong problems. Eczema, acne, and psoriasis are only a few of many conditions that require regular visits to the dermatologist. Managing the patient's condition often takes the form of control rather than cure.

Applications and Products

Dermatologist diagnose and treat many disorders

and diseases. The most common examples of disorders treated are infections, inflammatory diseases, papulosquamous diseases, and tumors.

Infections. Several categories of pathogens cause infections with cutaneous manifestations. *Staphylococcus aureus* and group A beta-hemolytic streptococci account for most skin and soft tissue infections, such as impetigo, folliculitis, cellulitis, and furuncles. Syphilis is an infectious disease caused by the bacterium *Treponema pallidum.* Primary syphilis, acquired by direct contact with a skin or mucosal lesion, manifests with a cutaneous ulcer (chancre). Warts are benign epidermal tumors caused by numerous types of human papillomaviruses (HPVs). These viruses infect epithelial cells of the skin, mouth, and other areas, causing both benign and malignant lesions.

Herpesvirus infections are caused by herpes simplex virus 1 (HSV1) and herpes simplex virus 2 (HSV2), distinguishable by laboratory tests. HSV1 is generally associated with oral infections, and HSV2 causes genital infections. The lesions appear as grouped vesicles on a red base.

The agents that induce superficial fungal infections include dermatophytes (responsible for tinea, or ringworm) and *Candida* species yeasts.

Inflammatory Diseases. Eczema is the most common inflammatory disorder. It manifests with itchiness and exhibits three clinical stages: acute (redness and vesicles), subacute (redness, scaling, fissuring, and scalded appearance), and chronic (thickened skin). There are numerous types of eczemas, including atopic dermatitis (in patients with personal or family history of allergies) and contact dermatitis (allergy to a common material such as nickel or poison oak).

Acne is a common disorder with important psychosocial effects. It occurs in predisposed individuals when sebum production increases. Proliferation of the microorganism *Propionibacterium acnes* in the sebum alters it and causes pore clogging. Lesions are noninflammatory (comedones, also known as blackheads and whiteheads) or inflammatory (papules, pustules, or nodules). The extent and severity of the lesions varies, from a few comedones to the strongly inflammatory acne conglobata.

Papulosquamous Diseases. The group of disorders known as papulosquamous diseases are characterized by scaly papules and plaques. Psoriasis, an immune-mediated skin and joint inflammatory disease, develops when inflammation primes basal stem keratinocytes to proliferate excessively. Initial red, scaling papules coalesce to form round-oval plaques. The scales are adherent, silvery white, and show bleeding points when removed (Auspitz sign). Inflammatory arthritis is present in some patients.

Tumors. The two most common skin cancers are basal cell carcinoma (BCC) and squamous cell carcinoma (SCC). Approximately 80 percent of nonmelanoma skin cancers are the basal cell type, and 20 percent are the squamous cell type.

Basal cell cancer, the most common invasive malignant skin tumor in humans, represents more than 90 percent of skin cancers in the United States. The patient typically has a bleeding or scabbing sore that heals and subsequently recurs. The tumor advances by direct extension and destroys normal tissue but rarely metastasizes. The cells of basal cell carcinoma resemble those of the basal epidermal layer. They have a large nucleus and develop an orderly line around the periphery of tumor nests (palisading).

Squamous cell carcinoma is the second most common cancer among light-skinned individuals. The relationship to ultraviolet radiation is stronger and the chances of metastasis much higher than for basal cell carcinoma. Actinic keratosis, the most common precursor of squamous cell carcinoma, begins on sun-exposed skin as isolated or multiple flat, pink-brown, rough lesions. Abnormal squamous cells originate in the epidermis from keratinocytes and proliferate indefinitely.

Malignant melanoma originates from melanocytes. Skin melanoma either begins on its own or develops from a preexisting lesion, such as a mole (nevus). One of the most aggressive tumors, melanoma can metastasize to any organ, including the brain and heart. Individuals who sunburn easily or who experienced multiple or severe sunburns have a twofold to threefold increased risk for developing skin melanoma. The goal of specialists and patients alike is to recognize melanomas as early as possible in their development. Compared with common acquired melanocytic nevi, malignant melanoma tend to have four characteristics: asymmetry, border irregularity, color variation, and diameter enlargement (ABCD). These four characteristics are the primary criteria for clinical melanoma recognition. Changes in the shape and color of a mole are important early signs and should always arouse suspicion. Ulceration and bleeding are late signs; at this stage, the chance of cure diminishes greatly.

Important Treatment Modalities. Common ways of dealing with dermatological problems are topical treatments (such as ointments and creams) and oral treatments (drugs taken by mouth). Any bodily injury, irritation, or trauma that eliminates water, lipids, or protein from the epidermis compromises its function. Restoration of the normal epidermal barrier can often be accomplished using mild soaps and emollient creams or lotions. The often-cited dermatologic adage is "If it is dry, wet it; if it is wet, dry it." Consequently, wet compresses are a frequently used remedy. A multitude of other topical treatments are available, from antibiotic, antiviral, or steroid ointments applied to treat infectious diseases or eczema to vitamin D derivative creams for psoriasis and retinoid creams for acne. Drugs can also be taken orally to treat a variety of conditions such as acne and autoimmune disorders.

Surgical and Cosmetic Procedures. Dermatologists use several techniques to obtain skin biopsies. Most procedures are done in the doctor's office, and each technique has specific indications. Punch biopsies are employed for most superficial inflammatory diseases and skin tumors (except melanoma). Shave biopsies are used for superficial benign and malignant tumors. Deep inflammatory diseases and malignant melanoma benefit from excisions. Electrodesiccation and curettage (ED&C; also known as scrape and burn) is an important technique for removing a variety of superficial skin lesions, such as cancerous growths and genital warts. The physician uses a sharp dermal curette to cut away the growth and a needle-shaped electrode that delivers an electric current to remove any remaining material and to stop the bleeding.

Blunt dissection is a fast, elementary, usually nonscarring surgical procedure used to remove warts and other epidermal tumors. Unlike ED&C and excision, it does not disturb normal tissue.

Small, superficial, nonmalignant lesions may be quickly and efficiently frozen with liquid nitrogen, administered with a spray or sterile contact probe.

Cryosurgery for malignant lesions, however requires experience and sophisticated equipment with thermocouples that measure the depth of freeze. This minimally invasive technique is also successfully

---

Fascinating Facts About Dermatology and Dermatopathology

- Under normal conditions, the top layer of skin on an adult human sheds every twenty-four hours, and the skin completely renews itself in three to four weeks.

- Throughout the centuries, people with leprosy have been ostracized by their communities. Although modern medicine has made diagnosis and treatment of leprosy easy, the stigma associated with the disease remains and presents an obstacle to self-reporting. About one hundred patients are diagnosed each year in the United States.

- The use of botulinum toxin is not limited to the treatment of wrinkles. It has also been used as a remedy for muscle spasms, migraines, strabismus (lazy eye), and other conditions.

- Vitiligo, a skin disorder that affects one in every two hundred people, causes patches of skin that lack pigment and are prone to sun damage but not to skin cancer. A gene mutation responsible for increasing the risk of developing vitiligo also decreases the risk of skin malignancy.

- Researchers are studying noninvasive techniques for removing adipose tissue that could help eliminate localized fat deposits in individuals of average weight. These include exposing fat cells beneath the skin to ultrasound waves or low temperatures.

- Scientists have created artificial skin with biomechanical properties similar to real skin using biomaterials such as fibrin (from blood), agarose (from seaweed), chitosan (from crustacean shells), and collagen.

---

employed for common lesions, including genital warts, actinic keratoses, and certain infectious conditions.

An important surgical breakthrough occurred in the 1930's, when physician Frederic Mohs developed a microscope-guided method of tracing and removing basal cell carcinomas. These—and other tumors—may not grow in a well- circumscribed fashion but instead extend in fingerlike projections. Thin layers of tissue are removed, and all margins of the specimen are mapped to determine whether any tumor remains. This tissue-sparing technique has high cure rates.

Chemical peeling of facial skin uses a caustic agent to achieve a controlled, chemical burn of the epidermis and the outer dermis. Skin regeneration results in a fresh and orderly epidermis with ablation of fine wrinkles and pigmentation reduction.

In liposuction surgery, fat is removed through half-inch incisions using small-diameter cannulae. Multiple to-and-fro movements mechanically disrupt the fat and create tunnels. The loosened fat is removed by strong suction.

Photothermolysis is based on the property of a chromophore (melanin, hemoglobin, tattoo ink) in a target tissue to strongly absorb a selected laser wavelength and generate heat. It removes the target tissue while producing only a local thermal injury, resulting in less injury to the surrounding tissue and lowered risk of scarring. Vascular lesions, for example, can be treated in this manner, including port-wine stains, benign tumors, and spider veins in legs. In vascular lesions, the targeted chromophore is hemoglobin.

Specific types of lasers can be used to treat benign pigmented lesions with a predominant epidermal component such as freckles and tattoos. In addition, numerous laser-based devices can remove unwanted hair.

Other common techniques and devices include the use of intense pulsed light for resurfacing (to treat vascular lesions and acne) and light-activated drugs in photodynamic therapy (for precancerous and cancerous cells, acne, rosacea, or skin enhancement).

One of the most popular nonsurgical cosmetic procedures is injections of botulinum toxin (Botox). This neurotoxin blocks the release of the chemical messenger acetylcholine, effectively causing chemical denervation. The injections reduce facial lines caused by hyperfunctional muscles.

IMPACT ON INDUSTRY

Government and University Research. The etiology, pathogenesis, and optimal therapeutic strategies for many skin disorders are still unclear. Understanding the biology of skin tumors, especially melanoma, has become imperative. The collaboration between academic dermatologists and basic scientists—geneticists, immunologists, and molecular biologists—in the United States and abroad is growing stronger.

Sources of research funds include grants and scholarships from governmental institutions, foundations, associations, and corporate partners. In the United States, at least two branches of the National Institutes of Health (NIH)—the National Cancer Institute and the National Institute of Arthritis and Musculoskeletal and Skin Diseases—are involved in patient education and research regarding cutaneous disorders. The American Academy of Dermatology and the Society for Investigative Dermatology also are committed to supporting the development of a strong skin research environment.

The mission of the European Academy of Dermatology and Venereology includes promoting excellence in research, education, and training. The European Society for Dermatological Research has focused on advancing basic and clinical science and has facilitated the exchange of information relevant to investigative dermatology among clinicians and scientists worldwide.

Industry and Business. Cosmetic dermatology has become vastly popular. More and more dermatologists devote a significant part of their time to nonsurgical and surgical cosmetic procedures. The financial gains are undeniable. The global market for cosmetic surgery is predicted to exceed $40 billion in revenues by 2013. A survey conducted by the American Academy of Cosmetic Surgery (AACS) found that more than 17 million cosmetic surgery procedures were performed in the United States in 2009. Nonsurgical procedures, such as injections of botulism toxin and acid peels, and prescription drugs that are used to treat conditions associated with perceived deficits in physical appearance, such as retinoids, are also in great demand. The aging baby boomers are strongly driving the demand for facial rejuvenation products. Although for dermatologists, cosmetic procedures are generally more lucrative than medical treatments, pharmaceutical companies find drugs that act on the skin to be a significant source of revenue. The psoriasis market is forecast to grow the most aggressively, doubling from $3.4 billion in 2009 to $6.8 billion in 2019 in the United States, France, Germany, Italy, Spain, the United Kingdom, and Japan.

Teledermatology and teledermatopathology use telecommunication technologies to exchange medical information for diagnostic, therapeutic, and educational purposes. Digital dermoscopy is an evolving computer-based version of traditional dermoscopy. It provides a reliable way to capture and store images, send them to pathologists, and compare them over time. The technique is quickly becoming an essential tool in dermatology practice. Numerous companies are developing hardware and software products for these advanced technologies; even more are in the business of providing cosmetic surgery products (lasers, instruments), phototherapy equipment, and other devices essential to modern dermatologic practices.

CAREERS AND COURSE WORK

Those considering a career in dermatology will require a strong foundation in biology, chemistry, and physics, followed by a decade of training and a lifetime of learning. Medical school requires four years of intense preparation in an accredited school. Subsequent specialization in dermatology requires a one-year internship, followed by a three-year residency in an accredited program. A one- or two-year fellowship can be undertaken after residency; examples of possible paths are dermatopathology, pediatric dermatology, immunodermatology, and dermatologic surgery. Veterinary colleges may also offer a three-year residency program in dermatology.

To pursue a career in dermatopathology, a physician also can specialize in surgical pathology via a three-year residency, then undergo further training in a dermatopathology fellowship program. Competency in dermatopathology is of utmost importance for both pathology and dermatology residents.

Opportunities to practice exist all over the United States and worldwide. Dermatopathologists and dermatologists can be self-employed, partner with others, or be employed by hospitals, clinics, and governmental agencies. Both categories of specialists teach in universities and colleges that have degree programs in these disciplines. Research opportunities are available in universities and in laboratories operated by corporations, pharmaceutical companies, or governmental agencies.

SOCIAL CONTEXT AND FUTURE PROSPECTS

The burden of skin diseases on society is significant. According to a 2004 study by the American Academy of Dermatology Association and the Society for Investigative Dermatology, the annual cost of skin diseases in the United States is about $39 billion;

direct medical costs account for $29 billion and indirect costs related to lost productivity make up the remaining $10 billion. At any given time, one in three people in the United States suffers from an active cutaneous condition. The most prevalent disorders are herpes simplex, shingles, sun damage, eczema, warts, and hair and nail conditions. The incidence of melanoma is on the rise. The main reasons for this high level of skin disease are increased exposure to the sun during recreational activities and the atmospheric changes brought on by pollutants that result in increased radiation. Understanding the biology of skin tumors, especially melanoma, has become a priority of research efforts worldwide.

New therapeutic agents such as antibodies and immunomodulators offer hope for stubborn medical conditions such as psoriasis, still an incurable disease in need of good long-term therapeutic approaches. Biological treatments are on their way to bringing relief. Stem cells hold promise for tissue regeneration.

Advances in understanding the pathogenesis of various disorders have led to improved management and to a reduced risk of incorporating nonevidence-based components into dermatological practice. The close cooperation between dermatologists, pathologists, rheumatologists, and surgeons enhances the quality and efficiency of care.

The ever-increasing preoccupation with young, healthy skin has fueled an unprecedented explosion in the popularity of cosmetic procedures. More important, skin diseases with significant aesthetic, psychological, and social consequences have prompted dermatologists to implement and refine numerous cosmetic techniques involving peeling, botulinum toxin, hyaluronic acid, and lasers. These techniques have enabled many categories of patients with skin disorders to lead a normal social life.

Mihaela Avramut, M.D., Ph.D.

FURTHER READING

Bickers, D. R., et al. "The Burden of Skin Diseases, 2004: A Joint Project of the American Academy of Dermatology Association and the Society for Investigative Dermatology." *Journal of the American Academy of Dermatology* 55, no. 3 (September, 2006): 490-500. Summary of the well-documented study assessing the prevalence and economic burden of skin diseases and how they effect quality of life.

Bolognia, Jean, et al., eds. *Dermatology.* 2d ed. 2 vols.

St. Louis, Mo.: Mosby Elsevier, 2008. A basic text-book that covers nearly all aspects of dermatology, from cancers to cosmetic procedures.

Ferri, Fred. *Ferri's Fast Facts in Dermatology: A Practical Guide to Skin Diseases and Disorders.* Philadelphia: Saunders/Elsevier, 2011. A handbook for the diagnosis of dermatological disorders.

Habif, Thomas P. *Clinical Dermatology.* 5th ed. St. Louis, Mo.: Mosby Elsevier, 2010. Leading manual with excellent photographs, online access, multiple appendixes, and an online differential diagnoses (DDX) mannequin for lesion localization.

Hall, Brian J., and John C. Hall. *Sauer's Manual of Skin Diseases.* 10th ed. Philadelphia: Lippincott, Williams & Wilkins, 2010. Accessible textbook includes numerous color photographs, diagnostic algorithms, and a dictionary-index. Has an accompanying Web site.

Pilla, Louis. "Cosmetic Versus Medical Dermatology: A Widening Gap?" *Skin and Aging* 11, no. 6 (June, 2003). Analysis of the interplay between medical and cosmetic dermatology in modern practices.

WEB SITES
American Academy of Dermatology
http://www.aad.org

European Academy of Dermatology and Venereology
http://www.eadv.org

European Society for Dermatological Research
http://www.esdr.org

Society for Investigative Dermatology
http://www.sidnet.org

See also: Geriatrics and Gerontology; Pathology; Pharmacology; Surgery.

DIFFRACTION ANALYSIS

FIELDS OF STUDY

Physics; chemistry; geology; X-ray diffraction; electron diffraction; neutron diffraction; materials science; crystallography; mechanical engineering; physical chemistry; quantum mechanics; organic chemistry; molecular biology; fiber diffraction; mineralogy; metallurgy; differential equations; partial differential equations; Fourier analysis; optics; spectroscopy.

SUMMARY

Diffraction analysis is a general term used to describe various methods of scattering beams of X rays, electrons, or neutrons from targeted materials to generate diffraction patterns from which atomic arrangements in gases, liquids, and solids can be precisely determined. Diffraction techniques are useful in identifying and characterizing both natural materials, such as minerals, and engineered materials, such as ceramics. X-ray diffraction played an important role in the discoveries of the three-dimensional structures of molecules such as proteins and deoxyribonucleic acid (DNA), and neutron diffraction has allowed researchers to investigate stresses in automobile and aerospace constituents.

KEY TERMS AND CONCEPTS

- **Bragg Diffraction:** Scattering of X rays from a three-dimensional periodic structure such as a crystal, caused by waves reflecting from different crystal planes.
- **Crystal Lattice:** Three-dimensional array of points, each of which represents a unit cell.
- **Crystalline Material:** Chemical substance in which the atoms or ions are arranged in an orderly pattern.
- **Diffraction:** Directional change of a wave group after it encounters an obstacle or passes through an aperture.
- **Electron Diffraction:** Technique used to study the atomic structures of various substances through the interference patterns resulting when electrons are directed at a sample.
- **Fiber Analysis:** Study of biological or artificial filaments by X-ray-, electron-, or neutron-diffraction techniques.
- **Neutron Diffraction:** Analytic technique in which a beam of neutrons is scattered from gaseous, liquid, crystalline, or amorphous materials to determine their atomic arrangements.
- **Powder-Diffraction Analysis:** Technique in which a coherent beam of monochromatic X rays is directed at a powdered sample to obtain a diffraction pattern used to determine the structure of a particular substance.
- **Texture Analysis:** Use of diffraction analysis to determine the nature of components in such materials as textiles, food ingredients, or soil samples.
- **Unit Cell:** Basic building block, or the smallest assemblage of atoms or ions, the repetition of which generates the overall pattern of a crystal.
- **X-Ray Diffraction:** Analytic method in which X rays are reflected from the lattice of a crystal to determine the crystal's structural arrangement of atoms or ions.

DEFINITION AND BASIC PRINCIPLES

In general, diffraction denotes a change in the directions and intensities of waves when they encounter obstacles. All waves, be they water, sound, or light, or even such particles as electrons and neutrons that exhibit wavelike properties, are subject to diffraction. When X rays, the wavelengths of which range from 0.001 nanometer to 10 nanometers (a nanometer is one billionth of a meter), interact with matter such as a crystal, a pattern is generated that can be photographed and analyzed. This X-ray diffraction analysis has led to the discovery of the precise atomic arrangements in an enormous number of substances, from a simple crystal such as sodium chloride (whose sodium and chloride ions are cubically arranged) to such complex crystals as vitamin B12 (cyanocobalamin), the gigantic molecular structure of which was worked out by Dorothy Crowfoot Hodgkin, helping her to win the 1964 Nobel Prize in Chemistry.

The discovery that moving electrons exhibit wave properties led to a new kind of diffraction analysis employing electron beams. This technique proved to be particularly valuable in studies of the structures of gaseous substances, adsorbed gases, and surface layers. It could also be used to study liquids and solids, and it was often used in tandem with X-ray diffraction analyses. For example, X-ray diffraction studies confirmed that liquid cyclohexane existed in a chair form, and electron diffraction showed that cyclohexane oscillated among two chair forms and a boat form. This led scientists to the discovery of the basic principles of conformational analysis, which proved essential to an understanding of a variety of molecules.

Although X-ray and electron-diffraction methods are powerful in elucidating the structures of various substances, they are limited in studying elements that are close together in the periodic table. However, such elements vary greatly in their neutron-scattering ability, and so neutron-diffraction analysis can easily distinguish such elements as carbon, nitrogen, and oxygen. Certain details of molecular structure such as hydrogen bonding can be more precisely observed by neutron diffraction than by X-ray or electron techniques. Because of this, neutron scattering has proved advantageous in structural studies of colloids, membranes, dissolved proteins, and viruses.

BACKGROUND AND HISTORY

The term "diffraction" owes its origin to Francesco Maria Grimaldi, who derived it from the Latin root *diffringere*, meaning "to break apart," in the seventeenth century. Others had observed diffraction patterns when a light beam was broken up by a bird feather, and in the early nineteenth century the English physician and physicist Thomas Young performed his famous double-slit experiment in which he proved that light is a wave phenomenon, since interfering waves constituted the sole explanation of the striped light and dark bands that he observed when light managed to traverse the two slits.

The discovery that X rays also produced interference patterns was made by the German physicist Max von Laue in 1912, when he showed that X rays passing through a crystal produced a diffraction pattern on a photographic plate. The next year the British father-and-son team of William Henry Bragg and William Lawrence Bragg used the X-ray diffraction technique to determine the structures of such crystals as sodium chloride and diamond, and William Lawrence Bragg formulated a mathematical equation relating the wavelength of the X rays, the angle between

the incident X rays and the crystal's parallel atomic layers, and its interplanar spacing.

The Braggs' work led to the flourishing new field of X-ray crystallography, in which the three-dimensional arrangements of a wide variety of materials were determined, for example, fluorite (calcium fluoride) and calcite (calcium carbonate). In 1916, the Dutch physical chemist Peter Debye showed that X-ray diffraction could be extended to powdered solids, and he used this technique to figure out the structure of graphite. During the 1920's and 1930's, the American physical chemist Linus Pauling used X-ray diffraction to determine the structures of more than thirty minerals, including some important silicates. Pauling learned about the electron diffraction of molecules in the gaseous state from its discoverer, Herman Mark, and Pauling as well as others determined the structures of many substances that could be studied in the gaseous and liquid states. The techniques of X-ray and electron diffraction were also helpful throughout the second half of the twentieth century in elucidating the structures of proteins and DNA.

HOW IT WORKS

X-Ray Diffraction. Crystals are orderly arrangements of atoms or ions, and crystals constitute about 95 percent of all solid materials. The scattering of X rays from a crystal is actually due to its electron densities (atomic nuclei have a negligible contribution). Every crystal bombarded by X rays gives a unique pattern, and so each pattern acts as a distinctive identifier of the substance. This pattern is a result of secondary waves emanating from the electrons, which are called scatterers, and the phenomenon itself is known as elastic scattering. In a crystal these scatterers are regularly arrayed, but the reflected X rays can interfere destructively as well as constructively. The Bragg equation describes how these waves add constructively in certain directions. The reason X rays can do this is that their wavelength is generally similar to the spacings between crystal planes. When a researcher changes the angle between the X-ray beam and a crystal face, the diffraction pattern changes, and a series of diffraction photographs taken at different angles allows the investigator to formulate the three-dimensional atomic structure of the crystal.

Powder X-Ray Diffraction. In this technique

experimenters use fine grains of a crystalline substance instead of a single crystal. It is extensively employed for identifying such materials as minerals and chemical compounds, or such engineered materials as ceramics. It is also used for characterizing materials, for studying particles in liquid suspensions or thin films, and for elucidating the structures of components of polycrystalline materials. Because the tiny crystals in the powder sample are randomly oriented, researchers can collect diffraction data either by reflecting X rays from the sample or by transmitting them through it. Powder X-ray diffractometer systems allow scientists to obtain a diffraction pattern for the substance under investigation, which can then be used to calculate the unit cell of the substance.

Electron Diffraction. This technique depends on the wave nature of electrons, but because electrons are negatively charged they interact much more intensely with the electromagnetic environment of samples than neutral X rays or neutrons. Like X rays, electron beams can be scattered by atoms in a sample, producing patterns that can be registered on a photographic plate or fluorescent screen. To determine the positions of atoms or ions in solids, electron-diffraction instruments that can generate high-energy electrons in a thin beam are required. Nevertheless, electrons do not have the penetrating power of X rays, which leads researchers to use thin slices of solids. Relative to solids, molecules in a gas are far apart, which means that electron diffraction readily generates patterns that allow molecular dimensions to be determined. Because air molecules scatter electrons, these measurements have to be made in a vacuum. The electron-diffraction technique has enabled scientists to study films on solid surfaces. The new field of quasicrystals, which exhibit fivefold symmetry in violation of the traditional principles of crystallography, has benefited from electron-diffraction analyses.

Despite its usefulness, electron diffraction has its limitations. In electron-diffraction experiments, interplanar spacings can be discovered to accuracies in the range of one to one hundred parts per thousand, whereas in powder X-ray diffraction, interplanar spacings can be found to precisions of one to one hundred parts per million.

Neutron Diffraction. For many scientists and engineers, neutron diffraction, when compared with X-ray and electron diffraction, is the least utilized technique for structural determinations and other applications. Nevertheless, as instruments become more sophisticated and techniques more refined, this method has been increasing in popularity. Neutron diffraction differs from X-ray diffraction in terms of scattering points—electrons for X rays and nuclei for neutrons. This makes neutron diffraction useful in distinguishing certain isotopes and helpful in studies of such compounds as the metal hydrides (hydrogen is difficult to determine by X-ray diffraction). Disadvantages of this technique include the requirement of large crystals and an efficient neutron source. Large crystals can often be difficult to grow, and neutron sources such as nuclear reactors are not as generally available as diffractometers.

Despite these problems, neutron diffraction studies have been successfully performed to measure precisely the carbon-carbon distances in graphite as well as to determine the absolute configuration of atoms in several complex chiral structures (those in which it is impossible to superimpose a configuration on its mirror image). Chemists have also found neutron diffraction studies helpful in illuminating the function of hydrogen bonds in inorganic and organic compounds, the role of magnetism in such classes of compounds as the ferrites and rare-earth nitrides, and the nature of such condensed inert gases as helium, argon, neon, and krypton.

APPLICATIONS AND PRODUCTS

X-Ray Diffraction. German physicist Wilhelm Conrad Röntgen discovered X rays in 1895. The power of X rays to reveal structures within the human body was quickly recognized, leading to many applications. With the discovery of the power of X rays to uncover the previously hidden atomic structures of crystals in the second decade of the twentieth century, physicists, chemists, and geologists enthusiastically embraced the X-ray diffraction technique as an essential tool in their disciplines. Starting with such simple substances as sodium chloride and diamond, scientists began working out the structures of increasingly complex inorganic materials such as the silicates, and the technique was also applied to organic crystals (hexamethylenetetramine was the first to be determined). As instrumentation and methods became more sophisticated, X-ray diffraction, in both its single-crystal and powder forms, was most notably applied to specify the molecular structures in living

Fascinating Facts About Diffraction Analysis

- By the early twenty-first century, the Cambridge Structural Database contained precise information for more than one-quarter million small-molecule crystal structures, and most of the new data had arrived electronically.

- Nicolaus Steno, a Danish anatomist and the father of modern geology, and René Just Haüy, a French mineralogist, made important contributions to crystallography. Steno discovered the first law of crystallography, which states that every crystal of a particular kind has angles between the faces that are the same. Haüy found that every crystal face can be explained by the stacking of basic building blocks having constant angles and sides interrelated by simple integral ratios, now known as unit cells.

- Max von Laue was a Privatdozent and Paul Peter Ewald was a doctoral student at the University of Munich in 1912 when Ewald told Laue about his thesis on crystal models. In what he later called the biggest mistake of his scientific life, Ewald did not realize that the spacings between layers of his model suggested the possibility of X-ray diffraction, which Laue quickly went on to discover and prove.

- Laue won the Nobel Prize in Physics in 1914 for his discovery of X-ray diffraction, but during World War II he decided to have his gold medal dissolved in aqua regia to prevent its falling into the hands of the Nazis. After the war, the Nobel Society recast his medal in its original gold.

- The young English physicist Henry Moseley, influenced by the discoveries of Laue and the Braggs, used X-ray diffraction to discover atomic number, which led to important improvements in the periodic table, but his promising scientific career was cut short when he was killed in action in the Gallipoli campaign during World War I.

- In 1952, the English physical chemist Rosalind Franklin took what has been called the most famous X-ray diffraction photograph ever made. It was of the B form of DNA, and, without her permission, it was shown to James Watson and Francis Crick, who used it in making their monumental discovery of the double-helix structure of DNA.

things, such as proteins and DNA. Once the three-dimensional structure of DNA was determined by James Watson and Francis Crick in 1953, an explosion of applications followed, leading to the flourishing field of biotechnology as well as to applications in criminology, genetically modified foods, and medicine.

At the beginning of diffraction analysis the instruments and other products associated with this technique (X-ray diffractometers, photographic film, and various crystals) were relatively simple and inexpensive. As the field evolved, instruments became more complex and expensive. In the early years analyses of diffraction photographs were time-consuming and labor-intensive (some complex structures took years to figure out). With the development of efficient mathematical techniques and ever-more-powerful computers, complex structures could be determined in days or even hours.

A major application of X-ray diffraction analysis has been the study of defects in metals, alloys, ceramics, and other materials. These studies can be done more quickly and efficiently, and methods have become so refined that they can be applied to defects in nanocrystalline materials. Although the use of diffraction techniques in such industries as aerospace, iron, and steel is well known, not-so-well known is their use in the manufacture of such common household products as cleansers. Diffraction techniques enable researchers to monitor the effectiveness of the abrasive minerals these cleansers.

Electron- and Neutron-Diffraction Analysis. Like X-ray diffraction, electron and neutron analyses have pplications in the areas of characterizations of materials, substance identification, measurement of purity, texture description, and so on. Electron diffraction is better than X-ray analysis in studying of purity, texture description, and so on. Electron diffraction is better than X-ray analysis in studyinguch substances as membrane proteins, because, unfortunately, X rays pass through these thin layers without forming a diffraction pattern. Electron crystallographic analysis has been used to determine the atomic arrangements in certain proteins. More common is determining structures of inorganic crystals, including complex materials such as zeolites. As with X-ray analysis, the instruments and other products associated with electron-diffraction methods have become more advanced, effficeint , and expensive, but certain applications, such as the use of electron diffraction in electron microscopy, have led many scientists, from mineralogists to biologists, to expand and deepen their discoveries.

Applications of neutron analysis developed much later than those found in X-ray and electron diffraction. Like electron diffraction, neutron analysis has often been used in conjunction with X-ray techniques. For example, researchers were able, in such a combined study, to elucidate the internal dynamics of protein molecules. Independently, neutron diffraction has allowed investigators to study the details of atomic movements in substances. This basic scientific knowledge has helped others to develop better products, including window glass, semiconductors, and other electronic devices. In industry, neutron diffraction has been used to study the stresses, strains, and textures of various building materials. For example, metal alloys and welds often exhibit cracks or expansion as well as shrinkage, which limits the value of the respective products. Indeed, this practice is so prevalent that it has been named "engineering diffraction."

IMPACT ON INDUSTRY

Many thousands of structure determinations being done at a variety of institutions—governmental, academic, and industrial—together with a rough estimate of the average cost per determination indicate that this scientific and technical field's valuation is most likely in the billions of dollars. Data from various professional organizations reveal that the growth or decline of different segments of this field, which is often tied to the market value of various materials, is sensitive to the economic state of particular countries. Viewed from a historical perspective, though, the growth of this field since the discovery of X-ray diffraction in 1912 has been truly remarkable.

Government and University Research. Diffraction analysis thrives in most advanced industrialized societies, and government agencies have played an important role in funding structural studies and the development of new equipment. Governmental agencies in the United States such as Sandia National Laboratories, part of the Department of Energy's National Nuclear Security Administration, and the U.S. Patent Office either fund their own projects or provide contracts for proposals and programs

at academic institutions and industrial research centers. A representative example is the X-ray Diffraction Center in the Department of Chemistry at the University of Missouri-Columbia.

Industry and Business Sectors. The mining, metallurgical, biochemical, and pharmaceutical industries make extensive use of diffraction-analysis technologies. As chemists create new compounds, alloys, and drugs, knowledge of the three-dimensional structures of these substances is essential in understanding how they function in nature, the human body, and machines. Diffraction techniques are important in new materials development, especially in the growing field of nanomaterials. Forensic scientists use these techniques for identifying and characterizing crime-scene materials. X-ray diffraction even plays a role in the imaging industries, for example, in evaluating existing and future photographic materials. Large companies such as Eastman Kodak have supported the research and development of new imaging chemicals, sensors, and hybrid substances. Other companies, such as Matco Services, use diffraction techniques in their corrosion investigations.

CAREERS AND COURSE WORK

Associations such as the American Crystallographic Association provide detailed information on the education needed to pursue a career in diffraction analysis. Courses in physics, chemistry, and higher mathematics form the basis for later study in specialized diffraction techniques, mineralogy, or metallography. If, for example, a person wishes to pursue a career in failure analysis, courses in fracture mechanics, fractography, and corrosion testing and engineering form part of the curriculum. On the other hand, a student desiring a career in diffraction analysis as it applies to biotechnology would take courses in molecular biology, molecular biophysics, molecular genetics, and biostatistics.

Many careers are possible for students who complete programs in diffraction analysis, from being an X-ray diffraction technician to becoming the head of an integrated imaging facility. Some companies want applicants with interdisciplinary expertise, for example, someone with a mastery of mineralogy, X-ray diffraction, and modern computer-modeling methods. Many universities have X-ray diffraction facilities and will hire research assistants as well as tenured professors. The increasing number of forensic science laboratories hire large numbers of scientists and technicians, many of whom should have expertise in diffraction techniques.

SOCIAL CONTEXT AND FUTURE PROSPECTS

If the past is a prelude to the future, then the twenty-first century will be a time when the determination of structures by diffraction techniques will become faster, more efficient, and more accurate than determinations in the twentieth century. Another established trend has been the application of these techniques to more complex structures, and this, too, should continue. Many prognosticators predict that nanotechnology will flourish in the twenty-first century, and diffraction techniques have already been applied to smaller and smaller crystals. The environmental concerns that intensified in the latter part of the twentieth century will most likely continue throughout the twenty-first, and diffraction techniques provide powerful tools for identifying, characterizing, and developing detailed structural understanding of pollutants and also of the chemical compounds that make up the life-forms that are increasingly threatened by these pollutants.

Another trend that is likely to continue is the improvement of traditional techniques and instruments and the discovery of new ones. In the late twentieth century computer-controlled diffractometers came into wide use because they facilitated the collection and processing of digitized diffraction patterns. According to Moore's law, the computing power of integrated circuits doubles about every eighteen months, and, if this law remains valid, the computerization of diffraction technologies should become more powerful and less expensive than ever before. As more and more structures are determined by X-ray, electron, and neutron diffraction, larger and larger databases will be created, and more sophisticated computer software will be written to mine this treasure trove for useful information. Diffraction analyses have been largely concentrated in educationally and industrially well-developed societies, but as the less developed countries evolve, diffraction techniques will undoubtedly spread to these nations.

Robert J. Paradowski, M.S., Ph.D.

FURTHER READING

Bacon, G. E., ed. *X-Ray and Neutron Diffraction.* New York: Pergamon Press, 1966. This book, part of a series intended to introduce undergraduates to important topics in physics through original sources, treats diffraction techniques as the culmination of a long tradition in crystallography.

Chung, Frank H., and Deane K. Smith, eds. *Industrial Applications of X-Ray Diffraction.* New York: Marcel Dekker, 2000. This well-illustrated book features extensive coverage of X-ray diffraction applications for a large number of industries.

Hammond, Christopher. *The Basics of Crystallography and Diffraction.* 3d ed. New York: Oxford University Press, 2009. The author's intention is to provide the essentials of crystallography and diffraction to beginning students from a wide variety of fields. He emphasizes making difficult topics crystal clear and using history and biography to bring out the human element in his subjects.

McPherson, Alexander. *Introduction to Macromolecular Crystallography.* 2d ed. Hoboken, N.J.: Wiley-Blackwell, 2009. Intended for students interested in pursuing careers in such areas as pharmacology, protein engineering, bioinformatics, and nanotechnology, this well-illustrated book provides an easily understood introduction for readers without a background in science and mathematics.

Scott, Robert A., and Charles M. Lukehart, eds. *Applications of Physical Methods to Inorganic and Bioinorganic Chemistry.* Hoboken, N.J.: John Wiley & Sons, 2007. The editors intend this book as a practical introduction to those physical methods, including diffraction techniques, that have been used to gather relevant structural information about inorganic and organic materials. Includes an extensive list of abbreviations.

Warren, B. E. *X-Ray Diffraction.* Mineola, N.Y.: Dover Publications, 1990. This inexpensive reprint is a well-illustrated introductory text that is an excellent resource for learning X-ray diffraction from the ground up.

WEB SITES

American Crystallographic Association
http://www.amercrystalassn.org

American Physical Society
http://www.aps.org

Elmer O. Schlemper X-Ray Diffraction Center
http://www.chem.missouri.edu/x-ray

International Union of Crystallography
http://www.iucr.org

Pittsburgh Diffraction Society
http://www.pittdifsoc.org

See also: Biophysics; Ceramics; Mineralogy; Optics.

DNA ANALYSIS

FIELDS OF STUDY

Biochemistry; biotechnology; molecular biology; molecular genetics; population genetics; forensic science; statistics.

SUMMARY

DNA analysis involves the use of scientific tools to access the information found in DNA to identify its source, whether some infectious agent, another organism of interest, or a particular individual, such as in forensic applications. Medical applications of this technology include the search for mutations associated with genetic disorders and the design of probes that are able to diagnose these disorders in a timely fashion.

KEY TERMS AND CONCEPTS

- **Base Pair:** Single unit of double-stranded DNA.
- **Combined DNA Index System (CODIS):** Federal Bureau of Investigation (FBI) database that contains the DNA profiles of more than 8 million convicted violent offenders.
- **DNA (Deoxyribonucleic Acid):** Nucleic acid that is the genetic component of all living cells. Nucleic acids are polymers of nucleotides.
- **Nucleotide:** Chemical composed of a nucleoside (a nitrogen-containing purine or pyrimidine base linked to a five-carbon sugar such as ribose or deoxyribose) and a phosphate or polyphosphate group.
- **Polymerase Chain Reaction (PCR):** Technique used to produce a large amount of DNA from very small quantities.
- **Restriction Fragment Length Polymorphism (RFLP):** Early method of DNA analysis involving the cleavage and separation of DNA.
- **Short Tandem Repeat (STR):** Repeating element in DNA that is from one to six base pairs long; also known as microsatellite.
- **Single Nucleotide Polymorphism (SNP):** Point mutation in DNA that differs among individuals.
- **Southern Blotting:** Process used to detect the presence of a particular DNA sequence in a sample. DNA fragments are separated using gel electrophoresis, then transferred, or blotted, onto a membrane, where they are exposed to a hybridization probe (a fragment of DNA that will hybridize to a complementary sequence).
- **Variable Number of Tandem Repeats (VNTR):** Repeating element in DNA that is tens to hundreds of base pairs long; also known as minisatellite or long tandem repeat.

DEFINITION AND BASIC PRINCIPLES

DNA analysis is, in the strictest sense of the term, an actual observation of the length or sequence of a portion of DNA. The length of a fragment of DNA can be determined using gel electrophoresis. This technique involves placing DNA onto a semisolid support, or gel, and applying an electric current to the gel so that DNA migrates toward the positive pole. The migration of DNA in gel electrophoresis is proportional to its mass, which is, in turn, proportional to its length. Determining the actual sequence of bases in a strand of DNA is much more complicated.

Although DNA analysis has at times been equated with genetic testing, the two are not always the same. Certain types of genetic testing developed in the 1960's did not technically involve DNA analysis. Amniocentesis, which allowed for Down syndrome testing, actually involved chromosomal analysis following the creation of a karyotype (organized profile of a person's chromosomes). Similarly, genetic testing for phenylketonuria (PKU) and Tay-Sachs disease originally involved enzyme assays, not an analysis of the defective genes themselves. Thus, actual DNA analysis did not begin in earnest until the mid-1970's.

BACKGROUND AND HISTORY

Although the double helical structure of DNA was first described in 1953 by American molecular biologist James D. Watson and British biophysicist Francis Crick, more than twenty years passed before scientists developed methods of comparing DNA for the purpose of identification. In 1974, British molecular biologist Joseph Sambrook described the differentiation of human tumor viruses following cleavage by a restriction endonuclease (an enzyme that cleaves DNA at a specific nucleotide sequence). He noticed that different-sized bands of DNA were visible following their separation on a gel. This discovery formed the basis for what has become known as restriction fragment length polymorphism (RFLP). Although the DNA of viruses and even bacteria could be analyzed directly by RFLP, the restriction enzyme cleavage patterns of higher organisms were of sufficient complexity that only a subset of the bands that were produced could be analyzed. DNA analysis of higher organisms was made possible in 1975 by the development of the Southern blotting technique by British biochemist Edwin Southern.

The following decade saw two ideas that would revolutionize the field of DNA analysis. In 1983, American biochemist Kary Mullis developed the polymerase chain reaction (PCR), a process that enabled the amplification across several orders of magnitude of small amounts of starting sample DNA in the laboratory. Then, in 1985, British geneticist Alec Jeffreys realized that human DNA was peppered with regions of repeating sequences, or variable number tandem repeats (VNTRs), and that comparisons of these regions could create a unique DNA fingerprint for any given individual.

How It Works

Probes and Primers. Most of the methods of DNA analysis take advantage of DNA's natural tendency to form a double helix. RFLP analysis has long been paired with Southern blotting. This procedure involves binding a small synthetic fragment of DNA to a region of interest that is contained within one or more of the bands of DNA that have been separated by gel electrophoresis and then blotted onto some type of membrane. This binding is made possible by the complementary nature of the DNA bases, the fact that adenine forms hydrogen bonds to pair with thymine and that cytosine pairs with guanine, a concept called Watson-Crick base pairing after the discoverers of DNA structure. The synthetic DNA, called a probe, is designed to contain about twenty complementary nucleotides to the sequence of interest. Binding of this probe to the blotted DNA is called hybridization. Originally, DNA probes were labeled with a radioactive marker to enable the detection of their position on a membrane, but later probe labeling has included nonradioactive alternatives such as fluorescent dyes.

The polymerase chain reaction (PCR) also involves the binding of small synthetic fragments of DNA to a region of interest, but in this process, two such fragments bind to opposite strands of the target DNA. These fragments, although identical in structure to the probes described earlier, are called primers because they are used to prime a DNA synthesis reaction. Also, the binding reaction, which involves the same process of complementary bases coming together to form hydrogen bonds, is referred to as annealing. It had been previously discovered that DNA could be made in the laboratory by taking a given single strand of DNA and adding a specific primer, the four types of nucleotides, and purified DNA polymerase (the enzyme normally involved in the polymerization process), but Mullis's insight in the early 1980's was that this process could be converted into a chain reaction that produced large amounts of DNA. By adding two primers instead of one and by using double-stranded DNA as a target, twice as many molecules of DNA could be created, but this necessitated a step in which the DNA had to be heated to near-boiling temperatures to separate, or melt, the two strands of the double helix. Mullis reasoned that a DNA polymerase that had been purified from a thermophilic microbe would be able to survive this heating step. This would allow the stringing together of a number of cycles with three different temperatures—one each for annealing, polymerization, and DNA melting—without having to add more DNA polymerase enzyme. Thus, after each cycle, the amount of DNA double helix would be doubled, resulting in more than a billion molecules of DNA after thirty cycles, even if the cycle started with only one strand of DNA.

DNA Polymorphism. DNA analysis takes advantage of the intrinsic variability that exists among organisms as well as among members of the same species. Polymorphism, a word derived from the Greek for "many forms," is used to describe this variability, the simplest form of which is a single nucleotide polymorphism (SNP). Single nucleotide polymorphisms, which are also referred to as point mutations in genetics, are detectable by RFLP analysis only when they occur within the recognition sequence for a particular restriction enzyme because the enzyme fails to cleave the altered sequence. RFLP analysis also readily detects deletions or insertions of DNA sequences that have occurred between restriction enzyme cleavage sites. What Jeffreys realized in the

1980's was that most restriction fragment length variation in humans was not caused by large insertions or deletions of unique DNA sequences but by a variation in the number of repetitive DNA elements that were found in tandem with one another. He did not, however, use the PCR method that was being developed at the time because a practitioner of PCR must know the precise sequences that flank a site of interest to design the primers used in this procedure. Instead, Jeffreys performed Southern blotting using a probe designed to hybridize with the about fifteen-nucleotide-long sequence that he was studying. This probe specifically labeled the regions of the membrane that contained these variable number tandem repeats (VNTRs). For this contribution, Jeffreys has been called the father of DNA fingerprinting.

Subsequent analysis of various regions in human DNA has taken advantage of PCR to produce results, focusing on even smaller tandem repeats with repeating units that are only one to six nucleotides in length. Discovered in 1989, these short tandem repeats (STRs) were eventually found to outnumber variable number tandem repeats by nearly one hundredfold, being found at more than 100,000 sites in human DNA. As more and more of these STR sites were characterized over time, primers that annealed to their flanking sequences were designed to amplify the repeat area in question.

APPLICATIONS AND PRODUCTS

Of Microbes and Man. Although the tools involved in DNA analysis are often used in basic research such as determining the evolutionary relationships between organisms, much of the application of this technology involves analysis for the purposes of identification. Although identification could potentially include any organism of interest, the primary focus of DNA analysis has been disease-causing viruses and microorganisms along with humans. Ever since Sambrook and colleagues first applied RFLP analysis to differentiate between two strains of viruses, viral epidemiology has remained an important application for tools such as PCR. For example, around the beginning of the twenty-first century, nucleic acid amplification testing (NAAT) was developed to detect the viral load of the human immunodeficiency virus (HIV). The procedure is a faster and more effective way to test for the presence of HIV in a person. NAAT has also been applied as a diagnostic test for certain bacterial infections. Other PCR-based methods have been adapted to test for bacterial contamination of foods as well as of hospital areas and supplies. In most cases, the identification of the precise strain of virus or microbe present is unnecessary because the physical presence of an infectious agent, not its detailed classification, is of interest. Tandem-repeat-based methods of identification are largely useless when analyzing such infectious agents because these agents tend to lack such repetitive DNA sequences. Because the DNA sequence of the entire genome (the complete set of DNA found in a particular organism) of most known infectious agents has been determined, it is possible to design primers that will specifically amplify DNA from a given target species.

In some cases, as in life-threatening illnesses, potential epidemics, and acts of bioterrorism, the speed at which an infectious agent is identified is critical to saving lives. For such applications, a type of PCR called real-time PCR has been developed. Rather than waiting to run gel electrophoresis after the full thirty or so cycles of a traditional PCR reaction have been completed, real-time PCR measures the production of a fluorescent-tagged product in real time, during the early phases of the reaction. This allows for an agent to be detected in minutes rather than hours.

Crime Scenes and Beyond. The best-known use of DNA analysis is probably in the area of forensics. The first case in which Jeffreys applied DNA fingerprinting was an immigration dispute. In 1983, British authorities had denied a thirteen-year-old boy entry into the country, claiming that his passport was forged and that his stated mother, a British subject, was not his biological mother. The dispute continued until 1985, when Jeffreys was able to apply his new technique to prove that the maternal relationship stated on the passport was indeed correct. Since that time, maternity tests have been vastly outnumbered by paternity tests, but the principle used in both types of parental testing remains the same.

The first use of DNA fingerprinting in a criminal case occurred in 1986, when it was used to exonerate a suspect accused of the rape and murder of a teenage girl near Leicester, England. Later, the same technique was used to identify the real killer. Since this early case, evidence from DNA fingerprinting has helped convict thousands of criminals. The source of DNA is blood in about half of all cases;

other common sources are semen and hair. DNA analysis also plays an important role in the identification of human remains following disasters, acts of terrorism, and war.

Limitations of PCR. Following the advent of PCR, the amount of forensic sample required for analysis was reduced significantly. The original DNA fingerprinting procedure developed by Jeffreys required a blood sample about the size of a quarter, but later methods needed only a few cells swabbed from a person's cheeks to perform an analysis. Although PCR requires much less starting material than RFLP analysis and is also a more rapid procedure to perform, it does have a number of limitations. The first limitation, that flanking DNA sequences must be known ahead of time, was largely overcome as more and more human short tandem repeats were characterized along with the DNA that surrounded them. A second limitation is that the method is so sensitive that it is prone to contamination by outside sources. Because even a single fragment of DNA can be amplified into large amounts on a gel, care must be taken not to introduce foreign DNA from an investigator's hair or fingertips. A third limitation is that only a single area, or locus, of DNA can be analyzed at one time. To overcome this limitation, a procedure called multiplex PCR has been developed. This method simultaneously employs a number of primers that have been labeled with fluorescent tags. These can be identified during the subsequent gel electrophoresis step based on their specific labels.

Medical Applications. Besides using DNA analysis to identify infectious agents, the medical community has begun to use this technique to study genetic disorders. However, common methods of DNA analysis cannot identify most genetic disorders, with the exception of a class of disorders called trinucleotide repeat expansion disorders. This rare class of disorders, which includes Huntington's disease as well as fragile X syndrome, is readily detectable using PCR amplification of the short tandem repeats that contribute to the disorders in question. A more common class of genetic disorders results from point mutations in genes and can therefore be linked to particular single nucleotide polymorphisms in the human genome. Unfortunately, single nucleotide polymorphisms are not detectable by PCR and will show up in RFLP analysis only if they occur in the restriction enzyme recognition site itself, which is a rare occurrence. The

identification of genetic disorders is therefore largely dependent on determining the actual sequence of the DNA, still a technically challenging and expensive undertaking despite progress that has been made since the inception of the Human Genome Project DNA sequencing program in the 1990's.

Methods involved in DNA sequencing include many of the same principles as other forms of DNA analysis. A single primer is labeled with a fluorescent dye and mixed with a target sequence in the presence of a thermostable DNA polymerase. This procedure does not amplify the DNA as in PCR but results in primer extension for a certain length along the target sequence. Another difference from PCR is that modified nucleotides are added to this mixture so that the primer extension is halted whenever these particular nucleotides are incorporated into a growing DNA strand. Four separate tubes are used in this method, one for each of the four DNA bases. Once these four reactions are separated by electrophoresis, the order of bases can be determined using computer software that monitors the relative migration of the bands that occurs from each of the four reaction tubes.

IMPACT ON INDUSTRY

Perhaps because the United States and the United Kingdom rank first and second, respectively, for the most crimes committed within an industrialized nation, or perhaps because they are world leaders in biomedical research, these two nations have played a leading role in the development of tools for use in DNA analysis from the very beginning. Soon after Jeffreys developed the DNA fingerprinting technique, the Cellmark Diagnostics division of the British company Imperial Chemical Industries (ICI) began offering the first commercially available DNA testing kit. Since then, molecular diagnostics/genetic testing has become a billion dollar industry, with many of the companies based in the United States. In addition to producing reagents required to perform genetic testing, some companies have focused on providing kits for the extraction of DNA from a variety of sources, including bacteria, soil, water, blood, tissue, and bone. Because the basic steps in DNA extraction usually include cell lysis, ribonucleic acid (RNA) removal, and the separation of DNA and proteins based on their differing solubilities in salt- and alcohol-containing solutions, many of these kits have a number of reagents in common, although the procedure used in extraction can be adapted to the specific source of the DNA.

The governments of the United States and the United Kingdom also assumed leading roles in DNA analysis. In 1990, the Federal Bureau of Investigation established the combined DNA index system (CODIS), a database to contain the information gained from DNA fingerprinting of convicted offenders. The numerical data that correspond to the migration of bands obtained from DNA fingerprinting are also referred to as an individual's DNA profile. Early on, the FBI chose thirteen specific STR loci to include in its system. As of 2010, CODIS contained more than 8 million DNA profiles. In 1995, the United Kingdom's Forensic Science Services established a similar system to catalog DNA profiles, including many of the same STR loci. In addition to law enforcement, the governments of both the United States and the United Kingdom were significantly involved in the Human Genome Project. On June 26, 2000, President Bill Clinton and Prime Minister Tony Blair made a simultaneous announcement that a first draft of the human genome had been completed ten years into the project.

CAREERS AND COURSE WORK

Scientists in general have traditionally chosen from three career paths, industry, academics, and government, with industry providing the most jobs and government the fewest. This general principle is true for those interested in the science of DNA analysis. Industry leads the way in the design of diagnostic tests and DNA extraction kits, academics is the domain of basic research, and government dominates the field of forensic science. Most of of the forensic analysis performed in the United States occurs at the governmental level. About half of the 400 publicly owned crime laboratories at the municipal, county, state, and federal level are capable of performing DNA analyses; the rest rely on a government laboratory higher up in their jurisdiction or on a private laboratory for this service.

Those who are interested in a career in forensic science should bear in mind a few facts. First, the hybrid job of police officer/detective/forensic scientist depicted on television dramas does not exist in reality. Forensic scientists either work in the field or in the laboratory but rarely in both . In large laboratories, the forensic analyst will tend to specialize in a

---

Fascinating Facts About DNA Analysis

- In addition to Southern blotting, two other forms of blotting are performed in molecular biology. These involve blotting of either RNA or protein for analysis and are named Northern blotting and Western blotting, respectively, as a humorous homage to Edwin Southern.
- Ironically, Kary Mullis, who developed the method used to screen for the viral load of HIV in humans, is among the handful of scientists who reject the scientific evidence that HIV is the cause of acquired immunodeficiency syndrome (AIDS).
- Colin Pitchfork, the first criminal convicted of murder based on DNA fingerprinting evidence, initially evaded arrest by telling a friend that he was terrified of needles and paying that friend to submit a blood sample for him.
- DNA extracted from blood actually comes from the white blood cells, not the red, even though the latter outnumber the former by a ratio of 700:1. Human red blood cells lose their nuclei, and therefore their DNA, during development.
- The first human genome sequence produced was actually a mosaic of DNA sequences from various anonymous donors. In 2007, J. Craig Venter, the head of a private company involved in the Human Genome Project, became the first individual to have his entire genome sequenced. Venter said his company had largely used his own DNA in the sequencing efforts that they had contributed to the project.
- According to the latest estimates, the Bureau of Justice Statistics at the U.S. Department of Justice reports that tens of thousands of requests for DNA analysis are backlogged at any given time because of the high demand for this service. This represents the highest percentage of backlogged requests for any type of analysis performed by crime laboratories under their jurisdiction.

---

particular area, and in small laboratories, the analyst will be more of a generalist but also will lack most of the resources and equipment shown on television. Second, although dozens of colleges have introduced degree programs in forensics to keep up with the increased demand, any bachelor's degree that gives its holder a solid background in science and mathematics and excellent communication skills is sufficient to work in this area. Third, students should realize that such a bachelor's degree prepares a person to begin working as a technician performing largely support functions such as preparing reagents for analysis. A master's degree or certification program is most likely required to specialize in a subfield of forensic science and to perform more of the scientific analysis of evidence, while a doctoral degree in an associated field is preferred for advancement to administrative positions such as laboratory director.

SOCIAL CONTEXT AND FUTURE PROSPECTS

Single nucleotide polymorphisms (SNPs), although not used extensively in forensic applications, potentially contain valuable information that can be of use to crime scene investigators. For example, the presence of particular SNPs may indicate a perpetrator's race, while others could indicate hair color. One disadvantage of SNPs, besides the relative difficulty of identifying them, is that many more of them are needed to provide a unique identification (compared to the number of short tandem repeats needed for PCR). Because most SNPs are biallelic, they contain one base or another but generally not all four possible bases, and it is estimated that as many as fifty would have to be analyzed to obtain the same level of confidence as provided by the thirteen STR loci contained in CODIS. This may not prove as difficult as it sounds because it is estimated that there are probably about 10 million SNP sites scattered throughout the human genome. If accurate, that would mean that SNPs outnumber short tandem repeats to the same degree that short tandem repeats outnumber variable number tandem repeats.

DNA sequencing in some form or another is likely to continue to play an increasing role in DNA analysis. The cost of DNA sequencing probably will drop as it becomes more prevalent and increasingly automated. Although the first human genome sequence was produced at a cost of billions of dollars, scientists have set a goal of reducing the cost of DNA sequencing to about one thousand dollars. At the same time, scientists are developing a number of methods that allow SNPs to be determined without first finding the sequence of the 99.7 percent of DNA bases that do not exist as SNPs. These methods include directed hybridizations, ligations, primer extensions, or nuclease cleavages that specifically involve SNPs while leaving the rest of the DNA alone.

With any increase in the involvement of DNA sequencing in forensics comes the likelihood that debate will intensify concerning privacy issues regarding the use of sequence information. Unlike commonly used methods of PCR analysis, SNP determination will reveal certain details about suspects that could be open to abuse. Ethical issues involving the use and dissemination of DNA data will have to be resolved as the methods of DNA analysis continue to evolve.

James S. Godde, Ph.D.

FURTHER READING

McClintock, J. Thomas. *Forensic DNA Analysis: A Laboratory Manual.* Boca Raton, Fla.: CRC Press, 2008. Examines the various methods of DNA analysis and DNA fingerprinting.

Nakamura, Yusuke. "DNA Variations in Human and Medical Genetics: Twenty-five Years of My Experience." *Journal of Human Genetics* 54 (2009): 1-8. A historic perspective on the progression of DNA analysis techniques with a particular emphasis on human disease characterization.

Pereira, Filipe, Joao Carneiro, and Antonio Amorim. "Identification of Species with DNA-Based Technology: Current Progress and Challenges." *Recent Patents on DNA and Gene Sequence* 2 (2008): 187-200. Contains an excellent table comparing methods of DNA analysis, along with a helpful flowchart. Also contains clear descriptions of each method, including diagrams.

Roper, Stephan M., and Owatha L. Tatum. "Forensic Aspects of DNA-Based Human Identity Testing." *Journal of Forensic Nursing* 4 (2008): 150-156. A straightforward description of all pertinent methods and applications of DNA analysis, including simple diagrams as well as a glossary of terms.

Rudin, Norah, and Keith Inman. *An Introduction to Forensic DNA Analysis.* 2d ed. Boca Raton, Fla: CRC Press, 2002. Discusses forensic DNA analysis from both the medical and legal standpoints. Examines the advantages and limitations of the various techniques.

Watson, James D., and Andrew Berry. *DNA: The Secret of Life.* New York: Alfred A. Knopf, 2006. This comprehensive introduction to DNA has the famous biologist Watson as one of its authors.

WEB SITES

Association of Forensic DNA Analysts and Administrators
http://www.afdaa.org

The DNA Initiative
Advancing Criminal Justice Through DNA
Technology
http://www.dna.gov

Federal Bureau of Investigation
CODIS
http://www.fbi.gov/hq/lab/html/codis1.htm

International Society for Forensic Genetics
http://www.isfg.org

See also: Criminology; DNA Sequencing; Genetic
Engineering; Human Genetic Engineering.

DNA SEQUENCING

FIELDS OF STUDY

Biology; genetics; population genetics; genomics; forensics; bioinformatics; microbiology; biotechnology; biological systematics; chemistry; biochemistry; medical genetics; clinical biochemical genetics; clinical molecular genetics; clinical cytogenetics; genetic counseling; molecular oncology; pharmaceuticals; pharmacogenomics; anthropology; archaeology; genealogy; agriculture; bioethics; engineering; computer science; mathematics; bioremediation.

SUMMARY

DNA (deoxyribonucleic acid) sequencing is a technique used to determine the order of the nitrogenous bases (adenine, guanine, cytosine, and thymine) that make up a gene, DNA molecule, or entire genome. Genome sequencing has been completed for many organisms, including animals, plants, and humans. Applications of this technology can advance the understanding, diagnosing, and treatment of disease; enable personalized health care; and produce innovative techniques that can be used in forensics, agriculture, and archaeology.

KEY TERMS AND CONCEPTS

- **Deoxynucleoside Triphosphate (dNTP):** Nucleotide used to synthesize strands of DNA during sequencing.
- **Dideoxynucleoside Triphosphate (ddNTP):** Nucleotide lacking the 3'-hydroxyl function on its deoxyribose sugar required for formation of bonds between nucleotides.
- **DNA Polymerase:** Enzyme functioning as a catalyst in the connection of deoxynucleoside triphosphates.
- **Electrophoresis:** Technique that exposes molecules in a medium to an electric field in order to separate them by some feature.
- **Nitrogenous Base:** Nitrogen-containing base, such as adenine, guanine, cytosine, and thymine (DNA) or uracil (RNA). Represented as A, G, C, T, and U.
- **Nucleotide:** Basic unit of DNA consisting of a five-carbon sugar, a phosphate group, and a nitrogenous base.

- **Oligonucleotide:** Short strand of DNA synthesized and used as a primer for a polymerase chain reaction.
- **Polymerase Chain Reaction (PCR):** Technique in which target sequences of DNA are amplified to large amounts for use in molecular and genetic analyses.

DEFINITION AND BASIC PRINCIPLES

DNA sequencing is a laboratory technique that allows scientists to determine the structure of DNA at its highest level of resolution. DNA is a double-stranded helix made of building blocks called deoxyribonuceotides. Deoxyribonuceotides are nucleotides that contain the deoxyribose sugar as well as a phosphate group and a nitrogenous base. The information in DNA is stored as a code made up of the nitrogenous bases adenine (A), guanine (G), cytosine (C), and thymine (T).

DNA sequencing has a number of applications that may revolutionize medicine, agriculture, anthropology, and archaeology. The prospective uses of these applications include personalized medicine, the decoding of genes, and the identification of mutations causing genetic diseases. Sequencing, coupled with genetic engineering and agriculture, has produced plants with improved nutritional quality, greater resistance against insects, and better ability to withstand poor soil and drought. Sequencing of entire genomes has been completed for several organisms, including extinct species. This has provided an extensive insight into human migration and the evolution of all living organisms.

Despite the prospective and generated benefits of DNA sequencing, the use of this technique has raised ethical, legal, and social questions. Concerns exist regarding genetic determinism and discrimination, the manipulation of an individual's attributes, the loss of genetic privacy, and the modification of food. Programs have been created in response to these concerns.

BACKGROUND AND HISTORY

In 1953, James D. Watson and Francis Crick proposed the double-helical structure of DNA. This discovery has since yielded revolutionary insights into the genetic code and protein synthesis. However, because of certain properties of DNA, it took about

191

fifteen years before the first sequencing experiments were completed.

The first nucleic acid to be sequenced was yeast alanine transfer RNA (tRNA) because of its size and availability for purification. Following this accomplishment, scientists began to purify genomes of bacteriophages and pursue sequencing. However, whole genome sequencing did not become an actuality until after the discovery of restriction enzymes by Hamilton Smith, Werner Arber, and Daniel Nathans in 1970 and the development of more modern methods of DNA sequencing by Frederick Sanger and Alan R. Coulson in 1975. In 1977, the first genome, belonging to the bacteriophage phiX174, was sequenced.

The increasing amount of information created a need for computer programs capable of compilation and analysis of DNA, followed by databases with rapid searching programs (such as Genbank, created in 1982). These developments, as well as several advances in laboratory techniques (such as automated sequencing), led to the completion in 1998 of the first genome for an animal, a nematode called *Caenorhabditis elegans*, and ultimately the complete mapping of the human genome in 2003.

How It Works

Plus and Minus Method. Before the development of direct DNA sequencing, DNA had to be converted into RNA, sequenced, and then decoded. In 1975, Sanger and Coulson introduced plus and minus sequencing, the first direct DNA sequencing method. Plus and minus sequencing begins with several asynchronous cycles of DNA synthesis with radioactively labeled deoxynucleoside triphosphates (dNTPs). The asynchronous cycles lead to an array of DNA fragments varying in nucleotide length (1, 2, 3, . . . 100). Products are separated into eight containers and dNTPs added; however, each container receives either one of the dNTPs (the plus reactions) or three of the four dNTPS (minus reactions). This allows for termination of synthesis in a sequence-specific manner. The products are separated via electrophoresis on a polyacrylamide gel. Subsequently, the gel is exposed to X-ray film that results in a series of bands corresponding to the radiolabeled DNA fragments, allowing the sequence to be constructed. Although this method revolutionized how sequencing was completed, it was inefficient, and therefore other techniques were developed.

Maxam and Gilbert Method. In 1977, Allan Maxam and Walter Gilbert developed a sequencing method that replaced the plus and minus method. This method was similar, as it required a radioactive label, gel electrophoresis for fragment separation, and the use of X-ray autoradiography for product visualization and inference of the sequence. However, the method differed in that it allowed for the direct analysis of purified double-stranded DNA and used another way of creating products ending in a specific nucleotide.

In the Maxam and Gilbert method, the double-stranded DNA is radioactively labeled, cut with restriction enzymes, and denatured. Subsequently, the DNA is treated chemically in four separate reactions, which cut DNA at different nucleotides. The first reaction, called the A+G reaction, cuts the nucleotides adenine (A) and guanine (G). The second reaction, called the G reaction, cuts at G. The third and fourth reactions are similar, but involve cytosine (C) and thymine (T) in the C+T and the C reaction. The products of these four reactions are run through gel electrophoresis in four adjacent wells and analyzed for the sequence. The G reactions determine the placement of G, the A+G reactions determine the location of A, and so forth.

Sanger Sequencing Method. Named after Sanger, this method was developed in December, 1977, by Sanger and Gilbert. Because of its overall efficiency and limited use of chemicals and radioactivity, it has become the most widely used technique. This method takes advantage of dideoxynucleoside triphosphates (ddNTPs), analogues of the dNTPs. The ddNTPs are nucleotides lacking the 3'-hydroxyl function on its deoxyribose sugar required for the formation of phosphodiester bonds between two nucleotides of a developing DNA strand. Therefore, the ddNTPs are used during the synthesis of DNA strands to terminate DNA extension, resulting in products of different lengths.

Originally, this method required four separate reactions. Each aliquot contained a template DNA primed with an oligonucleotide, DNA polymerase to extend the sequence, and dNTPs. Each reaction received one of the chain-terminating ddNTPs labeled radioactively for detection. Later, sequencing could be completed in one reaction by substituting the radioactive label with four unique fluorescent dyes corresponding to each ddNTP. Further progress was made in 1983, when Kary Mullis introduced polymerase chain reactions (PCR), which allowed target sequences of DNA to be amplified in a fraction of the time. After PCR, the strands are separated and analyzed with automated sequencers, resulting in a chromatogram with a series of four-colored peaks representing each of the DNA bases. Computers are used to assemble sequences and analyze them for a variety of characteristics.

APPLICATIONS AND PRODUCTS

Genetic Diagnostics. The increased knowledge of genes and the organization of the genome have had a significant impact on medicine. Any disorder is caused by a combination of the environment and genetics; however, the role of genetics may be large (as in Huntington's disease) or small (as in diabetes). Sequencing has helped elucidate the genetic variation and mutations responsible for predisposing a person to disease, modifying the course of a disease, or causing the disease itself. Understanding the molecular mechanisms of disease allows for the development of tests, diagnoses, treatments, and even cures and preventative options.

Personalized Medicine. Medicine is moving in the direction of using specific treatments based on patients' individual attributes, a development termed as personalized medicine. Although personalized treatments are being developed in many medical fields, the most striking examples are in oncology. For example, physicians are measuring the levels of human epidermal growth factor receptor 2 (HER2) in patients with breast cancer, and if the test is positive, the person is treated with trastuzumab.

Agriculture. Sequencing the genomes of plants and animals has made it possible to create transgenic organisms, which incorporate desired characteristics from other organisms. For example, genes from the bacterium *Bacillus thuringiensis* have been successfully transferred to crops such as rice, cotton, corn, and potatoes, thereby producing plants that are protected from insects. The alteration of genomes has also produced plants that resist drought and disease. Golden Rice, a genetically modified strain of rice, contains high levels of beta-carotene, which is converted to vitamin A in the human body. This rice has the potential to fight vitamin A deficiency in less-developed countries. Interestingly, bananas have also been modified to produce human vaccines against

---

Fascinating Facts About DNA Sequencing

- A completed sequence of the human genome was revealed in the spring of 2003, fifty years after James D. Watson and Francis Crick described the structure of DNA.

- DNA has been successfully amplified from several sources, including hair, blood, saliva, bone marrow, body tissue, and urine. DNA can be extracted from seeds, plant leaves, vegetables, fruits, bacteria, and any other material containing DNA.

- DNA sequencing is not limited to modern samples. Scientists have amplified DNA from ancient human remains and extinct species, including DNA that was more than 64,800 years old.

- In 1977, Frederick Sanger sequenced the first DNA genome, the bacteriophage phiX174, which had only 5,386 base pairs. Thirty years elapsed before the sequencing of the human genome, which has about 3 billion base pairs.

- There is considerable genetic variation within a species. Any two humans differ by about 3 million, or 1 out of every 1,000 base pairs. Most variation (about 90 percent) is within a population belonging to a specific continent; the additional variation is found among populations.

- The human genome contains coding and noncoding DNA. The coding DNA, the genes that encode for proteins, makes up only 2 percent of the human genome. The remaining 98 percent does not code for proteins; instead, it may function in regulation of expression and maintenance.

- DNA sequencing has shown that genomes vary by size, but the significance of this remains unclear. The human genome contains about 3 billion base pairs, whereas the protozoan *Amoeba proteus* has 290 billion base pairs, which makes it one hundred times larger.

- Analysis of genetic variation has confirmed that humans are remarkably similar to other life-forms. Humans share 98 percent of their genetic material with chimpanzees, 90 percent with mice, 23 percent with yeast, and 12 percent with *Escherichia coli*.

diseases such as hepatitis B.

Comparative/Evolution. The ability to sequence DNA has led to the study of the evolution of all forms of life, including humans. Since sequencing began, genomes have been mapped for innumerable organisms, including chimpanzees, mice, fish, fruit flies, plants, yeasts, bacteria, and viruses. The Human Genome Project, completed in 2003, identified 20,000 to 25,000 genes and about 3 billion bases in humans. DNA sequencing has also been completed on ancient DNA from clinical, museum, archaeological, and paleontological specimens. These data have greatly increased knowledge of genetic variation, thus increasing understanding of human differences, similarities, evolution, and origins.

Microbial Genomics. In 1994, the U.S. Department of Energy (DOE) launched the Microbial Genome Project. Scientists realized that these organisms, with their ability to withstand extremes of temperature, radiation, acidity, and pressure, provided an excellent resource for the development of applications of renewable energy production, environmental cleanup of toxic waste, management of environmental carbon dioxide, and industrial processing of antibiotics, insecticides, and enzymes. Although this project ended in 2005, the DOE has continued its research in the Genomic Science Project and the Joint Genome Institute's Community Sequencing Program.

Biological Weapons. Although highly controversial, the use of genetic sequencing has led to development of materials for biowarfare. One example is invisible anthrax, which was developed by introducing a gene that altered the immunological properties of the microorganisms *Bacillus anthracis*. Access to the DNA of virulent agents and strains is regulated and restricted, thus preventing the introduction of genes to create novel properties. However, with the advancement of microbiology it is becoming increasingly possible to synthesize agents artificially. In 2002, the poliovirus was synthesized using only the information of the genetic sequence.

IMPACT ON INDUSTRY

Biotechnology. The Biotechnology Industry Orga-Organization reports that the development oDNA nization reports that the development of DNA sequencing and other techniques has enabled the field to boom into a $30 billion a year industry creating more than 1,400 companies in the United States and employing more than 200,000 people. Biotechnology has been responsible for creating vaccines, hundreds of medical diagnostic tests, and treatments for cancer, diabetes, autoimmune disorders, and human immunodeficiency virus (HIV) infection. Additionally, biotechnology has dramatically affected the agricultural industry, one of the world's largest employers, by increasing yields and decreasing the need to apply pesticides.

Government and University Research. One of the largest effects of the development of DNA sequencing and other techniques has been the expansion of research. The goals of the Human Genome Project (1990-2003) were to identify all the genes in human DNA, determine the sequences of the base pairs that make up this DNA, record the information in databases, improve methods of analyzing data, transfer technologies to the private sector, and address all legal, ethical, and social issues stemming from the project. The cost for the project was about $2.7 billion. This project opened the door for additional research opportunities, including the International HapMap Project, the Genographic Project, and the Human Proteome Project.

The International HapMap Project is aimed at creating a catalog of common genetic variants in human beings. It records the differences and similarities in human DNA, where in the DNA they occur, and distribution patterns within and among human populations. The most common genetic variant, a difference in individual bases, is known as a single nucleotide polymorphism, or SNP. The project is identifying most of the 10,000 SNPs that commonly occur in the human genome. These SNPs are not being linked to specific genetic diseases but rather are being used in genetic association studies to estimate people's risk of developing many common diseases.

The Genographic Project, led by the National Geographic Society, IBM, geneticist Spencer Wells, and the Waitt Family Foundation, is using computer analysis of DNA submitted by people around the world in an effort to shed light on human migratory patterns and origins. The project seeks to collect DNA from 100,000 indigenous and traditional peoples as well as from the general public. Started in 2005, the project was to last five years but was extended to 2011.

The Human Proteome Organisation's Human Proteome Project, initiated in January, 2010, aims to characterize all the genes of the human body and generate a map of the protein-based molecular architecture of the human body to answer the question of how the genetic code can guide the enormous intricacies of human anatomy, physiology, and biochemistry. This research will aid in the diagnosis and treatment of disease. The field of proteomics has enabled many complexities of the regulation and structure of the sequenced DNA to be elucidated.

Direct-to-Consumer Marketing. Genetic testing has traditionally been available only through health care providers such as genetic counselors and doctors, who interpret test results for patients. However, some companies have begun marketing genetic tests directly to consumers. These companies send a DNA testing kit to the customer's home. Typically, the customer swabs the inside of his or her cheek and returns the DNA sample to the company. The company processes the sample and communicates the results through regular mail, telephone, or e-mail.

Although direct-to-consumer marketing of genetic testing kits promotes awareness of genetic disease, allows patients to take an active role in their health care, and helps people learn about their ancestral origins, it is highly controversial. The main concern is that guidance from health care providers regarding whether to have a test performed and how to interpret its results is missing. Patients may not fully understand the implications of a positive or negative test result or the predictive ability of these tests.

Regardless of the controversy, companies such as 23andMe, deCODE genetics, Navigenics, and GeneleX have begun offering home-based genetic testing for tracing ancestral origins and determining carrier status of genetic diseases and predisposition to physical and mental illnesses. These kits are priced between $300 and $1,000. In early 2010, Walgreens stated its intent to begin carrying over-the-counter genetic tests produced by Pathway Genomics; however, the company changed its mind after the U.S. Food and Drug Administration began an investigation of the supplier and product. In July, 2010, the Government Accountability Office (GAO) issued a report focusing on the direct-to-consumer kits marketed by four companies. It found the test results to be misleading and contradictory. The report noted "questionable marketing claims, serious quality control and privacy concerns, and questions about the accuracy of information provided to consumers."

CAREERS AND COURSE WORK

Students interested in pursuing careers involving genomic research must take a cross-disciplinary approach. Students should develop a solid background in science, including biology, chemistry, physics, and mathematics, at the undergraduate level. Pursuing higher degrees in the basic sciences or combining studies of journalism, law, computer science, anthropology, archaeology, bioethics, medicine, and pharmaceuticals can result in further specialization. Although higher education, including a master's degree, a medical degree, or doctorate is required for many advanced research positions in academic institutes or industries, opportunities are available for individuals without advanced degrees.

Careers may take several paths, including medicine, public health, pharmaceuticals, agriculture, computer science, engineering, business, law, history, archaeology, and anthropology. Individuals interested in medicine can pursue careers as genetic counselors, medical geneticists, or genetic nurses. Pharmacy students with backgrounds in genetics can pursue innovative research and development of personalized medicine and pharmacogenomics. Scientists with an interest in agriculture may be involved in the genetic modification of food. Other possibilities include programming and maintaining DNA databases, marketing and promoting new technologies, paternity testing, and forensic science. Many of the positions in genetics are likely to be in research into evolution, diagnostic testing, and development of the proteomes and HapMaps.

SOCIAL CONTEXT AND FUTURE PROSPECTS

DNA sequencing has come a long way since its inception in 1975. Enormous accomplishments have been made, creating a future full of many new careers and innovative applications. The knowledge gained from DNA sequencing is a starting point for years of additional research and developments. Armed with the blueprint of life, scientists have begun working to unlock some of biology's most intricate and complex processes, including determining how a human develops from a single cell, how genes regulate the functions of organs and tissues, and what is involved in the preposition of disease.

The completion of the Human Genome Project and the initiation of projects such as the Human Proteome Project and International HapMap Project have demonstrated the commitment of government and society to understanding the nature and role of genetics. However, DNA sequencing, coupled with developments in genetic engineering, have raised profound ethical and social concerns. The heightened ability to determine an individual's genetic profile has raised concerns about confidentiality and privacy, possible stigmatization, negative consequences in the areas of employment and insurance, and the psychological effects of knowing one's predisposition to diseases and conditions. The commercialization of DNA products is another area of concern.

In response, many U.S. governmental agencies such as the National Institutes of Health have created bioethics programs. For example, the Ethical, Legal and Social Implications (ELSI) Research Program, part of the National Human Genome Research Institute, supports research on the ethical, legal, and social implications of genetics research. Other programs include a bioethics component.

Amber M. Mathiesen, M.S.

FURTHER READING

Brown, T. A. *DNA Sequencing.* New York: IRL Press at Oxford University Press, 1995. A basic book examining the technique of DNA sequencing.

Hummel, Susanne. *Ancient DNA Typing: Methods, Strategies, and Applications.* New York: Springer, 2003. A manual for the analysis of ancient and degraded DNA. Includes information on extraction, techniques, and applications, including identification of objects, kinship, and population genetics.

Hutchison, Clyde A., III. "DNA Sequencing: Bench to Bedside and Beyond." *Nucleic Acids Research* 35, no. 18 (August, 2007): 6227-6237. Reviews the history and development of DNA sequencing from the discovery of the structure of DNA to modern times.

Janitz, Michal. *Next-Generation Genome Sequencing: Towards Personalized Medicine.* Weinheim, Germany: Wiley-VCH, 2008. Provides the reader with a comprehensive overview of next generation sequencing techniques, highlighting their implications for research, human health, and society's perception of genetics.

Jones, Martin. "Archaeology and the Genetic Revolution." In *A Companion to Archaeology,* edited by John Bintliff, Timothy Earle, and Christopher S. Peebles. Malden, Mass.: Blackwell, 2004. Reviews the expanding field of archaeogenetics by discussing the history of how DNA entered the field, the study of ancient DNA, human evolutionary studies, existing practices, and future prospects.

Lynch, April, and Vickie Venne. *The Genome Book: A Must-Have Guide to Your DNA for Maximum Health.* North Branch, Minn.: Sunrise River Press, 2009. Provides an explanation of the growing medical benefits provided from decoding the human genome. Discusses several health topics, including cancer, behavior, and heart conditions.

Meyers, Robert A., ed. *Genomics and Genetics: From Molecular Details to Analysis and Techniques.* Weinheim, Germany: Wiley-VCH, 2007. Covers the basics of genomics and genetics and discusses techniques such as DNA sequencing.

WEB SITES

The Genographic Project
https://genographic.nationalgeographic.com/genographic/index.html

The Human Genome Project
http://www.ornl.gov/sci/techresources/Human_Genome/home.shtml

The International HapMap Project
http://hapmap.ncbi.nlm.nih.gov

Microbial Genomics at the U.S. Department of Energy Microbial Research Programs: Past and Present
http://microbialgenomics.energy.gov/researchprograms.shtml

National Human Genome Research Institute
http://www.genome.gov

See also: Animal Breeding and Husbandry; Archaeology; DNA Analysis; Genetically Modified Organisms; Genetic Engineering; Human Genetic Engineering; Pathology.

ECONOMIC GEOLOGY

FIELDS OF STUDY

Geology; chemistry; physics; structural geology; mineralogy; petrology; geochemistry; geochronology; geologic occurrence; mining engineering; geophysics; stratigraphy.

SUMMARY

Economic geology is the study of the origin and distribution of mineral deposits of metals, other useful materials (such as building stone and salt), and fossil fuels (petroleum, natural gas, and coal). The economic geologist explores for these materials, predicts the mineral reserves of known deposits, and assesses the economics for the extraction of the ore. For instance, for an iron ore deposit, the geologist needs to assess the kinds and amount of iron minerals present, the depth of the deposit, the location of the deposit, and the ownership of the land so that estimates for the possibility of mining the iron ore profitably may be made.

KEY TERMS AND CONCEPTS

- **Basalt:** Dark, fine-grained igneous rock composed mostly of plagioclase (a calcium-sodium silicate mineral) and pyroxene (a magnesium iron silicate mineral).
- **Clay Minerals:** Very small minerals that result from the weathering of other silicate minerals.
- **Granite:** Igneous rock composed of coarse crystals (greater than about 2 millimeters) of plagioclase (calcium-sodium silicate mineral), alkali feldspar (potassium-sodium silicate mineral), and quartz (all silicate with no other plus ions), and a few dark silicate minerals.
- **Hydrothermal:** Ore deposit in which the minerals were deposited by "water" vapor.
- **Magma:** Molten silicate rock below the surface with suspended minerals that have crystallized out of the molten material.
- **Mineral:** Naturally occurring element or compound that has a definite ordered arrangement of atoms.
- **Ores:** Minerals that may be economically and legally extracted from the earth.
- **Sandstone:** Sedimentary rock containing mostly quartz, feldspar, or rock fragments with grain sizes mostly in the range of 0.0625 to 2 millimeters.
- **Shale:** Very fine-grained sedimentary rock often consisting mostly of clay minerals and quartz.
- **Subduction Zone:** Boundary where an oceanic plate (oceanic crust and part of the upper mantle) slowly slips below another oceanic or continental plate, creating earthquakes and magma.

DEFINITION AND BASIC PRINCIPLES

Economic geologists explore for nonrenewable resources that formed so slowly, often over millions to even billions of years, that the processes of formation are much slower than the speed at which the resources can be extracted. For example, much of the oil and gas were formed by the slow, deep burial of marine, organic plankton that were alive about 66 million to 144 million years ago. These organisms need to be gradually buried to a certain depth long enough so that they gradually change to petroleum. If they are buried too deeply, the petroleum may break down to form natural gas (methane) or the mineral graphite (pure carbon); if they are not buried deep enough, then useful petroleum or natural gas will not form.

Nonrenewable resources include the abundant metals (such as iron and aluminum), scarce metals (gold and copper), materials used for energy (fossil fuels and uranium minerals), building materials (limestone, crushed stone, sand, and gravel), and other miscellaneous minerals (halite or natural salt). Running water, wind power, and solar power are not included among materials used for energy because they are renewable sources.

Economic geologists use their knowledge of geology to interpret where certain economic deposits might form. For instance, copper and some other associated elements may occur in what are called copper porphyry deposits. These deposits formed as hot "water" vapor carried dissolved copper out of granite magma and upward along rock fractures in which the copper precipitated to form a variety of copper minerals. The porphyry copper deposits occur only in association with subduction zones. Therefore, the economic geologist knows to search for them only where ancient or existing subduction zones occur, such as along the west coasts of North and South America.

BACKGROUND AND HISTORY

The first use of natural resources was to obtain water, salt, and other natural materials to make tools and weapons. Larger cities were located where there was a source of water such as a river. Salt was used as a flavoring and to preserve food. A major discovery around 9000 b.c.e. was that clay minerals could be heated to make pottery so that food and water could be stored and transported much more easily.

The first metals used were those such as gold and copper, which sometimes were found as native metals (not combined with other elements). These native elements could be shaped into useful materials by hammering or cutting. By 4000 to 3000 b.c.e., minerals of copper, zinc, lead, and tin were heated by burning charcoal in a very hot flame so that these elements could be separated from the ore. Much later, this process was applied to iron ore. The purification of iron had to wait until the development of blast furnaces, in which oxygen was blown through molten iron to purify it.

German physician Georgius Agricola wrote about the ways that ore minerals might form in *De re metallica* (1556; English translation, 1912). He divided the origins of ore deposits into those formed in streams and those formed in place. However, Latin was the language of scholars, not those actually working with ore deposits, so his writings were largely ignored.

Up to the nineteenth century, most ores were discovered accidentally because no one understood how ores and the rocks that contained them were geologically formed. Indeed, many people believed that mysterious celestial powers formed the ore minerals. During the period from the latter part of the eighteenth century to the twentieth century, the origin of rocks and ores gradually became understood so that people had a much better understanding of how to search for ores.

HOW IT WORKS

Selection of the Potential Ore Region. An economic geologist may first examine a geologic map, an aerial photograph, or a satellite photograph of a region, usually where other economic deposits have been found, to see if any geographical features provide clues as to where other deposits might be located. Petroleum and natural gas, for instance, usually occur in sedimentary rocks where certain geologic structures may be seen on photographs or maps. There must be shales that contain organic matter that can potentially be converted to oil or gas if buried to the right depth below the surface. Then there must be an overlying sedimentary rock, such as a permeable sandstone, sandwiched between two impermeable shales, and the sandstone must have spaces between mineral grains through which the oil may move so that it can flow to where it can be trapped. A variety of traps can be used, but one kind uses an impermeable rock to stop the oil from migrating so that it can collect. Once geologists find these circumstances, they can drill below the ground along the trap to see if any petroleum is present. Narrowing the range of possibilities to find oil is important because drilling a single well may cost over $1 million.

The method for exploration differs depending on the type of deposit. For example, copper porphyry deposits often occur as veins in igneous rocks that formed along subduction zones; therefore, the exploration takes place along subduction zones. Some of the large iron ore deposits called banded iron formations found in the Lake Superior region formed only in shallow oceans from 1.8 billion to 2.6 billion years ago. Therefore, geologists search for new iron deposits only in other sedimentary rocks of similar ages.

Geophysical Methods. Remote methods of discovering an economic mineral deposit use instruments that can detect variations in magnetism, electrical conductivity, gravity, and radioactivity in rocks at or close to the surface of the Earth. For example, some iron-rich minerals such as magnetite (Fe_3O_4) and ilmenite ($FeTiO_3$) will produce magnetic attractions that suggest that a region may be hydrothermally mineralized. The variations in magnetism may be detected fairly quickly by flying a magnetic detector over a region in a regular pattern. A Geiger counter may be used to quickly assess if there are any radioactive elements such as uranium or thorium present at the Earth's surface. Also instruments that detect variations in electrical conductivity can be used to determine if certain sulfide minerals might be present.

Petroleum geologists may explode small charges along the surface of the Earth so that sound waves travel through the ground for some distance. The sound waves, which can be detected at a receiver station, travel at varying speeds through different kinds of rocks and geologic structures. A computer program can then be used to construct a cross-section of the characteristics of the rocks below the surface. Sedimentary rocks were deposited in horizontal layers, but in some areas, they may be folded into arches (anticlines), warped downward (synclines), or fractured along faults that displace the sedimentary rock layers. The top portion of an anticline or the side of some fault may provide a possible trap where oil and natural gas can accumulate. Geologists can drill a well into these traps to see if oil or gas is present.

Geochemical Methods. Ore minerals may be weathered out of ore deposits and move into streams, where they slowly move downstream. To search for such deposits, the geologist may collect a series of sediments, often at stream junctions, and look for ore minerals in the sediments or analyze the sediments to see if any metals are present in abnormally high concentrations. This technique was used to discover a number of metal ores, such as the gold and copper porphyry ores in Papua, New Guinea.

Once a suspected ore deposit has been found in a certain region, soil samples may be collected in a systematic grid over the area. The soil samples are analyzed for the elements most likely to be present in the potential ore deposit. Areas that have much higher concentrations of these elements may indicate the presence of ore directly below the surface. Geologists may drill directly into the areas with high concentrations of these elements to obtain subsurface samples to confirm the kinds of minerals and the elemental concentrations. This method, for example, found zinc deposits near Queensland, Australia.

The results of such a drilling program are used to evaluate whether mining the ore prospect might be worthwhile. The decision of whether to mine the prospect depends on many factors, such as the size and location of the deposit, the concentration and types of elements present, the finances of the company, the price obtained for the ore, the ownership of the land, and whether open-pit or underground mining can be used. For example, a gold deposit in northern Canada may not be profitable because of the costs of transporting equipment and miners to the area and of transporting the ore to a production facility. If gold prices drastically increase, then a previously uneconomical mine might be able to show some profit. Also large companies sometimes abandon smaller deposits and try to sell them to smaller companies that might still profitably mine the ore.

APPLICATIONS AND PRODUCTS

The mineral resources discovered by economic geologists are used by industries to produce the abundant metals, the scarce metals, fossil fuels, and other natural materials. Economic geologists are not directly concerned with the use of these materials, but they must be aware of the demand for these materials so that they know the amount of money that a company may receive for them.

Abundant Metals and Their Uses. The abundant metals are those found in the highest concentrations within the Earth. Iron minerals are used to make steel, which is used to make automobiles, buildings, roads, bridges, major appliances, and construction materials such as nails. The major iron ore minerals, hematite and magnetite (both iron oxides), are converted to steel by heating them to a high temperature with coke and limestone to form molten steel, which contains up to 2 percent carbon to harden the iron.

Aluminum is another abundant metal that is obtained from the ore bauxite, a mix of the minerals boehmite (aluminum oxyhydroxide) and gibbsite (aluminum hydroxide). The production of aluminum is very expensive because the bauxite must be heated with the compound cryolite to make a molten solution in which the aluminum is concentrated by an electric current. Aluminum is used in products such as cans, foil, windows, vehicles, and household items and mixed with other metals in alloys.

Titanium occurs in the minerals rutile (titanium oxide) and ilmenite (iron titanium oxide). Like aluminum, titanium is expensive to produce from its ore. Most titanium used is in the form of titanium oxide, which produces an intense white color in paint and paper. Titanium is also alloyed with aluminum, vanadium, and iron and used in ships, airplanes, and missiles. Magnesium oxide is often alloyed with aluminum to provide corrosion-resistant cans and materials used in vehicles. Magnesium metal is expensive to produce because like aluminum, electricity is used in its production.

Silicon is obtained by melting the mineral quartz

(silicon oxide) with iron and coke in an electric furnace. Silicon is often alloyed with iron, aluminum, and copper because it improves the strength of these alloys and guards against corrosion. Silicon is also used in transistors in electronic devices.

Scarce Metals and Uses. Scarce metals occur in the Earth in much lower concentrations than the abundant metals. Some minerals, however, may locally concentrate these metals, making them potentially economic to mine. Some scarce metals may be added to steel to give it certain characteristics. For instance, chromium, molybdemun, tungsten, or vanadium may be added to make steel harder, especially at higher temperature. Stainless steel contains more than 11 percent chromium combined with nickel to help keep steel from rusting. Chromium and nickel may improve the strength of steel. Vanadium added to steel decreases the weight of steel, its strength, ductility, and ease of welding.

In addition, chromium is also used in chromium compounds to produce paint and ink pigments with green, yellow, and orange colors. Molybdemun is used in catalysts, pigments, and lubricants. Tungsten is often combined with carbon to produce a very hard compound, tungsten carbide, which is nearly as hard as diamond.

Copper, lead, zinc, tin, mercury, and cadmium are often grouped together simply because they are not used to alloy with iron. Copper metal conducts electricity well and can be shaped into wires, so it has been used mainly in electric lines and electric motors. Lead and mercury are used much less than in the past because of their toxicity. Lead is still used in batteries for vehicles, and mercury is used in some batteries, electrical switches, and for some chemical compounds. Zinc or tin are used in protective coatings for steel to keep it from rusting. Zinc oxide is added to paint to produce a white color and to a variety of lotions to prevent sunburn. Many so-called tin cans contain tin plated on other metals.

The precious metals are gold, silver, and the platinum metals. These metals are often used in jewelry because of their beauty. Gold is also used for money, and in industrial materials such as electronic connectors and dental fillings. Silver is also used in some electrical equipment. Platinum is used as a catalyst, for example, in catalytic converters in cars.

A variety of other metals, such as the rare earth el-

Fascinating Facts About Economic Geology

- The desire to find gold and silver motivated the Spanish and Portuguese to explore Central America and South America from the fifteenth to seventeenth centuries.
- In the 1400's, Native Americans dug pits to mine the crude oil at Oil Creek near Titusville, Pennsylvania, where Edwin L. Drake would later drill for oil.
- In spring, 1859, Edwin L. Drake, the general agent of Seneca Oil Company, began drilling for oil in Titusville, Pennsylvania. On August 27, he hit oil, and the U.S. oil industry was born.
- The East Texas Oil Museum commemorates the discovery of oil in the 1930's near Kilgore. On October 3, 1930, a well drilled by wildcatter Columbus Marion Joiner, then seventy years old, produced a gusher. Two other wells were drilled, on the Crim family farm and the J. K. Lathrop lease in Gregg County.
- The commercial mining of coal began in Kanawha County, West Virginia, in 1817. West Virginia would be the site of many mining disasters and union unrest that produced the Matewan Massacre (May 18, 1920) and the Battle of Blair Mountain (summer, 1921).
- The California Gold Rush (1848-1855) drew prospectors who panned for gold in the stream sediment of the American River and significantly raised the population of California.
- Gold Maps Online used a U.S. Bureau of Land Management database listing abandoned and active gold mining claims to create gold maps in Google Earth, which it sells to modern prospectors.
- Oil companies often lease surface and mineral rights from landowners, who are given cash bonuses and paid rental fees. When production begins, the landowner receives a percentage royalty of the gas or oil extracted.

---

ements, are used for industrial purposes. Rare earth elements have been used as catalysts and to color glass and ceramics and to provide some colors in television screens.

Chemical Minerals and Fertilizers. A number of

nonmetallic minerals such as halite, baking soda, and sylvite are used in food. Halite and sylvite are salts used for flavoring. Halite is also used to soften water and to melt ice on roads. Borox is a boron compound that is used in some detergents and cosmetics.

A variety of potassium, nitrogen, and phosphorous minerals are used to fertilize crops. Commonly used nitrate compounds are sodium nitrate and potassium nitrate. Ammonium phosphate and apatite are examples of phosphate minerals.

Gypsum (calcium sulphate) is the main mineral in wallboard. Sulfur compounds derived from petroleum are used to make many industrial compounds such as sulphuric acid.

Building Materials. Building materials include building stones, crushed rocks, sand, gravel, cement, plaster, bricks, and glass. Common building stones are granite, limestone, and sandstone. Granite, for example, can be polished to form an attractive surface for building facings and countertops.

Huge quantities of crushed rocks, sand, and gravel are used for roads, building foundations, and concrete. In the United States, more than 100 billion tons of these materials are used yearly.

Gemstones. The most important gems are diamonds, sapphires, rubies, and emeralds. Gems must be harder than most other minerals, and they must be beautiful. Diamonds are composed of the element carbon, and rubies and sapphires are gem-quality varieties of the mineral corundum. Emeralds are gem-quality varieties of the mineral beryl.

Fossil Fuels. Petroleum, natural gas, and coal are the main fossil fuels. Much of the coal is burned to provide electrical power. Much of the natural gas is burned to provide heat in homes and industry. Petroleum products such as gasoline are used in providing power for automobiles and trucks.

IMPACT ON INDUSTRY

Most of the most easily found minerals and fossil fuels have already been discovered, so economic geologists must continue to devise new ways to search for deposits. The demand for most metals, fossil fuels, and industrial minerals is likely to increase in the future.

Industry and Business. Private companies control the search for metals, fossil fuels, and industrial minerals. The worldwide supply and demand of many of these commodities controls the prices. Many metals are easy to transport, and many are in high demand, so the prices for them continue to increase. Large groups that control the supply may dictate the prices of some commodities such as diamonds and oil. The Organization of Petroleum Exporting Countries (OPEC) is a cartel that controls the amount of oil exported from it members and hence its price. The DeBeers Group controls much of the supply of diamonds and therefore their price.

Governments. Governments do not usually explore for minerals or fuels, but they may, for instance, periodically stockpile them, thus driving up the prices. For instance, the United States has a stockpile of petroleum, to which it periodically adds. Sometimes it sells part of the petroleum to counteract rapid increases in petroleum prices if OPEC cuts back on the supply.

Governments will usually tax the income from private industry. They may also pass laws to increase the safety in mines and production facilities for a given commodity such as coal.

Universities. Geology and engineering departments in universities provide instruction in economic geology. Academic researchers in geology typically function on how to better explore for mineral resources, and those in engineering concentrate on how to mine minerals and improve on extraction methods.

CAREERS AND COURSE WORK

A person interested in exploring for metals, fossil fuels, and other industrial minerals should major in geology in college with a concentration in economic geology and exploration geology. Courses such as mineralogy, petrology, economic geology, and structural geology are required to obtain a bachelor's degree. A variety of mathematics, chemistry, and physics courses should also be taken. Further study for master's and doctoral degrees would enable the student to carry out research on a specific problem in economic geology. Those interested in solving geochemical aspects of exploration should take many supporting courses in chemistry; those interested in exploring using geophysical techniques should take supporting courses in mathematics and physics. Advanced degrees enable an individual to be given a lot of responsibility if employed by a company searching for natural resources.

Those interested in developing mining operations

once a potential site has been discovered can major in mining engineering with course work on how to develop mines and supporting courses in geology, physics, and chemistry. Those interested in solving environmental problems associated with an economic site should take a variety of courses in geology, chemistry, water chemistry, and hydrology.

SOCIAL CONTEXT AND FUTURE PROSPECTS

The increased demand for most metals, oil, natural gas, coal, and other raw materials such as sand and gravel is likely to continue indefinitely. Construction materials such as sand, gravel, and limestone, which are used in large quantities, need to be sourced close to where they will be used because of the high cost of transporting them. Most metals are used in smaller quantities, and they can be shipped economically from sites around the world to where they will be used. In industrial societies such as the United States, the amount of money spent on industrial minerals is much greater than that spent on the metals.

The use of iron, manganese, aluminum, copper, zinc, gold, graphite, nickel, silver, sulphur, vanadium, and zinc have increased substantially from 1950 to the twenty-first century. The use of some metals, such as lead, increased from the 1950's to the 1980's but subsequently declined because of associated environmental problems.

Worldwide, the distribution of many metals and fossil fuels is very uneven, which means that many countries must import these resources. For example, the United States uses much more petroleum than it can produce so it must obtain it from countries such as Saudi Arabia and Venezuela, which are rich in petroleum. The United States must also import a great deal of aluminum, platinum, manganese, tantalum, and tungsten to meet its demand. The United States, however, has an excess of salt, lead, copper, and iron.

Robert L. Cullers, B.S., M.S., Ph.D.

FURTHER READING

Craig, James R., David J. Vaughan, and Brian Skinner. *Earth Resources and the Environment.* Upper Saddle River, N.J.: Prentice Hall, 2010. Gives a basic overview of natural resources with terms that are well defined.

Evans, Anthony M. *An Introduction to Economic Geology and Its Environmental Impact.* 1997. Reprint. London: Blackwell Science, 2005. Has sections on the basics of economic geology and the types and distribution of ores through time.

Guilbert, John M., and Charles Frederick Park. *The Geology of Ore Deposits.* 1986. Reprint. Long Grove, Ill.: Waveland Press, 2007. Based on *Ore Deposits* (3d ed., 1975), by Park and Roy A. MacDiarmid. Gives some history of the development of the theory of finding ore deposits as well as more modern theories for their formation.

Klein, Cornelis, and Barbara Dutrow. *Manual of Mineral Science.* Hoboken, N.J.: John Wiley & Sons, 2008. Gives the methods to identify minerals and the properties of minerals.

Robb, H. G. *Introduction to Ore Forming Processes.* Oxford: Blackwell Science, 2005. Gives a detailed summary of how the major ores are formed.

Singer, D. A., and W. D. Menzie. *Quantitative Mineral Resource Assessments: An Integrated Approach.* New York: Oxford University Press, 2010. Draws on quantitative analyses including deposit density models and frequency distributions to enable a better determination of the likelihood of a mineral deposit and its size.

Wellmer, Friedrich-Wilhelm, Manfred Dalheimer, and Markus Wagner. *Economic Evaluations in Exploration.* New York: Springer, 2008. Gives the reader a feeling for how ore deposits are evaluated in the earlier stages of development.

WEB SITES

American Association of Petroleum Geologists
http://www.aapg.org

Society of Economic Geologists
http://www.segweb.org

U.S. Geological Survey
http://www.usgs.gov

See also: Mineralogy

ELECTROCHEMISTRY

FIELDS OF STUDY

Physical chemistry; thermodynamics; organic chemistry; inorganic chemistry; quantitative analysis; chemical kinetics; analytical chemistry; metallurgy; chemical engineering; electrical engineering; industrial chemistry; electrochemical cells; fuel cells; electrochemistry of nanomaterials; advanced mathematics; physics; electroplating; nanotechnology; quantum chemistry; electrophoresis; biochemistry; molecular biology.

SUMMARY

Electrochemists study the chemical changes produced by electricity, but they are also concerned with the generation of electric currents due to the transformations of chemical substances. Whereas traditional electrochemists investigated such phenomena as electrolysis, modern electrochemists have broadened and deepened their interdisciplinary field to include theories of ionic solutions and solvation. This theoretical knowledge has led to such practical applications as efficient batteries and fuel cells, the production and protection of metals, and the electrochemical engineering of nanomaterials and devices that have great importance in electronics, optics, and ceramics.

KEY TERMS AND CONCEPTS

- **Anode:** Positive terminal (or electrode) of an electrochemical cell to which negatively charged ions travel with the passage of an electric current.
- **Battery:** Electrochemical device that converts chemical energy into electrical energy.
- **Cathode:** Negative electrode of an electrochemical cell in which positively charged ions migrate under the influence of an electric current.
- **Electrolysis:** Process by which an electric current causes chemical changes in water, solutions, or molten electrolytes.
- **Electrolyte:** Substance that generates ions when molten or dissolved in a solvent.
- **Electrophoresis:** Movement of charged particles through a conducting medium due to an applied electric field.
- **Electroplating:** Depositing a thin layer of metal on an object immersed in a solution by passing an electric current through it.
- **Faraday's Law:** Magnitude of the chemical effect of an electrical current is directly proportional to the amount of current passing through the system.
- **Fuel Cell:** Electrochemical device for converting a fuel with an oxidant into direct-current electricity.
- **Ion:** Atom or group of atoms carrying a charge, either positive (cation) or negative (anion).
- **Nanomaterials:** Chemical substances or particles the masses of which are measured in terms of billionths of a gram.
- **pH:** Numerical value, extending from 0 to 14, representing the acidity or alkalinity of an aqueous solution (the number decreases with increasing acidity and increases with increasing alkalinity).

DEFINITION AND BASIC PRINCIPLES

As its name implies, electrochemistry concerns all systems involving electrical energy and chemical processes. More specifically, this field includes the study of chemical reactions caused by electrical forces as well as the study of how chemical processes give rise to electrical energy. Some electrochemists investigate the electrical properties of certain chemical substances, for instance, these substances' ability to serve as insulators or conductors. Because the atomic structure of matter is fundamentally electrical, electrochemistry is intimately involved in all fields of chemistry from physical and inorganic through organic and biochemistry to such new disciplines as nanochemistry. No matter what systems they study, chemists in some way deal with the appearance or disappearance of electrical energy into the surroundings. On the other hand, electrochemists concentrate on those systems consisting of electrical conductors, which can be metallic, electrolytic, or gaseous.

Because of its close connection with various branches of chemistry, electrochemistry has applications that are multifarious. Early applications centered on electrochemical cells that generated a steady current. New metals such as potassium, sodium, calcium, and strontium were discovered by electrolysis of their molten salts. Commercial production of such metals as magnesium, aluminum, and

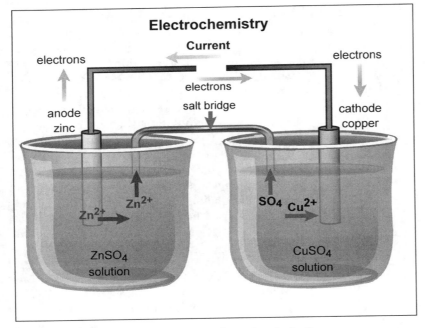

Electrochemistry

Current

electrons

electrons

electrons

salt bridge

anode
zinc

cathode
copper

Zn^{2+}

Zn^{2+}

SO_4 Cu^{2+}

$ZnSO_4$
solution

$CuSO_4$
solution

Connecting copper and zinc creates an electrochemical cell.

quality and length of human lives. Much research and development are being devoted to increasingly sophisticated electrochemical devices for implantation in the human body, and some even predict, such as American inventor Ray Kurzweil, that these "nanobots" will help extend human life indefinitely.

BACKGROUND AND HISTORY

Most historians of science trace the origins of electrochemistry to the late eighteenth and early nineteenth centuries, when Italian physician Luigi Galvani studied animal electricity and Italian physicist Alessandro Volta invented the first battery. Volta's device consisted of a pile of dissimilar metals such as zinc and silver separated by a moist conductor. This "Voltaic pile" produced a continuous current, and applications followed quickly. Researchers showed that a Voltaic pile could decompose water into hydrogen and oxygen by a process later called electrolysis. English chemist Sir Humphry Davy used the electrolysis of melted inorganic compounds to discover several new elements. The Swedish chemist Jöns Jacob Berzelius used these electrochemical studies to formulate a new theory of chemical combination. In his dualistic theory atoms are held together in compounds by opposite charges, but his theory declined in favor when it was unable to explain organic compounds, or even such a simple molecule as diatomic hydrogen.

Though primarily a physicist, Michael Faraday made basic discoveries in electrochemistry, and, with the advice of others, he developed the terminology of this new science. For example, he introduced the terms "anode" and "anion," "cathode" and "cation," "electrode," "electrolyte," as well as "electrolysis." In the 1830's his invention of a device to measure the quantity of electric current resulted in his discovery of a fundamental law of electrochemistry—that the quantity of electric current that leads to the formation of a certain amount of a particular chemical substance also leads to chemically equivalent amounts of other substances. Even though Faraday's discovery of

zinc were mainly accomplished by the electrolysis of solutions or melts. An understanding of the electrical nature of chemical bonding led chemists to create many new dyes, drugs, plastics, and artificial plastics. Electroplating has served both aesthetic and practical purposes, and it has certainly decreased the corrosion of several widely used metals.

Electrochemistry played a significant part in the research and development of such modern substances as silicones, fluorinated hydrocarbons, synthetic rubbers, and plastics. Even though semiconductors such as germanium and silicon do not conduct electricity as well as copper, an understanding of electrochemical principles has been important in the invention of various solid-state devices that have revolutionized the electronics industries, from radio and television to computers. Electrochemistry, when it has been applied in the life sciences, has resulted in an expanded knowledge of biological molecules. For example, American physical chemist Linus Pauling used electrophoretic techniques to discover the role of a defective hemoglobin molecule in sickle-cell anemia. A grasp of electrochemical phenomena occurring in the human heart and brain has led to diagnostic and palliative technologies that have improved the

the relationship between quantity of electricity and electrochemical equivalents was extraordinarily significant, it was not properly appreciated until much later. Particularly helpful was the work of the Swedish chemist Svante August Arrhenius, whose ionic theory, proposed toward the end of the nineteenth century, contained the surprising new idea that anions and cations are present in dilute solutions of electrolytes.

In the twentieth century, Dutch-American physical chemist Peter Debye, together with German chemist Erich Hückel, corrected and extended the Arrhenius theory by taking into account that, in concentrated solutions, cations have a surrounding shell of anions, and vice versa, causing these ions' movements to be retarded in an electric field. Norwegian-American chemist Lars Onsager further refined this theory by taking into account Brownian motion, the movement of these ionic atmospheres due to heat. Other scientists used electrochemical ideas to understand the nature of acids and bases, the interface between dissimilar chemicals in electrochemical cells, and the complexities of oxidation-reduction reactions, whether they occur in electrochemical cells or in living things.

HOW IT WORKS

Primary and Secondary Cells. The basic device of electrochemistry is the cell, generally consisting of a container with electrodes and an electrolyte, designed to convert chemical energy into electrical energy. A primary cell, also known as a galvanic or Voltaic cell, is one that generates electrical current via an irreversible chemical reaction. This means that a discharged primary cell cannot be recharged from an external source. By taking measurements at different temperatures, chemists use primary cells to calculate the heat of reactions, which have both theoretical and practical applications. Such cells can also be used to determine the acidity and alkalinity of solutions. Every primary cell has two metallic electrodes, at which electrochemical reactions occur. In one of these reactions the electrode gives up electrons, and at the other electrode electrons are absorbed.

In a secondary cell, also known as a rechargeable or storage cell, electrical current is created by chemical reactions that are reversible. This means that a discharged secondary cell may be recharged by circulating through the cell in a quantity of electricity equal to what had been withdrawn. This process can be repeated as often as desired. The manufacture of secondary cells has grown into an immense industry, with such commercially successful products as lead-acid cells and alkaline cells with either nickel-iron or nickel-cadmium electrodes.

Electrolyte Processes. Electrolysis, one of the first electrochemical processes to be discovered, has increased in importance in the twentieth and twenty-first centuries. Chemists investigating electrolysis soon discovered that chemical reactions take place at the two electrodes, but the liquid solution between them remains unchanged. An early explanation was that with the passage of electric current, ions in the solution alternated decompositions and recombinations of the electrolyte. This theory had to be later revised in the light of evidence that chemical components had different motilities in solution.

For more than two hundred years, the electrolysis of water has been used to generate hydrogen gas. In an electrolytic cell with a pair of platinum electrodes, to the water of which a small amount of sulfuric acid has been added (to reduce the high voltage needed), electrolysis begins with the application of an external electromotive force, with bubbles of oxygen gas appearing at the anode (due to an oxidation reaction) and hydrogen gas at the cathode (due to a reduction reaction). If sodium chloride is added to the water, the electrochemical reaction is different, with sodium metal and chlorine gas appearing at the appropriate electrodes. In both these electrolyses the amounts of hydrogen and sodium produced are in accordance with Faraday's law, the mass of the products being proportional to the current applied to the cell.

Redox Reactions. For many electrochemists the paramount concern of their discipline is the reduction and oxidation (redox) reaction that occurs in electrochemical cells, batteries, and many other devices and applications. Reduction takes place when an element or radical (an ionic group) gains electrons, such as when a double positive copper ion in solution gains two electrons to form metallic copper. Oxidation takes place when an element or radical loses electrons, such as when a zinc electrode loses two electrons to form a doubly positive zinc ion in solution. In electrochemical research and applications the sites of oxidation and reduction are spatially separated. The electrons produced by chemical processes can be forced to flow through a wire, and this electric current can be used in various applications.

Electrodes. Electrochemists employ a variety of electrodes, which can consist of inorganic or organic materials. Polarography, a subdiscipline of electrochemistry dealing with the measurement of current and voltage, uses a dropping mercury electrode, a technique enabling analysts to determine such species as trace amounts of metals, dissolved oxygen, and certain drugs. Glass electrodes, whose central feature is a thin glass membrane, have been widely used by chemists, biochemists, and medical researchers. A reversible hydrogen electrode plays a central role in determining the pH of solutions. The quinhydrone electrode, consisting of a platinum electrode immersed in a quinhydrone solution, can also be used to measure pH (it is also known as an indicator electrode because it can indicate the concentration of certain ions in the electrolyte). Also widely used, particularly in industrial pH measurements, is the calomel electrode, consisting of liquid mercury covered by a layer of calomel (mercurous chloride), and immersed in a potassium chloride solution. Electrochemists have also created electrodes with increasing (or decreasing) power as oxidizing or reducing agents. With this quantitative information they are then able to choose a particular electrode material to suit a specific purpose.

APPLICATIONS AND PRODUCTS

Batteries and Fuel Cells. Soon after Volta's invention of the first electric battery, investigators found applications, first as a means of discovering new elements, then as a way to deepen understanding of chemical bonding. By the 1830's, when new batteries were able to serve as reliable sources of electric current, they began to exhibit utility beyond their initial value for experimental and theoretical science. For example, the Daniell cell was widely adopted by the telegraph industry. Also useful in this industry was the newly invented fuel cell that used the reaction of a fuel such as hydrogen and an oxidant such as oxygen to produce direct-current electricity. However, its requirement of expensive platinum electrodes led to its replacement, in the late nineteenth and throughout the twentieth century, with the rechargeable lead-acid battery, which came to be extensively used in the automobile industry. In the late twentieth and early twenty-first centuries many electrochemists predicted a bright future for fuel cells based on

Fascinating Facts About Electrochemistry

- Alessandro Volta, often called the father of electrochemistry, invented the world's first battery, but he did not understand how it worked (he believed in a physical-contact theory rather than the correct chemical-reaction theory).
- English physicist Michael Faraday, with only a few years of rudimentary education, later formulated the basic laws of electrochemistry and became the greatest experimental physicist of the nineteenth century.
- Swedish chemist Svante August Arrhenius, whose ionic theory of electrolytic solutions became fundamentally important to the progress of electrochemistry, was nearly drummed out of the profession by his skeptical doctoral examiners, who gave him the lowest possible passing grade.
- The electrochemical discovery that charged atoms or groups of atoms in solution have a distinct electric charge (or some integral multiple of it) led to the idea that electricity itself is atomic (and not a fluid, as many believed), and in 1891, Irish physicist George Johnstone Stoney gave this electrical unit the name, "electron."
- Charles M. Hall in the United States and Paul L. T. Héroult in France independently discovered the modern electrolytic technique of manufacturing aluminum in 1886; furthermore, both men were born in 1863 and both died in 1914.
- Many molecules in plants, animals, and humans have electric charges, some with two charges (called zwitterions), others with many (called polyelectrolytes).
- In the second half of the twentieth century, many academic institutions in the United States taught courses in electroanalytical chemistry and electrochemical engineering, but no American college, university, or technical institute taught any courses in applied electrochemistry.

hydrogen and oxygen, especially with the pollution problems associated with widespread fossil-fuel use.

Electrodeposition, Electroplaying, and Electrorefining. When an electric current passes through a solution (for instance, silver nitrate), the precipitation of a material (silver) at an electrode (the cathode) is called electrodeposition. A well-known category of electrodeposition is electroplating, when a thin layer of one metal is deposited on another. In galvanization, for example, iron or steel objects are coated with rust-resistant zinc. Electrodeposition techniques have the advantage of being able to coat objects thoroughly, even those with intricate shapes. An allied technique, electrorefining, transforms metals contaminated with impurities to very pure states by anodic dissolution and concomitant redeposition of solutions of their salts. Some industries have used so-called electrowinning techniques to produce salable metals from low-grade ores and mine tailings.

Advances in electrochemical knowledge and techniques have led to evermore sophisticated applications of electrodeposition. For example, knowledge of electrode potentials has made the electrodeposition of alloys possible and commercial. Methods have also been discovered to provide plastics with metal coatings. Similar techniques have been discovered to coat such rubber articles as gloves with a metallic layer. Worn or damaged metal objects can be returned to pristine condition by a process called electroforming. Some commercial metal objects, such as tubes, sheets, and machine parts, have been totally manufactured by electrodeposition (sometimes called electromachining).

Electrometallurgy. A major application of electrochemical principles and techniques occurs in the manufacture of such metals as aluminum and titanium. Plentiful aluminum-containing bauxite ores exist in large deposits in several countries, but it was not until electrochemical techniques were developed in the United States and France at the end of the nineteenth century that the cost of manufacturing this light metal was sufficiently reduced to make it a commercially valuable commodity. This commercial process involved the electrolysis of alumina (aluminum oxide) dissolved in fused cryolite (sodium aluminum fluoride). During the century that followed this process's discovery, many different uses for this lightweight metal ensued, from airplanes to zeppelins.

Corrosion Control and Dielectric Materials. The destruction of a metal or alloy by oxidation is itself an electrochemical process, since the metal loses electrons to the surrounding air or water. A familiar example is the appearance of rust (hydrated ferric oxide) on an iron or steel object. Electrochemical knowledge of the mechanism of corrosion led researchers to ways of preventing or delaying it. Keeping oxidants away from the metallic surface is an obvious means of protection. Substances that interfere with the oxidizing of metals are called inhibitors. Corrosion inhibitors include both inorganic and organic materials, but they are generally categorized by whether the inhibitor obstructs corrosive reactions at the cathode or anode. Cathodic protection is used extensively for such metal objects as underground pipelines or such structures as ship hulls, which have to withstand the corrosive action of seawater. Similarly, dielectric materials with low electrical conductivity, such as insulators, require long-term protection from high and low temperatures as well as from corrosive forces. An understanding of electrochemistry facilitates the construction of such electrical devices as condensers and capacitors that involve dielectric substances.

Electrochemistry, Molecular Biology, and Medicine. Because of the increasing understanding of electrochemistry as it pertains to plant, animal, and human life, and because of concerns raised by the modern environmental movement, several significant applications have been developed, with the promise of many more to come. For example, electrochemical devices have been made for the analysis of proteins and deoxyribonucleic acid (DNA). Researchers have fabricated DNA sensors as well as DNA chips. These DNA sensors can be used to detect DNA damage. Electrochemistry was involved in the creation of implantable pacemakers designed to regulate heart beats, thus saving lives. Research is under way to create an artificial heart powered by electrochemical processes within the human body. Neurologists electrically stimulate regions of the brain to help mitigate or even cure certain psychological problems. Developments in electrochemistry have led to the creation of devices that detect various environmental pollutants in air and water. Photoelectrochemistry played a role in helping to understand the dramatic depletion of the ozone layer in the stratosphere and the role that chlorofluorocarbons (CFCs) played in exacerbating this problem. Because a large hole in the ozone layer allows dangerous solar radiation to damage plants, animals, and humans, many countries have banned the use of CFCs.

Nanomaterials in Electrochemistry. Miniaturization of electronic technologies became evident and important in the computer industry, where advances have been enshrined in Moore's law, which states that transistor density in integrated circuits doubles every eighteen months. Electrodeposition has proved to be a technique well-suited to the preparation of metal nanostructures, with several applications in electronics, semiconductors, optics, and ceramics. In particular, electrochemical methods have contributed to the understanding and applications of quantum dots, nanoparticles that are so small that they follow quantum rather than classical laws. These quantum dots can be as small as a few atoms, and in the form of ultrathin cadmium-sulfide films they have been shown to generate high photocurrents in solar cells. The electrochemical synthesis of such nanostructured products as nanowires, biosensors, and microelectroanalytical devices has led researchers to predict the ultimate commercial success of these highly efficient contrivances.

IMPACT ON INDUSTRY

Because electrochemistry itself an interdisciplinary field, and is a part of so many different scientific disciplines and commercial applications, it is difficult to arrive at an accurate figure for the economic worth and annual profits of the global electrochemical industry. More reliable estimates exist for particular segments of this industry in specific countries. For example, in 2008, the domestic revenues of the United States battery and fuel cell industry were about $4.9 billion, the lion's share of which was due to the battery business (in 2005, the United States fuel cell industry had revenues of about $266 million). During the final decades of the twentieth century, the electrolytic production of aluminum in the United States was about one-fifth of the world's total, but in the twenty-first century competition from such countries as Norway and Brazil, with their extensive and less expensive hydropower, has reduced the American share.

Government and University Research. In the United States during the decades after World War II, the National Science Foundation provided support for many electrochemical investigations, especially those projects that, because of their exploratory nature, had no guarantee of immediate commercial success. An example is the 1990's electrochemical research on semiconducting nanocrystals. The United States' Office of Naval Research has supported

projects on nanostructured thin films. It is not only the federal government that has seen fit through various agencies to support electrochemical research but state governments as well. For example, the New York State Foundation for Science, Technology and Innovation has invested in investigations of how ultrathin films can be self-assembled from polyelectrolytes, nanoparticles, and nanoplatelets with the hope that these laboratory-scale preparations may have industrial-scale applications.

Government agencies and academic researchers have often worked together to fund basic research in electrochemistry. The Department of Energy through its Office of Basic Energy Sciences has supported research on electrochromic and photochromic effects, which has led to such commercial products as switchable mirrors in automobiles. Researchers at the Georgia Institute of Technology have contributed to the improvement of proton exchange membrane fuel cells (PEMFCs), and scientists at the University of Dayton in Ohio have shown the value of carbon nanotubes in fuel cells. Hydrogen fuel cells became the focus of an initiative promoted by President George W. Bush in 2003, which was given direction and financial support in the 2005 Energy Policy Act and the 2006 Advanced Energy Initiative. However, a few years later, the Obama administration chose to de-emphasize hydrogen fuel cells and emphasize other technologies that will create high energy efficiency and less polluting automobiles in a shorter period of time.

Industry and Business. Because of the widespread need for electrochemical cells and batteries, companies manufacturing them have devoted extensive human and financial resources to the research, development, and production of a variety of batteries. Some companies, such as Exide Technologies, have emphasized lead-acid batteries, whereas General Electric, whose corporate interest in batteries goes back to its founder, Thomas Alva Edison, has made fuel cells a significant part of its diversified line of products. Some businesses, such as Alcoa, the world's leading producer of aluminum, were based on the discovery of a highly efficient electrolytic process, which led to a dramatic decrease in the cost of aluminum and, in turn, to its widespread use (it is second only to steel as a construction metal).

Careers and Course Work

Electrochemistry is an immense field with a large variety of specialties, though specialized education generally takes place at the graduate level. Undergraduates usually major in chemistry, chemical engineering, or materials science engineering, the course work of which involves introductory and advanced physics courses, calculus and advanced mathematics courses, and elementary and advanced chemistry courses. Certain laboratory courses, for example, qualitative, quantitative and instrumental analysis, are often required. Because of the growing sophistication of many electrochemical disciplines, those interested in becoming part of these fields will need to pursue graduate degrees. Depending on their specialty, graduate students need to satisfy core courses, such as electrochemical engineering, and a certain number of electives, such as semiconductor devices. Some universities, technical institutes, and engineering schools have programs for students interested in theoretical electrochemistry, electrochemical cells, electrodeposition, nanomaterials, and many others. For doctoral and often for a master's degree, students are required to write a thesis under the supervision of a faculty director.

Career opportunities for electrochemists range from laboratory technicians at small businesses to research professors at prestigious universities. The battery business employs many workers with bachelor of science degrees in electrochemistry to help manufacture, service, and improve a variety of products. Senior electrochemical engineers with advanced degrees may be hired to head research programs to develop new products or to supervise the production of the company's major commercial offerings. Electrochemical engineers are often hired to manage the manufacture of electrochemical components or oversee the electrolytic production of such metals as aluminum and magnesium. Some electrochemists are employed by government agencies, for example, to design and develop fuel cells for the National Aeronautics and Space Administration (NASA), while others may be hired by pharmaceutical companies to develop new drugs and medical devices.

Social Context and Future Prospects

Even though batteries, when compared with other energy sources, are too heavy, too big, too inefficient, and too costly, they will continue to be needed in the twenty-first century, at least until suitable substitutes are found. Although some analysts predict a bright future for fuel cells, others have been discouraged by their slow rate of development. As advanced industrialized societies continue to expand, increasing demand for such metals as beryllium, magnesium, aluminum, titanium, and zirconium will necessarily follow, forcing electrochemists to improve the electrolytic processes for deriving these metals from dwindling sources. If Moore's law holds well into the future, then computer engineers, familiar with electrochemical principles, will find new ways to populate integrated circuits with more and better microdevices.

Some prognosticators foresee significant progress in the borderline field between electrochemistry and organic chemistry (sometimes called electro-organic chemistry). When ordinary chemical methods have proved inadequate to synthesize desired compounds of high purity, electrolytic techniques have been much better than traditional ones in accomplishing this, though these successes have occurred at the laboratory level and the development of industrial processes will most likely take place in the future. Other new fields, such as photoelectrochemistry, will mature in the twenty-first century, leading to important applications. The electrochemistry of nanomaterials is already well underway, both theoretically and practically, and a robust future has been envisioned as electrochemical engineers create new nanophase materials and devices for potential use in a variety of applications, from electronics to optics.

Robert J. Paradowski, M.S., Ph.D.

Further Reading

Bagotsky, Vladimir, ed. *Fundamentals of Electrochemistry.* 2d ed. New York: Wiley-Interscience, 2005. Provides a good introduction to this field for those unfamiliar with it, though later chapters contain material of interest to advanced students. Index.

Bard, Allen J., and Larry R. Faulkner. *Electrochemical Methods: Fundamentals and Applications.* 2d ed. Hoboken, N.J.: John Wiley & Sons, 2001. For many years this "gold standard" of electrochemistry textbooks was the most widely used such book in the world. Its advanced mathematics may daunt the beginning student, but its comprehensive treatment of theory and applications rewards the extra effort needed to understand this field's fundamentals; index.

Brock, William H. *The Chemical Tree: A History of Chemistry*. New York: Norton, 2000. This work, previously published as *The Norton History of Chemistry*, was listed as a *New York Times* "Notable Book" when it appeared. The development of electrochemistry forms an important part of the story of chemistry; includes an extensive bibliographical essay, notes, and index.

Ihde, Aaron J. *The Development of Modern Chemistry*. New York: Dover, 1984. Makes available to general readers a well-organized treatment of chemistry from the eighteenth to the twentieth century, of which electrochemical developments form an essential part. Illustrated, with extensive bibliographical essays on all the chapters; author and subject indexes.

MacInnes, Duncan A. *The Principles of Electrochemistry*. New York: Dover, 1961. This paperback reprint brings back into wide circulation a classic text that treats the field as an integrated whole; author and subject indexes.

Schlesinger, Henry. *The Battery: How Portable Power Sparked a Technological Revolution*. New York: HarperCollins, 2010. The author, a journalist specializing in technology, provides an entertaining, popular history of the battery, with lessons for readers familiar only with electronic handheld devices.

Zoski, Cynthia G., ed. *Handbook of Electrochemistry*. Oxford, England: Elsevier, 2007. After an introductory chapter, this book surveys most modern research areas of electrochemistry, such as reference electrodes, fuel cells, corrosion control, and other laboratory techniques and practical applications.

WEB SITES

Electrochemical Science and Technology Information Resource (ESTIR)
http://electrochem.cwru.edu/estir

The Electrochemical Society
http://www.electrochem.org

International Society of Electrochemistry
http://www.ise-online.org

ELECTRON MICROSCOPY

FIELDS OF STUDY

Physics; cell biology; chemistry; morphology; materials science; nanotechnology; medicine; atomic theory; toxicology; virology; nanometrology; forensic engineering; geology; engineering; microanalysis.

SUMMARY

Electron microscopy is the use of electrons, instead of light, to study biological specimens and other nonliving materials under much greater resolution than the conventional light microscope. The electron microscope, first built in the 1930's, saw a number of technological advances, including better lenses and higher voltages for accelerating electrons for increased resolution and imaging. As technology and sample preparation techniques improved, electron microscopy has found multiple applications in the natural sciences, particularly in areas such as medicine, cell biology, morphology, materials sciences, and engineering.

KEY TERMS AND CONCEPTS

- **Contrast:** Distinction between an object and its background, or between two adjacent objects.
- **Field Of View:** What the microscope user sees when he or she looks through its lens.
- **Fixative:** Chemical (usually a solution) that will kill the cells to preserve the tissue for viewing under the microscope. A good fixative will keep the tissue as close to its living state as possible.
- **Light Microscopy:** Earliest form of microscopy, also called optical microscopy, it uses light to illuminate small objects under a series of magnifying glass
- **Magnification:** Amount by which an object can be enlarged.
- **Resolution:** Measure of the sharpness and quality of an image.
- **Resolving Power:** Ability to distinguish the difference between two adjacent objects as distinct.
- **Section:** Slice of a specimen thin enough to be illuminated.
- **Vacuum:** Environment devoid of air.

- **Whole Mount:** Placing an entire specimen or organism under a microscope.

DEFINITION AND BASIC PRINCIPLES

In contrast to traditional microscopes, which use light as an illuminating source, electron microscopes use a beam of electrons to visualize objects. A typical electron microscope is made of a long, hollow cylinder in which the electron beam (the illuminating source) passes through or is reflected off a specimen. At the top of the column is a cathode (an electrode or terminal), usually a heated tungsten filament, that provides electrons. After a high voltage is applied between the cathode and anode, the electrons are accelerated as a thin beam. It is also necessary to create a vacuum within the cylinder by pumping out the air—otherwise, the electrons would collide with gas molecules present in the air and be scattered before the beam reaches the specimen. The beam of electrons is focused using electromagnetic lenses located along the side of the column; the strength of the magnets is controlled by the current that is applied to them by the user. Depending on the current, the object can be magnified from 1,000 to 250,000 times.

The high resolving power of the electron microscope is due to the wave properties of electrons. The resolution of a microscope is limited by the wavelength of the light source in a proportional manner; in other words, the longer the wavelength, the poorer the resolution. The wavelengths of photons (of light) are constant, whereas wavelengths of electrons depend on the speed at which they are traveling, which in turn depends on the voltage that is applied. There are two main types of electron microscopy: transmission electron microscopy (TEM), and scanning electron microscopy (SEM). Both methods provide much greater resolution of the specimen than the light microscope does.

BACKGROUND AND HISTORY

The electron microscope was invented by Max Knoll and Ernst Ruska in 1931. The technology, developed at the Technical University of Berlin, was important because it was the first major improvement on the resolving power of microscopes. Six years later, the first version of the electron microscope was ready for the market. Ruska received the 1986 Nobel Prize in Physics for his design and creation of the first electron microscope.

When the electron microscope was first conceptualized, the goal was the ability to visualize individual atoms. Although this was not achieved for a few decades, the early electron microscopes were able to use electron beams to view objects. By the late 1930's, electron microscopes, which had theoretical resolutions of 10 nanometers (nm), were being produced. By 1944, the resolution was decreased to 2 nm—a vast improvement on the light microscope, which had a theoretical resolution of 200 nm. Other parts of the instrument were also improved; for example, the voltage accelerations were increased, which resulted in better resolution. Also, better technology in the electron lens decreased the amount of optical aberrations, and the vacuum systems and the electron guns were refined.

Other researchers helped further electron microscopy technology. Ladislaus L. Marton of Brussels assembled the first micrograph of a biological sample, while Manfred von Ardenne of Berlin constructed the first scanning electron microscope in Berlin in 1937. At the University of Toronto, Cecil Hall, Albert Prebus, and James Hillier built a model of the electron microscope that would later be used by the Radio Corporation of America (RCA) to build its own Model B, the first commercially produced electron microscope in North America. The first electron microscope for commercial use was produced by Siemens in Germany.

In the 1940's, electron microscopy was used to study materials and particles such as carbon black (a material that gives strength to car tires), paints, and pigments. Over the next ten years, the technology improved slightly, with lenses being stabler and brighter electron guns being produced.

In the 1960's, electron microscopy moved forward to include commercialization of the first scanning electron microscope and ultrahigh-voltage transmission electron microscopy. In addition, specimen stages that were able to rotate and tilt allowed more angles of the specimen to be viewed. In the 1980's and 1990's, environmental electron microscopes were developed to examine samples under more natural temperature and pressure conditions.

HOW IT WORKS

Transmission Electron Microscopy. In transmission electron microscopy (TEM), electrons are transmitted through the specimen. The electron beam is evenly illuminated across the entire field of view, and an image is generated by the specimen, which will scatter the electrons.

Scanning Electron Microscopy. In scanning electron microscopy (SEM), the principles of using electrons as an illuminating source are the same as in TEM. However, in SEM, the electrons are reflected off the surface of a specimen, making it the better tool for studying the external structure and surface of objects. SEM generally provides magnification in the range of 15 to 150,000 times, and its resolving power is related to the diameter of the electron beam.

Specimen Preparation. A number of methods are used to prepare specimens for electron microscopy. Because electron microscopy is much more powerful than traditional light microscopy, the steps and care required to fix and prepare specimens are significant to viewing success. Any damage sustained during the preservation process will be apparent under the microscope. Chemical fixing or rapid freezing can be used to preserve biological specimens quickly and prevent degradation. Commonly, chemical fixatives such as glutaraldehyde are used to link cellular proteins so that they can be removed. Next, water is removed by a dehydration process using alcohol, and the specimen is sectioned into extremely thin slices (often of less than 0.1 micrometer).

For SEM, the specimens (particularly biological ones) have to be prepared so that the surface morphology is preserved but completely devoid of liquid. Because water consists of a large portion of living cells, it has to be removed in a way that will not be destructive to the structure. The specimens cannot be air dried, because surface tension between air and water will disfigure the surface. Therefore, specimens or SEM are usually fixed, passed through a series of dehydrating alcohols, and dried using a process called critical-point drying. Briefly, the specimens are placed in a liquid such as carbon dioxide, which slowly replaces the water in the sample over a period of time. The carbon dioxide can then be vaporized into gas when the pressure is increased, leaving the surface of the specimen intact. When the specimen is dried using the method described above, it has to be coasted with a thin layer of metal (usually gold) to

Fascinating Facts About Electron Microscopy

- Cellular structure is typically measured using the angstrom (Å). One angstrom is equal to 1x10-10 meters. While the practical limit of most transmission electron microscopes is 3-5 Å, a range of 10-15 Å is required to observe most cellular structures.

- Cryosections, or sections of tissue that are rapidly frozen (using a liquid with a very low boiling point such as propane), can be used to study enzymes, which are proteins that catalyze reactions in cells. Cryosections can be a quicker process than methods such as using chemical fixatives and subsequently dehydrating the specimen.

- The use of electron microscopy in materials science began with examining carbon and plastics in the 1940's, to metals and semiconductors and minerals in the 1960's, to essentially all types of materials in the 1980's. The Microscopy Society of America, previously known as the Electron Microscopy Society of America, held meetings beginning in 1942 for microscopists and manufacturers to discuss the latest technology in electron microscopy.

- Ernst Ruska, winner of the 1986 Nobel Prize in Physics, realized that the focal length of waves could be decreased by using an iron cap. The polschuh lens was built using this concept and was eventually used in all magnetic electron microscopes.

- The 1938 Model Toronto Electron Microscope in the Physics Department of the University of Toronto was constructed in six sections, each joined by vacuum-tight seals. It was 6 feet tall (with the electron beam standing vertically). This design still exists in almost all commercially produced electron microscopes.

reflect the electrons from the electron beam.

APPLICATIONS AND PRODUCTS

Cellular Biology. When one thinks of microscopy, one often thinks of observing two-dimensional images through a narrow lens. An electron microscope allows scientists to scan a series of two-dimensional, high-resolution images to create a three-dimensional

reconstruction of the specimen. The final result is called a tomogram, and it has changed the way that scientists study structures within cells that have not been fixed or dehydrated with chemicals.

Material Science and Engineering. An application of scanning electron microscopy involves using an energy-dispersive X-ray spectroscopy system, which examines the phases of minerals and metals, as well as determines the size, shape, and distribution of particles. Using high-resolution electron microscopy, the molecular arrangements of polymers (such as plastics) can be viewed.

Medicine. If a patient is having a tumor removed, any tissue removed during the surgery can be examined immediately using either light or electron microscopy. The specimen can be rapidly frozen and examined by pathologists rather than undergoing more time-consuming techniques that require the tissue to be fixed and dehydrated.

IMPACT ON INDUSTRY

Major Corporations. Germany's Siemens and Japan's JEOL are two companies that began developing electron microscopy technology and applications early in the field's history, and they remain important manufacturers of these instruments. JEOL produced its first electron microscope in 1948. By 1956, the company had entered the overseas market, selling an electron microscope to the Atomic Energy Commission in France. The Japanese company Hitachi also produces electron microscopes.

Government and University Research. In the United States, the National Institutes of Health and the National Science Foundation award grants to fund research related to electron microscopy research and technology. Many university laboratories deal directly with electron microscopy. They include the biological electron microscopy facilities at the University of Illinois and the University of Hawaii; the Center for Cell Imaging at Yale University; and the Penn Regional Nanotechnology Facility at the University of Pennsylvania.

CAREERS AND COURSE WORK

Electron microscopy is often used to examine biological specimens and other materials on a magnification level that cannot be approached by light microscopy. Microscopists typically have a bachelor's degree in biology and have taken courses in cell and molecular biology concepts, as well as in the principles and techniques of electron microscopy, including specimen preparation and research methods.

Microscopists often work as technicians at a university, in government, or in private engineering firms, where they will analyze samples for various clients. The study of electron microscopy also leads to careers in the pharmaceutical industry and in manufacturing, medical research, and environmental agencies.

SOCIAL CONTEXT AND FUTURE PROSPECTS

The increasing resolution of electron microscopy will enable more scientists and engineers to observe structures on the molecular scale. For example, in 2009, JEOL, in partnership with the Japan Science and Technology Agency, the National Institute of Advanced Industrial Science and Technology in Japan, and the National Institute for Materials Science in Japan, developed a new electron microscope capable of analyzing individual atoms and molecules. This new technology, with which they were able to observe a single atom of calcium, makes use of a new correction mechanism such that the accelerating voltage can be decreased. An important application of this new electron microscopic technique is that a biological or organic specimen can easily be altered by high-voltage electron beams.

Jessica C. Y. Wong, B.Sc., M.Sc.

FURTHER READING

Chandler, Douglas E., and Robert W. Roberson. *Bioimaging: Current Concepts in Light and Electron Microscopy.* Sudbury, Mass.: Jones and Bartlett, 2009. Begins with the history of electron microscopy and examines topics such as specimen preparation, transmission and scanning electron microscopes, and fluorescence microscopy.

Kuo, John, ed. *Electron Microscopy: Methods and Protocols.* 2d ed. Totowa, N.J.: Humana Press, 2007. Presents clear instructions on how to process biological specimens. Protocols are described by experienced experts. Both transmission and scanning electron microscopy are covered.

McIntosh, J. Richard, ed. *Cellular Electron Microscopy.* Boston: Elsevier/Academic Press, 2007. Experts discuss various aspects of electron microscopy, including specimen preparation, analysis of data, imaging from frozen-hydrated samples, and three-dimensional imaging.

Slayter, Elizabeth M., and Henry S. Slayter. *Light and Electron Microscopy.* New York: Cambridge University Press, 2000. Examines the principles behind the electron microscope and how understanding these basics helps researchers interpret results.

WEB SITES

Microscopy Society of America
http://www.microscopy.org

Museum of Science
Scanning Electron Microscope
http://www.mos.org/sln/sem

National Center for Electron Microscopy
http://ncem.lbl.gov/index.html

See also: Electronics and Electronic Engineering; Histology; Microscopy; Optics; Pathology; Spectroscopy.

ELECTRON SPECTROSCOPY ANALYSIS

FIELDS OF STUDY

Mathematics; physics; chemistry; biology; materials science; biochemistry; biomedical technology; metallurgy

SUMMARY

Electron spectroscopy analysis is a scientific method that uses ionizing radiation, such as ultraviolet, X-ray, and gamma radiation, to eject electrons from atomic and molecular orbitals in a given material. The properties of these electrons are then interpreted to provide information about the system from which they were ejected.

KEY TERMS AND CONCEPTS

- **Auger Electron:** An electron emitted from a valence orbital as a result of an energy cascade initiated by the photoemission of an electron from a core orbital.
- **Balmer Series:** A set of absorption lines in the electromagnetic spectrum of the hydrogen molecule, corresponding to the specific frequencies of the energy differences between molecular orbitals in the H_2 molecule.
- **Bond Strength (or Bond Dissociation Energy):** The amount of energy required to overcome the bond between two atoms and separate them from each other.
- **Electron-Volt:** The energy acquired by any charged particle with a unit charge on passing through a potential difference of one volt, equal to 23,053 calories per mole.
- **Ionization Potential:** The amount of work required to completely remove a specific electron from an atomic or molecular orbital.
- **Paramagnetism:** A measurable increase in the strength of an applied magnetic field caused by alignment of electron orbits in the material.
- **Photoelectron Emission:** The emission of an electron from an orbital caused by the impingement of a photon.

DEFINITION AND BASIC PRINCIPLES

The quantum mechanical theory of matter describes the positions and energies of electrons within atoms and molecules. When ionizing radiation is applied to a sample of a material, electrons are ejected from atomic and molecular orbitals in that material. The measured energies of those ejected electrons provide information that corresponds to the chemical identity and molecular structure of the material.

The analytical methods that employ this technique, such as mass spectrometry, typically study the properties of the molecular ions themselves rather than the electrons that were removed. The two processes are related, however, because the energies observed for one technique are often identical to those observed for the other. This can be understood at a rudimentary level by considering the law of conservation of energy as it must apply to the overall process of rearrangement. Electrons move from one orbital to another after one has been removed from an inner orbital and rearrangement of the electron distribution takes place to "fill in the hole."

Electron spectroscopic methods require that the electron emission process be carried out under high vacuum and with the use of sensitive electronic equipment to capture and measure the emitted electrons and their properties. Each technique utilizes unique methods, but similar devices, to carry out its tasks.

BACKGROUND AND HISTORY

The beginning of the science of electron spectroscopy can only be equated to the experiments of British physicist and Nobel laureate J. J. Thomson in 1897. These experiments first identified electrons and protons as the electrically charged particles of which atoms were composed, according to the atomic model propounded by British chemist and physicist Ernest Rutherford. Thomson's experiments were also the first to measure the ratio of the charge of the electron to the mass of the electron. This feature must be known to utilize the interaction of electrons and electromagnetic fields quantitatively.

In 1905, Albert Einstein identified and explained the photoelectron effect, in which light is observed to provide the energy by which electrons are ejected from within atoms. This work, one of only a handful of scientific papers actually published by Einstein, earned him the Nobel Prize in Physics in 1921.

German physicist Wilhelm Röntgen's discovery of X rays in 1895, and the subsequent development of the means to precisely control their emission, provided an important way to probe the nature of matter. X rays are designated in the electromagnetic spectrum as intermediate between ultraviolet light and gamma rays. High-vacuum technology and, most recently, digital electronic technology, all combine in the construction of devices that permit the precise measurement of minute changes in the properties of electrons in atoms and molecules.

HOW IT WORKS

Photoelectron Spectroscopy. Two general categories of photoelectron spectroscopy are commonly used. These are ultraviolet photoelectron spectroscopy (UPS) and X-ray photoelectron spectroscopy (XPS). Both methods function in precisely the same manner, and both utilize the same devices. The difference between them is that UPS uses ultraviolet radiation as the ionizing method, while XPS uses X rays to effect ionization.

A typical photoelectron spectrometer consists of a high-vacuum chamber containing a sample target; both are connected to an ionizing radiation emitter and a detection system constructed around a magnetic field. In operation, the vacuum chamber is placed under high vacuum. When the system has been evacuated, the sample is introduced and the emitter irradiates the sample, bringing about the emission of electrons from atomic or molecular orbitals in the material. The emitted electrons are then free to move through the magnetic field, where they impinge upon the detector.

The ability to precisely control the strength of the magnetic field allows an equally precise measurement of the energy of the emitted electrons. This measured energy must correspond to the energy of the electrons within the atomic or molecular orbitals of the target material, according to the mathematics of quantum mechanical theory, and so provides information about the intimate internal structure of the atoms and molecules in the material. The methodology has been developed such that measurements are obtainable using matter in any phase as a solid, liquid, or gas. Each phase requires its own modification of the general technique.

The direct measurement of emitted electron energies through the use of photomultiplying devices is displacing more complex methods based on magnetic field because of the inherent difficulties of providing adequate magnetic shielding to the ever more sensitive components of the devices.

XPS is also known as electron spectroscopy for chemical analysis, or ESCA. The use of this identifier, however, is becoming less common in practice and in the chemical literature.

Auger Electron Spectroscopy (AES). The Auger electron process is a secondary electron emission process that begins with the normal ejection of a core electron by ultraviolet or X-ray radiation. In the Auger process, electrons from higher energy levels shift to lower levels to fill in the gap left by the emission of the core electron. Excess energy that accrues from the difference in orbital energies as the electrons shift then brings about the secondary emission of an electron from a valence shell. The overall process is in accord with both quantum mechanics and the law of conservation of energy, which requires the total energy of a system before a change occurs to be exactly the same as the total energy of the system after a change occurs.

Unlike UPS and XPS, AES generally utilizes an electron beam to effect core electron emission. Detection of emitted electrons is entirely by direct measurement through photomultiplying devices rather than by any magnetic field methods. The methodology of the technique is otherwise similar to that of UPS and XPS. It is amenable to the study of matter in all phases, except for hydrogen and helium, but is generally valuable for use only with solids as a surface analysis technique. This is true because sample materials must be stable under vacuum at pressures of 10-9 Torr. Also, AES is known to be highly sensitive and capable of fast response.

Electron Spin Resonance (ESR). The principles of ESR are based on an entirely different physical property of electrons in their atomic or molecular orbitals. In quantum mechanics, each electron is allowed to exist only in very specific states with very specific energies within an atom or molecule.

One of the allowed states is designated as "spin." In this state, the electron can be thought of as an electrical charge that is physically spinning about an axis, thus generating a magnetic field. Only two orientations are allowed for the magnetic fields generated in this way, and according to the Pauli exclusion principle, pairs of electrons must occupy both states in opposition. This requirement means that ESR can be used only with materials that contain single or unpaired electrons, including ions. When placed in an external magnetic field, the magnetic fields of the single electrons align with the external magnetic field.

Subsequent irradiation with an electromagnetic field fluctuating at microwave frequencies acts to invert the magnetic fields of the electrons. Measurement of the frequencies at which inversion takes place provides specific information about the structure of the particular material being examined. The precise locations of inversion signals depend upon the atomic or molecular structure of the material, as these environments affect the nature of the magnetic field surrounding the electron.

APPLICATIONS AND PRODUCTS

In application, electron spectroscopy is strictly an analytical methodology, and it serves only as a probe of material composition and properties. It does not serve any other purpose, and all applications and products related to electron spectroscopy are the corresponding spectroscopic analyzers and the ancillary products that support their operation.

Spectroscopic analyzers come in a variety of forms and designs, according to the environment in which they will be expected to function, but more with respect to the nature of the use to which they will be put. These range from machines for routine analysis of a limited range of materials and properties at the low end of the scale, to the complex machines used in high-end research that must be capable of extreme sensitivity and finely detailed analysis.

The applications of electron spectroscopic analysis are, in contrast, wide ranging and are applicable in many fields. In its roles in those fields, the methodology has enabled some of the most fundamental technology to be found in modern society.

One application in which electron spectroscopy has proven unequaled in its role is submicroscopic surface analysis. Both XPS and AES are the methods of choice in this application, because each can probe to a depth of about 30 microns below the actual surface of a solid material, enabling analysts to see and understand the physical and chemical changes that occur in that region.

The surface of a solid typically represents the point of contact with another solid, and physical interaction between the two normally effects some kind of change to those surfaces because of friction, impact, or electrochemical interaction. A good example of this is the tribological study of moving parts in internal combustion engines. In normal operation, a piston fitted with sealing rings moves with a reciprocating motion within a closely fitted cylinder. The rings physically interact with the wall of the cylinder with intense friction, even though well lubricated, under the influence of the high heat produced through the combustion of fuel. At the same time, the top of the piston is subjected to intense pressures and heat from the explosive combustion of the fuel. At an engine revolution rate of 2,400 revolutions per minute (rpm), each cylinder in a four-cylinder internal combustion engine goes through its reciprocating motion six hundred times each minute, or ten times each second.

In turbine and jet engines, for example, parts are subjected to such stress and friction at a rate of hundreds and even thousands of times per second. Engine and automobile manufacturers and developers must understand what happens to the materials used in the corresponding parts under the conditions of operation. Both XPS and AES are used to probe the material effects at these surfaces for the development of better formulations and materials, and for understanding the weaknesses and failure modes of existing materials.

ESR, on the other hand, is used entirely for the study of the chemical nature of materials in the liquid or gaseous phase. In this role, researchers and analysts use the methodology to study the reactions and mechanisms involving single-electron chemical species. This includes the class of compounds known as radicals, which are essentially molecules containing their full complement of electrons but not of atoms. The methyl radical, for example, is basically a molecule of methane (CH_4) that has lost one hydrogen atom. The remaining CH_3 portion is electrically neutral, because it has all of the electrons that would normally be present in a neutral molecule of CH_4, but with one of its electrons free to latch on to the first available molecule that comes along.

Radical reactions are understood to be responsible for many effects: aging in living systems, especially humans; atmospheric reactions, especially in the upper atmosphere and ozone layer; the detrimental effects of singlet oxygen; combustion processes; and many others. In biological systems, specially designed molecules are used to tag other nonparamagnetic molecules so that they can be studied by ESR. Such molecules often include a "nitroso" functional group in their structures to provide a paramagnetic radical site that can be monitored by ESR. The production and testing of these specialty chemicals is another area of application.

Fascinating Facts About Electron Spectroscopy Analysis

- The photoelectron effect, one of the basic principles of electron spectroscopy, was first explained by Albert Einstein in 1905, for which he received the 1921 Nobel Prize in Physics.
- Spinning electrons generate a magnetic field around themselves in the same way that moving electrons through a wire produces a magnetic field around the wire.
- X rays and ultraviolet light can both eject electrons from the inner orbitals of atoms.
- Auger electrons are electrons emitted by the extra energy released when electrons cascade into lower orbitals to replace electrons that have been ejected by X rays or ultraviolet light.
- Electron spectroscopy can measure the chemical and physical properties of materials up to 30 microns below the surface of a solid.
- Electron spin resonance measures the frequencies required to switch the orientation of the magnetic fields of single electrons.
- Scanning Auger microscopy can produce detailed maps of the distribution of specific metal atoms at the surface of an alloy, allowing metallurgists to see how the material is structured.

IMPACT ON INDUSTRY

The impact on industry of electron spectroscopy is not obvious. The methodology plays very much a behind-the-scenes or supportive role that is not apparent from outside any industry that uses electron spectroscopy. Although this is true, the role played by electron spectroscopy in the development and improvement of products and materials has been

valuable to those same industries.

Many advances in metallurgy and tribology, the study of friction and its effects, have been made possible through knowledge obtained by electron spectroscopy, especially AES and XPS. Given the untold millions of moving parts–bearings, pistons, slides, shafts, link chains–that are in operation around the world every single minute of every single day, lubrication and lubricating materials represents a huge world-wide industry. With the vast majority of lubricating materials (oils and greases) being derived from nonrenewable resources, the need to enhance the performance of those materials through better understanding of material interactions has been one driving force behind the application of electron spectroscopy in industry.

Other industrial processes require that materials undergo a chemical process called passivation, which is essentially the rendering of the surface of a material inert to chemical reaction through the formation of a thin coating layer of oxide, nitride, or some other suitable chemical form. With its ability to accurately measure the thickness and properties of thin films such as oxide layers on a surface, electron spectroscopy is uniquely appropriate to use in industries that rely on passivation or on the formation of thin layers with specific properties. One such industry is the semiconductor industry, upon which the computer and digital electronics fields have been built. AES and XPS are commonly used to monitor the quality and properties of thin layers of semiconductor materials used to construct computer chips and other integrated circuits.

CAREERS AND COURSEWORK

Electron spectroscopy is used to examine the intimate details of the atomic and molecular structure of matter, making electron spectroscopy an advanced career. Students who look to a career in this field will undertake highly technical foundation courses in mathematics, physics, inorganic chemistry, organic chemistry, surface chemistry, physical chemistry, chemical physics, and electronics. The minimum requirement for a career in this field is an associate's degree in electronics technology or an honors (four-year) bachelor's degree in chemistry, which will allow a student to specialize in the practice as a technician in a research or analytical facility. More advanced positions will require a postgraduate degree (master's or doctorate).

As the field finds more application in materials research and forensic investigation, general opportunities should further develop. These applications will require, however, that those wishing a career involving electron spectroscopy have specialist training. The vast majority of opportunities in the field are to be found in such academic research facilities as surface science laboratories and in materials science research. Forensic analysis also holds a number of opportunities for electron spectroscopists.

SOCIAL CONTEXT AND FUTURE PROSPECTS

Electron spectroscopy analysis is a methodology with an important role behind the scenes. The field is neither well known nor readily recognized. Nevertheless, it is a critical methodology for advancing the understanding of materials and the nature of matter. As such, electron spectroscopy adds to the general wealth of knowledge in ways that permit the development of new materials and processes and to advancing the understanding of how existing materials function.

One development of electron spectroscopy, known as scanning Auger microscopy (SAM) has the potential to become an extremely valuable technique because of its ability to generate detailed maps of the surface structure of materials at the atomic and molecular level. By tuning SAM to focus on specific elements, the precise distribution of those elements in the surface being examined can be identified and mapped, providing detailed knowledge of the granularity, crystallinity, and other structural details of the material. This is especially valuable in such widely varied fields as metallurgy, geology, and advanced composite materials.

XPS and AES have been applied in a variety of different fields and are themselves becoming very important surface analytical methods in those fields. These areas include the aerospace and automotive industries, biomedical technology and pharmaceuticals, semiconductors and electronics, data storage, lighting and photonics, telecommunications, polymer science, and the rapidly growing fields of solar cell and battery technology.

Richard M. Renneboog, M.Sc.

FURTHER READING

Chourasia, A. R., and D. R. Chopra. "Auger Electron Spectroscopy." In *Handbook of Instrumental Techniques for Analytical Chemistry*, edited by Frank Settle. New York: Prentice Hall Professional Reference, 1997. This chapter provides a thorough and systematic description of the principles and practical methods of Auger spectroscopy, including its common applications and limitations.

Kolasinski, Kurt W. *Surface Science: Foundations of Catalysis and Nanoscience.* 2d ed. Chichester, England: John Wiley & Sons, 2008. Provides a complete study of the utility of electron spectroscopy as applied to the study of processes that occur on material surfaces.

Merz, Rolf. "Nano-analysis with Electron Spectroscopic Methods: Principle, Instrumentation, and Performance of XPS and AES." In *NanoS Guide 2007.* Weinheim, Germany: Wiley-VCH, 2007. A lucid and readable presentation of the principles, capabilities, and limitations of XPS and AES based on actual applications and comparisons with methods such as scanning electron microscopy.

Strobel, Howard A., and William R. Heineman. *Chemical Instrumentation: A Systematic Approach.* 3d ed. New York: John Wiley & Sons, 1989. This book provides a valuable resource for the principles and practices of electron spectroscopy and for many other analytical methods. Geared toward the operation and maintenance of the devices used in those practices.

WEB SITES

Farach, H. A., and C. P. Poole "Overview of Electron Spin Resonance and Its Applications" http://www.uottawa.ca/ publications/interscientia/inter.2/spin.html

Molecular Materials Research Center, California Institute of Technology "Overview of Electron Spin Resonance and Its Applications"
http://mmrc.caltech.edu/SS_XPS/XPS_PPT/ XPS_Slides.pdf

See also: Electrochemistry; Electronics and Electronic Engineering; Electron Microscopy; Spectroscopy.

ELECTRONIC COMMERCE

FIELDS OF STUDY

Business information systems; computer science; computer programming; database administration; retailing; marketing; Web development; software engineering; Web design.

SUMMARY

Electronic commerce (e-commerce) refers to the buying, selling, and transfer of products and services using the Internet. It offers enormous advantages for supply-chain management and the coordination of distribution channels, in addition to being convenient for consumers. Convenience is a big selling point: In 2007, online retail generated $175 billion and is expected to climb to $335 billion by 2012.

KEY TERMS AND CONCEPTS

- **Bandwidth:** Difference between highest and lowest number of data that a medium can transmit, expressed in cycles per second or bits per second. The amount of data that can be transmitted increases as the bandwidth increases.
- **Browser:** Software program that is used to view Web pages on the Internet.
- **Download:** Copy digital data and transmit it electronically.
- **E-Book:** Electronic book that can exist with or without a printed form of the same content. E-books are downloaded from a Web site to an electronic reader (e-reader).
- **Encryption:** Any procedure used in cryptography to convert plaintext into ciphertext to prevent anyone but the intended recipient from reading the data.
- **Enterprise Resource Planning (ERP):** Software integration of projects, distribution, manufacturing, and employees.
- **Firewall:** Hardware device placed between the private network and Internet connection that prevents unauthorized users from gaining access to data on the network.
- **Hacking:** Process of penetrating the security of other computers by using programming skills.

- **Lead Time:** Time required for a product to be received from a supplier after an order has been placed.
- **Operating System:** Computer program designed to manage the resources (including input and output devices) used by the central processing unit and that functions as an interface between a computer user and the hardware that runs the computer.
- **Overhead Costs:** Daily operating expenses.
- **Server:** Computer that is dedicated to managing resources shared by users (clients).

DEFINITION AND BASIC PRINCIPLES

E-commerce refers to the communications between customers, vendors, and business partners over the Internet. The term e-business incorporates the additional activities carried out within a business using intranets, such as communications related to production management and product development. Many also view e-business as referring to collaborations with partners and e-learning organizations.

In addition to online purchases of goods and services, e-commerce involves bill payments, online banking, e-wallets, smart cards, and digital cash. E-commerce depends on secure connections to the Internet. Many precautions to ensure security are necessary to maintain successful e-commerce, including public key cryptography, the Secure Sockets Layer (SSL), digital signatures, digital certificates, firewalls, and antivirus programs. New technology is constantly being developed to protect against worms, viruses, and other cyber attacks.

BACKGROUND AND HISTORY

In 1969, the Advanced Research Projects Agency (now the Defense Advanced Research Projects Agency) of the U.S. Department of Defense proposed a method to link together the computers at several universities to share computational data via networks. This network became known as ARPANET, which was the precursor of the Internet. As a result, electronic mail (e-mail) was developed, along with protocols for sending information over phone lines in packets. The protocols for the transmission of these packets of data came to be known as Transmission Control

Protocol (TCP) and Internet Protocol (IP). Together these two protocols, known as TCP/IP, are still in use and are responsible for the efficient communication conducted through the network of networks referred to as the Internet.

Oxford University graduate Tim Berners-Lee initially created the World Wide Web in 1989 while working at CERN, the European Council for Nuclear Research, and made it available in 1991. During this time, Cisco Systems was growing to become the first company to produce the broad range of hardware products that allowed ordinary individuals to access the Internet. In 1993, Marc Andreessen and Eric Bina, employees at National Center for Supercomputing Applications (NCSA), created Mosaic, the first Web browser that supported clickable buttons and links and allowed users to view text and images on the same page. New software and programming-language developments rapidly followed, allowing ordinary consumers easy access to the Internet. As a result, companies saw the opportunity to gain customers, resulting in the creation of online businesses, including Amazon in 1994, eBay in 1995, and PayPal and Priceline in 1998.

HOW IT WORKS

The engineers who developed ARPANET created the use of digital packets for transmitting data via packet switching. The general idea was that it would be faster and cheaper to transmit digital data using small packets that could be sent, or routed, to their destination in the most efficient way possible, even if the original message had to be split up into smaller packets that were then joined back together at their destination. In order to accomplish the packaging of data and transmission via the best routes, the engineers who developed ARPANET also developed Transmission Control Protocol (TCP). The first and most common access method to the Internet was through the wiring that has transmitted telephone calls. However, wireless Internet connections can now be made much faster from many locations—even from cell phones. The economy has become dependent on digital communication, and the companies that sell the most goods and services have a strong Web presence. A great deal of planning goes into the maintenance of an effective Web site, and some of the most important steps in establishing an e-commerce company are discussed next.

E-Commerce Business Establishment. After first developing a practical business plan, the process for establishing an online business that will be able to compete for sales successfully could be overwhelming. One way to start an e-business is by using a turnkey solution, which is essentially a prepackaged type of software specifically for a new business. An alternative is to use the services of an Internet incubator, which is a company that specializes in e-business development. Both eToys and NetZero used Internet incubators to help them get started. An Internet incubator typically obtains ownership of at least 50 percent of the business and may also enlist funding help from venture capitalists to get started. Web-hosting companies sell space on a Web server to customers and maintain enough storage space for the Web site and to provide support services as well. A domain name for the Web site must be chosen and registered. This domain name is to be used in the URL (uniform resource locator) for the Web site. The URL, or Web site's Internet address, consists of three parts: the host name, which is shown by the www for World Wide Web; the domain name, which is usually the name of the company; and lastly, the top-level domain (TLD), which describes the type of organization that owns the domain name, such as .com for a commercial organization, or .gov for a government organization. An initial public offering (IPO) of stock to assist with funding usually follows for enterprises that achieve a certain level of success.

Design of Markets and Mechanisms of Transactions. Initially the business-to-business (B2B) types of transactions were the primary e-commerce activities. These activities quickly expanded to include sales to consumers via electronic retailing (e-tailing), often called business-to-consumer (B2C). Since the late 1990's e-commerce has expanded to include consumer-to-consumer (C2C) Web sites, including eBay, and consumer-to-business (C2B), such as Priceline, where several airlines or hotels will compete for the purchase dollars of consumers. Each of these types of transactions can be completed within the general structure of one of the many different types of e-commerce models to generate revenue.

Automated Negotiation and Peer-to-Peer Distribution Systems. Auction models allow an Internet user to assume the role of a buyer using either the reverse-auction model (where the buyer sets a price and sellers have to compete to beat that price) or the reverse-price model (where the seller sets the minimum price that will be accepted). Auction sites, such as eBay, update listings, feature items, and earn submission and commission fees, but they leave the processes of payment and delivery up to the actual buyers and sellers.

Dynamic-pricing models include the name-your-price companies, such as Priceline.com, that use a shopping bot to collect bids from customers and deliver these bids to the providers of services to see if they are accepted. A shopping bot is a computer program that searches through vast amounts of information, then collects, summarizes, and reports the information. This is one example of intelligent agents, software programs that have been designed to gather information, used by e-businesses. Priceline's immense success is due in part to its use of this technology.

Network Resource Allocation: Electronic Data Interchange (EDI). Portal models present a whole variety of news, weather, sports, and shopping all on one Web page that allows a visitor to see an overview and then choose to obtain more in-depth information. Vertical portals are specific for a single item, while horizontal portals function as search engines with access to a large range of items. Storefront models require a product line to be accessible online via the merchant server so that customers can select items from the database of products and collect them in the order-processing technology called a shopping cart. Businesses use EDI as a standardized protocol for communication to monitor daily inventory, shipments, and payments. Standardized forms for invoices and purchase orders are routinely accessible via the use of extensible markup language (XML). Companies such as Commerce One and TIBCO Software were created with the sole purpose of helping companies to move their businesses to the Web via B2B techniques. The transition of traditional brick-and-mortar stores to click-and-mortar stores has helped to decrease lead time and has caused an increase of just-in-time (JIT) inventory management. JIT inventory management allows e-businesses to save money because the companies do not overbuy goods and create an inventory surplus that they then have to worry about storing and selling. JIT in turn decreases overhead costs.

APPLICATIONS AND PRODUCTS

The development of computer technologies and the Internet has given rise to e-commerce, which, in turn, has spawned a variety of applications and products that facilitate e-business as well as enhance people's lives.

Consumer Products. Several tablets were among the most popular consumer digital purchases in 2010. Apple's iPad, Barnes & Noble's NookColor, Samsung's Galaxy Tab, Sony's Reader, and Amazon's Kindle are all capable of accessing the Internet and downloading magazines and books. The iPad is the most expensive of these tablets, but it is also capable of downloading video and audio files, while the Kindle is the least expensive and functions exclusively as an e-reader.

Both contact and contactless "smart cards," which resemble credit cards, have been developed to store much more information (banking, retail, identification, health care) because of a microprocessor embedded in the card. The E-ZPass, which is used by New York and New Jersey commuters to pay tolls, is an example of a contactless smart card. Smart cards are more secure than credit cards because they are encrypted and password protected.

Security Applications and Products. Companies have been created to help merchants accept credit card payments online, which are called card-not-present (CNP) transactions. These companies, such as CyberCash and PAYware, offer services to facilitate the authentication and authorization processes through the Secure Socket Layer (SSL) using Secure Electronic Transaction (SET) technology to minimize fraud. Additional security features include firewalls, encryption, and antivirus software. Visa and other major credit card companies have introduced e-wallets that allow customers to save their shipment address and payment information securely in an online database so that purchases can be made with one click of the mouse, instead of having to reenter the same information each time. In 1999, the Electronic Commerce Modeling Language (ECML) emerged as the protocol for e-wallet usage by merchants. PayPal can be used to transfer payments between consumers securely by simply creating an account using an e-mail address and a credit card, which is used to pay for goods and services. PayPal is ideally suited for use on an auction site, such as eBay. PayPal is especially secure because credit card information is checked before the transaction actually begins. This allows for payment to take place in real time, minimizing the opportunity for fraud.

Applications Using Wireless Transactions. Mobile business (m-business) made possible by wireless technology will continue to grow in importance. The third generation, called 3G technology, is allowing wireless devices to transmit data more than seven times faster than the 56K modem, and 4G technology began to replace it in 2011. Sprint PCS provides access to the Internet using the Code Division Multiple Access (CDMA) technology. CDMA technology assigns a unique code to each transmission on a specific channel, which allows each transmission to use the entire bandwidth available for that channel, greatly decreasing the time it takes to complete an e-commerce transaction.

Fascinating Facts About Electronic Commerce

- In 2010, e-commerce sales increased by 15.4 percent over 2009 e-commerce sales.
- E-book purchases in 2010 increased 150 percent over e-book purchases in 2009.
- By the year 2015 it is estimated that there will be more than 29.4 million e-books. Some believe that traditional book stores will eventually cease to exist. Borders bookstore is the largest national bookstore chain, and its financial difficulties (closing stores) could be an indication of this trend.
- The Apple iPhone is essentially a miniature, handheld computer, because it can download application programs ("apps") directly from the Internet, in addition to functioning as a phone. There are apps designed to perform just about any task, from accessing files on an office server to making the iPhone function as a flashlight. As of 2011 there were more than 100,000 apps available on the Apple App Store Web site.
- Almost 8 million Kindles were sold in 2010.
- Online sales for the 2010 holiday shopping season were estimated to total more than $36 billion.

IMPACT ON INDUSTRY

The developments in technology that have made the Internet more easily accessible to the average consumer have been made primarily by American companies. These companies have continued to grow and expand to reach consumers all around the world. Other nations have grown economically because of the explosion in e-commerce, and much of the e-commerce growth is expected to involve the BRIC nations (Brazil, Russia, India, and China).

Amazon.com was founded in Jeff Bezos's Seattle garage in 1994. Bezos was named *Time* magazine's Person of the Year in 1999, and the company that he started in his garage continues to grow at an amazing rate. In addition to books, products available on Amazon have come to include jewelry, sporting goods, shoes, digital downloads of music, videos, games, software, health and beauty aids, and just about everything else possible, including groceries. The company reported net sales of $7.56 billion for the third quarter of 2010, which was a 39 percent increase over the same period in 2009.

eBay.com has become the largest online auction site in the world. Visitors to the site can buy or sell just about anything, including iPods, laptops, digital cameras, tickets to concerts and sporting events, jewelry, books, antiques, crafts, sporting goods, pet supplies, and clothes. eBay is headquartered in San Jose, California, and revenue for the third quarter of 2010 was $2.2 billion, an increase of 1 percent over the same period of 2009. Its PayPal business, acquired in 2002, has grown by at least an additional 1 million accounts every month, and it can accept twenty-four different types of currencies worldwide.

PayPal was initially founded in 1998 in Palo Alto, California, and like many other original U.S.-based tech companies, it has long since expanded its operations worldwide. PayPal has locations in Berlin, Tel Aviv, Dublin, Luxembourg, and China. Its purpose is to facilitate the processing of online payments for various e-commerce businesses.

Priceline.com, located in Norwalk, Connecticut, developed its name-your-own-price system for its online auction type of business and has grown since it emerged in 1998. Visitors to the Web site can list the price they want to pay for travel-related services and items, including hotels, vacation packages, cruises, airplane tickets, and car rentals. Although it briefly tried to expand on its e-commerce activities to include home loans, long-distance telephone service, and cars, it discontinued these ventures in 2002 and has continued to excel in travel-related services. Its chief e-commerce competitors are Expedia.com, Travelocity.com, Orbitz.com, and Hotwire.com. As of 2011, Priceline.com was the leader. The company had more than three hundred employees, and its third-quarter 2010 revenue was reported to be $1 billion, a 37.1 percent increase over 2009.

FedEx is the leader in air and ground transportation and won recognition in *Fortune* magazine's "World's Most Admired Companies" in 2006, 2007, 2008, 2009, and 2010 and in *Business Week*'s "50 Best Performers" in 2006. In 2010, its second-quarter revenue was $9.63 billion, an increase of 12 percent from the same period in 2009. FedEx continues its commitment to innovation with its FedEx Institute of Technology at the University of Memphis, in the company's hometown. The institute focuses on nanotechnology, artificial intelligence, biotechnology, and multimedia arts.

While an undergraduate at Yale University in 1965, Fred W. Smith wrote a term paper describing the implementation of an airfreight system that could transport computer parts and medications in a timely fashion. Although at the time his idea was not viewed as feasible by Yale faculty, Smith later bought an interest in Arkansas Aviation Sales, which eventually grew to become FedEx. E-commerce has continued to grow because of the rapid and efficient transportation of goods made possible by FedEx and the United Parcel Service (UPS).

CAREERS AND COURSE WORK

The job titles, career paths, and salaries vary a great deal within the field of e-commerce. Some of the typical job titles include Web site developer, Web designer, database administrator, Web master, and Web site manager. Jobs related to Web content require skills in Web development tools and software languages, including hypertext markup language (HTML), extensible markup language (XML), Java, JavaScript, Visual Basic, Visual Basic Script, and Active Server Pages (ASP). Database administrators focus less on these Web tools and languages and more on database-related tools, such as structured query language (SQL), Microsoft Access, and Oracle. Knowledge of computer networks and operating systems is also very helpful.

SOCIAL CONTEXT AND FUTURE PROSPECTS

Due in part to the easy accessibility of goods and services via the Internet, online businesses have become increasingly competitive, with more made-to-order goods being produced by companies such as Dell and corresponding decreases in the costs associated with the maintenance of a large inventory. Since so much more data are exchanged digitally via stock trades, mortgages, purchases of consumer goods, payment of bills, and banking transactions, it is conceivable that eventually digital cash and smart cards could replace traditional cash. Because consumers enjoy the comparison shopping among goods offered by companies all over the world, as well as the twenty-four-hour-a-day, seven-days-a-week convenience of online shopping, sales at the traditional brick-and-mortar stores will no doubt continue to decline, or at least migrate toward those goods that consumers prefer not to purchase online (such as those they wish to consider physically and those they wish to obtain immediately).

Hacking, identity theft, and other types of cyber theft have become problems, which have only increased the need for better security tools, such as digital certificates and digital signatures. Increased security needs will continue to fuel the ever-expanding Internet security industry.

Jeanne L. Kuhler, B.S., M.S., Ph.D.

FURTHER READING

Byrne, Joseph. *I-Net+ Certification Study System.* Foster City, Calif.: IDG Books Worldwide, 2000. This review guide provides technical information regarding hardware for networks, as well as software, important for e-commerce.

Castro, Elizabeth. *HTML, XHTML, and CSS, Sixth Edition: Visual Quick Start Guide.* Berkeley, Calif.: Peachpit Press, 2007. An introductory text describing how a novice can set up an individual Web site and includes plenty of helpful screen shots.

Deitel, Harvey M., Paul J. Deitel, and Kate Steinbuhler. *E-Business and E-Commerce for Managers.* Upper Saddle River, N.J.: Prentice Hall, 2001. This introductory textbook describes e-commerce in fairly nontechnical terms from the business perspective.

Longino, Carlo. "Your Wireless Future." *Business 2.0.* May 22, 2006. http://money. c n n . c o m / 2 0 0 6 / 0 5 / 1 8 / t e c h n o l o g y /

business2_wirelessfuture_intro/. Discusses the future of wireless technology in all sectors—business, entertainment, and communications.

Turban, Efraim, and Linda Volonino. *Information Technology for Management: Improving Performance in the Digital Economy.* 7th ed. Hoboken, N.J.: John Wiley & Sons, 2010. This introductory textbook provides both technical and nontechnical information related to e-commerce.

Umar, Amjad. "IT Infrastructure to Enable Next Generation Enterprises." *Information Systems Frontiers* 7, no. 3 (July, 2005): 217-256. Describes advances in network protocols and design.

Web Sites
E-Commerce Times
http://www.ecommercetimes.com

National Retail Foundation's Digital Division
http://www.shop.org

ELECTRONIC MATERIALS PRODUCTION

FIELDS OF STUDY

Mathematics; physics; chemistry; crystallography; quantum theory; thermodynamics

SUMMARY

While the term "electronic materials" commonly refers to the silicon-based materials from which computer chips and integrated circuits are constructed, it technically includes any and all materials upon which the function of electronic devices depends. This includes the plain glass and plastics used to house the devices to the exotic alloys and compounds that make it possible for the devices to function. Production of many of these materials requires not only rigorous methods and specific techniques but also requires the use of high-precision analytical methods to ensure the structure and quality of the devices.

KEY TERMS AND CONCEPTS

- **Biasing:** The application of a voltage to a semiconductor structure (transistor) to induce a directional current flow in the structure.
- **Czochralski Method:** A method of pulling material from a molten mass to produce a single large crystal.
- **Denuded Zone:** Depth and area of a silicon wafer that contains no oxygen precipitates or interstitial oxygen.
- **Epi Reactor:** A thermally programmable chamber in which epitaxial growth of silicon chips is carried out.
- **Gettering:** A method of lowering the potential for precipitation from solution of metal contaminants in silicon, achieved by controlling the locations at which precipitation can occur.
- **Polysilicon (Metallurgical Grade Silicon):** A form of silicon that is 99 percent pure, produced by the reaction of silicon dioxide (SiO_2) and carbon (C) to produce silicon (Si) and carbon monoxide (CO) at a temperature of 2000 degrees Celsius.

DEFINITION AND BASIC PRINCIPLES

Electronic materials are those materials used in the construction of electronic devices. The major electronic material today is the silicon wafer, from which computer chips and integrated circuits (ICs) are made. Silicon is one of a class of elements known as semiconductors. These are materials that do not conduct electrical currents appreciably unless acted upon, or "biased," by an external voltage. Another such element is germanium.

The construction of silicon chips requires materials of high purity and consistent internal structure. This, in turn, requires precisely controlled methods in the production of both the materials and the structures for which they are used. Large crystals of ultrapure silicon are grown from molten silicon under strictly controlled environmental conditions. Thin wafers are sliced from the crystals and then polished to achieve the desired thickness and mirror-smooth surface necessary for their purpose. Each wafer is then subjected to a series of up to five hundred, and sometimes more, separate operations by which extremely thin layers of different materials are added in precise patterns to form millions of transistor structures. Modern CPU (central processing unit) chips have between 107 and 109 separate transistors per square centimeter etched on their surfaces in this way.

One of the materials added by the thin-layer deposition process is silicon, to fill in spaces between other materials in the structures. These layers must be added epitaxially, in a way that maintains the base crystal structure of the silicon wafer.

Other materials used in electronic devices are also formed under strictly controlled environmental conditions. Computers could not function without some of these materials, especially indium tin oxide (ITO) for what are called transparent contacts and indium nitride for light-emitting diodes in a full spectrum range of colors.

BACKGROUND AND HISTORY

The production of modern electronic materials began with the invention of the semiconductor bridge transistor in 1947. This invention, in turn, was made possible by the development of quantum theory and the vacuum tube technology with which electronic devices functioned until that time.

The invention of the transistor began the development of electronic devices based on the semiconducting character of the element silicon. Under the influence of an applied voltage, silicon can be induced to conduct an electrical current. This feature allows silicon-based transistors to function somewhat like an on-off switch according to the nature of the applied voltage.

In 1960, the construction of the functional laser by American physicist and Nobel laureate Arthur Schawlow began the next phase in the development of semiconductor electronics, as the assembly of transistors on silicon substrates was still a tedious endeavor that greatly limited the size of transistor structures that could be constructed. As lasers became more powerful and more easily controlled, they were applied to the task of surface etching, an advance that has produced ever smaller transistor structures. This development has required ever more refined methods of producing silicon crystals from which thin wafers can be cut for the production of silicon semiconductor chips, the primary effort of electronic materials production (though by no means the most important).

How It Works

Melting and Crystallization. Chemists have long known how to grow large crystals of specific materials from melts. In this process, a material is heated past its melting point to become liquid. Then, as the molten material is allowed to cool slowly under controlled conditions, the material will solidify in a crystalline form with a highly regular atomic distribution.

Now, molten silicon is produced from a material called polysilicon, which has been stacked in a closed oven. Specific quantities of doping materials such as arsenic, phosphorus, boron, and antimony are added to the mixture, according to the conducting properties desired for the silicon chips that will be produced. The polysilicon melt is rotated in one direction (clockwise); then, a seed crystal of silicon, rotating in the opposite direction (counterclockwise), is introduced. The melt is carefully cooled to a specific temperature as the seed crystal structure is drawn out of the molten mass at a rate that determines the diameter of the resulting crystal.

To maintain the integrity of the single crystal that results, the shape is allowed to taper off into the form of a cone, and the crystal is then allowed to cool completely before further processing. The care with which this procedure is carried out produces a single crystal of the silicon alloy as a uniform cylinder, whose ends vary in diameter first as the desired extraction rate was achieved and then due to the formation of the terminal cone shape.

Wafers. In the next stage of production, the nonuniform ends of the crystal are removed using an inner diameter saw. The remaining cylinder of crystal is called an ingot, and is then examined by X ray to determine the consistency and integrity of the crystal structure. The ingot then will normally be cut into smaller sections for processing and quality control.

To produce the rough wafers that will become the substrates for chips, the ingot pieces are mounted on a solid base and fed into a large wire saw. The wire saw uses a single long moving wire to form a thick network of cutting edges. A continuous stream of slurry containing an extremely fine abrasive provides the cutting capability of the wire saw, allowing the production of many rough wafers at one time. The rough wafers are then thoroughly cleaned to remove any residue from the cutting stage.

Another procedure rounds and smooths the edges of each wafer, enhancing its structural strength and resistance to chipping. Each wafer is also laser-etched with identifying data. They then go on to a flat lapping procedure that removes most of the machining marks left by the wire saw, and then to a chemical etching process that eliminates the marking that the lapping process has left. Both the lapping process and the chemical etching stage are used to reduce the thickness of the wafers.

Polishing. Following lapping and rigorous cleaning, the wafers move into an automated chemical-mechanical polishing process that gives each wafer an extremely smooth mirror-like and flat surface. They are then again subjected to a series of rigorous chemical cleaning baths, and are then either packaged for sale to end users or moved directly into the epitaxial enhancement process.

Epitaxial Enhancement. Epitaxial enhancement is used to deposit a layer of ultrapure silicon on the surface of the wafer. This provides a layer with different properties from those of the underlying wafer material, an essential feature for the proper functioning of the MOS (metal-oxide-semiconductor) transistors that are used in modern chips. In this process, polished wafers are placed into a programmable oven and spun in an atmosphere of trichlorosilane gas. Decomposition of the trichlorosilane deposits silicon atoms on the surface of the wafers. While this produces an identifiable layer of silicon with different properties, it also maintains the crystal structure of the silicon in the wafer. The epitaxial layer contains no imperfections that may exist in the wafer and that could lead to failure of the chips in use.

From this point on, the wafers are submitted to hundreds more individual processes. These processes build up the transistor structures that form the functional chips of a variety of integrated circuit devices and components that operate on the principles of digital logic.

APPLICATIONS AND PRODUCTS

Microelectronics. The largest single use of silicon chips is in the microelectronics industry. Every digital device functions through the intermediacy of a silicon chip of some kind. This is as true of the control pad on a household washing machine as it is of the most sophisticated and complex CPU in an ultramodern computer.

Digital devices are controlled through the operation of digital logic circuits constructed of transistors built onto the surface of a silicon chip. The chips can be exceedingly small. In the case of integrated circuit chips, commonly called ICs, only a few transistors may be required to achieve the desired function.

The simplest of these ICs is called an inverter, and a standard inverter IC provides six separate inverter circuits in a dual inline package (DIP) that looks like a small rectangular block of black plastic about 1 centimeter wide, 2 centimeters long, and 0.5 centimeters thick, with fourteen legs, seven on each side. The actual silicon chip contained within the body of the plastic block is approximately 5 millimeters square and no more than 0.5 millimeters thick. Thousands of such chips are cut from a single silicon wafer that has been processed specifically for that application.

Inverters require only a single input lead and a single output lead, and so facilitate six functionalities on the DIP described. However, other devices typically use two input leads to supply one output lead. In those devices, the same DIP structure provides only four functionalities. The transistor structures are correspondingly more complex, but the actual chip size is about the same. Package sizes increase according to the complexity of the actual chip and the number of leads that it requires for its function and application to physical considerations such as dissipation of the heat that the device will generate in operation.

In the case of a modern laptop or desktop computer, the CPU chip package may have two hundred leads on a square package that is approximately 4 centimeters on a side and less than 0.5 centimeters in thickness. The actual chip inside the package is a very thin sheet of silicon about 1 square centimeter in size, but covered with several million transistor structures that have been built up through photoetching and chemical vapor deposition methods, as described above. Examination of any service listing of silicon chip ICs produced by any particular manufacturer will quickly reveal that a vast number of different ICs and functionalities are available.

Solar Technology. There are several other current uses for silicon wafer technology, and new uses are yet to be realized. Large quantities of electronic-grade silicon wafers are used in the production of functional solar cells, an area of application that is experiencing high growth, as nonrenewable energy resources become more and more expensive. Utilizing the photoelectron effect first described by Albert Einstein in 1905, solar cells convert light energy into an electrical current. Three types are made, utilizing both thick (> 300 micrometers [µm]) and thin (a few µm) layers of silicon. Thick-layer solar cells are constructed from single crystal silicon and from large-grain polycrystalline silicon, while thin-layer solar cells are constructed by using vapor deposition to deposit a layer of silicon onto a glass substrate.

Microelectronic and Mechanical Systems. Silicon chips are also used in the construction of microelectronic and mechanical systems (MEMS). Exceedingly tiny mechanical devices such as gears and single-pixel mirrors can be constructed using the technology developed for the production of silicon chips. Devices produced in this way are by nature highly sensitive and dependable in their operation, and so the majority of MEMS development is for the production of specialized sensors, such as the accelerometers used to initiate the deployment of airbag restraint systems in automobiles. A variety of other products are also available using MEMS technology, including biosensors, the micronozzles of inkjet printer cartridges, microfluidic test devices, microlenses and arrays of microlenses, and microscopic versions of tunable capacitors and resonators.

Other Applications. Other uses of silicon chip technology, some of which is in development, include mirrors for X-ray beams; mirrors and prisms for application in infrared spectroscopy, as silicon is entirely transparent to infrared radiation; and the material called porous silicon, which is made electrochemically from single-crystal silicon and has itself presented an exceptionally varied field of opportunity in materials science.

As mentioned, there are also many other materials that fall into the category of electronic materials. Some, such as copper, gold, and other pure elements, are produced in normal ways and then subjected to methods such as zone refining and vapor deposition techniques to achieve high purity and thin layers in the construction of electronic devices. Many exotic elements and metallic alloys, as well as specialized plastics, have been developed for use in electronic devices. Organic compounds known as liquid crystals, requiring no extraordinary synthetic measures, are normally semisolid materials that have properties of both a liquid and a solid. They are extensively used as the visual medium of thin liquid crystal display (LCD) screens, such as would be found in wristwatches, clocks, calculators, laptop and tablet computers, almost all desktop monitors, and flat-screen televisions.

Another example is the group of compounds made up of indium nitride, gallium nitride, and aluminum nitride. These are used to produce light-emitting diodes (LEDs) that provide light across the full visible spectrum. The ability to grow these LEDs on the same chip now offers a technology that could completely replace existing CRT (cathode ray tube) and LCD technologies for visual displays.

IMPACT ON INDUSTRY

Electronic materials production is an entire industry unto itself. While the products of this industry are widely used throughout society, they are not used in the form in which they are produced. Rather, the products of the electronic materials industry become input supplies for further manufacturing processes. Silicon chips, for example, produced by any individual manufacturer, are used for in-house manufacturing or are marketed to other manufacturers, who, in turn, use the chips to produce their own particular products, such as ICs, solar cells, and microdevices.

This intramural or business-to-business market aspect of the electronic materials production industry, with its novel research and development efforts and especially given the extent to which society now relies on information transfer and storage, makes ascribing an overall economic value to the industry impossible. One has only to consider the number of computing devices produced and sold each year around the world to get a sense of the potential value of the electronic materials production industry.

Ancillary industries provide other materials used by the electronic materials production industry, many of which must themselves be classified as electronic materials. An electric materials company, for example, may provide polishing and surfacing materials, photovoltaic materials, specialty glasses, electronic packaging materials, and many others.

Given both the extremely small size and sensitivity of the structures created on the surface of silicon chips and the number of steps required to produce those structures, quality control procedures are stringent. These steps may be treated as part of a multistep synthetic procedure, with each step producing a yield (as the percentage of structures that meet functional requirements). In silicon-chip production, it is important to understand that only the chips that are produced as functional units at the end of the process are marketable. If a process requires two hundred individual construction steps, even a 99 percent success rate for each step translates into a final yield of functional chips of only 0.99^{200}, or 13.4 percent. The majority of chip structures fail during construction, either through damage or through a step failure. It is therefore imperative that each step in the construction of silicon chips be precisely carried out.

To that end, procedures and quality control methods have been developed that are applicable in other situations too. Clean room technology that is essential for maximizing usable chip production is equally valuable in biological research and medical treatment facilities, applied physics laboratories, space exploration, aeronautics repair and maintenance facilities, and any other situations in which steps to protect either the environment or personnel from contamination must be taken.

CAREERS AND COURSEWORK

Electronic materials production is a specialist field that requires interested students to take specialist training in many subject areas. For many such careers, a university degree in solid state physics or electronic engineering is required. For those who will specialize in the more general field of materials science, these subject areas will be included in the overall curriculum. Silicon technology and semiconductors are also primary subject areas. The fields of study listed here are considered prerequisites for specialist study in the field of electronic materials production, and students can expect to continue studies in these subjects as new aspects of the field develop.

Researchers are now looking into the development of transistor structures based on graphene. This represents an entirely new field of study and application, and the technologies that develop from it will also set new requirements for study. High-end spectrometric methodologies are essential tools in the study and development of this field, and students can expect to take advanced study and training in the use of techniques such as scanning probe microscopy.

SOCIAL CONTEXT AND FUTURE PROSPECTS

Moore's law has successfully predicted the progression of transistor density that can be inscribed onto a silicon chip. There is a finite limit to that density, however, and the existing technology is very near or at that limit. Electronic materials research continues to improve methods and products in an effort to push the Moore limit.

New technologies must be developed to make the use of transistor logic as effective and as economic as possible. To that end, there exists a great deal of research into the application of new materials. Foremost is the development of graphene-based transistors and quantum dot technology, which will drive the level of technology into the molecular and atomic scales.

Richard M. Renneboog, M. Sc.

FURTHER READING

Akimov, Yuriy A., and Wee Shing Koh. "Design of Plasmonic Nanoparticles for Efficient Subwavelength Trapping in Thin-Film Solar Cells." *Plasmonics* 6 (2010): 155-161. This paper describes how solar cells may be made thinner and lighter by the addition of aluminum nanoparticles on a surface layer of indium tin oxide to enhance light absorption.

Askeland, Donald R. *The Science and Engineering of Materials.* London: Chapman & Hall, 1998. A recom-

---

Fascinating Facts About Electronic Materials Production

- Large single crystals of silicon are grown from a molten state in a process that literally pulls the molten mass out into a cylindrical shape.
- About 70 percent of silicon chips fail during the manufacturing process, leaving only a small percentage of chips that are usable.
- Silicon is invisible to infrared light, making it exceptionally useful for infrared spectroscopy and as mirrors for X rays.
- Quantum dot and graphene-based transistors will produce computers that are orders of magnitude more powerful than those used today.
- The photoelectric effect operating in silicon allows solar cells to convert light energy into an electrical current.
- Semiconductor transistors were invented in 1947 and integrated circuits in 1970, and the complexity of electronic components has increased by about 40 percent each year.
- Copper and other metals dissolve very quickly in liquid silicon, but precipitate out as the molten material cools, often with catastrophic results for the silicon crystal.
- Porous silicon is produced electrochemically from single crystals of silicon. Among its other properties, porous silicon is highly explosive.

---

mended resource, this book provides a great deal of fundamental background regarding the physical behavior of a wide variety of materials and processes that are relevant to electronic materials production.

Falster, Robert. "Gettering in Silicon: Fundamentals and Recent Advances." *Semiconductor Fabtech* 13 (2001). This article provides a thorough description of the effects of metal contamination in silicon and the process of gettering to avoid the damage that results from such contamination.

Zhang, Q., et al. "A Two-Wafer Approach for Integration of Optical MEMS and Photonics on Silicon Substrate." *IEEE Photonics Technology Letters* 22 (2010): 269-271. This paper examines how photonic and micro-electromechanical systems on two different silicon chips can be precisely aligned.

Zheng, Y., et al. "Graphene Field Effect Transistors with Ferroelectric Gating." *Physical Review Letters* 105 (2010). This paper discusses the experimental development and successful testing of a graphene-based field-effect transistor system using gold and graphene electrodes with SiO2 gate structures on a silicon substrate.

WEB SITES

SCP Symposium (June 2005) "Silicon Starting Materials for Sub-65nm Technology Nodes."
http://www.memc.com/assets/file/technology/papers/SCP-Symposium-Seacrist.pdf

University of Kiel "Electronic Materials Course."
http://www.tf.uni-kiel.de/matwis/amat/elmat_en/index.html

See also: Electrochemistry; Electronics and Electronic Engineering; Surface and Interface Science.

ELECTRONICS AND ELECTRONIC ENGINEERING

FIELDS OF STUDY

Mathematics; physics; electronics; electrical engineering; automotive mechanics; analytical technology; chemical engineering; aeronautics; avionics; robotics; computer programming; audio/video technology; metrology; audio engineering; telecommunications; broadcast technology; computer technology; computer engineering; instrumentation

SUMMARY

A workable understanding of the phenomenon of electricity originated with proof that atoms were composed of smaller particles bearing positive and negative electrical charges. The modern field of electronics is essentially the science and technology of devices designed to control the movement of electricity to achieve some useful purpose. Initially, electronic technology consisted of devices that worked with continuously flowing electricity, whether direct or alternating current. Since the development of the transistor in 1947 and the integrated circuit in 1970, electronic technology has become digital, concurrent with the ability to assemble millions of transistor structures on the surface of a single silicon chip.

KEY TERMS AND CONCEPTS

- **Cathode Rays:** Descriptive term for energetic beams emitted from electrically stimulated materials inside of a vacuum tube, identified by J. J. Thomson in 1897 as streams of electrons.
- **Channel Rays:** Descriptive term for energetic beams having the opposite electrical charge of cathode rays, emitted from electrically stimulated materials inside a vacuum tube, also identified by J. J. Thomson in 1897.
- **Gate:** A transistor structure that performs a specific function on input electrical signals to produce specific output signals.
- **Operational Amplifier (Op-Amp):** An integrated circuit device that produces almost perfect signal reproduction with high gains of amplification and precise, stable voltages and currents.
- **Sampling:** Measurement of a specific parameter such as voltage, pressure, current, and loudness at a set frequency determined by a clock cycle such as 1 MHz.
- **Semiconductor:** An element that conducts electricity effectively only when subjected to an applied voltage.
- **Zener Voltage:** The voltage at which a Zener diode is designed to operate at maximum efficiency to produce a constant voltage, also called the breakdown voltage.

DEFINITION AND BASIC PRINCIPLES

The term "electronics" has acquired different meanings in different contexts. Fundamentally, "electronics" refers to the behavior of matter as affected by the properties and movement of electrons. More generally, electronics has come to mean the technology that has been developed to function according to electronic principles, especially pertaining to basic digital devices and the systems that they operate. The term "electronic engineering" refers to the practice of designing and building circuitry and devices that function on electronic principles.

The underlying principle of electronics derives from the basic structure of matter: that matter is composed of atoms composed of smaller particles. The mass of atoms exists in the atomic nucleus, which is a structure composed of electrically neutral particles called neutrons and positively charged particles called protons. Isolated from the nuclear structure by a relatively immense distance is an equal number of negatively charged particles called electrons. Electrons are easily removed from atoms, and when a difference in electrical potential (voltage) exists between two points, electrons can move from the area of higher potential toward that of lower potential. This defines an electrical current.

Devices that control the presence and magnitude of both voltages and currents are used to bring about changes to the intrinsic form of the electrical signals so generated. These devices also produce physical changes in materials that make comprehensible the information carried by the electronic signal.

BACKGROUND AND HISTORY

Archaeologists have found well-preserved Parthian

relics that are now believed to have been rudimentary, but functional, batteries. It is believed that these ancient devices were used by the Parthians to plate objects with gold. The knowledge was lost until 1800, when Italian physicist Alessandro Volta reinvented the voltaic pile. Danish physicist and chemist Hans Christian Oersted demonstrated the relationship between electricity and magnetism in 1820, and in 1821, British physicist and chemist Michael Faraday used that relationship to demonstrate the electromagnetic principle on which all electric motors work. In 1831, he demonstrated the reverse relationship, inventing the electrical generator in the process.

Electricity was thought, by American statesman and scientist Benjamin Franklin and many other scientists of the eighteenth and nineteenth centuries, to be some mysterious kind of fluid that might be captured and stored. A workable concept of electricity was not developed until 1897, when J. J. Thomson identified cathode rays as streams of light electrical particles that must have come from within the atoms of their source materials. He arbitrarily ascribed their electrical charge as negative. Thomson also identified channel rays as streams of massive particles from within the atoms of their source materials that are endowed with the opposite electrical charge of the electrons that made up cathode rays. These observations essentially proved that atoms have substructures. They also provided a means of explaining electricity as the movement of charged particles from one location to another.

With the establishment of an electrical grid, based on the advocacy of alternating current by Serbian American engineer and inventor Nikola Tesla (1856-1943) , a vast assortment of analogue electrical devices were soon developed for consumer use, though initially these devices were no more than electric lights and electromechanical applications based on electric motors and generators.

As the quantum theory of atomic structure came to be better understood and electricity better controlled, electronic theory became much more important. Spurred by the success of the electromagnetic telegraph of American inventor Samuel Morse (1791-1872), scientists sought other applications. The first major electronic application of worldwide importance was wireless radio, first demonstrated by Italian inventor Guglielmo Marconi (1874-1937). Radio depended on electronic devices known as vacuum

tubes, in which structures capable of controlling currents and voltages could operate at high temperatures in an evacuated tube with external contacts. In 1947, American physicist William Shockley and colleagues invented the semiconductor-based transistor, which could be made to function in the same manner as vacuum tube devices, but without the high temperatures, electrical power consumption, and vacuum construction of those analogue devices.

In 1970, the first integrated circuit "chips" were made by constructing very small transistor structures on the surface of a silicon chip. This gave rise to the entire digital technology that powers the modern world.

APPLICATIONS AND PRODUCTS

Electronics are applied in practically every conceivable manner today, based on their utility in converting easily-produced electrical current into mechanical movement, sound, light, and information signals.

Basic Electronic Devices. Transistor-based digital technology has replaced older vacuum tube technology, except in rare instances in which a transistorized device cannot perform the same function. Electronic circuits based on vacuum tubes could carry out essentially the same individual operations as transistors, but they were severely limited by physical size, heat production, energy consumption, and mechanical failure. Nevertheless, vacuum tube technology was the basic technology that produced radio, television, radar, X-ray machines, and a broad variety of other electronic applications.

Electronic devices that did not use vacuum tube technology, but which operated on electronic and electromagnetic principles, were, and still are, numerous. These devices include electromagnets and all electric motors and generators. The control systems for many such devices generally consisted of nothing more than switching circuits and indicator lights. More advanced and highly sensitive devices required control systems that utilized the more refined and correspondingly sensitive capabilities available with vacuum tube technology.

Circuit Boards. The basic principles of electricity, such as Ohm's resistance law and Kirchoff's current law and capacitance and inductance, are key features in the functional design and engineering of analogue electronic systems, especially for vacuum-tube control systems. An important application that facilitated the general use and development of electronic systems of all kinds is printed circuit board technology. A printed circuit board accepts standardized components onto a nonconducting platform made initially of compressed fiber board, which was eventually replaced by a resin-based composite board. A circuit design is photo-etched onto a copper sheet that makes up one face of the circuit board, and all nonetched copper is chemically removed from the surface of the board, leaving the circuit pattern. The leads of circuit components such as resistors, capacitors, and inductors are inserted into the circuit pattern and secured with solder connections.

Mass production requirements developed the flotation soldering process, whereby preassembled circuit boards are floated on a bed of molten solder, which automatically completes all solder connections at once with a high degree of consistency. This has become the most important means of circuit board production since the development of transistor technology, being highly compatible with mechanization and automation and with the physical shapes and dimensions of integrated circuit (IC) chips and other components.

Digital Devices. Semiconductor-based transistors comprise the heart of modern electronics and electronic engineering. Unlike vacuum tubes, transistors do not work on a continuous electrical signal. Instead, they function exceedingly well as simple on-off switches that are easily controlled. This makes them well adapted to functions based on Boolean algebra. All transistor structures consist of a series of "gates" that perform a specific function on the electronic signals that are delivered to them.

Digital devices now represent the most common (and rapidly growing) application of electronics and electronic engineering, including relatively simple consumer electronic devices such as compact fluorescent light bulbs and motion-detecting air fresheners to the most advanced computers and analytical instrumentation. All applications, however, utilize an extensive, but limited, assortment of digital components in the form of IC chips that have been designed to carry out specific actions with electrical or electromagnetic input signals.

Input signals are defined by the presence or absence of a voltage or a current, depending upon the nature of the device. Inverter gates reverse the sense of the input signal, converting an input voltage (high input) into an output signal of no voltage (low output), and vice versa. Other transistor structures (gates) called AND, NAND, OR, NOR and X-OR function to combine input signals in different ways to produce corresponding output signals. More advanced devices (for example, counters and shift registers) use combinations of the different gates to construct various functional circuits that accumulate signal information or that manipulate signal information in various ways.

One of the most useful of digital IC components is the operational amplifier, or Op-Amp. Op-Amps contain transistor-based circuitry that boosts the magnitude of an input signal, either voltage or current, by five orders of magnitude (100,000 times) or more, and are the basis of the exceptional sensitivity of the modern analytical instruments used in all fields of science and technology.

Electrical engineers are involved in all aspects of the design and development of electronic equipment. Engineers act first as the inventors and designers of electronic systems, conceptualizing the specific functions a potential system will be required to carry out. This process moves through the specification of the components required for the system's functionality to the design of new system devices. The design parameters extend to the infrastructure that must support the system in operation. Engineers determine the standards of safety, integrity, and operation that must be met for electronic systems.

Consumer Electronics. For the most part, the term "electronics" is commonly used to refer to the electronic devices developed for retail sale to consumers. These devices include radios, television sets, DVD and CD players, cell phones and messaging devices, cameras and camcorders, laptops, tablets, printers, computers, fax and copy machines, cash registers, and scanners. Millions such devices are sold around the world each day, and numerous other businesses have formed to support their operation.

IMPACT ON INDUSTRY

With electrical and electronic technology now intimately associated with all aspects of society, the impact of electronics and electronic engineering on industry is immeasurable. It would be entirely fair to say that modern industry could not exist without electronics. Automated processes, which are ubiquitous, are not possible without the electronic systems that control them.

The transportation industry, particularly the automotive industry, is perhaps the most extensive user of electronics and electronic engineering. Modern automobiles incorporate an extensive electronic network in their construction to provide the ignition and monitoring systems for the operation of their internal combustion engines and for the many monitoring and control systems for the general safe operation of the vehicle; an electronic network also informs and entertains the driver and passengers. In some cases, electronic systems can completely take control of the vehicle to carry out such specific programmable actions as speed control and parallel parking. Every automated process in the manufacture of automobiles and other vehicles serves to reduce the labor required to carry out the corresponding tasks, while increasing the efficiency and precision of the process steps. Added electronic features also increase the marketability of the vehicles and, hence, manufacturer profits.

Processes that have been automated electronically also have become core components of general manufacturing, especially in the control of production machinery. For example, shapes formed from bent tubing are structural components in a wide variety of applications. While the process of bending the tubing can be carried out by the manual operation of a suitably equipped press, an automated process will produce tube structures that are bent to exact angles and radii in a consistent manner. Typically, a human operator places a straight tube into the press, which then positions and repositions the tube for bending over its length, according to the program that has been entered into the manufacturing system's electronic controller. Essentially, all continuous manufacturing operations are electronically controlled, providing consistent output.

Electronics and electronic engineering make up the essence of the computer industry; indeed, electronics is an industry worth billions of dollars annually. Electronics affects not only the material side of industry but also the theoretical and actuarial side. Business management, accounting, customer relations, inventory and sales data, and human resources-management all depend on the rapid information-handling that is possible through electronics.

XML (extensible markup language) methods and

Fascinating Facts About Electronics and Electronic Engineering

- In 1847, George Boole developed his algebra for reasoning that was the foundation for first-order predicate calculus, a logic rich enough to be a language for mathematics.
- In 1950, Alan Turing gave an operational definition of artificial intelligence. He said a machine exhibited artificial intelligence if its operational output was indistinguishable from that of a human.
- In 1956, John McCarthy and Marvin Minsky organized a two-month summer conference on intelligent machines at Dartmouth College. To advertise the conference, McCarthy coined the term "artificial intelligence."
- Digital Equipment Corporation's XCON, short for eXpert CONfigurer, was used in-house in 1980 to configure VAX computers and later became the first commercial expert system.
- In 1989, international chess master David Levy was defeated by a computer program, Deep Thought, developed by IBM. Only ten years earlier, Levy had predicted that no computer program would ever beat a chess master.
- In 2010, the Haystack group at the Computer Science and Artificial Intelligence Laboratory at the Massachusetts Institute of Technology developed Soylent, a word-processing interface that lets users edit, proof, and shorten their documents using Mechanical Turk workers.

the corresponding applications and databases. XML is an application that promises to facilitate information exchange and to promote research using large-applications are being used (and are in development) to interface electronic data collection directly to physical processes. This demands the use of specialized electronic sensing and sampling devices to convert measured parameters into data points within the corresponding applications and databases. XML is an application that promises to facilitate information exchange and to promote research using large-scale databases. The outcome of this effort is expected to enhance productivity and to expand knowledge in ways that will greatly increase the efficiency and effectiveness of many different fields.

An area of electronics that has become of great economic importance in recent years is that of electronic commerce: the exclusive use of electronic communication technology for the conduct of business between suppliers and consumers. Electronic communications encompasses interoffice faxing, e-mail exchanges, and the Web commerce of companies such as eBay, Amazon, Google, and of the New York and other stock exchanges. The commercial value of these undertakings is measured in billions of dollars annually, and it is expected to continue to increase as new applications and markets are developed.

The fundamental feature here is that these enterprises exist because of the electronic technology that enables them to communicate with consumers and with other businesses. The electronics technology and electronics engineering fields have thus generated entirely new and different daughter industries, with the potential to generate many others, all of which will depend on persons who are knowledgeable in the application and maintenance of electronics and electronic systems.

CAREERS AND COURSEWORK

Many careers depend on knowledge of electronics and electronic engineering because almost all machines and devices used in modern society either function electronically or utilize some kind of electronic control system. The automobile industry is a prime example, as it depends on electronic systems at all stages of production and in the normal operation of a vehicle. Students pursuing a career in automotive mechanics can therefore be expected to study electronic principles and applications as a significant part of their training. The same reasoning applies in all other fields that have a physical reliance on electronic technology.

Knowledge of electronics has become so essential that atomic structure and basic electronic principles, for example, have been incorporated into the elementary school curriculum. Courses of study in basic electronics in the secondary school curriculum are geared to provide a more detailed and practicable knowledge to students.

Specialization in electronics and other fields in which electronics play a significant role is the province of a college education. Interested students can expect to take courses in advanced mathematics, physics, chemistry, and electronics technology as part of the curriculum of their specialty programs. Normally, a technical career, or a skilled trade, requires a college-level certification and continuing education. In some cases, recertification on a regular schedule is also required to maintain specialist standing in that trade.

Students who plan to pursue a career in electronic engineering at a more theoretical level will require, at minimum, a bachelor's degree. A master's degree can prepare a student for a career in forensics, law, and other professions in which an intimate or specialized knowledge of the theoretical side of electronics can be advantageous. (The Vocational Information Center provides an extensive list of careers involving electronics at http://www.khake.com/page19.html.)

SOCIAL CONTEXT AND FUTURE PROSPECTS

It is difficult, if not impossible, to imagine modern society without electronic technology. Electronics has enabled the world of instant communication, wherein a person on one side of the world can communicate directly and almost instantaneously with someone on the other side of the world. As a social tool, such facile communication has the potential to bring about understanding between peoples in a way that has until now been imagined only in science fiction.

Consequently, this facility has also resulted in harm. While social networking sites, for example, bring people from widely varied backgrounds together peacefully to a common forum, network hackers and so-called cyber criminals use electronic technology to steal personal data and disrupt financial markets.

Electronics itself is not the problem, for it is only a tool. Electronic technology, though built on a foundation that is unlikely to change in any significant way, will nevertheless be transformed into newer and better applications. New electronic principles will come to the fore. Materials such as graphene and quantum dots, for example, are expected to provide entirely new means of constructing transistor structures at the atomic and molecular levels. Compared with the 50 to 100 nanometer size of current transistor technology, these new levels would represent a difference of several orders of magnitude. Researchers suggest that this sort of refinement in scale could produce magnetic memory devices that can store as much as ten terabits of information in one square centimeter of disk surface. Although the technological advances seem inevitable, realizing such a scale will require a great deal of research and development.

Richard M. Renneboog, M.Sc.

FURTHER READING

Gates, Earl D. *Introduction to Electronics.* 5th ed. Clifton Park, N.Y.: Cengage Learning, 2006. This book presents a serious approach to practical electronic theory beginning with atomic structure and progressing through various basic circuit types to modern digital electronic devices. Also discusses various career opportunities for students of electronics.

Mughal, Ghulam Rasool. "Impact of Semiconductors in Electronics Industry." *PAF-KIET Journal of Engineering and Sciences* 1, no. 2 (July-December, 2007): 91-98. This article provides a learned review of the basic building blocks of semiconductor devices and assesses the effect those devices have had on the electronics industry.

Petruzella, Frank D. *Introduction to Electricity and Electronics 1.* Toronto: McGraw-Hill Ryerson, 1986. A high-school level electronics textbook that provides a beginning-level introduction to electronic principles and practices.

Platt, Charles. *Make: Electronics.* Sebastopol, Calif.: O'Reilly Media, 2009. This book promotes learning about electronics through a hands-on experimental approach, encouraging students to take things apart and see what makes those things work.

Robbins, Allen H., and Wilhelm C. Miller. *Circuit Analysis Theory and Practice.* Albany, N.Y.: Delmar, 1995. This textbook provides a thorough exposition and training in the basic principles of electronics, from fundamental mathematical principles through the various characteristic behaviors of complex circuits and multiphase electrical currents.

Segura, Jaume, and Charles F. Hawkins. *CMOS Electronics: How It Works, How It Fails.* Hoboken, N.J.: John Wiley & Sons, 2004. The introduction to basic electronic principles in this book leads into detailed discussion of MOSFET and CMOS electronics, followed by discussions of common failure modes of CMOS electronic devices.

Singmin, Andrew. *Beginning Digital Electronics Through Projects.* Woburn, Mass.: Butterworth-Heinemann, 2001. This book presents a basic introduction to electrical properties and circuit theory and guides readers through the construction of some simple devices.

Strobel, Howard A., and William R. Heineman. *Chemical Instrumentation: A Systematic Approach.* 3d ed. New York: John Wiley & Sons, 1989. This book provides an exhaustive overview of the application of electronics in the technology of chemical instrumentation, applicable in many other fields as well.

WEB SITES

Institute of Electrical and Electronics Engineers
http://www.ieee.org

See also: Electronic Commerce; Electronic Materials Production; Electronics and Electronic Engineering.

ENDOCRINOLOGY

FIELDS OF STUDY

Bariatrics; diabetes medicine; internal medicine; laboratory medicine; neuroendocrinology; obstetrics and gynecology; pediatric endocrinology; radiology reproductive endocrinology; thyroid medicine.

SUMMARY

Endocrinology is a medical field focused on the diagnosis and treatment of abnormalities of the endocrine system. The endocrine system consists of glands that produce hormones: the adrenal gland, hypothalamus, ovaries, pancreas, parathyroid glands, pituitary gland, testes, and thyroid gland. These hormones control metabolism (utilization of food by the body), reproduction, and growth. Endocrinology is practiced by medical doctors with specialized training in that field. Physicians in other fields (such as obstetricians, gynecologists, and internists) may devote some of their practice to endocrinology. Some endocrinologists specialize in one area of endocrinology (such as neuroendocrinology or pediatric endocrinology). Conditions treated by endocrinologists include diabetes, hypertension (high blood pressure), inadequate growth, infertility, obesity, osteoporosis (weak bones), menopause, metabolic disorders, and thyroid disorders.

KEY TERMS AND CONCEPTS

- **Adrenal Glands:** Glands that are situated above each kidney and produce hormones, which respond to stress, such as cortisol and adrenaline.
- **Hypothalamus:** Portion of the brain that contains a variety of specialized cells; an important function of the hypothalamus is the linkage of the brain to the endocrine system through the pituitary gland.
- **Ovaries:** Paired organs adjacent to the uterus, which release eggs (ova) for reproduction and produce a variety of hormones, primarily estrogen and progesterone.
- **Pancreas:** Organ located in the upper abdomen, which produces hormones that regulate blood sugar levels (insulin, glucagon, and somatostatin) and secretes digestive enzymes, which pass into the small intestine.

- **Parathyroid Glands:** Small glands, usually located within the thyroid gland, which produce parathyroid hormone; this hormone regulates calcium levels in the bloodstream and bones.
- **Pituitary Gland:** Small gland, located at the base of the brain, which is sometimes referred to as the master gland because it produces hormones that stimulate or suppress the secretions of other endocrine glands.
- **Testes:** Male reproductive organs, which produce sperm and male sex hormones, such as testosterone.
- **Thyroid Gland:** Butterfly-shaped organ located in the neck; it controls the metabolic rate, protein production, and the sensitivity of the body to other hormones.

DEFINITION AND BASIC PRINCIPLES

Endocrinology is a medical field dealing with the endocrine system, which is a complex system of organs that secrete hormones into the bloodstream. Hormones are chemicals that are released from cells in one location and affect cells located elsewhere in the body. To respond to these chemical messengers, a cell must possess a receptor to the hormone. Hormones control many bodily functions, including metabolism, reproduction, and growth. Diseases of the endocrine system often involve the abnormal production of a hormone or a cell's resistance to the effects of a hormone. For example, excess thyroid hormones produce a condition known as hyperthyroidism in which metabolism is increased. Patients with hyperthyroidism have increased nervousness, irritability, tremors, and a rapid heart rate. Individuals with hypothyroidism (inadequate level of thyroid hormones) have fatigue, poor muscle tone, constipation, and dry skin. Either thyroid condition may be caused by an abnormal functioning of the thyroid gland. These conditions may also be secondary to an abnormal functioning of the pituitary gland or hypothalamus, both of which regulate the thyroid gland. The hypothalamus produces a hormone, thyrotropin-releasing hormone (TRH), which causes the pituitary gland to release thyroid-stimulating hormone (TSH), which signals the thyroid gland to release thyroid hormones. Patients with diabetes usually have inadequate levels of insulin, which controls glucose (sugar) metabolism. Some cases of diabetes are caused by insulin resistance, a condition in which the cells do not respond well to insulin circulating in the bloodstream.

BACKGROUND AND HISTORY

Endocrinology is derived from the Greek words *endo* (within), *krīn* (to separate), and *logia*, which supplies the suffix "-ology" (referring to a field of knowledge). Endocrinology originated in 200 b.c.e. in China, when pituitary and sex hormones were isolated from the urine for medicinal purposes. In the Western world, an organ basis for pathology did not develop until the nineteenth century. In 1841, the German physician Friedrich Henle described "ductless glands" that secrete products directly into the bloodstream. In 1902, William Bayliss and Ernest Starling discovered secretin, which they described as a hormone. They defined a hormone as a chemical that is produced in an organ, then travels via the bloodstream to a distant organ and exerts a specific function on it. The field of endocrinology is based on the replacement of inadequate levels of a hormone with purified extracts. In cases in which hormone levels are unusually high, treatment involves lowering the hormonal level through surgical removal of a portion of the gland or destruction of some of the gland's cells by radiation. For example, hyperthyroidism is commonly treated with radioactive iodine, which concentrates in the thyroid gland and destroys some of the cells that produce hormones.

HOW IT WORKS

Endocrinology is a medical specialty, and usually patients are referred from other physicians (such as an internist or family physician) for evaluation. The endocrinologist examines the patient and makes a diagnosis. If the referring physician has made a preliminary diagnosis, the endocrinologist often orders further tests to confirm it. The endocrinologist must be well versed in biochemistry and clinical chemistry to properly interpret these tests.

The endocrinologist frequently relies on the radiologist for diagnosis and treatment of an endocrine disorder. This involves the use of imaging equipment such as ultrasound and scintigraphy. Scintigraphy is a two-dimensional visualization of a radionuclide in the body. A radionuclide is a radioactive substance that is taken up by an endocrine gland. Computed tomography (CT) and magnetic resonance imaging (MRI) are also used to visualize organs such as the thyroid and adrenal glands. After diagnosing an endocrine condition, the radiologist might be called on to treat the condition with the injection of a radionuclide. The treatment levels of radiation are much higher than the diagnostic level and are designed to destroy cells producing excessive amounts of a hormone.

Once a diagnosis is made, a course of treatment must be developed. Initial treatment might include referral to a surgeon for excision of a tumor or an abnormally functioning organ. Radiotherapy is employed to treat certain endocrine conditions such as hyperthyroidism (overactive thyroid) and Cushing's syndrome (overactive adrenal glands). Drug therapy is an option in some cases. Many diseases of the endocrine system are chronic and require lifelong treatment. A classic example is diabetes, which can develop in children and young adults (type 1 diabetes) and in adults (type 2 diabetes). Milder forms of type 2 diabetes can often be treated with medication; however, type 1 diabetes and severer forms of type 2 diabetes require insulin injections. Patients requiring insulin must be educated as to the importance of controlling their blood sugar level with self-administered insulin injections and of frequent monitoring of their blood glucose (sugar) through blood sampling. Diabetics are more prone to many conditions such as cardiovascular disease and loss of vision. Diabetics whose condition is under good control are less likely to develop serious health problems. Obesity greatly increases the risk of developing type 2 diabetes; therefore, a weight-loss program can sometimes return blood glucose levels to normal levels.

Some diseases are congenital (present at birth) or occur at a young age; the pediatrician plays a crucial role in diagnosing an endocrine problem in young patients. Inadequate levels of many hormones can severely affect a child's development, both physically and mentally. Children with endocrine abnormalities are often referred to a pediatric endocrinologist.

Couples with an infertility problem may seek the help of a reproductive endocrinologist. The problem may be caused by either male or female factors. Some of these problems are nonendocrine in origin (for example, Fallopian tubes blocked from an infection); however, many infertile women can greatly increase their chances of becoming pregnant with assisted reproductive technology (ART), which involves the administration of hormones and other medications to stimulate and regulate the ovulation process.

Gynecologists practice endocrinology related to the female reproductive system. A common problem dealt with by gynecologists is a menstrual irregularity. When women approach the menopause, which is caused by a drop in the level of female hormones (estrogen and progesterone), they often develop distressing symptoms, such as hot flashes, dry skin, and depression. Many women do not consult an endocrinologist at that time. Instead, they seek the advice of a gynecologist, family physician, or an internist. That physician will often diagnose and treat the condition; however, the patient may occasionally be referred to an endocrinologist.

APPLICATIONS AND PRODUCTS

Most hormones can be administered orally or by injection, skin patch, vaginal cream, or nasal inhalation. Some hormones cannot be administered orally. Hormones can be derived from animal or human sources. Some can be retrieved from urine, and others are obtained from animal organs harvested at slaughterhouses. In many cases, hormones have been synthesized in the laboratory.

Hormones derived from animal or human sources must be subjected to bioassay.

Bioassay involves administering a hormone sample to an animal and measuring its effect. By this process, the pharmaceutical manufacturer can adjust the hormonal level of that batch to a standardized level. Thus, an individual ingesting a hormone can be assured that the dose is uniform from day to day. Synthesized hormones are easier to standardize; however, they still may require a bioassay. They are less likely to contain impurities, which could produce an adverse effect, including an allergic reaction.

Many products on the market are used for diabetes, thyroid disease, osteoporosis, infertility, and the menopause. A significant market exists for performance-enhancing products, although athletes are banned from using hormones in this manner.

Specialized surgical procedures exist for medical conditions with an endocrine component, such as morbid obesity. Surgery for morbid obesity is known as bariatric surgery. An increasing proportion of surgical procedures are being done with laparoscopy, which has the advantage of a small incision and quicker recovery.

Medical Laboratory. Endocrinology is a field that uses laboratory services to a greater degree than many other specialties. Some medical laboratories contain specialized equipment to accurately measure specific hormonal levels. Inasmuch as this equipment can be quite expensive and may measure only one specific hormone, these services are usually found in specialty laboratories that analyze samples from a large geographic area.

Diabetes. Many products are marketed for diabetic patients. Meters that can reliably measure blood glucose levels with minimal discomfort are a necessity for all diabetics. Syringes for injection are also essential for management of the condition. Also available are insulin pumps that can be programmed to administer the appropriate insulin dosage throughout a twenty-four-hour period. The patient can adjust the rate at any time depending on any variance from the normal routine or an abnormal blood glucose reading.

Thyroid Disease. In addition to laboratory tests to measure thyroid levels, diagnosis of thyroid disease often involves radioactive iodine (iodine 131), which is manufactured in specialized laboratories. A small amount of the substance is given by injection or in tablet form for diagnosis. Diagnosis is facilitated by radiologists using a specialized device, a gamma camera. Treatment of hyperthyroidism involves the use of a much higher dosage of iodine 131, which destroys thyroid cells. Hyperthyroidism can also be treated with antithyroid drugs, such as methimazole and propylthiouracil. These drugs become concentrated in the thyroid gland and block production of thyroid hormones. For the immediate treatment of the symptoms of hyperthyroidism, beta-blockers such as propranolol, atenolol, and metoprolol can be used. These medications lower metabolism; however, they do not alter thyroid hormone levels in the circulation. Surgical removal of a portion of the thyroid gland is sometimes done. It is usually reserved for pregnant patients, children with an adverse reaction to antithyroid medications, and patients with a very large thyroid gland.

Osteoporosis. Osteoporosis is a weakening of the bones that can be triggered by an imbalance of a number of hormones. It commonly occurs in women at the time of the menopause. Osteoporotic bone is susceptible to fracture and disfiguring deformities. Specialized equipment, using specialized X-ray equipment or other imaging modalities, can measure the degree of osteoporosis. These devices are marketed not only for radiology facilities but also for physician offices and other health care facilities. Dual energy X-ray absortiometry (DEXA) scanning is the most common procedure used to measure the amount of calcium and other minerals present in bone. The amount of minerals present is known as the bone mineral density (BMD). The most commonly scanned areas are the hip and spine. If a patient is diagnosed with osteoporosis, hormonal and nonhormonal products are available to lessen the severity of or reverse osteoporosis. For women, estrogen preparations can slow or halt the progression of osteoporosis; however, they cannot reverse it. Bisphosphonates, pharmaceuticals that can reverse osteoporosis, are available.

Infertility. Reproductive endocrinologists are major users of specialized products. Assisted reproduction technology (ART) involves stimulating the ovaries to produce ova (eggs), extracting the ova, fertilizing the ova in the laboratory, culturing the ova (multiple cell stage or embryonic stage), then implanting the ova in the uterus. Medications such as clomiphene are used to stimulate the ovaries to release eggs; hormones, such as human menopausal hormone (HMG) and gonadotropins are also used to stimulate the ovaries. The menstrual cycle is often regulated with hormones such as estrogen and progesterone. Specialized equipment is used to extract the ova from the patient's ovaries. The process involves inserting a needle through the vaginal wall and into an ovarian follicle; it is conducted under the guidance of ultrasound. The extracted ova are placed in a culture medium and fertilized with sperm supplied by the husband or donor. The fertilized ova are then placed in specialized incubators for growth to the desired embryonic stage. The ova are then inserted into the uterine cavity for development. Ova and embryos are placed in vials and frozen with liquid nitrogen. At a future date, the embryos are thawed and inserted into the uterus.

Menopause. A number of products are on the market for replacement of hormones lost after the menopause. They include oral medication, injections, and transdermal skin patches (the hormones are absorbed through the skin and pass into the

Fascinating Facts About Endocrinology

- Gigantism is a condition marked by excessive growth because of the secretion of a growth hormone by a pituitary tumor. Trijntje Keever, born in 1616 in the Netherlands, suffered from the condition. She was the tallest woman in recorded history. When she died at age seventeen from cancer, her height was 8 feet, 4 inches.
- A deficiency of growth hormone results in dwarfism. General Tom Thumb (the stage name of Charles Sherwood Stratton), born in 1883, probably suffered from the condition. His height at the age of eighteen was 2 feet, 8.5 inches. Stratton became wealthy as a performer in P. T. Barnum's circus.
- Within a year of giving birth, 5 to 10 percent of women develop hypothyroidism secondary to postpartum thyroiditis. Initially, thyroid hormone levels may rise, then either return to normal or drop to hypothyroid levels. Of those women who become hypothyroid, about 20 percent will require lifelong treatment.
- Premarin is an estrogen preparation that is sometimes prescribed to women at the time of the menopause as hormone replacement therapy. The name is derived from "pregnant mare's urine," the source of the hormonal preparation.
- Girls stop growing in height at puberty because estrogen closes the epiphyses (growth plates), located in the arm and leg bones.
- Newborn girls sometimes have vaginal bleeding; this is caused by a drop in the level of estrogen, which the fetus was exposed to while in the uterus.

bloodstream). Many nonhormonal products are also available to treat menopausal symptoms.

Performance-Enhancing Hormones. Performance-enhancing hormones have been used by many athletes, both professional and amateur. Although most of these products are banned because they confer an unfair advantage on their user, their use continues, supported by a thriving black market enterprise. Hormones that can enhance athletic performance are anabolic steroids (male hormones), which promote muscle growth and strength, and erythropoietin (EPO), which increases red blood cell production. Stronger muscles allow a baseball player to hit more home runs and increases a cyclist's speed and endurance. Increased red cell production from EPO raises the blood's oxygen-carrying capacity.

The athlete who uses these hormones may not only be disqualified but also experience adverse effects. Anabolic steroids change cholesterol levels. They increase the level of low-density lipoprotein (LDL, or bad cholesterol) and decrease the level of high-density lipoproteins (HDL, or good cholesterol), which can raise a person's risk of developing cardiovascular disease and having a heart attack. Furthermore, these drugs can cause direct damage to the heart and liver. A prominent example of an athlete who was disqualified is cyclist Floyd Landis (although he continues to deny using anything). Landis was the overall leader of the 2006 Tour de France, a grueling multistage bicycle race, but in stage 16, he lost eight minutes to the second-place rider, and most experts believed he could not make up the time. The following day, he made a dramatic comeback and went on to win the event. A mandatory urine test taken after stage 17 revealed high levels of testosterone, and he was stripped of his title.

IMPACT ON INDUSTRY

Endocrinology plays a significant role in laboratory medicine, medical imaging, radiotherapy, assisted reproductive technology, and pharmacology. The medical facilities and industries associated with these areas derive significant revenue from endocrinology. Laboratory procedures for detecting and measuring hormone levels are complex and highly technical, and accurate measurement is imperative for proper diagnosis and treatment. Diagnostic and therapeutic radiology for endocrine conditions requires sophisticated equipment, which is often updated frequently as improved models come on the market. Both laboratory medicine and radiology require a team of skilled physicians, supervised by physicians with specialized training. Assisted reproductive technology is a high-ticket item and often is not covered by insurance; however, for many of those with moderate incomes, the hope of achieving a pregnancy is enough to justify the expense. The field of reproductive endocrinology requires a team of skilled technicians, supervised by one or more physician specialists.

The manufacturing of hormonal products and medications to treat hormonal disorders is a significant segment of the pharmaceutical industry. Patients with chronic conditions, such as diabetes and hypothyroidism, are lifelong consumers. Many women take hormonal medications to deal with menopausal symptoms or dysmenorrhea (painful menstruation), and many use hormonal contraceptives (pills, patches, or vaginal rings). In addition to hormonal products, many pharmaceuticals are designed to treat endocrine problems. These include medications to boost energy, aid in weight loss, induce sleep, and lower blood pressure.

Research. Significant endocrine research is conducted by the government and universities. A branch of the National Institutes of Health, the National Institute of Diabetes and Digestive and Kidney Diseases, funds research in many fields, including diabetes, digestive diseases, genetic metabolic diseases, immunologic diseases, and obesity. The institute also provides health information for the public in these fields. Virtually all developed nations have extensive endocrine research programs. Beyond government and university programs, many practicing endocrinologists devote a significant amount of their time to research in their field.

CAREERS AND COURSE WORK

To become an endocrinologist, one must first graduate from college and then complete a four-year course of medical training. Initial specialty training, typically a three- or four-year residency program, focuses on internal medicine, pediatrics, or gynecology. Subsequently, the physician completes a fellowship (two or more years) in the field of general endocrinology, pediatric endocrinology, reproductive endocrinology, or neuroendocrinology.

Most endocrinologists will locate their practice in a densely populated urban area, and many will join a group of specialists. A large number of endocrinologists practice in the medical university setting. Most endocrinologists are certified in the specialty of internal medicine, pediatrics, and gynecology. They may ultimately be board certified in endocrinology or a subspecialty of endocrinology.

Most endocrinologists become a member of one or more professional organizations, which provide continuing education and forums for physician members and educational material for the general public. In the United States, the main professional organizations for endocrinologists are the

American Association of Clinical Endocrinologists, the American Diabetes Association, the American Thyroid Association, the Pediatric Endocrine Society, and the Society for Reproductive Endocrinology and Infertility. In the United Kingdom, the two principal organizations are the British Society for Paediatric Endocrinology and Diabetes and the Society for Endocrinology. The world's largest professional organization for pediatric endocrinology is the European Society for Paediatric Endocrinology. Most developed nations throughout the globe have similar organizations.

SOCIAL CONTEXT AND FUTURE PROSPECTS

Research in endocrinology examines how hormones work in the body and how they relate to diseases and conditions. This research aims at improving the medical treatment of endocrine conditions. Some researchers are focusing on the development of synthetic hormones, which do not have the problems associated with animal-derived products. Others are looking at possible cures, in the form of transplanted or regrown organs or other ways for the body to manufacture hormones.

Organ transplantation has made significant progress since the 1960's. Research continues on transplanting endocrine organs, such as the pancreas for the treatment of diabetes. The major problem faced by transplanted organs is rejection of the organ by the recipient's body. Therefore, research is focusing on developing better medications to combat rejection and also on the creation of artificial endocrine organs. Although an artificial pancreas is still a distant possibility, insulin pumps have been developed to continuously administer insulin to diabetics. These devices are likely to become more sophisticated, possibly with sensors to monitor glucose levels and administer the proper insulin dosage.

A controversial topic, the subject of much debate on medical, political, and religious grounds, is embryonic stem cell research. These cells, derived from early embryos, have the potential to develop into any organ within the human body. For example, in theory, a diabetic could grow a new pancreas, which would produce insulin. Proponents of stem cell research allude to a future in which a patient could grow a new adrenal gland, a paraplegic could walk again, and a child with cystic fibrosis could be cured. Opponents of the research claim that it involves the destruction of an embryo and therefore a human life. Proponents counter that researchers use only surplus embryos—those that are destined for destruction. Treatment with stem cells is still in its infancy, but the ability to grow a replacement endocrine organ would cure many endocrine diseases.

Robin L. Wulffson, M.D., F.A.C.O.G.

FURTHER READING

American Diabetes Association. *American Diabetes Association Complete Guide to Diabetes.* 5th ed. Alexandria, Va.: Author, 2011. Provides information to help diabetics manage their disease. Begins with a discussion of the causes and effects of diabetes. Contains a glossary, an appendix on self-monitoring and injection techniques, and a list of resources and organizations.

Borer, Katarina T. *Exercise Endocrinology.* Champaign, Ill.: Human Kinetics, 2003. Looks at the role of hormones in exercise and athletic performance. Topics include regulation of hydration and fuel use during exercise, gender and performance, biological rhythms, and exercise as a stressor.

Gardner, David, and Dolores Shoback. *Greenspan's Basic and Clinical Endocrinology.* 9th ed. New York: McGraw-Hill Medical, 2011. Examines the molecular biology of endocrine glands and discusses metabolic bone disease, pancreatic hormones and diabetes mellitus, hypoglycemia, obesity, geriatric endocrinology, and many other diseases and disorders.

Hadley, Mac E., and Jon E. Levine. *Endocrinology.* 6th ed. Upper Saddle River, N.J.: Prentice Hall, 2007. Presents explanations of basic concepts and applications. Focuses on how glands and hormones control physiological processes.

Lebovic, Dan I., John D. Gordon, and Robert N. Taylor. *Reproductive Endocrinology and Infertility: Handbook for Clinicians.* Arlington, Va.: Scrub Hill Press, 2005. A ready reference for endocrinologists treating conditions and disorders related to reproduction. Information from textbooks, articles, and endocrinologists was gathered and analyzed to provide evidence-based approaches and strategies.

Potter, Daniel A., and Jennifer S. Hanin. *What to Do When You Can't Get Pregnant: The Complete Guide to All the Technologies for Couples Facing Fertility Problems.* New York: Marlowe, 2005. A thorough guide for couples with fertility problems.

Skugor, Mario, and Jesse Bryant Wilder. *The Cleveland Clinic Guide to Thyroid Disorders.* New York: Kaplan, 2009. Skugor, an endocrinologist, teamed with writer Wilder to present detailed information on thyroid diseases and treatment options.

Thacker, Holly. *The Cleveland Clinic Guide to Menopause.* New York: Kaplan, 2009. Thacker, a physician at the Center for Specialized Women's Health at the Cleveland Clinic, offers safe treatments for the menopause and explains myths and facts regarding hormonal replacement therapy.

WEB SITES
American Association of Clinical Endocrinologists
http://www.aace.com

American Diabetes Association
http://www.diabetes.org

American Thyroid Association
http://www.thyroid.org

Pediatric Endocrine Society
http://www.lwpes.org/index.cfm

Society for Reproductive Endocrinology and Infertility
http://www.socrei.org

See also: Cell and Tissue Engineering; Geriatrics and Gerontology; Obstetrics and Gynecology; Pediatric Medicine and Surgery; Pharmacology; Reproductive Science and Engineering.

ENVIRONMENTAL CHEMISTRY

FIELDS OF STUDY

Chemistry; chemical engineering; bioengineering; physics; physical chemistry; organic chemistry; biochemistry; molecular biology; electrochemistry; analytical chemistry; photochemistry; atmospheric chemistry; agricultural chemistry; industrial ecology; toxicology.

SUMMARY

Environmental chemistry is an interdisciplinary subject dealing with chemical phenomena in nature. Environmental chemists are concerned with the consequences of anthropogenic chemicals in the air people breathe and the water they drink. They have become increasingly involved in managing the effects of these chemicals through both the creation of ecologically friendly products and efforts to minimize the pollution of the land, water, and air.

KEY TERMS AND CONCEPTS

- **Anthrosphere:** Artificial environment created, modified, and used by humans for their purposes and activities, from houses and factories to chemicals and communications systems.
- **Biomagnification:** Increase in the concentration of such chemicals as dichloro-diphenyl-trichloroethane (DDT) in different life-forms at successively higher trophic levels of a food chain or web.
- **Carcinogen:** Chemical that causes cancer in organisms exposed to it.
- **Chlorofluorocarbon (CFC):** Organic chemical compound containing carbon, hydrogen, chlorine, and florine, such as Freon (a trademark for several CFCs), once extensively used as a refrigerant.
- **Dichloro-Diphenyl-Trichloroethane (DDT):** Organic chemical that was widely used for insect control during and after World War II but has been banned in many countries.
- **Ecological Footprint:** Measure of how much land and water a human population needs to regenerate consumed resources and to absorb wastes.
- **Greenhouse Effect:** Phenomenon in which the atmosphere, like a greenhouse, traps solar heat, with the trapping agents being such gases as carbon dioxide, methane, and water vapor.
- **Hazardous Waste:** Waste, or discarded material, that contains chemicals that are flammable, corrosive, toxic, or otherwise pose a threat to the health of humans or other organisms or a hazard to the environment.
- **Heavy Metal:** Any metal with a specific gravity greater than 5; usually used to refer to any metal that is poisonous to humans and other organisms, such as cadmium, lead, and mercury.
- **Mutagen:** Chemical that causes or increases the frequency of changes (mutations) in the genetic material of an organism.
- **Ozone:** Triatomic form of oxygen produced when diatomic oxygen is exposed to ultraviolent radiation (its presence in a stratospheric layer protects life-forms from harmful solar radiation).
- **Pesticide:** Chemical used to kill or inhibit the multiplication of organisms that humans consider undesirable, such as certain insects.

DEFINITION AND BASIC PRINCIPLES

Environmental chemistry is the science of chemical processes in the environment. It is a profoundly interdisciplinary and socially relevant field. What environmental chemists do has important consequences for society because they are concerned with the effect of pollutants on the land, water, and air that humans depend on for their life, health, and proper functioning. Environmental chemists are forced to break down barriers that have traditionally kept chemists isolated from other fields.

Environmental chemistry needs to be distinguished from its later offshoot, green chemistry. As environmental chemistry developed, it tended to emphasize the detection and mitigation of pollutants, the study of the beneficial and adverse effects of various chemicals on the environment, and how the beneficial effects could be enhanced and the adverse eliminated or attenuated. Green chemistry, however, focuses on how to create sustainable, safe, and nonpolluting chemicals in ways that minimize the ecological footprint of the processes. Some scholars define this field simply as sustainable chemistry.

245

The work of environmental chemists is governed by several basic principles. For example, the prevention principle states that when creating chemical products, it is better to minimize waste from the start than to later clean up wastes that could have been eliminated. Another principle declares that in making products, chemists should avoid using substances that could harm humans or the environment. Furthermore, chemists should design safe chemicals with the lowest practicable toxicity. In manufacturing products, chemists must minimize energy use and maximize energy efficiency; they should also use, as much as possible, renewable materials and energy resources. The products chemists make should be, if possible, biodegradable. Environmental chemists should use advanced technologies, such as computers, to monitor and control hazardous wastes. Finally, they must employ procedures that minimize accidents.

BACKGROUND AND HISTORY

Some scholars trace environmental chemistry's roots to the industrial revolutions in Europe and the United States in the eighteenth and nineteenth centuries. The newly created industries accelerated the rate at which chemicals were produced. By the late nineteenth century, some scientists and members of the general public were becoming aware of and concerned about certain negative consequences of modern chemical technologies. For example, the Swedish chemist Svante August Arrhenius recognized what later came to be known as the greenhouse effect, and a Viennese physician documented the health dangers of asbestos, which had become an important component in more than 3,000 products.

Modern industrialized societies were also generating increasing amounts of wastes, and cities and states were experiencing difficulties in discovering how to manage them without harm to the environment. World War I, often called the chemists' war, revealed the power of scientists to produce poisonous and explosive materials. During World War II, chemists were involved in the mass production of penicillin and DDT, substances that saved thousands of lives. However, the widespread and unwise use of DDT and other pesticides after the war prompted Rachel Carson to write *Silent Spring* (1962), which detailed the negative effects that pesticides were having on birds and other organisms. Many associate the start of the modern environmental movement with the publication of Carson's book.

The warnings Carson issued played a role in the establishment of the Environmental Protection Agency (EPA) in 1971 and the EPA's ban of DDT in 1972. Reports in 1974 that CFCs were destroying the Earth's ozone layer eventually led many countries to halt, then ban their production. From the 1970's on, environmental chemists devoted themselves to the management of pollutants and participation in government policies and regulations that attempted to prevent or mitigate chemical pollution. In the 1990's, criticism of this command-and-control approach led to the formation of the green chemistry movement by Paul Anastas and others. These chemists fostered a comprehensive approach to the production, utilization, and termination of chemical materials that saved energy and minimized wastes. By the first decade of the twenty-first century, environmental chemistry had become a thriving profession with a wide spectrum of approaches and views.

HOW IT WORKS

Chemical Analysis and the Atmosphere. Indispensable to the progress of environmental chemistry is the ability to measure, quantitatively and qualitatively, certain substances that even at very low concentrations, pose harm to humans and the environment. By using such techniques as gravimetric and volumetric analysis, various types of spectroscopy, electroanalysis, and chromatography, environmental chemists have been able to accurately measure such atmospheric pollutants as sulfur dioxide, carbon monoxide, hydrogen sulfide, nitrogen oxides, and several hydrocarbons. Many governmental and nongovernmental organizations require that specific air contaminants be routinely monitored. Because of heavy demands on analytic chemists, much monitoring has become computerized and automatic.

Atmospheric particles range in size from a grain of sand to a molecule. Nanoparticles, for example, are about one-thousandth the size of a bacterial cell, but environmental chemists have discovered that they can have a deleterious effect on human health. Epidemiologists have found that these nanoparticles adversely affect respiratory and cardiovascular functioning.

Atmospheric aerosols are solid or liquid particles smaller than a hundred millimicrons in diameter.

These particles undergo several possible transformations, from coagulation to phase transitions. For example, particles can serve as nuclei for the formation of water droplets, and some chemists have experimented with particulates in forming rain clouds. Human and natural biological sources contribute to atmospheric aerosols. The CFCs in aerosol cans have been factors in the depletion of the Earth's ozone layer. Marine organisms produce such chemicals as halogen radicals, which in turn influence reactions of atmospheric sulfur, nitrogen, and oxidants. As marine aerosol particles rise from the ocean and are oxidized in the atmosphere, they may react with its components, creating a substance that may harm human health. Marine aerosols contain carbonaceous as well as inorganic materials, and when an organic aerosol interacts with atmospheric oxygen, its inert hydrophobic (water-repelling) film is transformed into a reactive hydrophilic (water-absorbing) layer. A consequence of this process is that organic aerosols serve as a conduit for organic compounds to enter the atmosphere.

Carbon dioxide is an example of a molecular atmospheric component, and some environmental chemists have been devoting their efforts to determining its role in global warming, but others have studied the Earth's prebiotic environment to understand how inorganic carbon initially formed the organic molecules essential to life. Using photoelectrochemical techniques (how light affects electron transfers in chemical reactions), researchers discovered a possible metabolic pathway involving carbon dioxide fixation on mineral surfaces.

Water Pollution. Because of water's vital importance and its uneven distribution on the Earth's surface, environmental chemists have had to spend a great deal of time and energy studying this precious resource. Even before the development of environmental chemistry as a profession, many scientists, politicians, and citizens were concerned with water management. Various governmental and nongovernmental organizations were formed to monitor and manage the quality of water. Environmental chemists have been able to use their expertise to trace the origin and spread of water pollutants throughout the environment, paying special attention to the effects of water pollutants on plant, animal, and human life.

A particular interest of environmental chemists has been the interaction of inorganic and organic matter with bottom sediments in lakes, rivers, and oceans. These surface sediments are not simply unreactive sinks for pollutants but can be studied quantitatively in terms of how many specific chemicals are bonded to a certain amount of sediment, which in turn is influenced by whether the conditions are oxidizing or reducing. Chemists can then study the bioavailability of contaminants in sediments. Furthermore, environmental chemists have studied dissolution and precipitation, discovering that the rates of these processes depend on what happens in surface sediments. Using such techniques as scanning polarization force microscopy, they have been able to quantify pollutant immobilization and bacterial attachment on surface sediments. Specifically, they have used these methods to understand the concentrations and activities of heavy metals in aquatic sediments.

Hazardous Waste. One of the characteristics of advanced industrialized societies has been the creation of growing amounts of solid and liquid wastes, an important proportion of which pose severe dangers to the environment and human health. These chemicals can be toxic, corrosive, or flammable. The two largest categories of hazardous wastes are organic compounds such as polychlorinated biphenyls (PCBs) and dioxin, and heavy metals, such as lead and mercury. Environmental chemists have become involved in research on the health and environmental effects of these hazardous substances and the development of techniques to detect, monitor, and control them. For example, they have studied the rapidly growing technology of incineration as a means of reducing and disposing of wastes. They have also studied the chemical emissions from incinerators and researched methods for the safe disposal of the ash and slag produced. Because of the passage of the Resource Recovery Act of 1970, environmental chemists have devoted much attention to finding ways of reclaiming and recycling materials from solid wastes.

Pesticides. The mismanagement of pesticides inspired Carson to write *Silent Spring*, and pesticides continue to be a major concern of industrial and environmental chemists. One reason for the development of pesticides is the great success farmers had in using them to control insects, thereby dramatically increasing the quantity and quality of various agricultural products. By 1970, more than 30,000 pesticide products were being regularly used. This expansion in pesticides is what alarmed Carson, who was not in favor of a total ban of pesticides but rather their reduction and integration with biological and cultural controls.

Because pesticides are toxic to targeted species, they often cause harm to beneficial insects and, through biomagnification, to birds and other animals. Pesticide residues on agricultural products have also been shown to harm humans. Therefore, environmental chemists have become involved in monitoring pesticides from their development and use to their effects on the environment. They have also helped create pesticide regulations and laws. This regulatory system has become increasingly complex, costing companies, the government, and customers large amounts of money. The hope is that integrated control methods will prove safer and cheaper than traditional pesticides.

APPLICATIONS AND PRODUCTS

Anthrosphere. Environmental chemists are concerned with how humans and their activities, especially making and using chemicals, affect the environment. Building homes and factories, producing food and energy, and disposing of waste all have environmental consequences. Whereas some environmentalists study how to create ecologically friendly dwellings, environmental chemists study how chemical engineers should design factories that cause minimum harm to the environment. Specific examples of applications of environmental chemistry to industry include the creation of efficient catalysts that speed up reactions without themselves posing health or environmental hazards.

Because many problems arise from the use of hazardous solvents in chemical processes, environmental chemists try to develop processes that use only safe solvents or avoid their use altogether. Because the chemical industry depends heavily on petroleum resources, which are nonrenewable and becoming drastically diminished, environmental chemists study how renewable resources such as biomass may serve as substitutes for fossil fuels. They are also creating products that degrade rapidly after being discarded, so that their environmental impact is transient. Following the suggestions of environmental chemists, some companies are developing long-lasting, energy-saving batteries, and others are selling their products with less packaging than previously.

Hydrosphere. Throughout history, water has been essential in the development of human civilizations, some of which have declined and disappeared because of deforestation, desertification, and drought. Water has also been a vehicle for the spread of diseases and pollutants, both of which have caused serious harm to humans and their environment. Environmental chemists have consequently been involved in such applications as the purification of water for domestic use, the monitoring of water used in the making of chemicals, and the treatment of wastewater so that its release and reuse will not harm humans or the environment. For example, such heavy metals as cadmium, mercury, and lead are often found in wastewater from various industries, and environmental chemists have developed such techniques as electrodeposition, reverses osmosis, and ion exchange to remove them. Many organic compounds are carcinogens and mutagens, so chemists want to remove them from water. Besides such traditional methods as powdered activated carbon, chemists have used adsorbent synthetic polymers to attract insoluble organic compounds. Detergents in wastewater can contribute to lake eutrophication and cause harm to wildlife, and some companies have created detergents specially formulated to cause less environmental damage.

Atmosphere. In the twentieth century, scientists discovered that anthropogenic greenhouse gases have been contributing to global warming that could have catastrophic consequences for island nations and coastal cities. Industries, coal-burning power plants, and automobiles are major air polluters, and environmental chemical research has centered on finding ways to reduce or eliminate these pollutants. The Clean Air Act of 1970 and subsequent amendments set standards for air quality and put pressure on air polluters to reduce harmful emissions. For example, chemists have helped power plants develop desulfurization processes, and other scientists developed emission controls for automobiles.

The problem of global warming has proved difficult to solve. Some environmental chemists believe that capturing and storing carbon dioxide is the answer, whereas others believe that government regulation of carbon dioxide and methane by means of energy taxes will lead to a lessening of global warming. Some think that the Kyoto Protocol, an international agreement that went into effect in 2005, is a small but important first step, whereas others note that the lack of participation by the United States and the omission of a requirement that such countries as China and India reduce greenhouse gas emissions seriously weakened the agreement. On the other hand, the Montreal and Copenhagen Protocols did foster global cooperation in the reduction of and phasing out of CFCs, which should lead to a reversal in ozone layer depletion.

Agricultural and Industrial Ecology. Agriculture, which involves the production of plants and animals as food, is essential in ministering to basic human needs. Fertilizers and pesticides developed by chemists brought forth the green revolution, which increased crop yields in developed and developing countries. Some believe that genetic engineering techniques will further revolutionize agriculture. Initially, chemists created such highly effective insecticides as DDT, but DDT proved damaging to the environment. Activists then encouraged chemists to develop biopesticides from natural sources because they are generally more ecologically friendly than synthetics.

Industrial ecology is a new field based on chemical engineering and ecology; its goal is to create products in a way that minimizes environmental harm. Therefore, environmental chemical engineers strive to build factories that use renewable energy as much as possible, recycle most materials, minimize wastes, and extract useful materials from wastes. In general, these environmental chemists act as wise stewards of their facilities and the environment. A successful example of ecological engineering is phytoremediation, or the use of plants to remove pollutants from contaminated lands. Artificially constructed wetlands have also been used to purify wastewater.

IMPACT ON INDUSTRY

Industries are involved in a wide variety of processes that have environmental implications, from food production and mineral extraction to manufacturing and construction. In some cases, such as the renewable energy industries, the environmental influence is strong, but even in traditional industries such as utilities and transportation companies, problems such as air and water pollution have become corporate concerns.

Government and University Research. Since the beginning of the modern environmental movement,

---

Fascinating Facts About Environmental Chemistry

- At life's beginning, single-celled cyanobacteria made possible the evolution of millions of new species; however, what modern humans are doing to the atmosphere will result in the extinction of hundreds of thousands of species.
- More than 99 percent of the total mass of the Earth's atmosphere is found within about 30 kilometers (about 20 miles) of its surface.
- Although 71 percent of the Earth's surface is covered by water, only 0.024 percent of this water is available as freshwater.
- According to a National Academy of Sciences study, legally permitted pesticide residues in food cause 4,000 to 20,000 cases of cancer per year in the United States.
- From 1980 to 2010, the quality of outdoor air in most developed countries greatly improved.
- From 1980 to 2008, the EPA placed 1,569 hazardous-waste sites on its priority list for cleanup.
- According to the U.S. Geological Survey, even though the population of the United States grew by 16 percent from 1980 to 2004, total water consumption decreased by about 9 percent.
- Because of global warming, in February and March, 2002, a mass of ice larger than the state of Rhode Island separated from the Antarctic Peninsula.
- The United States leads the world in producing solid waste. With only 4.6 percent of the world's population, it produced about one-third of the world's solid waste.
- Each year, 12,000 to 16,000 American children under nine years of age are treated for acute lead poisoning, and about 200 die.

---

state and federal governments as well as universities have increased grants and fellowships for projects related to environmental chemistry. For example, the EPA's Green Chemistry Program has supported basic research to develop chemical products and manufacturing techniques that are ecologically benign. Sometimes government agencies cooperate with each other in funding environmental chemical projects; for instance, in 1992, the EPA's Office of Pollution Prevention and Toxics collaborated with the National Science Foundation to fund several green chemical proposals. These grants were significant, totaling tens of millions of dollars. Besides government and academia, professional organizations have also sponsored green research. For example, the American Chemical Society has established the Green Chemistry Institute, whose purpose is to encourage collaboration with scientists in other disciplines to discover chemical products and processes that reduce or eliminate hazardous wastes.

Industry and Business. Environmentalists and government regulations have forced leaders in business and industry to make sustainability a theme in their plans for future development. In particular, the U.S. chemical industry, the world's largest, has directly linked its growth and competitiveness to a concern for the environment. Industrial leaders realize that they will have to cooperate with officials in government and academia to realize this vision. They also understand that they will need to join with such organizations as the Environmental Management Institute and the Society of Environmental Toxicology and Chemistry to minimize the environmental contamination that has at times characterized the chemical industry of the past.

Major Corporations. Top American chemical companies, such as Dow, DuPont, Eastman Chemical, and Union Carbide, have gone on record as vowing to use resources more efficiently, deliver products to consumers that meet their needs and enhance their quality of life, and preserve the environment for future generations. Nevertheless, these promised changes must be seen against the background of past environmental depredations and disasters. For example, Dow is responsible for ninety-six of the worst Superfund toxic-waste dumps, and Union Carbide shared responsibility for the deaths of more than 2,000 people in a release of toxic chemicals in Bhopal, India. Eastman Chemical, along with other industries, is a member of Responsible Care, an organization devoted to the principles of green chemistry, and the hope is that the member industries will encourage the production of ecologically friendly chemicals without the concomitant of dangerous wastes.

CAREERS AND COURSE WORK

Students in environmental chemistry need to take many chemistry courses, including general chemistry, organic chemistry, quantitative and qualitative analysis, instrumental analysis, inorganic chemistry, physical chemistry, and biochemistry. They also should study advanced mathematics, physics, and computer science. Although job opportunities exist for students with a bachelor's degree, there are greater opportunities for those who obtain a master's degree or a doctorate. Graduate training allows students to specialize in such areas as soil science, hazardous-waste management, air-quality management, water-quality management, environmental education, and environmental law. Because of increasing environmental concerns in industries, governments, and academia, numerous careers are possible for environmental chemistry graduates. They have found positions in business, law, marketing, public policy, government agencies, laboratories, and chemical industries. Some graduates pursue careers in such government agencies as the EPA, the Food and Drug Administration, the Natural Resource Conservation Services, the Forest Service, and the Department of Health and Human Services. After obtaining a doctorate, some environmental chemists become teachers and researchers in one of the many academic programs devoted to their field.

SOCIAL CONTEXT AND FUTURE PROSPECTS

In a world increasingly concerned with environmental quality, the sustainability of lifestyles, and environmental justice, the future for environmental chemistry appears bright. For example, analysts have predicted that environmental chemical engineers will have a much faster employment growth than the average for all other occupations. Environmental chemists will be needed to help industries comply with regulations and to develop ways of cleaning up hazardous wastes. However, other analysts warn that, in periods of economic recession, environmental concerns tend to be set aside, and this could complicate the employment forecast for environmental chemists.

Some organizations, such as the Environmental Chemistry Group in England, have as a principal goal the promotion of the expertise and interests of their members, and the American Chemical Society's Division of Environmental Chemistry similarly serves its members with information on educational programs, job opportunities, and awards for significant achievement, such as the Award for Creative

Advances in Environmental Chemistry. These organizations also issue reports on their social goals, and documents detailing their social philosophy emphasize that environmental chemists should be devoted to the safe operation of their employers' facilities. Furthermore, they should strive to protect the environment and make sustainability an integral part of all business activities.

Robert J. Paradowski, M.S., Ph.D.

FURTHER READING

Baird, Colin, and Michael Cann. *Environmental Chemistry.* 4th ed. New York: W. H. Freeman, 2008. A clear and comprehensive survey of the field. Each chapter has further reading suggestions and Web sites of interest. Index.

Carson, Rachel. *Silent Spring.* 1962. Reprint. Boston: Houghton Mifflin, 2002. Originally serialized in *The New Yorker* magazine, this book has been honored as one of the best nonfiction works of the twentieth century. Its criticism of the chemical industry and the overuse of pesticides generated controversy, and most of its major points have stood the test of time.

Girard, James E. *Principles of Environmental Chemistry.* Sudbury, Mass.: Jones and Bartlett, 2010. Emphasizes the chemical principles undergirding environmental issues as well as the social and economic contexts in which they occur. Five appendixes and index.

Howard, Alan G. *Aquatic Environmental Chemistry.* 1998. Reprint. New York: Oxford University Press, 2004. Analyzes the chemistry behind freshwater and marine systems. Also includes useful secondary material that contains explanations of unusual terms and advanced chemical and mathematical concepts.

Manahan, Stanley E. *Environmental Chemistry.* 9th ed. Baca Raton, Fla.: CRC Press, 2010. Explores the anthrosphere, industrial ecosystems, geochemistry, and aquatic and atmospheric chemistry. Each chapter has a list of further references and cited literature. Index.

Schwedt, Georg. *The Essential Guide to Environmental Chemistry.* 2001. Reprint. New York: John Wiley & Sons, 2007. Provides a concise overview of the field. Contains many color illustrations and an index.

WEB SITES

American Chemical Society
Division of Environmental Chemistry
http://www.envirofacs.org

Environmental Protection Agency
National Exposure Research Library, Environmental Sciences Division
http://www.epa.gov/nerlesd1

Royal Society of Chemistry
Environmental Chemistry Group
http://www.rsc.org/membership/networking/interestgroups/environmental/index.asp

Society of Environmental Toxicology and Chemistry
http://www.setac.org

ENVIRONMENTAL MICROBIOLOGY

FIELDS OF STUDY

Environmental microbiology; molecular biology; environmental science; biology; chemistry; biotechnology; botany; entomology; plant pathology; biochemistry; genetics; toxicology; environmental monitoring; chemical and biomolecular engineering; microbial ecology; soil science; biogeochemistry; modeling.

SUMMARY

The majority of the Earth's biomass consists of microorganisms, and everything consumed and created by microorganisms has a very significant impact on the surrounding environment and all other living organisms. Environmental microbiology focuses on the role of these microorganisms in both causing environmental deterioration and rectifying ecological degradation, while also considering microbial ecology in both natural and polluted environments. Environmental microbiology technology harnesses the natural ability of microorganisms to remove pollutants, such as crude oil and industrial waste, from the environment.

KEY TERMS AND CONCEPTS

- **Bioassessment:** Measurement of living organisms (their presence, condition, and number) as determination of water quality.
- **Biocriteria:** Narrative or numerical standards assessing the qualities required to support a desired condition, particularly in a body of water; also known as biological criteria.
- **Bioindicator:** Bacterium used to detect and respond to surrounding environmental conditions.
- **Bioremediation:** Use of microorganisms, such as fungi and bacteria, and their enzymes to return a contaminated environment to its original condition.
- **Biotechnology:** Exploitation and manipulation of living organisms to produce valuable and useful products or results.
- **Culture:** Artificial cultivation of microorganisms, usually under laboratory conditions.
- **Fossil Fuel:** Deposit within the Earth's crust of either solid (coal), liquid (oil), or gaseous (natural gas) hydrocarbons produced through the natural decomposition of organic material (plants and animals) over many millions of years.
- **Genomic Analysis:** Examination of an organism's complete DNA.
- **Metagenome Technology:** Field of microbiology that applies genomic analysis to entire microorganism communities, thus creating a culture-independent approach that avoids the need to isolate and culture individual microbes; also known as metagenomics.

DEFINITION AND BASIC PRINCIPLES

Environmental microbiology is the study of microbial community structure and physiology within the environment, whether soil, water, or air. At a fundamental level, environmental microbiology is the study of the role of microorganisms as both the cause and the remediation of environmental pollution and ecological degradation.

Environmental microbiology is a multidisciplinary science, blending environmental science and the biology of microscopic organisms. It involves the development, application, adaptation, and management of biological systems and microorganisms to repair and prevent environmental damage caused by pollution. This discipline investigates how microorganisms can be both detrimental and beneficial to the environment and human society, particularly in relation to the restoration and remediation of the world's natural environment (air, water, and soil). The field has benefited from advancements in genetic engineering and modern microbiological concepts, which have encouraged the development of traditional and innovative solutions to environmental contamination.

BACKGROUND AND HISTORY

Any discussion of environmental microbiology must touch on the history of microbial discovery. Although the technology involved in environmental microbiology is evolving and expanding, the first steps toward the establishment of this field began hundreds of years ago with the discovery of microorganisms. In the mid-seventeenth century, both Antoni van Leeuwenhoek and Robert Hooke observed non-living and living microorganisms under self-made microscopes, thereby discovering the reason behind one of nature's conundrums—the decomposition of plants and animals. Although the identification of microbes (living things not visible to the naked eye) in 1675 was instrumental in advancing biological science, the true significance of the role of microbes in disease was not understood until some two hundred years later.

By the late nineteenth century, research conducted by Dutch microbiologist Martinus W. Beijerinck and Russian microbiologist Sergei Winogradsky had greatly advanced the understanding of microorganisms and their possible role in ecological processes. Beijerinck developed a method for isolating microorganisms called the enrichment culture technique, and Winogradsky enhanced scientific awareness of microbial diversity and discovered the autotrophic (self-feeding) ability of bacteria. He went on to develop the Winogradsky column culture technique, which showed that water in a lake contains a number of organisms performing various processes at different levels. The discipline of environmental microbiology was made possible by the work of these early scientists. Although traditional microbe research focused on the role of microbes in causing disease, Beijerinck and Winogradsky began to examine their role in disease prevention.

The roots of environmental microbiology also lie in urban waste management and treatment. The field originally focused on monitoring the movement of pathogens and treating them within natural and urban environments to protect municipal water quality and public health. As the world became more urbanized in the late nineteenth century, the incidence of communicable diseases such as typhoid fever and cholera increased. To combat the spread of diseases, cities and communities began to treat water with various filtration and disinfectant methods. For the most part, such approaches to water treatment were instrumental in the elimination of waterborne bacterial diseases in developed countries, and disinfection processes continue to be widely used.

As research continued into the 1960's, however, it became apparent that the viruses and protozoa parasites that caused waterborne diseases were much more resistant to the process of disinfection than were bacteria. Already treated urban water supplies were still plagued by uncontrolled outbreaks of giardia (which causes giardiasis), cryptosporidium (which causes cryptosporidiosis), and norovirus (which causes gastroenteritis). Although serious outbreaks within the developed world are rare, the continued occurrence of water pathogens has meant that the field of environmental microbiology still has a very strong focus on water quality and the treatment and control of water pathogens. Perhaps of most concern is the fact that some 10 to 50 percent of diarrheal illness is caused by waterborne microbial organisms not yet identified by science.

Before long, the effects of water quality on human health were not the only area of concern for environmental microbiology. By the 1960's, concern over the effects of chemicals in the natural environment had increased, and it became obvious that poor human health was not the only issue. Chemicals in the soil and water found their way into the human food chain through groundwater use and consumption, affecting not only human health but also animal and plant species using those same soil and water supplies. Significant chemical contamination, such as the massive *Exxon Valdez* oil spill in 1989, highlighted the need to investigate the potential for microorganisms in bioremediation. In the twenty-first century, the use of microorganisms in the treatment and control of pathogens and in bioremediation processes is a key feature of environmental microbiology.

How It Works

Microorganisms are found in all areas of the biosphere, even environments of extreme temperatures, acidity, salinity, and darkness that are inhospitable to most other organisms. They account for the vast majority of the Earth's biomass. The entire number of microorganisms living on Earth is almost immeasurable, but it is estimated that more than 1 billion microorganisms live in just 1 gram of soil. Microorganisms are vital to the health and function of the planet and are responsible for a vast number of natural life processes. They are the most significant players involved in the synthesis and degradation of important molecules. They provide energy to other organisms, are responsible for most of the planet's photosynthesis, are able to fix nitrogen and recycle nutrients in ecosystems, and are also capable of causing lethal diseases for humans, flora, and fauna.

Fundamentally, environmental microbiology is about harnessing the natural ability of microbes and their relationships with other microbes to address environmental issues and solve environmental problems. The main focus of environmental microbiology remains the use of microorganisms in the treatment and control of pathogens and in bioremediation processes. Modern-day technological advancements in genetic engineering and biotechnology have, however, greatly increased the applications of this scientific field. The increased applications for environmental microbiology are important in view of the ability of pathogens such as bacteria and viruses to evolve and emerge rapidly in an environment under increasing stress from the human population.

APPLICATIONS AND PRODUCTS

The application of genomics and molecular biology to environmental microbiology has allowed scientists to discover the vast diversity and complexity that exists in natural microorganism communities. Despite advanced human-developed technology, however, it is estimated that less than 2 percent of microorganisms have actually been described by science or have been grown in laboratory conditions. This means that scientists not only are unsure of all the possible effects of such organisms on human health and the environment but also have not even begun to understand or determine the many potential applications and products of such organisms.

New discoveries and principles of environmental microbiology can have vast potential in other areas of science such as biotechnology, pharmaceutics, biomedical research and engineering, the chemical and textile industries, bioremediation and wastewater treatment, pathogen control, mineral recovery, sustainable agriculture, and resource conservation. However, the majority of research in the applications for bioremediation, water quality, and the biotreatment of waste material and wastewater.

Bioremediation. Human society is increasingly exploiting the flexible appetite of microbes (particularly bacteria) to remediate environments containing contaminants such as industrial waste, crude oil, and polychlorinated biphenyls (PCBs). Bioremediation is usually classified as either in situ or ex situ and is defined as the use of microorganisms, such as fungi and bacteria, and their enzymes to return a contaminated environment to its original condition. In situ bioremediation entails treating the contamination in place and relies on the ability of the microorganisms to metabolize or remove the contaminants inside a naturally occurring system. Ex situ remediation involves removing the polluted material from the site and treating it elsewhere, and it relies on some form of artificial engineering and input. Although the process of bioremediation can occur naturally, it can be promoted through artificial human stimulus through the use of environmental microbiology technology.

For example, environmental microbiology technology is used in the remediation of crude oil spills in coastal areas. Although oil is a naturally occurring fossil fuel, it can cause major ecological damage when accidentally introduced into marine (and terrestrial) environments during oil spills. Large spills can be difficult to contain and collect, and the longer a spill remains in the environment, the greater the potential damage. Environmental microbiology techniques, specifically hydrocarbonoclastic bacteria (HCB), can be used to clean up oil spills. These bacteria are able to degrade hydrocarbons (in this case oil) and are therefore beneficial as bioremediators.

Waste Biotreatment and Water Quality Treatment. Biotreatment processes involve the environmentally friendly treatment of waste material, including wastewater, using living microorganisms. Scientific research has demonstrated that while wastewater processes are one of the most important functions of environmental microbiology and use a significant diversity of microorganisms, many of these microorganisms are yet to be classified or cultured. This indicates that the potential of microbes in biotreatment is not yet fully understood. Research on metagenome technology (metagenomics), however, is highlighting the diversity, structure, and functions of microorganisms. This research has looked at the role of microbes in processes such as nitrogen cycling, anaerobic ammonium oxidation, and methane fermentation in the treatment of wastewater, with a particular focus on the use of bacterial biofilms and bioreactors.

The World Health Organization (WHO) reported that 10 percent of the global health burden is related to poor-quality water and could be avoided. One of the most important applications of environmental

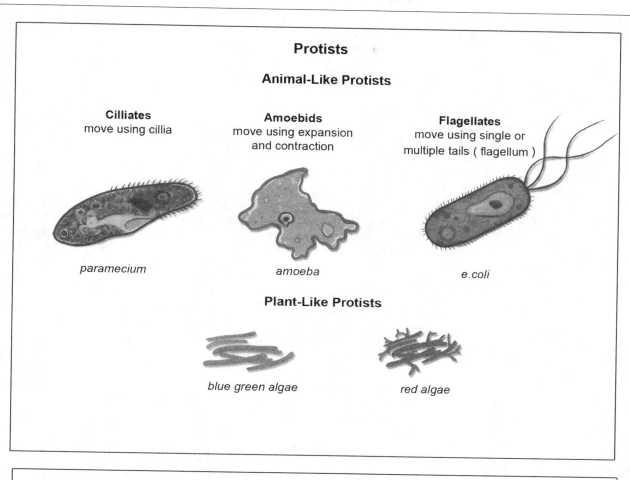

Protists

Animal-Like Protists

Cilliates
move using cillia

Amoebids
move using expansion
and contraction

Flagellates
move using single or
multiple tails (flagellum)

paramecium

amoeba

e.coli

Plant-Like Protists

blue green algae

red algae

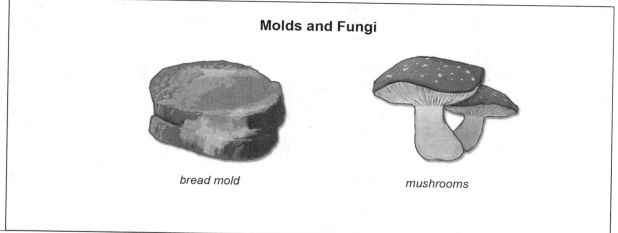

Molds and Fungi

bread mold

mushrooms

The microbial interactions of microorganisms such as protists and fungi are a primary focus of microbiologists.

microbiology, therefore, is in controlling waterborne pathogens and diseases. The treatment of drinking water is, of course, not new. Water disinfection processes were key in the fight against bacterial waterborne pathogens. The introduction of environmental microbiology techniques, however, has assisted in controlling pathogens and contaminants other than bacteria. In particular, microorganisms are used to decompose contaminants such as organic matter, nitrates, and phosphate in wastewater and to improve water quality. The removal of organic material is done through both bacterial aerobic and anaerobic decomposition, but the removal of ammonium and nitrates is much more complex. It requires both aerobic and anaerobic conversion to remove the pollutants and involves the bacterial conversion of ammonia to nitrite and nitrate to nitrogen gas, which can be safely released into the atmosphere.

Bioassessment and Bioindicators. In environmental microbiology, bioindicators are bacteria that are used to detect and respond to surrounding environmental conditions. Bioindicators can be any organism and are used in numerous industries and fields of science, including environmental microbiology. Bacteria, however, are considered to be superior precursors of human-caused environmental degradation, as they possess the highest surface area to volume ratio of all organisms.

IMPACT ON INDUSTRY

Major Organizations. The American Society for Microbiology is the world's largest professional life science organization. Although the society covers all areas of microbiology and releases numerous publications, including *Applied and Environmental Microbiology*, it is focused on advancing "microbiological sciences as a vehicle for understanding life processes and to apply and communicate this knowledge for the improvement of health and environmental and economic well-being worldwide." The International Society for Microbial Ecology is also one of the principal scientific societies involved in the emerging field of microbial ecology and its related disciplines. Organizations such as the World Health Organization are also concerned with environmental microbiology, particularly as it relates to environmental water quality and human health.

Government and University Research. A great many universities around the world are involved in environmental microbiology either directly or indirectly through biotechnology and microbial ecology. Significant research into environmental microbiology has been undertaken by universities such as Cornell and Stanford in the United States, the University of Ottawa in Canada, the University of Oxford and University of Birmingham in the United Kingdom, the University of New South Wales in Australia, and Peking University in China.

Many of the world's governments have been pursuing and implementing environmental microbiology research in an attempt to rectify and control environmental pollution and wastewater management. For example, the U.S. Environmental Protection Agency, particularly through the Microbiological and Chemical Exposure Assessment Research Division, has promoted significant research and remediation of environmental contaminants. Canada's National Research Council (NRC) has researched and developed environmental genomics, bioremediation of contaminated soil, and molecular techniques to assess environmental impacts and determine the diversity of microorganism species. Australia's leading scientific research center, the Commonwealth Scientific and Industrial Research Organization, has investigated bioremediation of contaminated sites, molecular detection and monitoring of microbial communities in natural and engineered environments, and the isolation and description of novel microorganisms. Additionally, the Chinese Academy of Sciences, China's prominent academic institution and broad research and development center in natural and technological science innovation, has undertaken significant research in the field of environmental microbiology, particularly in relation to water quality and bioremediation.

CAREERS AND COURSE WORK

Students who wish to pursue a career in environmental microbiology can come from a diverse number of fields but must have a strong grounding in basic microbiology, molecular biology, and environmental science. A solid grasp of genetic engineering technology and genetically modified organisms is also helpful.

Many universities provide undergraduate and postgraduate degrees in environmental microbiology. Upon course completion, students should have a solid understanding of the interactions

Fascinating Facts About Environmental Microbiology

- Scientists with the U.S. Geological Survey added nutrients to contaminated soil to stimulate natural microbes that were actively consuming fuel-derived toxins in Hanahan, South Carolina, in 1992. After a year, contamination in residential areas was reduced by 75 percent.
- In 2010, the United Nations reported that worldwide, annually, 2 million tons of industrial and agricultural waste and sewage are discharged into waterways and more than 1.8 million children under the age of five die from water-related diseases.
- According to estimates, nearly 99.8 percent—or less than 1 percent—of microorganisms in the environment are nonculturable or not readily culturable.
- Deep in the Black Sea, scientists from the Max Planck Institute for Marine Microbiology have found bacteria that thrive on methane.
- Microorganisms exist in the human mouth. For instance, *Actinomyces naeslundii* is a bacterium that is associated with good oral hygiene.
- In the 1960's, Thomas Brock found *Thermus acquaticus,* an archaea organism, in the hot springs at Yellowstone National Park. The microbe's polymerase enzyme can duplicate DNA at high temperatures and has been used in genetic studies.

between microorganisms in both natural and artificial environments, particularly contaminated aquatic and terrestrial ecosystems. Most courses focus teaching and research on both fundamental and applied features of biodegradation, bioremediation strategies, and the use of biosensors for toxicity assessment.

Graduates with environmental microbiology degrees who have done postgraduate research can enter careers in environmental and microbiology consulting, in water treatment and resource management, and bioremediation and bioassessment in the private sector, specialized government organizations and agencies, and universities and institutions undertaking teaching and research.

SOCIAL CONTEXT AND FUTURE PROSPECTS

The growing global population and its impact on the environment, particularly the supply of fresh water, make water treatment and bioremediation increasingly important. Therefore, future research will continue to focus on these aspects and applications of environmental microbiology. Less than 2 percent of the world's microbial species have been classified and cultured, which means there is significant unrealized potential for human uses of microorganisms. Many believe, however, that the emergence of metagenome technology holds the key to rapid advancement in the field of microbiology and in human understanding of life. Heralded as the most important advancement since the invention of the microscope, metagenomics applies genomic analysis (the examination of an organism's complete DNA) to entire microorganism communities, thereby avoiding the need to isolate and culture individual microbes.

Research has begun on applications of environmental microbiology to remove heavy metal pollution, destroy specific xenobiotics (foreign chemicals found in a living organism), treat water and air pollution caused by carbon dioxide and sulfur dioxide, and develop biodegradable plastics and other useful material.

Christine Watts, Ph.D., B.App.Sc., B.Sc.

FURTHER READING

Cummings, Stephen, ed. *Bioremediation: Methods and Protocols.* New York: Humana Press, 2010. Experts in the fields of environmental biotechnology and microbiology examine innovative and imaginative bioremediation techniques in pollution removal.

Hurst, Christon J., et al., eds. *Manual of Environmental Microbiology.* 3d ed. Washington, D.C.: ASM Press, 2007. A detailed examination of the role microbes play in planetary environments, with a focus on the basic principles of environmental microbiology, general methodologies, detection and impact of microbial activity within the environment, and detection and control of pathogens.

Illman, Walter, and Pedro Alvarez. "Performance Assessment of Bioremediation and Natural Attenuation." *Critical Reviews in Environmental Science and Technology* 39, no. 4 (April, 2009): 209-270. Reviews state-of-art performance assessment methods and discusses future research directions in bioremediation, natural attenuation, chemical fingerprinting, and molecular biological tools.

Liu, Wen-Tso, and Janet K. Jansson, eds. *Environmental Molecular Microbiology*. Norfolk, England: Caister Academic Press, 2010. Highlights the concepts and technology of environmental molecular microbiology, with contributions from international experts describing various technologies and their applications in environmental microbiology.

Maier, Raina M., Ian L. Pepper, and Charles P. Gerba, eds. *Environmental Microbiology*. 2d ed. Boston: Elsevier/Academic Press, 2009. Examines the fundamental concepts and principles related to environmental microbiology, with the addition of case studies to highlight relevant issues and solutions.

Mitchell, Ralph, and Ji-Dong Gu, eds. *Environmental Microbiology*. 2d ed. Hoboken, N.J.: Wiley-Blackwell, 2010. This revision of one of the most successful books on environmental microbiology takes a comprehensive look at the role of microbiological processes related to environmental deterioration, with a focus on the detection and control of environmental contaminants.

Rochelle, Paul A., ed. *Environmental Molecular Microbiology: Protocols and Applications*. New York: Springer-Verlag, 2001. This book provides a comprehensive collection of laboratory protocols, techniques, and applications in the field of environmental microbiology.

WEB SITES

American Society of Microbiology
http://www.asm.org

International Society for Microbial Ecology
http://www.isme-microbes.org

U.S. Environmental Protection Agency
EPA Microbiology
http://www.epa.gov/microbes

See also: Environmental Chemistry; Soil Science.

EPIDEMIOLOGY

FIELDS OF STUDY

Immunology; internal medicine and other medical specialties; demography; public health; sociology; biochemistry; bioethics; environmental science; genetics; microbiology; molecular biology; toxicology.

SUMMARY

Epidemiology is the branch of medical science that studies the occurrence of disease in human populations: How many people in the population have the disease? Are there common factors that distinguish people with the disease from those who are free of it? How is it spread? Why does it occur in a particular place? Are there associated risk factors? What causes the disease? What can be done to prevent or control it? Epidemiology is tasked with answering these questions, which will always have important public-health implications.

KEY TERMS AND CONCEPTS

- **Bias:** Flaw in a study's design or selection of subjects that distorts the results.
- **Confounding Variable/Factor:** Variable or factor that is related to both the risk factor and the disease or condition being studied and therefore distorts the relation between them.
- **Control Group:** Group of people who are similar to those in the study group but who do not have the disease or condition being studied or who did not receive the intervention being studied.
- **Dose-Response Curve:** Graph that plots changes in the dose of an agent against its effect.
- **Epidemic:** Occurrence of a disease that is clearly in excess of normal numbers.
- **Exposure:** Condition of being subjected to something that can be considered a determinant of disease.
- **Incidence:** Rate of occurrence of new cases of a disease during a specified time.
- **Morbidity:** Diseased state or condition.
- **Mortality Rate:** Death rate, usually discussed in relation to a particular morbidity.
- **Pandemic:** Epidemic that affects several countries or continents.

- **Pathogenic:** Having the ability to cause disease.
- **Prevalence:** Number of cases of a disease at a specific time.
- **Reservoir:** Carrier of a disease-causing organism that is not itself harmed by it.
- **Risk Factor:** Variable associated with a disease that increases the likelihood of developing that disease.

DEFINITION AND BASIC PRINCIPLES

Epidemiology is the study of variations in disease within population groups. The variations are rarely of only theoretical interest; rather they serve as the data source for devising measures to control or prevent health-related problems. Ideally, once a disease or health anomaly is identified, an investigation links it to risk factors and a cause is inferred. A range of activities can follow: design of biomedical research, communication of findings, and collaboration with other disciplines and public-health agencies. Epidemiologic studies can be expected to influence the policy and practice of public health.

Epidemiology derives from the term "epidemic," which traditionally described a rapidly spreading infectious disease. Epidemics have come instead to mean health-related conditions that occur in excess of expected numbers and may be related to environmental, socioeconomic, and other factors. The science of epidemiology considers a host of modern health issues within its purview.

BACKGROUND AND HISTORY

The history of epidemiology has been both consistent with and limited by the biological knowledge of the time, but its roots lie in the nineteenth century. In Victorian England and elsewhere, the explanation of choice for the cause of disease was miasma, a collective term, traceable to the ancient Greek physician Hippocrates, for foul atmospheric poisons. In the nineteenth century, a few pioneers attempted to make inroads into scientific methods in the face of widespread ignorance of disease causation. For example, the London Epidemiological Society, chartered in 1850, consisted of a group of physicians whose mission was to investigate the nature of disease and prevent epidemics.

Several historical events—cholera in London,

clinical study in Paris, and measles in the Faeroe Islands—are considered milestones in the development of epidemiologic studies.

Cholera in London. Miasma was believed responsible for London's periodic cholera epidemics, which occurred between 1831 and 1866. One expert of the time actually advocated removing noxious smells from the home by directing household waste to the water supply. John Snow, a physician widely known for administering chloroform to Queen Victoria during childbirth, reasoned that if symptoms of cholera were largely gastrointestinal, then ingesting something must cause it. He ultimately related variations in household and neighborhood mortality rates to the sources of their water supply. His resulting 1854 treatise indicting the infamous Broad Street pump is an epidemiologic classic.

Clinical Study in Paris. The early nineteenth century also saw the beginnings of clinical epidemiology studies. Pierre Louis recorded and published meticulous observations of the contemporary practice of bloodletting. His methods were mathematical, based on careful studies of patients with inflammatory disease. Among his conclusions: Bloodletting did not benefit patients with pneumonia, although it shortened the duration of the disease if done early and the patient survived.

Measles in the Faeroe Islands. Peter Panum was a young Danish doctor dispatched by his government in 1846 to study an outbreak of measles in the isolated Faeroe Islands, in the North Atlantic. Panum's deductions about the natural course of measles were derived largely from studying the islanders' way of life. In writing about his observations in a journal of the time, Panum emphasized that the physician who works in an unaccustomed environment must first study hygienic conditions that can affect the health of its population.

HOW IT WORKS

Hill's Postulates. The tools of epidemiology are methodology, investigation, and inference. Sir Austin Hill was a professor of medical statistics at the University of London. His 1965 address to London's Royal Society of Medicine enumerated six criteria—strength, consistency, specificity, temporality, biologic gradient, and plausibility—that should be used to infer cause from an association. None is absolutely required, but together they build a substantial case

for establishing causation. First, Hill examined the strength of an association. His example was the high prevalence of scrotal cancer among chimney sweeps. Second, he looked at whether the association was consistently observed by separate investigators in different places, times, and circumstances. Third, he examined whether the association was limited to specific groups, sites, and disease, and whether there were any associations between those variables and another disease. Fourth, he looked at temporality, noting that the cause of an association had to precede the effect. The fifth element was the biological gradient, or the dose-response curve, formed by the data. For example, the greater the number of hours of unprotected sun exposure, the higher the risk of cancer, if there is a causal relation. The sixth criterium was plausibility, whether the theorized cause made biological sense. This determination, of course, depends on the biological knowledge at the time it is made.

The tools of epidemiology are various types of studies and investigations. All studies depend heavily on biostatistics in their design and in testing the strength of the associations that they find. Most epidemiologic studies are observational; they look at diseases in their settings, estimate their prevalence, and point to risk factors. Data are collected and analyzed, outcomes are evaluated, and causal inferences are drawn when the evidence warrants it. The classic study types are case-control studies, cohort studies, cross-sectional studies, meta-analyses, and random controlled trials.

Case-Control Studies. Case-control studies are retrospective in that they look back in time, often involving the review of hospital charts of cases. The strategy compares people with a disease—cases—with a control group free of the disease but otherwise as similar as possible to the people in the cases. The goal is to identify disease risk factors. These studies are inexpensive and useful for studying rare diseases or those that take years to develop. They often rely on participants' memory, however, and bias in selecting cases or controls can mislead.

The first notable case-control study was reported in 1926 by Janet Lane-Claypon, who compared a large group of women with breast cancer with a group of healthy women. She found important differences, such as estrogen exposure, between the two groups.

Cohort Studies. Cohort studies are observational

studies in which subjects, with and without an exposure of interest, are followed forward in time to determine the outcome. Investigators do not control the exposure or the treatment. Participants are chosen, sampled, or classified according to whether they were exposed to a treatment or risk factor.

The strengths of cohort studies lie in their ability to compute incidence rates and relative risks. They are the best way to examine the incidence and natural history of a disease. Confounding factors are a particular problem, however. For example, one study showed an association between ice cream consumption and an increased risk of drowning. In this instance, the confounding factor was proximity to a beach community.

Cross-Sectional Studies. Often conducted through questionnaires or interviews, cross-sectional studies measure the presence or absence of a condition in a representative population at one particular time. The population consists of people of different ages and ethnic and socioeconomic groups, and questions include a number of possible exposures.

The studies cannot determine the timing of an exposure because the disease outcome and potential causes are ascertained at the same time; whether risk factor or outcome came first cannot be confirmed. Market research and political organizations frequently use this research method.

Meta-Analysis. Meta-analysis is a widely used method that employs statistical analysis to synthesize the findings of a number of independent studies. Meta-analyses assume that random variation accounts for differences in results from one study to another. These analyses require both statistical expertise and extensive clinical knowledge of the topic. Weaknesses of meta-analyses are implicit in the process. The included studies may have measured slightly different outcomes, they may have used different research designs, and there may have been selection bias in the criteria used to include a study.

Randomized Controlled Trials. Randomized controlled trials are large clinical trials, considered the gold standard for evaluating the effects of a medical intervention—often a single drug for a specific condition. A group of patients, the experimental group, is randomly assigned to receive the intervention; the control group receives a placebo or alternative treatment. The statistical assumption is that the groups differ only in whether they have received the

intervention. The trial follows patients over time and compares the occurrence of prespecified outcomes in each group.

APPLICATIONS AND PRODUCTS

The application of epidemiology is public health—the control or prevention of disease in populations through some organized community-based or governmental effort. Products are also produced through privately run organizations, albeit under some governmental control. Most of the thousands of vaccines and prescription drugs that are listed on hospital formularies and stocked in American pharmacies owe their existence to epidemiologic research.

Vaccination. Vaccines are the quintessential product of applied epidemiology. Vaccination against smallpox has been one of the great public-health successes of the modern era. The last reported case of smallpox transmitted through human contact occurred in a Somali village in 1977.

Children in the United States are vaccinated against a growing list of diseases: pertussis (whooping cough), diphtheria, tetanus, measles, mumps, rubella, varicella (chicken pox), polio, hepatitis B, and pneumococcal disease. Annual influenza vaccination, depending on the prevalent viral strain, is recommended for most age groups. The H1N1 influenza pandemic of 2009 saw the rapid development of an effective vaccine; however, a vaccine for HIV remains ellusive.

Research is being directed to vaccines in other disease areas. A vaccine has been developed that prevents breast cancer in genetically predisposed mice, but much research must be done before a vaccine is ready for women. Because some strains of *Staphylococcus aureus* are resistant to methicillin, research may begin on a vaccine against this staph-causing bacterium. A study of the economic benefit of an *S. aureus* vaccine for newborns found it to be cost-effective.

The Framingham Heart Study. No better illustration exists of an epidemiologic study that has profoundly affected public health and generated a plethora of products in the process than the Framingham Heart Study, which is still operating..

The *Physician's Desk Reference* (PDR), which lies

Fascinating Facts About Epidemiology

- In the 1300's, the plague, or the Black Death, spread from the Middle East to Europe, where it killed nearly 30 million people, one-third of the population. Caused by the bacterium *Yersinia pestis*, the plague typically spreads from rodents or fleas to humans.

- The smallpox virus brought to the Americas by European settlers had a devastating effect on the Native American population, who had not developed any immunity to the disease. The mortality rate among Europeans was around 30 percent compared with about 50 percent for the Cherokee and the Iroquois, 66 percent for the Omaha and the Blackfeet, 90 percent for the Mandan, and 100 percent for the Taino Indians.

- In the 1918-1919 influenza pandemic, about 500 million people worldwide (one-third of the world's population) were infected, and more than 50 million people died.

- A large sample of California men and women whose scores on a standard test indicated major depression ate more than twice as much chocolate as adults whose scores were in the normal range. The higher the score, the deeper the depression and the greater the amount of chocolate consumed. Researchers did not speculate whether self-medication or chocolate-induced depression were factors.

- A unique form of social withdrawal termed *hikikomori*, known in Japan since the 1970's, has been described in English medical literature. Affected adolescent and young adult males, numbering in the thousands, withdraw into their parental homes for months or years, have no social contact, and neither attend school nor hold a job. No basis exists for any specific treatment.

- A cross-sectional study of more than seven hundred men and women, with the average age of sixty, found that visceral (abdominal) fat was associated with lower brain volume (as measured by brain scans). Generalized body fat had a weaker association with brain volume, and brain size was unrelated to any cardiovascular risk factors.

---

within easy reach of any physician who prescribes drugs, lists eighteen different classes of cardiovascular medication, each with a different mechanism of action or combination of mechanisms. An additional twelve classes of antihypertensive medication have a separate heading. Although no well-marked route exists between these drugs and the Framingham study, much of what is known about cardiovascular disease can be traced to it, and few epidemiology textbooks fail to mention the study.

Initiated in 1949 in Framingham, Massachusetts, a small town 20 miles from Boston, the study served as a model for all the cohort studies that came after it. The study enlisted and followed more than 5,000 men and women between the ages of thirty and sixty-two who were examined and found free of coronary heart disease. The study served as a source of insight into the development of cardiovascular disease when infectious disease was the major epidemiologic concern.

Cardiovascular disease was a puzzle. The causes were unclear, the onset was not clinically observable, and the outcome could be suddenly lethal. Hypertension was considered a benign condition that was safely ignored. Gradually, serum cholesterol, high blood pressure, and cigarette smoking began to emerge as important associations with the development of heart disease. The very concept of "risk factor" comes from the Framingham study.

From the earliest reports, an elevated serum-cholesterol level differentiated those who developed heart disease from those who remained free of it. As a result, millions of people take drugs that lower cholesterol; at least one drug has been approved by the U.S. Food and Drug Administration to be administered to people with normal cholesterol levels. Research on new lipid-lowering agents is also under way.

The Diabetes Epidemic. Whether the incidence of type 2 diabetes is referred to as an epidemic or pandemic, it is increasing at an alarming rate. Among people who are at least twenty years old, the prevalence of diagnosed and undiagnosed cases has been estimated at 12.9 percent and rising.

This epidemic has spurred drug development. Some thirty brands of antidiabetic agents and insulin formulations are listed in the *Physician's Desk Reference*. Pharmaceutical companies are working to develop drugs that can compensate for loss of beta-cell function (insulin-secreting cells). Diabetes is expected to remain a major public-health concern for the foreseeable future.

IMPACT ON INDUSTRY

Wedded as it is to public health, epidemiology is largely a government enterprise, and consequently, its effects on industry are largely government mediated. For example, in the twentieth century, the research that conclusively linked cigarette smoking and lung cancer prompted increased government regulation of cigarette advertising.

The Food Industry. As the epidemiology of food-borne illness evolves, regulatory concern is increasingly directed at industries involved in food production. The traditional villains, undercooked meat and unpasteurized milk, have given way to contaminated eggs, fruits, and vegetables. New food-borne pathogens have emerged with new transmission routes, some international. From the farm to the table, keeping food safe requires industry to collaborate with regulatory agencies.

Better control of disease-causing *Escherichia coli* reservoirs in cattle, for example, might include improving the hygienic conditions in slaughterhouses, immunizing cattle against *E. coli* infection, and altering feed to make cattle more resistant to *E. coli* residence. Extensive use of antibiotics to promote animal growth is implicated in the prevalence of antibiotic-resistant pathogenic bacteria. Changing this practice would doubtless require government intervention.

Another pressure on the food industry is the epidemic of obesity. Epidemiologic studies have quantified a relationship between consumption of sugar-sweetened beverages (carbonated soft drinks, sports drinks, sweetened tea, and the like), long-term weight gain, and diabetes. Calls for changes in processed foods are increasing. The American Heart Association, the American Dietetic Association, and other associations have recommended limiting consumption of trans fats.

U.S. Governmental Agencies. The U.S. government contains many bureaus, institutes, and agencies that deal with epidemiology. A number of individual states also have departments or programs in this area. All are concerned with aspects of public health, and their activities and research findings filter down to affect a variety of industries.

The Centers for Disease Control and Prevention (CDC), founded in 1946, and part of the U.S. Department of Health and Human Services, is concerned with infectious and chronic diseases, injuries, and disabilities in and out of the workplace, as well as environmental health hazards. The CDC's Foodborne Diseases Active Surveillance Network (FoodNet) collaborates with industry, state health departments, and federal food regulatory agencies to monitor outbreaks of food-borne illness.

The National Institute of Environmental Health Sciences is one of twenty-seven institutes within the National Institutes of Health. The Epidemiology Branch of this institute applies epidemiologic methods to the study of environmental effects on human health. Its research into the long-term effects of lead exposure resulted in the removal of lead from paint and gasoline.

The Division of Cancer Epidemiology and Genetics, a branch of the National Cancer Institute, conducts research into the genetic and environmental determinants of cancer. Its projects include investigating municipal and privately supplied drinking water. Exposure to nitrates, arsenic, and chlorination by-products have been linked to several cancers, and studies are under way in a number of regions within the United States.

Worldwide Agencies. The primary international agency involved in epidemiology is the World Health Organization (WHO), which benefits more than 190 United Nations member states. The agency tracks global health trends, organizes and coordinates research efforts, provides technical support, and disseminates information on disease outbreaks. The WHO is geared to respond to pandemics and was actively involved in the 2009 H1N1 influenza pandemic. Under the guidance of the WHO, member states devised plans and stockpiled antiviral drugs.

Severe acute respiratory syndrome (SARS) holds the distinction of being the twenty-first century's first global epidemic of a new, life-threatening, and easily transmissible disease. The first appearance of the disease was in Guangdong Province, in southern China, in November, 2002. Because of the disease's rapid spread and severity, the WHO convened a global epidemiology conference just six months after its outbreak to gather together what was then known about SARS and to identify what still needed to be determined. Throughout the epidemic, the WHO coordinated a network of laboratories that shared information on developing a rapid diagnostic test and determining likely transmission routes. They determined that the virus has an animal origin, which leaves open the potential for another outbreak. Animals within the Guangdong Province still harbor the precursor virus, and animal markets in southern China still provide a venue for the initial animal-human contact, followed by human-human transmission. The WHO continues to monitor and guide efforts to develop a SARS vaccine.

CAREERS AND COURSE WORK

Epidemiology accommodates many interests. A college degree is a minimal requirement to work in the field, and some background in biostatistics is helpful. More than seventy colleges and universities across the United States have departments or entire schools devoted to epidemiology or public health. The Bloomberg School of Public Health at The Johns Hopkins University in Baltimore, Maryland, consists of ten departments, one of which is Epidemiology. Others are Biostatistics, Health Policy and Management, and Molecular Microbiology and Immunology.

Virtually all the government agencies concerned with epidemiology and public health (including the WHO) list job opportunities and training programs on their Web sites. For example, in addition to epidemiologists, the CDC employs a variety of specialists, including microbiologists, behavioral scientists, and public health advisors. Other positions, described as "mission support," resemble those found in any large corporation—accountants, public relations personnel, budget administrators, and Web developers.

Positions at the CDC and at other government agencies structure their requirements according to civil-service grades. Starting at the lowest grade, the minimum requirement for a position (for example, as a behavioral scientist) would be a bachelor's or higher degree in a related discipline from an accredited college or university, an equivalent combination of education and experience, or four years of appropriate experience. As the civil-service grade goes up, so do the requirements: education at the graduate level and beyond, specialized experience, and specific skills.

SOCIAL CONTEXT AND FUTURE PROSPECTS

The future focus of epidemiology is a frequent topic in epidemiology journals. Critics of purely academic epidemiology claim that it ignores the influence of society on population health and undervalues the part that community action can play in promoting health. This view stresses science's public-health mission, and it considers health and disease in the context of the entire human environment: society, economics, and ecology. Social epidemiology covers a range of topics that blend epidemiology and sociology.

Molecular/Genetic Epidemiology. The term "molecular epidemiology" was first used in the early 1980's. Rapidly emerging technology enabled molecular biomarkers to be added to traditionally collected data, first in the context of cancer research. The strategy was to identify molecular events and modify risk in vulnerable populations before cancer became a clinical entity, thus potentially preventing cancer.

The Human Genome Project extended the reach of research still further, with genome-wide association (GWA) studies. These studies examine common genetic variants that occur across the human genome to identify associations with observable disease traits (phenotypes).

Genome-wide association studies often take the form of molecular case-control studies, which compare differences in alleles between cases with the disease and controls. This genome-wide strategy has identified four genetic loci with variants that increase susceptibility to type 2 diabetes.

Epigenetics. The next frontier in epidemiology is likely to be epigenetics; the term "epigenetic epidemiology" has already appeared in the scientific literature. Epigenetics refers to changes in gene function or expression that do not change DNA sequences and that are heritable because the changes remain intact through cell division. For epidemiologists and other scientists, epigenetics provides another research pathway to environmental mechanisms that affect gene expression and cause disease.

Epigenetics has particularly excited the interest of cancer researchers, who see epigenetic mechanisms as contributing to cancer development. In contrast to genetic changes, epigenetic changes can potentially be reversed. Anticancer agents that counter epigenetic effects are already being tested in clinical trials.

Judith Weinblatt, M.A., M.S.

FURTHER READING

Byrne, Joseph P., ed. *Encyclopedia of Pestilence, Pandemics, and Plagues.* 2 vols. Westport, Conn.: Greenwood Press, 2008. An alphabetized treasure trove of information about anything and anyone connected with epidemiology, public health, and epidemic disease. Each entry has references.

Cockerham, William C. *Social Causes of Health and Disease.* Malden, Mass.: Polity Press, 2008. Highlights the confluence of epidemiology, public health, and medical sociology. Lifestyle, class, and ethnicity are considered in the context of health.

Desowitz, Robert S. *Who Gave Pinta to the Santa Maria? Torrid Diseases in a Temperate World.* San Diego, Calif.: Harcourt Brace, 1998. A renowned epidemiologist and lively writer tracks the spread of lethal tropical diseases into the Americas.

Dickerson, James L. *Yellow Fever: A Deadly Disease Poised to Kill Again.* Amherst, N.Y.: Prometheus Books, 2006. Focuses on yellow fever, which had a surprisingly virulent history in the early United States.

Gerstman, Burt B. *Epidemiology Kept Simple: An Introduction to Traditional and Modern Epidemiology.* 2d ed. Hoboken, N.J.: Wiley-Liss, 2003. Takes a comprehensive approach toward epidemiology and makes good use of concrete examples.

Gordis, Leon. *Epidemiology.* 4th ed. Philadelphia: Saunders Elsevier, 2009. Presents basic principles and important concepts with generous use of illustrations and examples.

Lock, Margaret, and Vinh-Kim Nguyen. *An Anthropology of Biomedicine.* Oxford, England: Wiley-Blackwell, 2010. Examines the assumption that human bodies are the same everywhere and presents the ways in which culture, history, politics, and environmental biology interact to change human biology.

WEB SITES

American Epidemiology Society
http://www.acepidemiology.org/societies/AES.shtml

Centers for Disease Control and Prevention
Epidemiology Program Office
http://www.cdc.gov/epo

National Cancer Institute
Division of Cancer Epidemiology and Genetics
http://dceg.cancer.gov

National Institute of Environmental Health Sciences
Epidemiology Branch
http://www.niehs.nih.gov/research/atniehs/labs/epi/index.cfm

Society for Epidemiologic Research
http://www.epiresearch.org

World Health Organization
Epidemiology
http://www.who.int/topics/epidemiology/en

See also: Immunology and Vaccination; Parasitology; Pathology.

ERGONOMICS

FIELDS OF STUDY

Biomechanics; rehabilitation engineering; sports engineering; prosthetics; mechanical engineering; industrial engineering; psychology; physiology; anthropometry; industrial design; kinesiology; engineering psychology; industrial hygiene; human factors engineering; physiology; psychology; environmental science; computer science.

SUMMARY

According to the International Ergonomics Association, "Ergonomics draws on the physical and life sciences, applying data on human physical and psychological characteristics to the design of machines, devices, systems, and environments as a means of improving the practicality, efficiency, and safety of human-machine relationships."

KEY TERMS AND CONCEPTS

- **Anthropometrics:** Statistical information about human body dimensions and product design applied to ergonomics.
- **Biomechanics:** Study of human body structure and function and the forces that affect the human body.
- **Cognitive Ergonomics:** Mental processes of humans such as memory, motor response, and reasoning as related to interactions with elements of the system such as mental workload, performance, and reliability.
- **Engineering Psychology:** Application of psychological factors related to design and use of equipment.
- **Human Factors Engineering:** Application of human characteristics, such as arm length, to design equipment or products to fit the human while enhancing work efficiency and limiting stress.
- **Human Performance Engineering:** Approach to solving problems by individuals or teams that analyzes performance gaps, defines interventions, and evaluates outcomes.
- **Job Risk Factors:** Conditions and job demands in the workplace that pose a risk to the worker.

- **Macroergonomics:** Broad system review of human worker and technology interface with a goal of optimal worker productivity, job satisfaction and commitment, and health and safety.
- **Occupational Health and Safety:** Programs, including ergonomics, that support and foster health and safety in the workplace.
- **Organizational Ergonomics:** Optimization of human-technology systems and organizational policies and procedures to include teamwork, cooperative participation, and community ergonomics.
- **Participatory Ergonomics:** Active participation of workers in the development and evaluation of a work-related ergonomics programs.
- **Physical Ergonomics:** Physiological, anatomical, and biomechanical characteristics of the human body related to physical activity.
- **Work-Related Musculoskeletal Disorders:** Physical conditions that result from work-related stress or injury.

DEFINITION AND BASIC PRINCIPLES

Ergonomics, a holistic multidisciplinary science, draws from the fields of engineering, psychology, physiology, and computer and environmental sciences. These sciences define ergonomics as optimizing effectiveness of human activities while improving the quality of life with safety, comfort, and reduced fatigue.

The term ergonomics dates to the mid-1800's, but credit for applying the term generally goes to Hywel Murrell, a British chemist. Ergonomics derives from the Greek word *ergon*, meaning "work," and *nomos*, meaning "law." Ergonomics studies work within the natural laws of the human body.

The International Ergonomics Association promotes a systematic approach to the ergonomic process, to incorporate human factors and human performance engineering and address problems in design of machines, environments, or systems. This can improve efficiency and safety of the human-machine relationship. The basic steps in the ergonomic process include organization of the process, identification and analyzation the problem, development of a solution, implementation of the solution, and evaluation of the result.

BACKGROUND AND HISTORY

The types of work and settings have changed over the centuries. Humans have consistently been aware of the need for a good fit between work tools and the human body. While he was a medical student at Parma University in Italy, Bernardino Ramazzini recognized that workers suffered certain diseases. In 1682, he focused on worker health concerns. His scholarly collection of observations, *De Mortis Artificum Diatriba* (*Diseases of Workers*), published in 1700, detailed conditions associated with specific work environments and factors such as prolonged body postures and repetitive motion. His work earned him the title "Father of Occupational Medicine."

The term "ergonomics" is attributed to Hywel Murrell, a chemist who worked with the British Army Operational Research Group during World War II. In 1949, he served as leader of the Naval Motion Study Unit. He invited people with like interests in human factors research to meet with him, forming the Ergonomics Research Society. He remained active in academia until his death in 1989 at age seventy-six. Murrell's specialties included skill development and use and fatigue and aging. He was interested in applications of psychology and ergonomics in day-to-day situations. Murrell authored the first textbook on ergonomics, *Ergonomics: Man in His Working Environment.*

In the industrial era, tools and machines were developed to increase productivity. These put a new strain on the relationship between work and the human body. Between World War I and II, classic work was accomplished by the British Industrial Fatigue Research Board on the impact of environmental factors and human work performance. By the time World War II had begun, worker safety became a primary concern, leading the way for the science of ergonomics.

HOW IT WORKS

Ergonomics, the science of adapting the workplace environment to the work and workers, seeks to maintain worker safety. The goal of industry employers is to keep workers well and comfortable while functioning efficiently on the job. This can be best accomplished by providing safe working conditions to prevent work-related injuries.

National Institute for Occupational Safety and Health's Seven-Step Approach. Businesses can assume a reactive or proactive approach. The National Institute for Occupational Safety and Health (NIOSH) has defined a seven-step program for evaluating and addressing potential musculoskeletal problems in the workplace. First is finding worker complaints of pains or aches and defining jobs that require repetitive movement or forceful exertion. Management must then commit to addressing the problem with input from the worker. Participatory ergonomics is important in encouraging workers to help define problems and solutions to work-related stress. Key is education and training about the potential work-related risks and musculoskeletal problems from defined jobs. Using attendance, illness, and medical records, management should investigate high-risk jobs, where injury is most common. Leadership must analyze job descriptions and functions to see if risky work-related tasks can be eliminated. Management should support health care intervention that emphasizes early detection and treatment to avoid work-related impairment and disability. Finally, management should use this information to minimize work-related musculoskeletal risks when creating new jobs, policies, and procedures.

Physical Ergonomics. Ergonomics can be subdivided into several disciplines: physical, cognitive, and organizational. Physical ergonomics is the body's response to physical workloads. It addresses physiological and anatomical characteristics of humans as related to physical activity. Biomechanics and anthropometrics fall into this category. This discipline is concerned with safety and health and encompasses work postures, repetitive movements, vibration, materials handling, posture, workspace layout, and work-related musculoskeletal disorders. Common injuries in an office setting result from computer use (keyboard, mouse, and viewing the monitor).

Cognitive Ergonomics. Cognitive ergonomics deals with human mental processes and capabilities at work, such as reasoning, perception, and memory, as well as motor response. Topics related to cognitive ergonomics include work stress and mental workload, decision making, performance, and reliability. Computer-human interaction and human training are sometimes listed here.

Organizational Ergonomics. Organizational ergonomics addresses sociotechnical systems of the organization and its policies, procedures, processes, and structures. Concepts in this subdiscipline could include work design and hours, job satisfaction, time management, telecommuting, ethics, and motivation, as well as teamwork, cooperation, participation, and communication.

APPLICATIONS AND PRODUCTS

Ergonomics can be applied to work in any setting with the goal of achieving efficiency and effectiveness while maintaining worker comfort and safety. The principles of ergonomics have been applied to many industries, including aerospace, health care, communications, geriatrics, transportation, product design, and information technology.

Office Workers. Global industry requires office workers to use computers every day. Product orders are taken by workers via phone and the Internet. Office workers spend time tied to phones and computers, while sitting in one place. These workers are subject to work-related injury and stress created by continuous computer and phone use.

Ergonomic experts have taken the principles of human factors engineering to improve the work environment for computer users. The placement and maintenance of the computer monitor will affect the user's eyes and musculoskeletal system. The monitor should be clean with brightness and contrast set for the comfort of the user. Placing the monitor directly in front of the user will minimize neck strain. The monitor should be set one arm's length away, tilted back by 10 to 20 degrees, and positioned away from windows or direct lighting to reduce glare.

Office workers often sit for extended periods while working, which is stressful on legs, feet, and the intervertebral discs of the spine. Pooling of the blood in the feet and ankles can cause swelling and place stress on the heart. Employers should encourage workers to alternate between standing and sitting. Ergonomic chairs are designed to relieve the pressure placed on the back while sitting for extended periods. Arm rests should be adjusted so arms rest at the side of the body, allowing the shoulder to drop to a natural, relaxed position.

Many ergonomic ailments occur in the soft tissues of the wrist and forearm, as continuous computer use subjects workers to repetitive motions and sometimes awkward positioning. Computer mouses are ergonomically designed to minimize worker injury, and the no-hands mouse uses foot pedals to navigate. Ergonomically friendly computer keyboards are also

---

Fascinating Facts About Ergonomics

- October is National Ergonomics Month.
- Ergonomics is sometimes called human engineering.
- In 2007, the National Institute for Occupational Safety and Health (NIOSH) stated that 32 percent of the workers' compensation claims in the construction industry could have been prevented by proper ergonomic procedures.
- Repetitive strain injuries cost U.S. businesses more than $1 billion annually.
- Five common aspects of ergonomics include safety, comfort, productivity or performance, ease of use, and aesthetics.
- Some ergonomically designed keyboards are available with a rest-time indicator, which encourages the user to take a break from the computer.
- Although laptop computers are portable and convenient, work-related musculoskeletal injuries can still occur if one does not observe proper ergonomic principles while using a laptop.
- Cumulative trauma disorders, caused by repetitive strain or motion injuries and work-related musculoskeletal disorders, are the largest cause of occupational disease in the United States.

---

and the no-hands mouse uses foot pedals to navigate. Ergonomically fiendly computer keyboards are also available.

Health Care. In the health care field, ergonomics is useful in designing products for conditions such as arthrtis, carpal tunnel syndrome, and chronic pain. Ergonomic applications for persons with arthritis, some 46 million adults in the United States, include appliances with larger dials that can be grasped more easily, levers rather than door knobs, and cars with keyless entry and ignition. Larger controls on the dashboard and thicker steering wheels can be more easily grasped.

Health care workers are at risk for work-related musculoskeletal injuries, such as back or muscle strain, without adequate ergonomics. This is true in nursing homes where nursing assistants must lift patients with impaired mobility. These workers can benefit from ergonomically designed patient-handling equipment and devices such as belts and portable hoists to lift patients.

Dentists are at risk for work-related musculoskeletal disorders. They experience repetitive hand movements, vibrating tools, and fixed and awkward posturing. Neck, back, hand, and wrist injuries are common. Ergonomic equipment is available for dentists, including specially designed hand instruments, syringes and dispensers, lighting, magnification tools, and patient chairs.

Transportation. Ergonomics has applications useful to anthropometry. A 2001 study in the United Kingdom found the airline industry did not provide adequate space for passengers in the economy-class sections. The study focused on seating standards and the passengers' ability to make a safe emergency exit. They found that the economy-class seats did not have enough space to brace for an emergency landing and that even the seats themselves could delay a safe exit. The study also stated that the existing seating would accommodate only up to the 77 percentile of European travelers based on the buttock-knee length dimension.

Other applications include ergonomic food carts and passenger delivery, crew rest seats, and ergonomic design for first-class, business, and economy passenger seats. Cockpit design is important for pilot comfort and safety and to minimize fatigue.

Many competitive manufacturers in the automotive industry have employed ergonomics in designing cars for comfort, safety, and efficiency. Examples include options for driver seat position to accommodate variation in body size and allow the steering wheel and backrest to be ergonomically positioned. Also noted are passenger-seat comfort and safety, placement of controls, and an option for cell phone placement. These considerations can lessen work stress considerably when someone drives as part of his or her job.

Communication Technology. Cell phones have been a plus for industry and individuals but can come at a price. Shoulder and neck pain may be related to cell phone use. Many people will cradle the phone between their head and shoulder when talking. Ergonomic solutions exist to decrease user strain and pain. Headsets keep hands and head free of awkward posturing. Frequently changing sides can help if a headset is not available. Keying into the phone's address book one's most-dialed numbers can decrease repetitive movement of the fingers. Using cell phone technology with ergonomic design can reduce the daily and cumulative stress of cell phone use.

Aging. With an increasing aging population, industries are applying ergonomic solutions to meet senior needs. Gerontechnology addresses the need for work and leisure, comfort and safety in older adults. Some automobiles have larger and simple dashboard controls. Many tools for use in kitchen or garden have special adaptive handles for less strain on the hands and muscles. Phones are equipped with different levels of tone for varying hearing issues; digits are larger and easier to push. Bathrooms are equipped with safety handles and equipment that allow independence.

Many seniors still hold jobs and make good employees. Employers need ergonomically designed workplaces to accommodate the physical and cognitive changes of normal aging. Seats may need to be firmer and higher to allow for decreased joint flexibility. Good lighting is important for safe work. Restrooms with modifications may be necessary. Flexible-schedule availability will assist with worker fatigue. By redesigning the work environment for aging workers, the risk of illness and injury can be diminished and performance improved.

IMPACT ON INDUSTRY

Worker safety and health is critical for employers in all industries. The science of ergonomics impacts the quality of life and job satisfaction for workers while affecting the financial success of the employer. Many organizations have dedicated programs and research to support this growing field.

Government and University Research. National Institute for Occupational Safety and Health (NIOSH) offers copies of ergonomic-related studies in downloadable files. This government-agency Web site contains links to many published research articles on various aspects of ergonomics. The landmark 1997 study about musculoskeletal disorders and workplace factors is one that is available to any business or individual. Others include the research of violence in the workplace and musculoskeletal pain with regard to nursing-home workers published in the journal *Occupational and Environmental Medicine*. This was the first study of the hazard of workplace violence as linked to work-related musculoskeletal disorders. The results showed that the incidence of musculoskeletal pain increased from 40 percent in those workers that were not assaulted to 70 percent in victims of workplace violence.

T. H. Tveito of the Harvard School of Public Health published findings on workers and low-back pain in the November 2010 issue of *Disability and Rehabilitation*. This study indicated that employers who wanted to retain workers with low-back pain needed to focus on worker communication, pacing work, and options for altered job routines.

Cornell University has a unique Web site called Cornell University Ergonomic Web, where visitors can read research findings from students and professors who participate in the Cornell Human Factors and Ergonomics Research Group. Topics range from computer workstation guidelines to hospital ergonomics.

Industry and Business. Most industries and businesses address worker safety as required by law. However, the Material Handling Industry of America (MHIA) went further by publishing an article entitled "Ergonomics = Good Business Practice." This work describes many reasons why industries should be proactive in addressing ergonomic concerns to improve worker safety and morale. The author provides startling statistics, such as the fact more than 5,000 American workers died on the job in 2008. The U.S. Bureau of Labor Statistics recorded about 3.7 million cases of worker illness and injury that same year, mostly strains and sprains. The article promotes ergonomics as the answer to many of these industry-specific worker events, including cranes, ergonomic shelving, lift tables, workbenches, and adjustable platforms to handle materials safely without unnecessary bending or twisting.

The MHIA has a subgroup of its industry dedicated to the study and implementation of ergonomics. The Ergonomic Assist Systems and Equipment (EASE) reviews standards and serves as a resource to other industry groups.

Major Corporations. In 1991, the Intel Corporation, known for computers and computer chips, recognized that a significant amount of its workers' injuries and illnesses was related to ergonomics. They hired ergonomic professionals to provide leadership for a interdisciplinary team of occupational health professionals, ergonomists, and management. They worked at all levels to provide a companywide training program and employed twenty-eight full-time ergonomists globally. Much of the training is Web based, and the company credits its success to proactive leadership and dedicated workers who practice ergonomic safety. The lost-day case rate has dropped 95 percent.

Intel has continued its dedication to worker safety through ergonomics. In 1996, the company won the Outstanding Achievement Award from the Institute of Industrial Engineers for their work in ergonomics and productivity. Three years later they were given the Outstanding Office Ergonomics Award from the Center for Office Technology, and in 2001 Intel received the National Safety Council's Green Cross for Safety.

CAREERS AND COURSE WORK

The health and safety of workers continues to be a primary concern for employers. The job opportunities for persons interested in ergonomics are varied and depend on the role desired. As with other professions, some jobs require formal education, while others provide on-the-job training. Other jobs require certification or special training in the area of interest.

Jobs related to ergonomics include the role of ergonomist, who has special knowledge and skills in the science of ergonomics, designing the workplace to fit the worker. Ergonomists typically have the minimum of a bachelor's degree in industrial or mechanical engineering, industrial design, psychology, or health care sciences and often a master's or doctoral degree in a related area such as human factors engineering. The International Ergonomics Association (IEA) encourages all ergonomists to become board certified.

Health care professionals such as occupational therapists have also become interested in the growing field of ergonomics.

SOCIAL CONTEXT AND FUTURE PROSPECTS

The IEA attests to the fact that ergonomics is an international concern that affects the global economy. IEA is composed of forty-two organizations worldwide run by a council with representatives from these groups. IEA supports ergonomic efforts in developing countries and keeps a directory of educational programs in some forty-five countries. IEA produced the standard guidelines for industry ergonomics and established a certification program called Ergonomics Quality in Design (EQUID).

The goals of IEA include advancing ergonomics to the international level and enhancing the contribution of the discipline of ergonomics in a global society.

Marylane Wade Koch, M.S.N., R.N.

FURTHER READING

Bhattacharya, Amit, and James D. McGlothin, *Occupational Ergonomics: Theory and Applications.* New York: Marcel Dekker, 1996. Provides a comprehensive look at basic ergonomic principles, including physiology of body movement; practical applications for the workplace; medical implications; and case studies in various industries.

Dul, Jan, and Bernard Weerdmeester. *Ergonomics for Beginners: A Quick Reference Guide.* 3d ed. Boca Raton, Fla.: CRC Press, 2008. A reference for basic ergonomic principles with updated applications for the growing communications-technology workplace, International Organization for Standardization (ISO) ergonomics standards, and human-centered workplace design.

Eastman Kodak Company. *Kodak's Ergonomic Design for People at Work.* 2d ed. Hoboken, N.J.: John Wiley & Sons, 2003. Written for people who may not be trained in ergonomics but want to reduce the incidence of workplace injury through understanding basic principles and reviewing ISO standards; illustrations.

Marras, William S., and Waldemar Karwowski, eds. *Interventions, Controls, and Applications in Occupational Ergonomics.* 2d ed. Boca Raton, Fla.: CRC Press, 2006. A complete resource book to help the reader understand every aspect of ergonomics from the basic ergonomic processes to the future of ergonomics and human work.

National Institute for Occupational Safety and Health. *Elements of Ergonomics Programs.* Cincinnati, Ohio: Alexander L. Cohen, et al. Available at http://www.cdc.gov/niosh/docs/97-117. Looks at what components are necessary for ergonomics programs.

Stanton, Nelville A., et al., eds. *Human Factor Methods: A Practical Guide for Engineering and Design.* Burlington, Vt.: Ashgate Publishing, 2005. Designed to serve as an ergonomics how-to manual for students and practitioners, with examples, flowcharts, and case studies.

WEB SITES

Cornell University Ergonomics Web
http://ergo.human.cornell.edu

Human Factors and Ergonomics Society
http://www.hfes.org/web/Default.aspx

International Ergonomics Association
http://www.iea.cc

National Institute for Occupational Safety and Health
Ergonomics and Musculoskeletal Disorders
http://www.cdc.gov/niosh/topics/ergonomics

Office of Research Services: Division of Occupational Health and Safety
Ergonomics for Computer Workstations
http://dohs.ors.od.nih.gov/ergo_computers.htm

See also: Biomechanics; Kinesiology.

G

GASTROENTEROLOGY

FIELDS OF STUDY

Biology; chemistry; mathematics; physiology; health; nutrition; anatomy; nephrology; neurology; hematology; internal medicine; endocrinology and metabolism; oncology; nuclear medicine; infectious disease; geriatric medicine; pediatric medicine.

SUMMARY

The medical specialty known as gastroenterology focuses on conditions and diseases involving the digestive tract and related organs. The esophagus, stomach, duodenum, and large and small intestines are part of the gastrointestinal system. Related organs include the liver, gallbladder, and pancreas. Conditions addressed by the gastroenterologist include gastric ulcers, heartburn, abdominal discomfort, nausea and vomiting, constipation, diarrhea, inflammatory bowel diseases, hepatitis, nutritional deficiencies, and gastric cancers. Endoscopy, a diagnostic tool that allows visualization of the digestive tract, is an integral component of patient care. Gastroenterologists must be skilled in all aspects of the digestive process, and they must be caring, compassionate, and excellent team leaders.

KEY TERMS AND CONCEPTS

- **Colon:** Large intestine.
- **Colonoscopy:** Procedure using a thin tube to project images of the interior of the large intestine and rectum to a monitor.
- **Endoscopy:** Procedures using a thin, flexible fiber-optic tube with a camera to project images of the gastrointestinal tract back to a monitor.
- **Gastrointestinal Tract:** Organs associated with the digestion and elimination of food, including the esophagus, stomach, duodenum, small intestine, large intestine (colon), rectum, and anus.
- **Hepatitis:** Inflammation of the liver caused by several different types of viruses.
- **Motility:** Movement of food through the intestines.
- **Physiology:** Study of the function of organs.
- **Sigmoidoscopy:** Procedure to inspect the lining of the last third of the colon, or descending colon.

DEFINITION AND BASIC PRINCIPLES

Gastroenterology is a medical specialty focused on the diagnosis and treatment of conditions involving the digestive tract and associated organs. Gastroenterologists have a thorough understanding of the normal function of the gastrointestinal tract, which encompasses the esophagus, stomach, small and large intestines, and the rectum. This specialty also includes other organs associated with the gastrointestinal tract, such as the pancreas, gallbladder, bile ducts, and liver. Training includes knowledge of the normal movement of food through the digestive tract (motility), the absorption of nutrients as foods move through the intestines, and the removal of waste products. A detailed knowledge of normal digestive processes allows gastroenterologists to evaluate abnormal conditions and plan a course of treatment.

Gastroenterological training stresses diagnostic procedures, such as endoscopies, combined with thorough patient interviews regarding signs and symptoms. Colorectal cancer screenings are an important part of the practice. Gastroenterologists may specialize further and narrow their practice to children or the elderly, or they may become board certified in surgery. Not all gastroenterologists are surgeons, but most gastroenterologists perform minimally invasive procedures, such as removal of colon polyps. Gastroenterologists specializing in bariatric surgery generally perform procedures such as gastric bypass or lapband surgery for weight loss.

BACKGROUND AND HISTORY

Gastroenterology is a relatively new medical specialty. Until the mid-1980's, gastroenterology was considered a subspecialty of internal medicine and required only an additional one or two years of training. The introduction of endoscopy, which uses a flexible fiberoptic tube to explore the digestive tract, transformed gastroenterology and allowed the field to emerge as a separate subspecialty. The addition of hepatology, or the study of liver conditions, to gastroenterology and the rise in hepatitis further defined the field. Although gastroenterologists are first trained as internists, several more years of training in gastroenterology are required. Gastroenterologists may specialize further in gastric conditions particular to children.

The field of gastroenterology has developed unevenly across developed and developing countries. Standardized training is lacking, with nations requiring different curricula. Whereas gastroenterology has emerged as a separate subspecialty in North America, the field is still considered part of internal medicine in Europe. Fellowships are required, but they last only one or two years and consist of on-the-job training. Some developing countries have no formal training guidelines or firm requirements regarding entrance and exit examinations.

How It Works

Patient Care. First and foremost, gastroenterologists are concerned with patient care. Most gastrointestinal conditions are chronic, allowing gastroenterologists to develop long-term relationships with their patients. Detailed medical histories are often the key to diagnosis, so gastroenterologists must be compassionate and able to listen closely to the patient's complaints.

The physical examination is an important aspect of diagnosis and treatment. Procedures such as colonoscopies and sigmoidoscopies are uncomfortable and may be embarrassing for patients, so the gastroenterologist must be skilled in relaxing patients and providing appropriate sedation.

Setting. Gastroenterologists may work in private practice or hospital settings. They also may work in nursing homes or long-term care facilities, hospices, or outpatient surgery centers.

Conditions Treated. Although they focus on the gastrointestinal system, gastroenterologists diagnose and treat a wide range of symptoms and conditions. Gastric and peptic ulcers and gastroesophageal reflux disease (GERD) are diagnosed and treated by gastroenterologists, as are gallstones. Gastroenterologists

are called on to diagnose and manage abdominal pain and discomfort, hemorrhoids and bloody stools, constipation, diarrhea, nausea, and vomiting. They treat intestinal conditions, such as inflammatory bowel disease, ulcerative colitis, Crohn's disease, and diverticulitis. Gastroenterologists screen for colon and rectal cancer and remove rectal polyps. Gastroenterologists detect reasons for unexplained weight loss and poor absorption of nutrients. They also treat liver diseases such as hepatitis and jaundice.

Management Duties. Regardless of the setting, gastroenterologists are expected to handle a number of management duties in addition to their clinical caseload. The gastroenterologist in private practice manages staff and ensures that the office is run efficiently. All gastroenterologists consult with patients and other medical professionals, who offer their medical expertise. Experienced gastroenterologists train students.

Imaging Techniques. Imaging techniques are the cornerstone of the field. Endoscopy is the use of a thin, flexible, lighted tube that sends real-time video to a monitor. For upper endoscopy, also called an upper GI, the patient is sedated and the endoscopy tube is fed down through the throat into the stomach and duodenum. The gastroenterologist examines the lining of the esophagus and stomach for ulcers or other abnormalities. Tissue samples may be taken for biopsy, or the gastroenterologist can remove a foreign object that has been swallowed.

Colonoscopy, or a lower GI, uses endoscopy to examine the colon and rectum for abnormalities. Patients must prepare for the procedure by cleansing their bowels of waste materials, usually by drinking water mixed with a substance that results in rapid elimination of the contents of the bowels. Preparation is usually done at home. For the colonoscopy itself, the patient is sedated. The imaging instrument, called a colonoscope, is inserted into the rectum through the anus. As the gastroenterologist guides the tube up through the intestine, carbon dioxide gas is used to inflate the colon to allow for better imaging. As in an upper endoscopy, the gastroenterologist examines the intestinal wall for ulcers, polyps, or inflammation. If polyps are present, the gastroenterologist removes the growths by inserting miniature cutting tools through the colonoscope.

Applications and Products

Endoscopy. Endoscopy has become the foundation

of gastroenterology. Endoscopy is an imaging technique that uses a thin, flexible fiberoptic tube. As the tube is fed through the patient's gastrointestinal tract, a small camera transmits the images to a screen for viewing. The gastroenterologist manipulates the tube and camera to visually inspect the esophagus, colon, or intestines for abnormalities, lesions, or ulcers. Gastroenterologists receive extensive training on endoscopy procedures, including proper sedation of patients, and on interpretation of the images. Basic endoscopy training includes upper endoscopy, sigmoidoscopy, and colonoscopy.

Upper endoscopy examines the esophagus, stomach, and duodenum for ulcers, precancerous growths, foreign bodies, and other conditions causing pain, nausea or vomiting, bleeding, unexplained weight loss, or anemia. Gastroenterologists also receive training on using endoscopy to diagnose and dilate a narrow esophagus and to stop bleeding along the gastrointestinal tract.

Colonoscopy and sigmoidoscopy allow the gastroenterologist to visualize different portions of the colon. Colonoscopy encompasses the entire colon whereas sigmoidoscopy is limited to the last section of the colon and the rectum. Although sigmoidoscopies are quicker to perform, colonoscopies are preferred for cancer screening because they examine the entire colon. Both procedures require the patient to use a colon-cleansing product beforehand. During the procedure, the gastroenterologist examines the lining of the colon and removes polyps, or precancerous lesions.

Colorectal Cancer Screening. Colorectal cancer is the second leading cause of cancer death in the United States, according to the National Cancer Institute. Early detection through screening is an important tool in successfully treating the disease. Colonoscopy and sigmoidoscopy, along with tests measuring the presence of blood in the stool, are the tools used to screen for colon and rectal cancers. By examining the intestinal walls, the gastroenterologist detects and removes polyps, or precancerous lesions, and identifies abnormalities.

Pediatric Gastroenterology. Pediatric gastroenterology is an established subspecialty of pediatrics, whereas gastroenterology branches off from internal medicine. As such, pediatric gastroenterologists have a residency in pediatrics, and they take a separate subboard certification exam. Gastroenterologists in this subspecialty focus on gastrointestinal and nutrition-based conditions in infants, children, and adolescents. With additional training, pediatric gastroenterologists may narrow their specialization to liver transplantation, motility disorders, pancreatic diseases, endoscopic techniques for children, or nutrition.

All gastroenterologists form long-term relationships with their patients, but this is especially important for those specializing in pediatrics. Many pediatric gastric conditions are chronic and will follow the child throughout life. After the patient reaches adolescence, the pediatric gastroenterologist transfers care of the patient to a colleague; therefore, the pediatric specialist must have excellent communication skills. Pediatric gastroenterologists must consider the effect of a patient's condition on normal growth and development. Pediatric conditions often involve specialists from several disciplines, which means that the pediatric gastroenterologist must function well in teams.

Gastroesophageal Reflux Disease. The regurgitation of stomach acids into the esophagus, which causes a burning sensation in the chest known as heartburn, can lead to gastroesophageal reflux disease (GERD). Normally, a round muscle closes off the opening at the base of the esophagus leading to the stomach. Occasionally, the muscle fails to close completely, and the stomach contents wash up into the esophagus, burning the lining and causing the characteristic burning sensation. GERD may occur in adults or children.

GERD may be treated with over-the-counter antacids or other agents, but severe cases with persistent symptoms are best treated by a gastroenterologist. Upper endoscopy is useful for diagnosing GERD and evaluating treatment options. Prolonged GERD may damage the lining of the esophagus and cause bleeding ulcers.

Gastric and Peptic Ulcers. An erosion of the lining of the stomach is called a gastric ulcer; an erosion on the duodenum is called a peptic ulcer. They are caused by infection with the bacterium *Helicobacter pylori* or the use of certain medications, such as nonsteroidal anti-inflammatory drugs (NSAIDs) but are not believed to be caused by stress. Smoking may worsen ulcers. Gastroenterologists played an important role in discovering *H. pylori* as a cause of ulcers.

Gastroenterologists diagnose ulcers by their

symptoms, which include a dull or burning pain, weight loss, or vomiting, and a blood test for *H. pylori*. Upper endoscopy may be used if the patient has bleeding or if the ulcer blocks food from leaving the stomach. The gastroenterologist prescribes antibiotics and antacids to treat ulcers.

Parenteral Nutrition. Patients with serious medical conditions may be unable to eat or tolerate feedings introduced into the digestive tract. Parenteral nutrition—the injection of nutrients intramuscularly, intravenously, or subcutaneously—helps avoid complications of malnutrition. However, parenteral nutrition may be harmful in some instances, and it is up to the gastroenterologist to determine when the technique should be used. For example, parenteral nutrition has little effect in most patients following surgery, although it reduces postoperative complications in patients following surgery for esophageal or stomach cancer. Infants who are unable to eat and adults with prolonged malabsorption conditions benefit from parenteral nutrition.

Long-Term and End-of-Life Care. Patients in long-term care facilities and those with terminal illnesses receiving end-of-life care often suffer from digestive difficulties. Patients may be unable to eat properly and become malnourished. Patients receiving opioid medications for pain relief may develop constipation as a side effect of the drugs. Opioid-induced constipation causes additional pain and discomfort to patients already suffering from end-stage illnesses.

Gastroenterologists are skilled at finding alternatives to taking patients off the analgesic medications, which would relieve abdominal discomfort but would leave the patients in severe pain.

Food Allergies. Gastroenterologists may be helpful in determining specific food allergies or intolerances (for example, lactose intolerance) that lead to gastric disorders. Patients with food allergies or intolerances are often seen by primary care physicians, who consult with gastroenterologists and allergists to determine the cause of the symptoms. Gastrointestinal signs of food allergy include constipation, colic, or gastroesophageal reflux in infants, or a severe reaction immediately after eating the food.

Obesity. Gastroenterologists are uniquely positioned to assist overweight patients because many conditions that require gastroenterological care result from excess weight. The gastroenterologist faces the challenge of recommending a realistic diet

---

Fascinating Facts About Gastroenterology

- As many as 70 million people in the United States are affected by a digestive disease, and about 9 percent of these people require hospitalization. An estimated 234,000 Americans die annually from digestive diseases, including gastric cancers.
- The United States has 11,704 certified gastroenterologists.
- Peptic ulcers affect 14.5 million Americans, resulting in more than 875,000 physician visits and 2 million prescriptions. Up to 20 percent of the population suffers from heartburn, or gastroesophageal reflux, at least once weekly.
- Colon polyps are small growths on the interior wall of the colon, some of which may become cancerous. Adenomas, or precancerous polyps, take about ten years to change from a polyp to a cancerous tumor.
- Women tend to have a higher incidence of constipation than men, perhaps because the colon empties slower in women than in men. Women also have less pressure in the anal sphincter compared with men, which enables men to withstand the urge to defecate longer than women.
- Colon cancer has few symptoms in the early stages, making screening important. Warning signs that may indicate colon cancer are blood in the stool, change in bowel habits (including formation of the stool), or abdominal pain.

---

and exercise plan that the patient will follow. After obtaining a detailed patient history that includes a psychiatric evaluation, the gastroenterologist must determine the amount of weight the patient must lose to achieve a healthy body mass index, balanced with the patient's level of commitment to changing lifestyle habits and losing weight.

Several treatment options are available to the gastroenterologist depending on the patient's present weight, the desired amount of weight loss, and the patient's outlook. Gastroenterologists can assist patients with dietary plans using portion-controlled servings of low-fat foods and in developing realistic exercise plans to increase physical activity and energy use. The gastroenterologist helps patients develop weight-loss goals and provides support and guidance. The doctor may prescribe medications such as sibutramine (Meridia) or orlistat (Xenical). If the patient has been unable to lose weight through diet and exercise, has no unusual surgical risks, and is at risk for obesity-related complications, such as heart disease or diabetes, the doctor may perform bariatric surgery.

Research. Gastroenterology is a fertile field for research. Gastroenterologists with an interest in research perform their own studies on the cause and diagnosis of gastrointestinal disorders. They also work with pharmaceutical companies developing medications for GERD and ulcers, Crohn's disease, ulcerative colitis, inflammatory bowel disease, and other disorders, including gastrointestinal cancers.

IMPACT ON INDUSTRY

Global Standards. Gastroenterology has become a major medical specialty all over the world, but gastroenterologists are called on to perform different roles depending on the country. In developed nations, gastroenterologists follow structured educational paths to achieve certification. Advanced skills and specialized care are the primary focus. The field is complex, with emphasis on diagnostic tools such as endoscopy and medical management of gastric conditions, which include gastroenterological cancers and immunology. The addition of hepatology, which includes all types of hepatitis and cirrhosis, has greatly expanded the field, added to the complexity, and necessitated more intense education.

In developing countries, the field of gastroenterology is less well defined and the training requirements vary widely. Provision of tightly structured and focused curricula is less important than the need to quickly educate potential gastroenterologists to enable them to treat growing numbers of people with chronic liver and gastric conditions. In many developing countries, postgraduate education is limited to two additional years of training in endoscopic procedures, compared with the five years of gastroenterological training required in the United States. In an effort to establish a global standard for gastroenterological training, the World Gastroenterology Organization (WGO) outlined educational standards in developed and developing nations, highlighting shortcomings and areas for improvement.

The WGO recommends that all gastroenterologists have a basic foundation in specific areas, including physiology, pharmacology, epidemiology, and the anatomy and development of the gastrointestinal tract. The organization also recommends that basic training cover nutrition and metabolism, diagnosis, prevention, treatment, bioethics, and cost containment.

The WGO outlines specific skills in which gastroenterologists should be proficient, regardless of the nation in which they practice or their specific patient population. These skills include endoscopy techniques and the ability to interpret laboratory data, as well as general skills such as leadership and team management, communication, and professionalism.

Research. Researchers in gastroenterology may work in academic settings conducting basic research or for pharmaceutical companies focusing on specific conditions or drug side effects.

Gastroenterological conditions are an active area of pharmaceutical development. Medications for GERD, gastric and peptic ulcers, and inflammatory bowel diseases are in demand, and active research into treatments for gastroenterological cancers is expanding as new therapies are developed.

Gastroenterologists are valuable as consultants for medications that treat conditions outside the gastroenterological tract but have side effects such as nausea, vomiting, or abdominal pain. Furthermore, they can often discern the appropriate treatment for various conditions having gastrointestinal distress as a symptom without compromising the action of medications designed to address the patient's primary condition.

Education. The gastroenterological training process in the United States requires students to train with seasoned professionals. Experienced gastroenterologists demonstrate proper interaction and communication with patients and other specialists, along with proper diagnostic techniques. They observe the students and critique their approach. The practical, personalized training provided by experienced practitioners is an important part of the educational process.

CAREERS AND COURSE WORK

Students planning a career in gastroenterology often study biology or chemistry as an undergraduate, making sure to fulfill any medical school requirements. In addition to aptitude in the life sciences, potential gastroenterologists need to demonstrate excellent listening and communication skills and compassion for their patients. After completing medical school and earning their license to practice, students must become certified in internal medicine. Internal medicine certification requires a three-year postgraduate residency, demonstrated clinical competency, and successful completion of a comprehensive final exam. To become certified in gastroenterology, students must complete three years of fellowship training exclusively in gastroenterology, complete eighteen months in clinical practice, and successfully pass a comprehensive subspecialty exam. By the time gastroenterologists become certified, they have had an additional five years of specialized training beyond medical school.

Students interested in pediatric gastroenterology must complete their three-year postgraduate residency and take the board exam in pediatrics rather than internal medicine. Students interested in specializing in treating liver problems or gastric cancers undergo additional training with specialists in those areas.

The specialized gastroenterology training is supervised by four professional societies: the American College of Gastroenterology, the American Gastroenterological Association, the American Society of Gastrointestinal Endoscopy, and the American Board of Internal Medicine. The American Board of Internal Medicine administers the final comprehensive competency exam.

During their fellowships, students become proficient in procedures essential to the gastroenterological field, including endoscopy, colonoscopy, sigmoidoscopy, and esophageal dilation. Students learn to remove colon polyps as well as foreign bodies lodged in the esophagus. Students are observed as they interact with patients and evaluated on their ability to examine patients and take complete health histories.

SOCIAL CONTEXT AND FUTURE PROSPECTS

Gastroenterology is an active field that continues to expand. As the population ages, more people will require cancer screenings and polyp removal. Conditions such as ulcers, GERD, and hepatitis are diagnosed with increasing frequency. The general population is not only growing older but also heavier. Obesity is a growing health concern. Gastroenterologists are integral in managing complications of obesity such as gallstones and GERD, and they serve as important partners in teaching overweight patients proper nutrition and weight-loss strategies, including gastric banding procedures.

In developing countries and regions with unsanitary conditions, gastroenterologists are needed to combat health concerns such as diarrhea and malnutrition. Gastroenterologists can provide basic nutritional counseling and assist patients in fighting parasitic or infectious diseases that compromise the absorption of nutrients.

Cheryl Pokalo Jones, B.A.

FURTHER READING

Butcher, Graham. *Gastroenterology: An Illustrated Colour Text.* Philadelphia: Elsevier Health Sciences, 2003. Full-color clinical photographs and detailed line drawings illustrate gastroenterological and liver diseases.

Collins, Paul. *Gastroenterology: Crash Course.* Philadelphia: Elsevier Health Sciences, 2008. Offers basic definitions and explanations of all aspects of gastroenterology in an easily understandable manner.

Grendell, James H., Scott L. Friedman, and Kenneth R. McQuaid. *Current Diagnosis and Treatment in Gastroenterology.* 2d ed. New York: Lange Medical Books, 2003. This comprehensive reference discusses all gastroenterological conditions, including hepatic, pancreatic, and biliary conditions.

Travis, Simon P. L., et al. *Gastroenterology.* 3d ed. Malden, Mass.: Blackwell, 2005. A concise, informative manual on gastroenterology with a global perspective.

WEB SITES

American College of Gastroenterology
http://www.acg.gi.org

American Gastroenterological Association
http://www.gastro.org

American Society of Gastrointestinal Endoscopy
http://www.asge.org

National Institute of Diabetes and Digestive and Kidney Diseases

National Digestive Diseases Information
Clearinghouse
http://digestive.niddk.nih.gov

See also: Epidemiology; Geriatrics and Gerontology;
Nutrition and Dietetics; Pediatric Medicine and
Surgery.

GEMOLOGY AND CHRYSOLOGY

FIELDS OF STUDY

Geology; mineralogy; chemistry; economic geology; crystallography.

SUMMARY

Gemology is the study of minerals that are attractive enough to be worn as jewelry after being cut and polished. Gemstones often are transparent and internally reflect light, are colored, and have a brilliant luster. Sometimes the term "gemstone" is used to refer to attractive aggregates of minerals (such as turquoise), organic materials (such as amber), rocks (such as marble), or synthetically made gemstones. Only about seventy of the many thousands of known minerals have varieties that are of gem quality. Diamond (carbon) is the most important gemstone. Chrysology refers to the study of precious metals such as gold, silver, and platinum. Precious metals are opaque and often have colors (such as the yellow of gold) that make them attractive in jewelry.

KEY TERMS AND CONCEPTS

- **Carat:** Measure of weight used for gemstones, equal to 0.2 gram.
- **Cleavage:** Tendency of a mineral or gemstone to break along smooth, plane surfaces.
- **Density:** Mass per volume of a substance.
- **Hardness:** Resistance of a mineral surface to scratching by another mineral. The hardness scale ranges from 1 to 10, with diamond having a hardness of 10 and the common mineral quartz having a hardness of 7.
- **Luster:** Appearance of a gem's surface in reflected light; some gemstones appear vitreous (reflect like glass), pearly, silky, adamantine (refracts light strongly like diamond), and metallic.
- **Mineral:** Naturally occurring element or compound that has a definite composition and arrangement of atoms that is reflected by its physical characteristics.
- **Refractive Index:** Ratio of the speed of light in air (300,000 kilometers/second) to the speed of light in a mineral (much less speed in the mineral).

- **Silicate:** Mineral that consists of silicon-oxygen negative ions combined with other positively charged metals.
- **Transparent:** Describes a gemstone that transmits light so that an object may be seen through it.

DEFINITION AND BASIC PRINCIPLES

Gemstones must be attractive to viewers. Their attractiveness depends on the color, transparency, and luster of the gem and the way it is cut. The most important gemstones are diamonds, sapphires, rubies, and emeralds. Diamonds are often yellowish and transparent, with a brilliant adamantine luster. Some of the most valuable diamonds, however, contain more intense green, red, blue, or black colors; their rarity often makes them more valuable.

Gemstones ideally should be harder than the common mineral quartz and should not fracture easily so that the gemstone is not easily scratched or broken when worn as jewelry. About ten minerals have these characteristics. Diamond is the hardest known mineral; however, diamonds can be fractured along cleavage planes if excessive force is applied to them. Gemstones softer than quartz will retain their luster and beauty as long as they are not scratched. Valuable gemstones should not contain many small impurities such as minerals or fractures that might degrade the beauty of the gem.

Gemstones are often cut in certain ways to enhance their beauty. The choice of cutting method depends in part on the characteristics of the stone. Diamond, for instance, has four major planes of cleavage (octahedral cleavage) along which the diamond may be broken. Diamonds are cut along these octahedral cleavage planes so that light is internally reflected within the diamond, producing many flashes of light as the gem is viewed.

In contrast to gemstones, precious metals such as gold and silver are opaque and have a metallic luster. Their beauty comes from their color in reflected light.

BACKGROUND AND HISTORY

Gemstones and precious metals are mentioned in ancient writings from cultures around the world, and they have been found during many archaeological digs, suggesting that they have been used for thousands of years. For instance, ancient Babylonians had gemstones of jasper (red, fine-grained quartz) and lapis lazuli (blue, the mineral lazulite, a sodium calcium aluminum silicate sulfate, mixed with other minor minerals).

A number of gemstones were found in Egyptian tombs dating from 5000 to 3000 b.c.e. These gems include lapis lazuli, agates (silica-rich mineral), jasper, and emeralds (dark green, transparent beryl, a beryllium aluminum silicate). Some of these gems may be traced to specific geologic sources, suggesting that trading occurred over large areas. An ancient Egyptian papyrus contains a map showing the location of several gold mines and the miners' quarters. Gold was likened to the skin of the gods, was used in funerary art, and worn only by royalty.

Many gemstones were worn not only because of their beauty but also because many people believed that the gems could protect them from evil forces and diseases and make them wise. Astrologers claimed that certain gems gave the wearer special powers in certain months. The Bible mentions a variety of gemstones, such as emeralds, sapphires (blue, transparent corundum, and aluminum oxide mineral), diamonds, agates, and amethysts (purple, transparent quartz), and gold and silver, which were sources of wealth and used for decoration.

How It Works

Traditional Identification Methods. Most gemstones and precious metals are minerals or are found in minerals, so the classic methods of identifying minerals using physical properties may be used. These physical properties are crystal forms (such as cubes and octahedrons), color, hardness, luster (metallic or nonmetallic), cleavage, density, and microscopic characteristics. Experienced gemologists may be able to identify a gemstone if they can determine its luster, color, and other features by visual inspection, using no more than a magnifying glass. Some properties such as hardness, however, may be of limited use to identify a valuable gemstone because checking for hardness could damage the specimen.

Some specialized techniques may be used to identify a gemstone. Determining the gemstone's density can be very useful if the gemstone is not mixed with other minerals and is not mounted in jewelry. Gemologists might place the gem in a series of liquids of known densities. If the gemstone is denser than the liquid, it will sink; if it is less dense, it will float. Most minerals have very different densities, so this process can confirm the identity of the material.

Identification Using Instruments. A binocular microscope with a ten- to sixty-power magnification may be used to observe a gemstone using light reflected from the specimen or light transmitted through the specimen. Imperfections within a gemstone, for instance, can be seen using transmitted light.

The refractive index of a gemstone may be obtained using a refractometer. Most minerals have a unique refractive index, so this determination can also help confirm the identity of the gemstone. Refractive indices commonly range from a low of 1.4 to a high of 3.0. A gemstone typically should have one flat face that is placed on the stage of the refractometer in immersion oil with an index of refraction of 1.81. The mineral and light source are moved relative to each other until the light is internally reflected, and the refractive index is read directly on a scale. Some minerals have only one refractive index, so the readings are straightforward; others have more than one refractive index in different directions, so determining these values is more challenging.

Another approach to determining the refractive index is to place a gemstone in a series of liquids of known and different refractive indices and observe them under a microscope. If the refractive index of the gem is the same as the refractive index of the liquid, then the mineral nearly disappears; if the refractive index of the gem is substantially different from the refractive index of the oil, then the mineral edges stand out in bold relief.

If these methods fail to identify a gemstone, then other more destructive techniques such as X-ray diffraction, electron microprobe analysis, and internally coupled plasma mass spectrometry might be used for identification. Often these techniques use material that has been scraped off parts of the gemstone that will not be exposed to a viewer. X-ray diffraction techniques can be especially useful in confirming the identity of a mineral because each mineral presents a unique pattern.

Applications and Products

The most important gemstones are varieties of diamond, corundum (aluminum oxide; sapphires and rubies), beryl (beryllium aluminum silicate;

especially emeralds), tourmaline (sodium calcium, lithium silicate borate hydroxide mineral), spinel (magnesium aluminum oxide), topaz (aluminum silicate hydroxide), and quartz (silica, such as amethyst).

Diamond, because of its hardness, high refractive index, numerous colors, rarity, and ease of cutting, is the most important gemstone. Diamonds are the most common gemstones in jewelry stores. The majority of diamonds, however, are not of gem quality, but they can be used for industrial purposes such as in cutting materials. Most diamonds have a yellow tinge, but some rare diamonds with few imperfections may have intense green, blue, gold, red, and black colors that make them very valuable. The largest diamond ever found was larger than 3,100 carats. It was cut into smaller diamonds, the most famous of which is the colorless 530-carat Star of Africa diamond.

Rubies are the red gemstone of corundum; sapphires can be blue, yellow, green, or violet. Only diamond is harder than rubies or sapphires. Rubies and sapphires are difficult to cut because they have no cleavage and are brittle. Some of the most famous rubies and sapphires are much more expensive per carat than diamonds. One of the most famous sapphires is the Star of India (53 carats), and one of the most famous rubies is the Edward ruby (167 carats).

Beryl has many colors, but the dark green, transparent emeralds are the most expensive. Blue beryls of gem quality are called aquamarines. Some of the biggest and most beautiful specimens of emeralds are larger than 2,200 carats.

Tourmaline, spinel, and topaz occur in many colors. Pink and green varieties of tourmaline are the most popular, and the red varieties of spinels are the most valued. Good specimens of pink topaz are the most expensive.

Quartz is usually transparent and colorless, but it may be violet, black, blue, yellow, pale green, and pink. Amethyst is the name for violet quartz, which is generally more valuable than the other varieties of quartz.

Lesser Gemstones. Many other gemstones of lesser importance are commonly used. For example, spodumene (lithium aluminum silicate) can occur as transparent yellow, green, and colorless specimens. Turquoise (calcium aluminum silicate) is a blue to green, opaque mineral that is usually mixed with other dark impurities or some copper minerals; it has a hardness less than quartz and a good cleavage, so care must be taken not to damage it.

Artificial Gemstones. Artificial gemstones have been produced since the early part of the twentieth century. They are often difficult to distinguish from natural gemstones, thereby somewhat diminishing the demand for some natural gemstones. Artificial gemstones can be made in several ways. Some are produced by precipitation of the gems from a hot, steam solution; others are created by melting, at the appropriate temperature and pressure, a mixture of components in a ratio that approximates that of the natural gem they will resemble.

Since the 1950's, artificial diamonds have been created by heating carbon to a very high temperature and pressure. Most of these diamonds are not of gem quality but can be used as abrasives or for other industrial purposes. Artificial diamonds can be distinguished from natural diamonds by the presence of certain impurities, such as nitrogen or boron used in their manufacture.

Precious Metals. Gold and silver have been used for thousands of years as both a medium of exchange and for ornamentation. Gold, a malleable metal, and silver were used to make plates, cups, drinking vessels, ornamental objects, religious icons, and jewelry. Gold and silver have also been used in applications such as dentistry, electronics, and engineering. The refining of platinum was not perfected until the nineteenth century, so fewer applications have been developed. Platinum began to be used more commonly in jewelry in the 1960's, and it is also used as a catalyst in some chemical reactions.

IMPACT ON INDUSTRY

The cost of gemstones is driven by supply and demand. The most desirable gemstones—diamonds, rubies, emeralds, and sapphires—have the highest cost per carat. The price of a particular gemstone depends on the rarity of its color, its luster, the number of imperfections, and the desirability of its cut. The revenue from diamonds accounts for about 90 percent of that from all gemstones.

Similarly, the prices of precious metals such as gold, silver, and platinum are driven by supply and demand. In the twentieth century, the cost of gold ranged from as little as about $35 per ounce from 1935 to 1967 to as high as $615 per ounce in 1980. In

2010, the price of gold rose to more than $1,300 per ounce, while silver sold for around $23 per ounce, and platinum for more than $1,600 per ounce.

Industry and Business. The discovery and distribution of gemstones is complicated and depends on the source and type of the gemstone. The De Beers Group has dominated the diamond market since the early part of the twentieth century, controlling most aspects of supply, distribution, and price. However, the discovery of diamonds in Russia, Australia, and Canada and the creation of artificial diamonds for the industrial market has somewhat altered the market.

Fascinating Facts About Gemology and Chrysology

- Platinum is used in drugs such as cisplastin to fight cancer.
- The addition of silver halide crystals to eyeglass lenses allows them to rapidly change from clear to dark. Ultraviolet light hits the crystals, causing a chemical change that blocks light and darkens the lenses.
- Near Barstow, California, 1,926 silver-coated mirrors direct sunlight to nitrate-filled tubes that are used to run generators and provide electric power for 10,000 homes.
- One treatment for prostrate cancer involves inserting three grains of gold into the prostrate, using ultrasound. This allows physicians to accurately determine the position of the prostrate during radiation treatments.
- In Japan, sake, tea, and other foods containing thin gold flakes are believed to be beneficial to one's health.
- Ordinary household switches uses silver. Silver does not corrode, which would result in overheating and pose a fire hazard.
- Several companies produce artificial diamonds from the cremated remains of people or pets.
- Because the origin of amethyst was believed to be related to Bacchus, the Greek god of wine, it was once believed that wearing amethyst would prevent a person from becoming drunk.
- Ancient Greeks believed that topaz made its wearer stronger and invisible in case of an emergency.

Diamonds are sorted and valued by De Beers and sent to London, where the diamonds are sold in groups to a few select individuals at a certain price. These groups of diamonds are then sold to diamond exchanges in Belgium, the Netherlands, New York City, and Israel. These diamond exchanges can then sell diamonds to other places such as wholesale jewelry stores. As the diamonds all come from De Beers, prices tend to remain steady. However, the diamonds emerging from Russia, Canada, and Australia do not go through these distribution channels and their prices are governed by market forces.

Although the markets for other gemstones are not dominated by a single producer, those gemstones that come from limited sources may become scarce, and their prices may rise rapidly. For instance, tanzanite (calcium aluminum silicate) can be mined only at one site in Tanzania, and few new specimens are being found. This scarcity has driven up the price of tanzanite.

Precious metals are refined from ores from mines or produced from recycled materials. The firms involved in gold production generally are involved in exploration, development, or production, and are not vertically integrated. Silver is recovered from silver mines and as a by-product of gold mining. Firms are generally involved in mining, refining, fabrication, and manufacture. Most of the world's platinum comes from Russia and South Africa. Platinum production is capital intensive and dominated by large, vertically integrated firms. Gold is used primarily for jewelry but also in electronic products. Silver is used for jewelry, silverware, photography, and industrial uses, such as in batteries and electronics. Platinum is commonly used as an autocatalyst and also as jewelry and in electronic products.

Governments and Universities. Governments are primarily involved in the regulation of the gemstone and precious metals industry. Regulation can include measures to ensure the safety of workers in mines or refineries, to deal with environmental issues arising from mining or refining, or to control how gemstones and precious metals are bought and sold. Smuggling of gemstones, often involving rebel forces, organized crime, or drug cartels, remains a problem in many nations. Unstable governments have had negative effects on the supply of gemstones. For example, in 1969, the government of Burma, the main supplier of rubies, closed all ruby mines within its borders.

Universities conduct research on the composition and uses of gemstones, mineral identification techniques, and all aspects of mining operations. They also are involved in research regarding the possible uses of precious metals in medicine, engineering, and manufacturing. They also provide training in mineralogy, geology, and mining operations.

CAREERS AND COURSE WORK

A career in gemology or chrysology begins with a bachelor's degree in geology or mining engineering, with course work in mineralogy and gemology. A doctorate in one of these fields is required for conducting research on gemstones or precious metals. Additional course work in chemistry and physics may be helpful. Research involving applications that use gemstones or precious metals requires course work and degrees in those areas, such as medicine or electronics.

Those desiring to work as a gemstone identifier and grader in a jewelry store or as a jewelry maker or designer can obtain specialized training in these areas. For example, the Gemological Institute of America offers a graduate gemologist degree or a jewelry manufacturing arts degree, both six-month programs.

SOCIAL CONTEXT AND FUTURE PROSPECTS

The demand for gemstones is likely to continue, rising and falling with upturns and downturns in the economy. Gemstones, like many minerals, are found in only a limited number of locations worldwide; therefore, the supply of a given gemstone can vary considerably. A major mine may become depleted, or a new source of the gemstone may be discovered. Diamond prices are the most stable because the market is dominated by De Beers; however, as other diamond producers increase their presence in the market, prices may begin to fluctuate. The total worldwide diamond production exceeded $19 billion in 2009.

The best rubies and sapphires are so rare that the price per carat for individual specimens may exceed that for diamonds. Burma is the source for the best rubies and sapphires, and Colombia produces the finest emeralds. In 2006, total world emerald production was 5,400 kilograms, total ruby production was 10,000 kilograms, and total sapphire production was 32,500 kilograms. These levels of production are expected to be maintained.

According to the U.S. Geological Survey, in 2009, worldwide production of gold was 2,350 metric tons; silver, 21,400 metric tons; and platinum metals, 178,000 kilograms. The demand for gold and its price was increasing, while the demand for platinum had fallen because of the decline in the automotive industry. However, precious metals as a whole were predicted to increase in demand, especially as sources become depleted.

Robert L. Cullers, B.S., M.S., Ph.D.

FURTHER READING

Babcok, Loren E. *Gemstones and Precious Metals.* Rev. ed. Hoboken, N.J.: Wiley Custom Services, 2009. Examines both gemstones and precious metals, looking at history, applications, and their value.

Crowe, Judith. *The Jeweler's Directory of Gemstones.* Buffalo, N.Y.: Firefly Books, 2006. Examines how to appraise and use gemstones in jewelry making.

Desautels, Paul E. *The Gem Kingdom.* New York: Random House, 2000. Discusses gemstone quality, history, cutting, and the making of jewelry.

Gasparrini, Claudia. *Gold and Other Precious Metals: Occurrence, Extraction, Applications.* 3d ed. Toronto: Space Eagle, 2000. Examines where precious metals are found and how they are mined and used.

Klein, Cornelis, and Barbara Dutrow. *Manual of Mineral Science.* Hoboken, N.J.: John Wiley & Sons, 2008. Describes how to identify minerals using physical properties and modern analytical techniques. It has a chapter on gemstones.

Rutland, E. H. *Gemstones.* New York: Hamlyn Publishing, 1974. Describes the origin, identification, jewelry, history, and collection of gemstones.

Schumann, Walter. *Gemstones of the World.* New York: Sterling, 2009. Contains many photographs of cut and uncut gemstones. Discusses the origin, properties, deposits, cutting, polishing, and identification of gemstones.

WEB SITES

International Colored Gemstone Association
http://www.gemstone.org

International Gem Society
http://www.gemsociety.org

International Platinum Group Metals Association
http://www.ipa-news.com/en

The Silver Institute
http://www.silverinstitute.org

World Gold Council
http://www.gold.org

See also: Mineralogy.

GENETICALLY MODIFIED ORGANISMS

FIELDS OF STUDY

Biology; chemistry; genetics; biotechnology; recombinant DNA technology; reproductive science; genetic engineering; molecular biology; botany; entomology; plant pathology; agricultural science; environmental science; medical science.

SUMMARY

Genetically modified organisms are produced through genetic engineering and biotechnology and basically involve genetic modifications in which genetic material is added or removed to alter the genetic structure of the organism. Many organisms have undergone genetic modification, including bacteria and viruses, plants and animals, and even human beings. The majority of genetically modified organisms are created for therapeutic reasons, such as medicine and food for human consumption. Such organisms have the potential to affect all members of human society and their surrounding environment and have therefore become one of the most controversial ethical and ecological issues of the early twenty-first century.

KEY TERMS AND CONCEPTS

- **Crop Yield:** Amount of plant crop harvested, as opposed to grown, in a given area for a given time.
- **DNA (Deoxyribonucleic Acid):** Nucleic acid found in a cell that contains all the genetic material and instructions used for growth and development of living organisms.
- **Gene:** Basic unit of heredity; occupies a precise position on a chromosome within an individual organism's DNA.
- **Genetically Modified Organisms (GMOs):** Organisms whose genetic makeup has been manipulated by genetic engineering techniques (gene technology), through either the addition or removal of genetic material; also known as genetically engineered organisms (GEO) and bioengineered organisms.
- **Natural Selection:** Process whereby a species evolves over time as individual organisms possessing the most practical and useful characteristics survive and reproduce; their offspring will, in turn, possess those same positive characteristics so as to survive and continue breeding.
- **Outcrossing:** Transfer of genes from genetically modified plants into crops or related plant species in the wild.
- **Plasmid Technology:** Technology related to plasmids, which are circular, double-stranded units of DNA. Plasmids are able to replicate inside a cell independently of the cell's chromosomal DNA and are thus useful in recombinant DNA technology and research in the transfer of genes between different cells.
- **Recombinant DNA Technology:** Techniques in genetic biology that alter an organism's genes by removing a specific gene from the cell of one organism and inserting it into the cell of another. This splicing together of gene fragments from different species produces a new organism that would not occur naturally.

DEFINITION AND BASIC PRINCIPLES

Humans have selectively bred and crossbred plants and animals for desired traits since almost the dawn of agriculture, but advances in genetic technology have given people novel ways in which to manipulate plants and animals. These advances are motivated by the desires to develop new medical treatments for genetic diseases and disorders and to increase food production to satisfy the world's growing population. Most advances involve recombinant DNA technology, in which an organism's genes are altered by removing a specific gene from the cell of one organism and inserting it into the cell of another. This splicing together of gene fragments from different species produces a new organism that would not be produced through natural reproduction processes or would not be feasible because of the impossibility of interspecies breeding. This new organism is defined as a genetically modified organism (GMO).

The advancement of genetic technology and the introduction of GMOs into the human food chain has prompted controversy over the ethics of manipulating nature and the potential for GMOs in worldwide agricultural production and medicine.

Although many experts state that GMOs are safe for human consumption and offer myriad benefits to humankind, others claim that the production and consumption of GMOs is unethical and untested, which means that GMOs involve unknown consequences, which potentially could be dangerous.

BACKGROUND AND HISTORY

The process of natural selection, first described by Charles Darwin in 1859 in his seminal *On the Origin of Species by Means of Natural Selection*, states that species evolve over time. Individual organisms that possess the most desired and useful characteristics survive, reproduce, and give birth to offspring; these offspring, in turn, possess the same positive characteristics. For many hundreds of years, people have manipulated the process of natural selection though traditional agricultural selection and crossbreeding to create or eliminate specific characteristics in plant and animal species, producing a wide variety of cereal crops, livestock animals, and pets.

Human interference has altered many plants and animals through crossbreeding or selection, but the desirable traits initially appeared through naturally occurring genetic variation. Because the desired traits were already in existence, human interference in the breeding process was often viewed as relatively benign and within natural bounds. Although humans have manipulated the breeding of plants and animals based on phenotypic characteristics for a long time, the ability to directly manipulate the genotype developed much later. Specifically, to feed a growing and hungry world population and develop medical treatments, medical and agricultural scientists have researched and advanced genetic modification technology.

Although genetic engineering is a phenomenon of the late twentieth century, the building blocks for such technology began with the first isolation of DNA in 1869 and the subsequent awareness of its relevance to heredity in 1928. The first accurate double-helix model of DNA was developed in 1953 by James D. Watson and Francis Crick, and the first gene sequence and recombinant DNA was created in 1972 by researchers from Stanford University. The latter discovery truly heralded the beginning of the biotechnological industry and the development of GMOs.

Genetic engineering research continued during the 1970's, and the first publicly and commercially available GMO, a form of human insulin produced by bacteria, was developed in the United States in 1982. However, for the most part, the majority of commercial GMOs sold and used in the twenty-first century are found in agriculture and food production. GMO research scientists believe that genetic engineering is the only method that will guarantee global food production, particularly as predications regarding global climates have indicated that traditional agriculture practices will fail to meet demand.

HOW IT WORKS

Fundamentally, the development and manufacture of GMOs is the replacement of natural selection processes with artificial genetic manipulation. At its most basic, GMO technology relies on a sound understanding of DNA and involves the subtraction of specific genetic material or substitution of material from one species with that from another. Genetic engineering, a complex endeavor, deals with the most fundamental building blocks of an organism. Within a cell are tiny strandlike structures called chromosomes, which contain a nucleic acid called DNA. This molecule contains all the genetic material required for inheritance and thus is the basis for genetic manipulation technology.

Initially, the term genetic manipulation referred to a vast array of techniques for the modification of organisms through reproduction and gene inheritance. Later, however, the definition became more restricted and refers specifically to recombinant DNA technology, a form of genetic engineering in which the genome of a cell or organism is artificially modified. The fundamental concept of this technology is that genetic material from different species is combined to create a new species or organism. That is, molecules of DNA from more than one source are united together inside a cell, which is then inserted into a new organism or host, where it is able to reproduce. Because the genome is passed on to offspring, the modification is considered to be self-perpetuating. An organism's biological activities and physical characteristics are controlled by its genome, so modification of the genome can significantly influence the organism's biological functions and traits. The objective behind such technology is to advance the fields of medicine and agriculture to develop more effective medical treatments and to improve crop yield and disease resistance.

Fascinating Facts About Genetically Modified Organisms

- Genetically modified seeds are used in the agricultural industry for the production of cereal crops, such as soybeans and canola.
- The first commercial GMO, a synthetic human insulin, was introduced in 1982.
- GMOs are being heralded as the solution to the world's continuing increase in population and food demand.
- GMOs offer great potential in the fight against the 3,000 genetic diseases and disorders that adversely affect people.
- In 2007, an estimated 12 million farmers in more than twenty countries were growing genetically modified crops. Roughly 90 percent of these farmers were categorized as resource-poor and were mostly in twelve developing countries.
- Bt-corn, a genetically modified version of corn, incorporates a gene from the bacterium *Bacillus thuringiensis* that acts as an insecticide.
- Cheeses and canola oil are just two of the genetically modified foods commercially available.

APPLICATIONS AND PRODUCTS

The possible applications and products of genetic engineering are vast, perhaps limited only by the imagination. For the most part, however, the major function of genetic engineering, and hence the development of GMOs, is related to their potential in agricultural, medical, and environmental applications.

Agricultural Applications. The continuing rapid expansion of the human population is necessitating an increase in the supply of food. Providing adequate food for a hungry world has become a significant issue for science. The need for food has been instrumental in promoting and advancing genetic modification techniques to produce new and improved organisms, particularly those that, for example, have higher yields or are drought and disease resistant. Through recombinant gene technology, it has become possible to create plant species that are capable of surviving in extreme temperatures and with low rainfall, that can convert atmospheric nitrogen into a useable form (thereby eliminating the need for nitrogen fertilizer), and that have the ability to produce their own resistance to pests and pathogens

(thereby eliminating the need for chemical pesticides). Versions of soybeans, canola, corn, potatoes, sugar beets, and cotton that have been genetically modified to increase herbicide tolerance and resist insects are all available for purchase.

With some specific exceptions, research and the development of genetically engineered animals has proven to be less straightforward than the genetic modification of plants and certainly more ethically problematic. In addition, although the public definitely shows some resistance to the idea of introducing genetically modified plants into the human food chain, most people express much greater resistance to the idea of directly consuming genetically modified animals. Therefore, research on genetically modified animals for use in agriculture has stayed in a relatively early stage of development. However, there has been some research into and experimentation with the genetic manipulation of animals to increase production and meat yield; such experimentation is most promising in fish species rather than in hoofed farm animals.

Medical Applications. Although agricultural applications are very important, the potential medical applications of GMOs are perhaps even more significant. The world's first commercial applications of genetic engineering were, in fact, medically oriented and included synthetic human insulin, approved for public sale in the United States by the Food and Drug Administration (FDA) in 1982, and a human hepatitis-B vaccine, approved by the FDA in 1987. Before the 1980's, synthetic human insulin (produced from animals) was available only in relatively limited quantities. Since the 1980's, research into medical applications for GMOs has rapidly advanced. Of particular benefit is the ability of genetic engineering to produce GMOs on a previously unavailable scale.

Perhaps the most significant potential application of genetic engineering and GMOs is in the treatment and possible cure of genetic diseases. Human society is plagued with both serious diseases and mild disorders, more than 3,000 of which are genetic in origin and therefore difficult to cure using conventional medicine. Although this technology is still in its infancy, the potential of gene therapy to assist people with genetic disease is perhaps limitless.

Environmental Applications. Increasing human populations are important not only in relation to agricultural food production but also in terms of their impact on the environment. Genetic engineering and GMOs could potentially solve some of the world's most serious ecological problems. Research has produced genetically modified viruses that can be used to create ecologically friendly lithium batteries, modified bacteria that can produce biodegradable plastic, and genetically manipulated bacteria that have been encoded for use in bioremediation. Genetic modification technology may even be of use in the fight for survival of some of the world's most vulnerable and endangered species.

IMPACT ON INDUSTRY

Major Organizations. Many agricultural organizations, the largest and most influential being the U.S.-based agricultural biotechnology corporation Monsanto, are directly involved in the promotion and advancement of GMOs for food—specifically cereal crops and meat production. Monsanto is the world's largest supplier of modified seeds (some 90 percent of the market). Genetically modified seeds for soybeans, canola, corn, potatoes, sugar beets, and cotton are commercially available.

Many organizations are involved in both the production of GMOs and their regulation. There are many environmental and ethical concerns regarding the creation of GMOs and the technologies involved in the genetic modification of organisms, particularly animals, for food and medicine. According to the World Health Organization, there are three basic concerns regarding GMOs and their possible impact on human health: increased antibiotic resistance because of the use of antibiotics in GMO manufacture; increased human allergic reactions to foods and medicines; and gene transfer.

To regulate these possible effects, the United Nations established the Convention on Biological Diversity, an international treaty that has been signed by some 190 nations. The United Nations was also responsible for the establishment of the Protocol on Biosafety, which aims to conserve biodiversity through a reduction in the potentially harmful consequences of biotechnology, specifically living GMOs, and to monitor and regulate GMO movement and international trade across borders and between countries.

Government Regulation. A number of countries, such as Australia and New Zealand, have distinguished between in-country production of GMO crops and the sale of imported GMOs. Both countries allow the sale of genetically modified soybean, canola, corn, potato, sugar beets, and cotton seeds or plants, which have been altered mostly to increase herbicide tolerance and minimize damage from insects. Australia and New Zealand have been quite measured in their approval of growing genetically modified crops in comparison with countries such as Canada and the United States. In Canada, for example, genetically modified foods are nutritionally assessed and treated the same as unmodified foods. Additionally, both the Canadian and the United States governments do not require genetically modified foods to be labeled as such.

Numerous governments around the world, however, have banned the domestic growing of GMOs for fear of contamination with native plant species and because of the possible unknown side effects of consumption, particularly by livestock. Many environmental organizations have lobbied the world's governments and suggested a total moratorium on GMOs, claiming they will have an adverse effect on the environment and cause loss of biodiversity through unintended crossbreeding with native plants and conventional crops.

CAREERS AND COURSE WORK

Students who wish to pursue a career in genetic engineering usually obtain an undergraduate degree in science or medicine. Typical majors include molecular biology, biomedical engineering, and genetics, although some universities offer specific undergraduate courses in genetic engineering. Graduate studies are essential for those wishing to pursue a career in genetic engineering. Following graduation, students studying genetically modified organisms will understand methods and processes involved in recombinant DNA technology and techniques, including DNA cloning, recombining genes, nucleic acid hybridization, gel transfers, and DNA sequencing.

Genetic engineering, while controversial, certainly offers potential as a significant tool in solving problems in agriculture, medicine, environmental science, and basic biology. Genetic technology is likely to play a central role in almost all areas of innovative biological sciences. Rapid advances in the field correspond to significant career potential and multiple avenues. Students involved in genetic engineering research and the application of GMOs can pursue careers in medical diagnosis, treatment, and gene therapy, agricultural food production, environmental bioremediation, and resource management within the private sector, nongovernmental organizations, specialized government organizations and agencies, and universities undertaking teaching and research.

SOCIAL CONTEXT AND FUTURE PROSPECTS

Genetically modified organisms potentially could be very advantageous to people. Specifically, scientists claim that GMOs will be vital to the future of food production and therapeutic medicine. Given the possibility of climate change due to global warming, crops that can produce higher yields, resist pests and pathogens, and better tolerate drought are very attractive. The use of GMOs in the treatment of genetic disorders makes them potentially life-saving. Supporters of GMO technology argue that genetic engineering has become an economic and environmental necessity in regard to agriculture, environmental bioremediation, and medicine.

Despite their obvious benefits, GMOs also hold many potential dangers. In addition, GMOs are not well received by the public. Surveys in some countries have revealed that the majority of people are actually against the creation and production of genetically modified foods, animals in particular. This opinion is shared by many environmental organizations, which claim that the undeniable benefits of GMOs are far outweighed by their possible effects on ecosystems, native flora and fauna, and human health. Of particular concern is that many of the potential risks of GMOs are as yet unknown. Opponents of GMO technology have stated that imposing GMOs onto the public without long-term rigorous testing is irresponsible and that more research is required.

Christine Watts, Ph.D., B.App.Sc., B.Sc.

FURTHER READING

Howe, Christopher. *Gene Cloning and Manipulation.* 2d ed. New York: Cambridge University Press, 2007. A comprehensive look at advances in recombinant DNA techniques, with both a broad and a concise examination of the concepts and principles involved.

Nelson, Gerald C., ed. *Genetically Modified Organisms in Agriculture: Economics and Politics.* San Diego, Calif.: Academic Press, 2001. Examines and analyzes the economic, ecological, and social factors involved in the production of GMOs for agriculture.

Nicholl, Desmond S. T. *An Introduction to Genetic Engineering.* 3d ed. New York: Cambridge University Press, 2008. Introduces basic molecular biology, the methods used to manipulate genes, and the technology's applications.

Primrose, Sandy B., Richard M. Twyman, and Robert W. Old. *Principles of Gene Manipulation.* 6th ed. Oxford, England: Blackwell Scientific, 2003. Discusses the genetic engineering of plants, animals, and microbes; the use of nucleic acids as diagnostic tools; and modern plant breeding.

Watson, James D., et al. *Recombinant DNA: Genes and Genomes—A Short Course.* 3d ed. New York: W. H. Freeman, 2007. One of the landmark texts of recombinant DNA technology, this work presents the fundamental concepts of genetics and genomics, the Human Genome Project, bioinformatic and experimental techniques, and a survey of epigenetics and RNA interference.

Young, Tomme R. *Genetically Modified Organisms and Biosafety: A Background Paper for Decision-Makers and Others to Assist in Consideration of GMO Issues.* Gland, Switzerland: International Union for Conservation of Nature, 2004. A detailed look at biosafety and genetically modified organisms, from species conservation, to sustainable livelihoods, to sociocultural policy.

WEB SITES

Union of Concerned Scientists, Food and Agriculture
Engineered Foods Allowed on the Market
http://www.ucsusa.org/food_and_agriculture/
science_and_impacts/science/engineered-foods-allowed-on.html

U.S. Food and Drug Administration
Genetically Engineered Foods
http://www.fda.gov/NewsEvents/Testimony/
ucm115032.htm

World Health Organization
Twenty Questions on Genetically Modified (GM) Foods
http://www.who.int/foodsafety/publications/
biotech/en/20questions_en.pdf

See also: Agricultural Science; Animal Breeding and Husbandry; Bioengineering; Genetic Engineering.

GENETIC ENGINEERING

FIELDS OF STUDY

Genetics; biology; molecular genetics; pharmacology; botany; cell biology; ethnobotany; ecology; developmental biology; evolutionary biology; microbiology; molecular biology; xenobiology; soil science; geology; chemistry; parasitology; zoology; biophysics; agroecology; agronomy; agricultural engineering; biological systems engineering; food engineering; food science; animal husbandry; agrology; plant science; bioengineering; environmental engineering; experimental evolution; biotechnology.

SUMMARY

Genetic engineering, also known as genetic modification, is an interdisciplinary scientific technique using molecular techniques to directly alter the basic genetic blueprint (DNA) of bacteria, plants, animals, humans, and other living organisms to achieve or enhance a specific trait or useful characteristic. Genetic engineering is used in diverse areas, including medicine and agriculture, to diagnose and treat diseases, produce industrial products, neutralize pollutants, create hardier crops, and perform scientific research. The genetic engineering process uses the tools of molecular genetics to explore and change living systems on a fundamental level and has revolutionized scientists' ability to understand, modify, and enhance the natural world.

KEY TERMS AND CONCEPTS

- **Biotechnology:** Use of modified living organisms in industrial and production processes. This technology has applications in industrial, agricultural, medical, and other production arenas.
- **Classic Selection:** Selective breeding of plants or animals with a desired feature so that the feature becomes more common or dominant in the subsequent generations.
- **Cloning:** Creating an identical copy of a gene, cell, or organism.
- **DNA (Deoxyribonucleic Acid):** Material within a cell that contains the genetic instructions used in the growth, development, replication, and functioning of living organisms. DNA forms the basic building blocks of genes.
- **Gene:** Basic unit of inheritance that contains information and instructions for the creation and maintenance of a living organism; consists of a segment of DNA.
- **Genetically Modified Food (GMF):** Food directly produced by genetic engineering or containing material from a genetically engineered plant or animal.
- **Genetically Modified Organism (GMO):** Organism whose genes have been purposefully modified through genetic technologies for a particular purpose.
- **Genome:** Collection of all genes and DNA present in a particular organism.
- **Insertion:** Addition of one or more pieces of genetic material into a particular DNA sequence or gene.
- **Isolation:** Process of removing a desired DNA segment or gene and placing it into a carrier organism (vector) for later insertion into particular gene or DNA sequence.
- **Ligase:** Type of enzyme that can be used to link or glue together DNA segments.
- **Plasmid:** Circular form of DNA used frequently as a vector to carry a desired gene segment into the organism being modified.
- **Recombinant DNA Technology:** Procedure used to move specific genetic information from one organism into another.
- **Restriction Enzyme:** Bacterial chemical or enzyme that cuts DNA at a specific site; used to remove a particular gene or gene segment from an organism.
- **Transformation:** In genetic engineering, the genetic alteration of a cell as a result of inserting new genetic material into an organism's standard genetic code.
- **Transgene:** Gene or genetic material that has been transferred from one organism into another organism of a different species.
- **Vector:** Virus or other chemical carrier that is used to deliver genetic material to a cell.

DEFINITION AND BASIC PRINCIPLES

Genetic engineering is the direct and purposeful alteration of an organism's DNA, the basic genetic blueprints of a bacterium, plant, animal, human, or other living organism to add or enhance a specific characteristic or trait. Although genetic engineering is most often discussed in the controversial arenas of crop production or theoretical human genetic manipulation, genetic engineering is used in diverse areas such as medicine, industry, and agriculture to treat disease, diagnose problems, produce industrial products, convert industrial waste, create hardier crops, and perform better scientific research.

The focus of genetic engineering is the gene. Genes are the basic units of inheritance that contain information and instructions for the creation, maintenance, and reproduction of living organisms. Genes are composed of DNA, a highly organized molecule located in almost every cell of an organism's body. In genetic engineering, scientists add very specific pieces of useful genetic material to another organism's genes to change an organism's natural characteristics.

Genetic engineering was made possible by the development of new molecular genetic procedures, often called recombinant DNA technology, that can identify the particular DNA sequence of a gene or an entire genome, allow scientists to find the genetic material that codes for useful or desired features, and then insert the new material into the correct place in another organism's genetic code.

Organisms that have had new genetic material inserted into their code are referred to as genetically modified organisms (GMOs) or genetically engineered organisms (GEOs). Examples of GMOs range from corn that has been engineered to produce an innate insecticide to cows that produce milk containing human insulin.

BACKGROUND AND HISTORY

Before modern genetic engineering was possible, farmers had long selected for desired traits by breeding plants and animals with the desirable traits. Brewers and bakers also changed grains and flour into preferred products such as beer and bread through the use of small organisms called yeast and the process of fermentation.

By the early twentieth century, plant scientists had begun to use the work done by Gregor Mendel in the nineteenth century on the inheritance patterns of specific plant features to more formally introduce improvements in a plant species in a process called classic selection. However, the features of the basic unit of inheritance were not known until James D. Watson and Francis Crick identified the structure of DNA in 1953.

The nature of DNA and the technology to manipulate and modify the genetics of an organism was not available until twenty years later, when the first successful recombinant organism was created by Herbert Boyer and Stanley Cohen. Boyer and his laboratory had isolated an enzyme that could precisely cut segments of DNA in an organism, and Cohen found a way to introduce antibiotic-carrying plasmids into bacteria and a way to isolate and clone the genes in the plasmids. They combined their knowledge to create a way to clone genetically engineered molecules in foreign cells. Their discoveries led to the creation of a quick and easy way to make chemicals such as human growth hormone and synthetic insulin.

After Boyer and Cohen, many other scientists worked with recombinant DNA techniques to improve the procedures and develop a variety of genetically modified organisms (GMOs) designed to meet specific scientific, agricultural, industrial, and medical needs. Over time, these techniques and applications in genetic engineering spawned the multibillion-dollar biotechnology industry.

As the biotechnology industry grew and genetically modified organisms became more widespread, it became important to define which organisms were genetically modified organisms and which were products of classic selection. It also became necessary to determine if living organisms produced through genetic engineering could be patented by the companies and universities designing them. In 1980, the U.S. Supreme Court ruled that genetically altered life-forms can be patented.

In 1982, the U.S. Food and Drug Administration (FDA) approved the first consumer product developed through modern genetic engineering: a biosynthetic human insulin, sold under the trademark Humulin. The bacterially produced insulin created by Genentech and marketed by Eli Lilly revolutionized the treatment of diabetes, as it produced fewer immune reactions and its supply no longer depended on the availability of animals.

In 1996, Genzyme Transgenics (which in 2002 became GTC Biotherapeutics) created a transgenic goat that produced milk containing a cancer-fighting protein. It soon created additional transgenic animals that could produce specific human proteins to treat human disease. The ability to produce human hormones, enzymes, and other therapeutic products has decreased the risk of disease transmittal from donors to recipients of human products, increased supply, decreased immune reactions, and decreased the variability between medication batches that had been seen in the past.

In the 1990's, scientists sought to develop genetically modified plants and crops. By 1992, the first plant designed for human consumption (the Flavr Savr tomato) was approved for commercial production by the U.S. Department of Agriculture. In 1994, the European Union approved genetically modified tobacco in France. After these genetically modified crops gained approval, genetically engineered plants and other organisms became more widespread in the United States. It has been estimated that 60 to 70 percent of food products on store shelves may contain at least a small quantity of genetically engineered crops.

The Human Genome Project (1990-2003), a collaborative international scientific research initiative spearheaded by the National Institutes of Health, advanced genetic engineering by its publication of human DNA sequencing data. These data allowed scientists to learn more about the physical and functional aspects of genes and DNA. By its completion, the Human Genome Project had provided a basic genetic road map for scientists to find human DNA segments of interest.

During the 1900's, scientists also used genetic engineering technology to develop numerous varieties of investigational organisms with very specific characteristics for use in research. Some genetically modified organisms were used to learn more about the natural progression of particular diseases. Others were created to test experimental therapies before moving to humans. These genetically modified organisms have helped scientists learn more about genetic disease, cancer, aging, and other chronic diseases.

In academic and industry laboratories, modern genetic engineering continues to solve problems related to health, disease, industry, and agriculture. Additional applications in humans and human disease have been assisted by government-funded initiatives such as the Human Genome Project. Although controversial at times, genetic engineering is a modern tool to be used in addressing a wide range of problems.

HOW IT WORKS

Although the types of organisms modified vary substantially in genome size and structure, all genetic engineering involves several general steps: identifying the desired feature or end application, isolating the gene segment that codes for the feature, inserting the gene segment into a vector, and adding it to the target organism, a process called transformation.

Identification. To create genetically modified organisms that will meet a specific need or solve a particular issue, the best way to engineer a solution must be determined. For example, if a large oil spill required cleanup, the first step would be to determine the type of organism and the desired features that would be most effective at removing the spilled oil. Issues to consider in solving a problem through genetic engineering include the desired size and type of organism to be modified, the availability of desired characteristics or features with a known DNA segment, and the possible positive and negative environmental impact resulting from a modified organism.

Isolation of the Proper Gene Segment. To modify an organism through insertion of a gene or DNA segment, the specific DNA segment in the donor organism must be known and be able to be effectively removed from its host genome. In some cases, the DNA segment coding for the desired characteristic is known and available because of previous scientific work. In other cases, this step can be very labor intensive and require long-term research.

After the DNA segment is identified, a particular recombinant technique, often a restriction enzyme, is used to cut the desired gene or DNA segment out of the donor organism and move it into a vector. Depending on the genetic engineering requirements, other techniques such as polymerase chain reaction or agarose gel electrophoresis can be used to isolate a gene or gene segment.

Insertion. Insertion is the genetic engineering step during which the desired gene or DNA segment is integrated into the vector. In this step, restriction enzymes are used to cut the vector open in a particular place so that the desired gene or DNA segment can attach itself to the vector. Then a special enzyme glue called ligase is used to attach the DNA segment to the vector. The most commonly used vectors in genetic engineering are circular form of DNA called bacterial plasmids; however, the type of vector used for a particular application depends on the size of the gene or segment being moved and the organism being modified.

Transformation. The next step of genetic engineering is called transformation. During this step, the desired gene or DNA segment is introduced and successfully added to the organism being modified. A variety of methods can be used to send the vector containing the desired DNA segment into the organism being modified in such a way that the new genetic information is added to the organism's standard genes. These methods include microinjection, use of a gene gun, electroporation, or use of viruses. In each of these methods, the vector carrying the desired genetic segment is forced into the new cells of the organism being modified. Completion of the transformation step relies on testing that determines whether the inserted DNA segment is producing the desired effect or trait in the newly modified organism.

APPLICATIONS AND PRODUCTS

Genetic engineering has far-reaching applications in food production, industry, medicine, and research.

Crops. One of the most widespread but also most controversial uses of genetic engineering is in the creation of genetically modified crops and food. The goal of genetic modification varies from crop to crop. Soybeans have been modified with a DNA segment conveying resistance to herbicides sprayed over fields to kill weeds growing amid the soybeans. The Flvr Savr tomato was engineered to decrease ripening time and increase shelf life. Varieties of rice and corn have been engineered using DNA segments for other plant genomes to have increased levels of vitamins.

From the first commercially grown genetically engineered product for human consumption (the Flavr Savr tomato), adoption of genetically engineered crops in the United States has increased quickly. According to data from the U.S. Department of Agriculture's Economic Research Service, fields planted with genetically engineered cotton (herbicide-tolerant and insect-resistant cotton) reached 88 percent of the total cotton acreage in 2009. That

same year, herbicide-tolerant soybeans and biotech corn accounted for 91 percent and 85 percent, respectively, of their crop populations.

Supporters of genetically modified crops feel that the plants can increase food production to meet the world's needs using lower amounts of pesticides and increasing farmer profits. Opponents of genetically engineered crops are concerned about perceived safety issues regarding food produced from these crops, ecological issues around increased use of herbicides, contamination between genetically modified and naturally grown crops because of cross-pollination of fields, and economic difficulties dealing with patents on genetically modified crops.

Livestock. Although farmers have bred particular varieties of livestock such as cows, goats, chickens, and sheep for thousands of years to maximize desirable qualities, genetic engineering allows a more rapid introduction of specific qualities that may or not occur naturally in the animals. The benefits of genetically engineered livestock are numerous and affect the producers, environment, and consumers. Producers benefit by having disease-resistant, increasingly productive, or fast-growing animals. For example, the gene responsible for regulating milk production in cows can be modified to increase milk production. Also, if animals are engineered to have milder waste, the environment will benefit. The FDA is reviewing genetically modified pigs that are better able to digest and process phosphorus in ways that release up to 70 percent less phosphorus in their waste. Consumers benefit from more nutritious, vitamin-enriched meat, as in the case of pigs that are engineered to produce omega-3 fatty acids through the expression of a roundworm gene.

Many of the concerns that apply to crops also apply to livestock. As of 2010, the FDA had not approved any genetically engineered meat products for human consumption. In September, 2010, the FDA stated that a type of salmon called AquAdvantage, genetically modified to grow twice as fast as conventional Atlantic salmon, was safe for human consumption. The FDA has placed several genetically modified animals under review in a category called food-drug. The FDA considers the DNA segment to be like a drug and is regulating transgenic animals in the same way it oversees animals that receive growth hormones or antibiotics.

Diagnosis. Genetic engineering has allowed the

development of faster, cheaper, and more accurate diagnostic tests for certain diseases to be used both in the laboratory and in the body. The tests based on genetic engineering are used to identify infectious diseases, hormonal changes, pregnancies, cancer, and other diseases and conditions. For example, a series of faster and more accurate tests for the presence of the human immunodeficiency virus (HIV) have been developed based on genetically modified HIV antigens. Other tests can diagnose diseases by detecting particular substances in specific locations in the body. These exams rely on genetically modified antibodies with markers that can be injected into the body.

Medications. The use of genetically modified organisms to produce human hormones, enzymes, vaccines, and medications has revolutionized the pharmaceutical industry. Since 1982, when Genentech's biosynthetic human insulin was introduced, the ability to manufacture new products in a controlled environment instead of collecting similar substances from the limited supply of human and animal sources has led to more readily available, effective, and reliable medications. Products include human growth hormone to treat children with insufficient growth, plasminogen activator to dissolve blood clots, and erythropoietin to treat low blood iron (anemia). In 1994, genetic engineering also led to innovative treatments for rare genetic disorders such as Gaucher disease with the production of specific human enzymes in genetically modified Chinese hamster ovary cells. The point of this enzyme replacement therapy is to replace the enzyme that the affected individuals are missing through intravenous injections of genetically engineered human enzymes. In multiple situations, the use of genetically engineered organisms to create medications has saved lives and decreased the burden of disease in ways that could not be imagined before. Future applications of genetic engineering in medicine are likely to focus on the creation of better medications for life-threatening indications.

Disease Cures. Genetic engineering made it possible to develop gene therapy. Genetic diseases are inherited conditions that occur because of one or more genetic changes or mutations that prevent the correct functioning of a particular gene. Most genetic diseases do not have a treatment or cure. However, with genetic engineering techniques, scientists hope that they will be able to transform an affected individual's mutated gene into a working gene

---

Fascinating Facts About Genetic Engineering

- Biotechnology has created more than two hundred new therapies and vaccines, including products to treat genetic diseases, cancer, diabetes, acquired immunodeficiency syndrome (AIDS), and autoimmune disorders.
- Scientists successfully manipulated the genetic sequence of a rat to grow a human ear on its back.
- If all the DNA in one person were laid in a straight line, it would stretch to the Sun and back more than thirty times.
- Glow-in-the-dark zebra fish are available for purchase at some pet stores. The glowing fish was made by adding naturally occurring genes from fluorescent organisms such as jellyfish and sea coral to the zebra fish's genome.
- Genetically engineered plants are being developed to detoxify pollutants in the soil or absorb and accumulate polluting substances in the air.
- The Human Genome Project produced a complete catalog of the human genome. This catalog is the size of several hundred average-sized telephone books and contains nothing but the letters A, T, G, and C in various permutations and combinations.
- Scientists still do not know the function of more than 80 percent of human DNA.
- Cows have been genetically engineered to produce medically useful proteins in their milk.

---

by replacing it with a functional copy of the gene. Gene therapy has shown some success in helping individuals with severe combined immunodeficiency (SCID), hemophilia type B, and several other genetic diseases; however, this type of treatment is still under investigation to determine how to safely and permanently cure genetic conditions.

Research. Genetic engineering and genome sequencing have been used to improve investigative techniques through the ability to manipulate organisms on a basic, genetic level. In genetic research, genetic engineering techniques have allowed scientists to create mice and other organisms affected by a specific gene change for detailed study of a specific genetic disease. For example, a genetically modified mouse that lacks the gene to produce amyloids has been used to study Alzheimer disease. On a broader scale, genetic engineering allows detailed analysis of an organism's structure, function, and development. Through the insertion of a marker in or near a gene coding for a product of interest, scientists can track the location of that gene's product over time.

Industrial Applications. Genetically engineered organisms are used in several manufacturing arenas in production, processing, and waste removal. Most industrial applications of biotechnology are based on naturally occurring processes using modified bacteria, yeast, and other small organisms to digest, transform, and synthesize natural materials from one form into another. More specifically, genetically modified microorganisms have been used to produce industrial chemicals such as ethylene oxide (for making plastics), ethylene glycol (antifreeze), and alcohol. Bacteria have also been engineered to remove toxic wastes from the environment, for example, the varieties of genetically engineered bacteria that consume oil by chemically transforming its compounds into usable basic molecules. Future directions in industry include production of textile fibers, fuels, plastics, and other industrial chemicals out of industrial wastes or raw materials.

IMPACT ON INDUSTRY

In 2009, biotechnical companies in the United States, Europe, Canada, and Australia had an aggregate net profit of $3.7 billion. In their 2010 report, Ernst and Young found that biotechnology companies in the United States, Europe, and Canada raised $23.2 billion in capital for 2009. This means that there is plenty of capital to create and market new genetic engineering applications. The biotechnology companies also report rapid job growth, which has been fueled primarily by growth in research, testing, and medical laboratories. The biotechnology industry is an important area of economic growth in the world as major advancements and innovations in medicine, agricultural, and industry based on genetic engineering continue globally.

Government and University Research. Funding for basic genetic engineering research often begins with the government-funded National Institutes of Health (NIH). The NIH performs on-site genetic research and awards grants to researchers in academic and university centers.

Beyond direct funding, various sectors of the government also encourage and stimulate research in genetic engineering. For example, the FDA is responsible for regulating genetically engineered foods and requires research studies on the safety and efficacy of particular gene changes in plants and animals before approval for consumption. In addition, the U.S. Department of Agriculture tracks the use of genetically modified crops worldwide and routinely performs research in the United States and internationally on food, farming, and biotechnology in order to issue industry reports.

Because of the wide scope of genetic engineering, research has been conducted by multiple academic groups, including the Zinc Finger (ZF) Consortium, whose members have focused their research on one method of targeting specific regions of DNA, and the Human Genome Project, an international public consortium of research laboratories led by the United States that worked together to complete the first sequence of the human genome. Groups such as the Registry of Standard Biological Parts, a consortium database of standardized gene sequences in vectors with known characteristics ready to be used in genetic engineering, have been formed primarily to encourage data sharing.

Industry and Business. Genetic engineering has led to the creation of an entirely new industry called biotechnology. Under the umbrella of biotechnology are diverse groups of companies that apply genetic engineering principles in bacteria, plants, animals, and humans to research, manufacturing, health issues, and marketing. Because of the fast-paced nature of scientific discovery, these companies are designed to fund cutting-edge research and then quickly integrate their findings into production plans. The four main sections of the biotechnology industry are agricultural feedstock and chemicals, drugs and pharmaceuticals, medical devices and equipment, and research-testing-medical laboratories. The use of genetically engineered organisms in production, processing, and waste removal stimulates research and development internally in these companies as they develop more efficient, high-yield processes.

CAREERS AND COURSE WORK

Courses in molecular genetics, biochemistry, human genetics, developmental biology, cell biology, microbiology, biological systems engineering, and biotechnology are the foundational requirements for students interested in pursuing careers in genetic engineering. Depending on the student's desired area of study, courses in animal husbandry or medicine might be required. A bachelor's degree in biology, chemistry, applied biotechnology, or genetics is appropriate preparation for graduate work in genetic engineering. In most circumstances a master's degree, medical degree, or doctorate is necessary for the most advanced career opportunities in both academia and industry. However, technician and laboratory roles for those without advanced degrees are often available.

Other career possibilities include marketing and sales staff for pharmaceutical companies, doctors and nurses involved in prenatal genetic testing, genetic ethicists, or genetic counselors. Whether students pursue a scientific, academic, or socially oriented position, they should take a variety of courses beyond the natural sciences to be aware of cultural and societal issues surrounding genetic engineering.

SOCIAL CONTEXT AND FUTURE PROSPECTS

Genetic engineering has already altered the course of agriculture, industry, and medicine with its life-changing applications. Crops have been modified so that they are more nutritious and naturally produce pesticides. Life-saving medications made of human hormones integrated into bacteria are widely available in consistent and purified forms. Bacteria that convert toxic chemicals into harmless basic elements have been developed. Great strides have been made in using gene therapy to cure genetic diseases. However, these significant scientific strides also come with important ethical questions and safety concerns.

Environmentalists are concerned about the impact of genetically modified crops on ecosystems, in particular whether the genes introduced into genetically modified crops will be transferred to conventional crops through cross-pollination.

Advocacy groups such as Greenpeace and the World Wildlife Fund are concerned about the safety of genetically modified food and feel that the available data do not prove that there are no risks to human health from consumption. Despite statements from the Royal Society of Medicine and the U.S. National Academy of Sciences in support of the safety of such foods, these groups have called for additional and more rigorous testing before genetically engineered foods are marketed.

Other countries have significant concerns about the safety of genetically modified foods. The European Union regulates genetically modified food imported from other nations, including the United States. Venezuela has banned the growing of genetically modified crops, and India has issued a moratorium on the cultivation of genetically modified foods pending an investigation into safety concerns. Other counties such as Japan and Zambia also have registered concerns over the safety of genetically modified foods.

There is significantly less controversy over the use of genetically modified organisms in industrial production and medicines. However, the use of genetic engineering techniques for human gene therapy and related applications has touched off a firestorm of ethical debate.

Dawn A. Laney, M.S., C.G.C., C.C.R.C.

FURTHER READING

Avise, John C. *The Hope, Hype, and Reality of Genetic Engineering: Remarkable Stories from Agriculture, Industry, Medicine, and the Environment*. New York: Oxford University Press, 2004. Contains a series of well-written essays about more than sixty genetically engineered organisms. Each short essay discusses the procedures and challenges involved in engineering each organism or trait.

Hodge, Russ. *Genetic Engineering: Manipulating the Mechanisms of Life*. New York: Facts On File, 2009. Covers genetics from the beginning to genetic engineering. Examines both the science and the social issues.

Nicholl, Desmond S. T. *An Introduction to Genetic Engineering*. New York: Cambridge University Press, 2008. Describes the basic principles of genes, molecular biology, methods used in genetic engineering, and applications of genetic engineering.

Shanks, Pete. *Human Genetic Engineering: A Guide for Activists, Skeptics, and the Very Perplexed*. New York: Nation Books, 2005. A well-written discussion of genetic engineering and biotechnology that covers cloning, stem cells, gene therapy, and the genetic engineering of food. The author, who does not support genetic engineering, provides a good overview of genetic engineering debates.

Yount, Lisa. *Biotechnology and Genetic Engineering*. 3d

ed. New York: Facts On File, 2008. A compact overview of issues in the genetic engineering field. Examines law and genetic engineering and key individuals involved in genetic engineering. Contains an extensive time line and glossary.

WEB SITES
Union of Concerned Scientists
Impacts of Genetic Engineering
http://www.ucsusa.org/food_and_agriculture/
science_and_impacts/impacts_genetic_engineering/impacts-of-genetic.html

U.S. Department of Agriculture
Agricultural Biotechnology Website
http://www.usda.gov

World Health Organization
Twenty Questions on Genetically Modified Foods
http://www.who.int/foodsafety/publications/
biotech/20questions/en

See also: Animal Breeding and Husbandry; Bioengineering; Biosynthetics; Cloning; DNA Analysis; DNA Sequencing; Environmental Microbiology; Genetically Modified Organisms; Genomics; Human Genetic Engineering.

GENOMICS

FIELDS OF STUDY

Genetics; bioinformatics; molecular genetics; microbial genetics; epigenetics; behavioral genetics; neurogenetics; phylogenetics; population genetics; ancient DNA; genetic anthropology; paleogenetics; evolutionary biology; evolutionary biochemistry; archaeology; forensics; biology; cell biology; microbiology; molecular biology; biophysics; synthetic biology; systems biology; biochemistry; chemistry; systematics; taxonomy; virology.

SUMMARY

Genomics is the branch of biotechnology that focuses on the genome, the entire set of genes in an organism, as well as the interaction of individual genes with one another and the organism's environment. Broad applications include comparative and functional genomics.

KEY TERMS AND CONCEPTS

- **Ancient DNA:** DNA sequences from museum specimens, archaeological finds, fossils, mummified remains, and ancient microorganisms that were embedded in ice, rock, or amber.
- **Bioinformatics:** Discipline that uses computer programs to scan sequences of genomic DNA to retrieve and analyze specific genes.
- **Chromosome:** Structural unit of genetic material that consists of a lone, double-stranded molecule, found in eukaryotes.
- **Comparative Genomics:** Field of biological research that compares the genomes of different species.
- **DNA (Deoxyribonucleic Acid):** Long linear polymer found in the nucleus of a cell, consisting of nucleotides in the shape of a double helix.
- **Gene:** Segment of DNA that serves as the functional and physical unit of heredity.
- **Genome:** Entire set of DNA found in an organism.
- **Genomic Library:** Collection of bacteria that have been genetically engineered to hold the whole genome of an organism
- **Messenger RNA (mRNA):** Single-stranded molecule of RNA that functions as a template for the production of protein on the ribosomes of the cell.
- **Mitochondrial DNA (mtDNA):** Genetic material within the mitochondria that is passed from the female parent to her children.
- **Ribosomal RNA (rRNA):** RNA found in the ribosome, which translates mRNA and synthesizes proteins.
- **RNA (Ribonucleic Acid):** Single-stranded polymer consisting of ribose nucleotides formed by transcription of DNA or viruses from another RNA strand.
- **Sequencing:** Technique that elucidates the sequence of the molecules that form an organism's genomic content.

DEFINITION AND BASIC PRINCIPLES

Genomics studies organisms' entire genomes, focusing on structure, function, and inheritance. It examines complex diseases, including cancer, heart disease, and diabetes, in which both genetic and environmental factors play a part. In contrast, genetics focuses on individual genes and their role in inheritance and heritable genetic diseases such as phenylketonuria and cystic fibrosis. Of central importance in genomics is determining the sequences of DNA that make up an organism. DNA sequencing can be used to find mutations or variations that may be involved in causing a disease. Functional genomics examines how individual genes function and interact with each other within a given genome at the functional level, and comparative genetics examines the relationships between different species. Comparative genomics may be used to study evolutionary relationships between species by comparing their chromosomes.

BACKGROUND AND HISTORY

Scientists became aware of the existence of genomes in the late nineteenth century when they first viewed chromosomes under a microscope, although the word "genome" was not coined until 1920. In the twentieth century, scientists studied the frequency at which chromosomes exchanged parts during meiosis to map the genes on chromosomes. However, this technique was useful for mapping primarily genes that had mutant phenotypes, which accounted for only a small part of the total genome.

Numerous scientists discovered the base-pair sequences of many human genes, but mapping the entire human genome remained a challenge. Determining the entire human genome was generally regarded as a worthwhile endeavor, but the process would take billions of dollars, taking funding away from more established, mainstream biomedical research. Ethicists, scientists, and economists debated the merits of the research versus its exorbitant cost. Nevertheless, the Human Genome Project was launched in 1990, supported by the U.S. Department of Energy (Human Genome Program, directed by Ari Patrinos) and the National Institutes of Health (National Human Genome Research Institute, directed by Francis S. Collins). The Human Genome Project became an international effort, with scientists from around the globe joining the team. Newly developed computer software helped facilitate the process. In 1998, Celera Genomics, a private company headed by J. Craig Venter, began an independent effort to complete the mapping of the human genome using an unorthodox shotgun, or whole-genome, method of sequencing. In 2000, Collins, Venter, and Patrinos jointly announced the completion of a rough draft of the human genome. In 2003, fifty years after Francis Crick and James D. Watson discovered the double-helix structure of DNA, the human genome had been sequenced, and the project was declared complete, although final papers were published in 2006.

HOW IT WORKS

Determining an organism's genome requires the sequencing of its DNA, which involves finding the order of the bases in a given stretch of DNA. Several sequencing methods were developed in the 1970's, but sequencing was a very slow process.

One problem was that the total amount of DNA necessary for sequencing and analyzing a genome of interest could be several times the total amount of available DNA. To increase the amount of DNA, scientists turned to cloning. The DNA fragments are replicated inside a bacterial cell, then the cloned DNA is extracted and placed in a machine for sequencing. The polymerase chain reaction, developed in 1983, allows ancient DNA or degraded DNA to be sequenced using special protocols to amplify the sample before sequencing.

Genomic sequences are usually elucidated with automatic sequencers; shotgun, (whole-genome) sequencing, developed by Venter, is the most popular DNA sequencing technique. The DNA is randomly cut into fragments, and then the ends of each fragment are sequenced, resulting in two reads per fragment. The reads are used to reconstruct the original DNA sequence. Newer sequencing techniques generally follow this model but may use different strategies along the way. As the fragments are sequenced, the data are set aside. When enough sequences have been gathered, they are linked through sequence overlaps. The resulting genomic sequence is then deposited into a publicly accessible database.

The genomic sequence is analyzed to discover the individual genes and how they are regulated. Bioinformatics is the discipline that relies on computer programs to search for genes in DNA sequences, using what is known about the gene of interst. After scientists find the sequences that make up a gene, the structure and function can be compared with similar gene sequences in other organisms.

APPLICATIONS AND PRODUCTS

Genomics has many applications. Among the most important applications, because of their impact on society, are those that deal with evolution and the environment. Other areas, such as synthetic biology and personal genomics, may lead to innovations that can be of great benefit to individuals and populations and to public health.

Evolutionary Genomics. Until about 30,000 years ago, Neanderthals, the closest relatives of modern humans, inhabited Europe and Asia. Since the early twentieth century, anthropologists and paleontologists have attempted to demonstrate an evolutionary relationship between Neanderthals and modern humans, who emerged about 400,000 years ago. In 1997, Svante Pääbo, director of the Department of Genetics at the Max Planck Institute for Evolutionary Anthropology, furthered the understanding of the genetic relationship of Neanderthals and modern humans when he sequenced Neanderthal mtDNA. In 2009, scientists from the Max Planck Institute announced that they had generated a draft sequence equivalent to more than 60 percent of the complete Neanderthal genome and had begun comparing it to that of humans. In May, 2010, these researchers revealed that the human genome contains some of the same genes as were found in the Neanderthal genome. Their continued efforts may shed light on the origin of humankind as well as the evolutionary process.

Environmental Genomics. In 2009, the International Union for Conservation of Nature estimated that Earth is home to 8 million to 14 million animals and plants, of which only 1.8 million have been identified and classified. Naturalists such as Charles Darwin classified plants and animals using taxonomic systems. Later scientists compared the genetic differences among species; a species was defined as a group of organisms that could breed and produce fertile offspring. Determining where one species ends and another begins is difficult, as in the case of wolves and coyotes. Modern technology allows species to be identified and classified by comparing modern and ancient samples of genomic DNA and tracking how a species descended from an ancestor. The question of what constitutes a species is important in determining which species can be considered endangered. Global climate change has led to the extinction of many plant and animal species, and many scientists hope to be able to preserve the world's biodiversity using knowledge gleaned from genomics.

In 2004, Venter and his colleagues at the Institute for Biological Energy Alternatives (later part of the J. Craig Venter Institute), of Rockville, Maryland, applied whole-genome sequencing to microbial populations collected from the Sargasso Sea, close to Bermuda, where researchers expected a low diversity of species. They analyzed more than 1 billion base pairs and found 1.2 million new genes and 1,800 species of microbes (including 150 new species of bacteria). According to Venter and his colleagues, the number of species suggests that microbial life in the ocean is more plentiful and diverse than previously thought.

Synthetic Biology. One of the goals of synthetic biology, which creates artificial biological systems (not existing in nature) and redesigns existing biological systems, is intelligent design. Synthetic biology applies the principles of large-scale engineering to biology and is built on the premise that organisms can be divided into discrete parts. In 2010, researchers at the J. Craig Venter Institute created a self-replicating bacterial cell containing totally synthetic DNA. The cell's genome was designed in the computer. No natural DNA was used, and chemical synthesis was used to make a viable cell.

Synthetic biologists believe that someday it may be possible to program bamboo to grow into preformed chairs or to program trees to spew oil from their stems. Reprogrammed bacteria may be able to heal, not harm, humans and animals. Some experts, such as David Rejeski of the Woodrow Wilson International Center for Scholars in Washington, D.C., believe that synthetic biology may fundamentally change the way things are made within the next one hundred years, creating manufacturing shift as significant as the Industrial Revolution.

IMPACT ON INDUSTRY

Genome Decoding for Individuals. The cost of sequencing the first human genome was about $3 billion. However, advances in technology have caused the price of obtaining a complete genome to drop rapidly. In 2010, Ilumina, the market leader in DNA-sequencing machines, offered to provide people with their genome for $9,500 if they might benefit from the information (as is the case with people with rare cancers). Life Technologies, Ilumina's closest competitor, is working with a group of cancer research centers to see how genetic information might help in the treatment of cancer. It is also offering a sequencer that sells for around $100,000, in contrast to the typical $700,000. The DNA-sequencing industry saw the entrance of Pacific Biosciences, which offers sequencing in a matter of minutes rather than days, and Complete Genomic, which has created a sequencing facility for drug companies to use.

Although genome decoding had focused on healthy individuals, in March of 2010, two research teams independently decoded the entire genome of patients who had genetic diseases to find the exact genes that caused their diseases. The lower cost of DNA sequencing may revive the largely unsuccessful efforts to identify the genetic causes of major killers such as heart disease, diabetes, and Alzheimer's disease. For example, James R. Lupski, a medical geneticist who has a Charcot-Marie-Tooth, a neurological disease, used whole-genome sequencing on his own DNA in an effort to better understand his illness. He was able to find the genetic variation causing his condition, although earlier research had not been able to pinpoint it.

Genomic Ventures. Many genomics companies formed in the 1990's and 2000's, but many are still in the research and development process and have yet to market products. Most of these companies apply genomics to the creation of pharmaceuticals, although some use genomic studies for environmental purposes or in synthetic biology. Many are involved in cooperative ventures with pharmaceutical companies, and mergers and acquisitions are common.

In October, 2006, Venter (former head of Celera Genomics) brought together several affiliated organizations, The Institute for Genomic Research (TIGR), The Center for the Advancement of Genomics (TCAG), The J. Craig Venter Science Foundation, The Joint Technology Center, and the Institute for Biological Energy Alternatives to create the J. Craig Venter Institute, with offices in Rockville, Maryland, and San Diego, California. The institute, known for high-throughput DNA sequencing, has research groups working on human genomic medicine, microbial and environmental genomics, plant genomics, infectious diseases, and synthetic biology and bioenergy. Its approach toward infectious disease is to examine the genome of the microbes and viruses that cause disease and the microbial flora in various cavities of the human body. Another focus is its bioinformatic group, which works on the software, databases, and mathematics to analyze the data created by DNA sequencing.

Other smaller companies are also active in various aspects of genomics. Millenium: The Takeda Oncology Company, based in Cambridge, Massachusetts, was established in 1993 as a genomics company and has evolved into a biopharmacological company focused on cancer. Its leading product is injectable VELCADE (bortezomib) for the treatment of people with multiple myeloma and a type of lymphoma. Rockville, Maryland-based Human Genome Sciences (founded 1992) is developing drugs to treat hepatitis C and systemic lupus erythematosus. Incyte, based in Wilmington, Delaware, is developing a drugs to treat rheumatoid arthritis, psoriasis, and cancer.

Affymetrix, based in Santa Clara, California, developed the first microarray (for genome-wide analysis) in 1989 and the first commercial microarray in 1996. Its genomic analysis tools are invaluable for researchers and pharmaceutical companies. San Diego, California-based Verdezyne, switched from being a provider of synthetic genes to drug and industrial enzyme companies to focus on a fermentation process to produce renewable energy fuels and

chemicals, such as ethanol and adipic acid. It is using genomics technology to create proprietary metabolic pathways in yeast for enhanced conversion of hexose and pentose sugars to ethanol.

Government Agencies. One of the main governmental agencies involved in genomics is the National Human Genome Research Institute, part of the National Institutes of Health. It began as part of the Human Genome Project and continues to conduct research into the genetic basis of human disease. It is divided into seven branches: Cancer Genetics, Genetic Disease Research, Genetics and Molecular Biology, Genome Technology, Inherited Disease Research, Medical Genetics, and Social and Behavioral Research.

The U.S. Department of Energy is also involved in genomics research. Its Genomic Science Program focuses on using genomics to deal with issues regarding the environment, energy, and the climate. The agency is involved in the genome sequencing of microbes that can help cycle carbon from the atmosphere, clean up toxic waste, and create biofuels. The DOE Joint Genome Institute, operated by the University of California, Berkeley, concentrates on providing clean energy and environmental solutions.

CAREERS AND COURSE WORK

There are many pathways to a career in genomics. However, a combination of biology and mathematics s essential for genomics. High school students interested in genomics should take courses in the basic sciences—biology, chemistry, mathematics, and physics—and computer science. At the college level, recommended courses are genetics, physiology, biology, cell and molecular biology, microbiology, developmental biology, physics, linear algebra, probability and statistics, and computer science (in particular, programming). Suitable majors include biology, biochemistry, bioinformatics, genetics, and computer science. Laboratory experience is also necessary for the would-be researcher. A master's or doctorate in genetics or genomics is required for most research, whether in private industry, academic research centers, or governmental agencies.

SOCIAL CONTEXT AND FUTURE PROSPECTS

The U.S. Department of Energy and the National Institutes of Health have devoted between 3 and 5 percent of their annual genome project budgets

Fascinating Facts About Genomics

- In 2005, the dog genome sequence was published in *Nature* magazine. The DNA sequenced came from Tasha, a purebred female boxer. Scientists think that the genetic contribution to disease may be easier to determine in dogs than in humans. About 5 percent of the human genome is also present in dogs.
- Humans and chimpanzees share 96 percent of their genomes.
- The average difference in the genomes of two people is 0.2 percent, or one in five hundred bases.
- About 97 percent of the DNA in the human genome consists of so-called junk DNA—sequences with no known functions.
- Scientists at the J. Craig Venter Institute are studying the genome of the SARS coronavirus, which causes severe acute respiratory syndrome, in humans and animals in an effort to determine how the virus crosses the species barrier.
- In 2010, the complete draft genome sequence of the soybean was published in *Nature* magazine. It was expected to increase understanding of the nitrogen-fixing process and to help scientists develop better soybeans, including a more digestible version.

toward the study of ethical, legal, and social issues. Societal concerns include privacy and confidentiality issues, possible social stigma and discrimination, and how genetic information will be used by insurance companies, the legal system, and academia. Various government agencies have taken the lead in addressing the existing and potential social and ethical issues that may arise from genome-centered biology.

The Genetic Information Nondiscrimination Act of 2008 prohibits insurance companies and employers from discriminating against people based on the results of genetic testing. In addition, under the law, employers and insurers are prohibited from requesting or demanding that individuals take genetic tests.

In an interview in 2010, National Institutes of Health director and former Human Genome Project member Francis S. Collins expressed support for genomic research because it is likely to enable people to prevent and treat disease. He also noted that the patenting of human genes, once very controversial, has become more common, with more than 20 percent of all know genes having been patented. The patents, he argued, help private biotech companies fund costly research into genetic diseases. He also made a point of stating that his religious beliefs do not interfere with his work in evolutionary genetics.

Aware of the legal and social issues and of most people's limited understanding of genetics, the J. Craig Venter Institute has created a division to promote understanding of genomics among policy makers and the general public and to foster a positive image for the biotechnology industry.

The possible benefits of genomic research are immense. Some experts feel that better understanding of the genomics may lead to radical innovations in disease treatment and prevention, and environmental and synthetic genetics may produce ways to create renewable fuels. The decreasing cost of obtaining a person's genome may mean that physicians can take an individualized approach to medicine. Therefore, although controversy over genetics remains, this field is likely to remain an active area of research, yielding numerous applications, and providing numerous work opportunities to those interested in the field.

Cynthia F. Racer, M.A., M.P.H.

FURTHER READING

Davies, Kevin. *The $1,000 Genome: The Revolution in DNA Sequencing and the New Era of Personalized Medicine.* New York: Free Press, 2010. Looks at how less expensive, faster means of obtaining a person's genome will change medicine and make it more tailored to the individual.

DeSalle, Michael, and Michael Yudell. *Welcome to the Genome: A User's Guide to the Genetic Past, Present, and Future.* Hoboken, N.J.: Wiley-Liss, 2005. Starts with a brief history of genetics and description of the science before examining how the genome was sequenced. Analyzes the likely medical and agricultural applications.

Fairbanks, Daniel J. *Relics of Eden: The Powerful Evidence of Evolution in Human DNA.* Amherst, N.Y.: Prometheus Books, 2007. An examination of the field of evolutionary genomics that asserts that there is no dichotomy between religion and science.

Gee, Henry. *Jacob's Ladder: The History of the Human Genome.* New York: W. W. Norton, 2004. Examines what human genome sequencing reveals and how this information may be used in the future.

Gibson, D. G., et al. "Reation of a Bacterial Cell Controlled by a Chemically Synthesized Genome." *Science Express* (May 20, 2010). Announces the creation of a self-replicating bacterial cell governed by a synthetic genome.

Shreeve, James. *The Genome War: How Craig Venter Tried to Capture the Code of Life and Save the World.* New York: Alfred A. Knopf, 2004. Describes the competition between Venter and Collins to decode the human genome.

WEB SITES
DOE Joint Genome Institute
http://www.jgi.doe.gov

Genomic Science Program, U.S. Department of Energy
http://genomicscience.energy.gov/index.shtml

The Human Genome Project
http://www.ornl.gov/sci/techresources/Human_Genome/home.shtml

J. Craig Venter Institute
http://www.jcvi.org

National Human Genome Research Institute
http://www.genome.gov

See also: Bioengineering; Bioinformatics; Biosynthetics; Cloning; DNA Sequencing; Genetically Modified Organisms; Genetic Engineering; Human Genetic Engineering; Metabolic Engineering; Paleontology.

GERIATRICS AND GERONTOLOGY

FIELDS OF STUDY

Medicine; nursing; physical therapy; occupational therapy; psychology; sociology; political science; anthropology; public policy; education; statistics; dentistry; biology; pharmacy; social work.

SUMMARY

Examining life-span development, particularly the later years, has never been so important. Servicing an aging population requires diverse teams of medical doctors, biologists, psychologists, and other professionals, and training such individuals has produced dedicated schools and degrees focused exclusively on the aging process. Students interested in the study of the aged have never before had so many opportunities to expand their careers. Further, the need for teachers and specialists in geriatrics and gerontology is expanding at a rapid pace.

KEY TERMS AND CONCEPTS

- **Activities Of Daily Living:** Basic chores of everyday life, such as eating and bathing, which are often used to determine level of independence.
- **Alzheimer's Disease:** Type of dementia that shrinks the brain, destroying memory and, in later stages, functional ability.
- **Centenarian:** Person who is one hundred years old or older.
- **Chronic Conditions:** Ongoing diseases that account for the most deaths and disability in the United States, such as heart disease and cancer.
- **Compression Of Morbidity:** Belief that a lifetime of healthy activities produces a small period of disability at the end of life.
- **Executive Functions:** Higher-order cognitive abilities including attention, decision making, planning, adapting to change, and inhibition.
- **Preventable Diseases:** Leading causes of death in the United States—heart disease, stroke, cancer, rheumatoid arthritis, and diabetes—that are often the direct result of poor lifestyle choices.

DEFINITION AND BASIC PRINCIPLES

Before the first baby boomer turned sixty in 2006, geriatrics, the medical subspecialty of treating the aged, and gerontology, the comprehensive field of aging studies, both experienced a substantial growth of interest. Although old age might have been viewed as a period of disengagement, as of 2011 nothing is further from the truth. Aging is now often couched in terms of "successful" or "productive"; in fact, many seniors remain as busy after retirement with volunteering activities and the like as when they were employed. Reframing the negative language surrounding aging has changed the perception of aging from a period of consuming goods and services to a period of continued growth and productivity. With an estimated 69 million baby boomers turning sixty-five by 2029, the fields of geriatrics and gerontology have never been more in demand.

Maintaining the health and well-being of this graying section of society involves extensive education, research, and policy initiatives. Thus, the field of gerontology, by necessity, is interdisciplinary in nature. No one field encapsulates the varied systems of inquiry—especially when so many seniors now remain fit and active well into their nineties.

BACKGROUND AND HISTORY

Gerontology as an official field of inquiry first began in 1903, when Russian biologist Élie Metchnikoff coined the term "gerontology" or the "study of old men." Six years later, Austrian physician Ignatz Nascher created the field of geriatrics and in 1914 published the first book on geriatrics. The Social Security Act, enacted in 1935, helped to pull millions of seniors, and others, out of poverty. The first organizations solely dedicated to the study of the aging process were formed in the 1940's: the American Geriatrics Society, founded in 1942, and the Gerontological Society of America, founded in 1945. In 1957, the National Institutes of Health formed the Center for Aging Research. In 1963, physician Sidney Katz and colleagues published the seminal work on gauging the independence of the elderly, the index of Activities of Daily Living—a concept still widely used. Irving Rosow published *Social Integration of the Aged* in 1967, which laid the groundwork for later theories on aging. The White House Conference on Aging, begun in 1961, spawned a formal division on aging at the National Institutes of Health. In 1976, Robert Butler was appointed the first director of that new National Institute on Aging. The 1980's through the 2000's saw an explosion in interest on aging as the baby boomers began to reach retirement age. Countless university centers on aging and degrees in geriatrics and gerontology have been created, and at many universities students may now declare gerontology as their major field of study.

HOW IT WORKS

Theories of Aging. Scientific theories organize the how and why of empirical findings and provide a certain epistemology from which to examine new data. Using supported data-driven theory contextualizes known parameters, permitting a foundation of knowledge from which to build new projects and inquiries into the aging process. Previous data sets and explanations about interrelated phenomena streamline future interventions and research investigations. In geriatrics, biological theories of aging are the norm; however, the interdisciplinary nature of gerontology creates barriers to building comprehensive theories of aging. Although gerontology includes biological theories of aging, these two fields will be examined separately.

Geriatric theories on aging abound, and the more popular theories are briefly reviewed here. The free-radical theory of aging posits that self-multiplying free radicals cause damage to deoxyribonucleic acid (DNA) and healthy cells. The hypothesized antidote is consumption of antioxidants, commonly found in vegetables. The programmed theory of aging proposes that every organism has an expected life span. Similarly, the wear-and-tear theory likens the human body to a machine where constant use degrades the machine, eventually leading to failure. Lastly, the immune-system theory holds that, with age, the human body becomes less able to fight off infection. In a world where health is increasingly viewed as more than the mere absence of disease, geriatricians use biological theories of aging to dispense health-promotion advice: Exercise both mind and body daily, consume the daily requirements of fruits and vegetables, and remain socially engaged in meaningful activities.

Gerontological theories of aging, often referred to as social theories of aging, are diverse in scope.

Although largely refuted, disengagement theory suggests with increased age individuals slowly remove themselves from society. Alternatively and largely supported by early twenty-first-century research literature, activity theory holds a positive relationship between activity levels and happiness and health. Similarly, continuity theory maintains that new roles should be substituted for lost roles; for example, volunteering can replace retirement from a paying job. The life-course perspective on aging holds that aging is in fact a lifelong process and adaptation and change are continuous rather than enacted on by a specific age (sixty-five). These, and other social theories on aging, guide the development of new research endeavors and assimilate such new findings into the evolving language of aging research.

Geriatrics in the Field. Geriatricians have advanced training and certifications in medicine and aging and work under the broad field of geriatrics. Such individuals actively develop health plans, treat comorbidities, promote health and wellness, focus on the prevention of disease and disability, and perform clinical evaluations. Unlike other disciplines, such as cardiology, geriatrics is not focused on a single organ or disease. Therefore other medical personnel with specialty training in aging often make up a treatment team; such a team can include osteopathic physicians, nurses, social workers, physical and occupational therapists, psychologists, and others. Such a team can create a complete wellness portfolio and assess a patient's activities of daily living.

Geriatricians have moved from a prevention of disease model to a health-promotion model that uses the best available research evidence, clinical knowledge, and patient feedback to create a holistic model of well-being. Such evidence-based approaches are now much in demand; with rising health care costs insurance companies demand proven treatments. Geriatrics as a subspecialty of medicine is especially important as older adults can often differ from younger persons in symptoms related to illness and react differently to treatment methods. The success of geriatrics is important for measures of public health as well, where reducing disability and disease would have far-reaching effects on the overburdened medical system.

Generally a person should consider seeing a geriatrician when he or she turns sixty-five, although individual health complications could necessitate an earlier visit. Often by the age of seventy-five, many older adults have multiple chronic conditions, such as sensory and cognitive impairments, that require the services of a geriatrician. Because the life span is seen as continuous, rather than demarcated by specific ages, it is important that individual decisions made throughout one's life are well-informed, and geriatricians are an important piece of the puzzle. Because the rate of aging is determined by the interaction of genetic and environmental conditions, which differ for every individual, it is important that aging seniors see a specialist. Only a qualified medical doctor should make medical decisions.

Gerontology in the Field. Applied gerontologists examine, study, and directly train the aging population and those who work with them in a variety of ways. Such areas include: learning to operate hearing aids, using assistive devices such as canes or walkers, maintaining proper nutrition, adjusting driving habits, proper use of corrective visual aids, and any other area affected by the aging process. Research by gerontologists suggests that change is the key to successful aging; static lives produce static minds and bodies.

Gerontologists also teach and promote preventative interventions to ensure successful aging. Such interventions can retrain mental acuity, strengthen ailing muscles and skeletal structure, teach positive behavioral methods to cope with loss and grieving. Gerontologists have led the aging revolution, where seniors are living longer, healthier, and more engaged lives. Accordingly, seniors are contributing to a level of human capital never before seen. For example, in 2010 Senior Corps, a federal governmental agency, saw its largest increase in senior volunteers since 2004. Part of the reason for this increase in productivity is because technology is changing the way productivity is perceived. No longer is physical stamina required for ongoing employment; technology has permitted older workers to stay in the workforce, thereby increasing social contribution and delaying age-related functional declines.

Gerontologists are quick to point out that productive aging includes activities outside of standard market contributions, such as volunteering. Such a revolution has caused many to rethink the very concept of old age. Often, age is a mixture of chronological age, biological age, psychological age, and sociological age. Functional age is a good marker of the aging process and aids in determining between three age conditions: normal aging, pathological aging, and successful aging. Another method of categorizing the diverse array of seniors is simply via chronological age. Gerontologists see three subcategories of seniors, young old (sixty-five through seventy-four), middle old (seventy-five through eighty-four) and old-old (eighty-five and older). Whatever the method of categorization, grouping the vast growing senior population is an arduous process given the inherent wide variability in human aging. Further, such grouping permits large-scale comparisons of health and wellness.

Unfortunately, the United States lags behind other countries, especially Great Britain, and the World Health Organization (WHO) recommendations on prevention and restraint of chronic diseases. The WHO's guide to *Global Age-Friendly Cities* provides eight guidelines for communities aiming for improvement: outdoor spaces, transportation, housing, social participation, respect, civic participation, communication and community support, and health services. In London and other European cities, free exercise playgrounds for the elderly are the norm. Costa Rica; Sardinia, Italy; and Ikaria, Greece, are examples of locales that possess the right mix of cultural and social factors that permit many seniors to live healthy lives into their nineties and beyond. Generally, such countries have a culture of respect for the aged. For example, those who study and treat the aging population, in Great Britain in particular, are held in high esteem. Conversely, the United States has historically stigmatized the aging and those who work with them. A slow tide of change is occurring as baby boomers prominently age in the American society, but changing preconceived notions is a slow process and stereotypes of the aged still abound.

APPLICATIONS AND PRODUCTS

Baby boomers possess a higher level of education than any previous generation; thus, the expectation is this cohort will be savvy consumers desiring the best proven treatments. Where daily life choices are more predictive of health status than genetic composition alone, the previous niche areas of applications and products for the aging is now a rampant growth industry spanning every conceivable field.

Medicine. Medical implantation devices provide relief to ailing organs and prolong well-being. The left ventricle assist device aids the normal functioning of failing hearts while an implanted defibrillator prevents cardiac arrest by shocking the heart back into a normal rhythm. Cochlear implants are placed directly under the skin behind the ear and return the gift of sound to many hearing-impaired individuals. Cameras encapsulated in pill casings that the patient swallows take video of the intestinal track, eliminating the need for costly and invasive scoping procedures.

The increasing need for evidence-based medicine has produced some creative solutions to gathering information from patients. Wireless home-based transfer of medical information from accelerometers, glucose monitoring, and implanted devices, such as pacemakers, allows a patient to provide real-time health data while remaining independent. Cell phones with Global Positioning Systems (GPSs) track exercise regimens and allow for intermittent queries about self-perceived health and well-being. For example, patients newly released from the hospital can transmit responses to doctor-initiated questions eliminating the need and cost of in-person follow-ups. This trend in distance-based medicine includes genetic-testing kits, available at local drug stores, that allow the user to mail in his or her sample for analysis.

Although currently emerging technologies sound like science fiction, many of these products are closer to the marketplace than one might imagine. Thought-controlled mechanical limbs that receive feedback from the environment, for instance temperature and pressure, can closely mimic an individual's lost arm or leg. In development are microscopic cleaning robots, called blood bots, that can be guided to clean plaque-filled veins and take biopsies. Noninvasive blood, saliva, and urine tests for Alzheimer's disease, cancer, and other difficult-to-detect diseases are currently in development. There is even an experimental Breathalyzer test in development that may replace expensive blood and urine analysis. Semipermanent prescription tattoos might be able to respond to glucose levels in a diabetic's bloodstream by changing color when placed under a handheld infrared light, eliminating the need for painful blood monitoring.

Common memory-storage cards, like those found in a digital camera, are being used as a portable patient medical archive. The cards fit into a wallet and facilitate communication and accuracy between the various health professionals many seniors visit. A

computerized medical information system will likely soon replace inefficient paper-based records. Such an electronic system will permit comprehensive care while anywhere in the world and coordinate the spectrum of health care services seniors receive.

Pharmacology. Perhaps the field most engaged with the aging population is pharmacology. Clinical drug trials deliver numerous pharmaceuticals to the market each year—many designed to treat and extend wellness into advancing age. Drugs are being developed that may fight obesity and even change one's DNA. Drug encapsulation involves the coating of medicine either to delay activation or enter affected areas. For example, most oral medications are unable to pass the blood-brain barrier, which means they do not enter the brain. Encapsulating, or masking, the active drug compound could permit the body to pass the drug into the brain, eliminating the need for invasive surgeries.

Although, often with variable scientific evidence, herbs and supplements for the aged have expanded exponentially in the early twenty-first century. In 2011, a senior can take a pill that is purported to cure any ailment. However, geriatric researchers have found minimal scientific evidence for many of these claims. Ginkgo biloba was claimed to improve memory for many years; however, numerous large-scale clinical trials have found no such evidence. Large annual doses of vitamin D, thought to improve bone health, was found to increase fractures. Conversely, some herbs and supplements have proven effective. Omega-3 fatty acids have shown promise in improving heart health. Capsaicin, found in hot peppers, has recently been added to topical arthritis creams because of its analgesic properties. However, medicines work only when taken as directed. Medication non-adherence costs the health care industry millions of dollars per year. Patients frequently take the wrong dose or fail to fill the prescription, necessitating additional doctor visits. Accordingly, an industry has developed to correct this problem. Pill bottles with reminder alarms, automatic medication dispensers, Internet-based and cell-phone text reminders to assist with medication adherence are widely available.

Assistive Technology. Numerous home-based assistive devices are available to prolong independence: swing-down shelves in kitchens, easy-open door handles along with a bevy of structural changes to

Fascinating Facts About Geriatrics and Gerontology

- Globally, vast disparities exist in expected life span. The United States lags behind France, Sweden, Canada, and Japan in mean life expectancy. In some African countries people have a life expectancy of forty years.
- By 2050, 21 percent of the world's population will be sixty or older. Almost 1.5 million people are expected to be one hundred years old or older by 2050.
- In 2011, the first wave of baby boomers will reach age sixty-five and by this point boomers will be twice the population of the entire country of Canada.
- The aged do not suffer more crime than younger persons, although they are more fearful of crime.
- Many older adults enjoy a healthy sex life even into old age.
- Memory loss is not an inevitable part of the aging process. Most senior citizens do not get Alzheimer's disease.
- Men over fifty years old have the highest suicide rates; those eighty-five or older have the highest overall suicide rates.
- The majority of old people do not live alone.
- Religiosity does not increase with age.
- Although physical strength does decline, strength-training exercises maintain and improve flexibility and strength even into old age.

acommodate individuals with decreased strength, decreased stature, complications due to arthritis, and the like. Such assistive technologies have shown to decrease the need for outside personal assistance, further prolonging independence in the home. Motion-sensor systems eliminate wandering, which is often associated with later stages of dementia. Additional home-based applications for seniors include special bathtubs, mechanical chairs that climb stairs, motorized wheelchairs, and remote home-monitoring of health status.

Often advances in care for the aging have spill-over benefits for the rest of the population. For example, the physical and occupational therapy fields have created user-friendly work environments for employees of all ages. Accordingly, ergonomics is now a household name with companies and therapists building optimal sitting and standing work-stations that relieve pressure and support working and moving bodies. The automobile industry has responded with adjustable gas and brake pedals, backup cameras, audible turn-by-turn directions, and parallel-parking assistance.

Education. Lifelong learning colleges offer continuing education to seniors through a variety of formats. Online centers of learning, interactive CD and DVD training programs, and book and workbook training manuals abound that purport to increase memory, mental speed, and generally bolster one's brain power. Many traditional university and community colleges offer vacant classroom seats to seniors at a discount. Train the Trainer is an emerging public health program that trains seniors to educate other seniors, often in classroom-like settings. Select groups of seniors undergo an extensive program to learn the latest health-improvement information and how to deliver such information effectively. They each educate a room full of seniors, thereby increasing the scope of health-promotion programs.

Physical activity is the single best way to improve one's health at any age, and there are many products and services that promote an active lifestyle. The Nintendo Wii video game system has enjoyed popularity with seniors across the country. The Wii is a low-impact, hand-eye coordination system that is believed to increase balance, strength, and cognitive performance. Many senior centers have created Wii bowling leagues, and online goal-setting and health-improvement Web sites permit seniors from across the globe to post their scores to encourage other seniors to maintain healthy lifestyles.

IMPACT ON INDUSTRY

Across the world, governments, businesses, academia, and individuals are responding to the needs of the aging population. Elder rights, rural aging, poverty, retirement, and pensions are a few of the issues facing the unprecedented number of seniors in the first half of the twenty-first century. China, in particular, will have an elder population that outnumbers that in the United States by 2050, and almost 70 percent of China's aging population is not covered by a public pension system. The cost to the United Kingdom's economy because of elderly discrimination is estimated in the tens of billions of dollars. In

Latin American countries where a minimum pension exists, elderly poverty rates are low; World Bank research indicates that enacting similar programs in the remaining South American countries would drastically reduce elderly poverty.

Government and University Research. The National Institute on Aging (NIA) is solely dedicated to funding basic and applied research projects that address the aging process. The scope of their endeavor is impressive and important. For example, basic science research at the molecular level drives animal models of investigation, which in turn drives clinical trials that result in new drugs and products. The ultimate goal of such research is the increased health and wellness of the aging population. Research at the university level is a critical link in this process. University faculty are often the main recipients of NIA research grants. The large infrastructure of the university environment permits many highly qualified individuals to conduct cutting-edge aging research. For example, New York University recently discovered that a gene linked to Alzheimer's disease is involved in removing toxic proteins—excessive amyloid protein deposits in the brain are hallmark characteristics of Alzheimer's disease.

Industry and Business. As discussed above, the opportunity for new aging-related products has produced a boom in the manufacturing and service industries. Entire departments and companies cater solely to the needs of the aged. The burgeoning cognitive-training, physical fitness, nutrition, legal, political advocate, and construction industries have responded in-kind. The various industries that cater to seniors have begun to target middle-aged individuals as well as it is believed that lifelong behaviors often predict health and well-being in one's later years. Accordingly, memory- and attention-training products exist for any age. Although once confined to installation of grab rails in bathtubs, portions of the construction industry now specialize in senior-friendly housing where lower countertops and no-slip surfaces are commonplace.

Major Corporations. Founded in 1958, AARP is the largest corporation catering to the aging population. Formerly known as the American Association of Retired Persons, AARP struggled with the outdated concept of retirement and opted to go by AARP instead. Following in their footsteps, many view their later years as a period of being redirected over the static concept of retirement. The comparable academic organization to AARP is the Gerontological Society of America (GSA). The GSA provides an organizational structure for the numerous educators, researchers, and other individuals interested in aging. In addition, GSA holds a yearly conference where academics share their current research. The Cognitive Aging Conference is a biennial gathering of researchers specifically interested in cognition and the elderly. Because Alzheimer's is one of the most feared diseases, maintaining and improving cognitive skills, thought to buffer the disease's effects, is a multimillion-dollar industry. The Cogmed company, in particular, has produced Internet-based memory- and attention-training programs that have solid scientific evidence supporting their effectiveness.

CAREERS AND COURSE WORK

The rapidly aging baby boomers have already created numerous and varied career opportunities in geriatrics and gerontology. Educators and social workers have experienced tremendous growth opportunities in the last ten years as the need for coordination of health education and social services increases. The shortage of nurses is particularly prevalent in hospitals, nursing homes, and other retirement-oriented settings. By 2030, the projected ratio is 1 geriatrician for every 4,200 seniors, so future job prospects are excellent.

Geriatrics involves the completion of a bachelor's degree, four years of medical school, a three-year internship in general medicine and aging populations or subspecialty, such as immunology or cardiology, and often a one- to two-year program in geriatrics. Typically labeled a geriatrician, the individual spends his or her days compassionately caring for the medical and quality-of-life issues of the aged. Geriatric nurses have earned a specialized bachelor's or master's degree and often play the major role in the daily care of the infirmed. Presently, geriatric psychiatrists are in short supply, leaving many seniors with severe mental health issues searching for adequate care.

Conversely, gerontology accreditation varies with the specialty. Typically, employment-specific master's-level course work is required for most therapist-oriented careers, such as physical therapy or counseling. Completion of doctoral-level course work is required for clinical careers, clinical geropsychologist, educators, university professors, and other aging-related careers, such as dentistry, pharmacy, policy, and political science. Physical or occupational therapists spend their days retraining aging muscular and structural systems, and counselors commonly deal with depression, loss, grieving, and other aging-related mental health issues. Clinical geropsychologists commonly treat more extensive cognitive-impairment issues such as dementia and mild to moderate mental-health problems.

SOCIAL CONTEXT AND FUTURE PROSPECTS

In 2011, aging in America is viewed through the experience of the baby boomers—those individuals born between 1946 and 1964—who started turning sixty in 2006. The most educated of any senior cohort, baby boomers grew up during unprecedented economic growth and, accordingly, possess a unique view on the aging process. The boomers are not taking retirement lying down: This group is more redirected than retired. Traditional leisure activities, continuing education, volunteering, and often part-time employment now replace the full-time work-week. The baby boomers in particular expect an unprecedented retirement lifestyle. In response to this expectation, retirement communities now resemble theme parks with golf courses, activity centers, and staff solely dedicated to planning events.

Future prospects have become reality. In central Florida, an entire city was developed for the retired. Specialized golf cart highways and parking spaces connect shopping malls and doctor offices to homes. Daily activities can include concerts, speeches, exercise facilities, and college classes. If such retirement cities become the norm, a complete redefining of aging will likely take place. Active and engaged theories of aging will continue to replace outmoded theories of aging such as the concept of disengagement. Whether the stigma of aging and corresponding stereotypes will also be replaced is yet to be determined.

Biotechnology holds great promise for the future of aging research. Caloric-restrictive diets have extended longevity in mice, and such research has led investigators to explore the possibility of "turning off" mechanisms responsible for fat storage. Emerging research suggests B vitamins may decrease depression, phenolic compounds might kill certain cancer cells, a juice elixir may prevent the common cold, and capsaicin, mentioned earlier in connection with arthritis, may promote weight loss. Cortisone and prednisone, often found in anti-inflammatory medications, might have an unintended side effect of bone loss as recently reported in the journal *Cell Metabolism*. Perhaps the most visible biotechnology project involves using stem cells to regenerate aging cells.

Dana K. Bagwell, B.S.

FURTHER READING

Antonucci, Toni, and James Jackson, eds. *Annual Review of Gerontology and Geriatrics: Life-Course Perspectives on Late Life Health Inequalities.* New York: Springer, 2010. Yearly review of the vast field of aging studies. This volume's emphasis is on health disparities and aging.

Chodzko-Zajko, Wojtek, Arthur F. Kramer, and Leonard W. Poon, eds. *Enhancing Cognitive Functioning and Brain Plasticity.* Champaign, Ill.: Human Kinetics, 2009. A review of recent research supporting the notion that physical activity and brain exercises strengthen the brain. The fairly new concept of neural plasticity, that aging brains can grow and form new neural connections, is discussed extensively.

Halter, Jeffrey, et al. *Hazzard's Geriatric Medicine and Gerontology.* 6th ed. New York: McGraw-Hill, 2009. A compendium of evidence-based medicine and clinical applications for treating the aged. Includes 300 illustrations, numerous tables and figures, and additional online resources.

Palmore, Erdman B. *The Facts on Aging Quiz: A Handbook of Uses and Results.* New York: Springer, 1988. Contains several quizzes that test common misconceptions of the aging process.

Schaie, K. Warner, and Laura L. Carstensen, eds. *Social Structures, Aging, and Self-Regulation in the Elderly.* New York: Springer, 2006. Examines the evolution of personal and social roles in aging with particular emphasis on familial changes, immigration, and increased life span.

WEB SITES

AARP International
http://www.aarpinternational.org

American Geriatrics Society
http://www.americangeriatrics.org

American Psychological Association
Division 20 (Aging)
http://apadiv20.phhp.ufl.edu

Gerontological Society of America
http://www.geron.org

See also: Pharmacology; Surgery.

HEMATOLOGY

FIELDS OF STUDY

Anatomy and physiology; biology; immunology; immunohematology; microscopy; chemistry; statistics; pathophysiology; microbiology; molecular biology.

SUMMARY

Hematology is the study of blood and blood-related disorders. It involves the diagnosis and treatment of a wide variety of diseases including anemia, leukemia, cancer, and clotting and bleeding disorders. As knowledge of the human genome has grown, many genetic mutations have been linked to hematological disorders. This has resulted in new methods of detection and novel treatments for diseases once thought incurable.

KEY TERMS AND CONCEPTS

- **Anemia:** Decrease in hemoglobin or the volume of red blood cells, resulting in less oxygen being delivered to body tissues.
- **Automated Cell Counter:** Instrument that uses electric impedance or optical light scatter to rapidly count blood cells.
- **Flow Cytometer:** Instrument that can simultaneously measure multiple physical characteristics of a single cell as it flows in suspension through a measuring device.
- **Hematopoiesis:** Production, differentiation, and development of blood cells.
- **Hematopoietic Stem Cell:** Precursor cell that has the ability to replicate and proliferate into all the lymphoid and myeloid blood cell lines.
- **Hemoglobin:** Respiratory protein that is found in the cytoplasm of the red blood cells.
- **Hemoglobinopathy:** Genetic defect that results in either an amino acid substitution or diminished production of one of the protein chains of hemoglobin.
- **Hemostasis:** Process resulting in the balancing of bleeding and clotting within the circulatory system

- **Leukemia:** Disease characterized by the overproduction of immature or mature cells of various leukocyte types in the bone marrow or peripheral blood.
- **Lymphoma:** Disease characterized by malignant tumors of the lymph nodes and associated tissues or bone marrow.

DEFINITION AND BASIC PRINCIPLES

Hematology is the study of blood and its related disorders. The components of blood include the plasma (55 percent) and the formed elements, or cells (45 percent). Plasma is 91.5 percent water and 8.5 percent solutes, mainly the proteins, albumin, globulins, and fibrinogen. The cells can be subdivided into three types, erythrocytes (red blood cells), leukocytes (white blood cells), and platelets. The primary step in assessing hematologic function and disease processes is to examine the cellular elements, determining the percentage of each type of cell and its morphology.

Erythrocytes consist of a plasma membrane that surrounds a solution of proteins (mainly hemoglobin) and electrolytes. Mature red blood cells do not contain a nucleus. They are small, about 7-8 micrometers (μm) and shaped like a biconcave disk. Evaluation of red blood cells is important in the diagnosis and monitoring of anemia.

Platelets are small (1-4 μm) bits of cytoplasm that contain granules and no nucleus. Platelets are necessary for hemostasis. They have the ability to adhere, aggregate, and provide a surface for coagulation to occur. Platelets are important in bleeding and clotting disorders.

Leukocytes are white blood cells that are differentiated into five varieties, neutrophils, eosinophils, basophils, lymphocytes, and monocytes. Mature white blood cells contain a nucleus and often cytoplasmic granules. It is important to distinguish the different types of white blood cells and their relative amounts. Neutrophils have a segmented nucleus and are important in fighting infections. They also contain granules that secrete specific enzymes that aid in the killing of bacteria. Neutrophils play a key role in phagocytosis and inflammation. Eosinophils contain large granules that stain reddish orange. The number of eosinophils increases in cases of parasitic infections and some types of allergies. Basophils also contain granules; however, these stain a dark purple and contain the components heparin and histamine. Basophils play an important role in allergic reactions. Lymphocytes are the second most numerous white cells in the blood. They typically do not contain granules in the cytoplasm. Lymphocytes play an important role in immunity because of their ability to initiate an immune response and produce antibodies. Monocytes are the largest of the white blood cells. Often the monocyte's cytoplasm displays psuedopods. These cells are mobile and are important in fighting infections and removing foreign elements from the blood and tissues.

Hematopoiesis is the continual process of blood cell production and the development of the various cell lines. Blood cells are the progeny of a hematopoetic stem cell. Through the process of hematopoiesis, the body is supplied with ample blood cells of all lines. The hematopoetic stem cell can both replicate itself and differentiate into the mature cells found normally in the peripheral blood. The hematopoietic system includes the bone marrow, liver, spleen, lymph nodes, and thymus. Regulatory growth factors also play a role in hematopoiesis.

BACKGROUND AND HISTORY

The field of hematology came into existence in 1642 when Antoni van Leeuwenhoek visualized cells using a microscope that he invented. It was not until 1842 that French microbiologist Alexandre Donne discovered platelets. In 1845, German physician Rudolf Virchow discovered an excess of white blood cells in a patient who had died from a condition he called leukemia (meaning white blood). A staining technique developed in 1877 by American scientist Paul Ehrlich allowed the visualization and differentiation of blood cells. French anatomist Louis-Charles Malassez invented the hemocytometer in 1874 and was able to quantitatively measure blood cells. The following year, French physician Georges Hayem developed a method for quantitatively measuring platelets using a hemocytometer.

Many diseases of the blood and their treatments, cures, and causes were discovered in the twentieth century. In 1925, American pediatrician Thomas Cooley described Cooley's anemia (thalassemia major). In the 1970's, American geneticist Janet Davison Rowley demonstrated the translocation of chromosomes 8 and 21 in acute myelogenous leukemia and of chromosomes 9 and 22 in chronic myelogenous leukemia. In 1972, American immunologist and geneticist Leonard Herzenberg invented the fluorescence-activated cell sorter (FACS), which aided the study of cancer. Argentine immunologist César Milstein and German biologist Georges Köhler used hybridization to develop monoclonal antibodies in 1975. These were followed by many other discoveries and treatments, some of which were the result of knowledge about the human genome.

HOW IT WORKS

Hematology diseases involve disorders of red blood cells, white blood cells, and platelets and hemostasis.

Disorders of Red Blood Cells. For the red blood cell to survive and function properly in the body, it must maintain a proper membrane, possess structurally correct and appropriately functioning hemoglobin, and have properly working metabolic pathways. Problems or defects in any of these areas will result in the red blood cell having a shortened life. Anemia results when the circulating red blood cells are unable to provide an adequate supply of oxygen to the tissues of the body. The many causes of anemia can be classified as nutritional deficiency (vitamin B12 or folic acid deficiency), blood loss, accelerated red cell destruction, hemoglobin defects (hereditary or acquired), and enzyme deficiencies.

Disorders of White Blood Cells. White blood cells perform a variety of functions in the body, including the destruction of bacteria, mediation of the inflammatory process, and production of antibodies or immunoglobulins. White blood cell disorders include diseases that affect the number of cells (quantitative defects) and those that affect the functioning of the cells (qualitative defects). White blood cell disorders range from slight inflammation to acute leukemia. Leukemia is a malignant disease that involves the hematopoietic tissue. Abnormal cells can be found in both the bone marrow and the peripheral blood. Leukemia is classified as either chronic or acute and can affect any of the white blood cell types, red blood cells, or platelets. It is important to distinguish between the different types of leukemias to provide the proper and most effective treatment.

Disorders of Platelets and Hemostasis. Hemostasis is the process by which the body stops bleeding and maintains the fluid state of the blood. In other words, it is a balance between bleeding and clotting. It involves the platelets, blood vessels, and specialized coagulation proteins. Disorders of this system include qualitative and quantitative platelet disorders, such as idiopathic thrombocytopenic purpura (ITP) and von Willebrand's disease. Some disorders that involve problems with coagulation proteins include factor V Leiden mutation and hemophilia A and B. The interaction between the platelets, clotting factors, and blood vessels plays an immensely important role in the functioning of the cardiovascular system. The field of hematology has been integral in developing modern cardiovascular therapies, including heart catheterization and the use of stents.

Disseminated intravascular coagulation (DIC) is another hemostasis disorder. It is a complex disorder that involves the development of small clots within the blood vessels and the dissolution of these clots. Platelets and the clotting factors often are consumed during this disease process, and intensive therapy is necessary to resolve its occurrence.

APPLICATIONS AND PRODUCTS

Automated Cell-Counting Instruments. Until the mid-1950's, cell counts were performed manually using a diluted fluid and a hemocytometer, and blood smears were viewed microscopically. Modern automated instruments can perform a complete blood count (CBC), which includes red blood cells, white blood cells, platelets, hemoglobin, hematocrit, and a five-part differential. This testing accounts for the primary hematology testing performed for almost every disorder. The instruments use the principles of impedance and optical light scattering to enumerate and determine the characteristics of each cell. Using the impedance principle, the cells are passed through an electrically charged aperture. Because blood cells are poor conductors of electricity, they create a resistance that can be measured as a pulse. The number of pulses is equivalent to the number of blood cells. The height of each pulse is equal to the cell's volume. Using the optical light

scattering principle, each cell is passed through a beam of light (either optical or laser). Forward scatter and side scatter of the light is created by each cell. The forward scatter represents the size of the cell, and the side scatter represents the degree of complexity of the cells (cytoplasmic organelles can be assessed).

Flow Cytometry. Flow cytometry can detect molecules on the surface of a cell, making it possible to sort cells according to their surface composition. The molecule of interest is labeled using a fluorescent marker. Flow cytometry is used for immunophenotyping (identification of antigens on the cell surface), reticulocyte counting, and analysis of DNA. The information obtained about the subset of cells in flow cytometry is of critical use in the diagnosis, classification, and treatment of malignancies of mature lymphocytes, acute leukemia, and immunodeficiency disorders.

Cytogenic Analysis. In cytogenic analysis, cells are harvested and processed to visualize chromosomes in mitotically active cells. The cells are arrested in metaphase, and chromosome bands are visualized by various staining procedures. A similar technique, fluorescence in situ hybridization (FISH) may also be used. In this technique, fluorescent-labeled DNA probes are used to visualize specific chromosome centromeres (region on a chromosome joining two sister chromatids), whole arms, whole chromosomes, and individual genes. Chromosomes are visualized microscopically, and a karyotype can be constructed by using a video-computer-linked analysis system. Many chromosome aberrations are considered diagnostic or have significant prognostic implications for hematologic malignancies and solid tumors, such as chronic myelogenous leukemia, acute myelogenous leukemia, acute lymphoblastic leukemia, and lymphomas.

Chemotherapy. The goal of chemotherapy is to destroy all malignant cells within the bone marrow and allow the bone marrow to be repopulated with normal precursor cells. However, the drugs used for chemotherapy are not specific for leukemic cells, and many normal cells are also killed in the process. This results in the severe complications of bleeding, infections, and anemia.

Molecular-Targeted Therapy. Therapies that target specific genetic mutations can either silence the expression of a particular gene or reactivate a

Fascinating Facts About Hematology

- A mature red blood cell does not contain a nucleus and therefore possess no DNA. Red blood cells serve the vital function of supplying oxygen to the rest of the body.

- The shortest-lived cell in the blood, the polymorphoneutrophil, circulates for only seventy-two hours. Other cells in the blood, such as the monocyte/macrophages, can live for months or even years.

- Sickle cell anemia is a genetic defect that results in a single amino acid substitution in the protein chain of hemoglobin. Hemoglobin has four protein chains with more than 574 amino acids in a specific sequence.

- Iron-deficiency anemia is the most common anemia and affects an estimated 2 billion people worldwide, according to the World Heath Organization.

- Although leukemia can strike at any age, more than 50 percent of all cases of leukemia occur after the age of sixty-four.

- In 2010, an estimated 43,050 new cases of leukemia were diagnosed in the United States.

- Allogenic bone marrow transplants have been successful in curing chronic myelogenous leukemia. Long-term survival rates of as high as 78 percent have been reported. A majority of patients have successfully resumed their personal and professional lives.

- Hemophilia A is a bleeding disorder that affects men. This disorder was found in the Royal House of Stuart in Europe and Russia. Queen Victoria was later proved to be the carrier. The disorder, previously thought to be an absence of factor VIII, is actually a molecular defect in the factor that renders it unable to perform its clotting function.

silenced gene. These types of therapies are better tolerated by patients than traditional chemotherapy and have been developed for chronic myelogenous leukemia and acute promyelocytic leukemia.

Bone Marrow Transplant. For a bone marrow transplant, drugs and radiation are used to first eradicate all leukemic cells that may be present. Then, bone marrow from a suitable closely matched donor is transplanted into the patient to provide a source of normal stem cells that can then repopulate the patient's bone marrow. Autologous bone marrow transplants are also an option. Some of the patient's bone marrow is removed while the patient is in remission. This specimen is treated to remove any residual leukemic cells and preserved. The patient is then treated to eradicate all leukemic cells and given back his or her own bone marrow. Many bone marrow transplants have been successful, and this procedure is being performed on an increasing basis.

Stem Cell Transplant. Stem cells can be found in either the bone marrow or the peripheral circulating blood. They also can be found in umbilical cord blood and fetal marrow and liver. Before stem cells can be harvested from a donor, mobilization of stem cells into the peripheral blood is stimulated by the use of cytokines. A process called apheresis is used to collect the stem cells from the donor's blood. These harvested stem cells are then injected into the patient. As with a bone marrow transplant, it is also possible to perform an autologous stem cell transplant. The transplanted patient will usually begin to produce new blood cells ten to twenty-one days after receiving the harvested stem cells.

IMPACT ON INDUSTRY

Industry that involves the field of hematology is mainly centered on its application, which is in the field of medicine and treating patients with hematological or cancerous diseases.

Medicine. Physicians may practice in hematology or oncology in government, nonprofit, or private institutions of medicine. Hematology is considered a specialty area of medicine. Usually patients are referred to a hematologist by their general physician. Hematologists also specialize in the area of oncology, since many cancers involve the blood and circulatory system.

Most large medical center will have a hematology/oncology department that is staffed with hematologists, oncology nurses, medical laboratory scientists, and radiology technologists. These professionals work together as a health care team to diagnose and provide treatment for hematological disorders.

Research. Hematology research is conducted in a variety of settings. Research is ongoing to understand basic cellular and molecular mechanisms, hematopoiesis, and the regulation of genes that control normal blood cell maturation and function. Other areas of research are stem cells, growth factors, synthetic hemoglobin, and drugs for the treatment of all types of cancers. For example, the U.S. Department of Health and Human Services funds a multifaceted research program through the National Institute of Diabetes and Digestive and Kidney Diseases.

Another area of medicine that relies heavily on hematology research is cardiology, particularly cardiovascular surgery and the treatment of cardiovascular diseases. Because the blood vessels are in contact with the blood at all times, these two areas are heavily reliant on each other.

Industry. The medical laboratory instrument, reagent, and pharmaceutical industries are all essential to the diagnosis and treatment of hematology disorders. Revenues from these industries are approaching $4.5 billion.

The United States leads the medical laboratory instrument industry, with an estimated 38 percent of the market. Asia, Europe, and Germany follow closely behind. A large area of development involves point-of-care instruments, which allow the patient and caregivers to monitor the patient's laboratory results at home or at the bedside.

The development of drugs for treatment is a market that continues to grow. Some examples of drugs developed for hematology treatments are recombinant erythropoietin (EPO), granulocyte-macrophage colony stimulating factor (G-CSF), and recombinant blood clotting factors. The market for pharmaceuticals for treatment of leukemia, anemia, and other blood disorders continues to grow as more diseases are diagnosed and as research provides information about the causes of these diseases.

CAREERS AND COURSE WORK

Courses in biology, genetics, molecular and cell biology, chemistry, immunology, hematology, and immunohematology are core courses for those interested in pursing a career in hematology. There is a need for professionals with hematology knowledge at all levels of careers.

Medical Laboratory Scientists. Medical laboratory scientists, previously known as medical technologists or clinical laboratory scientists, work in all areas of the clinical laboratory, including hematology. They perform the necessary tests to diagnose and treat diseases of the hematological systems. To become a medical laboratory scientist, one must obtain a bachelor's

degree and complete an accredited medical laboratory science program. This can be done through a four-year course of study at most universities.

Cytotechnologists. Cytotechnologists are responsible for examining human cells under the microscope. They look for early signs of cancer and other diseases. A bachelor's degree and completion of an accredited cytotechnologist program are required.

Hematologists/Oncologists. Medical doctors specializing in hematology or oncology can diagnose and treat patients with diseases that affect the hematology system. They can work in private practice, for academic centers, or in governmental agencies. To become a hematologist requires a bachelor's degree and a medical degree. A residency and a fellowship in the area of hematology or oncology also are necessary.

Pathologists. Pathologists are physicians who examine tissues and are responsible for the accuracy of laboratory tests. The pathologist and the patient's other doctors consult on which tests to order, the interpretation of test results, and the appropriate treatments. A bachelor's degree, medical school, and a residency and a fellowship in pathology are required.

SOCIAL CONTEXT AND FUTURE PROSPECTS

Never in the history of medical science have the opportunities for advancement in the field of hematology been greater. Many genes have been characterized as disease specific or disease related. This creates the possibility of finding cures or treatments for many diseases that have plagued humankind for centuries.

With this newfound knowledge also comes the challenge of balancing the treatment of patients with new technologies and the ethical application of these treatments. Stem cell research has spurred some controversy as to how to ethically provide the stem cells needed for treatments. Another controversy arises when genetic testing for diseases or certain gene markers that are tested for as risk factors are used in the everyday practice of medicine.

Mary R. Muslow, B.S., M.H.S., M.T.(ASCP)S.C.

FURTHER READING

Carradice, Duncan, and Graham J. Lieschke. "Zebrafish in Hematology: Sushi or Science?" *Blood* 111, no. 7 (April, 2008): 3331-3342. Describes the potential of the use of zebrafish for the modeling of hematologic diseases.

Harmening, Denise M. *Clinical Hematology and Fundamentals of Hemostasis*. 5th ed. Philadelphia: F. A. Davis, 2009. Includes chapters on types of anemia, white blood cell disorders, hemostasis, and laboratory methods.

Herzenberg, Leonard A., and Leonore A. Herzenberg. "Genetics, FACS, Immunology, and Redox: A Tale of Two Lives Intertwined." *Annual Review of Immunology* 22 (2004) 1-31. The inventors of flow cytometry describe their lives and work.

McKenzie, Shirlyn B., and J. Lynne Williams. *Clinical Laboratory Hematology*. 2d ed. Upper Saddle River, N. J.: Pearson, 2010. Cover types of anemia, neoplastic hematologic disorders, hemostasis, and hematology procedures. Includes some excellent photomicrographs and illustrations.

Patlak, Margie. "Targeting Leukemia: From Bench to Bedside." *The Federation of American Societies for Experimental Biology* 16, no. 3 (March, 2002): 273. An interesting review of leukemia diagnosis and treatment.

WEB SITES

American Society of Hematology
http://www.hematology.org

American Society of Pediatric Hematology Oncology
http://www.aspho.org

Hematology/Oncology Pharmacy Association
http://www.hoparx.org

See also: Cardiology; Histology; Stem Cell Research and Technology.

HISTOLOGY

FIELDS OF STUDY

Microscopy; cell biology; embryology; anatomy and physiology; histopathology; histochemistry, immunohistochemistry; hematology, oncology, tissue transplantation.

SUMMARY

Histology is an interdisciplinary branch of science that focuses on the structure and function of normal and diseased tissues of the human body using microscopy and staining techniques. It studies comparative morphology of tissues, changes in tissues and organs during embryonic development, evolutionary changes of structure, and function of tissues in different species. In clinical medicine, histology is used as a diagnostic tool to understand and treat pathological developments in the body tissues. In forensics, histology is employed to understand the degenerative events in injured and dead tissues (autopsies) and to determine the cause of death in criminal investigations.

KEY TERMS AND CONCEPTS

- **Autopsy:** Medical test that involves the extraction of cells or tissues from a corpse to evaluate any disease or injury and determine a cause of death. It is usually performed by a specialized medical doctor called a pathologist.
- **Biopsy:** Medical test that involves the extraction of cells or living tissues for observation and analysis of structural and functional abnormalities by methods such as microscopy or histochemistry.
- **Cytology:** Science that focuses on the fine structure and function of cells.
- **Embryology:** Study of the development of an embryo from fertilization to the fetus stage.
- **Hematology:** Study of blood cell formation (hematopoiesis), the organs that form blood cells, and diseases of the blood.
- **Histochemistry:** Use of chemical reactions between specific laboratory reagents and components within tissue biopsy material for diagnostic purposes.
- **Histology Staining:** Use of particular staining reagents to stain tissue sections for structural characterization of cells and tissues.
- **Histopathology:** Subdiscipline of histology that focuses on development of disease at a tissue level. It uses biopsies to diagnose and evaluate disease progression.
- **Immunohistochemistry:** Use of fluorescent-labeled antibodies on tissue sections to identify the location of particular structures (antigens) to which such antibodies would specifically bind.
- **Microscopy:** Study of objects too small to be seen with the naked eye, using a magnifying instrument called a microscope.
- **Microtome:** Device that slices ultra-thin sections of tissues that can be used for microscopic observation and analysis.
- **Oncology:** Branch of medicine that studies tumors (cancers).

DEFINITION AND BASIC PRINCIPLES

Histology studies the morphology of cells and tissues of the human body. It is sometimes called microscopic anatomy because it looks at the structure and function of the human body at the microscopic level. In the body, individual cells are organized in tissues, which form organs and organ systems to perform complex functions to maintain homeostasis. Although histology as a discipline is descriptive in many ways, it also involves a great deal of analysis because it focuses on structure and function relationships under normal and pathological conditions. The human body is made up of epithelial, connective, muscle, and nervous tissue types, but there are structural variations within each of these four groups. Changes in the body homeostasis are directly reflected in the changes–from temporary to irreversible–of the structure and function of tissues. Signs of infection, autoimmunity, aging, and malignant growths can be observed in various types of tissue by histological analysis of tissue probes. A trained histologist is capable of drawing a diagnostic conclusion based on such assessment alone or in combination with other methods.

Histology is based on the preparation of stained samples of tissue (from autopsies, biopsies, and cell cultures) and their microscopic analysis. The development of various staining techniques and tools for specimen preparation and microscopy have strengthened histology as a diagnostic approach in biomedical research and clinical medicine. The microscope evolved from a simple set of magnifying lenses to highly sophisticated optical (light), scanning, and electron microscopes that can use computer control to create high-resolution digital images for researchers, doctors, clinicians, and educators.

BACKGROUND AND HISTORY

The word "histology" comes from the Greek *histos* (mast, or tissue) and *logie* (study) and means the study of tissues. Descriptions of tissues can be found in the works of the ancient Greek philosopher Aristotle, the eleventh-century Persian physician Avicenna, and the sixteenth-century German physician Andreas Vesalius, long before the microscope was invented. The English natural philosopher Robert Hooke was the first to introduce the term "cell" in 1665, while studying structures of a cork tissue using a magnifying device, a simple microscope. In 1674, Dutch scientist Antoni van Leeuwenhoek, using a microscope of his own invention, observed living organisms. He discovered that an animal cell is structurally organized and has a nucleus. The term "histology" was introduced by August Mayer in 1819, and in the nineteenth century, histology was established as an academic discipline. Studies of animal tissues were conducted by such prominent scientists as Marcello Malpighi, Camillo Golgi, Caspar Friedrich Wolff, and many others.

Histology developed much faster after around 1838, when microscopes with a greater magnification became available. Theodor Schwann developed the cell theory, which states that living organisms are made of similarly structured units, or cells, and that each individual cell exhibits characteristics of life. Rudolf Virchow, regarded as the father of modern pathology, contributed to the cell theory by stating that cells divide to produce daughter cells. In 1906, the Nobel Prize in Physiology or Medicine was awarded to Golgi for his outstanding work on the histology of the nervous system. Modern histology is a multidisciplinary branch of biomedical science. The results of histological observations contribute to the understanding of mechanisms of diseases and ways to treat diseases.

HOW IT WORKS

Preparation of Tissue Samples. Histology and histopathology examine tissue samples using microscopy; therefore, preparation of high-quality samples is an important step toward obtaining reliable results. For light microscopy, a sample of tissue is acquired, processed so that it can be thinly sliced or sectioned using a microtome, and stained with a specific dye. The tissue section is then analyzed using a microscope, and the results are interpreted, usually by comparison to normal tissue samples. A research scientist would examine how his experimental protocol affected the tissue, while a physician would analyze biopsied tissue to observe disease progression and determine the correct treatment. All steps of this process must be performed according to certain standards to avoid misinterpretation.

There are many methods of studying extracted tissues, and the choice of methodology depends on the type of tissue and the final objective. Because tissues degenerate quickly, it is important to preserve them as soon as possible, using a chemical process called fixation. The chemicals preserve the cells in the tissue without changing their structure. Individual protocols for fixation of different tissues vary and are the product of experimentation by many histologists. After fixation, tissue samples are treated to remove water so that they can be saturated with a waxy substance such as paraffin, which solidifies at room temperature. The paraffin-saturated tissue specimen is then sliced into sections using a microtome. The sections are mounted on a glass slide, treated to dissolve the paraffin, and stained with dye. The thickness of the tissue sections can vary from 0.002 to 0.02 millimeters (mm). Certain modifications to this process are made to prepare bone tissue, which is very hard because of its high mineral content, and adipose tissue, which has a high fat content. Staining is an important step, because stains have different chemical affinities for various cell and tissue components. For instance, the same kind of dye might stain the nucleus, cytoplasm, and membrane structures differently. These differing absorption rates make it possible to highlight particular structures of interest. The specimen must be thoroughly rinsed to remove any unbound stain and avoid nonspecific staining.

Electron microscopy employs completely different techniques for preparation of specimens and allows observation of fine cell and tissue structures under very high magnification. A glass or diamond knife is used with an ultramicrotome to produce ultra-thin sections less than 0.001 mm thick. In some situations, extracted tissues can be frozen and sliced immediately, then stained and analyzed without lengthy tissue processing. Such tissue sections can deteriorate rapidly but are used to quickly analyze biopsies during surgery or in applications that do not require a detailed structural analysis.

Microscopy. The most common tool in examining cells and cell structures in tissue samples is the microscope. Optical microscopy uses visible light, which passes thought the object (a tissue section mounted on a glass slide) and the lenses to create a magnified image. The maximum resolution that can be achieved with light microscopes is 0.2 micrometers (μm). Transmission electron microscopes send a beam of electrons through the sample and allow much higher resolution, around 0.05 namometers (nm), but they require complex tissue processing and good technical skills to produce usable specimens on an ultramicrotome. Scanning electron microscopes produce a three-dimensional image of the cells and tissues. Electron microscopes are not suitable for routine or rapid observations of tissue samples.

Tissue Culture. Cell and tissue cultures can be grown on an artificial growth medium in a laboratory by providing adequate growing conditions in an incubator. Cultures are used to study living cells and tissues under experimental conditions. Tissue cultures are used to study tissue physiology, mechanisms of cell differentiation and development, cell-cell interaction, and gene regulation. Cell cultures are also used as a testing system for drug development. For example, fibroblasts, a cell type of connective tissue, grows well in the laboratory, providing a homogenous, live tissue material for drug testing.

Fluorescent Staining. Certain methods of tissue staining differentially detect the presence of particular chemical components in cells. One approach is to use a fluorescent label to trace components inside the cell. Fluorescent staining not only attests to the presence or absence of certain components in cells but also provides their precise localization in a particular cell or tissue type. For example, immunochemistry uses fluorescent antibodies that specifically bind with the target structures in the tissue so that the labeled antibodies will highlight the presence of the structures of interest in the analyzed specimen.

Some applications of these techniques are staining of virus-infected cells in tissues and detection of immune-complex deposition in the kidneys and skin of patients with lupus.

APPLICATIONS AND PRODUCTS

Progress in histology has been dependent on the development of microscopy. Many companies are developing microscopes with various technical specifications because microscopes are being used not only in the biomedical field but also in fields such as engineering, optical physics, and biotechnology. Industry leaders such as Olympus and Carl Zeiss offer high-quality optics, superior construction, and high-image contrast to satisfy the growing demands of researchers. The companies develop and produce instruments, software, and accessories for microscope systems for use in industrial settings and clinical and research laboratories. Advances in technology have resulted in the creation of stereo microscopy, virtual microscopy, and total internal reflection fluorescence (TIRF) microscopy.

Fluorescence Microscopy. Improvements in fluorescence observation have resulted in images that are twice as bright as conventional fluorescence images. Fluorescence microscopy is used for staining and observation of tuberculosis bacteria and of tissues from pulmonary adenocarcinoma and breast cancer. The Carl Zeiss company, a member of the Stop Tuberculosis Initiative, provided the Primo Star iLED fluorescence microscope at a lower price to seventy-four countries with a high incidence of the disease. The microscope delivers up to four times faster detection of tuberculosis than can be achieved with traditional techniques.

NanoZoomer Digital Pathology is a system for scanning glass sides of tissue and rapidly converting them to high-resolution digital slides, known as virtual slides. By magnifying or shrinking a chosen area while changing the focus on a monitor screen, researchers can view intracellular structures in fine detail. This system of virtual microscopy can be employed for toxicology tests, gene-expression analysis, and protein localization. The great advantage of digital slides is that they eliminate the worry of the fluorescence fading and degrading the image quality. Digital slide technology enables researchers to archive entire slides at high resolution, copy and edit them, and easily share them with others, allowing

Fascinating Facts About Histology

- Antoni van Leeuwenhoek was the first person to observe human red blood cells. The preserved tissue samples that he used are part of the collections of the Royal Society of London. Leeuwenhoek was not a trained scientist but a merchant who made microscopes for a hobby.

- Olympus's VivaView fluorescence incubator microscope allows researchers to observe cultured cells growing in real time. This combination of a microscope and an incubator results in high-quality live-cell imaging with the freedom to control growth conditions of cells and use the cells for a significantly longer time than was previously possible.

- A transmission electron microscope allows researchers to see parts of individual human chromosomes, single proteins, and DNA.

- The life span of red blood cells is between 100 and 120 days, after which the cells undergo apoptosis (programmed cell death).

- There are about 100 trillion cells in the human body.

- A Pap smear (also known as a Papanicolaou test) involves microscopic examination of cells scraped from stratified squamous epithelium of the vagina and cervix of the uterus. It is performed for early detection of a precancerous condition or cancer. Collected cells are smeared on a microscope slide and sent to a laboratory for analysis.

- Tissue engineering is a technology in which living tissues are combined with synthetic materials to grow new tissues in the laboratory. Laboratory-grown versions of skin, cartilage, and bone are developed by using biodegradable synthetic materials or natural collagen fibers as a scaffolding system to immobilize and grow cells. The new tissue is then implanted into the patient.

them to be used as the focal point for discussions.

Stereo microscopy is used by cutting-edge biological and medical laboratories that require the most effective imaging and observation of a large quantity of live specimens. It achieves the world's highest zoom ratio of 1:16.4, enabling remarkably sharp three-dimensional imaging and considerably enhanced specimen manipulation.

Live microscopes, such as Olympus's VivaView

fluorescence incubator microscope, achieve the dual objectives of culturing cells under monitored conditions and simultaneously observing the cultured cells under a microsope. These microscopes allows cell analysis in the most optimized environment.

Total internal reflection fluorescence (TIRF) microscopes are employed to study a diverse phenomena, including cell transport, signaling, replication, motility, adhesion, and migration; cell membranes and transport; the structure of ribonucleic acid (RNA); neurotransmitters; and virology.

Industrial microscopy offers a wide range of both industrial microscopes and microimaging systems for metrology, semiconductor wafer manufacturing, quality assurance, advanced materials analysis, metallography, and other precision applications in various industrial sectors.

Sample Preparation. Isolation of single cells and cell groups, as well as biomolecules, from a heterogeneous tissue is crucial for preservation of the material. Therefore, the proper preparation and staining of high-purity samples is critical for production of reliable results. Laser microdissection with optical tweezers, a technology developed by Carl Zeiss, delivers absolute purity in sample isolation with maximum preservation of the material and without affecting the viability of live cells. The technology combines microdissection and advanced imaging and incorporates digital camera technology. Major achievements of this technology include instant viewing, editing, and sharing of images over a network from any remote location. It is employed for specimen capture and for isolation of DNA, RNA, and proteins. Leica Microsystems was the first company to automate microtomy (microtome sectioning) and the first to introduce an integrated workstation for staining and cover slipping. The system does not require constant supervision, which provides an enormous time-saving advantage.

Clinical and Educational Applications. Histology is used in clinical, research, and educational settings. Biomedical research and clinical testing centers use fluorescent tags in staining to identify components of interest. For instance, flow cytometry analysis uses fluorescent-labeled antibodies as markers to perform rapid counts of blood cell types for clinical analysis. Hospitals, universities, and research centers provide training, education, and workshops on various protocols for specimen preparation as well as digital slide scanning for researchers and medical professionals. Histology is used in the manufacture of prepared tissue slides for biology, anatomy, and physiology courses at high schools and colleges.

Diagnosis of Diseases. Microscopic analysis of tissue samples is used to diagnose and classify many types of cancer, including cervical, breast, prostate, pancreatic, skin, and blood cancers, as well as noncancerous diseases. Histology tests can reveal if a tumor is cancerous, and if it is, they can determine the stage of progression, which leads to appropriate clinical treatment and an overall prognosis. Hematology analysis (of the blood cells) is usually the first step in the diagnostic process for cancer and noncancerous diseases.

IMPACT ON INDUSTRY

Histology, a valuable tool in biomedical research and clinical diagnostics, is funded through federal and private sources. The governmental agencies that provide most of the funding for biomedical research in universities and research institutes are the National Institutes of Health (NIH) and the National Science Foundation (NSF). Biomedical research funding increased from \$37.1 billion in 1994 to \$94.3 billion in 2003 and doubled when adjusted for inflation. By 2007, biomedical research funding had increased to \$101.1 billion. Funding for this area of research continues to grow steadily.

Manufacturers that develop innovative microscopes, microtomes for tissue sectioning, staining reagents, and the glassware used by biomedical researchers and clinical laboratories are part of a multibillion-dollar industry. These manufacturers provide histology-related equipment to hospital and academic research centers, diagnostic laboratories, biotechnology companies, and other biomedical companies. Every major hospital has histology facilities that test and interpret samples.

CAREERS AND COURSE WORK

Specialists in histology can choose from numerous training programs. Histologists are in demand, and job prospects are excellent. To become a histologist, an individual must study microscopy, anatomy, physiology, microbiology, and chemistry, and master laboratory and diagnostic techniques. Some histology students obtain special training in molecular biology so that they can isolate DNA and RNA from human tissue.

Careers in histology can take different paths. Histology specialists can work in hospitals, medical laboratories, and universities as technicians, assistants, and researchers, or they can become instructors or professors at universities. As of 2010, the average yearly salary for a histology technician in the United States was about \$46,000. University professors receive between \$40,000 and \$120,000 per year. Histology technicians and assistants work in hospitals and laboratories helping with clinical and laboratory procedures. A bachelor's degree program in histology is required for histology technicians. For histology assistants, a two-year associate's degree in medical assisting is sufficient. Advanced degrees such as a M.D. or Ph.D. are necessary for becoming a university professor. At universities, histology specialists may divide their time between research and teaching.

SOCIAL CONTEXT AND FUTURE PROSPECTS

Histology is a field of continuous growth. Many areas of medicine–such as oncology and regenerative medicine–depend on and require knowledge of histology. Histological parameters such as tumor size and morphologic characteristics of tissues are the most important diagnostic factors for cancer. Tissue engineering, a subspecialty of regenerative medicine, deals with repairing and replacing tissues (such as skin, bone, cartilage, or blood vessels) in part or in their entirety. Examples of tissue engineering include artificial bladders and edible artificial animal muscle tissue (artificial meat). Many materials, both natural and synthetic (including carbon nanotubes), are being actively investigated for use in tissue engineering. Scientists are also pursuing stem cell development and manipulation for tissue regeneration. Interest in this area has expanded, leading to the establishment of the California Institute for Regenerative Medicine in 2004.

Elvira R. Eivazova, Ph.D.

FURTHER READING

Croft, William J. *Under the Microscope: A Brief History of Microscopy.* Hackensack, N.J.: World Scientific, 2006. Traces the microscope from early beginnings to modern instruments, discussing how each works.

Hewitson, Tim D., and Ian A. Darby, eds. *Histology Protocols.* New York: Humana Press, 2010. This laboratory manual looks at tissue preparation and staining, with explanations of complex procedures.

Ovalle, William K., and Patrick C. Nahirney. *Netter's Essential Histology*. Philadelphia: Saunders/Elsevier, 2008. This atlas covers cells and tissues and the major bodily systems. Features a great collection of images by Frank H. Netter.

Ross, Michael H., and Pawlina Wojciech. *Histology: A Text and Atlas–With Correlated Cell and Molecular Biology*. 6th ed. Philadelphia: Wolters Kluwer/Lippincott Williams & Wilkins Health, 2011. Contains great illustrations, easy to follow diagrams. Recommended for medical, health professions, and undergraduate biology students

Tortora, Gerard J., and Bryan Derrickson. *Principles of Anatomy and Physiology*. 12th ed. Hoboken, N.J.: John Wiley & Sons, 2009. Excellent anatomy and physiology textbook with great illustrations and clinical references.

WEB SITES
American Society for Clinical Pathology
http://www.ascp.org

California Institute for Regenerative Medicine
http://www.cirm.ca.gov

Loyola University Medical Education Network
Histology
http://www.lumen.luc.edu/lumen/MedEd/Histo/frames/histo_frames.html

National Society for Histotechnology
http://www.nsh.org

Olympus America
Microscopy Resource Center
http://www.olympusmicro.com

See also: Biosynthetics; Cell and Tissue Engineering; Electron Microscopy; Hematology; Microscopy; Pathology; Stem Cell Research and Technology.

HUMAN GENETIC ENGINEERING

FIELDS OF STUDY

Biotechnology; genetics; cytogenetics; molecular genetics; biochemical genetics; genomics; population genetics; developmental genetics; clinical genetics; genetic counseling.

SUMMARY

Human genetic engineering is a branch of genetic engineering focusing on the understanding of human genes to produce applications that can improve human life. Genes, formulated by DNA (deoxyribonucleic acid), determine genotype–the complete genetic information carried by an individual, even if not expressed. Visible human characteristics, by contrast, are formed as the result of human genes interacting with the environment and are called the phenotype. Human genetic engineering aims to alter genotypes to cause changes in phenotypes; also, and more often, the knowledge of human genetics is used to engineer products, such as medications, that can cure or improve the quality of human life by addressing genetic disorders. Many of the applications of what is now known as human genetic engineering arose out of the mapping of the human genome during the Human Genome Project, completed in 2003.

KEY TERMS AND CONCEPTS

- **Bioinformatics:** Science of compiling and managing genetic and other biology data using computers, requisite in human genome research.
- **Biologics:** Medicines produced using genes and genetic manipulations.
- **Clones:** Genetically identical living organisms produced via genetic engineering.
- **DNA (Deoxyribonucleic Acid):** Molecule, found in all living organisms, that by reproducing itself allows for the inheritance of characteristics from one generation to the next.
- **Dysmorphology:** Abnormal physical development resulting from a genetic disorder.
- **Forensic Genetics:** Application of genetics, particularly DNA technology, to the analysis of evidence used in criminal cases and paternity testing.

- **Gene:** Specific DNA sequence that codes for a specific protein.
- **Gene Therapy:** Use of a viral or other vector to incorporate new DNA into a person's cells with the objective of alleviating or treating the symptoms of a disease or condition.
- **Genetic Screening:** Use of the techniques of genetics research to determine a person's risk of developing, or his or her status as a carrier of, a disease or other disorder.
- **Gene Transfer:** Using a viral or other vector to incorporate new DNA into a person's cells; used in gene therapy.
- **Genetic Testing:** Process of investigating a specific individual or population of people to detect the presence of genetic defects.
- **Genomics:** Branch of genetics dealing with the study of the genetic sequences of organisms, including the human being.
- **Pharmacogenomics:** Branch of human medical genetics that evaluates how an individual's genetic makeup influences his or her response to drugs.
- **Proteomics:** Study of how proteins are expressed in different types of cells, tissues, and organs.
- **Recombinant DNA:** DNA that has been transferred from one cell to another. Genes are recombined from a human chromosome to another cell, usually from bacteria; if the transferred human genes code for insulin, the bacteria accepting the transferred genes will now produce human insulin.
- **Stem Cell:** Progenitor cell that has the capability to become a more specialized cell, such as a kidney, liver, or heart cell. Once a cell becomes a specific kind of cell, it cannot change to another type of cell.
- **Toxicogenomics:** Science of evaluating ways in which genomes respond to chemical and other pollutants in the environment.

DEFINITION AND BASIC PRINCIPLES

Human genetic engineering is the science and technology of manipulating or changing human genes to alter or control visible characteristics of a human newborn or adult. Genes, formulated by DNA (deoxyribonucleic acid), are called the genotype. Visible human traits or characteristics, formed from the interaction of genes and the environment, are called the phenotype. Human genetic engineering aims to change genotypes to cause change in the phenotype. To understand human genetic engineering capabilities, it is important to understand basic genetic principles.

BACKGROUND AND HISTORY

Human genetic engineering is a scientific endeavor, and as such this field builds on the information and knowledge gained from the decades of experimentation accomplished in years past. Without this foundation, human genetic engineering would not exist. This foundation brings the prospect of human cloning and the use of human genetics for therapeutic purposes.

A keystone event in modern genetics occurred in 1953, when American biologist James D. Watson and English physicist Francis Crick deduced the double-helical structure of DNA. This structural information enabled effective study of how genetic material codes for life. In 1968, the DNA genetic code was deciphered. Armed with this important genetic information, geneticists undertook the first recombinant DNA experiments on bacteria in 1973. The ambitious Human Genome Project started in 1990, with the goal of mapping out the entire human genetic sequence. In June, 2000, the first working draft of the human genetic sequence was produced from the efforts of this project. April, 2003, saw the announcement of the first complete human genetic sequence–breakthrough information for human genetic engineering.

Cloning, a subdiscipline of bioengineering, is the reproduction of genetically identical living organisms. In 1996, the first mammal was cloned, a sheep named Dolly. Other animals have been cloned since this pioneering event, including a bull in 1999 and a pig in 2000. The year 2003 saw the cloning of a mule, a horse, and a rat, followed by the cloning of a dog in 2005. Attempts at pet cloning have occurred: John Sperling, a wealthy and influential American educator, has funded pet-cloning projects, and researchers at Texas A&M University successfully cloned a cat in 2002. Commercial attempts at pet cloning started in April, 2004, with a company called Genetics Savings and Clone (now defunct) offering pet gene banking and cloning. Korean researchers published claims of successful human

embryonic cloning in 2004, but these claims were later retracted because of fabricated data and other problems with the research. In May, 2010, the journal *Science* reported that scientists J. Craig Venter, Clyde Hutchison III, and Hamilton Smith had created a living creature in the laboratory. This new life-form, a bacterium, was artificially produced using genetic-engineering techniques. It had no ancestor and it reproduced, a key ability of living organisms.

Milestones in other areas of human genetic engineering—with more practicable and practical results—occurred in the areas of medicine, pharmaceuticals, forensics, and even psychology, as identified below in Applications and Products. Many, if not most, of these blossomed shortly after the mapping of the human genome was completed in 2003. Many more will be developed as scientists and researchers continue to investigate the data that were gathered through that monumental accomplishment.

HOW IT WORKS

Until the middle of the eighteenth century, most biologists believed in spontaneous generation—that life arises from combinations of decaying matter, as if flies arose from garbage. It is now known that DNA genetically codes for many physical characteristics.

The story of genetics starts with DNA and ends with protein. DNA, the genetic material found in the nucleus of every cell, codes for (that is, creates instructions for the building of) various proteins by means of components of the DNA molecule called nucleotides, which form the building blocks of DNA. The nucleotides establish the code. Proteins make up the structural elements of the body, including collagen, ligaments, tendons, and muscles; some hormones, such as insulin, are made of protein as well. Perhaps most important, however, are the protein enzymes.

All the enzymes in the body are made up of protein and protein alone. Enzymes are key because they accelerate chemical reactions. Thousands of chemical reactions occur in human bodies all the time. Protein enzymes catalyze all these reactions. DNA, by dictating the production of enzymes, controls these chemical reactions.

DNA dictates which proteins are produced by living and staying in the cell nucleus during the entire protein-making process. Much like a general in a

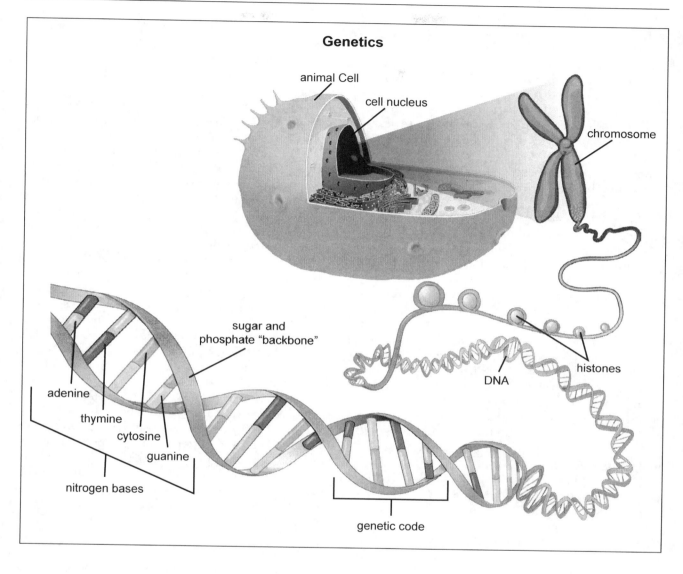

Genetics

animal Cell

cell nucleus

chromosome

sugar and phosphate "backbone"

histones

DNA

adenine

thymine

cytosine

guanine

nitrogen bases

genetic code

command center, the DNA sends out orders but does not leave the nucleus. DNA is made up of nucleotide bases, and the first step in protein production involves the reading of the nucleotide code in the DNA. When the nucleotides are read, a strand of messenger ribonucleic acid (mRNA) is produced and sent from the nucleus into the cytoplasm of the cell. The DNA stays in the nucleus, while the mRNA leaves the nucleus. This process of reading the DNA nucleotide code and producing mRNA is called transcription.

The next step involves reading the nucleotide code on the mRNA in the cytoplasm of the cell. The cytoplasm is the liquid environment inside the cell where all the cellular organelles float. Transfer RNA (tRNA) reads the code on the mRNA, and this process is called translation. tRNA is called transfer because it transfers a specific amino acid when it reads the appropriate code on the mRNA. The basic building blocks of protein are called amino acids. About twenty-two different amino acids build all the various proteins the body uses. It is like the English alphabet: Twenty-six letters and vowels comprise the English alphabet, and combining these various letters and vowels results in tens of thousands of different combinations and all the various words in the English language. Likewise, the body uses the different amino acids to form the tens of thousands of different proteins in the body.

During translation, a specific amino acid is coded for and carried by the transfer RNA to a ribosome. Ribosomes (along with the rough endoplasmic reticulum) are where the cell's proteins are produced. Along ribosomes, different amino acids are transported by tRNA and linked, forming a protein molecule. Some proteins may be only eighty or ninety amino acids long, whereas others, such as hemoglobin, may have more than 300 amino acids as their amino acid backbone.

The way DNA codes for all this involves the nucleotide bases that make up DNA. Four nucleotide bases make up DNA: adenine, cytosine, guanine, and thymine. Adenine will chemically bind with thymine, and cytosine always chemically binds with guanine. When DNA is transcribed to form mRNA, if the nucleotide sequence in the DNA reads cytosine-cytosine-guanine, these nucleotide bases will code for guanine-guanine-cytosine in the mRNA. Then when mRNA is translated by tRNA, the code goes back to the original DNA code, cytosine-cytosine-guanine. Cytosine-cytosine-guanine can code for a specific amino acid, and in that fashion DNA codes for the amino acid sequence of all protein molecules. The nucleotide base sequence in the mRNA is called the codon and the complimentary base sequence found in tRNA is called the anticodon.

The example of how Dolly the sheep was cloned demonstrates how genetic engineering in mammals works, and, hence, how human cloning could work. Cloning is an ultimate example of genetic engineering because cloning produces an entire living organism via genetic engineering. Dolly was a Finn Dorset sheep, which is all white. A Blackface ewe, named because of the distinctive black face these sheep have, was used as an egg donor and as a surrogate mother.

Cells taken from a Finn Dorset ewe were grown in a tissue culture. An egg cell, from the Blackface ewe, had the nucleus removed. The nucleus contained the genes and DNA. The nucleus and genetic information from the Finn Dorset ewe were placed in the enucleated Blackface ewe egg cell. The Blackface ewe egg cell, now containing genetic information from the Finn Dorset ewe, was placed in the uterus of the Blackface ewe after an electric pulse is applied to stimulate growth and duplication of the cells. The

Blackface ewe gave birth to Dolly, the all-white Finn Dorset ewe. The newborn Finn Dorset ewe was an identical genetic copy of the Finn Dorset ewe originally used to harvest the genetic information found in the nucleus.

Recombinant DNA refers to DNA transfer from one cell to another. In human genetic engineering, genes transfer from a human chromosome to another cell, usually bacteria. If the transferred human genes code for insulin, the bacteria accepting the transferred genes will now produce human insulin.

In this process, the desired genes are isolated and removed from the human cell. Bacterial cells have small, circular strips of DNA called plasmids. These circular plasmids are removed from the bacterial cells and opened up. Various enzymes are used to cut the human DNA and bacterial DNA sequences at specific points. Restriction enzymes cut the original DNA in specific locations. DNA ligase pastes strips of DNA together. Scientists mix isolated human genes with the opened bacterial plasmids, along with DNA ligase. The human genes are spliced into the bacterial plasmid and the circle of genetic information in the bacterial plasmid closes.

The bacterial plasmid with the spliced human genes is now called a vector. The plasmid vectors are taken up by the bacterial cells. Once inside the bacterial cells, the bacteria multiply and reproduce the spliced human genes. Whatever specific human genes were selected for splicing, for example, human insulin genes, are now functioning in the reproduced bacteria, and human insulin is harvested from the bacterial clones.

APPLICATIONS AND PRODUCTS

Early Medical Applications. Recombinant DNA techniques are remarkable biological life adaptations, and many medicines based on this technology are used. Medications generated through human genetic engineering techniques have been in use since 1982, with the production of human insulin using recombinant DNA techniques. Human genes are inserted into a bacterial host that then makes human insulin. Prior to this type of genetic manipulation, diabetics needing insulin had to rely on insulin harvested from pigs or cows. Genetic techniques produce human growth hormone, previously only available from human cadavers. A genetically engineered hepatitis B vaccine has been in use since 1987.

Since these first human genetic medicines and vaccines, many types of biological products have been introduced or are under current investigation and development. These new medicinal products are called biologics to distinguish them from chemically synthesized medicines. Genes and genetic manipulations produce biologics. Major types of biologics include hormones, antibodies, and cell-receptor proteins.

Insulin and human growth hormone, discussed above, are classic protein hormones produced with recombinant DNA technology. The immune system produces protein antibodies that attack disease causing-agents such bacteria and viruses. Genetic antibody production interferes with or attacks entities associated with diseases such as psoriatic arthritis and Crohn's disease. Recombinant DNA technology produces proteins binding with specialized white blood cells to reduce inflammation associated with rheumatoid arthritis.

Bioinformatics. The purpose of bioinformatics is to help organize, store, and analyze genetic biological information in a rapid and precise manner, dictated by the need to be able to access genetic information quickly. In the United States the online database that provides access to these gene sequences is called GenBank, which is under the purview of the National Center for Biotechnology Information (NCBI) and has been made available on the Internet. In addition to human genome sequence records, GenBank provides genome information about plants, bacteria, and animals other than humans.

Proteomics. Bioinformatics provides the basis for all modern studies of human genetics, including analyzing genes and gene sequences, determining gene functions, and detecting faulty genes. The study of genes and their functions is called proteomics, which involves the comparative study of protein expression. That is, it studies the metabolic and morphological relationship between the protein encoded within the genome and how that protein works. Geneticists are now classifying proteins into families, superfamilies, and folds according to their configuration, enzymatic activity, and sequence. Ultimately proteomics will complete the picture of the genetic structure and functioning of all human genes.

Toxicogenomics. Another newly developing field that relies on bioinformatics is the study of toxicogenomics, which is concerned with how human genes

Fascinating Facts About Human Genetic Engineering

- In 2011, researchers at Brigham and Women's Hospital in Boston identified a self-renewing human lung stem cell. This particular stem cell is also able to form and integrate a variety of biological lung structures, including bronchi, alveoli, and pulmonary vessels.
- DNA fingerprints, used in forensic genetics, are made by using enzymes splitting the genetic sequence up into patterns unique to an individual. DNA fingerprinting used to require test tubes of blood for analysis. Polymerase chain reaction technology now allows the reproduction of DNA (and subsequent analysis) from a sample as small as dried saliva from the back of a stamp.
- Synthetic biology aims to produce life-forms. Life-forms reproduce, by definition. An invention such the atomic bomb cannot reproduce itself, but biologics can.
- Dolly the sheep, the first cloned mammal, lived six years, half the normal life expectancy. The sheep originating the cloned DNA was six years old at the time of donation, raising speculation about the genetic age of the donor DNA.
- Genetic engineering techniques could resurrect life for extinct animals, such as mammoths, or extinct human species, such as Neanderthals. Although "farming" DNA that is viable for such use is difficult, theoretically such feats could be accomplished. A film that was based on this possibility is *Jurassic Park* (1993), based on a novel of the same name by Michael Crichton.

respond to toxins. As of 2011, this field is specifically concerned with evaluating how environmental factors negatively interact with mRNA translation, resulting in disease or dysfunction.

Gene Testing. In a gene-testing protocol, a sample of blood or body fluids is examined to detect a genetic anomaly such as the transposition of part of a chromosome or an altered sequence of the bases that comprise a specific gene, either of which can lead to a genetically based disorder or disease. As of 2011, more than 600 tests are available to detect malfunctioning or nonfunctioning genes. Most gene tests have focused on various types of human cancers, but other tests are being developed to detect genetic deficiencies that cause or exacerbate infectious and vascular diseases.

The emphasis on the relationship between genetics and cancer lies in the fact that all human cancers are genetically triggered or have a genetic basis. Some cancers are inherited as mutations, but most result from random genetic mutations that occur in specific cells, often precipitated by viral infections or environmental factors not yet well understood.

At least four types of genetic problems have been identified in human cancers. The normal function of oncogenes, for example, is to signal the start of cell division. However, when mutations occur or oncogenes are overexpressed, the cells keep on dividing, leading to rapid growth of cell masses. The genetic inheritance of certain kinds of breast and ovarian cancers results from the nonfunctioning tumor-suppressor genes that normally stop cell division. When genetically altered tumor-suppressor genes are unable to stop cell division, cancer results. Conversely, the genes that cause inheritance of colon cancer result from the failure of DNA repair genes to correct mutations properly. The accumulation of mutations in these "proofreading" genes makes them inefficient or less efficient, and cells continue to replicate, producing a tumor mass.

If a gene screening reveals a genetic problem several options may be available, including gene therapy and genetic counseling. If the detected genetic anomaly results in disease, then pharmacogenomics holds promise of patient-specific drug treatment.

Gene Therapy. The science of gene therapy uses recombinant DNA technology to cure diseases or disorders that have a genetic basis. Still in its experimental stages, gene therapy may include procedures to replace a defective gene, repair a defective gene, or introduce healthy genes to supplement, complement, or augment the function of nonfunctional or malfunctioning genes. Several hundred protocols are being used in gene-therapy trials, and many more are under development. As of 2011, trials are focusing on two major types of gene therapy, somatic cell gene therapy and germ-line gene therapy.

Somatic cell gene therapy concentrates on altering a defective gene or genes in human body cells in an attempt to prevent or lessen the debilitating impact of a disease or other genetic disorder. Some examples of somatic cell gene therapy protocols now being tested include ones for adenosine deaminase (ADA) deficiency, cystic fibrosis, lung cancer, brain tumors, ovarian cancer, and AIDS.

In somatic cell gene therapy a sample of the patient's cells may be removed and treated and then reintegrated into body tissue carrying the corrected gene. An alternative somatic cell therapy is called gene replacement, which typically involves insertion of a normally functioning gene. Some experimental delivery methods for gene insertion include use of retroviral vectors and adenovirus vectors. These viral vectors are used because they are readily able to insert their genomes into host cells. Hence, adding the needed (or corrective) gene segment to the viral genome guarantees delivery into the cell's nuclear interior. Nonviral delivery vectors that are being investigated for gene replacement include liposome fat bodies, human artificial chromosomes, and naked DNA (free DNA, or DNA that is not enclosed in a viral particle or any other "package").

Another type of somatic gene therapy involves blocking gene activity, whereby potentially harmful genes such as those that cause Marfan syndrome and Huntington's disease are disabled or destroyed. Two types of gene-blocking therapies being investigated include the use of antisense molecules that target and bind to the mRNA produced by the gene, thereby preventing its translation, and the use of specially developed ribozymes that can target and cleave gene sequences that contain the unwanted mutation.

Germ-line therapy is concerned with altering the genetics of male and female reproductive cells (gametes) as well as other body cells. Because germline therapy will alter the individual's genes as well as those of his or her offspring, both concepts and protocols are still very controversial. Some aspects of germ-line therapy now being explored include human cloning and genetic enhancement.

Clinical Genetics. Clinical genetics is that branch of medical genetics involved in the direct clinical care of people afflicted with diseases caused by genetic disorders. Clinical genetics involves diagnosis, counseling, management, and support. Genetic counseling is a part of clinical genetics directly concerned with medical management, risk determination and options, and decisions regarding reproduction of afflicted individuals. Support services are an integral feature of all genetic counseling themes.

Clinical genetics begins with an accurate diagnosis that recognizes a specific, underlying genetic cause of a physical or biochemical defect following guidelines outlined by the National Institutes of Health (NIH) Counseling Development Conference. Clinical practice includes several hundred genetic tests that are able to detect mutations such as those associated with breast and colon cancers, muscular dystrophy, cystic fibrosis, sickle-cell disease, and Huntington's disease.

Genetic counseling follows clinical diagnosis and focuses initially on explaining the risk factors and human problems associated with the genetic disorder. Both the afflicted individual and family members are involved in all counseling procedures. Important components include a frank discussion of risks, of options such as preventive operations, and of options involved with regard to reproduction. All reproductive options are described along with their potential consequences, but genetic counseling is a support service rather than a directive mode. That is, it does not include recommendations. Instead, its ultimate mission is to help both the afflicted individuals and their families recognize and cope with the immediate and future implications of the genetic disorder.

Pharmacogenomics. That branch of human medical genetics dealing with the correlation of specific drugs to fit specific diseases in individuals is called pharmacogenomics. This field recognizes that individuals may metabolically respond differentially to therapeutic medicines based on their genetic makeup. It is anticipated that testing human genome data will greatly speed the development of new drugs that not only target specific diseases but also will be tailored to the specific genetics of patients.

Forensic Genetics. Forensic genetics is the use of human genetics in criminal or paternity cases. For example, DNA testing on blood, saliva, or other tissue can be used to determine the source of evidence, such as blood stains or semen, left at a crime scene. Forensic DNA analysis is also used to determine paternity and other kinship. Finally, with the increasing use of forensic genetics since the 1990's, some incarcerated prisoners have been released after it was clearly determined that they could not possibly have been guilty of crimes they were convicted of, as DNA evidence eliminated them from suspicion.

Potential for Human Cloning. Human therapeutic cloning involves the production of cloned human embryos, with the idea of harvesting embryonic stem cells. The hope is that the stem cells can be grown into a wide variety of cells to replace or repair organs, such as liver, kidney, or heart cells. Although human cloning has not yet reached this potential, future applications could offer identically matched kidneys for people with failing kidneys or even a genetically duplicate heart for someone in severe heart failure.

IMPACT ON INDUSTRY

The many applications of our knowledge of human genetics have had a revolutionary impact on industry, spawning entirely new industries as well as new product lines. Three main areas on which these industries focus is research, or the analysis of the human genome to identify new and potentially profitable areas of investigation; pharmaceutical applications–many of the large pharmaceutical companies produce medications based on human genetic research, such as the aforementioned synthetic insulin hormones; and finally, genetic counseling, which the National Society of Genetic Counselors defines as the practice of "health care professionals with specialized graduate training in the areas of medical genetics and counseling" who "work as members of a health care team, providing information and support to families who have members with birth defects or genetic disorders and to families who may be at risk for a variety of inherited conditions."

Genentech, established in 1976, is considered the founder of the biotechnology industry, with special focus on human genetics. Its stated mission is "using human genetic information to discover, develop, manufacture, and commercialize medicines to treat patients with serious or life-threatening medical conditions . . . to create, produce, and market innovative solutions of high quality for unmet medical needs." Other leading companies in the field include Amgen, Genzyme, Gilead Sciences, Biogen Idec, Cephalon, and MedImmune in the United States; Merck Serono in Switzerland; CSL in Australia; and UCB in Belgium.

CAREERS AND COURSE WORK

A variety of career paths exist for people interested in genetics. Educational requirements vary from bachelor's to doctorate degrees. Research scientists, working at major universities, companies, and government agencies such as the NIH usually need a Ph.D.

or M.D. A clinical geneticist who treats patients typically completes medical school and specialty training. Geneticists or research scientists usually have Ph.D. training in fields such as molecular genetics, cytogenetics, or the burgeoning field of synthetic biology. Genetic counselors complete a master's degree program in genetic counseling. Research assistants in genetics laboratories usually have a master's degrees in genetics or biological sciences. Genetic laboratory technicians work in forensic or research labs with a bachelor's degree in science-related fields.

Select engineering schools, such as the University of Michigan, have biomedical engineering programs. Biomedical engineers interface with medical device and drug industries. Biomedical engineers will play an important role bringing genetic therapies into clinical practice. Tissue engineering involves the manipulation of genes to affect changes in phenotypes, the essence of genetic engineering.

SOCIAL CONTEXT AND FUTURE PROSPECTS

The Human Genome Project painstakingly mapped out the human DNA sequence in 2003, after a decade and a half of meticulous multicenter collaboration. Genetic databases are now rapidly filling with genetic detail because of technological advances in the speed of analyzing DNA sequences. While the speed of this analysis has increased considerably, the price of such investigations has dropped significantly. Genetic databases currently hold information on a wide variety of life-forms, and significant amounts of new generic information is added frequently.

The DNA sequencing found in genetic databases provides the burgeoning field of synthetic biology with important basic information needed for human genetic engineering. This information can be used for modeling and as supply depots for the mixing and matching of genes. As the speed of genetic analysis has increased significantly and the price of genetic investigations has dropped considerably, the process of DNA synthesis is much less expensive and faster than it was in the beginning of the twenty-first century.

More genetic information, faster artificial DNA synthesis, and significant technological cost savings result in more feasible human genetic engineering projects. Genes are the stuff of life, and the field is on the verge of changing life and even making new life-forms, via genetic engineering. How and what

changes are made will present significant bioethical and societal challenges, along with potentially fantastic and beneficial results.

Richard P. Capriccioso, M.D.

FURTHER READING

Andrews, Lori B. *The Clone Age: Adventures in the New World of Reproductive Technology.* New York: Henry Holt, 1999. A lawyer specializing in reproductive technology, Andrews examines the legal ramifications of human cloning, from privacy to property rights.

Baudrillard, Jean. *The Vital Illusion.* Edited by Julia Witwer. New York: Columbia University Press, 2000. A sociological perspective on what human cloning means to the idea of what it means to be human.

Capriccioso, Richard P. "Genetic Testing." In *Salem Health: Cancer,* edited by Jeffrey A. Knight. Pasadena, Calif.: Salem Press, 2009. A comprehensive overview of genetic testing covering different types of genetic tests with a review of the science behind the testing.

The Economist. "Artificial Lifeforms: Genesis Redux." 395, no. 8683 (May 20, 2010): 81-83. Informative article on synthetic biology and the creation of a new form of life in the laboratory.

Hartwell, Leland, et al. *Genetics: From Genes to Genomes.* 4th ed. New York: McGraw-Hill, 2011. A comprehensive textbook on genetics, including human genetics discussed in a comparative context.

Hekimi, Siegfried, ed. *The Molecular Genetics of Aging: Results and Problems in Cell Differentiation.* Berlin: Springer, 2000. Examines various genetic aspects of the aging process. Illustrated.

Jorde, Lynn B., John C. Carey, and Michael J. Bamshad. *Medical Genetics.* 4th ed. Philadelphia: Mosby, 2010. Provides both an introduction to the field of human genetics with chapters on clinical aspects of human genetics, such as gene therapy, genetic screening, and genetic counseling.

Lewis, Ricki. *Human Genetics: Concepts and Applications.* 9th ed. New York: McGraw-Hill, 2010. This textbook provides a broad overview of human genetics and genomics.

Pasternak, Jack J. *An Introduction to Human Molecular Genetics: Mechanisms of Inherited Diseases.* 2d ed. Hoboken, N.J.: Wiley-Liss, 2005. Discusses treatment advances, fundamental molecular mechanisms that govern human inherited diseases, the interactions of genes and their products, and the consequences of these mechanisms on disease states in major organ systems such as muscles, the nervous system, and the eyes. Also addresses cancer and mitochondrial disorders.

Rudin, Norah, and Keith Inman. *An Introduction to Forensic DNA Analysis.* 2d ed. Boca Raton, Fla.: CRC Press, 2002. An overview of many DNA typing techniques, along with numerous examples and a discussion of legal implications.

Shostak, Stanley. *Becoming Immortal: Combining Cloning and Stem-Cell Therapy.* Albany: State University of New York Press, 2002. Examines the question of whether human beings are equipped for potential immortality.

Wilson, Edward O. *On Human Nature.* 1978. Cambridge, Mass.: Harvard University Press, 2004. A look at the significance of biology and genetics for the way people understand human behaviors, including aggression, sex, and altruism, and the institution of religion.

WEB SITES

American Society of Human Genetics (ASHG)
http://www.ashg.org

Center for Genetics and Society
http://www.geneticsandsociety.org

Genetics Education Center
University of Kansas Medical Center
http://www.kumc.edu/gec

Human Genome Project
http://www.ornl.gov/sci/techresources/Human_Genome/home.shtml

National Institutes of Health
Stem Cell Information
http://stemcells.nih.gov/info/basics

National Society of Genetic Counselors
http://www.nsgc.org

See also: Bioengineering; Cell and Tissue Engineering; Cloning; DNA Analysis; DNA Sequencing; Genetic Engineering.

HYDROELECTRIC POWER PLANTS

FIELDS OF STUDY

Biology; civil engineering; earth science; electrical engineering; electronics; environmental science; geology; hydrogeology; mechanical engineering; physics; natural resource planning; seismology; urban planning

SUMMARY

Hydroelectric power plants produce electricity using water, a renewable resource. The power plants convert the energy in flowing water into electricity that can supply the needs of an entire city or supplement the power available for a region or other area on the power grid. To produce the electricity, water collects behind a dam before flowing through a turbine. As the water flows through the turbine, a generator uses magnets to create an electromagnetic field and then electricity. There is little pollution created in the process, but there is an impact on the ecosystem at the site of the plant.

KEY TERMS AND CONCEPTS

- **Dam:** Any barrier that curtails the flow of water.
- **Excitor:** Sends an electrical current to the rotor as the turbine turns.
- **Generator:** The part of the turbine that uses magnets to produce electricity.
- **Intake:** The penstock that brings water into the powerhouse.
- **Outflow:** The used water moved through pipes to reenter the river downstream from the dam.
- **Penstock:** Closed conduit or pipe that uses gravity to bring water to the powerhouse and into the turbine.
- **Powerhouse:** A facility for the generation of electricity.
- **Power Lines:** The wires that carry electricity from the powerhouse.
- **Reservoir:** A large natural or human-made lake used to contain water for drinking or power.
- **Rotor:** The series of electromagnets that spin inside the stator.
- **Stator:** The tightly wound coil of copper wire inside the generator.

- **Tailraces:** The pipelines that carry the outflow.
- **Transformer:** Located inside the powerhouse, it takes the alternating current and converts it to higher voltage currency.
- **Turbine:** A rotary engine that extracts energy from a fluid flow and converts it into usable energy.
- **Turbine Blades:** The parts of the turbine that spin when water passes over them as it falls over the dam.
- **Wicket Gate:** Controls the amount of water flowing from the penstock through the turbine.

DEFINITION AND BASIC PRINCIPLES

Hydroelectric power plants take the stored energy of water in a reservoir and convert it into electricity. Swiftly moving water is brought into the powerhouse and then moved through the turbine engine. As the water passes by the turbine, it causes the blades of the turbine to spin. These blades in turn cause a series of rotors with magnets mounted on rotors inside the generator to rotate past copper coils. The magnets and coiled copper wire act as giant electromagnets and produce an alternating current. The alternating current, or AC, is then converted into a higher voltage of current before exiting the powerhouse by way of the power lines that carry the power to the electric grid.

The amount of electricity generated depends upon the volume of water flow from the reservoir. The larger the reservoir, the greater the volume flow and the greater the amount of electricity produced, as the greater the volume flow, the more quickly the magnets will spin within the coil.

For a hydroelectric power plant to produce a steady flow of electricity, a steady source of water must flow into the reservoir. To ensure this, some hydroelectric power plants have an upper reservoir and a lower reservoir. Water from the upper reservoir is used to produce electricity and then channeled into a lower reservoir, rather than back into the river downstream of the dam. During off-peak hours, water from the lower reservoir is pumped back up to the upper reservoir to be used again as a source of flowing fluid for the turbines. In the absence of a lower reservoir, water is directed into the river downstream of the dam.

Current technology calls for large quantities of

water moving at a high rate of speed. Newer technologies that use the kinetic properties of water without building dams, like the waterwheels of years ago, are also being explored.

BACKGROUND AND HISTORY

Civilizations have been using water as a power source for thousands of years. From the ancient Greeks, who used water wheels to replace manual labor in the grinding of wheat into flower, to the Romans, who created floating mills when under siege by the Goths in 537 c.e., water has been used to get work done. Because they had fewer slave workers than the Romans, the ancient Chinese used waterpower to their advantage throughout the empire. Water, for example, was used to power the bellows used in iron casting.

During the medieval period in Europe, waterpower grew in prominence. Because of the Black Death (the plague) and the shortage of labor that resulted, it was critical to find an inexpensive way to grind wheat to sustain the surviving population. The number of mills increased dramatically, with some of the mills built to take advantage of tidal changes.

Another important development was the use of waterpower by a religious order. The Cistercian monastic order lived in rural areas, where they perfected the use of hydropower for milling, woodcutting, and olive crushing. They did not use dams, but rather placed their waterwheels in swiftly moving water without impeding the flow of that water. They also used the water for washing and for sewage disposal. As the order moved through Europe, the technology traveled with them, until waterpower next came to prominence during the Industrial Revolution. At that time, swiftly running streams were used to provide power for a variety of manufacturing processes before the widespread use of fossil fuel.

Hydroelectric power plants in the United States have often been constructed to take advantage of the force of moving water while also solving flooding or other water-related problems. Typically the dams have been among the largest concrete structures built in the area. They also have altered the ecosystem by controlling the flow of water downstream while flooding the area upstream of the dam. Because of this, a great deal of opposition to new construction accompanies any dam proposal; people debate issues of water rights and the environmental impact of dams in general.

HOW IT WORKS

The process of converting the energy from a flowing liquid, such as water, into electricity that can be used to supply the power needs of a city or to increase the amount of electricity available on a local or regional grid is actually simple, as long as the conditions are right.

Swift Water Flow. These conditions include the existence of a steady supply of swiftly moving water. The water can flow swiftly as a result of the pressure on the water from the reservoir as the water enters the intake, or penstock, area of the powerhouse located near the base of the dam. Alternately, it can flow swiftly as a result of being released over the top of the dam to spill into the penstock at a high speed. Either way, it is essential the water moves rapidly when it flows through the turbine, although it is not necessary that the water first be held in a reservoir created by a dam.

The turbine has blades that are turned by the flowing water. The more rapidly the water flows, the more rapidly the blades spin. The spinning motion of the blades in turn causes the magnets in the generator, mounted on rotors, to spin inside coils of copper wire. This spinning creates an electromagnetic field that produces alternating current that is then converted to a higher voltage current in the transformer. The higher voltage current can be stored but is typically transmitted over power lines to become part of the power grid serving a city or region.

A typical powerhouse will have water entering through multiple penstocks to flow through one of several turbines mounted under generators. Once the water has flowed past the turbine blades, it will reenter the flow of the river downstream of the dam. The water can also be channeled into a lower reservoir to be pumped back to the upper reservoir for reuse in generating additional electricity. Whether or not this will be done depends on the amount of water available in the reservoir and the expediency of pumping the water back to the higher reservoir.

The placement of the hydroelectric power plant is vitally important. Because a steady flow of water must be moving at a high speed, the dam needs to be located on a river with a reliable supply of water. Gravity also plays an important part in the generation of hydroelectric power because water picks up speed as it moves from a higher level to a lower level. Thus, dams are often built in areas where there is a natural downward flow.

Reservoirs. The construction of a dam often also includes building a reservoir that covers hundreds of acres of land. When that land is flooded, wildlife will lose their habitat and plant life will be ruined. Reservoir construction may also require the relocation of a significant number of people. Because of this, a study of the impact of the dam is often an important part of the planning process, and not all dams are built at the optimal site.

Land Integrity. Finally, the dam site must be one that can bear the weight of the dam and the water that will accumulate. It must also be a site without significant seismic activity or the likelihood of such activity. To ensure this, a thorough geological study of the site is necessary before construction begins.

APPLICATIONS AND PRODUCTS

Hydroelectric power plants supply power that is used for many purposes. Most plants supply power to an existing power grid. Once part of the grid, the power is allocated to the area of greatest need (along with power from other sources). It is possible, though not common, to have a dedicated hydroelectric power plant, one that is created specifically to meet the needs of an individual or factory.

The main application of hydroelectric power plants is to ensure a steady supply of electricity through a process that uses a renewable resource with little pollution. Hydroelectric power plants do have an environmental impact, however, and that needs to be taken into consideration.

Hydroelectric power plants have had a significant impact on industry through the ages. Originally a simple replacement for slave labor, the use of hydroelectric power peaked before the widespread use of fossil fuels as power sources. It is possible for a dedicated waterwheel to power a simple manufacturing process such as milling, on a small scale.

The largest hydroelectric power plant in the world is the Three Gorges Dam in the People's Republic of China. It has a capacity of 22,500 MW (megawatts) and is visible from space. The largest hydroelectric power installation in the United States, and the fifth largest in the world, is the Grand Coulee. Located in Washington State, the power plant produces electricity and is also used to irrigate the land around it. One of the largest concrete structures in the world, the Grand Coulee has a capacity of 6,809 MW. Brazil,

Venezuela, Russia, and Canada are also home to hydroelectric power plants with significant capacity.

IMPACT ON INDUSTRY

Hydroelectric power is considered a green power source. Water is a renewable resource that is plentiful on the North American continent. As a result, several government agencies are actively involved in managing and promoting hydroelectric power plants and projects.

The U.S. Department of the Interior's Bureau of Reclamation is responsible for water operations in the western United States. The region includes the Pacific Northwest, upper and lower Colorado River, and the mid-Pacific. Washington, Oregon, Idaho, California, Nevada, Utah, Arkansas, New Mexico, and portions of Wyoming, Colorado, and Texas are under their jurisdiction. The Hydropower Technical Services Group provides specialized technical knowledge to hydroelectric power engineers as they work to increase energy production at their own sites.

The U.S. Department of Energy's Water Power Program is designed to optimize existing hydroelectric power plant production while promoting construction of new plants and hydropower technologies. The agency maintains an online database with information and data about hydroelectric power installations, plans, and new technologies. The database also includes information about technologies in use or under development in the United States and abroad. It can be used by anyone and allows anyone to submit a technology or project idea for consideration.

The U.S. Geological Survey (USGS) also collects information about water resources throughout the United States. The USGA's Web site highlights water science by state and includes links to a full range of publications and reports on water resources, quality, and use. As governments worldwide move toward renewable sources of energy, hydroelectric power and associated technologies will remain at the forefront.

CAREERS AND COURSEWORK

Hydroelectric power and the technology behind existing and new forms of hydroelectric power are growing fields. Study in the fields of engineering, hydrogeology, civil engineering, environmental science, and seismology are just a few areas of study that will lead to careers in this industry.

Impact and feasibility studies must be performed

before construction can begin on a new facility. These studies include the expert opinions of geologists, seismologists, civil engineers, and environmental scientists about the quality of the sites under consideration and the type of facility that can be built and maintained there. The studies also detail the environmental impact of the facility under consideration. Decisions about the use of the water before and after the generation of power will also be made. The size of the facility will also be determined once the experts figure the desired output and uses of the water before and after the electricity is produced.

After approval, civil engineers oversee the construction of the dam and penstock while electrical engineers work to install the turbines, generators, and transformers. Specialists in the installation and operation of the turbines, generators, and transformers will be on hand throughout the construction phase. When construction is completed, professionals ensure that the hydroelectric power plant runs at optimal efficiency.

Engineers inspect the integrity of the dam at frequent intervals. Naturalists ensure that any fish ladders or other equipment function appropriately and are in operation at the times that are essential to the upriver journey of the fish. Naturalists also oversee the health of the fish population and the reservoir.

Social Context and Future Prospects

Hydroelectric power plants are important sources of renewable energy. Producing the electricity generated by these plants does not result in significant levels of pollution. It also does not consume resources that take centuries to replenish, does not require labor-intensive or costly processes, and does not damage the environment.

The dam site and its reservoir, however, do affect the local ecosystem. The land lost to the reservoir through flooding is likely already home to many species of animals and plants and may include towns or villages, all of which will be displaced or destroyed by the dam project. It is possible the reservoir will form a wetlands area at the shoreline. It also is possible that this will not occur, resulting instead in areas of stagnant water that are not hospitable to wildlife.

Furthermore, the dams built for hydroelectric power plants, for example, cut off access to the spawning grounds of anadromous, migratory salmon

---

Fascinating Facts About Hydroelectric Power Plants

- Hydroelectric power plants provide power from a renewable resource at low cost while generating little pollution.
- Construction of a dam for a hydroelectric power plant often alters, both positively and negatively, the surrounding ecosystem.
- The first hydroelectric power plant in the United States began operation in Appleton, Wisconsin, in 1882.
- Hydroelectric power is used around the world, providing almost 25 percent of the world's electricity.
- China generates more hydroelectricity than any other country.
- The largest hydroelectric power plant in the United States is located on the Columbia River in northern Washington State at the Grand Coulee Dam.
- The Three Gorges Dam project in China displaced 1.4 million people.
- The Three Gorges Dam is so large that it creates its own seismic activity.
- The Three Gorges Dam is one of only a few human-made structures visible from space.
- Hydroelectric power can be used anywhere there is falling water.
- Many dams incorporate fish lifts to help fish such as the American shad or the salmon return upstream to spawn.

---

or the American shad. These fish must return upriver to lay their eggs. To facilitate the return journey, fish lifts, ladders, or elevators are in place at many dams. These structures, which help fish move upstream, also help to avoid the disruption of the local habitat.

As the importance of renewable energy sources becomes incontrovertible, greater demand for hydroelectric power can be anticipated. As the cost in terms of loss of habitat gains greater appreciation, the call to protect existing wildlife and vegetation can also be expected. With existing technology, these goals are not easily met simultaneously.

The challenge for the next generation of hydroelectric power professionals will be to modify this existing technology or explore the use of alternatives that do not require such a large footprint. Using water as it moves, without impeding its flow, is one possible way that hydroelectric power plants can better coexist with the populations they serve.

Gina Hagler, M.B.A.

FURTHER READING

Gevorkian, Peter. *Sustainable Energy System Engineering.* New York: McGraw-Hill, 2007. Presents a variety of green energy solutions that includes hydroelectric power.

Hicks, Tyler. *Handbook of Energy Engineering Calculations.* New York: McGraw-Hill, 2011. Covers such topics as combustion of fossil fuels, alternative power sources, and hydroelectric power.

Monroe, James S., and Reed Wicander. *The Changing Earth: Exploring Geology and Evolution.* Belmont, Calif.: Thomson Higher Education, 2006. An excellent overview of the role of geology in general, with several chapters on energy sources such as hydroelectric power.

Nag, P. K. *Power Plant Engineering.* New Delhi, India: Tata McGraw-Hill, 2008. A thorough discussion of power generation from a variety of sources, including hydroelectric power. Also presents considerations for appropriate site selection.

WEB SITES

Tennessee Valley Authority "Hydroelectric Power."
http://www.tva.gov/power/hydro.htm.

U.S. Department of Energy, Water Power Program
http://www1.eere.energy.gov/water.

U.S. Department of the Interior "Reclamation: Managing Water in the West—Hydroelectric Power."
http://www.usbr.gov/power/edu/pamphlet.pdf.

U.S. Geological Survey "Water Resources of the United States."
http://water.usgs.gov.

See also: Landscape Ecology; Land-Use Management.

HYDROLOGY AND HYDROGEOLOGY

FIELDS OF STUDY

Hydrology; hydrogeology; physics; physical geography; geology; chemistry; biology; fluid mechanics; statistics; water resources; groundwater; water supply; precipitation; fluvial processes; earth sciences; aquifers; water quality; climatology; meteorology; evapotranspiration.

SUMMARY

Hydrology is a broad interdisciplinary science that includes the hydrologic cycle and global distribution of water in both solid and liquid form in the atmosphere, oceans, lakes, streams, and subsurface formations. Hydrogeology is a subset of hydrology that focuses on groundwater and related geologic factors governing its distribution and magnitude. The shortages of water and the increasing pollution in many countries have heightened concern about availability and quality of water for many people in the world.

KEY TERMS AND CONCEPTS

- **Aquifer:** Water-bearing geologic formation of saturated permeable materials.
- **Base Flow:** Groundwater discharge moving into receiving streams.
- **Discharge:** Amount of water flowing through a stream cross section, measured in cubic feet per second or cubic meters per second.
- **Evapotranspiration:** Combined loss of water from evaporation from soils, streams, and lakes, plus plant transpiration.
- **Exotic River:** Stream that flows in an arid region from water coming from distant well-watered uplands.
- **Groundwater:** Subsurface water in the saturated zone that varies with the type of geological formation.
- **Hydrologic Cycle:** Total transfer and storage of all of the readily available water on the Earth as it moves through the gaseous, liquid, and solid states.
- **Watershed:** Area of land that has a drainage network and is separated from other watersheds by a drainage divide; also known as a drainage basin.

DEFINITION AND BASIC PRINCIPLES

Hydrology is the science of water, a unique substance that affects all life on Earth. Although water could exist on Earth without life, life could not exist without water. Water is the most abundant liquid on Earth, covering 71 percent of the Earth's surface in its liquid and solid forms. In humans, water constitutes about 92 percent of blood plasma, 80 percent of muscle tissue, 60 percent of red blood cells, and more than half of most other tissues.

Hydrology is a part of many scientific disciplines. The study of water in the atmosphere involves the fields of climatology and meteorology. The study of the hydrosphere includes the fields of physical geography, potamology (rivers), glaciology, cryology (snow, ice, and frozen ground), and limnology (lakes). The study of the lithosphere (the topmost rock layer of the Earth) includes the fields of hydrogeology (groundwater location, movement, and magnitude), geomorphology (the science of surface processes and landforms on the Earth), and limnology (the science of lakes). Given the importance of water to plants and animals, hydrology includes the fields of silviculture (forestry), plant ecology, and hydrobiology (the biology of bodies of freshwater such as lakes).

Other fields related to hydrology by virtue of their strong connection to water resources include watershed management, potable water supply, wastewater treatment, irrigation, water law, political science (water policy), economics (costs of water projects), drainage, flood control, hydropower, salinity control and treatment, erosion and aspects of sediment control, navigation, lake and inland fisheries, and recreational uses of water.

BACKGROUND AND HISTORY

Credit for developing part of the hydrologic cycle, a fundamental component of hydrology, goes to Marcus Vitruvius Pollio, a Roman engineer and writer during the reign of the emperor Augustus, who developed a theory that groundwater is mostly recharged by precipitation infiltrating the ground. This early theory was buttressed by Leonardo da Vinci and Bernard Palissy during the sixteenth century. The seventeenth century was a period of measurement, when scientists studied precipitation, evaporation, and stream discharge in the Seine River in France. Hydraulic studies developed during the eighteenth century. In the nineteenth century, the active area of investigation was experimental hydrology, particularly in stream-flow measurement and in groundwater. The U.S. government created several important agencies, including the U.S. Army Corps of Engineers in 1802, the U.S. Geological Survey in 1879, and the U.S. Weather Bureau (now the National Weather Service) in 1891.

The nineteenth and twentieth centuries witnessed the increasing use of statistical and theoretical analysis in hydrologic studies. One example of this was the development of the bed-load function in sedimentation research in 1950 by Hans Albert Einstein, the son of Albert Einstein. Research has benefited from the increasing use of computers that can handle larger amounts of data in shorter periods of time. Several types of sophisticated statistical packages can assist in the analysis of increasingly complicated studies in hydrology and hydrogeology.

HOW IT WORKS

The Hydrologic Cycle. The never-ending circulation of water and water vapor over the Earth is called the hydrologic cycle. This continuous circulation affects all three parts of the global system: the water spreading over the Earth's surface (the hydrosphere), the gaseous envelope above the hydrosphere (the atmosphere), and the rock layer beneath the hydrosphere (the lithosphere). The Sun's energy and gravity power this circulation that has no beginning or end.

The oceans cover 71 percent of the Earth's surface and account for 86 percent of the moisture in the atmosphere. The evaporated water that is transported into the atmosphere can travel tens to hundreds of miles before it is returned to the Earth in the form of rain, hail, sleet, snow, or ice. When precipitation gets closer to the surface of the Earth, the water may be intercepted and transpired by vegetation, or it can reach the ground surface and eventually flow into streams or simply infiltrate into the ground. A large portion of the water that reaches plants and the runoff flowing in streams is evaporated back into the atmosphere. A portion of the water that infiltrates into the ground may penetrate to deeper layers in the Earth to become groundwater. In turn, this

groundwater may return to the streams as the baseflow component of runoff, which will eventually flow into the oceans and evaporate back into the atmosphere to complete the hydrologic cycle.

Urbanization and Stream Flow. The expansion of cities into open spaces outside the metropolitan area has strongly affected local streams. As impervious cover–in the form of houses, roads, driveways, and large parking areas for shopping malls and office buildings–increases, larger and larger areas of previously water-penetrable surfaces become impervious. Depending on local land-use regulations, impervious cover can easily approach or exceed 80 percent of the total area. The immediate effect of this high impervious percentage is to reduce the amount of water that can infiltrate into the ground and result in increased overland flow.

Storm drainage systems can also affect stream flow, as runoff is deliberately directed into nearby streams. This rapid exiting of water from the increased impervious area can quickly reduce the lag time between precipitation input and flood runoff. The resultant increase in the stream hydrograph invariably gives rise to peak discharge flows that result in local and regional flooding.

In the light of flooding problems associated with an increase in impervious cover, some counties and states have required new developments, particularly in suburban areas, to give up part of their site for detention basins. These structures are designed to detain, for varying amounts of time, the excess runoff that the new buildings and roads on the site will generate. The basins can reduce flooding and also provide an opportunity for sediment to settle out and thereby improve the quality of the water that moves downstream.

APPLICATIONS AND PRODUCTS

Flow-Duration Curves. One example of the type of analysis that is commonly employed in hydrogeologic research is to study the variability of stream flow in watersheds that have lithologic heterogeneity (differences in their rock formations). The physical attributes of watersheds affect stream-flow variability. Some formations found on coastal plains have large amounts of sand with high infiltration rates and thereby have high groundwater yields. Other geologic formations, such as basalt, diabase, and granite, have low infiltration rates and consequently have

Fascinating Facts About Hydrology and Hydrogeology

- Unlike most substances, which expand and decrease in density as they are heated, water reaches its highest density at 39.2 degrees Fahrenheit. Thus, ice, which is less dense than water, floats.

- Water in lakes and oceans freezes from the surface downward; this permits water circulation to continue under the ice so that fish can survive.

- Water has the highest specific heat of all of the common substances, and its huge heat capacity has an equalizing effect on the Earth's climate. Therefore, at comparable latitudes, coastal areas will have milder climates than interior areas.

- Water boils at 212 degrees Fahrenheit at sea-level pressure. It has one of the highest boiling points of any fluid on Earth.

- Almost all fluids experience an increase in viscosity as the pressure increases, but the viscosity of water decreases as the pressure increases. This explains why water under high pressure in water distribution systems is able to flow rather than dribble out of a kitchen tap.

- The surface tension of water is double or triple that of most common liquids because of hydrogen bonding. As a result, some insects can walk on water, and steel needles will float.

- Surface tension (called cohesion) and the natural tendency of water to wet solid surfaces (called adhesion) result in capillarity, which allows water to climb a tube or wall. If water had a weaker surface tension (and consequently weaker capillary forces), water in the soil would not be able to overcome gravity, and plant life would die.

- The oceans hold the vast majority of the world's water (97 percent). About 2 percent is frozen in icecaps and glaciers. This means that almost all the water in the world is either salty or frozen.

- The average volume of all of the rivers on Earth amounts to only 0.0001 percent of the total water on the planet.

order of magnitude.

One useful technique that developed in the twentieth century was to use the low-flow and high-flow ends of flow-duration curves to provide useful information about the hydrogeologic characteristics of any watershed. The flow-duration curve is a cumulative frequency curve that shows the percentage of time that specified stream discharges were equaled or exceeded. The values are plotted on logarithmic-probability graph paper with discharge in cubic feet per second on the y-axis (ordinate) and the frequency that specified discharges were exceeded in percent on the x-axis (abscissa). The slope of the curve provides a measure of temporal variability; the steeper the curve, the more variable the value plotted. Steeply sloping curves indicate a flashy stream, where the flow is mostly derived from direct runoff, indicating minimal groundwater storage. As a result, that watershed has limited potential for ample groundwater supplies.

Water-Quality Issues. Concern is growing about the release of pharmaceuticals from manufacturing facilities into surface waters. In a 2004-2009 study, the U.S. Geological Survey (USGS) found that the water released into surface waters by two wastewater-treatment plants in New York that received 20 percent of their wastewater from nearby pharmaceutical plants had concentrations of drugs that were ten to one thousand times higher than the water released from twenty-three other plants in the United States that were not treating any waste from pharmaceutical plants. A sampling of the maximum concentrations in the outflows from the two New York plants were 3,800 parts per billion (ppb) for the muscle relaxer metaxalone and 1,700 ppb and 400 ppb, respectively, for oxycodone and methadone, both opioid pain relievers. In stark contrast, the twenty-three plants that were not receiving any wastewater effluent from pharmaceutical facilities reported drug concentrations of less than 1 ppb for these drugs.

Treated effluent from wastewater-treatment plants is routinely discharged into streams that flow downstream to one or more water-treatment plants that distribute potable water to their service areas. A prime example is New Orleans, located close to the mouth of the Mississippi River, the largest drainage basin in North America, and downstream from many wastewater-treatment facilities.

The problem is that water containing a growing

low yields of groundwater. Indeed, the differgeologic formations, such as basalt, diabase, and granite, have low infiltration rates and consequently have low yields of groundwater. Indeed, the differences in water yield between formations can approach an number of pharmaceuticals is entering wastewater-treatment plants that are not equipped to remove them. The issue is compounded by the comparable lack of techniques available to water supply treatment facilities. This water-quality issue will most probably increase in importance.

Specific Capacity. The determination of specific capacity is a useful procedure to evaluate the magnitude of the expected yield of a well. It is obtained by simply dividing the tested well yield in gallons per minute by the drawdown of the well in feet (gpm/ft). Drawdown is a measure of how much the water table is lowered as a well is pumped. The purpose of this test is to ensure that the well can sustain a required minimum yield over the long term.

Pump tests for residential wells should take at least four hours, although some communities require six hours or more. Most states require a minimum pumping rate of 0.5 gpm/ft for residential use. In addition, the original static level of the water table should recover twenty-four hours after end of the test. For large-scale commercial and industrial users, many states have more stringent standards for pumping, such as a minimum testing period of at least forty-eight hours.

Specific capacity values can vary over several orders of magnitude, from less than 1 gpm/ft to more than 100 gpm/ft, depending on the type of geologic formations present. High permeability and porosity usually result in high specific capacity values, and the converse is expected if the formations are either poorly fractured or have low permeability and porosity.

Water Use. The U.S. Geological Survey (USGS) began estimating water use every five years beginning in 1950. Its reports present a large amount of data on a wide selection of water-use categories. The USGS collects information from all fifty states in the United States, as well as the District of Columbia, Puerto Rico, and the U.S. Virgin Islands.

Some water-use categories have changed over the years, but the overall data collected are very useful. For example, the public supply category pertains to water that is furnished to at least 25 people or that serves a minimum of fifteen connections. The distributed water category includes domestic, commercial, and industrial users and contains estimates about system losses, such as leaks and the flushing of pipes. About 258 million people (86 percent of the total population) depend on public water for household needs. Surface water (streams and lakes) accounts for about two-thirds of the public water supply, and the remaining one-third comes from groundwater sources. In New Jersey, the most densely populated state in the nation, 11 percent of the population uses their own wells for water. This relatively large percentage has remained about the same since about the 1980's. This situation presumably reflects large-lot zoning and the high cost of bringing in water from distant suppliers to isolated clusters of a few homes.

Water Disputes. Conflicts over the world's water resources have been numerous and lengthy. The list of water conflicts goes back several thousand years, and the growing disparity between well-watered and poorly watered countries means future conflicts are likely.

The hydrologic imbalance in water supplies is physically based, but social, economic, and political factors play an important role. Countries that have an abundance of headwater streams can build large dams and use the water for irrigation, resulting in less flow to downstream countries. For example, the headwaters of the Tigris and Euphrates rivers start in Turkey, and any diminishment in flow through Syria and Iraq en route to the Persian Gulf would obviously affect downstream agriculture. Another well-known example is the almost total dependence of Egypt on the Nile River, which begins in Ethiopia (Blue Nile), and the lakes Albert and Victoria (White Nile) in east-central Africa. A large diversion of the waters of the Nile by the upstream states would have a substantial impact on Egypt.

Continued growth in irrigation and population in the semi-arid portion of the southwestern United States has led to serious problems on the Colorado River. The drainage area of the Colorado is 246,000 square miles and flows for 1,450 miles from its headwaters in the Rocky Mountains in Colorado, Wyoming, and New Mexico through Utah, Arizona, Nevada, and California before emptying into the Gulf of California in Mexico. Although the seven states and Mexico have made various agreements pertaining to water allocation, numerous problems have developed and are likely to worsen in future years, as the initial allocation in the early twentieth century used an average flow that was based on precipitation values that were above normal. The allocations would have to be reduced when drier or even more normal precipitation cycles return, resulting in manifold problems to large users in the basin.

IMPACT ON INDUSTRY

The private and governmental organizations that deal with hydrology range from large-scale international and national government agencies and associations to small groups focused on local issues. The increase in both the size and the number of these organizations reflects the attention that must be given to the study of problems and issues related to water and water use.

Government and University Research. The International Association of Hydrological Sciences issues a series of publications that convey the results of hydrological research and practice to all interested parties. Research covers all water-related aspects from as many countries as possible.

The U.S. government has twelve agencies that deal with water issues. The oldest agency is the U.S. Army Corps of Engineers, established in 1802. Its major activities are flood control and the improvement of navigable waters by dredging and related means. Other activities include wetlands protection. The U.S. Geological Survey (established 1879) is charged with monitoring and examining the geological and mineral resources of the nation. In 1889, the agency established the first gauging station for measuring stream flow in New Mexico; it has come to manage, with some state participation, more than 7,600 sites in the United States. Additional responsibilities include hydrogeologic investigations, water-quality analyses, and water-use monitoring. The U.S. Bureau of Reclamation in the Department of the Interior was established in 1902 with a mission to build irrigation projects, such as the Hoover and Grand Coulee dams, in the drier lands west of the 100th meridian that include seventeen Western states. The Environmental Protection Agency was established in 1970 with the responsibility for protecting and improving water quality in the United States. It has the power to issue fines and set standards for drinking water, wastewater discharge, and levels of pollution in rivers.

Professional and Educational Organizations. The American Institute of Hydrology is a nonprofit scientific and educational organization that offers certification to professionals in all fields of hydrology, including groundwater, surface water, and water quality. The certification process includes examinations and information regarding the professional and academic experience of the applicant. The American Water Resources Association provides a forum for education, professional development, and information covering many related disciplines. It also sponsors meetings in different parts of the country on various hydrologic topics. The National Groundwater Association, which is focused on groundwater-related topics, holds technical meetings in different locations. The Association of American Geographers hosts professional meetings at varied locations that include specialty sessions in many water-related fields such as groundwater, water supply, and wastewater issues, and the application of new techniques in geographic information systems that pertain to water.

Trade Associations. The Water Quality Association is an international trade association for household and commercial water-quality improvement issues. The American Water Works Association, started in 1881, has become one of the largest organizations of water professionals, including scientists, manufacturers, and water-treatment plant personnel and managers. Its membership includes more than 4,600 treatment plants that are responsible for the delivery of potable water to 180 million people in North America. The Water Environment Federation, formed in 1928, is a nonprofit organization that provides educational meetings and technical material for wastewater-treatment plant operators and managers.

CAREERS AND COURSE WORK

Given the interaction that water has with a wide array of disciplines and subdisciplines, anyone entering the field of hydrology can pursue a variety of paths to become proficient in the subject. Although few academic institutions have a hydrology department, many offer courses in hydrology within academic departments such as civil engineering, geology, physical geography, and environmental science.

The course path for students is determined by their interests. For example, students interested in hydrogeology would most likely major in geology and possibly minor in geography, so that they could take courses in physical geography, cartography, and geographic information systems. Students interested in water-quality issues would be drawn to a major in environmental science, with chemistry and biology as suitable minors. Other useful courses include statistics, economics, civil engineering, mathematics, meteorology, and climatology.

Employment opportunities in hydrology include positions in federal, state, and local government, state water project associations (such as the Central Arizona Project), bi-state river basin commissions (such as the Delaware River Basin Commission), industry, professional associations, and nonprofit watershed associations that act as guardians for their local drainage area. Also, teaching and research positions are available at colleges and universities.

SOCIAL CONTEXT AND FUTURE PROSPECTS

Providing sufficient clean water for the growing population of the world is vital to human survival. The growing importance of an adequate water supply is demonstrated by the increasing number of books, articles, meetings, and commentaries on the topic. The Pacific Institute, established in 1987 in Oakland, California, focuses on water issues, including water and human health, controversies over large dams, freshwater conflicts, climate change and water resources, new water laws and institutions, and efficient urban water use. It conducts research; publishes reports, including the biennial series *The World's Water*; and works with stakeholders to develop solutions.

The future prospects for hydrologic research and technological change are good. For example, desalination plants are being considered for a variety of coastal locations in areas such as the Persian Gulf region, where alternative sources of water are very limited, and in the Los Angeles-San Diego area, where additional imports of water from the Colorado River would encounter opposition from other users. Another technologic advance is the invention in Israel of a type of drip irrigation designed to deal with water scarcity and salinity problems in water-short areas. This irrigation method is used in more than half of the irrigated land in Israel and its use is spreading in California, particularly for high-value crops such as orchards and vineyards.

Unlike most other resources on Earth, water is renewable. Technology and innovative thinking can play important roles in ensuring adequate supplies of clean water for a growing population. For example, the recognition that storm-water runoff, gray water (from dishwashers and washing machines), and reclaimed wastewater could be used for landscape irrigation and some industrial processes resulted in significant water conservation.

Robert M. Hordon, B.A., M.A., Ph.D.

FURTHER READING

Cech, Thomas V. *Principles of Water Resources: History, Development, Management, and Policy.* 3d ed. Hoboken, N.J.: John Wiley & Sons, 2010. A highly readable and very informative text containing a wide variety of useful and well-written chapters on water resources.

Fetter, Charles W. *Applied Hydrogeology.* 4th ed. Englewood Cliffs, N.J.: Prentice Hall, 2001. One of the standard texts in the field, it has good coverage of groundwater topics along with pertinent case studies.

Gleick, Peter H., et al. *The World's Water, 2008-2009: The Biennial Report on Freshwater Resources.* Washington, D.C.: Island Press, 2009. Contains a series of short informative chapters followed by numerous and detailed tables covering a wide range of useful data.

Glennon, Robert. *Water Follies: Groundwater Pumping and the Fate of America's Fresh Waters.* Washington, D.C.: Island Press, 2002. An interesting non-technical and readable discussion of a variety of groundwater problems in different areas of the United States resulting from misguided direction.

Powell, James L. *Dead Pool: Lake Powell, Global Warming, and the Future of Water in the West.* 2008. Reprint. Berkeley: University of California Press, 2010. A detailed but readable book on the water problems in the southwestern United States.

Spellman, Frank R. *The Science of Water: Concepts and Applications.* 2d ed. Boca Raton, Fla.: CRC Press, 2008. Contains many useful chapters on water biology, ecology, chemistry, and water pollution and treatment.

Strahler, Alan. *Introducing Physical Geography.* 5th ed. Hoboken, N.J.: John Wiley & Sons, 2011. An excellent textbook with diagrams, maps, and photographs that cover the fields of weather and climate, fresh water of the continents, and landforms that are made by running water.

WEB SITES

American Water Works Association
http://www.awra.org

International Association of Hydrological Sciences
http://iahs.info

National Groundwater Association
http://www.ngwa.org

U.S. Army Corps of Engineers
http://www.usace.army.mil

U.S. Bureau of Reclamation
http://www.usbr.gov

U.S. Geological Survey
Water Resources
http://www.usgs.gov/water

See also: Hydroelectric Power Plants; Meteorology; Oceanography.

HYPNOSIS

FIELDS OF STUDY

Psychology; humanistic psychology; psychoanalysis; neuroscience; philosophy of mind; placebo research; altered states of consciousness; complementary medicine; alternative healing; ethno-psychotherapy; shamanism.

SUMMARY

Hypnosis, a subfield of psychology, studies the influence of hypnotherapy (therapy undertaken in hypnosis) and hypnosis (as an altered state, social role, or response expectancy) on people, especially its impact on psychological and physical health. Hypnosis is believed to give access to the unconscious mind, and there is empirical evidence that in highly susceptible subjects, hypnosis is successful as a therapeutic treatment (or as adjunct treatment) against a variety of psychological and psychosomatic disorders, such as post-traumatic stress disorder and depression. Hypnotherapy has also proven to be helpful as a substitute for anesthetics, for pain reduction in general, and for relaxation. Finally, hypnosis is used in the context of sports, education, advertising, and marketing.

KEY TERMS AND CONCEPTS

- **Age Regression and Progression:** Attempts to focus the mental attention of a subject on an earlier or later age, which can be therapeutically constructive in the context of experiencing roles, elucidating expectations, and testing consequences but does not necessarily produce real-life memories.
- **Altered State of Consciousness (ASC):** Mental state different from wakefulness and sleeping, includes drug-induced and trance states, as well as those brought about by mental, physical, and holistic techniques; also known as altered state of awareness or altered state of mind.
- **Hypnotherapy:** Form of psychotherapy or counseling that employs hypnosis.
- **Induction:** First set of suggestions in the hypnotic process, inducing relaxation, concentration, and the establishment of hypnotic rapport, or the definition of the roles of the hypnotist and subject; also known as hypnotic induction.
- **Nocebo Effect:** Unpleasant or harmful effect brought about by a placebo or a medical intervention that simulates the actual procedures.
- **Placebo Effect:** Positive therapeutic effect caused by a simulated medical intervention.
- **Rapport:** Subjective experience of consciously or unconsciously feeling trust or an emotional affinity with another person; also known as hypnotic rapport.
- **Suggestion:** Guidance of thoughts, imaginations, feelings, and behaviors of one person by a different person (hetero-suggestion) or by the same person (auto-suggestion, self-suggestion); also known as hypnotic suggestion.
- **Susceptibility:** Ability to be hypnotized or to respond to hypnotic suggestion, as measured by such scales as the Stanford Hypnotic Susceptibility Scale and the Harvard Group Scale of Hypnotic Suggestibility.
- **Trance:** Altered state of consciousness achievable through hypnosis, meditation, prayer, shamanistic rituals, or drug use; also known as hypnotic trance.
- **Unconscious:** Domain of the mind that is not conscious, of which people are unaware during the normal waking state.

DEFINITION AND BASIC PRINCIPLES

Hypnosis is a wakeful state in which the attention is focused on one or several issues by diminished peripheral awareness and heightened suggestibility, usually induced by suggestions (hypnotic induction). Although the word "hypnosis" is derived from the Greek word for sleep (*hypnos*) and a hypnotized person might at times appear to be asleep, neurological research has revealed that brain waves during hypnosis do not resemble those of sleep. Hypnosis is an altered state of consciousness and a specific interactive situation with voluntarily assumed, defined roles in which a subject follows the suggestions of a hypnotist (in hetero-hypnosis) or the subject's own suggestions (in self-hypnosis). Some researchers think that a hypnotic state can occur without suggestion, as part of everyday life when people become extremely focused on a particular issue, for example, during concentrated learning or the creative process. The depth and the success of hypnosis are determined by psychological factors such as positive motivation, an appropriate attitude, expectations, susceptibility, and an active imagination.

Hypnotherapy is hypnosis in a psychotherapeutic or counseling setting, with goals such as stress management, pain reduction, and the modification of attitudes, habits, or behavior.

BACKGROUND AND HISTORY

Hypnosis as psychosocial phenomenon is as old as human culture; images from several cultures show trances in the context of what are probably religious rituals, in some cases, possibly induced for medical reasons. Hypnosis is similar to some forms of trance brought about by eastern meditative techniques and religious or shamanistic rituals in diverse traditional cultures.

In the eighteenth century, the German physician Franz Anton Mesmer (from whose name comes the word "mesmerize") invented a treatment dubbed "animal magnetism." The Scottish physician and surgeon James Braid developed a treatment known as neuro-hypnotism, or hypnotism, which shared some features with Mesmer's technique. Braid is the first advocate of hypnosis and hypnotherapy to gain scientific acceptance. In France, neurologists Jean-Martin Charcot and Hippolyte Bernheim conducted research and developed clinical forms of hypnosis. Austrian neurologist and psychiatrist Sigmund Freud was trained by Charcot and initially was enthusiastic about hypnotherapy, but he later abandoned it in favor of his own psychoanalytic approaches.

In the twentieth century, psychiatrist Milton H. Erickson, who founded the American Society of Clinical Hypnosis, was an advocate of hypnotherapy. Erickson's approaches were both innovative and controversial, according to his collaborator, André Muller Weitzenhoffer, one of the most prolific hypnosis researchers of the second half of the twentieth century.

HOW IT WORKS

In the first hypnosis or hypnotherapy session, the subject is informed about the basics of hypnosis. Each session usually begins with an introduction in which the subject, in most cases, will be asked to recline and to relax. This is followed by the first inductive suggestions (impressions of gravity, feelings of heaviness, and the like), followed by further suggestions that guide the subject toward becoming more relaxed but also more alert and focused toward his or her inner impressions, images, and imagination. The methods vary considerably depending on the therapist's training and philosophical worldview, the subject's aims or problems, and the respective circumstances. For example, induction can use the eye-fixation method, whereby the subject is told to keep his or her eyes fixed on a certain object such as the hypnotist's finger. Various other induction methods employ one or more senses and the imagination. According to the altered state theory, induction helps the subject transfer into an altered state of awareness or consciousness, and social role and response expectancy theories view the induction as a means of defining the roles of the client and hypnotist, increasing expectations, focusing attention, and increasing concentration. Posthypnotic suggestions given during the session are intended to trigger or support the subject's therapy goals, usually to change a behavior or alter an attitude in daily life. Hypnosis can be conducted with individuals and groups. Before the end of the session, the hypnotist conducts an exit procedure, in which any suggestions that are not posthypnotic are taken back and the subject is gradually brought back to a normal condition. The session can end with a review.

Altered State and Dissociation Theories. Braid, Erickson, and Weitzenhoffer, among many others, believed that hypnosis is an altered state of consciousness or an altered state of awareness. American psychologist Ernest Hilgard was of the opinion that consciousness is dissociated and that parallel streams of consciousness coexist and have certain degrees of autonomy (for example, one feeling pain and the other not). Therefore, the coordination and the emphasis of such streams of consciousness can be altered by suggestions. For example, a feeling of pain can be suggested to have less gravity, while a pleasant feeling can be emphasized. Additionally, many experts believe that hypnosis can give access to more remote, subordinated, or covered streams of the consciousness.

Social Role and Response Expectancy Theories. Social role theories (like those of American psychologist Theodore R. Sarbin), sociocognitive theories, and response expectancy theories emphasize the similarity of hypnosis and a placebo. Empirical evidence and meta-analyses suggest that the two main parameters that contribute to the effect of hypnosis in a significant way are the willingness to act socially compliant and an imaginative suggestibility (about one-third of subjects respond to imaginative suggestions and social pressure). Hypnotic suggestibility is not correlated with intelligence, social position, willpower, motivation, gender, introversion, extroversion, or credulity. Researchers such as psychology professor Irving Kirsch think that the effect of hypnosis, as well as that of placebos, is grounded in a kind of self-fulfilling prophecy, namely that largely subjects experience what they expect to experience. In patients with depression, the difference between a placebo and an antidepressant drug is not clinically significant.

Research in this area attempts to prove that the altered state theories are wrong, but that the effects of hypnosis are nevertheless real and that subjects do not fake the effects of hypnosis, as social role and response expectancy theorists believe. A number of researchers hold that both altered state and social role/response expectancy theories are correct and responsible for the effect of hypnosis. Additionally, it can be argued that a hypnotic state is a deeper form of hypnosis, following a nonaltered or less altered state in which suggestions are given and taken according to the social role/response expectancy theory.

APPLICATIONS AND PRODUCTS

Hypnotherapy, Psychotherapy, and Counseling. In clinical psychology, hypnosis, as an adjunct method, and hypnotherapy, as a stand-alone treatment, are successfully used to deal with pain reduction, psychosomatic symptoms, obsessive-compulsive disorder, post-traumatic stress disorder, anxiety disorders, and depression. Less successful is the use of hypnosis or hypnotherapy to treat problem habits such as excessive drinking, eating, and smoking, which are not manageable by self-control. Hypnosis is also used for pain and stress management, for self-improvement, and to change behavior and attitudes.

Hypnotic Analgesia. Hypnosis and shamanistic trance rituals have been used as analgesia. Hypnosis has been employed successfully to achieve relaxation and reduce anxiety, fear, discomfort, and pain before and during childbirth, in dental settings, and also during minor surgery. Hypnosis does not reduce the

physical reception of pain, but its perception can be manipulated by hypnotic suggestions, whether administered by a hypnotist or the self.

Nonclinical Applications. Hypnosis and hypnotic suggestions (self- and hetero-) have been successfully used to cope with stage fright, to reduce stress levels, and to increase the degree of concentration and focus. They can also intensify relaxation and concentration in the context of creative arts, education, and sports. Research has been conducted on applications for military intelligence, investigations, and forensics, but there is no scientific evidence that such applications are of value. A number of business applications, for advertising, marketing, and improving sales, have been created; however, such applications are ethically questionable. Interest in raising athletic performance levels, losing weight, or quitting smoking has resulted in the proliferation of self-hypnosis products, usually in the form of CDs, DVDs, or books. However, self-hypnosis is best learned from a qualified practitioner. Stage hypnosis is usually considered to be neither of therapeutic interest nor an overly important issue in academic research. Leaving aside stage hypnosis, most of the applications still take place in the medical or clinical field.

IMPACT ON INDUSTRY

Hypnosis Research and Clinical Organizations. Hypnotherapy is practiced and researched all over the world. Hypnosis and hypnotherapy organizations exist at the international, national, and regional levels. They include the International Society of Hypnosis, the American Psychological Association's psychological hypnosis division, and the European Society of Hypnosis. Established scholarly journals include the Society for Clinical and Experimental Hypnosis's *International Journal of Clinical and Experimental Hypnosis* and the American Society of Clinical Hypnosis's *American Journal of Clinical Hypnosis*. A number of medical schools and psychology departments at established universities such as Stanford University, Harvard University, and the University of California, Berkeley, deal with hypnosis in research and education.

Dubious Fields. Although the science-based hypnosis has advanced, an increasing number of institutes for hypnotherapy-training exist in more or less grey areas. These programs might promise anything from turning a person into a hypnotherapist after a weekend of training to turning an individual into a happy person in one evening. Therefore, anyone seeking training in hypnosis should seek out reputable organizations or institutes affiliated with them. For example, the American Society of Clinical Hypnosis offers professionals training courses that are approved by the American Psychological Association. Also, stage or show hypnosis generally is the work of actors with no actual hypnosis taking place, but if actual hypnosis were performed, it could endanger the health of participants, as it can pose, for example, cardiovascular or psychological risks.

CAREERS AND COURSE WORK

Students pursuing careers as hypnotherapists should study counseling, psychology, or medicine to become counselors, psychologists, physicians, or psychiatrists and specialize or subspecialize in hypnotherapy. Physicians, dentists, educators, social workers, nurses, counselors, psychologists, and psychiatrists can also complete diverse courses and programs to acquire the practical qualifications for using hypnosis as stand-alone or adjunct treatment.

Courses or programs in hypnosis should be part of or affiliated with accredited universities or reputable hypnosis societies. For higher positions in education, research, and the clinical field, a master's or even a doctoral degree is recommended. If such a career path is intended, the thesis or dissertation should focus on a particular issue very closely related to hypnosis, undertaken in a traditional field such as medicine, nursing science, psychology, human biology, anthropology, ethnology, philosophy, or any hypnosis, undertaken in a traditional field such as medicine, nursing science, psychology, human biology, anthropology, ethnology, philosophy, or any other field in which a scientific approach to hypnosis is possible and credible.

SOCIAL CONTEXT AND FUTURE PROSPECTS

Hypnosis still receives a lot of attention in the discourse concerning memory recovery. Debate exists as to in which circumstances, under which conditions, and how reliably forgotten memories of past events, especially traumatic experiences, can be "recovered" through hypnosis. The more research conducted on hypnosis and more empirical data collected, the more effectively and appropriately hypnotherapy can be used and the more it will

Fascinating Facts About Hypnosis

- In Greek mythology Hypnos, the god of sleep, is the brother of Thanatos, the god of death. Morpheus, the god of dreams, is Hypnos's son.
- The famous German writer Thomas Mann, in his 1930 novel *Mario und der Zauberer* (*Mario and the Magician*, 1930), explores how hypnosis relates to the mesmerizing power of Fascist political leaders.
- In a 1956 article, professor Frank A. Pattie claimed that Franz Anton Mesmer, one of the fathers of hypnosis, plagiarized his dissertation from a work by the English physician Richard Mead, an acquaintance of Sir Isaac Newton.
- American psychiatrist Milton H. Erickson (1901-1980), sometimes referred to as Mr. Hypnosis, was born tone-deaf and color-blind, and he attributed much of his heightened sensitivity to altered modes of sensory-perceptual functioning, body dynamics, and kinesthetic cues to his innate infirmities.
- Despite a lack of scientific evidence, many subjects claim to have past-life experiences under hypnosis; experiments undertaken by psychology professor Nicholas Spanos in the 1980's suggest that such past-life memories reflect social constructions.
- According to Guinness World Records, in 1987, the German stage hypnotist Manfred Knoke was able to hypnotize 1,811 people in the city of Bochum in a six-day period.
- In 1995, the German magazine *Der Spiegel* reported that International Society of Hypnosis president Walter Bongartz, appearing on a television show featuring Manfred Knoke and his feats, pleaded for a law to prohibit such stage-hypnosis spectacles, given the medical risks that they could pose.
- Dream Theater, a progressive metal band, released a concept album, *Metropolis Pt. 2: Scenes from a Memory* (1999), which tells the story of a man who undergoes hypnosis to experience the mystery of his past lives.

gain acceptance in mainstream medicine.

The question of whether hypnosis is a social role/expected response or a true altered state will not be conclusively answered until neuroscience advances and anthropologists use neurological tools to study shamanist-cultic and religious trance rituals. Anthropology and sociology have made it evident that the use of hypnosis as therapy is not a European invention but rather a phenomenon that can be traced back to therapeutic shamanistic rituals and religious trances in various parts of the world.

Another issue concerns the role of the subject's imagination in the curative powers of hypnosis. A subject in a hypnosis show is similar to the subject of a traditional cultic healing ritual in that the involvement and participation of the public is taken for granted. If both subjects experience a curative effect, then hypnosis is acting like a placebo, and its actual therapeutic benefits are questionable. The subject's own power of imagination and its neurological, biochemical, physiological, psychological, and holistic effects may be what is producing the curative effect.

Certain scholars hold that quite a number of everyday settings–such as intensive educational settings, artistic performances, and mass political events–have a hypnotic character. Research in this field will bring to light how business applications of hypnosis are possible. The ethical problem in such contexts is that subjects should not be hypnotized against their own will. Critics hold that this has already been done in the sphere of marketing.

From a feminist perspective, it can be argued that hypnosis cements and even perpetuates patriarchal structures, since most of the well-known hypnotherapists were men and the hypnotherapist exerts a kind of dominance over the patient or client, unlike as in a guided imagery setting in which the relationship is less hierarchical and less suggestive. Therefore, a unique feminist approach to hypnosis is also the subject of research.

Roman Meinhold, M.A., Ph.D.

FURTHER READING

Erickson, Milton H. *The Wisdom of Milton H. Erickson.* Compiled by Ronald A. Havens. 1985. Reprint. Williston, Vt.: Crown House, 2003. Erickson's thoughts on hypnosis and psychotherapy.

Kirsch, Irving, Steven J. Lynn, and Judith W. Rhue, eds. *The Handbook of Clinical Hypnosis.* 2d ed. Washington, D.C.: American Psychological Association, 2010. Broad introduction to hypnosis covering how it is used clinically to treat conditions and disorders.

Nash, Michael R., and Amanda J. Barnier, eds. *The Oxford Handbook of Hypnosis: Theory, Research and Practice.* New York: Oxford University Press, 2008. Presents both academic theory and practical approaches. Contains name and subject indexes.

Pattie, Frank A. *Mesmer and Animal Magnetism: A Chapter in the History of Medicine.* Hamilton, N.Y.: Edmonston, 1994. Mesmer specialist Pattie provides a thoughtful biography of Mesmer that includes the probable source of his thought.

Pintar, Judith, and Steven Jay Lynn. *Hypnosis: A Brief History.* Malden, Mass.: Wiley-Blackwell, 2008. Offers a compact but critical overview of the origins and the history of hypnosis, including accounts on debates within psychology, references and index.

Temes, Roberta. *The Complete Idiot's Guide to Hypnosis.* 2d ed. Indianapolis, Ind.: Alpha Books, 2004. Provides the basics of hypnosis. Contains a glossary, an index, and suggestions for further reading.

WEB SITES

American Psychological Association
Division 30: Society of Psychological Hypnosis
http://www.apa.org/about/division/div30.aspx

American Society of Clinical Hypnosis
http://www.asch.net

European Society of Hypnosis
http://www.esh-hypnosis.eu

International Society of Hypnosis
http://www.ish-hypnosis.org

Society for Clinical and Experimental Hypnosis
https://netforum.avectra.com/eWeb/StartPage.aspx?Site=SCEH

See also: Psychiatry; Somnology.

IMMUNOLOGY AND VACCINATION

FIELDS OF STUDY

Microbiology; general biology; cell biology; chemistry; molecular biology; biochemistry; biophysics; microbial genetics; immunology; genetic engineering.

SUMMARY

The function of vaccination is to induce immunity in humans and other animals for the purpose of providing protection against disease-causing organisms. Vaccination is generally carried out through the injection of attenuated or inactivated microorganisms such as bacteria or viruses, or the inactivated toxins produced by bacteria. The first vaccinations were directed against smallpox during the eighteenth century and involved the use of cowpox virus, similar but not identical to the virus that caused smallpox. Modern vaccinations often use purified components of the organism rather than the entire bacterium or virus, producing similar immunity without the danger of side effects or illness.

KEY TERMS AND CONCEPTS

- **Acellular Vaccine:** Vaccine prepared from subunits or genes from the agent rather than the whole bacterium or virus.
- **Antibody:** Protein produced by B lymphocytes in response to foreign molecules.
- **Antigen:** Commonly a protein, but any molecule perceived as being foreign to the body.
- **B Lymphocyte:** Type of white blood cell that produces and secretes antibodies.
- **Cellular Immunity:** Immunity based on activation of macrophages, natural killer cells, and antigen-specific cytotoxic T-lymphocytes.
- **Complement:** Pathway triggered by antigen/antibody complexes that consists of proteins that augment the immune response.
- **Humoral Immunity:** Immunity based on soluble proteins such as antibodies found in blood and body fluids.
- **Monoclonal Antibodies (mAB):** Antibodies produced by a clone of B lymphocytes, all of which are identical.
- **Phagocytosis:** Ingestion and digestion of material, including bacteria, by white blood cells.
- **T Lymphocyte:** White cell that matures in the thymus gland (indicated by the T) and that regulates the immune response.

DEFINITION AND BASIC PRINCIPLES

Immunology is the science that studies the reactions of immune cells within the body to foreign molecules referred to as antigens. The majority of immune cells are represented by populations of white blood cells, or leukocytes, found circulating within the bloodstream and lymphatic system. Although all immune cells originate and mature largely within the bone marrow, they undergo differentiation into highly specialized categories. Monocytes and neutrophiles are phagocytic cells circulating in the bloodstream and lymphatic system that function to ingest and digest both foreign antigens from outside the body and old or dying cells within the body. Monocytes mature within tissues and organs such as the spleen and lymph nodes into a class of cells called macrophage, transport the ingested antigens to the cell surface, and present the digested molecules to a second class of white cells called B and T lymphocytes. Only those lymphocytes that express a receptor specific for the presented antigen will respond, with B cells differentiating into plasma cells that secrete antibodies directed against the original antigen.

The principle underlying vaccination is that administration of a killed or attenuated form of bacterium, bacterial toxin, or virus will result in production of antibodies against the target, producing immunity in the individual against the bacterial or viral antigens. If the person is later exposed to the same microorganism or toxin, the presence of preexisting antibodies will protect against infection or poisoning.

BACKGROUND AND HISTORY

Immunitas originally referred to the freedom from taxes among ancient Romans. The medical concept of freedom from disease—immunity—was described by the Greek historian Thucydides, who in his description of the plague (actually probably a typhoid fever epidemic) in Athens in 430 b.c.e. noted that individuals who survived "were never attacked twice."

An understanding of the cellular basis for immunity was not reached until late in the nineteenth century. However, the principle of prior exposure to a disease resulting in immunity had been known since about 1000 c.e., when the Chinese carried out a practice called variolation. In this practice, a person inhaled dried crust from the pocks that developed on smallpox victims. If he or she developed only a mild form of the disease, the most common outcome, the person was immune for life. The practice traveled to Eastern Europe and then England by the early eighteenth century. Although variolation was generally successful, sometimes it resulted in the person contracting smallpox.

The first successful active immunization is ascribed to Edward Jenner, an English country physician who in the 1790's tested a belief common among local dairymaids—that prior exposure to a mild cowpox infection of the udder on a cow provided immunization against smallpox. Beginning in 1796, Jenner carried out tests in which he intentionally infected people by applying cowpox "lymph" obtained from a lesion to small slits cut in the arms of volunteers. During a subsequent epidemic, none of the inoculated individuals developed smallpox. Jenner called the practice "vaccination," from *vacca*, Latin for cow.

HOW IT WORKS

An understanding of the cellular mechanism underlying successful vaccinations did not begin until the late nineteenth century and was the outcome of both a scientific and a nationalistic rivalry between French and German scientists. The major proponent of a cellular theory of immunity was the Russian scientist Élie Metchnikoff. While studying the differentiation of cells in animals such as starfish larvae, Metchnikoff observed that insertion of a wooden splinter into the larvae resulted in the infiltration of both large and small white blood cells. He called these macrophage, "large eaters," and microphage, "small eaters." Microphage later became known as neutrophils. Metchnikoff subsequently joined the laboratory of French scientist Louis Pasteur, where he became a proponent of the cellular theory of immunity.

The competing theory was defined by the German school, and became known as humoral immunity. Robert Koch, Emil von Behring, and their associates noted that blood plasma obtained from animals previously exposed to etiological agents of disease or to bacterial toxins could directly kill bacteria or neutralize these toxins. Behring and Paul Ehrlich applied their discovery in the development of the first vaccines against diphtheria. Soluble proteins, including antibodies, became the basis for humoral immunity.

It was not until the mid-twentieth century that the basis for immunization was established as a combination of both cellular and humoral immunity. Phagocytosis is indeed carried out by several classes of white blood cells, while antibodies are produced by a class of white cells called B lymphocytes. The actual immune mechanisms involve a complex interaction between these classes of cells and their soluble products in which the phagocytic cell presents the digested antigen on its surface to the appropriate lymphocyte. The end result is that B lymphocytes mature and differentiate into an end-stage antibody-producing factory called a plasma cell. Each plasma cell produces a single type of antibody, selected on the basis of possessing a receptor specific for the antigen presented by the phagocyte.

Vaccine Production. Vaccine production is based on activation of lymphocytes through exposure to bacterial or viral antigens (proteins), the result of which is the production of antibodies or the stimulation of phagocytic cells. Vaccines have historically been produced by three major mechanisms: use of inactivated or killed microorganisms, use of attenuated or cross-reacting organisms, or use of purified portions of microorganisms in recombinant vaccines. The smallpox vaccine is an example of a cross-reacting organism; the cowpox virus is similar enough to smallpox that the immune response is protective against both.

Most viral vaccines have used attenuated strains of the original virus, selected either by passage through nonhuman animals or cells or by artificial selection on the basis of avirulence, an inability to cause disease. The strains of poliovirus vaccines developed by Albert Bruce Sabin, as well as vaccines against rabies, chickenpox, measles, and mumps all consist of attenuated viruses. The polio vaccine developed by Jonas Salk is a formalin-killed virus. A later generation of viral vaccines, those directed against viruses such as hepatitis B and human papillomavirus (which causes warts and cervical cancer), are subunit types consisting of surface proteins obtained from the virus, which through DNA recombination are linked to harmless carrier proteins. Vaccines directed against tetanus toxin are similar to those originally developed by Behring and Ehrlich against diphtheria toxin. The toxin is chemically modified and injected.

The principle behind all vaccinations is the same. Exposure to the agent results in an immune response within the individual. Antibodies are produced, and cellular immunity is activated. The response is already in place in the event of future exposure to the same organism or toxin. In most cases, immunity is long-lasting, though periodic boosters are recommended to ensure a proper level of immunity.

Autoimmune Disease. In principle the immune response is directed only against foreign agents that could potentially cause disease. However, in certain circumstances, alterations in immune regulation take place, and antibodies are produced against the person's own tissues. The precise molecular mechanism that triggers autoimmune function is unclear. Some diseases run in families or are gender specific (women are more likely to contract certain autoimmune diseases), and other illnesses may be triggered by cross-reaction with viral or bacterial antigens. The tissue involved depends on the type of autoantibody produced, but the mechanisms for damage are similar.

Autoimmune diseases are placed into two major categories, organ specific or systemic, reflecting the sites or systems involved. Examples of organ-specific diseases include type 1 diabetes, in which the B cells of the pancreatic islets of Langerhans are targeted; Crohn's disease, a form of inflammatory bowel disease; and multiple sclerosis, characterized by inflammation of tissue in the central nervous system. Systemic autoimmune diseases include systemic lupus erythematosus, in which antigen/antibody complexes lodge in different organs, and rheumatoid arthritis, characterized by immune complexes that lodge in joints or bone. Although the type of antibody may differ in autoimmune diseases, pathologies are similar in that each activates the complement pathway, components of which include degradative enzymes that contribute to inflammation and tissue destruction.

APPLICATIONS AND PRODUCTS

Vaccine Production. Historically, vaccines fell into two categories: live vaccines in which the agent was altered so as to be unable to cause disease but still able to replicate in the human host, triggering the immune response, and killed vaccines in which the organism was identical to the parent strain but unable to replicate. Each had advantages. Live vaccines produced a greater response and often a lifelong immunity, and killed vaccines would not result in reversion to the wild strain, causing disease. For example, before they were discontinued in 1990, the Sabin strains of attenuated poliovirus had a reversion rate of about one in one million persons inoculated in the United States, resulting in about ten vaccine-associated cases of polio per year.

Live, or attenuated, vaccines were originally created by passage in nonhuman animals or in cell cultures in a laboratory. This was particularly true for vaccines for viruses, including those against rabies, polio, measles, mumps, rubella, and chickenpox. Because viruses develop random mutations, variant strains were selected on the basis of sensitivity to pH (acidity-alkalinity) or to elevated temperatures (fever), or for their inability to infect certain tissues. The Sabin poliovirus strains represent a prototype of attenuated viruses, being both temperature sensitive and incapable of infecting tissue in the central nervous system. Later methods of developing attenuated strains have involved active modification of viral genetic material or creation of recombinant viruses in which those genes necessary for replication have been deleted.

Most viruses can be grown in cell culture for vaccine production. Animal cells are easy to maintain in the laboratory, and viruses for vaccines can be grown to necessary concentrations. Influenza viruses are exceptions, which is one reason quick production of yearly influenza vaccines has been difficult. The influenza virus genome consists of eight individual segments; coinfection of cells with two different strains, often involving viruses from two different species such as humans and birds, routinely creates a new recombinant strain that is not recognized by the human immune system. Influenza viruses do not grow well in cell culture, so vaccines must be produced using viruses grown in eggs. The lead time necessary to produce sufficient quantities of vaccine for the influenza season, which begins in the fall, is about six months. Therefore, health agencies such as the World Health Organization and the Centers for Disease Control and Prevention must decide in late winter which strains are most likely to produce an outbreak later that year.

Monoclonal Antibodies (mAB). Exposure to antigens such as those found on bacteria or viruses triggers the production of a large number of different antibodies, each of which is specific for a particular molecular determinant on an organism. In the 1960's, it was discovered that persons with multiple myeloma, a cancer of plasma cells, produced large quantities of homogeneous antibodies.

British scientists Georges J. F. Köhler and César Milstein found that because myeloma cells, like those of most cancers, are immortalized, they could artificially fuse myeloma cells with immune cells of known specificity to produce a clone of "immortal" cells producing identical antibodies; because these cells represented a clone, the product became known as monoclonal antibodies.

Because antibodies, in theory, can be generated in the laboratory against any target, monoclonal antibodies can be used as a probe for detection of any cellular molecule. Initial applications used monoclonal antibodies for detection of cell surface proteins for identification of cell types or the maturation stage of cells during differentiation. Because these surface proteins exhibited clustering, they became known as cluster of differentiation (CD) proteins. Nearly two hundred cluster of differentiation proteins are now known.

The ability of monoclonal antibodies to bind surfaces on specific cells has led to their use in the diagnosis or treatment of certain types of cancers. Immuno-conjugates are prepared by chemically attaching a toxin or radioisotope to a monoclonal antibody and injecting the molecule into a patient. Binding of the conjugated monoclonal antibody to the tumor cell results in killing of the target. Although in theory immunotherapy could be applied to many forms of cancers, most tumors do not express proteins unique to that type of cancer.

IMPACT ON INDUSTRY

Most basic research in immunology has been carried out in university and medical school laboratories, but only industry has the capability to carry out large-scale production and testing of vaccines. Several factors have had a significant impact on the ability of industry to maintain this work. Vaccine production represents a significant expense, ultimately reflected in the cost to the consumer. New or untested vaccines are particularly expensive, with no guarantee of a return on investment for the company that produces them.

The second factor that has had a negative impact on industry is the cost of litigation in response to real or imagined injury to the vaccine recipient. In court cases through the 1960's, the benefit associated with vaccines was generally considered to override the known risk. However, beginning in the 1970's, courts increasingly found vaccine manufacturers liable for damages that may or may not have been associated with their products. The vaccines administered in response to an outbreak of swine influenza in 1976 resulted in more than 4,000 claims and awards of damages exceeding $72 million. In this case, the government assumed liability, but in most cases, the manufacturer had to assume the loss, and payouts exceeded the profits associated with the sale of vaccines. For example, Lederle (later part of Pfizer) was forced to pay claims against its Sabin poliovirus vaccine and diphtheria-tetanus-pertussis (DTP) vaccine that were two hundred times greater than their sales. The cost of the DPT vaccine rose 10,000 percent by 1986.

The problem of cost has been directly reflected in the loss of American companies involved in vaccine production. In 1967, thirty-seven companies in the United States manufactured vaccines, but by 1984, less than half of those companies continued to manufacture vaccines, and by 2005, only three companies remained in the business. During the same period, the actual number of licensed vaccines fell from 380 to fewer than 40. The reluctance of companies to engage in vaccine production and the long production time involved has resulted in several shortages of influenza vaccines for the flu season in the twenty-first century. During the H1N1 flu (swine flu) pandemic of 2009, the vaccine was in short supply and was given only to those judged most in need of it. As of 2010, only ten companies worldwide were licensed to produce influenza vaccines, and only two, Medimmune-Avirion and Sanofi Pasteur, a French-owned company, had plants in the United States.

Careers and Course Work

As is true for most careers in medical science, students generally begin by earning a bachelor of science degree in a field such as biology or biochemistry. The undergraduate program should include courses in general biology, chemistry (particularly organic chemistry), microbiology, genetics, and biochemistry. An understanding of the human immune system is vital, so courses in immunology, virology, and pathogenic microbiology should be included.

The bachelor of science degree is sufficient for an entry-level position, but advanced training is necessary for someone wishing to be more than a technician. Historically, most research in the field of immunology has taken place in universities, often in association with medical schools or research institutes. A student wishing to pursue such research most commonly enters a graduate program in which faculty members are carrying out studies in a relevant area. The laboratory director may focus his or her research in development of a recombinant vaccine. The work often involves initial testing of efficacy and safety in nonhuman animals. A doctorate is necessary for working at the level of university faculty or laboratory director.

Development and marketing of any prospective vaccine requires a significant source of funding, which may be available through government grants but more likely involves funding from pharmaceutical companies. A master's degree in an area of science such as chemistry or biochemistry is the minimal requirement in industry for vaccine research and development, while a doctorate is preferred.

Social Context and Future Prospects

The effective control of most childhood infectious diseases by the end of the twentieth century has caused the fear of such diseases to all but disappear among most modern populations. As some segments of European, British, and American populations have grown up in a time in which childhood infectious diseases appeared to be a thing of the past, many of these people do not fully understand the

> ### Fascinating Facts About Immunology and Vaccination
>
> - Edward Jenner, the English physician who developed the first vaccine against smallpox, refused an offer by Captain James Cook of the HMS *Endeavour* to accompany the captain on his next trip to Pacific islands during the 1770's.
> - Bishop Cotton Mather of Boston is notorious for his condemnation of witchcraft, but in the 1720's, he was the primary proponent of variolation.
> - During the 1918 influenza epidemic, the bacterium *Haemophilus influenzae* was mistakenly thought to be the etiological agent of the disease. A vaccine was produced against it, which had no effect on the epidemic.
> - Louis Pasteur's rabies vaccine was first tested on nine-year-old Joseph Meister in 1885. In 1940, Meister, then a caretaker at the Pasteur Institute, committed suicide rather than open Pasteur's tomb for the Nazis.
> - The attenuated Edmonston strain of measles virus developed by physician John Enders was isolated in 1954 from an eleven-year-old boy, David Edmonston, son of a Bethesda mathematician. As an adult, Edmonston participated in the 1963 civil rights march on Washington and later became a schoolteacher in Mississippi.
> - The JL strains of the mumps virus vaccine were developed by physician Maurice Hilleman from an isolate obtained in 1963 from his six-year-old daughter, Jeryl Lynn.

devastating nature of these diseases and question the value of vaccines. Also, the sheer number of recommended vaccinations has created concern among parents, some of whom are afraid their children's immune systems could be overwhelmed, perhaps resulting in autism. Although no evidence for a link between autism and vaccination has been found, some parents still believe that such a link exists.

Future immunizations are likely to rely less on whole virus or bacterial vaccines and more on acellular or subunit vaccines. Side effects resulting from the pertussis vaccine, generally mild fever or inflammation but occasionally a more serious problem, led to the development of an acellular pertussis vaccine using only bacterial proteins. Similar vaccines, some containing only genetic information for production of viral proteins, are likely to be used against other diseases in the future.

Immunization against some agents such as influenza viruses, which undergo yearly changes, will probably involve some form of combination vaccines, incorporating proteins that are common to most major strains of the virus. The simplicity of world travel in the twenty-first century means scientists must take a worldview of new strains, as an outbreak in a few countries can rapidly develop into a worldwide pandemic.

The ability of the human immunodeficiency virus (HIV) to undergo rapid mutations, even within the same individual, means a vaccine against the acquired immunodeficiency syndrome (AIDS) remains unlikely in the foreseeable future.

Richard Adler, Ph.D.

FURTHER READING

Allen, Arthur. *Vaccine.* New York: W. W. Norton, 2007. Discusses the history of vaccine development, from eighteenth century variation to modern times, and the controversies that have surrounded vaccination.

Heller, Jacob. *The Vaccine Narrative.* Nashville, Tenn.: Vanderbilt University Press, 2008. Tells how four of the major vaccines of the twentieth century were developed.

Link, Kurt. *The Vaccine Controversy: The History, Use, and Safety of Vaccinations.* Westport, Conn.: Praeger, 2005. Short synopses of the history behind most major vaccines, from smallpox to acellular vaccines.

Plotkin, Stanley A., Walter A. Orenstein, and Paul A. Offit, eds. *Vaccines.* 5th ed. Philadelphia: Saunders/Elsevier, 2008. An excellent source for understanding the history of vaccine development against most major agents.

Tauber, Alfred. "Metchnikoff and the Phagocytosis Theory." *Molecular Cell Biology* 4 (November, 2003): 897-901. Discussion of Metchnikoff's discovery of phagocytosis. Includes original illustrations.

Williams, Tony. *The Pox and the Covenant.* Naperville, Ill.: Sourcebooks, 2010. Story of Cotton Mather and variolation in Boston during the 1720's.

WEB SITES

American Association of Immunologists
http://www.aai.org

Centers for Disease Control and Prevention
Vaccines and Immunization
http://www.cdc.gov/vaccines

World Health Organization
Immunization Surveillance, Assessment, and Monitoring
http://www.who.int/immunization_monitoring/en

See also: Epidemiology; Genetic Engineering; Human Genetic Engineering; Pathology; Pharmacology; Virology.

INTERNATIONAL SYSTEM OF UNITS

FIELDS OF STUDY

Metrology; agriculture; geography; history; sociology; ballistics; surveying; military applications and service; navigation; construction; civil engineering; mechanical engineering; chemical engineering; physics; biochemical analysis.

SUMMARY

The development of the metric system, which served as the basis of the International System of Units (Le Système International d'Unités; known as SI), occurred during the French Revolution in the mid-eighteenth century. This coincided with the beginning of the age of modern science, especially chemistry and physics, as the value of physical measurements in the conduct of those pursuits became apparent. As scientific activities became more precise and founded on sound theory, the common nature of science demanded an equally consistent system of units and measurements. The units in the SI have been defined by international accord to provide consistency in all fields of endeavor. The basic units are defined for only seven fundamental properties of matter. All other consistent units are derived as functions of these seven fundamental units.

KEY TERMS AND CONCEPTS

- **Base Unit:** Unit of measurement that is used for one of the seven fundamental properties; also known as defined unit.
- **Conversion Factor:** Ratio of one unit of measure in a particular system to the unit of measure of the same property in a different system of measurements.
- **Derived Unit:** Unit of measurement that is derived from the relationship of base units.
- **Dimensional Analysis:** Consideration and manipulation of the units involved in a measurement and their relationship to one another.
- **Fundamental Property:** Property ascribed to a phenomenon, substance, or body that can be quantified: length, mass, time, electric current, thermodynamic temperature, the amount of a substance, and luminous intensity.
- **International Standard:** Unit of measurement whose consistent value is recognized as a common unit of measurement by international agreement.
- **Measurement:** Ascribing a value to the determination of the quantity of a physical property.
- **Standard Temperature and Pressure (STP):** Value of the mean atmospheric pressure and mean annual temperature at sea level, chosen by international agreement to be designated as 1 standard atmosphere or 760 torr and 0 degrees Celsius/32 degrees Fahrenheit or 273.15 degrees Kelvin.
- **Unit:** Single definitive basic magnitude of portions used to measure various properties.

DEFINITION AND BASIC PRINCIPLES

The International System of Units (SI) is the internationally accepted standard system of measurement in use throughout the world. The units of the SI are ascribed to seven fundamental physical properties and two supplementary properties. These are length, mass, time, electric current, thermodynamic temperature, the amount of a substance, luminous intensity, and the magnitude of plane and solid angles.

Length is the extent of some physical structure or boundary in two dimensions, such as the distance from one point to another, how tall someone is, and the distance between nodes of a sinusoidal wave. The SI base unit associated with length is the meter.

Mass refers to the amount of material in a bulk quantity. The term is used interchangeably with weight, although the mass of any object remains constant while its weight varies according to the strength of the gravitational field to which it is subject. The SI base unit for mass is the gram.

Time is a rather more difficult property to define outside of itself, as it relates to the continuous progression of existence of some state through past, present, and future stages. The SI base unit of time is the second.

Electric current is the movement of electronic charge from one point to another, and it is ascribed by the SI base unit of the ampere. One ampere is defined as the movement of one coulomb of charge, as one mole of electrons, for a period of one second.

The amount of a substance is defined by an SI base unit called the mole. The term is essentially never used outside of the context of atoms, molecules, and certain subatomic particles such as electrons. A mole of any substance is the quantity of that substance containing the number of atoms or molecules as there are atoms in 12 grams of carbon-12; this number is referred to as Avogadro's number (or the Avogadro constant), which is 6.02214×1023 of atoms, molecules, or particles.

The term luminous intensity refers to the brightness, or the quantity, of light or other electromagnetic energy being emitted from a source. Initially luminous intensity was referred to by comparison to the light of a candle flame, but the variability of such a source is not conducive to standardization. The SI base unit for luminous intensity is the candela. This is about the amount of light emitted by a candle flame, but it has been standardized to mean an electromagnetic field strength of 0.00146 watts.

The derived unit known as the plane angle refers to the angular separation of two lines from a common point in a two-dimensional plane. The SI derived unit for plane angles is the radian. A complete rotation about a point origin is an angular displacement of 2ϖ radians. The extension of this into three dimensions is known as the solid angle and is measured by the SI derived unit called the steradian.

Measurement of any property or quantity is accomplished by comparing the particular amount of the property or quantity to the amount represented by the standard unit. For example, an object that is 3.62 times as long as the distance defined as 1 meter is said to be 3.62 meters long. Similarly, an object that is proportionately 6,486 times as massive as the quantity defined as 1 gram is said to weigh or to have a mass of 6,486 grams. The use of standard reference units such as the SI units ensures that measurements and quantities have the same meaning in all places where that system is used.

BACKGROUND AND HISTORY

Traditionally, systems of measurement have employed units that related to various parts or proportions of the human body. Ready and convenient measuring tools in earlier times included the hand, the foot, the thumb, and the pace. Because these human measures tend to vary somewhat from person to person, no work that depended on measurements could be repeated exactly, even if the same person carried it out. Specialization in tasks eventually led to

the realization that standard units of measurement would be beneficial, and determination of certain units were established by royal decree as long ago as in the signing of the Magna Carta.

The Industrial Revolution in Europe and the growth of science as a common international pursuit drove the need for a unified system of measurement that would be independent of human variability and consistent from place to place. The International System of Units was developed and first put to use in about 1799. It represents the first real standardized system of measurements. Prior to this time, a broad variety of measurement systems were in use because many countries had developed their own measuring standards for use internally and in any territories that it held.

A need for a standardized system of measurement had been recognized by various luminaries and was proposed in 1670 by French scientist Gabriel Mouton. The incompatibility and variability of the many measurement systems in use at that time often resulted in unfair trade practices and power struggles. In 1790, the French Academy of Sciences was charged with developing a system of measurement that would be fixed and independent of any inconsistencies arising from human intervention. On April 7, 1795, the French National Assembly decreed the use of the meter as the standard unit of length. It was defined as being equal to one ten-millionth part of one quarter of the terrestrial meridian, specified by measurements undertaken between Dunkirk and Barcelona. The liter was later defined as being a unit of volume equal to the volume of a cube that is one-tenth of a meter on a side, and the kilogram as being equal to the mass of one liter of pure water. Larger and smaller quantities than these are indicated by the use of increasing and decreasing prefixes, all indicating units that were some power of ten times greater or lesser than the base units.

Standard reference models of the metric units were made and kept at the Palais-Royale, and it was then required that measuring devices being used for trade and commerce be regularly checked for accuracy against the official standard versions.

HOW IT WORKS

All measurement practices relate the actual properties and proportions of something to a defined standard that is relevant to that property or proportion.

For example, two-dimensional linear quantities are related to a standard of length, and the mass of an object is proportionately related to a standard unit of mass. In the measurement of length, for example, the standard unit of the IS is the meter. An object that is determined to be, for example, 6.3 times longer than the base unit of one meter is therefore said to be 6.3 meters long. Similarly, an object that is, say, 5.5 times heavier than the base unit of one kilogram is said to weigh 5.5 kilograms, and something that has a volume of 22.4 times that of the base unit of one liter is said to have a volume of 22.4 liters.

The modern metric system is simple to use, especially when compared with its most common predecessor systems, the British Imperial and the U.S. customary systems. Whereas these predecessor systems use units such as inches, feet, yards, ounces, and pounds that relate to one another irregularly, the metric system has always used the simple relation of factors of ten for all of its units. For example, the decameter is equal to ten meters (10×1 meter), while the decimeter is equal to one-tenth of a meter (0.1×1 meter). A complete series of prefixes indicates the size of the smaller and larger units that are being used in a measurement. Thus, a decimeter is ten times smaller than a meter, a centimeter is a hundred times smaller than a meter, a millimeter is a thousand times smaller, and so on. Correspondingly, there are ten millimeters in a centimeter, ten centimeters in a decimeter, and ten decimeters in a meter. There are one thousand milliliters in a liter, one hundred liters in a hectoliter, one thousand grams in a kilogram, and one thousand milligrams in a gram. The uniformity of the system readily allows one to visualize and estimate relative sizes and quantities.

The basic units of the SI have always been defined with the intention of relating to some universal and unalterable standard. In some cases, the definitions have been changed to relate the unit to a more permanent universal feature or a more accurately known property. For example, with the high accuracy of time measurement that has become available, the definition of the meter has been changed from being a specific fraction of the distance between two fixed terrestrial points to "the length of the path traveled by light in vacuum during a time interval of $1/299792458$ of a second."

The units of measurement of other properties and quantities reflect the fact that those properties and quantities can be viewed as combinations of the seven fundamental properties. That is, for any dimensional property Q, the dimensions of Q are derived from the expression:

$$\dim Q = L^{\alpha} M^{\beta} T^{\gamma} I^{\delta} \theta^{\varepsilon} N^{\zeta} J^{\eta}$$

The exponents a, b, c, d, e, f, and g are integers representing the degree of involvement of the corresponding property of length (L), mass (M), time (T), electric current (I), thermodynamic temperature ( ), amount of material (N), and luminous intensity (J). (Remember that any quantity raised to the power 0 has the numerical value of 1.)

APPLICATIONS AND PRODUCTS

Applications and products related to the metric system essentially fall into two categories: educational devices and training to promote familiarity with the system and metric versions of existing products and the tools required for their maintenance.

Educational Devices and Training. The long history of independent measurement systems has served to entrench those systems in common usage. Therefore, a very large body of materials, products, and devices have been constructed using nonmetric measurements. More significantly, those independent systems have been so deeply entrenched in the education systems of many nations that several generations of people have grown up using no other system. The familiarity gained through a lifetime of using a particular system generally gives individuals the ability to visualize and estimate quantities in terms of that measurement system. Making the change to a different measurement system such as the metric system represents a paradigm shift that leaves many individuals unable to associate the new measurement units with even the most familiar dimensions. Such a change, however, can be accommodated in a number of ways as the world continues to adopt the metric system of measurement as its universal standard.

The most basic method of replacing one system with another is to incorporate the new system into the basic education system, teaching it in such a manner that it becomes the entrenched measurement system as children progress through school. It is, therefore, important not only that teachers are educated in the use of the metric system but also that they actively replace their own reliance on any former system that they have used. This guarantees that, in time, the older system is completely displaced from the public lexicon and the metric system becomes the primary measurement system over a period of essentially one generation.

Those who have already left the school system can learn the metric system through training programs. Training in the metric system can be incorporated into professional development programs at the workplace or take the shape of formal training programs offered by local educational institutions or third-party providers.

Metric Versions of Existing Devices. The vast majority of goods and devices produced in countries that have not adopted the metric standard must be maintained using their original component dimensions because metric values and nonmetric values are not generally interchangeable. The fundamental difference between the two bases of measurement ensures that any coincidence of size from one system to the other is exactly that: coincidental. Therefore, switching to metric goods and devices requires that complete new lines of products be made with metric rather than nonmetric dimensions. This requirement places an odd sort of constraint on the situation because nonmetric parts and tools must still be produced in order to maintain existing nonmetric devices. At the same time, new devices to replace those that fail are produced in metric dimensions. This means that tradespeople, such as automobile mechanics, maintenance workers, and engineers, must obtain double sets of tools, and supply stores must maintain double sets of components and supplies. In addition, tools for taking measurements must be capable of using both metric and nonmetric dimensions, although the incorporation of electronic capabilities into many tools has greatly minimized the difficulties that arise from this dual requirement.

IMPACT ON INDUSTRY

In some countries that have officially adopted the metric system to replace a system such as the Imperial system, the change has been sudden and dramatic. For example, when Canada adopted the metric system in the 1970's, the changeover was quick and, at times, dramatic. The Imperial system of measurement was phased out and replaced by the metric system in a relatively short time, with the key feature being that after a certain date, Imperial units would no longer be used for any quantities that were subject

---

Fascinating Facts About International System of Units

- On September 23, 1999, the Mars Climate Orbiter crashed on the planet Mars because engineers had neglected to take into account that some design data had been provided in metric units and some in U.S. customary units.
- The International System of Units is the first system of measurement based solely on scientific observation and physical characteristics that are not subject to change.
- Only three countries, the United States, Myanmar, and Liberia, still have not adopted the metric system.
- In the United States, the metric system is voluntary but preferred for trade and commerce.
- Demetrication (abandonment of the metric system) is taking place in some states and Canada. In 2005, the government of Ontario changed the secondary math curriculum to include Imperial measurements.

---

to government regulation. On that date, gas stations changed from price per gallon to price per liter, automobile speedometers were read in kilometers per hour, and Canadian schools taught using only the metric system. Average citizens were basically left on their own to learn and adapt to the new system. In many respects, it posed few hardships, because many goods were purchased by dollar amounts rather than unit quantities. Regulatory control of how goods were measured and delivered ensured that the consumer received the correct amount of what was purchased.

In other countries, such as the United States, the American democratic process has stalled conversion to the metric system. In 1800, the United States could have become the second nation in the world to adopt the SI because of its strong ties with France, where the SI was developed. It chose instead to adopt the U.S. customary system of measurement, which used units similar to the British Imperial system but sometimes varied in terms of quantities (an Imperial gallon held 160 fluid ounces, for example, while an American gallon held only 128 fluid ounces). The debate over adoption of the metric system in the United States continues, as many Americans oppose any official move to replace the customary system. As a result, the United States remains essentially the last major country in the world to maintain a system of measurement other than the metric system. In practical terms, the failure to completely adopt the metric system, even in the scientific and technical fields, has cost the country billions of dollars in lost trade and costly errors. For example, in 1999, a multimillion-dollar space probe crashed on Mars because the technical contractors had failed to notice that some of the data being used were in metric units while other data were in U.S. customary units.

CAREERS AND COURSE WORK

The International System of Units is not broad enough to provide a field of study or advanced course work and, by itself, is unlikely to form the basis of a career. However, during the process of converting from a nonmetric measurement system to a metric system, opportunities will arise for those familiar with SI to create and provide training to facilitate understanding and use of the metric system.

Anyone entering a technical or scientific field must become familiar with the metric system, as this has been the standard system of measurement in those areas since the SI was developed. As the metric system becomes more and more widely adopted and accepted, students and others should expect to carry out measurements and calculations using the appropriate metric system units.

Careers that depend specifically on measurement include quality-control engineering and most branches of scientific research and physical engineering. Specific examples of measurement-based careers include civil engineering, medical and biochemical analysis, analytical chemistry, metrology, mechanical engineering, and industrial chemical engineering.

SOCIAL CONTEXT AND FUTURE PROSPECTS

It seems likely that the metric system will ultimately become the standard system of measurement in the United States, as it already is in the rest of the world. As international trade and offshore manufacturing increase, the necessity for American industry to adopt the SI increases. In addition, educational systems have increased their focus on science and technology, and metric systems will become more familiar to children in American schools.

In the meantime, development of other measurement systems and hybrids of measuring systems progresses. For example, the manufacturing community has adopted a standard unit called the metric inch, which is the standard inch divided into hundredths, to be used on visual measuring devices. There is some logic to this adoption, as this measurement apparently corresponds with the limits of differentiation of which the human eye is capable to a better degree than does the millimeter division of the metric system. However, such developments are likely to delay and otherwise interfere with the adoption of the metric system.

Richard M. J. Renneboog, M.Sc.

FURTHER READING

Butcher, Kenneth S., Linda D. Crown, and Elizabeth J. Gentry. *The International System of Units (SI): Conversion Factors for General Use.* Gaithersburg, Md.: National Institute of Standards and Technology, 2006. A governmental guide to the International System of Units.

Cardarelli, Francois. *Encyclopedia of Scientific Units, Weights and Measures: Their SI Equivalences and Origins.* New York: Springer, 2003. A systematic review of the many incompatible systems of measurement that have been developed throughout history. It clearly relates those units to their modern SI equivalents and provides conversion tables for more than 19,000 units of measurement.

Finucane, Edward W. *Concise Guide to Environmental Definitions, Conversions, and Formulae.* Boca Raton, Fla.: Lewis, 1999. A comprehensive compilation of defined and derived units that clearly describes how each unit relates to the physical property that it quantifies.

Himbert, M. E. "A Brief History of Measurement." *European Physical Journal Special Topics* 172 (2009): 25-35. Describes measurement in a historical context.

Shoemaker, Robert W. *Metric For Me! A Layperson's Guide to the Metric System for Everyday Use with Exercises, Problems and Estimations.* 2d ed. South Beloit, Ill.: Blackhawk Metric Supply, 1998. A basic guide for the average person wanting to learn and understand the use of metric units.

WEB SITES
Bureau International des Poids et Mesures
http://www.bipm.org

National Institute of Standards and Technology
Weights and Measures
http://www.nist.gov/ts/wmd/index.cfm

See also: Measurement and Units.

K

KINESIOLOGY

FIELDS OF STUDY

Anatomy; physics; biomechanics; physiology; exercise physiology; chemistry; biochemistry; motor behavior; sport psychology; sport sociology; coaching; ergonomics.

SUMMARY

Kinesiology is a multidisciplinary field that specializes in the science of human movement. It can focus on improving health or performance or on preventing injuries. Traditionally, kinesiology concentrated on the structural anatomy and the mechanics of movement. Later, the field expanded to include the physiological and mental aspects of movement. Some common applications of kinesiology include proper running and jumping mechanics, correct weightlifting techniques, and perfecting the execution of sports skills. Kinesiology also can encompass the mechanics of work-related activities such as lifting, repetitive movements, and sitting at a desk.

KEY TERMS AND CONCEPTS

- **Action:** Type of movement made by a muscle contraction.
- **Adenosine Triphosphate:** Form of stored energy that can be used directly by muscle cells for contraction.
- **Aerobic Exercise:** Physical activity that is vigorous, continuous, and rhythmical.
- **Alveoli:** Lung structure in which oxygen is transferred from air to the blood.
- **Antagonist:** Muscle that has an opposite action from its paired muscle and limits its action.
- **Applied Kinesiology:** Diagnostic technique used in chiropractic medicine.
- **Hemoglobin:** Component in blood that attaches to and transfers oxygen.
- **Insertion:** Point of muscle attachment on the bone that moves.

- **Isokinetic Machine:** Instrument that measures the strength of muscles through the joint's full range of motion.
- **Metabolism:** Sum of all chemical reactions in a living organism.
- **Motor:** Relating to or involving movements of a muscle.
- **Origin:** Point of muscle attachment on the bone that remains stationary.

DEFINITION AND BASIC PRINCIPLES

Kinesiology is the study of human movement. Although it technically is not limited to humans, as an applied science, it almost always is. Kinesiology is primarily concerned with all kinds of physical activity, including competitive sports and activities designed to maintain and improve health. It is a major part of physical rehabilitation and injury prevention, and it can be used to design workstations that are ergonomic and minimize hazards.

A key component of kinesiology is to understand which muscles are involved in specific movements. Each major joint in the human body has identifiable planes of movement. Each movement in the plane is named according to its direction, and each movement has certain muscles that contribute to it. Muscles can be primary movers at the joint or can assist in the movement. Using this information, kinesiologists can observe movement patterns, identify which joints are involved, determine the movement at each joint, and ascertain which muscles contribute to the movement. Furthermore, they can develop training programs to work the appropriate muscles to improve the strength, movement, and muscle balance at the joint.

The field of kinesiology should not be confused with applied kinesiology, which is a diagnostic technique used in chiropractic medicine. The clinician applies a force to a muscle or muscle group, and the patient resists the force. Based on the patient's response to the force, the clinician makes a diagnosis.

This is not a generally accepted practice in medicine and not related to the field of kinesiology.

BACKGROUND AND HISTORY

The field of kinesiology is believed to have begun in ancient Greece. The term "kinesiology" comes from the Greek words *kinein* meaning "to move" and *logos* meaning "discourse." Aristotle is considered the father of kinesiology for his work using geometry to describe the movement of humans. It was not until the fifteenth century that Leonardo da Vinci helped expand the knowledge of human movement by studying the mechanics of standing, walking, and jumping.

One of the greatest contributors to kinesiology, Sir Isaac Newton, did not actually study human movement. In the late 1600's and early 1700's, Newton developed three laws of rest and movement that laid the foundation for the analysis of human movement in the following years. The development of photography in the 1800's enabled researchers to study how animals moved by taking a number of pictures in rapid succession. They first studied horses, then humans. Cinematography further enabled researchers to understand human movement.

In 1990, the American Academy of Physical Education (later the American Academy of Kinesiology and Physical Education) recommended that programs of study involving human movement be called kinesiology. This idea gained wide acceptance, and kinesiology as a field came to include many other specialized fields beyond the traditional anatomy and biomechanics. Kinesiology can include any area that relates to human movement, such as history, sociology, psychology, physiology, philosophy, and motor behavior.

HOW IT WORKS

Movement Analysis. Traditionally, courses in kinesiology focused on the anatomy and mechanics of human movement. Although more advanced courses sometimes take the same approach, introductory courses tend to be an overview of the broader field of kinesiology.

A thorough understanding of muscle and skeletal anatomy is required to understand movement. This includes the names of the bones and the names, origins, insertions, and actions of the major muscles of the human body. The origin and insertion of a muscle are the locations where it attaches to bones. These locations determine the action of the muscle, based on the angle of pull on the bones. This knowledge is important in understanding which muscles are involved in specific movements.

To describe movements of the body, planes and rotations are defined at the various joints. Some of these movements include flexion, extension, adduction, abduction, internal rotation, and external rotation. Kinesiologists watch a specific movement and determine the actions at the joint or joints being evaluated. They can also evaluate movements to determine if they are completed properly. Slow-motion cinematography is helpful when analyzing the very fast movements often found in sports.

After determining the movement at a joint, kinesiologists can use their knowledge of anatomy to determine which muscles are involved. Additionally, they can use this information to develop strength training programs to develop the specific muscles used in the activity. It is important to note that muscles that generate a movement (agonist muscles) have antagonist muscles that stop the movement. Therefore, strengthening an antagonist muscle is just as important as strengthening the muscle that initiates the movement.

Physiological Function. Human movement requires oxygen and energy beyond the levels needed to simply survive. Exercise physiology includes the study of how the body gets food and oxygen from the environment to the working muscles. Oxygen is very important for sustaining activity during intense exercise, during which the cardiovascular, respiratory, and muscular systems are primarily involved.

The respiratory system transfers oxygen from the atmosphere into the blood. Air, which is about 21 percent oxygen, is inhaled into the lungs, and much of it enters the alveoli at the ends of the airways. Oxygen diffuses from the alveoli into the blood, where it binds with the hemoglobin. The pumping action of the heart carries the blood to the muscles. When the oxygenated blood gets near muscle cells that are low in oxygen, the hemoglobin releases the oxygen, which goes into the cell. A strong, healthy heart and blood with normal amounts of hemoglobin are capable of delivering sufficient amounts of oxygen to support high levels of muscle movement and activity.

When oxygen enters the muscle cell, it is metabolized. It goes through a series of chemical reactions in which oxygen and energy are converted into adenosine triphosphate (ATP). Only ATP can be used by the muscle to make the fibers contract and the body move. When highly trained people exercise at a very high intensity, the amount of oxygen consumed can increase more than twenty times. It is the ability of the muscle cells to use oxygen to produce ATP that limits high-intensity human movement. The oxygen consumed and several other measures can be determined with a metabolic cart, which is an important type of testing equipment used in exercise physiology.

Behavioral Control. The areas of kinesiology that involve the brain and nervous system are motor development, sport psychology, and sport sociology. Motor development is concerned with skills that take stored movement patterns in the brain and communicates them to the muscles. Most skilled movements are an organized, synchronous set of smaller movements. Therefore, the series of muscle contractions needed to perform the skill must be stored and retrieved often. Practicing the movement patterns on a regular basis is required to refine the skills.

Sport psychology and sociology are the segments of kinesiology that relate to the mental aspects of human movement and performance. One of the largest fields is clinical sport psychology, in which psychologists assist athletes with aggression, stress, motivation, mood, adherence, and leadership as well as a number of other related issues. Sport sociology is a smaller element of kinesiology that focuses on social relationships in sport and how sports affect different segments of society and organizational structures.

APPLICATIONS AND PRODUCTS

Rehabilitation. Kinesiology techniques, especially those regarding the muscular and skeletal systems, are used in physical rehabilitation every day. Physicians, chiropractors, physical therapists, and athletic trainers use their knowledge of muscle origins, insertions, and actions to diagnose injuries and determine exercises that will help people recover from them. Isokinetic machines can be used to determine the muscular strength at any joint through the entire range of motion. Identifying points of weakness during movement helps determine which muscles need to be strengthened. A muscle and its antagonist can be tested to see if one is weaker than the other. Good muscle balance is needed to prevent injuries and reinjury. Limbs (arms and legs) can be

tested to see if the left and right sides are equally strong or if one side, usually the injured side, is weaker. Kinesiology is a very important component of muscular and skeletal rehabilitation.

Cardiovascular rehabilitation focuses on the area of kinesiology that includes the cardiovascular and respiratory systems. In this area, exercise is used to strengthen the heart muscle. When a person has a heart attack, some of the heart tissue dies and is replaced by connective tissue. With some of the muscle gone and not able to pump blood, the heart is weaker. In cardiac rehabilitation, patients perform aerobic exercises to increase the strength of the heart muscle that is left. Through kinesiology applications, patients can strengthen their hearts and improve their ability to engage in daily living activities.

Health Promotion and Injury Prevention. The principles of kinesiology are used to maintain good health and prevent injuries. For many years, exercise has been recognized as an important component of health promotion. Kinesiology studies have involved developing and researching the best types of exercise to improve health. Exercise specialists such as physical education instructors, fitness trainers, and sports coaches rely on kinesiology to help people exercise safely and efficiently. They create exercise programs or develop exercise sessions based on kinesiology principles. Exercise can take the form of group classes or individual instruction (personal training).

Cardiovascular exercise is very important for health and fitness. Exercise specialists are often charged with helping a person develop the endurance to engage in physical activity for extended periods. Based on an assessment of the person's health and fitness, exercise specialists use exercise physiology principles to write a prescription for activity. The prescription, or exercise plan, is often based on heart rate so that the individual can use his or her pulse to gauge whether the right level of effort is being attained. Another important type of exercise for health and fitness is strength training. Exercise specialists use kinesiology principles to demonstrate proper lifting techniques and help clients get stronger. Attention is paid to balancing muscle strength across the body. Movement mechanics are also used to improve flexibility. A stretching program is designed to improve or maintain flexibility throughout the body and help clients move more freely. A good training program will help clients stay healthier throughout their lives and enable them to perform the activities of daily living more easily and longer.

An overriding factor in exercise for health is adherence to the program. Health benefits are obtained only with regular participation. The psychological area of kinesiology studies provides information about getting and keeping exercisers motivated.

Coaching. Coaches use kinesiology in many of their activities. Sports skill development requires regular evaluation of movement to determine if the sports skill is being performed properly, which maximizes performance and reduces the chances of injuries. Coaches must consider all the involved joints, the type of muscle contractions, and the planes of movements. Of great importance is the synchronization of the movements around the involved joints. Energy transfer from one joint to the next is critical for superior performance. Coaches use their knowledge of kinesiology to teach proper sport skills.

Coaches also must use kinesiology for strength and conditioning. Weightlifting and other exercises must be performed with proper mechanics. The variety of available equipment makes a basic understanding of kinesiology imperative for teaching athletes effectively. Additionally, coaches must determine which exercises should be performed and which muscles are to be strengthened. Coaches also use sport psychology to motivate athletes to perform their best and to develop the leadership skills that are important for success. Athletic competition can be stressful, especially as most athletes must deal with demands on their time and concentration that stem from their social and academic or work-related obligation. Coaches often teach athletes stress management techniques to help them cope.

Ergonomics. Kinesiology can be applied at the workplace in the area of ergonomics, which science uses body mechanics concepts to design workstations that are more comfortable and minimize overuse injuries from repetitive movements. Any workstation can be analyzed and appropriately modified. Ergonomic solutions for people in desk jobs are relatively simple, but finding answers for people in jobs that require lifting and carrying require the use of more kinesiology principles. After the proper techniques for lifting and carrying have been determined, the worker must be trained to perform the movements properly.

Fascinating Facts About Kinesiology

- In 1972, University of Oregon track coach Bill Bowerman invented the Nike Waffle Racer, a shoe with extra traction, when he poured a liquid rubber compound into his wife's waffle iron.
- Exercise is believed to boost mental function, increase energy, reduce stress, build relationships, prevent disease, strengthen the heart, and allow a person to eat more. However, on an average day, only about 16 percent of Americans over the age of fifteen participated in sports or exercised.
- In 2007, there were 9.9 million health club members over the age of fifty-five, more than four times the number in 1990.
- In 2010, the Washington, D.C., area was ranked as the fittest metropolitan area in the United States for the third time in a row.
- A leading professional organization in the kinesiology field, the American College of Sports Medicine, was founded in 1954.
- There are 639 skeletal muscles and 206 bones in the human body. Numbers vary slightly depending on how the muscles and bones are counted.
- Indiana University psychologist Norman Triplett is considered the first sport psychologist for his work that found exercisers performed better in a group than when exercising alone. He concluded that people became more competitive in a group setting.
- Exercise physiology in the United States is believed to have originated at the Harvard Fatigue Laboratory in 1927.

IMPACT ON INDUSTRY

The health, fitness, and sports sector of the U.S. economy has been undergoing substantial growth, and this trend is also found in the international market. In 2009, gyms and health and fitness clubs (includes ice and roller rinks, tennis courts, and swimming facilities) generated $23.6 billion in revenues. Although economically developed countries spend the most on health and fitness, growth is found in most countries of the world. Kinesiology applications for the medical and business industries are also common.

Sports and Fitness. Traditionally, the field of kinesiology has been applied to sports skills and conditioning for performance. This is still the major application. When working with athletes, coaches and trainers use their knowledge in areas such as anatomy, biomechanics, exercise physiology, and sports psychology to help athletes maximize their performance. World records continue to be broken in most sports, so it appears that athletic performance can still be improved. The level of competition in many sports, whether professional or amateur, has increased. Much of the credit for these improvements goes to coaches, trainers, and others who have had education in kinesiology.

Higher levels of training and more intense competition has resulted in an increase in injuries. Kinesiology principles have been used to develop programs to rehabilitate injured athletes and to decrease the chances of injury. As sports become more popular and generate higher revenues, sports managers want to ensure that their best players remain active during the season and, if injured, return to play as quickly as possible. Kinesiology training enables rehabilitation professionals to facilitate the athlete's return to competition.

Workplace. Since the 1970's, the numbers of organizations offering fitness centers at their work sites has increased. Employees can more easily fit exercise into their workday at a facility that is convenient and often competitively priced. By offering group classes and quality supervision in the weight rooms, a company can be assured that its employees can exercise safely and obtain the health benefits. Workplace health programs have been shown to reduce health care costs and absenteeism and to improve productivity.

Kinesiology concepts—in the form of ergonomics—have been applied to workstations. Movement analysis can lead to better workstations and fewer overuse and acute injuries. Ergonomics can save business and industry millions of dollars in medical costs and workers' compensation.

Health. Kinesiology programs have placed a greater emphasis on recreational sports and health since the 1970's, when interest in fitness began to grow among the general population. People began running, bicycling, and lifting weights. Cardiovascular exercise, including aerobics, gained popularity. Gyms and clubs began offering places to exercise safety and obtain information about health and fitness. Helping participants reach and maintain healthy weights is one of the major benefits of these programs.

During this same time period, exercise began to be used to rehabilitate individuals with chronic diseases. The major program has been cardiac rehabilitation, in which patients who have had heart surgery or experienced a heart attack or other problem are placed in a supervised and monitored exercise program. Patients who participate in cardiac rehabilitation typically recover faster and are more likely to resume an independent lifestyle. Cardiac rehabilitation programs may include individuals with other chronic diseases such as diabetes, lung disease, and cancer. Sometimes specialized programs are created for those with other chronic diseases, but the common goal is to design an effective program that improves the improved health and function of participants.

CAREERS AND COURSE WORK

Students who are interested in a career in kinesiology need a basic science background in anatomy, physics, chemistry, and math. Kinesiology courses apply these basic science principles to human movement. Although a bachelor's degree provides an overview of the major areas within kinesiology, a master's degree allows for specialization. An advanced degree is not required for most entry-level positions but it does provide more opportunities, particularly in management. Some graduates may prefer to use their kinesiology education to pursue physical therapy, chiropractic medicine, or other clinical degrees. Those who want to do research in kinesiology will need a doctorate.

Careers in the kinesiology field are many and varied. Students can coach and train athletes. With the increase in the number of youth competing in sports and the parents' willingness to pay, opportunities for coaching are expanding. The most common job is that of a health and fitness trainer in a private or corporate fitness club or a community recreation facility. An experienced health and fitness trainer can become a personal trainer and work with clients individually or in small groups to prescribe training programs and help them meet their exercise goals.

Clinical settings such as hospitals offer positions in cardiac rehabilitation and sports medicine. Exercise physiologists can work in cardiac rehabilitation, and athletic trainers can work in sports medicine. These positions are for those who like the clinical setting and working with individuals and smaller groups.

Some opportunities exist for sport psychologists who use mental strategies with athletes to improve performance.

SOCIAL CONTEXT AND FUTURE PROSPECTS

Programs that use kinesiology professionals have traditionally been paid for by the participants. Kinesiology benefited those who could pay for good sport trainers, health and fitness clubs, and sports medicine. More programs, including youth sports programs, are emerging to provide access to exercise facilities and sports training to those who cannot afford to pay. Some insurance plans provide free or discounted gym membership, and many companies offer employee gyms or discounted gym memberships.

Kinesiology continues to be a growing field. Professional sports remain popular around the world, and the emergence of new competitive sports such as extreme sports results in more research and more need for people trained in kinesiology. Also, more people are interested in maintaining good health, and positions in fitness training are likely to increase to meet the demands of these exercisers. With more people exercising and participating in sports, the number of sports-related injuries is likely to grow. Kinesiology education will be needed to train rehabilitation professionals to research injuries, educate the public about injuries, and rehabilitate those injured.

Bradley R. A. Wilson, Ph.D.

FURTHER READING

Floyd, R. T. *Manual of Structural Kinesiology*. New York: McGraw-Hill, 2009. An introductory text that focuses on movements of the major joints.

Hoffman, Shirl J., ed. *Introduction to Kinesiology: Studying Physical Activity*. 3d ed. Champaign, Ill.: Human Kinetics, 2009. Includes an overview of the major areas of kinesiology and a discussion of professions.

Klavora, P. *Foundations of Kinesiology: Studying Human Movement and Health*. Toronto: Sport Books, 2007. An overview of basic anatomy and physiology, human performance, and the major components of kinesiology.

_____. *Introduction to Kinesiology: A Biophysical Perspective*. Toronto: Sport Books, 2009. A succinct text on basic kinesiology and human movement.

Kornspan, Alan S. *Fundamentals of Sport and Exercise Psychology*. Champaign, Ill.: Human Kinetics, 2009. Discusses basic opportunities and goals in the field.

Oatis, Carol A. *Kinesiology: The Mechanics of Human Movement*. Baltimore: Lippincott Williams & Wilkins, 2009. Covers biomechanics, movement at the major joints, and posture.

WEB SITES

American Academy of Kinesiology and Physical Education
http://www.aakpe.org

American Kinesiology Association
http://www.americankinesiology.org

Energy Kinesiology Association
http://www.energykinesiology.com

See also: Biomechanics; Ergonomics; Occupational Health; Rehabilitation Engineering.

L

LANDSCAPE ECOLOGY

FIELDS OF STUDY

Ecology; spatial ecology; conservation; forest ecology; forestry; agronomy; disturbance ecology; ecosystem management; environmental planning; natural resources policy and management; international development; regional and urban planning; landscape geography; landscape architecture; human ecology; animal behavior; plant biology; spatial statistics; modeling; remote sensing; geographic information systems (GIS).

SUMMARY

Landscape ecology is a relatively new field of science that focuses on spatial heterogeneity and is the study of how ecosystems, including the built environment, are arranged and how these arrangements affect the wildlife and environmental conditions that form them. In other words, it is the study of the abundance, distribution, and origin of elements within landscapes, coupled with their impact on the ecology of an area. Landscape ecology brings a spatial perspective to the integration of people with natural ecosystems. Because of this emphasis, landscape ecology has the potential to be of significant use in ecological and urban planning and management and in solving global scale problems.

KEY TERMS AND CONCEPTS

- **Connectivity:** Continuity, in a spatial sense, of habitat across an area (such as a landscape).
- **Corridor:** Narrow area, including line corridor, strip corridor, and riparian/stream corridor, that differs from the adjacent habitat on both of its sides.
- **Edge:** Outer boundary area of a landscape, which will often exhibit environmental conditions different from those within the landscape.
- **Fragmentation:** Division, separation, or breakup of a landscape into patches and corridors.
- **Heterogeneity:** Condition of diversity or consisting of dissimilar elements.
- **Landscapes:** Diverse land areas made up of repeating and comparable patterns of interconnected and interacting ecosystems.
- **Matrix:** Extensive and highly connected background area of a landscape, which surrounds and influences patches and corridors. Examples include forest or moorland.
- **Patch:** Broad and homogeneous area that differs in some respect from the surrounding landscape.
- **Scale:** Spatial and temporal dimensions of a landscape measured by grain (finest level of measurement) and extent (total area sample).

DEFINITION AND BASIC PRINCIPLES

Landscape ecology, also referred to as landscape science, is an interdisciplinary science. Ecology is the study of the relationships and interconnections between organisms and the environment in which they live in an attempt to determine how living (biotic) and nonliving (abiotic) patterns and processes influence organism abundance and distribution. A landscape is a diverse land area made up of repeating and comparable patterns of interconnected and interacting ecosystems. According to the International Association for Landscape Ecology, landscape ecology is the study of "spatial variation in landscapes at a variety of scales" that "includes the biophysical and societal causes and consequences of landscape heterogeneity."

One of the most fundamental concepts of landscape ecology is that all landscapes are heterogeneous, no matter whether the environment is natural or modified or large or small. Landscape ecology most significantly deviates from classic ecological studies (and more closely resembles geography) in its focus on spatial patterns and its emphasis on broad-scale research on the ecological consequences of spatial patterning on ecosystems. Because landscape ecologists study such spatial heterogeneity, landscape ecology theory and concepts are often considered to be somewhat outside the rigid structure of traditional science.

BACKGROUND AND HISTORY

Although landscape ecology is considered to be a young scientific discipline, the study of landscape ecology has a rich history in central and Eastern Europe and, as a concept, can be traced back to the 1930's. The German geographer Carl Troll first coined the term in 1939 based on the emergence of aerial photography as a tool of geography. Although he believed landscape ecology to be more of a philosophy than a branch of science, he also stated that it was the "perfect marriage between geography and biology." Over the following decades, landscape ecology gained acknowledgment as a branch of ecology in its own right but primarily in central and Eastern Europe and in strong connection with land planning.

The study of landscape ecology gained a measure of global recognition in 1981 following the first international landscape ecology congress in the Netherlands. The primary problems that emerged from this congress on landscape ecology differed widely depending on the region involved, and these differences were instrumental in the development of two distinct approaches, the European and the North American.

The concept of landscape ecology in Europe had been based on environments that had undergone significant human manipulation, and research was often focused on cultural landscapes. The North American approach, however, saw a much greater integration of ecological principles, and research tended to concentrate on natural and forest landscapes, studying the relationship of disturbance, spatial patterns, and change. Research in the theory and application of landscape ecology has developed rapidly since the 1980's, as landscape ecology has moved from a regional to a global science and has been significant in ecosystem management and planning.

HOW IT WORKS

The concept of landscape ecology is perhaps best clarified by its fundamental themes related to spatial heterogeneity, patterns, and the relationships between natural science and human beings. Specifically,

according to the International Association for Landscape Ecology, landscape ecology research tends to concentrate on "the spatial pattern or structure of landscapes, ranging from wilderness to cities; the relationship between pattern and process in landscapes; the relationship of human activity to landscape pattern, process and change; and the effect of scale and disturbance on the landscape." In other words, research in this area of science attempts to understand spatial heterogeneity and its influence on ecological processes, interactions within and between heterogeneous landscapes, and the management of such landscapes and spatial heterogeneity.

Landscape ecology differs from classical ecology in that it focuses on spatial patterns and broadscale research on how spatial patterning affects ecosystems. Theories of landscape ecology concentrate on the interaction and impact of people on landscape structure, function, and change. Landscape structure is determined by the spatial relationships between the elements common to all landscapes: patches, edges, boundaries, corridors, and matrices. Landscape function is defined by interactions between spatial elements and patterns, such as quantity of habitat, size of landscape structures, and landscape connectivity. Landscape change is determined by modifications in the structure and function of landscape over time.

Landscape ecology aims to establish the influence of human activity on the structure and function of landscapes. It aims to highlight the interactions between people and landscapes and to provide a basis for managing landscapes, restoring degraded landscapes, recovering from landscape disturbances, and planning land uses.

APPLICATIONS AND PRODUCTS

Landscape ecology is a young branch of ecology, incorporating theories from social, biological, and geographical sciences. Researchers and scientists in the field have stated that significant research is still required and an improved theoretical and conceptual basis needs to be developed. Once a more unified concept of landscape ecology has been established, further applications and products can be developed. Because of its emphasis on the interaction between natural landscapes and people, landscape ecology has the potential to be of significant use in ecological planning and management, and in solving global-scale problems.

Although the field of landscape ecology has been hampered by a lack of unified methodologies and theories, it has the potential to provide a true joining of people with the natural world. The practical application of landscape ecology has experienced rapid growth, especially in solving conflicts between human development and conservation of the natural landscape. The most successful applications of landscape ecology in relation to environmental and spatial planning rely heavily on effective translation and communication of base data, which have been garnered using such tools as modeling, remote sensing, and geographic information systems (GIS). As human impact on the environment increases, the demand grows for solid scientific data that will permit more effective management of entire landscapes and will further the identification and understanding of the effects of spatial heterogeneity in relation to land and resource management. Pressure is mounting for resources to be managed as a part of the entire ecosystem rather than in and of themselves and for the temporal and spatial sensitivity of certain management and utilization activities to be considered. Generally, therefore, applications of landscape

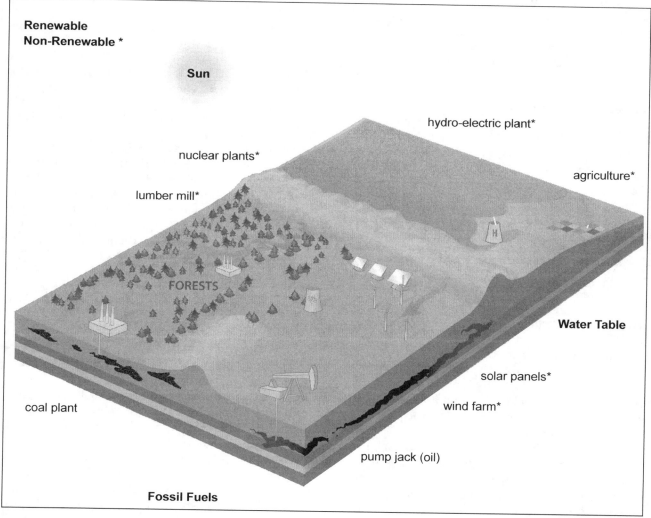

Landscape ecology concepts are implemented by most governments in their approach to environmental management and land use, including the harnessing of natural resources.

ecology can be categorized as being in nature conservation or in land management, which includes sustainable land-use decisions, forest management, ecological/environmental impact assessments, and broadscale environmental monitoring.

Landscape Planning and Land-Use Decisions. Human activity, particularly agriculture, mining, and urbanization, can fundamentally alter landscape structure by changing the dominant land cover and use from natural habitats to human-use areas. This changes the spatial patterns of natural landscapes and influences biodiversity. The study of such modifications in the structures of natural landscapes can be applied to decisions regarding the feasibility and preferred location of a development and the suitability of a specific land use. In agriculture and mining, for example, landscape ecology research tries to establish better options for the management of the environmental risks and hazards that accompany modern farming and mining practices.

Forest Management. A fundamental element of landscape ecology is the impact of forestry activity and natural processes on the spatial patterns of the forest landscape. Landscape ecology research can be applied to forest management to measure growth and decline of the world's forests, understand and minimize habitat fragmentation in forest landscapes, and determine harvesting processes and impacts on both the spatial and the temporal scale.

Ecological/Environmental Risk Assessments. Ecological and environmental risk assessments are fundamental to ecosystem management and policy development. They provide a basis from which to judge, estimate, and clarify the ecological risks associated with environmental developments, hazards, or specific land-use practices. As some hazards, such as climate change, affect large areas and can be influenced by landscape patterns, landscape ecology has been particularly applicable in addressing issues of long-term resource management over a wide range of time periods and distances.

Broadscale Environmental Monitoring. Examining and supervising large-scale landscapes is a significant application of landscape ecology, made more effective by the use of remote sensing and satellite imagery to determine landscape modifications. This application is particularly important because of large-scale problems such as climate change and its subsequent effects on biodiversity.

IMPACT ON INDUSTRY

Although landscape ecology is not a fully mature science, it has certainly experienced rapid development and interest since the 1980's. After its shift from a regional (mostly European) science to a global one, many government agencies, universities, industries, corporations, and organizations worldwide began researching and working in this field of applied science.

Major Organizations. One of the most significant organizations in the field of landscape ecology is the International Association for Landscape Ecology. This organization provides the infrastructure and the means for a global community of landscape ecologists to interact, collaborate, and network across disciplines. Many organizations also use landscape ecology approaches and research in large-scale conservation planning. For example, the World Wide Fund for Nature uses landscape ecology to "develop cost-effective, spatially-explicit strategies that meet the ecological needs of wildlife and habitats while minimizing human-wildlife conflicts and maximizing benefits to resident populations."

Government Research and Regulation. Most governments attempt to use landscape ecology concepts in their approach to environmental management. Such agencies and departments include environmental protection agencies and government or academic departments dealing with climate change, environmental issues, natural resource management, city planning, and the mining industry.

CAREERS AND COURSE WORK

Many universities offer science-based undergraduate and postgraduate degrees in landscape ecology. Most commonly, students who wish to pursue a career in landscape ecology will possess an ecological or geographical background. Because landscape ecology is widely recognized as an interdisciplinary field, students can also come from such varied backgrounds as the social sciences, architecture, animal behavior, plant biology, or conservation. Following graduation, students should have a solid understanding of spatial heterogeneity and its influences on natural processes, and the importance of landscape ecology to conservation management. In addition, because landscape ecologists provide the science-based data used for management and planning of areas at a variety of scales, students should gain knowledge and experience in the use of tools such as modeling, remote sensing, and geographic information systems. Students who study landscape ecology can pursue careers in environmental consulting, environmental advocacy, and other environmental fields in the private sector, nongovernmental organizations, and the government, particularly in the development of conservation proposals and environmental management plans.

SOCIAL CONTEXT AND FUTURE PROSPECTS

The world's human population is continuing to grow and the impact of human activity on the natural environment is increasing. To meet the challenge, landscape ecology must become a more holistic science, oriented toward solving problems on both the small and the large scale. It is important, particularly given the increasing global population, that concepts such as urban and natural sustainability are approached in an interdisciplinary and holistic way.

One of the most significant areas for future landscape ecology research is climate change. Models of climate change have predicted significant changes, particularly in relation to the frequency and timing of large-scale weather events. Climate change will affect the structure and function of landscapes and will alter land-use patterns and influence entire ecosystems. Scientific inquiry is playing an important role in the debate on climate change and in finding solutions. The strategies for combating global climate change must incorporate concepts and research in landscape ecology. Examples of how landscape concepts are influenced by climate change include land-use changes, such as modifying agricultural practices or using land to sequester carbon. The continued impact of greenhouse gases will fundamentally change landscapes for many years and perhaps forever. Landscape ecology can increase the scientific understanding of landscape resilience, recovery, and disturbance. Such information can provide an important framework for the sustainable management of landscapes and ecosystems.

Christine Watts, Ph.D., B.App.Sc., B.Sc.

FURTHER READING

Burel, Françoise, and Jacques Baudry. *Landscape Ecology: Concepts, Methods, and Applications.* Enfield, N.H.: Science Publishers, 2003. A comprehensive look into landscape structure, ecological processes, and dynamics, and how research can be applied to landscape management and planning.

Farina, Almo. *Principles and Methods in Landscape Ecology: Towards a Science of the Landscape.* Boston: Kluwer Academic, 2006. This holistic textbook is a useful tool for undergraduate and graduate students studying landscape ecology and coming from a variety of different disciplines.

Hong, Sun-Kee, et al., eds. *Landscape Ecological Applications in Man-Influenced Areas.* Dordrecht, the Netherlands: Springer, 2008. Concentrates on landscape ecology as it relates to urbanization, biodiversity, and land alteration. It provides a number of specific case studies for developing sustainable management through the examination of spatial analysis and landscape modeling.

Turner, Monica, Robert Gardner, and Robert O'Neill. *Landscape Ecology in Theory and Practice: Pattern and Process:* New York: Springer, 2001. Provides a fundamental framework regarding contemporary landscape ecology, while particularly focusing on the relationship and interaction between spatial patterns and ecological processes. Useful for both practicing ecologists and graduate students.

Wiens, John, et al., eds. *Foundation Papers in Landscape Ecology.* New York: Columbia University Press, 2007. Presents information on the origins and progress in the science of landscape ecology from a number of different standpoints, methodologies, and geographical areas.

Wu, Jianguo, and Richard Hobbs, eds. *Key Topics in Landscape Ecology.* New York: Cambridge University Press, 2007. Some of the most noted figures in landscape ecology discuss advances in the field and the concepts and methods used in understanding landscape patterns and in applying the science in a novel way.

WEB SITES

California Natural Resources Agency
Map Server
http://atlas.resources.ca.gov

International Association of Landscape Ecology
http://www.landscape-ecology.org

U.S. Environmental Protection Agency
Landscape Ecology
http://www.epa.gov/esd/land-sci/default.htm

See also: Climate Modeling; Ecological Engineering; Land-Use Management; Soil Science.

LAND-USE MANAGEMENT

FIELDS OF STUDY

Civil engineering; landscape architecture; regional and urban planning; design; water resource planning; agriculture; agricultural engineering; agronomy; environmental science; forestry; geography; social science; soil science.

SUMMARY

Land-use management concerns allocation of the landscape and its natural resources to appropriate uses for the purposes of providing food and shelter, ensuring adequate social and economic life-support systems, and protecting public health and safety, while preserving and sustaining the affected environment and natural resources for future generations. Another way to define land-use management is as a process that alters nature to benefit humanity and also strives to diminish the impact of the built and cultivated environment on natural resources such as soil, water, and air. The ultimate goal is to manage the competition for land and natural resources through integrated planning based on clearly stated objectives.

KEY TERMS AND CONCEPTS

- **Biodiversity:** Number and variety of plant and animal species in a particular area.
- **Built Environment:** Human-made features, such as buildings and structures, found in the environment.
- **Geospatial:** Relating to the relative position of things on Earth's surface.
- **Hydrology:** Study of water on the earth, underground, and in the atmosphere.
- **Land-Use Planning:** Integration of social, economic, and environmental objectives to ensure the best use of land and natural resources.
- **Mitigation:** Reduction of the negative impact of the built environment on the natural environment.
- **Stakeholders:** All those who have an interest in the success of a project or business.
- **Subdivision Control:** Regulation of land division, road construction, and infrastructure installation to serve specific uses.
- **Zoning:** Governmental regulation of buildings, structures, and land uses.

DEFINITION AND BASIC PRINCIPLES

Land-use management is a multidisciplinary science that integrates natural land systems with the built environment. Land-use managers assess, evaluate, analyze, and study geospatial data in setting public policy and proposing land-use controls that mitigate environmental impacts caused by development. The collaboration of multiple stakeholders and a high level of communication are necessary components in land-use management.

The graphic arts are often employed to prepare maps or plans that are used in communicating land-use management proposals. Land-use management plans provide viable support for a variety of management purposes. Maps and plans may depict existing and future uses for land such as transportation corridors, hazardous and solid-waste disposal sites, and historical districts, or water management solutions such as aquifers and watersheds. Maps of limiting factors, such as wetlands, floodplains, carrying capacity assessments, and the results of land suitability studies, help determine how land should be used and zoned. Soil and erosion mapping can aid in planning agricultural uses, and forest studies protect the habitats of flora and fauna from development.

When a lack of planning and land-use management creates haphazard and unsustainable growth, governments are often forced to take on the burden of correcting the situation.

BACKGROUND AND HISTORY

Around 8000 b.c.e., people turned from hunting to agriculture, and eventually, city-states began to spring up along rivers and bodies of water. In the fourth century b.c.e., the Greek philosopher Aristotle espoused a holistic view of city design, and laws began to be written to manage water resources for irrigation. People began constructing walls around cities to protect them from invaders, and they created geometric land patterns, including gardens that served both practical and aesthetic purposes. Ancient societies began to strategically locate buildings and select sites for monuments and religious structures based on purposes such as uniting the heavens with the earth. As people discovered metals and other resources that could be bartered, they developed transportation networks and began to exploit land with little regard for the environmental impact.

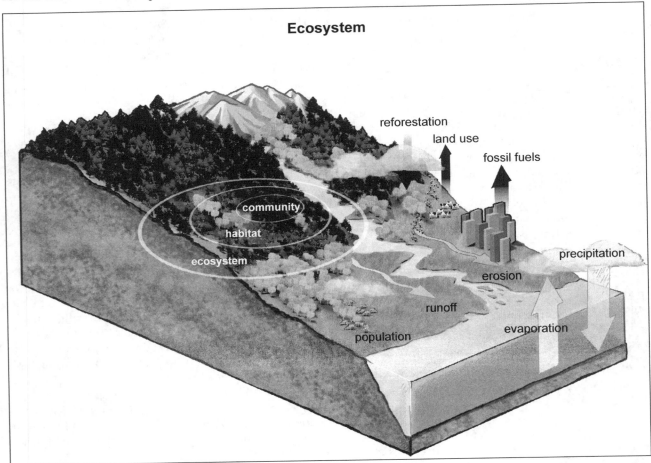

Ecosystem

Modern land-use management evolved from eighteenth century schools of thought such as Western classicism. In the United States, Thomas Jefferson is well known for his management of the landscape based on Renaissance traditions. Landscape architect Frederick Law Olmsted developed urban parks, and in partnership with architect Calvert Vaux, he planned communities such as Riverside Estate in Chicago. The depletion of natural resources and uncontrolled development that resulted from industrialization highlighted the need for planning and land

management. Pioneers in land-use planning include English town planner Sir Ebenezer Howard, the founder of the garden city concept, and American historian and philosopher Lewis Mumford, author of *The Story of Utopias* (1922) and cofounder in 1923 of the Regional Planning Association of America, a group devoted to planned development of urban areas.

In 1909, the first National Conference on City Planning and the Problems of Congestion was held in Washington, D.C., to address problems such as those uncovered by an extensive survey of Pittsburgh that had begun in 1907. Policymakers began to examine questions as to what American cities should provide for their citizens and how they could be economically and socially beneficial as well as aesthetically pleasing. This interest in land-use management resulted in the founding of organizations such as the American City Planning Institute (1917, later the American Planning Association), the creation of city planning boards (Boston, 1912), and the enacting of the first zoning ordinances (New York, 1916). Land use gradually became controlled by federal, state, and local government regulations and policies.

HOW IT WORKS

Land-use management begins with an inventory of past and existing land-use patterns for the purposes of proposing sustainable land uses and promulgating accompanying management policies. In preparing land-use plans, establishing public policy, and adopting comprehensive land-use laws and regulations, land-use managers make extensive use of technologies such as remote sensing and geographic information systems (GIS) to detect changes in land-use patterns and AutoCAD (two- or three-dimensional computer-aided design, or CAD, software) to depict existing and future features of selected areas.

Land-Use Controls. Those involved in managing land use must balance, evaluate, and integrate social, economic, and environmental concerns before arriving at a final management plan. Governmental regulators will usually seek public and private input as part of the decision-making process, and in many jurisdictions, land-use controls must be approved by legislative and administrative bodies as well as by the public. The types of land-use management controls are many and varied, but most include zoning and subdivision control regulations adopted by municipal and county governments based on comprehensive master plans. Other public land-use controls might include ordinances and bylaws that regulate water resources, earth removal, and building construction. Land use can also be managed through easements, both public and private, and proprietary regulations such as those found in real estate development covenants. Permitting and enforcement of land-use laws and regulations by local, state, and federal governmental entities also provide a form of land-use control, especially in the areas of environmental and transportation planning.

New Technologies. Many new technologies support modern land-use management. Private and public tracking systems are available to governments and others for the purposes of collecting and sharing geographic data. The U.S. Navy's electronic Land Use Control Information System (LUCIS) is a national digitized mapping system. Private companies also have created information gathering and tracking systems, such as Google Earth, which provides satellite images of the world as well as maps. Many of these new systems allow users not only to collect and track data on land-use controls and permitting but also to manage land-use activities by integrating with GIS and CAD systems.

Environmental modeling is another technology that has become popular as a land-use management tool, especially in the international community. Computer modeling allows experimentation that suggests an outcome when something occurs in nature. Modeling might be used to study the relationship between industrialization, air pollution, and environmental degradation or to improve agricultural production by suggesting the amount of surface-water runoff and erosion that can be expected based on soil and ground cover conditions.

Sustainable Land Use. Land-use management concerns the impact that the developed environment has on the natural environment and the strictures that the natural environment might place on land development. Climate, weather conditions, topography, soils, and hydrology are some of the physical features of the environment that land-use managers rely on in developing sustainable policies. Land-use managers must make difficult decisions such as whether developments should be allowed in areas subject to hurricanes or whether expansion into undeveloped areas should be limited so as to prevent urban overcrowding, sprawl, and unhealthy living conditions. The dilemma in achieving sustainable land-use management, therefore, is how to reach equilibrium between the built and natural environments and satisfy the multiple economic, social, and environmental interests.

APPLICATIONS AND PRODUCTS

Spatial Analysis. Geographic information systems (GIS) together with remote sensing technologies are used for spatial analysis of land uses. One application of spatial analysis is management of agricultural production. Remote sensing through agricultural satellites allows for monitoring of crop yields and distribution, atmospheric conditions that affect agricultural growth, moisture needs, pest and disease damage, and soil erosion. Planners also use spatial analysis to generate land-use maps and informational databases that promote sustainable development through evaluation of land-use change and adoption of regulatory schemes.

Land Cover Evaluation. In preparing land cover maps, land-use managers classify and quantify existing land uses into categories such as water resources, pavement, forest lands, agricultural lands, grasslands, and the built environment. This mapped information, together with database knowledge gleaned from new technologies and field experimentation, allows planners to gain an understanding of existing land-use conditions and monitor future changes throughout the world. Agricultural sustainability has benefited greatly from land cover analysis. Evaluation of land cover has also been used for predicting storm-water runoff to design drainage systems; urban planning, including integration of green space and recreational uses into the built environment; fire hazard mapping based on the amount of plant coverage that would provide fuel for a fire; and assessment of hydrology needs in arid and semi-arid regions of the world.

Land Surface Sensing. Remote-sensing technologies that use satellites to gather data enable land-use managers to study objects on the surface of the land, such as forest growth, buildings, pavement, and bodies of water, as well as the surrounding biosphere. Those involved in land-use management associated with natural resource conservation are likely to use land surface sensing together with field measurements to manage forests and ensure biodiversity by gathering data on plant growth, density of the canopy, and undercover species. Other applications for land surface sensing include meteorological studies, evaluation of soils and topography, collection of data concerning radiation from natural and non-natural sources, monitoring of ground temperature and snow and ice accumulation and decline, and water resources management.

Growth Management. Governmental regulation of growth allows for economic development in ways that make better use of existing land and natural resources. Land uses are managed by clustering the built environment around transportation networks, water resources, and other essential services, integrating open space into the developed landscape, and preserving natural land features. Smart growth is an example of growth management that promotes

---

Fascinating Facts About Land-Use Management

- Without implementation of land-use policies that preserve cultivable land and promote sustainable agriculture, global agricultural production is unlikely to be able to keep pace with the world population, which is increasing by more than 90 million people per year.
- The world's deserts are expanding. For example, the Sahara Desert is expanding at a rate of about 30 miles each year.
- The increase in the worldwide destruction of forests results in depletion of an energy source for more than three-quarters of the world's population, changes in rainfall patterns, increased river flooding, and extinction of flora and fauna.
- Every day, between 150 and 200 species of plants and animals become extinct because their habitats have been destroyed by the overdevelopment of land.
- Urban expansion has a major impact on land use: Each day, a city with a population of 1 million people uses about 625,000 metric tons of water and generates about 500,000 metric tons of wastewater and 2,000 metric tons of solid waste.
- More than 70 percent of the global population lacks access to clean water, and as a result, thousands of people die each day.

---

high density, compact projects in areas best suited for development based on topography, geographic features, water resources, and infrastructure. Smart growth consists of mixed-use developments, including residential, commercial, industrial, and recreational uses that support a self-sustaining community.

Environment Issues and Hazard Mitigation. Land-use managers can address environmental pollution issues by using modeling to predict the impact of various land uses on the environment and develop mitigation strategies. For example, planners use modeling to propose regulations to mitigate possible impacts of fertilizers and pesticides used in agriculture on water resources. In addition, GIS and related technologies enable land-use managers to locate and monitor hazards such as mine and hazardous-waste sites.

Natural Resources Protection. Those involved with the economic side of land-use management do not always provide the best protection for natural resources including water, air, soil, and plant and animal habitats. Ensuring that natural resources are preserved and open space is integrated into the built environment are important applications of land-use management. In addition, land-use managers are involved in land conservation through outright purchase of land or easements for conservation or through restricting land development in ways that preserve fragile ecosystems. Modern land-use management technologies provide the means for determining which land should be protected.

Habitat Modeling. The use of habitat modeling as a land-use management tool aids in protecting plant and animal habitats, including those that are rare; conserving biodiversity; monitoring ecosystems and their functionality; and predicting changes in habitats caused by nonnatural and natural disturbances. Often habitat modeling supports the preservation of specific species. Habitat modeling is used in conjunction with GIS, remote sensing, and fieldwork, including soil classification and plant coverage measurements. By using data from multiple sources, land-use managers are able to compare variables to make inferences concerning habitat health. Variables used in habitat modeling include tree types, soil classes, topography, and hydrology.

Open Land Management. Globally, as development continues to encroach on open land, governments have begun to regulate land uses to preserve rangeland, grazing land, grassland habitats, and agricultural areas. Land-use managers play a key role in monitoring land-use changes through new technologies and advising governments about which nonagricultural uses are compatible with open land uses. The preservation of open land uses is important for sustaining the future of life on the planet.

IMPACT ON INDUSTRY

The connections involving land use and development, climate change, the quantity and quality of water resources, and preservation of plant and animal habitats are some of the most important issues of the twenty-first century. The United Nations and international organizations such as the Earthwatch Institute, a nonprofit entity with offices in the United States, England, Australia, and Japan, are promoting

sustainable development. Through research and education, Earthwatch and similar organizations are attempting to balance the social and economic needs of society with protection of the environment.

Governmental and Nongovernmental Organizations. Many federal and state agencies are involved in land-use management. At the federal level, the U.S. Department of Agriculture addresses agricultural land uses and rural and community development. Within the department is the U.S. Forest Service (USFS), which manages federal forests and grasslands with the goal of preserving flora and fauna habitats, enhancing water resources, and responding to climate change, while allowing multiple uses of public lands (for recreation, as rangeland, and as a source of forest products). To manage the forest ecosystem, the agency evaluates human activity such as cutting and harvesting trees as well as natural systems including disease and insect infestations that affect the forest. The USFS employs several new technologies and tools, including integrated geographic information systems, simulators, and scale models to analyze how land-use management practices affect forest growth and watersheds. The U.S. Geological Survey provides scientific information for managing land-use resources including water and energy, and the Bureau of Land Management (in the U.S. Department of the Interior) manages public lands, especially those related to renewable energy resources, and is also responsible for habitat maintenance under the national wild horse and burro program. The Environmental Protection Agency (EPA) is the watchdog agency over all land uses that negatively affect the environment. The agency's mandate is accomplished through regulation, permitting, fines, and civil and criminal legal actions.

Colleges and Universities. Academic institutions, often in conjunction with the federal government, conduct research on land-use management. For example, the University of Washington and Yale University have worked with the USFS in analyzing forest ecosystems and generating database inventories of forest resources. Purdue University and the EPA have conducted hydrology research, including estimating changes in air and water resources as a result of existing and proposed land-use changes. This research is especially useful for localities that depend on groundwater for their public drinking water supply. Many universities with landscape architecture programs, often state universities that derived from land-grant colleges, conduct research and participate in public and private planning related to land-use management. For example, the University of Massachusetts brings together public and private stakeholders through its Citizen Planner Training Collaborative.

Scientific and Professional Organizations. The Interstate Technology Regulatory Council, whose members come from state and federal governmental agencies and private industries within the United States, conducts research and provides education on environmental public policy issues. The organization considers many of the land uses that affect the environment, including landfills, brownfields, and mining. Another scientific organization involved in land-use analysis is the Institute for Applied Ecology. Working with agencies and organizations including the U.S. Forest Service, Bureau of Land Management, and the Nature Conservancy, the institute's mission is preservation by evaluating the impact of land uses, such as transportation systems, energy generators, and landfills, on native plant and animal habitats. Internationally, an important scientific organization is the Global Observation of Forest and Land Cover Dynamics, which uses information on forest change obtained from satellites, aerial photography, and ground measurements to analyze the impact of multiple land uses on the environment.

Professional organizations include the International Federation of Landscape Architects, the American Planning Association, and the Urban Land Institute. The International Federation of Landscape Architects promotes not only design of the landscape but also worldwide land-use management and conservation through the profession of landscape architecture, while members of the American Planning Association are involved in land-use planning in all settings from urban to rural. The Urban Land Institute assists global leaders in the development of sustainable land use.

CAREERS AND COURSE WORK

An undergraduate or advanced degree in one of several planning disciplines, such as urban, regional, spatial, environmental, or transportation planning, is an excellent background for launching a land-use management career. Students also should take classes in planning theory, graphic arts and design,

geographic information systems, tracking and notification computer databases, economics, sociology, political science, public administration, botany, statistics, law, and natural sciences such as biology. Additional course work might include hydrology for a focus on water resource management or soil science and geology for students interested in soil management, especially as it relates to agricultural land use.

Job opportunities related to land-use management are best for students with advanced degrees that include training in modern technologies for geographical data collection, tracking, and sharing. Most advanced degrees are in one of the planning disciplines, and most planners also seek membership in professional organizations such as the American Institute of Certified Planners. Although doctorate degrees in the area of land-use management are becoming common, the typical advanced degree program culminates with a master's degree.

The majority of land-use management careers are in local, state, and federal governmental settings, often as a planner or public policy administrator. Opportunities also exist with international bodies, nongovernmental agencies, and nonprofit organizations in diverse fields associated with agriculture or environmental and natural resource management and conservation. Private companies that provide consultation services to governmental agencies also employ land-use managers. Some specialty careers in land-use management are related to oil and gas exploration, mining (including reclamation), and historic preservation.

SOCIAL CONTEXT AND FUTURE PROSPECTS

Although the ability to study and manage land use has been enhanced by geographic information systems, soil degradation, lack of quality water, and loss of biodiversity continue to be significant land-use management issues. Gaining global support to tackle these matters is not always viable, as less-developed nations do not have the economic resources or the social desire to develop policies that enable sustainable land use. In addition, some stakeholders in the most developed nations are unwilling to forgo immediate economic gain in exchange for future sustainability.

Individual nations and the international community, however, are working together to solve some of the most pressing land-use management concerns.

The greater availability of geographical data, including those collected by satellites, has led to the implementation of better agricultural practices in areas where soil degradation is obvious, the evaluation of water resources to solve issues regarding the quality and quantity of water, and better management of waste disposal. These data have also clarified land-use patterns and encroachment on the natural environment, including deforestation and loss of habitat for plants and animals. Refinements of existing technology and future advances are likely to add to the knowledge base that planners and policy makers can use and share in making informed land-use management decisions that affect the future of global life.

Carol A. Rolf, B.S.L.A., M.Ed., M.B.A., J.D.

FURTHER READING

Aspinall, Richard J., and Michael J. Hill, eds. *Land Use Change: Science, Policy, and Management.* Boca Raton, Fla.: CRC Press, 2007. Discusses land-use data analysis through remote-sensing technologies.

Birnbaum, Charles A., and Stephanie S. Foell, eds. *Shaping the American Landscape: New Profiles from the Pioneers of American Landscape Design Project.* Charlottesville: University of Virginia Press, 2009. Features articles authored by historic pioneers involved in diverse fields related to land-use management.

Butterfield, Jody, Sam Bingham, and Allan Savory. *Holistic Management Handbook: Healthy Land, Healthy Profits.* Chicago: Island Press, 2006. Provides step-by-step instructions for land management decision making using holistic practices.

Jellicoe, Geoffrey, and Susan Jellicoe. *The Landscape of Man: Shaping the Environment from Prehistory to the Present Day.* 3d ed. London: Thames and Hudson, 1995. Comprehensively covers the history of land-use management and the many schools of landscape design and planning.

Peterson, Gretchen N. *GIS Cartography: A Guide to Effective Map Design.* Boca Raton, Fla.: CRC Press, 2009. Practical guidebook for using GIS to create clear and informative maps that will be effective communication tools.

Randolph, John. *Environmental Land Use Planning and Management.* Chicago: Island Press, 2003. Considers advances in geographical data collection to mitigate environmental impacts, especially on water resources.

WEB SITES

American Planning Association
http://www.planning.org

Institute for Applied Ecology
http://appliedeco.org

International Federation of Landscape Architects
http://www.iflaonline.org

U.S. Department of Agriculture
U.S. Forest Service
http://www.fs.fed.us

U.S. Department of the Interior
Bureau of Land Management
http://www.blm.gov/wo/st/en.html

U.S. Geological Survey
Land Use History of North America
http://biology.usgs.gov/luhna/contents.html

See also: Agricultural Science; Agroforestry; Hydrology and Hydrogeology; Landscape Ecology; Soil Science.

M

MAGNETIC RESONANCE IMAGING

FIELDS OF STUDY

Radiology; diagnostics; radiofrequency; pathology; physiology; radiation physics; instrumentation; anatomy; microbiology; imaging; angiography; nuclear physics.

SUMMARY

Magnetic resonance imaging (MRI) is a noninvasive form of diagnostic radiography that produces images of slices or planes from tissues and organs inside the body. An MRI scan is painless and does not expose the patient to radiation, as an X ray does. The images produced are detailed and can be used to detect tiny changes of structures within the body, which are extremely valuable clues to physicians in the diagnosis and treatment of their patients. A strong magnetic field is created around the patient, causing the protons of hydrogen atoms in body tissues to absorb and release energy. This energy, when exposed to a radiofrequency, produces a faint signal that is detected by the receiver portion of the MRI scanner, which transforms it into an image.

KEY TERMS AND CONCEPTS

- **Artifact:** Feature in a diagnostic image, usually a complication of the imaging process, that results in an inaccurate representation of the tissue being studied.
- **Axial Slice:** Horizontal imaging plane that corresponds with right to left and front to back.
- **Claustrophobia:** Abnormal and persistent fear of closed spaces, of being closed in or being shut in.
- **Computed Tomography (CT) Scan:** Image of structures within the body created by a computer from multiple X-ray images.
- **Contrast Agent:** Dye used to provide contrast, for example, between blood vessels and other tissue.
- **Functional Magnetic Resonance Imaging (fMRI):**

Use of MRI to study physiological processes rather than just anatomy.
- **Gradient Coils:** Coils of wire used to generate the magnetic field gradients that are used in MRI.
- **Magnetic Resonance Angiogram (MRA):** Noninvasive complement to MRI to observe anatomy of blood vessels of certain size in the head and neck.
- **Sagittal Slice:** Vertical imaging plane that corresponds with front-to-back and top-to-bottom.
- **Scan:** Single, continuous collection of images.
- **Session:** Time that a single subject is in the magnetic resonance scanner; can be two hours for fMRI.
- **Tesla:** Unit of magnetic field strength; named for Nikola Tesla who discovered the rotating magnetic field in 1882.

DEFINITION AND BASIC PRINCIPLES

Magnetic resonance imaging (MRI), sometimes called magnetic resonance tomography, is a noninvasive medical imaging method used to visualize the internal structures and some functions of the body. MRI provides much greater detail and contrast between the different tissues in the body than are available from X rays or computed tomography (CT) without using ionizing radiation. MRI uses a powerful magnetic field to align the nuclear magnetization of protons of hydrogen atoms in the body. A radio frequency alters the alignment of the protons, creating a signal that is detectable by the scanner. The signals are processed through a mathematical algorithm to produce a series of cross-sectional images of the desired area. Image resolution and accuracy can be further refined through the use of contrast agents.

The detailed images produced are extremely valuable in detection and diagnosing of medical conditions and disease. The need for exploratory surgery has been greatly reduced, and surgical procedures and treatments can be more accurately directed by the ability to visualize structures and changes within the body.

BACKGROUND AND HISTORY

Magnetic resonance imaging is a relatively new scientific discovery, and its application to human diagnostics was first published in 1977. Two American scientists, Felix Bloch at Stanford University and Edward Mills Purcell from Harvard University, were both independently successful with their nuclear magnetic resonance (NMR) experiments in 1946. Their work was based on the Larmor relationship, named for Irish physicist Joseph Larmor, which stated that the strength of the magnetic field matched the radiofrequency. Bloch and Purcell found that when certain nuclei were in the presence of a magnetic field, they absorbed energy in the radiofrequency range of the electromagnetic spectrum and emitted this energy when the nuclei returned to their original state. They termed their discovery nuclear magnetic resonance: "nuclear" because only the nuclei of certain atoms reacted, "magnetic" because a magnetic field was required, and "resonance" because of the direct frequency dependence of the magnetic and radiofrequency fields. Bloch and Purcell were awarded the Nobel Prize in Physics in 1952.

NMR technology was used for the next few decades as a spectroscopy method to determine the composition of chemical compounds. In the late 1960's and early 1970's, Raymond Damadian, a State University of New York physician, found that when NMR techniques were applied to tumor samples, the results were distinguishable from normal tissue. His results were published in the journal Science in 1971. Damadian filed patents in 1972 and 1978 for a NMR system large enough to accommodate a human being that would emit a signal if tumor tissue was detected but did not produce an image. Paul Lauterbur, a physicist from State University of New York, devised technology that could run the signals produced by NMR through a computed back projection algorithm, which produced an image. Peter Mansfield, a British physicist from the University of Nottingham, further refined the mathematical analysis, improving the image. He also discovered echo-planar imaging, which is a fast imaging protocol for MRI and the basis for functional MRI. Lauterbur and Mansfield shared the 2003 Nobel Prize in Physiology or Medicine. Some controversy still exists regarding Damadian's exclusion from this honor.

HOW IT WORKS

The human body is made up of about 70 percent water. Water molecules are made up of two hydrogen atoms and one oxygen atom. When exposed to a powerful magnetic field, some of the protons in the nuclei of the hydrogen atoms align with the direction of the field. When a radio frequency transmitter is added, creating an electromagnetic field, a resonance frequency provides the energy required to flip the alignment of the affected protons. Once the field is turned off, the protons return to their original state. The difference in energy between the two states is called a photon, which is a frequency signal detected by the scanner. The photon frequency is determined by the strength of the magnetic field. The detected signals are run through a computerized algorithm to deliver an image. The contrast of the image is produced by differences in proton density and magnetic resonance relaxation time, referred to as T1 or T2.

An MRI scanning machine is a tube surrounded by a giant circular magnet. The patient is placed on an examination table that is inserted through the tube space. Some individuals experience claustrophobia when lying in the closed space of the scanning tube and may be given a mild sedative to reduce anxiety. Children are often sedated or receive anesthesia for an MRI. Patients are required to remain very still during the scan, which normally takes between thirty to ninety minutes to complete. During the scan, patients are usually provided with a hand buzzer or communication device so that they may interact with technicians. The magnetic field is created by passing electric current through a series of gradient coils. The strong magnetic fields are normally safe for patients, with the exception of people with metal implants such as pacemakers, surgical clips or plates, or cochlear implants, making them ineligible for MRI. During the scanning procedure, patients will hear a loud humming, beeping, or knocking noise, which can reach up to 120 decibels. (Patients are often provided with ear protection.) The noise is caused by the interaction of the gradient magnetic fields with the static magnetic field. The gradient coils are subject to a twisting force each time they are switched on and off, and this creates a loud mechanical vibration in the cylinder supporting the coils and surrounding mountings.

To enhance the images, contrast agents can be

injected intravenously or directly into a joint. MRI is being used to visualize all parts of the body by producing a series of two-dimensional images that appears as cross sections or slices. These slices can also be reconstructed to create three-dimensional views of the entire body or specific parts.

APPLICATIONS AND PRODUCTS

Research into the applications of magnetic resonance imaging technology beyond basic image generation has been progressing at a tremendous rate. Although the basic images are immensely valuable to physicians and scientists, the application of the scientific principles in the development of specialized scans is reaching far beyond original expectations and benefiting health care delivery and patient care.

Functional MRI (fMRI) is based on the changes in blood flow to the parts of the brain that accompany neural activity, and it provides visualization of these changes. This has been critical in detecting the brain areas involved in specific tasks, processes, or emotions. fMRI does not detect absolute activity of areas of the brain but it detects differences in activity. During the scan, the patient is asked to perform tasks or is presented with stimuli to trigger thoughts or emotions. The detection of the brain areas that are used is based on the blood oxygenation level dependent (BOLD) effect, which creates a variation signal, linked with the concentration of oxy-/deoxy-hemoglobine in each area. These scans are performed every two to three seconds over a period of minutes at a low resolution and do not often require additional contrast media to be used.

Diffusion MRI can measure the diffusion of water molecules in biological tissues. This is incredibly useful in detecting the movement of molecules in neural fiber, which can enable brain mapping, illustrating connectivity of different regions in the brain, and examination of areas of the brain affected by neural degeneration and demyelination, as in multiple sclerosis. Diffusion MRI, when applied to diffusion-weighted imaging, can detect swelling in brain cells within ten minutes of the onset of ischemic stroke symptoms, allowing physicians to direct reperfusion therapy to specific regions in the brain. Previously, computed tomography would take up to four hours to detect similar findings, delaying cerebral perfusion therapy to salvageable areas.

Interventional magnetic resonance imaging is

Fascinating Facts About Magnetic Resonance Imaging

- Magnetic resonance imaging (MRI) is based on nuclear magnetic resonance techniques, but it was named "magnetic" rather than "nuclear" resonance imaging because of the negative connotations associated with the word "nuclear" in the 1970's.
- An overwhelming majority of American physicians identified computed tomography (CT) and MRI as the most important medical innovations for improving patient care during the 1990's.
- In 1977, Raymond Damadian and colleagues built their first magnetic resonance scanner and named it "Indomitable." It is housed in the Smithsonian.
- MRI scanners were primarily developed for use in medicine but are also used to study fossils and historical artifacts.
- Functional MRI can create a video of blood flow in the brain and has been used to study monks serving under the Dalai Llama and the control they exert over mental processes through meditation.
- As of 2009, there were 7,950 magnetic resonance imaging systems in the United States, compared to only 266 in Canada.
- Tattoos received before 1990 may contain small amounts of metal in the ink, which may interfere with or be painful during MRI.
- While using functional magnetic resonance imaging (fMRI) to research children with attention deficit disorder, Pennsylvania psychiatrist Daniel Langleben discovered that deception activates regions in the prefrontal cortex, showing that fMRI could be used as a lie detector.

used to guide medical practitioners during minimally invasive procedures that do not involve any potentially magnetic instruments. A subset of this is intraoperative MRI, which is used during surgical procedures; however, most often images are taken during a break from the procedure in order to track progress and success and further guide ongoing surgery.

Magnetic resonance angiography (MRA) and venography (MRV) provide visualization of arteries and veins. The images produced can help physicians evaluate potential health problems such as narrowing of the vessels or vessel walls at risk of rupture as in an aneurysm. The most common arteries and veins examined are the major vessels in the head, neck, abdomen, kidneys, and legs.

Magnetic resonance spectroscopy (MRS) measures the levels of different metabolites in body tissues, usually in the evaluation of nervous system disorders. Concentrations of metabolites such as N-acetyl aspartate, choline, creatine, and lactate in brain tissue can be examined. Information on levels of metabolites is useful in determining and diagnosing specific metabolic disorders such as Canavan's disease, creatine deficiency, and untreated bacterial brain abscess. MRS has also been useful in the differentiation of high-grade from low-grade brain tumors.

Precise treatment of diseased or cancerous tissue within the body is a tremendous advance in health care delivery. Radiation therapy simulation uses MRI technology to locate tumors within the body and determine their exact location, size, shape, and orientation. The patient is carefully marked with points corresponding to this information, and precise radiation therapy can be delivered to the tumor mass. This drastically reduces excess radiation therapy and limits damage to healthy tissues surrounding the tumor. Similarly, magnetic resonance guided focused ultrasound (MRgFUS) allows ultrasound beams to achieve more precise and complete treatment and the ablation of diseased tissues is guided and controlled by magnetic resonance thermal imaging.

IMPACT ON INDUSTRY

In 1983, the Food and Drug Administration approved MRI scanners for sale in the United States. Magnetic resonance imaging is experiencing rapid growth on the global market and is expected to sustain this trend in the future as more clinics and health care centers use the imaging technique. The aging of the American population has increased the demand for efficient, effective, and noninvasive diagnosis, especially for neurological and cardiovascular diseases. Although a basic MRI system can cost more than $1 million, and the cost of construction of the suite to accommodate it can exceed $500,000, installation of these imaging systems in medical institutions still promises a positive return on investment for the health care providers. Some companies have begun offering refurbished MRI systems at reduced cost. MRI technology is an asset in challenging economic

times because it provides advanced and quick diagnoses, leading to faster patient care and greater patient turnover, leading to greater revenues for the institution. In the United States, health care insurers and the federal government provide very good reimbursement for the scan itself and a professional fee for a review of the resulting images by a radiologist.

Commercial development and industrial growth depends on advancements in research, which is being funded and conducted around the world. Some projects that are moving from experimental to commercial and driving the industry are integrated systems combining MRI with another modality, more portable MRI systems, improved contrast agents, advanced image-processing techniques, and magnetic coil technology.

Positron emission tomography (PET) is used in the diagnosis of cancer, but because it lacks anatomical detail, it is used in combination with CT scans. MRI scans provide superior contrast to CT for determining tumor structure and integrating PET and MRI into a single modality would be highly desirable and is in development.

Research is being done to improve the enhancement of the contrast agents used with MRI. Specifically, research involves targeted contrast agents that will bind only to desired tissue at the molecular and cellular level. Visualizing changes at a cellular level would make it possible to detect and diagnose disease sooner and provide opportunities for more focused treatment.

A growing niche market consists of MRI systems that allow patients to remain upright during the scan and open systems that can accommodate patients who previously were unable to have a scan because of their size or their claustrophobia.

Portable and handheld MRI technology has entered the international market. These small devices promise to deliver quality images at low cost, without the need for a special room to accommodate the device.

CAREERS AND COURSE WORK

The most common career choice in the field of magnetic resonance imaging is the MRI technician or technologist, individuals who operate the MRI system to effectively produce the desired images for diagnostic purposes while adhering to radiation safety measures and government regulations.

Researchers and government agencies are exploring potential occupational hazards to personnel because of prolonged and frequent exposure to magnetic fields. Technicians first explain the procedure to patients. Then, they ensure that patients do not have any metal present on their person or in their body and position them correctly on the examination table. Some technicians also administer intravenous sedation or contrast media to the patients. During the scan, the technologist observes the patient as well as the display of the area being scanned and makes any needed adjustment to density or contrast to improve picture quality. When the scan is complete, the technologist will evaluate the images to ensure that they are satisfactory for diagnostic purposes. The MRI training program may result in an associate's degree or certificate; some programs require prior completion of radiology or sonography programs and core competencies in writing, math, anatomy, physiology, and psychology. Once admitted to an accredited program, students receive training that includes patient care, magnetic resonance physics, and anatomy and physiology. The American Registry of Magnetic Resonance Imaging Technologists requires that students complete one thousand hours of clinical training. To satisfy this clinical training requirement, students are assigned to a specific hospital or are rotated through different hospitals. Becoming an MRI technician can take two to three years.

Diagnostic radiologists are physicians who have specialized in obtaining and interpreting medical images such as those produced by MRI. Becoming a radiologist in the United States requires the completion of four years of college or university, four years of medical school, and four to five years of additional specialized training.

MRI physicists are specialized scientists with a diverse background covering nuclear magnetic resonance (NMR) physics, biophysics, and medical physics, in combination with basic medical sciences, including human anatomy, physiology, and pathology. They also have a good understanding of engineering issues involving advanced hardware, such as large superconducting magnets, high-power radio frequencies, fast digital data processing, and remote sensing and control. Industrial MRI physicists often work in research and development for biotechnology companies or they implement new applications and provide support for equipment already installed in health care centers. Academic MRI physicists work in a university laboratory or in cooperation with a medical center involved in clinical research and training. Academic research may involve basic science in MRI spectroscopy, functional imaging, contrast media, or echo-planar imaging.

SOCIAL CONTEXT AND FUTURE PROSPECTS

Magnetic resonance imaging provides physicians with the ability to see detailed images of the inside of their patients to more easily diagnose and guide treatment. It provides researchers with valuable insight into the metabolism and physiology of the body. Still, there are drawbacks that make this scientific advance unavailable to some patients because of their economic circumstances or body shape. For patients who do not have health care insurance, the price of an MRI scan, which can range from $700 to $2500 depending on body part and type of examination, may be beyond what they can afford. Also, MRI systems have weight and circumference restrictions that make many people unsuitable candidates, limiting the quality of acute care for them. Typically, the weight limit is 350 to 500 pounds for the examination table, and the size of the patients is additionally limited by the diameter of the magnetic tube. As people tend to be larger in size, biotechnology companies are working on scanning systems, such as the upright or open concept scanner, that can accommodate larger people.

April D. Ingram, B.Sc.

FURTHER READING

Blamire, A. M. "The Technology of MRI—The Next Ten Years?" *British Journal of Radiology* 81 (2008): 601-617. Looks at the clinical status and future of MRI.

Filler, Aaron. "The History, Development and Impact of Computed Imaging in Neurological Diagnosis and Neurosurgery: CT, MRI, and DTI." *The Internet Journal of Neurosurgery* 7, no. 1 (2010). Provides a good history of the development of diagnostic imaging techniques.

Haacke, Mark, et al. *Magnetic Resonance Imaging: Physical Principles and Sequence Design.* New York: John Wiley & Sons, 1999. Explains the key fundamental and operational principles of MRI from a physics and mathematical viewpoint.

Simon, Merrill, and James Mattson. *The Pioneers of*

NMR and Magnetic Resonance in Medicine: The Story of MRI. Ramat Gan, Israel: Bar-Ilan University Press, 1996. Describes the history of MRI, from its development of scientific principles to application in health care.

Weishaupt, Dominik, Vitor Koechli, and Borut Marincek. *How Does MRI Work? An Introduction to the Physics and Function of Magnetic Resonance Imaging.* New York: Springer, 2006. A good resource for those who wish to familiarize themselves with the workings of magnetic resonance imaging and have some of the challenging concepts explained. It uses conceptual rather than mathematical methods to clarify the physics of MRI.

WEB SITES
Clinical Magnetic Resonance Industry
http://www.cmrs.com

International Society for Magnetic Resonance in Medicine
http://www.ismrm.org

MedlinePlus
Magnetic Resonance Imaging
http://www.nlm.nih.gov/medlineplus/mriscans.html

See also: Cardiology; Computed Tomography; Radiology and Medical Imaging; Ultrasonic Imaging.

MEASUREMENT AND UNITS

Metrology; agriculture; geographic information systems (GIS); geography; history; sociology; ballistics; surveying; military applications and service; navigation; construction; civil engineering; mechanical engineering; chemical engineering; physics; biochemical analysis.

SUMMARY

Measurement is the act of quantifying a physical property, an effect, or some aspect of them. Seven fundamental properties are recognized in measurements: length, mass, time, electric current, thermodynamic temperature, amount of a substance, and luminous intensity. In addition, two supplementary or abstract fundamental properties are defined: plane and solid angles. The base units for the seven fundamental properties can be manipulated to produce derived units for other quantities that are the effect of combinations of these fundamental properties. For instance, a Newton is a derived unit measuring force and weight, and a square meter is a derived unit used to measure area.

Historically, a number of units representing different amounts of the same properties have been used in various cultures around the world. For example, the United States traditionally uses the U.S. customary system (miles, cups, pints, ounces, and so on), while most other industrialized nations use the metric system (kilometer, milliliter, liters, grams, and so on). The metric system, developed in France in the late eighteenth century, represents the first true standard measurement system. The theory and physical practice of measurement is constant no matter what system of units is being used.

KEY TERMS AND CONCEPTS

- **Base Unit:** Unit of measurement that is used for one of the fundamental properties, also known as defined unit.
- **Calibration:** Process of ensuring that the measurements obtained with a specific measuring device agree with the value established by the base unit of that measured property or effect.
- **Conversion Factor:** Ratio of one unit of measure in a particular system to the unit of measure of the same property in a different system of measurements.
- **Derived Unit:** Unit of measurement that is derived from the relationship of base units.
- **Dimensional Analysis:** Consideration and manipulation of the units involved in a measurement and their relationship to one another.
- **Fundamental Property:** Property ascribed to a phenomenon, substance, or body that can be quantified: length, mass, time, electric current, thermodynamic temperature, the amount of a substance, and luminous intensity.
- **International Standard:** Unit of measurement whose consistent value is recognized as a common unit of measurement by international agreement.
- **Measurement:** Ascribing of a value to the determination of the quantity of a physical property.
- **Standard Temperature And Pressure (STP):** Value of the mean atmospheric pressure and mean annual temperature at sea level, chosen by international agreement to be designated as 1 standard atmosphere or 760 torr and 0 degrees Celsius/32 degrees Fahrenheit or 273.15 degrees Kelvin.
- **Unit:** Single definitive basic magnitude of portions used to measure various properties.

DEFINITION AND BASIC PRINCIPLES

Measurement has the purpose of associating a dimension or quantity proportionately with some fixed reference standard. Such an association is intended to facilitate the communication of information about a physical property in a manner that allows it to be reproduced in concept, or in actuality if needed. The function of an assigned unit that is associated with a definite dimension is to provide the necessary point of reference for someone to comprehend the exact dimensions that have been communicated. For example, a container may be described as having a volume of eight cubic feet. Such a description is incomplete, however, because it does not state the shape or relative proportions of the container. The description applies equally well to containers of any shape, whether they be cubic, rectangular, cylindrical, conical, or some other shape. At a more basic level, however, there is the assumption that the person who receives the description has the same understanding of what is meant by "cubic feet" as does the person who provided the description. This is the fundamental principle of any measurement system: to provide a frame of reference that is commonly understood by a large number of people and that indicates exactly the same thing to each of them.

A measurement system, no matter what its basic units, must address a limited group of fundamental properties. These are length, mass, temperature, time, electric current, amount of a substance, and luminous intensity. In addition, it must also be able to describe angles. All other properties and quantities can be described or quantified by a combination of these fundamental properties.

In practice, any of these fundamental properties can be defined relative to any randomly selected relevant object or effect. Logically, though, for a measurement system to be as effective as possible, the objects and effects that are selected as the defined units of fundamental properties must be readily available to as many people as possible and readily reproducible. If, for example, a certain king were to decree that the "foot" to be used for measurement was the human foot, a great deal of confusion would result because of the variability in the size of the human foot from person to person and from a person's left foot to right foot. Should he instead decree that the "foot" would correspond to his own right foot and no other, the unit of measurement becomes significantly more precise, but at the same time, the decree raises the problem of how to verify that measurements being taken are in fact based on the decreed length of the foot. A physical model of the length of the king's right foot must then be made available for comparison. The same logic holds true for any and all defined properties and units. All measurements made within the definitions of a specific system of measurement are therefore made by comparing the proportional size of an effect or property to the defined standard units.

BACKGROUND AND HISTORY

Historically, measurement systems were generally based on various parts of the human body, and some of these units have remained in use in the modern world. The height of horses, for example, is generally given as being so many hands high. In many languages, the word for "thumb" and the word for "inch" are closely related, if not identical: In French, both words are *pouce*; in Hungarian, *hüvelyk* and *hüvelykujj*; in Norwegian, *tomme* and *tommelfinger*; and in Swedish, *tum* and *tumme*. Although using the thumb as a basis for measurement is convenient in that almost everyone has one, in practice, the generally accepted thumb size tended to vary from city to city, making it impossible to interchange parts made in different locations by separate artisans.

Units of measurement traditionally have been defined by a decree issued by a political leader. In ancient times, units of length often corresponded to certain parts of the human body: The foot was based on the human foot and the cubit on the length of the forearm. Other units used to weigh various goods were based on the weight of commonly available items. Examples include the stone (still used widely but not officially in Britain) and the grain. Invariably, the problem with such units lay in their variability. A stone weight may be defined as equivalent to the size of a stone that a grown man could enclose by both arms, but grown men come in different sizes and strengths, and stones come in different densities. A grain of gold may be defined as equivalent to the weight of a single grain of wheat; however, in drier years, when grains of wheat are smaller and lighter, the worth of gold is significantly different from its worth in years of good rainfall, when grains of wheat are larger and heavier.

HOW IT WORKS

Measurements and units are most useful when they are standardized so that measurements mean the same thing to everyone and are comparable. One method of solving the problem of variability in measurement is to establish a standard value for each unit and to regularly compare all measuring devices to this standard.

Historically, many societies required that measuring devices be physically compared with and calibrated against official standards. Physical representations of measuring units were made as precisely as possible and carefully stored and maintained to serve as standards for comparison. In ancient Egypt, the standard royal cubit was prepared as a black granite rod and most likely kept as one of the royal treasures by the pharaoh's chief steward. That simple stone

Fascinating Facts About Measurement and Units

- More than three hundred distinct units of measurement have been defined in the British Imperial and U.S. customary systems of measurement but only seven in the metric system. All other metric units are either multiples of ten or combinations of these seven basic units.
- The U.S. customary system of measurement includes the "mark twain," indicating the two-fathom depth that was considered the minimum safe depth for steam-powered riverboats on the Mississippi River. The term "mark twain" was adopted as a pen name by the American writer Samuel Clemens, author of *The Adventures of Tom Sawyer* (1876) and *Adventures of Huckleberry Finn* (1884).
- A system of measurement used in Imperial Russia contained a unit of length called the "archine" or "arshin." It was equivalent to a distance of 28 inches and was in use until the mid-twentieth century. Many of the rifles used by the Soviet army in World War II and afterward had been fitted with sights scaled to that unit of measurement.
- The phrase "worth its weight in gold" comes from the use of a grain of wheat as the unit of measurement for the precious metal.
- The "carat" unit used to weigh gems comes from the use of the carob seed as the unit of weight.
- A design flaw in the programming of the Mars Climate Orbiter caused it to spin out of control and crash because part of the program called for data in metric units and another part called for data in U.S. customary units.

object would have been accorded such a status because of its economic value to trade and construction and because its helped maintain the pharaoh's reputation as the keeper of his kingdom.

Later nations and empires also kept physical representations of most standard units appropriate to the economics of trade. The British made and kept definitive representations of the yard and the pound, just as the French made and kept definitive representations of the meter and the kilogram after developing the metric system in the late 1700's. Early on, countries recognized that the representative of

the unit must not be subject to change or alteration. The Egyptian standard royal cubit was made of black granite; the standard meter and the kilogram, as well as the foot and the pound, were made from a platinum alloy so that they could not be altered by corrosion or oxidation.

France adopted the metric system as its official measuring system in 1795, and that system was standardized in 1960 as the International System of Units (Le Système International d'Unités; known as SI). The units of the metric system were defined based on unchanging, readily reproducible physical properties and effects rather than any physical object. For example, the standard SI unit of length, the meter, was originally defined to be one ten-millionth of the distance from the equator to the north polar axis along the meridian of longitude that passed through Paris, France. In 1960, for greater accuracy, the definition of the meter was based on a wavelengths of light emitted by the krypton-86 isotope. In 1983, it was changed to the length traveled by light in a vacuum during 1/299,792,458 of a second. Similarly, the length of a standard second of time had been defined as a fraction of one rotation of the Earth on its axis, until it was realized that the rate of rotation was not constant but rather was slowly decreasing, so that the length of a day is increasing by 0.0013 second per hundred years. In 1967, it was formally redefined as a duration of 9,192,631,770 periods of the radiation corresponding to the transition between the two hyperfine levels of the ground state of the cesium-133 atom. As technology develops, it becomes possible to measure smaller quantities with finer precision. This capability has been the principal that permits ever more precise definitions of the basic units of measurement.

The application of measuring procedures is of fundamental importance in the economics of trade, especially in the modern global economy. In manufacturing, engineering, the sciences, and other fields, the accuracy and precision of measurement are essential to statistical process control and other quality-control techniques. All such measurement is a process of comparing the actual dimensions or properties of an item with its ideal or design dimensions or properties. The definition of a standard set of measuring units greatly facilitates that process.

APPLICATIONS AND PRODUCTS

It is quite impossible to calculate the economic effects that various systems of measurement, both good and bad, have had throughout history. Certainly, commonly understood and accepted units of weight, distance, and time have played a major role in facilitating trade between peoples for thousands of years. In many ways, all human activity can be thought of as dependent on measurement.

The study of measurement and measurement processes is known as metrology. In essence, metrology is the determination and application of more precise and effective means of measuring quantities, properties, and effects. The value in metrology derives from how the information obtained is used. This is historically and traditionally tied to the concepts of fair trade and of well-made products. Measurement serves to ensure that trade is equitable, that people get exactly what they are supposed to get in exchange for their money, services, or other trade goods, and that as little goes to waste as possible.

By far, the largest segment of metrology deals with the design, production, and calibration of the various products and devices used to perform measurements. These devices range from the simplest spring scale or pan balance to some of the most sophisticated and specialized scientific instruments ever developed. In early times, measurements were restricted to those of mass, distance, and time because these were the foremost quantities used in trade. The remaining fundamental properties of electric current, amount of a substance, temperature, and luminous intensity either remained unknown or were not of consequence.

Weight Determination. Originally, weights were determined with relative ease by the use of the pan balance. In the simplest variation of this device, two pans are suspended from opposite ends of a bar in such a way that they are at equal heights. The object to be weighed is placed in one of the pans and objects of known weights are placed in the other pan until the two pans are again at equal heights. The precision of the method depends on the ability of the bar to pivot as freely as possible about its balance point; any resistance will skew the measurement by preventing the pans from coming into proper balance with each other.

Essentially, all balances operate on the principle of comparing the weight of an unknown object against the weight of an accurately known counterweight, or some property such as electric current that can in turn be measured very accurately and precisely. The counterweight may not actually be a weight but rather an electronic pressure sensor or something that can be used to indicate the weight of an object such as the tension of a calibrated spring or a change in electrical resistance. Scales used in commercial applications, such as those in grocery stores, grain depots, and other trade locations, are inspected and calibrated on a regular basis according to law and regulations that govern their use.

Length Determination. For many practical purposes, a linear device such as a scale ruler or tape measure is all that is needed to measure an unknown length. The device used should reflect the size of the object being measured and have scale markings that reflect the precision with which the measurement must be known. For example, a relatively small dimension being measured in a machine shop would be measured by a trained machinist against a precision steel scale ruler with dimensional markings of high precision. Training in the use of graduated scale markings typically enables the user to read them accurately to within one-tenth of the smallest division on the scale. High-precision micrometers are generally used to make more precise measurements of smaller dimensions, and electronic versions of such devices provide the ability to measure dimensions to extremely precise tolerances. Smaller dimensions, for which the accuracy of the human eye is neither sufficient nor sufficiently reproducible, are measured using microscope techniques and devices. Accurate measurements of larger dimensions have always been problematic, especially when the allowable tolerance of the measurement is very small. This has been overcome in many cases by the measuring machine, a semirobotic device that uses electronic control and logic programming to determine the distance between precise points on a specific object.

The 1960 definition of the meter was achieved through the use of a precision interferometer, a device that uses the interference pattern of light waves such that the number of wavelengths of a specific frequency of light can be counted. Using this device, the meter was precisely defined as the distance equal to 1,650,763.73 wavelengths of the $2p10$-$5d5$ emission of krypton-86 atoms. In 1983, however, the meter was defined as the distance traveled by light in vacuum in $1/299,792,548$ of a second.

Time Measurement. Essentially, the basic unit of time measurement in all systems is the second. This is a natural consequence of the fact that the length of a day is the same everywhere on the planet. The natural divisions of that period of time according to the observed patterns of stars and their motions almost inevitably results in twenty-four equal divisions. A natural result of the metric system is that a pendulum that is one meter in length swings with a period of one second. Pendulum clockworks have been used to measure the passage of time, coordinated to the natural divisions of the day, for thousands of years. More precise time measurements have become possible as better technology has become available. With the development of the metric system, the unit duration of one second was defined to correspond to the appropriate fraction of one rotation of the planet. Until the development of electronic methods and devices, this was a sufficient definition. However, with the realization that the rotation of the planet is not constant but is slowly decreasing, the need to redefine the second in terms that remain constant in time led to its redefinition in 1967. According to the new definition, a second is the time needed for a cesium-133 atom to perform 9,192,631,770 complete oscillations.

Temperature Determination. Of all properties, measurement of thermodynamic temperature is perhaps the most relative and arbitrary of all. The thermometer was developed before any of the commonly used temperature scales, including the Fahrenheit, Celsius, Kelvin, and Rankine scales. All are based on the freezing and boiling points of water, the most readily available and ubiquitous substance on the planet. The Fahrenheit scale arbitrarily set the freezing point of saturated salt water as 0 degrees. The physical dimensions of the scale used on Fahrenheit's thermometer resulted in the establishment of the boiling point of pure water as 212 degrees. The Celsius scale designated the freezing point of pure water to be 0 degrees and the boiling point of pure water to be 100 degrees. The relative sizes of Fahrenheit and Celsius degrees are thus different by a factor of 5 to 9. Conversion of Fahrenheit temperature to Celsius temperature is achieved by subtracting 32 degrees and multiplying the result by 5/9; to convert from Celsius to Fahrenheit, first multiply by 9/5, then add 32 degrees to the result. The temperature of -40 degrees is the same in both scales.

Each temperature scale recognizes a physical state called absolute zero, at which matter contains no thermal energy whatsoever. This must physically be the same state regardless of whether Fahrenheit or Celsius degrees are being used. Because of this, two other scales of temperature were developed. The Rankine scale, established as part of the school of British engineering, uses the Fahrenheit degree scale, beginning at 0 degrees at absolute zero. The Kelvin scale uses the Celsius degree scale, beginning at 0 degrees at absolute zero.

Temperature is accurately measured electronically by its effect on light in the infrared region of the electromagnetic spectrum, although less accurate physical thermometers remain in wide use.

Amount of a Substance. Of all the fundamental properties, amount of a substance is the least precisely known. The concept is intimately linked to the modern atomic theory and atomic weights, although it predates them by almost a hundred years. Through studies of the properties of gases in the early 1800's, Italian Amedeo Avogadro concluded that a quantity of any pure material equal to its molecular weight in grams contained exactly the same number of particles. This number of particles came to be referred to as the Avogadro constant. Thus, 2 grams of hydrogen gas (molecular weight = 2) contains exactly the same number of molecules as does 342 g of sucrose (common white table sugar, molecular weight = 342) or 18 milliliters of water (molecular weight 18, density = 1 gram per milliliter). Calculations have determined the value of the Avogadro constant to be about 6.02214×1023. Because all twenty-three decimal places are not known, the absolute value has not yet been determined, and therefore amount of a substance is the least precisely known unit of measurement.

The amount of material represented by the Avogadro constant (of atoms or molecules) is termed the "mole," a contraction from "gram molecular weight." The number is constant regardless of the system of measurement that is used, as a result of the indivisibility of the atom in modern chemical theory. Thus a gram molecular weight of a substance and a pound molecular weight of a substance both contain a constant number of molecules.

Electric Current. Of course, atoms are not indivisible in fact. They consist of a nucleus of protons and neutrons, surrounded by a cloud of electrons. The electrons can move from atom to atom through matter, and such movement constitutes an electric current. More generally, any movement of electronic charge between two different points in space defines an electric current. Although some controversial evidence suggests that electricity may have been known in ancient times, any serious study of electricity did not begin until the eighteenth century, after the discovery of the electrochemical cell. At that time, electricity was thought to be a mysterious fluid that permeated matter. With the discovery of subatomic particles (electrons and protons) in 1898 and the subsequent development of the modern atomic theory, the nature of electric currents came to be better understood. The ampere, named after one of the foremost investigators of electrical phenomena, is the basic unit of electric current and corresponds to the movement of one mole of electrons for a period of one second.

The development of the transistor—and the electronic revolution that followed—made it possible to measure extremely small electric currents of as little as 10-9 amperes, as well as corresponding values of voltage, resistance, induction, and other electronic functions. Basically, this made it possible to precisely measure fundamental properties by electronic means rather than physical methods.

Measurement of Luminous Intensity. Until it became important and necessary to know precisely the intensity of light being emitted from a light source, particularly in the fields of astronomy and physics, there was no need for a fundamental unit of luminous intensity. Light intensities were generally compared, at least in post-Industrial Revolution Europe, to the intensity of light emitted from a candle. This sufficed for general uses such as lightbulbs, but the innate variability of candle flames made them inadequate for precision measurements. The candela was set as the standard unit of luminous intensity and corresponds to the output of energy of 0.00146 watts. Modern lightbulbs typically are rated at a certain number of watts, but this is a measure of the electric power that they consume and not of their luminous intensity.

IMPACT ON INDUSTRY

Metrology is central to industry. The ability to measure quantities accurately and precisely is an absolute necessity given the scale of modern applications from nanotechnology to celestial mechanics. The tolerances of the dimensions required for modern devices and machines makes it imperative that measurement be on the same scale. If a dimension must be accurate to within a very narrow range of values, then the metrologist's limit of measurement of that dimension must also be within that range of values.

As analytical techniques and methods are designed to identify and quantify ever smaller amounts of materials, ever faster processes, ever finer distinctions in molecular structure, ever more distant stellar emissions, and so on, metrology must also progress in its capabilities. Specialized applications aside, the most important aspect of metrology in modern society is its applicability in the fields of medicine and manufacturing. Medical testing is carried out on a greater number of people than ever before. Routine analytical procedures must identify precisely and definitively the amount of a specific material that may be present in extremely small quantities in a sample. The analytical method used may require measurement or control of any of the properties of time, distance, luminous intensity, amount of material, electric current, weight, and temperature in the quantification of a specific material.

In industrial applications, metrology is central to the techniques of statistical process control (SPC). The basic goal of SPC is the elimination of product units that are outside of their respective design parameters. The units are closely monitored as they are produced and checked against their design standards. Interpretation of the nature and number of variations in the products is used to adjust and control the process by which they are produced. SPC and other quality-control methods are based on statistical analysis, which requires the measurement of several dimensions and aspects of product units. In a simple example, a machined part such as a fluid valve body, produced in a mass-production machining facility, may have been designed with two machined surfaces that must be parallel to each other within 0.01 millimeter and with three very precisely placed holes of a specific diameter. The quality-control program will call for measurement of those respective dimensions and comparison to the ideal design. The appearance of units with surfaces that are no longer suitably parallel or with holes that do not conform in size or in shape call for the adjustment of the machinery so that subsequent product units are again within design parameters. The value of this process is in the reduction of costs through the elimination of waste.

CAREERS AND COURSE WORK

Measurement and units, as they apply to metrology, are part of every technical and scientific field. The student planning on a career in any of these areas can expect to learn how to use the metric system and the many ways in which measurement is applied in a specific field of study. Because measurement is related to the fundamental properties of matter, a good understanding of the relationships between measurements and properties will be essential to success in any chosen field. In addition, the continuing use of multiple measuring systems, such as the metric, U.S. customary, and British Imperial systems, in the production of goods means that an understanding of those systems and how to convert between them will be necessary in many careers.

Specific courses that involve measurement include geometry and mathematics, essentially all of the physical science and technology courses, and design and technical drawing courses. Geometry, which literally means "earth measurement," is the quintessential mathematics of measurement and is essential for careers in land surveying, architecture, agriculture, civil engineering, and construction. Gaining an understanding of trigonometry and angular relationships is particularly important. The physical sciences, such as physics, chemistry, and geology, employ analytical measurement at all levels. Specializations in which measurement plays a prominent role are analytic chemistry and forensic research. Technology programs such as mechanical engineering, electronic design, and biomedical technology all rely heavily on measurement and the application of metrological techniques in the completion of projects that the engineer or technologist undertakes.

SOCIAL CONTEXT AND FUTURE PROSPECTS

Measurement and an understanding of the units of measurement are an entrenched aspect of modern society, taught informally to children from early childhood and formally throughout the course of their schooling. They are fundamental to the continued progress of technology and essential to the determination of solutions to problems as they arise, as well as to the development of new ideas and concepts. The need for individuals who are trained in metrology and who understand the relationship and use of measurements and units will be of increasing importance in ensuring the viability of both new and established industries.

Particular areas of growth and continuing development include the fields of medical research and analysis, transportation, mechanical design, and aerospace. The accuracy and precision of measurement in these fields is critical to the successful outcome of projects.

Richard M. J. Renneboog, M.Sc.

FURTHER READING

Butcher, Kenneth S., Linda D. Crown, and Elizabeth J. Gentry. *The International System of Units (SI): Conversion Factors for General Use.* Gaithersburg, Md.: National Institute of Standards and Technology, 2006. A guide to the International System of Units produced by the U.S. government.

Cardarelli, Francois. *Encyclopedia of Scientific Units, Weights and Measures: Their SI Equivalences and Origins.* New York: Springer, 2003. Systematically reviews the many incompatible systems of measurement that have been developed throughout history. It clearly relates those quantities to their modern SI equivalents and provides conversion tables for more than 19,000 units of measurement.

Kuhn, Karl F. *In Quest of the Universe.* 6th ed. Sudbury, Mass.: Jones and Bartlett, 2010. This excellent introduction to astronomy illustrates the uses and importance of measurement in that field.

Tavernor, Robert. *Smoot's Ear: The Measure of Humanity.* New Haven, Conn.: Yale University Press, 2007. An entertaining yet scholarly work that discusses measuring systems and the act of measurement in the context of the societies in which the systems were developed.

WEB SITES

Bureau International des Poids et Mesures
http://www.bipm.org

National Aeronautics and Space Administration
A Brief History of Measurement Systems
http://standards.nasa.gov/history_metric.pdf

National Institute of Standards and Technology
Weights and Measures
http://www.nist.gov/ts/wmd/index.cfm

See also: Barometry.

METABOLIC ENGINEERING

FIELDS OF STUDY

Biology; biochemistry; molecular biology; organic chemistry; analytical chemistry; microbiology; genetic engineering; biotechnology.

SUMMARY

Metabolic engineering is a new science that appeared in the 1990's. It is associated with biology and chemistry. Metabolic engineering allows the designing of biochemical pathways that do not exist in the natural world, as well as the redesign of existing biochemical pathways often with the use of genetic engineering. Metabolic engineers often modify biochemical pathways by reducing cellular energy use or waste production, by changing the nutrient flow to the cells, or improving the productivity and yield of a particular pathway. In addition, metabolic engineers may potentially design new organisms that are tailor-made for the desired chemicals and production processes. Many novel compounds of industrial and medical interest can be produced by metabolic engineering. In the twenty-first century, the main efforts of metabolic engineers are concentrated on biofuels and pharmaceuticals.

KEY TERMS AND CONCEPTS

- **Bioreactor:** Apparatus for cell growth with practical purpose under controlled conditions.
- **DNA Sequencing:** Determining of the precise order of nucleotides (such as adenine, guanine, cytosine, and thymine) in a DNA.
- **Enzymes:** Biological catalysts made of proteins.
- **Genetic Engineering:** Modification of genetic material to achieve specific goals.
- **Metabolism:** Sum of biochemical reactions within an organism.
- **Substrate:** Substance that is acted on (as by an enzyme).

DEFINITION AND BASIC PRINCIPLES

Metabolic engineering is a relatively new field that deals with the modification and optimization of metabolic pathways, mainly in microorganisms, by altering genes, nutrient uptake, or metabolic flow to allow production of novel compounds that are of industrial and medical interest. Metabolic pathways of living organisms are not optimal for specific practical applications, but they can be modified using the tools of modern biotechnology such as genetic engineering. The redesign of existing, natural metabolic pathways for useful purposes is a main objective of metabolic engineering. Metabolic engineering usually includes two phases: careful analysis of the metabolic pathway and genes involved in the pathway (analytical phase) and its modification (synthesis phase). Pathway analysis often includes the metabolic control analysis: determining which compounds can control the productivity and yield of particular pathway. Different tasks of metabolic engineering are as follows: improvements of productivity and yield of particular pathway; expansion of substrate range; elimination of waste; improvement of process performance; improvements of cellular activities; and extension of product array. Metabolic engineering is becoming one of the principal fields of biotechnology.

Production of many chemicals and fuels uses nonrenewable resources or limited natural resources. Metabolic engineering creates many alternatives to replace dangerous chemicals and petroleum-based transportation fuels with clean, green, and renewable chemicals and biofuels.

BACKGROUND AND HISTORY

The term "metabolic engineering" first appeared in the early 1990's. Since that time, the range of products that can be generated has increased significantly, partly because of remarkable advances in other fields related to metabolic engineering, such as DNA sequencing and genetic engineering. With DNA sequencing, scientists were able to identify the majority of metabolic genes and enzymes in many organisms. In the post-sequencing era, the obtained information is used for practical construction of biochemical pathways or whole organisms with optimized functions through metabolic engineering.

In the 1990's, scientists developed new genetic tools that gave metabolic engineers more precise control over metabolic pathways. They also created analytical tools that allowed the metabolic engineer to track metabolites in a cell to identify new biochemical pathways more precisely.

Earlier in the twenty-first century, metabolic engineers joined other scientists in their quest for alternative fuels, which are in high demand because of increasing oil prices and concern about climate change.

HOW IT WORKS

Metabolic engineering is based mainly on microbial metabolism. Microbes produce different kinds of substances that they use for the growth and maintenance of their cells. These substances can be useful for humans. The goal of metabolic engineering is to enhance the microbial production of useful substances. To achieve this goal, metabolic engineers must follow a particular route. They need to choose a friendly organism (host) for their metabolic manipulations. They need to find cheap and available substrates to use for modified metabolic pathways. Finally, metabolic engineers must be able to perform genetic manipulations of metabolic routes. Metabolic engineers can also alter nutrient uptake or metabolic flow. All these steps are dependent on each other. For example, genes cannot be manipulated in every organism; products or metabolic intermediates may be toxic to its host.

Host and Host Design. Generation of products by metabolic engineering has been achieved by transferring product-specific enzymes or entire metabolic pathways into so-called user-friendly microorganism hosts, which were used traditionally in industry. These industrial microorganisms grow rapidly on inexpensive culture media available in bulk quantities, are open to genetic manipulation (and genetic manipulation tools are available), and are nonpathogenic (do not cause disease). In addition, it is important that the host can survive (and thrive) under the desired process conditions (ambient versus extremes of temperature, pH). It is essential that the host is genetically stable (with the introduced pathway) and not susceptible to virus or another microbe's attack. Among the host microorganisms most widely used are *Saccharomyces cerevisiae* and *Escherichia coli*. *Saccharomyces cerevisiae*, or baker's yeast, has been used for making bread and alcohol for thousands of years. It is one of the earliest domesticated organisms. This organism has come to be used in a large number of different processes within the biotechnological and pharmaceutical industries. Comprehensive knowledge of *S. cerevisiae* has been accumulated over a long period of time. In addition, the complete genome sequence of yeast is available, and yeast is nonpathogenic. The well-established fermentation and process technology for large-scale production with *S. cerevisiae* in bioreactors makes this organism very attractive for several industrial purposes.

Escherichia coli, commonly known is *E. coli*, is a bacterium that is widely used as a research (model) organism. It is easy to grow and genetically manipulate this bacterium, and its genome sequence is available. Several important products such as interferon (flu-fighting drug), insulin, and growth hormone are manufactured by genetically modified *E. coli*.

In addition to *E. coli* and *S. cerevisiae*, several other microorganisms are widely used as hosts for metabolic engineering manipulations, including bacteria *Bacillus subtilis* and *Streptomyces coelicolor*.

Finally, in addition to redesigning particular metabolic processes, metabolic engineers may also design de novo artificial cells that will produce desired products.

Substrates. To make metabolically engineered products, chemical substrates are needed. To make these products economically viable, inexpensive sources of substrates are required. Substrates must contain different chemical components, such as carbon, nitrogen, oxygen, and hydrogen. For example, metabolic engineers are looking at sugars from cellulosic biomass as potential substrates for biofuel production. Cellulosic biomass is a very attractive biofuel feedstock because of its abundant supply. On a global scale, plants produce almost 100 billion tons of cellulose per year, making it the most abundant organic compound on Earth.

Genetic Manipulation of Metabolic Routes. Genetic manipulation of metabolic pathways by adding or deleting genes or modifying the expression of existing genes in the host can serve several useful purposes. It can extend the existing pathways or shifting metabolic route into a desired pathway or increase the rate-determined step of the particular metabolic route. Adding genes into the host consists of the following steps.

- The gene the for desired pathway is obtained from the non-host organism.
- The gene is inserted into the host cell.

- Host cells are induced to express (to cause the gene to manifest its effects) this "foreign" gene in order to produce the desired product.

One example of how gene manipulation is used in areas relevant to metabolic engineering is as follows: In the mold *Aspergillus terreus*, the producer of cholesterol-lowering drug lovastatin, genes were modified to increase their expression levels in order to change its metabolism in terms of drug production.

Another example is the introduction of bovine lactic acid pathway into *S. cerevisiae*. As a part of this, a gene responsible for speeding up removal of hydrogen, which participates in lactic acid production, was expressed in *S. cerevisiae*, and lactic acid was produced at rate of eleven grams per liter per hour.

Fascinating Facts About Metabolic Engineering

- In 2010, a team of scientists led by J. Craig Venter created the first synthetic cell. The team synthesized the artificial chromosome, which was then transplanted into the recipient cell. The artificial chromosome was able to take over the recipient cell. This research opens the door for creation of useful artificial cells to make products such as vaccines and biofuels.

- Antimalarial drug artemisinin has been produced from metabolically engineered laboratory yeast. The antimalarial comes from the *Artemisia annua* plant, which grows in Southeast Asia. Artemisinin could also possibly be used in cancer treatment.

- Scientists are able to synthesize large DNA molecules in the laboratory. Researchers have made artificial DNA containing all twenty-one genes encoding the small ribosomal subunit (cell protein factory) from *Escherichia coli* (*E. coli*).

- Using metabolic-engineering methods, researchers were able to modify *E. coli* bacterium to produce butanol. Among the types of biofuels that are on the road to commercialization, butanol has been the most promising. It is another alcohol fuel, but when compared with ethanol, it has higher energy content (roughly 80 percent of gasoline energy content). Butanol can also be stored and transported using existing infrastructure—and it does not occur naturally in *E. coli*.

Because it tolerates acid, yeast may serve as an alternative to bacteria, which is usually used in industry for lactic acid production. Lactic acid is widely used as a food preservative.

Altering Nutrient Uptake or Metabolic Flow. Alteration of nutrient uptake or metabolic flow can be done not only by genetic manipulation but also by using inhibitors—simple chemicals or physical factors such as light or temperature.

The alteration of molecular hydrogen (H_2) production in green algae using high-intensity light is an example of metabolic flow modification by physical factors. H_2 is one of the possible energy carriers of the future. Microscopic green algae produce H_2 in photosynthetic reactions from water using sunlight as an energy source, usually in anoxic (without oxygen) conditions. Oxygen (O_2) produced by photosynthesis in green algae is an inhibitor of H_2 production. Brief illumination of algal cells by high-intensity light was accompanied by rapid suppression of photosynthetic O_2 evolution. The decline in the rate of O_2 evolution was accompanied by stimulation of H_2 production in algal cells.

Production Systems. All of the above-mentioned considerations are very important in metabolic engineering, although it is also important to ensure that the production of desired compounds by modified cells can be reproduced. This can be achieved by using bioreactors, in which the important parameters such as pH, temperature, substrate supply, and other variables are controlled. It is even possible to modify cell metabolism by using bioreactors.

APPLICATIONS AND PRODUCTS

There are a wide range of metabolic engineering products and applications. Undoubtedly, a number of novel applications and products will arise in the future.

Pharmaceuticals. Metabolic engineering is most promising in the production of pharmaceuticals. These include pharmaceuticals from different classes of natural products: alkaloids, isoprenoids, and flavonoids. Biosynthesis of natural products is an emerging area of metabolic engineering that offers significant advantages over conventional chemical methods. Some pharmaceutical compounds are too complex to be chemically synthesized or extracted from biomass organisms inexpensively.

Alkaloids are mainly plant-derived compounds

that have been used as drugs such as morphine. Alkaloids are produced by simple extraction from plants. Studies show that alkaloids can be synthesized from amino acids by metabolic engineering in *E. coli* and *S. cerevisiae*.

Isoprenoids, organic compounds composed of two or more hydrocarbons, have a range of functions: pigments, fragrances, and vitamins. Isoprenoids are also the precursors to sex hormones. Many isoprenoids have been produced using microorganisms, including carotenoids and various plant-derived terpenes. Metabolic engineers are using *S. cerevisiae* as a cell factory for the biosynthesis of isoprenoids. One metabolic-engineering success is the production of Taxol, which is used to treat breast cancer. It is an isoprenoid that was first isolated in the bark of the Pacific yew (*Taxus brevifolia*). The demand for Taxol greatly exceeds the supply that can be obtained from its natural source. A partial Taxol biosynthetic pathway has been engineered in *S. cerevisiae*.

Another metabolic engineering success is the production of isoprenoids-carotenoids. Carotenoids are naturally occurring yellow, orange, and red pigments commonly found in plants such as carrots as well as in bacteria, algae, and fungi and play an important role in fighting disease. Metabolic engineers have successfully introduced carotenoid biochemical pathways into nonproducing carotenoid microbes such as *E. coli* and *S. cerevisiae*.

Flavonoids are a group of secondary plant metabolites. These compounds can be used as antioxidants or antiviral, antibacterial, and anticancer drugs. Many flavonoid biosynthetic pathways are known, and a wide array of flavonoid compounds from *S. cerevisiae* are expected to be produced by metabolic engineering in the near future.

Chemicals. Numerous chemicals, such as amino acids, organic acids, vitamins, flavors, fragrances, and nutraceuticals can be manufactured by metabolic engineering.

Glycerol (or glycerin) is a chemical produced by metabolic engineering. Glycerol is used to synthesize many products, ranging from cosmetics to lubricants. It is a by-product of soap or biodiesel manufacturing and its production is 1.2 billion liters annually. It can be also used a fuel. Metabolically engineered *S. cerevisiae* strain produced more than 200 grams of glycerol per liter of liquid medium.

Another example of chemicals produced with help of metabolic engineering are sterols. The most well-known sterol is cholesterol. Sterols are important for living organisms as they are a part of the cellular membrane, participate in the synthesis of several hormones, and are also nutrient supplements. Several sterols are being produced from metabolically engineered *S. cerevisiae*.

Fuels. Metabolic engineering can be used in the production of biofuels. Several scientific laboratories have demonstrated the feasibility of manipulating microorganisms to produce molecules similar to oil-derived products, although the yield is very low. Adjusting metabolic pathways of microbes to produce fuels similar to gasoline has the potential to save an enormous amount of money. These fuels can be used in existing engines, unlike other biofuels that require modified engines or fueling stations.

Several research groups are trying to metabolically engineer microorganisms to produce ethanol fuel using cellulose as substrate. Another example of the work of metabolic engineers is biodiesel production. Biodiesel is a diesel substitute primarily obtained from vegetable oils such as soybean. However, the production of this fuel is limited by the absence of sufficient vegetable oil feedstocks. Another problem is that in order to produce biodiesel, oils should be modified by transesterification, a chemical reaction with methanol, catalyzed by acids or bases (such as sodium hydroxide). *E. coli* has been metabolically engineered to produce biodiesel directly, using low-cost materials.

IMPACT ON INDUSTRY

Metabolic engineering may one day play a major role in a number of multibillion-dollar industries, including pharmaceutical, biotechnology, and biofuel. The United States maintains a dominant position in the world of metabolic engineering. The first large-scale industrial process of the human hormone insulin by metabolic engineering was developed in the United States. Other developed countries are researching the use of metabolic engineering to produce a variety of products, such as pharmaceuticals and biofuels.

Government and University Research. Governmental agencies such as the National Science Foundation (NSF) and the U.S. Department of Energy (DOE) provide funding for research in metabolic engineering. A vast majority of metabolic engineering research is concentrated in the areas of pharmaceuticals and biofuel generation. The Joint BioEnergy Institute (JBEI) is a major player in the metabolic engineering area. It is funded by the U.S. Department of Energy, and its main goal is to develop next-generation biofuels such as cellulosic ethanol. It is working in partnership with Sandia National Laboratories and the University of California, Berkeley. Other major players in metabolic engineering are University of Chicago and Rice University.

Industry and Business. Scientists in industry traditionally carry out a significant percentage of metabolic engineering research, a good portion of which is directed to health care products, such as pharmaceuticals. Successful scientific projects include creating of lysine, riboflavin, coenzyme Q-10, the aminoshikimate pathway, and beta-carotene by metabolically engineered microorganisms.

Major Corporations. At present, just a few companies are using metabolic engineering alone to achieve their goals, including Amyris, Integrated Genomics, and LS9. Some major biotechnological companies, such as Genentech, employ a metabolic engineering approach combined with traditional biotechnological techniques such as recombinant DNA technology.

CAREERS AND COURSE WORK

Biotechnology, pharmaceutical, and biofuel companies are the biggest employers of metabolic engineers. As new biology-based products move from research into production, more metabolic engineers will be needed in industry, universities, and government laboratories.

Metabolic engineering is an interdisciplinary science. Course work includes biochemistry, molecular biology, chemistry, genetic engineering, analytical chemistry, biochemistry, biochemical and bioprocess engineering, and microbiology. Most metabolic engineers have a bachelor of science in biology, biochemistry, genetic engineering, microbiology, or biotechnology. Advanced degrees (master's and doctorate) in molecular biology, biochemistry, biotechnology, or genetics are necessary for research and teaching positions. Some governmental institutions such as the Joint BioEnergy Institute help students to develop metabolic engineering educational paths by providing opportunities for internships.

SOCIAL CONTEXT AND FUTURE PROSPECTS

Though the redesign of life forms for the benefit of mankind is definitely an exciting career, metabolic engineers are paying particular attention to ethical, legal, and political issues. To continue in this work, the field as a whole will need sustained support from the public and government.

At present, metabolic engineering is more a collection of successful experiments than an established science. In the future, metabolic engineering may play a significant role in production of chemicals and fuels from inexpensive and renewable starting materials. Continued development of the techniques of metabolic engineering will be necessary to expand the range of products. The role of metabolic engineering in science is likely to expand in the future as a result of increasing needs for pharmaceuticals and biofuels.

Sergei A. Markov, Ph.D.

FURTHER READING

Bailey, James E., and David F. Ollis. *Biochemical Engineering Fundamentals.* 2d ed. New York: McGraw-Hill, 1986. Classic textbook on biochemical engineering.

Bourgaize, David, Thomas R. Jewell, and Rodolfo G. Buiser. *Biotechnology: Demystifying the Concepts.* San Francisco: Benjamin Cummings, 2000. Excellent introduction to biotechnology.

Lewin, Benjamin. *Genes VIII.* San Francisco: Benjamin Cummings, 2003. In-depth look at genes and molecular biology.

Madigan, Michael T., et al. *Brock Biology of Microorganisms.* 12th ed. San Francisco: Benjamin Cummings, 2008. Several chapters of this popular textbook describe microbial metabolism and the application of microorganisms in industry.

Marguet, Philippe, et al. "Biology by Design: Reduction and Synthesis of Cellular Components and Behavior." *Journal of the Royal Society Interface* 4, no. 15 (2007): 607-623. Review on metabolic engineering and synthetic biology written for the general public.

Ostergaard, Simon, Lisbeth Olsson, and Jens Nielsen. "Metabolic Engineering of *Saccharomyces cerevisiae*." *Microbiology and Molecular Biology Reviews* 64, no. 1 (2000): 34-50. Describes metabolic engineering techniques using *S. cerevisiae* as an example.

Stephanopoulos, Gregory N., Aristos A. Aristidou,

and Jens Nielsen. *Metabolic Engineering: Principles and Methodologies.* San Diego: Academic Press, 1998. Classic text on metabolic engineering.

WEB SITES
Biotech Career Center
http://www.biotechcareercenter.com/biotech.html

Biotechnology Industry Organization
http://www.bio.org

Nature Technology Corporation
http://www.natx.com

See also: Cloning; DNA Sequencing; Genetically Modified Organisms; Genetic Engineering.

METEOROLOGY

FIELDS OF STUDY

Climatology; hydrology; atmospheric physics; atmospheric chemistry, oceanography.

SUMMARY

Interdisciplinary study of physical phenomena occurring at various levels of the Earth's atmosphere. Practical applications of meteorological findings all relate in some way to understanding longer-term weather conditions. On the whole, however, meteorological weather forecasts are—in contrast to the research goals of climatology—mainly short term in nature. They concentrate on contributing factors, including temperature, humidity, atmospheric pressure, and winds.

KEY TERMS AND CONCEPTS

- **Cyclone:** Air mass closed in by spiraling (circular) winds that can become a moderate or violent storm, depending on factors of humidity, temperature, and the force, changing direction, and altitude of the winds.
- **Dewpoint:** Temperature at which air becomes saturated and produces dew.
- **Front:** Interface between air masses with different temperatures or densities.
- **Isotherms:** Graphically recorded lines connecting all points in a given region (large or more limited) having exactly the same temperature.
- **Jet Stream:** Narrow but very fast air current flowing around the globe at altitudes between 23,000 and 50,000 feet.
- **Ozone:** Variant form of oxygen made up of three atoms; forms a layer in the upper atmosphere that helps absorb ultraviolet rays from the Sun.
- **Relative Humidity:** Percentage value indicating how much water is in an air sample in relation to how much it can hold, or its saturation point, at a given temperature and pressure.
- **Saturation Point:** Point where the water vapor in the air is at its maximum for a given temperature and pressure; point where condensation occurs.
- **Stratosphere:** Atmospheric layer above the troposphere; temperatures in this layer increase as the altitude becomes greater.
- **Temperature Inversion:** Phenomenon that is the opposite of normal atmospheric conditions. When air close to ground level remains colder and denser than air at higher levels, the warmer air can form a sort of cover, trapping the colder air, with resultant increases in ground-level fog and smog.
- **Troposphere:** Lowest layer of the atmosphere; contains 80 percent of the total molecular mass of the atmosphere.

DEFINITION AND BASIC PRINCIPLES

Meteorology is the study of the Earth's atmosphere, particularly changes in atmospheric conditions. The three main factors affecting change in the atmosphere are humidity, temperature, and barometric pressure. Dynamic short-term interaction among these three atmospheric factors produces the various phenomena associated with changeable weather. Low barometric pressure conditions are generally associated with greater capacity for the atmosphere to absorb water vapor (resulting in various cloud formations), whereas high pressure prevents absorption of humidity.

Meteorological calculation of relative humidity, for example, reveals how much more moisture can be absorbed by the atmosphere at specific temperature levels before reaching the saturation point. Changes in temperature (either up or down) will affect this dynamic process. Rainfall occurs when colder air pushes warmer, moisture-laden air upward into higher altitudes, where the warmer air mass begins to cool. Cooler air cannot hold as much water as warmer air, so the relative humidity of the warmer air mass changes, resulting in condensation and precipitation. This phenomenon is closely associated with the presence of surface winds.

BACKGROUND AND HISTORY

Meteorology, like several other applied sciences that stem from observations of natural phenomena, has a long history. The term "meteorology" comes from the Greek word for "high in the sky." Aristotle's work on meteorology maintained that the Sun attracted two masses of air from the Earth's surface, one humid and moist (which returned as rain) and the other hot and dry (the source of wind currents). His student, Theophrastus, described distinct atmospheric signs associated with eighty types of rain, forty-five types of wind, and fifty storms.

During the Renaissance, Europeans developed instruments that could refine these ancient Greek theories. The Italian scientist Galileo, for example, used a closed glass container with a system of gauges that showed how air expands and contracts at different temperatures (the principle of the thermometer). The French philosopher Blaise Pascal developed what became the barometer, a device to measure surrounding levels of atmospheric pressure.

Although many important small-scale experiments would be carried out in the eighteenth and nineteenth centuries, a major breakthrough occurred in the first quarter of the twentieth century when the Swede Vilhelm Bjerknes and his son Jacob Bjerknes developed the theory of atmospheric fronts, involving large-scale interactions between cold and warmer air masses close to the Earth's surface. In the 1920's, a Japanese meteorologist first identified what came to be known as jet streams, or fast-moving air currents at altitudes of 23,000 and 50,000 feet.

The turning point for modern meteorology, however, came in April, 1960, when the United States launched TIROS 1, the first in a series of meteorological satellites. This revolutionary tool enabled scientists to study atmospheric phenomena such as radiation flux and balance that were known but had not been measured with high levels of accuracy.

How It Works

Meteorologists employ a variety of basic tools and methods to obtain the data needed to put together a comprehensive picture of local or regional atmospheric conditions and changes. At the most basic level, meteorologists direct their attention to three essential factors affecting the atmosphere: temperature, air pressure, and humidity.

Drawing on empirical data, meteorologists not only analyze the effects of temperature, pressure, and humidity in the area of the atmosphere they are studying but also apply their findings to ever-widening areas of the globe. From their analyses, they are able to predict the weather—for example, the direction and strength of the wind and the nature and the probable intensity of storms heading toward the area, even if they are still thousands of miles away.

Wind Strength and Direction. Wind, like rain and snowstorms, is a common weather phenomenon, but the meteorological explanations for wind are rather complicated. All winds, whether local or global, are the product of various patterns of atmospheric pressure. The most common, or horizontal, winds arise when a low pressure area draws air from a higher pressure zone.

To build a complete picture of the likely strengths and directions of winds, however, meteorologists must gather much more than simple barometric data. They must consider, for example, the dynamics of the Coriolis effect (the influence of the Earth's rotation on moving air). Except in the specific latitude of the equator, the Coriolis effect, which is greater near the poles and less near the equator, makes winds curve in a circular pattern. Normal curving from the Coriolis effect can be altered by another force, centrifugal acceleration, which is the result of the movement of air around high and low pressure areas (in opposite directions for high and low pressure areas). Tornadoes and hurricanes (giant cyclones) occur when centrifugal acceleration reaches very high levels.

When a near balance exists between the Coriolis effect and centrifugal acceleration, the resultant (still somewhat curved) wind pattern is called cyclostophic. If no frictional drag (deceleration associated with physical obstacles at lower elevations) exists, something close to a straight wind pattern occurs, especially at altitudes of about two-thirds of a mile and greater. This straight wind, called a geostrophic wind, is characterized by a balance between the pressure gradient and the Coriolis effect.

In some parts of the Northern Hemisphere, massive pressure gradient changes produced by differences in the temperature of the land and the ocean create monsoon winds (typically reversed from season to season), which in the summer are followed by storms and heavy rains. The best-known example of a monsoon occurs in India, where the rising heat of summer creates a subcontinent-wide thermal low pressure zone, which attracts the moisture-laden air from the Indian Ocean into cyclonic wind patterns and much needed, but sometimes catastrophic, heavy rainfall.

Atmospheric Absorption and Transfer of Heat. Meteorologists worldwide use various methods to determine how much heat from the Sun (solar radiation) actually reaches the Earth's surface. Heat values for solar radiation are calculated in relation to a universal reference, the solar constant. The solar constant (1.37 kilowatts of energy per square meter) represents the density of radiation at Earth's mean distance from the Sun and at the point just before the Sun's heat (shortwave infrared waves) enters the Earth's atmosphere. Not all this heat actually reaches Earth's surface. The actual amount is determined by various factors, including latitudinal location, the degree of cloud cover, and the presence in the atmosphere of trace gases that can absorb radiation, such as argon, ozone, sulfur dioxide, and carbon dioxide.

At the same time, data must be gathered to calculate the amount of heat leaving the Earth's surface, mainly in the form of (longwave) infrared rays. Meteorologists attempt to calculate, first on a global scale and then for specific geographic locations, ecologically appropriate energy budgets.

Cyclones. Cyclones include a number of forms of severe weather, the most violent of which is known as a tornado. Cyclones are characterized by circular or turning patterns of air centered on a zone of low atmospheric pressure. The direction of cyclone rotation is counterclockwise in the Northern Hemisphere but clockwise in the Southern Hemisphere. Frontal cyclones, the most common type, usually develop in association with low pressure troughs that form along the polar front, a front that separates arctic and polar air masses from tropical air masses. Typical cyclogenesis occurs when moisture-laden air above the center of a relatively warm low pressure area begins to rotate under the influence of converging and or diverging winds at the surface or at higher levels. Simply stated, the dynamic forces operating within a cyclone can pull broad weather fronts toward them, causing increasingly strong winds and precipitation. Anticyclones occur under opposite conditions, originating in areas of high pressure where air masses are pushed down from upper areas of the atmosphere.

APPLICATIONS AND PRODUCTS

Meteorology has both practical and professional, scientific applications. Everyday weather reports are probably the most common application of meteorology. Weather reports and forecasts are available through traditional sources such as newspapers, television, and radio broadcasts and can be obtained by calling the National Oceanic and Atmospheric Administration's National Weather Service, using a smartphone application, or checking one of the many Web sites devoted to weather. Real-time weather reports, including radar, and hourly, daily, and ten-day forecasts are available for the United States and other parts of the world.

Knowledge of existing and forecasted weather conditions is invaluable to companies that provide public transportation, such as airlines. Knowledge of weather conditions and forecasts is essential to ensuring the safety of passengers on airplanes, trains, buses, boats, and ferries. Motorists, whether traveling for pleasure or commuting, need to plan for weather conditions. People who participate in outdoor recreational activities or sports, such as hiking, biking, camping, hang gliding, fishing, and boating, depend on accurate forecasts to ensure that they are adequately prepared for the conditions and to avoid getting into dangerous situations. Those participating in outdoor activities and sports often purchase various meteorological instruments—such as lightning detectors, weather alert radios, and digital weather stations—designed to keep them informed and aware.

Weather Equipment. Meteorology involves the measurement of temperature, air pressure, and humidity, and many companies produce equipment for this purpose, ranging from portable products for outdoor use, to products for home use, to products for industry use, to weather balloons and satellites. Some companies, such as Columbia Weather Systems in Oregon, which provides weather station systems and monitoring, focus on the professional level, and others, including Oregon Scientific, Davis Instruments, and La Crosse Technology, concentrate on the many hobbyists who enjoy monitoring the weather. Equipment ranges from simple mechanical rain gauges to digital temperature sensors to complete home weather stations. Some companies offer software that provides specific weather information in a timely manner to companies or individuals who need it.

Scientific Applications. The World Data Center in Asheville, North Carolina, distributes meteorological data that it has gathered and processed. Its facilities are open on a limited basis to visiting research scientists sponsored by recognized parent organizations or international programs. A variety of meteorological data are readily available on an exchange basis to counterparts of the center in other countries.

One organization participating in the information exchange is the World Climate Programme, an international program headquartered in Switzerland devoted to understanding the climate system and using its knowledge to help countries that are dealing with climate change and variability.

IMPACT ON INDUSTRY

A wide range of industries and commercial concerns depend on meteorological data to carry out their operations. These include commercial airlines, the fishing industry, and agricultural businesses.

Aviation. Perhaps the most obvious industry that relies on the results of meteorology is the airline industry. No flight, whether private, commercial, or military, is undertaken without a clear idea of the predicted weather conditions from point of departure to point of arrival. Changing conditions and unexpected pockets of air turbulence are a constant concern of pilots and their crews, who are trained in methods of analyzing meteorological data while in flight.

Every day, the National Weather Service issues nearly 4,000 weather forecasts for aviation. Meteorologists at the Aviation Weather Center in Kansas City, Missouri; the Alaska Aviation Weather Center in Anchorage; and at twenty-one Federal Aviation Administration Air Route Traffic Control Centers across the nation use images from satellites as well as real-time information from observation units at major airports and from Doppler radar to create reliable forecasts.

Fishing Industry. Rapidly changing and violent weather on the seas and oceans can result in the loss of fishing vessels, their crews, and their catch, or it can cause severe damage to boats and ships, all of which hurt the fishing industry. In addition, when boats get caught in dangerous storms, rescuers must attempt to reach survivors, often endangering themselves. Therefore, accurate, up-to-the-minute forecasts are essential for the fishing industry, which also uses weather reports to help determine the areas where the fish are most likely to be found.

Agriculture. In agriculture, short meteorological predictions are of less importance than longer-term predictions, particularly for agricultural sectors that depend on rainfall rather than irrigation to grow their crops and that rely on predictable windows without rainfall to harvest and dry them. If the possibility of a prolonged drought menaces crops, both large- and small-scale agriculturalists turn to meteorologists to learn if changing weather patterns may bring them relief. However, short-term predictions are helpful, in that they help farmers deal with problems that could create catastrophic losses, such as too much rain in a short period, which could create flooding; extreme heat or cold, which, for example, can ruin the citrus crop; or strong winds and hail, which can flatten corn.

CAREERS AND COURSE WORK

Those seeking a career in meteorology can consider a wide range of possibilities. The American Meteorological Association requires at least twenty semester credits in the sciences, which include geophysics, earth science, physics, and chemistry, as well as computer science and mathematics. Specialized courses beyond such basic science courses (such as atmospheric dynamics, physical meteorology, synoptic meteorology, and hydrology) represent slightly over half the required courses. A number of governmental agencies and private commercial businesses employ full- or part-time meteorologists to provide technical information needed to carry out their operations.

Probably the most familiar job of a meteorologist is to prepare weather reports for broadcast on television or to be published in newspapers. The task of a weather forecaster for a local television station is more complex than it may seem judging from the briefness of the televised forecast. Local predictions are created from data gathered from diverse sources, often sources that gather information for an entire region or the whole country, including pulse-doppler radar operators, who receive special training to enter the profession.

Government Agencies. Opportunities for employment in agencies that gather meteorological data are to be found within the National Oceanic and Atmospheric Administration (NOAA), part of the U.S. Department of Commerce. The NOAA contains a number of organizations that deal with meteorology, including the National Environmental Satellite, Data, and Information Service, a specialized satellite technology branch that provides global environmental data and assessments of the environment. Another NOAA organization is the National Weather Service, which provides forecasts, maps, and information on water and air quality. The Office of Oceanic and Atmospheric Research contains the Climate Program Office, which studies climate variability and predictability.

The National Centers for Environmental Prediction, part of the National Weather Service, oversees operations by several key specialized service organizations run by professional meteorologists. The most important of these are the Hydrometeorological Prediction Center in Washington, D.C., the Tropical Prediction Center (includes the National Hurricane Center in Miami, Florida), and the Storm Prediction Center in Norman, Oklahoma, which maintains a constant tornado alert system for vulnerable geographical regions. Other operations include the Ocean Prediction Center, the Aviation Weather Center, the Climate Prediction Center, Environmental Modeling Center, and the Space Weather Prediction Center.

Private Commercial Operations. A number of private companies are involved in the development and production of instruments used for meteorological data gathering. The instruments range from simple devices to sophisticated electronic equipment, complete with software. The development of new weather satellites requires instruments that can be used by and can make the most of the gathered information.

SOCIAL CONTEXT AND FUTURE PROSPECTS

The worldwide need for accurate daily and short-term meteorological forecasts, delivered in various formats, will result in continuing development of more accurate instruments and better predictive models, as well as improved and additional methods of packaging and delivering the information to users. Accurate prediction of where extreme weather, such as hurricanes and tornadoes, will occur and how intense the storms will be has the potential to save many lives and possibly minimize economic damage, and meteorologists will continue to conduct research in this area. Data gathered by meteorologists also are gaining importance in analyzing global issues such as air pollution and changes in the ozone layer. The ability to gather information by satellites allowed meteorology to make significant advances, and future research is likely to focus on increasing satellites' data-sensing ability and the speed and quality of data transmission as well as on software to interpret and analyze the information.

Satellite Technology. Development of satellite technology continues to revolutionize the science of meteorology. Various types of polar-orbiting satellites have been devised since the early 1960's. The task of the satellite bus—the computer-equipped part of the

---

Fascinating Facts About Meteorology

- The 53rd Weather Reconnaissance Squadron, known as the Hurricane Hunters of the Air Force Reserve, has been flying into tropical storms and hurricanes since 1944. The squadron's planes fly into the storms and send data directly to the National Hurricane Center by satellite.

- Geology departments at some colleges and universities, such as Ball State University in Muncie, Indiana, have led storm chasing groups, which train students in basic storm knowledge and lead them in pursuit of tornadoes and other violent storms.

- The popularity of storm chasing, featured in a Discovery channel television series, has led to the creation of storm chasing commercial tours. One group warns potential customers that while storm chasing, the group may not be able to stop for dinner.

- Without the continual pull of gravity toward the center of the Earth, the atmosphere would gradually disperse into space. Because air is made up of molecules

- in a gaseous rather than solid material state, it clings to the Earth.

- The chemical content of sedimentary layers on ocean floors and gases trapped in Arctic ice reveal data concerning the composition of the Earth's atmosphere in earlier geological times.

- The amount of oxygen contained in the air in higher mountain environments is markedly less than at sea level or mid-range altitudes. People coming to the mountains from lower elevations can experience serious shortness of breath and even altitude sickness, which is characterized by headaches, loss of appetite, fatigue, and nausea.

- At times, the atmosphere cannot keep up with the Earth it surrounds. This is particularly true at higher levels of the atmosphere, especially near the equator. At the same time, some parts of the atmosphere travel faster than others because of strong jet stream winds.

satellite without sensitive recording instruments—is to transmit data gathered by an increasingly sophisticated variety of devices designed to collect vital data.

One such scanning device is the advanced very-high-resolution radiometer (AVHRR) used to measure heat radiation rising from localized areas on the Earth's surface. An AVHRR, very much like a telescope, projects a beam that is split by a set of mirrors, lenses, and filters that distribute the work of data recording to several different sensor devices, or channels. To obtain an accurate reading of radiation rising from a given target, the AVHRR must calibrate data received from these different sensors. AVHRR technology produces images that depict the horizontal structure of the atmosphere, and another radiation-recording instrument, the high-resolution infrared radiation sounder (HIRS), produces soundings based on the vertical structure of the atmosphere.

In zones of widespread cloud cover, the microwave sounding unit (MSU) may be used. The area scanned by an MSU is about a thousand times greater than that scanned by a device using infrared wavelengths. Other specialized sounding devices used by meteorological satellites include stratospheric sounding units and solar backscatter ultraviolet radiometers (SBUV), which measure patterns of reflection of solar radiation coming back from the Earth's surface.

The functions of SBUVs in particular became more and more critical as concern about changes in the composition of the ozone layer emerged in the last decades of the twentieth century. Ozone-depleting substances such as chlorofluorocarbons (used for air-conditioning systems and as a propellant for aerosol sprays) have had a negative effect on the protective ozone layer. Ozone depletion allows increased levels of ultraviolet radiation to reach the Earth, raising the incidence of skin cancer and damaging sensitive crops. Beyond the obvious utility of using satellites to predict global weather, these devices play a major role in helping meteorologists monitor the all-important radiation budget (the balance between incoming energy from the Sun and outgoing thermal and reflected energy from the Earth) that ultimately determines the effectiveness of the atmosphere in sustaining life on Earth.

Air Pollution. Although the presence of pollutants in the air has been recognized as an undesirable phenomenon since the onset of the Industrial Revolution, by the end of the twentieth century, the question of air quality began to take on new and alarming dimensions. As countries industrialized, the combustion of fossil fuels to power industry and later automobiles released ever-increasing amounts of solar-radiation-absorbing greenhouse gases into the atmosphere. The most alarming effects stem from carbon dioxide, but meteorologists are also concerned about other serious gaseous pollutants and a wide variety of chemical particles suspended in gases (aerosols). Even if industries do not pollute by burning fossil fuels, many of them release sulfur dioxide, asbestos, and silica in quantities large enough to seriously damage air quality. Natural phenomena, including massive volcanic eruptions, also can produce atmospheric chemical imbalances that challenge analysis by meteorologists.

Byron D. Cannon, M.A., Ph.D.

FURTHER READING

Budyko, M. I., A. B. Ronov, and A. L. Yanshin. *History of the Earth's Atmosphere.* Berlin: Springer-Verlag, 1987. Deals with various ways scientists can determine the probable composition of the atmosphere in earlier geological ages.

David, Laurie, and Gordon Cambria. *The Down-to-Earth Guide to Global Warming.* New York: Orchard Books, 2007. A nontechnical discussion of practical factors, many daily and seemingly inconsequential, that contribute to rising levels of carbon dioxide in the atmosphere.

Fry, Juliane L, et al. *The Encyclopedia of Weather and Climate Change: A Complete Visual Guide.* Berkeley: University of California Press, 2010. Covers all aspects of weather, including what it is and how it is monitored, as well as the history of climate change. Many photographs, diagrams, and illustrations.

Grenci, Lee M., Jon M. Nese, and David M. Babb. 5th ed. *A World of Weather: Fundamentals of Meteorology.* Dubuque, Iowa: Kendall/Hunt, 2010. Examines the study of meteorology and how it is monitored.

Hewitt, C.N., and Andrea Jackson, eds. *Handbook of Atmospheric Science.* Oxford, England: Blackwell, 2003. Examines research approaches to and data on the chemical content of the atmosphere.

Kidder, Stanley Q., and Thomas H. Vonder Haar. *Satellite Meteorology.* 1995. Reprint. San Diego: Academic Press, 2008. Covers ways in which satellite instruments can monitor meteorological phenomena at several levels of the atmosphere.

Williams, Jack, and the American Meteorological Society. *The AMS Weather Book: The Ultimate Guide to America's Weather.* Chicago: University of Chicago Press, 2009. Examines common weather patterns in the United States and discusses the science of meteorology.

WEB SITES
Air Force Weather Observer
http://www.afweather.af.mil/index.asp

American Meteorological Society
http://www.ametsoc.org

International Association of Meteorology and Atmospheric Sciences
http://www.iamas.org

National Oceanic and Atmospheric Administration
National Weather Service
http://www.weather.gov

World Meteorological Organization
http://www.wmo.int/pages/index_en.html

See also: Atmospheric Sciences; Barometry.

MICROSCOPY

FIELDS OF STUDY

Electronics; electrical engineering; physics; optical physics; atomic physics; mathematics; statistics; image analysis; materials science; photomicrography; interferometry; electromagnetics; quantum electrodynamics; computer science; nanotechnology; metallography; electron microscopy; optical microscopy; scanning probe microscopy; cell biology; chemistry.

SUMMARY

Microscopy is the science of creating, observing, analyzing, and capturing visible images of objects and their components that are too small to be seen by the naked eye. It also refers to research conducted with the aid of microscopes, or instruments used for visual magnification. Microscopy is an essential tool for conducting research in a large number of scientific disciplines, including chemistry, biology, and medicine. For example, microscopy enables biologists to examine, in fine detail, the structure and function of individual components of a cell. The field also has a variety of industrial, materials science, and other practical applications. Powerful microscopes are used, for instance, to inspect the composition of the tiny silicon crystals used to manufacture semiconductors and integrated circuits and to detect minute defects in glass.

KEY TERMS AND CONCEPTS

- **Cantilever:** Flexible bracket to which the tip of an atomic force microscope is attached.
- **Compound Microscope:** Optical microscope containing two lenses, the objective and the eyepiece (or ocular lens).
- **Condenser:** Lens in a microscope that focuses, or condenses, a beam of light or electrons onto the object being studied.
- **Magnifying Power:** Ability of a lens or lenses to enlarge an object; measured in the number of times the magnified image is larger than the object appears to the naked eye.
- **Numerical Aperture (NA):** Measure of how much light an objective lens in an optical microscope is capable of gathering.

- **Objective:** Lens located closest to the object being studied; an objective is the primary magnifying lens in an optical microscope.
- **Ocular Lens:** Upper lens, or eyepiece, in a microscope, through which the viewer looks.
- **Optical Aberrations:** Errors in the image produced by an optical microscope, caused by the inability of a lens to accurately focus rays of light.
- **Photomicrography:** Photography of magnified objects through a microscope.
- **Resolving Power:** Extent to which a particular lens or microscope is able to distinguish between two points (to see them as separate from each other) at a given distance.
- **Stereomicroscope:** Microscope with two optical systems, one for each eye. Objects viewed through a stereomicroscope appear three-dimensional.
- **Working Distance:** Distance between the objective lens and the object being studied—at the point when the magnified image is in clear focus.

DEFINITION AND BASIC PRINCIPLES

Microscopes are scientific instruments whose purpose is to create enlarged visual images of objects so tiny they cannot be seen by the unaided human eye. Microscopy is an applied science concerned with ways of developing and improving microscope technology and relies heavily on knowledge gained from physics, mathematics, and engineering. Different varieties of microscopes function in various ways, but it is useful to understand two basic principles that apply to how well a microscope performs.

First, it might seem that the fundamental purpose of a microscope is to magnify an object. In reality, however, photographic enlargements can always be used to further enlarge the image any given microscope creates. Therefore, for a microscope to be truly useful to scientists and other researchers, it must not only magnify the specimen being observed but also properly separate (or "resolve") the details of individual components within the image. In effect, the more resolving power a microscope has, the crisper and clearer the magnified image it can produce and the more information it can provide about the object in question. If the resolution of a microscope is not high enough, for instance, two tiny dots next to each other might be perceived as a single element, no matter how much the image of the specimen was magnified.

Second, the image a microscope creates must possess a high enough degree of contrast to allow the viewer to clearly distinguish the object from its background and to differentiate various details within the object. A mostly translucent specimen, for example, might be impossible to make out against a bright background. Special techniques are used to increase the contrast in microscopic images. Phase contrast microscopy takes advantage of differences in the refractive indexes (the extent to which a material bends light) of various components of the specimen. Microscopes using this technology translate these variations into differences in the amplitude of the light waves reflected from each component. This results in light and dark areas that can be seen by the viewer. Interferometry is another important technique used to increase contrast. It does so by creating two images of a single specimen, superimposed on top of each other.

BACKGROUND AND HISTORY

The first ground glass lenses that had the ability to magnify objects were created in the late Middle Ages by monks who used them for reading. In the sixteenth century, the first microscopes were created in the Netherlands by inventors Hans Janssen and Zacharias Janssen, who placed two lenses into a series of tubes to create a primitive compound microscope. The device was focused by drawing one of the tubes in and out of the other, and it had a magnifying power of about 10 times (10x). The seventeenth century saw a flurry of interest in microscope technology. The Dutch amateur scientist Antoni van Leeuwenhoek designed hundreds of microscopes in which a bi-convex lens was placed between two glass plates. He was the first person to ever observe bacteria and other single-celled organisms under a microscope. In the eighteenth century, cuff-style microscopes were created, whose design prefigured the modern laboratory optical microscope. These were instruments with two brass tubes, one fixed and one sliding. By sliding the assembly up and down and turning a small thumbscrew, the object under observation could be brought into fine focus. The whole mechanism was mounted on a solid wooden base and had a magnifying power of up to 100x.

In the nineteenth century, advances in optical science and lens production pushed microscope technology forward by leaps and bounds. Two people whose work was prominent in this era were the German physicists and engineers Ernst Abbe and Carl Zeiss; together they designed sophisticated lenses that cut down drastically on spherical aberration (which causes points of lights to look like discs) and chromatic aberration (which distorts the colors of objects). The twentieth and twenty-first centuries have seen major changes in the field of microscopy, with the development of new technologies such as electron and scanning probe microscopes and huge improvements in the design of optical microscopy. Digital imaging—the transformation of an image created in an optical microscope to digital form—is making it easier than ever to analyze magnified specimens.

HOW IT WORKS

Since the 1800's, hundreds of different varieties of microscopic technologies have been developed, each useful for performing certain types of observations on specific kinds of materials. Three broad categories of microscopy exist into which the majority of these technologies can be categorized.

Optical Microscopy. Optical microscopes, also known as light microscopes, create a magnified image by using a series of glass lenses to manipulate visible light, or light from the portion of the electromagnetic spectrum that can be seen by the naked eye. A condenser focuses light onto the specimen to be observed. As this light passes through or is reflected off the specimen, it is collected by one or more objectives. (Most microscopes have multiple objectives contained in a long tube with magnification powers ranging from 4x to 100x.) The objectives focus the light they have gathered into parallel rays; the result is a magnified image of the specimen. This image, however, is projected to a distance of infinity. To focus the image at a distance comfortable for the human eye, an ocular lens is required, through which the viewer looks. Most ocular lenses further magnify the image by another 10x.

In a conventional bright-field optical microscope, the source of this light is usually an incandescent or halogen lightbulb positioned directly below the specimen to be examined. Bright-field microscopy is useful for observing specimens that are either naturally dark or can be stained a dark color—such as cells or thin cross sections of biological material, usually placed on a glass slide. Images produced by a bright-field microscope appear dark against a bright white background.

Some specimens, such as living organisms, are difficult to see under bright-field microscopes. Other forms of optical microscopy, such as dark-field microscopy, have been developed to combat this problem. In dark-field microscopy, opaque material inside the condenser blocks the most central source of light, causing light to hit the specimen at oblique angles. When this angled light hits even the tiniest particle, the light scatters and makes the particle visible—like dust motes catching the angled light coming in from a window. Dark-field microscopes show specimens as bright points of light against a dark background.

Fluorescence microscopy is a special form of optical microscopy in which the specimen itself acts as the source of light. A fluorescence microscope irradiates the specimen with light of a certain wavelength, causing its atoms to become excited and emit energy as visible light. Some specimens, like chlorophyll, fluoresce naturally; others can be made to fluoresce through the use of chemicals. Other special forms of optical microscopy include phase contrast microscopy, in which small changes in the wavelength of light as it passes through transparent regions of the specimen are intensified so that they show up as areas of greater brightness, and confocal microscopy, in which light coming from out-of-focus regions of the specimen is filtered out of the final image through a pinhole aperture, eliminating blurry regions in the image.

Electron Microscopy. Electron microscopes operate using the same basic principles as optical microscopes—with one important difference. Where optical microscopy manipulates focused beams of visible light, electron microscopy manipulates focused beams of highly excited electrons, which have wavelengths much shorter than those of visible light. This technique enables objects to be magnified at far higher levels and resolved in far finer detail than optical microscopy. In addition, electron microscopes have larger depths of field, allowing a larger area of an object to be in focus at one time. The source of the electrons in an electron microscope is most often a thermionic electron gun—a device that shoots out a stream of electrons produced by heating a charged electrode. The path these electrons take is shaped by a series of lenses, just as in an optical microscope. However, rather than being made out of glass, the lenses in an electron microscope consist of coils of wire (solenoids). An electric current passed through a solenoid creates an electromagnetic field that can direct the flow of electrons and focus it into a thin beam that can be directed toward the object under study.

In a transmission electron microscope (TEM), the beam of electrons enters the specimen and passes through it. When electrons hit dense regions of the specimen, they bounce off and are not included in the resulting image. The remaining electrons travel through the object and then pass through more electromagnetic lenses that create a final, magnified image on a fluorescent screen. In a scanning electron microscope (SEM), the beam of electrons sweeps across the surface of the specimen in a back-and-forth pattern of parallel lines known as a raster. As the beam scans over the object, it causes atoms within it to become excited and emit electrons that escape from the object. These electrons, known as deflected secondary electrons, are collected, counted, and measured. The information from this analysis is then used to create a magnified pixelized image on a computer screen. Because the electron beam sweeps over the entire surface of the object under study, electron microscopes are able to produce an image of the specimen's structure in three dimensions. Electron microscopes generally require samples to be placed in a vacuum in order to operate because molecules of air might disturb the movement of the electrons used to form images.

Scanning Probe Microscopy. Scanning probe microscopy abandons lenses altogether and makes use of very fine mechanical tips, or probes, attached to a cantilever. The probes delicately scan back and forth over the surface of the specimen being studied in order to inspect it. Scanning probe microscopes can deliver information about not only the topography of an object but also its internal properties. Some can even map a specimen's properties on a nanoscale.

Scanning tunneling microscopes (STM) rely on a phenomenon discovered by quantum mechanics, in which electrons—which have wavelike properties—are able to "tunnel" outside of the surface of a solid object into surrounding space. Scanning tunneling microscopes have incredibly sharp metallic probes, often made of tungsten or an alloy of platinum and iridium, with tips that are a mere one or two atoms in size. The tip of the probe does not touch the surface of the specimen but is held very close to it. An electric current of low voltage is applied to the gap between the two. In response to the current, electrons from the object tunnel across the gap. Changes in the intensity of the tunneling electrons are analyzed to produce an image of the object that can then be magnified. Scanning tunneling microscopes can be used only to examine specimens that conduct electricity.

Atomic force microscopes (AFM), whose tips are typically made of silicon or diamond, can probe surfaces made of practically any material. As the tip of an atomic force microscope is dragged across the surface of a specimen, it is either deflected by or drawn toward the object, depending on whether the atoms in the object are repelled by or attracted to the microscope's tip. By measuring these forces, a magnified representation of the physical structure of the sample can be created. By changing the modes in which these microscopes operate, different properties such as magnetism, friction, and electrical conductivity can be assessed.

Near-field scanning optical microscopes (NSOM) use a probe that emits an incredibly fine beam of laser light very close to a specimen. These microscopes use the intensity of the reflected light to produce a magnified topographical image of the object. Scanning probe acoustic microscopes (SPAM) direct a focused, ultrasonic (high-pitched) sound wave toward the specimen being observed and form a magnified image of it based on how and how much the wave is reflected by the object's surface.

APPLICATIONS AND PRODUCTS

Scientific Research. At heart, all scientific research rests on the power of observation. Microscopes make it possible for scientists from fields such as biology, chemistry, metallurgy, mineralogy, and countless other disciplines to make more accurate and more complete observations of microscopic structures. Biologists, for example, use fluorescent microscopes to analyze the structure and function of minute intracellular organelles such as ribosomes, mitochondria, and even single strands of DNA. Using the microscope, researchers have been able to watch as individual cells undergo mitosis and viruses invade healthy cells to spread their own genetic material and

Fascinating Facts About Microscopy

- Microscopes need not be large themselves in order to enlarge other things. One of the world's smallest microscopes, the Cellvizio microscope, is less than one-tenth of an inch in diameter and can be inserted down a patient's throat to observe live cells.

- In 1986, no fewer than three recipients of the Nobel Prize in Physics were awarded their honors based on their work in improving microscopy technology. Ernst Ruska won for the invention of the electron microscope, and Gerd Binnig and Heinrich Rohrer were recognized for the invention of the scanning tunneling microscope.

- Microscopes have helped chemists at the University of California, Irvine, detect the presence of fat in strands of human hair. Their experiment seeks to determine whether fat is a natural component of hair or is deposited in hair by hair-care products.

- Atomic force microscopes, which probe the forces between atoms to produce an image of incredibly tiny particles, enable scientists to look at—and even pick up and move around—single strands of DNA or individual atoms.

- By examining either a rough or a cut-and-polished gem beneath a powerful microscope, a gemologist can easily tell whether it is an authentic natural stone or one that has been synthetically manufactured.

- Swiss inventor George de Mestral first got the idea for Velcro hook-and-loop fasteners when he used a microscope to examine the intricate hook-and-loop structure of the tiny burrs that had gotten firmly caught on his pant legs while he was walking through the forest.

- In 2008, a microscope attached to the National Aeronautics and Space Administration's Phoenix Mars Lander took a photograph of a single particle of the incredibly fine red dust that swirls around Mars and forms its soil. Dust particles on Mars are about 100 nanometers, about one-thousandth the width of a human hair—or even smaller.

also to identify the precise manner in which different kinds of proteins are folded. Analytical chemists and physicists use microscopes to conduct research at the scale of the atom or even on a nanoscale. For instance, researchers use scanning tunneling microscopy and atomic force microscopy to observe how peptides—organic compounds composed of two or more amino acids—interact with carbon and graphite nanotubes.

Electron microscopes are being used by botanical researchers to examine how leaves protect themselves from insects by forming crystals inside themselves, by ornithologists to figure out how minute structures on the surface of certain bird feathers create an iridescent effect, and by geoscientists to identify the weather-induced changes to geological features such as rocks. The field is even enabling complex scientific research to take place on other planets. Robotic space vehicles such as the Mars Exploration Rover are often equipped with autonomously operated scanning probe microscopes capable of studying the properties associated with the surfaces encountered by the vehicle. In all these applications, microscopy allows investigators to transcend the limitations of the human sense of sight and expand people's scientific understanding of the world.

Microscopy has a place in many applied sciences as well. It is a useful tool in food science, where it has provided a better understanding of the chemical properties of foods and how processing them alters their natural properties and also has helped isolate food contaminants. Microscopes enable materials scientists to analyze the three-dimensional structure of the plastics and polymers they are developing. Mechanical engineers use microscopes to develop sharper and more sophisticated edges on tools used for cutting.

Microscopy is indispensable in forensic science, where it is used to examine crime scene evidence such as blood, hair, dust, fingerprints, tiny shards of glass, and threads of fiber. Criminologists use high-powered microscopes to help them study the minute hand motions that were used to construct signatures on suspicious documents, looking for frequent stops and starts or other signs of possible forgery. Counterfeit currency makers are often foiled by microscopes, which help scientists detect very subtle discrepancies in the color and texture of the paper fibers used to manufacture counterfeit bills.

Medical Applications. Virtually all biomedical and bioengineering research projects make use of microscopy at some point. Scientists in the pharmaceutical industry, for example, need to closely examine the physical structure and dispersion characteristics of the active components (chemicals) used in the development of drugs, as well as the materials used to coat medical devices such as pacemakers or other implants. Fluorescence microscopy and confocal microscopy are commonly used for these purposes.

Besides their use in preclinical biomedical research, microscopes play a role in at least two other important areas of medical practice: diagnosis and surgery. In diagnosis, microscopes help physicians and laboratory technicians detect whether cells in a patient's tissue samples show signs of disease. When a female patient undergoes a Pap smear, for example, cells are taken from her cervix and analyzed under an optical or electron microscope. If the sample is cancerous or precancerous, a microscopic examination will show changes in the cervical cells that make them look flat or scaly. Microscopes are also used to detect the presence of pathogens in tissue samples. For example, blood cells from patients who have been infected by the malaria virus may appear enlarged or stippled; the malaria parasites themselves will also be visible under magnification of about 100x.

Surgery with the use of an operating microscope (sometimes called microsurgery) has become common in nearly all surgical fields, but it is vital for performing many brain, eye, and ear surgeries. Magnifying the sometimes minute biological structures involved in a procedure can help a surgeon perform delicate tasks that were practically impossible before the age of microscopes. One particularly significant microsurgery application, for example, is the ability to reattach limbs, fingers, or toes that have been severed from a patient's body. By magnifying the individual nerve fibers, blood vessels, and tendons both in the severed part and at the site of separation, a surgeon can connect them one by one.

The typical microsurgery setup involves a surgeon looking at an operating site through a microscope (or sometimes a television screen connected to the microscope) rather than facing the site directly. Many microsurgery procedures are minimally invasive, making use of instruments inserted through small cuts, or ports, in a patient's skin. Often, robotic instrumentation is used in conjunction with microsurgical tools to track a surgeon's hand motions and correct for tiny tremors in his or her movements. Using a combination of microscopes, remotely controlled tools, and large video screens, surgeons can conduct coronary artery bypasses without ever opening up a patient's chest.

Nanotechnology. Nanotechnology is an example of a scientific discipline whose very existence simply would not be possible without the use of extremely powerful microscopes. This emerging field takes advantage of the special ways in which molecules behave at the nanoscale to create nanoscale machinery such as tiny sensors that can detect and tally the number of specific types of molecules in a sample of chemicals or nanoparticles that systematically seek out and destroy cancerous cells within a patient's body. Optical microscopes do not have the magnifying power necessary to clearly resolve objects at the nanoscale (about 1-100 nanometers, or about one-thousandth the width of a strand of human hair), so electron microscopes and scanning probe microscopes serve as the foundational tools of nanoscientists.

Atomic force microscopes are particularly important in nanotechnology for several reasons. Like electron microscopes, they are capable of imaging structures that are incredibly small (including single atoms or molecules). Unlike electron microscopes, they do not have to operate in a vacuum, giving scientists the ability to work with a greater variety of samples, such as living biological cells. Most significantly, researchers can use the probes attached to atomic force microscopes not just for observing specimens but also for actually manipulating them.

Although most nanotechnology applications are still in the research and development stage, nanotechnology is already causing a transformation in manufacturing. With the help of microscopes that can characterize the behavior of nanoscale structures, scientists have created, for example, grease- and mildew-resistant paints, bacteria-killing storage containers, and nanoscale drug-delivery systems that introduce drugs directly into cells affected by disease.

Microscopy and Art. The applications of microscopy stretch far beyond the boundaries of science. Art historians make extensive use of microscopes to study paintings, sculptures, and other works of art in minute detail, using them to uncover insights about materials, artistic techniques, and what a piece has been through over the course of its history. For example, they often conduct microscopic analyses of the chemical and structural properties of the specific pigments used to create a painting. Museums also use microscopic inspections to authenticate artworks

and accurately date them. Using an optical microscope, art historians in Belgium were able to detect minute quantities of a cobalt blue pigment mixed in with an ultramarine pigment in a painting that had been attributed to the Dutch master painter Jan Vermeer. However, Vermeer lived and worked in the seventeenth century, and cobalt blue pigment was not developed until the nineteenth century. Without the ability to scrutinize the precise morphology and crystalline structure of the pigments involved, the historians would never have been able to determine that the painting was, in fact, a forgery.

IMPACT ON INDUSTRY

Manufacturing. In many manufacturing industries, microscopes are essential for various stages of production and inspection. By examining materials and finished products on a microscopic scale, manufacturers can catch flaws, remove contaminants, and ensure the quality and safety of their products. They can also easily assemble products whose individual components are too small to be seen by the naked eye. For example, high-powered microscopes are essential for the manufacture of the silicon wafer microchips—whose circuits can be as small as 0.001 millimeter—used in computers and other electronic devices. Microscopes are particularly important in metallurgical industries because incredibly tiny discrepancies in the crystal structure of a metal alloy can cause significant differences in its physical properties, including hardness, toughness (how likely it is that small fissures in a metal will expand and cause it to break), and tensile strength (how much the metal can stretch before it is unable to return to its original shape). Both transmission and scanning electron microscopes are commonly used in the metal industry to examine the microscopic crystals, or grains, in samples of metal under production as a quality-control measure.

Product Development. Microscopy is an important tool in the research and development stages of many different products. Optical, electron, and atomic force microscopy enable manufacturers of contact lenses and intraocular implants, for example, to characterize features such as the topography, adhesive quality, hardness, elasticity, and viscosity of a lens, as well as to determine how uniformly it has been made. Cosmetics firms use microscope technology to help them evaluate the effectiveness of products ranging from face creams to skin whiteners. Electron microscopes, for example, help biochemists determine how well shampoos smooth down the rough edges of hair, making it feel softer and look glossier.

Microscope Market. The sale of electron microscopes of all kinds is responsible for generating the biggest revenues in the market for microscopy products, largely because these machines are relatively large and expensive—the most sophisticated can cost up to $1 million—in comparison with other types of microscopes. Optical microscopes tend to dominate in contexts in which high-powered microscopes are not necessarily required, such as schools and smaller research laboratories. Scanning probe microscopes are less common than both electron and optical microscopes because they tend to be used in more specialized fields. The two biggest markets for microscopy applications are the life sciences research sector and the biomedical sector, including pharmaceutical companies and medical facilities. Other major consumers of microscope technology include manufacturers of semiconductors and textiles. Among the most important global corporations involved in the development, manufacture, and sale of microscopic equipment are Carl Zeiss, Nikon, Olympus, Leica Microsystems, JEOL, and Hitachi.

CAREERS AND COURSE WORK

A great number of career options exist for individuals with training in microscopy. Among the many professional settings into which microscopists are hired are scientific and clinical laboratories, medical facilities, the research departments of pharmaceutical firms, food processing companies, nanotechnology firms, metallurgy manufacturers, consumer product development laboratories, forensics departments, archaeological digs, and museums.

Ideally, preparation for a career in microscopy should begin with a comprehensive set of advanced high-school courses in physics, chemistry, biology, and mathematics. Students should follow this early training by pursuing a bachelor's degree in science. There are many appropriate areas of concentration for the budding microscopist at the undergraduate level, including physics, biology, chemistry, nanotechnology, and geology. A few colleges and universities in the United States have offered specialized degrees in applied microscopy, though this has not become common. Whatever one's major, important

topics to cover include optics, electromagnetism, and electronics. Practical laboratory experience with microscopes is also essential. An interest in photography and postprocessing of photographs would be a helpful addition to a microscopy student's list of qualifications, because these skills are relevant to photomicroscopy and the handling of microscopic images. A bachelor's degree alone (or an associate's degree with additional training or professional certification) is sufficient background for most laboratory technician positions involving microscopy, but those wishing to conduct independent research, either in microscopy itself or in a field such as cell biology or nanotechnology, will need a master's degree or a doctorate.

SOCIAL CONTEXT AND FUTURE PROSPECTS

Microscopy breaks down the barrier to knowledge created by the limitations of the human sense of sight. If knowledge is power, then microscopy represents one of science and technology's most powerful contributions to society. It enables researchers to discover more about the precise structure and behavior of healthy and diseased cells, pinpoint the mechanisms by which pathogens such as bacteria and viruses act in the body, and explore the chemical properties of potential pharmaceutical therapies. It even assists surgeons in performing difficult operations more safely and accurately, thereby saving countless lives. By providing scientists with an intimate knowledge of the way molecules, atoms, and subatomic particles interact, microscopy has propelled the formation of theories about the fundamental nature of the universe.

The growing needs of nanotechnology are inspiring further developments in microscopy. A scanning probe microscope built for the Argonne National Laboratory in Illinois allows researchers to "see" into an individual atom and observe its magnetic spin. The microscope, which cost $2 million, is itself very small—but it must be placed inside a machine 16 feet high and located in a soundproof room, so as to prevent even the tiniest vibration to throw off its focus.

On the other hand, microscopes intended for use in clinical settings in the developing world point the way toward ever smaller and cheaper instruments. A dime-sized microscope that sells for about $10 has no lenses but instead is made of a layer of metal set on top of an array of charge-coupled device (CCD) sensors arranged in a grid. The CCDs translate light into an electric signal; then, a great number of tiny channels are pierced into the metal. As a sample of blood, water, or other liquid flows over the channels, the particles in it block light from passing through to the CCDs in certain areas. The information about which channels are blocked and which remain open to light is used to create an image of the specimen.

M. Lee, B.A., M.A.

FURTHER READING

Cardell, Carolina, Isabel Guerra, and Antonio Sánchez-Navas. "SEM-EDX at the Service of Archaeology to Unravel Historical Technology." *Microscopy Today* 17, no. 14 (August, 2009): 28-33. An overview of the use of scanning electron microscopy to analyze archaeological materials. Includes diagrams and full-color photomicrographs.

Dykstra, Michael J., and Laura E. Reuss. *Biological Electron Microscopy: Theory, Techniques, and Troubleshooting.* 2d ed. New York: Kluwer Academic, 2003. A guide to using microscopic instrumentation in cytological research. Covers conventional light microscopy, transmission electron microscopy, scanning electron microscopy, and photomicroscopy.

Reitdorf, Jens, et al., eds. *Microscopy Techniques.* New York: Springer, 2005. A technical reference book designed for those with a biomedical background, including numerous tables and diagrams, plus appendixes detailing mathematical formulas.

Sluder, Greenfield, and D. E. Wolf, eds. *Digital Microscopy.* 3d ed. Boston: Elsevier Academic Press, 2007. A guide to coordinating microscopes with digital cameras to capture and analyze microscopic images. Includes detailed laboratory exercises to demonstrate principles in action.

Yao, Nan, and Zhong Lin Wang, eds. *Handbook of Microscopy for Nanotechnology.* New York: Kluwer Academic, 2005. An overview of microscopy applications in nanotechnology. Each of the twenty-two chapters contains a discussion of a specific microscopic instrument or technique by nanotechnology specialists working in different fields.

WEB SITES

Florida State University
Microscopy Primer
http://micro.magnet.fsu.edu/primer/index.html

Microscopy Society of America
http://www.microscopy.org/index.cfm

See also: Electron Microscopy; Histology; Mirrors and Lenses; Optics.

MILITARY SCIENCES AND COMBAT ENGINEERING

FIELDS OF STUDY

Physics; chemistry; weapons-systems engineering; civil engineering; materials science; metallurgy; electrical engineering; mechanical engineering; ocean engineering; oceanography; meteorology; aeronautical engineering; avionics; electronics; cryptography; medicine; medical technology; biology; biomechanics; mathematics; artificial intelligence; computer science; computer engineering; computer programming; naval architecture; geography; environmental engineering; nuclear engineering; systems engineering; aerospace engineering.

SUMMARY

Military sciences are scientific, engineering, and technical activities undertaken by those who identify, design, and produce innovative weapons, including such items as improved rifles for individual fighters and larger strategic weapons such as laser-guided missiles; equipment, ranging from improved clothing and night-vision goggles to armored tanks; and communications devices for use in warfare. Combat engineering includes activities such as building bridges, harbors, roads, temporary shelters, or improvised airfields used to assist combat troops, or removing obstacles from the battlefield.

KEY TERMS AND CONCEPTS

- **Asymmetric Warfare:** War between two opposing countries or armed forces in which one possesses significant advantages in technology, often forcing the weaker combatant to resort to unconventional means of warfare.
- **Ballistics:** Science that studies the characteristics of projectiles, including the mechanics of flight and effects on impact.
- **Electronic Warfare:** Using electronic energy to conduct offensive and defensive operations.
- **Information Warfare:** Collecting, managing, and using information to conduct military operations, either offensively or defensively.
- **Military-Industrial Complex:** Network of relationships between government and various industries involved in research, development, and delivery of weapons and equipment to the armed forces.
- **Network-Centric Warfare:** Activities aimed at gaining an advantage over an enemy through computer-based information systems.
- **Ordnance:** Exploding devices, or the weapons that propel explosive devices in combat.
- **Revolution in Military Affairs:** Significant shift in the conduct of warfare brought on by the emergence of new technologies.
- **Technology Transfer:** Process of sharing information about technology or manufacturing processes that allows governments or private businesses to benefit from the work done by other agencies.
- **Weapons Of Mass Destruction (WMD):** Armaments (often chemical, biological, or nuclear) that can inflict significant casualties, cause major property damage, and incite terror among military units or civilian populations.
- **Weapons Systems:** Integrated network of ordnance, delivery mechanisms, and control mechanisms (often computerized) that allows for effective employment of specific weapons in combat.

DEFINITION AND BASIC PRINCIPLES

The term "military sciences" is used to designate the broad scope of activities undertaken to develop weapons, weapons systems, and equipment for the military. This broad category includes basic research on the components used in fabrication of materials for weapons or weapons systems, including research in ballistics; design of new computers, programs, or electronic devices; engineering to fabricate armaments and ordnance, battlefield gear, or support equipment; and systems for transitioning new technologies and equipment into use by fighting forces.

"Combat engineering" is a term that describes the activities of military units that support the armed forces by performing a number of engineering functions in or near the battlefield. Many of these, such as bridge building, road building, and construction of temporary fortifications and camps for forces in the field, are handled improvisationally with materials on hand or brought into the theater of operations by military forces. By contrast, some military personnel engage in activities more akin to civil engineering.

This may involve construction of facilities at permanent bases, or in the case of the U.S. Army Corps of Engineers, projects that control the rivers and lakes within the boundaries of the United States.

It is important to distinguish "military sciences" from the closely related term, "military science." The latter refers to the broader study of war as an art and science and encompasses the study of tactics, strategy, and leadership.

BACKGROUND AND HISTORY

Technology has played an important role in the conduct of warfare since the dawn of civilization, when people first began to assemble crude weapons from natural materials. The discovery of techniques to forge metals led to improvements in weaponry, as bronze and then iron weapons made soldiers and sailors more effective fighters. The history of warfare, therefore, is inseparable from the history of technological advances that have influenced the development of weapons, equipment, fortifications, transportation, and logistics. At times, military sciences have simply been efforts to adapt existing technologies for military use; at other times, however, systematic efforts have been undertaken to design weapons of war or develop countermeasures to protect combatants and civilian populations.

It has become commonplace to describe the various epochs in the history of warfare as occurring in a series of "revolutions in military affairs," driven by advances in technology that brought about paradigmatic shifts in the conduct of warfare. Ways of waging war before the eleventh century did not change radically. Handheld weapons permitted armies to clash on the battlefield. Siege warfare was also commonplace, as invaders would surround cities and on occasion attempt to breach fortified walls. Engineers played a key role in both defensive and offensive operations, designing ever-more sophisticated fortifications from earth and stone to thwart enemy penetrations into cities (and later castles) and creating siege engines such as catapults for launching missiles over these ramparts or climbing devices for scaling them. The invention of the bow allowed armies to attack enemies from some distance. Mounted soldiers were used in limited fashion in open terrain. Shields and pikes to protect soldiers organized into tight battlefield formation were among the primary defensive weapons fashioned during these centuries. In naval battles, ships would get close with each other, grappling hooks would be tossed to secure the two vessels, and sailors would board with weapons similar to those used by infantry. Some ships carried mechanisms that allowed them to launch missiles such as fire grenades, but most naval warfare was conducted at close quarters.

The invention of gunpowder in the eleventh century revolutionized warfare around the globe. By the fourteenth century nations had learned to use the power of chemical explosions to launch projectiles (bullets, mortar, and artillery shells) that traveled greater distances and caused substantially more damage than muscle- or mechanically powered projectiles. Over the next four centuries, scientists devised more effective mixtures of chemical substances that allowed for controlled explosions, while engineers and gunsmiths designed more accurate and devastating weapons. By the eighteenth century, metallurgists had discovered ways to make artillery pieces lighter while improving their power and accuracy. Naval engineers adapted field artillery pieces for use aboard ships, providing navies new capabilities to attack other ships and provide fire support to forces onshore.

Another revolution in military sciences and engineering occurred in the late eighteenth and nineteenth centuries. Industrial-age technology led to numerous advances in military technology. The development of processes to standardize weapons production led to significant efficiencies in arming soldiers, since spare parts could be carried with troops and replaced easily. Innovations such as the invention of rifled barrels, breech-loading and repeating rifles, and the machine gun, gave armies more firepower. Though not developed specifically for military operations, rail transportation allowed troops to be deployed at greater distances in a shorter time. Advances in ship design led to the launch of ironclad warships capable of carrying cannons that could launch massive projectiles toward land-based targets from miles at sea. Clothing manufacturers and suppliers turned out gear that was more durable and better suited to soldiers' and sailors' needs.

During this time, chemists perfected a number of formulas for poisonous gases, which were deployed with devastating results during World War I. In response, protective gear was developed to counter the effects of the gas. Two additional inventions designed for peaceful purposes were quickly adapted for military use: the airplane and electronic communications devices. Planes gave field commanders opportunities for better surveillance and eventually provided platforms for delivering more sophisticated bombs over distant targets, or dropping troops behind enemy lines. Between World War I and II, naval engineers designed aircraft carriers that allowed navies to bring air power far from shore to attack enemy targets or defend friendly forces. World War II also saw the effective employment of electronic warfare, as both the Allied and Axis nations used newly created devices to intercept enemy communications, transmit messages over long distances, detect targets, or conduct countermeasures to neutralize the enemy's electronic devices.

The end of World War II saw the dawn of yet another revolution in military affairs: the introduction of atomic power into military conflict. This weapon of mass destruction allowed one combatant to inflict extensive damage on the enemy with minimal involvement of troops. The development of the atomic bomb was one of the great scientific achievements of the twentieth century, even if its deployment were morally questionable. Teams of physicists and engineers managed to harness the power of the atom to generate hitherto unseen explosive power. Atomic weapons became the signature armament of the ensuing Cold War between the Western allies and the Soviet Union. Even though no nuclear weapons were used in conflict, the threat they posed served to shape both military and political policy for four decades. Additionally, the proliferation of weapons of mass destruction (WMD) became an international concern during the second half of the twentieth century, as more stable countries grew fearful that such weapons might fall into the hands of fanatics and be used as instruments of terror. At the same time, however, nuclear-powered engines installed in surface ships and submarines gave naval vessels the ability to stay at sea for months without making port calls.

Beginning in the last decades of the twentieth century, the nature of warfare changed again with the introduction of sophisticated computer-based weapons systems, surveillance devices, and command-and-control networks. Laser-guided weapons, "smart bombs," and missiles capable of being guided to within ten meters of a target provided battlefield commanders more effective ways to hit enemy targets with minimal

collateral damage. New electronic devices permitted more sophisticated methods of gathering and processing intelligence, giving commanders better real-time data on which to base decisions. The growing presence of computers on the battlefield, networked to ones far away from the front lines, permitted commanders on both sides of a conflict to engage in network-centric war: Real-time exchange of information allowed combatants to gather intelligence and exploit weaknesses in an enemy's defenses. At the same time, weaker countries or groups engaged in asymmetric warfare often resorted to weapons using more primitive technology that could often produce casualties on combatants and civilians, often randomly, thereby creating terror among populations engaged in or living within the zone of conflict.

HOW IT WORKS

With few exceptions, the application of scientific and engineering work for military purposes is carried out under the direction of a nation's defense agency. That is not to say, however, that all research and development (R&D) is performed by government employees at state-owned facilities. It is common for socialist nations or totalitarian regimes to control the entire process, while democracies tend to follow the pattern used in the United States, where R&D is carried out through a complex arrangement that involves government agencies, private industry, and academic institutions.

The United States Department of Defense has an elaborate organization to oversee research and development. The Defense Advanced Research Projects Agency (DARPA) sponsors basic scientific research (focusing on physics and chemistry) and applied research that shows promise of producing new breakthroughs in designing military weaponry and equipment. The Army, Navy, and Air Force each has its own R&D agency, employing teams of scientists, engineers, and technicians to carry out projects funded through federal appropriations. Additionally, each agency engages in partnerships with private businesses to sponsor additional research and to underwrite engineering efforts to turn basic science into usable tools for the fighting forces. These agencies also control funding that can be used to support research at academic institutions across the country. While the military services direct much of the research, fabrication of end items—weapons and

Fascinating Facts About Military Sciences and Combat Engineering

- The first recorded instance of professional weapons systems development occurred in Syracuse on the island of Sicily in the fourth century b.c.e.
- The invention of the stirrup in the fifth century c.e. allowed armies to make more effective use of cavalry and eventually resulted in the development of combined-arms operations.
- The development of firearms made the armored knight on horseback obsolete as a fighter.
- In response to the French government's need to provision troops in the field, an enterprising inventor developed the process of canning food.
- The invention of the machine gun in the nineteenth century eventually eliminated the popular bayonet charge as an offensive maneuver in battle.
- The nuclear-powered propulsion system developed for use aboard submarines allows these vessels to go without refueling for twenty years or more.
- A program initiated in 1969 to link computers for improved information sharing and research resulted in the creation of the ARPANET, the forerunner of the Internet.
- The composite materials used to construct the U.S. Air Force B-2 Spirit Stealth bomber absorb radio waves, making the plane virtually undetectable on radar.
- The military's need for precise information on the locations of friendly and enemy positions and important targets led to the development of the Global Positioning System (GPS).

combat engineers handle tasks such as constructing or repairing roadways and temporary facilities and assembling prefabricated bridges. In the Air Force, combat-engineering functions include constructing temporary airfields, repairing existing airstrips, constructing roads and revetments, and providing general engineering support to field commanders. Navy construction battalions (Seabees) typically build wharves and harbor facilities, airfields, field hospitals, roads, and bridges.

APPLICATIONS AND PRODUCTS

Scientists and engineers engage in work to develop thousands of products for the military. Some are large, multimillion-dollar items such as aircraft carriers or supersonic planes; others may be small but of great importance to individual fighters, such as improved lenses for night-vision goggles. A brief outline of some of the major items used to conduct warfare in the twenty-first century suggests the scope and complexity of the work military scientists, engineers, and technicians are responsible for accomplishing.

Air Warfare. The design and construction of matériel for air warfare requires thousands of individual end items built using the latest technologies. At any time, a country like the United States is deploying new aircraft, maintaining older ones, and conducting research to create new planes that will be faster, lighter, and less susceptible to detection by the enemy. The composite materials developed by chemists, metallurgists, and engineers are often key components in the body designs of new planes, and many older ones are retrofitted to accommodate new equipment that enhances overall performance.

New weapons systems—missiles, bombs, and small-arms weapons such as machine guns—are designed to be carried on these platforms. Among those in use: air-to-air and air-to-surface missiles employing complex electronic guidance systems; bunker-buster bombs guided by laser systems or from satellites capable of penetrating as much as 20 feet of concrete; and "blackout" bombs that can knock out an enemy's electrical power grid.

The United States also builds and maintains satellites that provide secure voice and data transmission capability. Equipped with antijamming devices, these form the backbone of an elaborate satellite network that affords commanders from NATO countries a reliable system for worldwide command and control.

weapons systems, personal gear, and high-technology equipment such as radars and surveillance devices—is more often carried out by private industry.

Combat engineers are beneficiaries of the R&D that takes place within the various organizations and activities sponsored by the federal government. While the organization of combat engineering units varies by country, the operation of such units within the United States armed forces suggests how combat engineers make use of existing technologies and equipment to support operations in the field. Army

Other satellites are used to gather intelligence and serve as navigation aids to troops on the ground or at sea. Significant research is ongoing to improve the capabilities of unmanned aircraft, which can be remotely controlled and flown over enemy territory to deliver ordnance on targets or gather intelligence.

Ground Warfare. The changing nature of the battlefield and an enemy's capabilities make it necessary for armies to develop and maintain a host of new equipment to transport soldiers to the combat zone, provide them mobility once there, protect them from enemy attack, and arm them with weapons that offer sufficient firepower to subdue the enemy from close or medium range. The major rolling stock of most armies consists of tanks, armored personnel carriers, and self-propelled artillery. However, trucks used to haul supplies and transport personnel are also key components in an army's ability to remain mobile, and these are often equipped with armor and various detection devices to protect soldiers from enemy mines or other exploding devices. The weaponry designed for use by soldiers on the battlefield typically includes rifles, sidearms, and grenades. Sophisticated guidance systems and devices aid artillerists in launching ordnance accurately.

Soldiers are also equipped with a number of protective devices such as body armor (often made from composite materials), protective masks (commonly called gas masks), helmets, and special clothing that makes them less detectable by enemies. An array of products have been designed to aid in command and control, including sophisticated radios and computer devices (often handheld). Individual equipment and supplies are often subject to extensive research and design as well, so that rations, clothing, and personal gear carried for hygiene and comfort are carefully fabricated to make them usable in the difficult environment produced by combat.

Naval Warfare. Navies sail ships of various sizes and functions in conducting war at sea. Each is a floating platform for a variety of weapons systems that can project power at an enemy's navy or at targets onshore. The United States Navy employs aircraft carriers, cruisers, destroyers, submarines, and frigates as its principal fighting ships; amphibious assault ships are used to carry U.S. Marines to combat zones. The Navy also has a fleet of supply ships, hospital ships, and other support vessels. These are all equipped with guidance and navigation systems used for maneuvering and fighting. Naval ships carry missiles that can be launched from the deck, carried into flight by naval aircraft, or launched by submarines from beneath the surface. Ships are outfitted with conventional weapons ranging from medium-size machine guns used for defense against enemy air attack to large naval guns that can fire shells for several miles at enemy ship formations or onshore targets. Ships carry substantial amounts of equipment for surveillance, supply, and maintenance, much of it specifically designed by military scientists and engineers to withstand the rigorous conditions at sea. Aircraft used aboard ship are designed for short takeoffs and landings and carry armaments similar to those used by the air force for offensive and defensive operations. As with ground forces, those aboard ships are outfitted with personal gear, much of it specially designed to protect them in battle and provide comfort and hygiene.

Combat Engineering. Combat engineers carry equipment similar to that used by civilian construction and demolition firms, much of it modified for the specific needs of working in a combat zone. These include carpenter's and other construction tools (hammers, brush cutters, vises, shovels, posthole diggers), an array of power tools, and generators built to withstand the incidental damage caused by use in rough terrain. Bulldozers, earth movers, front loaders, and similar construction equipment is standard for many engineer units. Engineers also employ breaching vehicles to remove man-made obstacles such as barbed-wire fences or mines. Some combat-engineering units are equipped with amphibious vehicles to serve as ferries and bridging vehicles that allow engineers to transport and assemble portable bridges that allow the fighting forces to cross bodies of water.

IMPACT ON INDUSTRY

In countries where the state controls both research and production of military matériel, industries see only minimal impact from increases or decreases due to changing priorities in R&D or fabrication of armaments and equipment. Such is not the case in countries where there are close ties between the government and private sector. The situation in the United States presents a good example of how decisions regarding military R&D and procurement affect industry.

Historically, businesses that engaged in

manufacturing other items during peacetime (passenger cars and trucks, ball bearings, household chemicals) converted part or all of their facilities during wartime to manufacture items for the military (tanks and aircraft, ammunition, chemical weapons). Since World War II, however, many companies have specialized in doing work for the military—some exist solely for that purpose, creating what is in effect the defense industry. Certain firms have come to depend on government contracts for ships and aircraft, two items that require substantial expertise in science and engineering. These firms hire thousands of highly trained specialists to perform such work, hence employment is affected when the government increases or reduces orders for these items. Many technicians and tradespeople who work on construction of ships, planes, heavy armored equipment, and specialized items such as firearms or sophisticated surveillance equipment and guidance systems are also affected by government decisions to increase or decrease requirements for specific end items. In 2010, the United States allocated more than $82 billion for military R&D. Other countries have more modest budgets: In 2010, the United Kingdom allocated £4.6 billion (about $7 billion) for military R&D. Although some have questioned the wisdom of having this kind of permanent arrangement, the military-industrial complex has proven to be important in meeting the armed services' demands for more effective armaments that can be delivered to the fighting forces in a timely manner.

Additionally, much of the basic research undertaken in military laboratories, or by civilian researchers working on government-funded contracts, has proven to have significant benefit outside the defense industry. This is particularly true of research conducted in physics, chemistry, and basic electronics. The opportunity for technology transfer permits the adaptation or use of R&D from military agencies or defense contractors for civilian use. That transfer has been most evident to the general public in the adaptation of computer technology for the development of the Internet and Global Positioning System (GPS) technology for use by private citizens. One area where significant benefit is being realized is the medical field. The high number of incidents beginning in 2001 in Afghanistan and Iraq involving injury to extremities led to accelerated research in prosthetics. The military's willingness to share that technology has led to notable advances in the development of artificial limbs that provide patients increased mobility.

CAREERS AND COURSE WORK

Those with an interest in science and the military will find opportunities for work in government, private industry, and the academic world. A number of individuals involved in research and development for the military are members of the armed forces, often with specialized training that permits them to supplement their battlefield knowledge with classroom and laboratory preparation to carry out sophisticated scientific inquiry. Since the eighteenth century, governments have established and sponsored military academies to prepare officers for practical applications of military science and engineering, but individuals educated in other institutions can often receive commissions in the services and perform these roles.

Regardless of the institution one chooses to attend, obtaining a bachelor's degree in basic sciences (chemistry, biology, or physics), mathematics, or engineering is often adequate qualification for jobs as technicians working on an array of projects. Occasionally, those with associate's degrees in applied sciences can find work as technicians as well. By far the greatest opportunities for making significant contributions to the military sciences are available to those with advanced degrees, especially in the physical sciences, computer sciences, engineering, or medicine and medical research.

Technicians serving on active duty in military forces frequently receive on-the-job training or attend special schools established by the armed forces, although individuals wishing to pursue careers in military specialties such as avionics, ordnance disposal, medical technology, electronics equipment repair, or similar technical fields will find it helpful to have a sound foundation in mathematics and some understanding of the specific science or technology in which they plan to specialize.

Employment in military science and technology varies by country, but in almost every country opportunities for individuals to pursue their interests in these disciplines is available through commissioning or enlistment in the active service. In the United States, employment is also available with the Department of Defense at DARPA or one of the military service's laboratories and with contractors that provide products and services to the armed forces. Additionally, those with an interest in basic research that might have applications for military use can find rewarding work at a number of universities where government contracts provide funding for significant research in fields that show promise for military application.

Combat engineering is handled almost exclusively by members of the uniformed services. Those interested in working in that field must first enlist or receive a commission in one of the armed forces, and then select combat engineering as a career specialty. The academic qualifications for combat engineers are less stringent than those required of laboratory scientists or design and manufacturing specialists. Often, combat engineers are given specific instruction in the tasks they will perform as part of their military training and participate in refresher courses or advanced training to keep their skills up to date.

SOCIAL CONTEXT AND FUTURE PROSPECTS

If history is any guide, the inevitability of future conflict somewhere in the world suggests that there will be a continuing need for new technologies to wage warfare more effectively. Developments in military sciences are always carried out, however, in a social and political climate that affects both the budgets of those engaged in research and the constraints that are imposed on the kinds of weapons and equipment that may be developed. Working within those real-world parameters, military scientists, especially in countries that enjoy political stability and the financial wherewithal to support major research efforts, continue to explore new applications for existing technologies or work to create new ones that will enhance a country's ability to fight when necessary.

Many military strategists believe that the greatest prospects for advancing a country's ability to fight more effectively lie in the development of more sophisticated tools for information warfare. Devices already available, such as GPS, surveillance satellites, and radar have proven effective in combat; however, refinements to improve their accuracy and reliability, especially in the face of electronic countermeasures, will continue to be required. Electronic command and control tools—instruments such as the Internet and handheld devices that rely on satellites for broadcast capability—will also require constant updating or replacement with yet-undiscovered technologies that

can give commanders improved ability to communicate with subordinates or superiors to direct activities and provide necessary support on the battlefield.

Several areas of research suggest the variety of tasks in which military scientists may be engaged. One is the construction of hypersonic aircraft. The potential to create unmanned vehicles that can travel at Mach 5 (five times the speed of sound) or more has great military value for the development of missiles that can strike with exceptional speed against high-value targets, particularly enemy soldiers and their leaders that have the potential to move about. The United States and its allies are also developing more sophisticated weapons that rely on laser technology. Directed-energy weapons, as these devices are called, are being designed using both solid-state and chemical lasers. When operational, these weapons will provide even more accurate platforms from which to engage and neutralize enemy combatants. At the same time, research in neuroscience is ongoing to produce early-warning devices that will monitor soldier's brainwave activities and alert them when their heightened subconscious senses danger.

Significant medical research continues to develop better methods of prevention, treatment, and rehabilitation for members of the armed forces engaged in combat. Of special note is work in prosthetics. Continuing research in biomechanics is leading to improvements in devices that mimic human extremities, and researchers continue to devise ways to link artificial limbs to nerve endings to provide better mobility and control. At the same time, basic research into diseases most commonly associated with battlefield conditions, as well as those that exist in potential battle zones, is under way to create more effective prophylactics that will permit military personnel to ward off disease or recover from illnesses and return to duty more quickly.

In the twenty-first century, however, a major factor influencing military R&D is the increase in incidents of asymmetric warfare worldwide. Countries with large arsenals of sophisticated weapons are finding it necessary to defeat forces with considerably less technological capability. While scientists work to create better defensive armaments and offensive weapons with greater precision to minimize collateral damage, combat engineers and explosive-ordnance-disposal specialists are facing challenges on the front lines to create effective fortifications and remove hazards from battle areas where civilians and combatants are often indistinguishable.

Laurence W. Mazzeno, B.A., M.A., Ph.D.

FURTHER READING

Amato, Ivan. *Pushing the Horizon: Seventy-Five Years of High-Stakes Science and Technology at the Naval Research Laboratory.* Washington, D.C.: Government Printing Office, 1998. Describes the activities of the Naval Research Laboratory in developing weapons systems, communications technology, and equipment during the twentieth century.

Boot, Max. *War Made New: Technology, Warfare, and the Course of History, 1500 to Today.* New York: Gotham Books, 2006. Traces the impact of new technologies on the conduct of war from the gunpowder revolution in the sixteenth century through the development of nuclear armaments and other weapons of mass destruction in the twentieth.

Evans, Nicholas D. *Military Gadgets: How Advanced Technology Is Transforming Today's Battlefield and Tomorrow's.* Upper Saddle River, N.J.: Prentice Hall, 2004. Provides an overview of existing technologies being used by armed forces and ones in development.

Langford, R. Everett. *Introduction to Weapons of Mass Destruction: Radiological, Chemical, and Biological.* Hoboken, N.J.: John Wiley & Sons, 2004. Describes elements of weapons of mass destruction, explains their capabilities, and explains technological countermeasures used to assure individual and community survival.

Levis, Alexander H., ed. *The Limitless Sky: Air Force Science and Technology Contributions to the Nation.* Washington, D.C.: Air Force History and Museums Program, 2004. Essay collection describing several significant projects in which technological developments led to significant improvements in Air Force readiness and capability.

Lewer, Nick, ed. *The Future of Non-Lethal Weapons: Technologies, Operations, Ethics and Law.* Portland, Oreg.: Frank Cass, 2002. Includes several essays on technologies used to develop and assess nonlethal weapons and evaluates claims made for their effectiveness.

O'Hanlon, Michael. *Technological Change and the Future of Warfare.* Washington, D.C.: Brookings Institution, 2000. Describes ways new technologies influence changes in military doctrine and strategy and affect budgeting and international relations.

Price, Alfred. *War in the Fourth Dimension: U.S. Electronic Warfare, from Vietnam to the Present.* Mechanicsburg, Pa.: Stackpole, 2001. Traces the history of electronic warfare and its impact on combat operations.

Richardson, Jacques. *War, Science and Terrorism: From Laboratory to Open Conflict.* Portland, Oreg.: Frank Cass, 2002. Examines the interaction of scientific research and combat, explaining the roles of scientists, engineers, and those involved in production and manufacture of new technologies.

Waltz, Edward. *Information Warfare: Principles and Operations.* Norwood, Mass.: Artech House, 1998. Explains ways advances in information technology influence offensive and defensive military operations and describes a number of technologies in some detail.

WEB SITES

Defense Advanced Research Projects Agency
http://www.darpa.mil

National Defense University
Center for Technology and National Security Policy
http://www.ndu.edu/inss/index.cfm?secID=53&pageID=4&type=section

U.S. Air Force Research Laboratory
http://www.afrl.af.mil

U.S. Army Corps of Engineers, Engineer Research and Development Center
http://www.erdc.usace.army.mil

U.S. Naval Research Laboratory
http://www.nrl.navy.mil

See also: Biomechanics; Meteorology.

MINERALOGY

FIELDS OF STUDY

Geology; chemistry; geochemistry; physics; petrology; experimental petrology; environmental geology; forensic mineralogy; medical mineralogy; gemology; economic geology; geochronology; descriptive mineralogy; crystallography; crystal chemistry; mineral classification; geologic occurrence; optical mineralogy; mining; chemical engineering.

SUMMARY

Mineralogy is the study of the chemical composition and physical property of minerals, the arrangement of atoms in the minerals, and the use of the minerals. Minerals are naturally occurring elements or compounds. The composition and arrangement of the atoms that make up minerals is reflected in their physical characteristics. For example, gold is a naturally occurring mineral containing one element that has a definite density of 19.3 grams per milliliter and a yellow color and is chemically inactive. Sometimes mineral resources are broadened to refer to oil, natural gas, and coal, although those materials are not technically minerals.

KEY TERMS AND CONCEPTS

- **Clay Mineral:** Any of a group of tiny silicate minerals (less than 2 micrometers in size) that form by varied degrees of weathering of other silicate minerals.
- **Cleavage:** Breaking of a crystallized mineral along a plane, leaving a smooth rather than an irregular surface.
- **Hardness:** Resistance of a mineral to being scratched by another material.
- **Hydrothermal Deposit:** Mineral deposit precipitated from a hot water or gas solution.
- **Igneous Rock:** Rock formed from molten rock material.
- **Ion:** Atom or group of atoms with a positive or negative charge.
- **Lava:** Molten rock material at the Earth's surface.
- **Magma:** Molten rock material and suspended mineral crystals below the Earth's surface.

- **Major Elements:** Elements that make up the bulk of the chemical composition of a mineral or rock.
- **Petrology:** Study of the origin, composition, structure, and properties of rocks and the processes that formed the rocks.
- **Sedimentary Rock:** Rock formed from particles such as sand by the weathering of other rocks at the surface or by precipitation from water.
- **Silicate Mineral:** Mineral that contains silicon bonded with oxygen to form silicate groups bonded with positive ions; silicates make up 90 percent of the Earth's crust.
- **Trace Element:** Element present only in tiny amounts (in quantities of parts per million or less) in a mineral or rock.

DEFINITION AND BASIC PRINCIPLES

Minerals are solid elements or compounds that have a definite but often not fixed chemical composition and a definite arrangement of atoms or ions. For instance, the mineral olivine is magnesium iron silicate, with the formula of $(Mg, Fe)_2SiO_4$, meaning that it has oxygen (O) and silicon (Si) atoms in a ratio of 4:1 and a total of two ions of magnesium (Mg) and iron (Fe) in any ratio. The magnesium and iron component can vary from 100 percent iron with no magnesium to 100 percent magnesium with no iron, and all variations in between. Other minerals, however, such as gold, have nearly 100 percent gold atoms. Minerals are usually formed by inorganic means, but some organisms can form minerals such as calcite (calcium carbonate, with the formula $CaCO_3$).

Minerals make up most of the rocks in the earth, so they are studied in many fields. In geochemistry, for example, scientists determine the chemical composition of the minerals and rocks to derive hypotheses about how various rocks may have formed. Geochemists might study what rocks melt to form magmas or lavas of a certain composition. Environmental geologists attempt to solve problems regarding minerals that pollute the environment. For instance, environmental geologists might try to minimize the effects of the mineral pyrite (iron sulfide, with the formula FeS_2) in natural bodies of water. Pyrite, which is found in coal, dissolves in water to form sulphuric acid and high-iron water, which can kill some organisms. Forensic mineralogists may determine the origin of minerals left at a crime scene. Economic geologists discover and determine the distribution of minerals that can be mined, such as lead minerals (galena), salt, gypsum (for wall board), or granite (for kitchen countertops). Geophysicists may study the minerals below the Earth's surface that cannot be directly sampled. They may, for example, study how the seismic waves given off by earthquakes pass through the ground to estimate the kinds of rocks present.

BACKGROUND AND HISTORY

Archaeological evidence suggests that humans have used minerals in a number of ways for tens of thousands of years. For instance, the rich possessed jewels, red and black minerals were used in cave drawings in France, and minerals such as gold were used for barter. Metals were apparently extracted from ores for many years, but the methods of extraction were conveyed from person to person without being written down. In 1556, Georgius Agricola's *De re metallica* (English translation, 1912) described many of these mineral processing methods. From the late seventeenth century into the nineteenth century, many people studied the minerals that occur in definite crystal forms such as cubes, often measuring the angles between faces.

In the nineteenth century, the fields of chemistry and physics developed rapidly. Jöns Jacob Berzelius developed a chemical classification system for minerals. Another important development was the polarizing microscope, which was used to study the optical properties of minerals to aid in their identification.

During the late nineteenth century, scientists had theorized that the external crystal forms of minerals reflected the ordered internal arrangement of their atoms. In the early twentieth century, this theory was confirmed through the use of X rays. Also, it became possible to chemically analyze minerals and rocks so that chemical mineral classifications could be further developed. Finally, in the 1960's, the use of many instruments such as the electron microprobe allowed geologists to determine variations in chemical composition of minerals across small portions of the minerals so that models for the formation of minerals could be further developed.

HOW IT WORKS

Mineral and Rock Identification. A geologist may tentatively identify the minerals in a rock using characteristic such as crystal form, hardness, color, cleavage, luster (metallic or nonmetallic), magnetic properties, and mineral association. The rock granite, for instance, is composed of quartz (often colorless, harder than other minerals, with rounded crystals, no cleavage, and nonmetallic luster) and feldspars (often tan, softer than quartz, with well-developed crystals, two good nearly right-angle cleavages, and nonmetallic luster), with lesser amounts of black minerals such as biotite (softer than quartz, with flat crystals, one excellent cleavage direction, and shiny nonmetallic luster).

The geologist then slices the rock into a section about 0.03 millimeters thick (most minerals are transparent). The section is examined under a polarizing microscope to confirm the presence of the tentatively identified minerals and perhaps to find other minerals that could not be detected by eye because they were too small or present in very low quantities. The geologist can determine other relationships among the minerals, such as the sequence of crystallization of the minerals within an igneous rock.

Identification Using Instruments. The minerals in a rock can be analyzed a variety of other ways, depending on the goals of a given study. For instance, X-ray diffraction may be used to identify some minerals. One of the most useful applications of X-ray diffraction is to identify tiny minerals such as clay minerals that are hard to identify using a microscope. The wavelengths of the X rays are similar to the spacing between atoms in the clay minerals, so when X rays of a single wavelength are passed through a mineral, they are diffracted from the minerals at angles that are characteristic of a particular mineral. Thus, the mixture of clay minerals in a rock or soil may be identified.

The electron microprobe has enabled the analysis of tiny portions of minerals so that changes in composition across the mineral may be determined. The instrument accelerates masses of electrons into a mineral that releases X rays with energies that are characteristic of a given element so that the elements present can be identified. The amount of energy given off is proportional to the amount of the element in the sample; therefore, the concentration of the element in the mineral can be determined when the results are compared with a standard of known concentration. The electron microprobe can also be used to analyze other materials such as alloys and ceramics.

The scanning electron microscope uses an electron beam that is scanned over a small portion of tiny minerals and essentially takes a photograph of the mineral grains in the sample. Some electron microscopes are set up to determine qualitatively what elements are present in the sample. This information is often enough to identify the mineral.

Other Analytical Techniques. Many instruments and techniques are used to analyze the major elements, trace elements, and the isotopic composition of minerals and rocks. Commonly used methods are X-ray fluorescence (XRF), inductively coupled plasma mass spectrometer (ICP-MS), and thermal ionization mass spectrometry. X-ray fluorescence is used to analyze bulk samples of minerals, rocks, and ceramics for major elements and many trace elements. A powdered sample or a glass of the sample is compressed and is bombarded by X rays so that an energy spectrum that is distinctive for each element is emitted. The amount of radiation given off by the sample is compared with a standard of known concentration to determine how much of each element is present in the sample.

The inductively coupled plasma mass spectrometer is used to analyze many elements in concentrations as low as parts per trillion by passing vaporized samples into high-temperature plasma so that all elements have positive charges. A mass spectrometer sorts out the ions by their differing sizes and charges in a magnetic field, which permits a determination of the elemental concentrations when the results are compared with a standard of known concentration. Up to seventy-five elements can be rapidly analyzed in a sample at precisions of 2 percent or better.

Thermal ionization mass spectrometers can be used to analyze the isotopic ratios of higher mass elements such as rubidium, strontium, uranium, lead, samarium, and neodymium, which may be used to interpret the geologic age of a rock. The mineral or rock is placed on a heated filament so that the isotopes are ejected into a magnetic field in which the ions are deflected by varied amounts depending on the mass and charge of the isotope. The data may then be used to calculate the amount of a certain isotope in the sample and eventually the isotopic age of the sample.

Other specialized instruments are also available. For instance, gemologists use specialized instruments to study and cut gemstones.

APPLICATIONS AND PRODUCTS

Abundant Metals and Uses. The most abundant metals are iron, aluminum, magnesium, silicon, and titanium. Iron, which is mostly obtained from several minerals composed of iron and oxygen (hematite and magnetite), accounts for 95 percent by weight of the metals used in the United States. Much of the hematite and magnetite is obtained from large sedimentary rock deposits called banded-iron formations that are up to 700 meters thick and extend for up to thousands of square kilometers. The banded-iron formations formed 1.8 billion to 2.6 billion years ago. They are abundant, for instance, around the Lake Superior region in northern Minnesota, northern Wisconsin, and northwestern Michigan. The banded-iron formations have produced billions of tons of iron ore deposits. The ores are made into pellets and are mixed with limestone and coke to burn at 1,600 degrees Celsius in a blast furnace. The iron produced in this process is molten and can be mixed with small amounts of scarce metals (called ferroalloy metals) to produce steel with useful properties. For instance, the addition of chromium gives steel strength at high temperatures, and it prevents corrosion. The addition of niobium produces strength, and the addition of copper increases the resistance of the steel to corrosion.

Much of the aluminum occurs in clay minerals in which the aluminum cannot be economically removed from the other elements. Thus, most of the aluminum used comes from bauxite, which is a mixture of several aluminum-rich minerals such as gibbsite ($Al(OH)_3$) and boehmite ($AlOOH$). Bauxite forms in the tropics to subtropics by intense chemical weathering from other aluminum-rich minerals such as the clay minerals. The production of aluminum from bauxite is very expensive because the ore is dissolved in molten material at 950 degrees Celsius and the aluminum is concentrated by an electric current. Electricity represents about 20 percent of the total cost of producing aluminum. Recycling of aluminum is economical because the energy expended in making a new can from an old aluminum can is about 5 percent of the energy required to make a new aluminum can from bauxite.

Scarce Metals and Uses. At least thirty important trace metals occur in the Earth's crust at concentrations of less than 0.1 percent. Therefore, these elements are trace elements in most minerals and rocks until special geologic conditions concentrate them enough so that they become significant portions of some minerals. For instance, gold is present in the Earth's crust at only 0.0000002 percent by weight. Gold often occurs as gold uncombined with other elements and precipitated from hydrothermal solutions as veins. Gold may also combine with the element tellurium to form several kinds of gold-tellurium minerals. Also gold does not react very well chemically during weathering, so it may concentrate in certain streams to form placer deposits. Gold has been used for thousands of years in jewelry, dental work, and coins. In the past, much world exploration has been motivated by the drive to find gold.

Gold is a precious metal. Other precious metals are sliver and the platinum group elements (for example, platinum, palladium, and rhodium). Silver, like gold, has been used for thousands of years in jewelry and coins. In modern times, silver also finds uses in batteries and in photographic film and papers because some silver compounds are sensitive to light. Silver occurs in nature as silver sulfide, and it substitutes for copper in copper minerals so that much of the silver produced is a side product from copper mining. The platinum metals occur together as native elements or combined with sulfur and tellurium. They concentrate in some dark-colored igneous rocks or as placers. The platinum group metals are useful as catalysts in chemical reactions, so they are, for example, used in catalytic convertors in vehicles.

Another group of scarce metals are base metals. The base metals are of relatively low economic value compared with the precious metals. The base metals include copper, lead, zinc, tin, mercury, and cadmium. Copper minerals are native copper and various copper minerals combined with sulfate, carbonate, and oxygen. Copper minerals occur in some igneous rocks, including some hydrothermal deposits, mostly as dilute copper-sulfur minerals. These minerals may be concentrated during weathering, often forming copper minerals combined with oxygen and carbonate. Copper conducts electricity very well, and much of it is used in electric appliances and wires. Lead and zinc tend to occur together in

Fascinating Facts About Mineralogy

- In 1822, Friedrich Mohs developed a hardness scale for minerals, using ten minerals. From hardest to softest, they are diamond, corundum, topaz, quartz, potassium feldspar, apatite, fluorite, calcite, gypsum, and talc.
- Diamonds are the hardest natural material on Earth and are the most transparent material known. Ultra-thin diamond scalpels retain their sharp edge for long periods and can transmit laser light to cauterize wounds.
- Blood diamonds, also known as conflict diamonds, are produced or traded by rebel forces opposed to the recognized government in order to fund their activities. Most of the blood diamonds come from Sierra Leone, Angola, the Democratic Republic of Congo, Liberia, and the Ivory Coast.
- The mineral kaolin is used to create a hard clay used in rubber tires for lawn equipment, garden hoses, rubber floor mats and tiles, automotive hoses and belts, shoe soles, and wire and cable.
- In China, ownership of Imperial jade, a fine-grained jade with a rich, uniform green color, was once limited to nobility. Jade is believed to have powers to ward off evil and is associated with longevity.
- The average automobile contains about 0.9 mile of copper wiring. The total amount of copper used in an automobile ranges from 44 to 99 pounds.
- The manufacture of jewelry accounts for 78 percent of the gold used each year. However, because gold conducts electricity and does not corrode, it is used as a conductor in sophisticated electronic devices such as cell phones, calculators, television sets, personal digital assistants, and Global Positioning Systems.

hydrothermal deposits in combination with sulfur. Lead is used in automobile batteries, ammunition, and solder. Lead is very harmful to organisms, which restricts its potential use. Zinc is used as a coating on steel to prevent it from rusting, in brass, and in paint.

Fertilizer and Minerals for Industry. Fertilizers contain minerals that have nitrogen, phosphorus, potassium, and a few other elements necessary for the growth of plants. In Peru, deposits of guano (seabird excrement), which is rich in these elements, have been mined and used as fertilizer. Saltpeter (potassium nitrate) has also been mined in Peru. A calcium phosphate mineral, apatite, occurs in some sedimentary rocks, so some of these have been mined for the phosphorus.

Some minerals are a source for sodium and chlorine. Halite, commonly known as rock salt, is used as a flavoring and to melt ice on roads. Halite is produced in some sedimentary rocks from the slow evaporation of water in closed basins over long periods of time, a process that is occurring in the Great Salt Lake in Utah.

A variety of rocks or sediment—including granite, limestone, marble, sands, and gravels—are used for the exteriors of buildings, construction materials, or for countertops. Sands and gravels may be mixed with cement to make concrete.

IMPACT ON INDUSTRY

The materials of the Earth interact, which means that all societies depend on mineral resources. For example, crops grow in soil containing clay minerals formed from the weathering of rocks. Fertilizers composed of nitrogen, phosphorus, and potassium minerals must be added to the soil to grow enough food to support the growing population of the world. Machinery such as tractors, trains, ships, and trucks made of steel and other materials obtained from processing metal ores must be used to fertilize fields and to move food to markets. Electricity, much of which is obtained by burning coal, must be used to process food, whether in factories or in homes. Most of these mineral resources cannot be renewed quickly, so many are being rapidly used up.

Industry and Business. The largest industries directly related to mineralogy are public or private corporations engaged in the exploration, production, and development of minerals. As of 2009, the United States was the world's third largest producer of copper, after Chile and Peru, and of gold, after China and Australia. In 2009, net exports of mineral raw materials and old scrap materials was $10 billion. In 2009, Freeport-McMoRan Copper and Gold's Morenci open-pit mine in Arizona produced about 460,000 pounds of copper (about 40 percent of its total North American production), Barrick Gold Corporation's Goldstrike Complex in Nevada produced 1.36 million ounces of gold, and Teck's Red

Dog Mine in Alaska produced 79 percent of the total U.S. production of zinc.

Government Regulation. Although most governments do not directly control the mineral resources in their countries, they often regulate mining to ensure the workers' safety and support mining through subsidies and funding. In the United States, government agencies regulating mining include the Mine Safety and Health Administration of the U.S. Department of Labor and the mining division of the National Institute for Occupational Safety and Health. The U.S. Geological Survey maintains statistics and information on the worldwide supply of and demand for minerals.

Academic Research. Engineering departments in colleges and universities engage in mineralogy research such as improving the safety and efficiency of mining operations and searching for better, cleaner ways to process ores. For instance, some engineers analyze ways to break up ores and separate the ore minerals or develop ways to use more of the metals in the ores.

CAREERS AND COURSE WORK

Anyone interested in pursuing a career in the minerals industry should study geology, chemistry, physics, biology, chemical engineering, and mining engineering. A bachelor's degree in one of these subjects is the minimum requirement for working in oil exploration or as a mining technician or water-quality technician. A master's degree or a Ph.D. is required for more challenging and responsible jobs in geochemistry, geophysics, environmental geology, geochronology, forensic mineralogy, and academics. Geochemists should have a background in geology and chemistry. Geophysicists use physics, mathematics, and geology to remotely tell what kinds of rocks are below the surface of the earth. Environmental geologists should have backgrounds in geology, chemistry, and biology. Geochronologists use natural radioactive isotopes to give estimates of the time when some kinds of rocks formed. They should have a background in geology, physics, and chemistry. Forensic mineralogists should have a good background in mineralogy and criminology. Economic geologists should major in geology, with a concentration in courses concerned with ore mineralization. Those interested in an academic career should have a Ph.D. in geology, mining engineering, or chemical engineering. They will be expected to teach and do research in their subject area.

SOCIAL CONTEXT AND FUTURE PROSPECTS

Mineral resources are not evenly distributed throughout the world. In the nineteenth century, the United States did not import many mineral resources, but it has increasingly become an importer of mineral resources. The need to import is driven by many factors, including an increase in the minerals used. U.S. production of final products, including cars and houses, using mineral materials accounted for about 13 percent of the 2009 gross domestic product. Other factors include the depletion of some U.S. sources of minerals and the increased use of minerals that are not naturally available in the United States. In 2009, foreign sources supplied more than 50 percent of thirty-eight mineral commodities consumed in the United States. Nineteen of those minerals were 100 percent foreign sourced. The United States must import much of the manganese, bauxite, platinum group minerals, tantalum, tungsten, cobalt, and petroleum that it uses. In contrast, the United States still has abundant resources of gold, copper, lead, iron, and salt. Japan, however, has few mineral resources, so it must import most of them.

This need to import and export mineral resources has forced countries to cooperate with one another to achieve their needs. The mineral resources traded in the largest quantities, in descending order, are iron-steel, gemstones, copper, coal, and aluminum. The total annual world trade in mineral resources is approaching $700 billion and is likely to rapidly increase as countries such as China begin to use and produce more mineral resources.

As the supply of minerals decreases worldwide and environmental and safety concerns raised by mining gain in importance, it is likely that mineralogy research may turn to improving extraction and processing methods and looking for ways to recycle materials, capture minerals remaining in wastes, and restore mining areas after the minerals have been depleted.

Robert L. Cullers, B.S., M.S., Ph.D.

FURTHER READING

Craig, James R., David J. Vaughan, and Brian Skinner. *Earth Resources and the Environment.* Upper Saddle River, N.J.: Prentice Hall, 2010. Gives an overview of mineral resources in the Earth and describes environmental problems regarding fossil fuels, metals, fertilizers, and industrial minerals. Appendix, glossary, index, and many illustrations.

Guastoni, Alessandro, and Roberto Appiani. *Minerals.* Buffalo, N.Y.: Firefly Books, 2005. This guide describes the common minerals, gives their chemical formulas, and describes where they are found. Contains illustrations.

Kearny, Philip, ed. *The Encyclopedia of the Solid Earth Sciences.* London: Blackwell Science, 1994. Contains an alphabetical list of defined geologic terms, minerals, and geologic concepts.

Klein, Cornelis, and Barbara Dutrow. *Manual of Mineral Science.* Hoboken, N.J.: John Wiley & Sons, 2008. Describes how to identify minerals using physical properties and modern analytical techniques.

Pellant, Chris. *Smithsonian Handbook: Rocks and Minerals.* London: Dorling Kindersley Books, 2002. This identification guide for minerals and rocks uses many pictures combined with text.

Sinding-Larsen, Richard, and Friedrich-Wilhelm Wellmer, eds. *Non-renewable Resource Issues: Geoscientific and Societal Challenges.* London: Springer, 2010. Looks at minerals as nonrenewable resources and discusses the challenges that societies face.

WEB SITES

Geology and Earth Sciences
http://geology.com

Mineralogical Society of America
http://www.minsocam.org

National Institute for Occupational Safety and Health
Mining
http://www.cdc.gov/niosh/mining

U.S. Department of Labor
Mine Safety and Health Administration
http://www.msha.gov

U.S. Geological Survey
Minerals Information
http://minerals.usgs.gov/minerals

See also: Diffraction Analysis; Electronic Materials Production; Gemology and Chrysology; Land-Use Management; Soil Science.

MIRRORS AND LENSES

FIELDS OF STUDY

Physics; astronomy; mathematics; geometry; chemistry; mechanical engineering; optical engineering; aerospace engineering; environmental engineering; medical engineering; computer science; electronics; photography.

SUMMARY

Mirrors and lenses are tools used to manipulate the direction of light and images, mirrors by using flat or curved glass in the reversion, diversion, or formation of images, and lenses by using polished material, usually glass, in the refraction of light. Scientific applications of mirrors and lenses cover a broad spectrum, ranging from photography, astronomy, and medicine to electronics, transportation, and energy conservation. Without mirrors and lenses, it would be impossible to preserve memories in a photograph, view far-away galaxies through a telescope, diagnose diseases through a microscope, or create energy using solar panels. Miniaturized mirrors are also implemented in scanners used in numerous electronic devices, such as copying machines, bar-code readers, compact disc players, and video recorders.

KEY TERMS AND CONCEPTS

- **Achromatic Lens:** Lens that refracts light without separating its components into the various colors of the spectrum.
- **Aperture:** Opening through which light travels in a camera or optical device, determining the amount of light delivered to the lens, and affecting the quality of an image.
- **Catadioptric Imaging:** Imaging system that uses a combination of mirrors and lenses to create an optical system.
- **Catoptric Imaging:** Imaging system that uses only a combination of mirrors to create an optical system.
- **Diffraction:** Spreading out of light by passing through an aperture or around an object that bends the light.
- **Diopter:** Unit of measurement that determines the optical power of a lens by calibrating the amount of light refracted by the lens, in correlation with its focal length.
- **Fresnel Lens:** Flat lens that reduces spherical aberration and increases light intensity due to its numerous concentric circles, often creating a prism effect.
- **Negative Lens:** Concave lens with thick edges and a thinner center, used most often in eyeglasses for correcting nearsightedness.
- **Positive Lens:** Convex lens with thin edges and a thicker center, used most often in eyeglasses for correcting farsightedness.
- **Refraction:** Bending of light waves, used by a lens to magnify, reduce, or focus images.
- **Spherical Aberration:** Distortion or blurriness of image caused by the geometrical formation of a spherical lens or mirror.

DEFINITION AND BASIC PRINCIPLES

Mirrors and lenses are instruments used to reflect or refract light. Mirrors use a smooth, polished surface to revert or direct an image. Lenses use a piece of smooth, transparent material, usually glass or plastic, to converge or diverge an image. Because light beams that strike dark or mottled surfaces are absorbed, mirrors must be highly polished in order to reflect light effectively. Likewise, since rough surfaces diffuse light rays in many different directions, mirror surfaces must be exceedingly smooth to reflect light in one direction. Light that hits a mirrored surface at a particular angle will always bounce off that mirrored surface at an exactly corresponding angle, thereby allowing mirrors to be used to direct images in an extremely precise manner.

Images are reflected according to one of three main types of mirrors: plane, convex, and concave. Plane mirrors are flat surfaces and reflect a full-size upright image directly back to the viewer but left and right are inverted. Convex mirrors are curved slightly outward and reflect back a slightly smaller upright image, but in a wider angled view, and left and right are inverted. Concave mirrors are curved slightly inward and reflect back an image that may be larger or smaller, depending on the distance from the object, and the image may be right-side up or upside down.

BACKGROUND AND HISTORY

The ancient Egyptians used mirrors to reflect light into the dark tombs of the pyramids so that workmen could see. The Assyrians used the first known lens, the Nimrud lens, which was made of rock crystal, 3,000 years ago in their work, either as a magnifying glass or as a fire starter. Chinese artisans in the Han dynasty created concave sacred mirrors designed for igniting sacrificial fires, and Incan warriors wore a bracelet on their wrists containing a small mirror for the focusing of light to start fires.

It was not until the end of the thirteenth century, however, when an unknown Italian invented spectacles, that lenses subsequently became used in eyeglasses. Near the end of the sixteenth century, two spectacle makers inadvertently discovered that certain spectacle lenses placed inside a tube made objects that were nearby appear greatly enlarged. Dutch spectacle maker Zacharias Janssen and his son, Hans Janssen, invented the microscope in 1590. In 1608, Zacharias Janssen, in collaboration with Dutch spectacle makers Hans Lippershey and Jacob Metius, invented the first refractory telescope using lenses. However, it was Galileo Galilei who, in 1609, took the first rudimentary refracting telescope invented by the Dutch, drastically improved on its design, and went on to revolutionize history by using his new telescopic invention to observe that the Earth and other planets revolve around the Sun. In 1668, Sir Isaac Newton invented the first reflecting telescope, which, using mirrors, conveyed a vastly superior image than the refractory telescope, since it greatly reduced distortion and spherical aberrations often conveyed by the refractory telescope's lenses.

Almost two hundred years later, in 1839, with the invention of photography by Louis Daguerre, the convex lens became used for the first time in cameras. In 1893, Thomas Edison patented his motion-picture camera, the kinetoscope, which used lenses, and by the 1960's, mirrors were being used in satellites and to reflect lasers.

HOW IT WORKS

The law of reflection states that the angle of incidence is equal to the angle of reflection, meaning that if light strikes a smooth mirrored surface at a 45-degree angle, it will also bounce off the mirror at a corresponding 45-degree angle. Whenever light hits an object, it may be reflected, or it may be absorbed, which is what happens when light hits a dark surface. Light may also hit a rough shiny surface, in which case the light will be reflected in many different directions, or diffused. Light may also simply pass through an object altogether if the material is transparent.

When light travels from one medium to another, such as air to water, the speed of the light slows down and refracts, or bends. Lenses take advantage of the fact that light bends when changing mediums by manipulating the light to serve a variety of purposes. The angle that light is refracted by a lens depends on the lens's shape, specifically its curvature. A glass or plastic lens can be polished or ground so that it gathers light toward its edges and directs it toward the center of the lens, in which case the lens is concave, and the light rays will be diverged. Conversely, a lens which is convex, because it is thicker in the middle and has thinner edges, will cause light rays to converge.

Both concave and convex lenses rely on the light rays' focal point either to magnify, reduce, or focus an image. The focal point is simply the precise point where light rays come together in a pinpoint and focus an image. The distance from the center of the lens to the exact point where light rays focus is called the lens's focal length. Light that passes through a convex lens is refracted so that the rays join and focus out in front of the lens, creating a convergence. Light rays passing through a concave lens are refracted outward and create a divergence because the focal point for the light rays appears to be originating from behind the lens. Aberrations such as blurriness or color distortion may sometimes occur if lenses are made using glass with impurities or air bubbles or if the lens is not ground or polished properly to make it precisely curved and smooth.

APPLICATIONS AND PRODUCTS

Electronics. Lenses and mirrors are used extensively in consumer electronics, primarily as part of systems to read and write optical media such as CDs and DVDs. These optical disks encode information in microscopic grooves, which are read by a laser in much the same manner as a turntable's needle reads a record. Good-quality lenses are essential to focus the laser beam onto the disk and to capture the reflection from the disk's surface.

Cameras also make use of both mirrors and lenses. In single-lens reflex (SLR) cameras, the film or

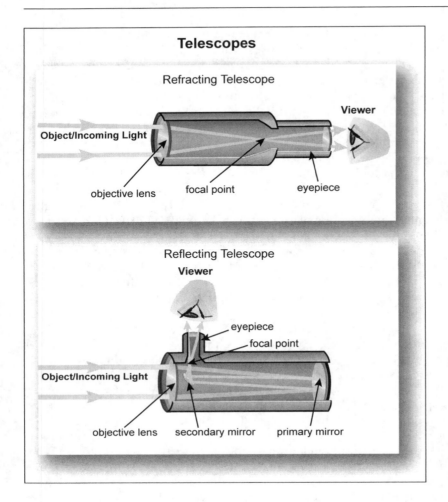

Telescopes

Refracting Telescope

Viewer

Object/Incoming Light

objective lens

focal point

eyepiece

Reflecting Telescope

Viewer

eyepiece

focal point

Object/Incoming Light

objective lens

secondary mirror

primary mirror

image sensor is protected from light by a flip mirror. When a photograph is taken, the flip mirror rotates, directing light onto the film or digital sensor to expose the image to a size that will fit onto the film. Modern camera lenses are actually composed of multiple simple lenses, which helps to improve the overall image quality.

Science. Astronomy has long been the driving force behind advances in optics, and many varieties of lenses and mirrors were developed specifically to enable astronomical observation. Large modern telescopes are limited not by the optical quality of their lenses and mirrors but by distortion caused by turbulence in the atmosphere. To address this, state-of-the-art telescopes use adaptive optical systems, which can change the shape of the telescope's primary mirror in response to fluctuations in the atmosphere, which are measured with a powerful laser.

Lens and mirror systems are also indispensable as tools in a larger research apparatus. Microscopes, high-speed cameras, and other digital equipment are commonplace in biological, chemical, and physical science laboratories, and many experimental setups include custom-made imaging equipment all based on combinations of mirrors and lenses.

Retail. JC Penney and Macy's department stores have installed full-size digital touch-screen mirrors in their stores, revolutionizing the retail experience. The mirrors enable consumers to model clothes, makeup, and jewelry digitally, without actually having to try on the products. Because an extraordinary amount of time and energy is saved by the interactive retail mirrors, customers have more flexibility to shop creatively by sampling a broader range of merchandise. The interactive digital mirrors also offer those shopping for friends or relatives the added convenience of being able to gauge sizes correctly, greatly reducing the number of returns. The increased ease and efficiency of touch-screen digital mirrors has generated more customers and sales for both retailers.

IMPACT ON INDUSTRY

Military. There are numerous military applications for lens- and mirror-based systems, and these have had an enormous influence on industry. Relatively simple applications include binoculars, range-finding systems, and weapon scopes. More complex optics are employed in guidance systems for aircraft and missiles, which often pair telescopic systems with image-recognition software. Reconnaissance satellites are also an important use for mirror and lens systems, and the general concept of a spy satellite is similar to a large space-based telescope, with multiple lenses and mirrors serving to magnify an image before projection onto an imaging sensor. Cutting-edge corporations such as Raytheon, Elbit Systems of America, and Christie were recently awarded multimillion-dollar defense contracts for military-based research involving lens and mirror applications in radar, lasers, and projectors.

Medicine. Lens technology has, of course, revolutionized the practice of optometry, and contact lenses in particular have become a multibillion-dollar industry worldwide. Because eyeglasses are essentially a single-lens optical that corrects for abnormalities in the shape of the eye, the basic geometry of the eyeglass lens has not changed for many years. However, major progress has been made in lens coatings and materials, which improve the aesthetics, durability, weight, and cost of eyeglasses. Additionally, modern manufacturing techniques have allowed greater lens diversity with more complex designs, creating progressive lenses that have a continuously variable amount of focus correction across the lens wearer's field of vision (as opposed to traditional bifocals, which provide only two levels of correction). Bausch + Lomb is at the forefront of engineering progressive lenses and adjustable-focus lenses, which allow the lens wearer to manually adjust the focus correction depending on the distance of the object being viewed.

Government and University Research. Undoubtedly, the greatest ongoing government research projects involving lenses and mirrors are undertaken by the National Aeronautics and Space Administration (NASA). The Hubble Space Telescope, launched by NASA in 1990, marks the culmination of decades of scientific research and the apex of mirror engineering and design. NASA's impact on the mirror and lens industry at large is paramount, since the majority of lens and mirror innovations have been a direct result of lens and mirror engineering for telescopes largely developed by NASA scientists. Although the Hubble Space Telescope contains the most advanced and sophisticated mirrors ever created, Hubble's successor, the James Webb Space Telescope, promises to surpass even Hubble by being able to see farther into space by monitoring infrared radiation.

Composite Mirror Applications, in Tucson, Arizona, has worked closely with NASA to build mirrors used in space, most notably building a mirror used by space shuttle Endeavor on its final mission to search for dark matter and anti-matter in the universe. Steward Observatory at the University of Arizona has also worked in close conjunction with

Composite Mirror Applications in the development of special optical projects with applications for the telescopic industry.

CAREERS AND COURSE WORK

Courses in physics, astronomy, advanced mathematics, geometry, optical engineering, mechanical engineering, aerospace engineering, computer science, and chemistry provide background knowledge and training needed to pursue a future working with various applications utilizing mirrors and lenses. A bachelor's degree in any of the above fields would assist in the basic applications of mirrors and lenses, but careers involving research or development of lenses in the corporate domain or academia would almost certainly necessitate a graduate degree in one of the above disciplines—ideally a doctorate. Although employment at any university as an educator and researcher is one potential career path, with an emphasis in physics and astronomy, working at an observatory is also a career possibility. Having an additional concentration of course work in aerospace engineering is also outstanding preparation for employment with NASA and government space projects.

Additional career opportunities working with lenses abound in the medical field, especially as an optical engineer. Either a master's or doctorate in physics or optics is a prerequisite for a career as an optical engineer; however, an associate's degree is all that is necessary to become trained to work as a contact lens technician. To work as an optometrist, one must graduate from a college of optometry after completing a bachelor's degree, and to work as an ophthalmologist, one must obtain a medical degree.

Additional career opportunities in the "green" industry using mirrors in solar panels requires a background in environmental engineering, and work in the photographic industry requires extensive knowledge of photography as well as lenses.

SOCIAL CONTEXT AND FUTURE PROSPECTS

Although the increased necessity of using mirrors in green technology (such as solar panels) has become universally recognized, mirrors may conceivably play a key role in another environmental technology in the future. Scientists researching the long-term impact of global warming, increasingly alarmed by the rapid escalation of carbon dioxide

Fascinating Facts About Mirrors and Lenses

- In the sixteenth century, mirror makers were held prisoner on the Venetian island of Murano, and they were sentenced to death if they left the island, for fear their secret mirror-making techniques would be revealed to the outside world.
- In 2010, French artist Michel de Broin created the world's largest disco mirror ball. Measuring almost twenty-five feet across and containing more than 1,000 mirrors, the mirror ball was suspended above Paris using a skyscraper crane.
- Other than humans and a handful of other higher primates, such as chimpanzees, orangutans, and bonobos, scientists studying self-awareness have discovered only four species that are able to recognize their own reflection in a mirror: bottlenose dolphins, killer whales, Asian elephants, and European magpies.
- In 1911, at the inaugural race of the Indianapolis 500, winner Ray Harroun used the first known automobile rearview mirror in his race car, the Marmon Wasp.
- The mirrors in the Hubble Space Telescope were ground to within 1/800,000 of an inch of a perfect curve and were the most highly polished mirrors on Earth.
- Roman emperor Nero, who was nearsighted, watched gladiator games at the Roman Colosseum through a large handheld emerald lens centuries before the invention of spectacles.

weight can be measured by the mirror and sent to a computer. Other mirrors may project a person's possible future appearance if certain negative lifestyle habits, such as smoking, overeating, excessive drinking, and lack of exercise, are practiced. Cameras placed throughout the home may record an individual's personal habits and then relay the information to an interactive home mirror linked to a computer. The mirror, in turn, may then portray possible physical results of a negative lifestyle to someone, traits such as obesity, discolored teeth, wrinkled skin, or receding hairline, turning mirrors into modern-day oracles.

Mary E. Markland, B.A., M.A.

FURTHER READING

Andersen, Geoff. *The Telescope: Its History, Technology, and Future.* Princeton, N.J.: Princeton University Press, 2007. A comprehensive survey of telescopes since their invention more than four hundred years ago, plus provides step-by-step instructions describing how to build a homemade telescope and observatory.

Burnett, D. Graham. *Descartes and the Hyperbolic Quest: Lens Making Machines and Their Significance in the Seventeenth Century.* Philadelphia: American Philosophical Society, 2005. Fascinating account of Descartes's lifelong dream of building a machine to manufacture an aspheric hyperbolic lens to end spherical aberration in lenses. Contains extensive drawings of Descartes's designs.

Conant, Robert Alan. *Micromachined Mirrors.* Norwell, Mass.: Kluwer Academic, 2003. Using profuse illustrations, examines how the miniaturization of mirrors has transformed modern technology, especially their use in scanners and in the electronics industry.

Kingslake, Rudolph. *A History of the Photographic Lens.* San Diego: Academic Press, 1989. Documents the development of photographic lenses from the beginning of photography in 1839 to modern telescopic lenses, providing ample pictures and sketches illustrating each lens type.

Pendergrast, Mark. *Mirror Mirror: A History of the Human Love Affair with Reflection.* New York: Basic Books, 2003. An exhaustive examination of mirrors and their influence on science, astronomy, history, religion, art, literature, psychology, and advertising, with an abundance of biographical

buildup in the Earth's atmosphere, have begun seriously investigating the potential for mirrors to help lower the temperature of the Earth's atmosphere. Scientists are studying the possibility of launching a series of satellites to orbit above the Earth, each one containing large mirrors, which could be controlled, like giant window blinds, to reflect back out into space as much of the Sun's rays as desired, thereby regulating the temperature of the Earth's atmosphere like a thermostat.

Another futuristic trend foresees the transformation of mirrors in the home, programming bathroom mirrors, for example, to monitor an individual's health by analyzing a person's physical appearance. Vital signs, such as heart rate, blood pressure, and material, supported by photographs, about scientists and mirror innovators.

Zimmerman, Robert. *The Universe in a Mirror: The Saga of the Hubble Space Telescope and the Visionaries Who Built It.* Princeton, N.J.: Princeton University Press, 2008. The behind-the-scenes story of the building of the Hubble Space Telescope, from its first conception in the 1940's, to its launch, repair, and triumph in the 1990's. Contains spectacular photographs of outer space sent back to Earth by the Hubble.

WEB SITES
American Astronomic Society
http://aas.org

The International Society for Optics and Photonics
http://spie.org

National Aeronautics and Space Administration
The James Webb Space Telescope
http://www.jwst.nasa.gov

N

NEPHROLOGY

FIELDS OF STUDY

Dialysis; transplantation; pediatric nephrology; proteomics; genetics; electrolyte physiology; hypertension; plasmapheresis; mineral metabolism; pharmacology; internal medicine; nephrolithiasis.

SUMMARY

Nephrology is a division of medical science associated with internal medicine, concentrated on the kidneys and the associated anatomy, physiology, diseases, and disorders. The word "nephrology" is of Greek origin: *nephros*, meaning kidney, and *logos*, meaning word, reason, thought, or discourse. Kidney diseases may include electrolyte-balance disorders and hypertension. Therapies associated with kidney disease manage limitations or failure of the renal system and may include dialysis or kidney transplant. Disorders of the kidney are often a result of systemic or congenital disorders, affecting more than one system or organ in the body, which can make treatment complicated and often challenging. A medical doctor who specializes in the diagnoses and treatment of the kidneys is a nephrologist.

KEY TERMS AND CONCEPTS

- **Calyx:** Cuplike urine-collection cavity located at the tip of each pyramid within the kidney.
- **Cortex:** Outer layer of the kidney that contains millions of microscopic nephrons.
- **Dialysis:** Process of removing toxic materials from the blood and maintaining fluid, electrolyte, and acid balance often using automated equipment when kidneys have become damaged or have been removed.
- **Kidney:** One of a pair of reddish brown, bean-shaped organs located on either side of vertebral column in the retroperitoneal area. Their function is to form urine from blood plasma by regulating water, electrolyte, and acid balance of the blood.
- **Nephrolithiasis:** Presence of calculi in the kidney, commonly called a kidney stone.
- **Nephron:** Functional unit of the kidney, consisting of a glomerulus, Bowman's capsule, renal tubule, and peritubular capillaries. Urine is formed by nephrons by a process of filtration, reabsorption, and secretion.
- **Pyramid:** Triangular tissue in the medulla (the inner region) of the kidney, which contains the loops and collecting tubules of the nephrons.
- **Renal Artery:** Blood vessel that carries blood from the aorta into the kidney. Inside the kidney, the renal artery branches into kidney tissue until the smallest artery (arteriole) leads to a glomerulus, where filtration can begin.
- **Renal Vein:** Blood vessel that carries blood from the kidney, back to the heart.
- **Ureter:** Muscular tube with a mucosal lining that leads from each kidney to the urinary bladder.
- **Urine:** Fluid created by the kidney that consists of 95 percent water and 5 percent dissolved solids (salts and nitrogen-containing wastes), which is eliminated by the body.

DEFINITION AND BASIC PRINCIPLES

Nephrology is a branch of medical science, a specialty of internal medicine, concerned with the structure and function of the kidneys. Any pathology within this system and the management of many systemic diseases affecting the kidneys are key responsibilities of a nephrologist. A nephrologist will determine the stage or degree of kidney disease, treat any associated complications (hypertension, bone disorders, vitamin imbalances), manage anemia, educate the patient about nutrition, risk factors, treatments, and transplantation. The nephrologist also commonly provides the vascular access placement for dialysis and coordinates treatment.

The purpose of the kidneys is to regulate water,

electrolytes, and acid-base content of the blood and, indirectly, all other body fluids. Filtration by the kidneys is a continuous process and the rate is affected by blood flow through the kidneys and daily fluid intake. Blood enters the kidney and passes through the glomerulus, where water and dissolved substances are filtered through capillary membranes, and the inner layer of Bowman's capsule, where it becomes glomerular filtrate. The blood cells and larger protein cells stay in the capillaries during this time. The filtrate travels through a series of renal tubules, where useful substances such as water, glucose, amino acid, minerals, and vitamins are reabsorbed into the capillaries to be used in the body. The amount of water that is reabsorbed is regulated by antidiuretic hormone and indirectly by aldosterone. The products that are not reabsorbed, such as metabolic products of medications, are considered waste and remain in the filtrate and become part of urine. The collecting tubules join together to form papillary ducts that empty urine into the calyxes and eventually into the ureter and bladder. The process from the ingestion of a large amount of liquid to the production of urine takes about forty-five minutes for most well-hydrated people. The average volume of urine produced each day is about 1,500 milliliters, but it depends on many factors—age, climate, activity, diet, and blood pressure. Patient pathology or disease may affect the urine production; however, people who have had part of a kidney removed or have only one kidney can have normal renal function.

There are five recognized stages of kidney disease.

- **Stage 1:** Slightly diminished function. Kidney damage with normal or relatively high glomerular filtration rate: >90 milliliters per minute per 1.73 m^2.
- **Stage 2:** Mild reduction in kidney function. Glomerular filtration rate: 60-89 milliliters per minute per 1.73 m^2.
- **Stage 3:** Moderate reduction in kidney function. Glomerular filtration rate: 30-59 milliliters per minute per 1.73 m^2.
- **Stage 4:** Severe reduction in kidney function. Glomerular filtration rate: 15-29 milliliters per minute per 1.73 m^2.
- **Stage 5:** Established kidney failure. Glomerular filtration rate: <15 milliliters per minute per 1.73 m^2.

BACKGROUND AND HISTORY

English physician Richard Bright first established the relationship between the symptoms and pathology of renal failure in 1827. It was not until 1854 when Scottish chemist Thomas Graham described osmotic force. In 1861, Graham went on to explain the process of dialysis using a hoop form dialyzer. In the late 1800's and early 1900's, several scientists began performing dialysis and kidney transplants on animals. Austrian surgeon Emerich Ullmann performed a kidney autotransplant on a dog, and in 1914, pharmacologist John Jacob Abel and his colleagues Leonard Rowntree and Benjamin Turner discovered that salicylic acid could be removed from the blood of rabbits using dialysis. In 1924, German physician George Haas performed the very first dialysis procedure on a human. Dutch physician Willem Kolff treated sixteen patients with acute kidney failure between 1943 and 1944 but with limited success. The first success came in 1945 with the seventeenth patient, a sixty-seven-year-old woman in uremic coma due to acute renal failure from gram-negative sepsis. After eleven hours of hemodialysis, the patient regained consciousness and began to produce urine. She went on to live seven more years.

The Scribner shunt, a U-shaped Teflon tube, is the creation of Chicago-born Belding Scribner. The shunt, inserted between an artery and vein in a patient's forearm, could be opened and connected to the artificial kidney machine during dialysis. Teflon was relatively new to the biomedical community at the time, and its nonstick properties made it less likely to clot. Before Scribner's shunt, a patient could receive only a few dialysis treatments before doctors would run out of places to connect the machine to the patient. The shunt was first used on March 9, 1960, on Clyde Shields, who was dialyzed repeatedly for eleven years. Another patient was dialyzed for thirty-six years, undergoing 5,700 cycles of hemodialysis, before his death. In 1962, Scribner and American physician James Haviland developed the first free-standing dialysis center in the world, the three-bed Seattle Artificial Kidney Center.

HOW IT WORKS

Nephrology is the science that concentrates on the kidneys and the associated anatomy, physiology, diseases, and disorders. The improper functioning of the kidney may disrupt electrolyte balance or lead to hypertension (high blood pressure) and is often related to other systemic or congenital conditions.

Diagnosing Kidney Disease. Treating the underlying condition may be complicated by reduced kidney function, and treatment of the kidney may be a challenge because of the underlying condition. Deteriorating kidney function has very unspecific symptoms, which makes kidney disease difficult to diagnose. Patients may feel generally unwell or have a reduced appetite. Quite often, kidney disease is not recognized until a major complication such as anemia, pericarditis, or cardiovascular disease has been detected. People diagnosed with high blood pressure or diabetes often have their kidney function assessed as part of normal screening procedure.

Chronic Kidney Disease Detection. Chronic kidney disease is detected by a blood analysis for levels of creatinine. Higher levels of creatinine indicate a decreased glomerular filtration rate resulting in a decline in normal kidney function. A glomerular filtration rate of less than 60 milliliters per minute per 1.73 m2, for a period of three months, is classified as having chronic kidney disease. Red blood cells or excess protein detected in urinalysis may cause a physician to investigate more thoroughly.

Kidney Stone Detection. Patients may develop nephrolithiasis or renal calculus, also known as a kidney stone, which is a solid mass normally composed of mineral salts. In the kidney, a calculus can block the ureter and urine flow. Symptoms may include a severe, sudden pain, chills, fever, and appearance of blood in the urine. If the blockage cannot be passed on its own by relaxation of surrounding smooth muscle, it should be removed surgically or disintegrated ultrasonically.

APPLICATIONS AND PRODUCTS

Dialysis, the common name for hemodialysis, is the procedure used to treat end-stage kidney failure, transient kidney failure, and some poisoning or drug-overdose situations. Other indications for dialysis include: hyperkalemia, uremia, uremic pericarditis, acidosis, and fluid overload. It is used to act as an artificial kidney, outside the body when a disorder is causing fluid, acids, electrolytes, and some drugs from being effectively eliminated.

Dialysis. Dialysis involves a series of five steps:

- Establishment of access to the patient's circulatory system via an arteriovenous fistula, graft, or catheter.
- Anticoagulating the patient's blood in order to prevent clotting during its circulation outside the body.
- Pumping of the patient's blood to a dialysis membrane.
- Adjusting the diffusion of solutes from the blood into a buffered dialysis solution.
- Returning the cleaned and buffered blood to the patient's circulation.

Typically, the entire process takes between three and four hours to complete and must occur several times per week. Even with regular dialysis, mortality rates remain quite high for end-stage renal disease patients. Most deaths are attributed to stroke, heart disease, or complications from diabetes. Complications of dialysis include hypotension (low blood pressure), infection at access site, sepsis, air embolism, bleeding, anemia, and muscle cramping.

Dialysis is the only treatment option for most renal disease patients, and the procedure that has not changed much since it was first performed. Unfortunately, during the days off of treatment, patients experience toxin buildup leaving them feeling bloated, tired, and uncomfortable.

There are two common types of dialysis, the most common being hemodialysis. Hemodialysis removes wastes and water by circulating blood outside the body through an external dialyzer that acts as a filter and contains a semipermeable membrane. Peritoneal dialysis uses a peritoneal membrane inside the body to filter wastes and water from the blood.

Home Dialysis Machines. In an effort to meet the growing need for dialysis treatment and provide some convenience to patients, home-treatment machines were created. The first home-treatment units were very large, very expensive, and cleaning and maintenance were difficult, but things are improving. Modern home dialysis machines are portable, and after rigorous training sessions, patients can treat themselves at home in six shorter sessions per week, rather than in several three- or four-hour sessions at a facility. Only about 1 percent of dialysis patients are being treated at home. Machines cost almost $25,000, are the size of a microwave, and perform the sophisticated filtration using a disposable cartridge.

Anemia Treatments. Anemia is an independent predictor of mortality in chronic kidney disease patients and is also associated with worsening of cardiovascular morbidity and accelerated rate of kidney damage. The administration of recombinant human erythropoietin (rHuEpo) has greatly reduced anemia in patients with chronic kidney disease. Unfortunately, almost 15 percent of patients show limited or no response to rHuEpo.

Genomic and Proteomic Research. Genomics (the study of genomes or DNA sequence of organisms) and proteomics (the study of structure and function of proteins) in nephrology are still in the early phases of research. The application of these approaches has recently produced promising new urinary biomarkers for kidney injury and chronic kidney disease, which may provide better understanding of renal physiology and assist in the development of new therapeutic strategies.

IMPACT ON INDUSTRY

Industry has stepped in to help manage the growing number of patients requiring dialysis treatment for end-stage renal disease. More than 70 percent of patients receiving dialysis in the United States are doing so at corporate-owned facilities. Many companies are public, and concern has been raised that decisions regarding patient health are managed based on corporate policy, rather than the physician's advice. Some physicians act as health care providers in these facilities as well as medical director, which may lead to complex conflicts of interest regarding patient care and corporate responsibility.

United States Renal Data System database requires that physicians provide information regarding their role in a facility and patient outcomes. Research is being done to investigate the trends of corporate involvement in dialysis treatment when compared with government or hospital facilities with regard to patient outcomes, costs, and transplantation referral rates.

A recent study concluded that kidney patients can improve their health by undergoing twice as many dialysis sessions. Industry has developed options for patients and health care providers to explore different ways to provide additional treatment while not overwhelming the health care system or the time commitment of the patient. Home-dialysis equipment makers are developing and seeking Food and Drug Administration (FDA) approval for home-based.

Fascinating Facts About Nephrology

- The World Health Organization (WHO) reports that more than 68,300 kidney transplants are performed every year worldwide.
- Kidney disease affects more than 600 million people globally.
- The 1990 Nobel Prize in Physiology or Medicine was awarded to Joseph E. Murray, the surgeon who performed the first-ever kidney transplant between identical twins, which demonstrated that previous failures were due to immunologic incompatibilities rather than surgical methods.
- The average kidney is about 4.5 inches long and weighs between four and six ounces.
- The kidneys have a higher blood flow than the heart, liver, or brain.
- World Kidney Day started in 2006 and is now an annual event held every March in more than one hundred countries.
- Most people are born with two kidneys but can survive with a single healthy kidney.
- The Voluntary Health Association of India estimates that about 2,000 Indians sell a kidney every year.
- The largest reported kidney stone weighed just more than thirty-one pounds.
- The United States has one of the highest incidences of end-stage renal disease in the world, 363 per million people, compared with Iceland, which has less than 60 per million.
- Chinese medicine believes that kidney stones may be caused by blockage or imbalance of chi (vital energy) in the kidney and acupuncture can restore positive energy flow.

---

devices and programs to keep up with the rising demand for treatment

CAREERS AND COURSE WORK

The most well-known career choice in nephrology is a nephrologist, which is a medical doctor (M.D.) who goes on to specialize in diseases and disorders of the kidney as a subspecialty of internal medicine. Becoming a nephrologist in the United States requires the completion of a bachelor's degree, four years of medical school, a three-year residency in internal medicine followed by an additional two-year fellowship in nephrology.

Nephrology social workers provide support and maximize the psychosocial functioning and adjustment of patients who are experiencing end-stage renal disease. They assist patients with the management of social and emotional stresses, which may include shortened life expectancy; altered lifestyle with changes in social, financial, vocational, and sexual functioning; and the complex, rigorous and time-consuming treatment required.

Dialysis technicians, also referred to as renal dialysis technicians, hemodialysis technicians, or nephrology technicians, operate the dialysis machines in hospitals and clinics, under the supervision of physicians. They may also travel to a patient's home to provide home treatment. Dialysis technicians must have a high school diploma or GED and have completed an additional training program at a technical school, community college, or hospital.

Renal scientists normally hold a master's or doctoral degree and have conducted years of research, as well as written and defended a thesis. Much of their work is funded by research grants from government or foundation sources or by academic institutions. Renal scientists normally find work in academic institutions, biotechnology laboratories, or government agencies. Renal scientists often collaborate with nephrologists to bring new findings and technology to the clinical phase of development.

SOCIAL CONTEXT AND FUTURE PROSPECTS

Patients requiring nephrology treatment are increasing at an alarming rate. In 2008, there were more than 1.64 million people on dialysis. This growth in end-stage renal disease patients is five times the world population growth (1.3 percent). Diabetes alone is expected to grow by 165 percent by 2050; therefore it is estimated that one out of every eight United States residents will have some level of kidney disease. Improving the facilitation of dialysis treatment to patients is a key concern as well as managing the growing number of people requiring treatment. The average cost to treat each patient with end-stage renal disease is about $85,000 annually.

Research has found that daily treatment provides better outcomes for patients. These findings are not accommodated by Medicare coverage, which, on average, pays for three treatments per week. Home-dialysis programs are being established to meet the growing need for treatment and allow patients the flexibility to accommodate treatment into their lives. Patients receive training on how to operate the machines as well as basic dialysis knowledge.

About one-third of patients who have a friend or family member willing to become a living kidney donor will not be able to receive the donor kidney because of an incompatible blood type or cross match. Kidney exchange programs are becoming more popular and are being facilitated by major medical centers and health providers. A kidney exchange increases the pool of available donors through an exchange between incompatible donor-recipient pairs, or a donor chain, managed by a specialized computer program that matches donors and recipients.

April D. Ingram, B.Sc.

FURTHER READING

Goldsmith, David, Satish Jayawardene, and Penny Ackland, eds. *ABC of Kidney Disease.* Malden, Mass.: Blackwell Publishing, 2007. This book contains excellent illustrations and clearly written information to provide a greater understanding of renal disease.

Greenberg, Arthur, ed. *Primer on Kidney Diseases.* Philadelphia: Saunders Elsevier, 2009. This book is endorsed by the National Kidney Foundation and provides very good background and basics of renal anatomy and physiology. Defines and classifies common kidney disorders and outlines treatment protocols and modalities.

Jörres, Achim, Claudio Ronco, and John A. Kellum, eds. *Management of Acute Kidney Problems.* Berlin: Springer-Verlag, 2010. This excellent reference contains information regarding the definition, epidemiology, pathophysiology, and clinical causes of acute kidney failure.

Lai, Kar Neng, ed. *Practical Manual of Renal Medicine: Nephrology, Dialysis and Transplantation.* Hackensack, N.J.: World Scientific, 2009. This manual provides practical information about dialysis, transplantation, and general nephrology in straightforward language. It describes treatment rationale and kidney disease management.

Schrier, Robert W. *Manual of Nephrology.* 7th ed. Philadelphia: Lippincott, Williams & Wilkins, 2009. An excellent clinical reference guide for the advanced student.

WEB SITES

American Board of Internal Medicine
http://www.abim.org

American Society of Nephrology
http://www.asn-online.org

International Society of Nephrology
http://www.isn-online.org

National Institute of Diabetes and Digestive and Kidney Diseases
http://www2.niddk.nih.gov

See also: Pharmacology; Surgery.

NEURAL ENGINEERING

FIELDS OF STUDY

Biomedical engineering; computational neuroscience; medicine; neural prostheses; neuroscience; chemistry; neural implants; neural interfacing; bioelectrical engineering; biology; biomaterials; neurosurgery; electrical engineering; materials science; nanotechnology; neural imaging; neural networks; tissue engineering.

SUMMARY

Neural engineering is an emerging discipline that translates research discoveries into neurotechnologies. These technologies provide new tools for neuroscience research, while leading to enhanced care for patients with nervous-system disorders. Neural engineers aim to understand, represent, repair, replace, and augment nervous-system function. They accomplish this by incorporating principles and solutions derived from neuroscience, computer science, electrochemistry, materials science, robotics, and other fields. Much of the work focuses on the delicate interface between living neural tissue and nonliving constructs. Efforts focus on elucidating the coding and processing of information in the sensory and motor systems, understanding disease states, and manipulating neural function through interactions with artificial devices such as brain-computer interfaces and neuroprosthetics.

KEY TERMS AND CONCEPTS

- **Cochlea:** Coiled part of the inner ear where the hearing receptors reside.
- **Electrode:** Solid conductor through which electrical current enters or leaves a medium.
- **Motor Cortex:** Area of cerebral cortex (outer brain layer) that processes motor information and control movement.
- **Photodiode:** Semiconductor component with light-sensitive electrical characteristics.
- **Retina:** Light-sensitive layer lining the inner eyeball.
- **Thalamus:** Mass of neural tissue situated deep in the brain.
- **Vagus Nerve:** Tenth and longest cranial nerve, which passes through the neck and thorax into the abdomen.
- **Visual Cortex:** Area of cerebral cortex that processes visual information.

DEFINITION AND BASIC PRINCIPLES

Neural engineering (or neuroengineering, NE) is an emerging interdisciplinary research area within biomedical engineering that employs neuroscientific and engineering methods to elucidate neuronal function and design solutions for neurological dysfunction. Restoring sensory, motor, and cognitive function in the nervous system is a priority. The strong emphasis on engineering and quantitative methods separates NE from the "traditional" fields of neuroscience and neurophysiology. The strong neuroscientific approach distinguishes NE from other engineering disciplines such as artificial neural networks. Despite being a distinct discipline, NE draws heavily from basic neuroscience and neurology and brings together engineers, physicians, biologists, psychologists, physicists, and mathematicians.

At present, neural engineering can be viewed as the driving technology behind several overlapping fields: functional electrical stimulation, stereotactic and functional neurosurgery, neuroprosthetics and neuromodulation. The broad scope of NE also encompasses neurodiagnostics, neuroimaging, neural tissue regeneration, and computational approaches. By using mathematical models of neural function (computational neuroscience), researchers can perform robust testing of therapeutic strategies before they are used on patients.

The human brain, arguably the most complex system known to humankind, contains about 10^{11} neurons and several times more glial cells. Understanding the functional neuroanatomy of this exquisite device is a sine qua non for anyone aiming to manipulate and repair it. The "neuron doctrine," pioneered by Spanish neuroscientist Santiago Ramón y Cajal, considers the neuron to be a distinct anatomical and functional unit. The extension introduced by American neuroscientist Warren S. McCullogh and American logician Walter Pitts asserts that the neuron is the basic information-processing unit

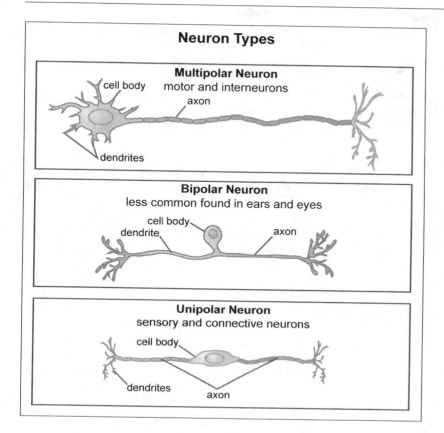

Neuron Types

Multipolar Neuron
motor and interneurons

cell body

axon

dendrites

Bipolar Neuron
less common found in ears and eyes

cell body

dendrite

axon

Unipolar Neuron
sensory and connective neurons

cell body

dendrites

axon

of the brain. For neuroengineers, this means that a particular goal can be reached just by manipulating a cell or group of cells. One argument in favor of this view is that stimulating groups of neurons produces a regular effect. Motor activity, for example, can be induced by stimulating the motor cortex with electrodes. In addition, lesions to specific brain areas due to neurodegenerative disorders or stroke lead to more or less predictable clinical manifestation patterns.

BACKGROUND AND HISTORY

Electricity (in the form of electric fish) was used by ancient Egyptians and Romans for therapeutic purposes. In the eighteenth century, the work of Swiss anatomist Albrecht von Haller, Italian physician Luigi Galvani, and Benjamin Franklin set the stage for the use of electrical stimulation to restore movement to paralyzed limbs. The basis of modern NE is early neuroscience research demonstrating that neural function can be recorded, manipulated, and mathematically modeled. In the mid-twentieth

century, electrical recordings became popular as a window into neuronal function. Metal wire electrodes recorded extracellularly, while glass pipettes probed individual cells. Functional electrical stimulation (FES) emerged with a distinct engineering orientation and the aim to use controlled electrical stimulation to restore function. Modern neuromodulation has developed since the 1970's, driven mainly by clinical professionals. The first peripheral nerve, then spinal cord and deep brain stimulators were introduced in the 1960's. In 1997, the Food and Drug Administration (FDA) approved deep brain stimulation (DBS) for the treatment of Parkinson's disease. An FES-based device that restored grasp was approved the same year.

In the 1970's, researchers developed primitive systems controlled by electrical activity recorded from the head. The U.S. Pentagon's Advanced Research Projects Agency (ARPA) supported research aimed at developing bionic systems for soldiers. Scientists demonstrated that recorded brain signals can communicate a user's intent in a reliable manner and found cells in the motor cortex the firing rates of which correlate with hand movements in two-dimensional space.

Since the 1960's, engineers, neuroscientists, and physicists have constructed mathematical models of the retina that describe various aspects of its function, including light-stimulus processing and transduction. In addition, scientists have made attempts to treat blindness using engineering solutions, such as nonbiological "visual prostheses." In 1975, the first multichannel cochlear implant (CI) was developed and implanted two years later.

HOW IT WORKS

Neuromodulation and Neuroaugmentation. Neural engineering applications have two broad (and sometimes overlapping) goals: neuromodulation and neuroaugmentation. Neuromodulation (altering nervous system function) employs stimulators and infusion devices, among other techniques. It can be applied at multiple levels: cortical, subcortical, spinal, or peripheral. Neural augmentation aims to amplify neural function and uses sensory (auditory, visual) and motor prostheses.

Neuromuscular Stimulation. Based on a method that has remained unchanged for decades, electrodes are placed within the excitable tissue that provide current to activate certain pathways. This supplements or replaces lost motor or autonomic functions in patients with paralysis. An example is application of electrical pulses to peripheral motor nerves in patients with spinal cord injuries. These pulses lead to action potentials that propagate across neuromuscular junctions and lead to muscle contraction. Coordinating the elicited muscle contractions ultimately reconstitutes function.

Neural Prosthetics. Neural prostheses (NP) aim to restore sensory or motor function—lost because of disease or trauma—by linking machines to the nervous system. By artificially manipulating the biological system using external electrical currents, neuroengineers try to mimic normal sensorimotor function. Electrodes act as transducers that excite neurons through electrical stimulation, or record (read) neural signals. In the first approach, stimulation is used for its therapeutic efficacy, for example, to alleviate the symptoms of Parkinson's disease, or to provide input to the nervous system, such as converting sound to neural input with a cochlear implant. The second paradigm uses recordings of neural activity to detect motor intention and provide input signal to an external device. This forms the basis of a subset of neural prosthetics called brain-controlled interfaces (BCI).

Microsystems. Miniaturization is a crucial part of designing instruments that interface efficiently with neural tissue and provide adequate resolution with minimal invasiveness. Microsystems technology integrates devices and systems at the microscopic and submicroscopic levels. It is derived from microelectronic batch-processing fabrication techniques. A "neural microsystem" is a hybrid system consisting of a microsystem and its interfacing neurons (be they cultured, part of brain slices, or in the intact nervous system). Technologies such as microelectrodes, microdialysis probes, fiber optic, and advanced magnetic materials are used. The properties of these systems render them suitable for simultaneous measurements of neuronal signals in different locations (to analyze neural network properties) as well as for implantation within the body.

APPLICATIONS AND PRODUCTS

Some of the most common applications of NE methods are described below.

Cochlear Implants. Cochlear implants (CI), by far the most successful sensory neural prostheses to date, have penetrated the mainstream therapeutic arsenal. Their popularity is rivaled only by the cardiac pacemakers and deep brain stimulation (DBS) systems. Implanted in patients with sensorineural deafness, these devices process sounds electronically and transmit stimuli to the cochlea. A CI includes several components: a microphone, a small speech processor that transforms sounds into a signal suitable for auditory neurons, a transmitter to relay the signal to the cochlea, a receiver that picks up the transmitted signal, and an electrode array implanted in the cochlea. Individual results vary, but achieving a high degree of accuracy in speech perception is possible, as is the development of language skills.

Retinal Bioengineering. Retinal photoreceptor cells contain visual pigment, which absorbs light and initiates the process of transducing it into electrical signals. They synapse onto other types of cells, which in turn carry the signals forward, eventually through the optic nerve and into the brain, where they are interpreted. Every neuron in the visual system has a "receptive field," a particular portion of the visual space within which light will influence that neuron's behavior. This is directly related to (and represented by) a specific region of the retina. Inherited retinal degenerations such as retinitis pigmentosa (RP) or age-related macular degeneration (AMD) are responsible for the compromised or nonexistent vision of millions of people. In these disorders, the retinal photoreceptor cells lose function and die, but the secondary neurons are spared.

Using an electronic prosthetic device, a signal is sent to these secondary neurons that ultimately causes an external visual image. A miniature video camera is mounted on the patient's eyeglasses that captures images and feeds them to a microprocessor, which converts them to an electronic signal. Then the signal is sent to an array of electrodes located on the retina's surface. The electrodes transmit the signal to the viable secondary neurons. The neurons process the signal and pass it down the optic nerve to the brain to establish the visual image.

Several different versions of this device exist and are implanted either into the retina or brain. Cortical visual prostheses could entirely bypass the retina, especially when this structure is damaged from diseases such as diabetes or glaucoma. Retinal prostheses, or artificial retinas (AR), could take advantage of any remaining functional cells and would target photoreceptor disorders such as RP. Two distinct retinal placements are used for AR. The first type slides under the retina (subretinal implant) and consists of small silicon-based disks bearing microphotodiodes. The second type would be an epiretinal system, which involves placing the camera or sensor outside the eye, sending signals to an intraocular receiver. In addition to challenges related to miniaturization and power supply, developing these systems faces obstacles pertaining to biocompatibility, such as retinal health and implant damage, and vascularization.

Functional Electrical Stimulation (FES). Some FES devices are commercialized, and others belong to clinical research settings. A typical unit includes an electronic stimulator, a feedback or control unit, leads, and electrodes. Electrical stimulators bear one or multiple channels (outputs) that are activated simultaneously or in sequence to produce the desired movement. Applications of FES include standing, ambulation, cycling, grasping, bowel and bladder control, male sexual assistance, and respiratory control. Although not curative, the method has numerous benefits, such as improved cardiovascular health, muscle- mass retention, and enhanced psychological well-being through increased functionality and independence.

Brain-Controlled Interfaces. A two-electrode device was implanted into a 1998 stroke victim who could communicate only by blinking his eyes. The device read from only a few neurons and allowed him to select letters and icons with his brain. A team of researchers helped a young patient with a spinal cord injury by implanting electrodes into his motor cortex that were connected to an interface. The patient was able to use the system to control a computer cursor and move objects using a robotic arm.

Brain-controlled interfaces (BCIs), a subset of NP, represent a new method of communication based on brain-generated neural activity. Still in an experimental phase, they offer hope to patients with severe motor dysfunction. These interfaces capture neural

Fascinating Facts About Neural Engineering

- Even though neuroengineering is still in its infancy, ethical questions are already arising. Will it affect human identity? Could it be used in the future to control thought processes? This is just the beginning.
- Cochlear implants are a great achievement of modern medicine and represent the most successful of all neural prostheses developed to date.
- Cell-containing polymer implants that release therapeutic factors hold promise for treating retinal disorders.
- An exciting new development in antiepilepsy therapy, "closed-loop" devices record electroencephalograph (EEG) signals, process them to detect imminent seizures, and deliver stimuli to stop them.
- The limb prostheses of the future will be equipped with multichanneled sensors that send tactile and proprioceptive feedback to the brain, continuously informing it about the effector's function. This approach will improve the patient's "sense of ownership" of the artificial limb.
- Scientists developed neuroprostheses that restore urinary bladder function by stimulating the spinal cord or nerves controlling the lower urinary tract.
- Advances in miniaturization and biosensors are expected to facilitate noninvasive monitoring of neuronal signaling and intracellular environment, thus greatly improving the diagnosis and treatment of nervous-system disorders.
- In a quest to replace the conventional, inadequate brain stimulation methods, scientists developed neural cells that become active when exposed to light and implemented carbon nanotube-based stimulators.

activity mediating a subject's intention to act and translate it into command signals transmitted to a computer (brain-computer interface) or robotic limb. Independent of peripheral nerves and muscles, BCI have the ability to restore communication and movement. This exciting technological advance is not only poised to help patients, but it also provides insight into the way neurons interact.

Every BCI has four main components: recording

of electrical activity, extraction of the planned action from this activity, execution of the desired action using the prosthetic effector (actuator), and delivery of feedback (via sensation or prosthetic device).

Brain-controlled interfaces rely on four main recording modalities: electroencephalography, electrocorticography, local field potentials, and singe-neuron action potentials. The methods are noninvasive, semi-invasive, or invasive, depending on where the transducer is placed: scalp, brain surface, or cortical tissue.

The field is still in its infancy; however, several basic principles have emerged from these and other early experiments. A crucial requirement in BCI function, for example, is for the reading device to obtain sufficient information for a particular task. Another observation refers to the "transparency of action" in brain-machine interface (BMI) systems: Upon reaching proficiency, the action follows the thought, with no awareness of intermediate neural events.

Deep Brain Stimulation (DBS) and Other Modulation Methods. Deep brain stimulation of thalamic nuclei decreases tremors in patients with Parkinson's disease. It may alleviate depression, epilepsy, and other brain disorders. One or more thin electrodes, about 1 millimeter in diameter, are placed in the brain. An external signal generator with a power supply is also implanted somewhere in the body, typically in the chest cavity. An external remote control sends signals to the generator, varying the parameters of the stimulation, including the amount and frequency of the current and the duration and frequency of the pulses. The exact mechanism by which this method works is still unclear. It appears to exert its effect on axons and act in an inhibitory manner, by inducing an effect akin to ablation of target area, much like early Parkinson's treatment. One major advantage of DBS over other previously employed methods is its reversibility and absence of structural damage. Another valuable neuromodulatory approach, the electrical stimulation of the vagus nerve, can reduce seizure frequency in patients with epilepsy and alleviate treatment-resistant depression. Transcutaneous electrical nerve stimulation (TENS) represents the most common form of electrotherapy and is still in use for pain relief. Cranial electrotherapy stimulation involves passing small currents across the skull. The approach shows good results in depression, anxiety, and sleep disorders.

Transcranial magnetic stimulation uses the magnetic field produced by a current passing through a coil and can be applied for diagnostic (multiple sclerosis, stroke), therapeutic (depression), or research purposes.

IMPACT ON INDUSTRY

Neural engineering is a fast-developing bioengineering specialty that is expected to grow tremendously. The increasing societal burden of neurological disorders, and the demand for more sophisticated medical devices, will drive an increase in new careers and employment. A global industry, with cutting-edge research under way in the United States, Europe, and Asia, neural engineering concentrates talent and capital in a network of neurotechnological innovation.

Government and University Research. Research in this field is funded through universities and various organizations such as National Science Foundation (NSF) and National Institutes of Health (NIH), including National Institute of Biomedical Imaging and Bioengineering (NIBIB). The Whitaker International Fellows and Scholars Program awards funds to emerging biomedical engineering leaders to conduct projects worldwide. Neurotechnology industry also supports some parts of academic research.

Therapeutic approaches approved by the FDA include spinal cord stimulation for pain, DBS for Parkinson's disease and essential tremor, and vagus nerve stimulation in epilepsy and depression. Techniques still at the investigational stage include DBS for depression, epilepsy, headache, Tourette's syndrome, and pain; cortical stimulation in Parkinson's disease, tremor, pain, depression, and stroke rehabilitation; and peripheral nerve stimulation for headache and tinnitus.

Industry and Business. With the promise of new treatments for billions of people suffering from nervous-system disorders, neurotechnology is fast becoming the leading recipient of life science venture capital worldwide. The NE and neurotechnology industry includes firms that manufacture neuromodulation devices, neural prostheses, rehabilitation systems, neurosensing devices including electroencephalograph (EEG) systems, magnetic sensing systems, sleep-monitoring equipment, neurosurgical monitoring equipment, and analytical tools. According to *Neurotech Reports*, the industry revenue is expected to grow to about $8.8 billion by 2012. Examples of prominent companies include Medtronic, Cyberonics, NeuroPace, and Trifectas Medical. Research endeavors at universities and clinical institutions frequently lead to start-up firms, such as Cyberkinetics Incorporated, a manufacturer of BCI devices.

The cochlear implant industry has progressed significantly in the first decade of the twenty-first century, achieving hundreds of millions of dollars in revenue. As hearing devices become more and more affordable and socially acceptable, the industry is expected to develop rapidly. Key players in this industry are Advanced Bionics, Cochlear, and Med-El.

CAREERS AND COURSE WORK

Most careers in NE require a bachelor's degree in engineering, biology, neuroscience, physics, or computer science. Neuroengineers often undergo formal training in mechanical or electronics engineering, combined with biomedical training. Educational programs in bioengineering, including NE, are growing rapidly in the United States. Typical course work integrates engineering and life sciences studies. Students may receive instruction in neuroengineering fundamentals; chemistry; fluid mechanics; engineering electrophysiology; diagnostic imaging physics; and drug development. In addition to core courses, students take electives related to their ultimate career goals. Subsequently, many pursue a master's or doctoral degree, sometimes followed by one or more years of postdoctoral training. As in other biomedical engineering fields, some graduates go on to obtain a medical, law, or business degree.

Neuroengineering researchers are employed by universities, medical institutions, industry (medical devices, pharmaceutical, biotechnology), and governmental agencies. They work as physicians, clinical engineers, product engineers, researchers, managers, or teachers.

SOCIAL CONTEXT AND FUTURE PROSPECTS

Bioelectrodes for neural recording and neurostimulation are an essential part of neuroprosthetic devices. Designing an optimal, stable electrode that records long-term and interacts adequately with neural tissue remains a priority for neural engineers. The implementation of microsystem technology opens new perspectives in the field.

More than 200 million people around the world suffer from hearing loss, mainly because sensory hair cells in the cochlea have degenerated. The only efficient therapy for patients with profound hearing loss is the CI. Improvements in CI performance have increased the average sentence recognition with multichannel devices. An exciting new development, auditory brainstem implants, show improved performance in patients with impaired cochlear nerves.

Millions of Americans have vision loss. The need for a reliable prosthetic retina is significant, and rivals the one for CI. Technological progress makes it quite likely that a functioning implant with a more sophisticated design and higher number of electrodes will be on the market soon. The epiretinal approach is promising, but providing interpretable visual information to the brain represents a challenge. In addition, even if they prove to be successful, retinal prostheses under development address only a limited number of visual disorders. Much is left to be discovered and tested in this field.

The coming years will also see rapid gains in the area of BCI. Whether they achieve widespread use will depend on several factors, including performance, safety, cost, and improved quality of life.

The advent of gene therapy, stem cell therapy, and other regenerative approaches offers new hope for patients and may complement prosthetic devices. However, many ethical and scientific issues still have to be solved.

Implanted devices are changing the way neurological disorders are treated. An unprecedented transition of NE discoveries from the research to the commercial realm is taking place. At the same time, new discoveries constantly challenge the basic tenets of neuroscience and may alter the face of NE in the coming decades. People's understanding of the nervous system, especially of the brain, changes, and so do the strategies designed to enhance and restore its function.

Mihaela Avramut, M.D., Ph.D.

FURTHER READING

Blume, Stuart. *The Artificial Ear: Cochlear Implants and the Culture of Deafness.* New Brunswick, N.J.: Rutgers University Press, 2010. Historical study of implant development and implementation.

DiLorenzo, Daniel J., and Joseph D. Bronzino, eds. *Neuroengineering.* Boca Raton, Fla.: CRC Press,

2008. Essential review of neuroengineering developments written by leaders in the field.

Durand, Dominique M. "What Is Neural Engineering?" *Journal of Neural Engineering* 4, no. 4 (September, 2006). Written by the editor in chief of the journal, who defines NE and its scope.

He, Bin, ed. *Neural Engineering.* New York: Kluwer Academic/Plenum Publishers, 2005. Introductory overview of research in neural engineering.

Katz, Bruce F. *Neuroengineering the Future: Virtual Minds and the Creation of Immortality.* Hingham, Mass.: Infinity Science Press, 2008. Fascinating introduction to this field, describing the state of the art and speculating on long-term developments.

Montaigne, Fen. *Medicine By Design: The Practice and Promise of Biomedical Engineering.* Baltimore: The Johns Hopkins University Press, 2006. Bioengineering (including neuroengineering) applications made accessible to the nonspecialist through vignettes and portraits of researchers.

WEB SITES
Engineering in Medicine and Biology Society
http://www.embs.org/index.html

International Functional Electrical Stimulation Society (IFESS)
http://ifess.org

National Institute of Biomedical Imaging and Bioengineering
http://www.nibib.nih.gov

Whitaker International Fellows and Scholars Program
http://www.whitaker.org

See also: Biomechanical Engineering; Bionics and Biomedical Engineering; Cell and Tissue Engineering; Computer Science; Electrical Engineering; Nanotechnology; Neurology.

NEUROLOGY

FIELDS OF STUDY

Internal medicine; neurosurgery; neuroscience; biology; molecular biology; biochemistry; pharmacology; neurophysiology; electrophysiology; neurobiology; neuroanatomy; neuroendocrinology; geriatric neurology; pediatric neurology; interventional neurology; veterinary neurology; psychology; behavioral neurology; cognitive neurology; neuroimaging; sleep medicine; movement disorders; neuromuscular disorders; cerebrovascular disorders; neurodegenerative disorders.

SUMMARY

Neurology is a rapidly developing branch of medicine dedicated to the diagnosis and treatment of disorders involving the brain, spinal cord, nerves, and muscles. The nervous system is crucial to human life. Understanding its operation under normal and pathological conditions is the focus of a group of disciplines that includes neurology, neuroscience, neurosurgery, and psychiatry. Considerable overlap and interaction exist between these specialties.

Among the most important and challenging neurological disorders are neurodegenerative conditions (such as Parkinson's disease, Alzheimer's disease, and amyotrophic lateral sclerosis), stroke, brain tumors, epilepsy, migraine, and multiple sclerosis.

KEY TERMS AND CONCEPTS

- **Anticholinergic:** Agent that blocks the neurotransmitter acetylcholine.
- **Ataxia:** Lack of muscle coordination during voluntary movements.
- **Basal Ganglia:** Group of nuclei deep within the cerebral hemispheres, functioning in movement coordination.
- **Blood-Brain Barrier:** Protective physical and functional barrier formed by a specialized brain capillary blood vessel structure; serves to keep the brain environment stable.
- **Brainstem:** Lower part of the brain, structurally continuous with the spinal cord, that plays important roles in consciousness, arousal, breathing, cardiovascular function, and digestion.
- **Central Nervous System:** Main processing center of the nervous system, consisting of the brain and spinal cord.
- **Cerebrospinal Fluid:** Clear liquid that surrounds the brain and spinal cord and fills their cavities.
- **Cortical:** Pertaining to the cerebral cortex, the outer part of the cerebrum, which is responsible for high brain functions.
- **Dopaminergic:** Related to the neurotransmitter dopamine.
- **Levodopa (L-Dopa):** Metabolic precursor of dopamine used in the treatment of Parkinson's disease.
- **Meninges:** Membranes that cover the central nervous system.
- **Myelin:** Insulating lipid and protein sheath covering nerve fibers that increases the speed of impulse transmission.
- **Peripheral Nervous System:** Part of the nervous system outside the brain and spinal cord, consisting of nerves that connect with target organs, muscles, blood vessels, and glands.
- **Synapse:** Specialized point of connection between two neurons or a neuron and an effector cell.

DEFINITION AND BASIC PRINCIPLES

Neurology is the medical discipline dedicated to the diagnosis, treatment, and management of diseases affecting the central nervous system, peripheral nerves, muscle effectors, and corresponding blood vessels. Its mission is thus distinct from that of other disciplines that focus on nervous system disorders, although some degree of overlap exists. Neuroscience studies the nervous system from a structural, functional, and molecular point of view, while neurosurgery is the surgical specialty involved in the treatment of nervous system disorders. Psychiatry deals with the diagnosis and treatment of mental illness. It is often taught that neurological diagnosis is based on answering two distinct questions: Where the lesion is located and what the nature of the disease is.

The concept of localization is central to the practice of neurology. A disorder can affect the cerebral hemispheres, the brainstem, the spinal cord, the peripheral nerves, and the muscles. Malfunctions in these territories fall into several categories:

motility disturbances, pain, disordered sensorium, seizures, altered consciousness, disturbed intellect or behavior, and changed speech and language. Alterations in mood or hormonal function are sometimes encountered. How the disorder manifests depends on the localization of the pathological process. Brainstem disease, for example, is suggested by cranial nerve palsies, ataxia of gait or limbs, or tremor. Involuntary movements often indicate that the basal ganglia might be affected. Identifying the precise nature of the disease process (for example, tumor, vascular malformation, neurodegeneration, infection) is facilitated by a variety of diagnostic tests.

BACKGROUND AND HISTORY

The structure and function of the human brain have preoccupied physicians and philosophers alike since the dawn of history. Prominent Greek physicians such as Alcmaeon of Croton, Hippocrates, and Galen correctly considered the nervous system to be the source of sensations, emotions, and cognitive faculties. In the sixteenth and seventeenth centuries, the anatomy of the brain was described in detail by Flemish physician Andreas Vesalius and English physician Thomas Willis. The term "neurology" was first used by Willis to describe the study of the brain.

Despite these efforts, the function of the nervous system remained obscure until the early 1800's, when studies of functional localization were performed. Using a microscope, Czech anatomist Jan Evangelista Purkyn described brain cells that he called "neurons." French physician Pierre Paul Broca's studies of speech localization and French anatomist Jean Cruveilhier's analysis of stroke-induced lesions laid the foundations of modern neuroscience and neurology. English physician James Parkinson described the "shaking palsy" later known as Parkinson's disease. In the second half of the nineteenth century, clinicians with exceptional observational abilities such as French neurologists Jean-Martin Charcot and Joseph Babinski made crucial contributions to the advancement of clinical neurology.

In the early twentieth century, Spanish histologist Santiago Ramón y Cajal and Italian physician Camillo Golgi refined the description of neurons as separate cells, which led to an improved understanding of synapses and interneuronal communication. German psychiatrist Alois Alzheimer gave a lecture in 1906 in which he first described what became known as

Alzheimer's disease. The complex processes occurring within a single neuron became accessible. Imaging techniques first introduced in the 1970's have been continuously improved and constitute essential diagnostic and research tools.

All these advances, and countless others, have shaped neurology into a discipline that is strongly anchored in scientific reasoning. Neurology remains a fast-growing medical field, rapidly incorporating advances in basic neuroscience, molecular genetics, and imaging technologies.

HOW IT WORKS

Understanding of the pathogenetic mechanisms in neurological disorders has improved tremendously, mainly because of significant progress in imaging, electrophysiology, and molecular genetic techniques. The diagnostic, however, still relies heavily on clinical history and neurological examination. Ancillary tests are frequently employed.

Clinical History. It is important for neurologists to know as much as possible about their patients' backgrounds and socioeconomic status. The tempo and duration of the disease are essential. Some conditions, such as stroke, occur suddenly, while others, such as tumors or dementia, have a gradual onset. Many disorders manifest continuously; others are characterized by remissions and exacerbations (multiple sclerosis, myasthenia gravis) or bouts (migraine headaches).

Neurological Examination. Neurological exams target multiple areas and are performed in an organized, step-wise manner. First, the patient's vital signs and general appearance are evaluated, with special focus on posture, motor activity, and potential signs of meningeal involvement. Alertness, speech, and language are assessed, usually in conjunction with a Mini Mental Status Exam. The integrity of cranial nerves 1 through 12 is evaluated. Motor system examination includes an evaluation of muscle tone and strength and possible muscular atrophy. Reflexes, coordination, and gait are also tested. Sensory function analysis involves testing the senses of touch, pain, temperature, vibration, and position.

Imaging Studies. Computed tomography (CT) and magnetic resonance imaging (MRI) are the core neurological imaging methods. In CT, the image is reconstituted with high speed from sets of X-ray

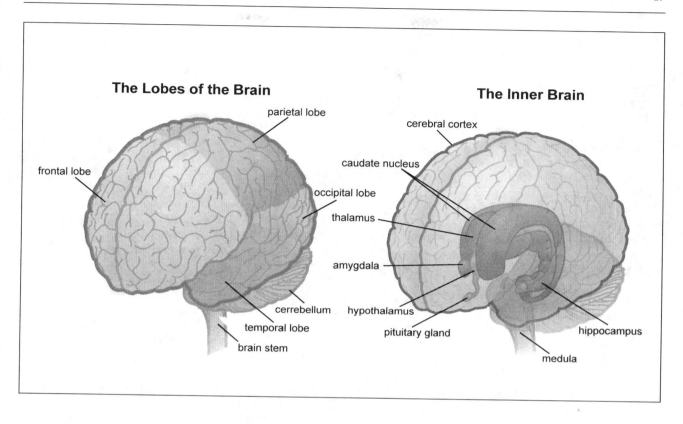

The Lobes of the Brain

- parietal lobe
- frontal lobe
- occipital lobe
- cerrebellum
- temporal lobe
- brain stem

The Inner Brain

- cerebral cortex
- caudate nucleus
- thalamus
- amygdala
- hypothalamus
- pituitary gland
- hippocampus
- medula

measurements. The method proves especially useful in evaluating acute strokes, head injuries, and acute infections, as well as in analyzing blood flow and identifying hemorrhages. It is also used for medically unstable or uncooperative patients and for patients with pacemakers or metallic implants, who cannot undergo an MRI. Magnetic resonance imaging is based on the ability of protons in water molecules to align to the direction of a strong magnetic field; on subsequent exposure to radio waves, the protons spin and emit signals that are detected by a receiver and processed by a computer. The technique plays an important role in detecting and delineating cerebral and spinal lesions. One advantage is the fact that it can be used when details on tissue physiology and biochemistry are needed.

Electrophysiological Studies. The arsenal of electrophysiological methods includes measures of brain electrical activity (such as electroencephalography and evoked potentials), electromyography, and nerve conduction studies. Electroencephalograms (EEGs) are noninvasive tests that display the electrical activity of the outer cortical layer neurons and serve as sensitive indicators of focal or generalized disturbances in neuronal activity. They can also reflect paroxysmal disturbances in neuronal function associated with seizure disorders. Evoked potentials (EPs) test the functional integrity of the sensory system and reflect the activity of the central nervous system in response to various stimuli. These measures of nervous system function are often important for differential diagnosis, especially in disorders that are not accompanied by morphological changes detectable with a CT or MRI scan.

Electromyograms (EMs) evaluate the electrical properties of the muscle. Nerve conduction studies (NCSs) assess the speed and strength of the electric impulse conduction along a peripheral nerve. The information derived from these studies allows the physician to establish if the patient's condition is of peripheral origin and whether the muscle or nerve is involved.

Lumbar Puncture and Cerebrospinal Fluid Studies. Studies of cerebrospinal fluid (CSF) for pressure, cellularity, pigments, protein, or glucose are undertaken after a thorough clinical evaluation and consideration of the value and risks of the lumbar puncture. They are especially valuable in the differential diagnosis of infections (such as meningitis and encephalitis), bleeding, stroke, malignancies, and demyelinating diseases. It is not undertaken when intracranial hypertension and brain herniation are suspected.

Other Diagnostic Tools. Additional techniques that can be employed for neurodiagnosis include ultrasonography, muscle and nerve biopsy, and DNA methods.

APPLICATIONS AND PRODUCTS

Neurological assessment aims to render an accurate diagnosis, provide therapy (including pain management), and lead rehabilitative efforts. At present, several categories of nervous system disorders are at the forefront of clinical and research efforts worldwide.

Movement Disorders. Many diseases with various pathological features and manifestations are included in this category. Patients diagnosed with this type of disorder often have difficulty walking, speaking, and performing basic daily tasks. They may exhibit involuntary movements, tremors, rigidity, and muscle spasms. The cause of these manifestations is incompletely understood but often involves the degeneration of neurons in central nervous system areas responsible for motor function. Examples of diagnostic entities commonly encountered in movement disorder clinics are Parkinson's disease, Huntington's disease, dystonia, myoclonus, and Gilles de la Tourette's syndrome. Therapeutic choices vary for different conditions and include medications (such as dopaminergic and anticholinergic drugs) and surgery that severs neural pathways, restores neuronal population, or stimulates certain brain areas. Physiotherapy represents an important component of disease management. Parkinson's disease treatment in particular is the object of an enormous basic and translational research effort, stemming from the inadequacy of long-term levodopa (L-dopa) administration and the absence of a treatment that addresses the degenerative process. Stem cell and gene therapy are examples of novel therapeutic approaches under investigation for Parkinson's disease.

Vascular Diseases. Stroke ranks in the top four causes of death in many countries and accounts for a large proportion of the neurological disease burden. It is caused by a reduction in cerebral blood flow (ischemia, usually induced by a blood clot, or thrombus) or bleeding (hemorrhage). The main clinical characteristic is a sudden onset of clinical signs of abnormal cerebral function, such as weakness, difficulty walking, sensory loss, vision and speech disturbances, and cognitive impairment. Treatment is directed at minimizing and preventing blood vessel occlusion, using antithrombotic drugs and surgical procedures. Extensive research studies focus on possible neuroprotective agents that can minimize the effect of nutrient deprivation on brain cells. Rehabilitation is essential and requires a team that includes neurologists and physical therapists.

Dementias. These disorders are diagnosed based on the development of multiple cognitive deficits that cause impairment in occupational and social functioning. Various forms of dementia have been identified. They can be grouped into four major categories, according to their pathogenesis: degenerative, vascular, metabolic, and infectious disorders.

The most frequently encountered degenerative dementia is named after Alzheimer, the German psychiatrist who first described its clinical features. Alzheimer's disease evolves progressively, with memory impairment (initially, for new information) and gradual cognitive loss, terminating in incapacitation and death. Neurological examination may show rigidity and postural changes, but primary motor and sensory functions are usually spared. Pathological changes in the brain include a striking atrophy, with loss of neurons; senile plaque microscopic lesions outside the cells, consisting of beta amyloid peptide; and tangle structures inside the neurons composed of filaments of a protein called tau. The exact causes and mechanisms of neuronal death need to be clarified and are under intense scrutiny by neuroscientists and neurologists all over the world. Medications that decrease the degradation of the neurotransmitter acetylcholine, thus increasing its levels, have been approved for treatment of this disorder. A drug that protects cells from the detrimental effects of the chemical messenger glutamate is used in intermediate to late stages. None of these agents, however, has disease-modifying effects. Alzheimer's disease remains a large part of the neuroscience and neurology research environment. Several strategies aimed at inhibiting disease progression have advanced to clinical trials. Among these, approaches targeting the production and clearance of the beta amyloid peptide are the most advanced. Attempts to modulate the abnormal aggregation of tau filaments and alleviate neuronal metabolic dysfunction are also being evaluated at the clinical level.

Demyelinating Diseases. Multiple sclerosis is a chronic disease characterized by multiple areas of inflammation in the white matter of the nervous system, associated with the loss of myelin sheath (demyelination) and scarring. It usually begins in young adults and has a relapsing and remitting course. Clinical manifestations are as variable as the distribution of the lesions; they often include unusual sensations, impaired vision, fatigue, muscle weakness, spasticity, ataxia, tremor, and bladder dysfunction. Cerebrospinal fluid examination, MRI, and evoked potentials are of great diagnostic value. The management of the condition is extremely challenging, and no cure has been developed. Treatment modalities include corticosteroids, beta interferons, glatimer acetate, mitoxantrone, and natalizumab. Symptomatic treatment and physical therapy are an important part of disease management. The use of transplanted stem cells that release immunomodulatory and neuroprotective factors is being explored in clinical trials.

Spinal Cord Diseases. Amyotrophic lateral sclerosis (ALS), a disease of unknown pathogenesis, is characterized by progressive muscle weakness and atrophy because of the selective degeneration of motor neurons and pathways in the brain and spinal cord. Muscle cramps and weight loss are characteristic symptoms. Breathing is compromised because of paresis of respiratory muscles. Treatment is symptomatic, as no effective drug treatment exists. In the twenty-first century, ALS research has seen a dramatic expansion, comparable to that of Parkinson's disease and dementia. Studies that hold significant promise are investigating mechanisms of disease and testing growth factors and chemical messenger receptor antagonists, as well as gene and stem cell therapies.

Paroxysmal Disorders. Migraines are paroxystic, intense hemicranial pain episodes, accompanied by vomiting and light sensitivity. The cause of these episodes is still unclear. Acute treatment of migraine employs drugs such as sumatriptan and ergotamine, while prophylactic therapy uses various agents such as propranolol, verapamil, amitriptyline, and valproate.

Epilepsy is a chronic condition characterized by recurrent seizures that are typically unprovoked and

---

Fascinating Facts About Neurology

- The human brain consists of about 200 billion neurons and many more supportive cells.
- The adult brain weighs around 1,500 grams and represents only 2 percent of total body weight; however, its oxygen supply needs (72 liters per day) account for 20 percent of the body's total oxygen consumption.
- Synesthesia is a neurological phenomenon in which a stimulus directed at a certain sense modality triggers sensations in a different sense territory. Individuals with this synesthesia can, for example, see yellow or smell cinnamon when they hear a G-sharp note.
- Locked-in syndrome sufferers have intact cognitive function and can see and hear; however, they cannot move, speak, or otherwise communicate except through coded eye movements.
- Prions are nonbacterial, nonviral agents that do not contain genetic material. They are responsible for some of the most intriguing disease entities in neurology, such as mad cow disease, Creutzfeldt-Jakob disease, fatal insomnia, and kuru, which affects tribe members in Papua New Guinea who eat the tissue of affected people during cannibalistic funeral rituals.
- In 1969, neurologist Oliver Sacks administered L-dopa (then a new miracle drug) to his institutionalized, lethargic patients suffering from Parkinson's disease symptoms after the mysterious World War I encephalitis epidemic. The drug gave most of these patients a spectacular, sudden awakening and allowed them to temporarily experience active life.
- Interventional neurology, also known as neuroendovascular therapy, is a neurology subspecialty that deals with blood vessel lesions of the brain and spine in a minimally invasive manner. A microscopic catheter is inserted through a nick in the leg, then advanced via the femoral artery to the site of the lesion.

---

seldom predictable. It affects more than 40 million people worldwide. Seizures are the result of a temporary perturbation in the function of cortical neurons that causes a self-limited, hypersynchronous electrical discharge. The electrophysiological disturbance can be detected on a scalp EEG recoding. Several distinct forms of epilepsy (such as partial, generalized tonic-clonic, myoclonic, and absence) have been identified, each of which has its own behavioral changes, EEG activity, natural history, and response to treatment. Therapy is aimed at eliminating or reducing seizures, avoiding side effects of long-term treatment, and restoring the patient's psychosocial function. Many antiepileptic drugs are in use; phenytoin, carbamazepine, phenobarbital, and valproate are some of the mainstays. Surgical treatment becomes necessary when seizures cannot be controlled pharmacologically, and they affect the patient's quality of life.

Other Disease Categories. Tumors, infections, genetic and developmental diseases, birth injuries, trauma, peripheral neuropathies, neuromuscular diseases, and myopathies are also the focus of neurological assessment.

IMPACT ON INDUSTRY

The burden of neurological disease affects every age group, every segment of society, and every country in the world; therefore, neurology has become a strong focus of academic and industrial research worldwide. The future of drug discovery and development is likely to rely on extensive cooperation among academic, governmental, nonprofit, and industry organizations.

Government and University Research. The basic sciences section of academic neurology is expanding, both in the United States and abroad. Basic scientists usually spend most of their time performing research, with some clinical responsibilities and teaching obligations. Patient-oriented researcher positions have been created in some institutions to optimize the translation of scientific advances into delivery of health care.

Funding mechanisms include grants and scholarships from governmental institutions such as the National Institutes of Health in the United States, associations, and private donors. Partnerships with industry represent an increasingly used avenue for research support. These sources fund numerous research studies every year in a collaborative but extremely competitive environment.

Within the National Institutes of Health, at least two branches, the National Institute of Neurological Disorders and Stroke (NINDS) and the National Institute of Mental Health (NIMH), are dedicated to researching nervous system disorders. Other major government agencies where neurology expertise is needed include the Environmental Protection Agency and the Food and Drug Administration.

Industry and Business Sectors. According to the Neurotechnology Industry Organization, more than five hundred international companies are active in the field of neurotechnology. These firms provide drugs, biological therapies, medical devices, diagnostic apparatus, and surgical equipment.

The pharmaceutical industry's keen interest in neurological disorders is evident from the rapid growth in central nervous system (CNS) drug development. In 2010, the global central nervous system drug market exceeded $60 billion. It is set to expand even further, especially as the number of people over the age of sixty-five increases, leading to a higher demand for more effective and safe medicines for neurological disorders.

Despite this impressive increase in drug discovery, the major challenge remains drug delivery across the blood-brain barrier, which limits the access of drugs to brain cells. Patients with fatal or debilitating central nervous system diseases such as tumors, HIV, encephalopathy, epilepsy, stroke, and neurodegenerative disorders far outnumber those dying of all types of systemic cancers or heart diseases. The presence of the blood-brain barrier is often the sole reason for the clinical failure of highly potent neurotherapeutic agents.

A booming neurology devices sector concentrates its efforts on perfecting neurostimulation devices, neurovascular stents, and diagnostic electrophysiology apparatus. Companies that produce these tools are involved in many major clinical trials.

CAREERS AND COURSE WORK

A career in neurology is demanding and extremely rewarding. Becoming a neurologist requires a solid foundation in the basic sciences and extensive training. Undergraduate education usually involves numerous science courses, including biology and chemistry. Many colleges offer a premedical degree option, designed to create the foundation for medical education. The next step is obtaining a medical degree (M.D. or D.O.), which requires four years of training. After completing a one-year internship with a strong internal medicine focus, physicians undergo specialty training in neurology in an accredited residency program. Doctors of veterinary medicine

(D.V.M.) who seek board certification can also opt for a neurology residency program.

General (three-year) residency training in neurology is sufficient for practicing adult neurology. Specializing in child neurology necessitates additional pediatrics and child neurology residency training. After completing the residency requirements, a physician is eligible for board certification by the American Board of Psychiatry and Neurology. Numerous fellowship opportunities (with a one- or two- year duration) provide further specialization in a variety of subfields, such as movement disorders, dementia, stroke, multiple sclerosis, neuro-oncology, clinical neurophysiology, neuroimaging, geriatric neurology, and interventional neurology. Subsequently, a physician can opt for one of many career pathways, in an academic setting, private practice, or pharmaceutical company.

Research experience accumulated during undergraduate, medical school, and residency training often leads to a position in a large university or medical center. Three potential options are available for the physician who pursues a career in academic neurology: basic neuroscientist, teacher-clinician neurologist, or patient-oriented researcher.

SOCIAL CONTEXT AND FUTURE PROSPECTS

Neurology is expected to be one of the most rapidly advancing disciplines of the twenty-first century. Its realm is expanding to encompass new areas of genetics, metabolism, sleep disorders, vascular diseases, and neuroimmunology.

With an improved understanding of the nervous system's function in health and disease, the pressure to correctly diagnose neurologic disease increases. The neurologist may function as a consultant or as principal physician for patients with primary nervous system disorders. These diseases represent an important area of medicine because of their high prevalence, disabling outcomes, and high costs for health systems.

The World Health Organization's Global Burden of Disease report shows that while neurological and mental disorders are responsible for about 1 percent of deaths, they account for almost 11 percent of disease burden the world over. As life expectancy increases and the general population ages, it is likely that the incidence of age-related neurological disorders will increase. The World Health Organization

is implementing programs aimed at prevention, diagnosis, and treatment of neurological disorders that are of public health importance. These include epilepsy, headache, dementia, multiple sclerosis, Parkinson's disease, stroke, pain syndromes, and brain injury. It is vital to ensure that an appropriate range of care is made available to all people with neurological disorders in every country of the world.

Identifying biochemical markers of disease for early detection implementing preventative measures are fast becoming central goals of worldwide research efforts. Scientists are investigating the use of agents that open the blood-brain barrier to allow drug penetration, as well as various alternative routes of drug delivery. They are also testing novel therapies based on gene replacement, growth factors, and stem cells that have the potential to promote neuronal repair and regeneration.

Mihaela Avramut, M.D., Ph.D.

FURTHER READING

Evans, Randolph W., ed. *Common Neurological Disorders*. Philadelphia: Saunders, 2009. Describes the most common nervous system disorders and their treatment.

Martin, Joseph B. "The Integration of Neurology, Psychiatry, and Neuroscience in the Twenty-first Century." *American Journal of Psychiatry* 159 (May, 2002): 695-704. Examines the historical basis for the divergence of neurology and psychiatry and discusses prospects for a potential convergence.

Mumenthaler, Marco, and Heinrich Mattle. *Fundamentals of Neurology: An Illustrated Guide*. New York: Thieme, 2006. A well-written, logically ordered textbook for the novice in neuroscience and neurology.

Rowland, Lewis P., Thomas A. Pedley, and H. Houston Merritt, eds. *Merritt's Neurology*. 12th ed. Philadelphia: Lippincott Williams & Wilkins, 2010. Provides the essential facts about common and rare diseases. Includes chapters on neurological symptoms, diagnostic tests, and rehabilitation.

Sacks, Oliver. *Awakenings*. 1973. Reprint. New York: Vintage Books, 1999. Exceptional case history and literary work, describing the outcome of treating encephalitis lethargica patients institutionalized since World War I.

_____. *The Man Who Mistook His Wife for a Hat*. 1970. Reprint. New York: Simon and Schuster,

2007. Sacks explores fascinating cases of excesses and losses in neurological conditions.

WEB SITES
American Academy of Neurology
http://www.aan.com

National Institute of Mental Health
http://www.nimh.nih.gov

National Institute of Neurological Disorders and Stroke
http://www.ninds.nih.gov

Neurotechnology Industry Organization
http://www.neurotechindustry.org

World Health Organization
Neurology
http://www.who.int/topics/neurology/en

See also: Computed Tomography; Epidemiology; Geriatrics and Gerontology; Magnetic Resonance Imaging; Pathology; Speech Therapy and Phoniatrics; Stem Cell Research and Technology.

NURSING

FIELDS OF STUDY

Chemistry; biology, microbiology, epidemiology, pharmacology; human anatomy and physiology; health assessment; surgery; medicine; community health; obstetrics; psychiatry; pediatric medicine; gerontology; emergency medicine; oncology; health care; nursing administration; nursing education.

SUMMARY

Nursing promotes and protects the health and well-being of people in a community through prevention of illness and injury, alleviation of pain and suffering through treatment of disease, and advocacy for the care of individuals, families, or populations. Nurses apply scientific knowledge, principles, and procedures to the treatment and care of people who are ill or seek preventive care. The art of nursing includes therapeutic relationships with people in the community and provision of compassionate patient care.

KEY TERMS AND CONCEPTS

- **Assessment:** Systematic and dynamic collection of physiological, psychosocial, economic, cultural, spiritual, and lifestyle data about the patient, with analysis of findings, in order to plan patient care.
- **Care Plan:** Plan that guides all involved in the patient's care.
- **Diagnosis:** Clinical determination of a patient problem, used to prioritize and plan patient care.
- **Evaluation:** Review of the care plan and outcomes with input from the patient and the health care team to determine which goals were met and what goals are to go into the next care plan.
- **Implementation:** Actualization of the care plan by communicating with the patient, caregivers, and interdisciplinary health care team.
- **Licensed Practical Nurse (LPN):** Nurse who has completed a one-year vocational or technical program and is licensed by the state to practice; also known as a vocational nurse, or LVN.
- **Nurse Practitioner:** Registered nurse who has received education in the diagnosis and treatment of medical conditions, has national board certification, and is licensed to practice by the state; usually works with a physician.
- **Nursing Process:** Five-step process for planning patient care and setting goals for health improvement; consists of assessment, diagnosis, planning, implementation, and evaluation.
- **Nursing Theory:** Consistent concepts and definitions within professional nursing that guide nursing research and provide the framework for patient care.
- **Planning:** Assimilation of the care plan, including measurable and achievable patient goals to improve health status and cope with illness.
- **Registered Nurse (RN):** Professional nurse who has completed a two-year associate degree (A.D.) in nursing, received a three-year diploma in nursing from a hospital, or completed a four-year baccalaureate degree in nursing (B.S.N.); passed a state board examination; and is licensed by the state under its Nurse Practice Act.
- **Standard of Care:** Generally accepted and consistent practice of patient care.

DEFINITION AND BASIC PRINCIPLES

Nursing is the term used to describe the practice of professional registered nurses (RNs) and licensed practical or vocational nurses (LPN, LVN). The International Council of Nurses defines nursing as the autonomous and collaborative practice of caring for people of all ages in a community, whether these people are sick, well, disabled, or dying. The goal of nursing is to promote health and prevent disease and illness. Key roles for nurses include patient and professional advocacy, active participation in health policy, promotion of a safe environment and prevention of injury, leadership in health care systems management, and patient and nursing education.

Nursing and Science. Nurses study the sciences that relate to the functions of the human body such as biology, microbiology, and chemistry. They must understand normal human anatomy and physiology as well as pathophysiology, the changes that can occur in the body as the result of disease. Pharmacology is important, as many diseases are treated with medications. To protect patient safety, these drugs must be administered using what are termed the five rights: the right medication and the right dosage, given at the right time by the right route to the right patient. Epidemiology applies to nursing in that nurses can assess the cause, distribution, and control of disease in the community.

Nursing Code of Ethics. Some basic foundational principles of nursing are contained in the Code of Ethics for Nurses. The code states that nurses must respect patients and treat them with dignity, regardless of the patient's social or economic status or health problem. It says that nurses' primary commitment is to patients and their needs, and nurses should promote the health of patients and protect their safety and rights. Further, nurses must take responsibility and be accountable for their nursing judgments and are obligated to provide high-quality patient care; they are also responsible for maintaining competency through continued professional learning and growth. Nurses are charged with establishing and maintaining a positive health care environment and workplace. Nurses are to advance the profession through research and involvement with health care policy, to collaborate with other health care professionals to promote the health needs of the public, and to be involved in social reform related to health.

Standards of Care. Nurses provide patient care using professional standards of care, which define the appropriate use of the nursing process in patient care as well as the standards of performance. The standards of performance address quality of patient care, nursing education, collegiality, collaboration, ethics, performance appraisal, research, and resource utilization. The American Nurses Association has also established position papers that address key concerns of nurses, such as access to universal health care or global health.

BACKGROUND AND HISTORY

Historically, care of the sick was provided by women who had little training. Mothers and grandmothers passed down various folks remedies to their daughters, who continued to care for the sick at home or in the community. In the 1600's, charitable organizations and churches began to organize women to deliver patient care. These "nurses" delivered babies and assisted new mothers as an outreach of the church or charity. A well-known group, the Sisters of Charity of France (1633), helped on the battlefield and in the home.

The Crimean War (1853-1856) increased the need for nurses. During the war, Florence Nightingale, as supervisor of the nursing staff, devloped a new approach to patient care. She published a description of her approach in *Notes on Nursing: What It Is and What It Is Not* (1860).

The Nursing Society of Philadelphia was established in 1836 and opened a training school for nurses in 1850. During the American Civil War, nurses were needed to care for the wounded, so the U.S. Army Nurse Corps was established. In 1881, Clara Barton founded the American Red Cross and served as president. By 1893, Lillian Wald had started a visiting nurse service in New York, providing patient care in the home. In 1899, the International Council of Nurses joined nurses in countries around the world.

By the early 1900's, U.S. nurses were being licensed and credentialed by their state governments. In 1909, Jane Delano, a member of the American Red Cross, established the Nursing Service to support the Army Nurse Corps. When the need for nurses increased during World War I and World War II, the federal government funded nursing education and provided stipends for students. Nursing school enrollment dropped during the 1980's, when nurses pursued expanded roles, and this resulted in a national shortage of nurses. In the twenty-first century, more than half of all health profession students are nursing students.

In 1992, Texan Eddie Bernice Johnson became the first nurse to be elected to the U.S. Congress. She has continued to serve, focusing on health care issues and human rights. Many nurses have embraced political activism to improve their working conditions and reduce nurse-to-patient ratios to improve patient care.

HOW IT WORKS

Nurses attend accredited schools or state-approved training programs depending on the type of nursing pursued. Some choose a one-year vocational program to become a licensed vocational (LVN) or licensed practical nurse (LPN). Others become a registered nurse (RN) through two-, three-, and four-year educational programs. Some registered nurses with master's degrees practice as nurse practitioners, clinical nurse specialists, nurse midwives, or nurse anesthetists. Nurses with a master's or doctoral degree in nursing may teach students in nursing schools.

State Board Examinations and Licensure. After graduating from an accredited nursing school, nurses take state board examinations. After passing these tests, they register with the state board of nursing to become licensed. Nursing licenses are renewed according to rules set by the state. Nurses must be licensed in the state in which they reside and also in the state where they work. For nurses to live in one state and work in another, the state boards of nursing in the involved states must have a compact agreement that recognizes multistate licensure.

Nurse Practice Acts. The Nurse Practice Acts are laws established by the individual states to delineate the scope of practice for nurses and protect the health and safety of the public. Nurses work under their individual state's Nurse Practice Act, which includes the requirements for licensure and defines which skills and services nurses can perform. The legal guidelines for nurses vary based on the nurse's education and licensure requirements. LPNs and RNs have different license and education requirements as well as different roles in health care delivery. Nurse practitioners, RNs with additional graduate-level education and skills, function under expanded legal parameters as defined by the laws of their states. Nurses are bound by the rules and regulations of the state in which they live and practice. The law authorizes the state's board of nursing to enforce the rules and regulations and to provide disciplinary actions as needed.

Dependent, Independent, and Collaborative Interventions. Nurses practice dependent, independent, and collaborative interventions based on the setting, institutional policies, and the laws of their state. Certain dependent nursing interventions require physician orders to complete, and others are independent of the physician. For example, when staff nurses administer medications in a hospital, they practice dependent interventions that require an order from the physician to prescribe the drug, the dosage, frequency of administration, and route of administration. Independent interventions must uphold the standard of care but do not require a physician's order. An example is listening and assessing lungs sounds or the patient's pulses to monitor the patient's health status. The nurse is responsible for

Fascinating Facts About Nursing

- Writer Walt Whitman worked as a volunteer nurse during the American Civil War. His work in more than forty hospitals inspired some of his poetry.
- Before her success in the musical group The Judds, Naomi Judd supported herself and two daughters as an emergency room nurse.
- Research has found that nurse practitioners can provide up to 80 percent of primary care services as well as physician services for less cost.
- In Gallup polls from 2005 to 2009, nurses ranked number one for their ethical standards and honesty.
- National nurses week is celebrated each year starting May 6 and ending May 12, the birthday of Florence Nightingale.
- As of 2004, 81.8 percent of registered nurses were non-Hispanic whites, and only 5.7 percent were men.
- In 2007, researchers found that after one year, 13 percent of new registered nurses had changed jobs and 37 percent wanted to change jobs.
- In 2008, the average age of registered nurses was nearly forty-seven years. Of those with active licenses, 83 percent were working.

notifying the physician or health care provider of any adverse changes. Another independent intervention is patient education. Collaborative interventions require work with other health care team professionals. An example is when nurses use teaching tools designed by physical therapists or dietitians.

APPLICATIONS AND PRODUCTS

Nursing is a service industry where workers provide direct or indirect patient care. Basic roles of nursing include care provider, educator, patient advocate, counselor, researcher, change agent, leader, and manager. Staff nurses deliver direct patient care. Nurses provide patient teaching to promote wellness, reducing the impact of costly disease. Nurse educators prepare student nurses to become professionals to provide patient care. Nurse administrators secure resources necessary to deliver patient care. Nurse practitioners develop private or clinical practices to meet the health care needs of the community. Nurse researchers examine practice to determine best practices for positive patient outcomes.

Nursing provides diverse opportunities in various settings. The hospital or acute care setting offers many nursing positions. For example, nurses work in hospital emergency centers, intensive care units, and various other units. Hospitals employ infection-control nurses who conduct studies on hospital-acquired diseases. Quality management nurses assess care provided against industry standards to improve the quality of patient care. Nurses may also work in community settings such as in physicians' offices or clinics. School nurses work within primary schools to improve the health of students. Occupational health nurses promote the health of industry or business workers. Home care nurses visit patients in their homes or residences to provide care, education, and support. Hospice nurses care for terminally ill patients in their homes or in hospices. Nurses may work as telephone triage professionals for physicians' offices or emergency rooms to help assess and evaluate patient needs and make appropriate referrals for care options. Still others work as flight nurses, who help transport high-risk patients safely, or as travel nurses, who offer their services to geographically diverse facilities as needed. Churches and places of worship sometimes have parish nurses to help congregation members choose healthy lifestyles. An expanding role of nursing is the informatic nurse specialist, who assists with data management and decision making.

Nurses with advanced degrees may work as nurse practitioners, clinical nurse specialists, certified registered nurse anesthetists, and certified nurse midwives. Nurse administrators or nurse executives work on the business side of nursing practice. Regardless of the education level or setting, the goal of nursing practice remains the same: to provide quality patient care.

IMPACT ON INDUSTRY

Nursing interfaces with many industries. Global industry is affected by the use of international nurse staffing agencies. Recruitment agencies contract with local hospitals and clinics to provide nurses to fill nursing shortages. This supports local industry as well as international industry.

Governments also are affected by nursing. Each state has a board of nursing to enforce regulations governing nursing practice and licensure. State schools and universities provide jobs for nurse educators to meet the growing demand for nurses. Public school nurses work with limited resources to improve student health and attendance. Community and veterans hospitals require nurses for patient care.

As the largest health care occupation, nurses are a major labor cost for hospitals and other health care agencies. These costs affect industry when passed on through the cost of insurance and other health care benefits. Occupational health nurses manage employee illness to minimize the use of expensive health care services. At the same time, corporations develop products that nurses need to provide patient care, so many companies benefit from the large number of nurses.

Communities receive many economic benefits from nursing. Nurses work in local industries to provide services needed by the community. In turn, they aid their communities by spending much of their salaries on housing, food, clothing, and personal and professional services. Nurses provide preventive care services as well as surveillance and early intervention, decreasing the overall costs of patient care within the community.

CAREERS AND COURSE WORK

Nurses may choose the path they take to become qualified to practice in the health care system. Nurses are educated through accredited schools or state-approved training programs. The specific course work required depends on the type of program attended and specific practice pursued.

Registered Nurses. Registered nurses are considered professional practitioners and perform their roles under the Nurse Practice Act of the state where they work. RNs must graduate from an accredited school of nursing and pass a state board examination. RNs are licensed by the state in which they reside. RNs can expect to make a median salary of $62,450.

Course work for RNs varies depending on the program. However, most schools of nursing offer courses in chemistry, biology, microbiology, pharmacology, human anatomy and physiology, pathophysiology, health assessment, research/evidence-based practice, ethics, nursing theory, informatics, and leadership or management. Special clinical studies include medical-surgical, community health or public health, maternal/newborn, psychiatric, child health/pediatric, perioperative/surgical, gerontological, emergency/trauma, and oncology nursing. Additional nursing courses may cover issues and trends, cultural diversity, health care policy, and quality management.

At the master's level, courses include evidence-based research, health care finance, human resource management, informatics or information management, nursing administration, and nursing education. Nurse practitioners have additional specialized courses based on their chosen practice (family nurse practitioners, pediatric practitioners, clinical nurse specialists, nurse midwives, or nurse anesthetists).

Licensed Practical and Vocational Nurses. Licensed practical and vocational nurses (LPNs or LVNs) attend a one-year state-approved training program in a community college or vocational school, pass the National Council Licensure Examination-Practical Nursing, and become licensed by the state. LPNs provide much of the hands-on patient care provided in hospitals, clinics, and doctor's offices under the supervision of an RN or physician. Most LPNs are considered generalists but may specialize in nursing home, home care, or in physicians' office practices. The skills of LPNs are defined by the laws of the state in which they works. Some institutions restrict certain functions such as intravenous medication administration or care of ventilator-dependent patients. However, in some states, LPNs can secure certification in these areas to practice in prescribed settings. Many LPNs continue their educations and become registered nurses. The median salary for LPNs is $39,030.

Desired Qualities and Challenges. Certain qualities are needed to be a nurse. Nurses should be caring and empathetic. Emotional stability is a plus, as nursing is a high-stress area. Nurses need good decision-making and communications skills. They must often multitask to care for a caseload of patients.

Nursing can be physically and emotionally demanding. Some nurses spend a lot of time on their feet providing patient care and must protect their physical well-being. Lifting patients can cause back strain or injuries, so nurses must be educated to use safety devices and techniques. Working with sick people can expose nurses to contagious diseases, so precautions must be taken to avoid infections. Nurses should take care of themselves to continue an effective practice.

SOCIAL CONTEXT AND FUTURE PROSPECTS

The job potential for registered nurses continues to be excellent. According to the U.S. Bureau of Labor Statistics, about 2.9 million registered nurses work in the United States. The bureau predicts that the number of RNs will grow to 3.2 million. The greatest growth will be in physicians' offices and home health care. Hospitals employ about 60 percent of RNs. In some areas, there is a shortage of nurses. Top pay for RNs exceeds $92,000.

The overall job outlook for licensed practical nurses is very good; the bureau predicted a 21 percent increase in jobs between 2008 and 2018. Settings where LPNs will be needed include physicians' offices and outpatient care centers. About 753,600 LPNs work in the United States, and the top pay for LPNs is about $55,000.

One major advantage of nursing is the many different opportunities available. Nurses have diverse choices when selecting the specialty or setting where they will work. They can work with specific populations, such as older adults, children, or terminally ill patients, or specialize in caring for patients with certain types of illnesses. Forensics is another growing practice opportunity for nurses. Some enjoy work as sales representatives for pharmaceutical companies or become legal nurse consultants. The possibilities for varied practice options continue to grow each year.

Marylane Wade Koch, M.S.N., R.N.

FURTHER READING

Andrist, Linda C., Patrica K. Nicholas, and Karen Wolf, eds. *A History of Nursing Ideas.* Sudbury, Mass.: Jones & Bartlett, 2006. A collection of essays on nursing issues and concepts.

Cherry, Barbara, and Susan R. Jacob. *Contemporary Nursing: Issues, Trends, and Management.* 5th ed. St. Louis, Mo.: Mosby Elsevier, 2011. Explains the complex issues of professional nursing practice, provides a historical perspective on nursing, and addresses career management.

Guido, Ginny Wacker. *Legal and Ethical Issues in Nursing.* 5th ed. Upper Saddle River, N.J.: Prentice Hall, 2009. Covers the legal and ethical aspects of nursing practice.

Nightingale, Florence. *Notes on Nursing and Other Writings.* New York: Kaplan, 2008. Includes her well-known book and other writings. Provides a historical perspective.

Taylor, Carol, et al. *Fundamentals of Nursing: The Art and Science of Nursing Care.* 7th ed. Philadelphia: Wolters Kluwer/Lippincott Williams & Wilkins, 2011. Aims to teach students to combine scientific knowledge and technical skills with responsible care for their patients.

WEB SITES

American College of Nurse Practitioners
http://www.acnpweb.org

American Nurses Association
http://www.nursingworld.org

International Council of Nurses
http://www.icn.ch

See also: Cardiology; Dermatology and Dermatopathology; Endocrinology; Gastroenterology; Geriatrics and Gerontology; Neurology; Obstetrics and Gynecology; Occupational Health; Orthopedics; Otorhinolaryngology; Pediatric Medicine and Surgery; Pulmonary Medicine; Surgery.

NUTRITION AND DIETETICS

FIELDS OF STUDY

Chemistry; physiology; medical nutrition therapy; management; sociology; economics.

SUMMARY

Nutrition and dietetics, a multidisciplinary field for acquiring and using nutrients from food sources to sustain life and growth, incorporates the science of economically producing and making available foods to a global society. Illnesses may result from inadequate or poor food choices, and certain health conditions require modifying foods in the diet to maintain or restore health.

KEY TERMS AND CONCEPTS

- **Body Mass Index (BMI):** Measure of body fat based on a person's height and weight.
- **Diet:** Food and beverage consumed by a person.
- **Dietary Reference Intake (DRI):** System of recommended nutrients developed by the National Academy of Science's Institute of Medicine for planning and assessing diets.
- **Dietitian:** Person trained in nutrition, food science, and areas related to food consumption.
- **Food Insecurity:** Inability to acquire safe, nutritious food to sustain and nourish life.
- **Hunger:** Lack of sufficient food as a result of food insecurity.
- **Malnutrition:** Condition of inadequate nutrient intake.
- **Metabolism:** Chemical reaction involved in obtaining and expending energy from food.
- **Nutrient:** All substances obtained from food to provide energy, growth, and regulation of cell activity.
- **Obese/Overweight:** Having excessive body fat.

DEFINITION AND BASIC PRINCIPLES

Nutrition is the science of providing and processing nutrients from food for survival and growth. It involves assimilating nutrients and turning food consumed into energy. Approximately fifty nutrients are essential for human life and health.

Dietetics is the science and art of procuring, planning, and preparing foods to supply nutrients in a palatable, pleasing, and economical way. It incorporates principles of nutrition and of the interactions among nutrients. In a broader scope, dietetics includes the social, cultural, and psychological aspects of food acquisition and consumption.

BACKGROUND AND HISTORY

Antoine-Laurent Lavoisier, the father of nutrition, recognized the relationship between food and respiration in the 1700's. In 1830, Dutch chemist Gerard Johann Mulder classified proteins. W. O. Atwater, the father of American nutrition, published the first table of food values in 1896. Two decades later, American biochemist E. V. McCollum referred to vitamins and minerals as "protective foods."

Most vitamins were discovered in the early 1900's, although their presence in certain foods had been noted. Scientists first recognized vitamin A in 1913. Since the Middle Ages, an unidentified substance in cod liver was known to prevent rickets, and in the 1920's, this substance, vitamin D, was isolated from vitamin A. As early as 1753, English physician James Lind had noted that sailors fed citrus fruits avoided scurvy. It was not until 1928, however, that vitamin C was first isolated. Deficiencies of substances, later named B vitamins, were known to result in beriberi, pellagra, and other diseases. Scientists classified B vitamins by numbers (B1, B2, B3, and so on) as they isolated specific types.

In 1917, Lenna F. Cooper founded the American Dietetic Association (ADA) to help the government conserve food and improve the nation's health and nutrition during World War I.

HOW IT WORKS

Energy Metabolism. Calories are units of heat measure used for body energy. One pound of body weight equals 3,500 calories. Body metabolism and activity levels determine daily calorie needs. Most adult women need 1,900 to 2,100 calories, while adult men need about 2,100 to 2,400 calories each day. Calorie requirements decrease with age because metabolism naturally slows about 5 percent per decade after age forty. Only the energy nutrients—carbohydrates, fats, and proteins—provide calories to the body.

Carbohydrates, also known as sugars and starches, are primary sources of calories. Each gram of carbohydrate yields about 4 calories. The simple sugar glucose is the major fuel source and the only sugar found in the body. Two other simple sugars are fructose, found mostly in fruits, and galactose, a component of the double sugar lactose. Sucrose, a double sugar known as table sugar, is the most common sugar in the diet. Complex glucose molecules, called starches, come primarily from potatoes and grains.

Lipids or fats yield about 9 calories per gram. Fats act as carriers for fat-soluble vitamins. Dietary fat improves palatability and provides satiety. Two common types of fats significant in health are saturated and unsaturated fatty acids. Saturated fats, primarily from animal sources, are generally solid at room temperature. These fats may cause buildup of fatty plaques in the blood and contribute to blood clots.

Unsaturated fats, found in plants, are better choices for a healthful diet. These fats, usually soft or liquid at room temperature, vary from monounsaturated (the major fatty acid in olive oil) to polyunsaturated fatty acids (found in cottonseed, soybean, corn, and canola oils). Hydrogenation, a process that changes fatty acids from liquid to solid, creates trans fats. Trans fats raise blood cholesterol levels and may contribute to heart disease.

Protein comes from the Greek word meaning "to take first place." Proteins, found in every living cell, are composed of amino acids. Of the twenty-two known amino acids, nine are essential in the diet. Nonessential (dispensable) amino acids are derived from the essential ones or manufactured in the body. Protein, which yields about 4 calories per gram, is inefficient as a source of energy. However, with insufficient carbohydrate, the body converts protein into glucose for energy. Protein builds and repairs body tissue. Amino acids may function as precursors for transport substances such as lipoprotein. Complete proteins, found in meats, eggs, and milk, contain all essential amino acids. Plant sources such as legumes and nuts lack at least one essential amino acid and are incomplete proteins.

Fat-Soluble Vitamins. Vitamins, organic substances needed by the body in minute quantities, are categorized into fat soluble and water soluble. Each vitamin has specific functions. Generally, vitamins regulate cell metabolism in conjunction with enzymes and contribute toward construction of body tissue. The fat-soluble vitamins—A, D, E, and K—can be stored in fatty tissues in the body.

Vitamin A (retinoid), necessary for growth and development of skeletal and soft tissues and maintenance of normal epithelial structures, is called the anti-infective vitamin. Major sources are dark green and deep yellow fruits and vegetables. Few sources other than cod liver oil provide significant amounts of vitamin D. Therefore, most all forms of milk are fortified at a level of 400 international units (IU) per quart. Vitamin D helps form normal bones and teeth and provides other functions. Vitamin E (tocopherol) is an antioxidant important in protecting cells from oxidation. Sources include vegetable oils, margarine, whole-grain products, seeds, and nuts. Vitamin K, important in blood clotting, is found in most green leafy vegetables.

Water-Soluble Vitamins. Vitamin C (ascorbic acid) functions in the formation of collagen, wound healing, metabolic functions, and other roles. Foods high in vitamin C include citrus fruits, strawberries, cantaloupe, and cruciferous vegetables. B vitamins are important in energy metabolism. Thiamin (B1) is called the antineuritic vitamin. Riboflavin (B2), rarely deficient in the diet, is found most abundantly in milk and dairy products. Niacin (B3) is prevalent in meats, poultry, fish, peanut butter, and other foods. Other major B vitamins include folic acid (B9), B6, and B12.

Minerals. Minerals have varied functions in building tissue. The major minerals, found in larger quantities in the body, include calcium, phosphorus, magnesium, and iron. More than 99 percent of calcium and 85 percent of phosphorus in the body is in the bones. Calcium, essential in blood coagulation, is involved in nerve, enzyme, and hormone functions and other activities. Iron is primarily involved in oxygen transport within the blood. Although many trace elements have been identified as essential, the best known are copper, zinc, manganese, cobalt, selenium, chromium, and molybdenum.

APPLICATIONS AND PRODUCTS

Even with sufficient global food supplies, much of the world fails to acquire appropriate nutrition to sustain good health. Food insecurity (a lack of nutritious food) or excessive weight problems often result

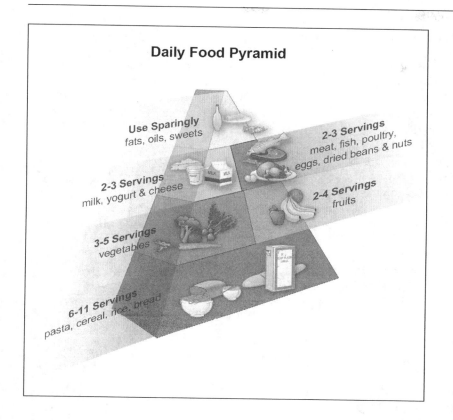

Daily Food Pyramid

Use Sparingly
fats, oils, sweets

2-3 Servings
meat, fish, poultry,
eggs, dried beans & nuts

2-3 Servings
milk, yogurt & cheese

2-4 Servings
fruits

3-5 Servings
vegetables

6-11 Servings
pasta, cereal, rice, bread

Department of Agriculture (USDA) programs, expansion of technical assistance, increasing public awareness of causes for food insecurity, and finding solutions.

Hunger and malnutrition affect individuals, nations, and the world community, affecting labor productivity and economic development. Multifaceted factors contributing to hunger include greed, overpopulation, unemployment, political and civil unrest, and limited productive resources. Elimination of worldwide hunger requires efforts from agriculture, development of human capital, and improved infrastructures. Advanced seed varieties increase crop production in developing countries, creating a need for additional workers who then acquire food from their incomes. However, a lack of costly irrigation systems, fertilizers, and pesticides make advanced technologies unsuitable for many poor farmers.

from limited knowledge, skills, or the financial means to procure more healthful foods.

Food Insecurity. Food insecurity affects more than 1.02 billion people worldwide. Almost 15 percent of Americans (nearly 50 million people) are food insecure. Of these people, about 8 percent have very low food security, causing reduced food intake and disruption of normal eating patterns. The remainder use a variety of coping strategies, including participation in various food programs.

Food insecurity is more prevalent in households with incomes near or below the federal poverty line, single parent households, and black and Latino households. In the United States, numerous agencies exist to alleviate food insecurity. The three largest federal food and nutrition assistance programs are the National School Lunch Program, the Supplemental Nutrition Assistance Program (formerly the Food Stamp Program), and the Special Supplemental Nutrition Program for Women, Infants, and Children. Efforts to reduce hunger 50 percent by 2015 depend on forming new partnerships at every level. That involves improved coordination among existing U.S.

Worldwide hunger is linked to price fluctuations in world markets and changing agricultural policies. Prices of imported items, such as fuel and manufactured goods, rise faster than exported food products, causing international debt to accumulate in developing countries. The economic crisis of 2009 differed from previous food crises. Simultaneous effects on large portions of the world hindered traditional coping strategies. The 2006 to 2008 food and fuel crisis intensified economic difficulties. Although food commodity prices decreased, they remained high by historical standards. The impoverished spent as much as 40 percent of their income on food.

The United Nations considers access to food a basic human right. Ethnic and political conflicts intertwined with cultural, religious, economic, and social systems remain problems in abating hunger.

The Weight Epidemic. Worldwide, 1 billion people are overweight, and 300 million of these people are obese. In the United States, the percentage of obesity remained relatively stable from 1960 to 1980. Since then, however, the prevalence of obesity has more than doubled. Nearly 70 percent of American are considered overweight, and about 33 percent are obese.

For adults, excessive weight ranks second only to smoking as a lifestyle choice affecting health and longevity. Overweight and obesity intensify and increase risks for chronic illnesses, including heart disease, stroke, certain cancers, and type 2 diabetes. Excessive weight has been linked to other health conditions and increased morbidity. Children and teens with weight problems show increased susceptibility to potential forerunners of heart disease. Additionally, excessive weight in children leads to lower self-esteem.

One recognized standard for determining a person's healthy weight is the body mass index (BMI). A BMI score between 19 and 24 suggests a healthy weight. A score under 19 indicates that the person is underweight. Scores between 25 and 29 are classified as overweight and scores over 30 signify obesity. Values above 35 indicate stage-one obesity and scores of 40 or greater, held by nearly 6 percent of the population, reflect stage-three obesity.

Weight management constitutes a balance between calories taken in and calories used. Overweight and obesity occur when more calories are consumed than the body uses. Being sedentary and exercising less can lead to weight gain. The appeal and convenience of high-calorie fast foods have caused them to replace more nutritious fruits and vegetables in people's diets. Oversized portions worsen the problem.

The expense and limited availability of more wholesome foods make the food insecure more vulnerable to becoming overweight or obese. Making nutritious foods and beverages more affordable and readily available in all communities may help prevent obesity. Ways to accomplish this include offering healthier choices in different settings, competitively priced foods in low income areas, and incentives for food purveyors and retailers to service those areas.

IMPACT ON INDUSTRY

The health of Americans affects the overall economy. Government, industry, and business sectors support education and research to combat health care problems related to poor nutritional habits.

Research and Education. Government support of medical research comes primarily through clinical trial or grants from the National Institutes of Health (NIH) of the U.S. Department of Health and Human

---

Fascinating Facts About Nutrition and Dietetics

- Cholesterol, a fat sterol found only in animal sources, can be high-density lipoprotein (HDL) or low-density lipoprotein (LDL). LDL, the "bad" cholesterol, can clog arteries and lead to heart attacks, but HDL, the "good" cholesterol, can keep it from sticking to artery walls.
- Ultraviolet rays from sunlight can form vitamin D with the aid of cholesterol in the skin.
- The amount of an element in soil determines the mineral content found in plant foods grown in that soil.
- Nutrient-dense foods provide substantial amounts of vitamins and minerals with relatively few calories.
- Stevia, an artificial sweetener derived from a South American plant, was declared safe in December, 2008, by the Food and Drug Administration. By early 2009, more than 110 beverages and health care products sold in the United States contained stevia.
- Among the obese, those who carry their weight in the abdominal area (waist size greater than 35 inches for women and greater than 40 inches for men) are more susceptible to disease than those who carry the excess weight in their butts, hips, and thighs.
- Free radicals, resulting from certain body functions and unhealthy lifestyle choices such as smoking, compromise cell integrity and cause greater susceptibility to cardiovascular disease and some cancers. Antioxidant properties in vitamins E and C and carotenoids (lutein, lycopene, and beta-carotene) neutralize free radicals and protect the cells.

Services (HHS). Other sponsors include individuals, organizations, or companies related to or interested in nutrition and health.

Coordinated efforts of the USDA and HHS provide several levels of education and initiatives to improve health. The Healthy People 2010 initiative focused on ensuring quality of life and eliminating disparities in health care among individuals of different races, ethnic groups, and economic levels. The initiative suggests including nutrition and weight status as an area of emphasis to assist Americans in making wiser food choices.

Since 1980, the USDA and the HHS have provided dietary guidelines for Americans to promote optimum health and reduce the risk of chronic diseases for those aged two years and older. These agencies use scientific research and findings to update the content of their guidelines every five years. Earlier versions targeted the general public. The 2005 report, which targeted policy makers, nutrition educators, nutritionists, and health care providers, summarized individual nutrients and foods information into a suggested pattern for the general public.

In 2005, My Pyramid replaced the Food Guide Pyramid to allow consumers and educators to adapt guidelines to individual tastes and lifestyles. Different calorie levels recommend specific numbers of servings in six color-coded categories of food: fruits, vegetables, grains, meat and beans, milk, and oils.

The dietary approach to stop hypertension (DASH) diet focuses on consuming more fruits, vegetables, low-fat dairy foods, whole-grain products, fish, poultry, and nuts. It recommends eating less red meat, limiting concentrated sweets to less than five servings per week, and limiting foods high in saturated fat, cholesterol, or total fat.

In 1987, dietary reference intake (DRI) replaced the recommended dietary allowance (RDA) established by the Food and Nutrition Board in 1941 to evaluate nutritional intakes of large populations. RDA data were based on average nutrients needed to prevent deficiencies. The DRI, determined by the Institute of Medicine and government funded, shifted the focus from preventing deficiencies to decreasing risks of chronic illnesses through nutrition. Four criteria for DRI include estimating average requirements of nutrients to meet the needs of 50 percent of the population, assessing nutrient values of foods without an established RDA, considering upper intakes or nutrient levels without causing health risks, and the RDA standard.

Industry and Business. Consumers choose from more than 45,000 items in the average supermarket. The food industry recognizes that consumers desire nutritional food choices and uses the latest health trends to market products, often prominently noting on the packaging that products are low fat, contain no trans fats, or are whole grain. For example, as people began to turn to products containing probiotics, which are believed to help the digestive system, Dannon began to market a yogurt containing probiotics. However, it came under fire for its marketing claims, revised the wording on its products, and settled a false-advertising lawsuit. The food industry's influence on diet and nutrition has come under criticism, especially in regard to its marketing of food products to children.

Under the 1990 Nutrition Labeling and Education Act, most food products were required to include a nutrition facts label listing nutritional information about the product, including serving size and calories, as well as the amount of fat, sodium, cholesterol, protein, carbohydrates, and vitamins and minerals as a percent of daily values. The government has launched campaigns to make consumers more aware of these labels and teach people how to read them. This raised awareness and scrutiny of nutrition labels has resulted in food companies developing low-fat, reduced-carbohydrate, whole-grain, and low-sodium versions of products. To develop these products and create nutrition labels, the food industry has employed many dietitians and nutritionists.

Industries in which dietitians and nutritionists have important roles are hospitals, nursing homes, schools, and weight loss or fitness centers. They design diets and nutrition plans that address the special health needs of people and provide adequate and appropriate nutrition for young people, often within strict budget limits.

Careers and Course Work

Dietetics provides opportunities in multifaceted areas, including working to improve the nutritional state of both healthy and diseased people and dealing with societal issues related to food supply, distribution, and consumption. Career choices continue to expand in education, research, media, health care sites, and industry. Many dietitians work in corporate settings. Others opt to launch independent practices, working with individuals in fitness centers and businesses such as supermarkets and creating and conducting employee wellness programs.

A dietitian must have at least a bachelor's degree from an accredited institution, with approved course work and training from the American Dietetic Association. A registered dietitian (RD) has successfully completed a national examination for credentialing. Many states have their own licensing systems as well.

SOCIAL CONTEXT AND FUTURE PROSPECTS

Cultural attitudes and beliefs, social influences, marketing, media, and other factors affect food choices, but taste preference remains a major influence. Although people tend to eat primarily according to their personal preferences, 60 percent of Americans claim to make an effort to eat a healthful diet to avoid future health problems. Heart disease, high blood pressure, diabetes, obesity, and other diseases and disorders in which diet plays a role remain common in the United States. Dietitians can play a significant role in helping people attain better health and physical well-being. As the roles that diet and nutrition play in human health are better understood, dietitians and nutritionists can make dietary recommendations to improve the health of the general public as well as of people with special dietary needs. In addition, they can advise companies in the food industry so that these companies can prepare processed and packaged foods that are not only healthful but also appealing.

Linda R. Shoaf, B.S., M.S., Ph.D.

FURTHER READING

Kaufman, Mildred. *Nutrition in Promoting the Public's Health: Strategies, Principles, and Practice.* Sudbury, Mass.: Jones and Bartlett, 2007. An overview of the role of governmental agencies in assessing and providing for the nutritional health of the public.

Shiels, Maurice Edward, et al. *Modern Nutrition in Health and Disease.* 10th ed. Philadelphia: Lippincott Williams & Wilkins, 2006. A comprehensive textbook and reference on the science of nutrition. Covers major areas from basic nutritional sciences to prevention and treatment of diseases.

U.S. Department of Health and Human Service. *A Healthier You: Based on the Dietary Guidelines for Americans.* Rockville, Md.: U.S. Department of Health and Human Service, 2005. A guide designed to help the public eat a more healthful diet.

Whitney, Ellie, and Sharon Rady Rolfes. *Understanding Nutrition.* 11th ed. Belmont, Calif.: Thomson Learning, 2008. A comprehensive textbook of general nutrition featuring a glossary for each chapter, nutrition portfolio, summary, and related Web sites. Illustrations, diagrams, and appendixes.

WEB SITES

American Dietetic Association
http://www.eatright.org

National Institute of Child Health and Human Development
Diet and Nutrition
http://www.nichd.nih.gov/health/topics/diet_and_nutrition.cfm

U.S. Department of Agriculture
Food and Nutrition Information Center
http://fnic.nal.usda.gov/nal_display/index.php?tax_level=1&info_center=4

See also: Cardiology; Gastroenterology; Geriatrics and Gerontology; Pediatric Medicine and Surgery.

O

OBSTETRICS AND GYNECOLOGY

FIELDS OF STUDY

Women's health care; maternal-fetal medicine; reproductive endocrinology; genetic counseling; genetics; infertility; perinatology; postpartum care; prenatal care; urogynecology.

SUMMARY

Obstetrics and gynecology is a medical specialty focused on women's health care. Obstetrics entails the care of a woman and her developing fetus from the moment of conception, through delivery, and following delivery (postpartum care). Gynecology encompasses the evaluation and treatment (both medical and surgical) of the female reproductive system (uterus, Fallopian tubes, ovaries, vagina, and external genitalia).

KEY TERMS AND CONCEPTS

- **Cesarean Section:** Surgical removal of a full-term infant from the uterus.
- **Hysterectomy:** Surgical removal of the uterus (womb).
- **Hysteroscopy:** Insertion of viewing device (hysteroscope) via the cervix (opening of the uterus) for visualizing and treating problems within the uterus.
- **In Vitro Fertilization:** Fertilization of an ovum (egg) in a laboratory for later introduction into the uterus for development.
- **Laparoscopy:** Surgical procedure that involves introduction of surgical instruments and a viewing device (laparoscope) into the abdomen to accomplish surgical procedures.
- **Maternal-Fetal Medicine:** Subspecialty of obstetrics and gynecology that focuses on high-risk pregnancies.
- **Pap Smear:** Scraping of cells from the cervix to check for abnormalities such as cancerous and precancerous conditions; also known as a Papanicolaou smear.
- **Reproductive Endocrinology:** Subspecialty of obstetrics and gynecology that focuses on the surgical and medical treatment of women with fertility problems.
- **Vaginal Delivery:** Delivery of an infant via the vaginal route; a normal delivery.

DEFINITION AND BASIC PRINCIPLES

Obstetrics and gynecology is a medical specialty limited to women's health care. Many obstetrician-gynecologists are generalists and provide both obstetrical and gynecologic care; however, some limit their practice to either obstetrics or gynecology. Some focus on problems of the menopause, and some focus on gynecologic care of children and adolescents. Some obstetrician-gynecologists receive additional training in a specific area of the field. The following subspecialties are recognized: gynecologic oncology, the medical and surgical treatment of cancers of the female genital tract (the ovaries, uterus, cervix, vagina, and external genitalia); reproductive endocrinology, the treatment of infertility in women; maternal-fetal medicine, the treatment of high-risk pregnancies); and urogynecology, the medical and surgical treatment of problems of the female urinary tract. Generalists often refer their more complicated and challenging cases to these subspecialists and, in many cases, comanage the patients' care.

As with other medical specialties, obstetrics and gynecology has become more sophisticated and technologically advanced. For example, infertility specialists are able to help women with certain conditions have children, although fifty years ago, women with those same conditions would not have been able to give birth. Endoscopy, the visualization of internal structures with a small viewing device, has enabled physicians to better understand their patients' conditions and provide appropriate treatment. Although a breast examination is a common part of the gynecologic exam, treatment of breast disease is usually referred to surgeons who specialize in breast disease.

BACKGROUND AND HISTORY

In Western civilization, obstetrics was first practiced by female midwives in the seventeenth century. In colonial America, about one in eight women would die in childbirth, and of every ten infants born, between one and three would die before reaching the age of five. Midwives would call male surgeons to the birthing room if they determined that the infant had died while in the womb. The surgeon's task was to use instruments to reduce the size of the baby's skull and extract the body. Toward the end of the seventeenth century, doctors began to perform cesarean sections. In the eighteenth century, forceps were developed to assist in the delivery process. In the nineteenth century, painkillers began to be used during childbirth, and in the 1940's and 1950's, the use of anesthesia increased to the point that women were often rendered semiconscious during the labor process. This excessive use of sedation produced a backlash, and in the 1970's, natural childbirth became popular. Advocates argued that childbirth was a natural process, not a disease needing treatment. Women began to use a variety of techniques to avoid using medication during labor and sought alternatives to hospital births, including using midwives rather than obstetricians.

In the nineteenth century, the direct viewing of a woman's genitals by a physician was regarded as immoral and immodest, and physicians had to rely on palpitation of the area. This made it difficult to diagnose and treat gynecologic problems. Gynecologic surgery, as well as surgery in general, did not develop until after the introduction of anesthesia in the nineteenth century. In the mid-1800's, J. Marion Sims developed a surgical treatment for vesicovaginal fistula, a complication of childbirth that caused urine leakage and discomfort, and he became known as the father of gynecology. Because Sims worked on slave women, he was able to directly observe their genitals. In 1869, Commander D. C. Pantaleoni performed the first operative hysteroscopy, paving the way for modern endoscopy (hysteroscopy and laparoscopy). In 1978, English gynecologist Patrick Steptoe announced the birth of a child through in vitro fertilization.

HOW IT WORKS

Some family practice doctors, typically in rural areas with a low population density, offer obstetrical services, but in urban areas, obstetricians generally handle pregnancy care. An obstetrician cares for a woman and her fetus/infant throughout pregnancy (antepartum), during labor (intrapartum), and following delivery (postpartum). An obstetrician (or obstetrical group) may also be associated with nurse practitioners and sometimes nurse midwives.

Obstetricians confirm a woman's pregnancy, then set up a schedule of visits, ranging from once a week to once a month, depending on the stage of pregnancy and whether complications are present. During a prenatal visit, the obstetrician listens for fetal heart tones, records the woman's weight and blood pressure, and checks her urine for abnormalities such as the presence of protein or sugar. Protein in the urine, accompanied by elevated blood pressure and increased reflexes, may indicate that the woman has toxemia (preeclampsia), which is life-threatening to both the woman and her fetus. The presence of sugar may indicate that the woman has developed gestational diabetes, which poses a threat to her fetus. Both conditions are treatable but require extra monitoring of the woman and fetus. Blood is drawn to check for anemia, screen for birth defects, and determine the blood type and Rh, or rhesus, factor. If a woman is Rh negative and her fetus is Rh positive, the risk of an immune reaction is present. This incompatibility may not cause problems during the woman's first pregnancy, but if a subsequent fetus is Rh positive, it could develop Rh disease, a life-threatening condition that could cause anemia or other serious problems. An obstetrician can also provide invaluable information on many aspects of pregnancy, such as nutrition, foods or medication to avoid, and exercise guidelines.

Some family practitioners and internists, especially those in rural areas, offer gynecologic services, although in urban areas, women often visit gynecologists for these services. Many women turn to gynecologists not only for problems with their reproductive system but also for other medical conditions; thus, their gynecologists function as primary care or family physicians. Many gynecologists also function as obstetricians and will continue to care for their patients if they become pregnant.

A gynecologic visit entails a breast examination

and a pelvic examination. During the pelvic examination, a gynecologist inserts a speculum (a two-bladed instrument) into the vagina to visualize the cervix and takes a Pap smear, which involves scraping some superficial cells from the cervix to check for cancerous or precancerous conditions. The gynecologist next places one hand in the vagina and the other on the lower abdomen to palpate (feel) the internal pelvic organs (uterus, Fallopian tubes, and ovaries). Unless a woman is unusually tense, the examination is relatively painless. The visit also includes a gynecologic history, which includes asking about any problems or concerns the patient may have in regard to her reproductive system. A woman may also be advised regarding birth control (contraception) or pregnancy.

APPLICATIONS AND PRODUCTS

Obstetrics and gynecology have developed markedly since the 1960's. Technological advances such as ultrasound and endoscopy have dramatically changed patient care. New medications have been developed to both prevent and increase the likelihood of pregnancy, as well as to treat numerous diseases and conditions. The understanding of disease processes has also grown. Old techniques have been replaced by more appropriate and beneficial ones.

Obstetrics. Most pregnancies progress uneventfully through delivery; however, complications can arise suddenly. The majority of women deliver vaginally; some of these deliveries are assisted by the use of forceps or a vacuum extractor (a suction device applied to the fetal head). Some deliveries are conducted via a cesarean section (operative delivery) because of problems that arise during labor or the intrapartum period. Common reasons for cesarean sections include lack of progress during labor or the development of fetal distress.

Modern-day obstetrical care involves a great deal of technology. Obstetricians visualize the development of the fetus via ultrasound; this modality does not involve X rays and is virtually harmless to the fetus and mother. Ultrasound can be used to evaluate fetal growth and to detect genetic abnormalities and the presence of multiple fetuses. About ten to twelve weeks into the pregnancy, chorionic villus sampling can be done to test for genetic defects in the fetus. In this procedure, ultrasound is used to guide a needle or catheter into the placenta and remove a small piece of tissue adjacent to the fetus. Amniocentesis is used for genetic testing at about fourteen to eighteen weeks into the pregnancy. In amniocentesis, ultrasound is used to guide a needle into the fetal sac (amnion) to withdraw fluid for genetic analysis.

Fetal monitoring involves recording the fetal heart rate on a continuous roll of graph paper and can be conducted during both pregnancy and labor. If a problem is noted, appropriate treatment or intervention can be initiated. For example, an obstetrician can induce (start) or augment (stimulate) labor with a variety of medications. A medication commonly used to promote labor is Pitocin (synthetic oxytocin), which acts directly on the uterus to stimulate contractions. The medication is administered by an intravenous drip, and the amount of medication administered is titrated (adjusted) so that contractions are strong enough for labor to progress but not excessively strong. Excessively strong contractions could cause harm to the fetus (even death) and possible rupture of the uterus. Instead of inducing labor, an obstetrician can also perform a cesarean section.

Gynecology. In an office setting, gynecologists offer contraceptive (birth control) advice and modalities. Oral contraceptives (birth control pills) and intrauterine devices (IUDs) are popular methods for preventing pregnancy. Gynecologists can also advise patients in regard to other contraceptive methods.

As women age, their ovarian function slows and stops. This condition is known as the menopause. Menstrual flow may change and then cease. Annoying symptoms, such as hot flashes, dry skin, and mood changes, may accompany the menopause and cause women to consult with gynecologists. Hormonal therapy or other medication may be prescribed to alleviate these symptoms.

Gynecologists often perform colposcopies when the results of a Pap smear are abnormal. The colposcope is a microscope that magnifies the cervix. If cervical abnormalities are found, a gynecologist can remove them by a cone biopsy, cryocautery (freezing), or the loop electrosurgical excision procedure (LEEP).

Gynecologists use ultrasound to aid in making diagnoses. Ultrasound can detect fibroid growths on the uterus and ovarian cysts. Cancerous cysts are more likely to have multiple cavities; therefore, ultrasound may detect the likelihood of a malignancy before surgery. Ultrasound can also measure the thickness of the endometrium (uterine lining); a thickened endometrium might indicate a cancerous condition. Fertility specialists commonly use ultrasound for in vitro fertilization. With ultrasound guidance, the specialist can insert a needle through the vaginal wall and into an ovarian follicle, which contains an ovum (egg). The ovum is aspirated into the needle and then fertilized with sperm. After an embryo develops, it is inserted into the uterine cavity to grow.

Gynecologic surgery is limited to the female reproductive system. A common gynecologic procedure is dilatation and curettage (D&C), which involves dilating the cervix (opening the womb) and scraping the uterine lining (endometrium); this procedure is done for diagnostic purposes and to control bleeding.

Some problems for which a D&C was previously performed are now diagnosed or treated by hysteroscopy (the insertion of a hysteroscope into the uterine cervix for viewing the interior of the uterus). A small piece of tissue (biopsy) can be removed for evaluation of an abnormality such as endometrial cancer (cancer of the uterine lining). Growths such as fibroid tumors (benign growths, which can cause pain or bleeding) can be removed, and abnormalities of the cavity can be corrected.

A hysterectomy involves removal of the uterus; sometimes this can be performed via the vagina (vaginal hysterectomy). Some hysterectomies are done because of fibroids that can cause pain or bleeding. Sometimes, a hysterectomy can be avoided by removing only the fibroids (myomectomy). Other common procedures are a salpingo-oophorectomy (removal of a tube and ovary) and vaginal repair surgery (vaginal damage is usually the result of childbirth).

A number of gynecologic procedures, including a hysterectomy, are done with a laparoscope. Laparoscopy is often done to coagulate a portion of the Fallopian tubes for women who do not wish to have children.

IMPACT ON INDUSTRY

Health care is a major portion of a nation's economy. The extent of governmental control varies, but in most countries, national, state, and local governments share control. Because women make up roughly half of the population, women's health care has a major impact on industry. Any health care delivery system must be cognizant of women's needs.

Industry and Business. Most larger hospitals have a section devoted to obstetrics, which includes a labor and delivery area as well as a postpartum area for recovery. These areas often feature a nonthreatening, comfortable environment with specialized equipment designed to blend in with the surroundings. A nursery is provided to care for the new infants. Some infants require intensive care; thus, many hospitals have a neonatal intensive care unit. If a hospital does not have such a unit, it typically transfers infants needing intensive care by ambulance or helicopter to another hospital.

Many hospitals have a separate floor or wing devoted to women with gynecologic conditions. Many surgical instruments and medical devices are designed specifically for gynecologic procedures. A large portion of the ultrasound equipment available on the market is designed specifically for obstetric-gynecologic uses. Colposcopy equipment has the sole purpose of viewing the cervix.

A significant market exists for items such as IUDs, diaphragms, condoms, and a variety of over-the-counter products used for contraception. In addition, many pharmaceutical products are focused on women's health care. These products include oral contraceptives, fertility-enhancing products, hormonal preparations, and medications to ease menstrual cramps.

In addition to the industries that provide health care and drugs for women, a number of companies have been founded to address women's other health care-related needs. These firms provide classes on pregnancy and breastfeeding, fitness programs for during and after pregnancy, products to make pregnancy and the menopause more comfortable, as well as books and audiovisual aids on pregnancy, childbirth, birth control, and the menopause.

CAREERS AND COURSE WORK

To become a specialist in obstetrics and gynecology, an individual must first obtain a bachelor's degree. As long as the course requirements for medical school are met, this degree can be in any field. Next, the individual must earn an M.D. from a four-year medical school, followed by an internship and residency program for specialty training. Some specialists limit their practice solely to either obstetrics or gynecology. Some specialists receive additional training in a subspecialty such as maternal-fetal

Fascinating Facts About Obstetrics and Gynecology

- In 1900, infant mortality in the United States was 1,650 per 100,000 births and maternal mortality was 900 per 100,000 births. In 2000, infant mortality had declined to 70 per 100,000 births and maternal mortality to 10 per 100,000 births.
- An estimated 350,000 to 500,000 women worldwide die in childbirth each year.
- Charles Clay performed the first abdominal hysterectomy in 1843. Although the operation was a success, the patient died fifteen days later of other causes. The first successful abdominal hysterectomy in which the woman lived was performed ten years later, by Ellis Burnham.
- Quintuplets are an obstetrical challenge even with modern technology. In 1934, the Dionne quintuplets were delivered by a country doctor with the assistance of two midwives. The infants were placed in a wicker basket containing heated blankets. They were brought into the kitchen and placed next to the open door of the oven to keep warm. All five infants survived into adulthood.
- Puerperal sepsis (severe infection occurring at the time of childbirth) became widespread in the nineteenth century when home delivery was replaced by hospital delivery. In 1843, Boston physician Oliver Wendell Holmes determined that the fever was carried from bed to bed on the unwashed hands of the physician. In 1847, Hungarian doctor Ignaz Semmelweis came to the same conclusion.
- In 1894, the use of rubber gloves became popular in surgery. The use was encouraged by Hunter Robb, a gynecologist at The Johns Hopkins University who had a special interest in wound contamination and sterile surgical technique.

medicine, infertility, or urogynecology (focused on the medical and surgical treatment of problems with the female urinary system).

The obstetrician-gynecologist has a variety of career opportunities. Entering private practice usually entails joining the practice of an established physician or group of physicians. Such groups may consist solely of obstetrician-gynecologists or contain specialists in a variety of fields. The trend is away from solo practitioners and small group practices. Larger groups have the benefit of sharing equipment and other resources. Also, being part of a group allows doctors to share weekend and night calls. Larger groups often employ physician assistants and nurse practitioners to handle less-complex problems. Opportunities exist for a career in academic medicine, including medical research and the training of medical students or resident physicians. Most obstetrician-gynecologists take written and oral examinations to become certified in the specialty. The specialty board in the United States is the American Board of Obstetrics and Gynecology.

SOCIAL CONTEXT AND FUTURE PROSPECTS

Any physician who deals with birth control and pregnancy may also be called on to deal with abortion. Abortion is a highly debated topic in many areas of the globe and an accepted procedure in others. Opinions range from the belief that life begins at the moment of conception and should not be disrupted under any circumstances to the view that it is a woman's right to terminate a pregnancy at any point up to full term. Some physicians will not perform an abortion under any circumstances, and some limit their practice to abortion and sterilization procedures. In the United States, this topic is likely to remain controversial for the foreseeable future.

Surgical techniques will continue to evolve, particularly in endoscopy (laparoscopy and endoscopy), where the use of robotics is likely to increase. As new equipment is developed, techniques will evolve. For example, during in vitro fertilization, traditionally, several embryos were implanted into the uterus because some were not expected to survive. However, this practice increases the likelihood of multiple births, which raises the risk of premature birth and miscarriage. Because of improvements in technology, a single embryo now has a good chance of survival, and many physicians have begun implanting only one embryo. Ultrasound and other imaging methods allow many birth defects to be diagnosed before birth, and intrauterine surgery is evolving to correct these problems.

Robin L. Wulffson, M.D., F.A.C.O.G.

FURTHER READING

Beckmann, Charles R. B. *Obstetrics and Gynecology.* 6th ed. Baltimore: Lippincott Williams & Wilkins, 2010. Covers all aspects of the specialty and is endorsed by the American College of Obstetricians and Gynecologists for its compliance with the organization's standards and procedures.

Carlson, Karen J., Stephanie A. Eisenstat, and Terra Diane Ziporyn. *The New Harvard Guide to Women's Health.* Cambridge, Mass.: Harvard University Press, 2004. Brings together doctors from Harvard Medical School, Massachusetts General Hospital, and Brigham and Women's Hospital to provide complete information on women's health.

Elit, Laurie, and Jean Chamberlain Froese, eds. *Women's Health in the Majority World: Issues and Initiatives.* New York: Nova Science Publishers, 2006. This first part of the book focuses on health issues that specifically affect women. The second part discusses how agencies such as governments, nongovernmental organizations, and professional societies can work together and improve standards for women.

Norwitz, Errol R., et al., eds. *Oxford American Handbook of Obstetrics and Gynecology.* New York: Oxford University Press, 2007. Covers all aspects of obstetrics and gynecology as well as diagnosis and treatment options.

Warsh, Cheryl Lynn Krasnick. *Prescribed Norms: Women and Health in Canada and the United States Since 1800.* Toronto: University of Toronto Press, 2010. Examines the history of medical health treatment of women in the United States and Canada, examining menstruation, the menopause, childbirth, and medical education, including nursing.

WEB SITES

American Congress of Obstetrics and Gynecology
http://www.acog.org

National Institutes of Health
Women's Health Initiative
http://www.nhlbi.nih.gov/whi

U.S. Department of Health and Human Services
National Women's Health Information Center
http://www.womenshealth.gov

See also: Anesthesiology; Geriatrics and Gerontology; Pharmacology; Reproductive Science and Engineering; Surgery.

OCCUPATIONAL HEALTH

FIELDS OF STUDY

Public health; environmental health; safety engineering; ergonomics; occupational hazards; psychology; medicine; nursing; epidemiology; health policy and management; industrial ethics; biostatistics; industrial hygiene; law; biomechanics; toxicology; risk sciences; sociology; chemical engineering; industrial chemistry.

SUMMARY

Occupational health takes a multidisciplinary approach to providing a positive work environment that minimizes work-related illness and hazards so that workers can perform their jobs safely. In the past, businesses focused on controlling the spread of disease; however, the chief goal of occupational health and safety has become the prevention of illness in workers. Prevention of work-related illness helps minimize worker absenteeism, enhances productivity, and promotes economic stability for both the worker and the business. Healthy and safe workers lead to productive businesses, healthy communities, and healthy families.

KEY TERMS AND CONCEPTS

- **Biomechanics:** Study of the structure and function of the human body, using mechanics, addressing internal and external forces and the effects of these forces on the body.
- **Epidemiology:** Study of factors that affect the health of certain populations.
- **Ergonomics:** Science of adapting the workplace to the needs and tasks of workers.
- **Industrial Toxicology:** Study of industrial chemicals and their effects on workers who are exposed to them in the workplace.
- **Occupational Accident:** Work-related accident that results in illness, injury, or death; also known as industrial accident.
- **Occupational Disease:** Sickness contracted from exposure to risk factors in the workplace that have demonstrated a strong cause-and-effect relationship.

- **Occupational Hazard:** Any condition in the workplace that most likely contributes to the illness, disability, or death of a worker.
- **Occupational Injury:** Any personal injury, disease, or death that results from a work-related accident.
- **Occupational Safety:** Processes that protect workers and foster a safe workplace.
- **Risk Assessment:** Process of identifying occupational hazards and workers affected by them, determining how workers are affected, evaluating the risks and implications, identifying actions to minimize the risk, and prioritizing the actions to be taken.

DEFINITION AND BASIC PRINCIPLES

"Occupational health" is the term used to describe the interdisciplinary approach to protecting the health and safety of workers. The World Health Organization and the International Labour Organization define occupational health as promoting and maintaining the highest level of well-being of workers, including their physical, social, and mental well-being. The work environment directly influences the ability of workers to perform their assigned duties. Overall occupational health strives to keep the worker well and on the job.

Occupational health is not a medical program for workers. However, the interdisciplinary team may include doctors and nurses. In the past, the occupational health program consisted of a designated physician who saw patients with work-related health issues who were referred by their employer. Companies often hired an occupational health nurse who operated a small on-site clinic, seeing and assessing workers' physical complaints and health issues. The nurse might also provide health education or screening for employees. The occupational health nurse could give flu and pneumonia shots or make community referrals for additional resources as needed.

The modern approach to occupational health and safety is much broader and more inclusive. Teams are multidisciplinary and address mental and social needs of workers in addition to their physical needs. Teams may include a physician, nurse, social worker, psychologist, occupational hygienist, and toxicologist. Programs on topics such as stress management may be available to support the worker in the work setting and at home. The value of a safe and healthy workforce is significant.

BACKGROUND AND HISTORY

Historically, work-related illnesses have not received much attention. The code of Hammurabi, the first code of law written about 1750 b.c.e., prescribed penalties that supported restitution for the loss of an eye by a worker. Early seeds of occupational health concerns can be seen in about 1500 b.c.e. when the famous Egyptian pharaoh Ramses provided a doctor to help quarry workers, recognizing that gold and silver fumes were detrimental to human health. The Greek physician Hippocrates, known as the father of medicine, instructed medical students that a harmful work environment, such as that of stonecutters, might negatively affect workers' health and well-being. Georgius Agricola, in his *De re metallica* (1556; English translation, 1912), wrote about the health hazards of mining, including mine dust that caused lung problems. In 1700, Italian professor of medicine Bernardino Ramazzini published *De morbis artificum diatriba* (*Diseases of Workers*, 1940), which described the diseases and health issues of those engaged in fifty-two occupations. Among his many observations, he noted the effects of lead and mercury on miners and how sitting, stress, and repetitive hand movements produced illnesses in clerks.

In 1775, London surgeon Percivall Pott discovered that chimney sweeps developed a cancerous growth in the skin of the scrotum, which the chimney sweeps called a "soot-wart," and he linked it to their working environment. Pott's work, along with that of many others, and the severe work conditions that accompanied the Industrial Revolution paved the way for legislation. In Great Britain, an initial health and safety act for workers was passed in the early 1800's. Subsequent legislation decreased the number of hours that women and children could work. The United Kingdom, with the passage of the Health and Safety at Work etc. Act of 1974, created a framework for protecting its workers.

In the United States, the advent of the Industrial Revolution created new workplace hazards from dangerous processes and materials and repetitive motions. Labor unions began lobbying for better working conditions to protect the health of their members. This resulted in the passage of regulations regarding mine workers and the implementation of safety methods in the railroad industry to decrease the loss of limbs by workers. In 1916, the U.S. Supreme Court legitimized workers' compensation laws to pay for financial loss and medical expenses that resulted from work-related injuries. After World War II, occupational health efforts addressed the effects of heavy physical labor, extreme levels of noise, excessive heat and cold, and psychological stressors on workers. The Occupational Safety and Health Act of 1970 created the Occupational Safety and Health Administration (OSHA) to ensure healthful working conditions.

HOW IT WORKS

In the past, many workers possessed skills in one key area and went to work for a company where they held full-time positions until retirement. If illness occurred during their employment, typically it was covered under health insurance. Company physicians or occupational health nurses also provided minor care in on-site clinics to keep workers on their jobs.

However, as the changing economy produced less job security and stability, workers have become more mobile, switching jobs several times during their careers and sometimes changing occupations. Workers must have multiple skills and embrace emerging technology. Many work part-time or take contract work without insurance or retirement benefits. They enter the workplace with preexisting health conditions that influence their work life. With these many changes in the workforce, occupational health and safety programs have expanded to include a multidisciplinary team to assist workers in organizations and industry.

The Multidisciplinary Approach. At one time, the company physician and occupational health nurse bore the primary responsibility for the health and safety of workers in their workplace. Although that physician-nurse team has expanded to include other disciplines, clinical services are still an important part of occupational health and safety programs. Some of the services provided by health care professionals include preemployment physicals, health assessments, and health surveillance through intermittent medical examinations and basic laboratory testing. Health education and counseling usually falls to occupational health professionals, as do treatment, rehabilitation, and referral.

The multidisciplinary team brings complementary skills and diverse experiences for more effective occupational health and safety programs. In addition to physicians and nurses, many other professionals have become specialists in occupational health and safety. These include psychologists, sociologists, toxicologists, hygienists, microbiologists, epidemiologists, health physicists, safety engineers, ergonomists, industrial chemists, and lawyers. This team provides businesses or agencies with the advantages of collaborative assessment and planning to ensure a safe and healthful workplace.

Protecting the Workforce. To have a healthy workforce that can fulfill job requirements, certain factors must be addressed by the business or organization. For example, the workplace must be assessed for risks and occupational hazards that pose threats to the health and safety of workers. Occupational hazards include threats within the physical environment, biological dangers, chemical exposures, and psychosocial conditions. After identification of any hazards, the actual impact of these hazards on the workers should be assessed. For example, determinations should be made as to whether these hazards have already resulted in disease, accidents, and injury (and if so, how often), and what the actual and perceived workplace risks are.

Managers are key to this process, as they are aware of all aspects of the business or workplace and understand the dynamics of their distinct workforce, such as rates of absenteeism. Managers may help evaluate the degree of risk associated with each hazard, adding their perceptions to those of skilled and experienced occupational health professionals, Together, they can develop and implement specific controls to minimize risks to workers. The occupational health and safety program will need regular review through audits and quality improvement processes to monitor its effectiveness. Changes in the business or workforce provide new opportunities to repeat this risk assessment process and improve the workplace.

APPLICATIONS AND PRODUCTS

Occupational health and safety professionals are service workers who either provide preventive services or support occupational health and safety programs. Each member of the interdisciplinary team contributes to the occupational health and safety program based on his or her skills and training. Some team members provide direct services to the workers, and others provide support services to the comprehensive occupational heath and safety program. Examples of applications of occupational health and safety include biomechanics, ergonomics, and occupational health psychology.

Biomechanics. Biomechanics studies the physics of human motion and can be useful in assisting workers with body mechanics and work-related movements. Biomechanics combines the knowledge of several fields of study, including physical, biological, and behavioral sciences and engineering. In the workplace, biomechanics is used to match job demands to workers in a way that minimizes the risk of musculoskeletal injuries. Biomechanics can help mitigate risk factors involved in jobs with repetitive or prolonged motion of the hands, such as in word processing or keyboarding, and jobs that require lifting, pulling, or pushing heavy objects or maintaining an unnatural body position for extended periods of time.

One common example of biomechanics in the workplace is designing the workspace for the millions of workers whose jobs require using a computer. The optimal workstation includes a computer monitor that is at or below eye level and a chair with lower back support and armrests close to the body. The worker should sit with relaxed shoulders and algin his or her head with the torso. If the worker makes extensive use of a mouse, an occupational health professional might suggest an ergonomically designed computer mouse to relieve strain from repetitive motion. Biomechanics studies how to design the workspace to enable the worker to minimize musculoskeletal stress and prevent injury. This leads to more efficient work and improved productivity.

Ergonomics. Ergonomics, which studies how people do their work, has been employed by occupational health and safety professionals and organizations to design workplaces and tools for worker health and safety. The demands of a job can be analyzed for possible problems by reviewing sources of information such as the company's workers' compensation claims. Hazard, such as high noise levels, and the associated possible problems, such as hearing loss, can be identified. OSHA provides information on specific ergonomic tools for certain at-risk workers in health care, sewing, printing, grocery warehousing, baggage handling, and beverage delivery as well as for electrical contractors and computer users. Some organizations have developed their own guidelines to support workers, including the American Apparel and Footwear Association, National Communications Safety Panel, and American Furniture Manufacturers' Association. OSHA offers free on-site evaluations and consultations so that small businesses can minimize ergonomic risks and maintain the health of their workers.

Occupational Health Psychology. The disciplines of health and industrial psychology combine with occupational health in occupational health psychology. Adverse workplace conditions may cause stress, anxiety, and burnout, with symptoms such as headaches, stomach problems, depression, and apathy. Interventions to avert employee illness or negative behavior include teaching workers to manage stress, use tools of conflict resolution, and deal with rudeness or uncivil behaviors in the workplace. Positive applications of occupational health psychology that can help organizations save health care dollars include motivational programs that encourage exercise, help decrease smoking, and promote compliant use of prescribed medications. Occupational health psychology can also be useful in minimizing workplace violence.

IMPACT ON INDUSTRY

Occupational health and safety is critical to the success of all industries. The effects of occupational health and safety programs extend beyond industry, reaching into the greater community. For example, most people agree that businesses cannot succeed unless workers are able to report to work and perform their jobs. However, other stakeholders are involved. Investors support businesses and expect a return on their investment. Customers want quality products and services provided as advertised. Suppliers want relationships with businesses that continue to use their products and services. Managers depend on workers to perform their jobs. Legislators introduce laws that affect workers, some of which increase the cost of doing business. Activist groups' protests and media coverage of problems can create pressure that encourages industry leaders and legislators to make changes that will improve the health and safety of workers. In the end, the well-being of the community and its businesses depends of the achievement of occupational health and safety.

OSHA. Several organizations provide leadership for occupational health and safety initiatives in industry. In the United States, the key agency is the Occupational Safety and Health Administration (OSHA), founded in 1970 under the U.S. Department of Labor. The purpose of OSHA is to make sure that workplace conditions ensure the health and safety of employees. OSHA sets and enforces safety and health standards for industry and provides education and training. Although OSHA is a federal agency, it monitors state programs to be sure they meet or exceed federal standards.

OSHA regulations apply to workers in private sector industries, the federal government, and state and local governments in OSHA-approved programs. However, self-employed workers, immediate family members of farm families without outside workers, and workers in some federal programs such as the Coast Guard or Federal Aviation Agency are exempt. Workers in OSHA-covered industries are encouraged to request that the agency make a site visit to their workplace and to report information about occupational hazards where they work with confidentiality.

Small business owners can request assistance from OSHA in identifying potential occupational hazards. This free consultation includes help developing an occupational heath and safety program. OSHA believes this can decrease workers' compensation claims. Better working conditions also improve worker morale, leading to increased productivity.

OSHA offers additional resources for the general public and industry. On its Web site, readers can access guidelines for preventing workplace violence, information on how to prepare for emergencies in the workplace and deal with chemical hazards, and fact sheets and information on personal protective equipment.

National Institute for Occupational Safety and Health. Another agency that provides support for workplace health and safety is the National Institute for Occupational Safety and Health, housed under the Centers for Disease Control and Prevention. This group provides health hazard evaluations to determine if workers have been exposed to hazardous materials in the workplace. Workers can request an evaluation be made at their workplace if, for example, workers acquire an illness from an unknown source or epidemiological studies demonstrate an inappropriately high incidence of an occupational disease among a particular group of workers.

International Labour Organization. Global industry is affected by occupational health and safety initiatives. The International Labour Organization was founded in 1914 and is part of the United Nations; it supports decent treatment of all workers. It provides international labor standards and oversees their implementation. The organization promotes social justice consistent with human rights in a safe and healthful workplace.

World Health Organization. The World Health Organization partners with the International Labour Organization to encourage companies worldwide to provide healthful and safe environments for their workers. Its model encourages companies to perform an assessment of work-related physical and psychological threats to their employees and to adopt healthful behaviors. In 2007, the organization adopted the Global Plan of Action on Workers' Health (2008-2017) to help national health systems deal with workers' health needs, establish global standards of workplace safety, and make sure that workers have access to preventive health care.

CAREERS AND COURSE WORK

Some of the most common careers in occupational health and safety are occupational health nurse, specialist, and technician.

Occupational health nurses are registered nurses employed to prevent work-related illness and injury, to provide emergency care at the workplace, and to implement health and safety regulations and standards. Basic nursing training can be in the form of an associate's or bachelor's degree. Most occupational health nurses have a bachelor's degree in nursing (B.S.N.) and make about $66,500 annually, depending on the industry size and setting. Registered nurses with a master's degree in nursing (M.S.N.) have annual salaries of about $75,000. Occupational health nurse practitioners earn about $106,000 each year. Over all, the demand for nurses is expected to increase, and the demand for occupational health nurse practitioners is growing because they are a less expensive way to provide quality one-site employee medical care.

The job potential for occupational health and safety specialists and technicians also is good. Occupational health and safety specialists try to ensure the safety of workers and to reduce work-related

---

Fascinating Facts About Occupational Health

- In 1902, Massachusetts passed the first compensation law for U.S. workers, but it was declared unconstitutional in 1904. In 1916, the U.S. Supreme Court declared workers' compensation laws to be constitutional.
- In 1970, President Richard Nixon signed the Occupational Safety and Health Act into law, and it became effective in 1971.
- The U.S. Department of Labor collects and publishes annual information on the number of work-related illnesses, injuries, and fatalities. The Injuries, Illnesses, and Fatalities program provides statistics by various industries, occupations, geographical locations, and other diverse parameters.
- The World Health Organization estimates that the cost of productivity lost because of work-related health concerns is about 4 to 5 percent of the gross domestic product.
- The American Association of Occupational Health Nurses estimates that poor worker health costs businesses $1 trillion each year.
- In 2009, 4,340 U.S. workers lost their lives on the job.
- Each year, more than 400,000 workers worldwide die from exposure to hazardous substances.
- About 10 percent of all skin cancers are thought to be related to exposure to hazardous materials in the workplace.

---

illnesses, thereby reducing absenteeism and downtime in industry, which can mean lower insurance premiums and fewer workers' compensation claims. They inspect workplaces to uncover hazards and then suggest improved workplace designs. They advise management on ways to avoid worker illness and injury. Most are required to have a bachelor's degree in health, safety, or a related field, and some positions require a master's degree. Certification is usually voluntary, and the average salary is about $62,000.

Occupational health and safety technicians enter the marketplace through on-the-job training, work experience, or a two-year associate's degree. Recommended high school courses include chemistry, math, biology, physics, and English. This job involves detail work, and employers usually encourage certification. The average annual income is about $45,000. Most technicians are employed in large private firms or work for the government.

The U.S. Bureau of Labor Statistics reports that about 41 percent of occupational health and safety specialists work for the government. The expected rise in employment is about 11 percent, an average rate. The overall job outlook for technicians is greater than average, with a 14 percent expected increase. Most technicians will work under the supervision of specialists, and the increase in jobs will be related to technological advances, additional regulations, and public interest.

SOCIAL CONTEXT AND FUTURE PROSPECTS

Since the early 1900's, occupational health and safety in the United States has altered in response to the changing work types and settings, as well as worker demographics. In the industrial era, workers were often factory employees who faced considerable physical, chemical, and mechanical hazards, often with no safeguards or standards. As the number of service workers—teachers, health care workers, social workers, and office workers—rose, the field of occupational health and safety focused on risks associated with these types of jobs, including biological, psychosocial, chemical, and musculoskeletal hazards (ergonomics) as well as physical agents such as noise, lighting, and air quality. The Centers for Disease Control and Prevention lists about twenty-five hazards stemming from the workplace setting.

The U.S. Department of Labor (DOL) is responsible for promoting the welfare of workers in the United States and improving working conditions. The department enforces laws related to workers and workplace safety and health in the public and private sector through OSHA. In 2010, OSHA announced the Severe Violator Enforcement Program, which will conduct concentrated inspections of those companies that have demonstrated significant indifference to the Occupational Safety and Health Act or have a history of repeated hazard violations. Noncompliance fees are as high as $70,000 for willful serious violation.

Occupational health and safety issues may change as the workplace develops, but they will continue to be primary concerns for employers, businesses, and the government in the global marketplace.

Maintaining worker safety and preventing injury and illness are economically sound and ethically correct business practices.

Marylane Wade Koch, M.S.N., R.N.

FURTHER READING

Aw, T. C., et al. *Occupational Health.* 5th ed. Malden, Mass.: Blackwell, 2007. A clinical guide covering hygiene and medicine, as well as legal and international issues.

Reese, Charles D. *Occupational Health and Safety Management: A Practical Approach.* 2d ed. Boca Raton, Fla.: CRC Press, 2009. A practical guide to identifying hazards and building and implementing a program to address them.

Smedley, Julia, Dick Finlay, and Steven S. Sadhra, eds. *Oxford Handbook of Occupational Health.* New York: Oxford University Press, 2007. Summarizes the theory and practice. Contains chapters on occupational hygiene, fitness for work, environmental medicine, laws, and diseases and hazards.

Stranks, Jeremy W. *Health and Safety at Work: Key Terms.* Oxford, England: Butterworth-Heinemann, 2002. A dictionary of all topics relevant to occupational health and safety.

Tompa, Emile, A. J. Culyer, and Roman Dolinschi, eds. *Economic Evaluation of Interventions for Occupational Health and Safety: Developing Good Practice.* New York: Oxford University Press, 2008. Examines ways to evaluate and assess the economic impact of programs designed to improve occupational health and safety.

WEB SITES

American Society of Biomechanics
http://www.asbweb.org/html/biomechanics/Biomechanics.html

Centers for Disease Control and Prevention
National Institute for Occupational Safety and Health
http://www.cdc.gov/niosh

Human Factors and Ergonomics Society
http://www.hfes.org/web/default.aspx

International Labour Organization
http://www.ilo.org

Occupational Safety and Health Administration
http://www.osha.gov

U.S. Bureau of Labor Statistics
Injuries, Illnesses, and Fatalities
http://www.bls.gov/iif

World Health Organization
Occupational Health
http://www.who.int/topics/occupational_health/
en

See also: Biomechanics; Epidemiology; Ergonomics; Immunology and Vaccination; Nursing.

OCEANOGRAPHY

FIELDS OF STUDY

Climatology; geography; geology; chemistry; marine biology.

SUMMARY

More than 70 percent of the Earth's surface is covered by saltwater oceans and seas. Oceanographers study a wide range of subjects, including major underwater geological formations, such as very high mountain ranges and very deep canyons, as well as the tectonic and volcanic movements that continue to change the submarine environment. They also study the plant and animal life in the oceans of the world and analyze the chemical bodies—gases and minerals—in the waters.

KEY TERMS AND CONCEPTS

- **Azoic Zone:** A depth (lower than 600 meters) at which it was thought marine life could not exist. This theory was put forth in 1840 but later disproved.
- **Bathyal Zone:** Marine environment associated with offshore continental slopes, with depths between 200 and 3,000 meters, characterized by many varieties of marine flora and fauna.
- **Equatorial Undercurrent:** Subsurface eastward-flowing current in the Atlantic beneath the westward-flowing Equatorial Current; originally observed in 1876.
- **Gravity Waves:** Larger sea surface waves that flatten (return to a state of equilibrium) because of what oceanographers call the restoring force of gravity.
- **Pycnocline:** Sudden change in sea-water density as depths increase.
- **Sedimentation:** Layered deposits of solid particles on the bottom of seas and oceans.
- **Sonar:** Sound navigation ranging; a method of recording the echo, or return of sound waves emitted by a sound device located in a submarine or similar vehicle.
- **Tsunami:** Very large ocean wave resulting from a sudden major displacement in the seafloor, either an earthquake, volcanic eruption, or a massive underwater landslide.

DEFINITION AND BASIC PRINCIPLES

Oceanography is not a single applied science but rather a combination of fields of study. Because the oceans and seas of the world contain both living creatures and inanimate physical components, the basic sciences that contribute to oceanographic knowledge are marine biology, chemistry, and geology. However, the ecological effects caused by phenomena such as oceanic tides, currents, and temperature variations must be studied before the marine biologist, chemist, or geologist can obtain meaningful results from research.

BACKGROUND AND HISTORY

Since early historical times, people have been interested in the oceans and seas. For many centuries, attitudes toward the ocean were strongly influenced by superstition and mythology. Even as late as the publication of Jules Verne's *Vingt mille lieues sous les mers* (1869-1870; *Twenty Thousand Leagues Under the Sea*, 1873), depictions of the ocean depths were dominated by fantastic and mysterious images rather than based on science. Exactly when the study of oceanography became a science is hard to say, but each time technology allowed scientists to study the ocean at a greater depth, knowledge advanced significantly. As scientific knowledge of oceans and seas increased, people began to realize that a knowledge of geology and, to some extent, astronomy, was valuable in understanding ocean phenomena.

The desire to develop submarines for use in warfare, particularly between World War I and World War II, spurred advances in ocean exploration. In the 1930's, Americans Otis Barton and William Beebe developed the bathysphere. A pressure-resistant spherical steel apparatus weighing 2,025 kilograms, the bathysphere could descend more than 900 meters, six times the depth that could be reached by helmeted divers. The bathysphere was connected to a ship on the surface by heavy cables and a hose that held electric and communication wires.

In 1951, Rachel Carson published *The Sea Around Us*, which popularized the study of oceanography. In her book, she focused on the vital interrelationships between submarine and surface phenomena and the delicate balance these relationships represent for the future of the world. Many of her concerns were adopted by later generations as the predominant issues confronting oceanography.

Another major figure who contributed to global recognition of the wonder of the oceans' secrets was the French oceanographer Jacques Cousteau, whose contribution to knowledge of the undersea world began with the 1943 film *Épaves* (shipwrecks), followed by his launching of the French navy's underwater research group. The navy group was the first of many to explore the ocean using methods developed by Cousteau and his team.

In 1953, Cousteau published *The Silent World*, which described his work as a diver and established the first scientific procedures for echolocation of large fish populations underwater. His accomplishments led to his appointment as director of the famous Oceanographic Institute and Museum in Monaco in 1957. Soon he developed the SP-350, known as the diving saucer, which became a prototype for descending—with significantly improved maneuverability—to ever-deeper reaches. The SP-30 reached a depth of 500 meters in 1965. Although early developers of the bathysphere had descended to almost 250 meters in 1930 and to more than 900 meters in 1934, the limited maneuverability of their invention could not compare with that of Cousteau's diving vessel. Before Cousteau died in 1997, he and his associates witnessed major developments in fields of oceanography that they and others had pioneered.

How It Works

Measuring Tides. Perhaps more than any other area of oceanography, the study of tides and waves requires knowledge from several fields of science. Principles of physics and astronomy, for example, help explain timing lapses and the strength of coastal tides. The rising and falling of tides are the result of gravitational pull from both the Moon and the Sun, and the strength and direction of tides are affected by forces connected with the Earth's rotation. Oceanographers may be involved in complicated mathematical calculations using the laws of physics to compare major tidal movements, while hundreds of relatively simple tidal gauge stations using float devices provide practical data for record keeping across the globe.

Systematic data gathering may not be enough, however, to predict potentially disastrous tsunamis or so-called monster waves in the middle of the ocean. Tsunamis are often caused by major earthquakes at fault points deep beneath the ocean surface. The earth movements unleashes swells that race toward coasts. The task of oceanographers in such cases is to measure the speed and forces in the tidal waves in an attempt to limit the damage when they reach the shore.

Oceanographers continue to investigate monster or rogue waves that appear rarely but suddenly on the high seas and pose major threats to shipping. Rogue waves, which can be tall enough to sink ships, are thought to be the result of several tidal movements that come together at one location in mid-ocean and combine their forces.

Mapping Ocean Currents. Ocean currents can be at the surface or much deeper. The best known of the surface currents is the Gulf Stream, which can reach speeds of 300 centimeters per second, followed by the Kuroshio off the east coast of Japan. Westward-flowing currents north and south of the equator (caused mainly by the trade winds over tropical regions of the Atlantic, Pacific, and Indian oceans) are separated by a 300-meter-wide countercurrent moving eastward. Most challenging for oceanographers are several major deep-ocean currents, which have both lateral and vertical patterns of movement. To study these currents, oceanographers use instruments to locate underwater strata that differ either in temperature or levels of saline (or other chemicals). They use these data to attempt to explain why massive volumes of water either drop or rise, displacing other bodies and creating currents in the depths of the ocean.

Surveying the Ocean Floor. Two obvious and spectacular applications of geology to oceanography are the study of deep trenches and mid-ocean mountain ridges. The Mariana Trench north of New Guinea, for example, is more than 2,542 kilometers long and, in places, 69 kilometers wide. At its deepest point, the Challenger Deep, the Mariana Trench plunges to more than 10,924 meters below sea level. The Mid-Atlantic Ridge (discovered in 1872) rises from more than 2,000 meters below sea level to form island mountains that are more than 2,000 meters high (notably in Iceland and the Azores).

Less spectacular but of considerable practical importance is the study of the continental shelves. Oceanographic studies of coastal ocean zones contribute increasingly to key economic development schemes, most notably (but with increasingly controversial environmental implications) in seeking zones for commercial exploitation of undersea petroleum reserves.

How El Niño Works. Although oceanographers are still investigating the phenomenon of El Niño (El Niño-Southern Oscillation, or ENSO), certain facts have been established. When there is an area-wide warming (El Niño conditions) or cooling (La Niña conditions) of surface waters in the Equatorial Pacific occurring at the same time that atmospheric pressure causes surface waters of the western Pacific to rise or fall (the Southern Oscillation), major climatic repercussions occur in the subtropical eastern Pacific. These conditions may bring extreme drought or excessive rain. The reasons for the ocean-wide coupling of this cause-and-effect situation are not clear. Oceanographers estimate that the phenomenon recurs in cycles of about five to seven years.

APPLICATIONS AND PRODUCTS

Harvesting Microscopic Organisms. Oceanographers have increasingly directed their attention to microscopic plant and animal organisms, most particularly to the smallest forms of plankton in the food chain and multiple varieties of algae. Extremely tiny and exponentially numerous holoplankton (as distinct from meroplankton, which are mainly tiny larvae that become larger independent organisms such as jellyfish, sea urchins, and starfish) never develop sufficient motor independence to propel themselves and instead drift like clouds in the world's seas and oceans. Concentrated masses of plankton provide nourishment to many varieties of fish, which in turn are eaten by larger fish. Therefore, oceanographers—as potential sources of information for the fishing industry—focus on studying not only the temperatures and saline/chemical content that contribute to plankton propagation but also the currents that can transport them to different areas of the ocean, creating concentrated feeding grounds for larger, commercially desirable fish species.

Commercial Uses of Algae. Oceanographers have contributed information to industries wishing to control the growth of certain species of algae.

Industrially organized algae cultivation is already used in producing fertilizers and feed for livestock, dyes, and some pharmaceuticals. In general terms, algae (larger forms of which are called seaweed) are plant plankton that, like all plants, use photosynthesis to convert carbon dioxide into various organic compounds and, most important for the environment, into oxygen. Algae's ecologically important role as an oxygen producer, along with its potential as a source for fuel, pose challenges for oceanographers. They are likely to be called on to identify the most relevant species for various industrial uses and the optimal conditions for their propagation under controlled conditions.

IMPACT ON INDUSTRY

Commercial Fishing. For many centuries, people have depended on the seas and oceans as a source of food, primarily through fishing. Fishing has evolved into a major industry, with both positive and negative implications for the global environment. As technology has allowed fishermen to capture more fish, a major concern is the possibility that overfishing will deplete populations of particularly sought-after species in critical geographical locations, including cod and tuna. Also, when large commercial fishing ventures catch fish, they snare various species of whales and dolphins, which die in the process. Many environmental groups have demanded that large-scale fishing operations be banned from or be heavily regulated in areas in which the whale and dolphin populations are dwindling. Seven species of whales and two species of seals (Caribbean and Hawaiian monk seals) have been placed on the endangered animals list.

Coral. Coral, once thought to be a plant, is made up of colonies of polyps. Its colorful skeletal remains have been widely used in jewelry, and coral reefs and atolls, known for their beauty, attract tourists. However, coral's popularity has led to excessive harvesting of coral and too frequent visitations to its reefs, which have threatened the existence of many highly sought-after species. Oceanographers, working with conservationists, must work to find ways to support coral habitats and to stop further destruction of the reefs.

Harvesting Chemicals and Minerals. The work of oceanographers often involves chemistry. Massive quantities of various atoms and molecules are

---

Fascinating Facts About Oceanography

- The ocean covers more than 70 percent of Earth's surface, and 65 percent of that is blue water (open water), or waters beyond the coastal shallows.
- An area on the floor of the South Pacific that is about the size of the state of New York is home to 1,133 volcanic cones and sea mounts.
- Most waves are caused by the friction of the wind on the water. How big the ripples become depends on the strength of the wind, how far it blows (fetch), and how long it blows (duration). Drops of water move forward with the crest (high spot of the wave) and back with the trough (low spot).
- The neritic zone, the first 200 meters of ocean water includes the seashore and most of the continental shelf. This is where phytoplankton and zooplankton live.
- The oceanic zone reaches from 200 meters deep to the ocean floor.

---

released into the atmosphere by the constant process of evaporation. The average person thinks automatically of the oxygen molecules that are released though evaporation from oceans (and inland waters) to become the source of rain. Scientists, however, study not only evaporation but also the huge amounts of various minerals and chemicals that are contained in ocean water. Some of these are released into the atmosphere in the form of gaseous bubbles that rise from the water's depths. Oceanographers work with private companies interested not only in capturing gases that can be useful to various industries but also in mining the oceans for minerals that have obvious economic value, such as barite, cobalt, copper, nickel, tin, and zinc. As technological methods advance, significant amounts of gold, platinum, and even diamonds may be extracted from the oceans on a commercial scale.

Transforming Ocean Energy. Because the movement of waves and tides is a (presumably) endless and pollution-free source of energy, environmentalists hope to use data gathered by oceanographers to convince private companies and governments to invest in rather uncomplicated but costly technology that can transform such forces into electric power. For example, certain spots off the northern shores

of Europe have been identified as good places to locate giant windmills that will capture the energy of offshore winds. Whether the energy of tides or ocean winds is tapped, oceanographic knowledge will be essential to guarantee a minimum of ecological disturbance.

CAREERS AND COURSE WORK

Individuals interested in preparing for a career in oceanography should not only plan their education carefully but also consider the employment paths likely to be open at various levels. It is generally recommended that undergraduates focus on the core sciences (such as biology and chemistry) and mathematics. For some entry-level jobs relating to oceanography, an undergraduate degree with a heavy emphasis on science may meet the basic requirements. Those seeking master's or doctoral degrees, however, must attend one of the approximately fifty universities offering graduate programs in specialized subareas of oceanography, such as biological, chemical, geological, or physical oceanography. A master's degree or doctorate in a subarea of oceanography is required to work in academic institutions (as teachers or researchers), private companies with involvement in various maritime economic pursuits, international or national organizations involved in oceanography, a variety of governmental agencies involved in environmental issues, and certain branches of the military.

A number of specifically defined areas are likely to attract the professional expertise of oceanographers in the future. Several of these are connected with growing environmental concerns, while others are more clearly tied to changing economic markets. Environmental concerns include global climate change issues and questions regarding the disposal of industrial and agricultural wastes. Increasingly, corporations are turning to oceanographers to develop ways to exploit the ocean waters. For example, fish farming based on scientifically grounded methods plays an ever-expanding role as world demand grows for common edible seafoods. Industrial extraction of minerals and the development of the oceans' biomedical and pharmaceutical resources will require a variety of oceanographic technical skills to guarantee proper responses to both ecological and health safety concerns.

The main U.S. government agency concerned with oceanographic issues is the National Oceanic and Atmospheric Administration (NOAA). There are also numerous private-sector research organizations and professional associations of interest to oceanographers. Three of the most famous research institutes in the United States are the Scripps Institution of Oceanography in La Jolla, California; the Woods Hole Oceanographic Institution in Woods Hole, Massachusetts; and the Lawrence Berkeley National Laboratory in Berkeley, California. Professionals in these institutes (together with their fellows in many other private research organizations) continue to play significant roles in defining the new frontiers of oceanography.

A variety of private companies that employ oceanographers (such as the Woods Hole Group, General Oceanics, and Sound Ocean Systems) share contracts and research projects with Scripps and Woods Hole, as well as similar research institutions in other countries. Those interested in career possibilities in oceanography should familiarize themselves with a number of professional associations or interest groups whose members are oceanographers. These include the Consortium for Oceanographic Research and Education in Washington, D.C.; the Reef Environmental Education Foundation in Key Largo, Florida; the Atlantic Coastal Zone Information Steering Committee in Halifax, Nova Scotia; and the National Geographic Society in Washington, D.C.

SOCIAL CONTEXT AND FUTURE PROSPECTS

A number of socially relevant issues are likely to involve professional oceanographers. Several of these are connected with growing environmental concerns, and others are tied to changing economic markets. In the first area, global climate change and questions of industrial-waste disposal rank very high. Rising concerns around the world over the effects of pollution on oceanic ecosystems suggest that very aggressive conservationist movements (such as Greenpeace International) are likely to be joined by an increasingly broad spectrum of environmental organizations focused on the world's oceans and seas.

Meanwhile, economically attractive activities involving controlled exploitation of the ocean waters is already involving oceanographers in key areas that seemed to be of only marginal importance before the twenty-first century. Scientifically based fish farming, for example, is growing in importance as demand for seafood increases. Conservationists hope that fish farms will decrease pressures on the natural habitat of many commercially attractive species. Also, oceanographers are likely to need to create better ways to extract minerals from the world's oceans and to develop marine pharmaceutical resources without damaging the environment.

Because ecological and climate issues are being recognized as vital for the future of the planet, oceanographic research projects are being conducted in these areas. Such programs include NOAA's Climate Variability and Predictability program, which seeks to find the causes of climate variability; the Integrated Ocean Drilling Program, an international organization that studies the history of the Earth through rocks and sediments beneath the ocean floor; and NOAA's Sustainable Seas Expeditions, which explores various marine ecosystems. High-tech environmental research projects include the Surface Heat Budget of the Arctic Ocean (SHEBA) of the National Center for Atmospheric Research's Earth Observing Laboratory, which is investigating the relationship between the Arctic ice pack and climate.

Development of remotely operated vehicles, greatly expanded since the 1980's, has made it possible to penetrate ocean depths that were unreachable by manned undersea devices. In May, 2009, the Woods Hole Oceanographic Institution used a remotely operated vehicle called the *Nereus* to revisit the depths of the Mariana Trench, which had been partially explored by a Japanese remotely operated vehicle in 1998. The small dimensions of the *Nereus* (8 feet wide and 14 feet long) enabled it to maneuver in tight spaces, and the vehicle combined several advanced technologies, most notably a remarkably thin fiber-optic tether.

Byron D. Cannon, M.A., Ph.D.

FURTHER READING

Carson, Rachel L. *The Sea Around Us.* 1951. Reprint. New York: Oxford University Press, 2003. A truly seminal work that introduced the general public to the extraordinary complexity and wonders of oceanic environments.

Denny, Mark. *How the Ocean Works.* Princeton, N.J.: Princeton University Press, 2008. Covers ten key technical areas studied by oceanographers.

Earle, Sylvia A. *The World Is Blue: How Our Fate and the Ocean's Are One.* Washington, D.C.: National Geographic Society, 2009. Emphasizes ecological issues in its discussion of the world's oceans.

Nichols, C. Reid, et al. *Recent Advances in Oceanography*. Westport, Conn.: Greenwood Press, 2003. Offers a chronology of research and statistics on budgets and funding, and details challenges posed by unanswered questions.

Seinfeld, J. H. *Atmospheric Chemistry and Physics of Air Pollution*. New York: John Wiley & Sons, 1986. Contains information on the effects of ocean water evaporation on the chemical content of the atmosphere.

Thurman, Harold V., and Alan P. Trujillo. *Essentials of Oceanography*. 10th ed. Upper Saddle River, N.J.: Pearson Education, 2009. Examines the basic concepts of oceanography and provides problems and exercises.

WEB SITES

National Geographic Society
The Ocean
http://ocean.nationalgeographic.com/ocean

National Oceanic and Atmospheric Administration
http://www.noaa.gov/ocean.html

The Oceanography Society
http://www.tos.org

Office of Naval Research
Oceanography
http://www.onr.navy.mil/focus/ocean

Scripps Institution of Oceanography
http://www.sio.ucsd.edu

Woods Hole Oceanographic Institution
http://www.whoi.edu

See also: Climatology; Coastal Engineering; Environmental Chemistry; Environmental Engineering; Fisheries Science; Marine Mining; Meteorology; Ocean and Tidal Energy Technologies; Sonar Technologies; Submarine Engineering; Water-Pollution Control.

OPHTHALMOLOGY

FIELDS OF STUDY

Laser refractive surgery; cataract surgery; visual neurophysiology; visual psychophysics; physiological optics; molecular biology; corneal surgery; oculoplastic surgery; retinal surgery; pediatric ophthalmology; strabismus and eye movements; glaucoma; age-related macular degeneration; diabetic retinopathy; neuro-ophthalmology; epidemiology; pathology; immunology; genetics; retinal cell biology.

SUMMARY

Ophthalmology is a division of medical science centered on the visual system and its associated anatomy, physiology, diseases, and disorders. An ophthalmologist is a medical doctor and surgeon who has specialized in diagnosis and treatment of the eyes. The science of ophthalmology is quite broad and can be separated into more specialized fields based on specific anatomy, such as the cornea, retina, or optic nerve. These structures can be further divided into functionality or pathology, down to the genetic or cellular level.

KEY TERMS AND CONCEPTS

- **Cataract:** Opacity of the lens of the eye occurring as the result of aging, trauma, or a metabolic or disease process.
- **Cornea:** Transparent front of the eye, covering the iris and pupil; the first layer of the eye to refract light.
- **Extracapsular Cataract Extraction:** Method of cataract surgery involving the removal of almost the entire natural lens, leaving the elastic lens capsule (posterior capsule) intact.
- **Low Vision:** Subspecialty of optometry or ophthalmology managing individuals who have reduced vision even with the use of corrective glasses or contact lenses.
- **Ophthalmoscope:** Instrument used to examine the interior of the eye, particularly the retina.
- **Presbyopia:** Loss of accommodation of the lens of the eye, causing the inability to maintain focus on objects near to the eye or at reading distance.

- **Pupil:** Contractile opening at the front of the eye, at the center of the iris.
- **Refraction:** Method to determine the amount of ocular refractive error; used to establish a prescription for corrective glasses or contact lenses.
- **Retina:** Innermost layer of the eye that contains receptors for vision, the rods and cones; receives images transmitted through the lens.
- **Retinopathy:** Any disease of the retina or retinal blood vessels, including diabetic, hypertensive, circinate, and solar retinopathy, as well as retinopathy of prematurity.
- **Visual Acuity:** Measure of the ability to read various sizes of letters or symbols at a standard distance. Normal visual acuity, 20/20, means that at a distance of 20 feet, the eye can see what an eye would normally see at that distance.

DEFINITION AND BASIC PRINCIPLES

Ophthalmology is the study of the eye and the visual system, which includes the eyes, the globe and orbit, the optic nerve connecting the eyes to the brain and visual cortex, and even the skin surrounding the eyes. The management of pathology within this system and the many systemic diseases affecting the eye are key responsibilities of ophthalmologists, requiring both medical and surgical proficiency. Symptoms that manifest in the eyes can be the first indication of numerous neurological, rheumatological, vascular, and cardiovascular medical conditions. In addition to ophthalmologists, vision scientists, usually with doctoral degrees, are the other major contributors to ophthalmic science. Most are employed by universities or medical schools, and many others are part of pharmaceutical and biotechnology companies. Some ophthalmology professionals opt for a combination of clinical practice and scientific research.

BACKGROUND AND HISTORY

Ophthalmology is one of the oldest of the medical sciences. Egyptian writings from as early as 1600 b.c.e. describe ophthalmic conditions such as cataracts, trachoma, and iritis; however, early treatments included things such as needles, sticks, lizard blood, and crocodile dung. Ophthalmic texts describing theories of how rays are transmitted and processed in the brain can be traced back to Galen in 200 c.e. In *De medicina* (c. 30 c.e.; *The Eight Books of Medicine*, 1830; better known as *De Medicina*, 1935-1938), Aulus Cornelius Celsus was the first to describe a surgical treatment for cloudiness of the eye, which most likely was a cataract. This early cataract removal procedure involved the use of a needle or stick to scrape cataracts into smaller pieces to be reabsorbed by the eye. The Indian surgeon Susruta performed the first cataract surgery more than 2,000 years ago by pushing the clouded lens back into the eye cavity. This technique was called couching or reclination and essentially remained the treatment for cataracts for centuries, despite high occurrences of infection and pain. In 1748, Jacques Daviel, eye doctor to Louis XV, was attempting to use the couching method on a wigmaker but was unsuccessful, so he decided to make an incision through the cornea and then used a spatula-like tool to scrape the cataracts out. The posterior lens capsule was left intact, resulting in the very first extracapsular procedure. Significant advances such as these and an increased understanding of refractive error correction and visual pathology have all contributed to centuries of advancement in ophthalmic science.

How It Works

Ophthalmology spans a broad range of sciences, from clinical applications to basic sciences on the cellular level. The goals of opthamology research are to better understand the visual system and disease processes and to develop more effective treatment methods for patients.

Physiological optics deals with the anatomy of the eye and its capacity to manage light and process images for interpretation by the brain. This field of ophthalmology measures the impact of anatomical changes due to disease or surgery on vision and visual processing.

Retinal cell biology is the key to understanding how the retina works and how certain conditions such as retinopathies or treatments such as laser photocoagulation affect its ability to function. Much of this work is done on a microscopic scale using retinal cells from laboratory animal models or human eyes from autopsies. Scientists look at experimental therapies and their impact on the retina at the cellular level. They can provide high-resolution imaging of how these cellular changes affect visual outcomes.

Genetics is becoming increasingly important in the diagnosis and treatment of eye disease. Molecular genetic testing can allow for precise diagnosis of genetic eye disease, possibly avoiding complex, expensive, or time-consuming diagnostic procedures. This is extremely important as it provides families with a risk profile for some diseases and, in some cases, allows for gene-specific treatment.

Visual neurophysiology and neuro-ophthalmology study the nerves and nerve cells that contribute to vision. A lot of study in this field is conducted in a controlled laboratory setting using animal subjects.

APPLICATIONS AND PRODUCTS

Millions of people each year visit ophthalmologists seeking a solution to problems they might be having with their eyes or vision. Technological advances in many fields of ophthalmology are offering patients more options and convenience than ever before. Many procedures can be done under local anesthetic and require minimal recovery time.

Laser refractive surgery to correct vision was approved for use by the Food and Drug Administration (FDA) in 1995, and since then, each year almost one million patients in the United States have had it performed. Refractive surgery can reduce nearsightedness, farsightedness, and even astigmatism. Surgeons use a guided laser to reshape the cornea at the front of the eye. Laser assisted in-situ keratomileusis (LASIK) requires that a thin flap be cut from the corneal stroma and hinged open to expose the portion of the cornea to be reshaped by the laser. Photorefractive keratectomy (PRK) is different in that the flap is not created; rather the laser reshapes the cornea, removing the thin front layer. Patients need to wear a bandage contact until the thin layer repairs itself, leading to longer recovery time. Laser assisted sub-epithelium keratomileusis (LASEK) similarly changes the shape of the cornea using the laser, but it first uses an alcohol solution to loosen then lift a thin front layer that can act as a natural bandage after the procedure.

Cataract surgeries are performed on millions of eyes around the world each year. Traditionally, ophthalmologists removed the clouded lens, and the patient had to wear very thick glasses to correct the resulting refractive error. However, ophthalmologist now replace the removed lens with a carefully selected artificial lens that can effectively correct the refractive error, resulting in near-perfect visual acuity. Multifocal lenses have been developed to correct eyes that required near as well as distance correction. The surgical procedure itself takes very little time and can be performed under local anaesthetic. The incision in the cornea is usually less than one-eighth of an inch. Ophthalmologists insert a very small probe that uses ultrasound waves to break up the cataract, then suck it out. The new corrective lens, an intraocular lens (IOL) is folded and injected through the small incision. The minimal size of the incision allows it to heal without the need for stitches.

Age-related macular degeneration (AMD) means that central vision becomes affected by blurriness, dark areas, or distortion as the macula, a small area at the back of the retina, degenerates. There used to be very little that could be done to slow the progression of macular degeneration, but studies have been done that conclusively show that nutritional supplementation, laser treatment, and drug injections provide promising options for patients. A large scientific study found that people at risk for developing advanced stages of AMD lowered their risk by about 25 percent when treated with a high-dose combination of vitamin C, vitamin E, beta-carotene, and zinc. A treatment called photodynamic therapy (PDT) uses a combinination drug and laser treatment to slow or stop leaking blood vessels that damage the macula. The latest generation of AMD treatments target a specific chemical, vascular endothelial growth factor (VEGF), which causes abnormal blood vessels to grow under the retina. Anti-VEGF drugs are injected into the eye and block the trouble-causing chemical, reducing the growth of abnormal blood vessels and slowing their leakage. These procedures may preserve sight, but they do not restore vision to normal. Despite advanced medical treatment, many people with macular degeneration still experience some vision loss.

Dry eye affects about 4.9 million people aged fifty and older in the United States. It may seem like a minor eye condition, but dry eye results in discomfort, changes in vision, tear film instability, and serious damage to the surface of the eye. Treatments for dry eye used to be limited to continued use of artificial tear products or surgical blockage of a tear duct in the eye, which decreased the amount of tears that could drain from the eye. In 2002, the FDA ap

---

Fascinating Facts About Ophthalmology

- The eye contains more than 2 million working parts.
- The average person blinks about 12 times per minute, or 10,000 times per day
- Human eyes are the same size from birth to death, although human noses and ears grow.
- Tears are made of salt, sugar, ammonia, urea, water, citric acid, and lysozomes.
- The first successful human-to-human transplant of any kind involved the transplantation of a cornea, performed in 1905 in what is now the Czech Republic.
- Normal vision includes a blind spot but people do not notice any hole in their vision because the brain fills in the gap.
- Glasses were so popular in the fourteenth century that painters and sculptors depicted biblical figures wearing them.
- The most common types of color blindness involve difficulty distinguishing between red and green and between blue and yellow.
- The National Eye Institute has designated 2010 to 2020 as the Decade of Vision. During this period, most of the 78 million baby boomers will reach the age of sixty-five and face a greater risk of developing eye diseases that affect older people.

---

proved cyclosporine ophthalmic solution (Restasis), an eye drop that, with continued use, helps eyes increase their tear production.

Ophthalmic science also develops products and techniques that overlap with the beauty and cosmetic industry. Bimatoprost ophthalmic solution (Latisse), a drug that promises to create longer, fuller eyelashes, is actually a reformulation of a common glaucoma drug, bimatoprost (Lumigan). Researchers noticed that patients being treated for glaucoma were experiencing the side effect of lash growth. Botulinum toxin (Botox), which dermatologists use to relax frown lines, has been injected by ophthalmologists into specific eye muscles, causing them to relax and provide improved eye alignment for strabismus patients.

IMPACT ON INDUSTRY

The aging population and advances in ophthalmic science and technology are likely to translate into increased demand for eye surgery and eye care to correct vision problems. Two common problems are age-related macular degeneration and glaucoma, which affects 1 out of every 200 people. The increased incidence of type 2 diabetes has heightened the chances of diabetic retinopathy. Key factors in the ophthalmic industry are new technologies, pharmaceutical products, and labor-saving devices. The global drug and medical device market for ophthalmology is expected to be a $36 billion industry by 2014. Finding solutions to clinical challenges drives the ophthalmic industry.

One such concern in ophthalmic surgery has been cross-contamination, which is being resolved with the development of single-use ophthalmic surgical products, a market that generated almost $2 billion in 2009. Refractive intraocular lenses provide the technology to correct patients' near, intermediate, and distance refractive errors with the surgical implantation of a lens during cataract surgery.

Ophthalmic pharmaceutical development is most often conducted at the laboratories of large biotechnology corporations or in conjunction with research at academic institutions. Drug development is highly competitive as corporations want to capitalize on the growing needs of the aging population. Popular and profitable drug treatments such as ranibizumab (Lucentis) for the treatment of age-related macular degeneration or latanoprost (Xalatan) for glaucoma are under constant threat of being replaced by better treatments, generic alternatives, or experimental off-label solutions.

The U.S. government is providing additional funding for ophthalmic research to address the needs and treatment of the aging population. In 2009, the National Eye Institute received more than $775 million from the National Institutes of Health in order to continue to fund ophthalmic research at academic institutions.

CAREERS AND COURSE WORK

Ophthalmic professionals and scientists are in demand, and the need is increasing annually, especially with the aging population and advancements in treatment modalities for eye disease. The well-known career choice in this field is to become an ophthalmologist, which is a medical doctor who specializes in the eyes and visual system. Becoming an ophthalmologist in the United States requires the completion of four years of college, four years of medical school, and four to five years of additional specialized training. Ophthalmology is reported to be one of the most satisfying medical professions because of the broad scope of ophthalmic science, encompassing medicine and surgery, serving pediatric to adult patients, and delivering primary care as well as highly specialized treatment.

Vision scientists normally hold a master's or doctorate degree and have conducted years of research, as well as written and defended a thesis. Much of their work is funded by research grants from government or foundation sources. Vision scientists normally find work in academic institutions, biotechnology laboratories, or governmental agencies. Ophthalmologists and vision scientists often collaborate to bring new technology to the clinical phase of development.

SOCIAL CONTEXT AND FUTURE PROSPECTS

Ophthalmology is experiencing increased growth and demand as a large portion of the population ages and begins to experience vision-related problems. Vision science and ophthalmic research is constantly evolving in the development of diagnostic and treatment options to preserve sight. The National Institutes of Health has identified priorities for research funding, and ophthalmic research addresses the top five: genomics, translational research, comparative effectiveness, global health, and the empowerment of biomedical enterprise. Accelerated research is occurring in an effort to predict, personalize, prevent, and preempt eye disease.

April D. Ingram, B.Sc.

FURTHER READING

Atchison, David A., and George Smith. *Optics of the Human Eye.* Boston: Butterworth-Heinemann, 2003. Looks at the physiological workings of the human eye.

Cavallotti, Carlo, and Luciano Cerulli. *Age-Related Changes of the Human Eye.* Totowa, N.J.: Humana, 2008. Discusses the effects of aging on people's eyes and how doctors deal with these changes.

Friedman, Neil, and Peter Kaiser. *Essentials of Ophthalmology.* Philadelphia. Saunders Elsevier, 2007. A general introduction to ophthalmology, organized by anatomical structure.

Ledford, Janice. *The Little Eye Book: A Pupil's Guide to Understanding Ophthalmology*. Thorofare, N.J.: Slack, 2008. Contains an easy-to-understand introduction to eye care. Includes photographs, drawings, tables, and charts.

Olver, Jane, and Lorraine Cassidy. *Ophthalmology at a Glance*. Malden, Mass.: Blackwell Science, 2005. A useful reference as an introduction to common eye problems and diseases. Provides an overview of knowledge and examination skills of ophthalmology and also includes the social and occupational aspects of vision.

WEB SITES

American Academy of Ophthalmology
http://www.aao.org

American Board of Ophthalmology
http://www.abop.org

National Institutes of Health
National Eye Institute
http://www.nei.nih.gov

See also: Geriatrics and Gerontology; Mirrors and Lenses; Optometry; Surgery.

OPTICS

FIELDS OF STUDY

Astronomy; geometrical optics; ophthalmology; optometry; photography; physical optics; physics; quantum optics; quantum physics.

SUMMARY

Optics is the study of light. It includes the description of light properties that involve refraction, reflection, diffraction, interference, and polarization of electromagnetic waves. Most commonly the word light refers to the visible wavelengths of the electromagnetic spectrum, which is between 400 and 700 nanometers (nm). Lasers use wavelengths that vary from the ultraviolet (100 nm to 400 nm) through the visible spectrum into the infrared spectrum (greater than 700 nm). Optics can be used to understand and study mirrors, optical instruments such as telescopes and microscopes, vision, and lasers used in industry and medicine.

KEY TERMS AND CONCEPTS

- **Coherence:** Light rays that are in synchronous phase.
- **Diffraction:** Change in direction of a light ray when it hits an obstruction.
- **Geometric Optics:** Use of ray diagrams and associated mathematical equations to describe light behavior.
- **Interference:** Interaction between light rays under specific circumstances that leads to an increase or decrease in light intensity.
- **Laser:** Light amplification by stimulated emission of radiation; monochromatic, coherent light used in various applications.
- **Ophthalmology:** Practice of examining eye health, diagnosing disease, prescribing medical and surgical treatment for eye conditions by a specialized medical doctor.
- **Optics:** Study of the behavior of light.
- **Optometry:** Practice of examining eye health and prescribing corrective lenses by a doctor of optometry.
- **Photoelectric Effect:** Release of electrons that occurs when light rays hit a metallic surface.
- **Photon:** Particle released because of the photoelectric effect.
- **Photonics:** Study of photons as they relate to light energy in applications such as telecommunications and sensing.
- **Quantum Optics:** Branch of quantum physics focused on the study of the dual wave and particle characteristics of light.
- **Refraction:** Bending of light waves that occurs when light traverses from one media to another, such as from air to water.

DEFINITION AND BASIC PRINCIPLES

Optics is the area of physics that involves the study of electromagnetic waves in the visible-light spectrum between 400 and 700 nm. Optics principles also apply to lasers, which are used in industry and medicine. Each laser has a specific wavelength. There are lasers that use wavelengths in the 100-to-400-nm range, others that use a wavelength in the visible spectrum, and some that use wavelengths in the infrared spectrum (greater than 700 nm).

Light behaves as both a wave and a particle. This duality has resulted in the division of optics into physical optics, which describes the wave properties of light; geometric optics, which uses rays to model light behavior; and quantum optics, which deals with the particle properties of light. Optics uses these theories to describe the behavior of light in the form of refraction, reflection, interference, polarization, and diffraction.

When light and matter interact photons are absorbed or released. Photons are a specific amount of energy described as the sum of Planck's constant h ($6.626 \times 10{-34}$) and the wavelength of the light. The formula to describe the energy of photons is $E = hf$. Photons have a constant speed in a vacuum. The speed of light is $c = 2.998 \times 108$. The constant speed of light in a vacuum is an important concept in astronomy. The speed of light is used in the measurement of astronomic distances in the unit of light-years.

BACKGROUND AND HISTORY

Optics dates back to ancient times. The

3,000-year-old Nimrud lens is crafted from natural crystal, and it may have been used for magnification or to start fires. Early academics such as Euclid in 300 b.c.e. theorized that rays came out of the eyes in order to produce vision. Greek astronomer Claudius Ptolemy later described angles in refraction. In the thirteenth century, English philosopher Roger Bacon suggested that the speed of light was constant and that lenses might be used to correct defective vision.

By the seventeenth century, telescopes and microscopes were being developed by scientists such as Hans Lippershey, Johannes Kepler, and Galileo Galilei. During this time, Dutch astronomer Willebrord Snellius formulated the law of refraction to describe the behavior of light traveling between different media such as from air to water. This is known as Snell's law, or the Snell-Descartes law, although it was previously described in 984 by Persian physicist Ibn Sahl.

Sir Isaac Newton was one of the most famous scientists to put forward the particle theory of light. Dutch scientist Christiaan Huygens was a contemporary of Newton's who advocated the wave theory of light. This debate between wave theory and particle theory continued into the nineteenth century. French physicist Augustin-Jean Fresnel was influential in the acceptance of the wave theory through his experiments in interference and diffraction.

The wave versus particle debate continued into the next century. The wave theory of light described many optical phenomena; however some findings such as the emission of electrons when light strikes metal can be explained only using a particle theory. In the early twentieth century, German physicists Max Planck and Albert Einstein described the energy released when light strikes matter as photons with the development of the formula $E = hv$, which states that the photon energy equals the sum of the wavelength and Planck's constant.

In the early twenty-first century it is generally accepted that both the wave and the particle theories are correct in describing optical events. For some optical situations light behaves as a wave and for others the particle theory is needed to explain the situation. Quantum physics tries to explain the wave-particle duality, and it is possible that future work will unify the wave and particle theories of light.

HOW IT WORKS

Physical Optics. Physical optics is the science of understanding the physical properties of light. Light behaves as both a particle and a wave. According to the wave theory, light waves behave similarly to waves in water. As light moves through the air the electric field increases, decreases, and then reverses direction. Light waves generate an electric field perpendicular to the direction the light is traveling and a magnetic field that is perpendicular both to the direction the light is traveling and to the electric field.

Interference and coherence refer to the interactions between light rays. Both interference and coherence are often discussed in the context of a single wavelength or a narrow band of wavelengths from a light source. Interference can result either in an increased intensity of light or a reduction of intensity to zero. The optical phenomenon of interference is used in the creation of antireflective films.

Coherence occurs when light is passed through a narrow slit. This produces waves that are in phase with the waves exactly lined up or waves that are out of phase but have a constant relationship with one another. Coherence is an important element to the light emitted by lasers and allows for improved focusing properties necessary to laser applications.

Polarization involves passing light waves through a filter that allows only wavelengths of a certain orientation to pass. For example, polarized sunglasses allow only vertical rays to pass and stop the horizontal rays, such as light reflected from water or pavement. In this way, polarized sunglasses can reduce glare.

Diffraction causes light waves to change direction as light encounters a small opening or obstruction. Diffraction becomes a problem for optical systems of less than 2.5 millimeters (mm) for visible light. Telescopes overcome the diffraction effect by using a larger aperture, however, for very large-diameter telescopes the resolution is then limited due to atmospheric conditions. Space telescopes such as the Hubble are unaffected by these conditions as they are operating in a vacuum.

Scattering occurs when light rays encounter irregularities in their path such as dust in the air. The increased scattering of blue light due to particles in the air is responsible for the blue color of the sky.

Illumination is the quantitative measurement of light. The watt is the measurement unit of light power. Light can also be measured in terms of the luminance of light as it encounters the eye. Units of luminance include lumens, candela, and apostilb.

The photoelectric effect that supports the particle theory of light was discovered by German physicist Heinrich Rudolph Hertz in 1887 and later by Albert Einstein. When light waves hit a metallic surface electrons are emitted. This effect is used in the generation of solar power.

Geometric Optics. Geometric optics describes optical behavior in the form of rays. In most ordinary situations the ray can accurately describe the movement of light as it travels through various media such as glass or air and as it is reflected from a surface such as a mirror.

Geometric optics can describe the basics of photography. The simplest way to make an image of an object is to use a pinhole to produce an inverted image. When lenses and mirrors are added to the pinhole a refined image can be produced.

Reflection is another optical situation where geometric optics applies. Reflections from plane (flat) mirrors, convex mirrors, and concave mirrors can all be described using ray diagrams. A plane mirror creates a virtual image behind the mirror. The image is considered virtual because the light is not coming from the image but only appears to because of the direction of the reflected rays. A convex mirror can create a real image in front of the lens or a virtual image behind the lens depending on where the object is located. If an object is past the focal point of the convex mirror then the image is real and located in front of the mirror. If the object is between the focal point and the convex mirror then the image is virtual and located behind the mirror. A convex lens will create a virtual image. Geometric optics involves ray diagrams that will allow the determination of image size (magnification or minification), location of the image, and if it is real or virtual.

Refraction of light happens when light passes between two different substances such as air and glass or air and water. Snell's law expresses refraction of light as a mathematic formula. One form of Snell's law is: $n_i \sin\,i = n_t \sin\,t$ where n_i is the refractive index of the incident medium, i is the angle of incidence, n_t is the refractive index of the refracted medium, and t is the angle of transmission. This formula, along with its variations, can be used to describe light behavior in nature and in various applications such as manufacturing corrective lenses. Refraction also occurs as

light travels from the air into the eye and as it moves through the various structures inside the eye to produce vision.

Magnification or minification can be a product of refraction and reflection. Geometric optics can be applied to both microscopes and telescopes, which use lenses and mirrors for magnification and minification.

Quantum Optics. Quantum optics is a division of physics that comes from the application of mathematical models of quantum mechanics to the dual wave and particle nature of light. This area of optics has applications in meteorology, telecommunications, and other industries.

APPLICATIONS AND PRODUCTS

Vision and Vision Science. There is a vast network of health care professionals and industries that study and measure vision and vision problems as well as correct vision. Optometrists measure vision and refractive errors in order to prescribe corrective spectacles and contact lenses. Ophthalmologists are medical doctors who specialize in eye health and vision care. Some ophthalmologists specialize in vision-correction surgery, which uses lasers to reduce the need for glasses or contact lenses. In order to perform vision-correction surgeries there are a number of optical instruments, including wave-front mapping analyzers, that may be used.

The industries that support optometry and ophthalmology practices include laser manufacturers, optical diagnostic instruments manufacturers, and lens manufacturers. Lenses are used for diagnosis of vision problems as well as for vision correction.

Development of new lens technology in academic institutions and industry is ongoing, including multifocal lens implants and other vision-correction technologies.

Research. Many areas of research, including astronomy and medicine, use optical instruments and optics theory in the investigation of natural phenomena. In astronomy, distances between planets and galaxies are measured using the characteristics of light traveling through space and expressed as light-years. Meteorological optics is a branch of atmospheric physics that uses optics theory to investigate atmospheric events. Both telescopes and microscopes are optimized using optical principles. Many branches of medical research use optical instruments in the investigations of biological systems.

Medicine. Lasers have become commonplace in medicine, from skin-resurfacing and vision-correction procedures to the use of carbon-dioxide lasers in general surgery.

Industry. As noted above, there is an industry sector that is dedicated to the manufacture and development of vision-correction and diagnostic lenses and tools. Optics is an important part of the telecommunications industry, which uses fiber optics to transmit images and information. Photography, from the manufacture of cameras and lenses to their use by photographers, involves applied optics. Lasers are also used for precision manufacturing of a variety of products.

IMPACT ON INDUSTRY

The twentieth century had several landmark advancements in the industrial applications of optics.

Fiber Optics. The reflective properties of light include the property of some materials to have total internal reflection. For specific materials of a certain size and with specific optical properties, a light ray will continue to propagate through the material through internal reflection without the light exiting the material. This feature of reflected light was investigated during the twentieth century, and was used in the invention of fiber optics by Corning scientists in 1970. By the mid-1970's, fiber optics was used to transmit telephone and computer communications. Millions of miles of fiber-optic cable are used worldwide.

Lasers. The term laser is an acronym for light amplification by stimulated emission of radiation. The possibility of lasers was postulated by Albert Einstein, and a microwave laser was developed in the 1950's. Some credit American physicist Gordon Gould with the invention of the first laser using light; however the ruby laser invented by American physicist Theodore Maiman in 1960 is considered to be the first laser to use light.

Lasers have since been developed using a wide range of wavelengths from the ultraviolet spectrum to visible light and into the infrared spectrum. In order to generate laser emissions, an energy source is used to excite atoms in the active medium, which then emits a particular wavelength of light. The active medium can be a gas or solid. This light is then amplified to increase coherence. Lasers emit monochromatic light that is a single wavelength or a narrow

spectrum of wavelengths. Some lasers also use polarizing filters to refine the beam characteristics further.

Since 1960, dozens of types of lasers have been developed and put to use in wide-ranging applications. From the laser pointer, which uses helium and neon, to medical and industrial lasers, the applications span all areas of industry. Lasers are used for measuring, cutting, shaping, cauterizing, printing, and numerous other applications. It is a remarkable revolution in the use of light for industrial applications.

Astronomy. The Hubble Space Telescope, launched by the National Aeronautics and Space Administration (NASA) in 1990, demonstrates the amazing progress in land- and space-based astronomy that has been made possible through the study of optics. Hubble has generated a large amount of data since its launch and is proposed to function until 2011. The success of the Hubble is in part because of the fact that as a space-based telescope it is not subject to optical interference from the atmosphere. The successor to the Hubble is the James Webb Space Telescope, which is scheduled to launch in 2014.

Photography. Cameras and color photography were refined over the twentieth century; however the biggest revolution in photography has been digital imaging. By the early twenty-first century, digital cameras had virtually replaced standard film. Digital photography takes advantage of the photoelectric effect described more than a century ago by Heinrich Rudolph Hertz.

Vision Care and Vision Science. The advances in optics that have changed industry have also been applied to changing how people are able to correct their vision. In 1949, English surgeon Sir Harold Ridley implanted the first intraocular lens following a cataract surgery. Before the intraocular lens was developed, a patient who had cataracts removed would have to wear very thick eyeglasses in order to have useful vision. As a result of advances in the ability to manufacture and use lightweight materials and an understanding of the optics of vision, surgeons and patients now have the choice to implant multifocal lenses, which can give the patient freedom from glasses. Similarly, advances in the ability to shape materials to achieve specific optical properties have led to advances in contact, bifocal, and progressive lenses, and to artificial corneal implants.

Perhaps the most widely known advancement since the 1990's is laser vision correction. With the

Fascinating Facts About Optics

- English surgeon Sir Harold Ridley decided on polymethylmethacrylate (PMMA) as suitable material for intraocular lens implants after observing Royal Air Force pilots with pieces of the PMMA airplane canopy in their eyes after accidents. He noticed that this material was not rejected in the eye, and it was used for subsequent decades to implant lenses after cataract surgery.

- The first images received from the Hubble Space Telescope in 1990 were blurry because of spherical aberration caused by a flaw the size of one-fiftieth of a sheet of paper in the focusing mirror. NASA scientists designed a series of small mirrors that were installed by a team of astronauts in 1993 to overcome this flaw. The subsequent Hubble images were free from the aberration and had the excellent resolution expected from a space-based telescope.

- The stereo images produced by the Mars Pathfinder's cameras functioned similarly to stereo vision produced by binocular vision in humans. Two sets of cameras produced individual images that were fused used prisms. The successor to the Pathfinder is NASA's Opportunity, which is still sending images from Mars.

- Geckos' eyes have 350 times more sensitivity to color in dim light than human eyes.

- Newer-generation excimer laser systems used for vision-correction surgery have the capacity to measure and correct higher-order optical aberrations of the human eye. Iris recognition with a rotational adjustment is also available on some lasers.

- The different colors of the northern lights are created when solar energy in the form of solar flares enter the Earth's magnetic sphere and collide with atmospheric gases. These collisions cause the gases to emit light. Collisions with oxygen will tend to cause a red color, while nitrogen or helium will produce blue or green colors.

Optics involves a combination of math and physics. An understanding of human eye anatomy is also essential for a career in vision care. For all of optics-related fields it is important to have a strong background in high school mathematics. For occupations in allied health care such as opticians or ophthalmic technicians, a high school diploma and technical training is required post high school. Photographers may pursue formal training through a university or art school or might develop skills through experience or an apprenticeship.

Many careers in physics, astronomy, and meteorology require at least a bachelor's degree and most require a master's or doctoral degree. University course work in these fields includes mathematics and physics. To become an optometrist a bachelor's degree plus a doctor of optometry degree is required. An ophthalmologist will need a bachelor's degree, a degree in medicine, and residency training.

SOCIAL CONTEXT AND FUTURE PROSPECTS

The advancements in optics theory and application have changed the fabric of life in industrialized countries, from the way people communicate to how the universe is understood. It is almost impossible to imagine what future advances will occur in optics, since the last fifty years has brought profound changes in the fields of photography, medicine, astronomy, manufacturing, and a number of other fields.

As wireless technology advances it seems possible that this technology may replace some of the millions of miles of fiber-optic telecommunications cables that currently exist. Because of their reliability, fiber optics will continue to be used for the foreseeable future. Existing lasers will continue to be to optimized, and most likely new lasers will be developed. With the launch of the James Webb Space Telescope in 2014, the understanding of the solar system and space will be further advanced.

Refinements in optical systems will aid in research in a variety of fields. For example, oceanographers already apply optics theory to the study of low-light organisms and to the development of techniques for conducting research in low light. Improved optical systems will likely have a positive impact on this and other research.

Quantum computers using photonic circuits are a possible future development in the field of optics. A quantum computer that takes advantage of the photoelectric effect may be able to increase the capacity of computation over conventional computers. Optics and photonics may also be applied to chemical sensing, imaging through adverse atmospheric conditions, and solid-state lighting.

Some scientists have commented that the wave and particle theories of light are perhaps a temporary solution to the true understanding of light behavior. The area of quantum optics is dedicated to furthering the understanding of this duality of light. It is possible that in the future a more unified theory will lead to applications of optics and the use of light energy in ways that have not yet been imagined.

Ellen E. Anderson Penno, B.S., M.S., M.D., F.R.C.S.C., Dip. A.B.O.

FURTHER READING

American Academy of Ophthalmology. *Clinical Optics.* San Francisco: American Academy of Ophthalmology, 2006. This volume of the American Academy of Ophthalmology basic science course covers the fundamental concepts of optics as it relates to lenses, refraction, and reflection. It also covers the basic optics of the human eye and the fundamental principles of lasers.

Meschede, Dieter. *Optics, Light, and Lasers: The Practical Approach to Modern Aspects of Photonics and Laser Physics.* 2d ed. Weinheim, Germany: Wiley-VCH, 2007. An undergraduate text that explains modern lasers and photonics using the fundamentals of optics theory. It includes chapters on everyday optics such as the human eye and telescopes as well as chapters on quantum optics.

Pedrotti, Frank L., Leno M. Pedrotti, and Leno S. Pedrotti. 3d ed. *Introduction to Optics.* Upper Saddle River, N.J.: Prentice Hall, 2007. Intended for the physics undergraduate and includes chapters on basic optics including wave optics, geometric optics, and modern optics.

Siciliano, Antonio. *Optics Problems and Solutions.* Singapore: World Scientific Publishing, 2006. Geared to physics and engineering undergraduates, this text includes chapters on basic optics with problems that can be used by students to improve understanding of optical concepts.

Tipler, Paul A., and Gene Mosca. *Physics for Scientists and Engineers.* 6th ed. New York: W. H. Freeman, 2008. A staple for introductory university physics courses for many years. Chapters cover basic physics concepts including optics and the dual wave and particle nature of light.

Wolfe, William J. *Optics Made Clear: The Nature of Light and How We Use It.* Bellingham, Wash.: SPIE, 2007. Contains basic optics chapters followed by a large number of chapters about applied optics, including industrial and environmental topics; may appeal both to students and general readers interested in this topic.

WEB SITES

American Academy of Ophthalmology
http://www.aao.org

American Academy of Optometry
http://www.aaopt.org

American Institute of Physics
http://www.aip.org

The Fiber Optic Association
http://www.thefoa.org

See also: Microscopy; Mirrors and Lenses; Ophthalmology; Optometry; Telescopy.

OPTOMETRY

FIELDS OF STUDY

Eye examination; low-vision rehabilitation; visual development; refractive correction; vision therapy; behavioral optometry; biomedical ocular research; electrodiagnostics/physiology.

SUMMARY

Optometry is a regulated and recognized branch of primary health care focused on the evaluation and management of eye health and vision. It is a key part of the eye health care team, which may also include ophthalmologists, orthoptists, and opticians. Doctors of optometry, called optometrists, are trained to provide comprehensive assessment of a patient's visual system, including refractive care (prescriptions for glasses and contact lenses) as well as disease detection, rehabilitation, and management. Optometrists do not typically provide surgical management of eye conditions but in most states are permitted to prescribe drugs for the treatment or management of eye diseases.

KEY TERMS AND CONCEPTS

- **Binocular Vision:** Ability to use both eyes together to enhance vision.
- **Cornea:** Transparent front of the eye, covering the iris and pupil, the first layer of the eye to refract light.
- **Low Vision:** Condition in which an individual has reduced vision even with the use of glasses or contact lenses.
- **Ophthalmologist:** Physician/surgeon who specializes in treatment of disorders of the eye.
- **Ophthalmoscope:** Instrument used to examine the interior of the eye, particularly the retina.
- **Optician:** Person who specializes in filling prescriptions for corrective lenses, such as glasses or contact lenses.
- **Orthoptist:** Person who specializes in assessing and treating defects in binocular vision resulting from abnormal optic musculature.
- **Pupil:** Contractile opening at the front of the eye, at the center of the iris.

- **Refraction:** Determination of the amount of ocular refractive error, used to determine a prescription for corrective glasses or contact lenses.
- **Retina:** Innermost layer of the eye. Receives images transmitted through the lens and contains receptors for vision, the rods and cones.
- **Tonometer:** Instrument used to measure pressure or tension in the eye, often in the detection of glaucoma.
- **Vision Therapy:** Nonsurgical techniques aimed at correcting and improving binocular, oculomotor, visual-processing, and perceptual disorders.
- **Visual Acuity:** Measure of the ability to read various sizes of letters at a standard distance. Normal visual acuity is 20/20, which means that at a distance of 20 feet, the eye can see what an eye would normally see at that distance.

DEFINITION AND BASIC PRINCIPLES

Optometry is an applied science of diagnosing, managing, and treating diseases and conditions of the eye. The word optometry comes from the Greek *optos*, which means seen or visible, and *metria*, meaning measurement. This measurement of vision is based on principles of optics (the relationship between light and vision) and is conducted using specialized devices such as refractors, lenses, and ophthalmoscopes.

Optometrists provide primary-level health care focused on the structure and function of eyes. An optometric eye exam usually involves testing a patient's visual acuity, color vision, depth perception, and binocular vision, and evaluating the structures of the eye to check for eye diseases such as cataracts, glaucoma, or retinopathy. The examination leads to a treatment plan for the patient and may involve referral to other eye care professionals, such as opticians, ophthalmologists, or orthoptists. In most states, optometrists can prescribe and administer drugs for the treatment of specific conditions; however, if more specialized or surgical intervention is required, they will refer the patient to an ophthalmologist. The optometrist will often remain involved in the patient's care after referral and provide low vision rehabilitation and preoperative and postoperative care for cataract or laser refractive surgery. Optometrists can diagnose potentially serious systemic diseases such as high blood pressure and diabetes, which would then necessitate referral to other health care professionals.

BACKGROUND AND HISTORY

The origins of optometry can be traced back to early studies of optics and image formation; however, the term "optometry" was not used until centuries later. The basic principles of optics, reflection, and light angles appear in writings by the Greek mathematician Euclid that date to 280 b.c.e. Around 150 c.e., Ptolemy described his observations of angles of incidence and reflection as light traveled through air to water. In the eleventh century, the Arab physicist Alhazen wrote *Kit b al-Man zir*, commonly known as *Optics*. His work was published in Latin in 1572 and partially translated into English in 1989. The use of a spectacle, although a rather primitive version, for magnification was first noted in 1260 in Italy.

Johannes Kepler published *Dioptrice* (1611; partial translation of the preface, 1880) describing mathematical concepts associated with mirrors, prisms, and lenses. In *Uso de los antojos y comentarios a propósito del mismo* (1623; *The Use of Eyeglasses*, 2004), Benito Daza de Valdés discusses eyes and refraction, spectacles, and clinical conditions. In the nineteenth century, professional opticians, either dispensing or refracting, provided vision correction. Refracting opticians were later termed optometrists.

Schools of optometry began to appear in the 1870's. The American Association of Opticians was formed in 1898, and by 1919, the group had changed its name to the American Optometric Association. In 1922, the American Academy of Optometry was organized and began to disseminate information through its journal, *Optometry and Vision Science*. By 1924, all states in the United States had passed laws regarding optometry licensing.

HOW IT WORKS

Optometry encompasses eye physiology and the measurement of vision, based on principles of optics. Examinations that measure how well a person can see include tests of visual acuity, visual field, and refraction. Those that focus on the physiology of the eye include tonometry, and slit lamp, ocular motility, and fundus examinations.

Visual Acuity Examination. Visual acuity is a measurement of sharpness or clarity of vision. It is often tested using a series of black letters or symbols on a white background to provide optimal contrast. The most well-known way to test distance visual acuity is to use the Snellen chart, which has the large "E" on the top and a series of progressively smaller letters below. Normal visual acuity is 20/20. This value indicates that 20 feet from the chart, the eye can clearly see what can be normally read by an eye with no refractive error at this distance. A visual acuity of 20/40 would reflect that the eye being tested can clearly see at 20 feet what a normal eye could see at 40 feet away, meaning a visual acuity of 20/40 is only half as good as 20/20. Testing can be done on each eye separately, both eyes together, or with corrective lenses. Near vision is tested similarly but using a modified Snellen chart, such as a Rosembaum chart, held at 15.7 inches from the eyes.

Ocular Motility Examination. There are six extraocular muscles that control eye movements. An ocular motility examination begins with an observation of the eyes together, as the patient is asked to fixate on an object. Some patients have eyes that are obviously misaligned. In these cases, one eye is typically able to fixate and the other eye is deviated. The deviations can be outward (exotropia), inward (esotropia), or upward (hypertropia). Cover/uncover testing determines if one eye dominates fixation or if fixation switches between the eyes. Misalignment of the eyes can be benign or a sign of something more serious, such as cranial nerve palsy or orbital floor fracture.

Refraction. The principles of optics are used in this clinical test to determine the eye's refractive error and prescribe corrective lenses to achieve the best possible visual acuity. Refractive errors can be spherical or cylindrical. Spherical errors mean that the optical power of the eye is either too large or too small to focus light on the retina, which causes blurry vision. Spherical errors can be further classified into myopia (nearsightedness) or hyperopia (farsightedness). Cylindrical errors occur when the optical power of the eye is too powerful or too weak across one meridian, known as astigmatism. A phoropter, retinoscope, or automated refractor can be used to measure refractive error.

Visual Field Examination. Visual fields are the areas in which objects can be seen with (peripheral) vision while the eyes are focused on a central point in front of the person. Visual field defects can be detected by an optometrist moving his or her hand or a target object to the side of a patient's head or with specialized devices such as Goldmann field examinations or automated perimetry.

Slit Lamp Examination. Also known as biomicroscopy, a slit lamp examination allows the optometrist to closely inspect the anatomical structures of the eye. If a biomicroscope is used, the optometrist can obtain a magnified and detailed view of the eye. The slit lamp is an instrument that can focus an intense, thin sheet of light into the eye. During this examination, the patient is asked to place his or her chin and forehead against a support or to keep his or her head and eyes steady.

Fundus Examination. Also known as ophthalmoscopy, a fundus examination is done to see the anatomical structures at the back of the eye, such as the fundus, macula, and retina. For the best view, the patient's pupils must be dilated with specialized eye drops that may cause light sensitivity and blurry vision for a few hours after the exam. When the pupil has sufficiently dilated, the optometrist, usually wearing a head-mounted lighting device, asks the patient to fixate on a distant point on a wall or ceiling. The optometrist then shines the light source into the patient's eye while looking through a handheld condensing lens that permits him or her to inspect the back of the eye.

Tonometry. Tonometry measures the pressure inside the eye, which is a common way to screen for eye diseases such as glaucoma. The two tonometry methods are contact and noncontact. In contact tonometry, a drop of anesthetic is placed in the eye, gentle contact is made with the cornea, and a measurement is determined. Normal eye pressure ranges from 10 to 21 millimeters of mercury. In the noncontact method, the patient feels a brief puff of air at the eye. The machine measures eye pressure by looking at how the light reflections change as the air hits the eye.

APPLICATIONS AND PRODUCTS

Specializations in optometry deal with the study of eye examination techniques, contact lenses, low vision rehabilitation, visual development, refractive correction, vision therapy, biomedical ocular research, and electrodiagnostics/physiology.

Diagnostic equipment in optometry has undergone major changes since the 1970's. Numerous automated and computerized testing devices—topographers, pupillometers, autorefractors, tonometers, pachymeters, and computerized acuity charts—have emerged, taking the place of numerous charts, rulers, and screens. Technology has increased the capacity to refine results and create an electronic record that can be transferred between practitioners. Computerization and enhanced communication has allowed for consultation with other practitioners and specialists to become routine and to take place in real time, even if the clinicians are practicing in different parts of the country or even the world. Digital fundus photography provides a digital record of the patient's eye, replacing hand drawings that were difficult to interpret and compare. These improvements in detailed diagnostics and record keeping also assist in billing, coding, and medical-legal issues.

New contact lens materials have been developed, and new diagnostic technology, such as corneal topography (mapping the surface of the cornea), has become integral to delivering care to contact lens patients. Modern consumers would find early contact lenses to be rather primitive. Those early lenses could be worn only for limited periods of time because they did not allow oxygen to permeate through to the cornea, and they were rigid and very expensive. The silicone hydrogel materials used in modern contact lenses allow oxygen to reach the cornea. Contact lenses can be worn for prolonged periods and are so inexpensive that patients can wear a new pair of disposable lenses every day. Further advances have allowed for the development of special toric contact lenses for people with astigmatism and contact lenses with multiple focuses for people who require both near and distance correction. Color and costume contact lenses are available with or without refractive correction for those who wish to temporarily change their eye color or to create a special effect.

Despite their name, eyeglasses rarely contain glass. The traditional glass lens has largely been replaced by plastic polymers. High-index, polycarbonate, and aspheric lenses are thinner and lighter than the traditional glass lenses. Photochromic lenses darken and lighten according to light conditions. Scratch-resistant and ultraviolet-blocking coatings are also available. The thick, heavy glasses of the 1950's have been replaced with thin functional lenses in stylish frames.

Low vision rehabilitation is still a significant part

of optometric practice. The devices and aids to daily living that have been developed assist patients with low vision and allow them to lead productive and functional lives with greater independence than ever before.

Vision therapy and behavioral optometry use nonsurgical methods to treat common visual problems such as strabismus, double vision, convergence insufficiency, and some learning disabilities. These methods are best described as physical therapy for the eyes and brain. Under the direction of an optometrist, the patient performs a series of exercises, some incorporating the use of specialized equipment such as patches, computer software, prisms, and optical filters. In ophthalmic and medical circles, controversy surrounds certain vision therapy practices because of doubts concerning their efficacy.

Biomedical ocular research encompasses a vast area of optometry. This work, which often takes place in a laboratory setting, examines the basic physiology of eye diseases or the application of technological advances in eye care, covering both structure and function of the eye as well as visual processing and vision assessment. These laboratories are found within academic institutions or as part of large eye and vision care corporations. Studies are ongoing in all areas of eye anatomy, physiology, microbiology, biomechanics, and immunology. Research into the treatment of conditions such as refractive error and amblyopia are spotlighted at annual academic meetings.

IMPACT ON INDUSTRY

Optometry is expected to experience higher than average growth as the population grows and ages. The aging population will require extensive screening and monitoring of diseases such as diabetic retinopathy, age-related macular degeneration, and cataracts. Optometrists do not provide surgical interventions for these diseases, but they do screen, manage, and provide care before and after surgery.

Government Research and Regulation. Governments often call on optometrists to advise on the science and status of vision in order to set safety standards. Optometrists act as consultants to determine the visual demands of a task and advise on vision standards for employment and for improvements in road safety. Legislation is constantly evolving and expanding the scope op optometric practice; for example, one law expanded the range

of pharmaceuticals that optometrists can prescribe. These changes have affected the pharmaceutical industry, expanding its markets. Although the increased popularity of laser surgery for refractive correction has lessened the need for glasses and contact lens prescriptions somewhat, optometrists are providing preoperative and postoperative care, thus creating collaborative relationships with ophthalmologists performing laser surgery.

Industry and Business. Industry has embraced optometry and respects the work being done in research on a consultative level, as well as the development of products and services to complement optometry practice. Many corporations have consulted optometrists in the development of their health and safety protocols in order to meet and exceed safe workplace guidelines. The motor vehicle industry uses optometry principles and experts to satisfy safety standards for motorists and apply these to vehicle design features.

Industry and business sectors, particularly pharmaceutical and biotechnology companies, are influenced by research and trends in optometric care. A vast new market opened up for drug companies after optometrists were allowed to prescribe pharmacologic agents to treat eye conditions. Biotechnology advances in optometric research aim to allow optometrists to employ comfortable, efficient, highly technical diagnostic tools and treatments in their practices. Eye care and vision companies have profited from scientific research, becoming multimillion-dollar corporations. For example, Bausch & Lomb, a small optical shop founded in 1853 in Rochester, New York, grew to employ more than 14,000 employees worldwide and to sell its products in more than one hundred countries.

Glasses are made with a variety of lens types, and contact lenses are available in a multitude of types. Fashion designers have leapt into the eye care industry, offering lines of designer frames, cases, and sunglasses. Children's glasses and amblyopia treatments are available in fashionable frames, some of which feature characters from television or films.

CAREERS AND COURSE WORK

The primary career choice in the field of optometry is to become a doctor of optometry (optometrist).

Fascinating Facts About Optometry

- The average eye blinks more than 10 million times per year.
- Sailors once believed that wearing a gold earring would help improve their eyesight.
- Prescriptions for glasses and contact lenses may be similar but are not interchangeable.
- In 1620, Peter Brown, a pilgrim, became the first person known to wear glasses on the North American continent.
- The cornea is the only living tissue in the human body that does not contain any blood vessels.
- The eyeball of a giant squid, at 18 inches in diameter, is the largest eyeball on the face of the Earth.
- Newborn babies may cry, but they do not produce any tears until they are one to three months old.
- Sam Foster sold the first pair of Foster Grant sunglasses at Woolworth's store on the Atlantic City Boardwalk in 1929.

Optometrists may specialize to areas such as pediatrics, geriatrics, vision therapy, low vision rehabilitation, cornea and contact lenses, refractive and ocular surgery, ocular disease, or vision science and optics research.

There are nineteen accredited colleges of optometry in the United States, two in Canada, and one in Puerto Rico. Admission to these programs requires at least three years of post-secondary study at a college or university, preferably in a field of science. To be accepted into an optometric education program, applicants must take a standardized examination, the Optometry Admissions Test (OAT), which covers biology, general chemistry, organic chemistry, reading comprehension, physics, and quantitative reasoning. Acceptance into an optometric program is highly competitive. In 2007, only one out of every three applicants was admitted.

The rigorous four-year optometric program covers study in visual optics, visual neurophysiology, and clinical techniques, along with diseases of the human visual system. Students also build a knowledge base in cell and molecular biology, pharmacology, genetics, epidemiology, clinical technology, ethics, and practice management. Following graduation from an accredited program, students must take licensing exams to ensure that their knowledge and techniques meet the requirements.

SOCIAL CONTEXT AND FUTURE PROSPECTS

Optometry has gone far beyond its beginnings as the fitting of spectacles, becoming, through continuing research and rigorous specialized training, a facet of primary health care. Optometry still provides patients with prescriptions for vision correction; however, the examinations are far more extensive. This broadened knowledge has allowed optometrists to work more closely with ophthalmologists, neurologists, and other health care practitioners. Some optometrists are specializing, focusing on areas such as contact lenses, vision therapy, pediatrics, low vision, sports vision, head trauma, learning disabilities, and occupational vision.

Optometry is expected to experience a high level of growth to meet the demand of the aging population. In 2008, there were more than 34,000 optometrists in the United States, about one-quarter of whom were self-employed in a private practice or in a group practice, while the remainder were employed by a larger facility, organization, or retail business. Optometry is considered to be very rewarding and interesting work, and according to the American Optometric Association, the average annual income for self-employed optometrists was $175,329 in 2007. The regular hours and good income make optometry an attractive and satisfying career choice.

Globally, expertise in optometry is lacking in many underserved areas. The World Health Organization (WHO) estimates that 500 million people who have refractive error do not have access to eye care services. In highly developed countries such as the United States, there is one optometrist for every 10,000 people. In contrast, the ratio in developing countries is one optometrist for every 600,000 people or more. The WHO's Vision 20/20 program is striving to provide more optometric expertise to combat the underserved areas of the world.

April D. Ingram, B.Sc.

FURTHER READING

Griffin, John R., and J. David Grisham. *Binocular Anomalies: Diagnosis and Vision Therapy.* 4th ed. Boston: Butterworth-Heinemann, 2002. Deals with vision therapy as a way to manage some disorders.

Keirl, Andrew William, and Caroline Christie. *Clinical*

Optics and Refraction: A Guide for Optometrists, Contact Lens Opticians, and Dispensing Opticians. Oxford, England: Butterworth-Heinemann, 2007. A guide to visual optics, providing straightforward information on clinical optics and refraction.

Kitchen, Clyde K. *Fact and Fiction of Healthy Vision: Eye Care for Adults and Children.* Westport, Conn.: Praeger, 2007. Starts with anatomy and proceeds to discuss how to maintain healthy eyes. Includes information on many conditions and the types of refractive surgery.

Millodot, Michel. *Dictionary of Optometry and Visual Science.* Oxford, England: Butterworth-Heinemann, 2008. Provides understandable definitions, tables, and illustrations.

Phillips, Anthony. *The Optometrist's Practitioner-Patient Manuel.* Oxford, England: Butterworth-Heinemann, 2008. Information and illustrations of procedures and conditions pertaining to eye health.

WEB SITES
American Optometric Association
http://www.aoanet.org

Association of Schools and Colleges of Optometry
http://www.opted.org

National Institutes of Health
National Eye Institute
http://www.nei.nih.gov

See also: Geriatrics and Gerontology; Mirrors and Lenses; Ophthalmology.

ORTHOPEDICS

FIELDS OF STUDY

Physiology; kinesiology; human anatomy; biology; cellular biology; chemistry; biochemistry; physics; mathematics; pathology; psychology.

SUMMARY

Orthopedics is a branch of medicine focusing on treating the skeleton, muscles, joints, ligaments, tendons, and nerves of the human body, collectively referred to as the musculoskeletal system. Orthopedic conditions may be treated by a family practice physician or a physician who specializes in treating disorders of the musculoskeletal system. Other medical specialists and health care providers such as physical and occupational therapists also treat orthopedic disorders and often play a part in the plan of treatment. A multidisciplinary team approach is important in managing the symptoms of an orthopedic condition, especially when symptoms are chronic and ultimately will change in severity.

KEY TERMS AND CONCEPTS

- **Arthritis:** Inflammation of one or more joints.
- **Arthroscope:** Type of surgical device inserted into the joint through a small incision.
- **Arthroscopy:** Examination or treatment of the inside of a joint using an arthroscope.
- **Fracture:** Break, rupture, or crack of a bone.
- **Joint Prosthesis:** Artificial joint that replaces structural elements within a joint to improve and enhance the function of the joint.
- **Magnetic Resonance Imaging (MRI):** Test that uses a magnetic field to create pictures of organs and structures inside the body.
- **Musculoskeletal System:** System encompassing the joints, ligaments, tendons, muscles, and nerves.
- **Occupational Therapist:** Allied health care professional specializing in assisting patients with various disabilities, from decreased motor skills to short-term memory loss.
- **Orthopedic Surgeon:** Physician educated in the workings of the musculoskeletal system and specializing in the surgical treatment of this system.
- **Physical Rehabilitation:** Process of returning function to an individual afflicted by conditions or disorders such as injury, chronic disease, or genetic dysfunction.
- **Physical Therapy:** Allied health care profession specializing in returning function to an individual.
- **X Ray:** Test that produces images of structures such as bones within the body.

DEFINITION AND BASIC PRINCIPLES

Orthopedics is the science that studies the bones, muscles, and joints in the human body, collectively referred to as the musculoskeletal system. Medical professionals who specialize in this branch of medicine treat a wide variety of conditions. Some of the many conditions that are treated by orthopedic physicians are arthritis, sports injuries, back pain, and leg and foot disorders. Orthopedic surgeons are specifically trained to deal with conditions and disorders associated with the musculoskeletal system and to perform surgical repair when necessary.

The care of bone fractures and the reconstruction or replacement of joints are common surgical procedures performed by orthopedic surgeons. Some orthopedic conditions do not require surgery and can be managed by medication, injections, or rehabilitation by medical professionals such as physical and occupational therapists. Some orthopedic surgeons specialize in a particular part of the body, such as the spine, hands, or feet. Some further specialize in orthopedics by exclusively treating children and are referred to as pediatric orthopedic surgeons. Pediatric orthopedic problems—such as curvature of the spine, hip joint disorders, and limb length discrepancies—can occur as a child grows and develops. In addition, active children often experience bone fractures, severe sprains or strains, and dislocated joints, which are treated by pediatric orthopedists. Orthopedics is a specialty that can be called on in emergency situations as well.

BACKGROUND AND HISTORY

Orthopedics can be traced as far back as prehistory, when humans underwent crude orthopedic procedures such as amputations of limbs and fingers in order to survive. Paintings and drawings on walls have suggested that early humans used forms of assistive walking aids such as crutches. The Greek historian Herodotus writes of a soldier who escapes from chains by cutting off his foot and later creates a prosthesis for himself. Other early writings mention wooden legs, iron hands, and artificial feet. The early writings of the Greek Hippocrates and his understanding of fractures and fracture care had an impact on orthopedics.

In the twelfth century and after the Dark Ages, universities and hospitals were beginning to be established in Europe. In many of these institutions, researchers were performing human dissections and gaining a better understanding of human anatomy. In addition, the ancient Greek texts were being translated from Arabic to Latin.

Orthopedics began to come into its own in the early 1900's. The discovery of the X ray was a major advancement in medicine, especially for orthopedics. Orthopedics was being seen as a true medical specialty. The British dominated orthopedic developments during this time; however, later in the century, Americans would make progressively more contributions.

HOW IT WORKS

Orthopedic Assessment. An individual with an orthopedic concern is often referred to a specialist through a primary care provider, often a family practice physician. An orthopedic assessment begins with the physician taking a thorough history. For example, he or she may ask where the problem is located, what (if any) event brought on the problem, and what type of activities increase or decrease the discomfort. The physician may order tests, such as X rays, magnetic resonance imaging (MRI) scans, or computed tomography (CT) scans. Once the problem has been identified, a plan of care is developed. This may consist of physical or occupational therapy, medication (including injections), and possibly surgery.

Fracture Care. Specialists in orthopedics are trained in the early management of fractures. This typically consists of the realignment of a broken limb, followed by the immobilization of the fractured extremity in a cast or splint. Nerve and blood vessels are assessed for possible injury associated with the fracture and documented before and after realignment.

Joint Pain and Dysfunction. Medical professionals involved in orthopedics treat joint pain and dysfunction on a regular basis. Orthopedic specialists often treat osteoarthritis (degenerative joint disease), with the primary goal of controlling pain. Physical therapy is used to decrease pain and swelling and increase muscle strength and joint motion. The orthopedic physician may also prescribe medications, including nonsteroidal anti-inflammatory drugs (NSAIDs). Surgery may be necessary if symptoms have not responded to conservative therapy such as medication and physical therapy. Surgical procedures include arthroscopy, which involves the removal of torn cartilage and the roughened joint surface (debridement), usually from the knee; hip resurfacing; and arthroplasty, replacement of the hip or knee. In arthrodesis, bones in a damaged joint (usually the spine, hand, ankle, or foot) are fused.

Developmental Disorders. Orthopedic physicians can become more specialized and treat a variety of developmental dysfunctions acquired at or before birth, such as cerebral palsy, a condition that describes a group of brain disorders affecting the communication between the brain and the muscles. The orthopedic team, which often includes physical and occupational therapists, addresses cerebral palsy and other developmental dysfunctions causing a permanent state of uncoordinated movement and posturing. Other developmental conditions seen and treated by the orthopedic physician are upper extremity misalignment and contractures, and joint deformities of the spine, hip, knee, ankle and foot.

Sports Medicine. Orthopedic specialists are trained to perform a comprehensive assessment, to treat, and to provide follow-up care to children, adolescents, and adults with sports-related orthopedic injuries. This is often accomplished with a team approach that includes sports medicine physicians and a physical therapy staff and results in the development of long-term treatment and activity plans.

Emergency Medicine. Orthopedic specialists serve as members of hospital-based emergency care teams, treating trauma-related injuries. In the emergency care setting, the physician assesses and treats acute illnesses and injuries that require immediate medical attention. Although emergency medicine physicians do not usually provide long-term or continuing care, those who specialize in orthopedics diagnose a variety of fractures and soft-tissue injuries and undertake acute interventions to stabilize patients.

APPLICATIONS AND PRODUCTS

Orthopedic Prosthesis. Arthroplasty (joint replacement) consists of the surgical removal of diseased or worn surfaces or parts of a joint and replacing them with a metal and plastic prosthesis. The prosthesis can be cemented in place or it can be attached by means of a porous coating designed to allow the bones to grow and adhere to the artificial joint. The decision made by the orthopedic surgeon to use bone growth or cement can depend on the age of the patient. The use of cement relieves pain more quickly, and the patient can bear weight much sooner, usually immediately after surgery. Although healing is slower with a porous coating and weight bearing is done in a much more progressive manner, attachment achieved through bone growth may last longer; therefore, a porous coating is used in younger patients. Joint replacements are available for joints such as the knee, hip, shoulder, elbow, wrist, ankle, and fingers. One problem that can arise with joint replacements is that the surfaces of the two parts of the implant can wear out, which leads to failure of the implant. Instead of traditional metal-on-plastic implants, orthopedic surgeons sometimes use ceramic-on-ceramic or metal-on-metal implants. Companies that design joint replacement implants are trying to improve wear characteristics.

Joint resurfacing is becoming a feasible alternative to complete joint replacement and is more prevalent in the hip. It is becoming more popular among younger and more active patients. This type of operation delays the need for the more traditional and less bone-conserving total hip replacement. However, this procedure does have a significant risk of early failure from fracture and bone death.

Spinal Stabilization. Since the early 1900's, orthopedic surgeons have performed spinal stabilization of the neck, truck, and lower back regions. Through trauma or progressive wear of each segment of the spine, tissues and nerves become aggravated and even compressed. The involved levels of the spine also become unstable, causing additional problems. This can result in pain in the arms or legs, reduced motion of the spine, and significant limitations in function. Orthopedic surgeons may use fixation devices, often a screw-and-rod plate system, to fuse the two levels of the spine. Orthopedic surgeons may also use a bone graft from another area of the body, such as the pelvis, rather than a fixation device. The bone graft is inserted between the unstable segments. Once the bone grafts heals and becomes solid, it fuses the unstable and worn segments. In both cases, surgery is followed by a relatively lengthy period of physical therapy before the patient is able to return to adequate function.

Arthroscopy is a procedure commonly used in orthopedic surgery, either to examine or treat a joint. By using a device called an arthroscope, the surgeon is able not only to see the inside the joint and examine the surfaces for damage but also to repair or remove floating debris and torn surface cartilage or perform reconstruction of ligaments. Arthroscopy has several advantages over traditional surgery because the joint does not have to be opened and exposed. Instead, several small incisions are made, one for the arthroscope and one or two for the surgical instruments used within the joint. This reduces recovery time and can improve the outcome because of reduced trauma to the surrounding tissues. It is a useful application in sports medicine, especially with college and professional athletes, who may require minimal healing time. Knee arthroscopy is one of the most common operations performed by orthopedic surgeons.

Pharmacotherapy and Medical Devices. Orthopedic surgeons prescribe low to moderate doses of simple analgesics and nonsteroidal anti-inflammatory medications such as acetaminophen, aspirin, ibuprofen, and naproxen. When these medications fail to relieve a patient's pain, alternative or additional pharmacologic agents are considered. Such pain and anti-inflammatory medications are carefully selected after the patient's risk factors are considered.

Orthopedic products are designed to improve surgical and rehabilitative outcomes, decrease pain, promote motion and strength, or simply improve overall comfort. Products in this medical field can range from an improved joint replacement device to strength training equipment for the physical therapist to an orthopedically designed pillow to help improve overall comfort while sitting or sleeping. Products made for the orthopedic industry are constantly changing in an effort to improve overall quality of life.

IMPACT ON INDUSTRY

Physical disabilities caused by musculoskeletal conditions are estimated to affect 4 to 5 percent of the global adult population. More than 150 diseases and conditions involve the musculoskeletal system. In the United States, the annual estimated direct and indirect cost for people with a musculoskeletal disease is $254 billion. The global orthopedic products market is about $36 billion per year. Between 3 and 4 percent of all practicing physicians are orthopedic surgeons, and there are about 20,400 actively practicing orthopedic surgeons and residents in the United States.

University and Government Research. Many renowned and respected hospitals are heavily involved in research conducted by orthopedic specialists. Other organizations such as the Orthopaedic Research Society are dedicated to the advancement of orthopedic research and to the transformation of basic and clinical research to clinical practice. Such organizations also advocate for increased resources for research and try to raise public awareness of the impact of orthopedic research.

About 70 percent of war wounds in the Afghanistan and Iraq conflicts are musculoskeletal injuries, and about 6 percent of casualties have had one or more limbs amputated. The U.S. Department of Defense has responded in several ways to meet the orthopedic needs of wounded soldiers, including improving battlefield care and providing rapid and orderly evacuation from the battle site. The military has created a state-of-the-art facility at Brooke Army Medical Center, San Antonio, Texas, dedicated to the care and rehabilitation of war-wounded soldiers. This facility focuses on prosthetic technology and physical and occupational therapies. It conducts orthopedic research to solve the major problems in military as well as civilian musculoskeletal health care.

Industry and Business. Some of the larger orthopedic product manufacturers conduct research in order to provide the world with a broad range of products that incorporate leading-edge medical technology. Such corporations work with respected medical professionals in orthopedics to help people lead more active and satisfying lives. Corporations in the field of orthopedics are important for introducing new materials that make joint replacements perform more and more like the natural joints. Businesses in the orthopedic sector strive to bring patients and physicians products that make orthopedic surgery and recovery more successful.

Fascinating Facts About Orthopedics

- The human body has more then 206 bones.
- In 2006, according to the American Academy of Orthopaedic Surgeons, musculoskeletal conditions were the number-two reason for visits to physicians.
- Exercising at least thirty minutes per day will reduce the risk of bone and joint injury.
- About 300,000 hip replacements and 500,000 knee replacements are performed in the United States each year.
- Every pound of weight gained places 3 pounds of additional stress on one's knees and six times the pressure on one's hips.
- Arthritis is the leading chronic condition reported by the elderly; however, more than half of those with arthritis are under the age of sixty-five.
- Many modern leg prostheses have microprocessors that allow the knee and ankle to adapt to changes in terrain or walking speed.
- The myoelectric arm is an electric prosthesis whose movements are controlled by electric signals produced by muscle contractions in the amputee's body.

CAREERS AND COURSE WORK

Careers in orthopedics can take numerous paths. The product industry is a large employer of engineers and technicians with and without a medical background. The first requirement for becoming a physical or occupational therapist or orthopedic nurse is a good background in the life and physical sciences such as biology, chemistry, physics, and mathematics at the high school level. Then, the student obtains a degree in nursing or physical or occupational therapy at an accredited college or university, which can take from four to six years.

Students interested in becoming an orthopedic physician will need to obtain a bachelor's degree from a college or university and meet course requirements for medical school. These include courses in biology, chemistry, physics, mathematics, and psychology. The next step is medical school, which generally takes four years. Orthopedic surgeons are specialists and have to undergo residency training in orthopedic surgery. The five-year residency consists of one year of general surgery training and four years of training in orthopedic surgery. Many orthopedic surgeons elect to do further training through fellowships. Examples of such orthopedic subspecialty training in the United States include hand surgery, pediatric orthopedics, orthopedic trauma, foot and ankle surgery, spine surgery, and surgical sports medicine.

SOCIAL CONTEXT AND FUTURE PROSPECTS

As the number of elderly people in the United States continues to rise, the incidence of age-related musculoskeletal conditions will increase, creating a greater demand for all health care professionals associated with orthopedics, including orthopedic surgeons. In addition, many soldiers in the conflicts in Afghanistan and Iraq need the services of orthopedic medical professionals to deal with musculoskeletal injuries that include amputations. Another area of growth is sports medicine, with orthopedic surgeons becoming trained in sports medicine or teaming with physicians specializing in the area to treat professional and amateur athletes.

As new technologies and advances in robotics and bone substitutes guide the way to less-invasive surgical interventions, health care experts predict that many more orthopedic procedures will be performed in outpatient settings or result in overnight hospital stays and that fewer procedures will involve extended stays in the hospital. As demand for orthopedic surgeons increases, hospitals are likely to form alliances and partnerships with other hospitals so that they may continue to provide orthopedic care at their facilities.

As with health care in general, among the challenges facing orthopedics are the cost of health care and the changes brought about by health care reform. Modern prostheses incorporate advanced technology so that they function more like the limbs they replace, but these devices are very expensive to produce and maintain. Adaptive devices, which range from grab bars for bathtubs to step-in bathtubs to leg prostheses designed for running to sports wheelchairs, are increasingly available, but they are not always covered by insurance and can be prohibitively expensive.

Jeffrey Larson, P.T., A.T.C.

FURTHER READING

Brotzman, S. Brent, and Kevin E. Wilk. *Clinical Orthopaedic Rehabilitation.* 2d ed. Philadelphia: Mosby, 2003. Includes chapters on rehabilitation of patients who have had a total knee replacement, lumbar fusion, and knee arthroscopy.

Magee, D. J. *Orthopedic Physical Assessment.* 5th ed. St. Louis, Mo.: Saunders 2008. Includes chapters on orthopedic assessments and rehabilitation of the cervical, thoracic, and lumbar spine.

Maxey, Lisa, and Jim Magnusson, eds. *Rehabilitation for the Postsurgical Orthopedic Patient.* 2d ed. St. Louis, Mo.: Mosby, 2007. Provides detailed descriptions of each orthopedic surgery and addresses patient rehabilitation and how to adapt therapy to geriatric, athletic, and pediatric populations.

Pagliarulo, Michael A., ed. *Introduction to Physical Therapy.* 3d ed. St. Louis, Mo.: Mosby, 2007. Examines the history of physical therapy, discusses financial and legal aspects, and contains a chapter on physical therapy for musculoskeletal conditions.

Scuderi, Giles R., and Peter D. McCann, eds. *Sports Medicine: A Comprehensive Approach.* 2d ed. Philadelphia: Mosby-Elsevier, 2005. Provides informative chapters on each body part with a focus on sports medicine injuries and rehabilitation.

Skinner, Harry B., ed. *Current Diagnosis and Treatment in Orthopedics.* 4th ed. New York: Lange Medical Books/McGraw-Hill Medical Publishing Division, 2006. An accessible resource for diagnosis of orthopedic conditions, covering trauma, sports medicine, oncology, surgery, amputations, and rehabilitation.

WEB SITES

American Academy of Orthopaedic Surgeons
http://www.aaos.org

American Association of Hip and Knee Surgeons
http://www.aahks.org

American Orthopaedic Association
http://www.aoassn.org

American Orthopaedic Foot and Ankle Society
http://www.aofas.org

American Orthopaedic Society for Sports Medicine
http://www.sportsmed.org

American Physical Therapy Association
Orthopaedic Section
http://www.orthopt.org

Orthopedic Surgical Manufacturers Association
http://www.osma.net/index.htm

See also: Geriatrics and Gerontology; Kinesiology; Occupational Health; Prosthetics; Rheumatology; Surgery.

OSTEOPATHY

FIELDS OF STUDY

Biology; cellular biology; human anatomy; chemistry; biochemistry; physics; mathematics; kinesiology; physiology; pathology; psychology.

SUMMARY

Osteopathic medicine is a branch of medicine that holds that the entire human being, not just the illness, should be treated. It emphasizes the relationship between the organs of the body and the musculoskeletal system. Osteopathic doctors use the body's self-healing abilities in their treatments. They are trained and licensed physicians with the ability to prescribe medication and to perform surgery. The majority of osteopaths are primary care physicians who focus on comprehensive family care and serving the medical needs of the community.

KEY TERMS AND CONCEPTS

- **Allopathic Medicine:** Traditional Western medicine.
- **Cerebrospinal Fluid:** Clear fluid in the spaces, passages, and cavities around and inside the brain and spinal cord.
- **Chiropractic Medicine:** Health care profession that diagnoses and treats disorders of the musculoskeletal system with additional emphasis on the spine.
- **Fascia:** Sheet or band of connective tissue surrounding and binding together structures such as the muscles.
- **Manipulation:** Hands-on therapeutic intervention performed on the body's joints, especially the spinal column.
- **Musculoskeletal System:** Human body system consisting of the muscles, bones, ligaments, and tendons that provide form, support, stability, and movement.
- **Primary Care Physician:** Medical doctor who provides patients with most of their routine care and acts as the initial contact in referrals to specialists.
- **Subluxation:** Incomplete dislocation of the bones in a joint so that they are no longer aligned.

DEFINITION AND BASIC PRINCIPLES

Osteopathy is a focused health care philosophy that emphasizes the musculoskeletal system of the human body and its relationship to the body's organs. Osteopaths strive to treat not just the disease but the whole person. Osteopaths practice medicine using the body's own healing power.

Osteopaths study patient history closely, often finding that elements of the body's structure–particularly the bones and muscles–are the underlying causes of illness and dysfunction. Osteopaths must have a highly developed sense of touch, allowing them to examine and often treat patients by palpitating the flow of fluids in their bodies, the texture and movement of soft tissue and muscles, and the body's overall structure.

Osteopathic medicine is not considered alternative health care, as this form of medicine has a 125-year history. It is recognized as a complete system of medical care founded on the philosophy of treating the whole patient–the body and the mind. The predominance of the musculoskeletal system within osteopathic medicine is also fundamental to chiropractic medicine, which is considered a separate but related health discipline.

BACKGROUND AND HISTORY

American physician Andrew Taylor Still founded the first school of osteopathic medicine in 1892, based on a theory he had developed in 1874. He theorized that the musculoskeletal system was the primary component of good health. Still believed that the structures that supported the nervous system—the skull and vertebrae–influenced the energy that flowed through it. This energy flow could be altered by any musculoskeletal defect, thereby producing disease. To cure the disease, therefore, the physician needed to restore the supporting structures to their natural state through adjustments.

In the early 1900's, osteopathic medicine increasingly incorporated the practices of allopathic medicine, while retaining the whole-person approach and manipulation techniques. In 1950, a court decision in Missouri recognized the basic equivalency of an osteopathic doctor (D.O.) and medical doctor (M.D.) and their right to practice and perform surgery in a county hospital. By 1973, all fifty states and the District of Columbia had granted full practice rights to doctors of osteopathy. Although osteopathic doctors and allopathic doctors perform the same functions, osteopaths have received education in the philosophy of osteopathy and training in manipulation techniques. About 5 percent of all American physicians are osteopaths. These 47,000 osteopaths handle about 100 million patient visits a year. The practice of osteopathy has spread from the United States to Canada and other nations.

How It Works

For osteopaths, treating a disorder involves not only returning a person's body to its normal alignment and functioning but also determining what caused the disorder. In the case of a joint injury, especially if a recent fall or mild blow to an area is the likely cause, osteopaths may treat the patient with manipulation.

Manipulation is a technique used by osteopaths, along with other medical professionals such as chiropractors and physical therapists. This particular hands-on approach is designed to ensure that the body's joints are moving properly, allowing the body's natural healing systems to function. Manipulation emphasizes joints, specifically those of the vertebrae and ribs. Osteopaths believe there may be two types of joint dysfunction. The joint members can be very slightly offset, which is referred to as a subluxation, or the joint surfaces may have a vacuum lock within them, so that movement of the joint squeezes out lubricants and produces abrasion of the joint surfaces. Some cases of joint pain may be the result of a postural misalignment caused by, for example, the patient's legs being of different lengths. Muscles can be tight or painful because of tension or stress. If the patient has experienced pain for a long time, some muscle deterioration may have occurred. Osteopaths look at all the possible causes and at the whole person when determining how to restore the patient's health.

Some controversy exists in regards to the efficacy of manipulation; however, it provides relief to millions of patients. Backs and necks are most commonly manipulated. Sometimes one manipulation or a short series is enough, but some patients require multiple treatments. Others fail to respond, and osteopaths must seek another form of treatment. Once a patient has been treated successfully by manipulation and

the cause of the illness or trauma has been identified, osteopaths can create exercise programs or prescribe other rehabilitative techniques.

Although osteopaths take a whole-person view of treatment, their methods of diagnosing and treating injuries and diseases are very similar to those of an allopathic doctor. Osteopaths treat patients with various forms of therapy and medicines as appropriate and perform surgery when necessary.

APPLICATIONS AND PRODUCTS

Osteopathic Manipulation. Osteopathic doctors look for restrictions in normal joint movement and try to correct the problem by moving the joint through the restricted area of motion yet not beyond the joint's normal range of motion. This is in contrast to chiropractic medicine, as this discipline may manipulate an affected joint beyond its normal range of motion. There are several approaches to osteopathic manipulation, and the techniques vary among osteopathic practitioners. For manipulation to be successful, the muscles must first be somewhat relaxed. To help relax muscles, osteopaths use forms of heat such as hot packs or ultrasound, medication, or gentle stretching. Articulation and thrust are commonly used in manipulation. Articulation (applying a gentle force against the restricted area) is used to improve motion and thrust (applying a sharp force against the restricted area) is used to regain range of motion.

Cranial Osteopathy. Some, but not all, practitioners believe that the skull bones can also be manipulated. Cranial osteopathy is a theory based on the premise that the bones of the skull permit small increments of movement. This application, which became popular for treating babies and children, is said to be based on a rhythm that can be felt by practitioners whose sense of touch is highly developed. Some osteopaths believe that improving the cranial rhythmic movement helps increase the flow of the cerebral spinal fluid that surrounds the brain and spinal cord. This increase in flow can then raise metabolic outflow and nutrition inflow.

Craniosacral Therapy. Craniosacral therapy (CST) is a hands-on approach to monitoring and mobilizing what osteopaths refer to as the craniosacral rhythm. In this therapy, the spine and the cranial junctions of the skull as well as soft tissue are gently manipulated to improve the flow of cerebrospinal fluid in the spinal cord and to open restricted nerve passages. Craniosacral techniques are used in the treatment of neck and back pain, mental stress, chronic pain, and migraines.

Craniosacral therapy and cranial osteopathy are two separate practices with different training backgrounds. Although the two are based on similar principles, craniosacral therapists are not doctors but rather an unlicensed group of individuals who perform such therapies. Cranial osteopaths, however, are fully licensed doctors of osteopathic medicine. They have graduated from a medical school and passed medical boards, so they are significantly more qualified than a craniosacral therapist.

Visceral Osteopathy. Osteopaths who propose the application of visceral osteopathy state that the internal organs and bodily structures (such as the digestive tract and respiratory system) and the body's motion are interconnected. For optimal health, connections among the organs and structures need to remain balanced and stable, despite the body's endless motion. Visceral osteopaths believe that manipulating the musculoskeletal system benefits the internal organs and vice versa. The effectiveness of visceral osteopathy remains controversial even within the osteopathic profession.

Muscle Energy Technique. Muscle energy techniques are also used by osteopaths to treat dysfunctional joints and the spine. Patients perform a combination of movements such as pushing against manual resistance, then relaxing and stretching the muscle at several points in the limb's motion. Such techniques can be used with various joints of the body such as the hip, knee, and shoulder. The technique can help decrease joint restriction and restore full range of motion.

Primary and Specialized Medical Care. Osteopathic doctors practice in many primary care areas, including family practice, internal medicine, obstetrics, gynecology, and pediatric medicine. They are licensed to prescribe pharmaceuticals and can specialize in areas such as surgery. The medical and surgical treatments that osteopathic doctors use on their patients are, for the most part, the same as those that allopathic doctors use.

Pain Management. Osteopathy often is used in pain management. In the treatment of pain, osteopaths and traditional doctors share the same goals and some of the same treatments, but osteopaths are

---

Fascinating Facts About Osteopathy

- One in 7 Americans has a musculoskeletal impairment.
- In 2000, more than 7 million people were hospitalized for musculoskeletal conditions.
- In addition to the United States, osteopathic medicine is recognized in most of Canada and forty-six additional countries.
- Of all osteopathic physicians, more than 45 percent are family doctors, and about 15 percent are specialists in fields such as pediatrics, internal medicine, surgery, obstetrics and gynecology, neurology, and psychiatry.
- Although the medical value of manipulation has been debated for years, it has helped millions of patients.
- A 1999 study showing that osteopathic manipulation was effective in treating lower back pain at a lower cost and with fewer side effects than medication alone appeared in *The New England Journal of Medicine*.

---

regarded as having a more in-depth understanding of the interrelationship between the body's own healing capabilities and the musculoskeletal system. Most doctors commonly prescribe drug therapies and surgical treatments to treat chronic pain, but osteopaths often take other approaches. In the treatment of acute or chronic pain, osteopaths place greater emphasis on how the organs and the musculoskeletal system are related and how to treat the whole individual not just the pain. For example, emotional problems can cause muscles to tighten, resulting in pain that produces more spasms and additional pain. The treatment is often drugs that relax muscle spasms and reduce pain. If the patient has been experiencing pain for a long time, his or her muscles may have deteriorated and lost strength, so osteopaths may recommend specific exercises or refer the patient to physical therapy. Osteopaths may investigate whether the patient has psychological problems that may need attention and medication. The osteopathic approach looks at the whole person, looking at multiple ways to decrease or eliminate the patient's pain.

Counterstrain Technique. Counterstrain techniques are used by osteopaths to treat joint dysfunction and relieve pain. These techniques place the affected joint in the position of the least discomfort, which is usually with the muscle at its shortest length. While monitoring the degree of discomfort at a nearby tender point, the position is held for ninety seconds and then the joint is slowly returned by the osteopath to a more neutral position.

Myofascial Release Techniques. Myofascial release techniques are similar to massage; however, osteopaths use their hands to apply a traction force to the alignment of muscles to stretch them and to the surrounding soft tissues to reduce tension. Related methods called soft tissue techniques basically involve massaging the muscles and the surrounding soft tissues. Osteopaths use their hands to stretch or relax dysfunctional soft tissue structures. Such techniques are applied almost anywhere throughout the body but are especially useful for the paraspinal muscles, which surround each vertebra. The ultimate goal of all soft tissue techniques is to relax tight muscles and stretch tight surrounding soft tissue structures called fascia.

IMPACT ON INDUSTRY

Osteopathic medicine has a great potential for growth. Because osteopaths focus on primary care and prevention, they are becoming the family physician for an increasing number of people. Osteopathic education emphasizes primary medical care, and more than half of all doctors of osteopathy are in family practice or internal medicine, often practicing in rural or underserved areas. Doctors of osteopathy perform 10 percent of all primary care visits in the United States. Most osteopathic doctors are actively practicing medicine but some have become educators or researchers. Research has suggested that osteopathic patients require less medication and less physical therapy compared with those who receive traditional medical care, and the total cost of care is less. As more research on osteopathy is being conducted, the field is increasingly affecting traditional medical practices.

CAREERS AND COURSE WORK

Individuals interested in osteopathy need to understand that those practicing osteopathic medicine must have a genuine concern for people. For that reason, prospective applicants should observe an osteopath at work and, if possible, perform volunteer work in a hospital or clinic to determine if they have or can develop the skills needed to interact with people.

To be admitted to an osteopathic school, applicants must have a bachelor's degree. Although they can chose almost any major, they must meet the minimum course requirements for osteopathic school. Some recommended courses include biology, genetics, general and organic chemistry, physics, mathematics, psychology, and English. Another requirement is a good score on the Medical College Admissions Test. Osteopathic medical colleges generally interview candidates, partly to determine the depth of their interest in osteopathy.

During the first two years of osteopathic medical school, students learn basic sciences, and the next two years, they gain experience in clinical work as well as study general medicine. Clinical rotations take students to urban, suburban, and rural settings, enabling them to gain exposure to all areas of medicine. Following graduation, osteopathic doctors complete an approved twelve-month internship, which provides exposure to hospital departments. At this point in their education, some students complete a two- to six-year residency program in a medical specialty. Osteopathic physicians can become certified in a specialty by completing a residency program and passing the pertinent exams.

Like all physicians, osteopaths must pass a national licensing exam. Osteopathic doctors are licensed to practice medicine and perform surgery in all fifty states and the District of Columbia, although state licensing procedures vary. Earning continuing medical education credits is a requirement for maintaining licensure.

SOCIAL CONTEXT AND FUTURE PROSPECTS

The demand for osteopathic doctors, like that for traditional medical physicians, is projected to grow faster than the average for all occupations. Contributing to this continued need are the growing and aging population and technological advances in tests and procedures that enable physicians to better treat patients. General and primary care practitioners most likely will have the best prospects and those pursuing specialties such as surgery will find the competition more intense. Compensation for osteopathic doctors depends on their specialty, the location and type of facility where they practice, and how long they have practiced.

Jeffrey Larson, P.T., A.T.C.

FURTHER READING

Gintis, Bonnie. *Engaging the Movement of Life: Exploring Health and Embodiment Through Osteopathy and Continuum.* Berkeley, Calif.: North Atlantic Books, 2007. Gintis, an osteopath, combines osteopathy with the Continuum Movement to describe an approach to life that uses movement, mindfulness, and breath to further mind and body health.

Hebgen, Eric. *Visceral Manipulation in Osteopathy.* Stuttgart, Germany: Thieme, 2011. Provides information on four concepts and techniques in visceral manipulation: the Barral concept, the Finet and Williame concept, the Kuchera and Kuchera technique, and the Chapman reflex points.

Liem, Torsten. *Cranial Osteopathy: A Practical Textbook.* Seattle: Eastland Press, 2009. Concentrates on the practice of cranial osteopathy, describes how it is done.

Möckel, Eva. *Textbook of Pediatric Osteopathy.* New York: Churchill Livingston Elsevier, 2008. Describes the challenges of treating children through osteopathy and provides methods of doing so.

Parsons, Jon, and Nicholas Marcer. *Osteopathy: Models for Diagnosis, Treatment, and Practice.* New York: Churchill Livingstone, 2006. Describes how an osteopath diagnoses and treats illnesses.

Richter, Philipp, and Eric Hebgen. *Trigger Points and Muscle Chains in Osteopathy.* New York: Thieme, 2009. Examines trigger points (sore points) which are essential in the manipulation that takes place in osteopathy. Examines myofascial release techniques and other hands-on methods.

WEB SITES

American Academy of Osteopathy
http://www.academyofosteopathy.org

American Association of Colleges of Osteopathic Medicine
http://www.aacom.org

American Osteopathic Association
http://www.osteopathic.org

See also: Kinesiology; Pediatric Medicine and Surgery; Rehabilitation Engineering.

OTORHINOLARYNGOLOGY

FIELDS OF STUDY

Audiology; biology; chemistry; ear, nose, and throat medicine; head and neck surgery; laryngology; medicine; neurotology; oncology; otolaryngology; otology; pediatric otolaryngology; physics; plastic and reconstructive surgery; rhinology; surgery.

SUMMARY

Otorhinolaryngology medicine and surgery practitioners diagnose and treat diseases of the ear, nose, throat, and adjacent head and neck structures. They are often called ear, nose, and throat (ENT) specialists. Hearing, balance, and voice disorders, as well as facial plastic and reconstructive surgery, constitute prominent areas of expertise for otorhinolaryngologists. Subspecialties focusing on these areas of the human body are among the most complex in modern medicine.

KEY TERMS AND CONCEPTS

- **Adenoids:** Masses of lymphoid tissue at the back of the nose.
- **Cochlea:** Coiled part of the inner ear where the hearing receptors reside.
- **Laryngology:** Branch of medicine focusing on disorders of the vocal apparatus.
- **Laryngoscopy:** Examination of the larynx by means of a laryngoscope.
- **Larynx:** Part of airway containing the vocal cords; also known as the voice box.
- **Mastoiditis:** Inflammation in the mastoid part of the temporal bone.
- **Oto-:** Prefix indicating "ear."
- **Otology:** Branch of medicine focusing on disorders of the ear (including hearing and balance).
- **Otoscopy:** Examination of the ear using an otoscope.
- **Pharynx:** Part of the digestive and respiratory systems situated behind the nose and mouth.
- **Rhino-:** Prefix meaning "nose."
- **Sinuses:** Air cavities in the cranial bones.
- **Tonsils:** Small masses of lymphoid tissue at the back of the mouth.

- **Tracheostomy:** Surgical procedure that creates an airway opening through the neck, into the trachea (windpipe).
- **Vestibular:** Pertaining to the vestibular apparatus—the part of inner ear and vestibulocochlear (auditory) nerve that contributes to balance and spatial orientation.

DEFINITION AND BASIC PRINCIPLES

Otorhinolaryngology, or otolaryngology, is the branch of medicine dedicated to the diagnosis, treatment, management, and prevention of diseases affecting the ear, nose, and throat, and structures of the head and neck. Otorhinolaryngologists are also known as ENT physicians. A characteristic of these physicians is that they are trained in both medicine and surgery. Despite considering ear, nose, and throat disorders as separate pathologic categories, general otolaryngologists treat patients with all of these disorders. The broad and challenging scope of otorhinolaryngology allows a choice of direction, while providing optimal patient care.

Treating ear disorders is a unique mission of otorhinolaryngologists. Both medical and surgical approaches are employed to care for patients with hearing impairment, ear and mastoid infections, balance disorders, tinnitus, and tumors. Most otolaryngologists employ audiologic equipment to test hearing and diagnose the cause of hearing loss. Some have a special interest in neurotology and disorders of balance such as vertigo.

Care of the nasal cavity and sinuses includes alleviating allergies, managing foreign bodies and nosebleeds, restoring the sense of smell and taste, and removing tumors.

The throat is a muscular tube leading into the respiratory and digestive tracts. It functions as a pathway for air and food, as well as in speech formation. It consists of the pharynx and the larynx. A feature of the throat is the epiglottis, which is a "flap" that protects the larynx during swallowing and prevents inhalation of food or liquids. Throat specialists make sure that patients can speak, sing, and eat. They treat a variety of disorders arising in the mouth and throat, including swallowing and breathing problems, infections, tumors, birth defects, speech or voice abnormalities, and trauma. Virtually all ENT doctors routinely handle tonsillectomy (tonsil removal) and adenoidectomy (adenoid removal).

Some esophageal (such as gastroesophageal reflux) and tracheal disorders also belong to the sphere of otorhinolaryngology.

In the head and neck area, otolaryngologists treat infectious diseases, trauma, and tumors (such as those arising in the aerodigestive tract, thyroid, and salivary glands). They may perform plastic and reconstructive surgery. Cranial nerve disorders (such as facial paralysis) also represent an area of focus for some otorhinolaryngologists.

BACKGROUND AND HISTORY

Early Egyptian and Hindu manuscripts mention ear and throat structures, including eustachian tube and tonsils. The first known rhinologist was an Egyptian named Sekhet'eananch, who lived around 3500 b.c.e. Ancient Hindus were probably the first to practice rhinoplasty. Hippocrates recognized the tympanic membrane as part of the hearing organ and pioneered a method for removing nasal polyps that remained in use for centuries. Manuscripts of prominent Byzantine physicians from the fourth to the fourteenth century frequently contained information on various otorhinolaryngology issues. Operations performed in the Middle Ages included removal of nasal polyps and tonsillectomy. Belgian anatomist Andreas Vesalius, Italian anatomist Bartolomeo Eustachio, and Italian physician Gabriel Fallopius described ear structures in the sixteenth century. The first published account of a successful tracheostomy dates back to this era. It was only much later, however, that the study of ENT disorders acquired a definite scientific basis.

During the first half of the nineteenth century, French physicians Jean-Marc-Gaspard Itard and Prosper Ménière systematically investigated ear physiology and diseases. French physiologist François Magendie published experiments on the physiology of the larynx and the purpose of the epiglottis. Surgeons started performing frontal sinus operations. Spanish singer and teacher Manuel Garcia, who started describing vocal cords movement in 1840, is recognized as the "father of laryngology." At this time, a better understanding of the connection between the ear and throat lead to the association of otologists and laryngologists. Ear hospitals started to merge with throat hospitals, to become de facto ear, nose, and throat centers. The British Rhino-Laryngological Association added otology to its focus in 1895.

The twentieth century saw great advances in otorhinolaryngology, along with significant progress in anesthesia and radiology. The American Board of Otolaryngology, founded in 1924, became the second oldest medical-certifying organization in the United States. The first cochlear implant received Food and Drug Administration (FDA) approval in 1984.

HOW IT WORKS

First, the otorhinolaryngologist obtains the patient's detailed medical and surgical history. This step is extremely important, because it constitutes the basis for the entire workup. Without it the evaluation and diagnosis are incomplete.

Clinical diagnosis involves the administration of a detailed physical examination and the utilization of tests. Equipment used for this purpose includes otoscopes, audiometers, endoscopes, microscopes, X-ray machines, MRI scanners, and many others. Frequently, tests serve to determine the extent of hearing loss due to ear injury or potential speech impairment that occurs following laryngeal damage. Many ENT disorders, however, can be diagnosed primarily on physical exam. Therefore, otolaryngologists can maintain a very hands-on approach to patient care.

After identifying the nature and extent of the disorder, the physician prescribes medications or performs surgery. The surgical approaches associated with the practice of otorhinolaryngology are as varied as the range of patients and subspecialties. The otolaryngologist can function as a microsurgeon, performing microvascular reconstruction or neurotologic procedures. He or she may be an endoscopic surgeon, using endoscopy to diagnose and treat sinus disease and laryngeal disease. Some surgical procedures are limited to relatively small areas, such as the temporal bone; others extend to the entire neck, in all its complexity.

APPLICATIONS AND PRODUCTS

Some of the more frequent diseases, conditions, and treatments encountered in the otorhinolaryngology practice are discussed below.

Otitis Treatment. Inflammatory conditions of the

middle ear (otitis media) affect all ages but occur mostly in children. Acute otitis media, caused by various bacterial pathogens, manifests with pain (otalgia), fever, irritability, nausea, and even fluid discharge. Physical examination reveals a thickened, red, bulging eardrum with limited mobility to pneumatic otoscopy. Otoscopy involves looking into the ear using an instrument called otoscope. The instrument consists of a handle, a head (with light source), and cone (inserted into the ear canal). Sometimes, air is insufflated via a pneumatic attachment to assess eardrum mobility. For otitis media, the physician prescribes antibiotics and analgesics. In some cases, he or she may place a hole in the eardrum (myringotomy) and insert drainage tubes.

Hearing-Loss Treatment. Hearing loss (HL) is the most common sensory impairment in humans. Its causes include aging (presbycusis), noise exposure, earwax accumulation, drugs, infections, tumors, and genetic abnormalities. To render an accurate diagnosis, otorhinolaryngologists and hearing specialists (audiologists) examine the ear with an otoscope, conduct screening tests using handheld tone generators, and perform tuning-fork and audiometer tests. Tuning-fork tests compare how well a patient hears sounds conducted by air versus those conducted by skull bones; they can differentiate between hearing loss caused by mechanical problems in the ear canal and hearing loss due to damaged sensory structures. Audiometric tests use an electrical device called audiometer that consists of a pure tone generator, a bone-conduction oscillator, an attenuator, as well as microphone and earphones. They identify the quietest tone a person hears in each ear and compare it with established standards. Speech threshold and discrimination tests may also be administered. Additional tests include auditory brain stem responses (using electrodes to determine the type of stimulus the brain receives from the ear) and electrocochleography (measuring cochlear and auditory nerve activity).

Treatment options vary according to the cause and severity of hearing loss and may involve earwax removal, hearing aids, or cochlear implants. Hearing aids are devices that deliver amplified acoustic signals to the ear. They can be placed behind the ear (BTE), in the ear (ITE), in the canal (ITC), or completely in the canal (CIC). Cochlear implants are complex electronic devices that send signals to the auditory nerve,

bypassing the damaged cochlear cells. Surgeons insert them–through an incision behind the ear–in selected people with severe and profound hearing loss, to provide them with a representation of the sounds in the environment.

Tinnitus Treatment. Tinnitus represents an auditory sensation originating in the head, without external stimulation. This common condition affects up to 10 percent of the U.S. general population. Patients describe it as ringing, whistling, hissing, roaring, buzzing, or clicking. The assessment of a patient with tinnitus includes physical examination, blood pressure measurements, audiometric profile, and brain imaging. Therapeutic steps vary according to the type of tinnitus and its affect on a patient's life. Specific medical or surgical treatment is provided for any underlying illness. Medication, hearing aids, noise-masking devices, and cognitive and behavioral therapy are sometimes recommended.

Vertigo Treatment. The term "vertigo" designates a specific type of dizziness characterized by a spinning sensation, which arises because of peripheral (vestibular) or central nervous system disorders. Common causes include benign paroxysmal positional vertigo, inner ear inflammation, Ménière's disease, migraine, and tumors. Diagnostic evaluation comprises audiologic, vestibular, and neurologic examination, complemented by imaging studies. Medical or surgical treatment is recommended, depending on the cause of the vertigo.

Allergy Treatment. Allergic rhinitis, an inflammation of nasal membranes induced by indoor or outdoor allergens, affects up to 20 percent of the adult population in the United States. It presents with itching, sneezing, and nasal discharge, often accompanied by eye and throat symptoms. The physician uses a nasal speculum or endoscope to visualize the nasal mucosa, which may be swollen and pale. Nasal endoscopy uses a slender optical instrument (telescope) that easily passes through the nostrils to visualize nasal cavity and sinuses. For allergic rhinitis, the physician recommends environmental control measures and prescribes oral antihistamines and decongestants.

Sinusitis Treatments. Sinusitis is an extremely prevalent inflammation of the sinuses that occurs following a bacterial, viral, or fungal infection. In the United States, tens of millions of people develop chronic sinusitis each year, making it one of the most common health complaints. People with allergies or other chronic nasal conditions may be particularly prone to sinus inflammation. The disorder has a significant impact on the quality of life of affected individuals. To diagnose it, ENT physicians rely on clinical presentation (such as nasal obstruction and facial pain), in addition to imaging studies and endoscopy. Medical therapy includes saline sprays, antibiotics, and decongestants. Procedures such as endoscopic sinus surgery or open sinus surgery can also be employed in certain circumstances. In the operating room, nasal endoscopy provides a brightly illuminated and magnified view of the inside of the sinuses and allows a thorough approach for any sinus disease.

Tonsillectomy. Patients with chronic tonsillitis have a persistent inflammation of the pharyngeal tonsils, caused by bacteria, such as streptococcal or staphylococcal. They may complain of throat pain, bad breath, and fatigue. Physical examination may show enlarged tonsils (mainly in young people) and enlarged tonsillar crypts filled with debris. Symptomatic patients are treated with tonsillectomy. This common major surgical procedure employs various dissection methods, such as cold knife (scalpel), electrocautery, harmonic (ultrasonic) scalpel, radiofrequency ablation, and laser ablation.

Voice Treatments. Voice disorders–such as hoarseness and whispered voice (aphonia)–are frequently encountered in an otorhinolaryngology practice. They arise because of a multitude of conditions, ranging from inflammation to neural impairment and tumors. To diagnose the condition, an otorhinolaryngologist examines the larynx and assesses vocal cord movements. Laryngoscopy (larynx examination) is performed indirectly using a mirror placed in the back of the throat or a thin, flexible fiberoptic endoscope inserted through the nose. This technique can be combined with the use of a special light called a stroboscope, to assess the vibratory capacity of the vocal cords. A neurological examination might be performed in patients with voice disorders. Speech-language pathologists evaluate voice production and quality. Medical, surgical, and rehabilitative options are available for voice restoration.

Head, Neck, and Throat Surgeries. Tumors occur at all levels of the ENT head and neck axis. They may manifest with voice changes, throat lumps, pain, and breathing or swallowing difficulties. The diagnosis

is reached using endoscopy, biopsy, and imaging studies. Laryngeal and laryngopharyngeal cancers are the most frequent malignancies of the head and neck. Treatment options depend on many factors, such as the type of tumor, location, size, and stage; generally, they involve surgical removal, chemotherapy, and radiation therapy.

Plastic and Reconstructive Surgery. These procedures include cleft lip and cleft palate repair, face and brow lifts, improving the appearance of the nose (rhinoplasty), changing the shape of the nasal septum (septoplasty), reshaping the ear (otoplasty), chin augmentation, and scar camouflaging.

IMPACT ON INDUSTRY

Ear, nose, throat, head, and neck disorders affect millions of people around the globe, significantly decreasing the quality of their life. For this reason, research programs in these fields are being increasingly developed in both academic and industrial environments. The future will likely see an extensive cooperation between academic, government, nonprofit, and industry organizations.

Government and University Research. Research in this field is funded through universities and various organizations such as the American Academy of Otolaryngology and the National Institutes of Health. The hearing-device manufacturing industry also supports some parts of academic research, such as product validation and testing.

Academic basic researchers in various fields (otorhinolaryngology, pathology, and neuroscience) focus on elucidating the cellular and molecular mechanisms of ENT disorders, particularly those pertaining to the vestibular and auditory systems. In addition, head and neck tumors constitute a complex and fascinating field of investigation.

Areas of active clinical research development in otorhinolaryngology include cochlear implants, microsurgery, and transplantation. The National Institute on Deafness and Other Communication Disorders is one of the institutions actively involved in auditory research. It sponsors studies that aim to improve the benefits of classical implants as well as efforts to create new brain-stimulating devices for patients with nerve damage.

Industry and Business. The hearing-devices industry is intimately linked with the practice of otorhinolaryngology and audiology. It consists of two main markets: hearing aids and hearing implants. The industry has progressed significantly over the past decade, achieving billions of dollars in revenue. After a brief decline during the economic downturn of 2008, the market resumed its growth. As more Americans reach an advanced age in relatively good health, with hearing devices more and more affordable and socially acceptable, the industry is expected to develop rapidly. Some manufacturers have created research laboratories in the United States that aim to develop their product on a scientific basis. Technological advances provide hearing-impaired patients with innovations such as disposable, programmable, and instant-fit hearing aids. The behind-the-ear (BTE) instruments represent more than 50 percent of the overall U.S. market. North America and Europe constitute the major hearing-aid markets in terms of sales. Worldwide, the hearing-aid industry is now progressing toward a shorter product life cycle, convergence with consumer electronics, and industry consolidation through mergers. Meanwhile, the relatively low rate of adoption for these novel instruments (especially hearing implants) leads to a tremendous potential for manufacturing growth. Advanced Bionics, Beltone, Cochlear Ltd., Hansaton, Med-El, and Phonak are some of the key players in this industry.

A distinct industrial sector specializes in ENT surgical instruments and disposables, endoscopic instruments and supplies, flexible endoscopes, otoscopes, audiometers, and hearing-screening equipment. These instruments are required in a wide range of medical and audiologic facilities. As patients increasingly request surgical options for conditions such as sleep apnea and sinusitis, the ENT surgical instruments market is expected to grow. New technologies are being developed in nasal and sinus packing, lasers, lighting, and flexible rhinolaryngoscopes. The U.S. specialty endoscopic-surgery products market (which includes revenues derived mainly from endoscopes and hand instruments) totaled more than $1 billion in 2009 and will approach $1.4 billion by 2014. Manufacturers and suppliers consist of multinational corporations as well as large, midsize, and small manufacturers. Most of them are located in North America, Western Europe (Germany), and Pakistan. Representative companies include Aztec Medical Products, Bausch + Lomb Instruments, Carl Zeiss Meditec, Jedmed Medtronic, Welch Allyn, Richard

Fascinating Facts About Otorhinolaryngology

- In William Shakespeare's time, some believed that poisons dropped in one's ear were as lethal as the ingested ones. This seems to be true in the case of Hamlet's father. In modern medical treatment, several drugs are applied directly to the inner ear for the treatment of ear disorders.

- Sir William Wilde, father of Oscar Wilde, set up a dispensary for eye and ear diseases in a disused stable in Dublin, which became St. Mark's Hospital. He was the first to teach otology in the United Kingdom. The surgical method to treat mastoiditis through a cut behind the ear bears his name: Wilde's incision.

- The advent of growth-factor products signals the beginning of a new era for wound healing. Bone morphogenetic protein and other recombinant growth factors will likely play important roles in the care of cleft lip and cleft palate.

- A robotic surgical system named da Vinci has been approved for minimally invasive throat cancer surgery. The robot uses pencil-like "arms" to reach the tumor through the patient's mouth. One of the arms bears a camera that enables the physician to visualize the surgical field.

- Cochlear implants are a great achievement of modern medicine. They represent the most successful of all neural prostheses developed to date.

- Obstructive sleep apnea syndrome is a complex chronic clinical syndrome, characterized by snoring, periodic apnea, reduced blood oxygen during sleep, and daytime somnolence. It affects 4 to 5 percent of the general population and appears to be linked to genetic factors.

Wolf Medical Instruments, and Summit Medical. Minimally invasive surgical devices are fast becoming the driving force in the industry.

Numerous pharmaceutical companies worldwide produce therapeutic agents used in the otorhinolaryngology practice. Examples include anti-inflammatory, antibiotic, and antihistamine medication.

CAREERS AND COURSE WORK

A strong foundation in biology, chemistry, and physics is required for a career in medicine–it supports years of training and a lifetime of learning.

Otolaryngologists can practice after completing up to fifteen years of college and postgraduate training. Upon graduation from college and medical school, an individual with an M.D. or D.O. degree undergoes at least five years of specialty training in a residency program accredited by the Accreditation Council for Graduate Medical Education (or equivalent). These programs are among the most competitive for residency training. One year is dedicated to performing general surgery and four years to specific otorhinolaryngology training. Next, the physician must pass the American Board of Otolaryngology examination. In addition, some otolaryngologists pursue a one- or two-year fellowship for in-depth training in an ear, nose, and throat subspecialty.

Major subspecialty areas include pediatric otolaryngology, otology-neurotology (ears, balance, and tinnitus), allergy, head and neck surgery, laryngology (throat), rhinology (nose), skull-base surgery, facial plastic and reconstructive surgery, and sleep medicine. Some otolaryngologists limit their practice to one of these areas. This multitude of options within the specialty generates significant freedom of choice.

Subsequently, ear, nose, and throat specialists can work as salaried physicians in hospitals, clinics, trauma centers, and government health departments. Some physicians work in the community, by establishing their own private practice or joining a group practice. Some otolaryngologists teach in medical schools and residency programs. They often conduct translational research in academic laboratories or pharmaceutical companies. The strength of this complex specialty lies in its diverse opportunity for medical and surgical practice.

SOCIAL CONTEXT AND FUTURE PROSPECTS

Otorhinolaryngology has enormous impact and widespread ramifications in medical practice. Ear, nose, and throat conditions affect patients of all ages and ethnicities. As of 2011, the burden of ear disorders in society is significant. According to the World Health Organization, moderate to severe hearing impairment affects more than 270 million people around the world.

According to the National Center for Health Statistics, by 2030, 32 percent of the population will be elderly, and 75 percent of this age group will have clinically significant hearing impairment. This will have a strong impact on the patients and on society.

Significant progress has been made in the performance of hearing aids and cochlear prostheses, but much room exists for improvements.

Other categories of patients will require specific otorhinolaryngology assistance as well. Treating ear and throat diseases in children, for example, is particularly important and requires special skills because of the complex developmental and psychosocial implications. The demand for pediatric otolaryngologists appears to be on the rise.

Technological advances such as robots and simulators enhance precision in surgical practice and are increasingly employed by otorhinolaryngology specialists. Advances in surgical techniques, radiation therapy, and chemotherapy have improved the survival and quality of life for patients with head and neck cancer. The outcomes of these treatments have allowed physicians to focus less on tumor removal and more on functional preservation and restoration. Preventative efforts, coupled with the study of tobacco use and occupational exposure as risk factors for carcinogenesis, have also become a priority.

Mihaela Avramut, M.D., Ph.D.

FURTHER READING

Blume, Stuart. *The Artificial Ear: Cochlear Implants and the Culture of Deafness.* Piscataway, N.J.: Rutgers University Press, 2010. Historical study of implant development and implementation.

Bull, Tony R., and John S. Almeyda. *Color Atlas of ENT Diagnosis.* 5th ed. New York: Thieme, 2010. Essential, concise overview with excellent illustrations.

Dhillon, R. S., and C. A. East. *Ear, Nose and Throat and Head and Neck Surgery.* 3d ed. New York: Elsevier, 2006. Concise introduction to this specialty; includes excellent illustrations.

Goldenberg, David, and Bradley J. Goldstein. *Handbook of Otolaryngology: Head and Neck Surgery.* New York: Thieme, 2010. Accessible, comprehensive textbook for a variety of practitioners.

Metson, Ralph B., with Steven N. Mardon. *Harvard Medical School Guide to Healing Your Sinuses.* New York: McGraw-Hill, 2005. Well-organized review for the general public, written by experts.

Møller, Aage R. *Hearing: Anatomy, Physiology, and Disorders of the Auditory System.* 2d ed. Burlington, Mass.: Academic Press, 2006. Textbook that covers the foundations of auditory anatomy and pathophysiology.

WEB SITES
American Academy of Otolaryngology
Head and Neck Surgery
http://www.entnet.org/index.cfm

American Speech-Language-Hearing Association
Voice Disorders
http://www.asha.org/public/speech/disorders/
voice.htm

National Institute on Deafness and Other Communication Disorders
Cochlear Implants
http://www.nidcd.nih.gov/health/hearing/coch.
asp

World Health Organization
Deafness and Hearing Impairment
http://www.who.int/mediacentre/factsheets/
fs300/en

See also: Audiology and Hearing Aids; Pediatric Medicine and Surgery; Surgery.

P

PALEONTOLOGY

FIELDS OF STUDY

Dinosaur studies; fossil record; evolution; extinction; taxonomy; vertebrate paleontology; invertebrate paleontology; micropaleontology; nanopaleontology; paleobiology; paleoecology; paleogeography; biology; biostratigraphy; climate change; functional morphology; geology; ichnology; stratigraphy; systematics; taphonomy.

SUMMARY

Paleontology is an interdisciplinary field that is concerned with the study of the record of life through time and the application of that information to solve scientific problems. Fossils are particularly used in the relative dating of rocks, which is important information for the minerals and petroleum industries. However, paleontology also provides valuable data toward an understanding of past life and therefore an understanding of extinction events and the effects of climate change. Modern technology is increasingly used to help solve problems, including computed tomography (CT) scanning to reconstruct the anatomy of extinct organisms.

KEY TERMS AND CONCEPTS

- **Biostratigraphy:** Use of fossil animals and plants in the dating of rocks.
- **Cladistics:** Computer-based method for delineating relationships among groups of organisms.
- **Evolution:** Change through time in successive generations of organisms.
- **Extinction:** Elimination of a taxon of organisms.
- **Fossil:** Remains of a once-living organism or a trace of its behavior.
- **Fossilization:** Process of preserving the remains of organisms within sediment.
- **Ichnology:** Study of trace fossils.
- **Paleobiology:** Approach to paleontology that

stresses the understanding of processes.
- **Paleoecology:** Study of the interaction between organisms and their environment through time.
- **Systematics:** Study of the relationships among organisms.
- **Taphonomy:** Study of the processes that affect organisms between death and burial.
- **Zone Fossil:** Organism that characterizes a particular period of time.

DEFINITION AND BASIC PRINCIPLES

Paleontology is the study of fossils, or once-living organisms preserved within sediment. The term "paleontology" is derived from the Greek *palaios* for "ancient," *ontos* for "being," and *logos*, for "study." Therefore, paleontology does not encompass the allied field of archaeology and overlaps only slightly with anthropology. Traditionally, the study of fossils has been aimed at reconstructing the organisms and analyzing them to develop an understanding of their evolutionary relationships. This has been enhanced in by the advent of cladistics, a computer-based analytical method that greatly increases the speed and rigor of such studies. In addition, the careful tabulation of fossils over several hundred years of study has enabled the development of a relative geological time scale that is the basic tool in all study of the crust of the Earth. Paleontology is an important industrial tool in that it helps extractive industries such as the petroleum and minerals industries understand where to find oil and minerals.

Increasingly, paleontology is being viewed as a potentially valuable predictive tool. The results of earlier climate changes are preserved in the fossil record and may help scientists understand the effects of global warming or other future climate changes.

BACKGROUND AND HISTORY

Fossils were probably picked up and admired long before they were studied, but the Greeks recognized that some of these objects represented marine organisms and could be used to determine where oceans had once been. In medieval times, fossils were often viewed as magical objects. For example, fossil shark teeth were termed *glossopetrae*, or "tongue stones," and were thought to provide protection from snakebite. They were viewed as structures that had "grown" within the rocks rather than the remains of actual organisms. In the seventeenth century, Nicolaus Steno demonstrated that *glossopetrae* were shark teeth by dissecting a modern shark and showing the similarity of its teeth to the fossils. At about the same time, Robert Hooke used early microscope technology to show that fossils and modern wood and shells had the same internal structure.

Once fossils were accepted as representing past life, they could be studied as such. In some sequences, scientists found clear evidence of extinction events in which whole faunas were replaced over short periods of time. In the early nineteenth century, work by French naturalist Georges Cuvier in the Paris basin showed that numerous mammals that lived when those rocks were laid down no longer existed. Studies of this type showed that long periods of time had been necessary for the evolution and extinction of groups of fossil organisms. At about the same time, English canal surveyor William Smith determined that fossils could be used as time markers, representing the period they were present on the surface of the Earth. His findings were developed into the geological time scale, which is still being refined but provides a basis for dating rocks in a relative sense.

How It Works

Dating. Understanding the spatial relationships of rocks developed from the need to predict where materials of economic value could be found. Initially, distinct rock beds were used to correlate between places, but this method was imprecise over long distances or in areas where the rocks were folded and faulted. Rock units could not be dated effectively until Smith realized that the geological succession corresponded to a regular, nonrepeating succession of fossils. Fundamental to this process of biostratigraphy is the use of zone fossils that represent a period of time equal to the duration of the species before it became extinct and that are widespread enough to enable useful correlations to be made. Increasingly, microscopic fossils (such as the marine, planktonic, and unicellular foraminifera) are used as they have

the advantage of being very abundant in sediments and of having short time ranges, which allows for accurate dating.

Taphonomy. The quality of the fossil record must be understood for it to be useful in applications such as dating or understanding mass extinctions or biodiversity loss. Taphonomy, which is the study of the processes that affect organisms from death to burial in sediment, is integral to such an understanding as it allows an evaluation of the extent to which the fossil record mirrors the original faunal situation. Most organisms do not become fossils, and those that do will first undergo decay of the soft tissues, then transport and breakage of the hard tissues, and finally their burial and modification. For example, soft-bodied organisms can make up 50 to 60 percent of the organisms in present-day marine settings, and in most cases, none of those organisms will be fossilized. Experimentation on modern forms shows that the presence of oxygen is vital to the initial breakdown of soft tissues, and its absence can lead to conditions of special preservation that result in unusually well-preserved faunas such as the Burgess Shale fauna (a Middle Cambrian age site in Canada). The breakage and disarticulation of organisms can be simulated in the laboratory or observed in nature. These studies show that organisms go through a number of stages as they progress toward fossilization. Factors that affect whether they are preserved include the presence of hard parts, how common the organisms are, and their life environment, as organisms living in areas of sedimentation (particularly aquatic environments) have a better chance of preservation. Processes within the rock, such as water filtering through, may result in replacement of the fossil or its total removal. Of course, before the information contained in the fossils can be used, the fossils must be found by a paleontologist and studied.

Taxonomy and Phylogeny. An understanding of the relationships of organisms enables scientists to appreciate and evaluate extinction events and the effects of climatic change, as well as to use fossils in dating. Naturalist Charles Darwin's views on natural selection led to an understanding not only of the evolutionary processes that result in the development of new species but also of phylogeny, the pattern of relationships of organisms, often shown as a branching diagram. Phylogeny has been greatly aided by the method termed cladistics, or phylogenetic analysis,

The Geological Timescale

CENOZOIC

PERIOD	EPOCH	MYA
QUATERNARY	HOLOCENE	
	PLEISTOCENE	0.01
	PLIOCENE	2.6
		5.3
NEOGENE	MIOCENE	
		23.0
	OLIGOCENE	
		33.9
PALEOGENE	EOCENE	
		55.8
	PALEOCENE	
		65.5

(TERTIARY)

MESOZOIC

PERIOD	EPOCH	MYA
CRETACEOUS		65.5
	LATE	
		99.6
	EARLY	
		145.5
JURASSIC	LATE	
		161
	MIDDLE	
		176
	EARLY	
		201.6
TRIASSIC	LATE	
		235
	MIDDLE	
		245
	EARLY	
		251

PALEOZOIC

PERIOD	EPOCH	MYA
PERMIAN	LATE	251
	MIDDLE	260
	EARLY	271
CARBONIFEROUS	PENNSYLVANIAN	299
	MISSISSIPPIAN	318
DEVONIAN	LATE	359
	MIDDLE	385
	EARLY	398
SILURIAN	LATE	416
	MIDDLE	423
	EARLY	428
ORDOVICIAN	LATE	444
	MIDDLE	461
	EARLY	472
CAMBRIAN	FURONGIAN	488
	SERIES 3	501
	SERIES 2	510
		521
	TERRENEUVIAN	542

PRE-CAMBRIAN

EON	ERA	MYA
PROTEROZOIC	NEOPROTEROZOIC	542
		1000
	MESOPROTEROZOIC	
		1600
	PALEOPROTEROZOIC	
ARCHEAN	NEOARCHEAN	2500
		2800
	MESOARCHEAN	
		3200
	PALEOARCHEAN	
		3600
	EOARCHEAN	
HADEAN		3850

developed by the German entomologist Willi Hennig in the mid-1900's. Hennig's method recognizes and orders characters in an objective fashion to show the most likely evolution of organisms through time. As it involves the processing of large amounts of data, this system lends itself well to the use of computer programs such as PAUP (Phylogenetic Analysis Using Parsimony). Results are presented as regularly branching diagrams, or cladograms, that show the relative sequence of speciation events.

APPLICATIONS AND PRODUCTS

Geological Dating. The ability of geological industries to accurately locate the materials they seek depends largely on an understanding of the structure of the Earth's crust. It is important initially to realize that rocks are normally laid down horizontally and that they become younger toward the top of any sequence. Detailed understanding, however, requires the ability to relate rocks of the same time period to one another, and this was not possible until fossils were recognized as providing unique temporal markers. Once this was understood, the sequences of

rocks that had been recognized in discrete outcrops could be related to one another. During the first half of the nineteenth century, most of the relative time scale was assembled, showing the age relationships of rocks and detailing the sequence of fossils that could be used. The oldest fossil-bearing rocks were described as from the Paleozoic (ancient life) era, followed by Mesozoic (middle life) and Cenozoic (new life) eras. Within these eras, the rocks were divided into periods, such as the Cambrian, Ordovician, and Silurian, which are part of the early Paleozoic era, and the Cretaceous period, the last in the Mesozoic era.

As the study of dating became more laboratory based and in response to the requirements of the oil industry, whose samples were typically drilling chips, which were too small to contain recognizable fossils, paleontologists moved toward using microfossils. These can be the remains of very small organisms or small parts of organisms, as long as they are common and widespread and show evidence of having evolved rapidly.

The field of applied micropaleontology started in

the late 1800's in Poland when paleontologist Józef Grzybowski described the foraminifera from the Carpathians in a series of monographs and then applied his findings to foraminifera from the Galician oil fields. He was soon able to demonstrate that it was possible to correlate subsurface strata in wells drilled for petroleum exploration using these fossils. The methods pioneered by Grzybowski continue to be used, and the study of microorganisms has been very much centered on the requirements of the petroleum industry.

Thermal Maturation. Organic microfossils such as spores and pollen can be used to indicate the temperature reached by the rocks (thermal maturity) that contain them because temperature increases as the rock is buried deeper. Organic microfossils are progressively altered by the loss of hydrogen and oxygen, and the resultant changes to physical properties such as color, reflectivity, and fluorescence can then be measured.

Reflectivity studies using vitrinite, an organic component, were initially carried out on coal to determine its rank, or thermal maturity. These studies were then applied to hydrocarbon generation, as hydrocarbons such as oil and gas are generated over time by the action of heat on fossil organic material. The reflectivity of vitrinite in the hydrocarbon source rocks reveals maturity and the likelihood of the presence of oil and gas in the sediments.

Spores and pollen change color from pale yellow to orange to dark brown as the temperature increases, and these changes can be used to show the level of organic maturity. A similar change is seen in conodonts, tiny jaw elements from primitive craniates that were abundant in the Paleozoic era, and an eight-point scale has been developed detailing a color change from pale yellow through black to colorless or clear. Below 60 degrees Celsius, organic material in rock is converted to kerogens and bitumen. Rock generates and expels most of its oil at temperatures between 60 and 160 degrees Celsius, and above 160 degrees Celsius, natural gas is formed. It is important for industry to understand the temperature level the rock has reached, and the application of thermal indices based on color or reflectivity provides that information.

Environment and Climate Change. Organisms are intimately connected to environments and therefore can be used to deduce past environments. This is particularly true of plants, which, unlike marine organisms, are directly exposed to the atmosphere. The size of leaves is related to temperature as well as to humidity and light levels; leaves decrease in size as the temperature or humidity drops. Similarly, the ratio of serrated leaf margins to smooth margins varies according to temperature, and a 3 percent change in this ratio is equivalent to a change of 1 degree Celsius in the mean annual temperature. In addition, the numbers of stomata (small openings on the surface through which plants lose oxygen and water and gain carbon dioxide) vary according to the carbon dioxide level. When carbon dioxide levels are high, plants need only a small number of stomata, but when the levels are low, plants will commonly have large numbers of stomata to enable them to get all the carbon dioxide they need. Because carbon dioxide is a greenhouse gas and traps radiation from the sun, the higher the atmospheric carbon dioxide levels, the higher the temperature. Therefore, a correlation can be made between high global temperatures and low numbers of stomata on fossil leaves. If this technique is applied to the end of the Triassic period extinction, the large drop in the number of stomata on leaves that occurred at this time indicates an increase in atmospheric carbon dioxide and a consequent increase in temperature. Calculations suggest that the temperature would have increased by 5 degrees Celsius globally and possibly up to 16 degrees Celsius locally. In the twenty-first century, carbon dioxide levels are rising because of the burning of fossil fuels, and these studies of carbon dioxide in past eras can provide information that will help scientists understand what changes in biodiversity are likely to occur because of higher carbon dioxide levels.

To obtain information about the temperature of ancient ocean waters, it is possible to use oxygen isotopes extracted from calcareous marine fossils. The ratios of oxygen 16 and oxygen 18 vary, with oxygen 18 increasing as the temperature decreases. Using isotope ratios from foraminifera from deep-sea sediments, it has been possible to show that over the last 100 million years, there has been a steady increase in the oxygen 18 level, indicating a decrease in ocean temperatures of up to 15 degrees Celsius over that period.

Evolution and Extinction. Studies on modern taxa have provided much information on how change in organisms takes place, but the fossil record is still the best evidence of evolution through time and also of its correlative, periods of extinction. Both of these areas are of interest because of their relevance to the changes in biodiversity that are taking place in the twenty-first century. To detail changes through time, it is necessary to develop a database of information on taxa and their longevity. That database is constantly gaining new entries, and therefore never complete; however, it has reached the point where information can be derived from it using statistical techniques aided by computers. Extinctions are always occurring, of course, but these studies show when the level of extinction exceeds that of the development of new species by a significant amount, a mass extinction has taken place. Although numerous extinction events have been recorded, there are seven major ones. The first occurred in the Middle Vendian period about 650 million years ago, followed by events at the end of the Cambrian, Ordovician, Devonian, Permian, Triassic, and Cretaceous periods, of which the end-Permian event is by far the most severe. Analysis of extinction rates of shallow marine organisms at the family level by University of Chicago paleontologists David M. Raup and Jack Sepkoski showed a 26-million-year periodicity to extinction events through the Mesozoic and Ceonozoic eras that was interpreted to indicate an extraterrestrial cause for the extinction events. Corroborative evidence is generally lacking for all but the Late Cretaceous event, however, leaving uncertainty as to the underlying cause of the periodicity.

IMPACT ON INDUSTRY

Fossil fuels will continue to be important to the development of economies in the foreseeable future, and the need for these fuels will increase as nations strive to increase their standards of living. Paleontology is important in the search for oil and natural gas, particularly in the search for coal. In addition, paleontologists are important in the development of a more complete understanding of the history of organisms on Earth, work that is carried on in museums and universities. The main centers for research in paleontology are in North America and Europe, with paleontologists being involved in projects worldwide.

Government and University Research. A considerable amount of paleontological research is carried out by the U.S. Geological Survey. Although some

---

Fascinating Facts About Paleontology

- Calcareous nannofossils in clay scraped from the shoes of a murder suspect in England were sufficiently distinctive to identify the crime scene.
- In India, ammonites (extinct shelled cephalopods) are identified with the god Vishnu and used in religious ceremonies.
- Ammonites were once believed to be petrified snakes, and snake heads were frequently carved on them. In Whitby, North Yorkshire, three ammonites with carved heads form part of the town coat of arms, commemorating the legend that Saint Hilda turned a plague of snakes into stone.
- The largest dinosaur, *Argentinosaurus,* was 120 feet long and weighed more than 100 tons.
- In Norway, which has no native chalk, calcareous nannofossils were used to determine the provenance of white chalk that was used to prepare medieval wooden panels and sculptures before painting.
- During the filming of *Jurassic Park* (1993), the digital artists held their arms next to their chests and ran along a stretch of road with obstacles in order to study the probable movements of the dinosaur *Gallimimus.*
- The largest known trilobite, with a length of 28 inches, is *Isotelus rex* from the late Ordovician period of Manitoba.
- Use of high-resolution computed tomography (CT) has shown that a specimen described as *Archaeoraptor* from Liaoning, China, in 1999 was not a link between birds and dinosaurs but a forgery in which the bones of a nonflying dinosaur and a primitive bird were combined.

---

of this research is related to the understanding of the evolution of organisms through time and the development of databases and collections that are available to answer broad evolutionary questions. A modern trend is for paleontologists to work as part of integrated teams that include members of other disciplines, such as geochemists or paleoclimatologists, to answer highly significant problems.

Industry and Business. The petroleum industry has been one of the largest employers of paleontologists and uses their expertise in deciding where to locate wells and in identifying the strata reached during drilling. However, the number of paleontologists employed at U.S.-based major oil companies has decreased by 90 percent since 1985, as other techniques, such as magnetostratigraphy and seismic information have partially supplanted microfossil dating. Generally, the major oil companies contract out their paleontology requirements to firms that exist to provide these services. Because of this, the demand for paleontologists in the petroleum industry will probably remain low.

The environmental industry has developed extensively since the 1980's as interest in the preservation of natural resources and the environment has become more focused. Paleontologists are employed in this area and are involved in assessment of construction projects and the preparation of environmental impact statements as well as the study of pollution levels using microfossils.

CAREERS AND COURSE WORK

Paleontologists may be trained initially as biologists or as geologists, but in either case, a solid foundation of basic science courses is required. For jobs in industry, a master's degree or a doctorate is required, and the field of study would normally have to be in some aspect of micropaleontology, as that is the area of interest to the petroleum industry. However, as positions in the petroleum industry decreased, universities eliminated or reduced the size of their training programs. In the mid-1980's, more than thirty universities in North America had programs in stratigraphic palynology (the use of spores in dating), but by the late 1990's, only two universities had such programs. Partnerships between industry and university programs are being touted as the way to maintain a flow of adequately qualified paleontologists. As of 2010, only one in ten graduating paleontologists was likely to get a job in industry.

Paleontologists are employed by universities and museums and by the U.S. Geological Survey. However, the number of university paleontologists in North America has not grown since the 1980's, and most of them are approaching retirement age. The trend is for these paleontologists to be replaced by faculty in other fields, particularly environmental geology, which became the top geology degree program in the 1980's and 1990's. The U.S. Geological Survey has actually reduced the number of its paleontology

staff since the 1990's, and although museums have responded to the popular interest in dinosaurs, this has not resulted in a significant increase in jobs for paleontologists.

SOCIAL CONTEXT AND FUTURE PROSPECTS

In 2007, 86.4 percent of world energy consumption came from fossil fuels, which includes oil, coal, and natural gas. Although there is a trend toward greater use of renewable resources, it will be many years before renewable resources provide a significant proportion of energy use. Therefore, paleontologists will still be needed to provide some of the basic data required in exploration to find fossil fuels. However, the numbers of paleontologists needed are likely to decline rather than to rise. Similarly, the need for paleontologists in academia and governmental agencies is declining.

One area in which paleontology could well have an effect in the future is in environmental studies and the study of climate change. As more emphasis is placed on the protection of natural resources, an understanding of past changes and an ability to monitor present-day changes using microorganisms will become increasingly important.

David Elliott, B.Sc., Ph.D.

FURTHER READING

Benton, Michael J. *Vertebrate Palaeontology.* 3d ed. Malden, Mass.: Blackwell, 2009. Covers the fossil record of vertebrates, from fish to humans. Contains lists of further readings and useful Web sites at the end of each chapter.

Benton, Michael J., and David A. T. Harper. *Introduction to Paleobiology and the Fossil Record.* Oxford, England: Wiley-Blackwell, 2009. Provides an overview of paleontology as a science and includes chapters on biostratigraphy, taphonomy, and mass extinctions.

Foote, Michael, and Arnold L. Miller. *Principles of Paleontology.* 3d ed. New York: W. H. Freeman, 2007. This revision of the classic text by David M. Raup and Steven M. Stanley provides good coverage of biostratigraphy and a number of multidisciplinary case studies in paleontology.

Lipps, Jere H. "What, if Anything, Is Micropaleontology?" *Paleobiology* 7, no. 2 (Spring, 1981): 167-199. Defines the subject, covers its history and application to industry, and makes suggestions for new applications of micropaleontological data.

Prothero, Donald R. *Bringing Fossils to Life: An Introduction to Paleobiology.* 2d ed. New York: McGraw-Hill Higher Education, 2004. Describes the major groups of fossils, including microfossils, and explains in some detail aspects of the fossil record, evolution, extinction, and biostratigraphy. Lists of further readings are provided for each chapter.

WEB SITES

The Palaeontological Association
http://www.palass.org

Paleontological Society
http://www.paleosoc.org

U.S. Geological Survey
Paleontology at the U.S. Geological Survey
http://geology.er.usgs.gov/paleo

See also: Gemology and Chrysology; Mineralogy.

PARASITOLOGY

FIELDS OF STUDY

Biology; internal medicine; ecology; infectious disease; public health; evolutionary biology; microbiology; pharmacology; zoology; bacteriology.

SUMMARY

Parasitology is the study of parasites and the relationship between parasites and host organisms. The primary interest in parasitology is to investigate the role of parasites in diseases that affect humans, livestock, and pets. Parasitologists work closely with medical professionals and pharmacologists to develop drugs that combat parasitic infections. Parasitology became a distinct branch of medicine and biology in the mid-nineteenth century after the development of microscope technology. Although medical parasitology dominates the field, some ecologists and zoologists are studying parasites to learn more about the role parasitism plays in evolution and ecology.

KEY TERMS AND CONCEPTS

- **Bacterium:** Single-celled organism that lacks a defined nucleus within its cell body; also called prokaryote.
- **Ectoparasite:** Parasitic organism that lives outside or on the surface of the host organism.
- **Endoparasite:** Parasitic organism that lives within the host organism.
- **Helminth:** Parasitic worm, such as a roundworm or tapeworm, in the intestines of vertebrates.
- **Host:** Organism that carries a parasite or organism and is the target organism for a parasite.
- **Protozoan:** Any of a diverse group of single-celled organisms from the eukaryotic group that has a nucleus contained within a discrete chamber inside the cell.
- **Vector:** Intermediary organism that delivers a parasite or parasitic reproductive stage to the ultimate host.

DEFINITION AND BASIC PRINCIPLES

Parasitology is the study of parasite organisms, their relationship with host organisms, and their role in disease. Parasitism is defined as a prolonged and intimate association between two organisms in which one organism (the parasite) benefits at the expense of the other organism (the host). Scientists have identified thousands of parasites from a variety of groups, including bacteria, protozoa, animals, plants, and fungi. The study of parasitic bacteria is covered under bacteriology, while parasitologists concentrate on parasites from the eukaryotic group, which includes protozoa, animals, fungi, and plants.

Parasites infect their hosts for the purpose of obtaining food or completing a portion of their reproductive cycle. They may infect their host by either attaching to the outside of the host's body, entering an existing opening such as the mouth or anus, or tunneling through the host's tissues. Some parasites are specialized to infect a single type of host organism, and others can parasitize organisms from a variety of species. Some parasites display complex life cycles that may include infecting hosts from different species during specific parts of their life cycles. Humans are vulnerable to infection by at least three hundred types of parasitic worms and more than seventy species of protozoa.

A key feature of parasitism is that it causes damage to the host. Some parasites reduce the host's ability to obtain or absorb nutrients, while others damage tissues directly. Some parasites are known to cause disease, such as malaria and Lyme disease.

BACKGROUND AND HISTORY

Early civilizations, including the ancient Greeks, kept records of patients suffering from infection by large, easily visible parasites such as the tapeworm. However, scientists were not able to examine parasites in detail until the invention of the microscope in the mid-seventeenth century. The next major advance was the discovery of bacteria in the late nineteenth century, which precipitated the discovery of the link between parasites and disease.

In the early twentieth century, scientists conducted the first detailed studies into the nature of malaria, giardiasis, and many other diseases caused by parasites. Although scientists have been unable to develop vaccines for malaria, sleeping sickness, and many other types of parasite-related diseases, researchers have been able to drastically reduce instances of infection because of a greater understanding of the life cycles of the parasites and the vectors (dispersal organisms) they use. In addition, cooperation between parasitologists and pharmacologists led to the development of medications that can effectively reduce the intensity of parasitic infections.

Although the medical study of parasitism remains the most active facet of the field, ecologists and evolutionary biologists began studying the role of parasites in nature in the 1980's, ushering in a new age of parasitology research. Studying the role of parasites in nature has also helped further the study of medical and agricultural parasitology.

How It Works

Parasitology is divided into three major branches of study: helminthology (parasitic worms), entomology (insect parasites), and protozoology (parasitic protozoa). Each field is further divided into specialties, of which the most important are medical, agricultural, and ecological parasitology.

Medical Parasitology. Medical parasitologists focus on parasites that cause disease and infection in humans. They work closely with pharmacologists in the development of vaccines and antiparasite drugs and with physicians to develop new therapeutic techniques.

Parasitologists also collect information that is used to create guidelines for diagnosis. This includes documenting symptoms common to each type of parasitic infection and cataloging the life stages of parasitic species. Intestinal parasites may cause diarrhea, intestinal pain, and the development of granulomas, or tumorlike masses. Other parasites can cause muscle and joint pain, fatigue, and various skin lesions and rashes.

In most cases, patients with parasitic infections are treated with drugs that function by differential toxicity, meaning that the chemicals in question are more toxic to the parasites than to the host. However, most antiparasite medications are also somewhat toxic to the host and can cause a variety of side effects.

Agricultural and Veterinary Parasitology. The subfields of agricultural and veterinary parasitology are extremely important from an economic perspective. A 2009 study from India estimated that crop parasites cause losses amounting to $200 billion annually worldwide, while an additional $3.5 billion is spent combating animal parasites. Some parasites, such as the parasitic roundworm, are capable of infecting plants, animals, and humans.

In many cases, parasites lead to the death of livestock or crop plants. In addition, research has shown that even minor infections can affect an animal's metabolism in such a way that the animal will never obtain full growth or development. Parasites therefore lead to lower yields from both crops and livestock.

Agricultural parasitologists perform a variety of laboratory experiments to develop antiparasite drugs and vaccines. Medications developed in agricultural laboratories are sometimes co-opted by medical parasitologists to create medications used to treat human patients.

Ecological Parasitology. Ecological parasitologists investigate the role of parasites in nature. This includes making field observations of the various methods parasites use to survive in their environments and their ultimate effect on the evolution of species.

Many evolutionary biologists believe that parasitism has had an important influence on the evolution of species. In a 2009 study of a species of snail from New Zealand, researchers found that in populations with high levels of parasites, the snails will switch from asexual reproduction (producing clones) to sexual reproduction, which produces genetically diverse offspring that are more resistant to infection. This discovery led the researchers to speculate that parasites may have been instrumental in the evolution of sexual reproduction.

Applications and Products

The development of antiparasite drugs involves research from all branches of parasitology as well as organic chemistry, pharmacology, and infectious disease medicine. Some antiparasite medicines have been found to have effects on nonparasitic illnesses, leading to greater integration between parasitology and other branches of medicine.

Malaria. Malaria is one of the most common and widespread diseases caused by parasite infection. According to the Centers for Disease Control and Prevention (CDC), between 350,000 and 500,000 cases of malaria are reported annually, and malaria has one of the highest fatality rates of all diseases. Malaria is caused by plasmodium, a parasite that is typically spread through mosquito bites.

The typical strategy for treating malaria involves using an array of antimicrobial drugs designed to kill plasmodium as it develops. Treatment is difficult because plasmodia have tremendous genetic diversity and many strains are resistant to existing treatments. In 2009, a team at the Monash University ARC Center announced the results of a set of experiments indicating that it may soon be possible to use a drug to deactivate an enzyme essential for the malaria parasite to digest nutrients. Deprived of this key enzyme, the malaria parasite will starve inside its host.

Parasite Prevention. In addition to treating parasites with drugs, researchers also focus on preventing infection through better hygiene and behavioral modification. In the case of malaria, controlling mosquito populations is an essential step toward preventing the spread of malaria. Parasitologists also contribute to the development of antibug sprays, mosquito netting, and other types of insect repellents in an effort to prevent parasite infection.

Another area of research in parasite prevention involves finding ways to prevent parasites from spreading through food or water. Ultraviolet (UV) light treatments have been effective in disinfecting food and water and killing parasites in various stages of life. Although UV radiation has often been used to disinfect foods, water treatment using UV radiation has just begun to become widespread.

In 2009, New York City became the first city in the United States to mandate UV light treatment for water processing in an effort to eliminate cryptosporidium, a parasitic microbe that causes a diarrheal infection known as cryptosporidiosis. Before the process was introduced in the United States, it had been used effectively in Britain to reduce instances of cryptosporidium infection.

Impact on Other Medical Fields. Discoveries from parasitology have filtered into other areas of research and development. For example, the physiological reaction to parasite infection gives important clues about the function of the immune system and the development of resistance. In some cases, medications developed to combat parasite infection have been found to have beneficial therapeutic effects for other types of diseases. In 2008, for instance, it was discovered that miltefosine, a drug used to treat patients suffering from protozoan infections, could potentially be effective in treating people infected with the human immunodeficiency virus (HIV). Miltefosine works by preventing the development of macrophage reservoirs, cells that, because of their long life, allow the virus to proliferate in secret before it becomes detectable.

IMPACT ON INDUSTRY

Parasitic diseases have a significant economic impact on society. Although diseases such as malaria are uncommon in developed nations, populations in developing countries still suffer from frequent outbreaks of malaria and other parasite-related diseases. In the East African nation of Burundi, economists estimate that malaria and other parasitic diseases stunt the nation's economic growth by at least 1.5 percent each year because of lost worker productivity. Because of the lack of financial incentive, large pharmaceutical companies invest relatively little in diseases that primarily affect developing nations.

Charitable organizations and medical institutions provide funding for research into many parasitic diseases. Organizations such as the Sandler Center for Basic Research in Parasitic Diseases, a collaborative effort between physicians representing several universities in the San Francisco Bay area, have formed in an effort to create drugs to treat parasite-related diseases affecting developing nations.

In the United States, the National Science Foundation and National Institutes of Health offer grants for parasitological research. In 2006, the National Institutes of Health awarded Yale University researchers $5.4 million to study cutaneous leishmaniasis, a parasitic disease spread by female sand flies. A variety of private organizations also support parasitology research. In 2009, the Bill and Melinda Gates Foundation announced a series of seventy-six $100,000 grants aimed at combating some of the major issues in society, at least one of which was slated for research into malaria prevention.

CAREERS AND COURSE WORK

For those interested in pursuing a career in parasitology, a strong background in biology is essential. However, individuals from many fields may contribute to parasitology. Because parasites are such a common feature in nature, numerous career opportunities are available for those wishing to study parasitism.

Most professional parasitologists have backgrounds in biology, zoology, and medicine. Ecological

Fascinating Facts About Parasitology

- There have been four Nobel Prizes awarded for work pertaining to the treatment of malaria.
- The parasitic isopod *Cymothoa exigua* infects fish by eating the fish's tongue. The worm then attaches to the fish's mouth and functions as a replacement tongue, living off scraps from the fish's meals.
- One of the methods being tested for fighting malaria involves using chocolate, which sticks to the blood fats on which the parasites feed and thereby weakens them.
- The filarial worm, a parasite that burrows into an animal's organ systems and causes damage, is recognized as the second leading cause of permanent and long-term disability in the world.
- In a 2006 study from the Imperial College of London, researchers found a potential link between toxoplasmosis, a parasite found in cat feces, and the family of mental disorders known as schizophrenia.
- A species of parasitic hairworm infects the brain of a grasshopper and influences the insect to commit suicide by leaping into open water. Once in the water, the parasite emerges from its host and morphs into its aquatic adult form.

parasitology involves specialists with backgrounds in evolutionary theory, taxidermy, and ecology. Medical parasitology invites contributions from those with backgrounds in infectious disease, public health, internal medicine, pharmacology, and medical statistics.

Alternatively, individuals may decide to study the effects of parasitism on society and culture. A full appreciation of parasitism involves economic analyses, sociological studies, and a variety of social and medical activities.

SOCIAL CONTEXT AND FUTURE PROSPECTS

With millions of people worldwide suffering from parasitic diseases and many more living with the threat of infection, the study of parasitism is vital to public health. Given the tremendous economic and social effect of parasitic diseases, research in parasitology has the potential to create major changes around the world.

Many parasitologists believe that the future of parasitology depends on developments in genomics and genetic medicine. Scientists have been examining the genetic components of parasitic organisms with the goal of finding more effective treatments for parasite-related diseases. The next generation of antiparasite medications may be genetically tailored to combat parasite organisms, thereby reducing the unpleasant side effects caused by differential toxicity.

Micah L. Issitt, B.S.

FURTHER READING

De Kruif, Paul. *Microbe Hunters*. 1926. Reprint. New York: Harcourt, Brace, 2006. A classic work that introduces some aspects of microbiology and bacteriology and contains some information about parasitology and eukaryotic microbes that cause disease as well as interesting coverage of malaria.

Esch, Gerald W. *Parasites, People and Places: Essays on Field Parasitology*. New York: Cambridge University Press, 2004. Anecdotal accounts of parasitology illustrate many aspects of the field and cover elements of the history of parasitology as well as medical and evolutionary investigations of parasite behavior.

John, David T., et al. *Markell and Voge's Medical Parasitology*. 9th ed. St. Louis, Mo.: Saunders Elsevier, 2006. Covers the basics of medical parasitology from treatments and pharmacology to genomic research. Written for readers with a strong medical or biological background.

Moore, Janice. *Parasites and the Behavior of Animals*. 2002. Reprint. New York: Oxford University Press, 2005. An interesting look at a variety of parasites and other animals, written for the general reader; contains information about medical, agricultural, and ecological parasitology.

Zimmer, Carl. *Parasite Rex: Inside the Bizarre World of Nature's Most Dangerous Creatures*. New York: Simon and Schuster, 2001. An account of interesting parasites, their ecology, evolution, and effect on culture and society; covers medical issues, parasite ecology, and other facets of the field.

Zuk, Marlene. *Riddled with Life: Friendly Worms, Ladybug Sex, and the Parasites That Make Us Who We Are*. Fort Washington, Pa.: Harvest Books, 2008. This introduction to parasites that affect humans in both negative and negligible ways focuses on the evolutionary aspect of parasitism and also contains information on medical and agricultural parasitology.

WEB SITES
Centers for Disease Control and Prevention
A-Z Index of Parasitic Diseases
http://www.cdc.gov/ncidod/dpd/parasites/index.
htm

U.S. Department of Agriculture, Food Safety and
Inspection Service
Parasites and Foodborne Illness
http://www.fsis.usda.gov/factsheets/parasites_and_
foodborne_illness/index.asp

See also: Agricultural Science; Genetic Engineering; Genomics; Immunology and Vaccination; Pharmacology; Virology.

PATHOLOGY

FIELDS OF STUDY

Biology; chemistry; biochemistry; immunology; chemical pathology; clinical pathology; forensic pathology; anatomic pathology; surgical pathology; blood banking; hematology; histology; cytology; genetics; microbiology; toxicology.

SUMMARY

Pathology is the scientific study of the nature of disease and its causes, processes, development, consequences, and resolution. Pathologists examine tissue and body fluids to determine if disease is present and, if so, its nature and extent. Many decisions on whether and how to treat disease are based on results of tests delivered by pathologists as laboratory reports. Pathologists also work to determine the cause of death in questionable or puzzling circumstances. This field is changing rapidly as new technologies that allow disease to be found in earlier stages are developed and implemented. Genetic mutations and diseases caused by them are also an emerging pathology field.

KEY TERMS AND CONCEPTS

- **Atrophy:** Wasting away of an organ or tissue because of disease or injury.
- **Biopsy:** Removal and examination of tissue.
- **Degeneration:** Deterioration or loss of function.
- **Hyperplasia:** Abnormal increase in number of cells.
- **Hypertrophy:** Abnormal increase in size of an organ or tissue.
- **Inflammation:** Response of tissue to injury or infection.
- **Positive Predictive Value:** Likelihood that the positive results of a test are reliably positive; a high positive predictive value means the test is reliable, while a low positive predictive value means further testing should be pursued.
- **Reagent:** Substance used to create a chemical reaction that detects, measures, or produces another substance.
- **Sensitivity:** Proportion of individuals who will correctly test positive when testing for a particular disease; if a test has high sensitivity, it will most likely recognize all those tested who truly have the disease.
- **Specificity:** Probability that a test will correctly identify those who do not have a specific disease; a test with high specificity can be used to rule out that disease.

DEFINITION AND BASIC PRINCIPLES

Pathology is the study of disease and involves examining biopsied tissues to determine whether atrophy, degeneration, hyperplasia, hypertrophy, or inflammation has occurred and, if so, to what extent. Pathologists also examine body fluids for signs of disease. They interpret the results of biopsies and other tests and send them to the practicing clinician who requested the test and will pass those results on to the patient. There are many subspecialties of pathology that focus on specific types of testing or specific body tissues or fluids.

Genetic testing is another area in which pathologists are involved. They examine a patient's genetic materials to determine whether there are any genetic mutations that are likely to be passed on to the patient's children or that could eventually result in a genetic disease. For example, examining a woman's genetic material for the presence of a gene associated with breast cancer could help that woman determine whether her personal risk of breast cancer is increased and decide whether she will make lifestyle changes that may help her avoid breast cancer.

Pathologists are often involved in research studies. Volunteers donate tissues or body fluids to help these scientists study the effects of long-term behaviors, such as smoking or exercise, on body tissues to help understand diseases and the disease process. Through research and development, pathologists may create better laboratory tests.

Pathologists are medical doctors and are very involved in determining how a patient's disease should be treated, although they are usually not involved in seeing patients in a clinical environment. They are much more likely to be involved in patient care through testing and research.

BACKGROUND AND HISTORY

Disease and its causes have been poorly understood throughout history. In many Western societies, autopsies were prohibited for religious reasons until the late Middle Ages, and scientists and doctors knew next to nothing about how death occurred and how diseases could affect body tissues and fluids at the cell level. In 1761, Italian anatomist Giovanni Battista Morgagni published the first book to discuss diseases in individual organs. Not until the mid-nineteenth century did German physician Rudolf Virchow's theories of cell-based disease replace the archaic notion that humors caused infections. That same century, German physician Robert Koch and French scientist Louis Pasteur theorized that bacteria caused some diseases.

Because pathology is based on the study of tissues and fluids, its history is closely tied to that of the microscope. Until the microscope was improved in the nineteenth century, it was impossible for scientists to closely examine tissues to determine whether disease was present and, if it was, what effect it had on the body. Advances in microscopic technology, including the electron microscope; new fields of study such as immunohistochemistry and molecular biology; and methods of preparing tissues such as staining and culturing have greatly improved pathologists' ability to study disease and disease processes. Scientists have come to understand that all diseases are reflected by changes that extend down to the molecular level, and testing methods such as polymerase chain reaction (PCR), fluorescence in situ hybridization (FISH), and mass spectrometry have changed the way laboratory tests are conducted.

HOW IT WORKS

The process of analyzing tissue or a body fluid depends on the type of tissue or fluid being examined and the reason for testing it. Usually a sample is taken in a clinical setting and prepared for delivery to a laboratory. For example, if the sample is blood, it may be drawn at a clinic, then placed in special tubes with a substance that keeps the blood from clotting.

When the sample arrives at a laboratory, it is prepared for the selected test. For example, tissue may be sliced very thinly, placed on a microscope slide, and stained so that a technician can examine it, or blood may be spun in a centrifuge to separate it into its components, then different reagents may be added to the blood to cause a chemical reaction that will help the technician determine whether the substance for which he or she is looking is in the blood sample. Another type of processing is that used for genetic material: A small sample of the genetic material is replicated many times to make it easier to determine if the sample contains errors. The sample is then processed, either automatically with specialized machines that determine the test results or by trained workers who examine the specimen and determine the results.

The test results are compared with values that have been determined to be normal, and if the results fall out of the normal range, they are further interpreted. For example, if a pregnant woman takes a standard blood test and some test values fall outside of the normal range for a woman who is not pregnant, the test values must be compared to what is normal for a woman in that particular stage of pregnancy.

After the results are interpreted, the laboratory sends a report containing the test values and an interpretation of the results to the clinician who ordered the test so that the results can be shared with the patient. The results often play a central role in determining the course of treatment.

APPLICATIONS AND PRODUCTS

Although pathology generally refers to laboratory testing, the science is usually broken down into the following specialties: Anatomic pathology, chemical pathology, forensic pathology, genetic pathology, hematology, immunology, and microbiology.

Anatomic Pathology. Sometimes called surgical pathology, anatomic pathology uses biopsied tissue from a person (living or dead) to diagnose disease (or a possible cause of death). A common example of this type of test is a Pap smear, which tests for cervical cancer. Another example is the examination of tissues for the presence of disease during surgery. A surgeon removes a tumor, and a pathologist examines its edges, or margins, to determine if the surgeon removed the entire tumor. If the pathologist does not see healthy tissue in all the margins, he or she informs the surgeon, who removes additional tissue. This process continues until the pathologist and surgeon are convinced that the tumor is fully excised. Subcategories of this specialty include histology (preparing tissues for examination), cytology (performing tests on tissues to determine if cancer exists), and forensic pathology (performing autopsies and analyzing tissues to determine the cause of death).

Chemical Pathology. Also called biochemistry, chemical pathology detects substances and changes in blood and bodily fluids. An example of this type of testing would be examining blood glucose levels to determine risk factors for diabetes or its progression. This type of testing also detects enzymes and proteins in blood that may change with the progression of illness or measures cancer tumor markers that show whether a tumor is increasing or in remission. A subspecialty is toxicology, which involves testing to look for poisons, drugs, or other toxins in the body.

Forensic Pathology. A subspecialty of anatomic pathology, forensic pathology focuses on investigating cases of sudden or unexpected death. Forensic pathologists perform autopsies to identify the cause of death and examine tissues and body fluids to reconstruct how the death occurred. This type of pathologist may be required to visit crime scenes or help law enforcement personnel in other ways.

Genetic Pathology. In genetic pathology, chromosomes and DNA are tested to diagnose diseases that may have genetic components. Examples include a chromosomal test to determine whether a fetus has Down syndrome and an evaluation of a DNA specimen to determine whether a woman has a gene associated with breast cancer. Subcategories include biochemical genetics (identifying specific genetic markers with biochemical testing), cytogenetics (performing an analysis of chromosomal abnormalities using a microscope), and molecular genetics (analyzing mutations in the DNA in genes).

Hematology. Pathological hematology focuses on diseases that affect blood and the organs that create blood. For example, blood could be tested to see if the patient has a clotting disorder. This specialty includes transfusion medicine, which involves performing blood typing and compatibility testing and managing the supply of blood products to ensure that blood is safe for transfusion into patients.

Immunology. Pathological immunology focuses on allergies, inflammations, and autoimmune diseases. For example, a patient's blood could be tested to determine if he or she is allergic to something and to identify the allergen. These tests can also determine if the immune system is malfunctioning, as in diseases such as lupus or multiple sclerosis, where the immune system targets normal systems as allergens and attempts to destroy the normal system's tissues.

Microbiology. Testing in pathological microbiology involves diseases caused by infectious agents. For example, a patient's urine sample could be tested to see if the individual has a urinary tract infection. Tests look for the presence of bacteria, fungi, parasites, and viruses. Pathologists may work with public health officials and be involved in efforts to control disease outbreaks, or they may attempt to solve problems related to drug-resistant bacteria.

Other Applications. Some pathologists practice as clinical pathologists. They may work in a rural area or with a community hospital and are usually trained in chemical pathology, hematology (including blood typing), and microbiology but usually do not practice anatomic pathology. General pathologists are trained in all areas of pathology (including anatomic) but at a lesser depth of knowledge, so if necessary, they refer cases to specialized pathologists. Another type of pathologist is a specialist in a body system, such as a dermatopathologist, who specializes in diseases of the skin, or a nephropathologist, who focuses on the renal system.

Pathologists are involved in research and development. They may develop laboratory tests that have more specificity, more sensitivity, better positive predictive value, or a faster turnaround time, or work on reducing the cost of testing. They also may conduct fundamental research, such as evaluating tissues or body fluid samples from patients with a specific disease and patients from a control group to determine similarities and differences.

IMPACT ON INDUSTRY

Clinical courses of treatment are often set into motion based on the results of a laboratory test; some estimates have determined that 80 percent of the decisions a clinician makes about a patient's treatment are based on laboratory test results. Therefore, pathologists and their interpretations of laboratory test results have a huge impact on the treatment that patients receive. This, in turn, has an enormous impact on health care in general. As health care becomes more and more personalized, pathologists must work toward creating and interpreting testing that generates results specific to each patient. This personalized type of medical treatment can improve outcomes for all patients by determining a course of care that has the highest likelihood of success.

Other ways in which pathologists influence the health care industry in general and costs of care in particular are in the research and development of new tests. These new tests can help improve the end results for the patient by continuous improvement of sensitivity, specificity, positive predictive value, and turnaround time. Pathologists can also help lower health care costs by working toward better use of laboratory testing. For example, they can make recommendations or educate clinicians to ensure that the right tests are ordered and at the appropriate times in the patient's treatment so as to ensure the most accurate result, which will enable the clinician to provide the best course of treatment.

CAREERS AND COURSE WORK

Becoming a pathologist requires graduation from medical school and the completion of a specialized course of study in a subfield. However, there are many jobs in this field that do not require an advanced degree.

Specimen processors prepare a sample for testing. This preparation may include adding reagents to a sample, preparing slides for examination, or spinning blood into components before testing and analysis. This position usually provides on-the-job training and may require only a high school diploma.

Medical laboratory technicians may perform moderately complex testing under the direction of a medical technologist. This position requires a two-year degree and a license in the field.

Medical technologists perform highly complex laboratory tests under the supervision of a medical director. They may specialize in various components of laboratory work, such as quality assurance, validation, or training, or they may specialize in a particular field, such as histotechnology or cytotechnology. This posiotion requires a bachelor of science degree in medical technology and a license in the field.

The relationship between a pathologist's assistant and a pathologist is similar to that between a physician's assistant and a physician. This position requires a master's degree in the field. A pathologist's assistant, with the pathologist's supervision, can perform many of the duties that were formerly performed only by a pathologist.

A medical director supervises and certifies laboratory testing. Like a pathologist, a medical director

---

Fascinating Facts About Pathology

- Forensic pathologists are said to speak for the dead. Their investigations tell the story of a person's death after that individual can no longer speak.
- The popular television series *CSI: Crime Scene Investigations*, which features the work of forensic evidence investigators, including pathologists, premiered in 2000 and spawned several spinoffs and imitators.
- Laboratories in the United States perform more than 10 billion tests each year.
- More than 2,000 laboratory tests can be performed on blood and body fluids alone.
- More than 99.9 percent of the samples that pathologists analyze are from living people, not from autopsies of corpses.
- In its 2006 annual report, the Los Angeles County Department of Coroner reported performing 4,401 complete and 450 partial autopsies. It also conducted 5,499 toxicology studies.
- Veterinary pathologists may participate in drug development as they monitor and describe the effects of drugs on laboratory animals.
- Environmental pathology deals with diseases resulting from environmental factors, such as chemicals.
- Phytopathology is the study of plant diseases.

---

must have a medical degree and specialized training in one of the subgroups of pathology. A researcher who creates new tests or performs traditional research usually must have a medical degree or a doctorate in a specialty in the field of pathology.

Blood banking is another area often associated with pathology and its subspecialty of hematology. Many laboratory tests must be performed on donated blood to ensure that it is safe for transfusion into patients. Jobs for phlebotomists, who draw blood from donors and patients, may be available at blood-banking facilities; these jobs often provide on-the-job training and require a high school diploma.

Another area of pathology that is growing quickly is veterinary pathology, which studies disease and its process in animals. Positions in veterinary pathology are very similar to those in human pathology. A pathologist in this field is usually a veterinarian who has specialized in pathology.

SOCIAL CONTEXT AND FUTURE PROSPECTS

The aging population in the United States is creating an increased need for laboratory testing to diagnose disease and determine an appropriate course of treatment. Pathologists can help reduce the rising cost of health care by identifying which tests are the most cost-effective to perform and which are unlikely to help determine a clinical path of treatment. For example, in men over the age of eighty-five, many studies have shown that the negatives (loss of continence and function) outweigh the positives of treatment for prostate cancer and, therefore, recommend no treatment for these men. Pathologists can help write and publicize guidelines for clinicians so that testing is not performed when the treatment does not vary depending on the outcome of the test. Pathologists can thus shape health care policies and practices that are cost-effective and improve patient care outcomes.

Genetic testing is another area in which pathologists can help guide informed decision making. This type of testing can help a couple with a history of family genetic problems determine their risk of having a child with genetic mutations. For example, a man and woman with a family history of cystic fibrosis can be tested to determine if they carry the gene that causes that disease. The results allow pathologists to calculate the couple's risk of conceiving a child who will carry that gene or who will develop the disease. Knowing the calculated risk, the couple can determine whether they want to have a child naturally or to investigate other options. Genetic testing on a fetus can show whether it has genes or genetic mutations that are likely to cause diseases, thus allowing the parents to make an informed decision as to whether to continue or terminate the pregnancy. If the decision is made to continue the pregnancy, the parents and health care workers can be better prepared to deal with the altered circumstances and health issues.

As health care becomes more personalized, a pathologist can interpret test data as it pertains to a particular patient and that patient's family history. For example, patients who have a family history of alpha-1-antitrypsin deficiency, a disease that has genetic components, could be tested to see whether they have inherited those genes, and if so, which level of lung disability they might be likely to have. This could affect lifestyle decisions, such as not smoking, not working in an industry that might negatively affect their lungs, or avoiding being treated for asthmalike symptoms with drugs that do not help this disease.

Another application of pathology to personalized health care is that of obtaining detailed information about the disease or condition that a patient has to determine the best course of treatment. For example, breast cancer is thought to be the result of several different disease processes, and the most effective treatment depends on the process involved. Identifying the type of breast cancer allows physicians to prescribe the most appropriate type of treatment. For example, if the patient's type of cancer does not respond well to chemotherapy, the doctor might recommend surgery.

Marianne M. Madsen, M.S.

FURTHER READING

Damjanov, Ivan. *Pathology for the Health Professions.* 4th ed. Maryland Heights, Mo.: Elsevier/Saunders, 2012. A basic overview of pathology with clear pictures and review questions; discusses normal pathology and diseases in various systems.

_____. *Pathology Secrets.* 3d ed. Philadelphia: Mosby/Elsevier, 2009. A basic review in question-and-answer format, focusing on practical knowledge.

Hayes, A. Wallace. *Principles and Methods of Toxicology.* 5th ed. Boca Raton, Fla.: CRC Press, 2008. Includes history of toxicology and discusses interpretation of data and problems that may arise. Comprehensive glossary.

Kemp, William L., Dennis K. Burns, and Travis G. Brown. *Pathology: The Big Picture.* New York: McGraw-Hill Medical, 2008. Focuses on broad pathology concepts. Contains full-color illustrations, summary tables and figures, and questions and answers.

Kumar, Vinay, et al. *Robbins and Cotran Pathologic Basis of Disease.* 8th ed. Philadelphia: Saunders/Elsevier, 2010. Regarded by some as the keystone book and used as the primary text in pathology at many medical schools. Includes illustrations and case studies.

Richards, Ira S. *Principles and Practice of Toxicology in Public Health.* Sudbury, Mass.: Jones & Bartlett, 2008. An overview of toxicology in public health practice, written in an easy-to-read style. Includes glossary and index.

Rubin, Raphael, and David S. Strayer, eds. *Rubin's Pathology: Clinicopathologic Foundations of Medicine.* 6th ed. Philadelphia: Lippincott Williams & Wilkins, 2011. Award-winning book taking a clinical approach to pathology. Includes full-color, enhanced illustrations.

Zaher, Aiman. *Pathology Made Ridiculously Simple.* Miami: MedMaster, 2007. A book of mnemonics and cartoons to aid in memory and retention.

WEB SITES

American Pathology Association
https://www.apfconnect.org

American Society for Clinical Pathology
http://www.ascp.org

Association for Molecular Pathology
http://www.amp.org

College of American Pathologists
Lab Tests Online
http://www.labtestsonline.org

See also: DNA Analysis; Hematology; Immunology and Vaccination; Parasitology; Veterinary Science; Virology.

PEDIATRIC MEDICINE AND SURGERY

FIELDS OF STUDY

Laparoscopic surgery; medicine; neonatology; pediatric endocrinology; pediatric medicine; pediatric nephrology; pediatric neurology; pediatric oncology; surgery.

SUMMARY

Pediatric medicine focuses on the diagnosis and medical treatment of diseases in infants and children. Doctors in this specialty are known as pediatricians. Pediatric surgery focuses on the surgical treatment of diseases in infants and children. Sometimes the diagnosis is made by a pediatrician who then refers the patient to a pediatric surgeon. On other occasions, the pediatric surgeon makes the diagnosis and then performs the surgery. Subspecialties exist within both the medical and surgical fields. For example, some pediatricians focus on pediatric endocrinology (endocrine glands) and some specialize in pediatric oncology (cancer treatment). Some pediatricians focus on a certain age group, such as adolescents, and others specialize in neonatology, the medical treatment of newborns.

KEY TERMS AND CONCEPTS

- **Incubator:** Specialized container for premature or sick infants that provides controlled temperature and humidity as well as an oxygen supply.
- **Laparoscopic Surgery:** Minimally invasive surgery that is accomplished with a laparoscope and the use of specialized instruments through small incisions and leads to a quicker recovery.
- **Neonatal Intensive Care Unit (NICU):** Intensive care unit for high-risk, often premature infants, who are suffering from life-threatening problems.
- **Neonatologist:** Pediatrician with specialized training in neonatology—the care of high-risk, often premature, infants.
- **Pediatric Endocrinologist:** Pediatrician with specialized training in endocrine conditions, such as diabetes, thyroid disorders, and problems with growth.
- **Pediatrician:** Physician who has specialized training in the medical care of infants and children.
- **Pediatric Neurologist:** Pediatrician with specialized training in neurologic disorders, such as epilepsy, brain tumors, and autism.
- **Pediatric Oncologist:** Pediatrician with specialized training in oncology— the treatment of malignant tumors.
- **Pediatric Surgeon:** Surgeon who limits his or her practice to surgical procedures on infants and children.
- **Ventilator:** Machine that mechanically moves air in and out of the lungs.

DEFINITION AND BASIC PRINCIPLES

Pediatric medicine is a medical specialty focused on the diagnosis and treatment of children from infancy through adolescence. Pediatric surgery is a surgical specialty focused on surgery for children. A pediatrician receives specialized training in pediatric medicine after completing medical school; a pediatric surgeon receives training in general surgery plus additional training in pediatric surgery.

Subspecialties exist in both the medical and surgical fields. For example, a pediatrician might specialize in adolescents and a pediatric surgeon might specialize in pediatric cardiothoracic surgery. Some medical and surgical problems are similar to those of adults (such as pneumonia and appendicitis); however, others are unique to children (such as a cardiac defect, which if not corrected will result in childhood death).

Preventive health care is a significant component of pediatric care. Immunizations are scheduled for diseases such as diphtheria, pertussis (whooping cough), and tetanus. The immunization is given in a combined injection known as DPT. Congenital disorders often appear at birth or within the first few years of life. Some, such as phenylketonuria, which causes mental retardation, can be corrected if recognized early. Others disorders, such as cystic fibrosis, cannot be cured; however, the patient can live longer and more healthily if the disease is recognized early and treatment is initiated.

An important and often underemphasized role of the pediatrician is that of educating parents so that their child can develop to his or her full potential—mentally, physically, and emotionally.

BACKGROUND AND HISTORY

Pediatric medicine arose as a medical specialty in the United States in 1861. Before that time, children's health care was included within fields such as general medicine, obstetrics, and midwifery. The German physician Abraham Jacobi emigrated to the United States in 1853 and established a training program at New York Medical College that focused on the diseases of infants and children. He published articles in medical journals and developed children's wards in several New York hospitals. In 1933, the American Board of Pediatrics was founded by a group of thirty-five pediatricians. Paralleling the United States in the development of pediatric medicine was Great Britain, where the Hospital for Sick Children was established in London in 1852.

Pediatric surgery arose as a specialty in the mid-twentieth century. Initially, it was focused on the correction of congenital defects, and as of 2011 birth-defect correction represents a significant portion of the specialty. One of the innovators in this field was C. Everett Koop, surgeon-in-chief at Children's Hospital of Philadelphia for a number of years. Koop performed many surgical milestones. Beginning in 1946, Koop and his team developed newer general anesthesia techniques that allowed for the surgical repair of previously untreatable congenital defects. In 1956, he had established the first neonatal surgical intensive care unit at the Children's Hospital of Philadelphia. In 1957, he and his team performed the first separation of conjoined ("Siamese") twins. Koop went on to become U.S. surgeon general under presidents Ronald Reagan and George H. W. Bush, from 1982 to 1989.

HOW IT WORKS

In the mid-twentieth century, pediatric care often began with a newborn examination shortly after birth. As of 2011, pediatric care often begins before conception. Women are advised via media sources or a health care provider to prepare for pregnancy and improve the chances for a healthy infant. This advice includes smoking cessation and supplementing the diet with folic acid, which decreases the incidence of neural tube defects, such as spina bifida.

During the pregnancy, ultrasound examinations are periodically performed, which can readily identify internal anatomical features, such as a kidney abnormality or a heart defect. If an abnormality is found, health care professionals can prepare for any special needs at the time of birth. Furthermore, if indicated, both medical and surgical treatment can be initiated. This treatment is coordinated between the obstetrician and other specialists, such as perinatologists, neonatologists, and pediatric surgeons. A perinatologist is an obstetrician with specialized training in high-risk pregnancy. His or her goal is to obtain a good outcome for both the mother and her developing infant. A neonatologist is a pediatrician with specialized training in the care of high-risk infants. A pediatric surgeon has specialized training in surgery on infants and children. In some cases, surgery is performed before birth. The uterus is incised, the surgery is performed, and the uterine incision is closed. This type of surgery is indicated when an abnormality is likely to cause death of the fetus before birth or delivery complications.

Pediatric Medicine. From the early twentieth century through 1975, an initial pediatric physical exam was conducted shortly after birth. With the introduction of ultrasound in the 1970's, the first newborn exam was conducted via a prenatal ultrasound examination. Most women receiving prenatal care in developed nations, such as the United States, undergo one or more ultrasound examinations during their pregnancy. For many women, this may be the most thorough exam she will ever have. A pediatrician or a neonatologist attends deliveries with increased risk (such as cesarean sections or multiple births). Most infants breathe spontaneously at birth; however, sometimes resuscitation is needed. If a pediatrician or neonatologist is present, that person will conduct the resuscitation. Sometimes an infant will unexpectedly need resuscitation. In that case, the obstetrician or delivery-room nurse (usually a registered nurse, RN) will perform the resuscitation. After birth, high-risk infants begin their pediatric care with a neonatologist. This care is given in a specialized intensive care unit known as a neonatal intensive care unit (NICU). The infant is placed in an incubator, which supplies controlled temperature and humidity as well as oxygen. Some infants require a ventilator, which mechanically assists their breathing and instills oxygen into the lungs.

In most cases, an infant is born healthy without

any special needs. All infants born in a hospital will receive a newborn examination before they are discharged. Follow-up care will be arranged with a pediatrician or family physician. Immunizations for childhood diseases will be scheduled, and at each visit, the child's health will be assessed. Usually, a growth chart is begun, which plots weight and height. This information not only allows the determination of proper growth and development but also can predict adult height. In many midsize and large hospitals, a separate section of the hospital is dedicated for pediatric patients. Metropolitan areas often have separate hospitals devoted to pediatric patients; however, some others do not offer pediatric care at all.

Pediatric Surgery. Since it arose as a specialty in the mid-twentieth century, a major portion of pediatric surgery is devoted to correction of congenital malformations. Some require immediate attention after birth, and others can be delayed for months or years. Those that require immediate attention include abdominal wall defects such as gastroschisis and omphalocele. With this condition, a portion of the abdominal contents, primarily intestines, protrude through a defect in the abdominal wall. Less serious abdominal wall defects in which repair can be postponed include hernias and undescended testes. Defects of the digestive tract vary in the degree of urgency; some require prompt treatment while others can be postponed to a time when the child is stronger and healthier. These defects include esophageal atresia (narrow or closed esophagus); pyloric stenosis (narrowing of the outlet from the stomach to the intestines); Hirschsprung's disease (blockage of the large intestine); and imperforate anus (no anal opening). Some types of malignancies typically appear in infants and children. These malignancies often require surgery. These tumors include neuroblastomas (the most common childhood brain tumor), rhabdomyosarcomas (muscle tumors), Wilms' tumor (kidney tumors), and teratomas. (Teratomas contain several types of tissue, such as bone, hair, and teeth). Although teratomas are thought to be present at birth, some are not diagnosed until adulthood. Conjoined twins require surgical correction after evaluation of shared organs and circulation. Surgery is usually postponed until a time when the twins' health permits. If an organ such as the heart is shared to some degree, separation may be difficult or impossible. In some cases, the surgical procedure focuses on the twin with the better chance for survival. A cleft lip and palate are readily diagnosed at birth, and many are diagnosed with a prenatal ultrasound. More urgent cases include those that interfere with nursing.

APPLICATIONS AND PRODUCTS

Products for pediatrics use include infant formulas, medications, and medical office, surgical, and endoscopic equipment. Many specialized products exist for use in a neonatal intensive care unit.

Infant Formulas. A wide variety of infant formulas are marketed that all attempt to approximate the composition of breast milk. Some are based on soy or other proteins for infants who are lactose intolerant or allergic to milk products. None of the products are identical to breast milk, which is a species-specific formulation for human babies. Each animal species has a unique milk formulation for its young. Also, breast milk can supply maternal antibodies, which can protect the infant from infection. For the foregoing reasons and others, many women choose to breast-feed. Some of these women will rely on breast pumps for regular or occasional use.

Medications. Pediatric medications include over-the-counter products such as vitamins and analgesics (painkillers); they also include a wide variety of prescription medications. Both prescription and over-the-counter medications are usually in a liquid or chewable formulation. They also are often flavored to make them more palatable to the child. Dosage must be based on the child's weight and age.

Medical Office Equipment. Medical equipment for a pediatric office is similar to adult equipment—it is modified for the "small people." Pediatric stethoscopes and otoscopes (ear scopes) are found in a pediatrician's office. The ophthalmoscope, which visualizes the inner eye, is identical to that used for adults, as the eye is adult size at birth. A scale and tape measure are used at each visit to record the child's height and weight.

Surgical Equipment. As in adult surgery, laparoscopic procedures are increasing in pediatric surgery. Laparoscopy, sometimes termed minimally invasive surgery, involves inserting a scope into the abdominal cavity through a small incision and then performing surgical procedures with specialized instruments. These instruments perform a variety of tasks, including manipulating, cauterizing, and suturing. Some of the equipment is similar to that used on adults while others are scaled down. Scaled-down or adult-size equipment is also used for traditional surgical procedures.

Endoscopic Equipment. Endoscopy involves the use of endoscopes, which are small scopes that can be passed in the body for visualizing internal structures. This equipment includes devices, which can biopsy, cauterize, and grasp internal structures. Endoscopes are designed for passage into the intestinal tract (esophagus, stomach, small intestines, large intestines, and the anus), the bronchi (lung tubes), and joints. It is not uncommon for a child to swallow an object that inadvertently finds its way into the bronchi. After the object (medically described as a foreign body) is located via a radiologic procedure, an endoscope can be passed to retrieve it.

Neonatal Intensive Care Unit Equipment. The neonatal intensive care unit (NICU) contains a great deal of high-tech equipment to facilitate growth and survival of high-risk infants. Many are premature; others have serious infections; and some have a combination of the foregoing problems. Common NICU equipment includes resuscitation equipment, ventilators, radiant warmers, incubators, cardiorespiratory monitors, pulse oximeters, and the basic, but essential, scales.

A cardiorespiratory monitor is attached to sensors on the infant and provides a continuous readout of heart rate and rhythm, respiratory rate, arterial or central venous pressure, as well as other useful information. Alarms can be adjusted to alert NICU staff when any of the vital signs go above or below a set limit. Some incorporate computer systems, which can filter out false alarms, record data over an extended period, and perform analysis of vital signs.

Incubators supply a controlled temperature and humidity to infants who are more stable than those requiring a radiant warmer. They also maintain a clean environment for the infant and protect it from drafts, noise, infection, and excessive handling.

The pulse oximeter monitors the oxygen saturation of the blood. This is accomplished by shining light through the infant's skin and measuring the color of the transmitted light. It works on the principle that blood, which is redder the higher oxygen content it possesses.

Radiant warmers supply warmth to unstable or extremely premature infants. Infants have a large surface area compared to their size and often possess little body fat; thus, they are unable to maintain their body temperature. Heat is radiated to the infant from an overhead element. Sensors on the baby's abdomen are attached to a thermostat, which adjusts the necessary amount of heat. The open design of the radiant warmer allows NICU personnel ready access to the infant from all sides.

A cardiac arrest is not an uncommon occurrence in a NICU. The unit is equipped with defibrillators, ventilators, and medication necessary for resuscitation.

The basic scale is an essential piece of NICU equipment. All feedings, intravenous-fluid administration, and medication administration is based on the infant's weight, so the weight must be accurate and current. The weight is carefully adjusted for the weight of the diaper and attached medical equipment. The weight is entered into a flow sheet on a daily basis.

Ventilators supply air to the infant's lungs when he or she is too ill or too weak to breathe on his or her own. Recent models of infant ventilators are highly computerized and feature diverse modes of operation, including "assist control," which allows for the infant to participate in the respiratory process. Some models incorporate real-time data on the infant's pulmonary function.

IMPACT ON INDUSTRY

Pediatrics has a significant impact on many medical fields: laboratory medicine; intensive care equipment; medical imaging (ultrasound, magnetic resonance imaging, computerized tomography scans, and scintigraphy); and the pharmaceutical industry (including the manufacture of infant formulas). These industries derive significant revenue from the field of pediatrics. Laboratory procedures include tests for genetic disorders such as cystic fibrosis, sickle-cell anemia, and Tay-Sachs disease. Both laboratory medicine and radiology require a team of skilled physicians, supervised by physicians with specialized training. The manufacturing of pediatric products and medications is a significant segment of the pharmaceutical industry. Patients with chronic conditions, such as diabetes and cystic fibrosis, are lifelong consumers.

Beyond medical fields, many products in the marketplace are devoted to infants and children. Until the latter part of the twentieth century, safety was not a major concern. For example, cribs were painted with lead-based paint and slats were spaced at a distance that could allow a child's head to become trapped. The safety of products for children is a major concern. Although manufacturers are responsive to this requirement, some products released on the market have safety issues. For this reason, various safety organizations, including the Food and Drug Administration (FDA), provide oversight. An example of a product that was found to have safety issues was the infant sleep positioner. These devices became popular because the manufacturers claimed that their products kept babies on their backs and reduced the chance of the sudden infant death syndrome (SIDS). However, in September, 2010, the FDA reported that they had received twelve reports of deaths over the past twelve years involving infants who had suffocated in a sleep positioner or became trapped and suffocated between a sleep positioner and the side of the crib or bassinet. The age of the infants ranged from one to four months, and most suffocated after rolling from their side to a stomach position. Some of the positioners were approved by the FDA in the 1980's to reduce symptoms of gastrointestinal reflux; however, these products were not approved for reducing the risk of SIDS. Furthermore, many of the products on the market as of 2011 have never received FDA approval for any purpose. Therefore, the FDA sent letters to eighteen manufacturers of sleep positioners requesting them either to stop manufacturing these products or to submit additional information to the FDA supporting the use of the products.

Significant pediatric research is ongoing by the government and universities. A branch of the National Institutes of Health (NIH), the National Institute of Diabetes and Digestive and Kidney Diseases (NIDDK) funds research in many fields, including diabetes, digestive diseases, genetic metabolic diseases, immunologic diseases, and obesity. The institute also provides health information for the public in these fields. Virtually all developed nations have extensive pediatric-research programs. Beyond government and university programs, some practicing pediatricians devote a significant amount of their time to research in their field.

CAREERS AND COURSE WORK

To become a pediatrician or to practice some

aspects of pediatrics, one must complete a bachelor's degree and a four-year course of medical training. Initial specialty training is in one of the following fields: pediatrics, surgery, family medicine, or psychiatry. This is typically a three- or four-year residency program. A pediatrics residency is the last training step for many pediatricians. However, some pediatricians will receive further training in neonatology or the following pediatric subspecialties: neurology, endocrinology, nephrology (kidney disease), neurology, oncology (cancer treatment), cardiology, or gastroenterology. It is also possible for a physician trained in a subspecialty who did not begin with a pediatrics residency program to transition to a pediatric subspecialty. In some cases, a physician will practice general pediatrics for a period of time before receiving additional training in a subspecialty. A surgeon who completes a surgical residency must then complete two or more years of training in pediatric surgery. Other areas of surgery also encompass pediatric specialties of their own, which require further training: pediatric cardiothoracic surgery, pediatric neurosurgery, pediatric orthopedic surgery, and pediatric urological surgery. For physicians desiring to practice child psychiatry, specific child-psychiatry residency programs exist as well as programs for general psychiatrists who wish to refocus on children.

Most pediatricians will become a member of one or more professional organizations. In addition to the benefits they supply, such as forums and continuing-education meetings to physician members, these organizations also provide educational material to the general public. In the United States, the main professional organization for pediatricians is the American Academy of Pediatrics. Pediatric surgeons and pediatric subspecialists often belong to professional organizations pertinent to their specialty.

SOCIAL CONTEXT AND FUTURE PROSPECTS

In 2010 in the United States, health care reform arose as a prominent topic. The expansion of health care brought up the topic of health care rationing. Two segments of the population consume the major portion of health care costs: infants and children with serious medical conditions and the elderly with serious medical conditions. Opponents of health care expansion expressed concerns that rationing could result in the loss of a child's life. For premature or seriously ill infants, a wealth of technology is

---

Fascinating Facts About Pediatric Medicine and Surgery

- Incubators were invented in France in the late 1800's and were modeled after poultry incubators, from which they inherited their name.
- The term orthopedics comes from the Greek words for "straight" (orthos) and "child" (paidion). This is because the specialty arose to treat skeletal deformities of children.
- Quintuplets (five developing infants) are an obstetrical challenge even in modern medicine. In 1934, the Dionne quintuplets were delivered by a country doctor with the assistance of two midwives. The infants were placed in a wicker basket that contained heated blankets and were brought into the kitchen and placed next to the open door of the oven to keep warm. All five infants survived.
- In January, 2009, Nadya Suleman, known as the "Octomom," delivered octuplets in Bellflower, California, after allegedly having six frozen embryos implanted. All eight survived. She had six children before the octuplets. In October, 2009, Suleman's physician, Michael Kamrava, was expelled from the American Society of Reproductive Medicine for a "pattern of behavior" detrimental to the industry after having been charged with gross negligence for his care of Suleman and two other patients.
- The conjoined twins Chang and Eng were born in Siam (now Thailand) in 1811; this was the origin of the term "Siamese twins." They were connected by a four-inch ligament at the chest. Physicians felt that separating them would be too risky.
- On December 12, 2006, Trishna and Krishna were born in Bangladesh and joined at the head. They were separated by a series of operations at the Royal Children's Hospital in Melbourne, Australia. The final surgery took place on November 16 and 17, 2009. They survived the separation; however, Krishna (the weaker of the two) required a tracheotomy and cannot talk.
- Girls stop growing in height at puberty because estrogen closes the epiphyses (growth plates), located in the arm and leg bones.
- Newborn girls sometimes have vaginal bleeding; this is due to a drop in estrogen level, which the fetus was exposed to while in the uterus.

---

available. Unfortunately, this therapy comes with a high price tag; the term "million-dollar baby" can be a truism. The health care costs for an infant with a serious condition can often exceed $1 million. In undeveloped and developing nations, a treatment dilemma does not exist—high-tech care is not available. In developed nations, very premature infants or infants with severe congenital abnormalities incompatible with life are given "comfort care". This involves family members or hospital personnel cradling the infant while life ebbs away. Many infants are born between the two extremes of perfectly healthy and gravely ill. It is an emotionally difficult decision on where to draw the line and predict which infants merit state-of-the-art treatment. With health care rationing, economics rears its ugly head in the decision process.

Medical technology, specifically assisted reproductive technology (ART), has created pediatric problems. Prior practice had surgeons implanting several embryos in the uterus to ensure that at least one survived. As the technology improved, the rate of multiple births rose. Multiple births—even twins—have a much higher incidence of complications, which may be severe. The trend is to implant a single embryo to reduce complications. In some cases of multiple births, the health of all the infants and sometimes the mother are threatened. Technology exists to inject potassium into an infant's heart, which results in death. Obviously, this is a ponderous medical, moral, and religious dilemma.

Most women strive for a healthy lifestyle when they become pregnant; however, some are nutritionally deprived, smoke cigarettes, and abuse alcohol or drugs. It is estimated that in the United States, about 5 percent of women use illicit drugs and 15 percent drink alcoholic beverages when pregnant. This results in a large number of infants with serious health problems as well as death. For example, a child born to an alcoholic mother can be born with the fetal alcohol syndrome, which is characterized by mental dling the infant while life ebbs away. Many infants are born between the two extremes of perfectly healthy and gravely ill. It is an emotionally difficult decision on where to draw the line and predict which infants merit state-of-the-art treatment. With health care rationing, economics rears its ugly head in the decision process.

Medical technology, specifically assisted reproductive technology (ART), has created pediatric problems. Prior practice had surgeons implanting several embryos in the uterus to ensure that at least one survived. As the technology improved, the rate of multiple births rose. Multiple births—even twins—have a much higher incidence of complications, which may be severe. The trend is to implant a single embryo to reduce complications. In some cases of multiple births, the health of all the infants and sometimes the mother are threatened. Technology exists to inject potassium into an infant's heart, which results in death. Obviously, this is a ponderous medical, moral, and religious dilemma.

Most women strive for a healthy lifestyle when they become pregnant; however, some are nutritionally deprived, smoke cigarettes, and abuse alcohol or drugs. It is estimated that in the United States, about 5 percent of women use illicit drugs and 15 percent drink alcoholic beverages when pregnant. This results in a large number of infants with serious health problems as well as death. For example, a child born to an alcoholic mother can be born with the fetal alcohol syndrome, which is characterized by mental retardation and physical deformities. Women who smoke cigarettes have an increased risk of delivering an infant who is premature, underweight, stillborn, or succumbs in infancy to the sudden infant death syndrome (SIDS).

Robin L. Wulffson, M.D., F.A.C.O.G.

FURTHER READING

American Academy of Pediatrics. *Caring for Your Baby and Young Child: Birth to Age Five.* 5th ed. New York: Bantam, 2009. A resource for parents that covers everything from preparing for childbirth to toilet training to nurturing a child's self-esteem.

Cavens, Travis. *Being a Pediatrician: The Struggles and Rewards of Caring for Children.* Longview, Wash.: Lake, 2000. A true story of the experiences of a female pediatrician.

Kalter, Harold. *Teratology in the Twentieth Century: Congenital Malformations in Humans and How Their Environmental Causes Were Established.* Amsterdam: Elsevier Science, 2003. Comprehensive reference on the environmental causes of congenital malformations.

Ketchedjian, Armen. *Will It Hurt? Parent's Practical Guide to Children's Surgery.* Southbury, Conn.: Warren Enterprises, 2008. Written by an anesthesiologist, this easy-to-read book provides parents and children reassurance to make the surgical experience as stress-free as possible.

Klass, Perri, ed. *The Real Life of a Pediatrician.* New York: Kaplan, 2009. Traces the careers of pediatricians and how they struggle to balance the conflicting needs of profession, self, and family.

Kliegman, Robert, et al. *Nelson Textbook of Pediatrics.* 18th ed. Philadelphia: Saunders, 2007. One of the classic pediatric texts; especially informative in the areas of cardiology and immunology.

Rennie, Janet, ed. *Robertson's Textbook of Neonatology.* 4th ed. Oxford, England: Churchill Livingstone, 2005. Covers all aspects of newborn care and includes contributions by a wide variety of neonatology practitioners.

WEB SITES

American Academy of Pediatrics
http://www.aap.org

American Congress of Obstetricians and Gynecologists
http://www.acog.org

National Association of Neonatal Nurses
http://www.nann.org

See also: Nephrology; Otorhinolaryngology; Surgery.

PENOLOGY

FIELDS OF STUDY

Sociology; criminology; criminal justice; law; psychology; law enforcement; forensic science; business management; psychiatry; behavior management; statistics; risk assessment; human rights; civil liberties; juvenile justice; victimology.

SUMMARY

Penology is the science and practice of prison management and criminal rehabilitation. Penologists study the ethics and effectiveness of various strategies for punishing crime, including incarceration and rehabilitation. The science of penology is especially relevant in the United States, a nation with less than 5 percent of the world's people but nearly 25 percent of its prisoners. More than 2 million individuals, or about 1 percent of the country's total population, is incarcerated.

KEY TERMS AND CONCEPTS

- **Correction:** Treatment of criminal offenders through a system of imprisonment, rehabilitation, probation, and parole.
- **Deterrence:** Prevention of criminal behavior by instilling a fear of punishment.
- **Incapacitation:** Prevention, rather than deterrence, of future criminal activity or offenses, often by the terms of a sentence handed down by a judge.
- **Incarceration:** Jailing or imprisonment of a person.
- **Mandatory Minimum Sentence:** Minimum prison term for a specific offense, established by law, which cannot be altered by a judge on an individual basis.
- **Parole:** Supervised release of a prisoner before his or her sentence has expired.
- **Penitentiary:** State or federal prison.
- **Probation:** Sentence imposed by a judge that allows a person who has been convicted of a crime to be released into the community, usually under specific conditions and close supervision.
- **Recidivism:** Tendency to relapse into criminal activity after having been incarcerated and released.
- **Rehabilitation:** Act of restoring a criminal offender to fulfill a useful role in society.

DEFINITION AND BASIC PRINCIPLES

Penology is the study and practice of prison management and criminal rehabilitation. The modern science of penology uses the principles of evidence-based research to evaluate and justify criminal punishments. Penologists use rigorously collected data, along with the tools of statistical analysis, to establish which criminal management strategies, under which circumstances, are the most effective in terms of their costs and benefits to both society and the individuals being punished. This approach is utilitarian, in the sense that it seeks solutions that maximize benefits for the greatest number of people. However, penologists are not merely concerned with numbers; they also study the daily lives and cultural habits of prisoners and correctional staff, examine the social and psychological effects of confinement, and address broader philosophical issues of power, control, and personal responsibility. In addition, they consider such issues as the rights of offenders, the rights of victims, and the ways in which laws are implemented by courts.

BACKGROUND AND HISTORY

Much of the history of penology is synonymous with the history of prisons, since for many years, incarceration was the only tool commonly used to combat criminal behavior in society. Between the twelfth and eighteenth centuries, most jails—which typically kept men, women, and children alike in deplorable conditions—were private rather than owned and operated by the state. In the late eighteenth and early nineteenth centuries, prison reform took place in Europe. The state became involved in managing prisons, and prisoners began to be separated into different categories. The first attempts at reforming, if not rehabilitating, offenders involved putting them to work at physically demanding jobs, intended to improve them both physically and mentally. In the late nineteenth century, the pendulum swung back, and prisons became harsher, with the emphasis on rigid rules and routines and strict discipline.

Penology in the twentieth century was characterized by a progressive movement that viewed criminal offenders as individuals with differing reasons for committing crimes. It called for a wider array of approaches to crime, including rehabilitation, parole, probation, and a juvenile justice system that treated children differently from adults. The study of penology began to be formalized during this time, as the strategies applied by governments to prevent or deal with criminal activity were increasingly based on principles derived from scientific research. In the late twentieth century, the pendulum once again swung back toward tougher measures and an increased reliance on incarceration and control to combat criminal offenses. Support for the death penalty also grew during this time. In the United States, penology has become a field whose major mission is to find ways to relieve the burden of what has become an overcrowded and dysfunctional prison system.

HOW IT WORKS

Penologists make a distinction between at least four major and potentially overlapping approaches to crime: retribution, deterrence, rehabilitation, and incapacitation. Retribution attempts to match the severity of a punishment with the severity of the crime, in effect making the offender repay his or her debt to society. Retribution is a moral rather than a rational model for formulating penal policies and has largely fallen out of favor within the field of penology as a primary basis for determining when and how punishments should be used. The desire for retribution continues to play an important role in public opinion about penal policy, however. For example, respondents to surveys most frequently cite retribution as their reason for supporting the use of capital punishment.

Deterrence attempts to formulate punishments that are so unwelcome that they will prevent individuals from committing crimes. For instance, many believe mandatory minimum sentences and the threat of capital punishment have a deterrent effect on crime. However, harsh punishments can be effective deterrents only if they are combined with a high probability of arrest and conviction. Therefore, the role of law enforcement and the courts in bringing offenders to justice is as important as the sentences themselves. The deterrence model of penal policy is based largely on a theory of criminal behavior called the rational choice theory, which holds that the decision to commit a crime is based on a reasoned examination of the potential costs and benefits of the act.

Rehabilitation also seeks to prevent future crimes but does so by working with offenders and helping them become useful members of society. Education, work training, counseling, or religious guidance are often components of rehabilitative programs. Rehabilitation is a model for determining penal policy that is compatible with emotional or social theories of criminal behavior. If criminal behavior is caused at least in part by personal factors such as depression or drug addiction or by social factors such as unemployment or poverty, rehabilitative programs that address those specific issues—through therapy, drug rehabilitation, or career development—should have a positive effect.

Incapacitation seeks to protect society by preventing (rather than simply deterring) future crimes by rendering the offender incapable of committing them. Examples of incapacitation include a long prison sentence or a death sentence. Incapacitation is a model for penal policy that is highly future oriented in that it assumes that offenders will commit additional crimes and attempts to take away any opportunity they may have to do so.

APPLICATIONS AND PRODUCTS

Incarceration. Perhaps the most basic application of penology as a science is in investigating questions related to incarceration. Penologists use empirical studies to answer questions such as when jail sentences are effective at reducing crimes for specific offenses or when alternatives should be used. They also ask how long sentences should be for maximum effectiveness. For example, a great deal of research has shone a spotlight on brief mandatory jail sentences for drunk driving offenses. Among other findings, penologists have shown that two-day sentences may help to reduce the risk of first-time offenders repeating their crime and that lengthier jail sentences may in fact have a counterproductive effect on recidivism. The relationship between incarceration and future crime is not straightforward. One study, for example, estimated that only about one-quarter of the drop in crime in the United States during the 1990's could properly be attributed to the use of jail sentences.

Penitentiaries. Penological research has had many

Fascinating Facts About Penology

- The incarceration rate is higher in the United States than in any other nation, with about 750 out of every 100,000 people behind bars, compared with an average of 166 out of every 100,000 in the rest of the world.
- Many prisons hold arts-based rehabilitation programs. For example, prisoners at Sing Sing Correctional Facility, a maximum security prison in New York state, produce plays, musicals, and other performance pieces. In 2003, a group of prisoners staged Reginald Rose's 1954 teleplay *Twelve Angry Men*, which is about a group of jurors who cannot agree on a verdict.
- In the early nineteenth century, penology was influenced by the pseudoscience of phrenology: the notion that one could analyze people's characters—or their tendencies toward criminal behavior—by examining the contours of their skulls.
- Sweden's unusual prison system allows incarcerated individuals to leave the prison temporarily for a few days at a time in order to visit their families.
- In the Stanford prison experiment of 1971, volunteers were randomly assigned to roles as prisoners or prison guards. The study showed that the same characteristics that emerge in prison culture—warden brutality, mental disturbances amongst prisoners, systems of negotiation and hierarchies—also emerge among people who are simply acting out those roles.
- The word "penology" comes from two Greek roots meaning "punishment" and "study."

important practical applications with regard to the design of penitentiaries. For example, studies have shown that exposure to pastel colors such as pink, blue, and yellow—as opposed to sterile white, for instance—can have a calming effect on incarcerated prisoners, reducing their tendencies toward violent behaviors. As a result, some penitentiaries have begun using paint color as a tool to reduce aggression among inmates. Other studies have shown that when prisoners are idle, they are more prone to violence and depression. Architects are able to use such evidence-based findings from penology to create penitentiaries that are more conducive to social interactions and productive activities—for example, prisons may include space dedicated to libraries or exercise facilities.

Rehabilitation. Penological studies have found that rehabilitation programs such as group therapy, religious counseling, and vocational training help reduce recidivism, especially with high-risk offenders, when it is directed specifically at the root causes of criminal behavior, is implemented with flexibility and altered to meet the needs of specific prisoners, and is followed by long-term support for the participants. Some examples of successful rehabilitation programs include educational classes in which prisoners work toward a general equivalency diploma (GED) and psychological therapy sessions for incarcerated sex offenders, in which prisoners are taught to control emotions such as anger and hostility and lessen their preoccupation with sex. Such programs have been shown to reduce recidivism rates.

Juvenile Justice. Many penologists help provide research and analysis that shapes the policies of the juvenile justice system. For example, researchers have investigated the hypothesis that trying adolescents in adult courts and making them eligible for harsher sentences has a deterrent effect on crime and a reductive effect on recidivism rates. One comparative study of juvenile offenders in the states of New York and New Jersey did not find this to be the case, however.

Parole. One particularly significant area of study for penologists is the effectiveness of alternatives to incarceration such as parole and probation. One study analyzed recidivism rates among released prisoners to clarify the circumstances under which parole is most effective. The study found that female offenders who did not have long prior criminal histories were more likely not to commit additional offenses than male offenders or those who had been incarcerated for crimes relating to drugs, property, or violence. The role of the science of penology is thus an eminently practical one—studies such as these can help inform government and state agencies as they decide whether to release a prisoner on parole.

The Death Sentence. One of the weightiest problems penologists attempt to tackle is determining whether the death sentence is an effective deterrent; that is, how well the threat of execution, or capital punishment, prevents crime. The many studies in this area have been inconclusive. Some, for example, have found no significant difference in the murder rate in states that impose the death penalty and those that do not, but one meta-analysis of death penalty studies found some evidence that the deterrence effect does exist. Another issue that concerns penologists with regard to the death penalty is that of wrongful conviction. In one study of prisoners on death row, nearly half were found to be eventually cleared of the crimes for which they had been found guilty, suggesting that the criminal justice system needs greater safeguards to prevent wrongful convictions, especially if they carry the harshest possible sentence.

IMPACT ON INDUSTRY

The businesses and organizations, both private and public, involved in constructing and maintaining a country's prison system are known collectively as the prison-industrial complex. The term is often used in a negative sense, as it implies a massive, self-sustaining system that relies on the constant supply of offenders for profit.

Government Research. Several U.S. government agencies are involved in penological research. The National Institute of Corrections sponsors studies on topics such as how to reduce the incarcerated population and how best to train and maintain effective prison staff. Smaller government organizations also conduct penological research. A county in Texas, for example, conducted a massive data-collection experiment to assess and improve its probation program. Because the annual budget spent on law enforcement and corrections by all levels of government runs into the billions, penological research is an essential tool for helping suggest ways to reduce these costs.

Industry and Business. The large number of private companies involved in the prison-industrial complex provide a wide array of services, including the design, construction, and management of prisons. The biggest business sector in United States industry is private prisons. Two of the largest and most influential corporations involved in private prisons are Corrections Corporation of America (CCA) and the Wackenhut Corporation, which collectively control the majority of the private prison market. CCA maintains its own penological research institute, which conducts independent studies on the viability of prison privatization.

CAREERS AND COURSE WORK

Penology is an interdisciplinary field. It is possible to approach it from a variety of academic and professional backgrounds. Common undergraduate and graduate concentrations for this career include psychology, criminology, sociology, and law. Because penology is concerned with the reasons behind human behavior and the ways to effectively control that behavior, many courses from business or management studies are also appropriate. Many penological studies involve synthesizing and analyzing trends across large amounts of data, so it is important for penologists to have a basic grounding in the mathematics of statistical analysis. After graduation, students will often find it beneficial to serve an internship at a state or federal prison system to obtain practical experience within the penitentiary system. Career paths related to penology include psychologist, criminologist, police officer, Federal Bureau of Investigation agent, correctional facility guard or administrator, and probation or parole officer.

SOCIAL CONTEXT AND FUTURE PROSPECTS

Penological research affects the lives not only of incarcerated prisoners and criminal offenders but also of members of the broader society. Research into the effectiveness of various deterrents to crime or attempts to reduce recidivism can help protect society from the harmful effects of criminal activity. In addition, when incarceration rates are high, there is an unseen cost to society in the form of lost man-hours because prisoners cannot participate in the economy and contribute to its growth. Penology has the potential to reduce the effects of crime on society by suggesting ways to reduce incarceration rates and recidivism, possibly through improving the effectiveness of alternatives such as probation and rehabilitation.

M. Lee, B.A., M.A.

FURTHER READING

Blomberg, Thomas G., and Karol Lucken. *American Penology: A History of Control.* 2d ed. New Brunswick, N.J.: Aldine Transaction, 2010. Addresses milestone moments in the history of American penal reform. Organized chronologically from colonial times to the 1990's and beyond. Also includes a section on ancient and medieval prisons.

Clear, Todd R., George F. Cole, and Michael D. Reisig. *American Corrections.* Belmont, Calif.:

Thomson Wadsworth, 2010. A comprehensive and accessible introductory textbook on the science of corrections. Each chapter includes an outline, focus boxes, definitions of key terms, and study questions.

Gottschalk, Marie. "Money and Mass Incarceration: The Bad, the Mad, and Penal Reform." *Criminology and Public Policy* 8, no. 1 (April, 2009): 97-109. Evaluates a number of proposed sentencing policies and alternatives to incarceration.

Horton, David M., ed. *Pioneers in Penology: The Reformers, the Institutions, and the Societies, 1557-1900.* 2 vols. Lewiston, N.Y.: Edwin Mellen Press, 2007. A history of penology in Europe and the United States. Each chapter includes further reading suggestions.

Johnson, Robert. *Hard Time: Understanding and Reforming the Prison.* 3d ed. Belmont, Calif.: Wadsworth, 2002. A broad look at the experience of incarceration that includes personal stories from prisoners.

Simon, Jonathan, and Malcolm M. Feeley. "The Form and Limits of the New Penology." In *Punishment and Social Control,* edited by Thomas G. Blomberg and Stanley Cohen. 2d ed. New York: Aldine de Gruyter, 2003. A substantive introduction to trends in the science of penology that emerged in the late twentieth century.

Thistlethwaite, Amy B., and John Wooldredge. *Forty Studies That Changed Criminal Justice: Explorations into the History of Criminal Justice Research.* Upper Saddle River, N.J.: Prentice Hall, 2010. A concise abstract of forty seminal penology studies, plus commentary on the significance of each. Each abstract includes a citation, information about methodology and results, and further reading notes.

Williams, Virgil L. *Dictionary of American Penology.* Rev. ed. Westport, Conn.: Greenwood Press, 1996. Contains cross-referenced, alphabetically arranged entries covering all aspects of technical terminology in the field.

WEB SITES
American Correctional Association
http://www.aca.org

International Corrections and Prisons Association
http://www.icpa.ca

U.S. Department of Justice
Federal Bureau of Prisons
http://www.bop.gov

See also: Criminology

PHARMACOLOGY

553

FIELDS OF STUDY

Chemistry; biology; biochemistry; physiology; microbiology; pathology; anatomy; genetics; mathematics; physics.

SUMMARY

Pharmacology is the study of how drugs or other chemical compounds act in a living organism. The field of pharmacology examines the basic chemical properties, biological effects, therapeutic value, and potential toxicity of drugs. Scientists who study pharmacology are involved in medicine, nursing, pharmacy, dentistry, and veterinary medicine. They develop new drugs, improve the safety of existing drugs or chemicals, and prevent and treat countless diseases. Pharmacology is a highly integrated field, applying basic principles of chemistry, biology, and physics to the safe and effective use of drugs and chemical compounds in living systems.

KEY TERMS AND CONCEPTS

- **Absorption:** Active or passive transport of a drug or chemical compound into the bloodstream.
- **Adverse Effect:** Unwanted negative or harmful effect of a drug.
- **Bioavailability:** Rate and extent to which a drug reaches the bloodstream and is available to exert its effect.
- **Distribution:** Transport of a drug or chemical compound from the bloodstream to the body's tissues.
- **Drug:** Any chemical compound, other than food, that affects a living organism; used in the field of medicine to diagnose, prevent, cure, and treat disease.
- **Drug Interaction:** Process by which one substance alters the activity or effect of a drug.
- **Excretion:** Irreversible elimination of a drug or chemical compound from a living system.
- **Half-Life:** Time required for a biological system to eliminate half the amount of drug or chemical compound administered.
- **Mechanism of Action:** Chemical process by which

a drug or chemical compound exerts its effect.
- **Metabolism:** Breakdown or modification of a drug or chemical compound.
- **Pharmacodynamics:** Relationship between drug concentration and physiologic response.
- **Pharmacokinetics:** Mechanics of drug absorption, metabolism, distribution, and excretion.

DEFINITION AND BASIC PRINCIPLES

Pharmacology is the study of how drugs and chemical compounds affect living processes. Pharmacology is closely related to biochemistry and physiology in terms of substantive knowledge and experimental techniques. Pharmacology is also closely related to chemistry and physics because it relies on a basic understanding of the fundamental chemical and physical properties of drugs, as well as of living organisms. Pharmacology also involves mathematics for quantification of its basic principles. Pharmacology is not an autonomous field of study but unifies the study of chemicals and living organisms and the interactions between them, and uses the knowledge and techniques of several other science disciplines for its foundation.

Pharmacology can be broadly divided into four categories: pharmacodynamics and pharmacokinetics, toxicology, pharmacotherapy, and pharmacy. Pharmacodynamics and pharmacokinetics examine the biological effect produced by a drug, as well as how the drug gets to the site of action and the fate of the drug in the body. They also evaluate the safety and effectiveness of drugs. Toxicology analyzes the toxic or harmful effects of drugs and chemical compounds and the mechanisms and conditions by which they occur. It also focuses on the signs, symptoms, and treatments for poisoning. Toxicology studies not only therapeutic agents but also the harmful effects of chemicals in manufacturing, food, water, and the atmosphere. It is instrumental in defining the legal relationship between accidental or intentional chemical exposure and the subsequent harmful effects.

Pharmacy is closely related to pharmacology, but it is a clinical, patient-focused discipline responsible for the safe preparation and dispensing of therapeutic agents. The application of drugs for the prevention or treatment of disease is called pharmacotherapy. Pharmacists work closely with physicians and other health care providers and use their knowledge of drugs and drug preparations to optimize pharmacotherapy.

BACKGROUND AND HISTORY

Modern pharmacology dates to the sixteenth century, when Swiss physician Paracelsus, also known as the grandfather of pharmacology, began applying the use of chemistry to medicine. Paracelsus taught that illness was a disturbance of the body's chemical systems and stressed the curative powers of inorganic substances.

In 1628, William Harvey published his explanation of the circulation of blood, which prompted the intravenous administration of drugs. This made possible the recognition of the relationship between the dose of a drug, the time of the dose, the site of administration, and its biological effect. The remainder of the seventeenth century, as well as the eighteenth and nineteenth centuries, included advances in experimental techniques in chemistry and physiology and the publication of the first large-scale studies of pharmacological experiments on animals.

Pharmacology separated itself from chemistry and physiology in the early twentieth century, owing to the work of early professors of pharmacology in Europe, including Rudolf Bucheim and Oswald Schmiedeberg, who wrote the earliest pharmacology textbooks and defined the modern purpose of pharmacology.

John Jacob Abel is known as the father of American pharmacology. He was the first professor of pharmacology in the United States and founder of the first department of pharmacology at the University of Michigan. Abel also founded the American Society for Pharmacology and Experimental Therapeutics in 1908.

The field of pharmacology continued to evolve over the course of the twentieth century as advances in medicine and research practices necessitated a deeper understanding of drug actions and interactions in the effort to treat and prevent diseases.

HOW IT WORKS

The foundation of pharmacology lies in the understanding of how drugs interact with their target receptors. Though only a few basic types of drug-receptor interactions exist, the molecular details vary widely among different classes of drugs and receptors. Additionally, classifying drugs by their mechanism of action simplifies pharmacology because this basic knowledge of how a drug works translates to predictable actions in cells, tissues, and organs, and explains how a drug mediates its therapeutic and adverse effects. Without basic fundamentals of chemistry, biology, physics, and physiology, the simple interaction between drug and receptor could not be expanded to its broad physiological context.

The technology used in the study of drug-receptor interactions includes in vivo and in vitro experiments, as well as modern computer software for mathematical and statistical modeling. Such software systems allow researchers to create models of drug compounds and receptors, simulate the interaction between drugs and receptors, predict the bioavailability of drugs, and recreate physiological environments in which a drug can be absorbed and distributed. Molecular modeling of drugs can also predict drug metabolism and drug interactions, although the clinical and therapeutic relevance of such activity is difficult to ascertain from a computer-generated model.

The study of drug-receptor interactions encompasses biopharmaceutics, which evaluates the relationships among the chemical and physical properties of a drug, the dosage form of the drug, the route of administration, and the biological effect. The Biopharmaceutics Classification System (BCS) is a research tool that provides a framework for making decisions in the early stages of drug development. The BCS provides solubility and permeability information for known compounds to predict the actions and reactions of a new drug in the body.

Chemical assays are important in the initial analysis of drug products and metabolites. Assays may be performed on a variety of substances or on samples from a variety of species, including humans, rats, and primates. Samples may be obtained from the liver, the drug's target organ, or blood or another body fluid. Liquid chromatography is often used to identify and quantify drug molecules or metabolites based on size, charge, or polarity. A small volume of the sample to be tested is placed into a liquid (mobile phase) that is pushed through a column (stationary phase) by a pump. The sample interacts with the material in the column—usually saturated carbon chains—and a detection device captures the activity of the sample as it leaves the stationary phase. Every compound has a characteristic time that it stays in the stationary phase, called retention time; therefore, this retention time can be used to identify a compound.

Several types of detection devices may be used at the end of a liquid chromatography column. Ultraviolet and fluorescence spectroscopies detect the ability of organic compounds to absorb light in the electromagnetic spectrum. Each compound absorbs light characteristically, allowing for the identification of organic compounds, including drug products and metabolites.

Positron emission tomography (PET) scans are used to evaluate the pharmacokinetics of drug products. Highly sensitive and specific, they are often used in drug development. PET scans allow modeling of drugs in the blood and tissues and yield measurements of receptor density on a target tissue. They are also used to study drug distribution throughout the body and drug metabolism. A PET scan is a specialized type of nuclear imaging that produces three-dimensional pictures of the processes of the body. To conduct a PET scan, a radioactive substance is tagged to a drug molecule and introduced into the body. The scanner detects the gamma rays emitted by the radioactive tag throughout the body over time. A computer then collects the images and reconstructs the body, allowing visualization of the target molecule within the body over a period of time.

After a drug is developed and its pharmacokinetic and pharmacodynamic properties are characterized, it is taken through several phases of clinical testing to determine the actual response achieved in patients. Clinical trials are necessary to determine the safe and appropriate doses for patient populations and uncover drug interactions or safety issues that may not be apparent during drug development phases. Patients are administered test drugs under highly regulated conditions.

APPLICATIONS AND PRODUCTS

Pharmacology is a highly integrated and interdisciplinary field that is divided into specialized applications. However, many of these divisions overlap, and pharmacologists collaborate with scientists in other fields to perfect the safe and effective use of drugs and other chemicals. Pharmacology principles may be applied to a variety of disease states, as well as individual and community health and environmental applications.

Clinical Pharmacology and Therapeutics. Clinical pharmacology and therapeutics are related areas that study the effects of the pharmacokinetics and pharmacodynamics of a drug in humans. Clinical pharmacology applies the knowledge of how drugs work to how they can affect disease processes. It also examines how a disease process may alter the action of drugs in the body. Pharmacogenomics is a new field within clinical pharmacology that examines how individual genetic variations and genome components alter drug pharmacokinetics and pharmacodynamics.

Closely related to clinical pharmacology, therapeutics encompasses the use of drugs to treat, prevent, or cure a disease. Therapeutic applications of pharmacology are necessary to optimize drug effectiveness and safety and prevent medication errors. Clinical pharmacology and therapeutics are also essential to the design and management of clinical trials.

Behavioral Pharmacology. Behavioral pharmacology studies the effects that drugs have on behavior. This specialized field of pharmacology covers topics including the molecular foundation of drug activity on neurotransmitters and the therapeutic and clinical ramifications of drugs on behaviors. Behavioral pharmacology also encompasses the evaluation and treatment of substance abuse and addiction disorders.

Cardiovascular Pharmacology. Cardiovascular pharmacology examines the effects of drugs on the heart and vascular system, as well as on other body mechanisms that regulate cardiovascular functions. Cardiovascular pharmacology includes all aspects of pharmacology, from the understanding of the mechanism of action of cardiovascular drugs to the use of drugs in clinical and therapeutic settings.

Endocrine Pharmacology. Endocrine pharmacology studies drugs that are hormones or hormone derivatives and drugs that alter the function of hormones. Endocrine pharmacology is concerned with the use of drugs to treat or prevent diseases of metabolic origin.

Neuropharmacology. Neuropharmacology investigates drugs that affect the central and peripheral nervous system, including the brain, spinal cord, and nerves. Neuropharmacologists may analyze the causes of disease or define brain activity to apply

Fascinating Facts About Pharmacology

- One of the world's oldest medical documents, the Ebers papyrus, which dates to around 1550 b.c.e., details how Egyptians prepared and used more than seven hundred drugs and medicinal concoctions, some of which are still used.
- Spanish physician and chemist Mateu Orfila, sometimes called the father of toxicology, published the first comprehensive text on forensic toxicology in 1813; his work emphasized the need for identification and quality assurance in the pharmaceutical, clinical, environmental, and industrial fields.
- Between 1923 and 2000, sixteen pharmacologists received the Nobel Prize in Physiology or Medicine.
- Paul Ehrlich is credited with originating modern chemotherapy, as he is the first scientist who used synthetic compounds to combat infectious diseases; he was awarded the Nobel Prize in Physiology or Medicine in 1908 for outlining the principles involved in using chemicals to destroy cells.
- During World War II, the United States accelerated a wartime program to investigate antimalaria drugs, thereby providing a boost to pharmacology research.
- Only one out of every five drugs in clinical trial receive FDA approval and are eventually marketed.

pharmacologic principles to the treatment or prevention of nervous system disorders, addiction and abuse disorders, behavioral disorders, and psychological disorders. Neuropharmacology also applies to the understanding and use of anesthesia.

Biochemical and Cellular Pharmacology. Biochemical and cellular pharmacology works at the smallest level of drug action. This field integrates biochemistry, physiology, and cell biology to understand how drugs influence the chemical actions of a cell.

Chemotherapy. Chemotherapy is the application of pharmacological principles to the treatment of microbial infections and cancer. Advances in chemotherapy make it possible to administer a drug to kill or inhibit the growth of a tumor or infecting organism without significantly impairing the normal function of the host body.

Drug Discovery and Legal Aspects. The field of drug discovery and development and regulatory affairs encompasses discovery and validation of new drugs or chemical products, molecular modeling, clinical testing, drug regulation, and the legal and economic aspects of drug use.

Drug Metabolism and Disposition. Drug metabolism and disposition studies the pharmacokinetics and enzymatic metabolism of drugs. This field encompasses the identification of drug metabolites, the regulation of drug metabolism, the relationship between the structure and function of drugs and metabolites, and the classification of drug-metabolizing enzymes.

Molecular Pharmacology. Molecular pharmacology studies the chemical and physical properties of drugs to define how they interact with receptors at the molecular level. This field is deeply rooted in mathematical techniques and molecular biology. Molecular pharmacology deals with all classes of drugs and all types of receptors at the cellular and subcellular level.

Toxicology. Toxicology studies the adverse effects of drugs or other chemicals. This includes not only therapeutic drugs but also household, environmental, and industrial chemicals that may pose a hazard to people or other organisms. Toxicology applies the principles of pharmacokinetics and pharmacodynamics to the design and interpretation of drug safety studies. The principles of pharmacology and toxicology may also be applied to forensics to aid in the medical and legal identification and interpretation of drugs or toxic substances.

Veterinary Pharmacology. Veterinary pharmacology applies the principles of pharmacology to drugs and disease states of animals.

Systems and Integrative Pharmacology. The study of systems and integrative pharmacology involves drug action and toxicity in the whole animal. Closely related to therapeutics and clinical pharmacology, this specialty examines the big picture of drug action to optimize and individualize drug therapy.

IMPACT ON INDUSTRY

The pharmacology industry involves university researchers, pharmaceutical and health care corporations, and government agencies. Although it is impossible to estimate the economic impact of the science of pharmacology, the closely related pharmaceutical and medications industry, which is deeply rooted in the principles of pharmacology, had global revenues of $643 billion in 2006. The industries related to pharmacology are highly integrated, with significant collaboration and overlap among academic institutions, industries, medical institutions, and government agencies. Many large pharmaceutical companies are headquartered in the United States, but others are in Europe and Asia.

University Research and Training. The pharmacology industry and profession relies on universities and the academic sector to train pharmacologists in basic science, the principles of pharmacology, and laboratory and experimental techniques. Universities offer expert educators and essential learning experiences in core pharmacology curriculum. Academic pharmacology researchers are often involved in partnerships with companies in the pharmaceutical or health care industries in which they provide input or research for new drug development or conduct or evaluate clinical trials.

The Association of Medical School Pharmacology Chairs (AMSPC), which brings together pharmacologists from medical schools throughout North America, works to promote pharmacology as a discipline within graduate and medical school curriculums. The association endorses basic science research, research funding, partnerships with industry, and faculty development to improve pharmacology research and training of students.

Emory University in Atlanta, Georgia, was ranked the premier university in the world for pharmacology and toxicology research. Many students within the program are members of the Georgia Biomedical Partnership, a professional group representing scientists and executives within the pharmaceutical and biotechnology sectors, highlighting the collaboration between academia and industry.

Industry and Business. The pharmaceutical and medications manufacturing industry is the largest end user of pharmacology. The industry designs, develops, manufactures, distributes, and markets prescription and over-the-counter medications and health care products. Although most new drugs receive research and clinical input from academia, the majority of drug development and testing is completed by the pharmaceutical industry. The Pharmaceutical Research and Manufacturers of America represents the pharmaceutical and biotechnology companies that are involved in producing safe and effective medications.

Government and Regulatory Agencies. The discovery, manufacture, and prescription of drugs has economic and political consequences. In the United States, the Food and Drug Administration (FDA) regulates the manufacture, sale, and administration of drug products to ensure safety and effectiveness for the consumer. The FDA imposes strict regulations on the management of clinical trials and requires extensive applications and regulatory filings for each new drug product to be developed or marketed in the United States.

In 2004, the FDA launched the Critical Path Initiative to transform the way that FDA-regulated products are developed, evaluated, manufactured, and used. The initiative affects the design of trials, surveillance of postmarketing safety, and training of scientists and researchers in drug discovery and development. It also encourages the use of new technologies and discoveries to improve drug design. In 2010, the FDA expanded the initiative to include advances in personalized medicine, defenses against drug-resistant bacteria, and the availability of drugs to combat tropical diseases. The initiative's efforts are intended to meet the ongoing public health needs of the United States and the rest of the world.

The United States Pharmacopeia (USP) is a nongovernmental organization that sets public standards for prescription and over-the-counter medicines, health care products, food ingredients, and dietary supplements. The USP dictates the quality, strength, purity, and consistency of these products if they are sold in the United States. The United States Pharmacopeia-National Formulary (USP-NF) presents standards for all drug products, dietary supplements, and excipients. It also sets forth standards for packing, labeling, and storing drug products, as well as detailed information regarding testing quality and assurance measurements.

CAREERS AND COURSE WORK

An education in pharmacology requires a firm foundation in chemistry, biology, physics, and mathematics. Pharmacology is not offered as an undergraduate degree by most universities. Therefore, most people first obtain an undergraduate degree in a basic science such as biochemistry or physiology, then pursue a master's or doctorate degree in pharmacology. The study of pharmacology is essential to professional curriculums in pharmacy, medicine, dentistry, and veterinary science. Many other health professions, including nursing and allied health, include pharmacology course work in the basic curriculum.

Courses in writing and liberal arts are also essential to becoming a skilled pharmacologist. The ability to think creatively and communicate effectively are critical skills for all scientists. Hands-on experience is invaluable to a pharmacology education, and students interested in a career in pharmacology should obtain an internship, summer job, or fellowship in a pharmacology laboratory or setting.

There are many career paths students can choose after studying pharmacology. A pharmacology education prepares students for work in pharmacology research, toxicology, and biotechnology. Pharmacologists develop new drugs in the pharmaceutical industry, perform basic or applied research at academic institutions, conduct clinical research to establish the effectiveness of drugs in humans, prepare legal and regulatory documents for drug product registration, and work in drug information or pharmaceutical sales. Pharmacologists may work in a research laboratory, a hospital, a university, a pharmaceutical company, or other health care business.

Pharmacologists work with cutting-edge technology and strive to improve existing drugs and develop additional drugs. A career in pharmacology is rewarding for students who are inquisitive problem solvers and want to advance medical science.

SOCIAL CONTEXT AND FUTURE PROSPECTS

The advances in science and medicine during the twentieth century stimulated advances in pharmacology that continue into the next century. Throughout the ages, drugs and medicinal products have been used to create safer, healthier, more productive lives. The world's population is growing and aging, and the understanding of disease states is more advanced than ever before. Pharmacologists are in high demand as the requirement for safe and effective new drugs remains strong in the short- and long-term future.

Better medicines are needed globally, and pharmacology is poised to deliver the necessary researchers, educators, clinical and therapeutic practitioners, and experts in drug evaluation, safety, use and regulation. Pharmacologists of the future will work with epidemiologists, public health scientists, economists, and social scientists to drive innovation and clinical application of drug products.

Pharmacology is critical to the pharmaceutical and health care industries and has a major impact on individual and public health. It optimizes the understanding and use of existing drugs and chemicals, discovers and evaluates new drugs, and defines the variability in toxic and therapeutic responses to drugs. In the future, pharmacologists may be able to use the principles of genomics and genetics to individualize drug therapy. This could lead to the use of drugs that target one specific protein or receptor to guarantee a clinical response, while erasing the possibility of unwanted adverse effects.

Jennifer L. Gibson, B.S., Pharm.D.

FURTHER READING

Bertomeu-Sánchez, José Ramón, and Agustí Nieto-Galan, eds. *Chemistry, Medicine, and Crime: Mateu J.B. Orfila (1787-1853) and His Times.* Sagamore Beach, Mass.: Science History Publications, 2006. A collection of essays that present an overview of the life and work of Orfila and his contributions to pharmacology and toxicology.

Katzung, Bertram, Susan Masters, and Anthony Trevor, eds. *Basic and Clinical Pharmacology.* 11th ed. New York: McGraw-Hill, 2009. A comprehensive textbook that includes more than three hundred full-color illustrations, drug comparison charts, and case studies outlining the clinical applications of pharmacology.

Rubin, Ronald P. "A Brief History of Great Discoveries in Pharmacology: In Celebration of the Centennial Anniversary of the Founding of the American Society of Pharmacology and Experimental Therapeutics." *Pharmacological Reviews* 59, no. 4 (2007): 289-359. An overview of the great scientists and pharmacology discoveries of the nineteenth and twentieth centuries.

Vallance, Patrick, and Trevor G. Smart. "The Future of Pharmacology." *British Journal of Pharmacology* 147 (January, 2006): S304-307. A perspective on the future of pharmacology as an integrative science, as well as discussions of the important divisions within the field.

Walsh, Carol T., and Rochelle D. Schwartz-Bloom.

Levine's Pharmacology: Drug Actions and Reactions.
7th ed. Abingson, Oxfordshire, England: Taylor &
Francis Group, 2005. Outlines the history of pharmacology, general principles, and individual factors that influence drug action.

WEB SITES
*American Society for Clinical Pharmacology and
Therapeutics*
http://www.ascpt.org

*American Society for Pharmacology and Experimental
Therapeutics*
http://www.aspet.org

United States Pharmacopeia
http://www.usp.org

U.S. Food and Drug Administration
Drugs
http://www.fda.gov/Drugs/default.htm

See also: Anesthesiology; Herbology; Immunology
and Vaccination; Toxicology.

PHOTONICS

FIELDS OF STUDY

Optics; quantum physics; fiber optics; holography; information processing; electro-optics; laser optics; solar cells.

SUMMARY

Photonics is a rapidly emerging field that uses the quantum interpretation that light has both wave and particle aspects that generate, detect, and modify it. Photonics covers the full range of the electromagnetic spectrum, but most applications are in the visible and infrared. Photonic systems are replacing electricity in the transmission, reception, and amplification of telecommunication information. Photonic applications include lasers, photovoltaic solar cells, sensors, detectors, and quantum computers.

KEY TERMS AND CONCEPTS

- **Diffraction Grating:** Series of tiny slits that allow light to pass through.
- **Holography:** Image construction of an object using interference effects that produce an apparent three-dimensional image.
- **Laser:** Light amplification and stimulated emission of radiation; atoms with electrons in excited states that all transition to lower energy states by emitting the same frequency photons.
- **Optical Fiber:** Thin, flexible, transparent material that serves as a wave guide for transmitting light.
- **Optical Waveguide:** Structure with a high index of refraction that allows light to pass through without side transmission losses.
- **Photoelectric Effect:** Conversion of light energy into electric energy.
- **Photons:** Localized, massless bundles traveling at the speed of light with energy proportional to their frequency.
- **Q-Bit:** Basic unit cell in a quantum computer.

DEFINITION AND BASIC PRINCIPLES

Photonics is the application of the scientific idea that light (electromagnetic radiation), in all its forms—from radio to cosmic—exhibits both wave and particle behavior. However, these different behaviors cannot be observed simultaneously. Which is observed depends on the physical arrangement at the time of detection. Since this radical idea was presented, science has come to accept this dualism, and photonics is the practical use of this duality in instruments and measurement methodologies. Light traveling from source to destination follows the rules of wave motion, but at its emission and reception points it behaves as a particle. Particles emit light energy in localized bundles (called photons), and photons transfer their energy when they interact with other particles.

Photonics uses this property to develop instruments sensitive to the interaction of photons with particles in the transmission, detection, and modulation of light. Two properties of light enable it to replace electricity in applications involving information technology and power transmission. Light travels at the fastest speed possible in nature and through the use of optical wave guides in fiber-optic material, there is almost no loss in the signal.

BACKGROUND AND HISTORY

In 1905, Albert Einstein expanded German physicist Max Planck's idea of quantized energy units to explain the photoelectric effect. It was known that certain metals had the ability to produce an electric current (photocurrent) when radiated by light energy. The classical understanding posits if there was sufficient intensity of the light, it would provide enough energy to free bound electrons in the metal's atoms and produce electric current. What Einstein realized that instead of a wave of diffuse intensity interacting with the metal's electrons, a local bundle of light carrying one quanta of energy could release the electron if its energy exceeded that holding the electron in the atom. This bundle of light was named a photon and the reaction was viewed as a particle-particle effect instead of a wave-particle interaction.

Soon afterward, other phenomena such as Compton scattering, X-ray production, pair creation and annihilation could be interpreted successfully using a photon picture of light. Light still retains its wavelike properties as it travels through space. It assumes its photon or particle-like behavior only when it interacts with matter in a detector or at a target.

HOW IT WORKS

Generation and Emission. An atom's electrons are placed in excited energy states and as they drop to lower states, photons are emitted. In lasers, an amplification of light is achieved through stimulated emission in which electrons are excited to a particular higher-energy state, and then they all emit the same photon when they drop in energy. This produces a coherent source of light at a particular frequency.

Transmission. Transmission is the process of sending and receiving a signal from one point to another. In photonics, the signal is sent over an optical-fiber wave guide to ensure the integrity of the signal. A transmitted signal may be altered by digitization or modulation in coding for security or error control.

Modulation. Modulation is the varying of any time input signal (carrier wave) by an accompanying signal (modulating wave) to produce some information that can be processed. In photonics, the two types of modulations are digital and analogue. In digital modulation of a laser diode, the output signal is zero when the input current (bias) is at the minimum (threshold) frequency. When the input frequency is greater than threshold, a constant positive value is produced by the output. In analogue modulation, the output signal varies in step with the input frequency.

Signal Processing. Signal processing represents the operations performed on input waveforms that provide amplification, coding, and information. The inputs are either analogue or digital representations of time-varying quantities. In photonics, the modulation of a light signal determines the type of processing performed.

Switching. Fiber-optical switches are useful in redirecting the optical signal in an optical network. A two-position switch reroutes a signal to one of two output channels. Factors determining the efficiency of a two-position switch are speed, reproducibility, and cross talk. Speeds of a few milliseconds are possible with electromechanical switches. Reproducibility provides the same intensity in the signal every time a switch is made. Cross talk measures how uncoupled one output channel is from the other in a multichannel optical system.

Amplification. There exists an array of optical amplifiers to increase the signal. Optical communications use fiber optic and semiconductor amplifiers. For research, there are Raman and quantum-dot amplifiers.

Photodetection and Sensing. Photodetectors are devices that take light radiation and directly convert it to electrical signals varying the electric current or voltage to replicate the changes in the input light source. In one type, electrons are emitted from the surface of a metal using the photoelectric effect. Photodiodes and photomultipliers operate under this effect. Another type is made of junctions of semiconductors. Electrons or electron holes (positive current) are emitted on the device's absorption of radiant energy. The p-n junction photodiode, the PIN photodiode, and the avalanche photodiode work under this property. Most fiber-optic communication systems employ a PIN or an avalanche photodiode. The effectiveness of a photodetector is measured by the ratio of the output electric current (I) over the input optical power (P).

APPLICATIONS AND PRODUCTS

Fiber Optics. This includes all the various technologies that use transparent materials to transmit light waves. A traditional fiber-optic cable consists of a bundle of glass threads, each of which has the capability of transmitting messages in light waves that have been modulated in some fashion. The advantages to transmitting electrons through conducting wire include their travel at the speed of light, less signal loss due to optical waveguides, and greater bandwidth. The data can be transmitted digitally, and the fiber-optic materials are typically lighter weight than metal cable lines. Fiber optics is used most heavily for local-area networks in data and telecommunications.

The heavy reliance on fiber-optic technology makes necessary the continual development of more efficient optical fiber materials with ever-increasing switching speeds and more bandwidth to accommodate users' increasing video demands.

Quantum Optics. The peculiar properties of quantum systems using photons make them candidates for quantum computing devices. The basic q-bit state has the ability through the property of superposition to be more than one value simultaneously. This can lead to properties of a computing system that can perform certain tasks such as code breaking faster and more efficiently than existing binary computers. Using photons that can be polarized into two states, an optical computer can be designed to take advantages of these light quanta. Whether such systems will ever have the stability to serve as computational devices and have the speed and low power consumption of the electronic computers commonly being used remains an active research question.

Telecommunications. Optical telecommunication devices send coded information from one location to another through optical fibers. The astonishing growth of the Internet and the ever-increasing demands for more information delivered faster with more efficiency have spurred the development of optical transmission networks. There are optical networks laid underneath the Earth's vast oceans as well as extensive ground-based systems connecting continental communication systems.

Holography. Holography is used optically to store, retrieve, and process information. Its ability to project three-dimensional (3-D) images has allowed for such videos to be more accessible for public viewing. The use of holograms in data storage inside crystals or photopolymers has increased the amount of memory that can be encoded in these structures. Holographic devices are used as security scanners in assessing contents of packages, in determining the authenticity of art, and for examining material structures.

Micro-optics. Microphotonics uses the properties of certain materials to reduce light to microscopic size so that it can be used in optical networking applications. Light waves are confined to move in materials because of total internal reflection using wave guides. The materials have a high index of refraction decreasing the critical angle. This enhances the total reflection capabilities. A photonic crystal has several reflections inside the material. Optical waveguides, optical microcavities, and waveguide gratings represent different materials and geometries.

Biophotonics. Biophotonics encompasses all the various interactions of light with biological systems. It refers especially to the effect of photons (quanta of light) on cells, tissues, and organisms. These interactions include emission, detection, absorption, reflection, modification, and creation of radiation from living tissue and materials produced from biological organisms. Areas of application include medicine, agriculture, and environmental science.

Medicine. Medical uses for photonic technologies include laser surgery, vision correction, and endoscopic examinations.

Fascinating Facts About Photonics

- In 1990, only about 10 percent of all telephone calls in the United States were carried by optical fibers. In 2010, more than 90 percent employed optical cables.
- Continued advances in photovoltaic conversion of solar energy to electricity may enable solar energy to produce about 50 percent of the world's electricity by 2050.
- Nanoplasmonics is an emerging field that studies the effect of light on the edges of metal surfaces. Applications include reduction of tumors, solar energy conversion, and detection of potential explosive reactions.
- The term "photonic engineer" is derived from the word "photon," which refers to a quanta or particle of light.
- Most photonic engineers work for large telecommunications firms and optical-fiber manufacturers. Optical physicists work for research institutions and universities.
- The demand for better visual displays for mobile (handheld) communication devices fuels the growth in the photonic industry. It is anticipated that by 2015 there will be more than 7 billion such mobile devices, which will be equivalent to the world's population.
- By 2015, two-thirds of the mobile device business will be for video transmission, thus increasing the need for better video displays.

Laser surgery uses laser light to removed diseasesd tissue or to treat bleeding blood vessels. Lasers are also extensively used in correcting problems in human vision. Laser-assisted in situ keratomileusis (LASIK) is a technique that uses a microkeratome laser to cut flaps in the cornea and remove excess tissue to correct myopia (near-sightedness). An alternative procedure is photorefractive keratectomy (PRK), which uses an excimer laser to reshape the corneal surface. Other optical uses of lasers include the removal of cataracts and the reduction of excess ocular pressure in the treatment of glaucoma.

Using a fiber-optic flexible tube and a suitable light source, a physician can obtain visual images of internal organs without more invasive surgery or high-energy X rays.

Military. Photonic devices have found use in military operations in terms of sensors, particularly infrared. Through the use of light-emitting diodes (LEDs) and lasers, photonics technologies are being developed for the infantry soldier on the battlefield and the field officer in the command center. This technology is also utilized in diverse areas such as navigation, search and rescue, minelaying, and detection. Applications range from an optical scope that enables soldiers to see around obstacles during night operations using a flexible fiber-optic tube and to weapons such as low-, medium-, and high-power lasers in the millimeter (microwave) wavelength region.

IMPACT ON INDUSTRY

Photonics is a worldwide industry that cuts across the sectors of academia, government, and industry. Although the United States has been on the forefront developing technologies and instruments, there is intense competition throughout the world. European leaders in photonics include Germany and Switzerland, while in Asia, China and India have rapidly emerging photonic industries.

Annual trade fairs held in different countries enable groups to get together to share ideas and display the latest optical technologies. Experts from all sectors present developments and applications, showcasing new technologies and new approaches to problems in the industry. It is this threefold collaboration that has supported and spurred the rapid development in the photonics industry.

Government. Photonic applications include governmental interest in the area of security and defense. Light sensors and detectors are being developed for airport security and to safeguard government facilities. Defense-related technologies include new guidance systems and laser weapons. There are also applications for civilian aviation including stabilization systems.

The space industry also is involved heavily in photonics. National Aeronautics and Space Administration (NASA) scientists and engineers design instruments to be placed aboard satellites to capture faint light emissions from distant objects in all regions of the electromagnetic spectrum.

Government funding has lead to new research initiatives. For example, in North America the United States' National Science Foundation (NSF) and the Natural Sciences and Engineering Research Council of Canada (NSERC) provide financial support and encouragement to university and research institutions to develop new photonic technologies to move this industry forward.

University Research. Tucson, Arizona, and Ottawa, Ontario in Canada are centers of the photonics industry in their respective countries. Tucson is known as "Optics Valley," and Ottawa is called "Silicon Valley North." The Optical Sciences Center (OSC) on the campus of the University of Arizona is recognized as a magnet for high-level academic and research programs.

One focus of interest at research universities has been in biophotonics. A large number of innovations in medicine and life sciences would be impossible without optical technologies. There is a subfield emerging called biomedical optics. The Center for Biophotonics, Science and Technology, located at the University of California, Davis, is focused on the development of optical tools and technology in medicine and the life sciences. Super-resolution optical microscopes, light sources, molecular sensors, and fluorescent proteins are developed there.

Massachusetts Institute of Technology's Microphotonics Center (MPhC) specializes in microphotonics and nanophotonics. It develops circuit-board devices with increased speed, capacity, and bandwidth for telecommunications, computing, and sensing. Researchers at Georgia State University in Atlanta have been studying the illumination of ultrasmall metal surfaces (nanoplasmonics).

Industry and Business. There are photonics technologies being used in diverse areas of industry and business. Some sample sectors include automobile manufacturing, mechanical engineering, micro-production, toolmaking, job-order production, and the aerospace, photovoltaic, and electronics industries.

For example, the Rochester, New York, area boasts several leading optical and imaging companies. The Rochester Regional Photonics Cluster is a nonprofit organization founded to promote and enhance photonics in New York State. There are more than fifty businesses in the Rochester area focused on optics and imaging.

A growing field is "green photonics," which includes energy and the environment, photovoltaics, lighting, energy efficiency, and climate protection. These optical technologies can contribute to efficient production and energy conservation when they are used with renewable energies.

CAREERS AND COURSE WORK

Photonics is a multidisciplinary field. It has roots in physics through classical optics and quantum theory. The explosion in the applications has been driven by the use of engineering to develop instruments and devices utilizing the particular properties of photons for transmitting, sensing, and detecting.

Career paths are in optical engineering, illumination engineering, or optoelectronics. There are more than one hundred universities in the United States offering degree programs or conducting research in photonics. There are also a number of community colleges that offer associate's degrees for careers as laser technicians. The basic undergraduate major would be physics with some emphasis in optics. A typical master's or doctoral program would concentrate on physics and quantum optics or optical engineering with research work in lasers or photonics.

The number of industries using photonics technology is growing. Photonics is prevalent in telecommunications, medicine, industrial manufacturing, energy, lighting, remote sensing, security, and defense. Job titles include research physicist, optical engineer, light-show director, laser manufacturing technician, industrial laser technician, medical laser technician, and fiber-optic packaging and manufacturing engineer.

SOCIAL CONTEXT AND FUTURE PROSPECTS

The photonics industry is an important component in the ever-growing use of handheld devices for voice, video, and data. The job growth in photonics is anticipated to be in the design and manufacture of display screens for television sets, computer monitors, mobile phones, handheld video-game systems, personal digital assistants, navigation systems, electronic-book readers, and electronic tablets such as the iPad. These systems have traditionally used semiconductor light sources such as light-emitting and superluminescent diodes (LEDs and SLDs), fluorescent lamps, and cathode ray tubes (CRTs). Plasma display panels (PDPs) and liquid crystal displays (LCDs) are in great demand. Green photonics develops organic light-emitting diodes (OLEDs) and light-emitting polymers (LEPs).

The design and development of media such as

glass or plastic fibers for transmission is another career path in photonics. There is a need for engineers to develop new photonic crystals, photonic crystal fibers, and metal surfaces (nanoplasmonics).

There is also demand for better photodetectors that range from very fast photodiodes (PDs) for communications to charge-coupled devices (CCDs) for digital cameras to solar cells that are used to collect solar energy.

Joseph Di Rienzi, Ph.D.

FURTHER READING

Cvijetic, Milorad. *Optical Transmission: Systems Engineering.* Norwood, Mass.: Artech House, 2004. Investigates the optimization of optical networks. It starts with first principles and then works through complex system-design architectures.

Hecht, Jeff. *Beam: The Race to Make the Laser.* New York: Oxford University Press, 2005. Captures the excitement and the personalities involved in the laser's discovery.

Longdell, Jevon. "Entanglement on Ice." *Nature* 469 (January 27, 2011): 475-476. Summary of two technical papers in same issue on the development of crystals using quantum entanglement for telecommunication memory units.

Menzel, Ralf. *Photonics: Linear and Nonlinear Interactions of Laser Light and Matter.* 2d ed. Berlin: Springer-Verlag, 2007. For the advanced student, covers how light interacts with matter, both for linear and nonlinear cases. The bibliography contains an enormous set of references.

WEB SITES

American Society for Photobiology
http://www.pol-us.net/ASP_Home/index.html

Illuminating Engineering Society
http://www.ies.org

National Center for Optics and Photonics Education
http://www.op-tec.org

Optoelectronics Industry and Development Association
http://www.oida.org

See also: Optics

PLANETOLOGY AND ASTROGEOLOGY

FIELDS OF STUDY

Astronomy; astrophysics; astrobiology; artificial intelligence; aeronautics; astronautics; biology; evolutionary biology; microbiology; biochemistry; chemistry; isotope geochemistry; computer science; computer programming; electronics; electrical engineering; geology; glaciology; metallurgy; geomorphology; meteorology; meteoritics; mineralogy; petrology; planetary geology; vulcanology; mathematics; mechanical engineering; field robotics; space robotics; space science.

SUMMARY

Planetology and astrogeology are separate branches of science that examine the physical and chemical characteristics of the planets and minor bodies in the solar system. The principle difference between these two scientific disciplines is that planetology is inclusive of all planetary bodies, while astrogeology concentrates on those worlds that are basically similar to the Earth. Scientists within the field of planetology can study a variety of topics that include planetary atmospheres, interiors, orbital characteristics, the potential for life, and all aspects of planetary formation and evolution. In comparison, astrogeology essentially concentrates on the various surface features and geological processes of the Earth as seen on other worlds.

KEY TERMS AND CONCEPTS

- **Asteroid:** Small solid body composed of either ice or rock that is between 30 meters and 600 kilometers in diameter and is the parent body for most meteorites.
- **Asteroid Belt:** Concentration of hundreds of thousands of asteroids positioned roughly between the orbits of Mars and Jupiter.
- **Astrobiology:** Branch of biology that deals with the possible life-forms that may exist on other planets or in extreme environments.
- **Cosmochemistry:** Branch of chemistry that deals with the chemical composition of extraterrestrial materials and the composition of stars.

- **Dwarf Planet:** Planet such as Pluto or Eris that is relatively small, is mainly composed of exotic ices, and orbits at the extreme limit of the solar system.
- **Extremophile Life-Form:** Microbial life-form that exists under extreme conditions of acidity, temperature, or salinity.
- **Jovian Planet:** Planet that has a very large mass, large size with a low density, and is composed mainly of hydrogen and helium gas, much like Jupiter.
- **Kuiper Belt:** Region of the solar system beyond Neptune where a large number of comets and dwarf planets exist.
- **Meteorite:** Piece of rock or metal that originated in the asteroid belt and has survived its fiery plunge through Earth's atmosphere and reached the surface intact.
- **Near-Earth Object:** Comet or asteroid that follows orbits that cross the path of the Earth and is a candidate for a possible impact event.
- **Runaway Greenhouse Effect:** Effect produced when a planet builds up an intense surface temperature like that on Venus because its atmosphere tends to trap heat rather than allow it to escape to space.
- **Terrestrial Planet:** Relatively small, cold, solid planet composed of iron and silicate minerals with surface features similar to those of Earth.

DEFINITION AND BASIC PRINCIPLES

The terms "planetology" and "astrogeology" are used to describe the scientific disciplines that study the planets in the solar system and those objects, believed to be planets, that orbit other distant stars. "Planetology" is a more general term that includes the study of all planets in every respect.

"Astrogeology" is actually somewhat of a misnomer; during the space program in the 1960's, the term was used to describe geological situations on other planets. If taken in a literal sense, astrogeology refers to the geology of the stars, which is not the case. Geology is the science that deals with the history of the Earth and life as recorded in rocks. The prefix "astro" indicates a place of origin beyond the Earth and is used in conjunction with many other sciences such as astrobiology, astrophysics, and astronautics.

The unmanned spacecraft missions of the 1960's to Mercury, Venus, the Moon, and Mars literally created the field of astrogeology. Traditional geologists could compare their understanding of Earth processes to what they were seeing on these other planets. Geologists wondered how the craters had formed on the Moon, but during the 1960's, it could not be determined with any certainty. The scientific community was equally divided between whether the craters were the result of impacts or volcanic in origin. A definitive answer had to wait for the astronauts of the Apollo program, who brought back rock samples. For the first time, scientists had a verifiable piece of the Moon to compare with Earth rocks. Since the Apollo program, several spacecraft have landed on Venus, Mars, and Titan (one of Saturn's moons), but none of these has brought back any materials.

BACKGROUND AND HISTORY

The planets in the solar system have attracted the attention of humans since before recorded history. One of the first observations was the recognition that stars remain in fixed positions while planets move in the sky. With the development of writing and numbers, early astronomers were able to accurately calculate and predict planetary motion. This was the limit of planetary science until the invention of the telescope in 1607.

The telescope transformed the planets from bright little points of light into actual worlds with definable surface features and cloud formations. In 1610, Galileo's telescopic observations of Jupiter revealed a "miniature solar system" of revolving moons, and his discovery of the phases of Venus helped support the heliocentric model of the solar system (the plants revolving around the Sun). Subsequent improvements in telescope design and quality led to the discovery of Saturn's rings and numerous moons, as well as the planets Uranus and Neptune. Further discoveries were limited only by technology.

The next major breakthrough came in the 1950's with the development of rocket technology and spacecraft design. For the first time, scientists were able to extend their observations beyond Earth's atmosphere and send instruments to the Moon and to various planets. By 1989, all the gas giant planets had been visited by spacecraft, leaving only Pluto as an unexplored world. In addition to gathering an enormous amount of data on each of these planets, the spacecraft also examined all of their major moons. In fact, many of these moons turned out to be more interesting than their planets. When the New Horizons mission was launched in 2004, it was aimed at the ninth planet, Pluto, and beyond into the Kuiper belt. Since then, controversy broke out in the astronomical community over whether Pluto should be defined as a planet. It has since been recognized as a dwarf planet.

HOW IT WORKS

The scientific disciplines of planetology and astrogeology attempt to answer fundamental questions concerning the origin and evolution of the planets in the solar system. To observe the planets from Earth requires the cooperation of several different sciences and technologies. Modern planetology includes such disciplines as astronomy, astrobiology, astrogeology, astrophysics, and cosmochemistry, coupled with various technologies such as computer science, electronics, and mechanical engineering. Their approach to problem solving involves a combination of direct observation and data collection with various laboratory experiments and computer simulation models.

Remote Sensing. Before the electronic age, planetary studies were limited by the vision of astronomers and the optics of their telescopes. In virtually all aspects of astronomy-related science, researchers must depend on indirect observations through various electronic instruments, particularly if they are studying distant stars and galaxies. Astronomers must understand how the basic principles of light and other forms of electromagnetic energy affect what they are seeing. Planetary scientists, by employing a combination of Earth-based telescopic images and direct spacecraft observations, can actually see events happening on these planets in real time.

Interference from the Earth's atmosphere has always been a problem for optical astronomers, especially when attempting to view surface details on the terrestrial planets such as Mercury or Mars. Planets with dense atmospheres such as Venus also present a problem for astronomers because their thick cloud layers prevent direct surface observations. To overcome this difficulty, in the 1950's, astronomers developed a technique using radar imaging to reveal surface features. Radar signals can easily penetrate clouds and are reflected back by items they hit. By timing the rate of return for these signals, astronomers assembled computer-generated maps indicating the high and low elevations. Although the quality of these early surface radar images was quite poor, it gave astronomers an idea of the nature of the geology of Venus. Later, orbiting spacecraft provided much higher quality images, which were used to construct a complete geological surface map of Venus. A similar technique employing laser technology has been used to map the elevations of geological features on Mars with pinpoint accuracy.

Direct Observation. The geologist primarily depends on fieldwork to construct geological maps and determine the location of mineral resources. However, extraterrestrial fieldwork has been limited to the Moon. All twelve of the Apollo astronauts who went to the Moon were trained in geology, but only one, Harrison Schmitt, was a professional geologist. Supplied with detailed lunar maps and reports of surface materials, the Apollo astronauts were able to successfully land at six locations and collect more than 400 kilograms of rock. Scientists continue to examine these materials and make exciting new discoveries with technologies that did not exist at the time of the Moon landings.

Before the Apollo Moon landings, geologists did have the opportunity to study extraterrestrial materials in the form of meteorites. By studying the chemical and mineralogical composition of meteorites and comparing them with Earth rocks, scientists were able to confirm their extraterrestrial origin. Meteorites proved to be older than the Earth and are believed to have originated in the asteroid belt between Mars and Jupiter. They represent some of the oldest solid material in the solar system and are the building blocks of the terrestrial planets.

APPLICATIONS AND PRODUCTS

Global Resource Management. One of the major benefits derived from the study of the other planets in the solar system is the ability to turn that technology around and study the Earth. Observing the Earth from space offers scientists the opportunity to view the Earth as a single entity rather than as a collection of apparently unrelated components. The technique of multispectral imaging has been used to map the mineral composition of the Moon's surface and to search for evidence of water. Similar technology has also been employed in mineral exploration on Earth.

Vast regions of the Earth's surface, including parts of Siberia and central Australia, remain virtually unexplored and are believed to contain great mineral wealth. The use of satellite technology has enabled geologists to assess a site's potential without actually setting foot on the ground. Although determining a site's true value still requires fieldwork, remote-sensing data obtained from space certainly can make the work more efficient and reduce expenses.

Food production management is another area that can directly benefit from planetary monitoring technology. As of 2010, the world's population was more than 6.8 billion and was expected to reach 9 billion in 2042. Space-age technology, by providing information about global conditions, can help farmers increase food production capabilities and manage resources to meet these demands. Fisherman can employ satellite data to help track schools of fish to increase the efficiency and productivity of their efforts. They can also use this information to monitor their fishing grounds and preserve them.

Planetary monitoring can help address another concern, the availability of an abundant supply of drinkable water, by keeping track of global water resources. The problem of maintaining an adequate supply of water affects the world's population in that water is also used for many other purposes, including irrigating crops.

Meteorology. Meteorologists use part of the data that planetologists have derived from their studies of planets with dense atmospheres. Satellite technology originally designed to study the atmosphere of another planet has been adapted to monitor the Earth's dynamic weather. Studying Venus, with its extremely dense atmosphere and its runaway greenhouse effect, provides Earth scientists with working models to use when trying to understand the effects of greenhouse gases in the Earth's atmosphere. Observing the various weather systems in the atmospheres of the Jovian planets also helps meteorologists understand wind and weather patterns on the Earth. Jupiter's Great Red Spot is essentially a 350-year-old hurricane, and Neptune exhibits the highest velocity winds of any planet in the solar system. Meteorologists have benefited from studying the dust devils seen blowing across the surface of Mars. These small dust storms closely resemble tornados on Earth. Periodically, Mars also experiences global dust storms that lift huge quantities of dust high into the atmosphere and block out most of its surface features for months on end.

Climate Change. The effects of climate change—whether the shrinking of the polar ice caps or the expansion of the deserts—can be seen clearly from space. The deforestation of the Brazilian rain forest as well as the amount of sediment that a river carries into the ocean each year can be precisely measured from satellite observations. Oceanographers can use the data from satellite observations in their studies of ocean currents and the effects of pollution on surface water and coastlines. City planners can use satellite technology to monitor urban sprawl and help develop better methods of waste management and disposal. Climatologists can study the localized weather patterns that develop over major cities and how they affect the smaller communities down wind. In the United States, these localized weather patterns are most apparent in the Great Lakes region and on the eastern seaboard. Similarly, the inadequacy of environmental controls in emerging industrial countries such as India and China is apparent when viewed from space. The pollution generated affects not only populations in India and China but also those in neighboring areas.

Analytical Instrumentation. Before the Apollo Moon landings, meteorites were the only extraterrestrial materials available for astrogeologists to study. Usually the classification of a meteorite requires a certain amount of destructive analysis. In many cases, the most interesting and rare meteorites are available only in very small quantities, thereby limiting the amount of material available for analysis. Similarly, only small amounts of the rocks recovered from the Moon were available for analysis. The National Aeronautics and Space Administration (NASA) deliberately preserved a large quantity of lunar material for future scientists to study with instruments not yet invented. They realized that another trip to the Moon might not occur for many years.

To cope with the limited availability, astrogeologists had to develop techniques to gain the maximum amount of data from the least amount of material. They needed the help of technicians to invent the electronic equipment and to develop the procedures needed to analyze the material. Radiation became an important component in modern analytical technology. Geologists often employ X-ray diffraction and X-ray fluorescence to identify the minerals in a rock. By using a mass spectrometer, the half-life decay rates of certain radioactive isotopes can be measured to obtain the age of Moon rocks or to determine the cosmic-ray exposure age of meteorites. Other instruments such as the electron microprobe, an instrument that can analyze particles as small as a few microns in size, are the primary tools of the scientists who study extraterrestrial materials.

The scanning electron microscope is another valuable tool that gives the scientist the ability to magnify the object that they are studying to an incredibly high power. This instrumentation is especially useful to astrobiologists who are attempting to prove that fossils of microbial life-forms are present in certain types of meteorites. If their theory is correct, then these extremophile life-forms could have originated somewhere else in the solar system and have been brought to Earth by either comets or meteorites very early in Earth's history.

IMPACT ON INDUSTRY

The space programs of the 1960's challenged American industry with the task of creating new technologies that were capable of taking humans into space and bringing them back alive. The Moon soon became the focus of space exploration. In April, 1961, President John F. Kennedy set a goal for the nation of landing a man on the Moon and returning him safely to Earth by the end of the decade. From the start, Project Apollo was presented to the public as pitting American technology against that of the Soviets. Winning the race to the Moon would require the total cooperation of various government agencies, university scientists, the military, and a broad spectrum of industry. Because it was a matter of national pride, funding was not one of NASA's major problems. However, many people viewed Project Apollo as an extravagant use of taxpayer's money, as the project cost more than $26 billion. However, some of the space technology was applied in the development of commercial products, and the project generated many jobs, at one time employing more than 600,000 workers in manufacturing components for the various lunar vehicles.

Project Apollo ended in 1972, when President Richard M. Nixon decided that NASA should discontinue the lunar missions and develop a space shuttle. President George W. Bush announced his vision of returning to the Moon and eventually landing

Fascinating Facts About Planetology and Astrogeology

- Saturn's moon Enceladus has active geysers that spew giant plumes of water hundreds of kilometers out into space. Scientists have been able to detect the presence of various organic molecules in these plumes.
- The twelfth man to set foot on the Moon in December, 1972, was a scientist. The first eleven were military pilots, but Harrison H. Schmitt was a professional geologist who later learned to fly aircraft.
- On October 6, 2008, astronomer Richard Kowalski discovered that a small asteroid was headed for an impact with Earth. Later calculations predicted that it would hit within the next thirteen hours somewhere in the Sudan. It did, and by December, more than forty-seven meteorite fragments had been collected.
- In the search for extraterrestrial life, Mars always seemed the most promising place to look. However, strong evidence of warm oceans of water beneath the icy surface of Jupiter's moon Europa make it the more likely location for life.
- Scientists were certain that planets had been hit by comets and other objects, but they never expected to

see it happen. In July, 1994, more than twenty large fragments of comet Shoemaker-Levy 9 plunged into the atmosphere of Jupiter, leaving behind huge dark scars that marked the points of impact.
- Clyde William Tombaugh, the astronomer who discovered Pluto in 1930, died in 1997. A small portion of his ashes was later placed onboard the New Horizons spacecraft, which is bound for Pluto.
- The Cassini spacecraft successfully landed a small probe on the surface of Saturn's moon Titan in 2005. The probe discovered features that resemble lakes, rivers, and shorelines on Earth, but on Titan, they were formed by liquid methane instead of water.
- The Moon has long been thought of as a waterless world. Studies have provided evidence that water ice may be present at the bottom of deep craters near the Moon's south pole that are never exposed to sunlight.
- The Earth is not the only planet with meteorites on its surface. The rovers *Spirit* and *Opportunity* photographed meteorites sitting among the rocks they encountered as they traversed the surface of Mars.

humans on Mars. NASA was charged with developing a new heavy-lift rocket and a crew vehicle that could provide support for the International Space Station and also carry astronauts to the Moon. However, these programs were underfunded from the start and did not get very far before being canceled.

The Moon in View. The Moon is once again becoming a focal point for manned exploration, with the possibly of another race between nations to see who gets there first. Political prestige is likely to be the motivating factor again, but this time the race will not be between the United States and Russia, but perhaps between China and India or possibly Japan. Apparently, Europe and Russia have little interest in landing humans on the Moon. Perhaps the optimal way to explore the Moon is through international cooperation rather than competition. NASA and the European Space Agency (ESA) conduct most of the planetary space missions.

President Barack Obama canceled the Constellation program that was to return Americans to the Moon. However, NASA is continuing to develop heavy-lift rockets, deep-space vehicles, and the technology necessary for human space flight. The

space budget allows for unmanned Mars missions, but not manned. The budget also addresses the need for more scientific studies on the potential threat from near-Earth objects.

CAREERS AND COURSE WORK

Students who are interested in making either planetology or astrogeology their career must first complete an undergraduate degree program in one of the fundamental sciences, which include biology, chemistry, geology, and physics. Adding computer science, mathematics, or electrical or mechanical engineering as a double major would increase job prospects. Most students pursue graduate work in a specialty. A master's degree is usually the minimum requirement for most technicians, while a doctorate is more appropriate for a senior scientist position. There are many ways to become employed in space science research, and students should possess a variety of technical skills to compliment their academic training. The industrial job market for highly skilled scientists and technicians in space science is quite unpredictable, but positions are likely to be available for the best applicants.

University teaching positions present an opportunity for scientists with doctorates to find employment and still pursue their own individual research interests. Major universities expect their professors to conduct independent research and encourage collaborative efforts with government agencies or major museums. Postdoctorate positions are usually available to recent graduates so that they can work with and learn from senior scientists in their field. Federal grants are available to scientists to support their research, although such grants can be difficult to obtain. Governmental agencies such as NASA, the National Oceanographic and Atmospheric Administration (NOAA), National Science Foundation (NSF), and the U.S. Geological Survey (USGS) all employ Earth scientists and geologists in various positions. The USGS actually has an astrogeology branch in Flagstaff, Arizona, where maps are created from the data returned from many of the planetary missions.

SOCIAL CONTEXT AND FUTURE PROSPECTS

Historians have often stated that the twentieth century will most likely be remembered for its two world wars and for the realization of space travel. The wars and the weapons race are probably responsible for humankind's venturing into space. However, in addition to powerful rockets and brave astronauts, sending people into space required major advancements in technology to create the necessary hardware. Technology originally developed in connection with the space program can be found in almost every modern technological necessity, such as cell phones, personal computers, and medical diagnostic instrumentation.

Although the primary motivation for sending a man to the Moon was political, it momentarily opened the eyes of the world to something greater than nationalism. The images of Neil Armstrong and Buzz Aldrin on the Moon on July 21, 1969, show what humankind is capable of achieving. The five subsequent lunar landings drew little public interest, and humankind has remained fixed in Earth orbit since 1972. Although human exploration of the solar system appears to be on hold for the foreseeable future, many more robotic missions are planned. Planetary probes have provided unimaginable visions of worlds. Perhaps it will take a major discovery for humankind to once again become fascinated with space exploration.

Paul P. Sipiera, Ph.D.

FURTHER READING

Chyba, Christopher. "The New Search for Life in the Universe." *Astronomy* 38, no. 5 (May, 2010): 34-39. A well-illustrated article written for the nonscientist describing the fundamentals of the science of astrobiology and the search for extraterrestrial life.

Geotz, Walter. "Phoenix on Mars." *American Scientist* 98, no. 1 (January, 2010): 40-47. A comprehensive article covering the scientific results from NASA's Phoenix Mars Lander.

Greeley, Ronald, and Raymond Batson. *The Compact NASA Atlas of the Solar System.* New York: Cambridge University Press, 2001. A wealth of detailed information that includes geological maps, photographs, and illustrations of all the planets and minor bodies in the solar system.

Publications and Graphics Department-NASA. *Spinoff: Fifty Years of NASA-Derived Technologies, 1958-2008.* Washington, D.C.: NASA Center for Aero-Space Information, 2008. Comprehensive review of the commercial products derived from technologies developed for the U.S. space program.

Schmitt, Harrison H. *Return to the Moon: Exploration, Enterprise, and Energy in the Human Settlement of Space.* New York: Copernicus Books, 2006. Written by the only geologist-astronaut to walk on the Moon, this book deals with the future economic benefits that can be realized from the Moon.

Sparrow, Giles. *The Planets: A Journey Through the Solar System.* London: Quercus, 2006. A spectacular collection of images taken from the various space missions to the planets.

Talcott, Richard. "How We'll Explore Pluto." *Astronomy* 38, no. 7 (July, 2010): 24-29. A well-illustrated article written for the nontechnical reader describing NASA's New Horizons mission to Pluto.

WEB SITES

Jet Propulsion Laboratory
Planetary Exploration
http://www.jpl.nasa.gov/missions

National Aeronautics and Space Administration
National Space Science Data Center
http://nssdc.gsfc.nasa.gov

The Planetary Society
http://www.planetary.org/home

U.S. Geological Survey
Astrogeology Science Center
http://astrogeology.usgs.gov

The White House
Office of Science and Technology Policy
http://www.whitehouse.gov/administration/eop/
ostp

See also: Atmospheric Sciences; Meteorology; Space
Science; Space Stations; Telescopy.

PROPULSION TECHNOLOGIES

FIELDS OF STUDY

Physics; chemistry; thermodynamics; gas dynamics; aerodynamics; heat transfer; materials; controls.

SUMMARY

The field of propulsion deals with the means by which aircraft, missiles, and spacecraft are propelled toward their destinations. Subjects of development include propellers and rotors driven by internal combustion engines or jet engines, rockets powered by solid- or liquid-fueled engines, spacecraft powered by ion engines, solar sails or nuclear reactors, and matter-antimatter engines. Propulsion system metrics include thrust, power, cycle efficiency, propulsion efficiency, specific impulse, and thrust-specific fuel consumption. Advances in this field have enabled humanity to travel across the world in a few hours, visit space and the Moon, and send probes to distant planets.

KEY TERMS AND CONCEPTS

- **Cycle Efficiency:** Ratio of the useful work imparted by a thermodynamic system to the heat put into the system.
- **Delta-V:** Velocity increment, expressed in units of speed. Measure of the energy required for propulsion from one orbit or energy level to another.
- **Equivalent Exhaust Velocity:** Total thrust divided by the propellant mass flow rate.
- **Fuel-To-Air Ratio:** Ratio of the fuel mass burned per unit time to the air mass flow rate through the burner per unit time.
- **Geostationary Earth Orbit (GEO):** Type of geosynchronous orbit, 35,785 kilometers above the equator, in which the orbiting object appears to stay at a fixed point relative to the Earth's surface at all times.
- **Low Earth Orbit (LEO):** Orbit occurring between roughly 100 and 1,500 kilometers above the Earth, with the high point sometimes reaching 2,000 kilometers.
- **Mass Ratio:** Ratio of the initial mass at launch to the mass left when propellant is expended.
- **Overall Pressure Ratio:** Ratio between the highest and the lowest pressure in a propulsion cycle.
- **Propulsive Efficiency:** Ratio of kinetic energy imparted to the propulsion system of the vehicle to the net work done by the propulsion system.
- **Specific Impulse:** Equivalent exhaust speed, divided by the standard value of acceleration due to gravity, to give units of seconds.
- **Thermal Efficiency:** Ratio of work done by the propulsion system to heat put into the system.
- **Thrust:** The force exerted by the propulsion system on the vehicle.

DEFINITION AND BASIC PRINCIPLES

Propulsion is the science of making vehicles move. The propulsion system of a flight vehicle provides the force to accelerate the vehicle and to balance the other forces opposing the motion of the vehicle. Most modern propulsion systems add energy to a working fluid to change its momentum and thus develop force, called thrust, along the desired direction. A few systems use electromagnetic fields or radiation pressure to develop the force needed to accelerate the vehicle itself. The working fluid is usually a gas, and the process can be described by a thermodynamic heat engine cycle involving three basic steps: First, do work on the fluid to increase its pressure; second, add heat or other forms of energy at the highest possible pressure; and third, allow the fluid to expand, converting its potential energy directly to useful work, or to kinetic energy in an exhaust.

In the internal combustion engine, a high-energy fuel is placed in a small closed area and ignited by compression. This produces expanding gas, which drives a piston and a rotating shaft. The rotating shaft drives a transmission whose gears transfer the work to wheels, rotors, or propellers. Rocket and jet engines operate on the Brayton thermodynamic cycle. In this cycle, the gas mixture is compressed adiabatically (no heat added or lost during compression). Heat is added externally or by chemical reaction to the fluid, ideally at constant pressure. The expanding gases are exhausted, with a turbine extracting some work. The gas then expands out through a nozzle.

BACKGROUND AND HISTORY

Solid-fueled rockets developed in China in the thirteenth century achieved the first successful continuous propulsion of heavier-than-air flying machines. In 1903, Orville and Wilbur Wright used a spinning propeller driven by an internal combustion engine to accelerate air and develop the reaction force that propelled the first human-carrying heavier-than-air powered flight.

As propeller speeds approached the speed of sound in World War II, designers switched to the gas turbine or jet engine to achieve higher thrust and speeds. German Wernher von Braun developed the V2 rocket, originally known as the A4 for space travel, but in 1944, it began to be used as a long-range ballistic missile to attack France and England. The V2 traveled faster than the speed of sound, reached heights of 83 to 93 kilometers, and had a range of more than 320 kilometers. The Soviet Union's 43-ton Sputnik rocket, powered by a LOX/RP2 engine generating 3.89 million Newtons of thrust, placed a 500-kilogram satellite in low Earth orbit on October 4, 1957.

The United States' three-stage, 111-meter-high Saturn V rocket weighed more than 2,280 tons and developed more than 33.36 million Newtons at launch. It could place more than 129,300 kilograms into a low-Earth orbit and 48,500 kilograms into lunar orbit, thus enabling the first human visit to the Moon in July, 1969. The reusable space shuttle weighs 2,030 tons at launch, generates 34.75 million Newtons of thrust, and can place 24,400 kilograms into a low-Earth orbit. In January, 2006, the New Horizons spacecraft reached 57,600 kilometers per hour as it escaped from Earth's gravity. Meanwhile, air-breathing engines have grown in size and become more fuel efficient, propelling aircraft from hovering through supersonic speeds.

HOW IT WORKS

Rocket. The rocket is conceptually the simplest of all propulsion systems. All propellants are carried on board, gases are generated with high pressure, heat is added or released in a chamber, and the gases are exhausted through a nozzle. The momentum of the working fluid is increased, and the rate of increase of this momentum produces a force. The reaction to this force acts on the vehicle through the mounting structure of the rocket engine and propels it.

Jet Propulsion. Although rockets certainly produce jets of gas, the term "jet engine" typically denotes an engine in which the working fluid is mostly atmospheric air, so that the only propellant carried on the vehicle is the fuel used to release heat. Typically, the mass of fuel used is only about 2 to 4 percent of the mass of air that is accelerated by the vehicle. Types of jet engines include the ramjet, the turbojet, the turbofan, and the turboshaft.

Propulsion System Metrics. The thrust of a propulsion system is the force generated along the desired direction. Thrust for systems that exhaust a gas can come from two sources. Momentum thrust comes from the acceleration of the working fluid through the system. It is equal to the difference between the momentum per second of the exhaust and intake flows. Thrust can also be generated from the product of the area of the jet exhaust nozzle cross section and the difference between the static pressure at the nozzle exit and the outside pressure. This pressure thrust is absent for most aircraft in which the exhaust is not supersonic, but it is inevitable when operating in the vacuum of space. The total thrust is the sum of momentum thrust and pressure thrust. Dividing the total thrust by the exhaust mass flow rate of propellant gives the equivalent exhaust speed. All else being equal, designers prefer the highest specific impulse, though it must be noted that there is an optimum specific impulse for each mission. LOX-LH2 rocket engines achieve specific impulse of more than 450 seconds, whereas most solid rocket motors cannot achieve 300 seconds. Ion engines exceed 1,000 seconds. Air-breathing engines achieve very high values of specific impulse because most of the working fluid does not have to be carried on-board.

The higher the specific impulse, the lower the mass ratio needed for a given mission. To lower the mass ratio, space missions are built up in several stages. As each stage exhausts its propellant, the propellant tank and its engines are discarded. When all the propellant is gone, only the payload remains. The relation connecting the mass ratio, the delta-v, and specific impulse, along with the effects of gravity and drag, is called the rocket equation.

Propulsion systems, especially for military applications, operate at the edge of their stable operation envelope. For instance, if the reaction rate in a solid propellant rocket grows with pressure at a greater than linear rate, the pressure will keep rising until the rocket blows up. A jet engine compressor will stall, and flames may shoot out the front if the blades go past the stalling angle of attack. Diagnosing and solving the problems of instability in these powerful systems has been a constant concern of developers since the first rocket exploded.

APPLICATIONS AND PRODUCTS

Many kinds of propulsion systems have been developed or proposed. The simplest rocket is a cold gas thruster, in which gas stored in tanks at high pressure is exhausted through a nozzle, accelerating (increasing momentum) in the process. All other types of rocket engines add heat or energy in some other form in a combustion (or thrust) chamber before exhausting the gas through a nozzle.

Solid-fueled rockets are simple and reliable, and can be stored for a long time, but once ignited, their thrust is difficult to control. An ignition source decomposes the propellant at its surface into gases whose reaction releases heat and creates high pressure in the thrust chamber. The surface recession rate is thus a measure of propellant gas generation. The thrust variation with time is built into the rocket grain geometry. The burning area exposed to the hot gases in the combustion chamber changes in a preset way with time. Solid rockets are used as boosters for space launch and for storable missiles that must be launched quickly on demand.

Liquid-fueled rockets typically use pumps to inject propellants into the combustion chamber, where the propellants vaporize, and a chemical reaction releases heat. Typical applications are the main engines of space launchers and engines used in space, where the highest specific impulse is needed.

Hybrid rockets use a solid propellant grain with a liquid propellant injected into the chamber to vary the thrust as desired. Electric resistojets use heat generated by currents flowing through resistances. Though simple, their specific impulse and thrust-to-weight ratio are too low for wide use. Ion rocket engines use electric fields or, in some cases, heat to ionize a gas and a magnetic field to accelerate the ions through the nozzle. These are preferred for long-duration space missions in which only a small level of thrust is needed but for an extended duration because the electric energy comes from solar photovoltaic panels. Nuclear-thermal rockets generate heat from nuclear fission and may be coupled with

ion propulsion. Proposed matter-antimatter propulsion systems use the annihilation of antimatter to release heat, with extremely high specific impulse.

Pulsed detonation engines are being developed for some applications. A detonation is a supersonic shock wave generated by intense heat release. These engines use a cyclic process in which the propellants come into contact and detonate several times a second. Nuclear-detonation engines were once proposed, in which the vehicle would be accelerated by shock waves generated by nuclear explosions in space to reach extremely high velocities. However, international law prohibits nuclear explosions in space.

Ramjets and Turbomachines. Ramjet engines are used at supersonic speeds and beyond, where the deceleration of the incoming flow is enough to generate very high pressures, adequate for an efficient heat engine. When the heat addition is done without slowing the fluid below the speed of sound, the engine is called a scramjet, or supersonic combustion ramjet. Ramjets cannot start by themselves from rest. Turbojets add a turbine to extract work from the flow leaving the combustor and drive a compressor to increase the pressure ratio. A power turbine may be used downstream of the main turbine. In a turbofan engine, the power turbine drives a fan that works on a larger mass flow rate of air bypassing the combustor. In a turboprop, the power is taken to a gearbox to reduce revolutions per minute, powering a propeller. In a turboshaft engine, the power is transferred through a transmission as in the case of a helicopter rotor, tank, ship, or electric generator. Many applications combine these concepts, such as a propfan, a turboramjet, or a rocket-ramjet that starts off as a solid-fueled rocket and becomes a ramjet when propellant consumption opens enough space to ingest air.

Gravity Assist. A spacecraft can be accelerated by sending it close enough to another heavenly body (such as a planet) to be strongly affected by its gravity field. This swing-by maneuver sends the vehicle into a more energetic orbit with a new direction, enabling surprisingly small mass ratios for deep space missions.

Tethers. Orbital momentum can be exchanged using a tether between two spacecraft. This principle has been proposed to efficiently transfer payloads from Earth orbit to lunar or Martian orbits and even to exchange payloads with the lunar surface. An extreme version is a stationary tether linking a point on

---

Fascinating Facts About Propulsion Technologies

- Thirty-two launches of the Saturn V rocket system were conducted in the 1960's and early 1970's. All succeeded.

- The Galileo mission to Jupiter and beyond obtained nearly 5 of the required 9 kilometers per second delta-v from one flyby of Venus and two flybys of Earth. This is only slightly higher than the delta-v required to reach lunar orbit.

- The hydrogen airships of the 1930's carried 110 people across the Atlantic in three days, at a speed of around 135 kilometers per hour. The Airbus A380, the largest modern airliner, carries 500 to 800 people at 900 kilometers per hour.

- The turbopump of the space shuttle's main engines uses liquid hydrogen at -250 degrees Celsius, but at the end of combustion, the temperature climbs to more than 3,300 degrees Celsius. The turbine of the turbopump is driven by combustion gases and is connected to the impeller, where liquid hydrogen comes in and is pressurized.

- In the linear aerospike nozzle used in the X-33 experimental vehicle, the contoured nozzle surface is on the inside, while the outer flow boundary adjusts itself, thus avoiding the cost of a large variable geometry nozzle.

- In space, to change the plane or direction of an orbit, a spacecraft must add a velocity component sideways, enough to make the resultant velocity vector point along the new direction. This is best done at the apogee of an elliptical orbit, where the orbital speed is lowest.

- A spacecraft can harvest electricity from the potential variation in the Earth's electromagnetic field by extending an electrostatic tether down from its orbit. A current-carrying electrodynamic tether between two spacecraft can be used to gain propulsive thrust because of the Faraday force exerted by the magnetic field on the tether.

- Like all high-performance systems, propulsion systems often push the edge of stable operation. Some descriptive terms include the "pogo" instability of rockets, the "screech" instability of jet engines, the supersonic inlet "buzz," compressor "surge," and "sloshing" in satellite fuel tanks.

Earth's equator to a craft in geostationary Earth orbit, the tether running far beyond to a counter-mass.

The electrostatic tether concept uses variations in the electric potential with orbital height to induce a current in a tether strung from a spacecraft. An electrodynamic tether uses the force that is exerted on a current-carrying tether by the magnetic field of the planet to propel the tether and the craft attached to it.

Solar and Plasma Sails. Solar sails use the radiation pressure from sunlight bounced off or absorbed by thin, large sails to propel a craft. Typically, this works best in the inner solar system where radiation is more intense. Other versions of propulsion sails, in which lasers focus radiation on sails that are far away from the Sun, have been proposed. In mini magnetospheric plasma propulsion (M2P2), a cloud of plasma (ionized gas) emitted into the field of a magnetic solenoid creates an electromagnetic bubble around solenoid creates an electromagnetic bubble around 30 kilometers in diameter, which interacts with the solar wind of charged particles that travels at 300 to 800 kilometers per second. The result is a force perpendicular to the solar wind and the (controllable) magnetic field, similar to aerodynamic lift. This system has been proposed to conduct fast missions to the outer reaches of the solar system and back.

IMPACT ON INDUSTRY

The propeller, the jet engine, and the rocket are the inventions that powered flight of heavier-than-air machines. Advances in engine technology have been transferred to power plants, automobile engines, and many other applications. Extensive development occurred before, between, and during the two world wars and the Cold War, with large numbers of engines being built and used.

The hard-won experience gained from these engines enabled long-distance passenger-carrying airliners and ambitious missions to outer space. The requirements of ever better engines for combat aircraft and missiles drove intense scientific development in structures and materials, combustion, gas dynamics, aerodynamics, and heat transfer and controls, with the resulting algorithms, prediction codes, and diagnostic techniques and devices revolutionizing many other industries. Examples are the technologies for growing single-crystal turbine blades, powder metallurgy for single-piece cast turbine stages, rapid prototyping using stereolithography, tribology for bearing seals to withstand high pressures while ensuring low friction, magnetic and air bearings, plasma igniters, fuel injectors, turbochargers for cars, fluidic thrust deflectors, several alloys that withstand high temperatures, composite structures technology to reduce propellant tank mass, ultralight solar sails, and research into solid hydrogen and antimatter production. The need to accurately measure velocity, temperature, and gas composition inside the hostile environment of combustors has driven development of the field of optical diagnostics, which supports a specialized industry making various types of lasers and associated equipment.

CAREERS AND COURSE WORK

Propulsion technology spans aerospace, mechanical, electrical, nuclear, chemical, and materials science engineering. Aircraft, space launcher, and spacecraft manufacturers and the defense industry are major customers of propulsion systems. Workplaces in this industry are distributed over many regions in the United States and near many major airports and National Aeronautics and Space Administration centers. The large airlines operate engine testing facilities. Propulsion-related work outside the United States, France, Britain, and Germany is usually in companies run by or closely related to the government. Because propulsion technologies are closely related to weapon-system development, many products and projects come under the International Traffic in Arms Regulations.

Students aspiring to become rocket scientists or jet engine developers should take courses in physics, chemistry, mathematics, thermodynamics and heat transfer, gas dynamics and aerodynamics, combustion, and aerospace propulsion.

Machinery operating at thousands to hundreds of thousands of revolutions per minute requires extreme precision, accuracy, and material perfection. Manufacturing jobs in this field include specialist machinists and electronics experts. Because propulsion

systems are limited by the pressure and temperature limits of structures that must also have minimal weight, the work usually involves advanced materials and manufacturing techniques. Instrumentation and diagnostic techniques for propulsion systems are constantly pushing the boundaries of technology and offer exciting opportunities using optical and acoustic techniques.

SOCIAL CONTEXT AND FUTURE PROSPECTS

Propulsion systems have enabled humanity to advance beyond the speed of ships, trains, balloons, and gliders to travel across the oceans safely, quickly, and comfortably and to venture beyond Earth's atmosphere. The result has been a radical transformation of global society since the early 1900's. People travel overseas regularly, and on any given day, city centers on every continent host conventions with thousands of visitors from all over the world. Jet engine reliability has become so established that jetliners with only two engines routinely fly across the Atlantic and Pacific oceans.

Propulsion technologies are just beginning to grow in their capabilities. As of 2010, specific impulse values were at best a couple of thousand seconds; however, concepts using radiation pressure, nuclear propulsion, and matter-antimatter promise values ranging into hundreds of thousands of seconds. Air-breathing propulsion systems promise specific impulse values of greater than 2,000 seconds, enabling single-stage trips by reusable craft to space and back. As electric propulsion systems with high specific impulse come down in system weight because of the use of specially tailored magnetic materials and superconductors, travel to the outer planets may become quite routine. Spacecraft with solar or magnetospheric sails, or tethers, may make travel and cargo transactions to the Moon and inner planets routine as well. These technologies are at the core of human aspirations to travel far beyond their home planet.

Narayanan M. Komerath, Ph.D.

FURTHER READING

Faeth, G. M. *Centennial of Powered Flight: A Retrospective of Aerospace Research.* Reston, Va.: American Institute of Aeronautics and Astronautics, 2003. Traces the history of aerospace and describes important milestones.

Henry, Gary N., Wiley J. Larson, and Ronald W.

Humble. *Space Propulsion Analysis and Design.* New York: McGraw-Hill 1995. Combines short essays by experts on individual topics with extensive data and analytical methods for propulsion system design.

Norton, Bill. *STOL Progenitors: The Technology Path to a Large STOL Aircraft and the C-17A.* Reston, Va.: American Institute of Aeronautics and Astronautics, 2002. Case study on a short takeoff and landing aircraft development program. Describes the various steps in a modern context, considering cost, technology, and military requirements.

Peebles, C. *Road to Mach 10: Lessons Learned from the X-43A Flight Research Program.* Reston, Va.: American Institute of Aeronautics and Astronautics, 2008. Case study of a supersonic combustion demonstrator flight program authored by a historian.

Shepherd, D. *Aerospace Propulsion.* New York: Elsevier, 1972. A prescient, simple, and lucid book that has inspired generations of aerospace scientists and engineers.

WEB SITES
American Institute of Aeronautics and Astronautics
http://www.aiaa.org

Jet Propulsion Laboratory
http://www.jpl.nasa.gov

National Aeronautics and Space Administration
Marshall Research Center
http://www.nasa.gov/centers/marshall

See also: Space Science

PROSTHETICS

FIELDS OF STUDY

Human anatomy; kinesiology; physiology; ergonomics; biology; chemistry; mechanical engineering; pathology; mathematics; physics; psychology.

SUMMARY

Prosthetics is the branch of medicine focused on the replacement of missing body parts with artificial substitutes so that an individual can function and appear more natural. Prostheses are commonly used to replace hands, arms, legs, and feet; however, examples of other prosthetic devices developed to improve one's quality of life are heart valves, pacemakers, and components of the ear. Some prosthetic devices, including eye and breast implants, are developed primarily for cosmetic reasons. Several health care professions work together as a team in this process, and teams include a surgeon, a nurse, a prosthetist, and physical and occupational therapists.

KEY TERMS AND CONCEPTS

- **Biomedical Engineering:** Engineering specialty that applies engineering principles and methods to medical problems such as the development and manufacture of prosthetic limbs and organs.
- **Neuroprosthetics:** Trend in prosthetics that aims to integrate body, mind, and machine.
- **Occupational Therapy:** Health profession concerned with using work, play, and everyday activities in a therapeutic manner to increase a person's function and to enhance development.
- **Osseointegration:** Technology of attaching prosthetic legs to a titanium material directly in the bone to avoid problems with anchoring methods.
- **Physical Rehabilitation:** Process of the return to function of an individual afflicted by disorders such as injury, chronic disease, or genetic dysfunction.
- **Physical Therapy:** Health profession that seeks to restore maximum function and movement to patients whose physical abilities have been reduced by pain, disease, or injury.
- **Prosthesis:** Artificial device that replaces a missing body part or augments an injured part; also known as prosthetic device.

DEFINITION AND BASIC PRINCIPLES

"Prosthetics" is the science of developing and fitting substitute body parts. This branch of medicine is devoted to assisting patients in regaining as much function as possible after they have lost a body part from trauma, a birth defect, or illness. The replacement of a limb or other impaired or lost body part involves fitting an individual with an artificial leg, arm, or other body part to allow him or her to perform the activities of daily living.

A "prosthesis" is a device that replaces a missing body part or augments a partial one; an individual who measures, fits, and modifies the prosthesis is referred to as the "prosthetist." Legs, arms, feet, and hands are the most commonly known artificial devices.

Although closely related to field of prosthetics, the term "orthotics" is not identical in meaning. Orthotics usually focuses on the management of impairment, but the treatment may be more temporary, whereas prosthetics concerns permanent artificial replacements of body parts. An orthotist designs and fits surgical appliances; this process is referred to as "orthosis." Orthotic devices include braces, neck collars, and splints. Such devices are designed to support the patient's limbs or spine while relieving pain and helping movement. They also may be designed to restrict movement and provide an environment of protection and healing.

BACKGROUND AND HISTORY

The use of prosthetic devices can be traced back to sixteenth century knights. A German knight known as Götz of the Iron Hand may have been the first to apply the science of prosthetics. He developed and used an appliance with movable fingers to help him hold a sword (hence his nickname).

Until the twentieth century, most prosthetic devices were made of wood, but because of the large number of amputees produced by World Wars I and II, these devices began to be made of metals and fibers. The devices were designed to increase function and therefore incorporated mechanical devices and elastic materials to allow individuals to move their artificial limbs more effectively and easily.

In the late twentieth and early twenty-first centuries, advances in biomechanics and bioengineering resulted in prosthetic devices such as hydraulic knees and computer-programmable hands that sense the slightest muscle movement. Such technology has led to advances in prosthetic devices for other body parts, including the heart and the ear.

How It Works

Prosthetics is often defined as a branch of surgery involving a team approach comprising such professionals as surgeons, nurses, prosthetists, physical and occupational therapists, prosthetic technicians and assistants, rehabilitation counselors, and social workers. By combining medical science with technology, rehabilitation engineering assists in the design and development of devices to meet each individual's needs.

Interaction with the patient begins well before any surgery, with the physician determining if replacement of the natural body part is required. If a surgical procedure such as amputation is required, the physician, nurse, and social workers must prepare the patient emotionally and physically. The prosthetist, physician, and physical therapist consult with the patient to determine the size, shape, and material most appropriate for the appropriate device. The physical therapist evaluates factors such as strength and ability to wear the prosthesis and works with the patient to increase physical strength as appropriate for the device involved.

Prosthetists and technicians work with their hands and high-tech machinery to make molds or casts of the amputated area to create the desired device. This process may include casting molds, using sewing machines, and heating plastics in a special oven. Rehabilitation engineers apply their expertise as well; for example, depending on the needs of the individual patient, they may suggest a prosthetic foot that offers a more natural spring to help push off from the floor from a standing start, or they may help design a prosthetic knee to facilitate stair climbing that avoids rubbing the foot on the steps (which could result in a fall).

Most prosthetists and related health care team members often work in a combination of environments that are inspected and regulated by the American Board for Certification in Orthotics, Prosthetics, and Pedorthics. Later stages of rehabilitation engage an occupational therapist, who focuses on helping the patient complete everyday tasks independently with the new prosthesis and suggests activities to strengthen weakened muscles.

Applications and Products

The products of prosthetics are primarily the physical devices used to replace lost body parts. The most common prosthetic devices replace limbs, but other, less familiar devices are also considered to be prostheses. The users of these products include a broad range of individuals, from children born with missing limbs to military personnel who have been injured during battle. The largest population of amputees in the United States are those individuals who have lost a limb from either diabetes or peripheral vascular disease. Trauma victims, such as those who have experienced motor vehicular accidents, make up another group of users, since accidents account for many lost limbs.

Prosthetic Limbs. Over the years, prosthetic limbs consisted of combinations of springs and hinges to increase motion and function. Prosthetic limbs use a socket to fit over the remaining part of the limb and provide a link between the body and the prosthesis. Additional straps and belts are often used to attach the device to the body with soft, socklike material used in between to protect the area of contact from excessive pressure and friction. The main body of the prosthesis is often made from material such as carbon fiber, popular for its light weight, strength, and durability. These properties require less exertion of effort by the patient, and the device appears more natural.

Prosthetic legs are generally of two types: transtibial and transfemoral. A transtibial prosthesis replaces the leg below the knee, which allows the knee joint to remain functional. A transfemoral prosthesis replaces the entire leg, including above the knee joint. Traditionally, the force needed to move either type of device has come from the patient's remaining muscles, including the momentum from using his or her entire body. New technology has brought the use of myoelectric limbs, which respond by converting muscle movement to an electric signal to move the device. This technology has allowed patients with leg and arm prostheses to have better control of the limb.

Prosthetic devices for the hip are made from materials similar to those used for prosthetic legs to provide strength, comfort, and support. A prosthetic hip joint is designed to support and link the patient to the prosthetic leg by way of a socket fitted to the body's torso and pelvis using a system of straps. Some artificial hip joints use a roller system to convey forces from the socket directly to the prosthetic leg.

Prosthetic arms and hands have advanced, using stronger and lighter materials that also look more like skin. The biomechanics of these devices has improved: Once anchored to the opposite shoulder with straps across the back, these devices have come to use electrical signals from the patient's nearby muscles to move specific fingers. The future of hand and arm prostheses gives promise to a technique known as targeted muscle reinnervation (TMR), whereby the arm or hand will respond to signals from the brain to specific remaining muscles. Patients potentially will be able to manipulate a prosthetic hand as naturally as they once could move their own hand.

Prosthetic applications for the foot have been primarily rigid in design, with little if any movement. Traditionally prosthetic feet were made from leather, metal, plastic, or a combination of such materials. Modern foot prostheses has improved, with computer-controlled components designed to handle the user's weight and the return of his or her momentum. Such products have been reported to be comfortable enough for participation in recreational sports. Further improvements have occurred with the use of a carbon fiber, compression springs, and telescoping tubes that help the prosthetic foot move more naturally without inducing pain or discomfort.

Nonlimb Prosthetic Devices. Other prosthetic devices include artificial eyes, breasts, heart valves, and pacemakers. The body's natural heart valve may need to be surgically replaced if it no longer functions properly because of disease, aging, or a birth defect. This vital prosthetic heart component is made from plastic, metal, or pig tissue. Calcification of the prosthetic heart valves is the major cause of product problems, and efforts have been aimed at constructing artificial valves with surfaces that resist calcification. Technological advancements in the durability of the tissue heart valves are also an area of research and could be more applicable to younger patients.

Prosthetic eyes are traditionally made from hard materials such as acrylic, gold, ceramics, and glass. When an individual loses an eye, it is replaced with a temporary implant that is positioned toward the back of the eye socket to allow proper room for the prosthetic eye. With the use of an impression, a wax model is made, followed by a mold for casting the prosthetic eye. Components of the eye, such as the pupil and the iris, are painted on a round plastic base and eventually inserted on the prosthetic eye. Some prosthetic eyes can even be designed to allow for the attachment of eye muscles. One of the trends in design is to develop an ocular system that allows for more natural movement.

A popular product for women who have had a mastectomy, prosthetic breasts are made of various materials, including lightweight silicone, soft gel, and a variety of fabrics. Breast prosthetics are often engineered using a cast of the body shape and other parameters. The prosthetic breast needs to be lightweight to avoid strain on the back and shoulders. Cosmetics—shape and contour—is a concern in the design and development of this form of prosthesis, as is comfort. For example, one breast prosthesis has at its center a climate control pad made of a soft gel that absorbs body heat, creating a cooling sensation. Future breast prostheses are expected to be developed with some type of climate-control technology.

Pacemakers are a complex form of prosthetic devices whose design and development require high-tech electric and computer expertise. These units are manufactured by the biomedical industry and have progressed significantly to keep up with advances in medicine. Originally stimulating the heart to beat at a standard rate of around 70 beats per minute, pacemakers can now interpret signals from the patient to change the rate of the heart. The latest device can take the electrical impulses from the heart's natural pacemaker, the sinoatrial node, and increase the heart rate during activity as needed.

Less well known but gaining in popularity are ear components, specifically the cochlea. Artificial cochleas are now able to duplicate the function of converting sound waves into electronic chemical impulses.

Neuroprosthetics. Neuroprosthetics, a subspecialty of prosthetics, aims to integrate body, mind, and machine. One example is the development of a system that can decipher brain waves and translate them into computer commands. A young science, this specialty promises to allow quadriplegics to gain sufficient function to operate household electric appliances and computers by using their thoughts, transmitted by an implant.

Tissue Engineering. The concept of tissue engineering to complement prosthetics promises to play a key role in twenty-first-century prosthetic devices. Surgical techniques are being developed that could lengthen the bone in a residual limb to fit artificial limbs more effectively. Problems associated with anchoring methods can be solved with tissue engineering by way of the developing technology of attaching prosthetic legs to a titanium bolt directly in the bone, a process known as osseointegration.

IMPACT ON INDUSTRY

The prosthetics industry is growing rapidly. The sale of prosthetic devices increased from $340 million in 1996 to about $600 million by 2010. Although private businesses are experiencing acquisitions and mergers, most manufacturers and distributors in the prosthetics industry remain small to medium-sized and are adapting to growing business opportunities.

Many companies have received grants to develop better limbs that can be used not only by military personnel but also by civilian amputees. In 2005, for example, the U.S. Deparment of Defense contracted with prosthetics researchers and manufacturers to develop a mind-controlled arm at a cost of $30.4 million. Some of these prostheses incorporate cutting-edge bionic technology, sensor technology, artificial intelligence, and micromechatronics.

The market for prostheses grows continuously because of the increasing number of people with diabetes and peripheral vascular disease who undergo amputations. These two conditions together rank as the number-one cause of amputation in the United States, contributing significantly to the average of 140,000 amputations performed annually. Overall, about 1.5 million Americans are diagnosed with diabetes each year, and a growing number will be faced with the unfortunate and serious complication of limb amputation.

The design and development of prosthetic devices has been supported in government and nongovernment sectors alike. The Johns Hopkins Applied Physics Laboratory has developed prosthetic arms that are rechargeable and motor-driven for people unable to use prosthetics that depend on muscle activity

---

Fascinating Facts About Prosthetics

- In the United States, about 150,000 amputations are performed each year, and about 1.7 million Americans have lost a limb.
- Up to 80 percent of amputees experience phantom pain, intermittent severe pain in the missing limb. Once thought to be psychological in origin, this pain has been determined to most likely have a physiological cause.
- Between 2001 and September, 2010, more than 1,400 soldiers in Operation Iraqi Freedom and Operation Enduring Freedom experienced major limb or partial amputations.
- People with diabetes account for 60 percent of nontraumatic lower limb amputations.
- Of all amputated limbs, 82 percent were made because of poor circulation.
- Nearly 69 percent of amputations caused by trauma are of upper limbs.
- About 76 percent of cancer-related amputations are of lower limbs.
- In the future, artificial hands are likely to contain sensors that will allow their users not only to adjust how tightly they grip an item but to gauge its temperature.
- Chips that were developed to interface between the brain and a prosthetic limb may some day be incorporated in an artificial eye, helping blind people see.

Alatheia Prosthetics of Brandon, Mississippi, has developed an artificial skin that closely resembles human skin. At Alatheia, architects use medical-grade silicone to create realistic tissue to ensure every crease, pore, and fingerprint is perfect for each patient.

Leading U.S. prosthetics manufacturers, such as Becker Orthopedic and the Hanger Orthopedic Group, sell primarily to special laboratories and workshops. Such workshops are staffed by trained and certified prosthetists and orthotists. Physicians contact these workshops to place an order or fill a prescription for a prosthesis to be fitted on a patient.

Advances in Technology. Advances in prosthetics have made an enormous impact on the medical profession and positive differences in people's lives. In general, designers of state-of-the-art prostheses strive to improve the comfort, function, and appearance of the artificial limb. These three dimensions are the primary focus of research, development, and eventually clinical assessment by the medical profession. These factors also drive further research and development of the industry, which at times is supported through federal grants and research funding. For example, manufacturers are introducing an increased number of artificial limbs designed for older and overweight users rather than younger patients, who typically have been the focus of research and development. Such technology has introduced sensors in the foot region that can determine the walking speed of the individual, his or her exact weight, and whether the person's foot is on an incline, so that the prosthesis can be regulated appropriately.

CAREERS AND COURSE WORK

The need for prosthetists, prosthetics assistants, and technicians is expected to increase as a result of an aging population, increases in obesity and diabetes, and the medical demands of war-related amputees.

A bachelor's degree in prosthetics is usually required from a program accredited by the American Board for Certification in Orthotics, Prosthetics, and Pedorthics. Following a period of supervised clinical internship, college graduates are eligible to take examinations given by its governing board. Another route to the profession has been designed for other members of the health care team, such as surgeons, nurses, and physical therapists. These professionals may receive training in prosthetics while studying to achieve certification in their respective specialties.

Another avenue is to become certified by earning an associate degree in any field, then completing a certificate program in orthotics and prosthetics, followed by working for four years in the field and eventually passing certification exams. Programs for prosthetics assistants and technicians range from six months to two years of study, and internships and are offered by the American Academy of Orthotics and Prosthetics, which also offers continuing education courses and forums so that those in the prosthetics industry can learn about new developments.

Because some patients will require both prosthetic devices and orthotics, many programs offer degrees and certificates in both disciplines. Individuals with

education and experience in both disciplines will possess much more knowledge and therefore be more employable, compared with those with degrees or certificates in only one of the disciplines.

SOCIAL CONTEXT AND FUTURE PROSPECTS

Prosthetic devices can restore independence to people who have lost limbs or function through the impairment of body parts. With the use of a prosthesis, people can return to such fundamental activities as walking, writing with a pen, feeding themselves with a fork or spoon, receiving a handshake, holding a newborn, and even performing sports. Such abilities, which most people take for granted, are dramatic to the person who has lost function and may mean the difference between independent living and institutionalization.

Jeffrey Larson, P.T., A.T.C.

FURTHER READING

Lusardi, Michelle M., and Caroline C. Nielsen, eds. *Orthotics and Prosthetics in Rehabilitation.* 2d ed. St. Louis, Mo.: Saunders Elsevier, 2007. Examines the role of orthotic and prosthetic devices in physical rehabilitation of amputees.

May, Bella J., and Margery A. Lockard. *Prosthetics and Orthotics in Clinical Practice: A Case Study Approach.* Philadelphia: F. A. Davis, 2011. Contains numerous case studies of amputations, examining the causes of amputation, the postsurgical care, physical therapy, and the prosthetic devices used.

Ott, Katherine, David Harley Serlin, and Stephen Mihm, eds. *Artificial Parts, Practical Lives: Modern Histories of Prosthetics.* New York: New York University Press, 2002. Examines both cosmetic and functional prosthetic devices. Contains a chapter on World War II veterans.

Pitkin, Mark R. *Biomechanics of Lower Limb Prosthetics.* New York: Springer, 2010. Looks at the design of prosthetic devices from a biomechanical approach.

Shurr, Donald G., and John W. Michael. *Prosthetics and Orthotics.* 2d ed. Upper Saddle River, N.J.: Prentice Hall, 2002. A basic overview of prosthetic and orthotic devices, examining when they are used and how.

Smith, Marquard, and Joanne Morra, eds. *The Prosthetic Impulse: From a Posthuman Present to a Biocultural Future.* Cambridge, Mass.: MIT Press, 2006.

A collection of sixteen essays that examine the boundaries between prosthesis and human. Contains some history, particularly of war veterans.

WEB SITES

American Academy of Orthotists and Prosthetists
http://www.oandp.org

American Board for Certification in Orthotics, Prosthetics, and Pedorthics
http://www.abcop.org

American Orthotic and Prosthetic Association
http://www.aopanet.org

Amputee Coalition of America
http://www.amputee-coalition.org

National Association for the Advancement of Orthotics and Prosthetics
http://www.naaop.org

National Commission on Orthotic and Prosthetic Education
http://www.ncope.org

See also: Cell and Tissue Engineering; Orthopedics; Rehabilitation Engineering.

PSYCHIATRY

FIELDS OF STUDY

Biopsychiatry; clinical psychology; neurology; neuroscience; counseling; social work; suicidology; addiction medicine; endocrinology; geriatrics; internal medicine; neurosurgery.

SUMMARY

Psychiatry is a branch of medicine that specializes in the diagnosis and treatment of mental illness. To diagnose a mental illness, psychiatrists conduct mental status examinations and psychiatric histories. Psychiatrists can also order brain scans and laboratory tests to rule out medical causes for psychiatric symptoms. To treat a mental illness, psychiatrists prescribe medications and engage patients in psychotherapy. As licensed physicians, psychiatrists are the only mental health professionals—with a couple of exceptions—who can prescribe medications to treat psychiatric disorders; clinical psychologists, nurses, social workers, and counselors cannot. In the United States, about one in four adults, in any given year, meets the criteria for a diagnosable mental illness, having conditions that cause distress and impairment in functioning. Hence, the field of psychiatry is important in helping individuals and families live more enjoyable and productive lives.

KEY TERMS AND CONCEPTS

- **Anxiety Disorder:** Disorder in which anxiety (persistent feelings of apprehension, tension, dread, worry, or uneasiness) is the predominant symptom.
- **Bipolar Disorder:** Disorder marked by excessive mood swings (alternating cycles of depression and mania).
- **Depression:** Common mood disorder that involves severe and persistent sadness, lack of interest in pleasurable activities, suicidal thoughts, and physical symptoms, such as sleep disturbance, loss of appetite, and reduced sexual desire.
- **Mental Status Examination:** Assessment of a patient's level of functioning, which examines the individual's thoughts, feelings, and behaviors.
- **Psychosis:** Mental state defined by a failure to differentiate between what is real and what is not in terms of thoughts and perceptions.
- **Psychotherapy:** Set of techniques for treating mental illnesses that consists of a variety of talk therapies.
- **Psychotropic Medication:** Group of medications used to treat mental illnesses.
- **Schizophrenia:** Severe mental illness, formerly known as dementia praecox, characterized by delusions (false beliefs) and hallucinations (false perceptions).

DEFINITION AND BASIC PRINCIPLES

Psychiatry is a subspecialty of medicine that involves the diagnosis and treatment of mental disorders, as defined and classified in the fourth edition of the Diagnostic and Statistical Manual of the American Psychiatric Association (2000; known as DSM-IV-TR). Mental illnesses are biologically based brain diseases that are widespread in the general population and vary in duration and severity. The main burden of mental illness is concentrated in the 6 percent of Americans who suffer from a serious mental illness such as schizophrenia, bipolar disorder, and major depression.

The field of psychiatry is predicated on the assumption that serious mental disorders result from abnormalities in the structure or function of the brain. Although no specific brain anomalies have been identified as the definitive cause of mental illness, the latest neuroscientific studies suggest that imbalances in neurochemicals—also called neurotransmitters—or malfunctions in their transportation from nerve cell to nerve cell might be responsible for the symptoms of mental illness, such as anxiety, depression, hallucinations, and delusions. Neurotransmitters include dopamine, norepinephrine, serotonin, and gamma-amino-butyric acid. Most experts believe that the causes of mental illness stem from a combination of genes (nature) and experiences (nurture).

BACKGROUND AND HISTORY

Centuries-old artifacts and the remains of ancient human skulls have provided evidence for the existence of a primitive surgical procedure known as trepanning. Cave paintings suggest that trepanning (or trepanation) began during the Neolithic era, possibly signaling that people in ancient societies thought of mental illness as a brain affliction. The technique employed a special surgical tool designed to bore a hole, or burr, in the skull to relieve cranial pressure or to release the evil spirits that were thought to cause disturbed behaviors or distress. Trepanning was practiced widely throughout the eighteenth century and is still used to relieve intracranial pressure.

One of the oldest theories of the causes of mental illness, propounded by Hippocrates and other physicians from Greek and Roman societies, postulated that imbalances of substances in the body, called humors, were responsible for emotional maladies, many of which resembled the major mental disorders described in American Psychiatric Association's diagnostic manual. More specifically, Hippocrates posited that disease and mental illness were caused by surpluses or deficiencies in the four humors: phlegm, yellow bile, black bile, and sanguine. For example, people with too much black bile were thought to be melancholic (very sad); those with too much yellow bile were thought to be irritable. For nearly 2,000 years (from ancient times until the nineteenth century), bloodletting to restore humoral balance was a common treatment for mental illness and other ailments.

Modern psychiatry was not recognized as a bone fide area of medical practice until the mid-nineteenth century. Emil Kraepelin's treatise on psychiatry, *Compendium der Psychiatrie*, first published in 1883, set the stage for the establishment of psychiatry as an independent medical specialty. Kraepelin, a physician, is widely considered the father of modern scientific psychiatry and is credited with seminal discoveries and advances in the field. For example, he believed that mental illness was caused by biologically based, genetically transmitted diseases of the brain and should be treated with scientifically tested medical interventions. Furthermore, Kraepelin noted that mental illness could be traced back several generations in families.

Kraepelin codiscovered Alzheimer's disease and was the first physician to describe schizophrenia and manic-depressive disorder (later known as bipolar disorder) in the medical literature. His most important contribution was the development of the first psychiatric nomenclature, or diagnostic system, which classified mental illness according to patterns of symptoms or syndromes. Much of Kraepelin's classification system and his discussions about the causes and symptoms of psychiatric conditions remain relevant in the twenty-first century.

Kraepelin's view of mental illness as a brain disease dominated psychiatry throughout the early decades of the twentieth century; after that time, his theories and writings abruptly faded into obscurity. Kraepelin publicly acknowledged that psychiatry offered few effective medical treatments and no cures for most psychiatric disorders. A contemporary of Kraepelin, Sigmund Freud and his followers, adopted a psychodynamic approach to mental illness called psychoanalysis. They theorized that psychiatric problems arose from childhood conflicts that could be remedied by talk therapy, which includes various techniques such as dream interpretation and free association. Thus, biological psychiatry was supplanted by an emphasis on early experiences over brain chemistry as the cause of mental illness and on psychoanalysis over medical interventions as the preferred treatment of such illnesses.

In the 1960's and 1970's, the field of psychiatry fell into further disfavor because of the pervasive abuse and neglect of psychiatric inpatients confined to state-run facilities, where they languished with little hope for recovery or reintegration into the community. Changes in involuntary commitment laws, a policy of deinstitutionalization, and a number of legal precedents dictating that patients be treated in the least restrictive environment have greatly reduced the number of psychiatric patients in hospital settings, as well as the power and prestige of psychiatrists.

Books that have declared mental illness a myth and criticized psychiatry as a method of social control and oppression deeply eroded the public's confidence in psychiatric practices. However, toward the beginning of the twenty-first century, the medical model in psychiatry began to experience a renaissance.

HOW IT WORKS

Early Practices. In the initial years of modern psychiatric practices, physicians administered a variety of well-intended but ineffective treatments, some of which were harmful and even deadly to patients. During the field's early history, psychiatrists experimented with many types of interventions that were often barbaric or dangerous. For example, to help relieve patients' agitation, patients were restrained with a specially designed apparatus that kept them completely and uncomfortably immobilized for many hours at a time. In other attempts at treatment, psychiatric patients were spun rapidly in chairs, wrapped in cold blankets, or steeped in freezing water in a technique known as hydrotherapy. As its name implies, insulin-coma therapy involved injecting patients with insulin to induce a short-lived coma from which they were revived with glucose infusions. During this procedure, numerous patients lapsed into a lengthy coma or died.

Perhaps the most infamous and maligned psychiatric treatment in the history of psychiatry is the lobotomy—a crude brain surgery that severed a patient's prefrontal cortex from the rest of his or her brain. A lobotomy was a common treatment in the mid-twentieth century. This procedure destroyed healthy brain tissue, leaving patients with permanent brain damage that resulted in docility, confusion, blunted emotions, extreme lethargy, and disengagement from pleasurable and productive activities. Many people who received a lobotomy suffered from cognitive and emotional deficiencies for the rest of their lives. Between about 1940 and 1952, about 50,000 lobotomies were performed in the United States.

Based on the erroneous notion that mental illness and seizure disorders were antagonistic, psychiatrists induced seizures as a treatment for depression and other psychiatric problems. Convulsive therapy was first described clinically by Ladislas J. Meduna. In 1934, he injected camphor liniment to induce seizures in patients with schizophrenia, some of whom actually improved. His technique was recognized immediately for its therapeutic value and adopted worldwide within a few years of its introduction. Other chemicals, such as metronidazole (Metrozol), were administered under the mistaken belief that such treatments would consistently relieve the symptoms of mental illness. Pharmaco-convulsive therapy frightened patients and was fraught with shortcomings, such as the unreliable induction of seizures.

In 1938, Ugo Cerletti and Lucino Bini introduced a technique called electroconvulsive therapy (ECT), which induced seizures electrically; it quickly became the preferred method of treatment for mental illness. In its earliest applications, ECT was highly traumatic for patients; it elicited a full-blown seizure that could result in broken bones, muscle tears, and permanent memory loss. The threat, or actual administration, of ECT was commonly used to punish or control patients. In modern psychiatry, ECT is administered with a protocol that minimizes trauma, injury, and memory loss, and it is considered a proven, effective treatment for patients with intractable depression and those who cannot take medication because of medical conditions or pregnancy.

Diagnosis. As with any other medical specialty, psychiatric treatment begins only after a proper diagnosis has been rendered. However, unlike other physicians, psychiatrists have no valid medical tests with which to diagnosis mental illness. Instead, they use medical tests to rule out medical illnesses or infections that could account for psychiatric symptoms. For example, a blood test can determine whether depression is attributable to hypothyroidism, while magnetic resonance imagining (MRI) can detect brain tumors or arterial blockages, which might be causing hallucinations or disturbed thinking and memory. Psychiatrists depend largely on their ability to interview and observe patients and their families in order to render a diagnosis.

The mental status examination (MSE) involves an evaluation of a patient's psychiatric condition. The psychiatrist elicits information about the patient's symptoms and observes the patient's appearance and demeanor to identify any signs of pathology. The examiner also gathers biographical and social data, as well as details about the reasons for the contact and the patient's description of problems and complaints. Social and conversational skills are also important indicators of the patient's level of functioning. Furthermore, the mental status examination is designed to assess the patient's reality testing, alertness, memory, and orientation to person, place, and time.

Other major areas of exploration in the examination include the patient's cognitions, emotions, and behaviors. The psychiatrist draws inferences about the patient's thought processes by the organization, coherence, and clarity of his or her speech patterns and notes whether the patient expresses any false unshakable beliefs, known as delusions, or reports any false perceptions, known as hallucinations. A patient's overall emotional state (for example, anxious, elated, angry, or dysphoric) is referred to as his or her mood, whereas the expression of emotions in the interview and the congruence between these emotions and the content of the patient's verbalizations is referred to as affect. For example, a patient who smiles and laughs while reporting the recent death of a loved one is expressing inappropriate affect. A patient who shows no emotion throughout the interview is judged to have a flat affect. In addition, the patient's behaviors during the interview, such as sluggishness, hyperactivity, abnormal movements, unusual gesticulations, or other unexplainable actions can suggest an underlying psychiatric or neurological disorder or the side effects of psychiatric medications. The psychiatrist might also meet with a patient's family members to corroborate the patient's self-reports and gain other perspectives into the course of the person's illness and its effects on the lives of others.

The gathering of a thorough psychiatric history is another element used to arrive at a proper diagnosis. The psychiatrist questions the patient regarding previous episodes of psychiatric illness and treatment. Information about close family members' emotional problems is also critical because of the tendency for mental conditions to be passed on genetically.

Psychiatric Treatment. Psychiatrists treat mental illnesses by prescribing psychopharmacological agents or medications (psychotropic drugs), which change the levels of neurochemicals and enhance communication among the neurons in the brain. They also use psychotherapy and various techniques to help patients change their maladaptive behaviors and live more satisfying and successful lives. Psychiatric medications fall into three major classes: antipsychotics, anxiolytics, and antidepressants.

Psychiatrists prescribe antipsychotics to treat mental illnesses that cause patients to experience marked breaks with reality (psychosis). The most common of such disorders is schizophrenia, which is a chronic, disabling, persistent, and severe brain disease that significantly impairs brain functioning and affects 1 percent of the world's population, including 3 million people in the United States alone. Antipsychotic medications are referred to as typical or atypical. Psychiatrists prescribe anxiolytics (anti-anxiety medications) to treat anxiety disorders, which include panic disorder, generalized anxiety disorder, specific phobias, obsessive-compulsive disorder, social anxiety disorder, and posttraumatic stress disorder. Psychiatrists prescribe antidepressants and

mood stabilizers to treat the symptoms of mood disorders, the most common and severe of which are major depression and bipolar disorder.

An effective but much less widely used treatment for depression is electroconvulsive therapy (ECT), which leaves patients with few permanent side effects. Patients receive about four to ten treatments over a two-week period. Confusion and memory loss are minimized by the common practice of applying the electric stimulus to only the nondominant brain hemisphere, usually the right-brain hemisphere. The exact mechanism of ECT is unknown. However, experts are certain that the seizures, rather than the electric current itself, are the basis for the treatment's effectiveness. More specifically, seizures can enhance the functioning of neurotransmitters in the brain, including norepinephrine and serotonin, and increase the release of pituitary hormones, thereby relieving the symptoms of depression.

Psychiatrists engage patients in psychotherapy sessions, implementing a wide range of techniques and modalities designed to relieve the symptoms of mental illness. One of the oldest and best-known forms of psychotherapy is psychoanalysis, developed by Sigmund Freud. The prototype for talk therapies, psychoanalysis, is an in-depth, intensive, and time-consuming method of treatment that deeply examines patients' early childhood experiences and helps them gain insight into their behaviors through dream interpretation. They explore their unconscious through free association and understand the meaning of their relationship with their doctor through transference analysis.

Although psychoanalysis revolutionized the treatment of mental illness and changed popular culture, little scientific evidence has demonstrated its effectiveness, and it has long been replaced by shorter and more present-focused approaches that attempt to modify a patient's irrational thinking habits and help him or her adopt more effective strategies to be successful at work and school and improve his or her relationships with others. People with serious mental illness can also benefit from a case-management approach that focuses on psychiatric rehabilitation, enhancing basic living skills, medication adherence, and stable living opportunities.

APPLICATIONS AND PRODUCTS

Since the release of the first edition of the

Diagnostic and Statistical Manual of Mental Disorders (DSM-I) in 1952 to the publication of the fourth edition (DSM-IV-TR) in 2000, the number of diagnostic entities has increased 300 percent from 106 to 365. The fifth edition (DSM-V, scheduled for release in 2013) is expected to include even more diagnostic categories. Because of the increase in the number of psychiatric diagnoses and a greater awareness of such problems in the general population, partially attributable to pharmaceutical company advertising, the reach of psychiatry is likely to extend in the first half of the twenty-first century, leading to what some critics refer to as the medicalization of everyday problems (such as shyness, grief, and sadness).

Since 1990, unprecedented numbers of children and adolescents have been diagnosed with serious mental illness. These illnesses include attention deficit and hyperactivity disorders, autism, and bipolar disorder. For example, from 1994-1995 to 2002-2003, the rate of doctor's office visits for bipolar disorder among youth increased from 25 per 100,000 youths to 1,003. In addition, autism has become the fastest-growing developmental disability in the United States.

As the population continues to age and longevity increases, the need for psychiatrists to treat dementia is growing. Projections indicate that by 2030, nearly 8 million people in the United States aged sixty-five and older will be diagnosed with Alzheimer's disease.

IMPACT ON INDUSTRY

The neo-Kraepelin revolution in psychiatry, which began around 1990 with a reemphasis on the biological basis for psychiatric illness, greatly benefited from significant developments in neuroscience and genetics. The so-called decade of the brain (the 1990's) generated a wealth of intriguing knowledge about how the living brain functions and brought more attention to the need to find more medical treatments for mental disorders. The newest family of psychotropic medications, the selective serotonin reuptake inhibitors (SSRIs), promised not only to release people from the shackles of depression but also to transform their personalities and enhance their productivity at work and school.

Antidepressant medications have become among the most prescribed drugs in the United States. Between 2002 and 2005, the number of

---

Fascinating Facts About Psychiatry

- The specific medicines prescribed by psychiatrists are often influenced by patients, who have viewed television commercials and advertisements in magazines that are part of the pharmaceutical industry's strategy of direct-to-consumer marketing.
- The highest rates of off-label prescriptions (drugs prescribed to treat illnesses for which they were not originally intended) are for antidepressant, antipsychotic, and antiseizure medications.
- Psychiatrists have been found to be the least religious physicians among all medical specialties. Among psychiatrists who report a religious affiliation, a large number are Jewish; far fewer are Protestant or Catholic, the two most common religions reported by physicians overall. However, among all physicians, psychiatrists are the ones most interested in the religious and spiritual aspects of their patients' lives.
- In the United States, most prescriptions for antidepressant medications are written (as a group) by internists, family practitioners, and general medical practitioners, not by psychiatrists.

---

antidepressant prescriptions filled rose from 154 million to 170 million. Drug treatment lies at the core of modern psychiatric practice and has been spurred by an abundance of psychiatric research on the biological nature and causation of mental disorders. The field of psychiatry has therefore become a prime sales target for the pharmaceutical industry. Drug companies have launched sophisticated marketing campaigns and offer incentives (dinners, vacation packages, and continuing medical education credits) to psychiatrists to encourage them to write prescriptions for the latest pharmaceuticals. The industry spends billions of dollars on the development, testing, and approval of psychiatric medications and uses television, the Internet, and print publications to advertise medications directly to potential patients (consumers).

CAREERS AND COURSE WORK

Psychiatrists are medical doctors who receive the standard four years of medical training to earn a doctor of medicine degree (M.D.). They must also pass their medical boards. The first year of medical school consists of courses on the development and structure of the human body, such as anatomy, physiology, histology, biochemistry, embryology, and neuroanatomy. The second year focuses on human diseases and treatments and consists of courses on pathology, pharmacology, microbiology, and immunology. The third and fourth years of training provide students with more hands-on medical experience and patient contact.

After graduating from medical school, physicians begin advanced training in psychiatry by entering a three-year residency program. The first year is known as an internship; during this time students learn to practice psychiatry under the close supervision of a licensed attending physician, usually in a hospital or clinic. Physicians in their final year of residency, known as senior residents, are responsible for the supervision and training of interns and junior residents. A residency can be followed by a fellowship, during which the psychiatrist is trained in a subspecialty, such as child and adolescent psychiatry, addiction medicine, or geriatric psychiatry. Board certification is the process by which a psychiatrist passes a standardized written test, which demonstrates a mastery of the basic knowledge and skills that define the area as a medical specialization.

SOCIAL CONTEXT AND FUTURE PROSPECTS

The early diagnosis and treatment of mental illness is vitally important in the remediation of symptoms and prevention of relapses. By ensuring access to effective treatments and social supports, recovery is accelerated and further harm is minimized. The best (evidence-based) treatments for serious mental illness are highly effective. Between 70 and 90 percent of patients experience a significant reduction in symptoms and an improved quality of life following a combined regimen of pharmacological and psychosocial treatments and habilitation programming.

The key to successful treatment is the adoption of an illness management and recovery model. With appropriate and effective medication and a wide range of services tailored to meet their needs, most people who live with serious mental illness can significantly reduce its impact and find a satisfying measure of achievement and independence.

Despite the urgency and potency of prompt mental health care, less than one-third of adults with psychiatric disorders receive mental health care treatment. For example, about 2 million individuals with schizophrenia and bipolar disorder receive no treatment. Without treatment, the consequences of mental illness for individuals, families, and communities are staggering: unnecessary disability, unemployment, substance abuse, homelessness, incarceration, suicide, and unproductive lives. The economic cost of untreated mental illness is more than $100 billion each year in the United States alone. Cutting budgets and instituting draconian limits on needed treatments and services exacerbates human suffering and puts an additional strain on state economies through increased reliance on emergency services, correctional systems, and welfare programs.

Arthur J. Lurigio, Ph.D.

FURTHER READING

Bentall, Richard. *Doctoring the Mind: Is Our Current Treatment of Mental Illness Any Good?* New York: New York University Press, 2009. Bentall attempts to distill the mountain of evidence regarding the effectiveness of psychiatric treatment, especially the use of medication to treat mental disorders. The volume describes the historical development of mental health treatments and debunks myths about the nature of severe mental illness and its purported remedies.

Berrios, German E. *History of Mental Symptoms: Descriptive Psychopathology Since the Nineteenth Century.* 1996. Reprint. New York: Cambridge University Press, 2002. Provides a sweeping account of the treatment and diagnosis of mental illness since the early 1900's. Discusses the manifestations and interpretations of major symptoms from the perspective of the historical time period. Covers the historical roots of mental illness and provides descriptions of psychopathology, cognition and consciousness, mood and emotions, and volition and action. Contains a chapter on suicide.

Frances, Allen, and Michael B. First. *Your Mental Health: A Layman's Guide to the Psychiatrist's Bible.* New York: Scribner, 1999. Explains in intelligent but nontechnical language the symptoms of and treatments for all the major diagnostic categories that are defined as mental illness by the psychiatric profession.

Gorman, Jack M. *The New Psychiatry: The Essential Guide to State-of-the-Art Therapy, Medication, and Emotional*

Health. New York: St Martin's Press, 1996. A comprehensive and well-written volume on all aspects of psychiatric illness and treatment. Contains a section on special topics for people who might be considering mental health treatment.

Kelly, Timothy. *Healing the Broken Mind.* New York: New York University Press, 2009. Discusses the shortcomings of the country's mental health policies and treatment system. Based on his experiences as the former state commissioner of Virginia's Department of Mental Health, the author offers a series of recommendations to improve mental health services in the United States.

Kramer, Peter, R. *Listening to Prozac: The Landmark Book about Antidepressants and the Remaking of the Self.* New York: Penguin, 1997. Explores early views of the clinical effectiveness of Prozac and other medications in its class, which have become among the most prescribed medications in the United States and the world. The author, a psychiatrist, argues that such drugs not only are effective in the treatment of depression and anxiety but also can transform people's personalities, making them more confident and less shy.

Shorter, Edward. *A History of Psychiatry: From the Era of the Asylum to the Age of Prozac.* Hoboken, N.J.: John Wiley & Sons, 1998. Condenses two hundred years of psychiatric theory and practice into a compelling and coherent narrative. Through a series of gripping anecdotes, it chronicles the sometime harsh and heroic efforts of generations of scientists and physicians to ease the suffering of people with mental illness as well as society's changing attitudes toward psychiatric problems and their treatment.

WEB SITES

American Psychiatric Association
http://www.psych.org

American Psychological Association
http://www.apa.org

National Institute of Mental Health
http://www.nimh.nih.gov/index.shtml

See also: Neurology; Pharmacology; Surgery.

PULMONARY MEDICINE

FIELDS OF STUDY

Biology; organic chemistry; statistics and methodology; physiology; epidemiology; internal medicine.

SUMMARY

Pulmonary medicine is a medical specialty concerned with the diagnosis and treatment of diseases of the respiratory system. Pulmonologists complete several years of postgraduate medical training before they begin caring for patients who have respiratory diseases. Pulmonologists often specialize in a specific area of pulmonary medicine, such as asthma management or lung transplantation. Pulmonary medicine has also spawned areas of focus that have become specialties, notably critical care medicine and sleep medicine.

KEY TERMS AND CONCEPTS

- **Auscultation:** Evaluative technique in which the physician places a stethoscope on different areas of the chest and back and asks the patient to take deep breaths to identify irregularities in the lungs.
- **Crackles:** Discontinuous, interrupted explosive sounds in the lung that are indicative of pulmonary disease; also known as rales.
- **Critical Care Medicine:** Medical subspecialty concerned with the management and treatment of acutely ill patients in an intensive care unit.
- **Interstitial Lung Disease:** Lung disease occurring in the small areas between the lungs or parts of the lung.
- **Lungs:** Primary organs of respiration, found in all land mammals, including humans.
- **Percussion:** Evaluative technique in which the physician places a hand over certain parts of the chest or back directly over the lungs and, with the other hand, taps with the fingertips to identify irregularities in the lungs.
- **Pulmonary:** Of, relating to, or affecting the lungs.
- **Pulmonology:** Study of the basic anatomy, physiology, and function of the respiratory system.
- **Respiratory System:** System of organs used to take in and exchange oxygen for carbon dioxide in an organism.
- **Ventilator:** Device that helps pulmonary patients breathe when they cannot adequately respirate on their own.

DEFINITION AND BASIC PRINCIPLES

Pulmonary medicine is a branch of medicine that is concerned with the maintenance and function of the respiratory system. It deals with causes, diagnoses, and treatments of diseases that affect the lungs and related systems, such as sleep, that are strongly supported by an efficiently working respiratory system. Pulmonary medicine is often confused with pulmonology, which is the scientific study of the basic anatomy, physiology, and function of the respiratory system. Practitioners of pulmonary medicine are called pulmonologists.

The practice of pulmonary medicine has evolved to include a number of distinct areas of focus. Examples include obstructive airway diseases (such as asthma and chronic obstructive pulmonary disease, or COPD), occupational and environmentally caused diseases (such as asbestosis), congenital lung diseases (such as cystic fibrosis), interstitial lung diseases (such as sarcoidosis), neoplasms (such as small cell lung cancer, or SCLC), diseases of the pleural space (such as pneumothorax), vascular diseases (such as deep venous thrombosis), lung transplants, sleep disorders (such as obstructive sleep apnea), and critical care medicine. The last two areas of focus—sleep disorders and critical care medicine—have become distinct areas of subspecialization that have significant overlap with the general practice of pulmonary medicine.

BACKGROUND AND HISTORY

Pulmonary medicine came into being as a subspecialty of internal medicine in the early part of the twentieth century. However, interest in treating the underlying cause of diseases of the lungs can be traced back to Hippocrates, who described the symptoms, physical findings, and treatment of thoracic empyema. Interest in developing treatments for people suffering from ailments of the respiratory system increased in the nineteenth century with the introduction of resorts—often situated in mountainous areas—that claimed to offer a healing climate for sufferers of afflictions of the lungs. Some early pioneers in pulmonary medicine were scientist-practitioners who had a particular interest in the physiological structure of the lungs and the respiratory tract. The scientific study of pulmonology and its application to the diagnosis and treatment of lung diseases began systematically and earnestly in the early twentieth century and has progressed steadily since that time.

During the late twentieth century, advances in the treatment of respiratory diseases were numerous and dramatic. Some of the most significant advances include lung volume reduction surgery, in which the diseased parts of the lungs of emphysema or COPD patients are surgically removed to allow the remaining, healthy parts of the lung to expand and perform better, and lung transplantation, in which badly diseased lungs are partially or completely replaced by donor organs. Advances in mechanical ventilation technology have allowed critically ill patients to be stabilized for survival and patients with chronic respiratory conditions (such as sleep apnea) to achieve a much higher quality of life.

How It Works

The Pulmonary Examination. Pulmonologists practice medicine in both outpatient (ambulatory) and inpatient (hospital) settings. In the United States, most pulmonologists see patients only after they are referred from primary care physicians because of the complexity of the patient's respiratory condition. When a patient sees a pulmonologist, the physician's initial examination will include questioning about pulmonary symptoms. The physician may ask about breathing problems that the patient may be having and whether the problems change after exertion. The pulmonologist may also ask about the quality of the patient's sleep, because certain pulmonary issues may be particularly evident at rest.

Auscultation and Percussion. After questioning the patient about symptoms, the pulmonologist will perform the initial pulmonary physical examination. The complete pulmonary examination takes about ten minutes to complete. The primary component of this examination requires the physician to listen to the patient's lungs for the distinct sounds of different respiratory diseases. These lung sounds have been identified empirically by a computerized examination of differences in the lung sounds of normal patients versus patients who have various confirmed lung diseases. Two primary components of the pulmonary physical examination are auscultation and percussion.

Auscultation is the act of listening to the lungs in an effort to diagnose potential respiratory disorders. Successful auscultation requires substantial clinical experience and listening skill. The physician places the stethoscope on different areas of the chest and back and asks the patient to take deep breaths. The pulmonologist is listening for several potential indicators of pulmonary dysfunction, including crackles, rhonchi, wheezes, and rubs in both lungs. Crackles (also known as rales) are defined as discontinuous, interrupted explosive sounds in the lung. Crackles may sound low-pitched (coarse crackles) or high-pitched (fine crackles). The area of the lung in which crackles are heard may be indicative of different pulmonary conditions. Rhonchi are low-pitched rumbling or gurgling sounds that suggest that air is being forced through fluid obstructions in the airways. Wheezes are continuous and high-pitched hissing sounds. The physician may also listen for rubs—sounds that have been traditionally described as being similar to two pieces of leather being rubbed together. Successfully identifying the type and intensity of different lung sounds through listening with a stethoscope is a skill that requires several years of practice. Having a naturally strong listening ability may be one factor in choosing pulmonary medicine as a subspecialty.

Percussion is somewhat different from auscultation in the pulmonary examination, but the two techniques ultimately serve the same purpose—to identify some dysfunction of the respiratory system through audible observation. Percussion is a technique in which the physician places a hand over parts of the chest or back directly over the lungs and, with the other hand, taps with his or her fingertips. The physician then notices the pitch of the resonant sound from the tap (called percussed sounds). Percussion is particularly useful for the identification of pneumothorax, a pulmonary condition in which air or gas is trapped in the pleural cavity. In modern pulmonary medicine, percussion is almost always used in conjunction with auscultation.

Imaging Studies. Auscultation and percussion are integral to the initial examination of the pulmonary patient. The next step in evaluation is often imaging studies. The most common imaging studies in pulmonary medicine are radiographs of the chest (X rays), computed tomography (CT) scans, high-resolution CT scans, pulmonary angiography, and ventilation/perfusion lung scans. Pulmonologists will often start with less invasive and more economical chest radiography to look for clues about the cause of the patient's symptoms. They may be looking for many different signs depending on the results of the patient's physical examination. One of the most common findings on radiography of the chest or chest CT is opacities on the film of the lungs that indicate the presence of some kind of lung infiltrate. The clarity of the radiograph, the availability of other imaging studies, and the physician's clinical suspicion about the etiology of the patient's condition will determine whether any of the other imaging studies are used. Some practices may immediately image the patient's respiratory system with a chest CT and bypass the radiograph altogether, but getting the chest radiograph first is still the most common practice.

Pulmonary Function Tests. Beyond imaging techniques, pulmonologists often need to test the functional performance of the lungs. Pulmonary function tests measure the functional performance of the lungs by measuring how much air they can hold, how quickly a person can move air into his or her lungs, and how well a person's lungs can distribute oxygen to the bloodstream and remove carbon dioxide from it. Lung function is most commonly measured with a process called spirometry. Some of the most common spirometic measures of lung function are FEV1 forced expiratory volume), FVC (forced vital capacity), RV (residual volume), and TLC (total lung capacity). FEV1 is a measure of the amount of air that the patient can exhale forcefully in one second. FVC is a measure of the amount of air that the patient can exhale after inhaling deeply. RV is a measure of how much air is left in the lungs after the patient has completely exhaled. TLC is a measure of the amount of air deposited in the patient's lungs after he or she has inhaled deeply. Other common pulmonary function studies include arterial blood gas studies, which measure the amount of oxygen and carbon dioxide in the patient's bloodstream. These and other pulmonary function tests are often employed after the pulmonologist has developed some clinical suspicion of the patient's diagnosis.

APPLICATIONS AND PRODUCTS

Bronchodilators. Some of the most common problems of pulmonary patients are obstructive airway diseases, in which something is preventing the normal flow of air in and out of the lungs and, thus, affecting the normal function of the respiratory system. Two of the most common obstructive airway diseases are asthma and chronic obstructive pulmonary disease (COPD). Physicians have a variety of methods at their disposal to treat obstructive airway diseases. One of the most common treatments is bronchodilator therapy. Bronchodilators are medications that work by relaxing the muscles that surround the airways in the lungs. This allows more air to flow easier into the lungs and prevents attacks of labored breathing in obstructive airway disease patients. Short-acting bronchodilators help with rapid, acute problems such as asthma attacks and acute COPD exacerbations, while long-acting bronchodilators work to prevent underlying symptoms that can lead to acute attacks.

Ventilators. Pulmonary patients who have severe conditions often must be placed on machines that aid their lungs in the respiratory process. These machines are called ventilators. The proper use of ventilators for various lung conditions is a primary part of the knowledge base of pulmonologists. Ventilators may be used for management of acute conditions in the emergency department or, more commonly, in the intensive care unit (ICU) for conditions such as acute lung injury. They may also be used for more chronic conditions such as COPD. Ventilators are either negative-pressure or positive-pressure. Negative-pressure ventilation involves directing air directly into the lungs, and positive-pressure ventilation involves directing air into the trachea. Some ventilators require intubation, the placement of a tube into the trachea from the nose or mouth. Ventilation requiring intubation is typically used for patients who will require ventilation for a protracted period. Other ventilators work with a breathing mask that can be placed over the mouth and nose. With the increase in respiratory-related sleep disorders (such as obstructive sleep apnea), use of two positive airway pressure systems—continuous positive airway pressure (CPAP) and bilevel positive pressure ventilators (BiPAP)—has become very common.

Lung Volume Reduction Surgery. Lung volume reduction surgery (LVRS) is a procedure that has been successful for patients who have emphysema and severe COPD. The procedure involves surgically removing the disease-damaged part of the lung to allow the other (healthy) part of the lung to expand and compensate.

Lung Cancer Therapy. Many types of respiratory cancers can affect pulmonary patients. Common neoplasms are small cell lung or nonsmall cell lung cancers. Small cell lung cancers are less common and include subtypes such as sacromatoid carcinoma, salivary gland tumors, and carcinoid tumors. More common nonsmall cell cancers include lung carcinoma, mediastinal lymph node cancer, and pleural cancers. Lung cancers are primarily treated with chemotherapy, radiation therapy, or combination therapy. Less common lung cancer treatments include adjuvant chemotherapy and radiotherapy. Patients who have lung cancer may be treated by an oncologist, a pulmonologist who specializes in lung cancer, or a multidisciplinary team.

Lung Transplant. In some very severe cases of lung disease, the patient may require lung transplantation, in which a diseased lung is surgically replaced by a live donor organ. Lung transplantation may be performed by a skilled pulmonologist specializing in transplant or, more commonly, by a thoracic surgeon. A national registry lists people who are waiting for lung transplants. Many factors, including the severity of disease and lifestyle factors (for example, continuing to smoke cigarettes), can determine a person's eligibility and position on the list.

IMPACT ON INDUSTRY

In 2000, 6,734 physicians in the United States indicated their primary specialty as pulmonary medicine. Most pulmonologists are internal medicine physicians treating adults, but a smaller number practice as pediatric pulmonologists.

Pulmonary medicine as a discipline requires expertise in diagnosing, managing, and treating the full array of respiratory conditions. It also involves large portions of smaller, more specialized areas such as critical care medicine and sleep medicine. Workforce issues in pulmonary medicine are difficult to fully grasp because of the complexity of the history and evolution of the specialty and the strength of its subspecialty areas. One of the most visible examples of this is workforce issues in critical care medicine.

Most physicians who work in the intensive care unit (intensivists) are trained in pulmonary medicine. A smaller number are trained in surgery, emergency medicine, or general internal medicine only. The number of intensivists in the United States is considered to be alarmingly inadequate. The potential increase in demand for intensive care because of the nation's substantial aging population has prompted leaders in pulmonary and critical care medicine to take concrete steps to increase the number of physicians who choose to practice critical care medicine. The link between pulmonary and critical care medicine is strong and bilateral: Most critical care physicians are pulmonologists and most pulmonologists are critical care physicians. Most of the training programs in pulmonary medicine are combined pulmonary and critical care programs. The need for experts in pulmonary and critical care medicine is expected to increase.

A subspecialty that branched out from pulmonary medicine is sleep medicine. Training in sleep medicine is still an integral part of most pulmonary training programs because of the strong link between breathing and sleep issues. Sleep medicine physicians must also be well versed in pulmonary medicine to be able to adequately diagnose, manage, and treat respiratory-related sleep disorders. Some pulmonologists use their training in sleep medicine to run sleep laboratories or interpret sleep studies. Pulmonary-trained sleep medicine specialists are expected to remain in high demand.

CAREERS AND COURSE WORK

Demand for competent, trained practitioners of pulmonary medicine should increase as the population ages and advances are made in the understanding of respiratory disease and in pulmonary science. Courses in advanced mathematics, biology, microbiology, and biochemistry are essential preparatory work for a career in pulmonary medicine. The training pathway is very straightforward for most pulmonologists. The path to becoming a pulmonologist starts in high school. Admission standards for top American universities are tougher than ever. To be accepted to a top-tier school, students must often show promise as critical thinkers at the high school level. Additionally, they need to have an adequate understanding of biology and chemistry to be able to master science courses at the college level.

After being accepted to a university, students interested in careers in pulmonary medicine pick a major area of study (often a science) and make certain they fulfill any requirements for medical school. Some schools allow the students to designate themselves as headed for medical school by adding a "pre-med" tag to their major. This allows the student to receive academic counseling that is customized for the student bound for medical school. After graduating with a bachelor's degree (usually four years), students apply to medical school. Medical school admission is even more difficult than college admission. In medical schools across the United States, acceptance rates hover around 8 percent, with the more selective programs much lower. In addition to outstanding academic achievement, evidence of scholarship potential, and other factors, medical school applicants must have adequate performance on the standardized Medical College Admissions Test (MCAT).

Medical school is typically four years, with the first two years devoted primarily to classroom and laboratory instruction, and the second two years devoted primarily to hands-on clinical rotations. After the second year of medical school, students begin taking steps assessments of the United States Medical Licensing Examination (USMLE) in preparation for eventual medical licensure. After being granted the medical degree (M.D. or D.O.) and completing all licensure requirements, students become physicians.

The next step in pulmonary medicine training is the medical residency. As mentioned earlier, most pulmonologists come to the field from internal medicine. Internal medicine residency is, again, competitive. After being accepted to a residency program, medical residents become junior staff physicians (often called "house staff"). These student-practitioners will be trained in internal medicine for three years with progressive increases in responsibility. After completing the internal medicine residency and passing the initial certification examination in internal medicine of the American Board of Internal Medicine (ABIM), the internist may apply to a postgraduate fellowship program in pulmonary medicine.

Most pulmonary training fellowships are combined pulmonary-critical care programs. Training in pulmonary medicine typically takes two years for pulmonary medicine alone and three years for pulmonary-critical care combined training. After completion of the fellowship and an attestation about

---

Fascinating Facts About Pulmonary Medicine

- In 1928, the first rudimentary hyperbaric chamber, the Timken tank, was constructed in Cleveland, Ohio, at a cost of $1 million. The tank was 64 feet in diameter and five stories tall.
- According to a national health survey, as many as 24 million Americans are affected by chronic obstructive pulmonary disease (COPD).
- Lung cancer is the leading cause of cancer death in both men and women for all ethnic groups in the United States.
- According to the Asthma and Allergy Foundation of America, about 20 million Americans (1 in 15) suffer from asthma, and the prevalence of asthma has been steadily increasing since 1980 for both men and women in all ethnic groups in the United States.
- Human lungs are filled with tiny air-filled structures called alveoli. The average adult lungs contain 600 million of these alveoli—enough to cover an area the size of a tennis court.
- Human lungs inhale more than two million liters of air everyday.

---

competency from the fellowship director, the pulmonologist sits for the subspecialty examination in pulmonary disease also given by the ABIM. After successfully passing the examination, the pulmonologist may set up practice as an inpatient or outpatient pulmonologist or may choose to pursue additional fellowship training (for example, a sleep medicine fellowship). Pulmonary medicine physicians are in high demand and that demand is expected to increase as the population ages.

SOCIAL CONTEXT AND FUTURE PROSPECTS

A functioning respiratory system is vital to life, and the goal of pulmonary medicine is to ensure that the respiratory system is functioning properly and adequately. Pulmonary medicine is a vital medical specialty that offers many interesting potential research areas and a variety of respiratory-related subspecialties that are at the cutting edge of modern medicine (such as critical care and sleep medicine).

Training in pulmonary medicine is long and laborious but can ultimately be very rewarding. As demand for pulmonary and critical care medicine practitioners is at an all-time high and expected to increase, the prospects for a successful career in pulmonary medicine are very good.

Jeremy Dugosh, M.S., Ph.D.

FURTHER READING

Fishman, Alfred P., et al. *Fishman's Pulmonary Diseases and Disorders.* 4th ed. New York: McGraw-Hill Professional, 2008. A basic pulmonary textbook and a good source for basic information about procedures for maintaining respiratory health and the history of the science of pulmonary medicine.

Henschke, Claudia I., Peggy Mcarthy, and Sarah Wernick. *Lung Cancer: Myths, Facts, Choices—and Hope.* New York: W. W. Norton, 2002. This book is a nontechnical examination of the prognosis and treatment of lung cancer for patients and their caregivers.

Kovitz, Kevin L. "Pulmonary and Critical Care: The Unattractive Specialty." *Chest* 127, no. 4 (April, 2005): 1085-1087. A frank discussion of the workforce issues surrounding pulmonary and critical care medicine and an accurate description of the day-to-day activities of a pulmonologist.

Macnee, William, and Stephen I. Rennard. *Chronic Obstructive Pulmonary Disease.* London: Health Press, 2009. Spends equal time on the science and clinical treatment of COPD. It is a useful reference to understand the disease and its manifestations at a deeper level.

Mason, Robert J., Jay A. Nadel, and John F. Murray. *Murray and Nadel's Textbook of Respiratory Medicine.* 5th ed. Philadelphia: Elsevier Saunders, 2010. The quintessential textbook on pulmonary medicine. Includes chapters on every aspect of pulmonary disease management from basic anatomy of the lungs to lung transplantation.

Wilkins, Robert L., and James R. Dexter. *Respiratory Disease: A Case Study Approach to Patient Care.* Philadelphia: F. A. Davis, 2006. Describes different pulmonary conditions by examining specific case studies of different pulmonary patients.

WEB SITES

American Board of Internal Medicine
http://www.abim.org

American College of Chest Physicians
http://www.chestnet.org

American Thoracic Society
http://www.thoracic.org

Lung Cancer Alliance
http://www.lungcanceralliance.org

Society of Critical Care Medicine
http://www.sccm.org

See also: Geriatrics and Gerontology; Otorhinolaryngology; Pathology; Pediatric Medicine and Surgery; Radiology and Medical Imaging; Surgery.

R

RADIO ASTRONOMY

FIELDS OF STUDY

Astronomy; physics; electronics; mathematics; electrical engineering; computer science; radio technology.

SUMMARY

Radio astronomy is the branch of astronomy associated with studying the heavens using radio frequency electromagnetic radiation, generally those frequencies below 300 gigahertz. Visual light composes only a tiny portion of the electromagnetic spectrum. Many astrophysical systems emit more strongly in radio waves than in visual light, and other systems emit only radio waves. Radio emissions also carry different information about the radiating system than do visual light waves, so without using radio astronomy, astronomers are unable to fully study the universe. Radio astronomers use radio telescopes, antennas dedicated to radio astronomy, sometimes used individually and sometimes grouped in arrays. The techniques of radio astronomy have been adapted to other fields, including low-intensity radio communication and image analysis.

KEY TERMS AND CONCEPTS

- **Antenna Temperature:** Theoretical temperature at which a blackbody would emit thermal radiation equivalent to the electrical noise of the antenna.
- **Aperture Synthesis:** Technique to combine the signals of multiple smaller radio telescopes to simulate the characteristics of a much larger instrument.
- **Array:** Set of radio telescopes arranged to work together to yield more information than the individual telescopes working separately.
- **Beamwidth:** Angular size measuring the resolution of an antenna.
- **Brightness Temperature:** Theoretical temperature at which a blackbody would radiate thermal energy at the rate detected by a radio telescope.
- **Intensity:** Energy per unit time per unit area.
- **Interferometry:** Technique of combining signals between multiple radio telescopes to increase the angular resolution of the telescopes.
- **Jansky:** Unit of radio power flux used in radio astronomy; equal to 10-26 watts per square meter per hertz.
- **Radio Window:** Range of electromagnetic radiation at which the atmosphere is mostly transparent, extending roughly from 5 megahertz to 300 gigahertz.
- **21-Centimeter Radiation:** Wavelength of the characteristic emission of neutral hydrogen atoms undergoing a spin-flip transition.

DEFINITION AND BASIC PRINCIPLES

Radio astronomy is the study of the heavens using radio frequency radiation. Physical processes often produce electromagnetic radiation. Although visual light is a form of electromagnetic radiation, it is only a very small part of the electromagnetic spectrum. Many physical processes produce radio radiation in addition to or sometimes instead of visual light. Thus, to fully study the universe, astronomers must observe more than just visual light.

Radio astronomy uses techniques for detection of radio signals from space similar to those used by radio communication systems. The radio telescope is basically the antenna that focuses the radio flux onto a detector. Electronic circuitry amplifies the detected signal. As with radio communication, the signal from multiple antennas, or multiple radio telescopes, can be combined electronically to yield more information than that which could be detected by one radio telescope alone. The similarity between the two technologies has permitted many advances in radio astronomy to be used in the field of commercial communication and vice versa.

Visual telescopes can be fitted with cameras to

take images. That is not possible, however, with radio telescopes, which measure only the intensity of radio flux coming from whatever direction the radio telescope is pointing. Thus, images must be constructed from multiple measurements, and several techniques have been developed for creating images using different types of radio telescopes. This is far more time-consuming than simply taking a photograph and often involves powerful computer software. This software can be adapted to build images of other systems from a series of measurements, such as with medical imaging.

BACKGROUND AND HISTORY

In 1930, Karl G. Jansky, an engineer with Bell Telephone Laboratories, began work to isolate sources of interference with long distance radio telephone communications. By 1932, Jansky had determined that one source of interference, a steady hiss of static, originated external to Earth. The following year, he published a paper showing that the extraterrestrial radio interference originated in the Milky Way, with the most intense interference coming from the general direction of the center of the galaxy. After hearing about Jansky's discovery, Chicago radio engineer and amateur radio operator Grote Reber decided to build his own radio telescope in his backyard in 1937. Reber's antenna was the first radio antenna constructed solely to monitor signals from space for scientific purposes and thus is often regarded as the first true radio telescope.

World War II brought an end to many basic scientific investigations, including radio astronomy. However, radar operators during the war monitored interference with radar systems caused by solar activity. After the war, military surplus radar technology became readily available to radio astronomers, and there was a surge in radio astronomy activity. Radio telescopes, however, must be extremely large, and thus extremely expensive, to compete with optical telescopes in resolution and sensitivity. The cutting edge of radio astronomy was too expensive for individual institutions to pay for without government aid. The establishment of the Jodrell Bank Experimental Station, with its giant radio telescope, in the United Kingdom in 1945 helped spur the United States and other nations to build their own radio astronomy facilities, giving rise to America's National Radio Astronomy Observatory in 1956,

with its 300-foot-diameter radio telescope, and the 1,000-foot-diameter Arecibo Observatory radio dish in Puerto Rico in 1963.

HOW IT WORKS

The heart of a radio telescope is a radiometer, a device that measures the strength of radio signals gathered by the telescope. The output of the radiometer is measured and recorded. The output of a radio telescope is essentially analogue, so older systems used a chart recorder to record the signal strength. More modern systems have used magnetic tape or other similar media. The radio telescope itself is essentially an antenna that concentrates and amplifies the incoming signals for the radiometer. Though the dish shape similar to a satellite dish is most commonly portrayed in photographs of radio telescopes, any radio antenna can be used as a radio telescope. The most effective antenna design depends on the wavelength and type of observation. The advantage of the dish shape is that the dish acts as a reflector to focus radio radiation striking the entire dish area toward the receiver that is located at the focus of the dish. The radiometer connects to the receiver, and often the two parts are a unit subassembly. The receiver is designed to measure the intensity of radio radiation across a particular radio frequency band. Unlike with many communication radio receivers, changing the frequency band is more complicated with a radio telescope than just pressing a button or turning a dial. Often, the entire receiver subassembly must be replaced. Other shapes, such a Yagi antenna radio telescope (a long boom with crosspieces, a example is a typical very-high-frequency, or VHF, external television antenna), are more easily constructed and handled than a dish design but are typically constructed to operate best at only one set of radio frequencies. A dish system is more versatile.

Image Construction. Radio telescopes measure intensity only. The radio intensity is governed by the amount of radio power concentrated by the antenna onto the receiver. The most useful antennas tend to be directional, so that the antenna preferentially concentrates radio energy from a particular direction in the sky, determined by the orientation of the antenna. Thus, radio telescopes are often mounted in such a way that they can be readily pointed in different directions as needed. However, there is no such thing as a radio camera or a radio imaging system for the radio telescope. Therefore, radio astronomers must take many measurements from slightly different orientations of the radio telescope to construct an image. This image is constructed using sophisticated computer software. The data analysis technique required to produce an image depends on the type of radio telescope system, and much of the work of a radio astronomer is done on the computer analyzing the data collected by the radio telescopes.

Aperture Synthesis. One of the problems of radio telescopes is that radio waves are quite long compared with other forms of electromagnetic radiation. The longer the wavelength, the larger the instrument must be to achieve comparable resolution, which is the ability to distinguish between nearby objects. The higher the resolution, the higher quality the images that can be produced and the more precisely the source of radio signals can be determined. The resolution of the radio telescope is the beamwidth of its antenna and is determined by the frequency band being used and the shape of the antenna. For a single dish radio telescope to have a resolution comparable with the human eye, it would need to be nearly a mile across, which is not feasible.

However, because of the wave nature of electromagnetic radiation, the principle of interferometry can be used to achieve greater resolution than would be possible with a single radio telescope. The signals from two radio telescopes can be combined electronically. Radio waves have to travel different paths to get to the two antennas, and thus the signals are slightly out of phase, depending on the location of the source relative to the antennas. If they are out of phase by one-half wavelength, then the signals cancel. The farther apart the two antennas, the greater the effect, and thus the more precisely the position of the source can be determined. This effectively allows the two radio telescopes to have a resolution along one direction equal to that of a radio telescope the diameter of the distance between the two instruments. By extending this technique to an array of many radio telescopes and combining the signals in the proper manner, radio astronomers can effectively simulate the resolution of a radio telescope equal in size to the size of the array in a technique called aperture synthesis. The National Radio Astronomy Observatory's Very Large Array (VLA) near Socorro, New Mexico, is an example of a radio telescope array that uses aperture synthesis to produce high-resolution images.

APPLICATIONS AND PRODUCTS

Active Galaxies. Some early radio survey objects were identified with galaxies, giving rise to the term "radio galaxies" to describe these objects. In the 1950's, however, a few radio sources were identified with what appeared to be stars. Analysis of these radio stars showed that they were not stars at all. They became known as quasi-stellar objects, or quasars. Later research showed that quasars and radio galaxies, along with other active galaxies, are powered by supermassive black holes in their centers. Radio astronomy continues to play an important role in understanding galactic formation and dynamics.

Interstellar Medium. The galaxy is filled with a very thin hydrogen gas. Much of this gas is in the form of individual neutral hydrogen atoms and does not shine with any visual light. However, when an electron flips its spin to a lower energy state in neutral hydrogen, it can emit radio waves having a wavelength of 21 centimeters. These radio emissions can be studied using radio telescopes. Observations of the 21-centimeter radiation have yielded a great deal of information about the density of material between stars and have been used to map the Milky Way and other galaxies. Studies of the 21-centimeter radiation in other galaxies are important in determining the rotational curves of those galaxies, an important tool in the study of dark matter in the universe. Rotational velocity measurements in other galaxies are also an important tool in calibrating the distance measurements needed to measure the cosmological expansion of the universe.

Pulsars. Jocelyn Bell and Antony Hewish discovered pulsating radio sources in 1967. These objects, named pulsars, turned out to be the first observational evidence of neutron stars, collapsed remnants of massive stars remaining after supernova explosions. Radio astronomy continues to play an important role in the study of pulsars, and knowedge of neutron star formation and properties is important in understanding how massive stars die and heavy elements form and are distributed throughout the universe.

Search for Extraterrestrial Intelligence. By the twentieth century, scientists were beginning to seriously consider the possibility that intelligent life may have evolved elsewhere in the universe. In the 1950's, some scientists suggested that perhaps such life could be detected by its radio transmissions. The first search for radio signals from intelligent life beyond Earth was conducted by Frank Drake in 1960. Since then, radio telescopes have routinely been used to search for extraterrestrial intelligence, but not all searches have been through dedicated radio telescopes. Several instrument packages piggyback on other scientific observations, looking for coherently modulate signals while the radio telescope is otherwise engaged in scientific research.

Radar Observations. Radar observations within the solar system fall under radio astronomy. The Arecibo Observatory has a powerful radar transmitter that has been used to refine the orbits of a number of near-Earth asteroids. Furthermore, radar reflections have been used to map the surface of Venus and the topography of several nearby asteroids, tasks that are either impossible or extremely difficult to do with visual astronomy.

IMPACT ON INDUSTRY

Jansky's discovery of radio signals coming from outer space opened the doorway for an entirely new way to study the heavens. With the large radio telescopes built in the latter half of the twentieth century, radio astronomy has become a major tool of astronomical research. Some astrophysical processes produce radio emissions rather than visual light, and radio telescopes are able to peer through interstellar media that block other wavelengths of electromagnetic radiation. Thus, radio telescopes are able to observe things that are not visible in other ways. Comprehensive studies of any celestial object have come to include observations in multiple wavelengths: visual, infrared, ultraviolet, and radio.

Government and University Research. Radio astronomy's biggest users are astronomers conducting research. The majority of research astronomers work either as faculty members at universities or researchers at government-funded laboratories. Radio observatories run or funded by the government operate on a shared-user basis. Astronomers submit proposals to use the facilities, and a review board judges which proposals are most qualified to warrant use of the facilities. Those researchers are then awarded time on the radio telescopes. Sometimes the researchers go the observatory to collect data, but other times data are collected by technicians at the observatory according to instructions from the researchers.

Most of the funding to operate large radio astronomy facilities comes from government agencies. The largest source of funding for radio astronomy in the United States is the National Science Foundation. Some smaller facilities associated with individual institutions were constructed with grant money, primarily from the foundation, but were operated out of state or local university funds. Budget shortfalls early in the twenty-first century, however, put pressure on many institutions to close their radio telescope facilities.

Some private organizations, such as the Planetary Society and the SETI Institute have secured private funds through donations or grants to fund their own radio astronomy studies. Generally, these organizations work with other facilities to fund operations of an existing radio telescope otherwise slated for decommissioning or to build new facilities, such as the Allen Telescope Array built by the SETI Institute in cooperation with University of California, Berkeley, at the university's Hat Creek Radio Observatory.

Industry and Business. There is very little economic incentive for private industry to conduct radio astronomy research for scientific gain. However, a small specialized group of companies works with the large government-funded observatories to construct new radio telescopes or to refurbish older ones. The physical structure of such large instruments requires mechanical engineering expertise to properly construct them, and several companies have found an economic niche supplying this expertise.

The technology developed for radio astronomy is virtually the same as that used for commercial communication receivers. Therefore, many of the advances in radio astronomy have commercial applications in the form of smaller, more sensitive, and less expensive satellite communication equipment. Many satellite communication companies keep abreast of the developments in radio astronomy technology to adapt this technology to produce better communication systems.

CAREERS AND COURSE WORK

Amateur radio astronomy is possible with simply a familiarity with electronics and astronomy; however, a career in radio astronomy generally requires more. Radio astronomy is a branch of astronomy, which is associated with physics and related to various engineering fields. Therefore, radio astronomers need courses in advanced mathematics, physics, computer programming, electronics, and astronomy. Chemistry and engineering courses are also useful. A radio astronomer should have a graduate degree in physics or astronomy, typically a doctorate, to do research. However, the field is quite broad, and most astronomers do not specialize in radio astronomy. Large radio telescopes, such as those that professional astronomers use, require mechanical engineers to design the structures and electrical engineers to design the receivers. Computer programmers are required to design the software to operate the telescope and analyze the data. These tasks often do not require a doctorate in the field but do require familiarity with astronomy. Because many astronomers who use radio telescopes are not themselves radio astronomers, technicians typically operate the telescopes. Technician positions generally require a bachelor's or master's degree in physics, astronomy, or electrical engineering.

Many astronomers work in academic settings as college professors. These positions always require advanced degrees. Astronomers often use many resources to study the heavens, including radio astronomy. However, technicians work directly with radio telescopes, either in construction or operation of the instruments. Therefore, technicians typically are employed directly by the radio telescope operators. A few private research laboratories or universities operate radio telescopes; however, most are operated by government laboratories or government-funded laboratories. There are comparatively few jobs available for radio telescope technicians, but there are also even fewer people who specifically choose this career path. Most radio telescope technicians start off as engineers or students in related fields who migrate to the field of radio astronomy.

SOCIAL CONTEXT AND FUTURE PROSPECTS

Radio astronomy continues to be an important tool in understanding the universe. However, the sensitive radio receivers of the radio telescopes are increasingly subject to interference from commercial communication systems. Additional technologies are being developed to deal with this interference, but the interference will most likely continue to be a problem and become worse at established radio

Fascinating Facts About Radio Astronomy

- After hearing about the discovery of pulsars, a tabloid dubbed the objects "little green men," or LGMs, speculating that they were beacons placed in space by aliens.
- The giant radio telescopes at Arecibo have been featured in several motion pictures, including the James Bond film *GoldenEye* (1995), and several television programs, including an episode of *The X-Files* (1993-2002).
- The jansky, a unit of radio power flux used in radio astronomy, was named for Karl G. Jansky, who is first credited with discovering extraterrestrial radio signals.
- The largest fully steerable radio telescope dish, the Green Bank Telescope, has a diameter about as large as a football field.
- When Grote Reber built his radio telescope in Chicago, it was on the eve of World War II, and it was rumored that he was working on an antiaircraft death ray for the U.S. Army.
- Simple radio telescopes can be built very inexpensively by amateurs using commercially available parts without modification. An old analogue television set and an ultra-high-frequency (UHF) antenna can be used to detect the Crab Nebula, and amateur radio equipment can be used to detect Jupiter.

observatories. Plans to build new radio observatories in even more remote locations will probably forestall the problem of interference but will not end it for ground-based radio astronomy.

The National Aeronautics and Space Administration (NASA) and several other space agencies have studied the feasibility of space-based radio telescopes. Several radio astronomy satellites have been launched into Earth orbit, where interference from terrestrial interference is much less. These instruments have been quite successful, and even more capable radio astronomy satellites are likely to be launched throughout the twenty-first century. Furthermore, several feasibility studies have been done on the prospect of building a radio astronomy observatory on the far side of the Moon, where the radio telescopes would be completely shielded from terrestrial interference. As yet, there are no plans for such a facility, but it is likely to eventually be built, probably at some point within the century. Naturally, such developments will bring new challenges and new opportunities to the field of radio astronomy.

Raymond D. Benge, Jr., B.S., M.S.

FURTHER READING

Burke, Bernard F., and Francis Graham-Smith. *An Introduction to Radio Astronomy.* 3d ed. New York: Cambridge University Press, 2009. A very thorough but also quite technical overview of the theory of radio astronomy and applications to astronomical observations.

Carr, Joseph J. *Radio Science Observing.* 2 vols. Indianapolis, Ind.: Prompt Publications, 1999. Contains radio astronomy ideas, circuits, and projects suitable for amateur radio operators or advanced high school or college students.

Kellermann, Kenneth I. "Radio Astronomy in the Twenty-first Century." *Sky and Telescope* 93, no. 2 (February, 1997): 26-34. An excellent overview of the development of radio astronomy through the twentieth century with a look forward to prospects for developments through the twenty-first century.

Lonc, William. *Radio Astronomy Projects.* Louisville, Ky.: Radio-Sky Publishing, 1996. Detailed instructions on the construction of various radio telescope systems and information on projects using those systems for the amateur astronomer with a good understanding of electronics. Suitable for engineering students.

Malphrus, Benjamin K. *The History of Radio Astronomy and the National Radio Astronomy Observatory: Evolution Toward Big Science.* Malabar, Fla.: Krieger, 1996. A chronicle of radio astronomy history from the early days to the development of large national radio astronomy laboratories.

Shostak, Seth. "Listening for a Whisper." *Astronomy* 32, no. 9 (September, 2004): 34-39. A very brief overview of how radio telescopes are used in the search for extraterrestrial intelligence.

Sullivan, Woodruff T., III. *Cosmic Noise: A History of Early Radio Astronomy.* New York: Cambridge University Press, 2009. A very thorough historical record of the first decades of radio astronomy, the people and instruments involved, and the discoveries made.

Verschuur, Gerrit L. *The Invisible Universe: The Story of Radio Astronomy.* 2d ed. New York: Springer, 2007. An excellent introduction and overview of the history of radio astronomy and discoveries made by radio astronomers.

WEB SITES

Jet Propulsion Laboratory
Basics of Radio Astronomy
http://www2.jpl.nasa.gov/radioastronomy

National Astronomy and Ionosphere Center
Arecibo Observatory
http://www.naic.edu

National Radio Astronomy Observatory
Very Large Array
http://www.vla.nrao.edu

The Planetary Society
http://www.planetary.org/home

SETI Institute
http://www.seti.org

See also: Planetology and Astrogeology; Space Science; Telescopy.

RADIOLOGY AND MEDICAL IMAGING

FIELDS OF STUDY

Physics; mathematics; radiology; medicine; electrical engineering; biomedical engineering; electronics; anatomy; thermography; microscopy; nuclear medicine; fluoroscopy.

SUMMARY

Radiology is the field of medicine concerned with imaging patients for the purposes of diagnosing and treating diseases and other medical conditions. Advances in medical imaging have helped revolutionize the contemporary practice of medicine. Through medical imaging, doctors are able to obtain a detailed view of the inside of the human body and obtain relevant qualitative and quantitative information regarding body tissues. Common uses of medical imaging include detecting broken bones using X-ray imaging, following fetus development using ultrasound imaging, diagnosing cancer through computed tomography or magnetic resonance imaging, and measuring the body's metabolic activity using positron emission tomography.

KEY TERMS AND CONCEPTS

- **Image Processing:** Process of converting raw data from the scan of an object into a useful image that can be qualitatively and quantitatively interpreted.
- **Magnetic Resonance Imaging (MRI):** Imaging modality that relies on the use of magnets and radio-frequency signals to manipulate specific atoms within the imaged structure.
- **Mammography:** Form of X-ray imaging customized for detecting abnormalities in the breast tissue of women and commonly used as a screening tool for breast cancer.
- **Nuclear Imaging:** Wide range of imaging procedures that depend on the use of radioactive agents to image the body.
- **Picture Archiving And Computer System (PACS):** Computer-based system of managing radiological information, including patient-specific data and images. PACS is used in all imaging modalities to facilitate the transfer, retrieval, and presentation of images.

- **Tomography:** Method of X-ray photography through which a single plane is imaged and other planes are eliminated.
- **Ultrasound:** Imaging modality based on the use of ultrasound to obtain quantitative and qualitative information on the scanned object.
- **X Ray:** Form of electromagnetic radiation characterized by its ionizing potential and short wavelength, ranging from 0.01 to 10 nanometers.

DEFINITION AND BASIC PRINCIPLES

Medical imaging, which uses a wide diversity of imaging modalities and procedures, relies on the use of radiation—ionizing and nonionizing—to obtain information regarding the imaged subject. Electromagnetic radiation covers a wide range of frequencies. The electromagnetic spectrum can be divided into categories based on wave frequencies. From low to high frequency waves, the categories are radio waves, microwaves, infrared, visible light, ultraviolet, X rays, and gamma rays.

Radiation can be broadly categorized as ionizing and nonionizing. Ionizing radiation is electromagnetic radiation that has sufficiently high energy to remove electrons on interaction with specific atoms, thereby producing ionization in the substance through which it passes. Nonionizing radiation cannot change the atomic structure of the materials with which it interacts. Gamma rays and X rays are ionizing radiation; ultraviolet, visible light, infrared, microwaves, and radio waves are nonionizing types. Both ionizing and nonionizing forms of radiation are used in medical imaging. In addition to electromagnetic radiation, sound waves are commonly used in medical imaging.

BACKGROUND AND HISTORY

Historically, people regarded light as something they could see with their eyes. Invisible radiation was not a scientifically proven idea until the early part of the seventeenth century. In 1800, British astronomer William Herschel experimented with a prism that split sunlight into a color spectrum. By measuring the temperature of each color, he realized that the temperature increased past the red light. Herschel interpreted the results by proposing the presence of an invisible form of radiation past the red part of the spectrum. This was eventually termed infrared radiation.

Other types of radiation were discovered in a similar manner. By observing physical effects on various materials, scientists were able to trace the cause to different forms of invisible radiation. The relationship between electricity and magnetism was discovered by Scottish scientist James Clerk Maxwell, who concluded that light is an electromagnetic disturbance that propagates through an electromagnetic field. Maxwell laid down much of the theoretical foundation that was used in subsequent discoveries of other forms of radiation in the electromagnetic spectrum.

The first type of ionizing radiation to be discovered was X rays. In 1895, when experimenting with a cathode-ray tube, German physicist Wilhelm Conrad Röntgen noticed a form of light that was capable of penetrating through most materials, including body tissues, and could form an image on a special form of phosphorescent plates. Shortly after the discovery of X rays, French physicist Antoine-Henri Becquerel discovered another form of ionizing radiation. Becquerel discovered natural radioactivity by experimenting with fluorescent minerals, including uranium. Radioactive materials could disintegrate by emitting specific forms of ionizing radiation. In 1898, Marie Curie discovered another radioactive element, radium.

When ionizing radiation was discovered, it was not known that it could harm people, and many scientists and others who were exposed to the rays suffered forms of radiation sickness and cancer. The field of health physics, which is most concerned with the safety of radiation, began to develop in the early part of the twentieth century. By the 2000's, the benefits and risks of radiation had come to be clearly understood by physicians, nuclear energy workers, and most of the public. Federal laws were created to regulate the use of ionizing radiation in medicine and other industries.

How It Works

The basis of all medical imaging modalities is radiation, which is coupled with detectors that acquire the information collected after the radiation has interacted with matter. Image processing and viewing can be performed entirely with computer software and algorithms. The end result of any imaging modality is a qualitative image that can be used to assess the body's parts based on prior knowledge of anatomy and physiology. Quantitative information can often be obtained from the images, and the quantitative parameters are characteristic of the imaging modality used.

Ultrasound Imaging. In ultrasound imaging, sound waves are generated and propagated through tissues at frequencies beyond the frequency of sound that can be heard by the human ear. The intercepting tissues either scatter (reflect) or absorb the sound waves. Absorbed wave energy is dissipated as heat and cannot be recovered, whereas scattered sound waves are acquired by the detector. How sound waves are reflected depends on the tissue type, with different types producing varying signal intensities on the final image. Ultrasonic transducers are used to generate and receive the ultrasound signals. The transducer converts the acquired acoustic energy into electric signals that are then manipulated using computer algorithms to produce an intensity image that can be interpreted by a radiologist.

Magnetic Resonance Imaging (MRI). The type of radiation used in magnetic resonance imaging lies in the radio frequency (RF) range of the electromagnetic spectrum. RF waves have frequencies ranging from 3 hertz to 300 billion hertz and are widely used in many broadcasting and electronic devices. In MRI, magnets with field strengths ranging from 3,000 to 30,000 times the Earth's magnetic field are used. RF waves and strong magnetic fields produce a specific RF frequency, termed the resonance frequency, which causes a change in the alignment of specific atoms within the specified magnetic field. The resonant frequency for hydrogen atoms—which are abundant in the human body—at a magnetic field strength of 1.0 by-prod is 42.6 megahertz. (By-prod is a unit of magnetic flux density, equal to one weber per square meter.) The Earth's magnetic field strength is 0.00005 by-prod. Most clinical MRI scanners have field strengths of 1.5 by-prod. Magnetic fields and RF signals are manipulated in many ways using computer algorithms. A specific set of signal manipulations is called a pulse sequence. Different pulse sequences can provide images with varying contrast, resolution, and intensity, depending on the type of information required.

X-Ray Imaging. Planar X-ray imaging and computed tomography (CT) imaging rely on the use of ionizing radiation in the X-ray part of the electromagnetic spectrum. X-ray photons are generated when high-energy electrons bombard a target material, such as tungsten, that is placed in an X-ray tube. The resulting X-ray beam has a continuous energy spectrum ranging from low-energy photons to the highest-energy photons, which corresponds with the X-ray tube potential. However, since low-energy photons increase the dose to the body and do not contribute to image quality, they are filtered from the X-ray beam. After a useful filtered X-ray beam is generated, the beam is directed toward the subject, while an image receptor (film or detector) is placed in the beam direction past the subject to collect the X-ray signal and provide an image of the subject. As X rays interact with the body's tissues, the X-ray beam is attenuated to different degrees, depending on the density of the material. High-density materials such as bone attenuate the beam drastically and result in a bright signal on the X-ray image. Low-density materials such as lungs cause minimal attenuation of the X-ray beam and are darkened on the X-ray image because most of the X rays strike the detector.

In CT, X-ray images of the subject are taken from many angles and reconstructed into a three-dimensional image that provides exquisite contrast of the scanned subject. At each angle, X-ray detectors measure the intensities of the X-ray beam, which are characteristic of the attenuation coefficients of the material through which the X-ray beam passes. Generating an image from the acquired detector measurements involves determining the attenuation coefficients of each pixel within the image matrix and using mathematical algorithms to reconstruct the raw image data into cross-sectional CT image data.

Nuclear Imaging. Nuclear imaging involves the use of radioactive substances to obtain functional information regarding various processes in the body. Specific radioactive compounds, called radionuclides, are manufactured using nuclear reactors and packaged in a safe way for administration to a patient. The radionuclides are biodegradable and do not pose a significant risk to the patient. However, because the emitted radiation is ionizing in nature, there is a slight risk of cancer induction. The most commonly used type of radionuclide for nuclear imaging is fluorodeoxyglucose (FDG), which is analogous to glucose and is metabolized by the body in the same manner as glucose.

Radionuclides can be administered orally or intravenously. Each radionuclide has a characteristic rate of decay, and therefore, a limited amount of time is available between uptake of the radionuclide and imaging of the patient. The products of radioactivity are detected using gamma cameras. The process of acquiring the signal from the radionuclide can be performed using either positron emission tomography (PET) or single photon emission computed tomography (SPECT). PET provides higher resolution images but uses more expensive and sophisticated equipment relative to SPECT. The level of intensity of the signal on PET or SPECT images is representative of the activity of the radionuclide. Radiologists can infer from PET/SPECT images any unusual activity in the body and request further examination based on the findings.

APPLICATIONS AND PRODUCTS

Ultrasound. The most common application for ultrasound is fetal imaging during pregnancy. Ultrasound is used to detect defects in the development of the fetus and to determine the sex of the fetus at later stages of development. Ultrasound is also used to detect abnormalities in muscles, tendons, and internal organs and to obtain information regarding the size and functionality of the organs. Cardiologists often use ultrasound to detect heart defects and plaque development in major body vessels. Patients with high cholesterol levels often undergo ultrasound imaging on a regular basis to measure plaque development in blood vessels. The goal is to detect atherosclerosis at an early stage and provide treatment to prevent devastating consequences such as heart attacks.

Magnetic Resonance Imaging. Applications of MRI exist in numerous branches of medicine in which detailed anatomical views of the patient are required. MRI provides soft-tissue contrast and high-resolution images without the use of ionizing radiation. MRI is commonly used to diagnose cancer, hemorrhage, stroke, musculoskeletal disorders, infections, inflammatory disorders (multiple sclerosis), and degenerative disorders (Alzheimer disease). MRI can also be used to obtain measurements of blood flow in the body. Such measurements provide doctors with pertinent information on the condition of the body's blood vessels, thereby allowing early intervention when abnormal conditions are found. Furthermore,

Fascinating Facts About Radiology and Medical Imaging

- Nobel prizes were awarded to those who discovered nearly every medical imaging modality.
- Functional MRI can be used as a lie detector. Lying and honesty activate different parts of the brain on functional magnetic resonance images.
- Magnets used in MRI are permanently turned on and have 30,000 times the magnetic field strength of the Earth. If a person enters an MRI unit carrying a credit card, the magnetic field will erase the credit card.
- When X rays were first discovered, they were used for entertainment purposes and people were imaged many times. Such use of X rays provided the earliest information on cancer induction caused by X-ray imaging.
- Radionuclides, used in nuclear imaging, are produced by particle accelerators, which are lengthy tracks that accelerate particles to extremely high velocities. The longest accelerator in the world is the Stanford linear accelerator, which is 3 kilometers long.
- Many of the fundamental aspects of medical imaging can be applied as therapies to cure or treat certain disorders. For example, ultrasound is used to relieve joint and muscle pain, while high-energy X rays are used to kill tumor cells.

developments in MRI technology have shown the feasibility of using it as a functional brain-mapping tool. Functional magnetic resonance imaging (fMRI) can detect which parts of the brain are activated in response to various external stimuli. Functional magnetic resonance imaging has found applications in early detection of Alzheimer's disease and in assessing cognitive and intelligence capabilities in children and adults.

Nuclear Imaging. The power of nuclear imaging lies in its ability to provide functional information on cellular processes within the body. PET imaging using the radionuclide fluorodeoxyglucose (FDG), or FDG-PET, is used to assess the body's metabolic activity and infer information regarding the body's condition. The most common use for FDG-PET imaging is detecting the spread of cancer throughout the body. Cancer patients have a high risk of developing

secondary cancers in the body; through routine nuclear imaging tests, physicians are able to monitor cancer patients and detect the spread of cancer at an early stage and offer effective treatment or palliation of the disease.

X Rays. The most common application of planar X-ray imaging is in orthopedics, where it is used to assess bone integrity and detect fractures and breaks. Bone absorbs X-ray photons readily and therefore appears bright on X-ray film. In addition to planar X-ray imaging, there are a number of commonly used X-ray-based imaging modalities. Fluoroscopy is an imaging procedure that allows doctors to obtain real time information on the relevant area in the patient. By obtaining sequential X-ray images through the use of an X-ray source and a fluorescent screen, fluoroscopic imaging can provide an interior view of the patient. Fluoroscopy has found applications in gastrointestinal (GI) imaging, where it is commonly used with a fluorescent contrast agent to detect abnormalities in the GI tract; in imaging of blood vessels (angiography), where it is used to detect blood clots and plaque development; in surgical procedures, where it can be used to guide surgeons to the target; and in numerous other areas of medicine. Because fluoroscopic procedures are performed over relatively long periods of time, the amount of radiation that the patient and the operating radiologist receive is concerning. Reducing exposure to radiation from fluoroscopy is of high importance to achieve good medical care.

Mammography is another X-ray-based imaging procedure. Mammography equipment is customized for breast imaging in women. Mammography uses a slightly different type of X ray because of the differing density and composition of breast tissue. The most common application for mammography is early screening for breast cancer in high-risk groups of women, including women with a family history of cancer and women above the age of forty. In some countries, including Canada, women above the age of forty are required to undergo routine mammography for screening of breast cancer. Early detection and treatment of breast tumors has been shown to prevent recurrence in late life and has been associated with long-term survival. As with all ionizing radiation, mammography carries a slight risk of cancer induction. In addition, the rate of false-positive results in mammography is slightly higher than for other imaging modalities. This can be distressing to patients undergoing screening for breast cancer.

Computed tomography (CT) is used as an initial procedure for the evaluation of specific types of patients and assessment of surgical or treatment options. From CT images, radiologists are able to locate the presence of foreign bodies such as stones, cancers, and fluid-filled cavities. Radiologists can also analyze CT images for size and volume of body organs and infer diagnosis of diseases and medical conditions such as pancreatitis, bowel disease, aneurysms, blood clots, abnormal narrowing of vessels, infections, injury to organs, tuberculosis, abnormal bone density, and diseases involving changes in tissue density or size.

In addition to the use of CT for disease diagnosis—which is based on a patient's complaint of a specific symptom or visual detection of an abnormality—CT has been used in private radiological clinics where patients may choose to obtain a CT image of their entire body to ensure the absence of abnormalities. However, the use of CT imaging for screening of apparently healthy people is controversial, because CT uses X-ray radiation, and patients who undergo X-ray imaging have an associated risk of cancer induction. CT imaging deposits between fifty to two hundred times the dose deposited by a conventional X-ray image. Although the association between CT imaging and cancer induction is not well established, its unjustified use remains an area of considerable debate.

Advances in computed tomography have allowed doctors to easily perform tissue biopsies that otherwise would have been invasive and time-consuming. Doctors insert a needle into the patient under the guidance of real-time CT imaging. By observing the location of the needle within the patient in real time, doctors are able to accurately obtain a tissue biopsy in a relatively short time and without the need for invasive procedures.

IMPACT ON INDUSTRY

The radiology industry continues to grow dramatically. According to the American College of Radiology, there were more than 30,000 radiologists in the United States as of 2008 as compared with 26,800 in 2005. The expectations of patients continue to grow with the advent of advanced imaging technologies; three-dimensional imaging is becoming a standard procedure in all imaging modalities as patients demand accurate and concise information regarding their medical conditions.

Government and University Research. Various forms of medical imaging have been integrated into routine health care standards in many countries. For detection of breast cancer, mammographies are recommended for all women above the age of fifty in Canada and the United States, and in addition, ultrasound is commonly used. Radiological centers exist in clinics and hospitals, and numerous facilities dedicated to specific imaging modalities have appeared in regions where routine testing is endorsed by governments and paid for by insurance companies. Governments are constantly seeking cost-effective means of reducing health care expenses through innovative radiological equipment and state-of-the-art computer-assisted technology.

Many universities across North America and Europe have established centers dedicated to medical imaging research at the clinical and basic science levels. The United States, in particular, houses some of the top radiological departments in the world, including Stanford University and Washington University. Numerous clinical trials are conducted at institutional and national levels to test the efficacy of new diagnostic tools and mechanisms. Emerging areas of research include computer-assisted radiology and surgery, four-dimensional imaging, image registration between modalities, data storage and picture archiving, digital mammography, and the combination of multiple modalities into a single unit.

Industry and Business. Significant business opportunities exist in the area of picture archiving and communication systems (PACS), which are related to medical imaging. In 2007, Frost and Sullivan, a business research and consulting firm, estimated the PACS market to be worth $590 million and growing strong. Among imaging modalities, CT technology was dominant, accounting for about 68 percent of the total revenues in 2006. MRI represented 20 percent of revenues, and the remaining 12 percent came from other imaging modalities such as PET, X-ray angiography, advanced (three-dimensional, four-dimensional) image fusion, and other imaging modalities.

Major Corporations. The major players in the global medical imaging market include General Electric (GE), Hitatchi, Siemens, Toshiba, and

Philips on the diagnostic level, and Varian on the radiotherapy level. Most corporations offer their equipment as a complete package that includes computer systems, customer support, file management, PACS, technology solutions, and customized support as required.

CAREERS AND COURSE WORK

Careers in radiology and medical imaging are primarily in the health care sector. Medical imaging equipment exists in every major hospital and in most small hospitals in the developed world. X rays are by far the most common and widely distributed imaging modality across the globe. More sophisticated imaging modalities, such as computed tomography and magnetic resonance imaging, are also widely distributed in hospitals in the developing world, and experts in the operation and maintenance of these scanners have frequently been imported from the West.

Many routes exist for working with medical imaging. From the academic side, several roads can lead to medical imaging careers. A degree in physics or engineering with specialization in biomedical or electrical engineering provides a solid ground in the electronics and hardware of medical imaging equipment. Engineers can find work in hospitals or in industries that design and manufacture scanners. Degrees in mathematics and computer science provide the necessary background for working with image reconstruction algorithms and advancing software-related operation of the various imaging modalities. Computer programmers can find work in the research and development sector of most imaging industries. In particular, MRI and CT software and computer algorithms are always being developed and upgraded. A graduate degree in medical physics provides theoretical and practical experience in medical imaging from data acquisition to image reconstruction and troubleshooting. Medical physicists often work in imaging facilities and hospitals where they usually supervise the personnel operating imaging equipment.

A degree in medicine with specialization in radiology provides theoretical and practical experience in understanding human anatomy, pathology, and physiology, and in interpreting medical images and providing diagnosis of specific disease conditions. Radiologists undergo residency training for a period of four to six years after graduating from medical school.

Technical colleges can provide education in the operation of the various medical scanners. Graduates with a technical degree in medical imaging often work as technologists in hospitals, where they are responsible for patient scheduling and machine operation. Technical colleges provide a broad overview of all the imaging modalities, while allowing the student to specialize and gain experience in one or more imaging procedures.

Medical imaging, with all of its tools and procedures, is an invaluable field in the medical sector. Career prospects involving any of the medical imaging modalities are very promising and there has been a continual demand for experienced radiology workers. In particular, MRI is increasingly becoming the modality of choice for doctors because of its high-quality images and use of nonionizing radiation. In some countries, patient waiting lists for MRI procedures can be as long as six months because of the heavy demand on the scanners. The number of MRI scanners is expected to increase to meet this demand.

SOCIAL CONTEXT AND FUTURE PROSPECTS

Medical imaging has grown substantially since the fourth quarter of the twentieth century and has become a multibillion-dollar global industry. As each imaging modality emerged, it was heralded by medical professionals, but the focus has shifted from new developments to further enhancing the efficiency, quality, and safety of each modality. In addition, emphasis has been placed on combining imaging modalities to obtain a comprehensive understanding of disease pathology. In the 2000's, combined PET/CT scanners and MRI/PET scanners emerged, providing an enhanced understanding of the anatomical and functional aspects of disease. In addition, portable imaging modalities such as ultrasound have commonly been merged with CT and MRI to provide a detailed portrait of a patient's condition.

Concerns regarding the use of ionizing radiation continue to exist, and the issue has often been debated by the medical community, the media, and regulatory bodies. MRI is the preferred imaging modality for acquiring high-quality anatomical information because of its use of nonionizing radiation, but its high cost and lower availability make CT a more practical imaging modality in many situations. In addition, CT provides different quantitative parameters than MRI.

The dependency of PET and SPECT imaging on radionuclides makes them vulnerable to the limitations of nuclear reactors. The 2009 closure of the Chalk River reactor, which supplied more than a third of the world's supply of medical radionuclides, had a significant impact on nuclear imaging in many clinics. Such incidents can have a negative impact on health care if nuclear imaging is the sole provider of imaging data in a clinic.

Ayman Oweida, B.Sc., M.Sc.

FURTHER READING

Bushberg, Jerrold T., et al. *The Essential Physics of Medical Imaging.* 2d ed. Philadelphia: Lippincott Williams & Wilkins, 2002. A comprehensive overview of all medical imaging modalities and their applications in medicine that provides a general introduction on the biological effects of ionizing radiation.

Delso, Gaspar, and Sibylle Ziegler. "PET/MRI Systems." *European Journal of Nuclear Medicine and Molecular Imaging* 36, no. S1 (March, 2009): S86-92. A basic review of the principles and challenges associated with combining PET and MRI in one system.

Huda, Walter. *Review of Radiologic Physics.* 3d ed. Baltimore: Lippincott Williams & Wilkins, 2010. Examines the physical concepts underlying each imaging modality. Contains illustrations and tables.

Mawlawi, Osama, and David Townsend. "Multimodality Imaging: An Update on PET/CT Technology." *European Journal of Nuclear Medicine and Molecular Imaging* 36, no. S1 (March, 2009): S15-29. A historical account of the evolution of combined PET/CT scanners, their status, and clinical applications.

Radiological Society of North America. http://www.radiologyinfo.org. Accessed October, 2009. A comprehensive Web site providing an overview of all medical procedures requiring the use of medical imaging, it is mainly directed toward patients but contains invaluable general knowledge to all readers. Representative images from each imaging procedure are abundant in the Web site.

WEB SITES

National Institutes of Health
Diagnostic Imaging
http://health.nih.gov/topic/DiagnosticImaging/ProceduresandTherapies

Occupational Safety and Health Administration
Radiation
http://www.osha.gov/SLTC/radiation/index.html#electromagnetic

Radiological Society of North America
http://www.radiologyinfo.org

See also: Computed Tomography; Magnetic Resonance Imaging; Ultrasonic Imaging.

REHABILITATION ENGINEERING

FIELDS OF STUDY

Human anatomy; kinesiology; physiology; ergonomics; mechanical engineering; pathology; physics; biochemistry; cellular biology; chemistry; biology; mathematics; psychology.

SUMMARY

Rehabilitation (or rehabilitative) engineering is the application of engineering sciences to improve the capabilities and quality of life for people with physical and cognitive impairments. Functional areas addressed through rehabilitation engineering may include the development of devices not only to enhance mobility but also to improve communication, hearing, and vision. For many individuals, rehabilitation engineering, which is considered a subset of bioengineering, becomes an important aspect of returning to employment, independent living, and integration into the community. Rehabilitation engineers may design prosthetic devices and seating and positioning technologies, or plan modifications of homes, workplaces, and vehicles for the disabled.

KEY TERMS AND CONCEPTS

- **Audiology:** Scientific study of hearing, balance, and related disorders.
- **Biomedical Engineering:** Engineering specialty that applies engineering principles and methods to medical problems such as the development and manufacture of prosthetic limbs and organs.
- **Ergonomics:** Science of designing equipment, typically in the workplace, to reduce user fatigue and discomfort and thereby maximize productivity.
- **Magnetic Resonance Imaging (MRI):** Diagnostic test that uses a magnetic field to create an image of organs and structures inside the body.
- **Occupational Therapy:** Medical field that uses work, play, and everyday activities in a therapeutic manner to increase a person's function and to enhance development.
- **Physical Rehabilitation:** Process by which functionality is returned to an individual afflicted by an injury, chronic disease, or genetic dysfunction.

- **Physical Therapy:** Medical field that seeks to restore maximum function and movement to patients whose physical abilities have been reduced by pain, disease, or injury.
- **Prosthesis:** Artificial device that replaces a missing body part or augments an injured part; also known as prosthetic device.
- **Speech Therapy:** Medical field that treats disorders involving speech, fluency, swallowing, language, and communication.

DEFINITION AND BASIC PRINCIPLES

Rehabilitation engineering applies the techniques of engineering to improve the quality of life of people with disabilities. Disabilities are of varying severities and often affect functional capability, such as mobility, communication, hearing, vision, and cognition. Such functional impairments can affect a person's participation in activities associated with employment, independent living, and recreation. Rehabilitation engineering examines the nature and involvement of disabilities so that an appropriate medical device such as a prosthetic limb or component of a knee replacement can be designed. After engineers gain an understanding of a particular injury or disability, they develop a device, which is tested, evaluated, and adapted as necessary. Engineers work with several other medical professionals in the design and develop rehabilitation devices such as physical, occupational, and speech therapists, scientists, physicians, special education teachers, and industrial designers.

BACKGROUND AND HISTORY

Rehabilitation engineering originated in the United States shortly after World War II when the National Research Council established a committee on prosthetic devices. It brought together physicians, surgeons, and allied health professionals who were involved in medical engineering research. In 1954, an increased awareness of the need to combine engineering with rehabilitation led to the passage of important amendments to the Vocational Rehabilitation Act, authorizing the funding of research and development. Between 1960 and 1970, the number of rehabilitation research facilities grew, and the size and stature of the medical specialty of physical medicine increased. James Garrett, chief of research and development at the Social and Rehabilitation Service, coined the term "rehabilitation engineering" around 1970. In 1971, several rehabilitation engineering centers were established with federal government funding. The Rehabilitation Act of 1973 ensured that these centers would continue, and as of 2010, there were seventeen rehabilitation engineering research centers funded by the National Center for the Dissemination of Disability Research.

How It Works

Rehabilitation engineering improves the quality of life for the physically handicapped by taking a total approach to rehabilitation. Medical professionals such as physicians and physical, occupational, and speech therapists study and treat problems confronted by those with physical, functional, and communication disabilities and rely on equipment and devices in the treatment of their patients. Whether used in an acute hospital setting or in an outpatient clinic specializing in sports medicine, the equipment and devices used in rehabilitation are a collaborative effort by engineers and medical professionals such as physical therapists. For example, physical and occupational therapy relies on equipment and devices in the treatment of patients recovering from loss of motion, strength, and balance and overall mobility. Engineers work with these professions to develop the type of equipment that will help the patient recover and gain increased ability to perform the activities of daily living. For example, researchers from engineering and physical and occupational therapy can determine how propelling a wheelchair relates to upper-extremity pain in individuals who have traumatic spinal cord injures or multiple sclerosis.

Biomechanics laboratories are important sites where rehabilitation engineers, rehabilitation professionals, and representatives of the medical device industry can work together. In these laboratories, new products and techniques pass through all phases of research and development and eventually reach the patient. Rehabilitation engineers engage in a lengthy process before individuals can be helped. Through clinical experience and research, rehabilitation professionals (and sometimes patients) identify a specific rehabilitation problem. Researchers develop a possible solution, which may include the need for a medical device such as a specific knee brace or an artificial arm with computer sensors. Rehabilitation engineers begin by developing a product concept and design. In the initial design phase, they may use electronic tools to create an abstract form of the product. Once the initial design phase is completed, a limited number of physical prototypes of the rehabilitation device (such as knee brace or strengthening equipment) are created and used to prove that the initial design is practical, effective, and above all safe. During this phase, it is common for the engineering team to develop the product by hand. If the rehabilitation device is satisfactory, the production phase begins, and the original design concept becomes reality.

Applications and Products

Devices and Technology. Health care professionals who work directly with individuals with physical, functional, and communication disabilities provide vital information to rehabilitation engineers on what type of devices are needed to compensate for what type and degree of disability. Engineers work to develop equipment best suited not only for assisting in the patient's initial recovery but also for facilitating activities of daily living. Strength training equipment, for example, has become more high tech, with advances made in computer-assisted machines that can maintain constant torque or tension as muscles shorten or lengthen. This technique, called isokinetic strengthening, can make rehabilitation more efficient by allowing the therapist to take a more specific approach.

A collaboration of biomechanics, bioengineering, and physical therapy in the late twentieth century led to significant advances in the development of prosthetic devices. The fabrication of robotic prosthetic hands that can use remaining muscles to reproduce all the motions of the human hand and fingers is just one example. The development of assistive devices to address a loss or lack of mobility can significantly improve the quality of life for individuals with disabilities. Physical therapists assist engineers in producing adaptive equipment such as specialized wheelchairs, walkers, and canes as well as rehabilitation equipment for enhancing strength, range of motion, and balance.

Occupational Therapy. Occupational therapists are trained in the structure and function of the human body and the effects of illness and injury. They can recommend or develop devices to help avoid injury or illness at home or on the job. They also help people with physical or cognitive limitations return to work or relearn the activities of daily living, often with assistive devices. Occupational therapists and rehabilitation engineers often collaborate to determine how certain components of the workplace (such as a specific tool in an assembly line or a computer station) can be adapted to produce a healthful and efficient environment. For example, an occupational therapist may work with a rehabilitation engineer to develop a hand and wrist brace to be used by workers who engage in repetitive movements or to modify tools and equipment so that injury or illness can be avoided. Therapists and engineers can collaborate to find better working conditions for employees who must maintain a fixed or awkward posture for extended periods of time. They also produce adaptive equipment (such as seating and mobility aids) for children and adults with disabilities. Generally, rehabilitation engineering and occupational medicine work together to examine the complex physical relationships among people, equipment, job duties, and work methods.

Speech Therapy. Rehabilitation engineers work closely with speech therapists, who deal with people who have difficulty with speech, communication, fluency, and swallowing. Speech therapists initially try to maximize a person's existing speech capabilities, whether the individual's problems result from a congenital disorder, an illness, or an injury. If normal communication cannot be achieved, assistive devices are used for speech and hearing. Nonverbal patients or those with severe impairment of speech functions typically need alternative methods of communication. At this point, rehabilitation engineers become involved, whether directly or indirectly, through the evaluation and development of electronic augmentative communication devices.

A wide range of communication needs can be met with augmentative communication systems, most of which are designed to be portable. For example, words and messages stored within the memory of a small computer can be accessed by a user and conveyed through a synthesized voice. ZYGO Industries' LightWRITER is a portable text-to-speech communication aid with two displays, one that faces the user and a second that faces the other person, allowing audible communication in a natural face-to-face position. Multicomponent communication systems

consisting of a collection of techniques, aids, symbols, and strategies also have been designed and developed.

Audiology. Audiologists evaluate and develop solutions to problems confronting people with hearing loss. Rehabilitation engineers work closely with audiologists to design hearing aids and related devices. The design and application of each device depends on the nature of the individual's auditory disorder, and professionals must be familiar not only with the design of the hearing device but also with its operation as some units are programmable, digital, and able to measure performance.

Rehabilitation engineering is active in the research, design, and development of devices used in phoniatrics, a subfield of audiology that studies communication disorders and tries to determine their causes. For example, rehabilitation engineers were involved in the research and development of cochlear implants, a surgical implant that enables individuals with certain hearing disorders to hear speech and everyday sounds. Postsurgical care often includes intensive therapy from a multidisciplinary medical team to help patients fully develop their speaking and comprehension skills.

IMPACT ON INDUSTRY

Rehabilitation engineering's impact on industry is growing, partly because disabilities have become a major health issue for an increasing number of Americans. According to the U.S. Census Bureau, in 2005, 54.5 million Americans (18.7 percent) had some level of disability and 35.0 million (12 percent) were severely disabled. The prevalence of disabilities is increasing because biomedical advances have made it possible for more people to survive severe levels of injury or disease. Rehabilitation engineering works to increase the capabilities of those with physical and cognitive impairments and to improve their quality of life. Of those age fifteen and older, 14.7 million had difficulty seeing, hearing, or speaking; 27.4 had difficulty with lower-limb mobility (walking or climbing stairs); 19.0 million had upper-limb mobility problems; and 8.5 million had trouble with one or more activities of daily living. The Centers for Disease Control and Prevention published a report in 2009 listing the top ten causes of disability: arthritis or rheumatism (8.6 percent); back or spine problem (7.6 percent), heart trouble (3.0), mental or emotional problem (2.2), lung or respiratory problem (2.2), diabetes (2.0), deafness or hearing problem (1.9), stiffness or deformity of limb (1.6), blindness or vision problem (1.5), and stroke (1.1). The significance of rehabilitation engineering in improving the quality of life for those with disabilities is likely to be a part of any discussion of health care reform in the future.

University and Government Research. Many hospitals and government agencies conduct research in rehabilitation engineering. Several medical research organizations employ rehabilitation engineers to investigate ways to improve the health of people with physical and mental disabilities as well as musculoskeletal and neurological conditions. These research institutes often make their scientific findings readily available to other researchers and provide educational programs for consumers, health professionals, and scientists.

Organizations such as the Orthopaedic Research Society aim to advance orthopedic research and transform this research to clinical practice. Such organizations rely on the field of rehabilitation engineering to help develop medical devices and equipment used in orthopedic research.

Another organization that engages in rehabilitation research is the Disability and Rehabilitation Research Coalition, a coalition of nonprofit organizations engaged in rehabilitation and disability research. The coalition seeks to help Americans with disabilities function and live as independently as possible and takes the lead in attempting to increase and leverage federal funds for disability and rehabilitation research.

Private Businesses. Some of the larger biomedical, assistive device, and orthopedic product corporations provide the world with leading medical technology and a broad range of products. Such corporations work with respected medical professionals in orthopedics, neurology, pediatrics, speech, and audiology to help people live active, worthwhile lives. Corporations in the field of bioengineering and rehabilitation engineering introduce new materials that make rehabilitation products perform effectively and enhance lives.

CAREERS AND COURSE WORK

Rehabilitation engineering offers careers for not only engineers but also physicians, physician

---

Fascinating Facts About Rehabilitation Engineering

- The first metallic hip replacement surgery was conducted in 1940 by American surgeon Austin T. Moore. Hip replacement is regarded as the most successful orthopedic surgery, with more than 97 percent of patients reporting successful outcomes.

- In 1958, Swede Arne Larsson became the first person to have a fully implantable pacemaker. Although the first device failed after three hours, Larsson received more than twenty replacement pacemakers before he died at the age of eighty-six in 2001.

- In 2006, Claudia Mitchell, a twenty-six-year-old woman, received a thought-controlled prosthetic arm. The $60,000 device, which uses redirected nerves, allows her to pick up small objects such as beverage bottles.

- South African Oscar Pistorius, a double amputee who runs with carbon fiber prosthetic legs, narrowly missed qualifying for the 400-meter dash in the 2008 Olympic Games.

- The Hokoma Lokomat system is an automated gait orthosis used on a treadmill that helps patients with spinal core and neurological disorders to improve their ability to walk. It supports patients' body weight and guides the legs.

---

assistants, laboratory technicians, statisticians, and supporting research laboratory personnel. A career in rehabilitation engineering can be rewarding and stimulating because technology in this area is ever advancing. As a branch of biomedical engineering, the career opportunities in rehabilitation engineering are expected to increase twice as fast in comparison with those in most science and engineering fields in the 2010's. Biomedical and rehabilitation engineers are employed in industry, hospitals, and the research facilities of medical institutions and universities. Typically, the medical equipment and supplies industry employed about 20 percent of all biomedical engineers, and the scientific research and development services industry employed another 20 percent. Rehabilitation and biomedical engineering positions in government agencies can involve product testing and developing safety standards for specific devices. Some biomedical and rehabilitation engineers are consultants for marketing departments of medical device companies, and others are employed in management positions.

High school students who are interested in biomedical and rehabilitation engineering as a career should focus on sciences, taking courses in biology, chemistry, physics, and mathematics, and in cellular biology, human anatomy, and biochemistry if they are available. Students planning on entering the field usually first obtain a degree in engineering, then choose a discipline within engineering, such as biomedical engineering, and later specialize further in rehabilitation engineering. A master's or doctoral degree is required for a position with most research and development programs but not for the majority of entry-level rehabilitation engineering jobs.

Professional organizations and associations offer a wide range of resources for planning a career in rehabilitation engineering. They can help students determine the proper course work and keep them abreast of trends in the industry. Associations promote the interests of their members and provide a network of contacts that assists in finding jobs. They can also offer a variety of services, including job referral, continuing education courses, insurance, and travel benefits, and often publish periodicals and hold meetings and conferences. Associations serving biomedical and rehabilitation engineers include the American Institute for Medical and Biological Engineering and the American Society of Mechanical Engineers, Bioengineering Division.

SOCIAL CONTEXT AND FUTURE PROSPECTS

According to the U.S. Bureau of Labor Statistics, biomedical engineers (which include rehabilitation engineers) occupy about 14,000 positions, or 0.9 percent of the 1.5 million positions held by engineers. Biomedical engineers are expected to experience 21 percent employment growth into the mid-2010's. This rapid increase is attributed to the aging of the population and the focus on health issues. The demand for better medical devices and equipment designed by biomedical and rehabilitation engineers is likely to grow, and advances in rehabilitation product technology are continually being made. Rehabilitation engineers have harnessed the progress in computer technology to improve quality of life. For example, engineers are developing fully implanted and wireless systems that will detect muscle activity from electrodes implanted just under the skin or scalp. These systems mean that users will no longer have to wear detecting hardware on their heads or faces. Studies are examining whether jaw muscle contractions, detected by sensors on the scalp or just under it, can be used to activate other muscles. Ongoing research covers a range of robotics technology, including manipulator, sensor, computer vision, and multimedia applications, and autonomous robots.

Jeffrey Larson, P.T., A.T.C.

FURTHER READING

Brotzman, S. Brent, and Kevin E. Wilk. *Clinical Orthopedic Rehabilitation*. Philadelphia: Mosby, 2003. A thorough textbook on orthopedic rehabilitation, designed for medical students and professionals.

Chau, Tom, and Jillian Fairley, eds. *Pediatric Rehabilitation Engineering: From Disability to Possibility*. Boca Raton, Fla.: Taylor & Francis, 2010. Addresses the rehabilitation needs of children and how engineering is applied.

Cooper, Rory A., Hisaichi Ohnabe, and Douglas A. Hobson, eds. *An Introduction to Rehabilitation Engineering*. Boca Raton, Fla.: Taylor & Francis, 2007. Examines rehabilitation engineering at the clinical level. Discusses wheelchairs, protheses, and many other devices.

Domholdt, Elizabeth. *Rehabilitation Research: Principles and Applications*. 3d ed. St. Louis, Mo.: Elsevier Saunders, 2005. Educates the reader on general and specific aspects of research within the science of rehabilitation medicine.

Hung, George K. *Biomedical Engineering: Principles of the Bionic Man*. Hackensack, N.J.: World Scientific, 2010. Examines scientific bioengineering principles as they apply to humans.

Maxey, Lisa, and Jim Magnusson. *Rehabilitation for the Postsurgical Orthopedic Patient*. St. Loius, Mo.: Mosby, 2001. Includes chapters on rehabilitation of the postoperative total knee arthroplasty.

WEB SITES

American Academy of Physical Medicine and Rehabilitation
Disability and Rehabilitation Research Coalition
http://www.aapmr.org/hpl/legislation/drrc.htm

American Institute for Medical and Biological Engineering
http://www.aimbe.org

American Society of Mechanical Engineers, Bioengineering Division
http://divisions.asme.org/bed

Centers for Disease Control and Prevention
Disabilities
http://www.cdc.gov/ncbddd/disabilities.htm

National Center for the Dissemination of Disability Research
http://www.ncddr.org

Orthopaedic Research Society
http://www.ors.org

See also: Audiology and Hearing Aids; Bioengineering; Geriatrics and Gerontology; Occupational Health; Orthopedics; Pediatric Medicine and Surgery; Speech Therapy and Phoniatrics.

REPRODUCTIVE SCIENCE AND ENGINEERING

FIELDS OF STUDY

Cell biology; molecular biology; biochemistry; developmental biology; medicine; pediatrics; physiology; obstetrics; gynecology; andrology; embryology; endocrinology; surgery; genetics; biomedical engineering; chemical engineering; reproductive technology; pharmacology; neurobiology; urology; pathology; immunology.

SUMMARY

Reproductive science and engineering is concerned with the examination and regulation of the physiological mechanisms involved in human reproduction, such as conception and birth, and with diagnosing and treating disorders of reproduction, such as male and female infertility. Because the ability to successfully bear offspring is the core objective driving the success of all animal species, reproductive research is of deep importance on a purely scientific level. In addition, by providing methodologies that enable infertile couples to conceive, such as artificial insemination and in vitro fertilization, this discipline has profound effects on both individual human lives and the population trends of societies, nations, and the world as a whole.

KEY TERMS AND CONCEPTS

- **Artificial Insemination:** Any procedure, other than sexual intercourse, by which sperm is placed directly into a woman's cervix.
- **Assisted Reproductive Technology (ART):** Any technique, including in vitro fertilization, in which ova (also known as oocytes or eggs) are removed from a woman before being fertilized with sperm.
- **Cloning:** Process by which an embryo, either human or animal, can be generated from the genetic information of another individual, creating an identical copy, or clone.
- **Cryopreservation:** Process by which gametes (sex cells) and embryos are preserved by freezing.
- **Follicle-Stimulating Hormone (FSH):** Hormone produced by the pituitary gland that stimulates the production of eggs; it can be produced in pill form and prescribed to improve fertility.
- **Human Chorionic Gonadotrophin (hCG):** Hormone produced by the placenta; it helps preserve the uterine lining and its presence is used to confirm pregnancy.
- **Intracytoplasmic Sperm Injection (ICSI):** Procedure by which a single spermatozoa is inserted directly into an individual ovum in a laboratory.
- **Micromanipulation:** Procedure in which sperm, ova, or embryos are manipulated under a microscope; also known as microinsemination.
- **Preimplantation Genetic Diagnosis (PGD):** Procedure by which DNA is removed from a fertilized egg or embryo and examined for the presence of genetic abnormalities.
- **Sperm Bank:** Facility that collects and stores (usually via cryopreservation) human sperm from donors.

DEFINITION AND BASIC PRINCIPLES

Reproductive science and engineering is the study of the physical and chemical processes that underlie human reproduction. It involves the application of scientific technologies to treat disorders of reproduction, such as infertility. It also includes the development and use of methods to interfere with or prevent impregnation, such as contraception and sterilization. Among the many approaches reproductive scientists take to these issues, three of the most significant involve mechanical, chemical, and genetic strategies for engaging with reproduction. Micromanipulation is an umbrella term for any reproductive assistance technique that involves the physical handling of sperm, oocytes, or embryos on a microscopic scale, using specialized tools. Another of the key tools that is widely used by reproductive scientists is a set of pharmaceutical products that mimic or interfere with the chemical signals naturally produced by the body. These artificial hormones can be used to manipulate the human reproductive system in a variety of ways. Reproductive genetics is a rapidly expanding subfield of reproductive science that applies the tools of genetic research and DNA-based technologies to issues of conception, childbirth, and inheritance.

BACKGROUND AND HISTORY

Written records indicate that humans have struggled with infertility since ancient times. For example, infertility is mentioned in texts from ancient Greece and Persia. For much of European history, infertility was generally attributed to women, not men, and was considered a sign of impiety because bearing children was considered a blessing from God. The first experiments with artificial insemination were carried out in the middle of the nineteenth century by the American gynecologist J. Marion Sims, who also carried out surgical procedures designed to widen the cervix and thus facilitate the entry of sperm. In the late nineteenth century, scientists began to acknowledge the potential role of male infertility, using microscopes to test the potency of sperm.

The first half of the twentieth century witnessed the discovery of the three most important hormones involved in reproduction, the female hormones estrogen and progesterone, and the male hormone testosterone. Soon after, companies began to manufacture the first synthetic hormones for the treatment of infertility. In 1944, the first laboratory test showing that human oocytes could be fertilized in vitro was carried out. Thirty-four years later, the first so-called test-tube baby was born in England, and sperm banks became more common. The late twentieth century saw two more milestones in infertility treatment: the first successful implantation and pregnancy with an egg that had been cryopreserved and the development of intracytoplasmic sperm injection technology.

HOW IT WORKS

Micromanipulation. The basic setup of a micromanipulator is a microscope connected to robotic arms that are powered by electric motors and moved by hydraulic or pneumatic controls that may require foot pedals or joysticks, or both. The robotic arms are in turn connected to incredibly tiny glass tools. By looking through the microscope, which magnifies the cells hundreds of times, and manipulating the controls, the operator is able to tinker with the gametes and embryos with great precision—in essence performing a kind of microsurgery. Some micromanipulation techniques involve lasers, which can move segments of a cell from place to place and slice open the thick membrane around an oocyte. This membrane, known as the zona pellucida, is often slit open to facilitate the entry of sperm. Other techniques involve the use of electric currents that can cause the membranes of two different cells to join together or transfer genetic material from one cell to another. Intracytoplasmic sperm injection is the most common micromanipulation procedure.

Reproductive Pharmacology. Reproductive pharmacology makes use of synthetic hormones to produce a desired effect—whether contraceptive in nature or intended to increase fertility. For example, most birth control pills use some combination of artificial forms of the female hormones estrogen and progesterone. These substances interfere with the normal cycle of ovulation and menstruation, thus suppressing a woman's ability to conceive. For example, the chemical signals sent by the pill may prevent a woman's pituitary gland from releasing a chemical signal that induces ovulation, so that the ovaries do not release any oocytes. Or it may prevent the lining of the uterus from thickening, thus inhibiting the implantation of a fertilized egg. Other ways in which artificial hormones may prevent pregnancy include thickening the mucus found in the cervix so that it is more difficult for sperm to travel through it and reducing the rate at which oocytes migrate from the ovaries toward the uterus. Artificial hormones can also be used to treat infertility. For example, injections of follicle-stimulating hormone (FSH) and human chorionic gonadotropin (hCG) stimulate the process of ovulation, while gonadotropin-releasing hormone alters the timing of ovulation, making a woman's fertile period more regular and ensuring that an oocyte is not released into the uterus until it has developed properly. These and other pharmaceutical tools can help physicians correct problems associated with common female fertility disorders. The most common disorder leading to infertility in women is polycystic ovary syndrome (PCOS), which results in excessively high levels of androgen (a male hormone) and ovulation that is irregular or entirely absent.

Genetics. Some infertility disorders are associated with a genetic defect of one kind or another. Male infertility has been linked with microdeletions, or tiny missing parts, in the Y chromosome, and with mutations in particular genes. To identify these genetic components of infertility, scientists often conduct what is known as a genome-wide association study, or whole-genome association study. This is a technique by which the entire set of genetic material belonging to each member of a group of subjects is scanned and compared in order to pinpoint specific genetic variations that are more prevalent in people with a certain trait, such as infertility.

Screening Tests. Another common tool of reproductive genetics is the use of screening tests to identify genetic traits in embryos produced by in vitro fertilization before they are implanted in a woman's uterus. After a cell sample has been retrieved from the blastocyst or embryo, various techniques can be used to screen its DNA for possible genetic abnormalities associated with diseases such as Down syndrome or cystic fibrosis. For example, short pieces of DNA can be artificially produced that are specially designed to bind to and mark mutated DNA in the sample, if it exists. Alternatively, the DNA in the sample can be directly examined to look for known mutations. Tests can also be carried out that reveal enzymes and proteins produced by specific genes.

APPLICATIONS AND PRODUCTS

In Vitro Fertilization. In vitro fertilization (IVF) is one of the most common applications of assisted reproduction technology. The Latin term *in vitro* literally means "in glass." In vitro fertilization is a medical procedure in which egg cells are fertilized not inside the body (in vivo) but within an artificial laboratory environment. Because this procedure is both complex and expensive, it is often employed to help couples for whom other infertility treatments have already failed.

The first step in an IVF cycle involves the use of artificial hormones, such as follicle-stimulating hormone and human chorionic gonadotropin, in drug form. During normal ovulation, a woman's ovaries produce a single egg; this treatment, known as superovulation, causes the ovaries to produce multiple oocytes. Next, the oocytes, along with some follicular fluid, are retrieved from the patient's ovaries via a needle. They are allowed to incubate under controlled laboratory conditions, with sperm collected from the patient's partner or with donor sperm. After a day or two, the eggs are examined to determine which, if any, have been successfully fertilized. If necessary, intracytoplasmic sperm injection may be used to inseminate the eggs. This is a micromanipulation procedure that can be used to artificially induce fertilization when sperm have low motility (do not move well). In this process, a micromanipulator with

a thin glass pipette on the end is used to pick up and inject a single sperm directly into an oocyte. Next, the fertilized eggs—now known as embryos—are either frozen for later use or transferred into the uterus using a speculum and a catheter. Typically, multiple embryos are inserted into the uterus so as to increase the chances of at least one successful implantation. This also increases the possibility of a multiple birth.

Because the success of any given cycle of IVF treatments is by no means guaranteed, many couples choose to use cryopreservation techniques to freeze embryos produced in one cycle for future use. This streamlines the process a couple must go through if treatment does, in fact, have to be repeated. It also reduces the need for performing invasive procedures on the female patient.

Intrafallopian Transfer. Gamete intrafallopian transfer (GIFT) is a procedure that resembles in vitro fertilization. The main difference between the two applications is that with GIFT, fertilization takes place not within a laboratory setting but rather inside the female patient's body. GIFT is a minimally invasive surgical procedure in which a catheter is placed through a small keyhole incision. Oocytes and semen are inserted through the catheter into the Fallopian tubes. At this point, fertilization and implantation of the embryo may or may not occur.

Zygote intrafallopian transfer (ZIFT) is a procedure that combines elements of both in vitro fertilization and GIFT. First, gametes are extracted and fertilized under controlled laboratory conditions. Next, they are inserted into the Fallopian tubes using the same method as in GIFT. Because GIFT and ZIFT are more invasive and expensive than in vitro fertilization, they are much less commonly performed. They may be recommended for couples whose struggles with infertility are more severe or who have not responded positively to previous cycles of IVF treatment.

Artificial Insemination. Artificial insemination is a widely used, minimally invasive procedure used to help couples with a variety of infertility problems, such as a male partner with low sperm count or motility, the existence of natural antibodies in either the male or female partner that attack sperm, or characteristics of the cervix shape that make fertilization difficult. Artificial insemination can be carried out using either the intended father's sperm or that of a donor obtained from a sperm bank. In either case, once the sperm has been collected, it is physically inserted, via a catheter, into the woman's cervix. Though the technique is simple, timing is extremely important—artificial insemination must take place either just before or on the day of ovulation to be successful. In practice, it is often carried out on two consecutive days to increase the chances of fertilization.

Surrogacy. Surrogacy is an attractive option for women who are infertile because their uterus is abnormal or has been removed or who are believed to be at high risk for miscarriage or other complications of pregnancy. In gestational surrogacy, an embryo fertilized though in vitro fertilization is implanted into the uterus of a woman who is healthy and fertile, and has agreed to act as a surrogate. In many cases, the surrogate donates her own egg to be fertilized via in vitro fertilization with the biological father's sperm or that of a donor, then carries the baby to term. Some surrogates perform this service out of pure altruism or generosity; these women are usually close friends or relatives of the intended mother. Others are unrelated strangers who are compensated financially by the parents.

Contraception and Sterilization. Some applications of reproductive science are intended to prevent, not facilitate, impregnation. Most of the available contraceptive methods are designed for use by the female partner alone, although a few are meant for shared use or use by the male partner alone. Barrier methods, such as the sponge, the diaphragm, the male and female condoms, and the cervical cap, prevent sperm from reaching the egg inside the Fallopian tube. Hormonal methods, such as the pill, the vaginal ring, injected or implanted devices, and certain intrauterine devices, release artificial hormones into a woman's body to interfere with the process of ovulation and prevent eggs from being released into the Fallopian tubes. Emergency contraception, which can be used up to three days after intercourse, uses high doses of estrogen to prevent a fertilized egg from being implanted in the uterus.

Other contraceptive methods are less reliable. These include the rhythm method, in which partners carefully monitor the woman's body temperature and menstrual cycle to determine the date of ovulation and avoid intercourse during this time. Sterilization is the most decisive method of preventing impregnation. A vasectomy is a reversible surgical procedure that results in sterility for the

---

Fascinating Facts About Reproductive Science and Engineering

- In 1996, the first mammal, a sheep, was successfully cloned. Dolly had the exact same DNA as the animal from which she was cloned.

- Louise Joy Brown was known as the first test-tube baby, conceived in a laboratory rather than in her mother's body. Brown was conceived in 1978 by in vitro fertilization, which was then a new technology.

- Through a process known as gestational surrogacy, one woman's fertilized egg can be carried to term in the uterus of a surrogate. The resulting baby will not be related to the woman who gestated and gave birth to it.

- The longest a human embryo created through in vitro fertilization has been known to survive outside of the womb is twenty-nine days.

- Frequently using saunas or hot tubs can lead to infertility in men because elevated temperatures are known to impair the body's ability to produce sperm.

- In 2009, artificial insemination helped a man who had become sterile as a result of radiation treatments father a baby girl. He had frozen his sperm twenty-two years earlier.

- In 2009, American Nadya Suleman made the headlines when she gave birth to what was only the second set of octuplets to be born in the United States and the first to have all the babies survive.

---

male partner. It involves cutting or otherwise sealing both the right and the left vas deferens, the tubes through which sperm travel into the penis. A tubal ligation is a nonreversible surgical procedure that results in sterility for the female partner. It involves sealing the Fallopian tubes so that eggs are unable to pass from them into the uterus.

IMPACT ON INDUSTRY

Government and University Research. Because the fertility industry is so lucrative, many reproductive science and engineering applications have emerged from private companies. However, reproductive issues are at heart social issues, and they have national and political significance. Therefore, government institutes are also important players in the field. In the United States, for example, the National Institute of Child Health and Human Development, the National Institutes of Health, and the U.S. Agency for International Development are all major sources of public funding for research within this discipline.

Industry and Business. Besides the huge fertility industry—estimated to be worth $4 billion each year in the United States alone—two other industries on which reproductive science and engineering makes a significant impact are agriculture and conservation. Techniques such as artificial insemination, embryo implantation, and the use of artificial reproductive hormones to increase the fertility of animals such as cows and pigs are commonly used on large industrial farms to simplify or automate the process of animal husbandry and to allow farmers to selectively breed generations of animals with desirable traits. Similarly, in zoos across the world, techniques borrowed from human reproductive science are increasingly being applied to conservation efforts aimed at increasing captive populations of threatened or endangered species.

Major Corporations. A great deal of reproductive science and engineering research takes place in private companies. Merck Serono, a pharmaceutical company based in Switzerland, is the biggest global developer and seller of fertility drugs, a market that is estimated to be worth about $1 billion annually. Among its products are synthetic versions of human follicle-stimulating hormone (hFSH), human luteinizing hormone (hLH), and human chorionic gonadotropin (hCG), all produced by a genetic engineering technique known as recombination. IntegraMed, a corporation based in the United States, is the leader in providing services and treatment to infertile couples across the country through the chain of fertility clinics it operates.

CAREERS AND COURSE WORK

Preparing for a future career in reproductive science and engineering should begin with a thorough grounding in science and mathematics at the high school level. Students should be sure to advance through the highest available levels of biology and chemistry. The next step is to earn a bachelor of science at the undergraduate level, preferably with a concentration in biology or biomedicine. Courses with particular significance to this field include

developmental biology, andrology, embryology, endocrinology, urology, bioengineering, and genetics. In addition, it is important to spend some time acquiring hands-on laboratory research experience, which can come in the form of internships or independent studies. At this point, some might choose to complete a degree in medicine and become a practicing physician specializing in fertility or embryology. Alternatively, many professionals in the field of reproductive science and engineering are physician-investigators who have pursued training in both medicine and scientific research. This path involves obtaining a dual doctor of medicine (M.D.) and doctoral degree (Ph.D.) after completing a bachelor of science degree at the undergraduate level. Typical M.D.-Ph.D. candidates require about seven to eight years to complete this program. This particular path is ideal for anyone considering conducting original research in an academic setting, serving as a clinician-researcher at a hospital, or working in a private laboratory dedicated to assisted reproductive technologies.

Professional careers in reproductive science and engineering can take a variety of other shapes. For instance, many opportunities exist in the context of fertility clinics, including jobs for nurses, medical assistants, and technicians such as ultrasonographers.

SOCIAL CONTEXT AND FUTURE PROSPECTS

Perhaps the most significant social impact of reproductive science and engineering is the way it has transformed the opportunities available to women in the workplace and, more broadly, the shape of families themselves. Because technologies such as in vitro fertilization enable women to bear children successfully later in their lives, many choose to delay parenthood until they have established themselves fully within their careers. Artificial insemination not only has enabled infertile couples to fulfill their desire to bear children but also—because sperm can be acquired through donor banks—has facilitated the rise of the modern phenomenon of single parenthood by choice. Also, assisted reproductive technologies have provided a means for same-sex couples to become biological parents.

Some very useful technologies created by reproductive science and engineering have the potential to be turned into what some observers fear are unethical applications. For example, preimplantation genetic diagnosis has profound benefits because it enables couples to raise their chances of having healthy babies. However, it has also generated a certain amount of controversy because the same techniques could, in theory, be used to allow couples to select embryos with certain very specific traits. For example, embryos could be chosen or engineered to have genes encoding for eye or hair color or perhaps traits such as intelligence or physical beauty—the so-called designer baby concept. Other ethical questions provoked by assistive reproductive technologies include the question of what to do with leftover frozen embryos and whether and how much women should be compensated for egg or embryo donation. Finally, cloning is a highly controversial and often misunderstood area in biomedical research. Reproductive cloning, which involves creating a precise genetic copy of an existing organism through a process known as somatic cell nuclear transfer, is banned from being done with humans in most countries.

M. Lee, B.A., M.A.

FURTHER READING

Elder, Kay, Doris Baker, and Julie Ribes. *Infections, Infertility, and Assisted Reproduction.* 2004. Reprint. New York: Cambridge University Press, 2010. A detailed, illustrated examination of the microbiology of assisted reproductive technologies. Each chapter includes references, further reading suggestions, and frequently appendixes outlining procedures and protocols.

Green, Ronald Michael. *Babies by Design: The Ethics of Genetic Choice.* 2007. Reprint. New Haven, Conn.: Yale University Press, 2009. A bioethicist tackles moral dilemmas provoked by emerging genetic engineering technologies. Contains a glossary of relevant technical terms.

Jones, Richard E., and Kristin H. Lopez. *Human Reproductive Biology.* Rev. ed. London: Academic Press, 2010. A comprehensive introductory textbook, heavily illustrated with diagrams and photographs. Each chapter includes a summary, further reading, and advanced reading list.

Romundstad, Liv Bente, et al. "Effects of Technology or Maternal Factors on Perinatal Outcome After Assisted Fertilisation: A Population-based Cohort Study." *The Lancet* 372, no. 9640 (August, 2008): 737-743. Finds that adverse outcomes associated with births following assisted reproduction are not caused by technological factors. Includes several tables.

Spar, Debora. *The Baby Business: How Money, Science and Politics Drive the Commerce of Conception.* Boston: Harvard Business School Press, 2006. A critical overview of the fertility industry and how reproductive technologies are used in the marketplace. Includes numerous tables and extensive end notes listing sources.

WEB SITES

Centers for Disease Control and Prevention

Assisted Reproductive Technology
http://www.cdc.gov/art

National Institute of Child Health and Human Development
Reproductive Sciences Branch
http://www.nichd.nih.gov/about/org/cpr/rs

Society for Assisted Reproductive Technology
http://www.sart.org

See also: Animal Breeding and Husbandry; Bioengineering; Cloning; DNA Analysis; Genetic Engineering; Human Genetic Engineering; Obstetrics and Gynecology; Stem Cell Research and Technology.

RHEUMATOLOGY

FIELDS OF STUDY

Internal medicine; orthopedics; immunology; pharmacology; pediatrics; neurology; hematology; physical therapy and rehabilitation medicine; radiology.

SUMMARY

Rheumatology is a subspecialty of the medical specialties of internal medicine and pediatrics. Physicians who practice rheumatology are called rheumatologists. Rheumatologists treat conditions of the bones, joints, muscles, and connective tissues of the body, but they are not surgeons. The conditions that they treat include a wide range of complex diseases that are difficult to diagnose, including arthritic and autoimmune disorders.

KEY TERMS AND CONCEPTS

- **Antibody:** Immune cell produced by the body to fight bacteria, viruses, and other foreign substances.
- **Auscultation:** Listening to the body and interpreting the sounds.
- **Autoimmunity:** Process where the body makes antibodies to some of its own cells, in order to destroy them.
- **Connective Tissue:** Tissue that occurs throughout the body and serves as support for it.
- **Inflammation:** Body's response to injury, which is characterized by redness, swelling, pain, and heat.
- **Malaise:** Lack of energy.
- **Osteopathy:** Alternative medical practice in which the patient is treated as usual but with the addition of joint and spinal manipulation.
- **Palpation:** Act of touching the human body.
- **Percussion:** Tapping of body organs to determine their shape, size, and location and whether there is fluid within the organ.

DEFINITION AND BASIC PRINCIPLES

Rheumatology is a subspecialty of internal medicine and pediatrics. Rheumatologists are medical doctors who treat conditions of the bones, joints, muscles, and connective tissue of the body, but they are not surgeons, nor do they treat fractures or sprains. They treat complex conditions that, despite the symptoms of joint pain, generally affect the whole body. The condition most commonly associated with rheumatologists is arthritis, which is joint inflammation due to many causes.

The practice of rheumatology consists of caring for patients who have conditions such as, gout, rheumatoid arthritis (RA), osteoarthritis, ankylosing spondylitis, fibromyalgia, Lyme disease, and vasculitis. Many of these conditions are challenging to diagnose because they do not always appear with the classic symptoms. Some will progress rapidly and others may develop gradually. Frequently, the rheumatologist works with other medical specialists to finally make a diagnosis.

Rheumatologists care for patients who are chronically ill, have chronic pain, and have difficulty leading a normal life. These patients are often frustrated because it has taken so long for them to receive a diagnosis and treatment for their condition. They see many physicians and take many medications with serious side effects. These patients require emotional support as well as physical care. The rheumatologist cannot cure their illnesses but rather assists them in dealing with their symptoms.

BACKGROUND AND HISTORY

In the 1700's and 1800's, physicians began to identify rheumatoid arthritis and gout in patients or on postmortem examination. Further discoveries in the 1900's, assisted physicians in diagnosing joint conditions using blood testing. Before World War II, there were few medical specialists. Most physicians were in general practice and handled the majority of health problems. During the war, doctors were drafted into the military to treat the injured, which often required specialists. Doctors found that their prestige and their income increased as specialists. After the war, many used the G.I. Bill to return to training to become specialists. When Medicare became law in 1965, specialists were needed to care for the elderly. By the 1990's, the number of specialists had grown, to the point where some felt there were too many specialists. Health maintenance organizations (HMOs) attempted to limit access to specialists by requiring patients to get approval from their insurance before seeing a specialist. Consumers did not like this extra step, so the popularity of HMOs declined, and the number of specialists continued to grow.

HOW IT WORKS

A patient complaining of a joint problem is sent to a rheumatologist by his or her primary care physician. The rheumatologist performs an assessment of the patient that includes taking the patient's medical history; performing a physical examination of the whole body, focusing on the affected joints or muscles; and ordering blood tests and X rays. The rheumatologist uses this information to diagnose the patient's condition. If the diagnosis is not obvious, the rheumatologist may order additional blood tests, X rays, or a biopsy, or even refer the patient to another specialist.

Diagnosis. The medical history should include questions designed to determine all the patient's health issues as well as his or her physical complaints. The chief complaint, often pain, is discussed first. The rheumatologist asks questions about the chief complaint such as when it began, what stimulates it, and whether it is intermittent or constant. The history is organized by body system.

Then, the physician examines the patient's body, paying special attention to areas where the patient indicates a problem. The body is examined by observation, palpation, auscultation, and percussion. A joint examination includes looking for redness and swelling, touching the area around the joint to feel for swelling, and comparing the painful joint with the normal joint, unless both joints are affected. The joint is then assessed for stiffness, range of motion, and weakness. For some conditions, the rheumatologist may have to apply pressure to the painful area. The location of pain can provide a clue as to the diagnosis of the condition.

The rheumatologist may perform routine blood tests such as a complete blood count, kidney function tests, liver function tests (LFTs), and blood chemistry. Sometimes these blood tests can provide helpful information but not always. Most often, the rheumatologist will do blood tests that assist with identifying specific diseases. They include the tests for the human leukocyte antigen, the erythrocyte sedimentation rate, the uric acid level, the antinuclear antibody,

the level of c-reactive protein, the rheumatoid factor, the anti-DNA antibody, and the antineutrophil cytoplasmic antibody. These tests can demonstrate the presence of inflammation in the body, indicate an autoimmune condition, identify specific antibodies, and point toward a certain condition, such as systemic lupus erythematosis (SLE) or gout.

X-ray tests of the affected areas are performed. Sometimes regular X rays are not sensitive enough to demonstrate the signs of the condition. In these cases, either computed tomography (CT) or magnetic resonance imaging (MRI) is performed. These scans provide much more information.

Treatments. The most common treatment for rheumatologic conditions is medication, which is a critical part of the care of people with rheumatologic diseases. In addition, the rheumatologist uses other treatments such as corticosteroid injections, withdrawal of fluid from joints, physical therapy, occupational therapy, the application of heat or cold, elevation of an extremity, and the application of joint support devices.

APPLICATIONS AND PRODUCTS

The rheumatologist uses the following groups of medications for treatment of most rheumatologic conditions: pain relievers; nonsteroidal antiinflammatory drugs (NSAIDs); corticosteroids; disease-modifying antirheumatic drugs (DMARDs); biologics; specific drugs for fibromyalgia, gout, and Sjögren's syndrome; and bone-sparing osteoporosis medications. Pain relievers range from mild drugs, such as acetaminophen (Tylenol) to very strong opiods, such as morphine. NSAIDs, such as ibuprophen, are medications that decrease pain by decreasing inflammation. Corticosteroids are drugs that have a very strong antiinflammatory property. They resemble cortisone that is produced by the body. These drugs may be given by mouth, intravenously, injected, inhaled, or applied locally. DMARDs are a group of medications that are not otherwise related to each other. They are able to suppress the immune system in the body and, thus, to prevent or limit damage to the body. Biologics, like DMARDs, help to prevent or limit damage to the body, but their action is to block specific steps in the inflammatory process. Many medications were developed expressly to treat rheumatoid arthritis. They have turned out to be effective at treating many other rheumatologic conditions. With the advent of these medications, particularly, DMARDs and biologics, rheumatologists are able to limit the damages caused by rheumatologic conditions and improve the quality of life of the patient.

Despite the good that they do, many of these medications, particularly corticosteroids, DMARDs, and biologics do have a number of serious side effects. They can cause insomnia, stomach upset, moon face, cataracts, osteoporosis, mood changes, increased risk of infections, diarrhea, hair loss, stomach pain, nausea, vomiting, dizziness, ulcers in the mouth, fevers or chills, tiredness, blood in the urine, skin rash, blurred vision, liver problems, and lightheadedness. They can cause a reaction at the site of injection of the drug or an infusion reaction, characterized by chest pain, change in blood pressure, shortness of breath, and hives.

An integral part of caring for a person with a rheumatologic condition is limiting or preventing the major side effects of his or her medications. The most serious side effect is infection, particularly fungal and viral infection. Any signs of infection should be treated aggressively with antibiotics, antivirals, or antifungal drugs. Two infections of particular concern are shingles (herpes zoster) and *Pneumocystis carinii* pneumonia (PCP). Shingles, a viral infection, follows a nerve, and produces blisters and severe pain. It should be treated with an antiviral agent. PCP is a fungal lung infection. Persons who are taking immune suppressant drugs should be on a drug that will prevent this fungal infection. The patient should be told to avoid receiving vaccines that are made from live viruses. If the vaccine is made from a dead virus (for example, flu vaccine), the patient can receive it. The rheumatologist should closely monitor the patient's blood values. Initially, blood tests should be performed every two weeks. Some of these drugs can cause nausea and loss of appetite. Patients should be given a drug that decreases stomach acid production.

Fibromyalgia, gout, Sjögren's syndrome, and osteoporosis are treated with specific medications. Pain relievers and NSAIDs often do not decrease the pain of fibromyalgia, but some antidepressants do. The patient may receive a muscle relaxant or a pain medication called tramadol (Ultram). The deposits of uric acid crystals in gout can be very painful, so these patients will need pain reliever medication. Also, gout is treated with drugs that either decrease uric

acid production or increase its excretion. Sjögren's syndrome causes dry mouth and dry eyes, as well as fatigue and lung inflammation. It can be treated with DMARDs, but the patient also requires medication to increase the lubrication in his or her eyes and mouth. Osteoporosis leads to bone loss and increased risk for fractures. The drugs given for osteoporosis increase bone growth and decrease bone loss.

IMPACT ON INDUSTRY

In developed countries such as the United States, Canada, Great Britain, and Europe, the number of rheumatologists is less than the demand. As the populations of these areas age, many rheumatologic disorders are beginning to appear. In the United States, Medicare has provided medical care for the elderly, increasing the demand for rheumatologists. However, in some less-developed countries, general medical doctors may be caring for persons with these conditions.

Government and University Research. The more-developed countries have government or private agencies dealing with rheumatologic conditions. These groups raise funds to support research into the diagnosis and treatment of rheumatologic conditions. They may also develop standards of care for these conditions. Great Britain has an arthritis organization called the Arthritis Research Campaign. Much medical research on rheumatologic conditions is performed in Europe. Europe's arthritis organization is named the European League Against Rheumatism. In the United States, the Arthritis Foundation, a private organization, and the National Institutes of Health (NIH), a governmental organization, deal with arthritis. Some rheumatologic disorders are rare, so research on them is less common.

In the United States, in addition to physicians, doctoral students and postdoctoral researchers are responsible for most of the medical research that is performed. Those researchers, studying immunology or genetics, often study rheumatologic conditions. They have played a major role in the development of the biologic drugs used to treat rheumatologic conditions. Some conditions that are frequently studied are SLE, fibromyalgia, osteoporosis, and Sjögren's syndrome. Fibromyalgia research has resulted in diagnostic tests and drugs that are effective against this condition.

Industry and Business. The industries most

affected by rheumatologic conditions are the pharmaceutical industry, the medical device industry, and the health care industry. The pharmaceutical industry develops the drugs used to treat rheumatologic conditions and researches their effectiveness and safety. Another governmental agency, the U.S. Food and Drug Administration (FDA) is responsible for deciding whether a drug is safe and effective to use for medical treatment. It weighs the information that the pharmaceutical industry provides. Some of this information is provided by small businesses that perform research studies for drug companies. Medical devices that may be used by people with rheumatologic conditions are walking aids such as walkers and canes and artificial joint prostheses. Because most rheumatologic conditions are chronic, the patients tend to be relatively high users of health care services. These health care services include physician's visits, physical therapy, hospitalization, and laboratory and X-ray services.

Of course, most industries can be affected by chronic health conditions that strike their employees. Employees cannot be fired for developing a rheumatologic condition, but they may be unable to perform some tasks or they may be absent from work more than the average employee.

CAREERS AND COURSE WORK

The rheumatologist must be a graduate of a medical school or a school of osteopathy. To be accepted to medical school, the student must have an undergraduate degree and appropriate course work. After medical school, the doctor must complete a residency at an accredited hospital. In the United States, the first year of the residency is the internship, which provides general medical experience. After the internship, the doctor does a two-year residency in either internal medicine or pediatrics. Then, if the doctor who chooses to become a specialist, a two- to three-year fellowship in that specialty is required. Osteopaths follow a similar pathway to that of medical doctors and must complete a fellowship to qualify as a specialist.

Other positions in the field of rheumatology are research positions. These positions require an M.D. or a Ph.D. in a field such as immunity or genetics. Rheumatology research may focus on broad concepts such as autoimmunity or genetic mutations, or it may focus on specific diseases.

Fascinating Facts About Rheumatology

- Corticosteroids were discovered in 1948 and have been used since 1949 to treat many parts of the body.
- Corticosteroids are made from a chemical found in yams.
- Immune cells, called T cells, are responsible for determining whether cells are foreign.
- With autoimmune disease, the body attacks itself, often causing life-threatening conditions.
- Some rheumatic diseases can cause rashes or other skin lesions.
- Although many rheumatologic diseases affect the joints and organs of the body, they can cause very general symptoms, such as fatigue, malaise, fever, night sweats, weight loss, weakness, and loss of appetite.
- Systemic lupus erythematosis is named for a symmetric facial rash that occurs early in the disease process. The disease is called systemic because it affects the whole body, and erythematosis means redness of the skin. Lupus, derived from the Latin word for wolf, implies that the disease is a sneaky foe that overwhelms the human body.

Rheumatology is a rapidly growing field of medicine. Concepts such as autoimmunity were late to be identified and defined, so there is limited information about them. Because of this, there are many opportunities to study rheumatologic diseases. Also, there is opportunity for drug research to search for a cure for these devastating disorders.

SOCIAL CONTEXT AND FUTURE PROSPECTS

Many rheumatologic disorders do not appear until a person reaches fifty or sixty years of age. As the population ages and the baby-boom generation approaches Medicare age, the number of persons with rheumatologic diseases will increase. With the discovery that disorders of the immune system are a cause of many rheumatologic conditions, there are increasing opportunities for research in rheumatology. The development of new drugs is ongoing. The long-term goal is to cure rheumatologic diseases. Some rheumatologic conditions are genetic in origin. A comparison of the genes of patients with rheumatologic conditions with those of disease-free individuals may reveal the genetic changes that lead to these conditions. In the future, it may be possible to alter genes to cure these disorders.

Autoimmunity is an important topic in the discussion of rheumatologic diseases. It is thought that some of the autoimmune conditions may be caused by environmental pollution. Toxic chemicals are in the air, earth, and water and also in people's homes and the products they use every day. For example, new carpeting and upholstered furniture are treated with chemicals to prevent staining and to make them fire resistant. Plastic water bottles have been found to leach out toxic chemicals, such as bisphenol A (PBA), as they age. Research is necessary to determine the impact of environmental chemicals on people.

Christine M. Carroll, B.S.N., R.N., M.B.A.

FURTHER READING

Baron-Faust, Rita, and Jill P. Buyon. *The Autoimmune Connection: Essential Information for Women on Diagnosis, Treatment, and Getting on with Life.* New York: McGraw-Hill, 2003. Examines autoimmune disorders, many of which are also treated by rheumatology.

Hakim, Alan, Gavin P. R. Clunie, and Inam Haq. *Oxford Handbook of Rheumatology.* 2d ed. New York: Oxford University Press, 2006. A handbook for diagnosing and treating rheumatologic conditions.

Hiorwitz, Randy, and Daniel Muller. *Integrative Rheumology.* New York: Oxford University Press, 2011. A rational and evidence-based approach to autoimmune and rheumatologic diseases that integrates some complementary and alternative medicine practices.

Hochberg, Marc C., et al., eds. *Rheumatology.* Philadelphia: Mosby/Elsevier, 2010. Covers the basics of rheumatology and the treatment of arthritis.

Weisman, Michael H. *Rheumatoid Arthritis.* New York: Oxford University Press, 2011. Contains information on the diagnosis and treatment of rheumatoid arthritis.

WEB SITES

American College of Rheumatology
http://www.rheumatology.org

Arthritis Foundation
http://www.arthritis.org

MedlinePlus
Rheumatoid Arthritis
http://www.nlm.nih.gov/medlineplus/rheumatoid-arthritis.html

See also: Geriatrics and Gerontology; Immunology and Vaccination; Neurology; Pharmacology.

S

SOIL SCIENCE

FIELDS OF STUDY

Chemistry; physics; microbiology; agronomy; geology; agriculture; forestry; range science; plant science; environmental science; ecology; hydrology; landscape architecture; soil mechanics; mining engineering; civil engineering; land-use planning.

SUMMARY

Soil science is the multidisciplinary study of soils, which are composed of mineral, water, air, and organic matter. Soil supports the growth of terrestrial plants, controls the percolation of water, recycles elements and organic molecules, provides habitats for biota, alters dust and gaseous components in the atmosphere, and stabilizes the foundations of buildings and constructions.

KEY TERMS AND CONCEPTS

- **Cation Exchange Capacity:** Sum total of exchangeable cations that a soil can adsorb on negative sites.
- **Humus:** Organic matter, compost, or decomposed plant and animal residues.
- **Leaching:** Removal of materials in solution from the soil by percolating water.
- **Loam:** Textural class for soil having a mixture of moderate amounts of sand, silt, and clay.
- **Pedology:** Study of soil that deals with the formation, morphology, and classification of soil bodies.
- **Rhizosphere:** Region of the soil around the roots of vegetation.
- **Sinkhole:** Depression in the ground formed when the underlying limestone is dissolved by groundwater.
- **Soil Colloids:** Microscopic-sized clay and humus particles.
- **Soil Horizon:** Horizontal layer of soil with different properties from the layers below or above it.
- **Soil Profile:** Vertical slice of the soil with its horizons and parent material.
- **Soil Texture:** Relative proportions of sand, silt, and clay in the soil.

DEFINITION AND BASIC PRINCIPLES

Soil science is the study of the physical and chemical properties of soils. Soil is the end product of the weathering of rocks and minerals and the decomposition of living organisms. Soil solids include minerals (inorganic particles) and humus (organic matter). As soil solids cling together, pore spaces, or voids, are formed between inorganic and organic materials to accommodate air and water. A typical loam soil contains 45 percent minerals, 5 percent humus, 25 percent water, and 25 percent air. Among the three inorganic particles (sand, silt, and clay), pure sands have the largest particle size, the lowest surface area per unit volume, and the lowest percent of pores. Sands have the largest pores, which allow faster water percolation and drainage. Clay particles are the smallest and have the highest surface area per unit volume and the highest percent of pore spaces. Therefore, clays have the greatest water-holding and element-adsorbing capacity among inorganic particles. Organic humus is the lightest and has the greatest percent of pore spaces among the soil solids. Addition of humus increases the retention of water and elemental nutrients in sandy soils and improves the drainage of water in clay soils.

Soils have net negative charge because the surfaces of soil particles have more negative sites than positive sites. Negative sites attract positively charged ions (cations) of nutrient elements. The attracted cations are loosely adsorbed and can be easily replaced by other cations. For example, application of ammonium fertilizers in the soils releases ammonium cations ($NH4+$), which can replace other adsorbed cations such as the cations of potassium ($K+$), hydrogen ($H+$), magnesium ($Mg2+$), calcium ($Ca2+$), or sodium ($Na+$). This chemical property of the soils plays an important role in crop production and in retaining chemical wastes and other pollutants.

BACKGROUND AND HISTORY

Edaphologists, who studied soil as a medium for plant growth, could be considered the pioneers of soil science. The next group of scientists, the pedologists, studied soil as a geologic entity, examining its origin, formation, morphology, and classification.

Edaphology research began in the early 1600's, when Dutch scientist Jan van Helmont conducted an experiment involving soil. He planted a 5-pound willow tree in 200 pounds of soil and let it grow for five years. During those years, the plant received only rainwater, no nutrients or fertilizers. At the end of the experiment, the weight of the soil had decreased while that of the plant increased. Helmont concluded that the elements from the soil had contributed to the weight increase of the willow tree. In the 1800's, German chemist Justus von Liebig postulated that the growth of plant is limited by sixteen essential nutrient elements, a number that scientists later increased to seventeen.

In the early 1870's, Russian geomorphologist Vasily Vasilyevich Dokuchayev, the father of pedology, conceived of soils as natural bodies, each with its own specific characteristics resulting from a unique process of development. In the 1890's, Dokuchayev created a classification system for Russian soils. In the1920's, the Russian technique was adapted by soil scientist C. F. Marbut who applied it to the soils of United States. In 1941, Swiss-born scientist Hans Jenny published *Factors of Soil Formation: A System of Quantitative Pedology*. His work summarized and illustrated investigations into soil science and became a standard work in the field. In 1965, an official soil classification system was completed for the U.S. National Cooperative Soil Survey; this was later published in 1975 as *Soil Taxonomy: A Basic System of Soil Classification for Making and Interpreting Soil Surveys*.

HOW IT WORKS

Soil Formation. It takes hundreds of year for the soils to develop from parent materials. The first step in soil formation is the weathering of rocks and minerals by physical, chemical, or biological processes. Over time, the consolidated rocks turn into unconsolidated soil materials lying on top of the existing bedrock. Some unconsolidated materials can be carried by wind, water, gravity, or a glacier and then deposited as sediment in different locations. Bedrock and transported sediments become the parent materials of inorganic (mineral) soils. Some land areas, which are cold or inundated with stagnant waters, are subject to faster accumulations of plant and animals residues. The decomposed residues serve as the parent materials of organic soils.

As the soil develops on top of the parent materials, a soil profile is formed with distinct layers of soil horizons. Soil taxonomists use the unique properties of horizons in a profile as the basis for soil classification. Soils are classified into twelve orders.

Physical Properties. The physical properties of soil (such as texture, density, and structure) affect their role in agriculture, the environment, and engineering. Soil texture represents the percentages of sand, silt, and clay particles. The greater the proportion of clay, the more the soil responds to compaction, which increases bulk density, decreases porosity, and reduces aeration. As a result, compacted clays prohibit root penetration but provide better foundation for engineering projects.

Chemical Properties. The chemical properties of soil—cation exchange capacity (CEC), pH (acidity-alkalinity), and nutrient availability—depend solely on microscopic soil colloids, clays, and humus. Each colloid possesses more negatively charged surfaces than positively charged surfaces. The total amount of cations that can be adsorbed by the soil colloids is called the cation exchange capacity. Exchangeable cations held by soil colloids are not easily removed by percolating water but can be easily removed by root absorption and replaced by other cations. Other adsorption reactions bind the ions more snugly, preventing them from leaching into the environment.

The cation exchange capacity influences the development of acidic or alkaline soils. In rainy locations, $H+$ ions from rain replace most of the basic cations ($K+$, $Ca2+$, $Mg2+$, $Na+$) on the negative sites. As rainwater percolates through the soil, the basic cations leach with water to the deeper horizons, leaving the top soil acidic with pH less than 7 because of high $H+$ accumulation in the exchange sites. In arid regions, alkaline soils with pH greater than 7 are found because basic cations are retained in the exchange sites.

Soil Biota. Based on their feeding habits, soil-dwelling organisms (plants, animals, fungi, protists, bacteria, and archaea) play a role in the formation of soil, ccntribute to the accumulation of organic matter

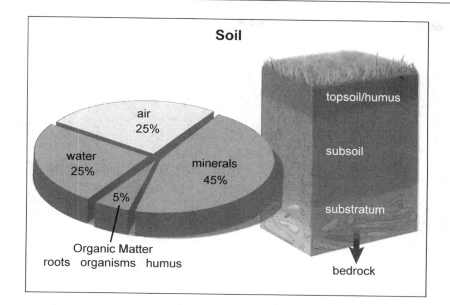

Soil

air 25%

water 25%

5%

minerals 45%

Organic Matter
roots   organisms   humus

topsoil/humus

subsoil

substratum

bedrock

conditions.

All higher plants require seventeen essential elements to complete their life cycle. Three elements—carbon (C), hydrogen (H), and oxygen (O)—mostly come from air and water. The remaining fourteen elements—nitrogen (N), phosphorus (P), potassium (K), calcium (Ca), magnesium (Mg), sulfur (S), boron (B), copper (Cu), chlorine (Cl), iron (Fe), manganese (Mn), molybdenum (Mo), nickel (Ni), and zinc (Zn)—are supplied by the soil. Macronutrients (N, P, K, Ca, Mg, and S) are the elements needed by the plants in large amounts; therefore, incorporation of these nutrients in the soil (as fertilizers) is often necessary. Micronutrients (B, Cu, Cl, Fe, Mn, Mo, Ni, and Zn) are required by plants in small amounts, such that the application of fertilizers containing micronutrients is usually not necessary.

Nutrient elements from the soil are absorbed by plant roots in the form of negatively charged ions (anions) or positively charged ions (cations). The forms of ions easily absorbed by plant roots are anions—nitrate (NO_3^-), hydrogen phosphate ($H_2PO_4^-$), sulfate (SO_4^{2-}), hydrogen borate ($H_4BO_4^-$), Cl^-, and molybdate (MoO_4^{2-})—on the positive sites and cations—NH_4^+, K^+, Ca^{2+}, Mg^{2+}, Cu^{2+}, Fe^{2+}, Mn^{2+}, Ni^{2+}, and Zn^{2+}—on the negative sites.

Soil amendments increase cation exchange capacity, improve the water-holding capacity and drainage of the soil, and supply more nutrients to the plants. Commercial potting mix for ground and container gardening contains a combination of fertilizer, perlite, vermiculite, and compost. Synthetic soils, which are made of potting mix plus additional amendments such as sludges, hygdrogels, and organic mulches, are commercially manufactured in large quantities and used for reclamation or remediation of abandoned mine sites, barren soils, or unproductive wetlands.

Water Storage and Filtration System. Water is stored in the soil but some water travels through soil pores and moves to streams, rivers, lakes, and aquifers. Contaminated waters that are spilled onto soils undergo purification and clarification processes that

and recycle nutrients in the soil. Large animals (such as badgers, mice, and prairie dogs) dig underground tunnels and aerate the soil. Earthworms strengthen the soil structure. As they burrow, they ingest plant and animals residues, including clays, and excrete them as granular aggregate. Most soil fungi (molds and mushrooms) are decomposers. They use the nutrients and organic molecules from the tissues of dead and living plants and animals. Protists include plantlike (algae), animallike (protozoa), and funguslike (slime molds) organisms. Algae, like plants and cyanobacteria, are not decomposers but rather primary producers because they are photosynthetic organisms. Protozoa (such as amoeba) are unicellular organisms that ingest small living organisms. Most bacteria and archaea, which decompose the tissues of all dead organisms, are found in the lowest part of the food chain.

APPLICATIONS AND PRODUCTS

Medium for Plant Growth. Soil anchors the roots of a plant (or shoot) and holds its above-ground parts in place. Pore spaces between soil solids supply oxygen to the roots of established plants or to germinating seeds during respiration. Soil pores also store water, which is supplied to the roots for absorption or to the seeds for imbibition. Moisture in the pores minimizes soil temperature fluctuations in the rhizosphere. As a result, soil serves as an insulator that protects plant roots from extremely hot and cold remove toxic ions and impurities before the cleansed water enters waterways. This filtration mechanism relies on the pores and exchange sites of the soil. As water moves through the soil, large molecules of pollutants are trapped within the microscopic pores, allowing only small molecules of water to pass through. Soil further removes ions of toxic wastes and heavy metals by electrostatic reaction with the ion exchange sites of the soil colloids.

Not all soils can store and filter water effectively. Sandy soils make ineffective water storage and filtration systems because of their large pores. Compacted clays have tiny pores that become impermeable to water. Torrential rains that pour into denuded mountains with impermeable clays result in flash floods of muddy water.

Nutrient and Waste Recycling. Soils are the major repositories of wastes created by humans. The capacity of soil to recycle nutrients from wastes depends on the living organisms that inhabit the soils. Organic wastes are decomposed into valuable humus by fungi and microbes. Mineral nutrients from humus are used by plants and animals. Plants acquire carbon as carbon dioxide from the atmosphere and release oxygen into the atmosphere through photosynthesis. Animals release carbon dioxide into the atmosphere through respiration. When plants and animals die and decompose, mineral nutrients are deposited in the soil.

Atmosphere Modification. Exposed soils that are dry and poorly structured are susceptible to erosion by wind, which can carry and disperse the soil particles as dust in the atmosphere. Suspended dust decreases the clarity of the air, increases the chance of developing respiratory problems, and cools the atmosphere and the ground. Soils that are covered with vegetation are less affected by erosion.

Soil alters the amounts of ammonia (NH_3) and dinitrogen gas (N_2) in the atmosphere. Decomposition of organic matter by bacteria and fungi in the soil generates NH_4^+ through ammonification. Some NH_4^+ is absorbed by plants and some volatilizes into the atmosphere as ammonia (NH_3). Soil bacteria convert NH_4^+ into nitrate (NO_3^-) by nitrification. Some nitrate is assimilated by plants and some is denitrified by soil bacteria into dinitrogen gas (N_2). Some N_2 in the soil pores is fixed by bacteria that live in the roots of the plants. Some atmospheric N_2 is converted into NO_3^- by lightning. Whatever form

of nitrogen is in the soil, nitrogen is subject to rapid loss. For this reason, soil nitrogen is the element most needed by plants. Nitrogen is often replaced by the addition of nitrogen-containing fertilizers or compost to soil.

Medium for Engineering Projects. Soil mechanics, an examination of the architecture and physical properties of the soil, determines which types of soil are best suited for construction. Soil, as compacted solid ground, serves as a stable and firm foundation on which to build roads, bridges, houses, towers, and other infrastructure. Most structures are erected on the soil surface, and many engineering projects (such as underground tunnels) involve digging deeper in the soil. Soil stability is based on its compressibility. Sands are difficult to compress, but compressed clays form a very stable compacted material. However, clays that shrink and form deep cracks when dry and expand when wet (such as montmorillonite) are not suitable sites for buildings. Highly weathered clays such as kaolinite do not shrink and swell, and are appropriate types of clays for engineering projects.

Soils are extensively used as materials in building houses. Walls of African huts are made of mud, and walls of modern Western houses use brick (baked clay), hollow cement blocks (a mixture of sand and cement), silica glass (made from sand), and other soil-cement mixtures. Green roofs are those that are partly or totally covered with soil and vegetation. They insulate the structures they cover and help to lower the air temperature within cities.

IMPACT ON INDUSTRY

Soil science contributes to the improvement of agriculture. Crop harvests in both developing and developed countries have increased markedly since 1950. Most of this increased output has been attributed to gains in input yields and not to the expansion of cultivated land. Scientific research on soils has brought these benefits to producers and consumers.

Government, University, and Corporation Research. The United States is one of the largest producers of food and fiber in the world. The Hatch Act of 1887 provided that the necessary basic and applied agricultural research would be conducted by agricultural experiment stations in state colleges of agriculture in cooperation with the U.S. Department of Agriculture (USDA). These agricultural experiment stations, in collaboration with USDA research

Fascinating Facts About Soil Science

- People who engage in geophagy, the practice of eating soil voluntarily as source of nutrients, are found in parts of Africa, the United States, and India. In Haiti and Uganda, clays are molded into pancakes or bars and are sold in the market.

- The color of hydrangea blossoms depends on the pH of the soil in which they are grown. Acidic soil (created by the application of ferrous sulfate, or $FeSO_4$) creates blue blossoms. Alkaline soil (created by the application of lime) creates pink blossoms.

- The Dust Bowl that struck the Great Plains in the 1930's was the result of wind erosion. Overcultivation, poor land management, and drought caused the top soil to turn to dust, which was lifted and carried off in severe winds and dust storms. Many farmers went bankrupt and left the land.

- In August, 2006, Fredrik Fransson and the crew of his yacht discovered a large area of floating pumice in the South Pacific Ocean near Tonga. The pumice was being emitted by an underwater volcano that created a new island.

- In the mid-fifteenth century, the Incas built Machu Pichu on top of a steep mountain, rising 760 meters above the valley floor. The ancient Peruvian city is surrounded by dozens of stepped agricultural terraces that prevent soil erosion.

- Florida has the most sinkholes of any state in the United States. In 1999, a sinkhole opened in Lake Jackson, a popular fishing lake in Tallahassee, allowing its water to drain into the aquifer. Within a few days, the lake had vanished. This lake has disappeared four times since 1907.

centers, conduct research programs in agriculture, plants, and soils. Two agencies within the USDA, Economic Research Service and the Agricultural Research Service, are dedicated to the creation of a safe, sustainable, competitive food and fiber system, as well as strong communities, families, and youth through integrated research, analysis, and education.

Business Sectors. The demand for the perlite and vermiculite (soil amendments) is growing. Perlite is a naturally occurring amorphous volcanic rock. Since 1950, vast deposits of perlite have been mined in New Mexico, Nevada, California, and other Western states, but the major producing countries are Greece and Turkey. When pulverized perlite is rapidly heated to 900 degrees Celsius, it softens, and the water inside the perlite evaporates. The escaping steam forms tiny pores and causes the perlite to expand up to twenty times its original volume. Because of its low density and porous nature, perlite has many applications for not only agriculture but also housing and infrastructure.

Vermiculite is a naturally occurring clay mineral. Vast vermiculite deposits exist in South Africa, Australia, Russia, China, and Brazil. In the United States, it is mined in Montana and the Carolinas. As of 2005, South Africa was the leading producer of vermiculite, followed by the United States and China. When finely ground vermiculite is quickly heated to 300 degrees Celsius in a furnace, water vapor escapes as steam, the mineral becomes porous and expands to about twenty times its original volume. Like the heat-treated perlite, expanded vermiculite is extremely light. Vermiculite furnaces are often used to heat perlite, as both are frequently sold together.

CAREERS AND COURSE WORK

Soil science has adopted tools and techniques from a wide variety of basic and applied sciences in the study of the soil. Courses include introductory soil science, soil fertility, soil chemistry, soil physics, soil microbiology, soil genesis, soil taxonomy, pedology, soil conservation, and soil management. The minimum requirement for a career in soil science is a bachelor of science degree in soil science or in agronomy, chemistry, physics, or environmental science with selected courses in soil science. For advanced positions, a master of science or doctorate in soil science is preferred.

Soil scientists may work in an office, in the field, or both. Soil scientists employed in agronomy and crop production provide recommendations to farmers regarding the correct amount of fertilizer for specific plants in large plots of cultivated land. Scientists at the USDA's Natural Resources Conservation Service map and describe the soil using digital and satellite imagery, take soil samples and chemically analyze them, classify soils, and evaluate the soil, identifying any problems. The USDA's Agricultural Research Service employs many soil scientists in numerous programs throughout the United States, including about seventy national programs in Natural Resources and Sustainable Agricultural Systems.

SOCIAL CONTEXT AND FUTURE PROSPECTS

Soils are highly used and abused natural resources. For example, farmers who employ conventional crop production methods cultivate the soils each year, and if they routinely apply more than the recommended amounts of nutrient fertilizers, this could lead to overfertilization. In rainy areas, the excess nutrients leach down to the aquifer and run off horizontally, polluting the surface water and groundwater. In arid regions, the excess nutrients facilitate the development of saline soils. Annual cultivation of soil removes organic matter and exposes the soil to wind and water erosion. Many modern farmers use windbreaks and no-till farming to reduce soil erosion.

The U.S. government has implemented many programs to combat soil problems. For example, the Natural Resources Conservation Service planted grasses on barren soils in Texas to counteract wind erosion. As a result, dust particles in the air were reduced drastically. Bioremediation projects operated by the Environmental Protection Agency have been able to identify plants that can extract toxic heavy metals from the soil and water. Although technologies are being continually developed to increase crop production and to improve soil and water quality, ultimately, it is up to humans to ensure that environmental quality is maintained for future generations.

Domingo M. Jariel, Jr., Ph.D.

FURTHER READING

Jariel, Domingo M., et al. "Plant and Soil Nutrient Concentrations in a Restored Cajun Prairie." *Proceedings of the Twenty-first North American Prairie Conference* 21 (2010): 95-107. Provides a discussion of nutrient recycling and dynamics in a prairie ecosystem. Includes graphs and tables.

Lal, Rattan, and Manoj K. Shukla. *Principles of Soil Physics.* New York: Marcel Dekker, 2004. Examines colloids, the mechanisms of aggregation, soil elasticity and plasticity, and hydrophobic soils.

Liu, Cheng, and Jack Evett. *Soil Properties: Testing, Measurement, and Evaluation.* 6th ed. Upper Saddle River, N.J.: Pearson/Prentice Hall, 2009. Discusses methods of field and laboratory testing for evaluating soil properties for engineering purposes.

Sposito, Garrison. *The Chemistry of Soils.* 2d ed. New York: Oxford University Press, 2008. Offers illustrations of silicate clays and explanations of biogeochemistry, hydrology, and the interaction between organic carbon and the soil.

Sylvia, David M., Jeffrey J. Fuhrmann, Peter G. Hartel, and David A. Zuberer, eds. *Principles and Applications of Soil Microbiology.* 2d ed. Upper Saddle River, N.J.: Pearson/Prentice Hall, 2005. Examines microbial metabolisms and genetics as well as the interactions of soil organisms with soil, air, and water. Contains many graphs, tables, and pictures.

WEB SITES

National Society of Consulting Soil Scientists
http://www.nscss.org/soil.html

Soil Science Society of America
https://www.soils.org

U.S. Consortium of Soil Science Associations
http://soilsassociation.org

U.S. Department of Agriculture
Natural Resources Conservation Service, Soils
http://soils.usda.gov

See also: Agricultural Science; Land-Use Management.

SOMNOLOGY

FIELDS OF STUDY

Sleep medicine; psychology; psychosomatics; psychiatry; pharmacology; placebo research; medicine; pediatrics; geriatrics; cardiology; pulmonology; human biology; chronobiology; molecular biology; endocrinology; neuroscience; neurology; oneirology; social psychology; epidemiology; anthropology; nursing science; nutritional science; lifestyle studies.

SUMMARY

Somnology is the scientific and clinical study of sleep, sleep disorders, and sleep-associated issues. Somnologists are interested in diverse physical as well as psychological processes of sleep and their correlations with wakefulness, lifestyle, and environment. Humans spend on average one-third of their lifetime sleeping, but one in two people experiences sleeping difficulties or disorders. The study of sleep is an important issue because many accidents involve lack of sleep. There is both high clinical and medical interest and a huge potential for services and products, ranging from medication to meditation, that will improve the quantity or quality of sleep.

KEY TERMS AND CONCEPTS

- **Circadian Rhythm:** Cycle of roughly twenty-four hours in behavioral, psychophysical, biochemical, and physiological processes in humans as well as in other animals, fungi, plants, and certain forms of bacteria.
- **Electroencephalogram (EEG):** Test that measures the electrical activity of the brain; used to determine the stages of sleep.
- **Insomnia:** Sleep disorder that involves difficulty falling or staying asleep, or sleep that is not refreshing; defined as transient if it occurs for less than one week, acute if it lasts one week to one month, and chronic if it lasts more than one month.
- **Nonrapid Eye Movement (Non-REM) Sleep:** Sleep in which no rapid eye movement occurs; it is divided into three stages, N1, N2, N3.
- **Parasomnias:** Sleep arousal disorders that involve nightmares or movements such as sleepwalking, teeth grinding, or restless leg syndrome.
- **Polysomnography (PSG):** Sleep test that records multiple types of biophysical data, including periods of REM and non-REM sleep, arousal, breathing, limb movement, and heart rhythm.
- **Rapid Eye Movement (REM) Sleep:** Stage of sleep characterized by rapid eye movement.
- **Sleep Medicine:** Field of medicine that provides diagnosis and treatment of insomnias and parasomnias.
- **Zeitgeber:** External cue, the most important being daylight, that serves to synchronize a living being's internal time-keeping systems.

DEFINITION AND BASIC PRINCIPLES

Somnology is the scientific and clinical study of sleep. The term is derived from the Latin *somnus*, "sleep," and the Greek *logos*, "word" or "reason." Somnologists study the processes—physical and psychological—that constitute sleep and examine how these processes relate to wakefulness, lifestyle, and environment. They conduct research on diverse scientific and medical problems such as insomnia or parasomnias.

Somnology is the science behind much of sleep medicine. Sleep medicine involves the diagnosis and treatment of sleeping disorders, disturbances, and problems. In sleep laboratories, neurological and biophysical data are recorded, collected, assessed, and interpreted with the help of polysomnography.

Somnology does not include oneirology, the scientific study of dreams, which includes cultural, arts-related, anthropological, psychoanalytical, psychological, and neurological approaches.

BACKGROUND AND HISTORY

In Greek mythology, sleep is viewed as having a close relationship with death (Hypnos, the god of sleep, is the brother of Thanatos, the god of death) and as being a negation of wakefulness. The Greek philosopher Aristotle believed that sleep was an active integrated aspect of the process of life, with the function of recreation and life maintenance.

In 1924, the German psychiatrist Hans Berger recorded the first human electroencephalogram

(EEG). In 1937, American physicist Alfred Lee Loomis and his colleagues used data from EEGs to first describe different stages of sleep. Nathaniel Kleitman, a professor of physiology at the University of Chicago, and Eugene Aserinsky, one of his graduate students identified rapid eye movement (REM) sleep in 1952 and defined it in 1953. In the 1960's, German biologist Jürgen Aschoff, one of the founders of chronobiology, first used the term "zeitgeber" to refer to an external cue that sets or resets an organism's biological clock in the scientific literature on sleep. William C. Dement, another of Kleitman's graduate students, was an important figure in the early study of REM sleep. Dement initiated a narcolepsy clinic in 1964 and founded the world's first sleep laboratory, the Sleep Research Center, at Stanford University in 1970.

In 1968, Allan Rechtschaffen and Anthony Kales standardized the criteria for the different stages of sleep in their Rechtschaffen and Kales (R&K) system. In 1975, Dement founded the Association of Sleep Disorders Centers, which later became the American Academy of Sleep Medicine. The association developed a diagnostic nomenclature of sleep disorders in 1979, which was revised in 1990. In 2007, the association published *The AASM Manual for the Scoring of Sleep and Associated Events: Rules, Terminology, and Technical Specifications*, a revision of the R&K system.

How It Works

Somnologists have divided sleep into non-REM and REM phases or stages. Three stages of non-REM sleep have been identified: N1, N2, and N3. Sleep normally consists of four to six cycles of non-REM and REM sleep. Each cycle usually follows the order of N1, N2, N3, N2, and REM and lasts between 90 and 120 minutes. In the N1 stage, also called drowsy sleep or somnolence, a person's conscious awareness of the environment as well as the tone of his or her muscles diminishes. In the N2 stage, conscious awareness of the environment disappears and muscular activity diminishes. Adults spend around one-half of their sleeping time in phase N2. In N3, known as deep sleep, delta sleep, or slow-wave sleep, parasomnias may occur. Usually N3 phases are longer earlier in the night, and the length of REM phases increases during late night and early morning. All four to six REM sleep phases added together amount to around 90 to 120 minutes (about one-quarter) of total sleeping time. During REM, heart and breathing rates are irregular, and normally clitoral or penile tumescence occurs. Although brain activity has been recorded in areas associated with sense perception, balance, and body movement during REM, in a healthy person, muscular activity is suspended, a phenomenon known as atonia. People who wake up during a REM phase can more easily remember their dreams; therefore, it is thought that most dreaming occurs during this stage of sleep and that little (or less) dreaming occurs in non-REM sleep. If REM sleep is repeatedly interrupted, the sleeper will compensate with longer REM phases at a later stage (rebound sleep).

Functions of Sleep. Sleep has diverse functions related to developmental, restorative, and memory processes. A certain quality and quantity of sleep (especially in the REM phase) is vital, not only for humans but also for mammals and certain other animals. The total sleeping time in adults ranges from six to nine hours per night; an average of seven hours is optimal for health. Newborn babies sleep up to eighteen hours per day, but as children grow older, they sleep fewer hours. Sleep (particularly REM) is important for brain development, especially in infants. Sleep deprivation (especially of slow-wave and REM sleep) negatively affects cognitive functions, complex processing capability, and memory. Sleep deprivation hampers the restorative processes of the organism and its immune system, which manifests, for example, as a lower white blood cell count and impeded wound healing. Anabolic hormones, which build tissues and organs, are secreted during sleep.

APPLICATIONS AND PRODUCTS

A polysomnogram can help physicians assess sleeping patterns and diagnose sleeping abnormalities. The quality or quantity of sleep can be improved through many tools and products, ranging from simple adjustments of behavior and the sleeping environment to significant medical interventions. If a sleep-related problem such as insomnia is not the symptom of a medical condition, often a patient's sleep can be improved through simple adjustments of the sleep environment or personal behaviors.

Polysomnography. Polysomnography is used to monitor and record biophysical and neurological data during sleep. It is used both in the scientific study of sleep and in diagnosing (or ruling out) parasomnias, insomnias, and other sleep-related problems. The test is usually performed during the night and monitored by a technician; the result is a polysomnogram. The test measures, records, and assesses brain waves by electroencephalography (EEG), eye movements by electrooculography (EOG), heart rhythm by electrocardiography (EKG), and skeletal muscle activity by electromyography (EMG). Additionally, depending on the purpose of the test, several other tests can be performed. Nasal and oral airflow can be measured to detect sleep apnea, a disorder in which a person stops breathing momentarily numerous times during sleep. Pulse oximetry can measure the oxygen saturation of the blood, which will drop in the case of sleep apnea. Microphones can pick up sounds, such as snoring or speaking during sleep. The sleeper is often videotaped during the test, partly to monitor the individual and partly to determine if parasomnias are present.

Behavioral Adjustments. Simple behavioral adjustments for those with sleep disorders include getting out of bed if the person is unable to get back to sleep after fifteen to twenty minutes and avoiding stimulants such as coffee, nicotine, chocolate, and sugary foods in the afternoon. Those with sleep disorders are also told to avoid alcohol, which initially might help induce sleep, because it can cause them to wake up more easily during the night. Sleep medicine professionals recommend that people eat a light meal two hours before going to bed or drink a glass of milk just before sleeping. Activities that encourage sound sleep are light exercise, sexual intercourse, walking, and taking a warm bath or shower. Sleeping in loose clothes made from natural materials, using guided imagery, and establishing sleeping rituals such as reading, listening to music, or meditating can improve sleep. Other suggestions include developing a positive attitude toward going to bed, limiting daytime naps, losing weight if overweight, and undergoing psychotherapy or counseling to address the root of the problem that keeps the person awake.

Environmental Adjustments. The environment can be modified so that it is conducive to falling and staying asleep. For example, the room can be darkened and made quieter through curtains and window shutters or special soundproof windows. The temperature and humidity of the sleeping environment can be adjusted with the help of air-conditioning, heating, fans, or air humidifiers. Other recommendations include removing all unnecessary items from the sleeping environment, using natural materials for bedding and bedroom furnishings, and limiting electronic sources of disturbance (such as televisions, computers, and cell phones).

Special Sleeping Devices. Many sleep products are available, although some of them are of questionable benefit. Products range from every-night products such as custom-made or weight-adjusted beds, contoured pillows, special bed or pillow covers, sleeping bags, and pajamas to items such as dawn-simulating devices (for shift workers), snoring aids, music to sleep by, white-noise machines, and bed-wetting alarms.

Sleeping Medications and Clinical Sleeping Aids. Sleep medications range from herbal, ayurvedic, homeopathic, and allopathic to placebo treatments. Some plants, such as lavender and hops, have a calming effect and are used in alternative treatments for sleeping problems. Most clinically prescribed allopathic treatments (sleeping pills) contain benzodiazepines or nonbenzodiazepines, which are sedative-hypnotic drugs. The side effects, which are various and can be severe, include reduction of the length of REM sleep, dizziness, confusion, anxiety, depression, drug dependency, cancer, and increased mortality. Circadian rhythm sleep disorders, related to jet lag, shift work, and delayed sleep phases can sometimes be treated successfully with melatonin. Melatonin has only a few short-term side effects but its long-term effects have not been thoroughly studied. Circadian rhythm sleep disorders can also be treated with bright light therapy, which exposes a person to a light of a specific bandwidth between 3,000 and 10,000 lux for a few minutes or up to several hours depending on the specific disorder. If sleep apnea is diagnosed, the use of a continuous positive airway pressure (CPAP) device might be prescribed. This device keeps the airway open during sleep so that breathing is not interrupted. In certain cases of obstructive sleep apnea, the surgical removal of throat tissue may be necessary. In many cases, weight loss can eliminate sleep apnea.

IMPACT ON INDUSTRY

Sleep Research and Sleep Medicine Organizations. Research related to sleep is in progress worldwide. The members of the World Sleep Federation include the American Academy of Sleep Medicine (AASM), the U.S.-based Sleep Research Society (SRS), the Canadian Sleep Society (CSS), the Federation of Latin American Sleep Societies (FLASS), the European Sleep Research Society (ESRS), the Asian Sleep Research Society (ASRS), and the Australasian Sleep Association (ASA). Most of these organizations offer training, publish academic journals, accredit sleep disorder centers and laboratories, and offer information to clinicians, patients, and the public. Some of them offer fellowships and scholarships for researchers, and a few (such as the ASRS and the ESRS) are umbrella societies with member nations. In the United States, where there are more than 2,000 sleep disorder centers and laboratories, the AASM acts as the national accrediting institution. It is also sponsors workshops and conferences and publishes the *Journal of Clinical Sleep Medicine*.

Industry. The pharmaceutical industry, which has a large stake in sleep medicine research, produces hypnotic drugs and sponsors education and symposia related to sleep. Some of the biggest corporations are Johnson & Johnson and Pfizer (based in the United States), GlaxoSmithKline (United Kingdom), Roche and Novartis (Switzerland), Sanofi-Aventis (France), and Bayer (Germany). These corporations engage in and sponsor research and education.

The medical equipment industry produces medical sleep equipment, including CPAP devices and nonmedical sleep aids such as special mattresses, pillows, air filters, and white-noise machines. The producers of polysomnography equipment are highly specialized and typically offer both software and hardware.

CAREERS AND COURSE WORK

The two most important professions related to sleep medicine are the sleep physician and the sleep technologist/polysomnographic technologist. Students who are interested in a career as a sleep technologist should study a program accredited by the Commission on Accreditation of Allied Health Education Programs. About thirty educational institutions offer accredited programs for polysomnographic technologists. In Canada, the Canadian Sleep Society is the accrediting body for educational programs for polysomnographic technologists. Sleep physicians are clinical specialists who diagnose and treat sleep disorders. They must graduate from medical school and should pursue a specialty in neurology, psychiatry, internal medicine, or pulmonology. Then, they can further specialize in sleep medicine and become board certified in the field of sleep medicine.

Fascinating Facts About Somnology

- According to an anecdote by the Greek-Roman philosopher Diogenes Laërtius, the Greek philosopher Aristotle tried to reduce his sleeping time and increase his time for research by going to sleep with holding an iron ball. The ball would drop into a bowl and wake him (probably immediately in or after the sleep stage N1).

- One of the discoverers of rapid eye movement sleep, Eugene Aserinsky, died because he fell asleep while driving. This suggests that conducting research on sleep does not guarantee long or good sleep.

- The famous German philosopher Immanuel Kant, renowned for accuracy in thinking and writing and also in everyday issues, went to bed every night at 10:00 p.m. and got up at 5:00 a.m. every day, no matter the circumstances. Kant instructed his servant to wake him up and get him out of bed, even if Kant resisted.

- In art, sleep and death frequently appear side by side, often based on the Greek myth that the god of sleep, Hypnos, was the brother of the god of death, Thanatos. In real life, sleep and death are linked in sudden infant death (SID), which occurs while babies sleep. Also, many elderly people hope to die in their sleep to avoid suffering.

- Both the American Albert Herpin (1862?-1947) and the Vietnamese Thai Ngoc (born 1942) reportedly did not sleep for decades, and many patients claim not to sleep for an entire night or several nights. Polysomnography can reveal whether a person really is not sleeping or just does not realize he or she is sleeping, a phenomenon called sleep state misperception.

- Sexsomnia, also known as sleep sex or clinically non-REM arousal parasomnia, is a rare sleeping problem that causes people to engage in sexual activities while they are asleep. Those with this condition, which can be diagnosed with polysomnography, rarely remember the sex acts. Charges of rape have been dropped after the alleged rapist was diagnosed with sexsomnia.

SOCIAL CONTEXT AND FUTURE PROSPECTS

Sleep patterns vary and depend on a person's culture. In areas with little or no artificial light, people have sleep patterns that tend to be more interrupted by wakefulness. In hunter-gatherer and nomadic cultures, people sleep for shorter periods but more often and also during the day. In such cultures, boundaries between wakefulness and sleep are at times less clear than in Western cultures. Cultural research suggests that sleeping for seven to eight hours at a stretch, which is regarded as normal in developed societies, is an ideal rather than a natural sleeping pattern. This research also sheds light on the definition of sleep abnormalities, as from this perspective, not all sleep abnormalities or disorders are necessarily unnatural. This is an area in which additional scientific study is needed.

Research, services, and products promising to improve the quality and quantity of sleep have to take a holistic approach to the individual sleeper and his or her environment. Highly industrialized societies have more sleep-disturbing factors, such as air and noise pollution, artificial light, and electronic appliances. Because prescription sleep medicines have side effects, including dependency, alternative methods of treating sleep disorders are in high demand. Those with sleep disorders are also ready consumers of information and products that they believe will help them sleep better. Treatment of sleep disorders may not only improve people's health but also produce significant profits for health care professionals and industries involved in health care.

Roman Meinhold, M.A., Ph.D.

FURTHER READING

Hirshkowitz, Max, and Patricia B. Smith. *Sleep Disorders for Dummies.* Hoboken, N.J.: John Wiley & Sons, 2004. An introduction to sleep disorders that focuses on practical solutions. Contains a foreword by William C. Dement, an index, and an appendix of resources and products.

Kryger, Meir H., ed. *Atlas of Clinical Sleep Medicine.* Philadelphia: Saunders Elsevier, 2010. Presents papers by experts in the field of somnology and sleep medicine covering a wide range of topics, including depictions of sleep in history, art, and literature. Contains color pictures of artwork, photographs, charts, and an extensive index.

Smith, Harold R., Cynthia L. Comella, and Birgit Högl, eds. *Sleep Medicine.* New York: Cambridge University Press, 2008. In-depth coverage of the phenomenon of sleep, its disorders and clinical sleep specialty areas, with index.

Summers-Bremner, Eluned. *Insomnia: A Cultural History.* London: Reaction Books, 2010. Historical-cultural studies approach to the phenomenon of sleep and insomnia; includes a bibliography.

WEB SITES

American Academy of Sleep Medicine
http://www.aasmnet.org

American Association of Sleep Technologists
http://www.aastweb.org

Sleep Research Society
http://www.sleepresearchsociety.org

Stanford University Center of Excellence for the Diagnosis and Treatment of Sleep Disorders
http://med.stanford.edu/school/psychiatry/coe

World Sleep Federation
http://www.wfsrsms.org

See also: Hypnosis; Neurology; Psychiatry; Pulmonary Medicine.

SPACE SCIENCE

FIELDS OF STUDY

Astronomy; physics; chemistry; geology; meteorology; space weather; computer science; communication technology; radio communications; electronics; aerospace engineering; electrical engineering; mechanical engineering; chemical engineering; biomedical sciences; medicine; environmental engineering; astrobiology; robotics; remote sensing.

SUMMARY

Space science is an all-encompassing term describing many different, often multidisciplinary fields focused on the study of anything beyond the Earth's atmosphere. At first, space science was limited to observational astronomy; however, developments in rocket technology have allowed the field to expand to experimental planetary studies and to astronautics. Astronomy has advanced to the point where several subfields, including astrophysics and cosmology, have become important fields of study in their own right. Applications of space technology have led to the exploitation of near space in the areas of communication, navigation, and space tourism. Space science is used to describe all these areas of study and technologies.

KEY TERMS AND CONCEPTS

- **Astrodynamics:** Study of orbits and trajectories.
- **Astronautics:** Engineering field associated with spacecraft engineering, especially manned spacecraft.
- **Booster:** Rocket used to lift off from the surface of a planetary body.
- **Cosmic Ray:** Ultra-high-energy particle radiation from space.
- **Exobiology:** Study of life, or the theoretical study of conditions for life, beyond Earth.
- **Geomagnetic Storms:** Disruptions in Earth's magnetic field caused by interactions with the Sun and particles emitted by the Sun.
- **Geosynchronous Orbit:** Orbit around Earth at a radius of 42,164 kilometers that has a period equal to the rotational period of the Earth.
- **Low Earth Orbit (LEO):** Orbit at an altitude of only a few hundred kilometers above the Earth's surface.
- **Micrometeoroid:** Tiny particle, often the size of a grain of sand, that moves through space at a very high speed.
- **Near-Earth Object (NEO):** Asteroid or comet whose orbit comes near or crosses the orbit of the Earth.
- **Satellite:** Natural or artificial object that orbits a planet.
- **Space Exploitation:** Use of space for monetary, commercial, or intellectual gain.
- **Space Weather:** Solar-terrestrial interactions, typically within Earth's magnetosphere, that can give rise to geomagnetic activity, radio blackouts, or radiation storms.

DEFINITION AND BASIC PRINCIPLES

Space science is not a single field of study. Rather, the term "space science" applies to any field that studies whatever is external to the Earth. These fields include planetary studies, stellar and galactic astronomy, and solar astronomy. Originally, astronomy involved simply looking at the skies, either with the naked eye or through telescopes, and recording what was observed. However, astronomy has become a branch of physics, with mathematical analysis and application of physical laws in an attempt to understand astrophysical processes. Additionally, robotic spacecraft have allowed planetary scientists to collect data on other planets that are useful to geologists and meteorologists. Discoveries of phenomena unknown on Earth brought many disciplines into the field of space science. Furthermore, theoretical investigations, such as models of potential planetary systems or the search for life in space, also are part of space science. The science of studying phenomena beyond Earth employs nearly every aspect, field, and subfield of science.

However, space science also encompasses the technology developed to further the aims of space science, such as rocketry and astronautics. Additionally, it includes the exploitation of space, with such technologies as satellites for communication and navigation, weather, and military surveillance. It also includes possible future technologies, such as mining asteroids and establishing space settlements and colonies.

BACKGROUND AND HISTORY

Throughout most of history, space was inaccessible to humans. The only way to study extraterrestrial objects was through observation. Astronomy began with these simple observations. In the seventeenth century, people began to use telescopes to view the sky, and physics and chemistry began to be important in interpreting their observations. In the nineteenth century, scientists discovered the existence of cosmic rays and learned the nature of meteorites, but the study of space remained largely limited to observations from the surface of the Earth.

In 1903, however, Russian scientist Konstantin Tsiolkovsky proposed that rockets could allow people to engage in space travel. The development of rocketry over the next half century eventually led to the creation of a rocket powerful enough to reach the edge of the Earth's atmosphere, allowing the edge of space to be studied directly. On October 4, 1957, the Soviet Union launched the first human-made Earth-orbiting satellite, *Sputnik*. The following month, the Soviets launched a larger satellite, *Sputnik II*, with a dog aboard to study the effect of space travel on a life-form. The United States launched *Explorer I* on January 31, 1958, the data from which helped scientists discover the Van Allen radiation belts surrounding Earth.

With the use of rockets, space exploitation became possible, leading to a flurry of interplanetary space missions by robotic spacecraft in the 1960's and 1970's, followed by much larger and more capable robotic spacecraft. In the 1960's, the first communication and weather satellites paved the way for the communication, navigation, and Earth-monitoring satellites that have become mainstays of modern society. Manned exploration of space began with simple capsules on top of converted missiles, evolved to more sophisticated craft that carried Apollo astronauts to the Moon, and then to reusable spacecraft such as the space shuttle. Manned space stations evolved from simple single-mission systems launched by the Soviet Union in the 1970's to the International Space Station, constructed during the first decade of the twenty-first century. Even telescopes took to space, with a number of astronomical satellites, including the Hubble Space Telescope, which was launched in 1990.

HOW IT WORKS

Astronomy, Astrophysics, and Cosmology. Astronomy is an observational science. Observations are made by a variety of instruments, including optical telescopes, radio telescopes, infrared and ultraviolet telescopes, and gamma-ray and X-ray telescopes. However, instead of simply taking photographs to study, modern astronomers measure spectra, intensities, and many other properties to understand the objects of their study. Some of the instruments they use are located at ground-based observatories with large telescopes, and others are located in orbit around Earth in space-based observatories, the most famous of which is the Hubble Space Telescope.

Astrophysics mathematically studies the physical processes of objects and develops theories based on physical laws. Cosmology involves the study of the universe and space itself. Both cosmology and astrophysics involve considerable computer modeling and mathematical analysis of astronomical observations.

Planetary Sciences. Until the 1960's, astronomers were limited to making observations of planets from Earth. However, the advent of interplanetary robotic spacecraft allowed scientists to study other planets in much the same way that they study Earth. Robotic spacecraft employ a battery of instruments, including cameras, spectrometers, neutron sensors, and magnetometers, to study planets from orbit. Spacecraft have landed on the Moon, Venus, Mars, and Saturn's moon Titan to study their surfaces.

Astrobiology/Exobiology. Since about the end of the sixteenth century, when astronomers were beginning to guess the nature of the planets, scientists have speculated on whether life could exist on other planets. Robotic spacecraft have searched for life on Mars, and rocks returned to Earth by the Apollo missions to the Moon have been studied for life. The basic essentials for life have been found in space, but as of the early twenty-first century, no definitive evidence of life has been found on other worlds. Biologists, working with astronomers, have studied the nature of life and probed its possible origins and the conditions necessary for life. This study was originally called astrobiology, although the term "exobiology" has slowly gained in popularity. Exobiology is largely theoretical.

Rocketry. Rockets are key to space study. Rockets are entirely self-contained and do not need to take in air to operate. They may be either liquid or solid fueled. Rockets operate on the principle of conservation of momentum. High-speed gases exiting the rocket carry momentum, so the rocket must have momentum in the other direction to conserve momentum. The rate of change of momentum is force and is the thrust of the rocket.

Astronautics. The space environment is harsh and hostile, not only to life but also to human-made devices. Thus, great care and redundancy must go into any spacecraft design. Aerospace engineers must design spacecraft that can operate in extremes of hot and cold, microgravity, and the vaccum of space, as well as under intense radiation exposure and despite repeated micrometeoroid impacts. These conditions are all difficult to reproduce on Earth. For manned spacecraft or space stations, the additional requirement is that a habitable environment must exist within the spacecraft. The spacecraft must be shielded from radiation as much as possible to protect its occupants, and systems must be reliable enough to keep the human occupants safe for the duration of the mission. This requires skill at mechanical and environmental engineering, as well as with advanced electronics systems.

Space Weather. Earth's magnetic field extends a long way above the surface of the planet. The region of the solar system in which Earth's magnetic field dominates is called Earth's magnetosphere. Particles trapped in the magnetosphere create regions of intense particulate radiation surrounding the planet. These regions were discovered in 1958 using instruments aboard the American *Explorer* spacecraft by physicist James Van Allen and are known as the Van Allen radiation belts.

Earth's magnetosphere is not static; it is constantly shifting and adjusting to the interplanetary space environment, which is constantly being affected by the Sun. The Sun continually emits particles that travel outward through the solar system. This stream of particles, called the solar wind, varies in intensity, density, and speed based on solar activity. Earth's magnetosphere adjusts accordingly.

Occasionally, very large bursts of energy, called solar flares, cause portions of the Sun's corona to detach and move across the solar system as a large bubble of plasma. When such a coronal mass ejection affects Earth's magnetosphere, a significant shift in the magnetic field occurs over a short period of time. Often solar flares are associated with significant elevations in solar cosmic rays, events called solar radiation storms. Passengers in aircraft and spacecraft can receive significant levels of radiation exposure during such radiation storms.

Variations in Earth's magnetic field are called geomagnetic storms. Large geomagnetic storms resulting from large rapid shifts in Earth's magnetic field can induce damaging electric currents in pipelines, power lines, and telephone lines. Geomagnetic storms are often associated with radio communication interference and even blackouts. Collectively, the study of the variations of Earth's magnetosphere and the solar-terrestrial interactions is called space weather.

APPLICATIONS AND PRODUCTS

Much of space science involves basic scientific research. Astronomers and planetary scientists seek to learn about stars, planets, and other celestial bodies and how they came to be the way that they are and why they have the properties that they have. Although many aspects of these studies have applications beyond academic studies, the goal of the scientist is to understand the system and to let other scientists or engineers apply the knowledge and technology to other purposes.

Weather and Remote Sensing. Being at high altitude permits an observer to get a bigger and broader picture than is available from the ground. Early researchers used balloons and then aircraft to better study the Earth. The advantages of observations from orbit were obvious to the very first space engineers in both the United States and the Soviet Union. Therefore, the first weather satellite, TIROS 1, was launched on April 1, 1960, less than three years after the first artificial satellite was launched into orbit around the Earth. Early weather satellites used film that was returned to Earth in canisters. Soon, engineers perfected the technology that enabled images to be transmitted electronically. In the twenty-first century, real-time or nearly real-time weather images have become essential tools for meteorologists. Modern weather satellites do not just take images but also contain instruments to make numerous measurements.

In addition to providing information on weather,

satellites can monitor land use, ocean currents, and atmospheric pollution. Collectively, this technology is called remote sensing. Originally, government agencies operated these remote-sensing orbital satellites; however, the demand for such data by researchers and private industry spurred several private companies to deploy their own remote-sensing satellites. Dozens of such satellites continually monitor the Earth from space.

Space Weather. Space weather storms have disabled satellites, navigation, and ground-based communication and power distribution grids; however, protective measures are sometimes possible given ample warning. The military, recognizing the dangers associated with communication blackouts during times of national tension or during armed conflicts, was the first to start monitoring the Sun for signs of solar flares that might trigger geomagnetic storms and communication blackouts. Both the United States and the Soviet Union established global networks of solar observatories for this purpose. However, as society became increasingly dependent on electronic technology that could be interrupted by such events, other government agencies have supplemented the military's observations of the Sun. By 2000, the Space Weather Prediction Center operated by the National Oceanographic and Atmospheric Administration (NOAA) had become one of the top clearinghouses for solar observation data and predictions of space weather events. Most of the center's data are available over the Internet. These data are important for satellite operators, electric utilities, airlines, and many other industries. The Space Weather Prediction Center issues space weather watches and warnings in much the same way that the National Weather Service, another NOAA division, issues thunderstorm and tornado watches and warnings. Many other technologically advanced nations have similar space weather centers.

Communication. When the first satellites were launched, microwave relay towers were being constructed to carry voice communications without the need for wires. These towers could be placed only so far apart because the curvature of the Earth would interfere with transmissions. Satellites, however, can act as ultra-high towers. The very first test of satellite communications occurred with the launch of the Project SCORE satellite in 1958. SCORE was not capable of real-time communication; it recorded an incoming signal and transmitted it later. Soon, however, technology was developed to permit satellites to relay signals in real time. By 1963, the first geosynchronous satellite, the Syncom 3, was launched. A geosynchronous satellite remains over a single position on the ground, making it easy to aim antennas and eliminating the need for expensive and complicated satellite-tracking systems. By 2010, several hundred geosynchronous communication satellites had been launched.

Although geosynchronous satellites have many advantages over low Earth orbit (LEO) satellites that track across the sky in a few minutes, LEO satellites have the advantage of being able to use less powerful signals. A number of LEO communication satellites exist, notably in the form of satellite telephones, systems similar to cell phones but using satellites instead of cell phone towers. Satellite phones are useful in remote regions where cell phone service is not available. The military often relies on satellite phones for battlefield operations.

Navigation. Celestial navigation, used for millennia, requires the ability of a skilled navigator to actually see the sky. Beginning in the 1930's, networks of radio transmitters were set up to aid in navigation. These systems become more advanced but were still limited to areas that had transmitters, typically the land areas of technologically advanced nations, and were unavailable at sea. Satellites overcome this problem in much the same way that they did for communication relay towers. The first satellite navigation satellite, Transit 1B, was launched in 1960.

Satellite navigation culminated with the U.S. Department of Defense's Global Positioning System (GPS). GPS technology, originally deployed as NAVSTAR strictly for military use in 1978, was eventually made available in civilian form in 1983 on the order of President Ronald Reagan. In 1993, GPS was made available to anyone free of charge, and it has been fully operational since 1995. Free use of the GPS signals for navigation spurred a number of private companies to sell commercial handheld GPS navigation systems. Software to display position on a map and to give audible turn-by-turn driving directions using maps and GPS position data made handheld GPS navigation systems very popular by the 2000's. Additional enhancements to GPS navigation are planned both by the Department of Defense and by private companies.

Space Tourism. Government space agencies have long been the only avenue for space exploration and manned space travel. However, there has been a gradual privatization of space technology, and many of the functions previously done only by government agencies are now carried out by private companies under government contract. A few private companies have formed to provide recreational space travel to wealthy individuals. Several people have purchased rides in the Russian Soyuz spacecraft to the former Mir and the International Space Station through Space Adventures. Although these private companies are working with the space agencies, other companies are developing their own spacecraft. Some of these companies aim to sell their services to the International Space Station, but others, such as Virgin Galactic, intend to provide rides into space for those willing and able to pay for a ticket. Other companies, such as Bigelow Aerospace, are planning to place private space stations into orbit as tourist destinations.

Military Applications. Even before the first satellite was launched, military leaders recognized the importance of space. Military reconnaissance satellites (often called spy satellites) were among the first satellites launched after the developmental flights. In the 1960's, many feared that orbital weapons platforms capable of dropping nuclear bombs anywhere on Earth would be deployed. This fear never materialized. Although long-range missiles lift their warheads into suborbital trajectories, they are not space-based weapons systems. The United States and the Soviet Union both began developing antisatellite weapon systems to counter the military advantage of reconnaissance satellites. The first test of such a system was conducted in 1963 by the Soviet Union. In 1985, the United States tested its first antisatellite weapon, a missile designed to destroy low Earth orbit satellites. In 2007, China conducted an antisatellite weapon test, and India and Russia are reportedly working on advancing such technology.

IMPACT ON INDUSTRY

Spending on space science grew from nearly nothing early in the twentieth century to nearly 4 percent of the gross domestic product of the United States during the height of the space race in the 1960's. By the beginning of the twenty-first century, it had fallen to a fraction of that sum, but the emphasis

Fascinating Facts About Space Science

- In September, 1859, one of the most powerful geomagnetic storms in history knocked out telegraph systems around the world.
- The word *sputnik* means "fellow traveler."
- Rockets do not need to push against anything to fly through space.
- Objects on the Moon weigh about one-sixth of what they do on Earth.
- The first communication satellite, Project SCORE, was launched December 18, 1958, and the first message broadcast was a holiday peace message from President Dwight Eisenhower.
- The United States' first space station, Skylab, was constructed from the third stage of a Saturn V rocket.
- Much of the universe is a vacuum, but other parts of space contain clouds of interstellar dust and tiny particles (the solar wind), many isolated particles and hydrogen atoms (which sometimes form nebulae), planetoids, asteroids, radiation bursting forth in solar flares, and beams of light, heat, and X rays radiating through space.

has changed from space exploration to space exploitation, and the funding has shifted from governments to private industry, primarily in the realm of communications.

Government and University Research. Governments still fund space exploration, but many researchers are employed as faculty members at universities. Space physics and astronomy research is mostly conducted by researchers, although the spacecraft and observatories used by the researchers are often operated or funded by government agencies. Knowledge of the space environment has been essential for private companies that want to develop technologies for space exploitation.

Industry and Business Sectors. Private industry, particularly in the United States, has always been involved in space science, particularly aerospace companies that designed and built components for spacecraft and satellites. In the 1980's and 1990's, however, companies began to build complete satellites, not just components, and private companies have begun building rockets to launch satellites for customers. As space travel became easier, the space tourism industry arose to offer wealthy individuals a chance to travel into space. The interest of the private sector in space has greatly increased the degree to which space has been exploited for human use.

Many nonspace industries, including banking and broadcasting, use space technology such as satellite communications. With the release of GPS technology to the public, an entire industry has emerged to produce GPS navigation units.

CAREERS AND COURSE WORK

The field of space science is really a great many different fields, with no single career path. Careers range from machinists building parts for rockets to assembly workers building rockets to engineers designing rockets and spacecraft. The field includes scientists studying the heavens, planetary scientists studying planets, and robotics engineers designing rovers to land on other worlds. It also includes businesses specializing in communications and remote sensing and the engineers designing the satellites to perform those functions. Some space businesses need employees with business degrees and experience rather than science backgrounds. Even astronauts do not all have the same qualifications. Some astronauts are pilots with extensive high-performance aircraft experience, and others are scientists with multiple graduate degrees. The growth of space science to include virtually every discipline means that there is no single career path.

However, space science is a very technical field, and therefore, a typical space science career generally requires many courses in science and mathematics. Physics and chemistry are needed for almost all space science careers. For applied space science work, a background in either electrical or mechanical engineering is probably necessary. For basic research, a background in physics or astrophysics is most common, but planetary scientists can have backgrounds in geology, meteorology, or any similar field. Exobiology would, of course, require a biology background. Almost all space science jobs that result in advancement require advanced degrees beyond the bachelor's degree.

SOCIAL CONTEXT AND FUTURE PROSPECTS

Space science has evolved from simply the science of astronomy to real, practical applications. Planetary science helps scientists understand how planets

function, yielding valuable insights about Earth as a planet. This knowledge gains in importance as issues such as global warming and ozone depletion in the stratosphere are more widely examined. Studies of stars help scientists understand how the Sun works, which is important in the realm of space weather.

The exploitation of space provides useful services for the inhabitants of Earth. Modern society relies very heavily on rapid communication and the transmittal of a vast amount of data across far distances on a regular basis. Satellite communication facilitates this need, and many common everyday activities, such as credit or debit card purchases or ATM withdrawals, use this technology. GPS navigation has become very common and probably will become even more common, as more and more automobiles come equipped with GPS systems as a standard feature. Accurate weather and storm forecasts would be impossible without weather satellite data.

Space travel used to be fanciful speculation, then evolved into expensive government endeavors. However, by 2010, several private companies were on the verge of providing space tourism at prices within the reach of many wealthy individuals. If the trend continues, space travel may eventually become almost as common and inexpensive as air travel.

Raymond D. Benge, Jr., B.S., M.S.

FURTHER READING

Angelo, Joseph A., Jr. *Space Technology*. Westport, Conn.: Greenwood Press, 2003. A good overview of spacecraft technology, with chapters on the history of development of the technology and on the people involved in its development.

Berinstein, Paula. *Making Space Happen: Private Space Ventures and the Visionaries Behind Them.* Medford, N.J.: Plexus, 2002. An overview of the emerging of private space venture capitalists and the emerging trend of the commercialization of space.

Carlowicz, Michael J., and Ramon E. Lopez. *Storms From the Sun: The Emerging Science of Space Weather.* Washington, D.C.: Joseph Henry Press, 2002. An excellent guide to space weather and solar-terrestrial interactions.

Comins, Neil F. *The Hazards of Space Travel: A Tourist's Guide.* New York: Villard, 2007. A nontechnical overview of the dangers associated with manned spaceflight and the technologies used to overcome those dangers.

Freedman, Roger, William J. Kaufmann, and Robert Geller. *Universe.* 9th ed. New York: W. H. Freeman, 2010. An excellent and very thorough college-level astronomy textbook.

Harra, Louise K., and Keith O. Mason, eds. *Space Science.* London: Imperial College Press, 2004. An advanced college level text and reference book about various areas of space science, including space weather, remote sensing, cosmology, and high-energy astrophysics.

Matloff, Gregory L. *Deep Space Probes: To the Outer Solar System and Beyond.* New York: Springer-Verlag, 2005. A thoughtful analysis of spacecraft propulsion and other issues, primarily with a focus on developing or possible future technology.

Swinerd, Graham. *How Spacecraft Fly: Spaceflight Without Formulae.* New York: Copernicus Books, 2008. Presents an overview of astronautics, including historical developments, modern technology, and developing technologies.

WEB SITES

National Aeronautics and Space Administration
http://www.nasa.gov

National Space Society
http://www.nss.org

The Planetary Society
http://www.planetary.org

Space Science Institute
http://www.spacescience.org/index.php

See also: Planetology and Astrogeology; Radio Astronomy; Space Stations; Telescopy.

SPACE STATIONS

FIELDS OF STUDY

Space engineering; aeronautics; astronautics; computer science; earth sciences; robotics and embedded systems.

SUMMARY

Space stations are Earth-orbiting vessels operated by national and international space agencies for the purposes of studying space and the Earth. Manned space stations are designed to be semipermanent habitats for working astronauts and scientists and are considered to be prototypes for human living environments in space. They do not have dedicated propulsion systems but are instead placed in space by other vehicles; this often means they are built over the duration of several missions. The International Space Station (ISS) is an orbiting space station inhabited by astronauts and operated by several nations, including the United States, Japan, Canada, Russia, and member nations of the European Space Agency (ESA).

KEY TERMS AND CONCEPTS

- **Expedition:** Term for a mission aboard the International Space Station.
- **Extravehicular Mobility Unit (EMU):** Enhanced and pressurized space suit or system that allows astronauts to conduct extravehicular activity in Earth's orbit. The Russian equivalent is the Orlan space suit. The EMU is slated to be replaced by the Constellation space suit system.
- **Life-Support System:** System developed to allow human and other biological life-forms to live in space environments.
- **Low Earth Orbit (LEO):** Refers to orbits occurring between roughly 100 and 1,500 kilometers, with the high point sometimes reaching 2,000 kilometers. Space stations and most satellites operate in low Earth orbit.
- **Modularized Space Station:** Space station whose overall structure is created by connecting modules and other operational pieces in orbit.
- **Prefabricated Space Station:** Space station whose overall structure is prefabricated before launch.
- **Space Engineering:** Science of engineering as applied to off-Earth environments; by their nature, these environments have different physical properties and therefore require knowledge specifically related to space science and physics.
- **Space Habitat:** Semipermanent or proposed permanent human living environment on an off-Earth planetary body (such as a satellite or planet) or aboard an orbiting spacecraft.

DEFINITION AND BASIC PRINCIPLES

Space stations are Earth-orbiting vessels operated by national and international space agencies to study space and the Earth. Space stations differ in design and purpose from other types of spacecraft, such as rockets and space shuttles, because they serve an entirely different purpose. Space stations are built for the purpose of long-term orbit around Earth, and they are intended to carry more crew members than other types of spacecraft. They are also intended to be laboratories for the study of space and earth sciences. An important aspect of space station research is the astronauts and scientists themselves. Their physical responses to being in zero-gravity environments and their psychological responses to living in space for long periods of time (far longer than typical space missions) help scientists understand the requirements for off-Earth human and animal habitats.

The design of space stations evolved since the first space station missions of the early 1970's. The space stations of the Salyut and other programs were monolithic structures, so called because they were designed as centralized spacecraft. Later space stations favored modular designs that allow diverse elements to be added as needed. For example, the ISS is made up of several modules added over several years. The most common uses of space stations by scientists are as laboratories for experiments in a range of life and material sciences.

BACKGROUND AND HISTORY

In science fiction, the space station has long held a place of special interest because it offers a glimpse into what life aboard an inhabited spacecraft might be like. Whereas sending rockets to the Moon and beyond answers questions about how far away from Earth and how fast humankind can travel, space stations answer the more fundamental questions of whether humans can live in space and for what length of time. The answers to these questions may provide a positive, long-term direction for the future of humans in space.

The first space station in orbit around Earth was the Salyut 1, the first of several Soviet space stations launched between 1971 and 1982. Despite its overall success, the mission ended in tragedy when a depressurization accident resulted in the deaths of its three crew members. Other early Salyut missions were plagued with system failures and equipment malfunctions, but later missions were large-scale successes, in part because of the increased duration of their time in orbit and the vast amount of data gathered about the effects of space environments on the human body and the overall viability of space stations as research laboratories.

The first U.S. space station was Skylab, which was launched in 1973 after several years of development by the National Aeronautics and Space Administration (NASA). Like Salyut and other Soviet missions, Skylab had its share of problems, such as the loss of a micrometeorite shield, but it was an important first step in establishing a U.S. space station program. In the years since these early missions, many improvements have been made to the overall concept and design of space stations, and space stations have continued to return vital information to the scientists and astronauts who have been given the task of designing the next generation of Earth-orbiting laboratories.

How It Works

Design and Uses. The design and uses of a space station differ from other types of spacecraft. Some vehicles, such as satellites, probes, and flyby craft, are unmanned vehicles whose flight may or may not be dedicated to a specific orbit and whose small size helps extend the duration of their missions. For example, the probe Voyager 1 has been regularly transmitting data about the Sun's heliosphere since it left Earth in 1977. The probe, which will not return to Earth, is believed to have enough power to last until 2020. Other vehicles, such as space shuttles and staged rockets, are built to accommodate human passengers. These vehicles are controlled by their

crew with aid from a mission control center and are launched into space from Earth. These vehicles commonly lose rocket systems after launch, and only the manned vehicular portions return to Earth. Space stations differ in their uses from other types of spacecraft. The spacecraft used in the Apollo program, for example, were built to be used as exploration craft on the lunar surface with provisions for temporary human habitation and were also built as return vessels. Like space stations, shuttle craft are used for scientific research but are not intended to be inhabited for long periods of time. Shuttle craft are also used as transportation vehicles between space stations and Earth.

Architecture. Manned space stations are designed to be used as semipermanent habitats and laboratories. Therefore, they are roomier and have more advanced life-support systems and are technologically more complex than other space vehicles. The ISS exemplifies this design consideration: It is 240 feet long and 356 feet wide, with more than 12,000 square feet of room in which to live and work. It is also modular, with solar arrays, pressurized living and research modules, connector nodes, storage modules, and other pieces all connected around a central truss system. Each piece of the ISS is launched into space (either aboard a shuttle craft or rocket) and locked into place. The last component of the ISS was scheduled to be added in 2011.

Mission Duration. An ever-evolving aspect of space stations is the duration of their missions in space. The Salyut 1 mission launched in April, 1971, and returned to Earth in October after 175 days. The Russian space station Mir launched in February, 1986, and returned in March, 2001, after 5,510 days in orbit, which remains the record for mission duration. The ISS was launched in 1998 and may remain in orbit until 2020. Future space station designs are expected to extend these periods, leading to semipermanent living and working facilities that will be operable for decades or more. The obstacles that must be overcome to make this happen involve funding and political issues, as well as the outcomes of the continuous research on the long-term effects of these environments on humans.

Applications and Products

Space stations have many applications, including serving as scientific research stations and prototypes for permanent Earth-orbiting human settlements and acting as tools for international diplomacy. Because they have been used for a relatively short period of time in any of these capacities, their potential is still mostly unrealized, at least in terms of how they were once envisioned by space scientists and science-fiction writers. However, if private entities continue to develop their plans for space stations, perhaps they will create uses that go beyond previously envisioned ideas.

Science and Research in Space. One of the benefits of the space station design is that it can provide room for humans to carry out scientific research. Scientific research has long been a primary reason for sending both manned and unmanned vehicles into space. However, depending on one's orientation, this is not always the most important reason for spaceflight. Consider the attitudes of the Mercury Seven, the first seven astronauts selected by NASA, many of whom thought of themselves as pilots first and scientists second (if at all). Although there is value in using space missions to test a nation's technological capabilities and explore the environs, there is no doubt that space exploration with the primary goal of scientific discovery has many benefits. For example, the effects of space weather and atmospherics on animals and plant life are important areas of knowledge if living in space is to be an option for the human race.

The science performed aboard the ISS illustrates the breadth of the research that can be accomplished. From 2007 to 2009, for example, crew members photographed portions of Earth from space, and these data were used to supplement studies of climate change. Other studies have focused on space farming, cellular growth, new treatment options for Duchenne muscular dystrophy (DMD) through the study of protein crystal growth, and the development of telemedicine strategies, which could be used to aid injured people across great distances.

Human Habitats in Space. Humans in space face dangers and problems such as space radiation, temperature extremes, decompression sickness, and the negative effects of a zero-gravity environment on humans. Designers and engineers have had some success in finding ways to mitigate and even eliminate the danger from radiation and extreme temperatures, but the negative effects of a zero-gravity environment may be harder to combat. It is possible that

humans may have to learn to adapt to zero gravity. However, as of 2009, only about 350 people had ever been in space, and therefore, it may take many more trips into zero gravity for scientists to catalog the effects on all physiological types. Other worries for astronauts include cardiovascular problems, loss of bone density, muscular atrophy, and many other issues that can lessen their productivity and even cause life-threatening conditions.

Research is helping scientists understand how to create better and more efficient habitats for humans in space. Other areas of study include determining what types of materials best shield astronauts from radiation, which recycling methods are most effective and provide the most long-term benefits to human-inhabited space settlements (whether large or small), and how to generate power from limited resources.

International Makeup. During the space race—from the 1950's to the mid-1970's—the United States and the Soviet Union competed for dominance in space. As the primary players in this worldwide drama, the diplomatic and political strength of either nation was tied to its success in space. For many years, "success" in this race meant the ability to launch men into space and return them safely to Earth. However, the lunar missions were only one point in this race—other measures of success were met handily by the Soviets, including their launch of the Salyut program in the early 1970's. In 1975, the space race effectively ended when both sides teamed up for the Apollo-Soyuz Test Project (ASTP). Although this mission yielded some scientific research results, its primary importance was in launching a scientific détente between the two nations.

There have been many international missions in the years since Apollo-Soyuz, including a joint program combining the resources of the U.S. space shuttle program with the Russian Mir program, a predecessor to the ISS. The ISS itself combines the resources of NASA, the Canadian Space Agency, the European Space Agency, the Japan Aerospace Exploration Agency, and the Russian Federal Space Agency. As space exploration and settlement become more relevant topics on the world stage, this spirit of international cooperation could prove beneficial. However, a number of nations, including China, India, and Japan, appear to be eager to develop their own programs, with planned missions to the Moon and beyond.

IMPACT ON INDUSTRY

Many areas of research and development are dedicated to the aeronautics, engineering, technology, and science related to space stations. Much of this is done by or in partnership with national and international space agencies, but many private and public organizations also are working in these fields. If the operation of space stations becomes a for-profit enterprise, it is likely that there will be more areas of development that are explicitly private, from concept and design to launch and management.

National Space Programs. The main agencies behind the ISS are its international partners—NASA, the Canadian Space Agency, the European Space Agency, the Japan Aerospace Exploration Agency, and the Russian Federal Space Agency. These agencies are responsible for transporting crew and materials to the ISS and managing various aspects of space station construction. Each of these agencies also actively oversees the research conducted by their crews in pursuit of scientific endeavors. However, given the large-scale efforts that go into space programs and the limitations of governments, it is unlikely that any space station mission could ever get off the ground purely through the efforts of a single national space agency. Instead, commercial partners and contractors must be brought in to help with various aspects of a mission, including construction of craft and materials and creation of hardware and software.

Public-Private Initiatives. Although the spirit of international cooperation exhibited by the ISS crews will most likely continue as more nations focus on the benefits and possibilities of space travel, management of the ISS may be shifted to nongovernmental organizations. These organizations would in turn be managed by a group or groups of institutions whose broad goals would be to uphold the missions of the space agencies involved in the program and possibly determine what types of research are done in space. Such public-private partnerships have often served as a template for contemporary space program development.

Private Space Initiatives. Private space companies, although rare, are already planning to venture into space. One such company, Bigelow Aerospace, has already begun experimenting with the unmanned space habitats Genesis I and Genesis II. Its mission is to develop "next-generation crewed space complexes to revolutionize space commerce and open up the final frontier to all of humanity." Although the Genesis modules are not yet part of any space station design, their design could eventually contribute to how space stations are built in future programs. Companies such as Bigelow are pioneers in what many believe will be the next waves in space travel—space tourism and entrepreneurial space endeavors.

Another pioneer in private space travel is the Russian Federal Space Agency, which came into being after the breakup of the Soviet Union in the early 1990's. Although the Russian Federal Space Agency is technically a national agency, it does offer commercial satellite services and is the sole transportation for space tourists traveling to and from the ISS. The services provided by the Russian agency are similar to those that would be provided by private companies when travel aboard space stations becomes more common.

University Research. University research is crucial to developing an understanding of astronautics and space science. Funding for such studies comes in great part from space agencies themselves. For example, in 2009, NASA awarded grants to several American universities for technological research and development related to space colonization. It is very common for NASA and other agencies to sponsor university research that will allow the agency to enhance space programs by broadening the uses of the ISS. Such initiatives are also helpful in attracting bright, talented students to work for national agencies onces they graduate. However, as private space initiatives become more common, they will provide another career path for these students.

CAREERS AND COURSE WORK

Compared with medicine or other areas of science, the history of space-station-related jobs is short. However, this short history has also revealed many areas where inventiveness and innovation are major assets for students. Anyone considering a career in space stations has much to consider regarding fields of study and potential careers.

Areas of focus for those wishing to build or design space stations are astronautics and space engineering, as well as robotics and embedded systems. Purdue University and the University of Washington offer schools of aeronautics and astronautics in which students work on next-generation technologies and

Fascinating Facts About Space Stations

- On its Web site, NASA publishes a list of dates and times when the International Space Station can be seen as it traverses the sky. Space station tracking and even video and audio feeds from the ISS are also available.
- The European Space Agency has eleven member countries: Belgium, Denmark, France, Germany, Italy, the Netherlands, Norway, Spain, Sweden, Switzerland, and the United Kingdom.
- The International Space Station flies 218 miles above the Earth at a speed of more than 17,000 miles per hour.
- Space stations are not designed to fly or land, so when they reach the end of their mission status, they are destroyed as they reenter Earth's atmosphere. When the Russian space station Mir reentered in March, 2001, its entry was controlled so that it would burn up over the South Pacific Ocean.
- Cosmonaut Valeri Polyakov holds the record for the most time in space. He flew aboard the space station Mir from January 8, 1994, to March 22, 1995.
- Seven people have visited the International Space Station. One such space tourist, a Canadian businessman named Guy Laliberté, may have paid about $35 million for the privilege.

techniques that will someday be part of space stations and other missions. Graduates from these and similar programs often go on to careers in the aerospace industry and NASA. NASA itself is heavily involved in educational programs for institutions of all sizes and levels, with the purpose not only of educating students about its missions but also of letting students know how they may become part of the space program as employees.

Those seeking to become astronauts aboard a space station should earn at least a bachelor's degree (a higher degree is desired) in engineering, biology, physical science, or mathematics. These types of degrees are available at most larger universities and colleges. To pilot a spacecraft, a person must accumulate at least one thousand hours as a pilot in command of jet aircraft. Astronauts also must have perfect vision and be between 62 and 75 inches tall. Actual training

for being an astronaut comes after acceptance into NASA's astronaut training program.

SOCIAL CONTEXT AND FUTURE PROSPECTS

The importance of space stations as research laboratories for both space and earth sciences was realized long before the first station was launched, which may account for the continuation of space station programs. However, space stations are important for many other reasons. First, space stations allow national and international space agencies to expand and improve on their knowledge of space environments and the potential of space stations as semipermanent or even permanent habitats for humankind. Second, space stations offer an unprecedented opportunity for nations to experience a form of diplomacy and sharing of resources that could serve as a future model of foreign relations. Third, if the quest for new ways to understand the environment is to continue, that search must be extended into space to maintain an essential perspective that had been unavailable. Space stations are invaluable tools for extending the ways humankind learns about itself and its environment.

Specific future applications of space stations are difficult to predict. They may someday be used as educational institutions, vacation getaways, or way stations between Earth and other spacecraft, or they may remain as they are while other space missions attempt to fly to and from Mars and beyond. Because of the expenses involved in building and maintaining space stations, their benefits must outweigh their costs. As of the start of the 2010's, space stations served as extensions of past discoveries in science, technology, and engineering. As their capabilities are further explored, probably humankind can derive greater benefit from them.

Craig Belanger, M.S.T., B.A.

FURTHER READING

Bond, Peter. *The Continuing Story of the International Space Station.* New York: Springer-Praxis, 2002. Comprehensive history and overview of the International Space Station, with extensive explorations into the international cast of characters involved in the development of the ISS. Appendixes detail the construction and staffing of expeditions.

Calhoun, Nicole M., ed. *International Space Station Research: Accomplishments and Challenges.* Hauppauge,

N.Y.: Nova Science Publishers, 2010. A collection of essays examining the research conducted on the space station and the physiological toll on astronauts.

Catchpole, John. *The International Space Station: Building for the Future.* New York: Springer, 2008. Examines the construction and uses of the station and discusses the political interactions involved.

Howe, A. Scott, and Brent Sherwood, eds. *Out of This World: The New Field of Space Architecture.* Reston, Va.: American Institute of Aeronautics and Astronautics, 2009. A collection of essays examining the demands of creating habitats for humans in space. Some essays look at aspects of the International Space Station, including the design of its crew quarters and the performance of its interior.

Linenger, Jerry M. *Off the Planet: Surviving Five Perilous Months Aboard the Space Station Mir.* New York: McGraw-Hill, 2000. Autobiographical book by an American astronaut who worked aboard the space station Mir in 1997. Valuable for its first-person description of an on-board emergency as well as the descriptions of the station itself and life as an astronaut.

Thirsk, Robert, et al. "The Space Flight Environment: The International Space Station and Beyond." *Canadian Medical Association Journal* 180, no. 12 (June 9, 2009): 1216-1220. Discusses the basic science behind the effects of space travel and extended periods of time in zero-gravity on the human body and details specific risks to astronauts aboard the International Space Station.

WEB SITES
National Aeronautics and Space Administration
International Space Station
http://www.nasa.gov/mission_pages/station/main/index.html

Russian Federal Space Agency
http://www.federalspace.ru/main.php?lang=en

See also: Space Science

SPECTROSCOPY

FIELDS OF STUDY

Analytical chemistry; physical chemistry; organic chemistry; biochemistry, molecular biology; immunology; cell biology; physics; astronomy; inorganic chemistry; geology; forensic science; materials science and engineering; chemical engineering; biomedical engineering; toxicology; agricultural engineering; electrochemistry; polymer chemistry; military science.

SUMMARY

Spectroscopy is the study of the interactions of electromagnetic radiation, or light, with matter in order to gain information about the atoms or bonds present within the system. There are many different types of spectroscopic techniques; however, most of the techniques are based on the absorption or emission of photons from the material being studied. The applications of spectroscopy span a variety of disciplines and can allow scientists to, among countless other things, determine the elemental composition of a nearby dwarf star, the chemical identity of an unknown white powder sample, whether a transfected gene has been expressed, or the types of individual bonds within a molecule.

KEY TERMS AND CONCEPTS

- **Absorption:** Process that occurs when an atom absorbs radiation.
- **Analyte:** Component of a system that is being analyzed.
- **Beer's Law:** Law that states that absorbance of a solution is proportional to its concentration.
- **Electromagnetic Spectrum:** Entire range of wavelengths of light.
- **Emission:** Process that occurs when an atom or molecule emits radiation.
- **Excitation:** Process that occurs when matter absorbs radiation and is promoted to a higher energy state.
- **Photon:** Massless, finite unit of electromagnetic energy that behaves as both a particle and a wave.
- **Scattering:** Process that occurs when light does not

bounce off a molecule at the same wavelength at which it started but instead undergoes a change in wavelength.
- **Synchrotron:** Particle accelerator that produces ultra-intense light.
- **Wavelength:** Distance a photon travels in the time that it takes for it to complete one period of its wave motion.

DEFINITION AND BASIC PRINCIPLES

Spectroscopy is the study of how light interacts with matter. It allows scientists in a broad array of fields to study the composition of both extremely large and extremely small systems. Each spectroscopic technique is unique; however, most of the widely used techniques are based on one of three phenomena: the absorption of light by matter, the emission of light by matter, or the scattering of light by matter. A photon can behave as both a particle and a wave. For most spectroscopic techniques, the wave nature of the photon is the most critical because the wavelength of light being emitted, absorbed, or scattered is where the information about the sample is contained.

Absorption spectroscopy involves the absorption of photons by matter and can give information about the types of atoms or bonds in a molecule. Typically, a given material will absorb specific wavelengths of light and will reflect or transmit all the other wavelengths. Emission spectroscopy involves the emission of photons from a sample on excitation. Scattering deals with light that is inelastically scattered from a sample, meaning that the wavelength of light bouncing off the sample is not the same as the wavelength of light that was shined on the sample.

Related disciplines include electron spectroscopy and mass spectroscopy. Electron spectroscopic techniques generally involve either irradiating a sample with light and causing the emission of photoelectrons or bombarding a sample with electrons and causing the emission of X rays. Either the photoelectrons or X rays can be detected to gain chemical information about a sample. Mass spectroscopy is unique in that various methods can be employed to liberate ions from a sample, the masses of which are then detected to gain a chemical fingerprint of the molecules that

651

are present.

BACKGROUND AND HISTORY

In 1666, Sir Isaac Newton became the first person to discover that ambient light could be separated into a continuous band of varying colors using a prism. This discovery was the beginning of all spectroscopic research. In 1803, Thomas Young performed his famous slit experiment demonstrating the wave nature of light. William Hyde Wollaston observed the first absorption bands in solar radiation in 1802; however, it was Joseph von Fraunhofer who assigned these lines (now known as Fraunhofer lines) significance in 1817, thus paving the way for the fields of spectroscopy and astrophysics. In 1848, French physicist Jean Bernard Leon Focault determined that the locations of the lines were characteristic of the elements that were present. Lord Rayleigh (John William Strutt) investigated the elastic scattering of light in 1871, and Albert Einstein followed in his footsteps in 1910. However, a wavelength shift caused by inelastic scattering of light was not observed until 1928, a discovery for which Sir Chandrasekhara Venkata Raman won the Nobel Prize in Physics. In 1960, Theodore Harold Maiman built the first successful optical laser, which would later prove essential for modern spectroscopic instruments.

HOW IT WORKS

Absorption Techniques. Fourier transform infrared spectroscopy (FTIR) is a vibrational spectroscopy technique that has the potential to elucidate what types of bonds are present within a sample. In transmittance mode, the sample is placed between an infrared light source and a detector and is irradiated with infrared light. Attenuated total reflectance (ATR) is a closely related technique that has minimal sample preparation. A sample is placed in contact with a crystal, such as diamond or germanium, and an evanescent infrared wave is sent through the crystal. Most bonds will absorb a specific wavelength of infrared light that causes them to vibrate or bend, and these wavelengths of light are absorbed in these processes and do not reach the detector. Polar bonds, such as carbonyls and ethers, are the bonds most easily detected using FTIR. The output of both FTIR and ATR is a spectrum with bands showing which wavelengths of light were absorbed by the sample. These wavelengths can then be interpreted to determine what types of bonds or functional groups were present in the sample.

Ultraviolet visible (UV-Vis) spectroscopy is similar to FTIR in that it involves shining light of a known wavelength range, in this case visible and ultraviolet, at a sample and determining which wavelengths are absorbed by the sample. UV-Vis does not provide as much specific information about bond character as FTIR; however, it is a very useful and fast technique for quantifying the concentration of a known analyte in solution according to Beer's law.

Atomic absorption spectroscopy (AAS) is a quantitative elemental analysis technique that relies on the reproducible absorption of specific wavelengths of light by a given element on excitation. A sample in solution is atomized by drawing the solution into a heat source and then irradiated with a specific wavelength of light selected to excite a particular element. The amount of light absorbed by the sample at that wavelength is used to calculate the concentration of the analyte in solution. AAS is very sensitive and is often used for detection of trace elements. Several types of heat sources, including an open flame, a graphite furnace, or an acetylene torch, can be employed to atomize the sample.

Astronomical spectroscopy is a method that astronomers use to determine the elemental composition of far-away celestial bodies that cannot be sampled in a laboratory. These bodies emit electromagnetic radiation across the entire electromagnetic spectrum; however, each element absorbs several characteristic wavelengths of light. So, by measuring which wavelengths of light are not being emitted by the very large sample in question, a fingerprint of all the elements present can be obtained.

Emission Techniques. Atomic emission spectroscopy (AES) is a technique that provides similar information to the information obtained using AAS but by essentially opposite means. In AES, a sample is atomized and excited by the same heat source, often a flame or a plasma, which is a highly ionized and energized gas. If a plasma is used, the technique is referred to as inductively coupled plasma AES (ICP-AES). The excited analyte will emit light at wavelengths that are characteristic of the elements that are present, so by detecting the emitted wavelengths of light and quantifying them, the elements in the sample can be identified and quantified. Multiple elements can be detected at once in AES.

Fluorescence spectroscopy is a family of techniques that uses higher energy radiation, such as ultraviolet light or X rays, to excite the atoms of a sample. The excited state is not stable, so atoms must release lower energy photons to relax back to a more energetically favored state. Depending on the type of incident radiation, information about the sample can be obtained to varying degrees of specificity. For some analyses, it is enough to induce fluorescence so that the emitted photons can be imaged, such as is the case in many biological applications. In other methods, such as X-ray fluorescence (XRF), specific elemental and chemical information can be obtained.

Scattering Techniques. Raman spectroscopy is a nondestructive vibrational spectroscopic technique based on inelastic light scattering. In Raman spectroscopy, a monochromatic laser is shined on the surface of a sample. Most of the light is elastically scattered, meaning that the light bounces off the sample at the same wavelength at which it entered; however, a very small fraction of the light is inelastically scattered, meaning that the light bounces off the sample at a different wavelength than that at which it entered because of interactions with the bonds in the sample. These interactions cause the bonds to vibrate and bend, similar to what happens in FTIR. The magnitude of the wavelength shift is characteristic of the bonds that caused it, and the resulting spectra are quite similar to FTIR spectra. However, whereas FTIR is most sensitive to polar bonds, Raman spectroscopy is most sensitive to polarizable bonds, such as carbon-carbon double bonds and aromatic rings.

Other Spectroscopic Methods. There are many other spectroscopic techniques that are in use across industry and academia; however, most are based on the same principles as the more common types. Examples of other techniques include circular dichroism, dynamic light scattering, and spectroscopic ellipsometry. One widely used spectroscopic method that is not based on conventional absorption, emission, or scattering methods is nuclear magnetic resonance (NMR). In organic chemistry, NMR is used to study the organization of bonds in a molecule and can give information about the number and location of hydrogens, carbons, or other elements being studied in a sample.

APPLICATIONS AND PRODUCTS

Spectroscopy is widely used in industrial and government-funded scientific endeavors for a variety of purposes, including pharmaceutical analysis, failure analysis, materials science, reverse engineering, and toxicology. Vibrational spectroscopy is widely used to aid in the identification of unknown materials. For example, an analytical chemist may use FTIR to identify foreign material found on a manufacturing line or in a finished product, materials used in a competitor's product, or unknown material recovered from a crime scene. Searchable libraries of vibrational spectra can be consulted in conjunction with spectral interpretation to determine the identity of the material in question. There are many commercially available libraries, but libraries can also be constructed from in-house samples and standard materials.

Raman spectroscopy is a nondestructive, surface-sensitive technique and therefore is often used to identify pigments in historical texts, art, and textiles to learn more about the people who created them and to verify their authenticity. Glass is largely transparent to Raman spectroscopy, making this technique ideal for identifying unknown materials found in glass containers without opening the container, thereby reducing the risks associated with exposure to an unknown material. This technique is especially useful in the identification of unknown flammable liquids. Ahura manufactures a portable Raman spectrometer for use in the field to allow emergency workers to evaluate potentially hazardous materials, such as chemical warfare agents or explosives. Techniques such as FTIR and Raman are used in airport screenings, as they are fast and reliable techniques for quickly identifying potentially harmful compounds.

Spectroscopic methods can also be used to quantify known materials. UV-Vis spectrometers can be used alone or as detectors in a chromatographic system in the pharmaceutical industry to quantify concentrations of drugs, biologics, and excipients in solution, as well as to verify the identity, purity, and stability of such compounds. Toxicologists use methods based on spectroscopy to quantify drugs in biological tissues and fluids.

Production facilities will often place spectroscopic instrumentation in line to continuously monitor products. For example, spectroscopic ellipsometry, which has the potential to measure thin films on the order of nanometers, can be placed in a production setting to monitor the thickness of vapor-deposited thin films.

NMR is typically used by synthetic chemists to determine or verify the structure of the compounds they synthesized, and it is often used in polymer science, drug chemistry, and materials science. In the medical community, NMR is used to image various types of soft tissue in situ and is better known as magnetic resonance imaging (MRI). Hospitals also use spectroscopy to measure blood counts, as well as coupling spectroscopy with immunoassays to screen for drugs or other compounds in urine.

Various elemental analysis techniques, such as AAS or AES, are often used along with Raman spectroscopy to analyze the composition of geological samples and meteorites. By understanding what elements and compounds are present in these samples, geologists can gain a better understanding of their formation and hypothesize about the processes that would create such a specimen. For example, researchers at Université de Lyon in France used Raman spectroscopy to identify two new types of ultrahard diamonds, one predicted and one unexpected, in the Havaro meteorite, which indicated that the parent asteroid had undergone an extremely violent collision.

Biologists use fluorescence spectroscopy to determine whether gene expression has occurred. By labeling the gene they wish to introduce into an organism with green fluorescent protein (GFP), they can visualize everywhere that the gene of interest was expressed. GFP can be visualized in cells, tissues, and organisms by exciting the GFP with blue light, which causes the emission of green light. Real-time polymerase chain reaction (PCR), a method for making many copies of DNA, often uses fluorescent probes that bind to the new double-stranded DNA. By using fluorescence spectroscopy to quantify the amount of fluorescence, biologists can monitor how much DNA they have synthesized.

Several spectroscopic techniques, such as Raman, FTIR, and NMR, can be used to chemically map out surfaces and volumes of materials. By taking individual spectra at various points across a sample, the intensity of a given signal or combination of signals can be tracked and mapped on an intensity scale. The resolution of these techniques, that is, the smallest features that can be resolved, depends on the wavelength of the light that is used. Thus, Raman

microscopy typically has much better resolution than FTIR because of the shorter wavelengths of light used in Raman spectroscopy. Chemical mapping can reveal details that would not otherwise be seen in a sample, such as where a specific protein may be expressed in a type of tissue or cell.

IMPACT ON INDUSTRY

Because of the wide variety of spectroscopic techniques and the even wider variety of applications, spectroscopy has had a far-reaching and diverse impact on industry. The majority of its impact comes from the power of the techniques to help scientists and engineers improve processes and solve problems rather than from marketing spectroscopic instrumentation to the general public.

Instrument Manufacturers. Many analytical instrumentation companies market spectroscopy systems and the publicly traded companies vary widely in terms of size and revenue. Most companies also market products that are not related to spectroscopy. For example, in 2009, Thermo Fisher Scientific employed about 35,000 people worldwide and had revenues of more than $10 billion, while Bio-Rad Laboratories employed 6,800 people worldwide and had revenues of nearly $1.8 billion. Bruker Corporation, which markets many spectroscopic systems, employed about 4,000 people in 2009 and earned revenue of $1.1 billion. Other large companies, such as Beckman Coulter, Agilent, Varian, Hitachi, and PerkinElmer, market spectroscopy-based systems. The HORIBA Group is a Japan-based global corporation with divisions that handle analytical equipment and medical diagnostic instruments and systems. Smaller private companies, such as Ocean Optics, an American company with between 51 and 200 employees in 2009, and WITec, a German company with 33 employees worldwide in 2009, have also successfully marketed spectroscopy systems. This list of companies is by no means exhaustive but is meant to provide a general sense of the companies involved in the spectroscopy world. Additionally, many academic and government groups choose to build their own state-of-the-art systems rather than purchase commercially available equipment.

Government Agencies. Most government agencies work with spectroscopic techniques. The national laboratories, such as Sandia, Lawrence Livermore,

Fascinating Facts About Spectroscopy

- Consumer products incorporating spectroscopy are becoming more prevalent, and one gadget Web site markets a keychain that detects and indicates the levels of ultraviolet light at a given moment.
- Environmental chemists often employ spectroscopy to check for levels of pollutants in water and air samples.
- Ocean Optics manufactures a line of handheld spectrometers that are ideal for performing analyses on the go.
- Any object or being that radiates heat, such as an animal, star, bonfire, or car engine, emits a spectrum of light.
- The wavelength range over the entire electromagnetic spectrum is 1,000 meters (longwaves) to 0.1 angstroms (gamma rays). The visible range of the spectrum lies between about 400 and 700 nanometers in wavelength.
- Sir Chandrasekhara Venkata Raman, the winner of the Nobel Prize in Physics in 1930, was so confident that his discovery of Raman scattering would win him the prize that he booked tickets to Stockholm even before the award winners were announced.
- Biologists amplify DNA in the presence of fluorescent markers, then use capillary electrophoresis with a fluorescence detector to determine the nucleic acid sequence of the DNA strand.
- Thomas Young's famous double slit experiment proved that light behaves as both a particle and a wave.

and Brookhaven, all perform research in spectroscopy. Brookhaven has a synchrotron facility and is using several beam lines for cutting-edge spectroscopic research. The National Institute of Standards and Technology (NIST) is involved in creating analysis methods and libraries of known compounds, as well as in performing research in related areas. Government agencies often collaborate with academic institutions on research and use commercial systems in everyday operations. For example, homeland security and military personnel use handheld Raman systems to identify possibly dangerous substances in the field.

Academic Research. Most graduate chemistry departments have at least one group focused on cutting-edge spectroscopy, and many other academic research groups use established spectroscopic methods to facilitate research in other areas. These programs can be found worldwide.

One interesting example of using established spectroscopic methods to assist in research in a field not related to spectroscopy involves researchers at the Saxelab Social Cognitive Neuroscience Laboratory and the Massachusetts Institute of Technology. At this laboratory, social cognitive researchers use a type of NMR known as functional magnetic resonance imaging (fMRI) to help them determine what areas of the brain are associated with various thought processes. These researchers are applying spectroscopy to investigate which areas of the brain are associated with the construction of abstract thoughts. One of their early findings was that one area of the brain specializes in thinking about what other people think.

Spectroscopy Societies and Conferences. Spectroscopy conferences and societies are a good way for spectroscopists to connect with one another to share ideas and research findings. Organizations such as the Society for Applied Spectroscopy have local chapters so that scientists can get to know other spectroscopists in their own geographical region.

Many conferences have sections on spectroscopy, including those held by the Materials Research Society, the American Chemical Society, the Federation of Analytical Chemistry and Spectroscopy Societies, and the Spectroscopy Society of Pittsburgh and the Society for Analytical Chemists of Pittsburgh. These conferences include presentations and posters on state-of-the-art research, as well as opportunities for networking with other people working in a similar field and interacting with instrument manufacturers and vendors. New instrumentation is often unveiled at larger meetings. Instrument vendors will often offer classes, Web seminars, and meetings for users of their instruments to network and share research.

CAREERS AND COURSE WORK

Students interested in spectroscopy have a wide variety of career paths from which to choose, and their final career goal should determine their course work. Completion of a bachelor of science degree in one of the core sciences such as chemistry or physics serves as preparation for graduate school, which, although not required in every case, greatly improves one's prospects of obtaining a career in spectroscopic research. A bachelor's degree qualifies an individual for a position at the level of a laboratory technician; such an employee is usually trained on the operation of a spectroscopic instrument and performs mainly routine tasks. A master's degree or doctorate and appropriate research experience qualify the holder for a position involving advanced use and maintenance of spectroscopic instruments; design of spectroscopy experiments; spectroscopic data analysis, interpretation, and communication; or possibly even the design of new instruments.

The careers that involve spectroscopy are as diverse as the spectroscopic methods themselves. Chemists, physicists, astronomers, materials scientists, chemical engineers, biologists, and geologists can all be spectroscopists. Careers involving spectroscopy can be found in industries such as pharmaceutical companies, in colleges and universities, and with the government, perhaps at a national laboratory or at a Federal Bureau of Investigation crime laboratory. Despite the diversity of these options, it is imperative that applicants possess a strong foundation in basic science, including chemistry and physics, as well as in mathematics. Course work in analytical and physical chemistry is highly pertinent to most careers involving spectroscopy.

SOCIAL CONTEXT AND FUTURE PROSPECTS

Spectroscopic techniques are powerful methods for identifying and sometimes quantifying components of a system, and applications of these techniques are likely to diversify further in the future. Handheld spectrometers probably will become commonplace in field forensic and military work because they allow personnel to identify the compounds with which they are dealing and to know what safety precautions to take. Spectrometers will also most likely find their way into everyday-use products and continue to play a significant role in space missions, such as looking for the presence of water or amino acids on distant planets.

An area of spectroscopy of interest is research using synchrotron light. Synchrotrons such as the Canadian Light Source and the European Synchrotron Radiation Facility use particle accelerators to produce ultra-high-intensity light of many wavelengths, which can then be diverted from the particle accelerator down a beam line and into a spectroscopy facility using a set of mirrors, gratings, attenuators, and lenses. Synchrotron light can penetrate deeper into materials and space, allowing information to be gained from farther away than before.

The resolution of spectroscopic chemical maps had been limited by physics, with maximum resolution being obtained using synchrotron light and operating right at the limit of diffraction. However, several commercial techniques are able to image below the diffraction limit of light. One example is the nanoIR by Anasys Instruments, which uses an atomic-force microscope tip to gain infrared information about very small areas of a sample. Typically, ATR resolution is on the order of 3 to 10 microns, but with nanoIR, resolution can be obtained at the submicron level. A great deal of research at universities, government-funded labs, and at spectroscopy companies is focusing on achieving better resolution so that smaller and smaller features can be accurately probed.

Lisa LaGoo, B.S., M.S.

FURTHER READING

Harris, Daniel C. *Quantitative Chemical Analysis.* New York: W. H. Freeman, 2003. An excellent resource for anyone interested in analytical chemistry in general; chapters 18 through 21 deal with spectroscopic techniques.

Hashemi, Ray H., William G. Bradley, and Christopher J. Lisanti. *MRI: The Basics.* 3d ed. Philadelphia: Lippincott, Williams & Wilkins, 2010. Written for the general public, this resource introduces the physics of MRI imaging.

Pavia, Donald L., et al. *Introduction to Spectroscopy.* 4th ed. Belmont, Calif.: Brooks Cole, 2009. Contains detailed information on interpreting spectra for basic techniques, such as FTIR, NMR, and UV spectroscopy.

Robertson, William C. *Light.* Arlington, Va.: National Science Teachers Association Press, 2003. Offers a solid background on the properties of light and electromagnetic energy.

Robinson, Keith. *Spectroscopy: The Key to the Stars—Reading the Lines in Stellar Spectra.* New York: Springer-Verlag, 2007. Presents information on astronomical spectroscopy in a nontechnical manner.

WEB SITES
Federation of Analytical Chemistry and Spectroscopy Societies
https://facss.org

MIT Spectroscopy
http://web.mit.edu/spectroscopy/index.html

Society for Applied Spectroscopy
http://www.s-a-s.org

See also: Magnetic Resonance Imaging; Radiology and Medical Imaging; Toxicology.

SPEECH THERAPY AND PHONIATRICS

FIELDS OF STUDY

Communication disorders; communication sciences; otorhinolaryngology; laryngology; logopedics; audiology; neurology; psychiatrics; pediatrics; dentistry; orthodontics; linguistics; acoustics; psychology; behavior sciences; pedagogics.

SUMMARY

Speech therapy and phoniatrics explore the therapeutic and medical aspects of improving the lives of people with communication disorders. Applications include games or activities designed to give a person more opportunities to practice and pronounce specific sounds correctly. Other communication disorders may require surgical intervention or botulinum toxin injections.

KEY TERMS AND CONCEPTS

- **Articulatory:** Pertaining to movement or contact of two or more articulators (lips, teeth, tongue, and palate).
- **Fluency:** Ability to combine sounds and words together naturally and without inappropriate pauses.
- **Intelligibility:** One's ability to speak in a comprehensible manner.
- **Intervention:** Plan to resolve or improve communication disorders.
- **Language Disorder:** Difficulty understanding or producing verbal communication.
- **Otorhinolaryngology:** Study and medical practice that treats the ears, nose, and throat.
- **Phoniatrics:** Medical evaluation and treatment of communication disorders.
- **Phonological:** Pertaining to sounds; a phonological communication disorder affects a person's ability to produce one or more speech sounds.
- **Speech Therapy:** Therapeutic specialty that encompasses various intervention methods used to improve speech and communication.

DEFINITION AND BASIC PRINCIPLES

Speech therapy and phoniatrics are two overlapping fields that evaluate, prevent, and treat communication disorders, which include any defect, disease, or deficit that compromises one's ability to use and understand spoken or written language. Communication disorders consist of any anatomical, physiological, or neurological disorder that negatively affects a person's voice, articulation, language fluency, respiratory system, vocal tract resonancy, use of language, quality of speech, or ability to hear and swallow.

Speech therapy is a method of improving speech and communication skills by altering physical, cognitive, and emotional behaviors and environments that influence the vocal tract during speech. The field has many specializations. Not all speech therapists work with people who have communication disorders. Some work with actors and singers to improve the quality and strength of their voices. Speech therapists, especially within the United States, are required to have a master's degree in the field.

Phoniatrics is the medical evaluation and treatment of any disease or disorder that affects communication. Phoniatricians are trained at medical facilities and have a comprehensive knowledge of many types of communication disorders. The field of phoniatrics is prevalent in Europe but not common in North America.

BACKGROUND AND HISTORY

The study of language and speech communication disorders originally derived from two medical disciplines: otorhinolaryngology (the study of the ears, nose, and throat) and internal medicine. A major breakthrough in understanding and exploring voice production occurred in the mid-1850's when Manuel Garcia, a Spanish voice teacher, became widely recognized as the first person to see his own vocal folds with the assistance of a dental mirror. Following Garcia's demonstration before the Royal Society of Medicine, medical interest in the larynx and pathology of the vocal folds increased. The ability to see the larynx and the vocal folds altered the world of laryngology and also general knowledge about the voice. Before that discovery, most diseases or disorders of the larynx had to be viewed and studied postmortem, after several such diseases had proven fatal.

Several decades later, in 1905, voice, speech, and language pathology was finally recognized as a separate academic discipline. One of the driving forces behind this movement was German physician Hermann Gutzmann, recognized as the father of voice and speech-language pathology.

The American Speech-Language-Hearing Association was founded in 1925. However, at the time, most American speech therapists were interested in communication disorders that dealt with aphasia, articulation, and fluency; there was little interest in disorders of the voice and larynx. In the late 1930's and early 1940's, the Nazi regime forced many European phoniatricians and voice experts to emigrate to the United States. These emigrants initiated U.S. interest in voice pathology, which became a more prominent field of study in the 1980's.

How It Works

Ingredients for Verbal Expression. Verbal communication requires both anatomical and neurophysiological processes used during respiration, phonation, resonation, articulation, and perception. To produce speech in English, air flows from the lungs and passes through the vocal tract; this is known as egressive airflow. Some languages use ingressive airflow (air external to the body flows into the oral cavity) to produce speech sounds. An example of this type of airflow is "tsk, tsk," which is used to communicate disapproval in English.

Changes in the vocal tract account for the different speech sounds humans can produce. These changes include positions of the vocal folds, the velum (soft palate), and the articulators (lips, teeth, tongue, and palate) within the oral cavity. The airstream flowing from the lungs is first modified in the larynx, where the vocal folds are located. If the vocal folds are open, the air will pass through freely; this occurs when a person breathes or pronounces a voiceless sound such as /p/ or /t/. If the vocal folds are pressed loosely together, air will pass through them, causing the vocal folds to vibrate and create a voiced sound such as /b/ or /d/. Complete closure of the vocal folds creates a glottal stop—the airflow is completely stopped, as heard in the pause between "uh" and "oh" when saying "uh-oh." Additional vocal fold alterations produce vocal qualities that are breathy or creaky. These vocal states are distinguishing speech features in some languages.

Next, the position of the velum or soft palate determines the nasality of an utterance. When the velum is pressed against the back wall of the pharynx, airflow is blocked from traveling up the nasal cavity. Air then flows out of the mouth, creating oral sounds such as /p/. Nasal sounds, such as /m/ and /n/, occur when the velum is lowered and air flows through both the oral and the nasal cavities but mainly exits out of the nostrils.

Finally, the placement of articulators modifies the speech airstream before it exits the vocal tract. To create the sounds /p/ and /b/, one presses the lips together and then quickly releases them. To create the sound /s/, one presses the tip (or the front of the tongue) up against the back of the lower teeth.

People rarely make single sounds separate from each other unless they are practicing speech exercises or telling someone to "Shhh." Verbal expression requires users to blend and sequence different sounds together to create words; those words are then sequenced to form larger utterances and sentences.

Being able to do all of this relies on language-specific cognitive abilities and many physiological systems, including the skeletal, articular, muscular, digestive, vascular, nervous, and respiratory systems. Because language is so complex, many things can go wrong. This is when a speech therapist or phoniatrician comes in.

Determining an Individual Intervention Plan. When there is a communication issue, the speech therapist or phoniatrician needs to create an individual intervention plan. Evaluation of a person's communication skills, including sounds that are present and absent, helps determine the type of speech problem, the anatomical location of the problem, and also any communication strengths. This information can then be used as a baseline to compare against future speech. It is important to note that the absence of performing a communication skill does not mean the skill cannot be performed.

Interviews, observations, and statistically sound tests give the therapist insight into the communication disorder being addressed. Information gathered from and collaboration with the individual and his or her family is important to determining intervention goals and creating a care plan that is culturally sensitive and unique to the person.

Developing and Implementing Interventions. Therapy and intervention largely strive to improve a person's intelligibility, accuracy in making speech gestures, ability to use communication correctly, and ability to monitor his or her own verbal behaviors (actions needed to formulate speech) and outputs.

The best interventions or care plans are based on an individual's strengths and goals to improve communication. In addition to knowledge of the client and his or her values or needs, therapists use knowledge of typical and disordered communication—how communication disorders affect typical communication, knowledge of scientific research on various treatment methods, and knowledge gained from previous experience working with people who have similar disorders.

Case studies, research from similar fields (including psychology, medicine, and linguistics), knowledge of child development, and knowledge of language processes help determine potential intervention methods.

APPLICATIONS AND PRODUCTS

Therapeutic Approaches to Intervention. Many approaches to intervention are based on different aspects of child development and theories on language processing. Three major approaches are used: behaviorist, linguistic-cognitive, and social-interactionist.

The behaviorist approach is based on operant conditioning, in which a behavior is modified by rewards or consequences. Rewards, such as stickers, toys, and praise, are used during speech therapy to encourage young clients to reach the target communication skill.

When a patient is taught general linguistic rules, the linguistic-cognitive approach is used. For example, past tense in English is most often marked by the suffix "-ed." This approach emphasizes learning through active interaction with the environment. The client takes an active role, and the therapist simply provides different environments for the person to practice and learn linguistic rules.

The importance of socialization is emphasized in the social-interactionist approach, particularly scaffolding or learning from a more experienced person. The therapist carefully monitors what the client can do on his or her own and what the person can do with the help of others to offer only necessary assistance.

Medical Approaches to Intervention. Medical intervention methods for communication disorders include prescription drugs, injections (usually to the vocal folds), and surgery performed by an otorhinolaryngologist or laryngologist.

Application of Therapeutic and Medical Approaches. Various therapies and medical approaches are used to aid communication disorders depending on the underlying conditions.

People affected by dementia often have trouble naming objects, people, and sounds. Medications given to improve memory can also improve this related speech disorder. Speech therapy is also beneficial; the client can practice communication skills, and family and friends can be taught how to give subtle verbal cues to the client without causing added frustration.

Treatment for reading disabilities emphasizes phonological awareness and development. Verbal, visual, and hands-on games and activities help the client break down both spoken and written words into syllables and individual sounds and letters. Educators, reading specialists, and speech therapists use phonics to improve the individual's ability to recognize different sounds and letters.

Therapies for children with phonological or articulatory communication disorders also emphasize phonological awareness. Speech and communication specialists will concentrate on making the speaker aware of minimal pairs or word pairs, such as "bad" and "tad," which have only one different sound, in this case the /b/ and /t/. The person realizes that the intended meaning can be easily misconstrued if he or she does not properly pronounce each individual sound. Verbal and intraoral tactile cues are given to assist those who cannot pronounce specific sounds.

Like all forms of therapy, voice therapy is based on setting and achieving short- and long-term goals. A few goals of voice therapy are to strengthen vocal behaviors that have or potentially can have a negative effect on the voice, to teach and promote appropriate vocal behavior such as breathing properly and speaking at a healthy volume, and to teach the person different vocal behaviors, for example, speaking at an audible volume. Exercises in breathing, loudness reduction, and relaxation teach people how to use their voices appropriately. Various computer programs can identify incorrect speech behaviors and also measure improvement over time.

Fascinating Facts About Speech Therapy and Phoniatrics

- Julie Andrews lost her singing voice in 1997 after a surgical attempt to remove vocal nodules damaged the elasticity of her vocal folds. Vocal nodules are a result of overusing one's vocal folds.
- Although language activity is predominantly found in the left hemisphere of the brain, positive emission tomography (PET) and single photon emission computed tomography (SPECT) studies have found that people who stutter show greater activity in the right hemisphere of the brain, even after the stutter has been resolved.
- About 12 percent of children have a language disorder that affects their ability to either understand or use language.
- Around 10 percent of children have an articulatory or phonological communication disorder, making it difficult for them to produce specific speech sounds.
- A child's speech becomes difficult to understand when a speech error is present more than 40 percent of the time.
- About 1 in every 700 babies is born with a cleft palate, a disorder in which the infant's palate is not fully fused together. Speech produced by children with a cleft palate often has a nasal quality because the child's velum or soft palate is often too small or weak to push against the back of the pharyngeal wall and block off the nasal cavity.

Medications. Spasmodic dysphonia negatively affects movement of the vocal folds. Routine botulinum toxin (Botox) injections are a common treatment for this disorder. These injections temporarily and partially relieve the problems by weakening the laryngeal muscles so that they have partial movement. The injections make it easier for the person to use verbal expression, although the voice may not sound "normal." On the negative side, the person has to receive an injection every few months.

Injections are also used to help people with unilateral or one-sided paralysis of the vocal folds. Because individuals who have vocal fold paralysis sometimes experience spontaneous recovery, injections are usually not given for the first six to nine months. In addition to botulinum toxin, specialists also inject gelfoam paste, fat, bovine collagen, or polydimethylsiloxane (a substance originally used in plastic surgery and urology). In the United States, laryngeal injections are performed by otorhinolaryngology or laryngology specialists.

Surgical Options. Phonosurgery is the general term used to describe surgical treatment of voice disorders. A common surgery, thryoplasty, reconstructs the thyroid cartilage to change the tension and position of the vocal folds. Another surgery, laryngectomy, involves the removal of the larynx, which is where the vocal folds are located. People who have had this surgery must use a device to speak. Another procedure, called a tracheosophageal puncture, connects the trachea to the esophagus with a small device. To speak, the person manipulates air within the esophagus instead of the larynx. Muscular action compresses the air in the esophagus and forces it through the top of the closed esophagus. As a result, the pharyngoesophageal region vibrates to create voicing.

IMPACT ON INDUSTRY

Worldwide Impact. The United States and various other countries, including Canada, Singapore, and New Zealand, emphasize speech therapy to treat communication disorders. Europe, however, remains the world leader in phoniatrics. Several other regions of the world—including Africa (Egypt); Asia (India, Indonesia, and Japan); Australia; Latin and South America (Brazil, Jamaica, and Mexico); and the Middle East (Israel)—are in the process of developing their own phoniatrics specialties, which may or may not include an interdisciplinary approach with otorhinolaryngologists and speech therapists.

Government Agencies and University Research Centers. The National Institute on Deafness and Other Communications Disorders (NIDCD) is under the umbrella of the National Institutes of Health. The NIDCD emphasizes public education on speech disorders in addition to research studies in areas such as the genetics of childhood language disorders, how the brain controls language, and computer-assisted devices. European funding for phoniatrics research occurs at both the national and the local levels. Research funding has focused on numerous categories including acoustic correlates of vocal function and dysfunction, student programs, and the needs of underdeveloped countries.

There are nearly seven hundred speech therapy and phoniatric training programs worldwide. The United States alone has three hundred programs at the university level. However, India has only twenty programs, and China has just a handful.

Commercial Endeavors. The field of biotech communications continues to evolve, and computer-assisted speech therapy aids are becoming more commonplace. The devices can help people with communication disorders understand sounds; develop perception of pitch, volume, or even voice; and assist in breath control or fluency during speech production. These devices, which have proven useful to people of all ages, are meant to be used with a trained therapist.

Professional Associations. The American Speech-Language-Hearing Association (ASHA) is the largest American organization for speech therapists. ASHA is part of the International Association of Logopedics and Phoniatrics (IALP), based in the Netherlands. IALP is a nonpolitical worldwide organization of professionals and scientists in communication voice, speech-language pathology, audiology, and swallowing. It holds a congress every three years and also heads dozens of other phoniatric and speech-language societies throughout the world.

Europe has several other professional societies overseeing the phoniatric field. The Union of the European Phoniatricians (UEP) meets annually to promote phoniatrics and the professional conditions of phoniatricians in public health service and private practice. The UEP is part of the European Academy of Otorhinolaryngology and Head and Neck Surgery. The European Laryngological Society is influential in promoting phoniatrics. The International Affairs Association and Communication Therapy International both emphasize the need to help countries with disadvantaged populations.

CAREERS AND COURSE WORK

To become a speech therapist in the United States, one needs to obtain a master's degree in the field. In many other countries, a bachelor's degree is accepted. Basic course work includes topics in child development, linguistics, psychology, physics and acoustics, physiology, anatomy, and neurology. Other qualifications vary by state. Many states require speech therapists to complete a minimum of four hundred hours of clinical experience, a thirty-six-week clinical

fellowship, and the speech therapy Praxis exam to earn a clinical competence certificate (C.C.C.).

Related professions include otorhinolaryngology, laryngology, neurology, augmentative and alternative communication, assistive technology, and audiology (ear and hearing specialist). Augmentative and alternative communication and assistive technology use communication devices for people who cannot express themselves intelligibly (comprehensively) through verbal speech or have severe difficulty doing so. Augmentative and alternative communication users include people with amyotrophic lateral sclerosis, autism spectrum disorder, or cerebral palsy.

SOCIAL CONTEXT AND FUTURE PROSPECTS

Like all medical specialists, speech therapists and phoniatricians are constantly looking for new, more efficient therapies and treatments. Scientists and researchers also regularly reevaluate the effectiveness of treatments.

However, intervention, especially with the vocal folds, does not make speech and communication perfect. Even after treatment, many people with communication and voice disorders will still have distinct and possibly unusual vocal qualities. However, as technology and science evolves, so will this field, thereby creating more opportunities for resolving communication hardships.

Holly Nyple, B.A. and Renée Euchner, B.S.R.N.

FURTHER READING

Bunning, Karen. *Speech and Language Therapy Intervention: Frameworks and Processes.* London: Whurr, 2004. Examines topics such as the construction, enactment, and processes of intervention. Discusses cultural aspects and how to enhance communication with the patient.

Colton, Raymond H., and Rebecca Leonard. *Understanding Voice Problems: A Physiological Perspective for Diagnosis and Treatment.* 4th ed. Philadelphia: Lippincott Williams & Wilkins, 2010. Examines the psychological aspects of speech problems, particularly the effect of behavioral and emotional factors.

Freeman, Margaret, and Margaret Fawcus, eds. *Voice Disorders and Their Management.* 3d ed. Philadelphia: Whurr, 2001. Each chapter is written by clinicians and looks at the physiological issues as well as the social and psychosocial consequences.

Gordon-Brannan, Mary E., and Curtis E. Weiss. *Clinical Management of Articulatory and Phonological Disorders*. 3d ed. Baltimore: Lippincott Williams & Wilkins, 2007. Textbook on articulation and phonological development that examines the treatment and management of disorders.

Haynes, William O., and Rebekah H. Pindzola. *Diagnosis and Evaluation in Speech Pathology*. 7th ed. Boston: Pearson Education, 2008. Provides information about different factors that may play a role in assessing communication disorders.

Justice, Laura M. *Communication Sciences and Disorders: A Contemporary Perspective*. 2d ed. Boston: Allyn & Bacon, 2010. Comprehensive discussion of different communication disorders, including basic information about the disorder, assessment methods, and treatment methods.

Ladefoged, Peter. *A Course in Phonetics*. 5th ed. Boston: Thomson Wadsworth, 2006. Offers a detailed and easy-to-follow description of phonetics and how individual sounds are produced.

Paul, Rhea, and Paul W. Cascella, eds. *Introduction to Clinical Methods in Communication Disorders*. 2d ed. Baltimore: Paul H. Brookes, 2007. Provides information on assessing and working with individuals who have communication disorders.

WEB SITES

American Speech-Hearing-Language Association
http://www.asha.org

European Academy of Otorhinolaryngology and Head and Neck Surgery
http://www.eaorl-hns.org

International Association of Logopedics and Phoniatrics
http://ialp.info/joomla

National Institute on Deafness and Communication Disorders
http://www.nidcd.nih.gov

Union of the European Phoniatricans
http://www.phoniatrics-uep.org

See also: Audiology and Hearing Aids; Occupational Health; Otorhinolaryngology; Pediatric Medicine and Surgery.

STEM CELL RESEARCH AND TECHNOLOGY

FIELDS OF STUDY

Biotechnology; cell and molecular biology; transplantation; genetic engineering; cell and tissue transplantation; cloning; cell-based therapies; regenerative medicine.

SUMMARY

Stem cell research is the field of science that examines specific cells that have the ability to divide indefinitely in culture and that give rise to specialized cells in order to provide therapy for diseases. There are two main types of stem cells: embryonic and somatic stem cells. Embryonic stem cells are formed in the early stages of embryonic development, and somatic stem cells are adult stem cells found in various tissues in the body. Stem cells have the potential to be used as therapy to replace or repair a person's cells or tissues that are damaged or dysfunctional in the treatment or cure of diseases.

KEY TERMS AND CONCEPTS

- **Blastocyst:** Embryo in a very early stage of development, produced by cell division of a zygote (fertilized egg); consists of about 150 cells in a spherical cell mass of two regions, the inner cell mass and the trophoblast.
- **Cloning:** Process of producing one or more genetically identical copies of a cell, tissue, or organism, by either natural means, such as cell division (mitosis), or artificial means, such as through an in vitro laboratory setting.
- **Differentiation:** Process of development in which cells change their complexity to acquire a specialized function.
- **Embryonic Stem Cell:** Cell derived from the inner cell mass at the blastocyst stage in the development of an embryo.
- **Induced Pluripotent Stem Cell:** Cell that is genetically reprogrammed to be induced to express genes and factors to maintain cells in a stem cell line state.
- **Multipotent Stem Cell:** Stem cell that has the potential to become many different cell types in an organism's body.

- **Pluripotent Cell:** Stem cell that has the potential to become any type of cell in an organism's body.
- **Somatic Cell Nuclear Transfer:** Technique that removes the nucleus of a somatic cell, which is then injected or transferred into an egg (that has had its nucleus removed), which will be implanted into the womb of another individual to be born as a clone.
- **Somatic Stem Cell:** Stem cell found in the non-germ-line (egg and sperm) tissues of an individual that remains undifferentiated and can give rise to specialized cell types of the tissue from which it is derived; also known as adult stem cells.

DEFINITION AND BASIC PRINCIPLES

Stem cells have the basic properties of being undifferentiated cells that can divide indefinitely and have the potential to develop into many different types of cells in a body during early embryogenesis and during growth of an individual. Stem cells are different from other cells in the body in that they can renew themselves through cell division, allowing them to act as a repair mechanism and to replenish cells that are damaged or that die. When each stem cell divides, it has the potential of either remaining a stem cell or becoming another cell type with a more specialized function.

There are two main types of stem cells: embryonic and somatic (also called adult stem cells). Embryonic stem cells are pluripotent, in that they have the capability of becoming any type of cell in the body. This is because these cells arise from the blastocyst, early in embryogenesis, making up the inner mass of cells. The inner cell mass gives rise to the entire body of an organism, including all the specialized cell types and organs, such as the heart, muscle, brain, skin, and other tissues. Somatic stem cells are considered to be multipotent and are found only in specialized tissues in the body, which are specific populations of cells that are used to generate replacements for cells that are damaged or die through the normal aging process of cells and because of injury or disease.

Stem cell therapy uses stem cells to replace or repair a patient's cells or tissues that are damaged or missing. Stem cell therapy is still experimental, in that it has not yet proven to be effective or safe, but stem cells have the potential to treat many diseases.

Background and History

Embryonic stem cells were first studied in the mouse in 1981, when scientists discovered ways to derive embryonic stem cells from mouse embryos. This led to the discovery, in 1998, of a method to derive stem cells from human embryos and grow them in the laboratory. However, the use of human embryonic stem cells—taken from embryos that were originally created for reproductive purposes—has been limited because a number of stem cell lines have been allowed to be grown in the laboratory for research purposes. This constraint led to further discoveries of how to derive stem cells from somatic tissues. In 2006, scientists discovered conditions that would allow these specialized tissue stem cells to be reprogrammed genetically to become pluripotent. These stem cells, reprogrammed to express certain genes or maintain these cells in a stem cell-like state, are called induced pluripotent stem cells. In March, 2009, the ban on generating new stem cell lines was lifted by President Barack Obama, making federal funding for embryonic stem cell research available without the previous limits on the stem cell lines generated.

How It Works

To identify stem cells, cells first have to be grown in the laboratory, or cultured. The first step in isolating stem cells is to transfer the inner cell mass of a blastocyst into a cultural medium in a laboratory dish. The culture medium contains nutrients that cells need to grow and divide. Stem cells do not always grow, but when the cells continue to grow and divide, they are then divided into other culture dishes, called subculturing, so that millions of copies of the same stem cell (cloning) can be used for research.

Embryonic Stem Cells. Embryonic stem cells are the easiest of stem cells to divide and reproduce in culture, and they have been shown to live for months without differentiating. When these cells continue in their stem cell state, they are considered pluripotent and have the same genetic makeup as the original stem cells from the inner cell mass. These cells are referred to as an embryonic stem cell line. These cells may be frozen and shipped to other laboratories for further culturing and experimentation.

Somatic Stem Cells. Somatic (adult) stem cells are undifferentiated cells that are found in a tissue or organ that can renew themselves and differentiate to become specialized cells of that tissue or organ. Adult stem cells are used to regenerate or repair the tissue in which they are located. Known somatic stem cells are located in the brain, bone marrow, peripheral blood and blood vessels, muscles, skin, teeth, heart, liver, ovarian epithelium, and testes. To be used as a somatic stem cell, these cells need to demonstrate that they can generate a line of genetically identical cells that can give rise to all the differentiated cell types of that tissue. Once these cells are identified, they can be used to regenerate and repair cells within that tissue. Experiments are ongoing in transdifferentiation, in which certain somatic stem cells are reprogrammed into other cell types or even to become like embryonic stem cells, called induced pluripotent stem cells, with the introduction of embryonic cells.

Applications and Products

There are several reasons why stem cells are important in science and the advancement of health care.

Cell Specialization and Development. Pluripotent stem cells help scientists understand the complexity of human development and how genes work to make decisions so that cells differentiate to become specialized cells. As development proceeds from an embryo to an individual human, genes turn on and off to give rise to protein expression and cell differentiation. These decision-making genes control the expression of pluripotent stem cells. Scientists know that certain diseases, such as cancer and birth defects, are caused by abnormal cell division and cell specialization. Understanding normal cell development will allow scientists to determine the errors that cause debilitating and often lethal diseases.

Medical Drug Testing. Stem cell research potentially may change the way new medical drugs are developed and tested for safety. These new drugs can be tested on stem cell lines first. Using pluripotent stem lines will expand the cell types that can be tested in the laboratory, before a drug is tested on animals and humans, streamlining the process for drug development.

Cell Therapies. Stem cells have the potential to be used to generate cells and tissues that could replace or regenerate damaged cells and tissues in humans. Such cell therapies could help treat disorders that disrupt cell function or destroy tissues, such as cancer, heart disease, diabetes, spinal cord injury, arthritis, Parkinson's disease, and Alzheimer's

Fascinating Facts About Stem Cell Research and Technology

- Clinical trials began in 2010 on a stem cell therapy to restore spinal cord function. Oligodendrocyte progenitor cells derived from human stem cells were to be injected directly into the patient's damaged spine.
- In 2010, the biotech company ACT was preparing to enter into clinical trials of retinal cells derived from stem cells in the treatment of Stargardt's macular dystrophy.
- Scientists in 2010 used a combination of three transcription factors to directly reprogram postnatal mouse heart or skin fibroblasts into differentiated cells that shared many features of heart muscle cells, including beating. Scientists hope to use such cells in heart repair.
- Qingdao University began conducting clinical trials in 2010 of umbilical cord mesenchymal stem cells to treat ulcerative colitis.
- Muscle stem cells have proved difficult to grow. However, in 2010, scientists reported that making the culture conditions mimic the physical properties of the muscle improved cell growth.
- University of Rotterdam scientists became the first to film the birth of blood stem cells in 2010. From this knowledge, the Dutch scientists hope to develop a technique to grow blood stem cells in the laboratory.
- Scientists used induced pluripotent stem cells (rather than embryonic stem cells) in 2010 to produce neural stem cells and dopaminergic neurons, which improved the behavior of mice with a condition similar to Parkinson's disease. Neurons from induced pluripotent and embryonic stem cells were similar in their gene expression but not identical.
- In human embryonic stem cells from people with Fragile X Syndrome, the *FMR1* gene is expressed normally until the cells begin to differentiate, when it is silenced. However, in induced pluripotent stem cells from people with Fragile X, the gene is silenced before differentiation, demonstrating that the two types of stem cells can have significant differences.

diseases. Modern medicine relies on donated organs and tissues to replace destroyed tissue in heart, bone marrow, and kidney transplants. However, the number of people suffering from these disorders far outnumbers the organs and cells available. Stem cells offer a unique opportunity to create a renewable source of replacement cells and tissues to treat these diseases. Another problem in the transplant process is that the recipient's body tends to reject the foreign cells from the donor. With stem cells, research could focus on developing modifications to these cells to minimize tissue incompatibility or to create tissue banks with common tissue type profiles that would be accepted by a large number of individuals.

Somatic Cell Nuclear Transfer. The technique called somatic cell nuclear transfer is still in the research stage, and no human stem cell lines have been created using it. In somatic cell nuclear transfer, the nucleus of virtually any somatic cell is taken from an individual patient and fused with a donor egg cell from which the nucleus has been removed. That cell is then stimulated to develop into a blastocyst and the inner cell mass is taken to create a culture of pluripotent stem cells. These stem cells can be stimulated to develop into specialized cells that are needed to repair damaged tissues or organs. Because the genetic information is taken from the individual patient, these cells theoretically would not be rejected by the patient as they are genetically identical to those of the individual. This type of transplantation would not require immune-suppressing drugs to be successful, and patients would have a far greater chance of survival.

Somatic Stem Cell Therapies. There are disadvantages and advantages to using somatic stem cells for therapies. One disadvantage is that these stem cells are multipotent but not pluripotent, and the types of cells that can be developed are limited. Previously, it was thought that somatic stem cells could develop into only the specialized cells from which they were derived, making it necessary to use only bone marrow stem cells for bone marrow transplantation, liver stem cells for liver diseases, and so on. However, experiments on mice have shown that, for example, when neural stem cells were placed into bone marrow, a variety of blood cell types were produced. So it is possible that even specialized stem cells may be manipulated to be wider reaching in their potential than previously thought. However, the biggest limitation of somatic stem cells to date is that they have not been isolated from all the tissues of the body. So far, it has not been possible to locate adult cardiac stem cells or pancreatic islet stem cells in humans, which would help in heart disease or diabetes, respectively.

Transplantation. One advantage of somatic stem cells is in transplantation. If these cells could be isolated from a patient and directed to divide and specialize in a manner that conveys normal cell function, they could then be transplanted back into the patient without immune rejection. This would also reduce or avoid the need for embryonic stem cells from human embryos or human fetal tissue. However, isolating somatic stem cells and growing them in culture has been difficult. Even if it becomes possible, growing and manipulating them quickly enough to correct a disease state may be impossible. Rigorous research will be required to overcome the obstacles of this type of cell therapy.

IMPACT ON INDUSTRY

Government. The National Institutes of Health has been a leader in stem cell research since the beginning in the 1980's. However, controversy over the use of human embryos and fetal tissue for generating stem cell lines led to a presidential executive order by President George W. Bush. The order, effective from August 9, 2001, to March 9, 2009, allowed federal funding for research using human embryonic stem cells only if the process of creating the stem cell line has begun before August 9, 2001, the stem cells were from an embryo that was created for reproductive purposes and was no longer needed, and informed consent was obtained for the donation of the embryo and that donation did not involve financial inducements.

To meet these criteria, research investigators from fourteen laboratories in the United States, India, Israel, Singapore, Sweden, and South Korea derived stem cells from seventy-one individual, genetically diverse blastocysts. These cell lines were the only ones that U.S. researchers were allowed to use if they received any National Institutes of Health funding for their research.

On March 9, 2009, President Barack Obama issued an executive order that permitted the National Institutes of Health to "support and conduct Institutes of Health to "support and conduct responsible, scientifically worthy human stem cell research, including human embryonic stem cell (hESC) research, to the

extent permitted by law." This order expanded U.S. policies to fully fund human embryonic stem cell research as long it was conducted in an ethically and scientifically responsible way.

The United States has developed stem cell research that has become quite extensive, including research in many institutes within the National Institutes of Health. These include the National Heart, Lung and Blood Institute; National Institue of Dental and Craniofacial Research; National Institute of General Medicine Sciences; National Institue of Diabetes and Digestive and Kidney Diseases; and the National Institute of Neurological Disorders and Stroke. All these institutes can potentially use stem cells in treating their respective diseases.

Other government agencies, such as the California Institute for Regenerative Medicine, the International Society for Stem Cell Research, and European and other international governmental agencies have been prominent in research and education regarding stem cells and their potential education regarding stem cells and their potential uses.

University Research. Most major universities throughout the world are involved in stem cell research, and researchers publish articles in scientific journals on a daily basis. Research uses both animal models and human embryonic and somatic stem cells to look at ways that these cells may improve the health of people with various diseases or conditions. Because most research is funded by governmental agencies or private institutes, it generally needs to be peer-reviewed and accepted by the scientific community. Such work is also presented at scientific meetings and subject to reproducibility by other researchers before it is accepted as plausible. This rigorous approach to scientific advancement has led to major breakthroughs in all fields of research and has the potential for reaching the overarching goals of stem cell research.

Business Sector. Products and services are now being offered by a wide range of for-profit public and private organizations to further develop the field of stem cell research. Certainly, to develop strategies for conducting stem cell research, many products must be developed, and some institutions rely on the business sector to do testing and research. This has led to further progress in the field and an extension of services to the scientific community.

CAREERS AND COURSE WORK

The field of stem cell research is growing, and careers in the field can be as diverse as medical doctors, doctorate level researchers, and laboratory technologists that have a bachelor's degree in biology or a related field. The number of laboratories that conduct stem cell research is getting larger, and funding for this type of research has expanded.

A bachelor's degree is required even for low-level positions in a research laboratory. A master's degree in a biological field may help in the competitive job market. A doctorate degree is essential to be involved in making decisions about the type of research and the funding received, as well as to direct a laboratory, including technologists. Medical doctors often participate in medical research at an institution and are also involved in the patient aspect of the research and how it is applied in the clinical setting.

In the United States, the National Institutes of Health supports short-term training courses in human embryonic stem cell culture techniques. These training courses include hands-on experience to improve the knowledge and skills of biomedical researchers to maintain, characterize, and use human embryonic stem cells in basic research studies. The courses are given at various locations throughout the United States.

SOCIAL CONTEXT AND FUTURE PROSPECTS

Acceptance of stem cell research has been greatly expanded with publicity regarding the potential benefits of this type of research. Also, there have been numerous scientific publications leading to advances in the field. However, considerable controversy remains regarding the ethical implications of using embryonic stem cells. In the United States, much debate has centered on the use of human embryos and fetal tissue created for reproductive use. Although these embryos are no longer needed, using them for research means they no longer can be used to produce a viable individual. The morals and ethics of this continue to be debated.

However, stem cell research is not limited by the availability of embryonic stem cells. Alternative stem cells, such as somatic stem cells and induced stem cells, have been developed, and researchers may be able to use these instead. The competitive nature of scientific endeavors has led to the advancement of all fields of science, and continued work in this field has produced further success in the use and potential of stem cells as a source of eliminating the threat of some of the most deadly human diseases.

Susan M. Zneimer, Ph.D.

FURTHER READING

Fox, Cynthia. *Cell of Cells: The Global Race to Capture and Control the Stem Cell.* New York: W. W. Norton, 2007. Looks at the competition that arises among researchers as they attempt to find applications for stem cell therapy.

Haerens, Margaret, ed. *Embryonic and Adult Stem Cells.* Detroit: Green Haven Press, 2009. Contains a collection of essays arguing the pros and cons of stem cell research.

Humber, James M., and Robert F. Almeder, eds. *Biomedical Ethics Reviews: Stem Cell Research.* Totowa, N.J.: Humana Press, 2004. A collection of objective essays reviewing the principle arguments for and against stem cell research and whether this type of work violates the rights of human embryos.

Panno, Joseph. *Stem Cell Research: Medical Applications and Ethical Controversy.* New York: Facts On File, 2004. Provides information on the technological advances, applications, and issues of stem cell research, including the use of stem cells to repair damaged nerve tissue and the ethical and legal implications of research in this field.

Sell, Stewart, ed. *Stem Cells Handbook.* Totowa, N.J.: Humana Press, 2007. Explains the origins of stem cells and describes how they function and how they can treat illness and disease. Emphasis is placed on the role of stem cells in development, tissue regeneration, repair mechanisms, and carcinogenesis. Also includes technical approaches to obtaining stem cells and manipulating them for therapeutic use.

Wobus, A. M., and K. R. Boheler, eds. *Stem Cells.* New York: Springer, 2006. Presents many novel aspects of stem cell biology, including existing and future applications in research and medicine, particularly uses in drug therapies.

WEB SITES

California Institute for Regenerative Medicine
http://cirm.ca.gov

International Society for Stem Cell Research
http://www.isscr.org/public

National Institutes of Health
Stem Cell Information
http://stemcells.nih.gov

See also: Bioengineering; Cloning; Genetic Engineering; Human Genetic Engineering.

SUPERCONDUCTIVITY AND SUPERCONDUCTING DEVICES

FIELDS OF STUDY

Electrical engineering; physics; solid-state physics; materials science and engineering; cryogenics; physical chemistry; thermodynamics; computer science and engineering; advanced mathematics; quantum mechanics; electricity and magnetism; X-ray crystallography; electronic ceramics; aeronautics, space science, and engineering; railway engineering; photolithography; military engineering.

SUMMARY

Superconductivity is an electrical phenomenon in which current flows without resistance in certain metals, alloys, and ceramics at very low temperatures. Low-temperature superconductors exhibit their characteristic zero electrical resistance and perfect diamagnetism at temperatures close to absolute zero, and high-temperature superconductors manifest these properties at temperatures from 23 Kelvin (K) to more than 135 K. Major applications of superconductors include generating the powerful magnetic fields for magnetic resonance imaging (MRI) and nuclear magnetic resonance (NMR) machines. They have been used in making extremely sensitive magnetometers that are able to measure magnetic fields a hundred billion times weaker than the Earth's. Superconducting magnets have appeared in transportation ("levitating" trains), particle accelerators, and a variety of industrial and military applications. Possible future applications include quantum computing, electric power generation and transmission, refrigeration, and various nanotechnology devices.

KEY TERMS AND CONCEPTS

- **BCS Theory:** Explanation of superconductivity first proposed by American physicists John Bardeen, Leon N. Cooper, and John R. Schrieffer (for which they won the 1972 Nobel Prize in Physics). This theory shows how interactions between electrons and atoms in a lattice result in the paired electrons responsible for zero electrical resistance.
- **Cooper Pair:** At very low temperatures, two electrons in an atomic lattice can consolidate and thus account for superconductivity; named for Leon Cooper.
- **Diamagnetism:** Phenomenon in which certain materials, including superconductors, exclude magnetic fields from their interiors.
- **Electron-Phonon Interaction:** Interplay between electrons and crystal-lattice vibrations that accounts for both ordinary resistance and the zero resistance of superconductivity.
- **Josephson Junction:** Insulating barrier separating two superconductors through which individual and paired electrons can travel in a process called tunneling.
- **Lattice:** Orderly arrangement of atoms in a crystal.
- **Meissner Effect:** Phenomenon of a material expelling magnetic lines of force when it transitions to the superconducting state. The magnetic flux in the superconductor then becomes zero; also known as perfect diamagnetism.
- **Perovskite:** A class of crystalline materials that often has the same formula and crystal structure. Two perovskites, strontium titanium trioxide and lanthanum aluminum trioxide, were central in the discovery of high-temperature superconductivity.
- **Phonons:** Based on an analogy between crystal lattice vibrations and those of an electromagnetic field, these particles of quantized vibrational energy were used by physicists to facilitate calculations of thermal and electrical conduction in solids.
- **SQUID:** Acronym for a superconducting quantum interference device; SQUIDs are supersensitive superconducting detectors of extremely weak magnetic fields, hence their use in magnetometers and interferometers.
- **Transition Temperature:** Temperature at which a substance becomes superconducting.
- **YBCO:** Also known as 1-2-3, YBCO is an acronym commonly used to define the high-temperature superconducting ceramic yttrium-barium-copper oxide, the chemical formula of which has one yttrium, two barium, and three copper atoms (hence 1-2-3).

DEFINITION AND BASIC PRINCIPLES

Unlike most natural phenomena, superconductivity

can be defined in terms of an absolute: Electric currents flow through superconductors with absolutely no resistance. This resistless flow is what warranted the name "superconductivity," because in traditional electrical behavior electrons traveling in wires lose energy in the form of heat to the atomic array of the wire. In superconductivity, in defiance of a long-held understanding, the resistance of certain metals and alloys did not simply decrease to a residual value as the material was cooled but precipitously fell to zero. This abrupt transition to the superconducting state took place at a specific temperature called the critical temperature, which is different for each superconducting material. The phenomenon of zero resistance applies only to superconductors through which direct electrical current flows. For alternating current, higher frequencies lead to greater resistance in the superconductor.

Unlike most substances, which allow magnetic field lines to pass through them (though certain other substances can become magnetized by strong fields), superconductors reject a magnetic field. For example, if either the north or south pole of a magnet is brought near a superconductor, each pole is repelled, which is what diamagnetism means. Studies have shown that a superconductor is not only a perfect conductor, but it is also a perfect diamagnet, and these two properties are often used to define superconductivity.

Superconducting devices have made use of these unique electrical and magnetic properties. For example, superconductors provide a way to circulate direct electric currents with no resistive loss. Even though alternating electric currents generate resistance in superconductors, careful choice of material and frequency for conveying these currents can be done with minor resistive losses. SQUIDs have become the principal achievement of superconductor electronics and have been used to precisely measure voltage, electrical currents, and gravity. Superconductors have had numerous applications in medicine, including magnetoencephalography, magnetocardiography, and magnetoneurography.

BACKGROUND AND HISTORY

Dutch physicist Heike Kamerlingh Onnes was the first to discover superconductivity as an outgrowth of his efforts to reach extremely low temperatures— he is sometimes called the "father of cryogenics."

In 1908 he succeeded in liquefying helium, but he also wanted to investigate how specific substances behaved at very low temperatures. Three years later he found, to his surprise, that mercury superconducted at 4 degrees above absolute zero (4 K). Scientists around the world were exuberant about his discovery, and Kamerlingh Onnes won the 1913 Nobel Prize in Physics. During the decades after this discovery many more superconductors were found; it turned out that about a quarter of the natural elements are superconductors. Further research revealed that hundreds of alloys and compounds also superconducted, but most did so only at very low temperatures. By the 1980's, despite seventy-five years of research, the highest temperature achieved for superconductivity was only 23 K.

Most physicists searching for the elusive high-temperature superconductor had studied metals and alloys, but in 1986 Swiss physicist Karl Müller and German physicist J. Georg Bednorz decided to study a ceramic material composed of lanthanum, barium, copper, and oxygen. They were working at the International Business Machines (IBM) research laboratory in Switzerland, where they found that their ceramic material superconducted at 35 K, a temperature much higher than any known substance. After they published their results, thousands of scientists in many countries began searching for new high-temperature superconductors. Within six months of the publication by Bednorz and Müller, more than 800 papers appeared on the chemical and physical properties of various new superconductors along with some theories to explain them. Particularly important was the discovery by American researchers of ceramic material YBCO, which superconducted at temperatures above 77 K. This meant that inexpensive liquid nitrogen could be used to study this superconductor rather than expensive and hard-to-handle helium. By the first decade of the twenty-first century physicists and chemists had created substances that superconducted at temperatures in excess of 135 K, and many scientists and engineers were racing to develop commercial applications of these new superconductors.

HOW IT WORKS

After a century of research on superconductivity, scientists have deepened their understanding about how this new and exciting phenomenon occurs, but a complete theory accounting for all superconductors has yet to be formulated to the satisfaction of a majority of scientists. The first theory to explain superconductivity actually drew from an explanation of the electrical properties of metals developed before Kamerlingh Onnes made his discovery. Dutch physicist Hendrik A. Lorentz proposed in 1900 that a crystalline metallic solid with no imperfections would actually conduct electricity without any resistance. However, real crystals have edges, faces, and missing atoms in their interiors, creating obstacles to passing electrons. Furthermore, high temperatures produce jiggling of the atoms in the lattice, thus impeding electron flow. Consequently, this old theory was unsatisfactory in its explanation of both conductivity and superconductivity.

Quantum Mechanics. By the mid-1920's physicists had developed quantum mechanics, a powerful new theory explaining the behavior of electrons in atoms. Different forms of quantum mechanics emphasized electrons as particles (matrix mechanics) and electrons as waves (wave mechanics), and these theories were eventually shown to be equivalent. Quantum mechanics proved very successful for understanding ionic and covalent crystals, organic chemical molecules, and many other physical and chemical phenomena, but it proved unable to unlock the mysteries of superconductivity. However, in 1933 German physicist Walther Meissner discovered a superconductor's ability to repel magnetism, which provided a clue to understanding superconductivity, since study of the Meissner effect showed how transitions from normal to superconducting states are thermodynamically reversible. Other studies helped explain some of the electromagnetic properties of superconductors. Nevertheless, theoretical physicists were still unable to explain superconductivity in terms of basic physical laws.

Josephson Effect. Another major development in understanding how superconductivity works came in 1962 when Welsh physicist Brian Josephson, a twenty-two-year-old graduate student, predicted the tunneling of electrons and Cooper pairs between linked superconductors. Within a year, experiments proved that pairs could travel across a barrier as easily as single electrons. In 1973 Josephson shared the Nobel Prize in Physics with physicists Leo Esaki and Ivar Giaever, who had also worked on tunneling.

High-Temperature Superconductivity. During the twenty-five years after the discovery of the first high-temperature superconductor by Bednorz and Müller, various physicists tried to develop an appropriate theory that explained the superconductivity of both low-temperature and high-temperature superconductors. BCS theory was able to explain low-temperature superconductivity, but, even with clever modifications, it failed to account for the properties of high-transition-temperature ceramics. One physicist remarked that there were nearly as many theories about these new superconductors as there were theorists. Some tried to explain particular categories of high-temperature superconductors, while others used superstring and gauge theories to try to resolve the mysteries. By 2011 some physicists were describing high-temperature superconductivity as one of the great, unexplained mysteries of condensed matter physics.

APPLICATIONS AND PRODUCTS

By 2011, superconductivity was a century old, and, from the beginning, Kamerlingh Onnes, its discoverer, foresaw practical applications for this unique and propitious phenomenon, particularly in the resistance-free generation and distribution of electricity. However, for the next seventy-five years commercial applications and products that he and other pioneers envisioned were few and far between. After the discovery of high-temperature superconductivity in 1986, practical uses for superconductors multiplied in a variety of fields, from scientific research and electronics to medicine and the military. Books, journals, newsletters, government programs, and professional organizations proliferated, urging the commercialization of low- and high-temperature superconductors for small-, medium-, and large-scale applications. Despite concrete evidence of multifarious applications, some critics point out that superconductivity, with its hundred years of research and development, has had much less commercial success than what followed: the discovery of electromagnetic waves, which spawned such phenomenally successful businesses as wireless telegraphy, radio, television, and myriad other electronic devices.

Scientific Research. Scientists have used superconductors to expand and deepen their understanding of the natural world, from the microcosm of nuclei, subatomic particles, and atoms to the macrocosm of stars and galaxies. High-energy physicists were among the first to grasp how superconducting magnets could facilitate their research on nuclei, protons, and electrons. Fermilab, the largest high-energy facility in the United States, changed from conventional electromagnets to superconducting magnets because they provided a superior means of controlling proton beams. Had the Superconducting Super Collider been realized (it was stopped because of its $8 billion cost), physicists believe it would have provided a way to discover new elementary particles and create new understanding about the universe's fundamental forces. Despite this setback, physicists have used superconducting techniques in such particle accelerators as the electron synchrotron and in plasma research, in the attempt to develop a practical nuclear fusion reactor. Research is also underway to develop superconductors for use in spacecraft, artificial satellites, and launch vehicles, because superconductors provide maximum performance with a minimum input of electrical power.

Metrology. Superconductors first became famous because of their electrical properties, but they first made money because of their magnetic properties. SQUIDs, the best known of the metrological devices, were the result of research and development in the mid-1950's at the Ford Motor Company Scientific Laboratory in Michigan. Because SQUIDs were extraordinarily sensitive to very weak magnetic fields, they led to several magnetometric applications. For example, they helped create detailed magnetic maps of geological formations, which were valuable in locating ore deposits. SQUID magnetometers have also helped oil geologists to penetrate rock layers to locate possible petroleum resources. Initially, field use of SQUID magnetometers was limited because of the necessity of working with liquid-helium-cooled devices, but with the advent of high-temperature superconductors, liquid-nitrogen-cooled SQUIDs have proved more convenient for use in geophysical explorations.

Electronics and Computers. Particularly after the discovery of high-temperature superconductivity, an explosion of research occurred as scientists in a variety of fields, from physics and chemistry to computer science and engineering, worked to find practical applications. Many researchers concentrated on developing superconducting wire, cable, and thin films. With officials in nations around the world aware of the growing "energy crisis," more efficient

ways of generating, storing, transmitting, and using electrical energy would be most welcome. During the period of research and development certain problems surfaced, such as the refrigeration costs involved in keeping transmission cables cool enough to superconduct. Nevertheless, companies in the twenty-first century began producing high-temperature superconducting wire for such applications as generators, motors, and cables.

Computer scientists and engineers have long recognized that a principal impediment to smaller, faster, and more energy-efficient computers has been waste heat produced by closely packed integrated circuits. Even before the discovery of high-temperature superconductivity IBM invested more than $100 million to develop a superconducting computer by using Josephson junctions for logical elements and superconducting rings for memory, but the program ended in the early 1980's. While the superconducting computer failed to become a reality, even after the appearance of high-temperature superconductivity, superconducting magnets have made it possible to grow silicon wafers more efficiently and economically than earlier methods. Furthermore, scientists at the Massachusetts Institute of Technology and several other institutions are intensely involved in creating quantum computers, which may (or may not) make use of superconductors.

Medicine. Superconducting devices have been used in medicine principally in assisting physicians to make faster, more detailed, more accurate, and less discomfiting diagnoses. For example, a superconducting magnetometric device has been used to detect the difficult-to-diagnose disease, hemochromatosis, which leads to excessive iron buildup in many tissues. Superconductors have led to new technologies that allow physicians to diagnose problems non-invasively in brains and hearts. For example, magnetoencephalography provides doctors with the means to pinpoint sites deeply within the brain responsible for epileptic seizures. Magnetocardiography has proved useful in mapping cardiac arrhythmias so that catheters can precisely ablate the source of the problem. As advanced high-temperature superconductors become part of various magnetic resonance imaging (MRI) systems, the ability to generate detailed pictures of structures (and their functioning) within the human body is expected to improve and costs decline.

Transportation. Trains based on traditional wheels-on-rails technologies have insurmountable limits on very fast speeds, but these limitations can be overcome by superconducting magnets, new kinds of track, and the levitation of the train. From the 1970's to the 1990's Japanese engineers developed the superconducting maglev (for *mag*netic *lev*itation) train, which was capable of speeds in excess of 300 miles per hour. With the success of the bullet train operating between Osaka and Tokyo, other countries, including the United States, have been exploring more advanced systems that will make use of high-temperature superconductors. These have also been used for sea transportation, as in the Jupiter II, a ship powered by superconducting motors.

Military. The U.S. Department of Defense and other agencies have been the chief supporters of the research and development of high-temperature superconducting devices for military uses. For example, the U.S. Air Force is developing superconductors that will increase the efficiency of electrical systems in jet engines for fighters and bombers. Other applications include superconducting magnetic detection systems that offer distinct advantages over sonar for submarines. Both surface ships and submarines will also be able to employ SQUID magnetometers for mine detection.

IMPACT ON INDUSTRY

The market research report,"Superconductors: Technologies and Global Markets," gave an estimate of the global market value for superconductors as $2 billion in 2010. Furthermore, the global superconductor market is expected to grow to $3.4 billion by 2015. The largest segment of the market has been superconducting magnets, and this segment is predicted to increase from $1.9 billion in 2010 to nearly $2.4 billion in 2015. Some analysts believe that a superconductor revolution is well on its way, and this will have a dramatic effect on many industries, their executives, researchers, workers, and investors.

Government and University Research. Throughout a large portion of the history of superconductors, research and development has taken place primarily at universities and technical institutes. Kamerlingh Onnes did his research at his Cryogenic Laboratory at Leiden University; Leon Cooper and John Schrieffer worked at universities, though John Bardeen worked for Bell Telephone Laboratories; and Karl Müller

and J. Georg Bednorz worked for IBM. Government agencies, such as the U.S. Department of Energy and the Defense Advanced Research Projects Agency, have funded superconducting research at universities and government laboratories. In the years after the discovery of high-temperature superconductivity, the U.S. government's response was rapid and substantial, leading to more funding for advanced superconductor research and development than any other country. The Department of Energy set up Superconductivity Pilot Centers at Argonne, Los Alamos, and Oak Ridge national laboratories to study applications of high-temperature superconductors, often in collaboration with researchers in universities and industries. The government agency most deeply involved in terms of investments has been the Department of Defense, because of the importance of superconductors for a wide variety of military technologies, from antennas to missiles.

Industry and Business. Industrial research and development of new applications of superconductors with the concomitant creation of new businesses cannot be isolated from government and university research, because federal funding has been distributed to relevant projects in government laboratories, academic institutions, and industrial facilities, with the goal of keeping the United States in a competitive position in superconductivity knowledge and applications. These programs have fostered strong American companies in such areas as wire and cable production, superconducting electronics, and superconducting magnets for medical and military devices. On the other hand, the United States finds itself in a weak competitive position in such applications as maglev transportation systems. Some analysts, aware of the decline in U.S. competitiveness in the steel, automobile, and electronics-products industries, have expressed concern that, unless U.S. government policies concerning the research, development, and commercialization of superconductors are insightfully formulated and implemented, the incipient American businesses dealing with superconductors may experience difficulties competing globally.

CAREERS AND COURSE WORK

Those advising students on how to prepare for jobs in superconductor technology emphasize matching desired careers with education. If a student wants to install and maintain superconductor cables, an associate's degree in electrical technology from a technical school would be sufficient. If a student wants to work in the research, development, and manufacture of superconductors, a bachelor's, master's, or doctoral degree in electrical engineering, physics, chemistry, or materials science may be required, depending on how advanced a rank in the profession one wishes to attain. Excellent programs in superconductor technology exist at such institutions as Stanford University; University of California, Berkeley; Massachusetts Institute of Technology; Florida State University; and the University of Houston. After taking introductory and advanced courses in physics, chemistry, mathematics, and electrical engineering, students may then specialize in such courses as engineering thermodynamics, fluid mechanics, cryogenics, and the theory and applications of superconductors.

Many opportunities exist for graduates with degrees in superconductivity. The field has been growing, and prognosticators predict accelerated future growth, so that successful careers can be pursued in government agencies, academic institutions, and a variety of well-established or new companies. After the discovery of high-temperature superconductors, some enthusiasts foresaw an accelerated need for those with expertise in this new field with an expected rapid introduction of applications in electric power, transportation, and medicine. Because of the modest increase in commercialized applications during the twenty-five years since this breakthrough discovery, early overly optimistic estimates of employment growth have had to be tempered with the more modest job opportunities that actually exist.

SOCIAL CONTEXT AND FUTURE PROSPECTS

Before 1986, most scientists were skeptical that a room-temperature superconductor would ever be discovered, but in the twenty-first century such a discovery seems increasingly likely. If such a superconductor could be efficiently and economically manufactured, then most industries of modern society would be affected, and a superconductor revolution would occur, similar to the computer revolution of the second half of the twentieth century. Room-temperature superconductivity would transform the electrical-power, transportation, and consumer-electronics industries. The era of copper's dominance would end, leading to a new generation of smaller and more efficient home appliances, such as refrigerators, washing machines, and air conditioners. Societies whose technologies make use of these room-temperature superconductors would be quieter, cleaner, and more energy efficient.

Though this idealistic view of the future motivates many researchers, realists caution that, based on the previous century of superconductivity research and development, the path to the elusive room-temperature superconductor will not be smooth and straight. Furthermore, those traveling this path should be prepared for unexpected difficulties as well as pleasant surprises.

Robert J. Paradowski, M.S., Ph.D.

FURTHER READING

Blundell, Stephen. *Superconductivity: A Very Short Introduction.* New York: Oxford University Press, 2009. Provides the general reader with a succinct history of the most important discoveries and theories of the field and concludes with an analysis of applications and prospects for the future. Index.

Hazen, Robert M. *The Breakthrough: The Race for the Superconductor.* New York: Summit Books, 1988. This popular account of the discoverers and discoveries of increasingly important high-temperature superconductors emphasizes the competitiveness of many of those involved in the race. Index.

Matricon, Jean, and Georges Waysand. Translated by Charles Glashausser. *Cold Wars: A History of Superconductivity.* New Brunswick, N.J.: Rutgers University Press, 2003. This book, translated from the original French version, emphasizes the physical principles, principal personalities, and the politics involved in the story of low- and high-temperature superconductivity. Index.

Mayo, Jonathan L. *Superconductivity: The Threshold of a New Technology.* Blue Ridge Summit, Pa.: Tab Books, 1988. John Schrieffer was one of the Nobel laureates who developed the BCS theory of superconductivity, and this introduction, considered as one of the best in the field, profits from his experience and expertise.

Schechter, Bruce. *The Path of No Resistance: The Story of the Revolution in Superconductivity.* New York: Simon & Schuster, 1989. This survey of early work on low-temperature superconductivity and particularly on the pivotal discoveries of high-temperature superconductivity also includes a discussion of applications and possible products.

Simon, Randy, and Andrew Smith. *Superconductors: Conquering Technology's New Frontier.* New York: Perseus Books, 1988. The authors, physicists with superconductor experience, wrote this book to bring the story of superconductivity to readers with no background in science or technology. They discuss the nature and history of the field, its most important theories, and various practical uses of superconductors, both realized and potential. Glossary, bibliography, and index.

Tinkham, Michael. *Introduction to Superconductivity.* 2d ed. New York: McGraw-Hill, 1996. Accessible to students with some background in science and mathematics, this paperback has been called "an amazing introduction to an amazing field." It contains references to many of the most important papers by the founders and developers of superconductivity.

WEB SITES

Aerospace Industries Association
http://www.aia-aerospace.org

American Society for Engineering Education
http://www.asee.org

Institute of Electrical and Electronics Engineers
http://www.ieee.org

See also: Ceramics

SURFACE AND INTERFACE SCIENCE

FIELDS OF STUDY

Materials science and engineering; chemistry; physics; inorganic chemistry; forensic science; electrical engineering; semiconductors; biomedical engineering; chemical engineering; toxicology; biology.

SUMMARY

Surface and interface science examines the properties of materials and products at the surface and at the locations where two or more materials meet. The discipline typically focuses on surface analysis, surface modification, and interface science. Surface modification changes the properties of a surface, such as the hydrophobicity, lubricity, chemistry, or topography, and surface analysis tells what those properties are for the material being studied. Interface science can include studying interactions between materials and how well the materials adhere to one another, as well as how to achieve desired interactions. The interface between a material and a solution or air can also be studied.

KEY TERMS AND CONCEPTS

- **Adhesion:** Attraction process that brings two or more materials together in direct contact.
- **Corrosion:** Oxidation of a metal surface; for example, rust on iron-containing metals.
- **Functionalization:** Attaching new functional groups to a surface to impart new properties to the surface.
- **Interface:** Region in a system where two or more materials meet.
- **Plasma:** Highly activated and ionized gas.
- **Surface:** Uppermost 0.5 to 3 nanometers of a sample or material.
- **Surface Chemistry:** Molecular and atomic species and bonds present at the surface of a material.
- **Topography:** Peaks and valleys of a surface and their relationship to each other in three-dimensional space.

DEFINITION AND BASIC PRINCIPLES

Surface and interface science encompasses several disciplines, including surface analysis, interface analysis, and surface modification.

Surface analysis is the study of the chemistry, crystal structure, and morphology of surfaces, using a combination of various surface sensitive analytical methods. Surface sensitivity can mean different things for different applications; however, a good working definition would be the uppermost 0.5 to 3 nanometers (nm) of a surface, or two to ten layers of atoms. Because surfaces are often modified with films in the tens to hundreds of nanometers range, the uppermost 100 nm can be considered the surface for some applications. Any surface thicker than 100 nm is typically considered bulk material.

Interface analysis is the study of the morphology and chemistry of internal interfaces, or regions where two or more materials meet. A good understanding of the interface between two materials will provide scientists and engineers with a wealth of information, including how well these materials are bonded, whether a certain amount of force is likely to separate the materials, whether any new species have been formed, or whether the two materials are diffusing into each other.

Surface modification is the process by which scientists and engineers change the chemistry, physical properties, or topography of a surface to improve the function of the material, such as changing the coefficient of friction, the lubricity, the hydrophobicity, the roughness, or bondability of the material.

Surface analysis and surface modification are both quite sensitive to contamination, so much care must be taken that the surfaces of interest do not come in contact with materials that could potentially transfer to the surfaces. However, even if every precaution is taken to prevent transfer, contamination can occur from simple exposure to air, which is why many surface modifications and analyses are conducted under vacuum. A vacuum is also essential for many of the surface techniques to prevent interactions between air molecules and the species being analyzed.

BACKGROUND AND HISTORY

Surface science is a relatively new area of science. The first X-ray photoelectron spectrum was collected in 1907 by P. D. Innes, and the first vacuum photoelectron spectrum was taken in 1930. American physicists Clinton Davisson and Lester Germer pioneered low energy electron diffraction (LEED) in 1927 and observed diffraction patterns on crystalline nickel. Ultra-high vacuum techniques for surface analysis and modification (specifically welding) were invented in the 1950's. In 1968, a surface science division of the American Vacuum Society was proposed, the same year that Auger electron spectroscopy began to be used. German scientist Alfred Benninghoven introduced time-of-flight secondary ion mass spectrometry (TOF-SIMS) in the 1980's, about the same time that scanning tunneling microscopy began to be used.

HOW IT WORKS

Surface Modification. Surface modification can be accomplished in many ways, including methods that modify the chemistry of a surface, the topography of a surface, or both. Films of material can be grown on a substrate in a vacuum or can be deposited using an ultrasonic spray-coating, spin-coating, or dip-coating process. Anodization is a process whereby an oxide layer is deposited on a metal using oxidation and reduction reactions. It has the potential to grow oxide films in a very controlled manner. Passivation is a process that changes the oxide layer on a metal to make it more resistant to oxidation and is typically carried out in an acid solution. Surface topography can be modified using laser or acid etching, which roughens the surface, or by using electropolishing, which uses electricity to smooth conductive surfaces.

Plasma treatment is a method of modifying the chemistry and often the topography of a surface. It uses a highly ionized, activated gas to react with the molecules of a surface. The plasma gas can vary from an inert gas, such as argon or helium, which would be expected to cause the species at the surface to react with one another, or a polymeric monomer, which could polymerize on the surface and create a thin plasma-treated layer. Plasmas can also be employed to clean surfaces before modification. Chemical vapor deposition is a technique in which the sample is exposed to a vapor that reacts with the surface to modify it.

Surface Analysis Techniques. There are many surface analytical techniques. The most common techniques are complementary to one another, and several are often used in combination to gain all the necessary information about a surface.

Scanning electron microscopy (SEM) is a technique that uses a focused electron beam that scans across the surface of a sample and allows for the construction of a very detailed map of surface topography. The surface sensitivity of this technique varies based on the sample and the instrument parameters, which can make the instrument range from simply surface sensitive to able to gain information up to microns within a sample.

Electron spectroscopy for chemical analysis (ESCA), also known as X-ray photoelectron spectroscopy (XPS), uses X rays to generate electrons from the surface of a sample. These electrons can then be detected and traced back to whatever elements are present at the surface. XPS gives quantitative elemental information about the uppermost 5 to 10 nm of a sample surface and can also give some information about the types of bonds present. Auger electron spectroscopy (AES) is a surface-sensitive technique that uses an electron beam to gain surface elemental information similar to that obtained by XPS. AES is less quantitative than XPS; however, higher resolution mapping is possible with AES than with XPS.

Time-of-flight secondary ion mass spectrometry (TOF-SIMS) is an extremely surface-sensitive analytical technique that uses a beam of liquid metal ions to probe the top one to three molecular layers of a sample, causing the emission of ions from the sample to be detected in a time-of-flight mass spectrometer, which gives detailed information about the specific molecules and atoms at a surface.

White light interferometry, sometimes called profilometry, uses white light interference patterns to construct detailed maps of surface topography very quickly. It can detect minute surface features in the nanometer range. Spectroscopic ellipsometry is often used to measure the thickness of surface films, such as oxides or plasma treatments on materials. Atomic force microscopy (AFM) and scanning tunneling microscopy (STM) both employ a very sharp tip called a cantilever that rasters across a sample and yields high-resolution images of surface topography. AFM is typically employed for nonconducting samples, and STM is useful for conducting samples.

Interface Science. The interface of two materials can be probed using several techniques, including some of the techniques used for surface analysis. Depth profiling is a method by which information is gained from within the sample. If the interface in question is near the surface, several techniques can be used. Both TOF-SIMS and XPS use beams of ions or carbon 60 (Buckministerfullerene) to sputter material away from the surface to obtain information about interfacial layers and boundaries. An advantage to these techniques is that they are performed under vacuum; therefore, the interface is not exposed to ambient conditions before analysis. Confocal Raman microscopy has the potential to probe interfaces up to 100 microns below the surface of a sample by focusing the laser on the interface and using a pinhole to eliminate out-of-focus data.

APPLICATIONS AND PRODUCTS

Surface Modification. Surface modification has a very wide range of applications. Surfaces are modified to impart desirable properties and often change properties such as lubricity, surface energy, the coefficient of friction, or the functional groups present on a surface. Self-assembled monolayers can be deposited, and crystals can be grown in a highly controlled manner. Sensors and catalysts can also be fabricated using surface modification. Surface modification is applied across a wide variety of industries, including the automobile, aerospace, medical device, textile, chemical, steel, and electrical industries.

A major application for surface modification is making materials more corrosion and wear resistant. Steel is used in everything from skyscrapers to motorcycles, so increasing the stability of this crucial material is a priority not only for many people in the steel industry but also for industries that use the steel to fabricate other products.

In the medical device community, the surface of a material is extremely important because it is what is in contact with the body. For example, the components of implants may be plasma treated to make their surfaces more biocompatible, which reduces cell adhesion and the formation of fibrous tissue around the implant. Implantable metal devices are often passivated to make the device resistant to corrosion when subjected to the aqueous environment inside the body.

In the automobile industry, surfaces of parts that must be bonded to other parts to provide structural integrity to the vehicle are sometimes plasma treated to promote better adhesion and reliability. The surfaces of exhaust systems are modified with a catalyst that greatly reduces toxic emissions.

Surface Analysis. Surface analysis can be employed in many situations, but it is particularly well suited for the analysis of contamination or surface damage. In many industries, contamination at the nanometer scale can spell disaster for a process. For example, semiconductor materials have very predictable conductive behavior, which is essential for designing microelectronics that work properly. The addition of contaminants to the system will change the behavior of the materials and can cause failures, so it is essential that surface analysis be employed in the development of a product and sometimes during manufacturing stages to ensure that the materials are clean and reliable.

Surface analysis is critical in the study of nanomaterials, as these materials and structures are often so minute that the majority of the mass of a sample is considered surface material. Less sensitive techniques would not be able to effectively study these extremely small materials.

Interface Science. The interface and interactions between two solid materials are very interesting to materials scientists; however, interface science also includes the study of the interactions between materials and a gas, such as air, or a liquid, such as water. Proteins such as insulin are often sold in solution form. Because proteins are unstable under most conditions, it is important for drug companies to understand the interactions between proteins and the surfaces of the containers in which they are stored.

Adhesion specialists often need to determine whether the two materials they are trying to bond are well bonded. Mechanical tests, such as peel tests or scratch tests, are often employed to determine the integrity of the bond in combination with a chemical analysis that has the ability to look for new bond types between the layers.

IMPACT ON INDUSTRY

Because of the wide variety of surface modification and analysis techniques and the even wider variety of applications, surface science has had a far-reaching but difficult to quantify impact on industry. The majority of the impact comes from the power of the techniques to help scientists and engineers improve processes and solve problems rather than from marketing surface science instrumentation to the general public.

Surface Analysis Instrument Manufacturers. The private and publicly traded companies that manufacture surface science equipment vary widely in terms of size and revenue. For example, the publicly traded Thermo Fisher Scientific, which manufactures several surface analysis systems among many other products, employed about 35,000 people worldwide and had revenues of more than $10 billion in 2009. Kratos Analytical is a surface analysis subsidiary of the Japan-based Shimadzu, which employed more than 9,600 people in 2009. Japan-based ULVAC-PHI and the German company ION-TOF are prominent surface science instrument manufacturers. Other companies that make surface analysis instruments include Hitachi, JEOL, and the Zygo Corporation.

Fascinating Facts About Surface and Interface Science

- Researchers at IBM inadvertently modified the tip of an atomic force microscope (AFM) with a carbon monoxide molecule, which ended up allowing them to image the structure of interlocking benzene molecules with more clarity than ever before.
- Ion scattering spectroscopy (ISS) has the potential to determine the structure of the uppermost layer of atoms of a sample.
- Textiles are sometimes plasma-treated to make the fabric resistant to wear, bacteria, fungi, and shrinkage.
- Varying the thickness of the oxide layer on titanium parts changes the color of the part.
- A single monolayer of contamination can obscure the results of surface analysis techniques.
- One of the most common surface contaminants is silicone oil.
- Attenuated total reflectance (ATR) and Raman spectroscopy are not true surface techniques, but both can be modified to give surface chemistry information about the specific types of functional groups present.
- Surface enhanced Raman spectroscopy (SERS) involves adsorbing molecules in a metal surface or depositing metal nanoparticles on a surface. The amount of Raman signal obtained can be enhanced by 1015, making what is normally a bulk technique extremely surface sensitive.

Surface Modification Companies. Numerous companies engage in surface modification, but some have more of a focus on surface modification than others. SurModics is a U.S.-based drug delivery and surface modification company. Surface Modification Systems (SMS), in California, focuses on protective coatings. March Plasma Systems, also based in California, specializes in plasma treatments and plasma cleaning.

Government Agencies and Academic Research. Because of the broad range of surface science applications, most government agencies and academic institutions are engaging in research involving surface science. The Surface and Interface Sciences Department of Sandia National Laboratories conducts advanced research in surface and interface science applications and has developed new surface techniques, such as interfacial force microscopy, a scanning probe technique similar to atomic force microscopy and scanning tunneling microscopy. The University of Washington's Department of Bioengineering operates the Surface Analysis Recharge Center to address problems and conduct research associated with the surfaces of biomedical devices, biomaterials, and surfaces that interact with biological materials.

CAREERS AND COURSE WORK

A wide variety of careers is available in surface science, and the desired career will determine the proper course work. To work in surface science requires a bachelor of science degree in a core science such as chemistry or physics, and having a graduate degree greatly improves a student's chances of finding work in the field. A bachelor's degree is adequate qualification for becoming a laboratory technician, which generally involves being trained on the operation of an instrument and performing mainly routine tasks. A master's or doctoral degree and appropriate research experience are the necessary qualifications for a position that involves the advanced use and maintenance of surface analysis instruments, experiment design, and data analysis, interpretation, and communication, as well as possibly the development of new instruments.

Careers involving surface science can be found in industry (such as at an automobile company), in academia, and in government laboratories. Surface and interface science is critical in materials science and engineering and is likely to play a major role in the development of the field. Regardless of the specific career choice, a strong foundation in basic science, including chemistry and physics, as well as in mathematics, is required.

SOCIAL CONTEXT AND FUTURE PROSPECTS

Surface science is a growing field that will continue to play a major role in the development of materials science and engineering, especially in light of the focus on nanomaterials and the potential of surface science to make a significant contribution to that field. In many industries, the focus is on making devices smaller and smaller, which naturally leads to the surface becoming a more significant part of the device because of the increase in the surface area to volume ratio.

New applications are being discovered for surface modification every day. Plasma treatment is becoming more common in many industries, such as the automobile and medical device industries, because it can make it easier to bond one material to another. The process increases quality and is relatively fast, which suits production in the fast-paced cultures of these industries.

Additionally, as governing agencies such as the Food and Drug Administration step up their scrutiny of medical devices and drug-device combination products with respect to biocompatibility and stability, surface analysis will probably play an increasing role in characterizing the surfaces that come into direct contact with a patient.

Lisa LaGoo, B.S., M.S.

FURTHER READING

D'Agostino, Riccardo, et al., eds. *Plasma Processes and Polymers*. Weinheim, Germany: Wiley, 2005. Brings together papers from a conference on plasma treatments for polymers.

Hudson, John B. *Surface Science: An Introduction*. New York: John Wiley & Sons, 1998. Contains information on surface analysis, interface interactions, and crystal growth.

Kim, Hyun-Ha. "Nonthermal Plasma Processing for Air-Pollution Control." *Plasma Processes and Polymers* 1, no. 2, (September, 2004): 91-110. Discusses the history and future of using plasma treatment to reduce air pollution.

Kolasinski, Kurt W. *Surface Science: Foundations of Catalysis and Nanoscience*. 2d ed. Hoboken, N.J.: John Wiley & Sons, 2009. A detailed, more technical resource for those interested in cutting-edge applications of surface science.

Rivière, John C., and S. Myhra, eds. *Handbook of Surface and Interface Analysis: Methods for Problem-Solving*. 2d ed. Boca Raton, Fla.: CRC Press, 2009. Discusses common surface analysis methods and problems; chapter 7 deals with synchrotron-based techniques.

Vickerman, John C., ed. *Surface Analysis: The Principal Techniques*. 2d ed. Chichester, England: John Wiley & Sons, 2009. Describes the techniques involved in surface analysis. The introduction contains a basic overview of surface analysis that is easy to read and informative.

WEB SITES

Materials Research Society
http://www.mrs.org/s_mrs/index.asp

Sandia National Laboratories
Surface and Interface Sciences Department
http://www.sandia.gov/pcnsc/departments/surfaceinterface.html

SURGERY

FIELDS OF STUDY

Medicine; pediatrics; geriatrics; ophthalmology; dermatology; otolargyngology; cardiology; obstetrics and gynecology; neurology; orthopedics; urology; laser surgery; plastic and reconstructive surgery.

SUMMARY

Surgery involves penetration of the body with specialized instruments to examine, repair, remove, or replace internal structures. General practitioners and specialists in a variety of fields perform surgical procedures ranging from simple to complex. For example, a dermatologist may excise a small growth on the skin and a cardiothoracic surgeon may perform a heart transplant. Some specialties (such as general surgery and neurosurgery) focus primarily on surgical procedures and preoperative and postoperative care; others (such as ophthalmology and gynecology) include surgery as part of a treatment plan. Surgery can be performed directly through incision into a body area or indirectly through instruments known as endoscopes.

KEY TERMS AND CONCEPTS

- **Aseptic Technique:** Use of sterile equipment and antiseptic agents to prevent infection.
- **Craniotomy:** Surgical entrance into the brain cavity to perform a neurosurgical procedure.
- **Hemostasis:** Cessation of bleeding by surgical techniques such as suturing or cauterization.
- **Laparoscopy:** Surgical procedure that involves introducing surgical instruments and a small viewing instrument (laparoscope) into the abdomen to accomplish surgical procedures.
- **Laparotomy:** Surgical entrance into the abdominal cavity to perform surgery.
- **Outpatient Surgery:** Surgery involving less complex procedures that is done within a hospital or freestanding facility. Patients remain in the facility for less than twenty-four hours, as opposed to inpatient surgery, when the patient is admitted to a hospital and stays more than a day.
- **Suture:** Thread composed of a variety of materials (such as silk, synthetic fiber, or catgut) and used to join body structures.
- **Thoracotomy:** Surgical entrance into the chest cavity to perform surgery on the heart or lungs.

DEFINITION AND BASIC PRINCIPLES

Surgical procedures are usually classified by the type of procedure (such as laparotomy or laparoscopy), the urgency (for example, emergency or elective), the body area involved, any special instrumentation used (such as microsurgery or laser surgery), and degree of invasiveness. Superficial surgery takes place near the surface of the skin (such as removal of a wart) and minimally invasive surgery is accomplished through small skin incisions; in invasive surgery, the physician makes incisions and works with instruments or devices inside the body. A common type of minimally invasive surgery is laparoscopy, originally used by gynecologists to view pelvic structures and perform simple procedures such as cauterizing the Fallopian tubes for sterilization. Later, gynecologists used laparoscopy for more complex procedures such as hysterectomies, and the procedures began to be used by physicians in other specialties. For example, the laparoscope is used by urologists to perform nephrectomies (kidney removals) and by general surgeons to perform cholecystectomies (gallbladder removals).

The basic principles of surgery are the use of aseptic technique, exposure, hemostasis, and closure. Aseptic techniques maintain sterility. For example, surgeons scrub their hands and don sterile gloves for surgery. The operative field is cleansed with a strong antiseptic, and great care is taken to prevent any contamination of the area. Exposure involves making an incision in the skin and manipulating internal structures with retractors and other instruments to view the area needing attention. Hemostasis, or controlling bleeding, is accomplished with sutures and electric coagulation (cauterizing) of bleeding areas. Closure involves removing any instruments used during surgery and closing the tissue layers. All surgery requires some form of anesthesia. Minor procedures on the skin can be accomplished with a local anesthetic. More involved procedures require a general or regional anesthetic. General anesthesia involves rendering a patient unconscious by administering an anesthetic drug. Regional anesthesia involves administering an anesthetic agent to nerves supplying a body area, so that the patient does not feel pain in that area. Examples of regional anesthesia are spinal and epidural anesthetics.

Background and History

Surgery has been performed since prehistoric times. Human skulls from the Neolithic era show evidence of trepanation, a procedure that involved drilling or scraping a hole in the skull. The purpose of trepanation is unknown; however, evidence exists that many of the individuals survived the procedure. Written evidence of surgery has been found on a 3,500-year-old papyrus in ancient Egypt. Surgery was performed by priests who specialized in medical treatment. Other ancient civilizations that had surgical knowledge were China, India, and Greece. In ancient Greece, citizens could obtain medical advice and treatment at temples dedicated to Asclepius, the god of medicine and healing. In the first century, the Greek physician Galen performed a number of complex surgical procedures including brain and eye surgery. In China, Hua Tuo, a Chinese physician during the Tan Dynasty (221 to 206 b.c.e.) performed surgery with a rudimentary form of anesthesia. After the fall of the Roman Empire, surgery declined in the West; however, it continued to survive and develop in the Middle East.

In medieval Europe (fifth through fifteenth centuries), surgery was often performed by barbers. These individuals cut hair, pulled teeth, and performed simple operations such as setting bones and amputations. In the twelfth and thirteenth centuries, surgical techniques were revived in the West. Surgical training was offered at universities at Montpellier, Padua, and Bologna. In the late nineteenth century, London universities began offering a bachelor of surgery degree (Ch.B.).

How It Works

Most surgeons have had specialty training in a surgical field; however, some family practitioners perform minor surgical procedures or assist surgeons in a hospital operating room. A general surgeon performs a wide range of procedures, primarily in the abdominal cavity (such as hernia repair and appendectomy). General surgeons also commonly operate on the breast. Some specialists, including plastic surgeons, begin their training in general surgery before being trained in procedures specific to their specialty. Cardiothoracic surgeons perform surgery involving the chest cavity (heart and lungs), while neurosurgeons specialize in brain and spinal surgery. Urologists specialize in surgeries of the bladder, kidney, and ureters (tubes connecting the kidneys to the bladder), and orthopedic surgeons perform surgery that involves the bones, joints, and muscles. Some surgeons further specialize within their field. For example, orthopedic surgeons might limit their practice to a specific body area such as the hand or hip. Some surgeons specialize in oncology, usually concentrating on specific types of cancer. For example, a gynecologic oncologist will operate on cancerous tumors of the female reproductive system.

Most surgeons have patients referred to them from other physicians. These physicians are often internists, family practitioners, or emergency room physicians; however, they can be other specialists. Surgeons may see patients in their offices, the emergency department, or the hospital bedside. Surgeons, especially specialists, are sometimes called to operating rooms when something requiring their expertise is encountered. When a patient visits a surgeon's office, the surgeon takes a medical history, performs a physical examination, and orders any necessary laboratory tests and X rays or other imaging studies. If surgery is determined to be the best option, the surgeon performs the procedure and provides postoperative care. Once the patient has recovered sufficiently, generally the patient's primary care provider provides followup care. In some cases, the followup care can last for months or years and may require further visits to the surgeon and additional surgery. Obstetricians/gynecologists perform surgery such as cesarean sections and hysterectomies and also provide long-term care to their patients. Most surgeons are in group practice with other surgeons or a multispecialty group.

If at all possible, a patient must be apprised of the risks and benefits of a surgical procedure and give written consent. Consent may also be given by a parent in the case of a minor child or by a family member if the patient is unable to give consent (as with an unconscious or comatose person). If time permits, the patient has the option of obtaining a second opinion from another surgeon.

Not all surgery is performed in a hospital operating room. Some is performed in outpatient surgery centers, and simple procedures can be carried out in the surgeon's office. Many inpatient surgeries require an assistant surgeon. This individual is often a surgeon with comparable training or training in another specialty. For example, a gynecologist and a urologist might join forces for a complex bladder repair procedure in which expertise in both specialties is necessary.

An inpatient procedure requires a surgical team made up of the surgeon, assistant surgeon, anesthesiologist, scrub nurse, and circulating nurse. The anesthesiologist administers the anesthetic and monitors the patient's vital signs. The surgeon, assistant surgeon, and anesthesiologist communicate with one another during the procedure. The scrub nurse is positioned at the operating table with the two surgeons and hands off instruments, sutures, swabs, and other material during the procedure. He or she may be asked to participate directly in the procedure (for example, to hold a retractor). The circulating nurse brings requested material (such as a particular instrument, a unit of blood, or medication) to the operating table. Other personnel may be present for more complicated procedures. For example, a heart procedure requiring a heart-lung machine will require technicians to run and monitor the equipment.

APPLICATIONS AND PRODUCTS

Since the mid-twentieth century, technology in the surgical arena has developed markedly. Technological advances in endoscopy—laparoscopy (abdomen), cystoscopy (bladder), hysteroscopy (uterus), and arthroscopy (joints)—have resulted in dramatic changes in patient care. Endoscopy has replaced traditional methods in many procedures. Many medical products have been designed for endoscopy and other surgical procedures. In addition, robotics is being developed to assist in endoscopic surgical procedures.

Laparoscopy. In laparoscopy, a needle is inserted into the abdominal cavity and the abdominal cavity is insufflated (filled) with carbon dioxide gas. This makes it possible to see abdominal organs with a small viewing instrument (laparoscope). One or more additional incisions are made to insert specialized instruments for grasping, cutting, and suturing internal organs. Gynecologists use laparoscopy to accomplish complex tasks such as removing fibroid tumors from the uterus and performing complete hysterectomies. General surgeons perform appendectomies with the laparoscope. The advantage of laparoscopic surgery is that the recovery time can be much less because the small incisions are much less painful. However, if complications arise, they are more difficult to control via the laparoscope and can result in serious consequences, even death. For example, if a hemorrhage occurs, visualization is lost and an immediate incision must be made to control it. Surgeons had to use eyepieces to view their work with the early laparoscopes, but video cameras were incorporated into the laparoscopes, and surgeons and all members of the surgical team can see the surgical area on a monitor. This not only increased the surgeons' comfort but also allowed the entire surgical team to watch and monitor the procedure.

Cystoscopy. Urologists and urogynecologists (gynecologists who specialize in the female urinary tract) perform cystoscopy, which involves passing an instrument (cystoscope) into the bladder to view its inner surface. The procedure also can visualize the opening of the ureters. If necessary, a urethroscope can be passed into the ureters and up into the kidney. This procedure can be used to remove bladder tumors or resect (reduce the size of) the prostate gland.

Hysteroscopy. In hysteroscopy, a hysteroscope is passed through the cervix into the uterus for viewing the endometrial cavity (uterine interior) and taking a biopsy of the uterine lining. The procedure can be used to remove fibroid tumors, polyps, and uterine abnormalities such as a septum (wall) in order to correct infertility problems. A small flexible video lens can be passed into the Fallopian tubes for viewing problems inside them and removing scar tissue.

Arthroscopy. Orthopedic surgeons use the arthroscope to view the interior of joints. As with other endoscopic procedures, abnormalities can be visualized and sometimes repaired. For example, a torn ligament can be rejoined using arthroscopy. Almost every joint in the human body can be examined with the arthroscope. The most common joints visualized are the knee, hip, ankle, foot, and shoulder.

Robotics. Robotic surgery is increasingly being used with endoscopes for a variety of surgeries, including cardiothoracic surgery, gastrointestinal surgery, gynecologic surgery, neurosurgery, pediatric surgery, and urologic surgery. Robotic surgery allows for superior visualization, precise manipulation, and improved dexterity. A commonly used robotic system is the da Vinci System. It consists of a console and a patient side cart with four arms controlled from the console. Three of the robotic arms hold and manipulate surgical instruments. The fourth arm is an endoscopic camera, which can visualize the surgical area in three dimensions. The surgeon sits comfortably at the console and manipulates the three arms with hand and foot pedals. The system allows the surgeon to make small, precise movements, which are beyond the capabilities and range of motion of the human arm and wrist.

Invasive Radiology. A rapidly evolving medical field is invasive radiology, which has replaced many complicated and difficult surgical procedures. Traditionally, the radiologist reviewed X-ray images to diagnose medical conditions. The invasive radiologist uses X-ray images or ultrasound to thread catheters inside the body to perform many surgical procedures. The procedures can be divided into three groups: neuroradiology procedures, vascular radiology procedures other than those that involve the brain and spinal cord, and nonvascular radiology procedures. Invasive neuroradiology procedures involve the brain and spinal cord. For example, to treat a brain aneurysm (weakened, dilated blood vessel that can rupture), special plugs are passed through blood vessels entering the brain and placed to seal off the weakened blood vessel. Invasive vascular radiology of other body areas involves either opening or closing a blood vessel. For example, a stent (tube) can be passed into a coronary artery to unblock it. Invasive nonvascular radiology involves passage of an instrument into the body cavity to obtain a biopsy. For example, an instrument may be passed through the chest wall into the lung to obtain a tissue sample.

Cutting Tissue. A surgeon in the mid-twentieth century almost exclusively used the scalpel for surgical procedure. Scalpels are small, sharp knife blades, which are fashioned in a variety of shapes. Since that time, a variety of devices have been developed for cutting tissue. These include electric current devices, which cut and cauterize tissue simultaneously; laser devices, which cut through a focused light beam; and the harmonic scalpel, in which ultrasound produces a vibration that can cut tissue.

Fascinating Facts About Surgery

- The tradition of placing a white-and-red pole outside a barber shop dates back to the Middle Ages. Many people were illiterate, so barber-surgeons used a white-and-red pole, representing blood and bandages, to help customers find their shops.
- In 1894, surgeons began to regularly wear rubber gloves during operations. Their use was encouraged by Hunter Robb, a gynecologist at The Johns Hopkins University who had a special interest in wound contamination and sterile surgical technique.
- In the sixteenth century, syringes were used to irrigate wounds with wine.
- Surgical procedures used by the Greek physician Galen in the first century were not employed again for almost 2,000 years.
- In 1878, American physician J. Marion Sims performed the first cholecystectomy (gallbladder removal), but the patient died. When French physician Philippe Mouret performed the first laparoscopic cholecystectomy in 1987, it was a success.
- Most American Civil War surgeons had a basic kit, which contained amputation tools: a saw, pliers, hook, and a few knives of different sizes. Less than 1 percent of the doctors on either side had ever performed surgery before going off to war. At the time, medical school training took two years and there were no licensing boards.
- In 2004, Connie Culp became the first person in the United States to undergo a face transplant. She endured a series of operations to restore a normal appearance after her husband shot her in the face with a shotgun.

IMPACT ON INDUSTRY

Health care—encompassing hospitals, medical device manufacturers, pharmaceutical companies, and insurance companies—is a major portion of a nation's economy. In most countries, national, state, and local governments exert some control over health care, although the extent of control varies. A large portion of health care is associated with surgical procedures. Surgery and the associated technology is expensive; therefore, surgery-related items have a major impact on industry.

Government and University Research. Many medical schools conduct research involving surgery, and that research is usually supported by funds from the private sector and the government. For example, the Trauma Research Center at the University of Texas, which focuses its research on the role of plasma in hemorrhagic shock, is funded by the National Institutes of Health. Yale Surgery Research, part of Yale University, collaborated with U.S. Surgical in research into the design and implementation of new surgical instruments and appliances for laparoscopic surgery.

Surgical experimentation and research is commonly conducted on animals before it is applied to human subjects. Many countries have formulated strict guidelines to ensure that the animals are treated humanely and that the people who are part of clinical trials are fully informed of possible benefits and risks before giving their consent. The World Health Organization has created a set of ethical standards and procedures that govern research on humans in an attempt to standardize practices worldwide.

Industry and Business Sectors. A significant portion of all hospitals is devoted to surgery. Facilities include operating rooms, post-surgery recovery rooms, and rooms and units exclusively for surgery patients. Surgery is often performed in hospital emergency units, and recovering patients are often treated in intensive care units. Some larger hospitals have special care wards (such as orthopedic and urology floors). Most hospitals also have areas devoted to outpatient surgery. The obstetrics unit usually contains one or more operating rooms, primarily used for cesarean sections.

Medical equipment and devices can cost millions of dollars, have a limited service life, and involve complex electronics and computers. Models with improved features are constantly being released into the medical marketplace. Some medical instruments, devices, and supplies—including surgical drapes, scalpels, and sutures—are consumable or disposable. Production of these items supports multimillion-dollar industries.

Major Corporations. Many hospitals are members of a hospital system, which can include any number of hospitals. Some of these hospital systems are health maintenance organizations, which bundle health care insurance, hospitals, doctors, and other medical personnel into a single entity. Medical corporations can consist of hundreds of surgeons. A solo practitioner can also form a corporation.

CAREERS AND COURSE WORK

Students interested in pursuing careers as surgeons must first graduate from a four-year college. Students usually major in one of the sciences, such as biology, but whatever the choice of major, they must fulfill the course requirements for medical school. The next requirement is a medical degree from a four-year medical school, followed by an internship and residency program for specialty training. Surgeons complete a general surgery training program, and some continue training in a surgical specialty. Most surgeons take written and oral examinations to become certified in their specialty.

A variety of career opportunities await the surgeon. The surgeon can enter private practice, which usually entails joining the practice of an established physician or group of physicians. These physician groups may consist solely of surgeons or contain specialists in a variety of fields. Physicians in larger groups can benefit from the sharing of equipment and other resources, and they can assist one another with weekend and night calls. Larger groups often employ physician assistants and nurse practitioners to handle less complex problems. Surgeons can also enter a career in academia, where they engage in medical research and train medical students and resident physicians.

SOCIAL CONTEXT AND FUTURE PROSPECTS

Surgery continues to play a major role in medical treatment. Surgical techniques and equipment continue to undergo research and development. Surgical research involves finding better techniques to repair injuries or remove diseased tissues, discovering devices that enable these body parts to regain or maintain function, and developing means to replace body parts that cannot be repaired. Surgeons will be involved in both the research behind and the implementation of new developments. Areas that are the focus of research include endoscopy (laparoscopy, hysteroscopy, arthroscopy, and cystoscopy) and its associated robotics, invasive radiology, implants, artificial organs, and organ transplantation.

Techniques for minimally invasive surgery have shortened recovery time and will continue to be refined and improved. To deal with failing body parts, researchers have developed devices such as artificial heart valves and heart assistance devices (pacemakers), although not an artificial heart. Organ transplant research involves not only surgical techniques but also pharmaceuticals and other ways to reduce rejection of the transplanted organ. Considerable research has been conducted in stem cells, which many scientists believe may someday enable a person to grow a new heart or kidney to replace a defective one. In the meantime, research has developed equipment that can take on the function of the lungs, heart, and kidneys, if for a limited period of time, and these devices will continue to evolve.

Robin L. Wulffson, M.D., F.A.C.O.G.

FURTHER READING

Bellomo, Michael. *The Stem Cell Divide: The Facts, the Fiction, and the Fear Driving the Greatest Scientific, Political, and Religious Debate of Our Time.* New York: AMACOM, 2006. An unbiased accounting of the benefits and risks of stem cell research.

Gharagozloo, Farid, and Farzad Nazam. *Robotic Surgery.* New York: McGraw-Hill Medical, 2008. Describes surgical robots and how they are employed during surgical procedures.

Gollaher, David. *Circumcision: A History of the World's Most Controversial Surgery.* New York: Basic Books, 2001. A comprehensive history of the most common surgical procedure performed in the United States

Kontoyannis, Angeliki. *Surgery.* 3d ed. London: Elsevier, 2008. Provides basic information on surgery for medical students, including diagnosis and techniques.

Nezhat, Camran, Farr R. Nezhat, and Ceana Nezhat. *Nezhat's Operative Gynecologic Laparoscopy and Hysteroscopy.* 3d ed. New York: Cambridge University Press, 2008. Contains information on how laparoscopy and hysteroscopy are used in gynecological procedures.

Panno, Joseph. *Stem Cell Research: Medical Applications and Ethical Controversy.* New York: Checkmark Books, 2006. Examines stem cell research with a focus on its medical applications.

Schwartz, Seymour, et al. *Schwartz's Principles of Surgery.* 9th ed. New York: McGraw-Hill, Medical Publishing Division, 2010. Provides the basic principles and methods of general surgery.

Sutton, Amy L. *Surgery Sourcebook.* 2d ed. Detroit, Mich.: Omnigraphics, 2008. Basic consumer information about common inpatient and outpatient surgeries.

WEB SITES

American Society of General Surgeons
http://www.theasgs.org

American Surgical Association
http://www.americansurgical.info

Centers for Disease Control and Prevention
Having Surgery? What You Should Know Before You Go
http://www.cdc.gov/Features/SafeSurgery

U.S. Food and Drug Administration
Medical Devices, Surgery and Life-Support Devices
http://www.fda.gov/MedicalDevices/ProductsandMedicalProcedures/SurgeryandLifeSupport/default.htm

World Health Organization
Surgery
http://www.who.int/topics/surgery/en

See also: Cardiology; Gastroenterology; Neurology; Orthopedics; Otorhinolaryngology; Pediatric Medicine and Surgery; Urology; Xenotransplantation.

T

TELESCOPY

FIELDS OF STUDY

Electronics; electrical engineering; systems engineering; aerospace engineering; mechanical engineering; physics; optical physics; atomic physics; particle physics; astrophysics; astronomy; astrometry; cosmology; mathematics; quantitative analysis; statistics; image analysis; interferometry; communications; electromagnetism; spectroscopy; robotics; quantum electrodynamics; meteorology; environmental science; earth science; computer science; photography.

SUMMARY

Telescopy is the science behind the creation and use of telescopes, devices that clearly render objects that are too dim or distant to be seen by the naked eye. The word "telescope" has its origins in two ancient Greek words meaning "to watch" and "from afar." By enabling people to study and accurately pinpoint stars such as Polaris and Sirius, planets such as Jupiter and Mars, and other structures in the sky, telescopy radically transformed humankind's ability to map locations and navigate from place to place. The use of telescopes in astronomy and cosmology has also profoundly deepened knowledge of the complexities of the universe and its origins.

KEY TERMS AND CONCEPTS

- **Adaptive Optics:** Technique of increasing the resolving power of a telescope by correcting for errors caused by atmospheric effects such as turbulence.
- **Angular Resolution:** For any given telescope, the smallest angle between two closely situated objects that can be clearly distinguished as being separate. Measured in arcseconds and also known as a telescope's diffraction limit.
- **Aperture:** Size of the opening in a telescope that collects light or other electromagnetic radiation.

- **Arcsecond:** 1/3600 of a degree.
- **Catadioptric Telescope:** Optical telescope that makes use of both reflective and refractive properties.
- **Interference:** Process in which multiple electromagnetic waves are superimposed and combine to form a single wave. Depending on whether the waves have the same or different amplitudes, they can either reinforce or cancel each other.
- **Interferometry:** Process in which the signals from multiple telescopes are combined to form a single image with a higher resolving power than that of any of the individual telescopes.
- **Optical Telescope:** Telescope that refracts or reflects light to create a clear, bright magnified image of a distant object.
- **Radio Telescope:** Telescope that bends or reflects radio waves to create a clear, bright magnified image of a distant object.
- **Spectrograph:** Instrument designed to record or photograph various properties of electromagnetic radiation; often used as an auxiliary device on astronomical telescopes.
- **Very Long Baseline Interferometry:** Technique in which astronomical observations are carried out by separate radio telescopes thousands of miles apart, then combined to form a single image.

DEFINITION AND BASIC PRINCIPLES

Telescopy is the field concerned with the development, improvement, and practical application of telescopes. A telescope is any device that enables the viewing or photographing of objects that are either too dim or too far away to be seen without aid. Although they are built to operate in many different ways, all telescopes make use of information gathered from various parts of the electromagnetic spectrum. The electromagnetic spectrum is made up of various forms of radiation—energy that comes from a particular source and travels through space or some other material in the form of a wave. Various types of radiation have different wavelengths and frequencies. The higher the frequency of a wave of radiation and the shorter its wavelength, the higher its energy as it travels.

The most common kinds of telescopes, known as refractors and reflectors, use systems of mirrors and lenses to gather and focus visible light. Both types of telescopes fall under the umbrella of optical telescopy. Radio telescopes gather information not from light but from radio waves, which have the longest wavelengths on the electromagnetic spectrum. X-ray telescopes and gamma-ray telescopes use the kinds of radiation with the shortest wavelengths. Specialized microscopes detect other types of electromagnetic radiation, including ultraviolet light and infrared light, and are used for specific purposes. For example, infrared telescopes are similar to reflecting optical telescopes in construction, but they are designed to collect radiation that is invisible to the naked eye. One reason these telescopes are so useful is that infrared radiation is able to travel through thick clouds of dust and gas in a way that visible light cannot. Thus, infrared telescopes allow scientists to gain insight into the phenomena taking place within hidden regions of space. With these tools and others, modern telescopy is able to detect every region of the electromagnetic spectrum. Researchers use telescopy to create clear images of stars, planets, galaxies, and other celestial objects.

No matter how it is constructed, a telescope is designed to serve three basic functions. First, it should effectively collect large amounts of electromagnetic radiation and focus, or concentrate, that radiation. This makes objects that would otherwise appear very dim seem much brighter and easier to see. Second, it should resolve, or clearly distinguish between, the small details of an image. This makes objects that would otherwise appear blurry seem sharp and focused. Third, it should magnify the image it creates, so that objects at a distance appear larger. Although many people think of magnification as being the primary purpose of a telescope, it is in fact the least important function—if an image is not bright or clearly resolved, no matter how much it is magnified, it will not be useful.

BACKGROUND AND HISTORY

Although it is difficult to trace the invention of the telescope to a single individual, the first person to

have tried to patent a basic telescope (in 1608) was a German-Dutch lensmaker named Hans Lippershey, who used a combination of two lenses separated by a tube to magnify objects by about three or four times. At about the same time, Italian mathematician Galileo built a very similar instrument using a combination of a concave (inwardly curving) lens and a convex (outwardly curving) lens. Galileo promptly showed his version of the telescope to the chief magistrate of Venice and became famous for using it to conduct astronomical observations showing, among other things, that the Earth revolved around the Sun, rather than the other way around. As a result of these well-known activities, Galileo is often wrongly credited as the inventor of the telescope. While Galileo did not invent the telescope, he may have been the first to call the instrument a telescope.

The devices made by Lippershey and Galileo were both refractor telescopes, which rely primarily on lenses that gather and focus light. In the second half of the seventeenth century, English physicist Isaac Newton was among the early pioneers of the reflecting telescope, which relies primarily on curved mirrors that bend light.

Over the next three hundred years, optical telescopes underwent vast technological improvements. For example, achromatic lenses were invented to compensate for the errors in color caused by older lenses that failed to treat all the colors of visible light in the same way. Radio telescopes were first developed in the twentieth century, based on the discovery that faraway celestial bodies were constantly emitting faint amounts of radiation in the form of radio waves. This new form of telescopy was soon applied to both military radar operations and astronomical research.

Other technological advances have affected the field of telescopy. The development of photography enabled astronomers to create permanent still images of the celestial bodies they were observing and to use light-sensitive plates to gather, over long periods of time, even more light than could be collected by lenses or mirrors. Similarly, the invention of increasingly sensitive electronic devices for capturing light, such as the charge-coupled device (CCD), revolutionized modern telescopy. In addition, advances in computer technology allow astronomers and the military to constantly monitor selected portions of the sky using computers that alert human overseers if anything unusual is detected.

HOW IT WORKS

Optical Telescopes. Optical telescopes are designed to gather and focus light that radiates from distant objects. The two main types of optical telescopes are reflectors and refractors. Each type is based on a different principle derived from the physics of light. A third type, a catadioptric telescope, is a hybrid of reflectors and refractors.

Refraction is a phenomenon by which light is bent as it travels from a medium of a certain density to a medium of a different density (such as moving from air into glass). Basic refracting telescopes use a combination of two lenses to refract light. As light rays from a distant object, such as a star, approach a telescope, they travel in nearly perfectly parallel lines. The first lens, known as the primary, bends or refracts these parallel rays of light so that they converge on a single point. This creates an intermediary image of the object that is both bright and in focus. The purpose of the second lens, known as the secondary, is to take that bright, focused image and magnify it by spreading the light rays once more, enabling them to form a larger image on the retina of the eye. In the course of this process, the rays of light cross (light from the top of the object is bent downward and light from the bottom of the object is bent upward), so the image is upside down. Many refracting telescopes use another pair of lenses to render the image right side up.

Reflection telescopes are based on the principle that if light waves meet a surface that will not absorb them, they are redirected away from the surface at the same angle at which they were originally traveling. The angles at which light meets and are deflected from a surface are called the angles of incidence and reflection. Basic reflecting telescopes use a combination of two mirrors rather than lenses to reflect light. The mirrors are usually coated with a thin film of a shiny metal such as aluminum, which makes them more reflective. As light enters the tube of a simple reflective telescope, it is reflected off the primary mirror and travels back in the direction from which it came to form a bright, focused image just as in a refraction telescope. A secondary mirror in a reflection telescope functions similarly to the secondary lens in a refraction telescope, creating a magnified image focused comfortably on the retina.

Some optical telescopes use a combination of reflecting and refracting techniques; these are known as catadioptric telescopes. One of the most common types of catadioptric telescopes is the Schmidt-Cassegrain telescope, which takes its name from two scientists whose work informed its design. This type of telescope contains a deeply curved concave primary mirror at the back of the tube, which reflects light toward a convex secondary mirror at the front of the tube. Schmidt-Cassegrain telescopes also contain a corrective lens that helps counteract the optical aberrations caused by the mirrors, such as making points of light look like disks.

Radio Telescopes. Rather than manipulating light, radio telescopes collect and focus radio waves—the same kinds of electromagnetic radiation that are used to transmit radio, television, and cell phone signals. The reason radio telescopes are useful for purposes such as astronomical observation is that faraway celestial objects, including stars and quasars (incredibly bright, high-energy bodies that resemble stars), are constantly emitting radio waves. There are many different kinds of radio telescopes, but each is made up of the same fundamental parts. The first is a radio antenna, which often looks like a huge, curved television satellite dish. The greater the surface area of the antenna, the more sensitive it can be to the relatively weak radio waves being transmitted from cosmic sources and the fainter and more distant the objects it can detect. The second basic part of a radio telescope is a radiometer, also known as an amplifier. This instrument is placed at the central focusing point of the antenna, and its purpose is to receive and amplify the signal produced by the antenna, convert it to a lower frequency, and transmit it via cable to an output device that charts or displays the information collected by the telescope.

Many radio telescopes use a technique known as interferometry to increase their angular resolving power, which is relatively weak compared with that of optical telescopes. (The reason for the relative weakness is that the angular resolving power of a telescope is defined by the wavelength of the radiation it measures divided by the telescope's diameter. Radio waves, with much longer wavelengths than those of visible light, require telescopes with very large diameters to achieve the same angular resolution as optical telescopes.) An interferometer is a device that takes advantage of the interference phenomenon to electronically combine the signals from multiple telescopes and create a single image. For instance, the

National Radio Astronomy Observatory's Very Large Array (VLA), a prominent radio astronomy observatory in New Mexico, has nearly thirty radio antennae each measuring 75 feet in diameter. By spacing the antennae far apart, the observatory has created an array that functions like a single telescope with a diameter as wide as the distance between the first and the last antenna. When the signals from the antennae are combined through interferometry, the array is able to resolve details in the sky at a much greater power.

Spectrographs. Spectrographs are important auxiliary instruments that are often attached to optical telescopes. Their primary function is to split up the light collected by a telescope and separate it into its individual wavelengths, thereby creating a spectrum. Spectrographs can be extremely complicated, but their basic construction involves an entrance slit, two lenses, a prism, and a charge-coupled device (CCD). The entrance slit is designed to reduce the interference of any background light not coming from the particular star being observed. The first lens is designed to direct the rays of light coming from the star into the prism, which then breaks up the light into its different wavelengths. After the light exits the prism as a spectrum, it is directed by the second lens onto the CCD, which produces a readout of how much light of each wavelength is coming from the star. This information can then be used to analyze various important characteristics of the object under observation. For example, spectrographs can help astronomers learn the chemical composition of a star as well as its temperature and rotation speed.

APPLICATIONS AND PRODUCTS

Astronomical Observations. Perhaps the most important scientific application of telescopy is its use in facilitating astronomical observations. Telescopes are the fundamental tools used by astronomers and astrophysicists to further their understanding of space, celestial objects, and the universe as a whole. Measurements produced with the aid of telescopes, for example, revealed the shape of the Milky Way galaxy and the location of Earth within it. Decades of careful observations through the Hooker optical telescope at the Mount Wilson Observatory in Pasadena, California, enabled astronomer Edwin Hubble to prove not only that the Galaxy is just one among many such systems in the universe but also

that the universe itself is expanding as these galaxies move farther apart. Without the help of telescopes, no human eye would ever have laid sight on such astonishing phenomena as the icy rings that surround Saturn, the gigantic high-pressure storm on Jupiter known as the Great Red Spot, the craggy craters on the far side of the Moon, or the brilliant azure of the atmosphere around Neptune caused by the reflection of blue light by methane gas.

The highest angular resolution achievable by ground-based telescopes is limited by the fact that radiation of some wavelengths does not travel well through the Earth's atmosphere but is absorbed by water vapor and carbon dioxide as it travels. Ground-based telescopes are also affected by atmospheric turbulence—small, irregularly moving air currents—which can cause blurry images. However, because space telescopes orbit the Earth at a high altitude, they are not affected by these problems. Therefore, some astronomical observations can be carried out only by telescopes located above the atmosphere of the Earth.

The Hubble Space Telescope, which was launched into orbit by the National Aeronautics and Space Administration (NASA) in 1990, is the most advanced space-based telescope system ever conceived. It is a large optical visible-ultraviolet reflecting telescope (its primary mirror is nearly 8 feet in diameter) that travels around the Earth several times a day, collecting images of star systems, planets, comets, galaxies, and other celestial bodies. The Hubble Space Telescope is also equipped with a wide-field planetary camera that can record images of space at resolutions several times higher than any telescope based on Earth. In addition, the telescope has a faint-object camera designed to detect extremely dim celestial objects, a faint-object spectrograph that collects information about the chemical composition of these objects, and a high-resolution spectrograph that gathers ultraviolet light from very distant objects.

Although telescopes are not generally used for navigation purposes, one of their most important early contributions was in helping sailors and explorers pinpoint their exact locations on the seas by finding the position of known stars or planets in the sky. Astronomers still rely on telescopes to pinpoint the positions of celestial bodies. In doing so, they are able to create detailed and systematic surveys, or maps, of the sky. For example, since 2000, the Sloan

Digital Sky Survey has been using a large reflecting telescope with charge-coupled devices to pinpoint the location of distant galaxies and quasars.

Military Surveillance. For centuries, telescopes have provided army and navy surveillance teams with an invaluable tool by enabling military personnel to detect the movements of hostile forces from a distance. Initially, only ground-based telescopes were used, but in the late eighteenth century, it became possible to greatly increase the visual range of telescopes by placing them on board hot-air balloons. In World War I, European military forces used refractor telescopes mounted on airplanes to perform aerial surveillance, and in World War II, surveillance airplanes were equipped with sophisticated telescopes with powerful lenses and cameras that produced high-resolution images of military bases and enemy territories far below. Modern aerial surveillance techniques typically involve telescopes with extraordinarily good angular resolutions mounted on unmanned aircraft systems (remotely piloted aircraft) such as the United States Air Force's Global Hawk. In addition, ground-based telescopes are still used by countries around the world to keep an eye on objects in the sky, such as enemy aircraft, missiles and other weapons, and satellite surveillance equipment belonging to other nations. For example, there are two optical surveillance sites in the United States, one in Hawaii and one in New Mexico, equipped with the latest in adaptive optics telescopes.

Increasingly, military surveillance is also conducted in space, with telescopes mounted on satellites. Satellite telescopes can move at vastly greater speeds than airplanes and have the advantage of being able to navigate to any region above the desired surveillance target without having to contend with national airspace boundaries. Although detailed information about the tools used by military surveillance units is closely guarded, it is generally thought that most satellite-based telescopes travel in low-Earth-orbit altitudes, about 62 to 310 miles above sea level. They probably conduct observations by collecting electromagnetic radiation with short wavelengths, such as infrared light and green light and are likely to be about 20 to 26 feet in diameter. Based on these parameters, experts have calculated that military satellite telescopes are most likely capable of distinguishing details that are less than an inch apart on the Earth—enough resolving power to read a newspaper headline.

Environmental Applications. Governments and other organizations also use satellite-based telescopes in surveillance applications. One of the most common uses is to monitor changes in the environment. NASA's Earth-observing system satellite *Terra* carries multiple sophisticated telescopes onboard that are used to detect such phenomena as volcanic activity, emerging forest fires, and floods. *Terra* also provides scientists with images so that they may track the effects of climate change on the Earth's surface; for example, scientists track the melting of the ice sheets in the Arctic over time. Brazil, a country whose rain forests have been reduced by centuries of cattle ranching and other agricultural activity, uses satellite-based telescopes to closely monitor the extent of deforestation and to evaluate how well its efforts to preserve the rain forests are working.

Communications. One application of telescopy that is still largely in the research-and-development stage is its potential use in laser communications between deep space and the Earth. A laser is a device that uses excited atoms or molecules to emit a powerful beam of electromagnetic radiation in a single wavelength (called monochromatic light). The light produced by lasers is intense and directed, meaning that the light rays do not spread out very quickly. This makes lasers a useful tool for transmitting a secure information-carrying signal directly to a receiver in a specific location. NASA, for example, has been looking into using laser signals to transmit data (including photographs, radar images, and analyses of space dust) collected by space probes such as the *Cassini*, which is orbiting Saturn. However, collection of the laser signal on Earth would require a huge telescope.

Recreational Applications. Telescopes are far from just a tool for scientists, military personnel, or government officials. Durable, lightweight general-purpose optical telescopes known as spotting scopes are used frequently in everyday life in a wide variety of recreational applications. These instruments have greater magnification and resolving powers than binoculars. Naturalists, for instance, use spotting scopes to identify plumage markings and observe the behavior of bird species at a distance, without alerting the birds as to the presence of humans . Airplane and train spotters use them to distinguish fine details in faraway vehicles. Long-range game hunters use spotting

---

Fascinating Facts About Telescopy

- Between 1990 and 2009, the Hubble Space Telescope traveled around the Earth more than 100,000 times, covering a total distance of nearly 3 billion miles.
- One story about the invention of the telescope claims that German-Dutch lensmaker Hans Lippershey was inspired to develop an early version of a refractor telescope by watching two children playing with lenses in his shop.
- Scientists are working on making a miniature telescope that can be implanted behind the iris of the eye. It projects magnified images onto the retina, helping patients with the type of blindness known as macular degeneration see once again.
- Satellites equipped with telescopes that capture images of the surface of the Earth are used by governments to spy on other nations, as well as to observe the effects of fires, flooding, and deforestation.
- The first planet to have been discovered with the aid of the telescope was the large blue-green orb known as Uranus.

- It is possible to build a very simple refractor telescope using two magnifying glasses and a cardboard tube.
- Since light travels at a constant, finite speed, the further away an object is, the longer it takes for light from that object to reach the Earth. To look at a very distant star or galaxy is to see it as it was thousands, millions, or even billions of years ago. In that sense, telescopes are a little like time machines.
- In the late nineteenth century, many telescopes contained spider webs. The fine, strong silk produced by spiders made the perfect material for crosshairs, a pair of crossed lines in the optical viewfinder that help the viewer position and focus a telescope.
- Some telescopes exist just to help other telescopes do their job. Astronomical transit instruments are small refracting telescopes that precisely determine the locations of specific stars and planets, enabling larger telescopes to focus directly on them.

---

scopes to view their prey as they take aim, and scopes to view their prey as they take aim, and sharpshooters use them to check on the position of their targets.

IMPACT ON INDUSTRY

Government and University Research. Most developed nations around the world devote a sizable portion of their annual budgets to supporting the development of both telescopy technology and astronomy research applications involving telescopes. In the United States, for example, NASA spends more than a billion dollars a year on astrophysics research, including the design and construction of new telescopes such as the one attached to the Kepler spacecraft launched in 2009 and charged with finding evidence of Earth-like planets. Other U.S. government agencies, such as the Energy Department, the National Science Foundation, and the Defense Advanced Research Projects Agency (DARPA), add millions more to research in telescopy.

One reason for these large budgets is the generally strong public support for astronomical observations, which are seen as being both inspiring and relatively untainted by scientific controversy. Another is that, much like spacecraft, large, state-of-the-art telescopes

function as a kind of global status symbol for the nations that house them. China, for instance, spent $34 million to construct the Large Sky Area Multi-Object Fiber Spectroscopy Telescope (LAMOST), which was completed in 2009. The LAMOST is a huge optical telescope containing two dozen hexagonal mirrors, each more than 3.5 feet in length. It is designed to collect information about the spectra of celestial objects, including their densities, atmosphere, degree of magnetism, and precise chemical composition. Because astronomical projects are so expensive, countries often collaborate on telescopy research and development. For example, the European Space Agency, NASA, and the Canadian Space Agency are sharing the cost of building the James Webb Space Telescope, set to be launched in 2014 to serve as the successor to the Hubble Space Telescope.

Major Corporations. For many years, a small handful of large corporations based in the United States has dominated the global amateur telescope market. These companies, the most significant of which are Meade, Celestron, and Orion, design, manufacture, market, and distribute mostly portable entry-level to high-end optical telescopes for home astronomers. Their products range from a few hundred dollars for a budget telescope to several thousand for an instrument at the high end of the range. The market for telescopes produced for serious amateurs is extremely competitive and largely driven by technological developments that take place almost as often as those in the computer industry. Even telescopes that are designed for use in the home rather than in industrial or research settings are generally equipped with the latest in advanced optics and electronic features, such as built-in Global Positioning System (GPS) capabilities that enable stargazers to quickly locate the portion of the sky in which particular celestial bodies can be found. Although the technology packaged into high-end telescopy products has developed rapidly, the prices of such products have not risen at a comparative pace. One reason is the entrance of East Asian telescope manufacturing companies into the global market. In particular, China has become a serious player in telescopy, with dozens of new Chinese telescope companies.

CAREERS AND COURSE WORK

Preparation for a career in any of the major fields related to telescopy, namely astronomy, astrophysics, meteorology, or military surveillance, should begin with a complete course of high school mathematics, up to and including precalculus. In addition, chemistry and physics are important subjects to cover in high school. Outside the academic environment, astronomy clubs or observatories are excellent places to gain practical experience using telescopes and to learn about the details of their operation. At the undergraduate level, a student interested in a career involving telescopy should work toward a bachelor of science degree with a concentration in a field such as physics, astronomy, mathematics, or computer science. No matter what major is chosen, additional course work in optics, electromagnetism, thermodynamics, mechanics, atomic physics, cosmology, statistics, and calculus provides essential background knowledge for further study—an important consideration, because almost all research positions in astronomy or related sciences, as well as any job involving the development of telescope technology itself, require the completion of a graduate degree, preferably a doctorate.

The typical career path for a student interested in telescopy involves pursuing work as an astronomer either in an academic or a government-based observatory such as the National Radio Astronomy Observatory or the Mauna Kea Observatories at the University of Hawaii. Most astronomers who work at universities teach in addition to conducting research. Because jobs for practicing astronomers can be relatively scarce, students may benefit from taking one or more short-term positions such as a paid internship or postdoctoral fellowship to gain experience and contacts in the field before seeking a more permanent appointment.

Another career option is to approach telescopy from the point of view of engineering rather than scientific research. Electrical engineers and other technicians are essential members of the teams at astronomical observatories. These jobs, which involve repairing, upgrading, testing, and maximizing the efficiency of high-powered telescopes, generally do not require graduate degrees. A bachelor's degree in engineering or electronics and a strong background in mathematics and physics are sufficient qualifications to pursue this kind of telescopy career.

SOCIAL CONTEXT AND FUTURE PROSPECTS

Over the four hundred years of its existence, the telescope has enabled humankind to transcend not just visual limitations but also mental ones. The telescope has been the impetus for a flood of astonishingly deep revelations (and further questions) about the origin of matter itself, the place of the Earth within the universe, and the future of the universe.

For the entire length of recorded human history, humans have constructed stories and mythologies about how the world came into being. Telescopes have provided a way to approach that issue from a scientific point of view. With the help of ever larger telescopes such as the Giant Magellan Telescope in Chile (scheduled for completion in 2018), physicists hope to be able to see what the universe looked like just a few hundred million years after the big bang and thereby to gain an understanding of how the very first stars, planets, and galaxies were formed. For millennia, humankind believed that Earth held a central place in the universe. Telescopes have turned that worldview on its head by showing that, in fact, there may be billions of Earth-like planets in the Milky Way galaxy alone and countless more across the entire universe. Although scientists once believed that the universe was unchanging, they have come to know—because of telescopy—that it is dynamic and expanding.

In an age in which the devastating effects of environmental pollution and the growing impact of climate change dominate the headlines, telescopy—by giving humans a holistic view from afar of the beautiful, vulnerable planet they inhabit—has a particularly important role to play in inspiring those who live on Earth to preserve and protect it for future generations.

M. Lee, B.A., M.A.

FURTHER READING

Andersen, Geoff. *The Telescope: Its History, Technology, and Future.* Princeton, N.J.: Princeton University Press, 2007. This guide to optical telescopes, heavily illustrated with photographs and diagrams, covers applications such as laser communications, surveillance, and astronomical observations. Includes an appendix explaining the mathematics of telescopy.

Burke, Bernard F., and Sir Francis Graham-Smith. *An Introduction to Radio Astronomy.* 3d ed. New York: Cambridge University Press, 2010. An advanced textbook of techniques and tools relating to radio telescopy, designed for the graduate student of astronomy and requiring at least a basic technical background.

Koupelis, Theo. *In Quest of the Universe.* 6th ed. Sudbury, Mass.: Jones and Bartlett, 2010. An undergraduate-level textbook dealing with both optical and radio telescopy and their applications in astronomy. Contains numerous photographs, diagrams, and highlighted definitions, as well as a study guide with practice questions.

Pugh, Philip. *The Science and Art of Using Telescopes.* New York: Springer, 2009. Discusses how to use telescopes to view celestial objects, including how to photograph images.

Schilling, Govert, and Lars Lindberg Christensen. *Eyes on the Skies: Four Hundred Years of Telescopic Discovery.* Chichester, England: John Wiley & Sons, 2009. The official book of the International Year of Astronomy (2009) contains a history of the telescope and the discoveries it enabled. Many illustrations.

Zirker, Jack B. *An Acre of Glass: A History and Forecast of the Telescope.* Baltimore, Md.: The Johns Hopkins University Press, 2006. A comprehensive, accessible introduction to the technological and scientific aspects of telescopy, complete with dozens of full-color and black-and-white photographs and diagrams. Contains a glossary and endnotes explaining important concepts.

WEB SITES

American Astronomical Society
http://aas.org

Giant Magellan Telescope
http://www.gmto.org

National Aeronautics and Space Administration
HubbleSite
http://hubblesite.org

National Radio Astronomy Observatory
http://www.nrao.edu

Sloan Digital Sky Survey
http://www.sdss.org

See also: Military Sciences and Combat Engineering; Mirrors and Lenses; Optics; Planetology and Astrogeology; Space Science.

TOXICOLOGY

FIELDS OF STUDY

Analytical chemistry; biochemistry; biology; chemistry; clinical chemistry; environmental science; forensics; mathematics; pharmacology; toxicology; veterinary medicine.

SUMMARY

Toxicology involves the study of toxicants, whether biological, chemical, or physical, and how they affect people, animals, and the environment. Toxicologists determine whether these chemicals are actually or potentially harmful by using their knowledge of chemistry and biology and help develop and implement strategies to eliminate, reduce, or control exposure to those harmful substances.

KEY TERMS AND CONCEPTS

- **Analytical Chemistry:** Study of the chemical composition of natural and artificial materials.
- **Biochemistry:** Study of the chemical substances and vital processes occurring in living organisms.
- **Chain of Custody:** Process used to maintain and document the chronological history and person responsible for evidence used in a criminal investigation.
- **Environmental Science:** Science of the interactions between the biological, chemical, and physical components in the environment including the effects of these interactions on all types of organisms.
- **Forensics:** Use of science, scientific methods, and technology to investigate a crime and establish facts that are admissible in a court of law.
- **Pharmacology:** Study of drugs and their sources, nature, and properties, and how an organism reacts to them.
- **Poison:** Toxicant that causes immediate death or illness when experienced in even a small amount.
- **Reagent:** Substance used in a chemical reaction to detect, measure, and examine a substance or to produce other substances.
- **Toxicant:** Substance that may produce adverse biological effects of any nature.

- **Toxicodynamics:** Study of the effects poison has on the body systems and structures.
- **Toxicokinetics:** Study of how the body processes poisons, including the body systems that are involved.
- **Toxin:** Specific protein produced by a living organism.
- **Xenobiotic:** Foreign substance taken into the body.

DEFINITION AND BASIC PRINCIPLES

Toxicologists study the adverse effects of biological, chemical, or physical agents on living organisms (humans, animals, and plants). Adverse effects can manifest in many forms, ranging from immediate death to subtle changes at a molecular level that do not become known until years later. These effects can also manifest themselves at various levels in the body. For example, some chemical agents affect a certain body organ, others damage a particular type of cell, and even others may interfere with a specific biochemical reaction in the body necessary for life to continue. As medical knowledge has progressed, the understanding of how toxic agents affect the body has changed. A body can be affected on a cellular level by unseen toxins, the damage of which will not be know for many years.

This realization has led to an expansion in the field of toxicology. Toxicologists are now tasked with examining the physical environment to determine whether, how, and at what levels environmental toxins affect humans and other living things. These types of examinations can affect many industries, such as those that emit toxins into the environment and even those that dispose of toxic and hazardous waste and develop agents for biological warfare. Other fields in which toxicology is key is that of animal science (veterinarians who determine treatment for animals who are affected by toxins) and drug development (scientists who determine how certain therapeutic drugs affect the human body and determine safe and effective dosages).

BACKGROUND AND HISTORY

Toxicology and the study of poisons has a long and interesting background, possibly beginning with early humans, who recognized poisonous plants and animals and used them in the process of killing, whether for food or in war. Writings as early as 1500 b.c.e. depict substances such as hemlock, opium, and certain metals that were used on arrows to kill animals or humans or even as agents in state execution processes. Stories are told of "poison maidens," beautiful young girls who were fed tiny amounts of poison on a daily basis, causing them to become immune to the effects of the poison, until they became poisonous themselves. They were then sent as gifts to rival kings who died when they touched the poisonous girl.

Poisoning as a method of assassination become more popular in the eighth century, when an Arab chemist discovered how to turn arsenic into an odorless, tasteless, nearly undetectable powder. This substance became an easily available murder weapon, and by the Renaissance period, poison rings, knives, letters, and lipstick were in use for those who wished to do away with a political or amorous rival easily and quickly.

Philippus Aureolus Theophrastus Bombastus von Hohenheim (known more commonly as Paracelsus), a sixteenth-century Swiss physician, was formulating ideas about poisons and toxicology that are still in use. He carefully studied plant and animal poisons and determined that specific chemical compounds, rather than the plant or animal itself, which was immune to the poison it carried, were responsible for toxicity. He documented how the human body responded to those specific chemical compounds and understood that doses of a particular compound could be beneficial or toxic, depending on the amount given (known as the dose-response relationship). A major concept of toxicology, credited to Paracelsus, is that "all substances are poisons; there is none which is not a poison. The right dose differentiates a poison and a remedy." Drug companies continue to use this idea, as many drugs, such as warfarin, were developed from substances that caused immediate death. In the case of warfarin, it began as a type of poison for rats that caused their blood to thin, and they would bleed to death. Therapeutic doses of warfarin help stroke victims (or possible victims) to keep from forming blood clots.

French toxicologist Mathieu Joseph Bonaventure Orfila is referred to as "the father of toxicology" and was the first major proponent of forensic toxicology.

In the nineteenth century, he prepared a systematic correlation between chemical and biological properties of poisons. He analyzed autopsy materials to show the effects of poisons on specific organs by showing tissue damage and made chemical analysis a routine part of forensic medicine. Orfila is credited as being one of the first to use a microscope to look for blood and semen stains, and he became an expert witness in the sensational murder trials of his time.

Poisoning is still a relatively major cause of death. In the United States, from 2001 to 2004, there were more than 147,000 deaths related to poison; of these 434 were considered homicides, though more may have been murders as it is sometimes difficult to distinguish a poisoning from a natural death or an accident.

HOW IT WORKS

Toxicologists work in laboratories, performing tests on substances of different types—often human tissue. They must be familiar with and know how to operate highly sophisticated laboratory equipment and understand the functioning of chemical reagents. They must understand and apply highly sophisticated and exact methodologies to determine reliably the presence or absence of a substance in a sample. Each step of every complicated process must be documented to ensure that procedures have been exactly followed, especially in circumstances involving a chain of custody for criminal cases.

Toxicologists must also make informed conclusions about the impact of a certain amount of a specific substance and what effect it would have on a certain individual (based on weight, for example) or what effect a substance would have on a particular environment. These educated opinions are often based on professional, educational, and scientific experience and are sometimes required in court testimony.

APPLICATIONS AND PRODUCTS

Toxicologists can focus their efforts in a variety of areas. Below are a few of the major areas of specialization.

Forensic Toxicology. These scientists usually work as part of a crime-scene team. They perform tests on bodily fluids and tissues to determine whether any drugs or chemicals in the body may have contributed to a crime, such as alcohol, chemicals, drugs (illegal or prescription), gases, metals, or poisons.

Alternatively, a forensic toxicologist may work in drug testing, trying to discover evidence of date-rape or performance-enhancing drugs, or in animal-tissue testing for evidence of wildlife crime or environmental contamination, such as chemical spills.

Environmental Toxicology. These professionals focus on the interaction of chemicals on living systems, including how areas and environments are affected by toxic waste or released industrial chemicals. They may also work with workplace exposure to chemicals and metals and understand principles of toxicodynamics.

Medical Toxicology. This type of toxicologist usually works in a laboratory performing tests on bodily fluid and tissue samples to determine whether there are chemicals present. Though their work is similar to that of a forensic toxicologist and may even involve criminal investigations, this type of toxicologist works more with medical cases, such as chemotherapy adjustments or accidental exposures, rather than criminal cases.

Pharmacological Toxicology. Drug companies use toxicologists to help determine the chemical toxicity of drugs under development. These professionals help determine therapeutic levels for drugs and evaluate whether the proposed drugs build up in tissues or are eliminated from the body to determine maximum safe dosages and durations. They may also help determine under what conditions certain drugs should be avoided by monitoring interactions of drugs with other drugs a patient may be taking or other conditions a patient may have. Their knowledge of toxicokinetics can be helpful in these situations. This knowledge also helps determine age-related effects of certain toxic agents, such as whether a drug affects children and the elderly differently than it affects adults.

IMPACT ON INDUSTRY

Environmental companies, toxic-waste-disposal industries, and drug development are just a few of the places that toxicologists prove their worth in industry. In the environmental field, toxicologists help determine whether chemicals released into the environment are likely to harm ecosystems, including animals and humans. They also test and monitor waste-disposal methods, such as factory effluent or toxic-waste disposal from energy production, that may be harmful to the environment.

Toxicologists are also key to pharmaceutical companies. As new drugs are developed, they monitor the effects on the human body to determine whether the efficacy of the drug outweighs the risks.

The United States Department of Health and Human Services houses a National Toxicology Program that focuses on shaping public health policy involving any toxicological agents of public concern. Toxicologists who work in this area develop and apply the tools of modern toxicology and molecular biology to evaluate toxicological substances, develop and validate tests to discover these agents, and communicate these tools and tests to public health agencies.

CAREERS AND COURSE WORK

A toxicologist may have an undergraduate degree in biology, chemistry, environmental science, or pharmacology. Most toxicologists have an M.D. or D.V.M. (Doctor of Veterinary Medicine) degree. Some universities, such as the University of Maryland, offer programs with master's or doctorate degrees in different aspects of toxicology. A good program will be accredited by the American Academy of Forensic Sciences. Certification as a toxicologist is available from the American Board of Forensic Toxicology, the American Board of Clinical Chemistry, and the American Board of Toxicology.

Toxicologists are employed by colleges and universities as teachers and researchers. They may also find work in government laboratories, where they investigate the safety and effectiveness of different types of chemicals. Another industry that employs toxicologists is the veterinary field. These toxicologists work in diagnostic laboratories examining the effects of chemicals on animals. This type of work may also be necessary in the drug-testing field.

SOCIAL CONTEXT AND FUTURE PROSPECTS

Toxicologists are necessary in many aspects of environmental industry. They are important in many ecological fields, and some professional societies of toxicologists focus exclusively on this area. For example, the Society of Environmental Toxicology and Chemistry concentrates efforts on the study and analysis of environmental problems.

It also focuses on environmental education and the management and regulation of natural resources. Its goal is to find soultions to environmental

---

Fascinating Facts About Toxicology

- Toxicology is sometimes called "the science of poisons."
- Most toxicologists, especially forensic toxicologists, work in labs that are part of law-enforcement agencies. Others work with medical examiners to determine cause of death. Private drug-testing facilities or poison-control centers are another source of employment for these scientists.
- Toxicologists work every day with body fluids and tissues. It can be messy, smelly work.
- Toxicologists must be mentally strong. They are often exposed to details of horrific crimes and must make judgments about whether a crime was committed.
- A forensic toxicologist is often called on to testify in court as to the effect a certain amount of a substance would have on a particular person. He or she must explain complicated testing methods in language that a jury can understand.
- Some famous victims of poisoning include Socrates (hemlock) and Cleopatra (snakebite). The Emperor Claudius (Tiberius Claudius Drusus Nero Germanicus) was said to have been poisoned by his wife. Some say she served him poison mushrooms. Another story says that he was suspicious of her and would only eat figs he himself had picked from the tree, so she went into the garden and poisoned figs still on the tree.
- In the fifteenth century, Lucrezia Borgia was one of the Borgia family members famous for poisoning rivals. She was said to have worn a ring that contained poison that she poured into drinks of men and women who were threatening to her family and its status.
- Viktor Yushchenko, a popular Ukrainian politician, was said to have been poisoned by government agents after announcing that he would run for president. After a dinner with Ukrainian officials, his face became pockmarked and disfigured. Toxicologists found that he had more than 1,000 times the normal amount of TCDD dioxin in his body.
- Toxicology is a constantly changing field. Successful toxicologists are constantly learning, keeping pace with new chemicals, methodologies, and technologies. A good toxicology candidate is someone who is fascinated by chemicals and the effect they can have on the human body.

---

problems that people can live with on a long-term, everyday basis that support sustainable environments and ecosystems.

As the field of health care expands, toxicologists have opportunities to become more and more involved. New drugs are constantly being developed, and toxicologists are heavily involved with drug testing, both on animals and on humans. Their knowledge of the human body and how chemicals interact with it is crucial in this field.

Marianne M. Madsen, M.S.

FURTHER READING

Evans, G. O., ed. *Animal Clinical Chemistry: A Practical Handbook for Toxicologists and Biomedical Researchers.* 2d ed. Boca Raton, Fla.: CRC Press, 2009. Covers pre-analytical and analytical variables along with information on specific-organ toxicity.

Fenton, John Joseph. *Toxicology: A Case-Oriented Approach.* Boca Raton, Fla.: CRC Press, 2002. Includes case studies and information about diagnosis, testing, and treatment.

Hayes, A. Wallace, ed. *Principles and Methods of Toxicology.* 5th ed. Boca Raton, Fla.: CRC Press, 2008. Discusses principles of absorption, distribution, metabolism, and excretion; helps with understanding and using basic experiments in toxicology.

Klaassen, Curtis D. *Casarett & Doull's Toxicology: The Basic Science of Poisons.* 7th ed. New York: McGraw-Hill, 2008. The "gold standard" of toxicology, includes detailed discussions of concepts, principles, and mechanisms of toxicology.

Nelson, Lewis S., et al. *Goldfrank's Toxicologic Emergencies.* 9th ed. New York: McGraw-Hill, 2011. Includes comprehensive references; begins with general principles and moves to detailed discussions of biochemical principles; discusses various exposures—drugs, plants, metals, household products, as well as occupational and environmental.

Osweiler, Gary D., et al., eds. *Blackwell's Five-Minute Veterinary Consult Clinical Companion: Small Animal Toxicology.* Ames, Iowa: Wiley-Blackwell, 2011. Overview of toxicology in veterinary practice; includes color photos and tables in an appendix to help with quick differential diagnoses.

Richards, Ira S. *Principles and Practice of Toxicology in Public Health.* Sudbury, Mass.: Jones and Bartlett, 2008. Introduction to the field of toxicology and its practice in the public-health environment.

Wright, David A., and Pamela Welbourn. *Environmental Toxicology.* Cambridge, England: Cambridge University Press, 2002. Overview of interaction of chemicals and the environment from molecular to ecosystem levels; includes case studies.

WEB SITES
American College of Medical Toxicology
http://www.acmt.net

Society for Environmental Toxicology and Chemistry
http://www.setac.org

Society of Toxicology
http://www.toxicology.org

United States Department of Health and Human Services
National Toxicology Program
http://ntp.niehs.nih.gov

See also: Pathology; Pharmacology; Veterinary Science.

TRIBOLOGY

FIELDS OF STUDY

Applied mathematics; physics; chemistry; material science; mechanical engineering; fluid mechanics; thermodynamics; rheology; polymer chemistry; biomechanics; biophysics

SUMMARY

All physical materials interact where their surfaces interface. The interaction is characterized by friction, abrasion, and the generation of heat. These effects have deleterious effects in every instance. Tribology, the study of those effects, works to eliminate negative effects and to find positive ways to harness them. Tribological effects, primarily friction, play a role in every mechanical aspect of existence. They are of particular significance in the ultrasmall devices of nanotechnology and in the biomechanics of living systems. The development of scanning probe microscopy has made it possible to acquire an understanding of tribology at the atomic scale.

KEY TERMS AND CONCEPTS

- **Abrasion:** Material deformation or removal that results from frictional contact.
- **Friction:** The resistance to lateral relative motion of two surfaces in contact with each other.
- **Interface:** The surface on which two differentiated materials make contact with each other.
- **Lubrication:** The interposition of a third material between two frictional surfaces for the purpose of minimizing the coefficient of friction between them.
- **Nanotribology:** The study of friction, abrasion, heat, and lubrication at the scale of the nanometer.
- **Oil Whirl:** A vibrational instability of the lubricating fluid in high-speed journal bearings.

DEFINITION AND BASIC PRINCIPLES

The word "tribology" means "study of rubbing." In every practical sense it applies to the study of the interactions of physical matter at an interface (that is, where one surface contacts another). These interactions are characterized by friction, abrasion, and the generation of heat, all of which affect the subsequent behavior of the surfaces and the dimensional characteristics of the material.

Friction can be described as the resistance to relative lateral motion between two surfaces, while abrasion describes the deformation and forcible removal of material from one surface by material of another surface. Both of these effects facilitate the release of energy, altering physical surface structure at the atomic and molecular level of the materials. The energy released by the alteration of surface structure by friction and abrasion becomes sensible as heat conducted through the mass of the material.

Tribology examines and quantifies the relationships of friction, abrasion, and heat as they relate to the physical performance of mechanical devices. An especially significant field of research in tribology is the study of the qualities of lubrication, as lubricating materials are used to counteract tribological effects. At the same time, however, the lubricating materials themselves become active contributors to tribological effects, and their study seeks to identify and quantify their corresponding effects.

BACKGROUND AND HISTORY

Friction is one of the oldest known technological effects. Archeologists have unearthed implements dating from the Paleolithic Age (early Stone Age) that had been fitted with pieces of antler or bone to act as antifriction bearings. Chariots found in tombs dating from ancient Egypt (from about five thousand years ago) contained the residue of animal fats in the axle-bearing surfaces of their wheels, indicating that the Egyptians of the time understood the value of lubrication. Tomb paintings also indicate that the use of lubrication was an essential component in the movement of large stone blocks used in construction. The physical concept of the coefficient of friction was deduced by Renaissance artist and thinker Leonardo da Vinci in the fifteenth century, but remained generally unknown because his notebooks were not published until some centuries later.

The rules of friction were rediscovered in 1699 by French physicist and inventor Guillaume Amontons and were later verified by French physicist

Charles-Augustin de Coulomb. These rules acquired great significance with the mechanization developed during the Industrial Revolution of the eighteenth century. In modern times, the precision with which mechanical devices are built demands that the effects of friction, abrasion, and heat be fully understood, from the atomic scale upward, so that their detrimental effects can be minimized or eliminated.

How It Works

Friction. The classical view of tribology is focused on the study of the causes and effects of friction. The simplest explanation comes from the view that no matter how smooth a surface may appear to be, as the scale of resolution becomes ever smaller, even the smoothest of surfaces becomes more and more irregular.

This process is exemplified by the examination of a billiard ball, a hard and extremely smooth spherical object. If one could expand the scale of the billiard ball to the size of the planet, maintaining the surface irregularities to scale, then the surface of the billiard ball would be covered with bumps and ridges higher than Mount Everest and depressions deeper than the Mariana Trench. In simplest of terms, friction results from the binding of the irregularities of one surface in those of another.

The processes of friction are much more complicated than this simple view, however. Since the development of the current atomic theory and quantum mechanics, it is now known that many other effects play a role in the causes of friction. Researchers in tribological phenomena are only now beginning to acquire an understanding of details of the process at the atomic level, where tribology begins. This new understanding has been made possible by the development of scanning probe microscopy and, particularly, of the atomic force microscope.

Scanning probe microscopes allow examination of surfaces at the atomic scale, with resolutions as fine as 10 picometers (10-11 meters). Even the most cursory examination of a surface image from a scanning probe microscope reveals that at the atomic level an assumed perfectly smooth surface consists instead of a series of bumps and depressions reminiscent of what one would observe in a layer of golf balls, marbles, or any other spherical object. Additionally, quantum effects such as van der Waals forces, magnetism, and electronic interactions, and the chemical nature of the material, are important components of friction at the atomic level. The accumulation of effects from the atomic level to the normal size of the object determines how friction is generated between material objects.

The basic principles of friction are deduced from empirical observation. First, the frictional force that resists the sliding of one surface against another is directly proportional to the normal load between them. In other words, the more pressure that is mutually exerted against the two surfaces, the harder it is to slide them across each other. Second, the amount of frictional force does not depend on the size of the area of contact between the two surfaces. This can be examined simply, and cursorily, by sliding an irregularly shaped object across a tabletop, using different surface areas each time. Third, once the sliding motion has begun, the frictional force is independent of the velocity. That is to say, sliding two surfaces against each other at a high velocity requires the same force as it does at a low velocity.

Abrasion. Abrasion is friction to the extreme, resulting in the deformation and displacement of material from one or both interacting surfaces. Abrasion is not the result of matter passing across the surface of other matter. Rather, abrasion is the result of matter physically passing through the same space occupied by other matter. The harder or tougher of the two materials will correspondingly force the other into a new relative position, to the point of separating from the main mass.

It is also possible for material abraded from one surface to transfer to the other surface in a chemical sense and in a physical sense. Research on friction between Teflon and aluminum surfaces, for example, has revealed the formation of a certain amount of aluminum trifluoride on the aluminum surface, a condition made possible only by chemical reaction between the aluminum metal and heat-induced breakdown products of the perfluorinated chemical structure of the Teflon surface.

Heat. Heat is the third major component of tribological effects, easily examined by rubbing one's hands together briskly, first dry and then wet. The heat produced through friction can be intense, leading to dimensional changes that in turn aggravate both friction and abrasion and perhaps lead to the failure of the mechanism.

APPLICATIONS AND PRODUCTS

Tribological effects can have both positive and negative effects, both of which are crucial to the functioning of modern machinery. The applied science of tribology is a multi-aspect study that seeks first to identify the genesis of tribological effects. It then seeks to identify the ways in which negative effects can be reduced or eliminated and positive effects used or enhanced.

Friction and wear (abrasion) occur simultaneously in all physical systems. Environmental erosion and skeletal joints obey the same principles of tribology as do steel bearings and internal combustion engines. Examples of positive applications of friction include braking and clutching systems; the drive wheels of trains, cars, and other vehicles; and bolts, nuts, and other devices whose proper function depends on the application of friction. Positive wear or abrasion includes such diverse applications as pencils, pens, and other writing or drawing materials; various machining and polishing techniques; and even a morning shave. Negative friction includes the resulting dimensional changes and physical damage that occur with internal combustion engines, gears, cams, bearings, and seals, and even such minor inconveniences as getting stuck halfway down a playground slide.

Lubrication and Lubricants. The essential principle of lubrication is simply to add a third material to a system to lower the coefficient of friction between them as much as possible. In the worst possible sense, lubrication can have disastrous results, as when water-soaked soil slides under the force of gravity as an avalanche or mudslide. Controlled lubrication, on the other hand, is essential to the long-term functioning of machines and other mobile structures, including biological and biomechanical systems.

Lubrication is a surprisingly complex system in its own right, because a lubricating material interacts somewhat differently with each other material in the system. For example, in a system in which an oil is used between steel and aluminum components, the oil molecules will have a different level of adhesion and adsorption to the aluminum surface than to the steel surface, resulting in a dynamic movement of material within the oil that may affect how the system functions over time.

Lubricating materials come in a variety of forms and viscosities, ranging from plain dry air to microgranulated solid particles, such as graphite powder. Typically, the selection of a lubricant depends on the amount of pressure that it must bear in application. Teflon is rather unique in this regard because it is a material that becomes more slippery as the pressure it bears increases; typical lubricants tend to lose their lubricating properties as the pressure placed upon them increases. There are literally as many possible lubricating materials as there are materials and material combinations, presenting an impossible challenge to tribological research. The vast majority of lubricants in common use therefore fall into a few general classes: liquids and semiliquids (such as oils and greases) and solid lubricants (such as graphite). Within these classes there are hundreds of variations.

A special class of liquid lubricants are those that function as abrasive carriers as they lubricate. The almost exclusive use of such lubricants is in deep boring operations; for instance, in petroleum and natural gas recovery. The extreme pressures encountered during deep well boring in rock formations demand the use of water as the lubricant, while simultaneously transporting abrasive and abraded material at the drill head to assist the boring process. Various cutting fluids and honing oils used in machine-shop operations for fine grinding and polishing procedures serve a similar function.

Tribological Research and Control Devices. At the lowest end of this technology are grease guns and oil cans for the crude application of lubricants. At the highest end are scanning probe microscopes, enabling researchers to examine the causes and processes of friction at the atomic level. Between the two ends are numerous specially designed devices that test and measure the properties and capabilities of lubricating materials and machine components under operating conditions likely to be encountered in the working environments of those devices. In most cases, such as for a synthetic oil blend in standard roller bearings, this is an almost trivial exercise. In other cases, the working environment is extreme, ranging from the deep ocean floor to deep space, demanding that the materials and designs function flawlessly the first time and for the lifetime of the device.

IMPACT ON INDUSTRY

It is estimated that fully one-third of all energy

Fascinating Facts About Tribiology

- Early Stone Age people used boring tools that had been fitted with pieces of bone or antler to act as bearings.
- Chariots found in Egyptian tombs dated to more than five thousand years ago have traces of animal fat, used as a lubricant, on the axle-bearing surfaces of their wheels.
- Leonardo da Vinci deduced the basic physical laws of friction in the late fifteenth century, but these laws remained unknown largely because da Vinci's personal notebooks were not published. The laws were rediscovered by Guillaume Amontons in 1699.
- Atomic force microscopes can examine the causes of friction between single atoms.
- If it were expanded to the size of the planet, the surface of a glass-smooth billiard ball, for example, would have peaks higher than Mount Everest and valleys deeper than the Mariana Trench.
- Images from scanning probe microscopes demonstrate that no surface is perfectly flat, but that all surfaces are made up of bumps and holes, like a layer of golf balls in a box.
- Ignorance of the effects of friction costs about 4 percent of a nation's gross national product, which amounts to about $200 billion per year in the United States.
- About one-third of all energy consumed around the world is used to overcome the force of friction.

consumed throughout the world serves no useful purpose other than to overcome the force of friction. From this standpoint alone, it is obvious that effective ways to reduce or eliminate frictional forces, through lubrication or design, can generate a large return on investment. Even simply raising awareness of tribological effects throughout the general population would be helpful. Current estimates place the economic cost of general ignorance in this area at about 4 percent of a nation's gross national product, or some $200 billion per year in the United States alone.

Tribological effects in industry are responsible for equipment and machinery wear that requires a certain level of continual maintenance and that inevitably results in the obsolescence or failure of the machinery. Enormous costs accompany the ongoing need to replace parts and machinery. Many types of replaceable parts are thus categorized as "consumables" rather as than primary components. In many cases this is unavoidable, as the cost of researching and developing materials to better withstand the rigors of tribological processes would greatly exceed the cost of disposable replacements and the downtime required for replacement.

A more subtle effect of tribological processes in industry is the gradual erosion of acceptable quality standards in the production of goods. Quality control has become an essential career position, as production facilities now work to closer tolerances and every rejected part that is out of specification represents an impact on the profitability and viability of a manufacturer. Gradual wear through friction and abrasion, and through chemical interaction between components, corresponds to gradual changes of dimension in the product, eventually reaching the point where too many units become unacceptable.

CAREERS AND COURSEWORK

Tribology is both a very practical and a highly theoretical field of study. Students interested in a career examining the interaction of material surfaces should develop a strong foundation in mathematics, physics, chemistry, and materials science as the foundation requirements for this field. A college or university degree with specialization in mechanical engineering and tribology are the minimum qualifications for a career in tribology. A more limited career option includes on-the-job experience in the hands-on aspects of the work involved. No special qualification is required for work in a quality inspection capacity, for example. However, a more advanced position in quality management will require specialist training in statistical process control and quality testing procedures. In such cases, training in electronics and metrology is essential.

At the highest career level, a graduate degree in applied physics or equivalent training for work with scanning probe microscopes will be required. Courses of study in advanced mathematical disciplines such as fluid dynamics will also be essential.

SOCIAL CONTEXT AND FUTURE PROSPECTS

In the general scheme of society, tribology plays a large economic role, in both positive and negative ways. Tribological effects are integral parts of the physical world. A world in which friction did not function would be a grand failure at a basic level, given that friction and frictional wear has, for example, enabled humans to walk upright and write meaningful information on materials, whose production, in turn, was possible only because of frictional processes. In modern times, however, devices that function with moving parts call for more ways to defeat the negative effects of friction, abrasion, and heat.

Richard M. Renneboog, M.Sc.

FURTHER READING

Bhushan, Bharat. *Principles and Applications of Tribology.* New York: John Wiley & Sons, 1999. The introduction to this book, also available online, provides an excellent account of the history of tribology.

Bhushan, Bharat, ed. *Measurement Techniques and Nanomechanics.* Nanotribology and Nanomechanics 1. Berlin: Springer, 2011. Provides a concise overview of the conceptual history and principles of tribology.

Donnet, Christophe, and Ali Erdemir. *Tribology of Diamond-like Carbon Films: Fundamentals and Applications.* New York: Springer Science, 2008. This book deals specifically with one specialized material aspect of tribology, demonstrating the breadth and depth of the field.

Gohar, Ramsey, and Homer Rahnejat. *Fundamentals of Tribology.* London: Imperial College Press, 2008. An introductory book for undergraduate engineering specialists that includes heavy mathematical descriptions and relationships for several common situations.

Sinha, Sujeet K., and Brian J. Briscoe. *Polymer Tribology.* London: Imperial College Press, 2009. This book addresses the relatively recent study of the tribology of polymeric materials with their increasing use as dynamic components such as nylon gears and Teflon-coated slides.

See also: Biomechanics; Biophysics; Surface and Interface Science.

U

ULTRASONIC IMAGING

Medical imaging; echocardiography; obstetrics and gynecology; radiology; reproductive endocrinology.

SUMMARY

Ultrasonic imaging is a medical diagnostic tool that visualizes internal structures of the body with a high-frequency sound beam. In contrast to X-rays, which produce harmful, ionizing radiation, ultrasound has no known harmful effects. The heart of the device is the transducer, which transmits the sound beam to and receives an echo from internal structures. Structures within the body reflect the sound beam to different degrees. Sound passes through liquid readily, is reflected to some degree by muscle, and is strongly reflected by bone. Ultrasound is commonly used for visualization of the developing fetus within the uterus. The amniotic fluid surrounding the fetus readily transmits the sound beam to the fetus and its internal structures.

KEY TERMS AND CONCEPTS

- **Acoustic Window:** Fluid-filled structure, such as the urinary bladder, which aids in visualization of structures beneath it.
- **Coupling Gel:** Gel placed on the skin to facilitate contact of the transducer with underlying structures.
- **Doppler Technique:** Technique that visualizes motion in blood vessels; color Doppler measures the degree and direction of blood flow and displays it by color variations; flow away from the transducer is in blue tones while flow toward the transducer is in red tones.
- **Echocardiogram:** Visualization of the internal structure of the heart via ultrasound.
- **Echogenicity:** Refers to the degree of reflection from a structure. A structure may be termed hyperechoic (high echogenicity), hypoechoic (low echogenicity), or anechoic (no echogenicity).
- **Piezoelectric Effect:** Production of electric voltage when pressure is applied to certain crystals; when placed in an electric field, these crystals become compressed.
- **Ultrasonic Transducer:** Device that transmits and receives high-frequency sound waves (above 20,000 hertz); also known as transceivers because they both transmit and receive sound.

DEFINITION AND BASIC PRINCIPLES

Ultrasound works on a principle similar to that of radar, which transmits then receives radio waves and converts them into an image. Structures within the body are differentiated by the varying degrees that they reflect a focused sound beam. Ultrasound is one of the most widely used diagnostic medical tools.

Compared with other imaging modalities, ultrasound is relatively less expensive and more portable. It can image many internal organs to visualize their size, structure, and any abnormal (pathological) lesions, such as a cancerous tumor, within them. It is used extensively in the field of obstetrics to observe the growth and internal organs of a fetus. Many abnormalities can be diagnosed through this modality. It is used by cardiologists to image the heart in real time; this technology is known as echocardiography. It is used by ophthalmologists to visualize the internal structure of the eye.

Transducers come in a variety of shapes and sizes, depending on their use. They also are designed to emit different frequencies. Higher frequencies produce a more detailed image; however, they do not penetrate as deeply. In addition, transducers are designed to focus at different depths, depending on their intended use. Transducers can be placed over the skin or within a body cavity, such as the vagina or rectum.

Ultrasound can be used to guide instruments

passed into the body. For example, an obstetrician can use ultrasound to guide a needle within the amniotic cavity (sac around the fetus) for an amniocentesis (withdrawal of amniotic fluid for analysis). Reproductive endocrinologists use ultrasound to guide needles passed through the vaginal wall and into the ovary for aspiration of ova (eggs) from the ovary.

BACKGROUND AND HISTORY

In 1841, Swiss physicist Jean-Daniel Colladon conducted experiments regarding sound transmission in Lake Geneva; he determined that sound traveled more than four times faster in water than in air. In 1881, French physicist Pierre Curie, who is well known for his work regarding ionizing radiation, discovered the piezoelectric effect, which later made the development of the ultrasound transducer possible.

In 1937, Karl Dussik, an Austrian physician, developed a technique that he termed "hyperphonography." His equipment purportedly aided in the diagnosis of brain tumors using heat-sensitive paper that recorded extremely rudimentary images of sound echoes generated from quartz crystals. Over the next decade, Dussik continued research on the use of ultrasound for differentiating body tissues. This type of ultrasound, later termed "A-mode," produced an echo spike on recording paper or an oscilloscope.

During the 1950's and 1960's, B-mode scanners were developed and improved. Using a linear array of transducers, these scanners produced a static, two-dimensional image of internal structures. Subsequently, real-time sonography was developed. The image was two-dimensional; however, it was continually updated in real time. Real-time ultrasound can display a beating heart and its internal structure; it also images fetal motion. Also, at this time, Doppler ultrasound, which allowed visualization of blood flow, was being devloped. In 1987, Olaf von Ramm and Stephen Smith of Duke University developed three-dimensional and four-dimensional ultrasound for imaging fetuses. Both techniques are three dimensional; however, the four-dimensional version adds real-time recording of movement to the three-dimensionsl image.

HOW IT WORKS

Many internal structures can be visualized with ultrasound. Compared with X rays, its main limitation is its inability to penetrate bone; structures behind a bone are obscured. An ultrasound examination consists of manipulating a transducer over a portion of the body. It is moved or angled over an area, and images of interest are recorded through film, a printer, or a videotape. Continuous recording of an ultrasound examination is often done for later review. A number of different transducers can be attached to an ultrasound machine. An examination might involve the use of more than one transducer.

Ultrasound is commonly used by obstetricians and gynecologists. Ultrasound is an excellent modality for imaging the fetus and charting its growth and development. It can visualize a gestational sac at about four weeks of gestation and can detect a fetal heart beat about two weeks later. Once a living fetus is visualized, various measurements can be made to determine its gestational age. In a number of instances, the fetal age does not coincide with the age calculated from the last menstrual period. Up to twelve weeks of gestation, the crown-rump length (distance from top of head to the buttocks) is used to calculate the gestational age. Later, the head diameters, abdominal diameters, and femur length can be used. Sequential ultrasounds can determine if the fetus is growing properly. Many fetal anomalies can be detected with ultrasound. Ultrasound can be used for conducting a biophysical profile, which can assess fetal well-being. It is a valuable diagnostic tool for imaging abnormalities of the ovaries and uterus. Obstetrical ultrasound employs two methodologies. For early pregnancy (twelve weeks or less), a vaginal transducer is covered with a condom containing conductive gel and placed in the vagina; it is then pressed against the upper vaginal wall. For the remainder of the pregnancy, conductive gel is placed on the abdomen and an abdominal transducer is manipulated to view the uterine cavity and its contents. The patient is asked to have a full bladder to provide an acoustic window for better visualization. Gynecologic ultrasounds are conducted with a vaginal or abdominal transducer.

Other Uses. The breast is scanned by placing the transducer over it. The liver and gallbladder are scanned by an abdominal transducer placed just under the ribs. The kidneys are imaged between the ribs on the back, and the heart is imaged between the ribs on the chest. In addition to the standard two-dimensional real-time image of the heart, M-mode is employed. This mode displays the motion of the heart in a linear display somewhat like an electro-cardiogram. M-mode is used for analyzing the function of the heart both in the uterus and after birth. Imaging of the hearts of adults and children is known as echocardiography. The vaginal transducer can also be inserted in the rectum for imaging of the prostate. Other than mild discomfort from a full bladder or internal probe, the procedure is painless.

Safety. Although, ultrasound is a far safer diagnostic modality than X-rays, it might have a slight risk; therefore, studies are ongoing to evaluate this possibility. In 2008, the American Institute of Ultrasound in Medicine published a report in which they stated that potential risks to an ultrasound exam might exist. These potential risks include "postnatal thermal effects, fetal thermal effects, postnatal mechanical effects, fetal mechanical effects, and bio-effects considerations for ultrasound contrast agents." Animal studies with long-term, high-intensity administration of ultrasound to cattle reported that it caused decreases in the diameter of red blood cells. To date, any harmful effect from diagnostic ultrasound is unknown. If ultrasound can be harmful, the following factors would come into play: duration of exposure, intensity of ultrasound waves, and the number of exams. A greater risk exists for three-dimensional ultrasound because the level of ultrasound energy is higher.

In any event, in most cases, the information gained from the procedure far outweighs any possible harmful effect. Ultrasound can identify many conditions that place the fetus in jeopardy and provide the opportunity for reduction of that risk. The U.S. Food and Drug Administration (FDA) limits the amount of ultrasound energy for obstetrics and gynecology. The limit (94 milliwatts per square centimeter) is the same regardless of the type of ultrasound being applied.

APPLICATIONS AND PRODUCTS

Since the 1970's, ultrasound has been an essential diagnostic tool for obstetricians; in fact, many obstetrician-gynecologists have ultrasound machines in their offices. This allows for rapid assessment of a problem. For example, if a woman complains that she has not felt the baby move for a period of time or a fetal heart beat cannot be heard, an ultrasound examination can either provide reassurance or identify a fetal death. Ultrasound is used to guide a needle for an amniocentesis to check for genetic defects. The condition of the placenta and fetal well-being at later stages of pregnancy can be evaluated. Ultrasound can image ovarian cysts; the internal structure of the cyst can help the gynecologist determine whether the cyst is malignant. Abnormalities of the uterus, such as fibroid tumors, can be imaged.

Ultrasound is used in many other medical applications. It is sometimes used as a supplement to mammography to image the breast because it can differentiate between solid and fluid-filled cysts and can produce a better image of areas near the chest wall. In cardiology, echocardiography serves as an essential diagnostic tool to evaluate cardiac abnormalities. It can also examine venous clots and arterial blockage or narrowing indicative of vascular disease. Neurologists can examine the carotid arteries in the neck for stenosis (narrowing) and can confirm the diagnosis of brain death. The aqueous humor (in the front of the eyeball) and the vitreous humor (filling most of the eyeball) readily conduct ultrasound for imaging inner structures. Trauma-induced bleeding into the abdominal cavity, chest cavity, or other regions can be promptly diagnosed. Ultrasound can also be used to image the pancreas, liver, gallbladder, bile ducts, and kidney.

An ultrasound machine can be connected to a wide variety of transducers, depending on the intended use. For example, in the field of obstetrics and gynecology, the machine is usually equipped with a vaginal transducer and one or more abdominal transducers. The head of a vaginal transducer is about an inch long, and the abdominal probes are several inches long. Transducers can be focused for optimum imaging at different depths. Ultrasound machines, in general, are portable; thus, they can readily be moved from one location to another. In many locations, mobile ultrasound is available. A sonographer travels between locations in a van containing an ultrasound machine. Small machines—about the size of a shoebox—also are available. Although not as full-featured as larger machines, they have the advantage of extreme portability.

IMPACT ON INDUSTRY

Because of the widespread usage of ultrasound equipment and the variety of specialized manufacturers of ultrasound equipment, manufactures of ultrasound equipment and associated products have a significant share of the medical equipment market. In 2009, the global market for ultrasound imaging equipment was estimated to be $4.9 billion, which was a 6 percent decrease from 2008. Analysts cite health care spending cuts and postponed equipment purchasing as the major reasons for the decline in Western Europe and North America. They predicted that sales would be slow for a few years and pick up when the economic situation improved. However, the need for ultrasound equipment is ongoing for many medical fields. Over time, equipment is subject to failure, which necessitates repair or replacement. Also, continued improvement in technology creates a need for a superior, newer model.

Subindustries. Ultrasound supports a number of subindustries, including maintenance and repair. Ultrasound equipment is connected to devices such as printers, photographic equipment, and video recorders. Consumables must be replenished. These include coupling gel, film, videotapes, and printer paper. Needles and syringes are necessary for performing an amniocentesis. A number of specialized instruments are manufactured for tissue biopsy or ovum extraction. Textbooks and other educational resources are purchased by practitioners. Other educational material is marketed for the public.

CAREERS AND COURSE WORK

Ultrasound is performed by physicians in many specialties as well as ultrasound technicians. Physicians who perform ultrasound must first graduate from college and then complete a four-year course of medical training. Initial specialty training can be in a number of fields, including radiology, obstetrics and gynecology, cardiology, nephrology, and ophthalmology. This is typically a three- or four-year residency program. The physician usually will receive training in ultrasound as part of the residency program. Many will receive additional training following their residency through continuing education courses. Others will take a fellowship of one or more years in which part of the course work is in ultrasound. An example is reproductive endocrinology (infertility), which is a heavy user of ultrasound for ovum (egg) harvesting. Physicians may use ultrasound to varying degrees, ranging from occasional to daily use.

For individuals desiring a career as an ultrasound technician, courses are available through sources

Fascinating Facts About Ultrasonic Imaging

- The film *ET: The Extra-Terrestrial* (1982) contained a scene in which ET's health was failing and physicians were at a loss as to what the problem was. One segment included a scene in which a physician suggested an ultrasound and one was subsequently performed. However, the ultrasound material was cut before the film's release.

- The actor Tom Cruise purchased a $200,000 ultrasound machine in 2005 to perform numerous examinations on his pregnant girlfriend, Katie Holmes. A year later, he promoted legislation to prohibit the sale of ultrasound equipment to anyone who is not a medical professional.

- Ian Donald, a professor of medicine at the University of Glasgow, Scotland, had a strong interest in ultrasound. He became famous in the 1950's when he performed an ultrasound examination on a patient suspected of having inoperable stomach cancer. His examination identified an ovarian tumor, which was surgically removed; the patient regained her health.

- Pro-life organizations have offered free three-dimensional ultrasounds to young pregnant women to influence them against choosing an abortion. The organizations receive charitable donations that defray the cost.

- The sinking of the *Titanic* gave a boost to the field of ultrasound. After the tragedy, French physicist Paul Langevin invented the first transducer, a hydrophone to detect icebergs. The hydrophone was later used for submarine detection in World War I.

- Some animals (including bats, mice, cats, dogs, and dolphins) can hear higher frequencies than humans; thus, they can hear ultrasound.

such as city colleges, technical schools, and hospitals. The minimum requirement for enrollment is a high school diploma. Physicians and nonphysicians interested in ultrasound may belong to one or more professional organizations. In the United States, physicians and registered nurses can join the American Institute of Ultrasound in Medicine. Another professional organization, the Society of Diagnostic Medical Sonography, has a broader membership base; it includes physicians, nurses, technicians, hospital administrators, and researchers. It is also international in scope. A major European organization is the European Federation of Societies for Ultrasound in Medicine and Biology (EFSUMB), based in London, England. Its membership is open to physicians, nurses, and technicians.

SOCIAL CONTEXT AND FUTURE PROSPECTS

Research and development is ongoing in ultrasound technology. Since ultrasound first emerged in the 1970's, the imaging quality has improved tremendously. It has progressed from blurry images lacking in detail to stunning three-dimensional images. Imaging quality is expected to continue to improve in the foreseeable future. Ultrasound has and will continue to be a significant component to the field of obstetrics. Seeing a living fetus within the uterus has a profound effect on the parents. It transforms a vague entity into a recognizable, living creature. Three-dimensional and four-dimensional ultrasound are very popular. In many instances, the technology does not aid in diagnosis of an abnormality; however, the high-quality images are extreme patient pleasers. A number of companies advertise production of three-dimensional and four-dimensional ultrasound images of fetuses for the public. Opponents of this practice state that this essentially nonmedical use of the technology should be discouraged until it can be proven that the fetus can suffer no harm from the procedure.

Robin L. Wulffson, M.D., F.A.C.O.G.

FURTHER READING

Callen, Peter W. *Ultrasonography in Obstetrics and Gynecology.* 5th ed. Philadelphia: Saunders/Elsevier, 2008. Starts with techniques, then proceeds to diagnosis. Includes information on three-dimensional ultrasound and ultrasound scanners.

Jacobson, Jon A. *Fundamentals of Musculoskeletal Ultrasound.* Philadelphia: Saunders/Elsevier, 2007. Presents the techniques of ultrasound in diagnosing musculoskeletal problems.

Rumack, Carol M., Stephanie R. Wilson, and J. William Charboneau, eds. *Diagnostic Ultrasound.* 3d ed. 2 vols. St. Louis, Mo.: Elsevier Mosby, 2005. Examines how ultrasound is used for imaging internal organs and how it is used in obstetrics.

Schmidt, Guenter, ed. *Ultrasound.* New York: Thieme, 2007. A manual that starts with the basic techniques and looks at the various applications and diagnoses.

Timor-Tritsch, Ilan E., and Steven R. Goldstein. *Ultrasound in Gynecology.* Philadelphia: Elsevier Churchill Livingstone, 2007. Looks at the use of ultrasound in gynecology for ovarian cancer, infertility, and inflammatory disease, among others.

WEB SITES

American Institute of Ultrasound in Medicine
http://www.aium.org

European Federation of Societies for Ultrasound in Medicine and Biology
http://www.efsumb.org

Society of Diagnostic Medical Sonography
http://www.sdms.org

See also: Cardiology; Computed Tomography; Gastroenterology; Magnetic Resonance Imaging; Nephrology; Neurology; Obstetrics and Gynecology; Ophthalmology; Radiology and Medical Imaging.

UROLOGY

FIELDS OF STUDY

Biology; chemistry; anatomy; physiology; biochemistry; endocrinology; embryology; developmental biology; neurology; pharmacology; microbiology; pathology.

SUMMARY

Urology is the study of the anatomy and physiology of the liquid waste removal system in men and women and the associated reproductive system in men. This system includes the kidneys, ureters, urinary bladder, and urethra in both sexes, and the testes, vas deferens, prostate, and penis in men. By understanding the normal structures and functions of the organs involved with the urinary tract, urology can address abnormalities and elective changes such as the correction of birth defects, implantation of a penile prosthesis for erectile dysfunction, sterilization by vasectomy, and sex reassignment surgery (both male-to-female and female-to-male). Urologists typically diagnose and treat more men than women because women generally bring urological concerns to their gynecologists. For men, urologists are comparable specialists for reproductive health.

KEY TERMS AND CONCEPTS

- **Benign Prostatic Hyperplasia:** Enlargement of the prostate gland.
- **Bladder:** Muscular organ that distends to hold urine until excretion.
- **Calculi:** Solid masses of mineral salts; also called kidney stones.
- **Catheter:** Hollow tube inserted up the urethra into the bladder to drain urine.
- **Erectile Dysfunction:** Inability to achieve or sustain an erection of the penis.
- **Kidney:** Organ that filters waste products and excess water from the blood.
- **Prostate:** Organ that secretes an alkaline fluid as a component of semen.
- **Ureter:** Tube that connects a kidney and the bladder.
- **Urethra:** Tube through which urine is excreted.
- **Vas Deferens:** Duct that propels sperm from the epididymis to the urethra.

DEFINITION AND BASIC PRINCIPLES

The urinary tract for the removal of liquid waste from the body consists of several organs. The kidneys, located in the lower back, filter metabolic waste products and excess water from the blood and convert them into urine. Urine passes through tubes called ureters into the bladder, where it is stored until it is excreted through the urethra. In males, the urethra passes through the penis. If metabolic waste products such as urea, uric acid, and creatinine were not removed, they would accumulate in the blood and poison the tissues throughout the body.

The male reproductive system consists of several organs. The testicles, or testes, produce sperm and testosterone as well as other male sex hormones. The sperm mature in the epididymis, which connects the testicle to the vas deferens. The vas deferens propels sperm toward the urethra. Along the way, the sperm mix with alkaline fluid from the prostate and fluid excreted from seminal vesicles. The sperm and semen are expelled through the urethra during ejaculation.

Urology studies the structure and function of these related organs. Urologists are physicians who are consulted for the treatment of urinary tract infections, kidney stones, prostate enlargement, and urological cancers; the correction of urogenital birth defects; and the management of stress incontinence (involuntary urination), male infertility, and erectile dysfunction.

BACKGROUND AND HISTORY

Some urological treatments have been in evidence since early times. Catheterization, for example, is one of the first therapeutic interventions. Early civilizations used catheters made of onion stalks, wooden tubes, and metal cylinders to drain painfully distended bladders.

It was not until the twentieth century, however, that urology came into its own as a medical specialization. In 1908, Nobel Prize laureate Paul Ehrlich discovered that salvarsan, an arsenic compound, was an effective treatment for syphilis (a venereal, or sexually transmitted, disease with devastating effects on the urological system) and less toxic than the mercury compounds that had been used previously to treat this disease. This therapy was significant to urologists because they typically treated venereal diseases.

In 1910, Hugh Hampton Young developed a novel technique for surgically treating benign prostatic hyperplasia (an enlarged prostate) using a perineal approach. As a result, he became known as the father of American urology. Then, in 1935, Frederick E. B. Foley introduced a rubber balloon catheter that did not require bandages or medical tape to keep it in place.

The overwhelming number of veterans with spinal cord injuries returning home from World War II created a need for advancements in urology and established it as a recognized and respected medical specialty. In 1973, when F. Bradley Scott developed a device to be implanted into a penis and pumped with saline to achieve an erection, urology began to expand into the area of sexual dysfunction as well as the treatment of disease.

In 1986, the U.S. Food and Drug Administration (FDA) approved the first prostate-specific antigen test for prostate cancer screening and monitoring. In 1997, the drug Flomax (tamsulosin) was introduced to improve urination for men with benign prostatic hyperplasia. In 1998, Viagra (sildenafil citrate) was introduced by Pfizer for the treatment of erectile dysfunction. The subsequent popularity of this drug and others designed to treat erectile dysfunction expanded not only the domain of urology but also the validation of disease states as a whole—particularly after major medical insurers began to cover these drugs and thereby to acknowledge the treatment as medically necessary.

HOW IT WORKS

Urology uses a broad range of technology and processes to diagnose urological conditions. The most important of these are endoscopy and several imaging techniques.

Endoscopy. The development of instruments that could help physicians visualize the urinary tract and thereby enable them to diagnose and treat patients was crucial to the advancement of urology as a medical specialty. In 1805, German physician Philip Bozzini created the Lichtleiter, a tube for viewing the inside of the body, using a wax candle as a light source. This instrument is sometimes referred to as the first laryngoscope.

In 1879, German urologist Maximilian Nitze and Viennese instrument maker Joseph Leiter introduced the cystoscope, a small endoscope for directly viewing urethras and bladders. It consisted of an incandescent platinum wire loop heated by electricity and cooled by ice water for internal illumination and a system of magnifying lenses.

In 1926, in New York City, urologist Maximilian Stern developed the resectoscope, an instrument containing an electrified tungsten wire loop that could be manipulated to clear urinary tract obstructions. The loop was electrified by high-frequency alternating current, and it vaporized cells as it passed through tissue. This design was later modified by American physician Theodore Davis, who used his engineering background to strengthen the parts, enhance the loop, and add insulation. He also introduced a dual-action foot pedal to control the current for either cutting or cauterizing.

The biggest development in this field was the Hopkins rod lens system, introduced in 1959. The inventor was Harold Hopkins, a British professor of applied optics. Hopkins had earlier developed zoom lenses, which can change their degree of magnification without having to be refocused. He used flexible glass rods to transmit light internally from an external source. These rods provided more illumination than a small lightbulb and were even smaller, so that an endoscope using the rod lens system could explore smaller spaces. The resulting endoscopes were maneuverable, sterilizable, and durable, and the light source did not burn the patient.

Imaging Modalities. When direct visualization is not sufficient or possible, urologists rely on other imaging modalities such as ultrasonography, computed tomography (CT), and magnetic resonance imaging (MRI). These imaging modalities use computer reconstruction software programs to generate three-dimensional information from two-dimensional images and thereby function as virtual endoscopies. Imaging provides a noninvasive manner in which to evaluate the urinary tract. It is especially advantageous for watching for the recurrence of bladder cancer as it does not have the risks of repeated cystoscopies: infection, trauma, scarring, and pain. However, imaging lacks the sensitivity to replace cystoscopies altogether.

Although still in trial, wireless capsule endoscopy

shows promise for use in urology. This technology consists of a wireless transmitter and a camera contained in a small capsule that the patient swallows. The capsule can be steered by external magnets. Using continuous real-time images, the capsule can be guided to its targeted destination. One potential application of a version of this technology is the delivery of chemotherapy directly to the bladder following bladder cancer surgery.

The application of technologies from seemingly unrelated fields has produced advancements in endoscopy. Miniaturization may lead to microendoscopy, and robotics may make remote control of internal scouts possible. These and further developments in endoscopy technologies are designed to reduce the adverse effects of diagnosis and treatment on patients and improve the course of recovery from urological diseases.

APPLICATIONS AND PRODUCTS

Advances in urology have resulted in a host of technologies and devices to treat urological conditions.

Prosthetic Devices. In urology, a prosthetic device almost always refers to a penile implant, which is a treatment option for men with erectile dysfunction for whom other treatments have failed. About 30 million men in the United States have erectile dysfunction, and about one-third of them do not respond to other treatments. Implantation requires a surgical procedure, which is why other options are tried first. Implants have several advantages. They do not require the doctor visits for prescriptions and monitoring that are necessary with pills, creams, suppositories, and injections, and they are available for immediate use and are aesthetically pleasing because they prevent the penis from contracting when cold.

There are two kinds of implants, flexible and inflatable. The less popular choice is a flexible prosthesis, consisting of two firm cylinders made from a silicone elastomer that is able to bend without breaking. These cylinders are surgically implanted within the two erection chambers (corpora cavernosa) of the penis. The implant allows erectile function to return by reducing the cavernosal volume to be engorged, inhibiting venous drainage, and increasing the intrapenile pressure. Although the prosthesis may be bent down for urination and up for sexual intercourse, its constant size and rigidity make it seem less natural.

The inflatable prosthesis consists of two fluid-filled cylinders that are surgically implanted within the two erection chambers of the penis. These cylinders are connected by tubing to a pump that is implanted within the scrotum. To achieve an erection, the pump transfers fluid to the cylinders, increasing the intrapenile pressure. Following ejaculation, the fluid returns to the pump from the cylinders. The adjustable size and rigidity make this prosthesis seem more natural. This type of implant may experience mechanical failure; there is a 10 to 15 percent failure rate in the first five years. If a cylinder leaks, the fluid inside is saline, which is absorbed by the body without harm. However, a failed prosthesis must be replaced or at least removed in a subsequent operation.

An implant procedure may be performed on a variety of patients, including young men who are trauma patients, men left impotent after having their prostate removed, and men with age-related erectile dysfunction. One of the biggest considerations is whether the man is otherwise healthy enough for general anaesthesia. Surgeons attempt to minimize the size of the incisions to preserve the sensitivity of the natural tissues. A penile prosthesis is not intended for penile enlargement of an otherwise perfectly functioning organ.

Although rarely used, another urological prosthetic is an artificial testicle, a saline-filled silicone ball that when implanted, mimics the appearance and movement of a natural testicle. It was developed for use in reconstructive surgery on men whose testicle had been removed in the treatment of cancer. It carries a risk of rupture from a sports injury, and one in three cases requires a surgical adjustment within the first year. This prosthetic is not medically necessary and serves only a cosmetic purpose. Perhaps a model will be developed that will provide long-term delivery of testosterone.

Neurourology. Another form of prosthesis under development is an implantable neuroprosthetic device for the treatment of urinary incontinence. An existing treatment for urinary incontinence is electrical stimulation of the nerves and muscles in an attempt to restart them so that they resume their normal function and regulation. However, this stimulation is from an external source and deliberately delivered. As research progresses, scientists hope to create an internal source of electrical stimulation that would be delivered on a regular schedule or in response to a particular signal. Trials have been conducted on subjects, most of whom were not paraplegic but experienced urinary incontinence. Research is also ongoing into rewiring the nerves of patients with spinal cord injuries to help them regain bladder control and eliminate their need for self-catheterization.

The application of artificial intelligence, in the form of artificial neural networks, has been successfully used in several fields of medicine and is being applied to urology. Many studies have applied artificial neural networking to the diagnosis and staging of prostate cancer with promising results. It is an analytical way to uncover nonlinear relationships among various clinical parameters, from which to determine optimal therapies and predict outcomes.

Drug Therapies. The pharmaceutical industry has developed numerous prescription drugs for the treatment of urological conditions. Drugs for the treatment of benign prostatic hyperplasia (enlarged prostate) include tamsulosin, Avodart (dutasteride), Jalyn (dutasteride and tamsulosin), and Uroxatral (alfuzosin hydrochloride). Drugs for the treatment of prostate cancer include Jevtana (cabazitaxel), Eulexin (flutamide), Zoladex (goserelin), and Gemzar (gemcitabine hydrochloride). Drugs for the treatment of overactive bladder include Toviaz (fesoterodine), Sanctura (trospium), Vesicare (solifenacin), Detrol LA (tolterodine tartrate), and Ditropan XL (oxybutynin chloride). Drugs for the treatment of cystitis and uncomplicated urinary tract infection include Cipro (ciprofloxacin), Elmiron (pentosan polysulfate), Levaquin (levofloxacin), and Doribax (doripenem). Drugs for the treatment of erectile dysfunction include sildenafil citrate, Cialis (tadalafil), and Levitra (vardenafil).

IMPACT ON INDUSTRY

Urologists and urology researchers are needed in various settings, from government agencies to hospitals to research institutions.

National Agencies. Prostate cancer is the most common cancer among men of all races in the United States. Researchers at the Centers for Disease Control and Prevention, Division of Cancer Prevention and Control, are studying the barriers to screening and early detection, including cost and patient education.

The National Institutes of Health oversees the National Institute of Diabetes and Digestive and Kidney Diseases (NIDDK), which funds urological diseases research. Areas of study include host-pathogen interactions in urinary tract infections, pediatric enuresis (bed-wetting), and kidney disease associated with human immunodeficiency virus (HIV) infection.

The NIDDK-sponsored Urologic Disease in America project, which has entered its second stage, tracks the changes in epidemiology, health care costs, and practice patterns of such diseases as cancers, kidney stones, and benign prostatic hypertrophy.

In 2010, the Food and Drug Administration approved the sale of generic tamsulosin capsules for the improvement of urination in cases of benign prostatic hyperplasia. The FDA also oversees the manufacturing of medical devices and takes action when diagnostic, prosthetic, or surgical devices are suspected of misfunctioning.

Research Hospitals. The urology research department at Beaumont Hospitals in southwest Michigan is pursuing projects in neurourology to better manage urinary incontinence. Researchers are rerouting nerves in the spinal cord with the goal of giving bladder control to people with spinal cord injuries or birth defects who otherwise practice self-catheterization.

Pediatric urology research at New York-Presbyterian Morgan Stanley Children's Hospital is focused on the genetics of pediatric renal disease and the developmental biology involved in the formation of the urogenital tract, including the kidneys.

Scientists at the John Duckett Center for Pediatric Urology, a research arm of the Children's Hospital of Philadelphia, are studying cryptorchidism (undescended testicles) from three angles: analysis of data collected from patients in whom this condition was surgically corrected, analysis of seminal parameters found in these patients, and gene analysis related to the etiology of cryptorchidism and the related potential infertility.

University Medical Centers. The urology department at Stony Brook University Medical Center in New York offers robotic prostate surgery using the da Vinci surgical system. The use of this less-invasive technique is likely to become more common throughout the country as the technology becomes less expensive and more urologists have access to training.

The physicians at Loma Linda University Medical Center in California use robot-assisted surgery to remove malignant tissue and organs in cases of

Fascinating Facts About Urology

- An adult's kidneys filter 50 gallons of blood and produce about 0.5 gallon of urine each day.
- Many believe that the children's song "Frère Jacques" was written about Frère Jacques Beaulieu, a seventeenth-century French monk who specialized in surgically removing urinary stones. He performed more than 5,000 lithotomies in thirty years.
- American inventor and statesman Benjamin Franklin designed a silver coil catheter for his brother in 1752.
- In 1831, American physician Philip Syng Physick removed more than 1,000 stones from the bladder of John Marshall, chief justice of the United States Supreme Court. The seventy-six-year-old Marshall recovered and returned to the bench.
- In the nineteenth and early twentieth centuries, men with chronic bladder outlet obstruction commonly self-catheterized with tubes they carried in hatbands, umbrella handles, and hollow walking sticks.
- Railroad magnate James Buchanan "Diamond Jim" Brady donated the funds to found a urological institute at The Johns Hopkins University after urologist Hugh Hampton Young surgically relieved Brady's chronically inflamed prostate on April 12, 1912.
- German aerospace engineer Claude Dornier noticed pitting on the surface of an airplane as it approached the sound barrier; the minute destruction was the result of the shock wave created in front of moisture. This led to the development in 1984 of extracorporeal shock wave lithotripsy, a noninvasive procedure to demolish kidney stones with intense shock waves.

The Urological Sciences Research Foundation is a nonprofit organization based in California. It solicits money to fund clinical trials, focusing on prostate disease, erectile dysfunction, overactive bladder, and incontinence. Many of its industry sponsors are well-known pharmaceutical and medical device companies.

The American Medical Association serves physicians of all specialties, including urologists. Although urologists have their own medical society and journals, they often present articles and book reviews related to urology in the *Journal of the American Medical Association* to educate and update other physicians. Articles have addressed drug treatments, discussed the controversy over prostate cancer screening, and offered advice on how to talk to patients about overactive bladder and urinary incontinence.

CAREERS AND COURSE WORK

As more people are living longer, the prevalence of age-related urogenital changes is increasing. Furthermore, as sex reassignment is becoming more accepted, more people are seeking specialized urological surgery. According to the U.S. Bureau of Labor Statistics, urologists will see greater than a 20 percent growth in their field by 2014.

To become a urologist, a person must complete an undergraduate college degree, majoring in a life science such as biology, chemistry, or biochemistry (or at least fulfilling medical school requirements). That graduate must then complete a medical degree, studying detailed anatomy, physiology, pathology, and pharmacology. That physician must then complete additional training in urology, through residencies and fellowships, and become board certified as a urologist. Urologists may further specialize in areas such as urological oncology, pediatric urology, and sex reassignment surgery. Urologists may be employed by a hospital or have a private practice with hospital privileges.

Urological researchers must complete an undergraduate college degree, majoring in a life science. That graduate may follow the same path as a urologist, but most choose instead to pursue additional education such as master's and doctoral programs in the sciences. They then do postdoctoral work and become published authors before they apply for and receive funding for their own research projects. Most urological research laboratories are associated with

prostate, bladder, and kidney cancer. Robot-assisted surgery is a less-invasive procedure that provides a greater level of surgical precision. The robot's camera system magnifies the surgical field by a power of ten, and its computer fuses images from two cameras to give the surgeon three-dimensional information.

Nonprofit Organizations. The American Urological Association is a national medical society that holds definite and informed positions on controversial issues such as circumcision, penile-augmentation surgery, and prostate cancer screening.

university medical schools, although some are associated with government agencies or private companies such as pharmaceutical manufacturers.

Jobs are also available in related fields, such as medical assisting, urological nursing, and laboratory diagnostic testing. Educational opportunities vary; some schools offer specific vocational programs with job placement, and some physicians prefer to train a college graduate who majored in a life science. Urological nurses have a bachelor's degree in nursing with additional training in urology.

SOCIAL CONTEXT AND FUTURE PROSPECTS

Urology will remain a necessary clinical specialty: The development of calculi is a common occurrence in industrialized nations, and prostate disease affects 75 percent of men fifty years of age or older. Neither of these can be prevented with a pharmacological intervention. Because urologists provide both surgical and nonsurgical treatments, they must remain knowledgeable about advances in physiology, pharmacology, and technology.

Urologists are continuing to develop their surgical skills to achieve improvements in appearance, function, and sensation. Internal penile pumps remain a popular choice versus flexible implants, vasodilator pills, or external pumps and creams; more than 250,000 men have already chosen to have such a device implanted. Surgeons will continue to correct birth defects and repair urogenital trauma. Their aim is to restore function or at least preserve it.

Scientists are using tissues harvested during surgery for ongoing cancer research, focusing especially on the genes underlying the disease. They are also trying to understand and block the mechanisms of cancer recurrence. Other research groups are studying kidney stones to determine the underlying factors of calculi formation to devise methods of prevention.

Diagnostic testing is advancing to facilitate earlier detection of prostate cancer and sexually transmitted diseases. Imaging equipment and technology will continue to develop, offering modalities with stronger clarity and contrast in real time. These will be used for earlier detection of abnormalities as well as closer monitoring of changes and better planning of surgical interventions.

New therapies will continue to be sought. For example, researchers are investigating the use of botulinum toxin (commonly known as Botox) in the treatment of urinary incontinence. New solutions for male infertility and enhancement of sperm conditions to increase the efficiency and effectiveness of fertilization are being studied.

Bethany Thivierge, B.S., M.P.H.

FURTHER READING

Chan, Evelyn C. Y., et al. "Informed Consent for Cancer Screening with Prostate-Specific Antigen: How Well Are Men Getting the Message?" *American Journal of Public Health* 93, no. 5 (2003): 779-785. Presents balanced information on why prostate-specific antigen (PSA) screening is controversial and yet beneficial.

Field, Michael J., Carol Pollock, and David Harris. *The Renal System: Systems of the Body Series.* 2d ed. New York: Churchill Livingstone, 2010. Covers the basic anatomy, physiology, and biochemistry of the renal and urogenital system.

Genadry, Rene, and Jacek L. Mostwin. *A Women's Guide to Urinary Incontinence.* Baltimore: The Johns Hopkins University Press, 2007. Provides information about a common yet embarrassing condition and encourages women and their families to seek compassionate treatment.

Gomella, Leonard G. *The Five-Minute Urology Consult.* 2d ed. Philadelphia: Lippincott Williams & Wilkins, 2009. Offers immediate, practical information on a broad range of urological topics.

Tanagho, Emil A., Jack W. McAninch, and Donald R. Smith. *Smith's General Urology.* 17th ed. New York: McGraw-Hill Medical, 2008. This comprehensive textbook contains visual aids, including clinical images.

Walsh, Patrick C., and Janet Farrer Worthington. *The Prostate: A Guide for Men and the Women Who Love Them.* New York: Warner Books, 1997. This comprehensive yet easy-to-read classic explains diagnostic tests and treatments in a reassuring manner suitable for patients and their families.

WEB SITES

American Board of Urology
http://www.abu.org

American Medical Association
http://www.ama-assn.org

American Urological Association
http://www.auanet.org

Kidney and Urology Foundation of America
http://www.kidneyurology.org

Society for Pediatric Urology
http://www.spuonline.org

Urological Sciences Research Foundation
http://www.usrf.org

See also: Endocrinology; Nephrology; Prosthetics; Reproductive Science and Engineering; Surgery.

V

VETERINARY SCIENCE

FIELDS OF STUDY

Biology; zoology; anatomy: chemistry; microbiology; genetics; mathematics; algebra; calculus; statistics; organic chemistry, physics; biochemistry.

SUMMARY

Veterinary science is a medical science dealing with the study, research, prevention, and treatment of disease in animals and the relation of animal disease to human health. It deals with the health of individual animals and groups of animals. Veterinary scientists include veterinarians who treat animal patients, research veterinary scientists who work in academic and private research laboratories, and technicians and technologists who assist veterinarians and veterinarian scientists. Veterinarians provide health care to companion animals and livestock. They play an important role in maintaining a wholesome food supply. Veterinary scientists involved in research are instrumental in the development of medicine, treatments, and surgical procedures applicable to both humans and animals.

KEY TERMS AND CONCEPTS

- **Anesthetic:** Drug used to induce loss of the sense of pain.
- **Anorexia:** Lack of the desire to eat.
- **Anthelmintic:** Medication that eliminates worms from the gastrointestinal tract.
- **Diagnosis:** Determination of a state of disease subsequent to examination.
- **Euthanasia:** Humane ending of the life of a terminally ill, severely injured, or unwanted animal, by chemical injection or other means.
- **Prognosis:** Expected outcome of illness and its treatment.
- **Respiratory Rate:** Breaths per minute.
- **Vaccine:** Serum injected subcutaneously or intramuscularly or substance given as an inhalant to stimulate the animal's immune system and prevent disease.

DEFINITION AND BASIC PRINCIPLES

Veterinary science is a medical science that deals with all aspects of health care provided to animals. It includes the areas of medical, surgical, dental, and ophthalmic treatment used to prevent and cure disease in animals as well as research in these areas. Biology is the base science from which veterinary science evolves; veterinary scientists must have an extensive knowledge of the structure of the tissues and organs of each animal species that they treat. Knowledge of organ function and homeostasis of each species from both a physiological and a biochemical basis is essential. Familiarity with disease-causing agents such as viruses, bacteria, and parasites is important in veterinary science. In addition to knowing the causes of disease in animals, veterinary scientists must have a thorough understanding of pathogenesis (how a disease develops), including immune and inflammatory responses and tumor development. Veterinary science also relies heavily on the principles of pharmacology and toxicology. Safe and effective treatment of disease is based on knowledge of the effects and interactions of drugs and their toxicity levels in various animal species. The toxicology of plants and poisonous substances found naturally in the environment and those manufactured by human beings is also essential to the working knowledge of the veterinarian.

An understanding of animal behavior and husbandry is an integral part of veterinary science. Veterinarians work in hands-on situations with animals and must understand the psychology of animals to ensure the safety of both the handlers and the animals. Proper restraint of animals during procedures that are often unpleasant is of the utmost importance. Animal husbandry plays an important role in the veterinary scientist's work with the herds and flocks of animals raised for food and other animal products. Thorough knowledge of animal nutrition, the physiology and endocrinology of reproduction, genetics, and housing requirements are at the base of successful management of herds and flocks.

BACKGROUND AND HISTORY

Veterinary science did not actually develop as a discipline until the eighteenth and nineteenth centuries. However, care of animals and concern for their well-being dates back to early civilizations in Egypt, India, and Rome. Throughout the Middle Ages and during the seventeenth century, "veterinary medicine" was performed primarily by farriers because horses were necessary for transportation, farming, and waging war. Treatment was primitive and often very ineffective.

Veterinary science as a discipline originated in 1762, when a veterinary school was founded in Lyon, France, by Claude Bourgelat. Before the end of the century, schools had been established in Germany, Austria, Denmark, and Sweden. The Royal College of Veterinary Surgeons was established in England in 1844. In the United States, the American Veterinary Medical Association was created in 1863, and the first college of veterinary medicine in 1879 at Iowa State University.

As long as the horse continued to play a major role in farming and transportation, veterinary science focused on equine care. However, advances in medical science and in the understanding of disease pathology in the early twentieth century fostered the field of public health veterinary medicine. In the 1920's, as farming became more mechanized and the role of the horse diminished, small-animal practice became a more important part of veterinary medicine. During the second half of the twentieth century, as the population in the United States became increasingly urban and more people began keeping companion animals, the number of small-animal practices increased. Specialization and technologically advanced procedures became major components of the field of veterinary practice.

In the twenty-first century, veterinary science has become a broad field in which specialization is playing an ever-greater role. Changing attitudes toward animals and their welfare have expanded the field to include more emphasis on animal nutrition, animal behavior, and areas of animal welfare such as environmental enrichment. Veterinarians also play a much greater role in disease prevention and public health as health care and disease prevention have become global issues involving both animals and human beings.

HOW IT WORKS

Examination. Veterinarians perform complete physical examinations of their animal patients to determine their state of health. The examination begins with observation of the animal for visible signs of illness or injury such as lameness, wounds, evidence of external parasites, and tumors. Female animals may also be examined for signs of pregnancy. Vital signs including temperature, respiratory rate, and heart rate are taken. Veterinarians use equipment such as stethoscopes, thermometers, otoscopes, weight scales, and florescent and ultraviolet lights during the examination. If warranted, the examination also includes the taking of samples of blood, urine, fecal matter, and skin. Aspirations of fluid from the lungs and from the abdomen and a transtracheal wash are other means of gathering samples that are available to the veterinarian. Collected samples are either examined in-house or sent to diagnostic laboratories. Ultrasound and X rays are important diagnostic tools in the modern veterinary clinic. More invasive examination processes include biopsy and exploratory surgery. The results of these investigations provide the information necessary for the veterinarian to make a diagnosis.

Diagnosis and Prognosis. Having gathered the evidence from the examination, the veterinarian assesses the state of health of the animal and makes a diagnosis of any disease, illness, or injury found. If disease, illness, or injury is found, the veterinarian states its possible outcome, along with available treatments and their efficacy. Treatments can be cures or palliative measures in cases of terminal or chronic illness. Referral to a specialist or veterinary college hospital may be made. In cases of terminal, incurable, or severely crippling or painful chronic conditions, euthanasia may be advised. In some circumstances, veterinarians advise postponing euthanasia until the animal's qualify of life is substantially decreased and instead prescribe palliative treatment that will provide an acceptable quality of life for some time.

Veterinarians have many different tools, products, and options for the treatment of disease. Medicines, such as antibiotics, urinary acidifiers, anthelmintics, steroids, and pain relievers, may be administered by mouth or injection. Intravenous fluids, oxygen therapy, surgery, and many other procedures are performed by veterinarians in small private clinics. More advanced treatments such as the implantation of pacemakers, hip replacements, or cataract removal are performed by veterinarians in university veterinary hospitals or specialized clinics.

Preventive and Cosmetic Treatment. Veterinarians also perform procedures and prescribe medicines to prevent the occurrence of disease and illness, to make animals more manageable, to prevent unwanted pregnancy, and to make cosmetic changes. Veterinarians in all types of practices establish schedules and administer vaccinations to their patients. Veterinarians prescribe medicines to prevent heartworm in small animals and to control fleas and other external parasites. They prescribe anthelmintics to control or eliminate internal parasites in both small and large animals. In small-animal practices, they perform spay and neuter operations, which help control the overpopulation of dogs and cats. Declawing of cats is also done. Tail docking, ear trimming, and dewclaw removal are some of the procedures done on dogs for cosmetic reasons. In a farm setting, veterinarians castrate, dehorn, and debeak animals and poultry. Veterinarians in private practice refer animals that require specialized treatment or highly technical surgeries to special clinics or university animal hospitals.

Experimentation. Veterinarians in academic research laboratories and other research facilities perform experimental studies and surgeries on animals to develop new medicines, treatments, and surgeries to improve the health care available to both human beings and other animals. They also conduct nutritional and management studies to improve the quality of meat produced from animals and poultry.

Euthanasia. Euthanasia, the humane termination of an animal's life, varies considerably by the type of practice and the situation encountered. In small-animal and equine practice, the veterinarian is most often dealing with an emotionally charged situation. Very strong bonds develop between these animals and their owners; many of the owners are emotionally dependent on their animal companion or equine partner. Therefore, the veterinarian and his staff must be prepared to counsel the owner and assist the

individual in the grieving process after the death of the animal. Often special arrangements are made to ensure privacy and give the owner time to cope with the actual euthanasia. Euthanasia is also performed on unwanted or unadoptable animals at animal shelters. This practice is disturbing to many people, and some cities have established no-kill shelters, and animal rescue groups have developed to help many of these animals.

In the euthanization of wildlife or livestock, where the major reason for euthanasia is eradication of disease from the herd, flock, or wildlife population, the veterinarian faces a very different situation. Euthanasia is also performed on wildlife or livestock to eliminate suffering in cases of terminal illness or severe injury. Economics and management are the primary concerns of livestock owners.

Zoo animals are often traded or sold to other zoos, so their handlers and keepers are usually more detached from them than pet owners but have more affection for them than owners of livestock. Therefore, if veterinarians are called on to euthanize zoo animals, they encounter a range of attitudes and varying depths of attachment.

Euthanasia of laboratory animals is part of the normal routine in an experimental laboratory. Often the animal must be destroyed for scientists to perform tests that will determine the effectiveness of medicines and procedures. Many or most of the animals cannot be used in other experiments, so it is also a means of controlling costs and saving space. Although many people are not troubled by this practice, especially when it involves animals such as rats or mice, the fate of primates, such as chimpanzees, used in experiments gave rise to a number of organizations that have created sanctuaries for these animals.

APPLICATIONS AND PRODUCTS

Companion Animals. In companion or small-animal practices, veterinarians apply their knowledge of animal physiology and function, of disease pathology, and of treatment of disease, illness, and injury to care for dogs, cats, and other small animals kept as pets or companions by individuals and families. They also use their knowledge of animal behavior and requirements for animal well-being to help pet owners solve behavior problems such as excessive barking, furniture scratching, aggression, and improper elimination. Because of the close bond that exists between many pet owners and their animal companions, small-animal veterinarians also use their understanding of these bonds to respond compassionately to the concerns and emotional needs of their clients.

Farm Animals. Veterinary science is an essential element in the raising of animals for food. Veterinarians who treat livestock and poultry travel from farm to farm and carry their equipment with them. These veterinarians are on call twenty-four hours a day and often work in difficult conditions. Veterinarians provide the knowledge and expertise necessary to maintain disease-free herds and flocks. They also advise livestock farms about genetics, breeding, nutrition, housing, and problems related to animal behavior. Veterinarians make regular scheduled visits to livestock operations to prevent outbreaks of disease in the herds or flocks. Through the health care that they provide to food animals, veterinarians contribute to the production of the meat, eggs, leather goods, wool goods, and other animal-derived products available in the marketplace.

Meat and Poultry Products Safety. Veterinary science also addresses the relationship and interaction of disease in animals with disease in human beings. There are about eight hundred diseases that can be transmitted from animals or animal products to human beings. Therefore, a number of veterinarians are employed in the public health sector at both the federal and the state levels. Public health veterinarians are rarely involved in actual treatment of animals but rather initiate and supervise programs of destruction of infected animals. They work as inspectors, advisers, and educators in regard to control of animal disease and food safety. They inspect meat and poultry plants throughout the process of bringing meat and poultry products to the consumer. They enforce government regulations regarding the humane transport and slaughter of animals, the handling of the meat or poultry during processing, and the transport of the products to the point of sale. They are also involved in work in pathology and epidemiology as well as playing a significant role in major occurrences of foodborne illness.

Equine, Zoo, and Wildlife Management. Veterinarians with a concentration in equine medicine deal primarily with prevention of infectious diseases such as encephalitis and West Nile virus, lameness, infertility, and gastrointestinal illness.

Veterinarians working with zoo and wildlife populations have special training. In zoos, they address issues of injury, nutrition, behavior, and environment. Wildlife veterinarians work with conservation officers to ensure the continuance and health of wildlife populations and to prevent the spread of disease from these populations to domestic animals and humans.

Biomedical Research. Veterinarians in charge of laboratory animals are responsible for the health and well-being of the animals used in the laboratory. They may oversee as many as 70,000 such animals, many of which are genetically engineered mice. They supervise teams of veterinary technicians and other trained personnel who take care of these animals on a daily basis. They also assess proposed experiments and evaluate how well they conform to regulations regarding laboratory animals.

Veterinarians with postgraduate training in specialized areas such as cardiology work as research scientists in academic laboratories and other research facilities. They perform experimental research using live animals or computer-generated programs to develop new antibiotics, vaccines, and surgical procedures that benefit both animals and humans. Those who are doing research at veterinary colleges usually teach classes as well.

IMPACT ON INDUSTRY

Concern for the environment and endangered species has focused people's attention on animal welfare and habitat preservation. Many cooperative programs between developed countries and developing countries have been put in place to aid in the preservation of endangered species and to improve the health of zoo animals. These programs are funded by various sources, including governments, charitable organizations, corporations, and private individuals. They provide financial support to veterinary projects in developing countries and enable international cooperation among veterinarians. For example, Elephant Care International has programs addressing tuberculosis in elephants in Asia, using contraceptives instead of culling in Africa, and distributing an elephant veterinary text throughout Asia.

At the same time, global travel and rise of imports and exports have made people aware of the possibility of global outbreaks of disease. Because 60 percent of infections contracted by human beings and

75 percent of emerging diseases have their origin in animals, maintaining a healthy animal population worldwide is essential to maintaining a healthy human population, which provides the workforce and consumer pool necessary to maintain an efficient, profitable, and successful world economy. The role that veterinarians play in protecting and improving both animal and human health contributes directly and indirectly to industry.

Agriculture and Animal Food Industry. Without availability of veterinary expertise, the livestock branch of the agricultural industry would not be able to operate on a sustained basis. Regular scheduled checkups by a veterinarian are vital to herd or flock health. Early detection of disease and its immediate eradication are vital to the industry. Diseases such as anthrax are capable of decimating a livestock producer's entire herd unless they are discovered in their early stages and appropriate measures are taken. Airborne diseases such as foot-and-mouth disease can easily and quickly spread from one livestock facility to another.

At companies such as Purina, Iams, Science Diet, and Pet Ag, veterinary science plays an essential role in the development of better animal food products. These companies offer foods formulated for each stage of an animal's life from birth to its geriatric years. Other animal food products are formulated to meet the nutritional needs of working animals and animals that are sedentary or have food health problems such as urinary calculi, renal failure, diabetes, or obesity.

Pharmaceutical Industry. Pharmaceutical companies such as Merck, Bayer, and Merial depend on veterinary science and research by veterinarians to develop their products and check them for safety. Practicing veterinarians are the main marketplace for products used in animal health. In addition to using the vaccines and medications, veterinary clinics are the major outlet for many animal health products that they sell to their clients.

CAREERS AND COURSE WORK

Preparation for a career in veterinary science requires a combination of undergraduate courses in biology, chemistry, physics, and mathematics. Courses in English and the humanities are also necessary for admission to a veterinary college. The completion of a graduate program leading to the

---

Fascinating Facts About Veterinary Science

- The United States has only twenty-eight veterinary colleges, located in twenty-six of the fifty states.
- Almost 80 percent of veterinarians are involved in private practice, and most of these practices treat small animals or a mixture of small and large animals. Only 15 percent of veterinary students go into large-animal practice.
- By 2007, women outnumbered men in the veterinary profession, and more than 75 percent of veterinary school students were women.
- Adult cats spend 30 to 50 percent of the time they are awake grooming. Grooming, because of enzymes in the cat's saliva, reduces stress and prevents infection in open wounds.
- Because of their excitability, horses are placed in slings, in a pool of water, or on an inflatable mattress when recovering from anesthesia after major surgery.
- Dogs with spinal injuries have been successfully treated at Purdue University. An injection of polyethylene glycol (PEG) within seventy-two hours of the occurrence of the injury prevents permanent spinal damage and the resulting paralysis. Two weeks after treatment, more than 50 percent of the dogs were able to stand and walk.
- An improper ratio of calcium to phosphorus in a horse's diet is a major cause of splints. This injury occurs often in yearlings placed on pastures that have been heavily fertilized with phosphorus.

---

doctorate of veterinary medicine (D.V.M.) and a passing score on the North American Veterinary Licensing Exam are required for a veterinarian to be admitted to practice. Individuals who wish to specialize or become research veterinarians or professors must complete a one-year internship and three- to four-year residency program. Careers in private practice, research, zoos, wildlife management, and public health are open to individuals trained in veterinary science.

Other careers in veterinary science requiring less study and training are those of veterinary technician (two-year program) or technologist (four-year program). Veterinary technicians and technologists work in clinics and in research facilities. They assist veterinarians with animals in the examination room and in the surgical room and are responsible for the daily care of hospitalized and laboratory animals. They also perform laboratory procedures, keep records, and assist with client education.

Social Context and Future Prospects

Veterinarians play a very important role in modern society and are well respected as vital members of the community. Livestock producers depend on them for successful management of their herds and flocks. Veterinarians in public health positions ensure the safety of the world's food supply. Many owners of companion animals depend on veterinarians and value them as much as they do the family doctor. Research veterinarians are highly valued by many segments of society; however, animal welfare and animal rights movements have surrounded the use of animals in research with emotionally charged controversy.

Only about one-third of those who apply to veterinary college are accepted; however, those who complete their studies have excellent prospects for employment. The increase in the number of companion animals, especially cats, and the significant role that they play as family members ensures a continuing need for small-animal hospitals and specialized clinics. Shortages of veterinarians in farm animal practice, in the public health sector, and in research facilities offer excellent opportunities for employment in these sectors. The field of biomedical research is constantly expanding, and with the new areas of stem cell research and cloning, the need for both veterinarians in charge of laboratory animals and research veterinarians should continue to increase.

Shawncey Jay Webb, Ph.D.

Further Reading

Hanie, Elizabeth A. *Large Animal Clinical Procedures for Veterinarian Technicians.* St. Louis, Mo.: Mosby, 2005. Explains restraint of animals, diagnosis of disease, and the treatment of horses, cows, sheep, and swine. Good illustrations.

Hosey, Geoff, et al. *Zoo Animals: Behaviour, Management and Welfare.* New York: Oxford University Press, 2009. Discusses all aspects of animals in zoos: nutrition, health, housing, breeding, behavior, and research.

Lawhead, James, and MeeCee Baker. *Introduction to Veterinary Science.* 2d ed. Clifton Park, N.Y.: Delmar, Cengage Learning, 2009. Good overview of veterinary medicine. Contains chapters on systems, nutrition, and principles of disease and a section on career opportunities.

McGavin, M. Donald, and James F. Zachary. *Pathologic Basis of Veterinary Disease.* 4th ed. St. Louis, Mo.: Elsevier Mosby, 2007. Somewhat technical but very comprehensive coverage of animal disease based on a systems approach. Good graphics and photographs.

Rollin, Bernard E. *An Introduction to Veterinary Medical Ethics: Theory and Cases.* 2d ed. Ames, Iowa: Blackwell, 2006. Discusses the veterinarian's responsibility to animals, clients, and society using case histories and considers animal welfare awareness.

Smith, Gary, and Alan M. Kelly, eds. *Food Security in a Global Economy: Veterinary Medicine and Public Health.* Philadelphia: University of Pennsylvania Press, 2008. Reviews the role of veterinarians in public health, assesses the changes ahead, and notes emerging health threats.

Web Sites

American Association of Public Health Veterinarians
http://www.acvpm.org

American Association of Wildlife Veterinarians
http://www.aawv.net

American Association of Zoo Veterinarians
http://www.aazv.org

American Physiological Society
Animal Research
http://www.the-aps.org/pa/policy/animals.htm

American Veterinary Medical Association
http://www.avma.org

American Veterinary Society of Animal Behaviors
http://www.avsabonline.org/avsabonline

Humane Society of the United States
http://www.humanesociety.org

See also: Agricultural Science; Animal Breeding and Husbandry; Pharmacology.

VIROLOGY

FIELDS OF STUDY

Microbiology; general biology; cell biology; chemistry; molecular biology; biochemistry; biophysics; microbial genetics; immunology; genetic engineering.

SUMMARY

Virology is a scientific field emphasizing the study of submicroscopic entities known as viruses. Because genetic material within viruses can be easily manipulated and viruses replicate at a relatively rapid rate, research in molecular genetics between the 1940's and 1970's largely consisted of the study of viruses. Scientific application within virology has led to understanding cell processes such as the molecular basis for cancer, as well as developments in the field of biotechnology. Genetic engineering, the process of altering the genetic makeup of organisms using viral vectors, has also provided a means for production of numerous pharmaceuticals.

KEY TERMS AND CONCEPTS

- **Antibody:** Protein produced by lymphocytes following exposure to an antigen or anything perceived by the body as foreign.
- **Cell Culture:** Animal or plant cells grown in the laboratory. Used for growth and study of viruses.
- **Host Range:** Animals or specific tissues and cells within the animal that a specific virus is capable of infecting.
- **Molecular Cloning:** Isolation of specific gene or fragment of DNA and its insertion into a viral vector.
- **Plaque Assay:** Method of quantifying viruses in which a viral solution is used to infect a layer of cells growing in a dish. Sites of infection by viral particles will appear as "holes" in the layer.
- **Restriction Enzymes:** Enzymes used to cut DNA to allow for insertion of genetic material.
- **Transfection:** Cellular uptake of "naked" DNA from the environment. Used to introduce DNA vectors without the use of viral capsids, bypassing the problem of host range.
- **Virus Particle:** Genetic material, either DNA or RNA, enclosed within a protein capsid or coat. The capsid may also be surrounded by a membrane or envelope.

DEFINITION AND BASIC PRINCIPLES

Viruses are intracellular parasites that infect all types of organisms, including bacteria, plants, and animals. Because viruses are commonly associated with human diseases, the public usually views these biological entities only in this context. However, the ability of viruses to infect cells and, in some cases, to integrate within the host genetic material has led to the development of new technologies for insertion or replacement of defective genetic material.

Although all cells are capable of being infected by viruses, viruses themselves often exhibit a specific host range. Bacterial viruses infect only specific strains of bacteria, and animal viruses usually are restricted to a specific species or tissues within the species. Viruses contain on their surfaces proteins that determine which cells can be infected. In a manner analogous to a lock and key, viruses attach to specific receptors expressed on the surface of the cell. Viral vaccines induce the body to produce neutralizing proteins, or antibodies, which bind to the surface of the virus and block its adsorption into the cell.

BACKGROUND AND HISTORY

Viral diseases have probably existed since the early years of human civilization, but viruses were not discovered until the late 1800's. The cause of many human diseases was unknown until the development of the germ theory of disease in the mid-nineteenth century. During the golden age of microbiology, from about 1875 to 1900, scientists were able to determine that microorganisms caused certain diseases. French scientist Louis Pasteur and German physician Robert Koch were prominent among these early microbiologists. The work of Pasteur, Koch, and their associates involved isolating and growing bacteria in the laboratory and demonstrating that these organisms were able to cause specific diseases in animals.

Pasteur, working with rabies in the early 1880's, noted that unlike bacteria, the agent that caused rabies did not grow on culture media and was capable of passing through minute filters. The agent was called a "virus," from the Latin word for poison. During the 1890's and early twentieth century, scientists demonstrated that viruses were capable of replicating, indicating that they were a life-form rather than a poison. A number of human diseases, including smallpox, influenza, and rabies, were shown to be caused by these agents.

The development of the electron microscope during the 1930's allowed viruses to be observed. The first viruses to be extensively studied were bacteriophages, viruses that infect bacteria. Because some bacteriophages kill bacteria, it was at one time thought that they could be used to treat bacterial infections; the plot of Sinclair Lewis's book *Arrowsmith* (1925), in which physician Martin Arrowsmith used "phage" to deal with an outbreak of plague, was predicated on that idea.

HOW IT WORKS

Much of the study of viruses has revolved around finding a way to treat viral diseases or prevent them through vaccines. However, as the mechanism by which they infect people became better understood, scientists began to develop ways in which they could use viruses to treat other diseases and conditions.

Viral Vaccines. The first human vaccine against a viral disease was created in the 1790's through a serendipitous discovery by British physician Edward Jenner. He observed that people previously infected with material from lesions on the udders of cattle (cowpox) became immune to smallpox. Although Jenner was unaware of why cowpox provided immunity to smallpox, scientists later determined that the cowpox virus contains proteins that cross-react with those of smallpox and produce immunity in those exposed to it. The process of immunization became known as vaccination, derived from the Latin word *vacca*, which means cow.

Although the smallpox vaccine uses a naturally cross-reactive virus, modern viral vaccines take advantage of the ability of artificially attenuated or killed viruses to provide immunity. Attenuated viruses, such as those used to immunize against rabies, polio (Sabin vaccine), measles, and mumps, are viruses that have had their disease-causing ability reduced through animal or cell culture passage, and they produce immunity to the disease although they are incapable of causing illness. The Salk vaccine against polio uses a virus that has been killed through chemical treatment.

The theory behind any immunization is that components of the vaccine induce the recipient to produce antibodies, which provide protection against any subsequent exposure to that agent. Viruses have their own means to avoid neutralization by antibodies. Influenza virus is prone to two forms of change: Antigenic drift, mutations in the surface hemagglutinin protein that is the target of the antibody response, and antigenic shift, the recombination of the human strain with animal influenza strains to create an entirely new variety of influenza virus. The 2009 swine flu (H1N1 flu) was the result of such a shift. Rhinoviruses, associated with most cases of the common cold, do not mutate significantly. However, more than one hundred strains of the virus exist, and immunity to one does not confer immunity to the other strains.

Genetic Manipulation. Genetic diseases generally have their origin in mutations that affect either the proper expression of specific genes or the function of the gene product. Although in most cases the mutation affects only a single gene, the effect may be significant because many gene products are pleiotropic, producing multiple effects in the organism. An example is the *FBN1* gene, which encodes fibrillin, a protein necessary for proper connective tissue formation. Mutations in the gene result in Marfan syndrome, a weakening in connective tissue throughout the body, which can affect the limbs, aorta, or particular organs.

The principle behind gene therapy is that replacement of the gene containing the mutation with a normal copy may reverse the effects of the genetic error. The idea applies in particular to genetic defects that are associated with a single gene, or monogenic diseases. The challenge has been how to introduce the correct gene into the tissues that are affected in a manner that minimizes side effects yet produces a permanent correction. Viruses are particularly useful in serving as gene vectors because some have the ability to infect multiple cells and multiple types of tissue and, in some cases, integrate into the host genome, becoming a permanent part of the genetic material.

Because viruses by their nature generally alter or kill the cells they infect, they must first be rendered harmless to be used in gene therapy. This is carried out by first deleting the genes necessary for viral replication. For example, to render human adenoviruses harmless, two specific genes, *E1* and *E3*, are deleted to block virus replication. Deletion of an additional gene called *F1*, which encodes the surface fiber that allows the virus to attach, reduces the danger of inflammation following the virus's introduction into a host. Deletion of the fiber also reduces the host range, limiting the variety of cells that can be infected. Cutting of the DNA with the proper restriction enzymes allows the desired therapeutic gene to be inserted at the site once occupied by the *E1* and *E3* genes, creating recombinant DNA. Cell cultures that can provide the necessary functions for duplication of the recombinant DNA are then transfected with the engineered viral DNA.

Anticancer Therapy. The limitation of host range for viral infection has been applied to try to develop anticancer therapies. The theory is that because viruses can infect only certain cells, viruses could be altered to express proteins detrimental to the survival of cancer cells. Gene therapy using viruses has taken a number of approaches. Several clinical trials have involved the introduction into cancer cells of retroviruses carrying genes for cytokines, proteins that stimulate an immune response directed against the neoplasm. A second approach uses viruses that infect only specific types of cells such as nerve cells. For example, viral agents such as the herpesviruses naturally target tissues found in the nervous system. Genetically altered herpesviruses could in theory be used to kill only tumor cells that arise in the brain or spinal cord.

APPLICATIONS AND PRODUCTS

Gene Introduction and Integration. The ability to insert specific genes into viral vectors provides a mechanism for introducing genetic material into individual cells or tissues within organisms. Several viruses, each with its own advantages and disadvantages, have served as such vectors. These viruses have in common the ability to infect a broad range of hosts, allowing for a wide range of applications in the field of genetic engineering. Replacement of defective genes in human cells remains the goal, but only a few therapies have been developed to the point where they could be tested in clinical trials.

The choice of virus is generally based on the size of the genetic material to be introduced and the likelihood of any immune or inflammatory response.

For larger segments of DNA, ranging between 7,500 and 35,000 base pairs, enough to encode a protein or proteins with about 2,500 to 10,000 amino acids, the primary choices as vectors are either vaccinia virus or adenovirus. Vaccinia, a large enveloped double-stranded DNA virus, has previously served a role as the vaccine against smallpox. When vaccinia is used as a vector, viral genes necessary for replication are deleted before inserting the desired genetic material, rendering the virus unable to replicate. Unfortunately, vaccinia frequently elicits a significant inflammatory response, limiting its usefulness.

Adenoviruses are relatively large nonenveloped DNA viruses with a protein capsid. Spikes or fibers attached to the capsid determine the host range for infection. There are about fifty serotypes of adenoviruses, some of which are associated with human respiratory and gastrointestinal infections. The most commonly used adenovirus strain for genetic studies is serotype 5. Recombinant adenovirus DNA has been used in limited clinical trials to treat patients with cystic fibrosis and ornithine transcarbamylase (OTC) deficiency. Cystic fibrosis, the most common inherited genetic disease in the West, is associated with a mutation in the gene that encodes a regulator protein necessary for proper transport of ions in and out of cells. OTC deficiency is a metabolic disorder in which urea metabolism is affected. About 25 percent of clinical trials using a viral vector have used adenoviruses.

Lentiviruses such as HIV have a vector capacity significantly smaller than that of adenoviruses, but they have the ability to integrate into the host genome, resulting in the recombinant gene becoming part of the host genetic material. In animal trials, lentiviruses have introduced growth factor genes into mouse cells as well as the gene encoding Factor VIII, which is lacking in the most common forms of human hemophilia. Preclinical trials have demonstrated that recombinant HIV can be used to replace defective genes in diseases such as cystic fibrosis and muscular dystrophy. About 25 percent of clinical trials have used retroviruses as the vector, including most of the original clinical trials that used viral vectors.

The ability of lentiviruses to integrate into the host genetic material, one of the desirable features of such viruses, does have its drawbacks. Integration is not random and frequently takes place within the introns (DNA regions in a gene that is not translated into protein) of preexisting genes, some of which are necessary for proper cell function. This can produce severe side effects, such as those that occurred during trials in France that attempted to insert an adenosine deaminase gene to cure a form of severe combined immunodeficiency syndrome (SCID). Although nine patients did show improved immune function, three developed T-cell leukemia, and one person died.

Adeno-associated viruses are naturally defective viruses that require a helper adenovirus to replicate normally. However, they are capable of infecting a range of cells and can integrate at specific sites in human chromatin. Adeno-associated viruses have been used experimentally to kill certain forms of breast, cervical, and prostate cancer cells. As viral vectors, they have been used to introduce a gene for production of insulin and the genes for both Factor VIII and Factor IX, lacking in humans with certain forms of hemophilia (hemophilia A and B, respectively), into mice genes that encode erythropoietin, a glycoprotein that induces the bone marrow to increase red blood cell production.

Adeno-associated viruses have been associated with only a small proportion of all clinical trials. Other vectors have used naked plasmid DNA, vaccinia, and other poxviruses, as well as other types of viruses.

IMPACT ON INDUSTRY

Vaccine production has historically involved a symbiotic relationship between university research centers and industry, with universities, often in association with medical centers, carrying out the initial development of a vaccine and industry carrying out the development and marketing of the vaccine. Since the 1960's, vaccine production has been affected by two cost-related issues. First, developing a vaccine costs hundreds of millions of dollars but does not guarantee a profitable return. Second, the cost of dealing with lawsuits arising from problems with vaccines and of insuring against possible litigation is considerable. For example, illnesses allegedly associated with the 1976 swine influenza vaccine resulted in a payout of $100 million, a cost assumed by the federal government.

University Research. Basic research involving the isolation and study of viruses has historically been carried out in universities. Two of the most prominent examples involved the development of poliomyelitis

vaccines by Jonas Salk and Albert Sabin. Salk, a faculty member of the University of Pittsburgh, had some research experience during World War II in the development of one of the early influenza vaccines, but he was a relative newcomer to the polio field when he began work on a polio vaccine in the late 1940's. By the early 1950's, he had developed a formalin-killed virus ready for a nationwide field trial. Working with the School of Public Health at the University of Michigan, Salk helped direct the largest field trial of a vaccine in history. During the summer of 1954, more than 400,000 children were inoculated with his vaccine. The vaccine proved to be more than 80 percent effective in reducing the incidence of polio.

Sabin, who carried out much of his research at Children's Hospital in Cincinnati, developed an attenuated version of the polio vaccine by the 1950's. The initial large-scale trial of Sabin's vaccine, which involved more than 100 million people, was carried out in Eastern Europe during the late 1950's. By the 1960's, it had largely replaced the Salk vaccine in the United States.

Industrial Production of Vaccines. Research and insurance costs have driven most manufacturers out of the vaccine business. During the early 1970's, more than a dozen companies were involved in production of polio vaccines, but by the mid-1980's, only a single company, Lederle (later Wyeth-Lederle), still produced the vaccine. Those companies that produce viral vaccines for measles and polio can do so because these are mature products for which costs are largely fixed.

The production and field testing of viral vectors face similar challenges. Trials in France using recombinant lentiviruses for amelioration of severe combined immunodeficiency syndrome (SCID) produced significant side effects in subjects, and one person died. Fear of litigation, often warranted, has made many companies conclude that the cost-benefit ratio is simply too high.

CAREERS AND COURSE WORK

Because virology is a biological science, students wishing to pursue a career in the field should emphasize undergraduate work in biology, working toward a bachelor of science degree. At the undergraduate level, students should elect course work in general biology and chemistry, with advanced courses in cell biology, microbiology, and biochemistry. Students should also enroll in virology and molecular biology if those courses are available.

Obtaining a graduate degree, such as a master's degree in biochemistry or in a biological science, is highly recommended. Graduate work should emphasize the molecular biology of viruses, and the student should try to obtain experience in molecular cloning or genetic engineering. A master's degree is adequate for performing laboratory work. However, either a medical degree or a doctorate is required for proposing or directing clinical trials. Much of the basic research in virology is carried out in university settings, although researchers with advanced degrees are also part of hospital or medical school staffs.

Vaccines or other virus-related treatments are usually tested first on animals to determine if there are any obvious problems with safety and to measure the treatment's efficacy. The treatments are then tested on humans in clinical trials, usually carried out by physicians associated with university or hospital health centers. The director of a trial, in conjunction with any associates, obtains the necessary approval of the appropriate government agency; funding may be provided by either a government agency or a private source. The phase or level of the clinical trial reflects the number of patients involved in the trial. Phase I trials involve fewer than one hundred volunteers, and Phase II trials involve upward of several hundred.

SOCIAL CONTEXT AND FUTURE PROSPECTS

Researchers who attempt to control viral disease must address both the evolving nature of viruses and the new or unusual viral diseases that develop. Many viruses, including influenza and HIV (human immunodeficiency virus), mutate at a rapid rate. Newly emerging viral diseases include Ebola hemorrhagic fever and Lassa fever (both in Africa) and hantavirus (in the United States).

Influenza and HIV provide particular challenges. The potential virulence of influenza was demonstrated during the 1918 worldwide pandemic in which an estimated 50 million people died. Any influenza vaccine that would provide protection against multiple strains would have to address the problem of both genetic drift and genetic shift. The inability to grow the virus in anything but embryonated chicken eggs rather than cell culture also significantly increases the lead time necessary for large-scale vaccine production.

Fascinating Facts About Virology

- Blossom was the name of the cow from which Edward Jenner isolated cowpox virus, which he used to immunize people against smallpox.
- The association of viruses and cancer was demonstrated by Peyton Rous in 1911, when he used cell-free extracts from a tumor extracted from a Plymouth Rock chicken to induce tumors in healthy chickens. Fifty-five years later, four years before his death at the age of ninety, Rous was awarded a Nobel Prize.
- The Raggedy Ann doll was created in 1915 by a New York illustrator, John Gruelle, for his daughter who became ill following a smallpox vaccination.
- The bacterium *Haemophilus influenzae* does not cause influenza. The species name resulted from its isolation from influenza patients.
- Restriction enzymes, critical to genetic engineering, were discovered during experiments studying the resistance of bacteria to viral infections.

HIV, the etiological agent for AIDS (acquired immunodeficiency syndrome), provides its own challenges. HIV mutates at a high rate immediately following infection of human lymphocytes, producing dozens of varieties in weeks. The high rate of mutation is the primary reason that no AIDS vaccine has thus far proven effective.

The encroachment of human civilization into previously isolated areas has exposed humans to new or unusual forms of viruses. In the 1980's, people in parts of western Africa were exposed for the first time to rodent viruses that produced Lassa fever in people. Luckily, the outbreaks were limited. The movement of people into new areas in both central and western Africa during the 1970's probably brought nonhuman primates infected with HIV in contact with people, and the virus jumped from one species to the other. Although vaccines against newly emerging viruses such as Ebola hemorrhagic fever or Lassa fever could probably be manufactured, the cost of addressing what so far have been localized outbreaks is prohibitive.

Richard Adler, Ph.D.

FURTHER READING

Allen, Arthur. *Vaccine.* New York: W. W. Norton, 2007. The story of vaccines and the controversies that have long surrounded their use.

Edelstein, Michael, Mohammad Abedi, and Jo Wixon. "Gene Therapy Clinical Trials Worldwide to 2007—An Update." *The Journal of Gene Medicine* 9 (August, 2007): 833-842. Summary of the results of more than 1,300 clinical trials using viral vectors for insertion of replacement genes.

Madigan, Michael, et al. *Brock Biology of Microorganisms.* 12th ed. San Francisco: Pearson Benjamin Cummings, 2009. Several chapters in this textbook of microbiology address the role of viruses in disease as well as in biotechnology.

Strauss, James, and Ellen Strauss. "Virus Vector Systems." In *Viruses and Human Disease.* New York: Elsevier, 2008. Examines the molecular biology of animal viruses and their role in human disease. Addresses the role of viruses as vectors for genetic engineering. Includes numerous illustrations of procedures.

WEB SITES

American Society for Virology
http://www.asv.org

Centers for Disease Control and Prevention
Vaccines and Immunizations
http://www.cdc.gov/vaccines

Pan American Society for Clinical Virology
http://www.virology.org/links.html

See also: Epidemiology; Immunology and Vaccination; Pathology; Pharmacology.

XENOTRANSPLANTATION

FIELDS OF STUDY

Cell biology; molecular biology; cell and tissue engineering; surgery; immunology; organ transplantation; biomedicine; genetics; genetic engineering; animal testing; medical ethics; public health; infectious disease; virology; infectious diseases; regenerative medicine; stem cell research.

SUMMARY

Xenotransplantation is a medical procedure in which live cells, tissues, or whole organs are surgically transplanted from one animal species to another, such as transplanting pig hearts or baboon livers into humans. It is a potential solution to the shortage of human organs and tissues available for transplant. Xenotransplantation also offers many other potential therapeutic applications, including the treatment of diabetes, spinal cord injury, and Parkinson's disease. However, the use of xenotransplantation has raised many ethical and safety concerns.

KEY TERMS AND CONCEPTS

- **Allotransplantation:** Transplantation of tissues, cells, or organs between two nongenetically identical members of the same species.
- **Graft:** Healthy tissue that replaces removed damaged tissue; an allograft is a graft between the same species, and a xenograft (or xenotransplant) is a graft between different species.
- **Histocompatibility:** Condition in which graft and recipient tissues can survive in each others' presence without provoking an immunological response.
- **Immunosuppression:** Act of lowering the body's natural protective response to invasive substances, often achieved by immunosuppressive drugs.
- **Rejection:** Any of a number of immune system reactions that destroy transplanted material that is not accepted by the host's body; the opposite of rejection is tolerance.
- **Retrovirus:** Virus that inserts a copy of its own genetic material into the genetic material (DNA) of a host cell in order to replicate itself.
- **Transgenic:** Describing an organism whose genetic material has been altered through the introduction of a gene or genes from another species.
- **Xenozoonosis:** Transmission of an infectious disease between members of two different species.

DEFINITION AND BASIC PRINCIPLES

Xenotransplantation is the science and technology of transplanting tissues, cells, or entire organs from a member of one animal species to a member of a different animal species. For example, a human patient with heart failure might have his or her own organ replaced with a living heart from a pig. Xenotransplantation does not include giving human recipients inert, nonliving animal tissues or cells, such as the common practice of replacing faulty human heart valves with cow or pig heart valves.

The basic principles of xenotransplantation are similar to those of allotransplantation, which has a long and successful history in medicine. In both procedures, healthy living tissues from another organism are implanted into a patient whose own tissues have failed because of disease or trauma. Xenotransplantation, however, is more complex, since the donor's cells are not likely to be histocompatible with the host's cells. The benefits of xenotransplantation assume that the long-term biological functionality of these foreign materials can be achieved in the recipient without rejection or infections from the donor's viruses or bacteria. Medical research has devised a number of immunosuppressive therapies to combat the rejection response and the risk of graft-versus-host disease associated with allotransplantation, but because xenotransplantation involves transplanting tissues from a completely different species, the risk of complications is greater.

For this reason, as well as for a number of social and cultural barriers, xenotransplantation is largely an experimental field.

Background and History

Attempts at cross-species transplantation date back at least as far as the seventeenth century and include attempts to transfuse sheep blood and graft dog bones into human recipients. In 1961, British immunologist and geneticist Peter Gorer proposed replacing the term "heterotransplantation," which was then used to refer to transplants from a different donor species, with the word "xenotransplantation." Beginning in the 1960's and continuing into the 1990's, a large number of whole-organ xenografts, including operations involving baboon and chimpanzee kidneys, as well as pig hearts, were attempted by several pioneering practitioners. However, the patients died within weeks or months of their surgeries, either from a rejection response that caused organ failure or from acute infections.

Later tests of xenotransplantation have concerned tissue and cell grafts, rather than whole-organ grafts. For example, a number of trials are proceeding in which liver and pancreatic cells from animals are transplanted into human patients to treat liver failure and diabetes. Although important strides have been made in suppressing graft rejection responses through techniques such as the use of encapsulation, the field has yet to entirely fulfill its promise.

How It Works

Broadly speaking, xenotransplantation techniques are similar to those associated with allotransplantation: healthy, properly functioning cells, tissues, and organs are introduced into a recipient's body to take the place of malfunctioning or damaged tissues. The biggest obstacle to both allotransplantation and xenotransplantation is the human immune system itself, which is programmed to attack foreign pathogens (harmful substances such as bacteria and viruses). It applies the same response to foreign cells and tissues, often destroying transplanted material. Various techniques are used to combat this response and reduce the risk of graft rejection.

Ex Vivo Xenotransplantation. The definition of xenotransplantation includes the method of giving a human recipient bodily fluids, tissues, or cells that have previously been exposed to the cells of another animal species ex vivo, or outside the body. For example, a patient may receive human cells and mouse cells that have been cultured together.

Immunosuppression. Immunosuppressive drugs, which may be given to xenotransplantation patients both before and after the transplantation process (sometimes their use is required for the rest of the recipient's life), are designed to inhibit the activity of the patient's immune system, reducing its ability to attack the foreign cells, tissues, or organs. Some classes of immunosuppressive drugs include corticosteroids and anticytokines. One of the drawbacks of immunosuppression is that inhibiting the patient's immune system renders the individual more vulnerable to infection.

Transgenesis. In transgenesis, a human gene is inserted into the genome, or genetic material, of another species, either by injecting it directly into the nucleus of a fertilized egg or by introducing a virus containing human DNA into animal cells. This technique is designed to produce animal tissues that are histocompatible with human recipients. Once the human DNA spliced into the animal's genome is successfully expressed by the animal's cells, its tissues or organs will no longer be identified as foreign pathogens to be destroyed by the human immune system. So far, attempts to create transgenic mice, sheep, goats, chickens, pigs, and small primates have been conducted, with varying degrees of success.

Encapsulation. Encapsulation is a technique that allows the cells or tissues from the donor animal to be separated from the human immune system. Usually it does not literally involve a capsule but rather droplets of a viscous gel derived from a nonanimal source such as seaweed. The gel allows the grafted cells to accept nutrients from the human recipient and to diffuse hormones, such as insulin, into the body. At the same time, it protects the grafted cells from attacking antibodies. Encapsulation is not a viable technique for whole organ xenografts.

Applications and Products

Organ Failure. The single most significant potential application of xenotransplantation is the replacement of organs that have failed, because there is a severe shortage of human organs available for such transplants. Organ failure is one of the leading causes of death in the developed world. In theory, countless numbers of patients with liver failure caused by hepa

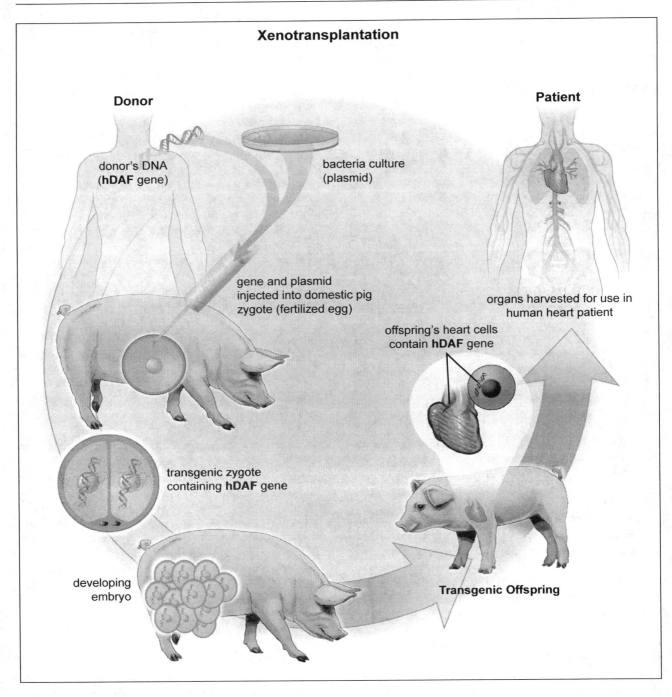

Xenotransplantation

Donor

donor's DNA (**hDAF** gene)

bacteria culture (plasmid)

Patient

gene and plasmid injected into domestic pig zygote (fertilized egg)

offspring's heart cells contain **hDAF** gene

organs harvested for use in human heart patient

transgenic zygote containing **hDAF** gene

developing embryo

Transgenic Offspring

titis, heart failure caused by cardiovascular disease, or lung failure caused by cystic fibrosis could be saved from premature death by xenografts harvested from transgenic animals such as pigs or baboons. Xenotransplantation could also benefit patients without life-threatening diseases. For example, it may one day be feasible to implant pig or baboon corneas into patients who are blind because of a cloudy or damaged cornea.

Gene Therapy. Potentially, xenotransplantation

Fascinating Facts About Xenotransplantation

- In the early twentieth century, thousands of testicle transplants were carried out on men all over the world using thin slices of testicular tissue from monkeys. Although the human patients survived these operations, later research showed that the transplanted material had in fact been totally rejected by the hosts' bodies.
- In 1984, a premature baby known as Fae lived for three weeks after receiving a heart transplant from a baboon.
- Scientists Peter Medawar and Sir Frank MacFarlane Burnet were honored with a 1960 Nobel Prize in Physiology or Medicine for their discovery that the rejection response to foreign cells is not present in animals in the embryonic stage.
- Usually, it is the recipient's own body that attacks a xenograft (or xenotransplant), but the grafted tissue itself can also launch an attack on the host's cells, a phenomenon known as graft-versus-host disease.
- In 1682, a Russian patient with a head injury had a piece of dog bone grafted onto his skull; the procedure was reported to be successful.
- The word "xenotransplantation" is derived from a Greek root meaning "foreign" or "strange."

could be used as a means of delivering targeted gene therapy. In other words, animal cells genetically programmed to express a particular gene could be implanted into a human patient for therapeutic purposes, such as to replace a defective gene with one that functions normally. For example, patients with genetic disorders such as Huntington's disease could receive cellular xenotransplantations in which the donated animal cells were engineered to express the normal human gene that is defective in the patient's own cells.

Chronic Diseases. Xenotransplantation has the potential to treat many chronic diseases that cause cell death. For example, in diabetes, pancreatic cells that produce insulin are destroyed. Encapsulation is being tested as a means of introducing porcine islet cells (pancreatic cell structures from pigs) into human patients with diabetes. The encapsulated porcine cells help these patients produce the insulin that

they would otherwise have to inject into themselves.

Ex vivo xenotransplantation can also be used to treat chronic diseases. For example, the blood of patients with liver failure may be passed through a pig liver, or a pump containing pig liver cells, before being returned to the patients. This process, which allows the human blood to be purified of its toxins, is known as xenoperfusion. In such cases, the human blood cells are separated from the pig liver cells by a semipermeable membrane (a barrier that allows some, but not all, materials to pass through). This reduces the chance of dangerous inflammation and damage to the human cells, which can occur if biochemical materials secreted by the animal cells become bound to the patient's blood.

Degenerative Diseases. Xenotransplantation trials are taking place with Parkinson's disease, a neurodegenerative disorder that causes progressive deterioration in the cells of the brain. The trials involve patients who have had fetal pig neurons implanted into their brains. The hope is that these cells will help reverse the damage caused by their disease. Xenotransplantation is therefore a potentially useful therapy in cases in which whole-organ transplantation is not suitable and for which no effective treatment exists.

Burn Treatment. Patients who have suffered severe burns may be treated with grafts of skin grown from their own cells. To promote the rapid growth of these grafts, the human skin cells are cultured in a laboratory together with mouse stem cells. This particular xenotransplantation application has already been tested for safety and effectiveness and is in clinical use.

IMPACT ON INDUSTRY

Government and University Research. A number of countries, including Australia and Canada, have placed moratoriums on cellular and whole-organ xenotransplantation trials. As a result, xenotransplantation trials sometimes take place in countries with more liberal guidelines. For example, in July 2009, Australian researchers began conducting trials on encapsulated porcine islet cells in Russia because they could not do so in their own country. In the United States, where xenotransplantation products are regulated by the Food and Drug Administration (FDA) and trials are subject to the same reviews and guidelines as other scientific research, the Mayo Clinic in

Bethesda, Maryland, has emerged as a major academic center for xenotransplantation research. The Mayo Clinic research focuses on the use of transgenic pigs for organ and tissue donation.

Industry and Business. The industries where xenotransplantation has the potential to make the greatest impact are medicine, biotechnology, and health care. However, should xenotransplantation become a more commonly used technology, it will also shape the direction in which animal agriculture is headed by sharply increasing demand for those animals whose tissues are most widely used in xenografts, such as pigs. Close collaborations between biotechnology companies and the agricultural industry would be required to produce transgenic animals for use in therapeutic applications.

Major Corporations. Because of the potential profits to be made if a xenotransplantation product were to be successfully tested and brought to market, many biotechnology companies are investing significant resources in this research. Among the major corporations involved in xenotransplantation trials are the large Swiss pharmaceutical company Novartis, the New Zealand company Living Cell Technologies, and the British company PPL Therapeutics.

CAREERS AND COURSE WORK

The academic path toward a career in xenotransplantation must begin with a solid foundation in the biological sciences, especially cellular biology and immunology. Other important course work would include genetics, cell and tissue engineering, and the study of infectious diseases. Many xenotransplantation researchers are trained physician-investigators who have pursued a dual M.D.-Ph.D. degree after obtaining a bachelor of science degree at the undergraduate level. The M.D.-Ph.D. program provides training in both medicine and scientific research and typically requires about seven to eight years to complete.

SOCIAL CONTEXT AND FUTURE PROSPECTS

As news that the porcine endogenous retrovirus (PERV) could infect human cells surfaced in the 1990's, xenotransplantation fell under a cloud of public suspicion and fear that has not yet fully lifted. Because of the potential for a xenozoonotic outbreak to spread rapidly across the world, there is an international dialog taking place among countries such as Canada, Sweden, Spain, France, and the United States about the safeguards that should be associated with xenotransplantation research.

Many believe harvesting animal organs for human use is a violation of animal cruelty laws. The choice of donor animal is also a subject of debate. Because of their genetic similarity to humans, nonhuman primates such as chimpanzees are in some ways ideal candidates. However, many object to the use of such animals for xenotransplantation purposes because many nonhuman primates are endangered species. Also, closely related species are more likely to pass pathogens to the recipients.

Despite the fears and ethical concerns about xenotransplantation, it is a technology with the potential for producing immense therapeutic benefits in many different clinical areas. It especially offers hope to the thousands of patients on waiting lists for transplants or living with chronic or degenerative diseases.

M. Lee, B.A., M.A.

FURTHER READING

Cooper, David K. C., and Robert P. Lanza. *Xeno: The Promise of Transplanting Animal Organs into Humans.* New York: Oxford University Press, 2000. An accessible overview of xenotransplantation technology by two physician-researchers. Includes a glossary of biomedical terms and a bibliography of international attempts to develop xenotransplantation guidelines.

McLean, Sheila, and Laura Williamson. *Xenotransplantation: Law and Ethics.* Burlington, Vt.: Ashgate, 2005. Explores the legal and ethical issues surrounding xenotransplantation, including the humane use of animals, public safety concerns, and the issue of informed medical consent.

Petechuk, David. *Organ Transplantation.* Westport, Conn.: Greenwood Press, 2006. The chapter "Xenotransplantation: Animal Use and Infectious Disease" addresses the costs and benefits of xenotransplantation and its philosophical implications.

Platt, Jeffrey L. "Biotechnology: Xenotransplantation." In *Encyclopedia of Animal Science,* edited by Wilson G. Pond and Alan W. Bell. New York: Marcel Dekker, 2005. Covers the basic science of xenotransplantation and includes a table showing how various animals express a compound provoking an immune response in humans.

Rothblatt, Martine Aliana. *Your Life or Mine: How*

Geoethics Can Resolve the Conflict Between Public and Private Interests in Xenotransplantation. Burlington, Vt.: Ashgate, 2004. This robust argument in favor of the benefits of xenotransplantation proposes solutions to the medical and ethical risks associated with it. Includes figures and tables showing survival rates for xenografts and stages of graft rejection

WEB SITES

Food and Drug Administration
Xenotransplantation
http://www.fda.gov/BiologicsBloodVaccines/
Xenotransplantation/default.htm

World Health Organization
Xenotransplantation
http://www.who.int/transplantation/xeno/en

See also: Animal Breeding and Husbandry; Bioengineering; Cardiology; Cell and Tissue Engineering; Hematology; Pulmonary Medicine; Veterinary Science.

Appendixes

BIOGRAPHICAL DICTIONARY OF SCIENTISTS

Alvarez, Luis W. (1911-1988): A physicist and inventor born in San Francisco, Alvarez was associated with the University of California, Berkeley, for many years. He explored cosmic rays, fusion, and other aspects of nuclear reaction. He invented time-of-flight techniques and conducted research into nuclear magnetic resonance for which he was awarded the 1968 Nobel Prize in Physics. He contributed to radar research and particle accelerators, worked on the Manhattan Project, developed the ground-controlled approach for landing airplanes, and proposed the theory that dinosaurs were rendered extinct by a massive meteor impacting Earth.

Archimedes (c. 287-c. 212 B.C.E.): A Greek born at Syracuse, Sicily, Archimedes is considered a genius of antiquity, with interests in astronomy, physics, engineering, and mathematics. He is credited with the discovery of fluid displacement (Archimedes' principle) and a number of mathematical advancements. He also developed numerous inventions, including the Archimedes screw to lift water for irrigation (still in use), the block-and-tackle pulley system, a practical odometer, a planetarium using differential gearing, and several weapons of war. He was killed during the Roman siege of Syracuse.

Babbage, Charles (1791-1871): An English-born mathematician and mechanical engineer, Babbage designed several machines that were precursors to the modern computer. He developed a difference engine to carry out polynomial functions and calculate astronomical tables mechanically (which was not completed) as well as an analytical engine using punched cards, sequential control, branching and looping, all of which contributed to computer science. He also made advancements in cryptography, devised the cowcatcher to clear obstacles from railway locomotives, and invented an ophthalmoscope.

Bacon, Sir Francis (1561-1626): A philosopher, statesman, author, and scientist born in England, Bacon was a precocious youth who at the age of thirteen began attending Trinity College, Cambridge. Later a member of Parliament, a lawyer, and attorney general, he rejected Aristotelian logic and advocated for inductive reasoning—collecting data, interpreting information, and carrying out experiments—in his major work, *Novum Organum* (*New Instrument*), published in 1620, which greatly influenced science from the seventeenth century onward. A victim of his own research, he experimented with snow as a way to preserve meat, caught a cold that became bronchitis, and died.

Baird, John Logie (1888-1946): A Scottish electrical engineer and inventor, Baird successfully transmitted black-and-white (in 1925) and color (in 1928) moving television images, and the BBC used his transmitters to broadcast television from 1929 to 1937. He had more than 175 patents for such far-ranging and forward-thinking concepts as big-screen and stereo TV sets, pay television, fiber optics, radar, video recording, and thermal socks. Plagued with ill health and a chronic lack of financial backing, Baird was unable to develop his innovative ideas, which others later perfected and profited from.

Bardeen, John (1908-1991): A Wisconsin-born electrical engineer and physicist, Bardeen worked for Gulf Oil, researching magnetism and gravity, and later studied mathematics and physics at Princeton University, where he earned a doctoral degree. While working at Bell Laboratories after World War II he, Walter Brattain (1902-1987), and William Shockley (1910-1989) invented the transistor, for which they shared the 1956 Nobel Prize in Physics. In 1972, Bardeen shared a second Nobel Prize in Physics for a jointly developed theory of superconductivity; he is the only person to win the same award twice.

Barnard, Christiaan (1922-2001): A heart-transplant pioneer born in South Africa, Barnard was a cardiac surgeon and university professor. He performed the first successful human heart transplant in 1967, extending a patient's life by eighteen days, and subsequent transplants—using innovative operational techniques he devised—allowed new heart recipients to survive for more than twenty years. He was one of the first surgeons to employ living tissues and organs from other species to prolong human life and was a contributor to the effective design of artificial heart valves.

Bates, Henry Walter (1825-1892): A self-taught

naturalist and explorer born in England, Bates accompanied anthropologist-biologist Alfred Russel Wallace (1823-1913) on a scientific expedition to South America between 1848 and 1852, which he described in his 1864 work, *The Naturalist on the River Amazons*. He collected thousands of plant and animal species, most of them unknown to science, and was the first to study the survival phenomenon of insect mimicry. For nearly thirty years he was secretary of the Royal Geographical Society and also served as president of the Entomological Society of London.

Becquerel, Antoine-Henri (1852-1908): A French physicist and engineer born into a family boasting several generations of scientists, Becquerel taught applied physics at the National Museum of Natural History and at the Polytechnic University, both in Paris, and also served as primary engineer overseeing French bridges and highways. He served as president of the French Academy of Sciences and received numerous awards for his work investigating polarization of light, magnetism, and the properties of radioactivity, including the 1903 Nobel Prize in Physics, which he shared with Pierre and Marie Curie.

Bell, Alexander Graham (1847-1922): A Scottish engineer and inventor whose mother and wife were deaf, Bell researched hearing and speech throughout his life. He began inventing practical solutions to problems as a child. His experiments with acoustics led to his creation of the harmonic telegraph, which eventually resulted in the first practical telephone in 1876 and spawned Bell Telephone Company. Bell became a naturalized American citizen and also invented prototypes of flying vehicles, hydrofoils, air conditioners, metal detectors, and magnetic sound and video recording devices.

Benz, Carl (1844-1929): A German engineer and designer born illegitimately as Karl Vaillant, Benz designed bridges before setting up his own foundry and mechanical workshop. In 1888, he invented, built and patented a gas-powered, engine-driven, three-wheeled horseless carriage named the Benz Motorwagen, which was the first automobile available for purchase. In 1895, he built the first trucks and buses and introduced many technical innovations still found in modern automobiles. The Benz Company merged with Daimler in the 1920's and introduced the famous Mercedes-Benz in 1926.

Berzelius, Jons Jakob (1779-1848): A physician and

chemist born in Sweden, Berzelius was secretary of the Royal Swedish Academy of Sciences for thirty years. He is credited with discovering the law of constant proportions for inorganic substances and was the first to distinguish organic from inorganic compounds. He developed a system of chemical symbols and a table of relative atomic weights that are still in use. In addition to coining such chemical terms as "protein, "catalysis," "polymer," and "isomer," he identified the elements cerium, selenium, silicon, and thorium.

Bessemer, Henry (1813-1898): The English engineer and inventor is chiefly known for development of the Bessemer process, which eliminated impurities from molten pig iron and lowered costs in the production of steel. Holder of more than one hundred patents, Bessemer also invented items to improve the manufacture of glass, sugar, military ordnance, and postage stamps, and built a test model of a gimballed, hydraulic-controlled steamship to eliminate seasickness. His steel-industry creations led to the development of the modern continuous casting process of metals.

Birdseye, Clarence (1886-1956): The naturalist, inventor, and entrepreneur was born in Brooklyn, New York. He began experimenting in the early 1920's with flash-freezing fish. Using a patented process, he was eventually successful in freezing meats, poultry, vegetables, and fruits, and in so doing changed consumers' eating habits. Birdseye sold his process to the company that later became General Foods Corporation, for whom he continued to work in developing frozen-food technology. His surname—split in two for easy recognition—became a major brand name that is still familiar.

Bohr, Niels (1885-1962): A Danish theoretical physicist, Bohr introduced the concept of atomic structure, in which electrons orbit the nucleus of an atom, and laid the foundations of quantum theory, for which he was awarded the 1922 Nobel Prize in Physics. He later identified U-235, an isotope of uranium that produces slow fission. During World War II, after escaping from Nazi-occupied Denmark, he worked as consultant to the Manhattan Project. Following the war, he returned to Denmark and became a staunch advocate for the nondestructive uses of atomic energy.

Bosch, Carl (1874-1940): The German-born chemist,

metallurgist, and engineer devised a high-pressure chemical technique (the Haber-Bosch process) to fix nitrogen, used in mass-producing ammonia for fertilizers, explosives, and synthetic fuels. He was awarded (along with Friedrich Bergius, 1884-1949) the 1931 Nobel Prize in Chemistry for his work. He was a founder and chairman of the board of IG Farben, for a time the largest chemical company in the world, but was ousted in the late 1930's for criticizing the Nazis.

Brahe, Tycho (1546-1601): A nobleman born of Danish heritage in what is modern-day Sweden, Brahe became interested in astronomy while studying at the University of Copenhagen. He made improvements to the primitive observational instruments of the day but never had access to the telescope. Nonetheless, he was able to study the positions of stars and planets accurately and produced useful catalogs of celestial bodies, particularly for the planet Mars, which helped Johannes Kepler (1571-1630) to formulate the laws of planetary motion. Craters on the Moon and on Mars are named in Brahe's memory.

Brunel, Isambard Kingdom (1806-1859): A British-born civil engineer and inventor, Brunel designed and built tunnels, bridges, and docks—many still in use—often devising ingenious solutions to problems in the process. He is best remembered for developing the SS *Great Britain,* the largest and most modern ship of its time and the first ocean-going iron ship driven by a propeller. Brunel was also a railroad pioneer, serving as chief engineer for Great Western Railway, for which he specified a broad-gauge track to allow higher speeds, improved freight capacity, and greater passenger comfort.

Burbank, Luther (1849-1926): Despite having only an elementary-school education, the Massachusetts-born botanist and horticulturist was a pioneer in the field of agricultural science. Working from a greenhouse and experimental fields in Santa Rosa, California, Burbank developed more than 800 varieties of plants, including new strains of flowers, peaches, plums, nectarines, cherries, peaches, berries, nuts, and vegetables, as well as new crossbred products such as the plumcot. One of his most useful creations, the Russet Burbank, became the potato of choice in food processing, particularly for French fries.

Calvin, Melvin (1911-1997): A Minnesota-born chemist of Russian heritage, Calvin taught molecular biology for nearly fifty years at the University of California, Berkeley, where he founded and directed the Laboratory of Chemical Biodynamics (later the Structural Biology Division) and served as associate director of the Lawrence Berkeley National Laboratory. He and his research team traced the path of carbon-14 through plants during photosynthesis, greatly enhancing understanding of how sunlight stimulates chlorophyll to create organic compounds. He was awarded the 1961 Nobel Prize in Chemistry for his work.

Carnot, Sadi (1796-1832): A French physicist and military engineer, Carnot was an army officer before becoming a scientific researcher, specializing in the theory of heat as produced by the steam engine. His *Reflections on the Motive Power of Fire* focused on the relationship between heat and mechanical energy and provided the foundation for the second law of thermodynamics. His work greatly influenced scientists such as James Prescott Joule (1818-1889), William Thomson (Lord Kelvin, 1824-1907), and Rudolf Diesel (1858-1913) and made possible more practically and efficiently designed engines later in the nineteenth century. Carnot's career was cut short by his death from cholera.

Carson, Rachel (1907-1964): A marine biologist and author born in Springdale, Pennsylvania, Carson worked for the U.S. Bureau of Fisheries before turning full-time to writing about nature. Her popular and highly influential articles, radio scripts and books, including *The Sea Around Us, The Edge of the Sea, Under the Sea-Wind,* and *Silent Spring* enlightened the public about the wonders of nature and the dangers of pesticides such as DDT, which was eventually banned in the United States. Carson is credited with spurring the modern environmental movement.

Celsius, Anders (1701-1774): A Swedish astronomer, Celsius studied the aurora borealis and was the first to link the phenomena to the Earth's magnetic field. He also participated in several expeditions designed to measure the size and shape of the Earth. Founder of the Uppsala Astronomical Observatory, he explored star magnitude, observed eclipses, and compiled star catalogs. He is perhaps best known for the Celsius international temperature scale, which accounts for atmospheric pressure in measuring the boiling and freezing points of water.

Clausewitz, Carl von (1780-1831): As a Prussian-born soldier and military scientist, Clausewitz participated in numerous campaigns, beginning in the early 1790's, and fought in the Napoleonic Wars. After his appointment in 1818 to major general, he taught at the Prussian military academy and helped reform the state army. His principal written work, *On War*, unfinished at the time of his death from cholera, is still considered relevant and continues to influence military thinking via its practical approach to command policies, instruction for soldiers, and methods of planning for strategists.

Colt, Samuel (1814-1862): An inventor born in Hartford, Connecticut, Colt designed a workable multishot pistol while working in his father's textile factory. In the mid-1830's he patented a revolver and set up an assembly line to produce machine-made weapons featuring interchangeable parts. The perfected product, the Colt Peacemaker, was used in the Seminole and Mexican-American wars and became popular during America's western expansion, and Colt became a millionaire. Colt's Manufacturing Company continues to produce a wide variety of firearms for civilian, military, and law-enforcement purposes.

Copernicus, Nicolaus (1473-1543): The Polish mathematician, physician, statesman, artist, linguist, and astronomer is credited with beginning the scientific revolution. His major work, published the year of his death, *De revolutionibus orbium coelestium* (*On the Revolutions of the Heavenly Spheres*), was the first to propose a heliocentric model of the solar system. The book inspired further research by Tycho Brahe (1546-1601), Galileo Galilei (1564-1642), and Johannes Kepler (1571-1630) and stimulated the birth of modern astronomy.

Cori, Gerty Radnitz (1896-1957): A biochemist born in Prague (now the Czech Republic), Cori came to the United States in 1922 and became a naturalized American citizen in 1928. She worked with her husband Carl at what is now Roswell Park Cancer Institute in Buffalo, New York, researching carbohydrate metabolism and discovered how glycogen is broken down into lactic acid to be stored as energy, a process now called the Cori cycle. She was awarded the 1947 Nobel Prize in Physiology or Medicine, the first American

woman so honored.

Cousteau, Jacques (1910-1997): A French oceanographer, explorer, filmmaker, ecologist, and author, Cousteau began underwater diving in the 1930's, and it became a lifelong obsession. He coinvented the Aqua-Lung in the 1940's—the precursor to modern scuba gear—and began making nature films during the same decade. He founded the French Oceanographic Campaigns in 1950 and aboard his ship *Calypso* explored and researched the world's oceans for forty years. In the 1970's he created the Cousteau Society, which remains a strong ecological advocacy organization.

Crick, Francis (1916-2004): An English molecular biologist and physicist, Crick designed magnetic and acoustic mines during World War II. He was later part of a biological research team at the Cavendish Laboratory. Focusing on the X-ray crystallography of proteins, he identified the structure of deoxyribonucleic acid (DNA) as a double helix, a discovery that greatly advanced the study of genetics. He and his colleagues, American James D. Watson (b. 1928) and New Zealander Maurice Wilkins (1916-2004), shared the 1962 Nobel Prize in Physiology or Medicine for their groundbreaking work.

Curie, Marie Sklodowska (1867-1934) and Pierre Curie (1859-1906): Polish-born chemist-physicist Marie was the first woman to teach at the University of Paris. She married French physicist-chemist Pierre Curie in 1895, and the couple collaborated on research into radioactivity, discovering the elements polonium and radium. She and her husband shared the 1903 Nobel Prize in Physics for their work; she was the first woman so honored. After her husband died, she continued her research and received the 1911 Nobel Prize in Chemistry, the first person to receive the award in two different disciplines. She founded the Radium (later Curie) Institute.

Daimler, Gottlieb (1834-1900): The German-born mechanical engineer, designer, inventor, and industrial magnate was an early developer of the gasoline-powered internal combustion engine and the automobile. He and fellow industrial designer Wilhelm Maybach (1846-1929) began a partnership in the 1880's to build small, high-speed engines incorporating numerous devices they patented—flywheels, carburetors, and cylinders—still found in modern engines. After creating the first motorcycle they founded Daimler Motors and began selling automobiles in the early 1890's. Their Phoenix model won history's first auto race.

Darwin, Charles Robert (1809-1882): An English naturalist and geologist, Darwin participated in the five-year-long worldwide surveying expedition of the HMS *Beagle* during the 1830's, observing and collecting specimens of animals, plants, and minerals. The voyage inspired numerous written works, particularly *On Natural Selection, On the Origin of the Species,* and *The Descent of Man,* which collectively supported his theory that all species have evolved from common ancestors. Though modern science has virtually unanimously accepted Darwin's findings, his theory of evolution remains a controversial topic among various political, cultural, and religious groups.

Davy, Sir Humphry (1778-1829): A chemist, teacher, and inventor born in England, Davy began conducting scientific experiments as a child. As a teen he worked as a surgeon's apprentice and became addicted to nitrous oxide. He later researched galvanism and electrolysis, discovered the elements sodium, chlorine, and potassium, and contributed to the discovery of iodine. He invented the Davy safety lamp for use in coal mines, was a founder of the Zoological Society of London, and served as president of the Royal Society.

Diesel, Rudolf (1858-1913): A French-born mechanical engineer and inventor of German heritage, Diesel designed an innovative refrigeration system for an ice plant in Paris and improved the efficiency of steam engines. His self-named, patented diesel engine introduced the concept of fuel injection. The efficient diesel engine later became commonplace in trucks, locomotives, ships, and submarines and, after redesign to reduce weight, in modern automobiles. Diesel disappeared while on a ship. His body was later discovered floating in the sea, but it is still unknown whether he fell overboard, committed suicide, or was murdered.

Edison, Thomas Alva (1847-1931): A scientist, inventor and entrepreneur born in Milan, Ohio, Edison worked out of New Jersey in the fields of electricity and communication and profoundly influenced the world. Credited with more than 1,000 patents, he is best known for creating the first practical incandescent light bulb, which has illuminated the lives of humans since 1879. Other inventions include the stock ticker, a telephone transmitter, electricity meters, the mimeograph, an efficient storage battery, and the phonograph and the kinetoscope, which he combined to produce the first talking moving picture in 1913.

Einstein, Albert (1879-1955): A German-born theoretical physicist and author of Jewish heritage, Einstein came to the United States before World War II. Regarded as a genius, and one of the world's most recognized scientists, he was awarded the 1921 Nobel Prize in Physics. He developed general and special theories of relativity, particle and quantum theories, and proposed ideas that continue to influence numerous fields of study, including energy, nuclear power, heat, light, electronics, celestial mechanics, astronomy, and cosmology. Late in life, he was offered the presidency of Israel but declined the honor.

Euclid (c. 330-c. 270 B.C.E.): A Greek mathematician who taught in Alexandria, Egypt, Euclid is considered the father of plane and solid geometry. He is remembered principally for his major extant work, *The Elements*—a treatise containing definitions, postulates, geometric proofs, number theories, discussions of prime numbers, arithmetic theorems, algebra, and algorithms—which has served as the basis for the teaching of mathematics for two thousand years. He also explored astronomy, mechanics, gravity, moving bodies, and music and was one of the first scientists to write about optics and perspective.

Everest, Sir George (1790-1866): A geographer born in Wales, Everest participated for 25 years in the Great Trigonometrical Survey of the Indian subcontinent—which surveyed an area encompassing millions of square miles while locating, measuring, and naming the Himalayan Mountains—and served as superintendent of the project from 1823 to 1843. Later knighted, he served as vice president of the Royal Geographical Society. The world's tallest peak, Mount Everest in Nepal (known locally as Chomolungma), was named in his honor.

Faraday, Michael (1791-1867): A self-educated British chemist and physicist, Faraday served an apprenticeship with chemist Sir Humphry Davy (1778-1829), during which time he experimented with liquefied gases, alloys, and optical glasses. He invented a prototype of what became the Bunsen burner, discovered benzene, and performed experiments that led to his discovery of electromagnetic induction. He built the first electric dynamo, the precursor to the power generator, researched the relationship between magnetism and light, and made numerous other contributions to the studies of electromagnetism and electrochemistry.

Farnsworth, Philo (1906-1971): An inventor born in Utah, Farnsworth became interested in electronics and mechanics as a child. He experimented with television during the 1920's, and late in the decade he demonstrated an electronic, nonmechanical scanning system for image transmissions. During the early 1930's, he worked for Philco but left to carry out his own research. In addition to significant contributions to television, Farnsworth held more than 300 patents and devised a milk-sterilizing process, developed fog lights, an infrared telescope, a prototype of an air traffic control system, and a fusion reaction tube.

Fermi, Enrico (1901-1954): An Italian-born experimental and theoretical physicist and teacher who became an American citizen in 1944, Fermi studied mechanics and was instrumental in the advancement of thermodynamics and quantum, nuclear, and particle physics. Awarded the 1938 Nobel Prize in Physics for his research into radioactivity, he was a member of the team that developed the first nuclear reactor in Chicago in the early 1940's and served as a consultant to the Manhattan Project, which produced the first atomic bomb. He died of cancer from sustained exposure to radioactivity.

Feynman, Richard (1918-1988): A physicist, author, and teacher born in New York, Feynman participated in the Manhattan Project and made numerous contributions to a diverse field of specialized scientific disciplines including quantum mechanics, supercooling, genetics, and nanotechnology. He shared the 1965 Nobel Prize in Physics—with Julian Schwinger (1918-1994) and Sin-Itiro Tomonaga (1906-1979)—for work in quantum electrodynamics, particularly for his lucid explanation of the behavior of subatomic particles. He was a popular and influential professor for many years at California Institute of Technology.

Fleming, Alexander (1881-1955): A Scottish-born biologist, Fleming served in the Royal Army Medical Corps during World War I and witnessed the deaths of many wounded soldiers from infection. He was a professor of bacteriology at a teaching hospital, and he specialized in immunology and chemotherapy research. In 1928 he discovered an antibacterial mold, which over the next decade was purified and mass-produced as the drug penicillin, which played a large part in suppressing infections during World War II. A major contributor to the development of antibiotics, he shared the 1945 Nobel Prize in Physiology or Medicine.

Forrester, Jay Wright (b. 1918): An engineer, teacher, and computer scientist born in Nebraska, Forrester built a wind-powered electrical system while in his teens. Associated with the Massachusetts Institute of Technology as a researcher and professor for many years, he developed servomechanisms for military use, designed aircraft flight simulators, and air defense systems. He founded the field of system dynamics to produce computer-generated mathematical models for such tasks as determining water flow, fluid turbulence, and a variety of mechanical movements.

Fourier, Joseph (1768-1830): A French-born physicist, mathematician, and teacher Fourier accompanied Napoleon's expedition to Egypt, where he served as secretary of the Egyptian Institute and was a major contributor to *Description of Egypt*, a massive work describing the scientific findings that resulted from the French military campaign. After returning to France, Fourier explored numerous scientific fields but is best known for his extensive research on the conductive properties of heat and for his theories of equations, which influenced later physicists and mathematicians.

Franklin, Benjamin (1706-1790): A Boston-born author, statesman, scientist, and inventor, Franklin worked as a printer in his youth and from 1733 to 1758 published *Poor Richard's Almanack*. A key figure during the American Revolution and a founding father of the United States, he established America's first lending library and Pennsylvania's first fire department, served as first U.S. postmaster general and as minister to France and Sweden. He experimented with electricity and is credited with inventing the lightning rod, bifocal glasses, an odometer, the Franklin stove, and a musical instrument made of glass.

Freud, Sigmund (1856-1939): An Austrian of Jewish

heritage, Freud studied neurology before specializing in psychopathology and conducted extensive research into hypnosis and dream analysis to treat hysteria. Considered the father of psychoanalysis and a powerful influence on the field, he originated such psychological concepts as repression, psychosomatic illness, the unconscious mind, and the division of the human psyche into the id, ego, and superego. He fled from the Nazis and went to London. Riddled with cancer from years of cigar smoking, he took morphine to relieve his suffering and hasten his death.

Frisch, Karl von (1886-1982): The Vienna-born son of a surgeon-university professor, von Frisch initially studied medicine before switching to zoology and comparative anatomy. Working as a teacher and researcher out of Munich, Rostock, Breslau, and Graz universities, he focused his research on the European honeybee. He made many discoveries about the insect's sense of smell, optical perception, flight patterns, and methods of communication that have since proved invaluable in the fields of apiology and botany. He was awarded the 1973 Nobel Prize in Physiology of Medicine in recognition of his pioneering work.

Fuller, R. Buckminster (1895-1983): The Massachusetts-born architect, philosopher, engineer, author, and inventor developed systems for lightweight, weatherproof, and fireproof housing while in his twenties. Teaching at Black Mountain College in the late 1940's, Fuller perfected the geodesic dome, built of aluminum tubing and plastic skin, and afterward developed numerous designs for inventions aimed at providing practical and affordable shelter and transportation. He coined the term "synergy" and advocated exploiting renewable sources of energy such as solar power and wind-generated electricity. He was awarded the Presidential Medal of Freedom in 1983.

Galen (129-c. 199): An ancient Roman surgeon, scientist, and philosopher, Galen traveled and studied widely before serving as physician to Roman emperors Marcus Aurelius (121-180), Commodus (161-192), Septimus Severus (146-211), and Caracalla (188-217). In the course of his education he explored human and animal anatomy, became an advocate of proper diet and hygiene, and advanced the practice of surgery by treating the wounds of gladiators and ministering to plague victims. His medical discoveries and healing methods, detailed in numerous written works, influenced medicine for more than 1,500 years.

Galilei, Galileo (1564-1642): A physicist, astronomer, mathematician, and philosopher born in Pisa, Italy, Galileo is known as the father of astronomy and the father of modern science. A keen astronomical observer, he made significant improvements to the telescope, through which he studied the phases of Venus, sunspots, and the Milky Way, and he discovered Jupiter's four largest moons. He risked excommunication and death championing the heretical Copernican heliocentric view of the solar system. He also invented a military compass and a practical thermometer and experimented with pendulums and falling bodies.

Galton, Francis (1822-1911): An anthropologist, geographer, meteorologist, and inventor born in England, Galton was a child prodigy. He traveled widely, exploring the Middle East and Africa, and wrote about his expeditions. Fascinated by numbers, he devised the first practical weather maps for use in newspapers. He also contributed to the science of statistics, studied heredity—coining the term "eugenics" and the phrase "nature or nurture"—and was an early advocate of using fingerprints in criminology. Galton is responsible for inventing a high-frequency whistle used in training dogs and cats.

Gates, Bill (b. 1955): An entrepreneur, philanthropist, and author born in Seattle, Gates became interested in computers as a teenager. He left Harvard in 1975 to cofound and to serve as chairman (until 2006) of Microsoft, which developed software for IBM and other systems before launching its own system in 1985. The result, Microsoft Windows, became the dominant software product in the worldwide personal computer market. Profits from his enterprise made Gates one of the world's richest people, and he has used his vast wealth to assist a wide variety of charitable causes.

Goddard, Robert H. (1892-1945): A physicist, engineer, teacher, and inventor born in Massachusetts, Goddard became interested in science as a child and experimented with kites, balloons, and rockets. He received the first of more than 200 patents in 1914 for multistage and liquid-fuel rockets, and during the 1920's he conducted successful test flights using liquid fuel. Goddard experimented with solid fuels and ion thrusters and is credited with developing tail fins, gyroscopic guidance systems, and many other basics of rocketry that greatly influenced the designs of rocket scientists who came after him.

Grandin, Temple (b. 1947): An animal scientist born in Massachusetts, Grandin was diagnosed with autism as a child. As an adult she earned advanced degrees before receiving a doctorate from the University of Illinois in 1989. A professor at Colorado State University, an author, and an autism advocate, she has made numerous humane improvements to the design of livestock-handling facilities that have been incorporated into meat-processing plants worldwide to reduce or eliminate animal stress, pain, and fear.

Haber, Fritz (1868-1934): A chemist and teacher of Jewish heritage (later a convert to Christianity) born in Germany, Haber developed the Haber process to produce ammonia used in fertilizers, animal feed, and explosives, for which he was awarded the 1918 Nobel Prize in Chemistry. At Berlin's Kaiser Wilhelm Institute (later the Haber Institute) between 1911 and 1933, he developed chlorine gas used in World War I, experimented with the extraction of gold from seawater, and oversaw production of Zyklon B, the cyanide-based pesticide that was employed at extermination camps during World War II.

Halley, Edmond (1656-1742): An English astronomer, mathematician, physicist, and meteorologist, Halley wrote about sunspots and the solar system while a student at Oxford. In the 1670's, he cataloged the stars of the Southern Hemisphere and charted winds and monsoons. Inventor of an early diving bell and a liquid-damped magnetic compass, a colleague of Sir Isaac Newton (1642-1727), and leader of the first English scientific expedition, Halley is best remembered for predicting the regular return of the comet that bears his name.

Heisenberg, Werner (1901-1976): A theoretical physicist and teacher born in Germany, Heisenberg conducted research in quantum mechanics with Niels Bohr (1885-1962) at the University of Copenhagen. There he developed the uncertainty principle, which proves it is impossible to determine the position and momentum of subatomic particles at the same time. Awarded the 1932 Nobel Prize in Physics for his work, he also contributed research on positrons, cosmic rays,

spectral frequencies, matrix mechanics, nuclear fission, superconductivity, and plasma physics to the continuing study of atomic theory.

Herschel, William (1738-1822) and Caroline Herschel (1750-1848): A German-born astronomer, composer, and telescope maker who moved to England in his teens, William spent the early part of his career as a musician, playing cello, oboe, harpsichord, and organ and wrote numerous symphonies and concerti. In the 1770's he began building his own large reflecting telescopes and with his diminutive (4 feet, 3 inches) but devoted sister Caroline spent countless hours observing the sky while cataloging nebulae and binary stars. The Herschels are credited with discovering two of Saturn's moons, the planet Uranus and two of its moons, and coining the word "asteroid."

Hersey, Mayo D. (1886-1978): A mechanical engineer born in Rhode Island, Hersey was a preeminent expert on tribology, the study of the relationship between interacting solid surfaces in motion, the adverse effects of wear, and the ameliorating effects of lubrication. He worked as a physicist at the National Institute of Standards and Technology (1910-1920) and the U.S. Bureau of Mines (1922-1926) and taught at the Massachusetts Institute of Technology (1910-1922). He was a consultant to the Manhattan Project and won numerous awards for his contributions to lubrication science.

Hippocrates (c. 460-c. 377 B.C.E.): An ancient Greek physician born on the island of Kos, Hippocrates was the first of his time to separate the art of healing from philosophy and magical ritual. Called the father of Western medicine, he originated the belief that diseases were not the result of superstition but of natural causes, such as environment and diet. Though his concept that illness was the result of an imbalance in the body's fluids (called humors) was later discredited, he pioneered such common modern clinical practices as observation and documentation of patient care. He originated the Hippocratic Oath, which for many centuries served as the guiding principle governing the behavior of doctors.

Hooke, Robert (1635-1703): A brilliant, multitalented British experimental scientist with interests in physics, astronomy, chemistry, biology, geology, paleontology, mechanics, and architecture, Hooke was instrumental as chief surveyor in rebuilding the city of London following the Great Fire of 1666. Among many accomplishments in diverse fields he is credited with inventing the compound microscope—via which he discovered the cells of plants and formulated a theory of fossilization—devised a balance spring to improve the accuracy of timepieces, and either created or refined such instruments as the barometer, anemometer, and hygrometer.

Howlett, Freeman S. (1900-1970): A horticulturist born in New York, Howlett was associated with the Ohio State University as teacher, administrator, and researcher for more than forty-five years and was considered an expert on the history of horticulture. His investigations focused on plant hormones, embryology, fruit setting, reproductive physiology, and foliation for a variety of crops, including fruits, vegetables, and nuts. He created five new varieties of apples popular among consumers. A horticulture and food science building at Ohio State is named in his honor.

Hubble, Edwin Powell (1889-1953): An astronomer born in Missouri, Hubble was associated with the Mount Wilson Observatory in California for more than thirty years. Using what was then the world's largest telescope, he was the first to discover galaxies beyond the Milky Way, which greatly expanded science's concept of the universe. He studied red shifts in formulating Hubble's law, which confirmed the big bang or expanding universe theory. The American space telescope launched in 1990 was named for him, and he was honored in 2008 with a commemorative postage stamp.

Huygens, Christiaan (1629-1695): Born in the Netherlands, Huygens was an astronomer, physicist, mathematician, and prolific author. He made early telescopic observations of Saturn and its moons and was the first to suggest that light is made up of waves. He discovered centrifugal force, proposed a formula for centripetal force, and developed laws governing the collision of celestial bodies. An inveterate inventor, he patented the pendulum clock and the pocket watch and designed an early internal combustion engine. He is also considered a pioneer of science fiction for writing about the possibility of extraterrestrial life.

Jacquard, Joseph Marie (1752-1834): A French inventor, Jacquard created a series of mechanical looms in the early nineteenth century. His experiments culminated in the Jacquard loom attachment, which could be programmed, via punch cards, to weave silk in various patterns, colors, and textures automatically. The labor-saving device became highly popular in the silk-weaving industry, and its inventor received royalties on each unit sold and became wealthy in the process. The loom inspired scientists to incorporate the concept of punch cards for computer information storage.

Jenner, Edward (1749-1823): A surgeon and anatomist born in England, Jenner experimented with cowpox inoculations in an attempt to prevent smallpox, a virulent infectious disease of ancient origin with a high rate of mortality that killed millions of people. In the early nineteenth century, Jenner successfully developed a method of vaccination that provided immunity from smallpox and late in life became personal physician to King George IV. The smallpox vaccination was made compulsory in England and elsewhere, and the disease was declared eradicated worldwide in 1979.

Jobs, Steven (b. 1955-2011): An inventor and entrepreneur of Syrian and American heritage born in San Francisco, Jobs worked at Hewlett-Packard as a teenager and was later employed at Atari designing circuit boards. In 1976, he and coworker Steve Wozniak (b. 1950) and others founded Apple, which designed, built, and sold a popular and highly successful line of personal computers. A multibillionaire and holder of more than 200 patents, Jobs continued to make innovations in interfacing, speakers, keyboards, power adaptation, and myriad other components related to modern computer science until his death in late 2011

Kepler, Johannes (1571-1630): A German mathematician, author and astronomer, he became interested in the cosmos after witnessing the Great Comet of 1577. He worked for Tycho Brahe (1546-1601) for a time and after Brahe's death became imperial mathematician in Prague. A major contributor to the scientific revolution, Kepler studied optics, observed many celestial phenomena, provided the foundation for Sir Issac Newton's theory of gravitation, and developed a set of laws governing planetary motion around the Sun—including the discovery that the orbits of planets are elliptical—that were confirmed by later astronomers.

Krebs, Sir Hans Adolf (1900-1981): A biochemist and

physician born the son of a Jewish surgeon in Germany, Krebs was a clinician and researcher before moving to England after the rise of the Nazis. As a professor at Cambridge University, he explored metabolism, discovering the urea and citric acid cycles—biochemical reactions that promote understanding of organ functions in the body and explain the cellular production of energy—for which he shared the 1953 Nobel Prize in Medicine or Physiology. He was also knighted in 1958 for his work.

Lawrence, Ernest O. (1901-1958): A South Dakota-born physicist and teacher, Lawrence researched the photoelectric effect of electrons at Yale. In 1928, he became a professor at the University of California, Berkeley, where he invented the cyclotron particle accelerator, for which he was awarded the 1939 Nobel Prize in Physics. During World War II, he was involved in the Manhattan Project. Lawrence popularized science and was a staunch advocate for government funding of significant scientific projects. After his death, laboratories at the University of California and the chemical element lawrencium were named in his honor.

Leakey, Louis B. (1903-1972) and Mary Nicol Leakey (1913-1996): Louis, born in Kenya, was an archaeologist, paleontologist, and naturalist who married London-born anthropologist and archaeologist Mary Nicol. Together and often with their sons, Jonathan, Richard, and Philip, they excavated at Olduvai Gorge in East Africa, where they unearthed the tools and fossils of ancient hominids. Their discoveries of the remains of Proconsul africanus, Australopithecus boisei, Homo habilis, Homo erectus, and other large-brained, bipedal primates effectively proved Darwin's theory of evolution and extended human history by several million years.

Leonardo da Vinci (1452-1519): An Italian genius considered the epitome of the Renaissance man, da Vinci was a superb artist, architect, engineer, mathematician, geologist, musician, mapmaker, inventor, and writer. Creator of such famous paintings as the *Mona Lisa* and *The Last Supper*, he is credited with imagining the helicopter, solar power, and the calculator centuries before their invention. His far-ranging mind explored such subjects as anatomy, optics, vegetarianism, and hydraulics, and his journals, written in mirror-image script, are filled with drawings, ideas, and scientific observations that are still closely studied.

Linnaeus, Carolus (1707-1778): A Swedish botanist, zoologist, physician, and teacher, Linnaeus began studying plants as a child. As an adult, he embarked on expeditions throughout Europe observing and collecting specimens of plants and animals and wrote numerous works about his findings. He devised the binomial nomenclature system of classification for living and fossil organisms—called taxonomy—still used in modern science, which provides concise Latin names of genus and species for each example. Linnaeus also cofounded the Royal Swedish Academy of Science.

Lippershey, Hans (c. 1570-c. 1619): A master lens grinder and spectacle maker born in Germany who later became a citizen of the Netherlands, Lippershey is credited with designing the first practical refracting telescope (which he called "perspective glass"). After fruitlessly attempting to patent the device, he built several prototypes for sale to the Dutch government, which distributed information about the telescope across Europe. Other scientists, such as Galileo, soon duplicated and improved upon Lippershey's invention, which became a primary instrument in the science of astronomy.

Lumière, Auguste (1862-1954) and Louis Lumière (1864-1948): The French-born brothers worked at their father's photographic business and devised the dry-plate process for still photographs. From the early 1890's, they patented several techniques—including perforations to guide film through a camera and a color photography process—that greatly advanced the development of moving pictures. From 1895 to 1896, they publicly screened a series of short films to enthusiastic audiences in Asia, Europe, and North and South America, demonstrating the commercial potential of the new medium and launching what would become the multibillion-dollar film industry.

Maathai, Wangari Muta (b. 1940-2011): An environmental and political activist of Kikuyu heritage born in Kenya, Maathai studied biology in the United States before becoming a research assistant and anatomy teacher at the University of Nairobi, where she was the first East African woman to earn a Ph.D. She founded the Green Belt Movement, an organization that plants trees, supports environmental conservation, and advocates for women's rights. A former member of the Kenyan Parliament and former Minister of Environment, she was awarded the 2004 Nobel Peace Prize for her work and is the first African woman to receive the award.

McAdam, John Loudon (1756-1836): A Scottish engineer, McAdam became a surveyor in Great Britain and specialized in road building. He devised an effective method—called "macadam" after its inventor—of creating long-lasting roads using gravel on a foundation of larger stones, with a camber to drain away rainwater, which was adopted around the world. He also introduced hot tar as a binding agent (dubbed "tarmac," an abbreviation of tarmacadam) to produce smoother road surfaces. Modern road builders still use many of the techniques he innovated.

Mantell, Gideon (1790-1852): A British surgeon, geologist, and paleontologist, Mantell began collecting fossil specimens from quarries as a child. As an adult, he was a practicing physician and pursued geology in his spare time. He discovered fossils that were eventually identified as belonging to the Iguanodon and Hylaeosaurus—which he named Megalosaurus and Pelorosaurus—and he became a recognized authority on dinosaurs. His major works were *The Fossils of South Downs: Or, Illustrations of the Geology of Sussex* (1822) and *Notice on the Iguanodon: A Newly Discovered Fossil Reptile* (1825).

Marconi, Guglielmo (1874-1937): An Italian-born electrical engineer and inventor, Marconi experimented with electricity and electromagnetic radiation. He developed a system for transmitting telegraphic messages without the use of connecting wires and by the early twentieth century was sending transmissions across the Atlantic Ocean. His devices eventually evolved into radio, and the transmitter at his factory in England was the first in 1920 to broadcast entertainment to the United Kingdom; he shared the 1909 Nobel Prize in Physics with German physicist Ferdinand Braun (1850-1918).

Maxwell, James Clerk (1831-1879): A Scottish-born mathematician, theoretical physicist, and teacher, Maxwell had an insatiable curiosity from an early age and as a teenager began presenting papers to the Royal Society of Edinburgh. He experimented with color, examined hydrostatics and optics, and wrote about Saturn's rings. His most significant work, however, was performed in the field of electromagnetism, in which he showed that electricity,

magnetism, and light are all results of the electromagnetic field, a concept that profoundly affected modern physics.

Mendel, Gregor Johann (1822-1884): Born in Silesia (now part of the Czech Republic), Mendel became interested in plants as a child. In the 1840's he entered an Augustinian monastery, where he studied astronomy, meteorology, apiology, and botany. Called the father of modern genetics, he is best known for his experiments in hybridizing pea plants, which evolved into what later were called Mendel's laws of inheritance. Though his work exerted little influence during his lifetime, his concepts were rediscovered early in the twentieth century and have since proven invaluable to the study of heredity.

Mendeleyev, Dmitri Ivanovich (1834-1907): A Russian chemist, teacher, and inventor, Mendeleyev studied the properties of liquids and the spectroscope before becoming a professor in Saint Petersburg and later serving as director of weights and measures. He created a periodic table of the sixty-three elements then known arranged by atomic mass and the similarity of properties (a revised form of which is still employed in modern science) and used the table to correctly predict the characteristics of elements and isotopes not yet found. Element 101, mendelevium, discovered in 1955, was named in his honor.

Meng Tian (259-210 B.C.E.): A general serving under Qin Shi Huang, first emperor of the Qin Dynasty (221-207 B.C.E.), Meng Tian led an army of 100,000 to drive warlike nomadic tribes north out of China. Descended from architects, he oversaw building of the Great Wall to prevent invasions, cleverly incorporating topographical features and natural barriers into the defensive barricade, which he extended for more than 2,000 miles along the Yellow River. After a coup following Emperor Qin's death, Meng Tian was forced to commit suicide. The Qin Dynasty fell just three years later.

Montgolfier, Joseph Michel (1740-1810) and Jacques-Etienne Montgolfier (1745-1799): Born in France to a prosperous paper manufacturer, the Montgolfier brothers designed and built a hot-air balloon, and in 1783 Jacques-Etienne piloted the first manned ascent in a lighter-than-air craft. The French Academy of Science honored the brothers for their exploits, which inspired further developments in ballooning. The Montgolfier brothers subsequently wrote books on aeronautics and continued experimenting. Joseph is credited with designing a calorimeter and a hydraulic ram, and Jacques-Etienne invented a method for the manufacture of vellum.

Morse, Samuel F. B. (1791-1872): An artist and inventor born in Massachusetts, Morse painted portraits and taught art at the City University of New York before experimenting with electricity. In the mid-1830's, he designed the components of a practical telegraph—a sender, receiver, and a code to translate signals into numbers and words—and in 1844 sent the first message via wire. Within a decade, the telegraph had spread across America and subsequently around the world. The invention would inspire such later advancements in communication as radio, the Teletype, and the fax machine.

Nernst, Walther (1864-1941): A German physical chemist, physicist, and inventor, Nernst discovered the Third Law of Thermodynamics—defining the chemical reactions affecting matter as temperatures drop toward absolute zero—for which he was awarded the 1920 Nobel Prize in Chemistry. He also invented an electric lamp, and developed an electric piano and a device using rare-earth filaments that significantly advanced infrared spectroscopy. He made numerous contributions to the specialized fields of electrochemistry, solid-state chemistry, and photochemistry.

Newton, Sir Isaac (1642-1727): The English physicist, mathematician, astronomer, and philosopher is considered one of the most gifted and scientifically influential individuals of all time. He developed theories of color and light from studying prisms, was instrumental in creating differential and integral calculus, and formulated still-valid laws of celestial motion and gravitation. He was knighted in 1705, the first British scientist so honored. From 1699 until his death he served as master of the Royal Mint and during his tenure devised anticounterfeiting measures and moved England from the silver to the gold standard.

Nobel, Alfred (1833-1896): A Swedish chemist and chemical engineer, Nobel invented dynamite while studying how to manufacture and use nitroglycerin safely. In the course of building a manufacturing empire based on the production of cannons and other armaments, he experimented with combinations of explosive components, also producing gelignite and a form of smokeless powder, which led to the development of rocket propellants. Late in his life, he earmarked the bulk of his vast estate for the establishment of the Nobel Prizes, annual monetary awards given in recognition of outstanding achievements in science, literature, and peace.

Oppenheimer, J. Robert (1904-1967): A brilliant theoretical physicist, researcher, and teacher born to German immigrants in New York City, Oppenheimer was the scientific director of the Manhattan Project, which developed the atomic bombs dropped on Japan during World War II. Following the war, he was primary adviser to the U.S. Atomic Energy Commission and director of the Institute for Advanced Study in Princeton, New Jersey. He contributed widely to the study of electrons and positrons, neutron stars, relativity, gravitation, black holes, quantum mechanics, and cosmic rays.

Owen, Richard (1804-1892): An English biologist, taxonomist, anti-Darwinist, and comparative anatomist, Owen founded and directed the natural history department at the British Museum. He originated the concept of homology, a similarity of structures in different species that have the same function, such as the human hand, the wing of a bat, and the paw of an animal. He also cataloged many living and fossil specimens, contributed numerous discoveries to zoology, and coined the term "dinosaur." Owen advanced the theory that giant flightless birds once inhabited New Zealand long before their remains were found there.

Paré, Ambroise (c. 1510-1590): A French royal surgeon, Paré revolutionized battlefield medicine, developing techniques and instruments for the treatment of gunshot wounds and for performing amputations. He greatly advanced knowledge of human anatomy by studying the effects of violent death on internal organs. He pioneered the life-saving practices of vascular ligating and herniotomies, designed prosthetics to replace amputated limbs, and was the first to create realistic artificial eyes from such substances as glass, porcelain, silver, and gold.

Pasteur, Louis (1822-1895): A chemist, microbiologist, and teacher born in France, Pasteur focused on researching the causes of diseases and methods for preventing them after three of his children died from typhoid. He proposed a germ theory, demonstrating that microorganisms affect food-stuffs. This ultimately led to his invention of pasteurization—a method of killing bacteria in milk, which was later applied to other substances. A pioneer in immunology, he also developed vaccines to combat anthrax, rabies, and puerperal fever.

Pauli, Wolfgang (1900-1958): An Austrian theoretical physicist of Jewish heritage who converted to Catholicism, Pauli earned a Ph.D. at the age of twenty-one. While lecturing at the Niels Bohr Institute for Theoretical Physics, he researched relativity and quantum physics. He discovered a new law governing the behavior of atomic particles and the characteristics of matter, called the Pauli exclusion principle, for which he was awarded the 1945 Nobel Prize in Physics. During World War II, he moved to the United States and became an American citizen but later relocated to Zurich.

Pauling, Linus (1901-1994): Born in Portland, Oregon, Pauling earned advanced degrees in chemical engineering, physical chemistry, and mathematical physics. A Guggenheim Fellow, he studied quantum mechanics in Munich, Copenhagen, and Zurich before teaching at the California Institute of Technology. He specialized in theoretical chemistry and molecular biology and greatly advanced understanding of the nature of chemical bonds. A political activist who warned of the dangers of nuclear weapons, he became one of a handful of scientists to receive Nobel Prizes in two fields: the 1954 prize in chemistry and the 1982 peace prize.

Pavlov, Ivan (1849-1936): A Russian physiologist and psychologist, Pavlov began investigating the digestive system, which led to experiments with the effects of behavior on the nervous system and the body's automatic functions. He used animals in researching conditioned reflex actions to a variety of visual, tactile, and sound stimuli—including bells, whistles, and electric shocks—to discover the relationship between salivation and digestion and was able to make dogs drool in anticipation of receiving food. He was awarded the 1904 Nobel Prize in Physiology or Medicine for his work.

Planck, Max (1858-1947): A German theoretical physicist credited with founding quantum theory—which affects all matter in the universe—Planck earned a doctoral degree at the age of twenty-one before becoming a professor at the universities of Kiel and Berlin. He explored electromagnetic

radiation, quantum mechanics, thermodynamics, blackbodies, and entropy. He formulated the Planck constant, which describes the proportions between the energy and frequency of a photon and provides understanding of atomic structure. He was awarded the 1918 Nobel Prize in Physics for his discoveries.

Ptolemy (c. 100-c. 178): A mathematician, astronomer, and geographer of Greek heritage who worked in Roman-ruled Alexandria, Egypt, Ptolemy wrote several treatises that influenced science for centuries afterward. His *Almagest*, written in about 150, contains star catalogs, constellation lists, Sun and Moon eclipse data, and planetary tables. Ptolemy's eight-volume *Geographia* (*Geography*) followed and incorporates all known information about the geography of the Earth at the time and helped introduce the concept of latitudes and longitudes. His work on astrology influenced Islamic and medieval Latin worlds, and his writings on music theory and optics pioneered study in those fields.

Pythagoras (c. 580-c. 500 B.C.E.): An ancient Greek philosopher and mathematician from Samos, Pythagoras traveled widely seeking wisdom and established a religious-scientific ascetic community in Italy around 530 B.C.E. He had interests in music, astronomy, medicine, and mathematics, and though none of his writings survived, he is credited with the discovery of the Pythagorean theorem governing right triangles (the square of the hypotenuse is equal to the sum of the squares of the other two sides). His life and philosophy exerted considerable influence on Plato (c. 427-347 B.C.E.) and through Plato greatly affected Western thought.

Reiss, Archibald Rodolphe(1875-1929): A chemist, photographer, teacher, and natural scientist born in Germany, Reiss founded the world's first school of forensic science at the University of Lausanne, Switzerland, in 1909. He published numerous works that greatly influenced the new discipline, including *La photographie judiciaire* (*Forensic photography*, 1903) and *Manuel de police scientifique. I Vols et homicides* (*Handbook of Forensic Science: Thefts and Homicides*, 1911). During World War I he investigated alleged atrocities in Serbia and lived there for the rest of his life. The institute he founded more than a century ago has become a major school offering numerous courses in various forensic sciences, criminology, and criminal law.

Röntgen, Wilhelm Conrad (1845-1923): A German physicist, Röntgen studied mechanical engineering before teaching physics at the universities of Strassburg, Giessen, Würzburg, and Munich. He experimented with fluorescence and electrostatic charges. In the process of his work he discovered X rays—and also discovered that lead could effectively block the rays—meanwhile laying the foundations of what would become radiology: the medical specialty that uses radioactive imaging to diagnose disease. He was awarded the first Nobel Prize in Physics in 1901. Element 111, roentgenium, was named in his honor in 2004.

Rutherford, Ernest (1871-1937): A chemist and physicist born in New Zealand, Rutherford studied at the University of Cambridge before teaching physics at McGill University in Montreal and at the University of Manchester. He made some of the most significant discoveries in the field of atomic science, including the relative penetrating power of alpha, beta, and gamma rays, the transmutation of elements via radioactivity, and the concept of radioactive half-life. His work, for which he received the 1908 Nobel Prize in Chemistry, was instrumental in the development of nuclear energy and carbon dating.

Sabin, Albert Bruce (1906-1993): A microbiologist born of Jewish heritage as Albert Saperstein in Russia, Sabin later became an American citizen and changed his name. Trained in internal medicine, he conducted research into infectious diseases and assisted in the development of a vaccine to combat encephalitis. His major contribution to medicine was an effective oral polio vaccine, which was administered in mass immunizations during the 1950's and 1960's and eventually led to the eradication of the disease worldwide. Among other honors, he received the Presidential Medal of Freedom in 1986.

Sachs, Julius von (1832-1897): A German botanist, writer, and teacher, Sachs made great strides in the investigation of plant physiology, morphology, heliotropism, and germination while professor of botany at the University of Würzburg. In addition to numerous written works on photosynthesis, water absorption, and chloroplasts that significantly advanced the science of botany, he also invented a number of devices useful to research, including an auxanometer to measure growth rates, and the clinostat, a device that rotates plants to compensate for the effects of gravitation on botanical growth.

Sakharov, Andrei (1921-1989): A Russian nuclear physicist and human rights activist, Sakharov researched cosmic rays, particle physics, and cosmology. He was a major contributor to the development of the hydrogen bomb but later campaigned against nuclear proliferation and for the peaceful use of nuclear power. He received the 1975 Nobel Peace Prize, and though he received several international honors in recognition of his humanitarian efforts, he spent most of the last decade of his life in exile within the Soviet Union. A human rights center and a scientific prize are named in his honor.

Scheele, Carl Wilhelm (1742-1786): A chemist born in a Swedish-controlled area of Germany, Scheele became a pharmacist at an early age. Though he discovered oxygen through experimentation, he did not publish his findings immediately, and the discovery was credited to Antoine-Laurent Lavoisier (1743-1794) and Joseph Priestly (1733-1804), though science later gave the Scheele recognition he deserved. Scheele also discovered the elements barium, manganese, and tungsten, identified such chemical compounds as citric acid, glycerol, and hydrogen cyanide, experimented with heavy metals, and devised a method of producing phosphorus in quantity for the manufacture of matches.

Shockley, William (1910-1989): A physicist and inventor born to American parents in England, Shockley was raised in California. After earning a doctoral degree, he conducted solid-state physics research at Bell Laboratories. During World War II, he researched radar and anti-submarine devices. Following the war, he was part of the team that invented the first practical solid-state transistor, for which he shared the 1956 Nobel Prize in Physics with John Bardeen (1908-1991) and Walter Brattain (1902-1987). He later set up a semiconductor business that was a precursor to Silicon Valley. His major work, *Electrons and Holes in Semiconductors* (1950), greatly influenced many scientists.

Sikorsky, Igor (1889-1972): A Ukrainian engineer and test pilot who immigrated to the United States and became a naturalized American citizen, Sikorsky was a groundbreaking designer of both airplanes and helicopters. Inspired as a child by the drawings of Leonardo da Vinci (1452-1519), he created and flew the first multi-engine fixed-wing aircraft and the first airliner in the 1910's. He built the first flying boats in the 1930's—the famous Pan Am Clippers—and in 1939 designed the first practical helicopter, which introduced the system of rotors still used in modern helicopters.

Spilsbury, Sir Bernard Henry (1877-1947): The first British forensic pathologist, Spilsbury began performing postmortems in 1905. He investigated cause of death in many spectacular homicide cases—including those of Dr. Crippen and the Brighton trunk murders—that resulted in convictions and enhanced the science of forensics. He was a consultant to Operation Mincemeat, a successful World War II ruse (dramatized in the 1956 film *The Man Who Never Was*) involving the corpse of an alleged Allied courier, which deceived the Axis powers about the invasion of Sicily. Spilsbury was found dead in his laboratory—a victim of suicide.

Stephenson, George (1781-1848): A British mechanical and civil engineer, Stephenson invented a safety lamp for coal mines that provided illumination without the risk of explosions from firedamp. He designed a steam-powered locomotive for hauling coal, which evolved into the first public railway line in the mid-1820's, running on his specified track width of 4 feet, 8.5 inches. This measurement became the worldwide standard railroad gauge. He worked on numerous rail lines, in the process making many innovations in the design and construction of locomotives, tracks, viaducts, and bridges that greatly advanced railroad transport.

Teller, Edward (1908-2003): An outspoken theoretical physicist born in Hungary, Teller came to the United States in the 1930's and taught at George Washington University while researching quantum, molecular, and nuclear physics. A naturalized American citizen, he was a member of the atomic-bomb-building Manhattan Project. A strong supporter for nuclear energy development and testing for both wartime and peacetime purposes, he cofounded and directed Lawrence Livermore National Laboratory and founded the department of applied science at the University of California, Davis.

Tesla, Nikola (1856-1943): Born in modern-day Croatia, the brilliant if eccentric Tesla came to the United States in 1884 to work for Thomas Edison's company and later for Edison's rival George Westinghouse (1846-1914). In 1891, Tesla became a naturalized American citizen. A physicist, mechanical and electrical engineer, and an inventor specializing in electromagnetism, he created fluorescent lighting, pioneered wireless communication, built an alternating-current induction motor, and developed the Tesla coil, variations of which have provided the basis for many modern electrical and electronic devices.

Vavilov, Nikolai Ivanovich (1887-1943): A Russian botanist and plant geneticist, Vavilov served for two decades as director of the Institute of Agricultural Sciences (now the N. I. Vavilov Research Institute of Plant Industry) in Leningrad (now Saint Petersburg). During his tenure, he collected seeds from around the world, establishing the world's largest seed bank—with more than 200,000 samples—and conducted extensive research on genetically improving grain, cereal, and other food crops to produce greater yields to better feed the world. Arrested during World War II for disagreeing with Soviet methods of agronomy, he died of complications from starvation and malnutrition.

Vesalius, Andreas (1514-1564): A physician and anatomist born as Andries van Wesel in the Habsburg Netherlands (now Belgium), Vesalius taught surgery and anatomy at the universities of Padua, Bologna, and Pisa in Italy. Dissatisfied at the inaccuracies in the standard texts of the day—based solely on the 1,400-year-old work of ancient physician Galen, since Rome had long discouraged performing autopsies—he dissected a human corpse in the presence of artists from Titian's studio. This resulted in the seven-volume illustrated work, *De humani corporis fabrica libri septem* (*On the Fabric of the Human Body*, 1543), which served as the foundation for modern anatomy.

Vitruvius (c. 80-c. 15 B.C.E.): A Roman architect and engineer, Vitruvius served in many campaigns under Julius Caesar (100-44 B.C.E.), for whom he designed and built mechanical military weapons, such as the ballista (a projectile launcher) and siege machines. His major written work, *De architectura* (*On Architecture*, c. 27 B.C.E.), set the standard for building structures solidly, usefully, and attractively. The book covers the construction of machines—including cranes, pulleys, sundials, and water clocks. It discusses construction materials and describes ancient Roman building innovations that greatly influenced later architects, particularly during the Renaissance.

Watt, James (1736-1819): A Scottish mechanical and civil engineer, Watt designed a steam engine to pump water out of mines. Refinements of his engine were used in grinding, milling, and weaving, and further improvements—including gauges, throttles, gears, and governors—enhanced the engine's efficiency and safety, making it the prime mover of the Industrial Revolution and the power source of choice for early trains and ships. Watt also devised an early copying machine and discovered a method for producing chlorine for bleaching. The unit of electrical power is named for him.

Wegener, Alfred (1880-1930): A German meteorologist, climatologist, and geophysicist, Wegener was one of the first to employ weather balloons. He was first to advance the theory of continental drift, proposing that the Earth's continents were once a single mass that he called Pangaea; his ideas, however, were not accepted until long after his death. From 1912, he worked in remote areas of Greenland examining polar airflows and drilling into the ice to study past weather patterns. He died in Greenland during his last ill-fated expedition.

Westinghouse, George (1846-1914): Born in Central Bridge, New York, Westinghouse was an engineer, inventor, entrepreneur, and a rival of Thomas Edison (1847-1931). He built a rotary steam engine while still a teenager and in his youth patented several devices—including a fail-safe compressed-air braking system—to improve railway safety. He developed an alternating-current power distribution network that proved superior to Edison's direct-current scheme, invented a power meter still in use, built several successful hydroelectric generating plants, and devised shock absorbers for automobiles.

Whittle, Sir Frank (1907-1996): Born the son of an engineer in England, Whittle joined the Royal Air Force as an aircraft mechanic and advanced to flying officer and test pilot before eventually rising to group captain. While in the Royal Air Force, he began designing aircraft engines that used turbines rather than pistons. In the mid-1930's, he formed a partnership, Power Jets, which produced the first effective turbojet design before the company was nationalized. He later developed a self-powered drill for Shell Oil and wrote a text on gas turbine engines.

Wiener, Norbert (1894-1964): A mathematician born in Missouri, Wiener was a child prodigy. He began college at the age of eleven, earned a bachelor's degree in math at fourteen, and a doctorate in philosophy from Harvard at the age of eighteen. During World War I, he researched ballistics at Aberdeen Proving Ground and afterward spent his career teaching mathematics at Massachusetts Institute of Technology. A pioneer of communication theory, he is credited with the development of theories of cybernetics, robotics, automation, and computer systems, and his work greatly influenced later scientists.

Woodward, John (1665-1728): An English naturalist, physician, paleontologist, and geologist, Woodward was an early collector of fossils, which served as the basis for his *Classification of English Minerals and Fossils* (1729), a work that influenced geology for many years. He also conducted pioneering research into the science of hydroponics. His collection of specimens formed the foundation of Cambridge University's Sedgwick Museum, and his estate was sold to provide a post in natural history, now the Woodwardian Chair of Geology at Cambridge.

Wright, Orville (1871-1948) and Wilbur Wright (1867-1912): The Wright brothers were American aviation pioneers who began experimenting with flight in their teens. In the early 1890's they opened a bicycle sales and repair shop, which financed their research into manned gliders. They soon progressed to designing powered aircraft. They eventually invented and built the first practical fixed-wing aircraft and piloted the world's first sustained powered flight—a distance of more than 850 feet over nearly a minute—in 1903 at Kitty Hawk, North Carolina. The Wright Company later became part of Curtiss-Wright Corporation, a modern high-tech aerospace component manufacturer.

Zeppelin, Ferdinand von (1838-1917): Born in Germany, Zeppelin served in the Prussian army and made a balloon flight while serving as a military observer in the American Civil War. After returning to Europe, he designed and constructed airships and devised a transportation system using lighter-than-air craft. He created a rigid, streamlined, engine-powered dirigible in 1900 and was instrumental in the creation of duralumin, which later led to lightweight all-metal airframes. By 1908, he was providing commercial air service to passengers and mail, which had an enviable record for safety until the *Hindenburg* disaster in 1937.

Zworykin, Vladimir (1889-1982): A Russian who emigrated to the United States after World War I, Zworykin worked at the Westinghouse laboratories in Pittsburgh. An engineer and inventor who patented a cathode ray tube television transmitting and receiving system in 1923, he later worked in development for the Radio Corporation of America (RCA) in New Jersey, where his inventions were perfected in time to be used to telecast the 1936 Olympic Games in Berlin. He also contributed to the development of the electron microscope.

Jack Ewing

GLOSSARY

absolute zero: The complete absence of thermal energy, resulting in a temperature of -273.15 degrees Celsius. This temperature is the basis for the Kelvin scale (starting at 0 Kelvin) developed by the British physicist, Lord Kelvin, in 1848. What living organisms feel as heat or warmth is a difference in temperature between two objects, which results in a transfer of thermal energy. Molecules at absolute zero have no thermal energy to transfer but can receive thermal energy from contact with a warmer object. *See also* cold, heat, temperature.

acid: A compound containing hydrogen ions (with a positive charge) in its molecules, which are released when the acid is dissolved in water. Acids include such familiar hazardous substances as sulphuric, nitric, and hydrochloric acid, essential nutrients such as ascorbic acid (vitamin C), and common flavorings or preservatives such as acetic or ethanoic acid (vinegar). Acids react chemically with substances known as bases. The balance of acids and bases in a solution is measured by the pH scale, from 0 (strongly acidic) to 7 (neutral) to 14 (strongly alkaline). *See also* alkali, basic chemical.

alkali: A base that is dissolved in water. Alkaline substances are identified by a measurement from 8 to 14 on the pH scale. *See also* basic chemical.

alpha particle: One of three common forms of radiation from the nuclei of unstable radioactive elements, consisting of two protons and two neutrons, identical to the nucleus of a helium atom, without its electron shell. It has a velocity in air of one-twentieth the speed of light. *See also* beta particle, gamma ray.

amino acids: Biological molecules that serve as the building blocks of proteins and enzymes. Amino acids are incorporated into proteins by transfer RNA, according to the genetic code contained in DNA. The majority of amino acids have names ending with -ine, and are complex arrangements of atoms of carbon, nitrogen, hydrogen, and oxygen. *See also* enzyme, protein.

animal husbandry: The art and science of breeding, raising, and caring for domesticated animals, primarily in small- or large-scale agriculture, as sources of food, leather, wool, and other products useful to humans. Husbandry skills are not only required for many jobs in agriculture but for zookeepers, maintaining rodent and amphibian populations in laboratories, and for large-scale veterinary and animal-vaccination practices.

antiseptic: Any chemical substance that kills or inhibits the growth of microorganisms causing sepsis—putrefaction, decay, or other infection—generally applied to surface tissues of human or other living organisms or to nonliving surfaces that may harbor microorganisms.

atmosphere: The layers of gas surrounding the solid or liquid surfaces of a planet. The atmosphere of the Earth is 78.08 percent nitrogen, 20.95 percent oxygen, less than one percent argon, and hundredths or thousandths of a percent neon, helium, and hydrogen. The amounts of water vapor, carbon dioxide, methane, nitrous oxide, and ozone vary with biological (and more recently industrial) processes. Water can rise to as high as four percent. The atmosphere has been divided by different studies into five to six distinct layers: the troposphere, tropopause, stratosphere, mesosphere, and ionosphere (or thermosphere), plus the very thin exosphere fading into interplanetary space. The ozone layer is in the upper level of the stratosphere.

atom: The smallest particle of matter that has the characteristics of an element, such as oxygen, iron, calcium, or uranium. Three subatomic particles are common to all atoms: protons, neutrons, and electrons. The characteristics of any atom are determined by the number of these particles, particularly the negatively charged electrons in the outer shell. *See also* compound, element, molecule, periodic table of the elements.

atomic number: The number of protons (positively charged particles) in the nucleus of an atom, also the number of electrons (negative charge) in the atom in its standard form. Ions of an atom have larger or smaller levels of electron charge. *See also* electron, ion, periodic table of the elements, proton.

atomic weight: The total mass of the protons and neutrons in an atomic nucleus, with a tiny addition for the weight of electrons. Uranium has the atomic number 92, for 92 protons and 92 electrons, but different isotopes such as U-235 (atomic weight 235, adding 143 neutrons to 92 protons) or U-238 (atomic weight 238, adding 146 neutrons to 92 protons).

ballistics: The science of propelling objects, from rocks and spears to spacecraft. Mastery of this field requires mathematical precision in determining the energy required to put a stationary object into motion in a desired direction, and adjust its course, considering friction from wind or water, or the absence of friction in relatively empty vacuum, and the effect of any body powerful enough to exert gravitational pull, such as the Earth, Moon, or Sun.

basic chemical: Any substance that reacts with an acid. Bases include some metals, such as sodium, calcium, zinc, and aluminum when not protected by an aluminum oxide coating. Other bases include carbonates, hydroxides, and metal oxides (compounds formed by burning metals in oxygen). When a base reacts with an acid, the result is a metal salt and water. *See also* acid, alkali.

battery: In electricity, any device for storing an electrical charge so that it can be used later to power a machine, heater, or light source. Common types of batteries include lead-acid batteries, used for internal combustion automobiles and backup power for industries and military bases; solid alkaline and carbon-zinc batteries, used for flashlights and portable radios; mercury oxide batteries, used in small electronic equipment such as hearing aids, rechargeable nickel-cadmium and nickel hydride batteries; and lithium-ion batteries, an advanced rechargeable type used in portable computers, iPods, and hybrid or electric motor vehicles. Every battery relies on an oxidation-reduction chemical reaction induced by passing a current through its component materials and a reverse reaction that gives off an electric current when plugged into a circuit.

beta particle: One of three common forms of radiation from the nuclei of unstable radioactive elements, carrying a negative charge, similar to an electron, but moving at a high rate of speed, formed when a neutron (neutral electrical charge) transforms into a proton (positive electrical charge) by ejecting a negative charge. *See also* alpha particle, gamma ray.

binary number system: A mathematical system having only two numerals, 0 and 1, most commonly used in computer hardware and software, because at its most basic, a computer can turn a series of switches on (value = 1) or off (value = 0). *See also* computer.

biosphere: First defined by Austrian geologist Edward Seuss in 1875 as "the place on Earth where life dwells." The concept has been expanded to include all living organisms on Earth, dead organic matter, and the biological component of dynamic processes such as the cycling of carbon, nitrogen, oxygen, phosphorous, and other elements.

British thermal unit (Btu): The heat required to raise one pound of water one degree Fahrenheit, equal to 252 calories. Also, the heat required to produce 779.9 foot-pounds of energy in a mechanical system. This is a common measure of the energy potential in fuels. *See also* heat.

calorie: The heat required to raise the temperature of one gram of water one degree Celsius. This is a common measure of the energy potential in food but can be applied to fuels and mechanical processes. *See also* heat.

carboniferous fuels: Any source of energy obtained from carbon-based compounds, particularly coal and oil. These two common fuel sources accumulated during the Carboniferous period of the Paleozoic era.

catalyst: A substance that makes a chemical reaction between two other substances proceed at a significantly faster rate, without being consumed in the reaction. (See also enzyme.)

ceramics: Inorganic, nonmetallic solids, particularly made from clay, that are processed at high levels of heat, then cooled. In addition to common use for pottery and tableware, ceramics have many industrial applications, such as ceramic-based thermocouples to measure high temperatures, ceramic insulators, laser components, heat storage and diffusion, and capacitors.

chromosome: A basic unit of heredity in living cells. Each chromosome is composed of proteins and DNA, which carry thousands of genes. In a healthy, normal, human cell, there are twenty-three pairs of chromosomes. In sexual reproduction, one chromosome in each pair comes from the father, the other from the mother. *See also* DNA, gene.

climate: Prevailing weather conditions in a specific region or area over the course of a year, characterized by a typical range of temperatures, seasonal or year-round humidity, precipitation, prevailing winds, and extremes of seasonal variation.

cold: The sensation felt when a living organism is in contact with a substance of a lower temperature. Thermal energy naturally flows from a warmer object to a colder object when they are in physical contact. *See also* absolute zero, heat, temperature.

compound: A chemical substance composed of molecules, containing atoms of two or more different elements, forming a single particle. For example, water is a compound, in which each water molecule contains two atoms of the element hydrogen and one atom of the element oxygen. *See also* atom, molecule, element, periodic table of the elements.

computer: Any device that can be programmed to perform mechanical or electrical computation, processing numbers. Since the middle of the twentieth century, the term is commonly used for equipment that can be programmed using a binary number system to perform a variety of work. For a computer to process letters, words, graphic images, or maps, human programmers have to encode nonnumerical data in a numeric form. *See also* binary number system.

cryptology: The science of encrypting or decrypting information, including creating codes for privacy or security and finding means to break a code by working from a message in an unknown code to learn the pattern. Any code in which a symbol is substituted for each letter of the alphabet is particularly vulnerable to decryption, because in a large sample, there is a probability for how often each letter will appear, with "e" being used more often than any other. *See also* encoding.

demography: Variations in human population that can be studied statistically, in defined groups, rather than individual behavior. Almost any characteristic can form the basis for demographic research: ethnicity, diet, religion, wealth, language, education, urbanization, occupation, marriage customs, class or caste distinctions.

desalination: Removing salt from water for use in drinking or agriculture. Ninety-seven percent of the water on earth is salt water, mostly in the oceans. The dissolved salt content is harmful to freshwater land plants and land animals. Removing the salt is energy intensive and therefore has a high cost, but

in places where freshwater is in short supply, it is sometimes considered worth the expense.

detergent: A type of surfactant that has the property of removing stains or particles from a surface, keeping it suspended in water, and allowing the suspended solids to be rinsed away. *See also* surfactant.

distillation: Isolating and purifying a liquid substance by heating a solution to the exact boiling point of the desired end product, producing a vapor, then capturing and condensing the vapor in a separate container. Water, perfumes, and alcohols are all common examples of liquids that can be distilled.

DNA (deoxyribonucleic acid): The complex molecule making up genes, encoding inheritance in all living species. It is known for a unique double-helix structure, with the code in an "alphabet" of four types of molecule: cytosine pairs with guanine, while adenine pairs with thymine. DNA is increasingly used for identification, particularly in crime scenes, to determine paternity of a child, and to study inheritance of both individuals and demographic groups. Study of DNA is also leading to new treatments for genetically inherited diseases. *See also* chromosome, gene.

ecology: A branch of biology that studies the complex relationships between living organisms, their environment, the manner in which a variety of plant and animal life forms a mutually interdependent ecosystem, and the competition between life forms within and between ecosystems.

electrical storage: See battery.

electrolysis: Applying an electric current to take apart (decompose) the molecules of a substance in solution. If an electric current is run through a container of water, the hydrogen and oxygen atoms in the water molecule will separate and form diatomic molecules (two atoms in each molecule) of oxygen and hydrogen gas. Electrolysis can be applied to many compounds. Industrial uses include separating chlorine and caustic soda from brine and refining metals such as sodium, calcium, magnesium, and aluminum from common ore compounds.

electromagnetism: A fundamental force of nature acting on all electrically charged particles. The previously separate studies of electricity, magnetism, and optics (initially visible light, which is only one spectrum of electromagnetic radiation) were unified by the work of English scientists Michael

Faraday and Robert Maxwell in the mid-nineteenth century.

electron: A negative charge in the outer shell of every atom. Electrons are sometimes described as particles, but within an atom they act more as electrical charges in shells, rather than particles in orbit. In a stable atom, the number of electrons exactly equals the number of protons in the nucleus. Ions have fewer or additional electric charges. *See also* atom, ion, neutron, proton.

element: A substance formed by atoms of a single type, found in the periodic table of the elements. Hydrogen, helium, and oxygen are all examples of elements. *See also* atom, compound, molecule, periodic table of the elements.

encoding: Representing information by a system of symbols or characters. Encoding processes exist in human memory, heredity, computer programming, and in written or verbal communication, including military or business communications intended to be secret. The most common use is to take a message in a plain language and convert it into a sequence of characters that can be read only by a person instructed in the code—or by a cryptologist who can break the code by mathematically analyzing the pattern. Natural encoding processes include the genetic code, which stores in long molecules of DNA the structure, physiology, and metabolism of a complete living organism. *See also* binary number system, computer, cryptology, DNA.

engineering: Practical application of the knowledge of pure science, and sometimes of art as well, not only to construct buildings, bridges, infrastructure, and engines, but to plan and organize industrial and community processes. There are many branches of engineering, including electrical, industrial, mechanical, civil, aeronautical, geotechnical, transportation, water management, disaster preparedness and management, and telecommunications.

entropy: The spontaneous direction of any natural process, tending to lose energy or to become more chaotic. All things naturally tend toward equilibrium: Two objects of different temperatures, placed in contact, will equalize to a common temperature as heat is transferred from the warmer object to the colder object. A solid dissolved in a solution spreads evenly throughout the solution, gas at two different pressures will equalize, but in none of these examples will objects spontaneously develop uneven temperatures, pressures, or concentrations in solution.

enzyme: A biological catalyst, any protein molecule within a living organism that speeds up biochemical reactions to a rate that will sustain life. The effect may speed up metabolic reactions by a factor of one million, compared with what would occur chemically outside the body. Names and classification of enzymes are regulated by the International Commission on Enzymes. Most enzymes are named by adding -ase to the root of a corresponding substrate, the molecule an enzyme acts upon. Sucrase catalyzes the hydrolysis of sucrose into glucose and fructose. A living cell has a unique set of 3,000 enzymes, each defined by the cell's DNA.

erosion: The process of wind or water wearing away soil or rock. The existence of soil is due in part to erosion of stone surfaces from the time solid rock first formed on Earth. Whole mountain chains have been worn down by erosion—the Appalachian Mountains were once as high as the Rockies. Sandstone is formed by compression of eroded rock under the oceans. Soil erosion can destroy cultivated land. Sheet erosion removes soil in a uniform layer, while rill erosion cuts small channels into the soil, and gully erosion forms deeper channels carrying away a large volume of soil.

eukaryotic cell: A complex cell on which all life more complex than a bacterium or yeast is based, characterized by a nucleus containing the cell's DNA and a number of specialized organelles, supplied with energy by mitochondria, which individually resemble the more primitive prokaryotic cell. *See also* prokaryotic cell.

exosphere: A layer of Earth's atmosphere from 500 kilometers above the surface to between 10,000 and 190,000 kilometers—or halfway to the Moon. At 190,000 kilometers, the force of solar radiation is more powerful than the force of Earth's gravity on the thinly distributed atmospheric molecules, but many scientists consider 10,000 kilometers to be the boundary with interplanetary space. Within the exosphere, a gas molecule can travel hundreds of kilometers before bumping into another gas molecule.

fermentation: A biological process for breaking down complex organic compounds into simpler compounds. One of the most familiar in human history is the conversion by yeast of sugar to carbon dioxide, alcohol, and water. Fermentation also occurs in cells, including animal muscle cells, breaking down glucose to produce lactic acid, lactate, carbon dioxide, and water, as well as adenosine triphosphate, a source of energy. It is less efficient than cellular respiration but occurs when muscles are short of oxygen. Many anaerobic bacteria ferment sugars: Lactobacillus ferment milk to produce yogurt. Fermentation also produces lactic acid in a variety of foods, such as sauerkraut and sourdough bread.

forensic: The application of science to legal concerns. The analysis of crime scenes, firearms, DNA, and the pathology of dead bodies are common subjects of forensic investigation, but dentists, toxicologists, psychiatrists, engineers, and practitioners in many other fields can also be called upon.

fuel cell: A source of electric current that operates in a manner similar to a battery, generating electricity as a by-product of a chemical reaction. The difference is that a fuel cell continues generating power as long as it is supplied with fuel. The hydrogen fuel cell, one of the most commonly known, generates electricity, heat, and water.

gamma ray: One of three common forms of radiation from the nuclei of unstable radioactive elements, which has no mass and no electrical charge. It is made up of photons, the fundamental particle of light, moving at the speed of light, emitted at a high energy level. X rays are similar but originate in electron fields rather than in the nucleus of atoms. *See also* alpha particle, beta particle.

gene: A unit of hereditary information found within a chromosome that determines the characteristics of an organism. Each gene is an ordered series of nucleotides, which are subunits of DNA, composed of a base molecule containing nitrogen, a phosphate molecule, and a pentose sugar molecule. *See also* chromosome, DNA.

Global Positioning System (GPS): A system owned by the United States government and operated by the Air Force for determining the exact position on the surface of the Earth of any user or defined landmark. By receiving signals from any of 24 orbiting satellites that are in direct line of site to a user's position, GPS devices can calculate latitude, longitude, altitude, and time.

gravity: As defined by Sir Isaac Newton's law of universal gravitation, a force that attracts any two objects in the universe. The strength of gravitational attraction depends on the mass of the two objects and decreases according to the distance between them, squared.

halogens: Elements fluorine, chlorine, bromine, iodine, and astatine in the periodic table of the elements. Being one electron short of filling the outermost shell of each atom, halogens form chemical bonds easily with other elements. They often form salts, including common table salt (sodium chloride), and calcium chloride—the salt applied to roads in winter to melt ice.

heat: The transfer of thermal energy, the kinetic energy of molecules, between two objects as the result of a temperature difference. Thermal energy is a vibration at the molecular level, which increases the volume of a substance, or increases the pressure of a gas or liquid in a closed space. Heat can be used to accomplish work mechanically; it is also produced by friction of moving parts, which wastes energy applied to accomplish work in a mechanical system.

husbandry: The cultivation of land to raise edible plant crops (and textile crops) or breed and raise domestic animals for food. *See also* animal husbandry.

hydrocarbons: Among the simplest of organic molecules, made up of a number of carbon and hydrogen atoms. Compounds with a benzene ring in the molecular structure are called aromatic hydrocarbons. Those without a benzene ring are called aliphatic hydrocarbons, which include alkanes (single carbon bond), alkenes (one double bond), and alkynes (one triple bond). There are a nearly unlimited number of derivative carbon compounds that can be formed by adding oxygen atoms to hydrocarbons.

hydrology: The study of water and, more specifically, of the way water cycles through lakes, streams, ponds, rivers, oceans, the atmosphere, and underground water flows and reservoirs, including evaporation, rain-, and snowfall. This includes study of contamination in water, as well as movement and distribution.

hydroponics: Growing plants in a solution of water and selected nutrients, without need for soil. Hydroponics can rely on sunlight or can be established in a closed, indoor environment using grow lights.

inflammable: An object or substance that catches fire easily, from the Latin *inflammare*, meaning to kindle a flame. Often confused, because "flammable" has the same meaning, and for English words, the prefix in- often means opposite, such as invisible. Safety officials have encouraged use of "flammable" to avoid misunderstanding and "non-flammable" to mean a substance that will not burn easily. Inflammability of any substance increases with higher concentration of oxygen in the surrounding atmosphere and decreases with lower oxygen concentration. Burning is an oxidation process—combining oxygen with the molecules of the burning substance—which can begin at lower temperatures with a higher concentration of oxygen. Once started, sufficient heat is given off to make the fire self-sustaining.

ion: An atom that has more or fewer electrons in its outermost shell than protons in its nucleus. *See also* atom, compound, element, molecule, periodic table of the elements.

ionosphere: The outermost layer of Earth's atmosphere, so named because solar radiation causes many atoms at this level to ionize, gaining or losing an electron. *See also* ion.

irradiation: Exposing any substance, or living organism, to radiation most commonly used to destroy bacteria, viruses, fungi (mold), and insects in food or on surfaces used for food preparation. The level of radiation used is not strong enough to disintegrate the nucleus of any atom making up the substance of any food item, so the food itself does not become radioactive. Irradiation is also used in treating cancer, checking luggage at airports, and sterilizing many items.

joule: A measure of heat or energy, used more commonly in scientific research than in industry; the energy required to accelerate a body with a mass of one kilogram using one newton of force over a distance of one meter. Equal to 0.2390 calories or 0.738 foot-pounds.

laser: A light source that emanates from a well-defined wavelength; originally an acronym for light amplified by stimulated emission of radiation.

light-emitting diode (LED): A diode constructed to provide illumination from the movement of electrons through a semiconductor, which is housed in a bulb that concentrates the light in a desired direction. Because there is no filament to warm up, as in a conventional electric light bulb, LEDs use ten percent or less electrical current to provide the same brightness of light and last up to twenty times longer. LEDs can be constructed to provide almost any desired color or hue of light.

lithography: A printing process that is the most common method of printing and publishing. The earliest process, invented by Alois Senefelder in Germany in 1798, is still used by hand-applying a greasy ink to a specially prepared block of limestone, which is moistened with water. The water is repelled by the ink, and in turn repels an oil-based ink, applied with a roller, from moistened areas, creating the desired image when paper is applied against the plate. Modern offset lithography burns a photosensitive metal plate through a negative film image to create a pattern of roughened areas—to which oil- or rubber-based inks will adhere—while a thin layer of water repels the ink from unexposed smooth metal surfaces.

magnetism: One aspect of electromagnetism, involving fields either generated by a current moving through a wire or a magnetized object, in which the molecules are aligned with magnetic north in one uniform direction and magnetic south in the opposite direction.

mean: Sometimes called the arithmetic mean, or average, it is the sum of a series of figures divided by the total number of figures.

median: The middle number in a group of numbers arranged in order from lowest to highest. Comparing the mean and the median can help to correct for the distortion of extreme highs or lows.

mesosphere: A layer of Earth's atmosphere, between the stratosphere and the ionosphere (or thermosphere), variously defined as beginning at an altitude from 30 to 55 kilometers above sea level and continuing to an altitude of about 80 to 90 kilometers.

metabolism: Physical processes and chemical reactions within a living body that convert or use energy, including those associated with digestion, excretion, breathing, blood circulation, growing and using muscle tissues, communication through the nervous system, and body temperature. *See also* enzyme, protein.

metals: A majority of known elements, generally

shiny in appearance and good conductors of heat and electricity. In ionic compounds, metals usually provide the positive ion. Many metals react with acids and therefore act as a base.

metric system: Scales of measurement in which each unit is one tenth of the next largest and ten times the next smallest, simplifying conversion, recording, and mathematical operations. The system is built around the unit of the meter (equivalent to 39.39 inches), which the system's inventors defined as one-ten millionth the distance from the North Pole to the equator.

microbe: Any microscopic form of life, also called a microorganism, particularly bacteria, protozoa, fungi, or virus. Most commonly, this term refers to pathogenic microscopic life—those that cause infection, disease, decay, sepsis, or gangrene. However, biologists are identifying an increasing number of microbes that are beneficial, even essential to life, including a variety of those found in the human intestine.

mitochondria: A type of organelle within eukaryotic cells, where oxygen and nutrients are converted into adenosine triphosphate, the molecule that stores chemical energy for the cell. This process, called aerobic respiration, is possible only in the presence of oxygen. Mitochondria are rod-shaped, have their own DNA, and reproduce independently within the cell, resembling some primitive prokaryotic cells. This suggests that prokaryotic cells were absorbed within the cell walls of evolving eukaryotic cells in a symbiotic relationship. Mitochondria enable cells to produce adenosine triphosphate fifteen times more efficiently than is possible by anaerobic respiration.

mode: The most frequent value in a series of numbers. For example, if the students in a class are all age sixteen, seventeen, or eighteen, and there are more sixteen-year-olds than either seventeen- or eighteen-year-olds, then sixteen would be the mode.

molecule: A particle formed by two or more atoms of the same element or of different elements. A water molecule is formed by one oxygen atom and two hydrogen atoms. Bonds holding molecules together are formed by the electrons in each atom's outer shell. *See also* atom, compound, element, periodic table of the elements.

navigation: The art and science of plotting a course

from a starting point to a desired destination, most commonly piloting a boat on water or travel in outer space. The term is sometimes used for travel on land as well, particularly in a desert or flatland without significant landmarks, or in recent years, using a GPS navigation device for ordinary driving. Historically, navigation has been accomplished by using certain stars as reference points for latitude. Longitude was often guesswork, based on the time of day at the starting point compared with the current location, until invention of the seagoing chronometer in 1764. Radio beacons, radar, the gyroscopic compass, accurate maps, and the satellite-based GPS have all provided increased precision in navigation.

neuron: A nerve cell, the basic unit of the spinal cord, nervous system, and brain of human beings and other mammals. Neurons communicate information in chemical and electrical forms throughout an organism. Sensory neurons provide information to the brain from receptors in every part of the body. Motor neurons transmit direction from the brain to muscles. Shortly after birth, neurons stop reproducing, while other cell types continue to do so. Neurons have specialized structures called axons and dendrites to send and receive signals at connections known as synapses.

neutron: A neutral particle within the nucleus of an atom, having neither a positive or negative electrical charge and a weight slightly more than that of a proton. *See also* atom, electron, proton.

noble gases: Elements helium, neon, argon, krypton, xenon, and radon in the periodic table of the elements. They are called "noble" because having their outer electron shell filled to capacity, they do not react with other elements or form compounds. Radon, however, is a radioactive element. All of these elements exist in the form of gases, forming liquids or solids only at extremely low temperatures, approaching absolute zero.

nucleus (atomic): The tightly packed protons and neutrons at the core of every atom, which account for nearly all of an atom's mass. The total size of an atom is defined by its electron shells.

nucleus (cell): An organelle within each living eukaryotic cell that acts as a control center, storing genes on chromosomes, producing messenger RNA molecules (which transfer code for essential proteins from genes in the chromosomes), producing ribosomes, and organizing replication of DNA, including complete copies for cell division.

optics: The study of light, including all systems for gathering, concentrating, and manipulating light, such as mirrors, spectacles, telescopes, microscopes, cameras, spectroscopes, lasers, fiber optic communications, and optical data storage and retrieval.

organic compounds: Compounds that are created in or by living cells, rather than as a result of spontaneous physical processes. The simplest organic compounds are hydrocarbons, composed of carbon and hydrogen. Sugars and some other compounds are composed of carbon, hydrogen, and oxygen. The most complex organic compounds are composed of carbon, hydrogen, oxygen, nitrogen, occasionally sulfur, and sometimes small traces of metals.

oxidation: The process of any other element forming chemical bonds with atoms of oxygen. Common examples include the formation of rust on metal, and the burning of wood or any other inflammable substance. Oxidation occurs at many points in human and animal metabolism, including oxygen binding to hemoglobin in the blood stream and many chemical reactions within each cell in the body.

oxygenation: Infusion of oxygen gas into a solution or organic process, such as the transfer of oxygen through the membranes in the lung to enter the bloodstream.

ozone layer: An outer layer of Earth's atmosphere made up of ozone, a molecule of oxygen containing three atoms, instead of the two atoms of the oxygen breathed by living things. While ozone is toxic to life, causing chemical burns at relatively low concentrations, the ozone layer absorbs much of the ultraviolet radiation in sunlight, protecting life at the planet's surface.

pasteurization: A process discovered by French microbiologist Louis Pasteur for rapidly heating milk to destroy disease-causing bacteria, while leaving the nutritional content of the milk unaffected. It is also used to destroy bacteria in wine and beer manufacturing.

periodic table of the elements: The table of chemical elements arranged according to their atomic number. *See also* atom, atomic number, element, valence.

photon: The particle of light. Since light is a wave of electromagnetic radiation, it is something of a paradox that it also travels in particles. Quantum mechanics is based on the observation that electromagnetic radiation travels in discrete quantities of energy, called quanta. The photon is also the particle that, moving at a high volume of energy, is called a gamma ray.

photosynthesis: A chemical process in which carbon dioxide and water are converted into carbohydrates and oxygen by the energy from a light source, generally the Sun. This reaction, which all plants and many bacteria rely on, is the source of the unnatural presence of oxygen in Earth's atmosphere and supplies the entire food chain upon which animal life depends for existence.

photovoltaic: Electricity from light—any process using the energy of light to generate electricity directly. The most common practical application is the photovoltaic cell, made of a thin semiconductor wafer, treated to form an electric field, with electrical conductors attached to its positive and negative sides. When these poles are connected by a circuit, an electrical current is generated by photons from incoming sunlight, knocking loose electrons in the semiconductor.

physiology: Study of the entire system of physical functions in a living body: its mechanical, biochemical, and bioelectrical processes, the purpose and operation of each organ, and the interaction of different organs and parts.

polarity: The existence of two opposite characteristics, such as the north and south poles of a magnet or the positive and negative poles of a battery.

prokaryotic cell: The earliest and most primitive type of cell, probably the first life form on Earth, lacking a cell nucleus. Most bacteria are prokaryotic. Most one-celled animals, such as paramecium, and simple plants such as algae have the more complex eukaryotic cell.

protein: A long chain of amino acids. There are thousands of different proteins in each cell of the human body, and since each species has slightly different proteins in its cells, there are millions of different proteins in the biosphere. A balance of all necessary proteins is essential to the continued life of any organism. Food consumption must either supply each complete protein that the human body cannot manufacture for itself or a wide variety of incomplete proteins that can be assembled into complete proteins.

proton: The positively charged particles in the nucleus of every atom. Along with neutrons, protons account for most of the weight in an atom, because electrons have very little weight. *See also* atom, electron, neutron.

protoplasm: The living substance of a cell, including the content of the cell membrane and the substance within the cell—a transparent gelatinous material composed of inorganic substances (90 percent water with mineral salts and gases such as oxygen and carbon dioxide), and organic substances (proteins, carbohydrates, lipids, nucleic acids, and enzymes). Protoplasm outside the cell nucleus is called cytoplasm.

radiation: Emission of subatomic particles or of photons at high energy levels from radioactive atoms. *See also* alpha particles, beta particles, gamma rays.

radio waves: A band of wavelengths in the electromagnetic spectrum that can be generated by a spark gap in an electrical circuit and are commonly used for broadcast communication. *See also* electromagnetism, ionosphere.

refrigerant: A compound that transfers thermal energy in cooling systems, including air conditioners, freezers, refrigerators, and low-temperature manufacturing processes. Releasing the refrigerant at low pressure, in tubes that are in physical contact with the area to be cooled, transfers heat to the refrigerant, which is then transferred to radiator coils outside the area to be cooled, by compressing the refrigerant to a liquid state. Common refrigerants include ammonia, dichlorodifluoromethane, propane, hydrochlorofluorocarbon (HCFC), and hydrofluorocarbon (HFC).

RNA (ribonucleic acid): A complex molecule similar to, but less complex than, DNA. RNA molecules transfer genetic information from genes in longer DNA molecules forming the chromosomes of living cells to the active metabolic proteins in a living cell. *See also* DNA.

scientific method: A process for investigating nature by observation and experiment, creating hypotheses to make sense of observations.

seismology: The study of earth movement, particularly the mechanism and causes of earthquakes.

semiconductor: A material that conducts electricity more efficiently than an insulator, but less efficiently than a conductor, useful in constructing diodes, which conduct electricity in only one direction. Common semiconductor materials include silicon, germanium, and selenium. Semiconduc-

tors are essential to construction and design of computers and most electronic equipment.

sewage: Waste, usually carried in water, which is either chemical industrial waste or organic waste. Industrial sewage often includes metals, other toxins, and complex molecules that are not naturally metabolized. In most industrialized economies, sewage goes through several stages of treatment to remove solid waste for disposal, and the water is returned to lakes and rivers in relatively clean condition.

spectroscopy: Study of the spectrum of wavelengths of electromagnetic radiation, particularly the wavelengths of light visible to the human eye. When light reflected off any substance is viewed or projected through a spectroscope, each element casts a unique pattern of lines on the visible spectrum. This is useful in astronomy for identifying the chemical composition of stars and planets, since physical samples cannot be obtained, as well as in analytical chemistry.

statistics: Methods of obtaining, organizing, analyzing, interpreting, and presenting numerical data, used in many areas of science and industry, as well as in demographic studies of human populations. *See also* mean, mode, median.

sterilization (of microbes and pathogens): Killing all or most microbes on a working surface, or on the surface of instruments to be used in medical care or food preparation, by means of heat (including pasteurization), irradiation, or chemical antiseptics.

sterilization (pertaining to reproduction): Surgically or chemically preventing an organism from producing offspring. This includes neutering or spaying of pets and domesticated farm or service animals, castration of male animals, irradiating male or female insects as a pest-control measure, and, in humans, severing and tying the Fallopian tubes to prevent pregnancy in a woman or vasectomy to prevent a man inseminating a woman.

stratosphere: A layer of Earth's atmosphere from 18 kilometers to 50 kilometers. Unlike the troposphere, airflow is mostly horizontal with no weather patterns. The ozone layer is in the upper level of the stratosphere.

surfactant: A chemical substance that reduces the surface tension of water. A common use of surfactants is in manufacture of detergents and soaps. Surfactants are complex molecules with one component attracted to water molecules (hydrophilic) and the other component repelled by water molecules (hydrophobic).

temperature: In theory, a measure of the average kinetic energy of molecules: The larger or denser an object is, the more heat is required to raise its temperature by one degree. The most common scales for measuring temperature select arbitrary fixed points and assign arbitrary numerical values. The Fahrenheit scale assigns a value of 32 degrees to the temperature at which water freezes and 212 degrees to the temperature at which water boils. The Celsius scale assigns zero to the freezing point and 100 degrees to the boiling point. The Kelvin scale begins with absolute zero, the temperature at which a substance lacks any thermal energy. *See also* absolute zero, cold, heat.

thermosphere: *See* ionosphere.

toxin: A poison; any substance that will have a toxic effect on organic life, particularly proteins produced by bacteria, plants, or in specialized glands of certain animals. Toxins may damage or paralyze without killing or have a caustic effect, but in high enough concentrations, exposure to most toxins will cause death. Some toxins, in low doses, can be used in medical treatment: Two of the seven types of botulinum toxin are used to inhibit muscle spasms, smooth wrinkles of the upper face (Botox), and treat cervical dystonia.

troposphere: The layer of Earth's atmosphere closest to the surface, up to 14 kilometers, where all weather takes place, with rising and falling air currents. Air pressure at the upper limit of the troposphere is about 10 percent of the pressure at sea level.

ultrasound: Vibrations or sound waves at a higher frequency than the human ear can detect.

vacuum: The absence of matter, including air or other gases, inside an enclosed space or in remote areas of outer space. No perfect vacuum is known, since interstellar space is estimated to contain at least one hydrogen atom per cubic meter.

valence: The capacity of the atoms of an element to combine with other atoms of the same element or another element. *See also* atom, molecule, periodic table of the elements.

Charles Rosenberg

TIMELINE

The Time Line below lists milestones in the history of applied science: major inventions and their approximate dates of emergence, along with key events in the history of science. The developments appear in boldface, followed by the name or names of the person(s) responsible in parentheses. A brief description of the milestone follows.

2,500,000 B.C.E.	**Stone tools:** Stone tools, used by Homo habilis and perhaps other hominids, first appear in the Lower Paleolithic age (Old Stone Age).
400,000 B.C.E.	**Controlled use of fire:** The earliest controlled use of fire by humans may have been about this time.
200,000 B.C.E.	**Stone tools using the prepared-core technique:** Stone tools made by chipping away flakes from the stones from which they were made appear in the Middle Paleolithic age.
100,000-50,000 B.C.E.	**Widespread use of fire by humans:** Fire is used for heat, light, food preparation, and driving off nocturnal predators. It is later used to fire pottery and smelt metals.
100,000-50,000 B.C.E.	**Language:** At some point, language became abstract, enabling the speaker to discuss intangible concepts such as the future.
16,000 B.C.E.	**Earliest pottery:** The earliest pottery was fired by putting it in a bonfire. Later it was placed in a trench kiln. The earliest ceramic is a female figure from about 29,000 to 25,000 B.C.E., fired in a bonfire.
10,000 B.C.E.	**Domesticated dogs:** Dogs seem to have been domesticated first in East Asia.
10,000 B.C.E.	**Agriculture:** Agriculture allows people to produce more food than is needed by their families, freeing humans from the need to lead nomadic lives and giving them free time to develop astronomy, art, philosophy, and other pursuits.
10,000 B.C.E.	**Archery:** Archery allows human hunters to strike a target from a distance while remaining relatively safe.
10,000 B.C.E.	**Domesticated sheep:** Sheep seem to have been domesticated first in Southwest Asia.
9000 B.C.E.	**Domesticated pigs:** Pigs seem to have been domesticated first in the Near East and in China.
8000 B.C.E.	**Domesticated cows:** Cows seem to have been domesticated first in India, the Middle East, and sub-Saharan Africa.
7500 B.C.E.	**Mud bricks:** Mud-brick buildings appear in desert regions, offering durable shelter. The citadel in Bam, Iran, the largest mud-brick building in the world, was built before 500 B.C.E. and was largely destroyed by an earthquake in 2003.
7500 B.C.E.	**Domesticated cats:** Cats seem to have been domesticated first in the Near East.
6000 B.C.E.	**Domesticated chickens:** Chickens seem to have been domesticated first in India and Southeast Asia.
6000 B.C.E.	**Scratch plow:** The earliest plow, a stick held upright by a frame and pulled through the topsoil by oxen, is in use.
6000 B.C.E.	**Electrum:** The substance is a natural blend of gold and silver and is pale yellow in color like amber. The name "electrum" comes from the Greek word for amber.

6000 B.C.E.	**Gold:** Gold is discovered—possibly the first metal to be recognized as such.
6000-4000 B.C.E.	**Potter's wheel:** The potter's wheel is developed, allowing for the relatively rapid formation of radially symmetric items, such as pots and plates, from clay.
5000 B.C.E.	**Wheel:** The chariot wheel and the wagon wheel evolve—possibly from the potter's wheel. One of humankind's oldest and most important inventions, the wheel leads to the invention of the axle and a bearing surface.
4200 B.C.E.	**Copper:** Egyptians mine and smelt copper.
4000 B.C.E.	**Moldboard plow:** The moldboard plow cut a furrow and simultaneously lifted the soil and turned it over, bringing new nutrients to the surface.
4000 B.C.E.	**Domesticated horses:** Horses seem to have been domesticated first on the Eurasian steppes.
4000 B.C.E.	**Silver:** Silver can be found as a metal in nature, but this is rare. It is harder than gold but softer than copper.
4000 B.C.E.	**Domesticated honeybees:** The keeping of bee hives for honey arises in many different regions.
4000 B.C.E.	**Glue:** Ancient Egyptian burial sites contain clay pots that have been glued together with tree sap.
3500 B.C.E.	**Lead:** Lead is first extruded from galena (lead sulfide), which can be made to release its lead simply by placing it in a hot campfire.
c. 3100 B.C.E.	**Numerals:** Numerals appeared in Sumerian, Proto-Elamite, and Egyptian hieroglyphics.
3000 B.C.E.	**Bronze:** Bronze, an alloy of copper and tin, is developed. Harder than copper and stronger than wrought iron, it resists corrosion better than iron.
3000 B.C.E.	**Cuneiform:** The method of writing now known as cuneiform began as pictographs but evolved into more abstract patterns of wedge-shaped (cuneiform) marks, usually impressed into wet clay. This system of marks made complex civilization possible, since it allowed record keeping to develop.
3000 B.C.E.	**Fired bricks:** Humans begin to fire bricks, creating more durable building materials that (because of their regular size and shape) are easier to lay than stones.
3000 B.C.E.	**Pewter:** The alloy pewter is developed. It is 85 to 99 percent tin, with the remainder being copper, antimony, and lead; copper and antimony make the pewter harder. Pewter's low melting point, around 200 degrees Celsius, makes it a valuable material for crafting vessels that hold hot substances.
2700 B.C.E.	**Plumbing:** Earthenware pipes sealed together with asphalt first appear in the Indus Valley civilization. Greeks, Romans, and others provided cities with fresh water and a way to carry off sewage.
2650 B.C.E.	**Horse-drawn chariot (Huangdi):** Huangdi—a legendary patriarch of China—is possibly a combination of many men. He is said to have invented—in addition to the chariot—military armor, ceramics, boats, and crop rotation.

2600 B.C.E.	**Inclined plane:** Inclined planes are simple machines and were used in building Egypt's pyramids. Pushing an object up a ramp requires less force than lifting it directly, although the use of a ramp requires that the load be pushed a longer distance.
c. 2575-c. 2465 B.C.E.	**Pyramids:** Pyramids of Giza are built in Egypt.
1750 B.C.E.	**Tin:** Tin is alloyed with copper to form bronze.
1730 B.C.E.	**Glass beads:** Red-brown glass beads found in South Asia are the oldest known human-formed glass objects.
1600 B.C.E.	**Mercury:** Mercury can easily be released from its ore (such as cinnabar) by simply heating it.
1500 B.C.E.	**Iron:** Iron, stronger and more plentiful than bronze, is first worked in West Asia, probably by the Hittites. It could hold a sharper edge, but it had to be smelted at higher temperatures, making it more difficult to produce than bronze.
1500 B.C.E.	**Zinc:** Zinc is alloyed with copper to form brass, but it will not be recognized as a separate metal until 1746.
1000 B.C.E.	**Concrete:** The ancient Romans build arches, vaults, and walls out of concrete.
1000 B.C.E.	**Crossbow:** The crossbow seems to come from ancient China. Crossbows can be made to be much more powerful than a normal bow.
1000 B.C.E.	**Iron Age:** Iron Age begins. Iron is used for making tools and weapons
700 B.C.E.	**Magnifying glass:** An Egyptian hieroglyph seems to show a magnifying glass.
350 B.C.E.	**Compass:** Ancient Chinese used lodestones and later magnetized needles mostly to harmonize their environments with the principles of feng shui. Not until the eleventh century are these devices used primarily for navigation.
350-100 B.C.E.	**Scientific method (Aristotle):** Aristotle develops the first useful set of rules attempting to explain how scientists practice science.
300 B.C.E.	**Screw:** Described by Archimedes, the screw is a simple machine that appears to be a ramp wound around a shaft. It converts a smaller turning force to a larger vertical force, as in a screw jack.
300 B.C.E.	**Lever:** Described by Archimedes, the lever is a simple machine that allows one to deliver a larger force to a load than the force with which one pushes on the lever.
300 B.C.E.	**Pulley:** Described by Archimedes, the pulley is a simple machine that allows one to change the direction of the force delivered to the load.
221-206 B.C.E.	**Compass:** The magnetic compass is invented in China using lodestones, a mineral containing iron oxide.
215 B.C.E.	**Archimedes' principle (Archimedes of Syracuse):** Archimedes describes his law of displacement: A floating body displaces an amount of fluid the weight of which is equal to the weight of the body.
200 B.C.E.	**Astrolabe:** A set of engraved disks and indicators becomes known as the astrolabe. When aligned with the stars, the astrolabe can be used to determine the rising and setting times of the Sun and certain stars, establish compass directions, and determine local latitude.

40 C.E.	**Ptolemy's geocentric system (Ptolemy):** A world system with the Earth in the center, and the Moon, Venus, Mercury, Sun, Mars, Jupiter, Saturn, and fixed stars surrounding it. The geocentric Ptolemaic system would remain the most widely accepted cosmology for the next fifteen hundred years.
90 C.E.	**Aeolipile (Hero of Alexandria):** The aeolipile—a steam engine that escaping steam causes to rotate like a lawn sprinkler—is developed.
105 C.E.	**Paper and papermaking (Cai Lun):** Although papyrus paper already existed, Cai Lun creates paper from a mixture of fibrous materials softened into a wet pulp that is spread flat and dried. The material is strong and can be cheaply mass-produced.
250 C.E.	**Force pump (Ctesibius of Alexandria):** Ctesibius develops a device that shoots a jet of water, like a fire extinguisher.
815 C.E.	**Algebra (al-Khwārizmī):** al-Khw{amacr}rizm{imacr} develops the mathematics that solves problems by using letters for unknowns (variables) and expressing their relationships with equations.
877 C.E.	**Maneuverable glider (Abbas ibn Firnas):** A ten-minute controlled glider flight is first achieved.
9th century	**Gunpowder:** Gunpowder is invented in China.
1034	**Movable type:** Movable type made of baked clay is invented in China.
1170	**Water-raising machines (al-Jazari):** In addition to developing machines that can transport water to higher levels, al-Jazari invents water clocks and automatons.
1260	**Scientific method (Roger Bacon):** Bacon develops rules for explaining how scientists practice science that emphasize empiricism and experimentation over accepted authority.
1284	**Eyeglasses for presbyopia (Salvino d'Armate):** D'Armate is credited with making the first wearable eyeglasses in Italy with convex lenses. These spectacles assist those with farsightedness, such as the elderly.
1439	**Printing press (Johann Gutenberg):** Gutenberg combined a press, oil-based ink, and movable type made from an alloy of lead, zinc, and antimony to create a revolution in printing, allowing mass-produced publications that could be made relatively cheaply and disseminated to people other than the wealthy.
1450	**Eyeglasses for the nearsighted (Nicholas of Cusa):** Correcting nearsightedness requires diverging lenses, which are more difficult to make than convex lenses.
1485	**Dream of flight (Leonardo da Vinci):** On paper, Leonardo designed a parachute, great wings flapped by levers, and also a person-carrying machine with wings to be flapped by the person. Although these flying devices were never successfully realized, the designs introduced the modern quest for aeronautical engineering.
1543	**Copernican (heliocentric) universe:** Copernicus publishes *De revolutionibus* (*On the Revolutions of the Heavenly Spheres*), in which he refutes geocentric Ptolemaic cosmology and proposes that the Sun, not Earth, lies at the center of the then-known universe (the solar system).

1569	**Mercator projection (Gerardus Mercator):** The Mercator projection maps the Earth's surface onto a series of north/south cylinders.
1594	**Logarithms (John Napier):** Napier's logarithms allow the simplification of complex multiplication and division problems.
1595	**Parachute (Faust Veranzio):** Veranzio publishes a book describing sixty new machines, one of which is a design for a parachute that might have worked.
1596	**Flush toilet (Sir John Harington):** Harington's invention is a great boon to those previously assigned to empty the chamber pots.
1604	**Compound microscope (Zacharias Janssen):** Janssen, a lens crafter, experiments with lenses, leading to both the microscope and the telescope.
1607	**Air and clinical thermometers (Santorio Santorio):** Santorio develops a small glass bulb that can be placed in a person's mouth, with a long, thin neck that is placed in a beaker of water. The water rises or falls as the person's temperature changes.
1608	**Refracting telescope (Hans Lippershey):** Lippershey is one of several who can lay claim on developing the early telescope.
1609	**Improved telescope (Galileo Galilei):** Galileo grinds and polishes his own lenses to make a superior telescope. Galileo will come to be known as the father of modern science.
1622	**Slide rule (William Oughtred):** English mathematician and Anglican minister Oughtred invents the slide rule.
1629	**Steam turbine (Giovanni Branca):** Branca publishes a design for a steam turbine, but it requires machining that is too advanced to be built in his day.
1642	**Mechanical calculator (Blaise Pascal):** Eighteen-year-old Pascal invents the first mechanical calculator, which helps his father, a tax collector, count taxes.
1644	**Barometer (Evangelista Torricelli):** Torricelli develops a mercury-filled barometer, in which the height of the mercury in the tube is a measure of atmospheric pressure.
1650	**Vacuum pump (Otto von Guericke):** After demonstrating the existence of a vacuum, von Guericke explores its properties with other experiments.
1651	**Hydraulic press (Blaise Pascal):** Pascal determines that hydraulics can multiply force. For example, a 50-pound force applied to the hydraulic press might exert 500 pounds of force on an object in the press.
1656	**Pendulum clock (Christiaan Huygens):** Huygens discovers that, for small oscillations, a pendulum's period is independent of the size of the pendulum's swing, so it can be used to regulate the speed of a clock.
1662	**Demography (John Graunt):** Englishman Graunt develops the first system of demography and publishes *Natural and Political Observations Mentioned in the Following Index and Made Upon the Bills of Mortality*, which laid the groundwork for census taking.
1663	**Gregorian telescope (James Gregory):** The Gregorian telescope produces upright images and therefore becomes useful as a terrestrial telescope.

1666	**The calculus (Sir Isaac Newton):** Newton (and independently Gottfried Wilhelm Leibniz) develop the calculus in order to calculate the gravitational effect of all of the particles of the Earth on another object such as a person.
1670	**Spiral spring balance watch (Robert Hooke):** Hooke is also credited as the author of the principle that describes the general behavior of springs, known as Hooke's law.
1672	**Leibniz's calculator (Gottfried Wilhelm Leibniz):** Leibniz develops a calculator that can add, subtract, multiply, and divide, as well as the binary system of numbers used by computers today.
1674	**Improvements to the simple microscope (Antoni van Leeuwenhoek):** Leeuwenhoek, a lens grinder, applies his lenses to the simple microscope and uses his microscope to observe tiny protozoa in pond water.
1681	**Canal du Midi opens:** The 150-mile Canal du Midi links Toulouse, France, with the Mediterranean Sea.
1698	**Savery pump (Thomas Savery):** Savery's pump was impractical to build, but it served as a prototype for Thomas Newcomen's steam engine.
1699	**Eddystone Lighthouse (Henry Winstanley):** English merchant Winstanley designs the first lighthouse in England, located in the English Channel fourteen miles off the Plymouth coast. Winstanley is moved to create the lighthouse after two of his ships are wrecked on the Eddystone rocks.
1700	**Piano (Bartolomeo Cristofori):** Cristofori, a harpsichord maker, constructs an instrument with keys that can be used to control the force with which hammers strike the instrument's strings, producing sound that ranges from piano (soft) to forte (loud)—hence the name "pianoforte," later shortened to "piano."
1701	**Tull seed drill (Jethro Tull):** Before the seed drill, seeds were still broadcast by hand.
1709	**Iron ore smelting with coke (Abraham Darby):** Darby develops a method of smelting iron ore by using coke, rather than charcoal, which at the time was becoming scarce. Coke is made by heating coal and driving off the volatiles (which can be captured and used).
1712	**Atmospheric steam engine (Thomas Newcomen):** Newcomen's engine is developed to pump water out of coal mines.
1714	**Mercury thermometer, Fahrenheit temperature scale (Daniel Gabriel Fahrenheit):** Fahrenheit uses mercury in a glass thermometer to measure temperature over the entire range for liquid water.
1718	**Silk preparation:** John Lombe, owner of the Derby Silk Mill in England, patents the machinery that prepared raw silk for the loom.
1729	**Flying shuttle (John Kay):** On a loom, the shuttle carries the horizontal thread (weft or woof) and weaves it between the vertical threads (warp). Kay develops a shuttle that is named "flying" because it is so much faster than previous shuttles.
1738	**Flute Player and Digesting Duck automatons (Jacques de Vaucanson):** De Vaucanson builds cunning, self-operating devices, or automatons (robots) to charm viewers.

1740	**Steelmaking:** Benjamin Huntsman invents the crucible process of making steel.
1742	**Celsius scale (Anders Celsius):** Celsius creates a new scale for his thermometer.
1745-1746	**Leiden jar (Pieter van Musschenbroek and Ewald Georg von Kleist):** Von Kleist (1745) and Musschenbroek (1746) independently develop the Leiden jar, an early type of capacitor used for storing electric charge.
1746	**Clinical trials prove that citrus fruit cures scurvy (James Lind):** Others had suggested citrus fruit as a cure for scurvy, but Lind gives scientific proof. It still will be another fifty years before preventive doses of foods containing vitamin C are routinely provided for British sailors.
1752	**Franklin stove (Benjamin Franklin):** Franklin develops a stove that allows more heat to radiate into a room than go up the chimney.
1752	**Lightning rod (Benjamin Franklin):** Franklin devises a iron-rod apparatus to attach to houses and other structures in order to ground them, preventing damage during lightning storms.
1756	**Wooden striking clock (Benjamin Banneker):** Banneker's all-wood striking clock operates for the next fifty years. Banneker also prints a series of successful scientific almanacs during 1790's.
1757	**Nautical sextant (John Campbell):** When used with celestial tables, Campbell's sextant allows ships to navigate to within sight of their destinations.
1762	**Marine chronometers (John Harrison):** An accurate chronometer was necessary to determine a ship's position at sea, solving the pressing quest for longitude.
1764	**Spinning jenny (James Hargreaves):** Hargreaves develops a machine for spinning several threads at a time, transforming the textile industry and laying a foundation for the Industrial Revolution.
1765	**Improved steam engine (James Watt):** A steam condenser separate from the working pistons make Watt's engine significantly more efficient than Newcomen's engine of 1712.
1767	**Spinning machine (Sir Richard Arkwright):** Arkwright develops a device to spin fibers quickly into consistent, uniform thread.
1767	**Dividing engine (Jesse Ramsden):** Ramsden develops a machine that automatically and accurately marks calibrated scales.
1770	**Steam dray (Nicolas-Joseph Cugnot):** Cugnot builds his three-wheeled fardier à vapeur to move artillery; the prototype pulls 2.5 metric tons at 2 kilometers per hour.
1772	**Soda water (Joseph Priestley):** Priestley creates the first soda water, water charged with carbon dioxide gas. The following year he develops an apparatus for collecting gases by mercury displacement that would otherwise dissolve in water.
1775	**Boring machine (John Wilkinson):** Wilkinson builds the first modern boring machine used for boring holes into cannon, which made cannon manufacture safer. It was later adapted to bore cylinders in steam engines.

1776	**Bushnell's submarine (David Bushnell):** Bushnell builds the first attack submarine; used unsuccessfully against British ships in the Revolutionary War, it nevertheless advances submarine technology.
1779	**Cast-iron bridge:** Abraham Darby III and John Wilkinson build the first cast-iron bridge in England.
1779	**Spinning mule (Samuel Crompton):** Crompton devises the spinning mule, which allows the textile industry to manufacture high-quality thread on a large scale.
1781	**Uranus discovered (Sir William Herschel):** Herschel observes what he first believes to be a comet; further observation establishes it as a planet eighteen times farther from the Sun than the Earth is.
1782	**Hot-air balloon (Étienne-Jacques and Joseph-Michel Montgolfier):** Shaped like an onion dome and carrying people aloft, the Montgolfiers' hot-air balloon fulfills the fantasy of human flight.
1782	**Oil lamp (Aimé Argand):** Argand's oil lamp revolutionizes lighthouse illumination.
1783	**Parachutes (Louis-Sébastien Lenormand):** Lenormand jumps from an observatory tower using his parachute and lands safely.
1783	**Wrought iron (Henry Cort):** Cort converts crude iron into tough malleable wrought iron.
1784	**Improved steam engine (William Murdock):** In an age when much focus was on steam technology, Murdock works to improve steam pumps that remove water from mines. He will go on to invent coal-gas lighting in 1794.
1784	**Bifocals (Benjamin Franklin):** Tired of changing his spectacles to see things at close range as opposed to objects farther away, Franklin designs eyeglasses that incorporate both myopia-correcting and presbyopia-correcting lenses.
1784	**Power loom (Edmund Cartwright):** Cartwright's power loom forms a major advance in the Industrial Revolution.
1785	**Automated flour mill (Oliver Evans):** Evans's flour mill lays the foundation for continuous production lines. In 1801, he will also invent a high-pressure steam engine.
1790	**Steamboat (John Fitch):** Fitch not only invents the steamboat but also proves its practicality by running a steamboat service along the Delaware River.
1792	**Great clock (Thomas Jefferson):** Jefferson's clock, visible and audible both within Monticello and outside, across his plantation, is designed to maintain efficiency. He also invented an improved portable copying press (1785) and will go on to invent an improved ox plow (1794).
1792	**Coal gas (William Murdock):** Murdock develops methods for manufacturing, storing, and purifying coal gas and using it for lighting.
1793	**Cotton gin (Eli Whitney):** Whitney's engine to separate cotton seed from the fiber transformed the American South, both bolstering the institution of slavery and growing the "cotton is king" economy of the Southern states. Five years later, Whitney develops an assembly line for muskets using interchangeable parts.

1793	**Semaphore (Claude Chappe):** Chappe invents the semaphore.
1796	**Smallpox vaccination (Edward Jenner):** Jenner's vaccine will save millions from death, culminating in the eradication of smallpox in 1979.
1796	**Rumford stove (Benjamin Thompson):** The Rumford stove—a large, institutional stove—uses several small fires to heat the stove top uniformly.
1796	**Hydraulic press (Joseph Bramah):** Bramah builds a practical hydraulic press that operates by a high-pressure plunger pump.
1796	**Lithography (Aloys Senefelder):** Senefelder invents lithography and a process for color lithography in 1826.
1799	**Voltaic pile/electric battery (Alessandro Volta):** Volta creates a pile—a stack of alternating copper and zinc disks separated by brine-soaked felt—that supplies a continuous current and sets the stage for the modern electric battery.
1800	**Iron printing press (Charles Stanhope):** Stanhope invents the first printing press made of iron.
1801	**Pattern-weaving loom (Joseph M. Jacquard):** Jacquard invents a loom for pattern weaving.
1804	**Monoplane glider (George Cayley):** Cayley develops a heavier-than-air fixed-wing glider that inaugurates the modern field of aeronautics. Later models carry a man and lead directly to the Wright brothers' airplane.
1804	**Amphibious vehicle (Oliver Evans):** Evans builds the first amphibious vehicle, which is used in Philadelphia to dredge and clean the city's dockyards.
1805	**Electroplating (Luigi Brugnatelli):** Brugnatelli develops the method of electroplating by connecting something to be plated to one pole of a battery (voltaic pile) and a bit of the plating metal to the other pole of the battery, placing both in a suitable solution.
1805	**Morphine (Friedrich Setürner):** Setürner, a German pharmacist, isolates morphine from opium, but it is not widely used for another ten years.
1806	**Steam locomotive (Richard Trevithick):** After James Watt's patent for the steam engine expires in 1800, Trevithick develops a working steam locomotive. By 1806 he has developed his improved steam engine, named the Cornish engine, which sees worldwide dissemination.
1807	**Internal combustion engine (François Isaac de Rivaz):** De Rivaz builds the first vehicle powered by an internal combustion engine.
1807	**Paddle-wheel steamer (Robert Fulton):** Fulton's steamboat becomes far more commercially successful than those of his competitors.
1808	**Law of combining volumes for gases (Joseph-Louis Gay-Lussac):** Gay-Lussac discovers that, when gaseous elements combine to make a compound, the volumes involved are always simple whole-number ratios.
1810	**Preserving food in sealed glass bottles (Nicolas Appert):** Appert answers Napoleon's call to preserve food in a way that allows his soldiers to carry it with them: He processes food in sealed, air-tight glass bottles.

1810	**Preserving food in tin cans (Peter Durand):** Durand follows Nicolas Appert in preserving food for the French army, but he uses tin-coated steel cans in place of breakable bottles.
1815	**Miner's safety lamp (Sir Humphry Davy):** Davy devises a miner's safety lamp in which the flame is surrounded by wire gauze to cool combustion gases so that the mine's methane-air mixture will not be ignited.
1816	**Macadamization (John Loudon McAdam):** McAdam designs a method of paving roads with crushed stone bound with gravel on a base of large stones. The roadway is slightly convex, to shed water.
1816	**Kaleidoscope (Sir David Brewster):** The name for Brewster's kaleidoscope comes from the Greek words *kalos* (beautiful), *eidos* (form), and *scopos* (watcher). "Kaleidoscope," therefore, literally means "beautiful form watcher."
1816	**Stirling engine (Robert Stirling):** The Stirling engine proves to be an efficient engine that uses hot air as a working fluid.
1818	**First photographic images (Joseph Nicéphore Niépce):** Niépce creates the first lasting photographic images.
1819	**Stethoscope (René-Théophile-Hyacinthe Laënnec):** Laënnec invents the stethoscope to avoid the impropriety of placing his ear to the chest of a female heart patient.
1820	**Dry "scouring" (Thomas L. Jennings):** Jennings discovered that turpentine would remove most stains from clothes without the wear associated with washing them in hot water. His method becomes the basis for modern dry cleaning.
1821	**Diffraction grating (Joseph von Fraunhofer):** Von Fraunhofer's diffraction grating separates incident light by color into a rainbow pattern. The various discrete patterns reveal the structure of specific atomic nuclei, making it possible to identify the chemical compositions of various substances.
1821	**Braille alphabet (Louis Braille):** Braille develops a tactile alphabet—a system of raised dots on a surface—that allows the blind to read by touch.
1821	**Electromagnetic rotation (Michael Faraday):** Faraday publishes his work on electromagnetic rotation, which is the principle behind the electric motor.
1822	**Difference engine (Charles Babbage):** Babbage's "engine" was a programmable mechanical device used to calculate the value of a polynomial—a precursor to today's modern computers.
1823	**Waterproof fabric is used in raincoats (Charles Macintosh):** Macintosh patents a waterproof fabric consisting of soluble rubber between two pieces of cloth. Raincoats made of the fabric are still often called mackintoshes (macs), especially in England.
1824	**Astigmatism-correcting lenses (George Biddell Airy):** Airy develops cylindrical lenses that correct astigmatism. An astronomer, Airy will go on to design a method of correcting compasses used in ship navigation and the altazimuth telescope. He becomes England's astronomer royal in 1835.

1825	**Electromagnet (William Sturgeon):** Sturgeon builds a U-shaped, soft iron bar with a coil of varnished copper wire wrapped around it. When a voltaic current is passed through wire, the bar becomes magnetic—the world's first electromagnet.
1825	**Bivalve vaginal speculum (Marie Anne Victoire Boivin):** Boivin develops the tool now widely used by gynecologists in the examination of the vagina and cervix.
1825	**"Steam waggon" (John Stevens):** Stevens builds the first steam locomotive to be manufactured in the United States.
1826	**Color lithography (Aloys Senefelder):** Senefelder invents color lithography.
1827	**Matches (John Walker):** Walker coats the ends of sticks with a mixture of antimony sulfide, potassium chlorate, gum, and starch to produce "strike anywhere" matches.
1827	**Water turbine (Benoît Fourneyron):** Fourneyron builds the first water turbine; it has six horsepower. His larger, more efficient turbines powered many factories during the Industrial Revolution.
1828	**Combine harvester (Samuel Lane):** Patent is granted to Lane for the combine harvester, which combines cutting and threshing.
1829	**Rocket steam locomotive (George Stephenson):** Stephenson builds the world's first railway line to use a steam locomotive.
1829	**Boiler (Marc Seguin):** Seguin improves the steam engine with a multiple firetube boiler.
1829	**Polarizing microscope (William Nicol):** Nicol invents the polarizing microscope, an important forensic tool.
1830	**Steam locomotive (Peter Cooper):** Cooper's four-wheel locomotive with a vertical steam boiler, the *Tom Thumb*, demonstrates the possibilities of steam locomotives and brings Cooper national fame. His other inventions and good management enable Cooper to become a leading industrialist and philanthropist.
1830	**Lawn mower (Edwin B. Budding):** Budding, an English engineer, invents the lawn mower.
1830	**Paraffin (Karl von Reichenbach):** Von Reichenbach, a German chemist, discovers paraffin.
1830	**Creosote (Karl von Reichenbach):** Von Reichenbach distills creosote from beachwood tar. It is used as an insecticide, germicide, and disinfectant.
1831	**Alternating current (AC) generator (Michael Faraday):** Faraday constructs the world's first electric generator.
1831	**Mechanical reaper (Cyrus Hall McCormick):** McCormick's reaper can harvest a field five times faster than earlier methods.
1831	**Staple Bend Tunnel:** The first railroad tunnel in the United States is built in Mineral Point, Pennsylvania.
1832	**Electromagnetic induction (Joseph Henry):** Henry discovers that changing magnetic fields induce voltages in nearby conductors.

1832	**Codeine (Pierre-Jean Robiquet):** French chemist Robiquet isolates codeine from opium. Because of the small amount found in nature, most codeine is synthesized from morphine.
1834	**Hansom cab (Joseph Aloysius Hansom):** English architect Hansom builds the carriage bearing his name.
1835	**Colt revolver (Samuel Colt):** The Colt revolver becomes known as "one of the greatest advances of self-defense in all of human history."
1835	**Photography (Joseph Nicéphore Niépce):** Niépce codevelops photography with Louis-Jacques-Mandé Daguerre.
1836	**Daniell cell (John Frederic Daniell):** Daniell invents the electric battery bearing his name, which is much improved over the voltaic pile.
1836	**Acetylene (Edmund Davy):** Davy creates acetylene by heating potassium carbonate to high temperatures and letting it react to water.
1837	**Electric telegraph (William Fothergill Cooke and Charles Wheatstone):** Wheatstone and Cooke devise a system that uses five pointing needles to indicate alphabetic letters.
1837	**Steam hammer (James Hall Nasmyth):** Nasmyth develops the steam hammer, which he will use to build a pile driver in 1843.
1837	**Steel plow (John Deere):** Previously, plows were made of cast iron and required frequent cleaning. Deere's machine is effective in reducing the amount of clogging farmers experienced when plowing the rich prairie soil.
1837	**Threshing machine (Hiram A. and John A. Pitts):** The Pitts, brothers, develop the first efficient threshing machine.
1838	**Fuel cell (Christian Friedrich Schönbein):** Schönbein's fuel cell might use hydrogen and oxygen and allow them to react, producing water and electricity. There are no moving parts, but the reactants must be continuously supplied.
1838	**Propelling steam vessel (John Ericsson):** Swedish engineer Ericsson invents the double screw propeller for ships allowing them to move much faster than those relying on sails.
1839	**Nitric acid battery (Sir William Robert Grove):** The Grove cell delivered twice the voltage of its more expensive rival, the Daniell cell.
1839	**Daguerreotype (Jacques Daguerre):** Improving on the discoveries of Joseph Nicéphore Niépce, Daguerre develops the first practical photographic process, the Daguerreotype.
1839	**Vulcanized rubber (Charles Goodyear):** Adding sulfur and lead monoxide to rubber, Goodyear processes the batch at a high temperature. The process, later called vulcanization, yields a stable material that does not melt in hot weather or crack in cold.
1840	**Electrical telegraph (Samuel F. B. Morse):** Others had already built telegraph systems, but Morse's system was superior and soon replaced all others.

1841	**Improved electric clock (Alexander Bain):** With John Barwise, Bain develops an electric clock with a pendulum driven by electric impulses to regulate the clock's accuracy.
1841	**First negatives in photography (William Henry Fox Talbot):** Talbot, an English polymath, invents the calotype process, which produces the first photographic negative.
1842	**Commercial fertilizer (John B. Lawes):** Lawes develops superphosphate, the first commercial fertilizer.
1843	**Rotary printing press (Richard March Hoe):** Patented in 1847, the steam-powered rotary press is far faster than the flatbed press.
1843	**Multiple-effect vacuum evaporator (Norbert Rillieux):** Rillieux develops an efficient method for refining sugar using a stack of several pans of sugar syrup in a vacuum chamber, which allows boiling at a lower temperature.
1845	**Suspension bridges (John Augustus Roebling):** A manufacturer of wire cable, Roebling wins a competition for an aqueduct over the Allegheny River and goes on to design other aqueducts and suspension bridges, culminating in the Brooklyn Bridge, which his son, Washington Augustus Roebling, completes in 1883.
1845	**Sewing machine (Elias Howe):** Howe develops a machine that can stitch straight, strong seams faster than those sewn by hand.
1846	**Neptune discovered (John Galle):** German astronomer Galle observes a new planet, based on irregularities in the orbit of Uranus calculated the previous year by England's John Couch Adams and France's Urbain Le Verrier.
1847	**Nitroglycerin (Ascanio Sobrero):** Italian chemist Sobrero creates nitroglycerin.
1847	**Telegraphy applications (Werner Siemens):** Siemens refines a telegraph in which a needle points to the alphabetic letter being sent.
1849	**Laryngoscope (Manuel P. R. Garcia):** Spanish singer and voice teacher, Garcia, known as the father of laryngology, devises the first laryngoscope.
1851	**Foucault's pendulum (Léon Foucault):** Foucault's pendulum proves that Earth rotates.
1851	**Sewing machine (Isaac Merritt Singer):** Singer improves the sewing machine and successfully markets it to women for home use.
1851	**Ophthalmoscope (Hermann von Helmholtz):** Helmholtz invents a device that can be used to examine the retina and the vitreous humor. In 1855, he will invent an ophthalmometer, an instrument that measures the curvature of the eye's lens.
1854	**Kerosene (Abraham Gesner):** Canadian geologist Gesner distills kerosene from petroleum.
1852	**Hypodermic needle (Charles G. Pravaz):** French surgeon Pravaz devises the hypodermic syringe.

1855	**Bunsen burner (Robert Wilhelm Bunsen):** Bunsen—along with Peter Desaga, an instrument maker, and Henry Roscoe, a student—develops a high-temperature laboratory burner, which he and Gustav Kirchhoff use to develop the spectroscope (1859).
1855	**Bessemer process (Sir Henry Bessemer):** Bessemer creates a converter that leads to a process for inexpensively mass-producing steel.
1856	**Synthetic dye (William H. Perkin):** British chemist Perkin produces the first synthetic dye. The color is mauve, which triggers a mauve fashion revolution.
1857	**Safety elevator (Elisha Graves Otis):** Otis's safety elevator automatically stops if the supporting cable breaks.
1858	**Internal combustion engine (Étienne Lenoir):** Lenoir's engine, along with his invention of the spark plug, sets the stage for the modern automobile.
1858	**Transatlantic cable (Lord Kelvin):** Kelvin helps design and install the under-ocean cables for telegraphy between North America and Europe, serving as a chief motivating force in getting the cable completed.
1859	**Signal flares (Martha J. Coston):** Coston's brilliant and long-lasting white, red, and green flares will be adopted by the navies of several nations.
1859	**Lead-acid battery (Gaston Planté):** French physicist Planté invents the lead-acid battery, which led to the invention of the first electric, rechargeable battery.
1860	**Refrigerant (Ferdinand Carré):** French inventor Carré introduces a refrigerator that uses ammonia as a refrigerant.
1860	**Electric incandescent lamp (Joseph Wilson Swan):** Swan produces and patents an incandescent electric bulb; in 1880, two years after Edison's light bulb, Swan will produce a more practical bulb.
1860	**Web rotary printing press (William Bullock):** Bullock's press has an automatic paper feeder, can print on both sides of the paper, cut the paper into sheets, and fold them.
1860	**Henry rifle (Tyler Henry):** American gunsmith Henry designs the Henry rifle, a repeating rifle, the year before the Civil War begins.
1860	**First mail service:** Pony Express opens overland mail service. The service eventually expands to include more than 100 stations, 80 riders, and more than 400 horses.
1861	**Machine gun (Richard Gatling):** Gatling develops the first machine gun, called the Gatling gun. It has six barrels that rotate into place as the operator turns a hand crank; the shells were automatically chambered and fired.
1861	**First color photograph:** Thomas Sutton develops the first color photo based on Scottish physicist James Clerk Maxwell's three-color process.
1861-1862	**USS Monitor (John Ericsson):** Ericsson develops the first practical ironclad ship, which will be use during the Civil War. He goes on to develop a torpedo boat that can fire a cannon from an underwater port.

1862	**Pasteurization (Louis Pasteur):** Pasteur's germ theory of disease leads him to develop a method of applying heat to milk products in order to kill harmful bacteria. He goes on to develop vaccines for rabies, anthrax, and chicken cholera (1867-1885).
1863	**Subway:** The first subway opens in London; it uses steam locomotives. It does not go electric until 1890.
1865	Pioneer **(Pullman) sleeping car (George Mortimer Pullman):** Pullman began working on sleeping cars in 1858, but the *Pioneer* is a luxury car with an innovative folding upper birth to allow the passenger to sleep while traveling.
1866	**Self-propelled torpedo (Robert Whitehead):** English engineer Whitehead develops the modern torpedo.
1866	**Transatlantic telegraph cable:** The first successful transatlantic telegraph cable is laid; it spans 1,686 nautical miles.
1867	**Dynamite (Alfred Nobel):** Nobel mixes clay with nitroglycerin in a one-to-three ratio to create dynamite (Nobel's Safety Powder), an explosive the ignition of which can be controlled using Nobel's own blasting cap. He goes on to patent more than three hundred other inventions and devotes part of the fortune he gained from dynamite to establish and fund the Nobel Prizes.
1867	**Baby formula (Henri Nestlé):** Nestlé combines cow's milk with wheat flour and sugar to produce a substitute for infants whose mothers cannot breast-feed.
1867	**Steam velocipede motorcycle (Sylvester Roper):** Roper spent his lifetime making steam engines lighter and more powerful in order to make his motorized bicycles faster. His velocipede eventually reaches 60 miles per hour.
1867	**Flat-bottom paper bag machine (Margaret E. Knight):** Knight designs a machine that can manufacture flat-bottom paper bags, which can stand open for easy loading.
1867	**Dry-cell battery (Georges Leclanché):** French engineer Lelanche invents the dry-cell battery.
1868	**Typewriter (Christopher Latham Sholes):** American printer Sholes produces the first commercially successful typewriter.
1869	**Periodic table of elements (Dmitry Ivanovich Mendeleyev):** The periodic table, which links chemical properties to atomic structure, will prove to be one of the great achievements of the human race.
1869	**Air brakes for trains (George Westinghouse):** In 1867, Westinghouse developed a signaling system for trains. The air brake makes it easier and safer to stop large, heavy, high-speed trains.
1869	**Transcontinental railroad:** The United States transcontinental railroad is completed.
1869	**Celluloid (John Wesley Hyatt):** American inventor Hyatt produces celluloid, the first commercially successful plastic, by mixing solid pyroxylin and camphor.
1869	**Suez Canal opens:** The canal, 101 miles long, took a decade to build and connects the Red Sea with the eastern Mediterranean Sea.

1871	**Fireman's respirator (John Tyndall):** The respirator grows from Tyndall's studies of air pollution.
1871	**Commercial generator (Zénobe T. Gramme):** Belgian electrical engineer Gramme builds the Gramme machine, the first practical commercial generator for producing alternating current.
1872	**Blue jeans (Levi Strauss):** Miners tore their pockets when they stuffed too many ore samples in them. Strauss makes pants using heavy-duty material with riveted pocket corners so they will not tear out.
1872	**Burbank russet potato (Luther Burbank):** Burbank breeds all types of plants, using natural selection and grafting techniques to achieve new varieties. His Burbank potato, developed from a rare russet potato seed pod, grows better than other varieties.
1872	**Automatic lubricator (Elijah McCoy):** McCoy uses steam pressure to force oil to lubricate the pistons of steam engines.
1872	**Vaseline (Robert A. Chesebrough):** Chesebrough, an American chemist, patents his process for making petroleum jelly and calls it Vaseline.
1873	**QWERTY keyboard (Christopher Latham Sholes):** After patenting the first practical typewriter, Sholes develops the QWERTY keyboard, designed to slow the fastest typists, who otherwise jammed the keys. The basic QWERTY design remains the standard on most computer keyboards.
1874	**Barbed wire (Joseph Farwell Glidden):** An American farmer, Glidden invents and patents barbed wire. Barbed-wire fences make farming and ranching of the Great Plains practical. Without effective fences, animals wandered off and crops were destroyed. At the time of his death in 1906, Glidden is one of the richest men in the country.
1874	**Medical nuclear magnetic resonance imaging (Raymond Damadian):** Damadian and others develop magnetic resonance imaging (MRI) for use in medicine.
1876	**Four-stroke internal combustion engine (Nikolaus August Otto):** In order to deliver more horsepower, Otto's engine compresses the air-fuel mixture. His previous engines operated near atmospheric pressure.
1876	**Ammonia-compressor refrigeration machine (Carl von Linde):** Breweries need refrigeration so they can brew year-round. Linde refines his ammonia-cycle refrigerator to make this possible.
1876	**Telephone (Elisha Gray):** Gray files for a patent for the telephone the same day that Alexander Graham Bell does so. While the case is not clear-cut, and Gray fought with Bell for years over the patent rights, Bell is generally credited with the telephone's invention.
1877	**Phonograph (Thomas Alva Edison):** Edison invents the phonograph—an unexpected outcome of his telephone research.
1878	**First practical lightbulb (Thomas Alva Edison):** Twenty-two people have invented lightbulbs before Edison and Joseph Swan, but they are impractical. Edison's is the first to be commercially viable. Eventually, Swan's company merges with Edison's.

1878	**Loose-contact carbon microphone (David Edward Hughes):** Hughes's carbon microphone advances telephone technology. In 1879, he will invent the induction balance, which will be used in metal detectors.
1878	**Color photography (Frederic Eugene Ives):** American inventor Ives develops the halftone process for printing photographs.
1879	**Saccharin (Ira Remsen):** Remsen synthesizes a compound that is up to three hundred times sweeter than sugar; he also establishes the important *American Chemical Journal*, serving as its editor until 1915.
1880	**Milne seismograph (John Milne):** Milne invents the first modern seismograph for measuring earth tremors. He will come to be called the father of modern seismology.
1881	**Improved incandescent lightbulb (Lewis Howard Latimer):** Latimer develops an improved way to manufacture and to attach carbon filaments in lightbulbs.
1881	**Sphygmomanometer (Karl Samuel Ritter von Basch):** Von Basch invents the first blood pressure gauge.
1882	**Induction motor (Nikola Tesla):** Tesla's theories and inventions make alternating current (AC) practical.
1882	**Two-cycle gasoline engine (Gottlieb Daimler):** Daimler builds a small, high-speed two-cycle gasoline engine. He will also build a successful motorcycle in 1885 and (with Wilhelm Maybach) an automobile in 1889.
1883	**Solar cell (Charles Fritts):** American scientist Fritts designs the first solar cell.
1883	**Shoe-lasting machine (Jan Ernst Matzeliger):** The machine sews the upper part of the shoe to the sole and reduces the cost of shoes by 50 percent.
1884	**Fountain pen (Lewis Waterman):** The commonly told story is that Waterman was selling insurance and lost a large contract when his pen leaked all over it, prompting him to invent the leak-proof fountain pen.
1884	**Vector calculus (Oliver Heaviside):** Heaviside develops vector calculus to represent James Clerk Maxwell's electromagnetic theory with only four equations instead of the usual twenty.
1884	**Roll film (George Eastman):** Roll film will replace heavy plates, making photography both more accessible and more convenient. In 1888, Eastman and William Hall invent the Kodak camera. These developments open photography to the masses.
1884	**Roll film (George Eastman):** Roll film will replace heavy plates, making photography both more accessible and more convenient. In 1888, Eastman and William Hall invent the Kodak camera. These developments open photography to the masses.
1884	**Steam turbine (Charles Parsons):** Designed for ships, Parsons's steam turbine is smaller, more efficient, and more durable than the steam engines in use.
1884	**Census tabulating machine (Herman Hollerith):** Hollerith's machine uses punch cards to tabulate 1890 census data. He goes on to found the company that later becomes International Business Machines (IBM).

1885	**Machine gun (Hiram Stevens Maxim):** Maxim patents a machine gun that can fire up to six hundred bullets per minute.
1885	**Bicycle (John Kemp Starley):** English inventor Starley is responsible for producing the first modern bicycle, called the Rover.
1885	**First gasoline-powered automobile (Carl Benz):** Benz not only manufactures the first gas-powered car but also is first to mass-produce automobiles.
1885	**Incandescent gas mantle (Carl Auer von Welsbach):** The Austrian scientist invents the incandescent gas mantle.
1886	**Dictaphone (Charles Sumner Tainter):** Tainter, an American engineer who frequently worked with Alexander Graham Bell, designs the Dictaphone.
1886	**Dishwasher (Josephine Garis Cochran):** Like modern washers, Cochran's dishwasher cleans dishes with sprays of hot, soapy water and then air-dries them.
1886	**Electric transformer (William Stanley):** Stanley, working at Westinghouse, builds the first practical electric transformer.
1886	**Gramophone (Emile Berliner):** A major contribution to the music recording industry, Berliner's gramophone uses flat record discs for recording sound. Berliner goes on to produce a helicopter prototype (1906-1923).
1886	**Linotype machine (Ottmar Mergenthaler):** Pressing keys on the machine's keyboard releases letter molds that drop into the current line. The lines are assembled into a page and then filled with molten lead.
1886	**Electric-traction system (Frank J. Sprague):** Sprague's motor can propel a tram up a steep hill without its slipping.
1886	**Hall-Héroult electrolytic process (Charles Martin Hall and Paul Héroult):** The industrial production of aluminum from bauxite ore made aluminum widely available. Prior to the electrolytic process, aluminum was a precious metal with a value about equal to that of silver.
1886	**Coca-Cola (John Stith Pemberton):** Developed as pain reliever less addictive than available opiates, the original Coca-Cola contains cocaine from cola leaves and caffeine from kola nuts. It achieves greater success as a beverage marketed where alcohol is prohibited.
1886	**Yellow pages (Reuben H. Donnelly):** Yellow paper was used in 1883 when the printer ran out of white paper. Donnelly now purposely uses yellow paper for business listings.
1886	**Fluorine (Henri Moissan):** French chemist Moissan isolates fluorine and is awarded the Nobel Prize in Chemistry in 1906. Compounds of fluorine are used in toothpaste and in public water supplies to help prevent tooth decay.
1887	**Radio transmitter and receiver (Heinrich Hertz):** Hertz will use these devices to discover radio waves and confirm that they are electromagnetic waves that travel at the speed of light; he also discovers the photoelectric effect.
1887	**Distortionless transmission lines (Oliver Heaviside):** Heaviside recommends that induction coils be added to telephone and telegraph lines to correct for distortion.

1887	**Olds horseless carriage (Ransom Eli Olds):** Olds develops a three-wheel horseless carriage using a steam engine powered by a gasoline burner.
1887	**Synchronous multiplex railway telegraph (Granville T. Woods):** Woods patents a variation of the induction telegraph that allows messages to be sent between moving trains and between trains and railway stations. He will eventually obtain sixty patents on electrical and electromechanical devices, most of them related to railroads and communications.
1888	**Cordite (Sir James Dewar):** Dewar, with Sir Frederick Abel, invents cordite, a smokeless gunpowder that is widely adopted for munitions.
1888	**Pneumatic rubber tire (John Boyd Dunlop):** Dunlop's pneumatic tires revolutionize the ride for cyclists and motorists.
1888	**Kodak camera:** George Eastman, founder of Eastman Kodak, introduces the first Kodak camera.
1889	**Electric drill (Arthur James Arnot):** Arnot's drill is used to cut holes in rock and coal.
1889	**Bromine extraction (Herbert Henry Dow):** Dow's method for extracting bromine from brine enables bromine to be widely used in medicines and in photography.
1889	**Rayon (Louis-Marie-Hilaire Bernigaud de Chardonnet):** Bernigaud de Chardonnet, a French chemist, invents rayon, the first artificial fiber, as an alternative to silk.
1889	**Celluloid film:** George Eastman replaces paper film with celluloid.
1890	**Improved carbon electric arc (Hertha Marks Ayrton):** The carbon arc produces an intense light that is used in streetlights.
1890	**Pneumatic (air) hammer (Charles B. King):** A worker with a pneumatic hammer can break up a concrete slab many times faster than can a worker armed with only a sledgehammer.
1890	**Smokeless gunpowder (Hudson Maxim):** Maxim (perhaps with brother Hiram) develops a version of smokeless gunpowder that is adopted for modern firearms; he goes on to develop a smokeless cannon powder that will be used during World War I.
1890	**Rubber gloves in the operating room:** American surgeon William Stewart Halsted introduces the use of sterile rubber gloves in the operating room.
1891	**Rubber automobile tires (André and Édouard Michelin):** The Michelin brothers manufacture air-inflated tires for bicycles and later automobiles, which leads to a successful ad campaign, featuring the Michelin Man (Bibendum).
1891	**Carborundum (Edward Goodrich Acheson):** Attempting to create artificial diamonds, Acheson instead synthesizes silicon carbide, the second hardest substance known. He will develop an improved graphite-making process in 1896.
1892	**Kinetoscope (Thomas Alva Edison):** Edison completes Kinetoscope; the first demonstration is held a year later.
1892	**Calculator (William Seward Burroughs):** Burroughs builds the first practical key-operated calculator; it prints entries and results.

1892	**Dewar flask (Sir James Dewar):** Dewar invents the vacuum bottle, a vacuum-jacketed vessel for storing and maintaining the temperature of hot or cold liquids.
1892	**Artificial silk (Charles F. Cross and Edward J. Bevan):** British chemists Cross and Bevan create viscose artificial silk (cellulose acetate).
1893	**Color photography plate (Gabriel Jonas Lippmann):** Also known as the Lippmann plate for its inventor, the color photography plate uses interference patterns, rather than various colored dyes, to reproduce authentic color.
1893	**Alternating current calculations (Charles Proteus Steinmetz):** Steinmetz's calculations make it possible for engineers to determine alternating current reliably, without depending on trial and error, when designing a new motor.
1894	**Cereal flakes (John Harvey Kellogg):** Kellogg, a health reformer who advocates a diet of fruit, nuts, and whole grains, invents flaked breakfast cereal with the help of his brother, Will Keith Kellogg. In 1906 Kellogg established a company in Battle Creek, Michigan, to manufacture his breakfast cereal.
1894	**Automatic loom (James Henry Northrop):** Northrop builds the first automatic loom.
1895	**Streamline Aerocycle bicycle (Ignaz Schwinn):** Through hard work and dedication, Schwinn develops a bicycle that eventually makes his name synonymous with best of bicycles.
1895	**Victrola phonographs (Eldridge R. Johnson):** Johnson develops a spring-driven motor for phonographs that provides the constant record speed necessary for good sound reproduction.
1895	**Cinématographe (Auguste and Louis Lumière):** The Lumière brothers' combined motion-picture camera, printer, and projector helps establish the movie business. Using a very fine-grained silver-halide gelatin emulsion, they cut photographic exposure time down to about one minute.
1895	**Antenna:** Aleksandr Stepanovich Popov demonstrated radio reception with a coherer, which he also used as a lightning detector.
1896	**Wireless telegraph system (Guglielmo Marconi):** Marconi is the first to send wireless signals across the Atlantic Ocean, inaugurating a new era of telecommunications.
1896	**Aerodromes (Samuel Pierpont Langley):** Langley's "Aerodrome number 6," using a small gasoline engine, makes an unmanned flight of forty-eight hundred feet.
1896	**Four-wheel horseless carriage (Ransom Eli Olds):** Oldsmobile patents Olds's internal combustion engine and applies it to his four-wheel horseless carriage, naming it the "automobile."
1896	**High-frequency generator and transformer (Elihu Thomson):** Thomson produces an electric air drill, which advances welding to improve the construction of new appliances and vehicles. He will also invent other electrical devices, including an improved X-ray tube.

1896	**X-ray tube (Wilhelm Conrad Röntgen):** After discovering X radiation, Röntgen mails an X-ray image of a hand wearing a ring and paving the way for the medical use of X-ray imaging—one of the most important discoveries ever made for medical science.
1896	**Better sphygmomanometer (Scipione Riva-Rocci):** Italian pediatrician Riva-Rocci develops the most successful and easy-to-use blood-pressure gauge.
1897	**Modern submarine (John Philip Holland):** Holland's submarine is the first to use a gasoline engine on the surface and an electric engine when submerged.
1897	**Oscilloscope (Karl Ferdinand Braun):** The oscilloscope is an invaluable device used to measure and display electronic waveforms.
1897	**Jenny coupler (Andrew Jackson Beard):** Beard's automatic coupler connects the cars in a train without risking human life. The introduction of automatic couplings reduces coupling-related injuries by a factor of five.
1897	**Escalator (Charles Seeberger):** Before Seeberger built the escalator in its now-familiar form, it was a novelty ride at the Coney Island amusement park.
1897	**Automobile components (Alexander Winton):** The Winton Motor Carriage Company is incorporated, and Winton begins manufacturing automobiles. His popular "reliability runs" helps advertise automobiles to the American market. He will produce the first American diesel engine in 1913.
1897	**Diesel engine (Rudolf Diesel):** Diesel's internal combustion engine rivals the efficiency of the steam engine.
1897	**Electron discovered (J. J. Thomson):** Thomson uses an evacuated tube with a high voltage across electrodes sealed in the ends. Invisible particles (later named electrons) stream from one of the electrodes, and Thomson establishes the particles' properties.
1898	**Flashlight (Conrad Hubert):** Hubert combines three parts—a battery, a light-bulb, and a metal tube—to produce a flashlight.
1898	**Mercury vapor lamp (Peter Cooper Hewitt):** Hewitt's mercury vapor lamp proves to be more efficient than incandescent lamps.
1899	**Alpha particle discovered (Ernest Rutherford):** Rutherford detects the emission of helium 4 nuclei (alpha particles) in the natural radiation from uranium.
1900	**Aspirin:** Aspirin is patented by Bayer and sold as a powder. In 1915 it is sold in tablets.
1900	**Dirigibles (Ferdinand von Zeppelin):** Von Zeppelin flies his airship three years before the Wright brothers' airplane.
1900	**Gamma ray discovered (Paul Villard):** Villard discovers gamma rays in the natural radiation from uranium. They resemble very high-energy X rays.
1900	**Brownie camera:** George Eastman introduces the Kodak Brownie camera. It is sold for $1 and the film it uses costs 15 cents. The Brownie made photography an accessible hobby to almost everyone.
1901	**Acousticon hearing aid (Miller Reese Hutchison):** Hutchison invents a battery-powered hearing aid in the hopes of helping a mute friend speak.

1901	**Vacuum cleaner (H. Cecil Booth):** Booth patents his vacuum cleaner, a machine that sucks in and traps dirt. Previous devices, less effective, had attempted to blow the dirt away.
1901	**String galvanometer (electrocardiograph) (Willem Einthoven):** Einthoven's device passes tiny currents from the heart through a silver-coated silicon fiber, causing the fiber to move. Recordings of this movement can show the heart's condition.
1901	**Silicone (Frederick Stanley Kipping):** English chemist Kipping studies the organic compounds of silicon and coins the term "silicone."
1902	**Airplane engine (Charles E. Taylor):** Taylor begins building engines for the Wright brothers' airplanes.
1902	**Lionel electric toy trains (Joshua Lionel Cowen):** Cowen publishes the first Lionel toy train catalog. Lionel miniature trains and train sets become favorite toys for many years and are prized by collectors to this day.
1902	**Air conditioner (Willis Carrier):** Whole-house air-conditioning becomes possible.
1903	**Windshield wipers (Mary Anderson):** At first, the driver operated the wiper with a lever from inside the car.
1903	**Wright Flyer (Wilbur and Orville Wright):** The Wright Flyer is the first heavier-than-air machine to solve the problems of lift, propulsion, and steering for controlled flight.
1903	**Safety razor with disposable blade (King Camp Gillette):** Gillette's razor used a disposable and relatively cheap blade, so there was no need to sharpen it.
1903	**Space-traveling projectiles (Konstantin Tsiolkovsky):** Tsiolkovsky publishes "The Exploration of Cosmic Space by Means of Reaction-Propelled Apparatus," in which he includes an equation for calculating escape velocity (the speed required to propel an object beyond Earth's field of gravity). He is also recognized for the concept of rocket propulsion and for the wind tunnel.
1903	**Ultramicroscope (Richard Zsigmondy):** Zsigmondy builds the ultramicroscope to study colloids, mixtures in which particles of a substance are dispersed throughout another substance.
1903	**Spinthariscope (Sir William Crookes):** Crookes invents a device that sparkles when it detects radiation. He also develops and experiments with the vacuum tube, allowing later physicists to identify alpha and beta particles and X rays in the radiation from uranium.
1903	**Crayola crayons (Edwin Binney):** With his cousin C. Harold Smith, Binney invents dustless chalk and crayons marketed under the trade name Crayolas.
1903	**Motorcycle:** Harley-Davidson produces the first motorcycle, built to be a racer.
1903	**Electric iron:** Earl Richardson introduces the lightweight electric iron.
1904	**Glass bottle machine:** American inventor Michael Joseph Owens designs a machine that produces glass bottles automatically.

1905	**Novocaine (Alfred Einkorn):** While researching a safe local anesthetic to use on soldiers, German chemist Einkorn develops novocaine, which becomes a popular dental anesthetic.
1905	**Special relativity (Albert Einstein):** At the age of twenty-six, Einstein uses the constancy of the speed of light to explain motion, time, and space beyond Newtonian principles. During the same year, he publishes papers describing the photoelectric effect and Brownian motion.
1905	**Intelligence testing:** French psychologist Alfred Binet devises the first of a series of tests to measure an individual's innate ability to think and reason.
1906	**Hair-care products (Madam C. J. Walker):** Walker trains a successful sales force to go door-to-door and sell directly to women. Her saleswomen, beautifully dressed and coiffed, are instructed to pamper their clients.
1906	**Broadcast radio (Reginald Aubrey Fessenden):** In broadcast radio, sound wave forms are added to a carrier wave and then broadcast. The carrier wave is subtracted at the receiver leaving only the sound.
1906	**Klaxon horn (Miller Reese Hutchison):** Hutchison files a patent application for the electric automobile horn.
1906	**Chromatography (Mikhail Semenovich Tswett):** Tswett, a Russian botanist, invents chromatography.
1906	**Freeze-drying (Jacques Arsène d'Arsonval and George Bordas):** D'Arsonval and Bordas invent freeze-drying, but the practice is not commercially developed until after World War II.
1907	**Sun valve (Nils Gustaf Dalén):** Dalén's device uses sunlight to activate a lighthouse beacon. His other inventions make automated acetylene beacons in lighthouses possible.
1907	**Mantoux tuberculin skin test (Charles Mantoux):** French physician Mantoux develops a skin-reaction test to diagnose tuberculosis. He builds on the work of Robert Koch and Clemens von Pirquet.
1908	**Helium liquefaction (Heike Kamerlingh Onnes):** Kamerlingh Onnes produces liquid helium at a temperature of about 4 kelvins. He will also discover superconductivity in several materials cooled to liquid helium temperature.
1908	**"Tin Lizzie" (Model T) automobile (Henry Ford):** Ford's development of an affordable automobile, manufactured using his assembly-line production methods, revolutionize the U.S. car industry.
1908	**Electrostatic precipitator (Frederick Gardner Cottrell):** The electrostatic precipitator is invaluable for cleaning stack emissions.
1908	**Geiger-Müller tube (Hans Geiger):** Geiger invents a device, popularly called the Geiger counter, that is a reliable, portable radiation detector. Later his student Walther Müller helps improve the instrument.
1908	**Vacuum cleaner (James Murray Spangler):** Spangler receives a patent on his electric sweeper, and his Electric Suction Sweeper Company eventually becomes the Hoover Company, the largest such company in the world.

1908	**Cellophane (Jacques Edwin Brandenberger):** Brandenberger builds a machine to mass-produce cellophane, which he has earlier synthesized while unsuccessfully attempting to develop a stain-resistant cloth.
1908	**Water treatment:** Chlorine is used to purify water for the first time in the United States, in New Jersey, helping to reduce waterborne illnesses such as cholera, typhoid, and dysentery.
1908	**Audion (Lee De Forest):** De Forest invents a vacuum tube used in sound amplification. In 1922, he will develop talking motion pictures, in which the sound track is imprinted on the film with the pictures, instead of on a record to be played with the film, leading to exact synchronization of sound and image.
1909	**Synthetic fertilizers (Fritz Haber):** Haber also invents the Haber process to synthesize ammonia on a small scale.
1909	**Maxim silencer (Hiram Percy Maxim):** The silencer reduces the noise from firing the Maxim machine gun.
1909	**pH scale:** Danish chemist Søren Sørensen introduces the pH scale as a standard measure of alkalinity and acidity.
1910	**Chlorinator (Carl Rogers Darnall):** Major Darnall builds a machine to add liquid chlorine to water to purify it for his troops. His method is still widely used today.
1910	**Bakelite (Leo Hendrik Baekeland):** Bakelite is the first tough, durable plastic.
1910	**Neon lighting (Georges Claude):** Brightly glowing neon tubes revolutionize advertising displays.
1910	**Syphilis treatment:** German physician Paul Ehrlich and Japanese physician Hata Sahachiró discover the effective treatment of arsphenamine (named Salvarsan by Ehrlich) for syphilis.
1911	**Colt .45 automatic pistol (John Moses Browning):** Commonly called the Colt Model 1911, an improved version of the Colt Model 1900, the Colt .45 is the first autoloading pistol produced in America. Among Browning's other inventions are the Winchester 94 lever-action rifle and the gas-operated Colt-Browning machine gun.
1911	**Gyrocompass (Elmer Ambrose Sperry):** Sperry receives a patent for a nonmagnetic compass that indicates true north.
1911	**Atomic nucleus identified (Ernest Rutherford):** Rutherford discovered the nucleus by bombarding a thin gold foil with alpha particles. Some were deflected through large angles showing that something small and hard was present.
1911	**Ductile tungsten (William David Coolidge):** Coolidge also invented the Coolidge tube, an improved X-ray producing tube.
1911	**Ochoaplane (Victor Leaton Ochoa):** In addition to inventing this plane with collapsible wings, Ochoa also developed an electricity-generating windmill.
1911	**Automobile electric ignition system (Charles F. Kettering):** Kettering invents the first electric ignition system for cars.
1912	**Automatic traffic signal system (Garrett Augustus Morgan):** Morgan also invents a safety hood that served as a rudimentary gas mask.

1913	**Gyrostabilizer (Elmer Ambrose Sperry):** Sperry develops the gyrostabilizer, a device to control the roll, pitch, and yaw of a moving ship. He will go on to invent the flying bomb, which is guided by a gyrostabilizer and by radio control.
1913	**Erector set (Alfred C. Gilbert):** Erector sets provide hands-on engineering experience for countless children.
1913	**Zipper (Gideon Sundback):** While others had made zipper-like devices but had never successfully marketed them, Sundback designs a zipper in approximately its present form. He also invents a machine to make zippers.
1913	**Improved electric lightbulb (Irving Langmuir):** Langmuir fills his lightbulb with a low-pressure inert gas to retard evaporation from the tungsten filament.
1913	**Industrialization of the Haber process (Carl Bosch):** Bosch scales up Haber's process for making ammonia to an industrial capacity. The process comes to be known as the Haber-Bosch process.
1913	**Bergius process (Friedrich Bergius):** Bergius develops high-pressure, high-temperature process to produce liquid fuel from coal.
1913	**Electric dishwasher:** The Walker brothers of Philadelphia produce the first electric dishwasher.
1913	**Stainless steel (Harry Brearley):** Brearley invents stainless steel.
1913	**Thermal cracking (William Burton and Robert Humphreys):** Standard Oil chemical engineers Burton and Humphreys discover thermal cracking, a method of oil refining that significantly increases gasoline yields.
1914	**Backless brassiere (Caresse Crosby):** The design of a new women's undergarment leads to the expansion of the U.S. brassiere industry. Caresse was originally a marketing name that Mary Phelps Jacob eventually adopted as her own.
1915	**Panama Canal opens:** The passageway between the Atlantic and Pacific oceans creates a boon for the shipping industry.
1915	**General relativity (Albert Einstein):** Einstein refines his 1905 theory of relativity (now called special relativity) to describe the theory that states that uniform accelerations are almost indistinguishable from gravity. Einstein's theory provides the basis for physicists' best understanding of gravity and of the framework of the universe.
1915	**Jenny (Glenn H. Curtiss):** The Jenny becomes a widely used World War I biplane, and Curtis becomes a general manufacturer of airplanes and airplane engines.
1915	**Pyrex:** Corning's brand name for glassware is introduced.
1915	**Warfare:** Depth-charge bombs are first used by the Allies against German submarines.
1916	**By-products of sweet potatoes and peanuts (George Washington Carver):** Carver publishes his famous bulletin on 105 ways to prepare peanuts.
1919	**Proton discovered (Ernest Rutherford):** After bombarding nitrogen gas with alpha particles (helium 4 nuclei), Rutherford observes that positive particles with a single charge are knocked loose. They are protons.
1919	**Toaster (Charles Strite):** Strite invents the first pop-up toaster.

1920	**Microelectrode (Ida H. Hyde):** Hyde's electrode is small enough to pierce a single cell. Chemicals can also be very accurately deposited by the microprobe.
1921	**Ready-made bandages:** Johnson & Johnson puts Band-Aids on the market.
1921	**Antiknock solution (Thomas Midgley, Jr.):** While working at a General Motors subsidiary, American mechanical engineer Midgley develops an antiknock solution for gasoline.
1921	**Insulin:** University of Toronto researchers Frederick Banting, J. J. R. Macleod, and Charles Best first extract insulin from a dog, and the first diabetic patient is treated with purified insulin the following year. Banting and Macleod win the 1923 Nobel Prize in Physiology or Medicine for their discovery of insulin.
1923	**Improved telephone speaker (Georg von Békésy):** Békésy's studies of the human ear lead to an improved telephone earpiece. He will also construct a working model of the inner ear.
1923	**Quick freezing (Clarence Birdseye):** Birdseye's quick-freezing process preserves food's flavor and texture better than previously used processes.
1924	**Coincidence method of particle detection (Walther Bothe):** Bothe's method proves invaluable in the use of gamma rays to discover nuclear energy levels.
1924	**Ultracentrifuge (Theodor Svedberg):** Svedberg's ultracentrifuge can separate isotopes, such as uranium 235 from uranium 238, from each other—a critical step in building the simplest kind of atomic bomb.
1924	**EEG:** German scientist Hans Berger records the first human electroencephalogram (EEG), which shows electrical patterns in the brain.
1925	**Leica I camera:** Leitz introduces the first 35-millimeter Leica camera at the Leipzig Spring Fair.
1925	**First U.S. television broadcast:** Charles Francis Jenkins transmits the silhouette image of a toy windmill.
1926	**Automatic power loom (Sakichi Toyoda):** Toyoda's loom helps Japan catch up with the western Industrial Revolution.
1926	**Liquid-fueled rocket (Robert H. Goddard):** A solid-fueled rocket is either on or off, but a liquid-fueled rocket can be throttled up or back and can be shut off before all the fuel is expended.
1927	**Aerosol can (Erik Rotheim):** Norwegian engineer Rotheim patents the aerosol can and valve.
1927	**Adiabatic demagnetization (William Francis Giauque):** Adiabatic demagnetization is part of a refrigeration cycle that, when used enough times, can chill a small sample to within a fraction of a kelvin above absolute zero.
1927	**All-electronic television (Philo T. Farnsworth):** Farnsworth transmits the first all-electronic television image using his newly developed camera vacuum tube, known as the image dissector. Previous systems combined electronics with mechanical scanners.
1927	**First flight across the Atlantic:** Charles Lindbergh flies the Spirit of St. Louis across the Atlantic. He is the first to make a solo, nonstop flight across the ocean.

1927	**Iron lung (Philip Drinker):** Drinker, a Harvard medical researcher, assisted by Louis Agassiz Shaw, devises the first modern practical respirator using an iron box and two vacuum cleaners. Drinker calls the device the iron lung.
1927	**Garbage disposal (John W. Hammes):** American architect Hammes develops the first garbage disposal to make cleaning up the kitchen easier for his wife. It is nicknamed the "electric pig" when it first goes on the market.
1927	**Adjustable-temperature iron:** The Silex Company begins to sell the first iron with an adjustable temperature control.
1927	**Analogue computer (Vannevar Bush):** Bush builds the first analogue computer. He is also the first person to describe the idea of hypertext.
1928	**Sliced bread (Otto F. Rohweddeer):** Bread that came presliced was advertised as "the greatest forward step in the baking industry since bread was wrapped." Today the phrase "the greatest thing since sliced bread" is used to describe any innovation that has a broad, positive impact on daily life.
1928	**First television programs:** First regularly scheduled television programs in the United States air. They are produced out of a small, experimental station in Wheaton, Maryland.
1928	**Link Trainer (Edwin Albert Link):** Link's flight simulator created realistic conditions in which to train pilots without the expense or risk of an actual air flight. Link also developed a submersible decompression chamber.
1928	**New punch card:** IBM introduces a new punch card that has rectangular holes and eight columns.
1928	**Radio network:** NBC establishes the first coast-to-coast radio network in the United States.
1928	**Pap smear (George N. Panpanicolaou):** Greek cytopathologist Panpanicolaou patents the pap smear, a test that helps detect uterine cancer.
1928	**Portable offshore drilling (Louis Giliasso):** Giliasso creates an efficient portable method of offshore drilling by mounting a derrick and drilling outfit onto a submersible barge.
1929	**Iconoscope (Vladimir Zworykin):** Zworykin claims that he, not Philo T. Farnsworth, should be credited with the invention of television.
1929	**Strobe light (Harold E. Edgerton):** Edgerton's strobe is used as a flash bulb. He pioneers the development of high-speed photography.
1929	**Dymaxion products (R. Buckminster Fuller):** Fuller's "Dymaxion" products feature an energy-efficient house using prefabricated, easily shipped parts.
1929	**Van de Graaff generator (Robert Jemison van de Graaff):** Van de Graaff invents the Van de Graaff generator, which accumulates electric charge on a moving belt and deposits it in a hollow glass sphere at the top.

1929-1936	**Cyclotron (Ernest Orlando Lawrence and M. Stanley Livingston):** Lawrence and Livingston are studying particle accelerators and develop the cyclotron, which consists of a vacuum tank between the poles of a large magnet. Alternating electric fields inside the tank can accelerate charged particles to high speeds. The cyclotron is used to probe the atomic nucleus or to make new isotopes of an element, including those used in medicine.
1930	**Schmidt telescope (Bernhard Voldemar Schmidt):** Schmidt's telescope uses a spherical main mirror and a correcting lens at the front of the scope. It can photograph large fields with little distortion.
1930	**Pluto discovered (Clyde Tombaugh):** Tombaugh observes a body one-fifth the mass of Earth's moon. Pluto comes to be regarded as the ninth planet of the solar system, but in 2006 it is reclassified as one of the largest-known Kuiper Belt objects, a dwarf planet.
1930	**Freon refrigeration and air-conditioning (Charles F. Kettering):** After inventing an electric starter in 1912 and the Kettering Aerial Torpedo in 1918 (the world's first cruise missile), Kettering and Thomas Midgley, Jr., use Freon gas in their cooling technology. (Freon will later be banned because of the effects of chlorofluorocarbons on Earth's ozone layer.)
1930	**Synthetic rubber (Wallace Hume Carothers):** Carothers synthesizes rubber and goes on to develop nylon in 1935. His work professionalizes polymer chemistry as a scientific field.
1930	**Scotch tape (Richard G. Drew):** After inventing masking tape, Drew invents the first waterproof, see-through, pressure-sensitive tape that also acted as a barrier to moisture.
1930	**Military and commercial aircraft (Andrei Nikolayevich Tupolev):** Tupolev emerges as one of the world's leading designers of military and civilian aircraft. His aircraft set nearly eighty world records.
1930's	**Washing machine (John W. Chamberlain):** Chamberlain invents a washing machine that enables clothes to be washed, rinsed, and have the water extracted from them in a single operation.
1931	**Electric razor (Jacob Schick):** Schick introduces his first electric razor, which allows dry shaving. It has a magazine of blades held in the handle.
1931	**Radio astronomy (Karl G. Jansky):** One of the founders of the field of radio astronomy, Janksy detects radio static coming from the Milky Way's center.
1932	**Positron discovered (Carl D. Anderson):** Anderson discovers the positron, a positive electron and an element of antimatter.
1932	**Neoprene (Julius Nieuwland):** The first synthetic rubber is marketed.
1932	**Neutron discovered (James Chadwick):** Chadwick detects the neutron, an atomic particle with no charge and a mass only slightly greater than that of a proton. Except for hydrogen 1, the atomic nuclei of all elements consist of neutrons and protons.

1932	**Phillips-head screw (Henry M. Phillips):** The Phillips-head screw has an X-shaped slot in the head and can withstand the torque of a machine-driven screwdriver, which is greater than the torque that can be withstood by the conventional screw.
1932	**Duplicating device for typewriters (Beulah Louise Henry):** Henry's invention uses three sheets of paper and three ribbons to produce copies of a document as it is typewritten. Henry also develops children's toys—for example, a doll the eye color of which can be changed.
1932	**Cockroft-Walton accelerator (John Douglas Cockcroft and Ernest Thomas Sinton Walton):** The Cockcroft-Walton accelerator is used to fling charged particles at atomic nuclei in order to investigate their properties.
1932	**Richter scale (Charles Francis Richter):** Richter develops a scale to describe the magnitude of earthquakes; it is still used today.
1932	**Neutron (Sir James Chadwick):** Chadwick proves the existence of neutrons; he is awarded the 1935 Nobel Prize in Physics for his work.
1933	**Nuclear chain reaction (Leo Szilard):** Szilard conceives the idea of a nuclear chain reaction. He becomes a key figure in the Manhattan Project, which eventually builds the atomic bomb.
1933	**Magnetic tape recorder (Semi Joseph Begun):** Begun builds the first tape recorder, a dictating machine using wire for magnetic recording. He also develops the first steel tape recorder for mobile radio broadcasting and leads research into telecommunications and underwater acoustics.
1933	**Electron microscope (Ernst Ruska):** Ruska makes use of the wavelengths of electrons—shorter than those of visible light—to build a microscope that can image details at the subatomic level.
1933	**Recording:** Alan Dower Blumlein's patent for stereophonic recording is granted.
1933	**Polyethylene (Eric Fawcett and Reginald Gibson):** Fawcett and Gibson of Imperial Chemical Industries in London accidentally discover polyethylene. Hula hoops and Tupperware are just two of the products made with the substance.
1933	**Modern airliner:** Boeing 247 becomes the first modern airliner.
1933	**Solo flight:** Wiley Post makes the first around-the-world solo flight.
1934	**First bathysphere dive:** Charles William Beebe and Otis Barton make the first deep-sea dive in the Beebe-designed bathysphere off the Bermuda coast.
1934	**Langmuir-Blodgett films (Katharine Burr Blodgett):** A thin Langmuir-Blodgett film deposited on glass can make it nearly nonreflective.
1934	**Passenger train:** The Burlington Zephyr, America's first diesel-powered streamlined passenger train, is revealed at the World's Fair in Chicago.
1935	**Frequency modulation (Edwin H. Armstrong):** Armstrong exploits the fact that, since there are no natural sources of frequency modulation (FM), FM broadcasts are static-free.

1935	**Diatometer (Ruth Patrick):** Patrick's diatometer is a device placed in the water to collect diatoms and allow them to grow. The number of diatoms is sensitive to water pollution.
1935	**Kodachrome color film (Leopold Mannes and Leopold Godowsky, Jr.):** Mannes and Godowsky invent Kodachrome, a color film that is easy to use and produces vibrant colors. (With the digital revolution of the late twentieth century, production of Kodachrome is finally retired in 2009.)
1935	**Physostigmine and cortisone (Percy Lavon Julian):** Julian synthesizes physostigmine, used to treat glaucoma, and cortisone, used for arthritis. He will hold more than 130 patents and will become the first African American chemist inducted into the National Academy of Sciences.
1935	**Mobile refrigeration (Frederick McKinley Jones):** Mobile refrigeration enables the shipping of heat-sensitive products and compounds, from blood to frozen food.
1935	**Radar-based air defense system (Sir Robert Alexander Watson-Watt):** Watson-Watt's technical developments and his efforts as an administrator will be so important to the development of radar that he will be called the "father of radar."
1935	**Fallingwater (Frank Lloyd Wright):** Wright designs and builds a showcase house blending its form with its surroundings. One of the greatest architects of the twentieth century, he will produce many architectural innovations in structure, materials, and design.
1936	**Field-emission microscope (Erwin Wilhelm Müller):** Müller completes his dissertation, "The Dependence of Field Electron Emission on Work Function," and goes on to develop the field-emission microscope, which can resolve surface features as small as 2 nanometers.
1936	**Pentothal (Ernest Volwiler and Donalee Tabern):** Pentothal is a fast-acting intravenous anesthetic.
1937	**Muon discovered (Seth Neddermeyer):** Neddermeyer, working with Carl Anderson, J. C. Street, and E. C. Stevenson discover the muon (a particle similar to a heavy electron) while examining cosmic-ray tracks in a cloud chamber.
1937	**Concepts of digital circuits and information theory (Claude Elwood Shannon):** Shannon's most important contributions were electronic switching and using information theory to discover the basic requirements for data transmission.
1937	**X-ray crystallography (Dorothy Crowfoot Hodgkin):** Hodgkin uses X-ray crystallography to reveal the structure of molecules. She goes on to win the 1964 Nobel Prize in Chemistry.
1937	**Model K computer (George Stibitz):** The model K, an early electronic computer, employs Boolean logic.
1937	**Artificial sweetener:** American chemist Michael Sveda invents cyclamates, which is used as a noncaloric artificial sweetener until it is banned by the U.S. government in 1970 because of possible carcinogenic effects.
1937	**First pressurized airplane cabin:** The first pressurized airplane cabin is achieved in the United States with Lockheed's XC-35.

1937	**Antihistamines (Daniel Bovet):** Swiss-born Italian pharmacologist Bovet discovers antihistamines. He is awarded the 1957 Nobel Prize in Physiology or Medicine for his work.
1938	**Teflon (Roy J. Plunkett):** Plunkett accidentally synthesizes polytetrafluoroethylene (PTFE), now commonly known as Teflon, while researching chlorofluorocarbon refrigerants.
1938	**Electron microscope (James Hillier and Albert Prebus):** Adapting the work of German physicists, Hillier and Prebus develop a prototype of the electron microscope; and in 1940 Hillier produces the first commercial electron microscope available in the United States.
1938	**Xerography (Chester F. Carlson):** Xerography uses electrostatic charges to attract toner particles to make an image on plain paper. A hot wire then fuses the toner in place.
1938	**Walkie-talkie (Alfred J. Gross):** Gross's portable, two-way radio allows the user to move around while sending messages without remaining tied to a bulky transmitter. Gross invents a pager in 1949 and a radio tuner in 1950 that automatically follows the drift in carrier frequency due to movement of a sender or receiver.
1937-1938	**Analogue computer (George Philbrick):** Philbrick builds the Automatic Control Analyzer, which is an electronic analogue computer.
1939	**Helicopter (Igor Sikorsky):** Sikorsky, formerly the chief construction engineer and test pilot for the first four-engine aircraft, tests his helicopter, the Vought-Sikorsky 300, which after improvements will emerge as the world's first working helicopter.
1939	**Jet engine (Hans Joachim Pabst von Ohain):** The first jet-powered aircraft flies in 1939, while the first jet fighter will fly in 1941.
1939	**Atanasoff-Berry Computer (John Vincent Atanasoff and Clifford Berry):** The ABC, the world's first electronic digital computer, uses binary numbers and electronic switching, but it is not programmable.
1939	**DDT (Paul Hermann Müller):** Müller discovers the insect-repelling properties of DDT. He is awarded the 1948 Nobel Prize in Physiology or Medicine.
1940's	**Solar technology (Maria Telkes):** Telkes develops the solar oven and solar stills to produce drinking water from ocean water.
1940	**Cavity magnetron (Henry Boot and John Randall):** Boot and Randall develop the cavity magnetron, which advances radar technology.
1940	**Penicillin:** Sir Howard Walter Florey and Ernst Boris Chain isolate and purify penicillin. They are awarded, with Sir Alexander Fleming, the 1945 Nobel Prize in Physiology or Medicine.
1940	**Blood bank (Charles Richard Drew):** Drew establishes blood banks for World War II soldiers.
1940	**Color television (Peter Carl Goldmark):** Goldmark produces a system for transmitting and receiving color-television images using synchronized rotating filter wheels on the camera and on the receiver set.

1940	**Paintball gun (Charles and Evan Nelson):** The gun and paint capsules, invented to mark hard-to-reach trees in the forest, are eventually used for the game of paintball (1981), in which people shoot each other with paint.
1940	**Audio oscillator (William Redington Hewlett):** Hewlett invents the audio oscillator, a device that creates one frequency (pure tone) at a time. It is the first successful product of his Hewlett-Packard Company.
1940	**Antibiotics (Selman Abraham Waksman):** Waksman, through study of soil organisms, finds sources for the world's first antibiotics, including streptomycin and actinomycin.
1940	**Plutonium (Glenn Theodore Seaborg):** Seaborg synthesizes one of the first transuranium elements, plutonium. He becomes one of the leading figures on the Manhattan Project, which will build the atomic bomb. While he and others urged the demonstration of the bomb as a deterrent, rather than its use on the Japanese civilian population, the latter course was taken.
1940	**Thompson submachine gun (John T. Thompson):** Thompson works with Theodore Eickhoff and Oscar Payne to invent the American version of the submachine gun.
1940	**Automatic auto transmission:** General Motors offers the first modern automatic automobile transmission.
1941	**Jet engine (Sir Frank Whittle):** Whittle develops the jet engine independent of Hans Joachim Pabst von Ohain in Germany. After World War II, they meet and become good friends.
1941	**Solid-body electric guitar (Les Paul):** Paul's guitar lays the foundation for rock music. He also develops multitrack recording in 1948.
1941	**Z3 programmable computer (Konrad Zuse):** Zuse and his colleagues complete the first general-purpose, programmable computer, the Z3, in December. In 1950, Zuse will sell a Z4 computer—the only working computer in Europe.
1941	**Velcro (Georges de Mestral):** Burrs sticking to his dog's fur give de Mestral the idea for Velcro, which he perfects in 1948.
1941	**Dicoumarol:** The anticoagulant drug dicoumarol is identified and synthesized.
1941	**RDAs:** The first Recommended Dietary Allowances (RDAs), nutritional guidelines, are accepted.
1942	**Superglue (Harry Coover and Fred Joyner):** After developing superglue (cyanoacrylate), Coover rejects it as too sticky for a 1942 project. Coover and Joyner rediscover superglue in 1951, when Coover recognizes it as a marketable product.
1942	**Aqua-Lung (Jacques-Yves Cousteau and Émile Gagnon):** The Aqua-Lung delivers air at ambient pressure and vents used air to the surroundings.
1942	**Controlled nuclear chain reaction (Enrico Fermi):** In 1926 Fermi helped develop Fermi-Dirac statistics, which describe the quantum behavior of groups of electrons, protons, or neutrons. He now produces the first sustained nuclear chain reaction.

1942	**Synthetic vitamins (Max Tishler):** After synthesizing several vitamins during the 1930's, Tishler and his team develop the antibiotic sulfaquinoxaline to treat coccidiosis. He also develops fermentation processes to produce streptomycin and penicillin.
1942	**Bazooka:** The United States military first uses the bazooka during the North African campaign in World War II.
1943	**Meteorology:** Radar is first used to detect storms.
1944	**Electromechanical computer (Howard Aiken and Grace Hopper):** The Mark series of computers is built, designed by Aiken and Hopper. The U.S. Navy uses it to calculate trajectories for projectiles.
1944	**Colossus:** Colossus, the world's first vacuum-tube programmable logic calculator, is built in Britain for the purpose of breaking Nazi codes.
1944	**Phased array radar antennas (Luis W. Alvarez):** Alvarez's phased array sweeps a beam across the sky by turning hundreds of small antennas on and off and not by moving a radar dish.
1944	**V-2 rocket (Wernher von Braun):** Working for the German government during World War II, von Braun and other rocket scientists develop the V-2 rocket, the first long-range military missile and first suborbital missile. Arrested for making anti-Nazi comments, he later emigrates to the United States, where he leads the team that produces the Jupiter-C missile and launches vehicles such as the Saturn V, which help make the U.S. space program possible.
1944	**Quinine:** Robert B. Woodward and William von Eggers Doering synthesize quinine, which is used as an antimalarial.
1945	**Automatic Computing Engine (Alan Mathison Turing):** While the Automatic Computing Engine (ACE) was never fully built, it was one of the first stored-program computers.
1945	**Atomic bomb (J. Robert Oppenheimer):** Oppenheimer, the scientific leader of the Manhattan Project, heads the team that builds the atomic bomb. On the side of military use of the bomb to end World War II quickly, Oppenheimer saw this come to pass on August 6, 1945, when the bomb was dropped over Hiroshima, Japan, killing and maiming 150,000 people; a similar number of casualties ensued in Nagasaki on August 9, when the second bomb was dropped. Japan surrendered on August 14.
1945	**Dialysis machine (Willem Johan Kolff):** Kolff designs the first artificial kidney, a machine that cleans the blood of patients in renal failure, and refuses to patent it. He will construct the artificial lung in 1955.
1945	**Radioimmunoassay (RIA) (Rosalyn Yalow):** RIA required only a drop of blood (rather than the tens of milliliters previously required) to find trace amounts of substances.
1945	**Electronic Sackbut (Hugh Le Caine):** Le Caine builds the first music synthesizer, joined by the Special Purpose Tape Recorder in 1954, which could simultaneously change the playback speed of several recording tracks.

1945	**ENIAC computer (John William Mauchly and John Presper Eckert):** The Electronic Numerical Integrator and Computer, ENIAC, is the first general-purpose, programmable, electronic computer. (The Z3, developed independently by Konrad Zuse from 1939 to 1941 in Nazi Germany, did not fully exploit electronic components.) Built to calculate artillery firing tables, ENIAC is used in calculations for the hydrogen bomb.
1945	**Microwave oven (Percy L. Spencer):** The microwave oven grew out of the microwave generator, the magnetron tube, becoming more affordable.
1946	**Tupperware (Earl S. Tupper):** Tupper exploits plastics technology to develop a line of plastic containers that he markets at home parties starting in 1948.
1946	**Carbon-14 dating (Willard F. Libby):** Libby uses the half-life of carbon 14 to develop a reliable means of dating ancient remains. Radiocarbon dating has proven to be invaluable to archaeologists.
1946	**Magnetic tape recording (Marvin Camras):** Camras develops a magnetic tape recording process that will be adapted for use in electronic media, including music and motion-picture sound recording, audio and videocassettes, floppy disks, and credit card magnetic strips. For many years his method is the primary way to record and store sound, video, and digital data.
1946-1947	**Audiometer (Georg von Békésy):** Békésy invents a pure-tone audiometer that patients themselves can control to measure the sensitivity of their own hearing.
1946	**Radioisotopes for cancer treatment:** The first nuclear-reactor-produced radioisotopes for civilian use are sent from the U.S. Army's Oak Ridge facility in Tennessee to Brainard Cancer Hospital in St. Louis.
1947	**Transistor (John Bardeen, Walter H. Brattain, and William Shockley):** Hoping to build a solid-state amplifier, the team of Bardeen, Brattain, and Shockley discover the transistor, which replaces the vacuum tube in electronics. Bardeen is later part of the group that develops theory of superconductivity.
1947	**Platforming (Vladimir Haensel):** American chemical engineer Haensel invents platforming, a process that uses a platinum catalyst to produce cleaner-burning high-octane fuels.
1947	**Tubeless tire:** B.F. Goodrich announces development of the tubeless tire.
1948	**Holography (Dennis Gabor):** Gabor publishes his initial results working with holograms in Nature. Holograms became much more spectacular after the invention of the laser.
1948	**Long-playing record (LP) (Peter Carl Goldmark):** Goldmark demonstrates the LP playing the cello with CBS musicians. The musical South Pacific is recorded in LP format and boosts sales, making the LP the dominant form of recorded sound for the next four decades.
1948	**Gamma-ray pinhole camera (Roscoe Koontz):** Working to make nuclear reactors safer, Koontz invents the gamma-ray pinhole camera. The pinhole should act like a lens and form an image of the gamma source.
1948	**Instant photography (Edwin Herbert Land):** Land develops the simple process to make sheets of polarizing material. He perfects the Polaroid camera in 1972.

1948	**Synthetic penicillin (John C. Sheehan):** Sheehan develops the first total synthesis of penicillin, making this important antibiotic widely available.
1949	**First peacetime nuclear reactor:** Construction on the Brookhaven Graphite Research Reactor at Brookhaven Laboratory on Long Island, New York, is completed.
1949	**Magnetic core memory (Jay Wright Forrester):** Core memory is used from the early 1950's to the early 1970's.
1950's	**Fortran (John Warner Backus):** Backus develops the computer language Fortran, which is an acronym for "formula translation." Fortran allows direct entry of commands into computers with Englishlike words and algebraic symbols.
1950	**Planotron (Pyotr Leonidovich Kapitsa):** Kapitsa invents a magnetron tube for generating microwaves. He becomes a corecipient of the Nobel Prize for Physics in 1978 for discovering superfluidity in liquid helium.
1950	**Purinethol (Gertrude Belle Elion):** Elion develops the first effective treatment for childhood leukemia, 6-mercaptopurine (Purinethol). Elion later discovers azathioprine (Imuran), an immunosuppressive agent used for organ transplants.
1950	**Artificial pacemaker (John Alexander Hopps):** Hopps develops a device to regulate the beating of the heart to treat patients with erratic heartbeats. By 1957, the device is small enough to be implanted.
1950	**Contact lenses (George Butterfield):** Oregon optometrist Butterfield develops a lens that is molded to fit the contours of the cornea.
1951	**Fiber-optic endoscope (fibroscope) (Harold Hopkins):** Hopkins fastened together a flexible bundle of optical fibers that could convey an image. One end of the bundle could be inserted into a patient's throat, and the physician could inspect the esophagus.
1951	**The Pill (Carl Djerassi):** The birth-control pill, which becomes the world's most popular and is possibly most widely used contraceptive, revolutionizes not only medicine but also gender relations and women's status in society. Its prolonged use is later revealed to have health consequences.
1951	**Field-emission microscope (Erwin Wilhelm Müller):** Müller develops the field-ion microscope, followed by an atom-probe field-ion microscope in 1963, which can detect individual atoms.
1951	**Maser (Charles Hard Townes):** The maser (microwave amplification by stimulated emission of radiation) is a "laser" for microwaves. Discovered later, the "laser" patterned its name the acronym "maser."
1951	**Artificial heart valve (Charles Hufnagel):** Hufnagel develops an artificial heart valve and performs the first heart-valve implantation surgery in a human patient the following year.
1951	**UNIVAC (John Mauchly and John Presper Eckert):** Mauchly and Eckert invent the Universal Automatic Computer (UNIVAC). UNIVAC is competitor of IBM's products.
1952	**Bubble chamber (Donald A. Glaser):** In a bubble chamber, bubbles form along paths taken by subatomic particles as they interact, and the bubble trails allow scientists to deduce what happened.

1952	**Photovoltaic cell (Gerald Pearson):** The photovoltaic cell converts sunlight into electricity.
1952	**Improved electrical resistor (Otis Boykin):** Boykin's resistor had improved precision, and its high-frequency characteristics were better than those of previous resistors.
1952	**Language compiler (Grace Murray Hopper):** Hopper invents the compiler, an intermediate program that translates English-language instructions into computer language, followed in 1959 by Common Business Oriented Language (COBOL), the first computer programming language to translate commands used by programmers into the machine language the computer understands.
1952	**Amniocentesis (Douglas Bevis):** British physician Bevis develops amniocentesis.
1952	**Gamma camera (Hal Anger):** Nuclear medicine pioneer Anger creates the first prototype for the gamma camera. This leads to the inventions of other medical imaging devices, which detect and diagnose disease.
1953	**Medical ultrasonography (Inge Edler and Carl H. Hertz):** Edler and Hertz adapt an ultrasound probe used in materials testing in a shipyard for use on a patient. Their technology makes possible echograms of the heart and brain.
1953	**Inertial navigation systems (Charles Stark Draper):** Draper's inertial navigation system (INS) is designed to determine the current position of a ship or plane based on the initial location and acceleration.
1953	**Heart-lung machine (John H. Gibbon, Jr.):** American surgeon Gibbon conducts the first successful heart surgery using a heart-lung machine that he constructed with the help of his wife, Mary.
1953	**First frozen meals:** Swanson develops individual prepackaged frozen meals. The first-ever meal consists of turkey, cornbread stuffing, peas, and sweet potatoes.
1954	**Geodesic dome ® (Buckminster Fuller):** After developing the geodesic dome, Fuller patents the structure, an energy-efficient house using prefabricated, easily shipped parts.
1954	**Atomic absorption spectroscopy (Sir Alan Walsh):** Atomic absorption spectroscopy is used to identify and quantify the presence of elements in a sample.
1954	**Synthetic diamond (H. Tracy Hall):** Hall synthesizes diamonds using a high-pressure, high-temperature belt apparatus that can generate 120,000 atmospheres of pressure and sustain a temperature of 1,800 degrees Celsius in a working volume of about 0.1 cubic centimeter.
1954	**Machine vision (Jerome H. Lemelson):** Machine vision allows a computer to move and measure products and to inspect them for quality control.
1954	**Hydrogen bomb (Edward Teller):** The first hydrogen bomb, designed by Teller, is tested at the Bikini Atoll in the Pacific Ocean.
1954	**Silicon solar cells (Calvin Fuller):** Silicon solar cells have proven to be among the most efficient and least expensive solar cells.

1954	**First successful kidney transplant (Joseph Edward Murray):** American surgeon Murray performs the first successful kidney transplant, inserting one of Ronald Herrick's kidneys into his twin brother, Richard. Murray shares the 1990 Nobel Prize for Physiology or Medicine with E. Donnall Thomas, who developed bone marrow transplantation.
1954	**Transistor radio:** The first transistor radio is introduced by Texas Instruments.
1954	**IBM 650:** The IBM 650 computer becomes available. It is considered by IBM to be its first business computer, and it is the first computer installed at Columbia University in New York.
1954	**First nuclear submarine:** The United States launches the first nuclear-powered submarine, the USS *Nautilus*.
1955	**Color television's RGB system (Ernst Alexanderson):** The RGB system uses three image tubes to scan scenes through colored filters and three electron guns in the picture tube to reconstruct scenes.
1955	**Floppy disk and floppy disk drive (Alan Shugart):** Working at the San Jose, California, offices of International Business Machines (IBM), Shugart develops the disk drive, followed by floppy disks to provide a relatively fast way to store programs and data permanently.
1955	**Hovercraft (Sir Christopher Cockerell):** Cockerell files a patent for his hovercraft, an amphibious vehicle. He earlier invented several important electronic devices, including a radio direction finder for bombers in World War II.
1955	**Pulse transfer controlling device (An Wang):** The device allows magnetic core memory to be written or read without mechanical motion and is therefore very rapid.
1955	**Polio vaccine (Jonas Salk):** Salk's polio vaccine, which uses the killed virus, saves lives and improves the quality of life for millions afflicted by polio.
1956	**Fiber optics (Narinder S. Kapany):** Kapany, known as the father of fiber optics, coins the term "fiber optics." In high school, he was told by a teacher that light moves only in a straight line; he wanted to prove the teacher wrong and wound up inventing fiber optics.
1956	**Scotchgard (Patsy O'Connell Sherman):** Sherman develops a stain repellent for fabrics that is trademarked as Scotchgard.
1956	**Ovonic switch (Stanford Ovshinsky):** Ovshinsky invents a solid-state, thin film switch meant to mimic the actions of neurons.
1956	**Videotape recorder (Charles P. Ginsburg):** The video recorder allows programs to be shown later, to provide instant replays in sports, and to make a permanent record of a program.
1956	**Liquid Paper (Bette Nesmith Graham):** Graham markets her "Mistake Out" fluid for concealing typographical errors.
1956	**Dipstick blood sugar test (Helen M. Free):** Free and her husband Alfred co-invent a self-administered urinalysis test that allows diabetics to monitor their sugar levels and to adjust their medications accordingly.

1956	**350 RAMAC:** IBM produces the first computer disk storage system, the 350 RAMAC, which retrieves data from any of fifty spinning disks.
1957	**Wankel rotary engine (Felix Wankel):** Having fewer moving parts, the Wankel rotary engine ought to be sturdier and perhaps more efficient than the common reciprocating engine.
1957	**Laser (Gordon Gould, Charles Hard Townes, Arthur L. Schawlow, Theodore Harold Maiman):** Having conducted research on using light to excite thallium atoms, Gould tries to get funds and approval to build the first laser, but he fails. Townes (inventor of the maser) and Schawlow of Bell Laboratories will first describe the laser, and Maiman will first succeed in building a small optical maser. Gould coins the term "laser," which stands for light amplification by stimulated emission of radiation.
1957	**Intercontinental ballistic missile (ICBM):** The Soviet Union develops the ICBM.
1957	**First satellite:** The Soviet Union launches Sputnik, the first man-made satellite.
1958	**CorningWare:** CorningWare cookware is introduced. It is based on S. Donald Stookey's 1953 discovery that a heat-treatment process can transform glass into fine-grained ceramics.
1958	**Integrated circuit (Robert Norton Noyce and Jack St. Clair Kilby):** The microchip, independently discovered by Noyce and Kilby, proves to be the breakthrough that allows the miniaturization of electronic circuits and paves the way for the digital revolution.
1958	**Ultrasound:** Ultrasound becomes the most common method for examining a fetus.
1958	**Planar process (Jean Hoerni):** Hoerni develops the first planar process, which improves the integrated circuit.
1960's	**Lithography:** Optical lithography, a process that places intricate patterns onto silicon chips, is used in semiconductor manufacturing.
1960	**Measles vaccine (John F. Enders):** Enders, an American physician, develops the first measles vaccine. It is tested the following year and is hailed a success.
1960	**Echo satellite (John R. Pierce):** The first passive-relay telecommunications satellite, Echo, reflected signals. The signals, received from one point on Earth, "bounce" off the spherical satellite and are reflected back down to another, far distant, point on Earth.
1960	**Automatic letter-sorting machine (Jacob Rabinow):** Rabinow's machine greatly increased the speed and efficiency of mail delivery in the United States. He also invented an optical character recognition (OCR) scanner.
1960	**Ruby laser (Theodore Harold Maiman):** Maiman produces a ruby laser, the world's first visible light laser.
1960	**Helium-neon gas laser (Ali Javan):** Javan produces the world's second visible light laser.
1960	**Chardack-Greatbatch pacemaker (Wilson Greatbatch and William Chardack):** Greatbatch and Chardack create the first implantable pacemaker.

1960	**Radionuclide generator:** Powell Richards and Walter Tucker and their colleagues at Brookhaven Laboratory in New York invent a short half-life radionuclide generator for use in nuclear medicine diagnostic imaging procedures.
1961	**Audio-animatronics (Walt Disney):** Disney established WED, a research and development unit that developed the inventions he needed for his various enterprises. WED produced the audio-animatronic robotic figures that populated Disneyland, the 1964-1965 New York World's Fair, films, and other attractions. Audio-animatronics enabled robotic characters to speak or sing as well as move.
1961	**Ruby laser:** The ruby laser is first used medically by Charles Campbell and Charles Koester to excise a patient's retinal tumor.
1961	**First person in space:** Soviet astronaut Yuri Gagarin becomes the first person in space when he orbits the Earth on April 12.
1962	**Soft contact lenses (Otto Wichterle):** Wichterle's soft contacts can be worn longer with less discomfort than can hard contact lenses.
1962	**Continuously operating ruby laser (Willard S. Boyle and Don Nelson):** The invention relies on an arc lamp shining continuously (rather than the flash lamp used by Theodore Maiman in 1960).
1962	**Light-emitting diode (Nick Holonyak, Jr.):** Holonyak makes the first visible-spectrum diode laser, which produces red laser light but also stops lasing yet remains a useful light source. Holonyak has invented the red light-emitting diode (LED), the first operating alloy device—the "ultimate lamp."
1962	**Telstar satellite (John R. Pierce):** The first satellite to rebroadcast signals goes into operation, revolutionizing telecommunications.
1962	**Quasar 3C 273 (Maarten Schmidt):** Schmidt shows that this quasar is very distant and hence very bright. Further research shows quasars to be young galaxies with active, supermassive black holes at their centers.
1962	**First audiocassette:** The Philips company of the Netherlands releases the audiocassette tape.
1962	**Artificial hip (Sir John Charnley):** British surgeon Charnley invents the low-friction artificial hip and develops the surgical techniques for emplacing it.
1963	**Learjet (Bill Lear):** The Learjet, a small eight-passenger jet with a top speed of 560 miles (900 kilometers) per hour, can shuttle VIPs to meetings and other engagements.
1963	**Self-cleaning oven:** General Electric introduces the self-cleaning electric oven.
1963	**Artificial heart (Paul Winchell):** Winchell receives a patent (later donated to the University of Utah's Institute for Biomedical Engineering) for an artificial heart that purportedly became the model for the successful Jarvick-7.
1963	**6600 computer (Seymour Cray):** The 6600 was the first of a long line of Cray supercomputers.
1963	**Carbon fiber (Leslie Philips):** British engineer Philips develops carbon fiber, which is much stronger than steel.

1964	**Three-dimensional holography (Emmett Leith):** Leith and Juris Upatnieks present the first three-dimensional hologram at the Optical Society of America conference. The hologram must be viewed with a reference laser. The hologram of an object can then be viewed from different angles, as if the object were really present.
1964	**Moog synthesizer (Robert Moog):** The Moog synthesizer uses electronics to create and combine musical sounds.
1964	**Cosmic background radiation (Arno Penzias and Robert Wilson):** Penzias and Wilson detect the cosmic background radiation, which corresponds to that which would be radiated by a body at 2.725 kelvins. It is thought to be greatly redshifted primordial fireball radiation left over from the big bang.
1964	**BASIC programming language (John Kemeny and Thomas Kurtz):** Kemeny and Kurtz develop the BASIC computer programming language. BASIC is an acronym for Beginner's All-purpose Symbolic Instruction Code.
1965	**Minicomputer (Ken Olsen):** Perhaps the first true minicomputer, the PDP-8 is released by Digital Equipment Corporation. Founder Olsen makes computers affordable for small businesses.
1965	**Aspartame (James M. Schlatter):** Schlatter discovers aspartame, an artificial sweetener, while trying to come up with an antiulcer medication.
1965	**First space walk:** Soviet astronaut Aleksei Leonov is the first person to walk in space.
1966	**Gamma-electric cell (Henry Thomas Sampson):** Sampson works with George H. Miley to produce the gamma-electric cell, which converts the energy of gamma rays into electrical energy.
1966	**Handheld calculator (Jack St. Clair Kilby):** While working for Texas Instruments, Kilby does for the adding machine what the transistor had done for the radio, inventing a handheld calculator that retails at $150 and becomes an instant commercial success.
1966	**First unmanned moon landing:** Soviet spacecraft Luna 9 lands on the moon.
1967	**Electrogasdynamic method and apparatus (Meredith C. Gourdine):** Gourdine develops electrogasdynamics, which involves the production of electricity from the conversion of kinetic energy in a moving, ionized gas.
1967	**Pulsars (Jocelyn Bell and Antony Hewish):** Pulsars, rapidly rotating neutron stars, are discovered.
1968	**Practical liquid crystal displays (James Fergason):** Fergason develops an liquid crystal display (LCD) screen that has good visual contrast, is durable, and uses little electricity.
1968	**Lasers in medicine:** Francis L'Esperance begins using the argon-ion laser to treat patients with diabetic retinopathy.
1968	**Computer mouse (Douglas Engelbart):** Engelbart presents the computer mouse, which he had been working on since 1964.
1968	**Apollo 7:** Astronauts on Apollo 7, the first piloted Apollo mission, take photographs and transmit them to the American public on television.

1968	**Interface message processors:** Bolt Beranek and Newman Incorporated win a Defense Advanced Research Projects Agency (DARPA) contract to develop the packet switches called interface message processors (IMPs).
1969	**Rubella vaccine:** The rubella vaccine is available.
1969	**First person walks on the moon:** Neil Armstrong, a member of the U.S. Apollo 11 spacecraft, is the first person to walk on the moon.
1969	**Boeing 747:** The Boeing 747 makes its first flight, piloted by Jack Waddell.
1969	**Concorde:** The Concorde makes its first flight, piloted by André Turcat.
1969	**Charge-coupled device (Willard S. Boyle and George E. Smith):** Boyle and Smith develop the charge-coupled device, the basis for digital imaging.
1969	**ARPANET launches:** The Advanced Research Projects Agency starts ARPANET, which is the precursor to the Internet. UCLA and Stanford University are the first institutions to become networked.
1970's	**Digital seismology:** Digital seismology is used in oil exploration and increases accuracy in finding underground pools.
1970's	**Mud pulse telemetry:** Mud pulse telemetry becomes an oil-industry standard; pressure pulses are relayed through drilling mud to convey the location of the drill bit.
1970	**Optical fiber (Robert Maurer and others):** Maurer, joined by Donald Keck, Peter Schultz, and Frank Zimar, produces an optical fiber that can be used for communication.
1970	**Compact disc (James Russell):** The compact disc (CD) revolutionizes the way digital media is stored.
1970	**UNIX (Dennis Ritchie and Kenneth Thompson):** Bell Laboratories employees Ritchie and Thompson complete the UNIX operating system, which becomes popular among scientists.
1970	**Network Control Protocol:** The Network Working Group deploys the initial ARPANET host-to-host protocol, called the Network Control Protocol (NCP), establishing connections, break connections, switch connections, and control flow over the ARPANET.
1971	**Computerized axial tomography (Godfrey Newbold Hounsfield):** In London, doctors performed the first CAT scan of a living patient and detected a brain tumor. In a CAT (or CT) scan, X rays are taken of a body like slices in a loaf of bread. A computer then assembles these slices into a detail-laden three-dimensional image.
1971	**First videocassette recorder:** Sony begins selling the first videocassette recorder (VCR) to the public.
1971	**Microprocessor (Ted Hoff):** The computer's central processing unit (CPU) is reduced to the size of a postage stamp.
1971	**Electronic switching system for telecommunications (Erna Schneider Hoover):** Hoover's system prioritizes telephone calls and fixes an efficient order to answer them.
1971	**Intel microprocessors:** Intel builds the world's first microprocessor chip.

1971	**Touch screen (Sam Hurst):** Hurst's touch screen can detect if it has been touched and where it was touched.
1972	**First recombinant DNA organism (Stanley Norman Cohen, Paul Berg, and Herbert Boyer):** The methods to combine and transplant genes are discovered when this team successfully clones and expresses the human insulin gene in the Escherichia coli.
1972	**Far-Ultraviolet Camera (George R. Carruthers):** The Carruthers-designed camera is used on the Apollo 16 mission.
1972	**Cell encapsulation (Taylor Gunjin Wang):** Wang develops ways to encapsulate beneficial cells and introduce them into a body without triggering the immune system.
1972	**Pioneer 10:** The U.S. probe Pioneer 10 is launched to get information about the outer solar system.
1972	**Networking goes public:** ARPANET system designer Robert Kahn organizes the first public demonstration of the new network technology at the International Conference on Computer Communications in Washington, D.C.
1972	**Pong video game (Nolan K. Bushnell and Ted Dabney):** Bushnell and Dabney register the name of their new computer company, Atari, and issue Pong shortly thereafter, marking the rise of the video game industry.
1973	**Automatic computerized transverse axial (ACTA) whole-body CT scanner (Robert Steven Ledley):** The first whole-body CT scanner is operational. Ledley goes on to spend much of his career promoting the use of electronics and computers in biomedical research.
1973	**Packet network interconnection protocols TCP/IP (Vinton Gray Cerf and Robert Kahn):** Cerf and Kahn develop transmission control protocol/Internet protocol (TCP/IP), protocols that enable computers to communicate with one another.
1973	**Automated teller machine (Don Wetzel):** Wetzel receives a patent for his ATM. To make it a success, he shows banks how to generate a group of clients who would use the ATM.
1973	**Food processor:** The Cuisinart food processor is introduced in the United States.
1973	**Air bags in automobiles:** The Oldsmobile Tornado is the first American car sold equipped with air bags.
1973	**Space photography:** Astronauts aboard Skylab, the first U.S. space station, take high-resolution photographs of Earth using photographic remote-sensing systems. The astronauts also take photographs with handheld cameras.
1974	**Kevlar (Stephanie Kwolek):** Kwolek receives a patent for the fiber Kevlar. Bullet-resistant Kevlar vests go on sale only one year later.
1975	**Ethernet (Robert Metcalfe and David Boggs):** Metcalfe and Boggs invent the Ethernet, a system of software, protocols, and hardware allowing instantaneous communication between computer terminals in a local area.

1975	**Semiconductor laser:** Scientists working at Diode Labs develop the first commercial semiconductor laser that will operate continuously at room temperature.
1976	**First laser printer:** IBM's 3800 Printing System is the first laser printer. The ink jet is invented in the same year, but it is not prevalent in homes until 1988.
1976	**Apple computer (Steve Jobs):** Jobs cofounds Apple Computer with Steve Wozniak.
1976	**Jarvik-7 artificial heart (Robert Jarvik):** The Jarvik-7 allows a calf to live 268 days with the artificial heart. Jarvik combined ideas from several other workers to produce the Jarvik-7.
1976	**Apple II (Steve Wozniak):** Wozniak develops the Apple II, the best-selling personal computer of the 1970's and early 1980's.
1976	**First Mars probes:** The National Aeronautics and Space Administration (NASA) launches Viking 1 and Viking 2, which land on obtain images of Mars.
1976	**Kurzweil Reading Machine (Ray Kurzweil):** Kurzweil develops an optical character reader (OCR) able to read most fonts.
1976	**Microsoft Corporation (Bill Gates):** Gates, along with Paul Allen, found Microsoft, a software company. Gates will remain head of Microsoft for twenty-five years.
1976	**Conductive polymers:** Hideki Shirakawa, Alan G. MacDiarmid, and Alan J. Heeger discover conductive polymers. They are awarded the 2000 Nobel Prize in Chemistry.
1977	**Global Positioning System (GPS) (Ivan A. Getting):** The first GPS satellite is launched, designed to support a navigational system that uses satellites to pinpoint the location of a radio receiver on Earth's surface.
1977	**Fiber-optic telephone cable:** The first fiber-optic telephone cables are tested.
1977	**Echo-planar imaging (Peter Mansfield):** British physicist Mansfield first develops the echo-planar imaging (EPI).
1977	**Gossamer Condor (Paul MacCready):** MacCready designs the Gossamer Albatross, which enables human-powered flight.
1978	**Smart gels (Toyoichi Tanaka):** Tanaka discovers and works with "smart gels," polymer gels that can expand a thousandfold, change color, or contract when stimulated by minor changes in temperature, magnetism, light, or electricity. This capacity makes them useful in a broad range of applications.
1978	**Charon discovered (James Christy):** Charon is discovered as an apparent bulge on a fuzzy picture of Pluto. Its mass is about 12 percent that of Pluto.
1978	**First cochlear implant surgery:** Graeme Clark performs the first cochlear implant surgery in Australia.
1978	**The first test-tube baby:** Louise Brown is born in England.
1978	**First MRI:** The first magnetic resonance image (MRI) of the human head is taken in England.
1979	**First laptop (William Moggridge):** Moggridge, of Grid Systems in England, designs the first laptop computer.

1979	**First commercially successful application:** The VisiCalc spreadsheet for Apple II, designed by Daniel Bricklin and Bob Frankston, helps drive sales of the personal computer and becomes its first successful business application.
1979	**USENET (Tom Truscott, Jim Ellis and Steve Belovin):** Truscott, Ellis, and Belovin create USENET, a "poor man's ARPANET," to share information via e-mail and message boards between Duke University and the University of North Carolina, using dial-up telephone lines.
1979	**In-line roller skates (Scott Olson and Brennan Olson):** After finding some antique in-line skates, the Olson brothers begin experimenting with modern materials, creating Rollerblades.
1980's	**Controlled drug delivery (Robert S. Langer):** Langer develops the foundation of controlled drug delivery technology used in cancer treatment.
1980	**Alkaline battery (Lewis Urry):** Eveready markets alkaline batteries under the trade name Energizer. Urry's alkaline battery lasts longer than its predecessor, the carbon-zinc battery.
1980	**Interferon (Charles Weissmann):** Weissmann produces the first genetically engineered human interferon, which is used in cancer treatment.
1980	**TCP/IP:** The U.S. Department of Defense adopts the TCP/IP suite as a standard.
1981	**Ablative photodecomposition (Rangaswamy Srinivasan):** Srinivasan's research on ablative photodecomposition leads to multiple applications, including laser-assisted in situ keratomileusis (LASIK) surgery, which shapes the cornea to correct vision problems.
1981	**Scanning tunneling microscope (Heinrich Rohrer and Gerd Binnig):** The scanning tunneling microscope shows surfaces at the atomic level.
1981	**Improvements in laser spectroscopy (Arthur L. Schawlow and Nicolaas Bloembergen):** Schawlow shares the Nobel Prize in Physics with Nicolaas Bloembergen for their work on laser spectroscopy. While most of Schawlow's inventions involved lasers, he also did research in superconductivity and nuclear resonance.
1981	**First IBM personal computer:** The first IBM PC, the IBM 5100, goes on the market with a $1,565 price tag.
1982	**Compact discs appear:** Compact discs are now sold and will start replacing vinyl records.
1982	**First artificial heart:** Seattle dentist Barney Clark receives the first permanent artificial heart, and he survives for 112 days.
1983	**Cell phone (Martin Cooper):** The first mobile (wireless) phone, the DynaTAC 8000X, receives approval by the Federal Communications Commission (FCC), heralding an age of wireless communication.
1983	**Internet:** ARPANET, and networks attached to it, adopt the TCP/IP networking protocol. All networks that use the protocol are known as the Internet.
1983	**Cyclosporine:** Immunosuppressant cyclosporine is approved for use in transplant operations in the United States.

1983	**Polymerase chain reaction (Kary B. Mullis):** While driving to his cottage in Mendocino, California, Mullis develops the idea for the polymerase chain reaction (PCR). PCR will be used to amplify a DNA segment many times, leading to a revolution in recombinant DNA technology and a 1993 Nobel Prize in Chemistry for Mullis.
1984	**Domain name service is created:** Paul Mockapetris and Craig Partridge develop domain name service, which links unique Internet protocol (IP) numerical addresses to names with suffixes such as .mil, .com, .org, and .edu.
1984	**Mac is released:** Apple introduces the Macintosh, a low-cost, plug-and-play personal computer with a user-friendly graphic interface.
1984	**CD-ROM:** Philips and Sony introduce the CD-ROM (compact disc read-only memory), which has the capacity to store data of more than 450 floppy disks.
1984	**Surgery in utero:** William A. Clewall performs the first successful surgery on a fetus.
1984	**Cloning:** Danish veterinarian Steen M. Willadsen clones a lamb from a developing sheep embryo cell.
1984	**AIDS blood test (Robert Charles Gallo):** Gallo and his colleagues identify the virus HTLV-3/LAV (later renamed human immunodeficiency virus, or HIV) as the cause of acquired immunodeficiency syndrome, or AIDS. Gallo creates a blood test that can identify antibodies specific to HIV. This blood test is essential to keeping the supply in blood banks pure.
1984	**Imaging X-ray spectrometer (George Edward Alcorn):** Alcorn patents his device, which makes images of the source using X rays of specific energies, similar to making images with a specific wavelength (color) of light. It is used in acquiring data on the composition of distant planets and stars.
1984	**DNA profiling (Alec Jeffreys):** Noticing similarities and differences in DNA samples from his lab technician's family, Jeffreys discovers the principles that lead to DNA profiling, which has become an essential tool in forensics and the prosecution of criminal cases.
1985	**Windows operating system (Bill Gates):** The first version of Windows is released.
1985	**Implantable cardioverter defibrillator:** The U.S. Food and Drug Administration (FDA) approves Polish physician Michel Mirowski's implantable cardioverter defibrillator (ICD), which monitors and corrects abnormal heart rhythms.
1985	**Industry Standard Architecture (ISA) bus (Mark Dean and Dennis Moeller):** Dean and Moeller design the standard way of organizing the central part of a computer and its peripherals, the ISA bus, which is patented in this year.
1985	**Atomic force microscope:** Calvin Quate, Christoph Gerber, and Gerd Binnig invent the atomic force microscope, which becomes one of the foremost tools for imaging, measuring, and manipulating matter at the nano scale.
1986	**Mir:** The Soviet Union launches the Mir space station, the first permanent space station.
1986	**Burt Rutan's Voyager:** Dick Rutan (Burt's brother) and Jeana Yeager make the first around-the-world, nonstop flight without refueling in the Burt Rutan-designed Voyager. The Voyager is the first aircraft to accomplish this feat.

1986	**High-temperature superconductor (J. Georg Bednorz and Karl Alexander Müller):** Bednorz and Müller show that a ceramic compound of lanthanum, barium, copper, and oxygen becomes superconducting at 35 kelvins, a new high- temperature record.
1987	**Azidothymidine:** The FDA approves azidothymidine (AZT), a potent antiviral, for AIDS patients.
1987	**Echo-planar imaging:** Echo-planar imaging is used to perform real-time movie imaging of a single cardiac cycle.
1987	**Parkinson's treatment:** French neurosurgeon Alim-Louis Benabid implants a deep-brain electrical-stimulation system into a patient with advanced Parkinson's disease.
1987	**First corneal laser surgery:** New York ophthalmologist Steven Trokel performs the first laser surgery on a human cornea. He had refined his technique on a cow's eye. Trokel was granted a patent for the Excimer laser to be used for vision correction.
1987	**UUNET and PSINet:** Rick Adams forms UUNET and Bill Schrader forms PSINet to provide commercial Internet access.
1988	**Transatlantic fiber-optic cable:** The first transatlantic fiber-optic cable is installed, linking North America and France.
1988	**Laserphaco probe (Patricia Bath):** Bath's probe is used to break up and remove cataracts.
1989	**Method for tracking oil flow underground using a supercomputer (Philip Emeagwali):** Emeagwali receives the Gordon Bell Prize, considered the Nobel Prize for computing, for his method, which demonstrates the possibilities of computer networking.
1989	**World Wide Web (Tim Berners-Lee and Robert Cailau):** Berners-Lee finds a way to join the idea of hypertext and the young Internet, leading to the Web, coinvented with Cailau.
1989	**First dial-up access:** The World debuts as the first provider of dial-up Internet access for consumers.
1990's	**Environmentally friendly appliances:** Water-saving and energy-conserving washing machines and dryers are introduced.
1990	**Hubble Space Telescope:** The Hubble Space Telescope is launched and changes the way scientists look at the universe.
1990	**Human Genome Project begins:** The U.S. Department of Energy and the National Institutes of Health coordinate the Human Genome Project with the goal of identifying all 30,000 genes in human DNA and determining the sequences of the three billion chemical base pairs that make up human DNA.
1990	**BRCA1 gene discovered (Mary-Claire King):** King finds the cancer-associated gene on chromosome 17. She demonstrates that humans and chimpanzees are 99 percent genetically identical.

1991	**Nakao Snare (Naomi L. Nakao):** The Snare is a device that captures polyps that have been cut from the walls of the intestine, solving the problem of "lost polyp syndrome."
1991	**America Online (AOL):** Quantum Computer Services changes its name to America Online; Steve Case is named president. AOL offers e-mail, electronic bulletin boards, news, and other information.
1991	**Carbon nanotubes (Sumio Iijima):** Although carbon nanotubes have been seen before, Iijima's 1991 paper establishes some basic properties and prompts other scientists' interest in studying them.
1991	**The first hot-air balloon crosses the Pacific (Richard Branson and Per Lindstrad):** Branson and Lindstrad, who teamed up in 1987 to cross the Atlantic, make the 6,700-mile flight in 47 hours and break the world distance record.
1992	**Newton:** Apple introduces Newton, one of the first handheld computers, or personal digital assistants, which has a liquid crystal display operated with a stylus.
1993	**Mosaic (Marc Andreessen):** Andreessen launches Mosaic, followed by Netscape Navigator in 1995—the first Internet browsers. Both Mosaic and Netscape allow novices to browse the World Wide Web.
1993	**Flexible tailored elastic airfoil section (Sheila Widnall):** Widnall applies for a patent for this device, which addresses the problem of being able to measure fluctuations in pressure under unsteady conditions. She serves as secretary of the Air Force (the first woman to lead a branch of the military) and also serves on the board investigating the space shuttle Columbia accident of 2003.
1993	**Light-emitting diode (LED) blue and UV (Shuji Nakamura):** Nakamura's blue LED makes white LED light possible (a combination of red, blue, and green).
1994	**Genetically modified (GM) food:** The Flavr Savr tomato, the first GM food, is approved by the FDA.
1994	**Channel Tunnel:** Channel Tunnel, or Chunnel, opens, connecting France and Britain by a railway constructed beneath the English Channel.
1995	**51 Pegasi (Michel Mayor and Didier Queloz):** Mayor and Queloz detect a planet orbiting another normal star, the first extrasolar planet (exoplanet) to be found. As of June, 2009, 353 exoplanets were known.
1995	**Saquinavir:** The FDA approves Saquinavir for the treatment of AIDS. It is the first protease inhibitor, which reduces the ability of the AIDS virus to spread to new cells.
1995	**iBot (Dean Kamen):** Kamen invents iBOT, a super wheelchair that climbs stairs and helps its passenger to stand.
1995	**Global Positioning System (Ivan A. Getting):** The GPS becomes fully operational.
1995	**Illusion transmitter (Valerie L. Thomas):** A concave mirror can produce a real image that appears to be three-dimensional. Thomas's system uses a concave mirror at the camera and another one at the television receiver.
1996	**LASIK:** The first computerized excimer laser (LASIK), designed to correct the refractive error myopia, is approved for use in the United States.

1996	**First sheep is cloned:** Scottish scientist Ian Wilmut clones the first mammal, a Finn Dorset ewe named Dolly, from differentiated adult mammary cells.
1997	**Robotic vacuum:** Swedish appliance company Electrolux is the first to create a prototype of a robotic vacuum cleaner.
1998	**PageRank (Larry Page):** The cofounder of Google with Sergey Brin, Page devises PageRank, the count of Web pages linked to a given page and a measure how valuable people find that page.
1998	**UV Waterworks (Ashok Gadgil):** The device uses UV from a mercury lamp to kill waterborne pathogens.
1998	**Napster:** College dropout Shawn Fanning creates Napster, an extremely popular peer-to-peer file-sharing platform that allowed users to download music for free. In 2001 the free site was shut down because it encouraged illegal sharing of copyrighted properties. The site then became available by paid subscription.
1999	**Palm VII:** The Palm VII organizer is on the market. It is a handheld computer with 2 megabytes of RAM and a port for a wireless phone.
1999	**BlackBerry (Research in Motion of Canada):** A wireless handheld device that began as a two-way pager, the BlackBerry is also a cell phone that supports Web browsing, e-mail, text messaging, and faxing—it is the first smart phone.
2000	**Hoover-Diana production platform:** A joint venture by Exxon and British Petroleum (BP), the Hoover-Diana production platform goes into operation in the Gulf of Mexico. Within six months it is producing 20,000 barrels of oil a day.
2000	**Clone of a clone:** Japanese scientists clone a bull from a cloned bull.
2000	**Minerva:** The Library of Congress initiates a prototype system called Minerva (Mapping the Internet Electronic Resources Virtual Archives) to collect and preserve open-access Web resources.
2000	**Supercomputer:** The ASCI White supercomputer at the Lawrence Livermore National Laboratory in California is operational. It can hold six times the information stored in the 29 million books in the Library of Congress.
2001	**XM Radio:** XM Radio initiates the first U.S. digital satellite radio service in Dallas-Ft. Worth and San Diego.
2001	**Human cloning:** Scientists at Advanced Cell Technology in Massachusetts clone human embryos for the first time.
2001	**iPod (Tony Fadell):** Fadell introduces the iPod, a portable hard drive-based MP3 player with an Internet-based electronic music catalog, for Apple.
2001	**Segway PT (Dean Kamen):** Kamen introduces his personal transport device, a self-balancing, electric-powered pedestrian scooter.
2003	**First digital books:** Lofti Belkhir introduces the Kirtas BookScan 1200, the first automatic, page-turning scanner for the conversion of bound volumes to digital files.
2003	**Aqwon (Josef Zeitler):** The hydrogen-powered scooter Aqwon can reach 30 miles (50 kilometers) per hour. Its combustion product is water.

2003	**Human Genome Project is completed:** After thirteen years, the 25,000 genes of the human genome are identified and the sequences of the 3 million chemical base pairs that make up human DNA are determined.
2004	**Stem cell bank:** The world's first embryonic stem cell bank opens in England.
2004	**SpaceShipOne and SpaceShipTwo (Burt Rutan):** Rutan receives the U.S. Department of Transportation's first license issued for suborbital flight for SpaceShipOne, which shortly thereafter reaches an altitude of 328,491 feet. Rutan's rockets are the first privately funded manned rockets to reach space (higher than 100 kilometers above Earth's surface).
2004	**Columbia supercomputer:** The NASA supercomputer Columbia, built by Silicon Graphics and Intel, achieves sustained performance of 42.7 trillion calculations per second and is named the fastest supercomputer in the world. It is named for those who lost their lives in the explosion of the space shuttle Columbia in 2003. Because technology evolves so quickly, the Columbia will not be the fastest for very long.
2005	**Blue Gene/L supercomputer:** The National Nuclear Security Administration's BlueGene/L supercomputer, built by IBM, performs at 280.6 trillion operations per second and is now the world's fastest supercomputer.
2005	**Eris (Mike Brown):** Working with C. A. Trujillo and D. L. Rabinowitz, Brown discovers Eris, the largest known dwarf planet and a Kuiper Belt object. It is 27 percent more massive than Pluto, another large Kuiper Belt object.
2005	**Nix and Hydra discovered (Pluto companion team):** The Hubble research team—composed of Hal Weaver, S. Alan Stern, Max Mutchler, Andrew Steffl, Marc Buie, William Merline, John Spencer, Eliot Young, and Leslie Young—finds these small moons of Pluto.
2006	**Digital versus film:** Digital cameras have almost wholly replaced film cameras. *The New York Times* reports that 92 percent of cameras sold are digital.
2007	**First terabyte drive:** Hitachi Global Storage Technologies announces that it has created the first one-terabyte (TB) hard disk drive.
2007	**iPhone (Apple):** Apple introduces its smart phone, a combined cell phone, portable media player (equal to a video iPod), camera phone, Internet client (supporting e-mail and Web browsing), and text messaging device, to an enthusiastic market.
2008	**Roadrunner:** The Roadrunner supercomputer, built by IBM and Los Alamos National Laboratory, can process more than 1.026 quadrillion calculations per second. It works more than twice as fast as the Blue Gene/L supercomputer and is housed at Los Alamos in New Mexico.
2008	**Mammoth Genome Project:** Scientists sequence woolly mammoth genome, the first of an extinct animal.
2008	**Columbus lands:** The space shuttle Atlantis delivers the Columbus science laboratory to the International Space Station. The twenty-three-foot long laboratory is able to conduct experiments both inside and outside the space station.

2008	**Retail DNA test (Anne Wojcicki):** Wojcicki (wife of Google founder Sergey Brin) offers an affordable DNA saliva test, 23andMe, to determine one's genetic markers for ninety traits. The product heralds what *Time* magazine dubs a "personal-genomics revolution."
2009	**Large Hadron Collider:** The Large Hadron Collider (LHC) becomes the world's highest energy particle accelerator.
2009	**Hubble Space Telescope repairs (NASA):** STS-125 astronauts conducted five space walks from the space shuttle Atlantis to upgrade the Hubble Space Telescope, extending its life to at least 2014.
2009	**AIDS vaccine:** Scientists in Thailand create a vaccine that seems to reduce the risk of contracting the AIDS virus by more than 31 percent.
2010	**Jaguar supercomputer:** The Oak Ridge National Laboratory in Tennessee is home to Jaguar, the world's fastest supercomputer, the peak speed of which is 2.33 quadrillion floating point operations per second.

Charles W. Rogers, Southwestern Oklahoma State University, Department of Physics; updated by the editors of Salem Press

GENERAL BIBLIOGRAPHY

Aaboe, Asger. *Episodes from the Early History of Astronomy.* New York: Springer-Verlag, 2001.

Abbate, Janet. *Inventing the Internet.* Cambridge, Mass.: MIT Press, 2000.

Abell, George O., David Morrison, and Sidney C. Wolff. *Exploration of the Universe.* 5th ed. Philadelphia: Saunders College Publishing, 1987.

Achilladelis, Basil, and Mary Ellen Bowden. *Structures of Life.* Philadelphia: The Center, 1989.

Ackerknecht, Erwin H. *A Short History of Medicine.* Rev. ed. Baltimore: The Johns Hopkins University Press, 1982.

Aczel, Amir D. *Fermat's Last Theorem: Unlocking the Secret of an Ancient Mathematical Problem.* Reprint. New York: Four Walls Eight Windows, 1996.

Adler, Robert E. *Science Firsts: From the Creation of Science to the Science of Creation.* Hoboken, N.J.: John Wiley & Sons, 2002.

Alberts, Bruce, et al. *Molecular Biology of the Cell.* 2d ed. New York: Garland, 1989.

Alcamo, I. Edward. *AIDS: The Biological Basis.* 3d ed. Boston: Jones and Bartlett, 2003.

Aldersey-Willliams, Hugh. *The Most Beautiful Molecule: An Adventure in Chemistry.* London: Aurum Press, 1995.

Alexander, Arthur F. O'Donel. *The Planet Saturn: A History of Observation, Theory, and Discovery.* 1962. Reprint. New York: Dover, 1980.

Alioto, Anthony M. *A History of Western Science.* 2d ed. Upper Saddle River, N.J.: Prentice Hall, 1993.

Allen, Oliver E., and the editors of Time-Life Books. *Atmosphere.* Alexandria, Va.: Time-Life Books, 1983.

Ames, W. F., and C. Rogers, eds. *Nonlinear Equations in the Applied Sciences.* San Diego: Academic Press, 1992.

Andriesse, Cornelis D. *Christian Huygens.* Paris: Albin Michel, 2000.

Angier, Natalie. *Natural Obsessions: Striving to Unlock the Deepest Secrets of the Cancer Cell.* Boston: Mariner Books/Houghton Mifflin, 1999.

Annaratone, Donnatello. *Transient Heat Transfer.* New York: Springer, 2011.

Anstey, Peter R. *The Philosophy of Robert Boyle.* London: Routledge, 2000.

Anton, Sebastian. *A Dictionary of the History of Science.* Pearl River, N.Y.: Parthenon Publishing, 2001.

Archimedes. *The Works of Archimedes.* Translated by Sir Thomas Heath. 1897. Reprint. New York: Dover, 2002.

Arms, Karen, and Pamela S. Camp. *Biology: A Journey into Life.* 3d ed. Philadelphia: Saunders College Publishing, 1987.

Armstrong, Neil, Michael Collins, and Edwin E. Aldrin. *First on the Moon.* New York: Williams Konecky Associates, 2002.

Arrizabalaga, Jon, John Henderson, and Roger French. *The Great Pox: The French Disease in Renaissance Europe.* New Haven, Conn.: Yale University Press, 1997.

Arsuaga, Juan Luis. *The Neanderthal's Necklace: In Search of the First Thinkers.* Translated by Andy Klatt. New York: Four Walls Eight Windows, 2002.

Artmann, Benno. *Euclid: The Creation of Mathematics.* New York: Springer- Verlag, 1999.

Asimov, Isaac. *Exploring the Earth and the Cosmos.* New York: Crown, 1982.

_____. *The History of Physics.* New York: Walker, 1984.

_____. *Jupiter, the Largest Planet.* New York: Ace, 1980.

Aspray, William. *John von Neumann and the Origins of Modern Computing.* Boston: MIT Press, 1990.

Astronomical Society of the Pacific. *The Discovery of Pulsars.* San Francisco: Author, 1989.

Audesirk, Gerald J., and Teresa E. Audesirk. *Biology: Life on Earth.* 2d ed. New York: Macmillan, 1989.

Aughton, Peter. *Newton's Apple: Isaac Newton and the English Scientific Revolution.* London: Weidenfeld & Nicolson, 2003.

Aujoulat, Norbert. *Lascaux: Movement, Space, and Time.* New York: Harry N. Abrams, 2005.

Aveni, Anthony F., ed. *Skywatchers.* Rev. ed. Austin: University of Texas Press, 2001.

Baggott, Jim. *Perfect Symmetry: The Accidental Discovery of Buckminsterfullerene.* New York: Oxford University Press, 1994.

Baine, Celeste. *Is There an Engineer Inside You? A Comprehensive Guide to Career Decisions in Engineering.*

2d ed. Belmont, Calif.: Professional Publications, 2004.

Baker, John. *The Cell Theory: A Restatement, History and Critique.* New York: Garland, 1988.

Baldwin, Joyce. *To Heal the Heart of a Child: Helen Taussig, M.D.* New York: Walker, 1992.

Barbieri, Cesare, et al., eds. *The Three Galileos: The Man, the Spacecraft, the Telescope: Proceedings of the Conference Held in Padova, Italy on January 7-10, 1997.* Boston: Kluwer Academic, 1997.

Barkan, Diana Kormos. *Walther Nernst and the Transition to Modern Physical Science.* New York: Cambridge University Press, 1999.

Barrett, Peter. *Science and Theology Since Copernicus: The Search for Understanding.* Reprint. Dorset, England: T&T Clark, 2003.

Bartusiak, Marcia. *Thursday's Universe.* New York: Times Books, 1986.

Basta, Nicholas. *Opportunities in Engineering Careers.* New York: McGraw-Hill, 2003.

Bates, Charles C., and John F. Fuller. *America's Weather Warriors, 1814-1985.* College Station: Texas A&M Press, 1986.

Bazin, Hervé. *The Eradication of Smallpox: Edward Jenner and the First and Only Eradication of a Human Infectious Disease.* Translated by Andrew Morgan and Glenise Morgan. San Diego: Academic Press, 2000.

Beatty, J. Kelly, and Andrew Chaikin, eds. *The New Solar System.* 3d rev. ed. New York: Cambridge University Press, 1990.

Becker, Wayne, Lewis Kleinsmith, and Jeff Hardin. *The World of the Cell.* New York: Pearson/Benjamin Cummings, 2006.

Berlinski, David. *A Tour of the Calculus.* New York: Vintage Books, 1997.

Bernstein, Jeremy. *Three Degrees Above Zero: Bell Labs in the Information Age.* New York: Charles Scribner's Sons, 1984.

Bernstein, Peter L. *Against the Gods: The Remarkable Story of Risk.* New York: John Wiley & Sons, 1996.

Bertolotti, M. *Masers and Lasers: An Historical Approach.* Bristol, England: Adam Hilger, 1983.

Bickel, Lennard. *Florey: The Man Who Made Penicillin.* Carlton South, Victoria, Australia: Melbourne University Press, 1995.

Bizony, Piers. *Island in the Sky: Building the International Space Station.* London: Aurum Press Limited, 1996.

Blackwell, Richard J. *Galileo, Bellarmine, and the Bible.* London: University of Notre Dame Press, 1991.

Bliss, Michael. *The Discovery of Insulin.* Chicago: University of Chicago Press, 1987.

Blumenberg, Hans. *The Genesis of the Copernican World.* Translated by Robert M. Wallace. Cambridge, Mass.: MIT Press, 1987.

Blunt, Wilfrid. *Linnaeus: The Compleat Naturalist.* Princeton, N.J.: Princeton University Press, 2001.

Bodanis, David. *Electric Universe: The Shocking True Story of Electricity.* New York: Crown Publishers, 2005.

Bohm, David. *Causality and Chance in Modern Physics.* London: Routledge & Kegan Paul, 1984.

Bohren, Craig F. *Clouds in a Glass of Beer: Simple Experiments in Atmospheric Physics.* New York: John Wiley & Sons.

Boljanovic, Vukota. *Applied Mathematics and Physical Formulas: A Pocket Reference Guide for Students, Mechanical Engineers, Electrical Engineers, Manufacturing Engineers, Maintenance Technicians, Toolmakers, and Machinists.* New York: Industrial Press, 2007.

Bolt, Bruce A. *Inside the Earth: Evidence from Earthquakes.* New York: W. H. Freeman, 1982.

Bond, Peter. *The Continuing Story of the International Space Station.* Chichester, England: Springer-Praxis, 2002.

Boorstin, Daniel J. *The Discoverers.* New York: Random House, 1983.

Bottazzini, Umberto. *The Higher Calculus: A History of Real and Complex Analysis from Euler to Weierstrass.* New York: Springer-Verlag, 1986.

Bourbaki, Nicolas. *Elements of the History of Mathematics.* Translated by John Meldrum. New York: Springer, 1994.

Bowler, Peter J. *Charles Darwin: The Man and His Influence.* Cambridge, England: Cambridge University Press, 1996.

_____. *Evolution: The History of an Idea.* Rev. ed. Berkeley: University of California Press, 1989.

_____. *The Mendelian Revolution: The Emergence of Hereditarian Concepts in Modern Science and Society.* Baltimore: The Johns Hopkins University Press, 1989.

Boyer, Carl B. *A History of Mathematics.* 2d ed., revised by Uta C. Merzbach. New York: John Wiley & Sons, 1991.

Bracewell, Ronald N. *The Fourier Transform and Its Applications.* 3d rev. ed. New York: McGraw-Hill, 1987.

Brachman, Arnold. *A Delicate Arrangement: The Strange Case of Charles Darwin and Alfred Russel Wallace.* New York: Times Books, 1980.

Bredeson, Carmen. *John Glenn Returns to Orbit: Life on the Space Shuttle.* Berkeley Heights, N.J.: Enslow, 2000.

Brock, Thomas, ed. *Milestones in Microbiology, 1546-1940.* Washington, D.C.: American Society for Microbiology, 1999.

Brock, William H. *The Chemical Tree: A History of Chemistry.* New York: W. W. Norton, 2000.

Brooks, Paul. *The House of Life: Rachel Carson at Work.* 2d ed. Boston: Houghton Mifflin, 1989.

Browne, Janet. *Charles Darwin: The Power of Place.* New York: Knopf, 2002.

Brush, Stephen G. *Cautious Revolutionaries: Maxwell, Planck, Hubble.* College Park, Md.: American Association of Physics Teachers, 2002.

Brush, Stephen G., and Nancy S. Hall. *Kinetic Theory of Gases: An Anthology of Classic Papers With Historical Commentary.* London: Imperial College Press, 2003.

Bryant, Stephen. *The Story of the Internet.* London: Pearson Education, 2000.

Buffon, Georges-Louis Leclerc. *Natural History: General and Particular.* Translated by William Smellie. Avon, England: Thoemmes Press, 2001.

Burger, Edward B., and Michael Starbird. *Coincidences, Chaos, and All That Math Jazz: Making Light of Weighty Ideas.* New York: W. W. Norton, 2005.

Burke, Terry, et al., eds. *DNA Fingerprinting: Approaches and Applications.* Boston: Birkhauser, 2001.

Byrne, Patrick H. *Analysis and Science in Aristotle.* Albany: State University of New York Press, 1997.

Calder, William M., III, and David A. Traill, eds. *Myth, Scandal, and History: The Heinrich Schliemann Controversy.* Detroit: Wayne State University Press, 1986.

Calinger, Ronald. *A Contextual History of Mathematics.* Upper Saddle River, N.J.: Prentice Hall, 1999.

Canning, Thomas N. *Galileo Probe Parachute Test Program: Wake Properties of the Galileo.* Washington, D.C.: National Aeronautics and Space Administration, Scientific and Technical Information Division, 1988.

Cantor, Geoffrey. *Michael Faraday: Sandemanian and Scientist: A Study of Science and Religion in the Nineteenth Century.* New York: St. Martin's Press, 1991.

Carlisle, Rodney. *Inventions and Discoveries: All the Milestones in Ingenuity—from the Discovery of Fire to the Invention of the Microwave.* Hoboken, N.J.: John Wiley & Sons, 2004.

Carlson, Elof Axel. *Mendel's Legacy: The Origin of Classical Genetics.* Woodbury, N.Y.: Cold Spring Harbor Laboratory Press, 2004.

Carola, Robert, John P. Harley, and Charles R. Noback. *Human Anatomy and Physiology.* New York: McGraw-Hill, 1990.

Carpenter, B. S., and R. W. Doran, eds. *A. M. Turing's ACE Report of 1946 and Other Papers.* Cambridge, Mass.: MIT Press, 1986.

Carpenter, Kenneth J. *The History of Scurvy and Vitamin C.* Cambridge England: Cambridge University Press, 1986.

Carrigan, Richard A., and W. Peter Trower, eds. *Particle Physics in the Cosmos.* New York: W. H. Freeman, 1989.

_____, eds. *Particles and Forces: At the Heart of the Matter.* New York: W. H. Freeman, 1990.

Cassanelli, Roberto, et al. *Houses and Monuments of Pompeii: The Works of Fausto and Felice Niccolini.* Los Angeles: J. Paul Getty Museum, 2002.

Caton, Jerald A. *A Review of Investigations Using the Second Law of Thermodynamics to Study Internal-Combustion Engines.* London: Society of Automotive Engineers, 2000.

Chaikin, Andrew. *A Man on the Moon: The Voyages of the Apollo Astronauts.* New York: Penguin Group, 1998.

Chaisson, Eric J. *The Hubble Wars.* New York: HarperCollins, 1994.

Chaisson, Eric J., and Steve McMillan. *Astronomy Today.* 5th ed. Upper Saddle River, N.J.: Pearson Prentice Hall, 2004.

Chandrasekhar, Subrahmanyan. *Eddington: The Most Distinguished Astrophysicist of His Time.* Cambridge, England: Cambridge University Press, 1983.

Chang, Hasok. *Inventing Temperature: Measurement and Scientific Progress.* Oxford, England: Oxford University Press, 2004.

Chang, Laura, ed. *Scientists at Work: Profiles of Today's Groundbreaking Scientists from "Science Times."* New York: McGraw-Hill, 2000.

Chant, Christopher. *Space Shuttle.* New York: Exeter Books, 1984.

Chapman, Allan. *Astronomical Instruments and Their Users: Tycho Brahe to William Lassell.* Brookfield, Vt.: Variorum, 1996.

Chase, Allan. *Magic Shots*. New York: William Morrow, 1982.

Check, William A. *AIDS*. New York: Chelsea House, 1988.

Cheng, K. S., and G. V. Romero. *Cosmic Gamma-Ray Sources*. New York: Springer-Verlag, 2004.

Christianson, John Robert. *On Tycho's Island: Tycho Brahe and His Assistants, 1570-1601*. New York: Cambridge University Press, 2000.

Chung, Deborah D. L. *Applied Materials Science: Applications of Engineering Materials in Structural, Electronics, Thermal and Other Industries*. Boca Raton, Fla.: CRC Press, 2001.

Clark, Ronald W. *The Life of Ernst Chain: Penicillin and Beyond*. New York: St. Martin's Press, 1985.

_____. *The Survival of Charles Darwin: A Biography of a Man and an Idea*. New York: Random House, 1984.

Cline, Barbara Lovett. *Men Who Made a New Physics*. Chicago: University of Chicago Press, 1987.

Clos, Lynne. *Field Adventures in Paleontology*. Boulder, Colo.: Fossil News, 2003.

Clugston, M. J., ed. *The New Penguin Dictionary of Science*. 2d ed. New York: Penguin Books, 2004.

Coffey, Patrick. *Cathedrals of Science: The Personalities and Rivalries That Made Modern Chemistry*. New York: Oxford University Press, 2008.

Cohen, I. Bernard. *Benjamin Franklin's Science*. Cambridge, Mass.: Harvard University Press, 1990.

_____. *The Newtonian Revolution*. New York: Cambridge University Press, 1980.

Cohen, I. Bernard, and George E. Smith, eds. *The Cambridge Companion to Newton*. New York: Cambridge University Press, 2002.

Cole, K. C. *The Universe and the Teacup: The Mathematics of Truth and Beauty*. Fort Washington, Pa.: Harvest Books, 1999.

Cole, Michael D. *Galileo Spacecraft: Mission to Jupiter: Countdown to Space*. New York: Enslow, 1999.

Collin, S. M. H. *Dictionary of Science and Technology*. London: Bloomsbury Publishing, 2003.

Connor, James A. *Kepler's Witch: An Astronomer's Discovery of Cosmic Order Amid Religious War, Political Intrigue, and the Heresy Trial of His Mother*. San Francisco: HarperSanFrancisco, 2004.

Conrad, Lawrence, et al., eds. *The Western Medical Tradition, 800 B.C. to A.D. 1800*. New York: Cambridge University Press, 1995.

Cook, Alan. *Edmond Halley: Charting the Heavens and the Seas*. New York: Oxford University Press, 1998.

Cooke, Donald A. *The Life and Death of Stars*. New York: Crown, 1985.

Cooper, Geoffrey M. *Oncogenes*. 2d ed. Boston: Jones and Bartlett, 1995.

Cooper, Henry S. F., Jr. *Imaging Saturn: The Voyager Flights to Saturn*. New York: H. Holt, 1985.

Corsi, Pietro. *The Age of Lamarck: Evolutionary Theories in France, 1790-1830*. Berkeley: University of California Press, 1988.

Coulthard, Malcolm, and Alison Johnson, eds. *The Routledge Handbook of Forensic Linguistics*. New York: Routledge, 2010.

Craven, B. O. *The Lebesgue Measure and Integral*. Boston: Pitman Press, 1981.

Crawford, Deborah. *King's Astronomer William Herschel*. New York: Julian Messner, 2000.

Crease, Robert P., and Charles C. Mann. *The Second Creation: Makers of the Revolution in Twentieth Century Physics*. New York: Macmillan, 1985.

Crewdson, John. *Science Fictions: A Scientific Mystery, A Massive Cover-Up, and the Dark Legacy of Robert Gallo*. Boston: Little, Brown, 2002.

Crick, Francis. *What Mad Pursuit: A Personal View of Scientific Discovery*. New York: Basic Books, 1988.

Crump, Thomas. *A Brief History of Science as Seen Through the Development of Scientific Instruments*. New York: Carroll & Graf, 2001.

Cunningham, Andrew. *The Anatomical Renaissance: The Resurrection of the Anatomical Projects of the Ancients*. Brookfield, Vt.: Ashgate, 1997.

Cutler, Alan. *The Seashell on the Mountaintop: A Story of Science, Sainthood, and the Humble Genius Who Discovered a New History of the Earth*. New York: Dutton/Penguin, 2003.

Dalrymple, G. Brent. *The Age of the Earth*. Stanford, Calif.: Stanford University Press, 1991.

Darrigol, Oliver. *Electrodynamics from Ampère to Einstein*. Oxford, England: Oxford University Press, 2000.

Dash, Joan. *The Longitude Prize*. New York: Farrar, Straus and Giroux, 2000.

Daston, Lorraine. *Classical Probability in the Enlightenment*. Princeton, N.J.: Princeton University Press, 1988.

Davies, John K. *Astronomy from Space: The Design and Operation of Orbiting Observatories*. New York: John Wiley & Sons, 1997.

Davies, Paul. *The Edge of Infinity: Where the Universe Came from and How It Will End*. New York: Simon & Schuster, 1981.

Davis, Joel. *Flyby: The Interplanetary Odyssey of Voyager 2.* New York: Atheneum, 1987.

Davis, Martin. *Engines of Logic: Mathematicians and the Origin of the Computer.* New York: W. W. Norton, 2000.

Davis, Morton D. *Game Theory: A Nontechnical Introduction.* New York: Dover, 1997.

Davis, William Morris. *Elementary Meteorology.* Boston: Ginn, 1894.

Dawkins, Richard. *The Ancestor's Tale: A Pilgrimage to the Dawn of Evolution.* New York: Houghton Mifflin, 2004.

_____. *River Out of Eden: A Darwinian View of Life.* New York: Basic Books, 1995.

Day, Michael H. *Guide to Fossil Man.* 4th ed. Chicago: University of Chicago Press, 1986.

Day, William. *Genesis on Planet Earth.* 2d ed. New Haven, Conn.: Yale University Press, 1984.

Dean, Dennis R. *James Hutton and the History of Geology.* Ithaca, N.Y.: Cornell University Press, 1992.

Debré, Patrice. *Louis Pasteur.* Translated by Elborg Forster. Baltimore: The Johns Hopkins University Press, 1998.

DeJauregui, Ruth. *100 Medical Milestones That Shaped World History.* San Mateo, Calif.: Bluewood Books, 1998.

De Jonge, Christopher J., and Christopher L. R. Barratt, eds. *Assisted Reproductive Technologies: Current Accomplishments and New Horizons.* New York: Cambridge University Press, 2002.

Delaporte, François. *The History of Yellow Fever: An Essay on the Birth of Tropical Medicine.* Cambridge, Mass.: MIT Press, 1991.

Dennett, Daniel C. *Darwin's Dangerous Idea: Evolution and the Meanings of Life.* New York: Simon & Schuster, 1995.

Dennis, Carina, and Richard Gallagher. *The Human Genome.* London: Palgrave Macmillan, 2002.

DeVorkin, David H. *Race to the Stratosphere: Manned Scientific Ballooning in America.* New York: Springer-Verlag, 1989.

Dewdney, A. K. *The Turing Omnibus.* Rockville, Md.: Computer Science Press, 1989.

Diamond, Jared. *The Third Chimpanzee: The Evolution and Future of the Human Animal.* New York: Harper-Collins, 1992.

DiCanzio, Albert. *Galileo: His Science and His Significance for the Future of Man.* Portsmouth, N.H.: ADASI, 1996.

Dijksterhuis, Eduard Jan. *Archimedes.* Translated by C. Dikshoorn, with a new bibliographic essay by Wilbur R. Knorr. Princeton, N.J.: Princeton University Press, 1987.

Dijksterhuis, Fokko Jan. *Lenses and Waves: Christiaan Huygens and the Mathematical Science of Optics in the Seventeenth Century.* Dordrecht, the Netherlands: Kluwer Academic, 2004.

Dimmock, N. J., A. J. Easton, and K. N. Leppard. *Introduction to Modern Virology.* 5th ed. Malden, Mass.: Blackwell Science, 2001.

Dore, Mohammed, Sukhamoy Chakravarty, and Richard Goodwin, eds. *John Von Neumann and Modern Economics.* New York: Oxford University Press, 1989.

Drake, Stillman. *Galileo: Pioneer Scientist.* Toronto: University of Toronto Press, 1990.

_____. *Galileo: A Very Short Introduction.* New York: Oxford University Press, 2001.

Dreyer, John Louis Emil, ed. *The Scientific Papers of Sir William Herschel.* Dorset, England: Thoemmes Continuum, 2003.

Duck, Ian. *One Hundred Years of Planck's Quantum.* River Edge, N.J.: World Scientific, 2000.

Dudgeon, Dan E., and Russell M. Mersereau. *Multidimensional Digital Signal Processing.* Englewood Cliffs, N.J.: Prentice Hall, 1984.

Dunham, William. *The Calculus Gallery: Masterpieces from Newton to Lebesgue.* Princeton, N.J.: Princeton University Press, 2005.

_____. *Euler: The Master of Us All.* Washington, D.C.: Mathematical Association of America, 1999.

_____. *Journey Through Genius.* New York: John Wiley & Sons, 1990.

Durham, Frank, and Robert D. Purrington. *Frame of the Universe.* New York: Cambridge University Press, 1983.

Easton, Thomas A. *Careers in Science.* 4th ed. Chicago: VGM Career Books, 2004.

Edelson, Edward. *Gregor Mendel: And the Roots of Genetics.* New York: Oxford University Press, 2001.

Edey, Maitland A., and Donald C. Johanson. *Blueprints: Solving the Mystery of Evolution.* Boston: Little, Brown, 1989.

Edwards, Robert G., and Patrick Steptoe. *A Matter of Life.* New York: William Morrow, 1980.

Ehrenfest, Paul, and Tatiana Ehrenfest. *The Conceptual Foundations of the Statistical Approach in Mechanics.* Mineola, N.Y.: Dover, 2002.

Ehrlich, Melanie, ed. *DNA Alterations in Cancer: Genetic and Epigenetic Changes.* Natick, Mass.: Eaton, 2000.

Eisen, Herman N. *Immunology: An Introduction to Molecular and Cellular Principles of the Immune Responses.* 2d ed. Philadelphia: J. B. Lippincott, 1980.

Espejo, Roman, ed. *Biomedical Ethics: Opposing Viewpoints.* San Diego: Greenhaven Press, 2003.

Evans, James. *The History and Practice of Ancient Astronomy.* New York: Oxford University Press, 1998.

Fabian, A. C., K. A. Pounds, and R. D. Blandford. *Frontiers of X-Ray Astronomy.* London: Cambridge University Press, 2004.

Fara, Patricia. *An Entertainment for Angels: Electricity in the Enlightenment.* New York: Columbia University Press, 2002.

_____. *Newton: The Making of a Genius.* New York: Columbia University Press, 2002.

_____. *Sex, Botany, and the Empire: The Story of Carl Linnaeus and Joseph Banks.* New York: Columbia University Press, 2003.

Farber, Paul Lawrence. *Finding Order in Nature: The Naturalist Tradition from Linnaeus to E. O. Wilson.* Baltimore: The Johns Hopkins University Press, 2000.

Fauvel, John, and Jeremy Grey, eds. *The History of Mathematics: A Reader.* 1987. Reprint. Washington, D.C.: The Mathematical Association of America, 1997.

Feferman, S., J. W. Dawson, and S. C. Kleene, eds. *Kurt Gödel: Collected Works.* 2 vols. New York: Oxford University Press, 1986-1990.

Feldman, David. *How Does Aspirin Find a Headache?* New York: HarperCollins, 2005.

Ferejohn, Michael. *The Origins of Aristotelian Science.* New Haven, Conn.: Yale University Press, 1991.

Ferguson, Kitty. *The Nobleman and His Housedog: Tycho Brahe and Johannes Kepler—The Strange Partnership That Revolutionized Science.* London: Headline, 2002.

Ferris, T. *Coming of Age in the Milky Way.* New York: Doubleday, 1989.

Ferris, Timothy. *Galaxies.* New York: Harrison House, 1987.

Field, George, and Donald Goldsmith. *The Space Telescope.* Chicago: Contemporary Books, 1989.

Field, J. V. *The Invention of Infinity: Mathematics and Art in the Renaissance.* New York: Oxford University Press, 1997.

Fincher, Jack. *The Brain: Mystery of Matter and Mind.* Washington, D.C.: U.S. News Books, 1981.

Finlayson, Clive. *Neanderthals and Modern Humans: An Ecological and Evolutionary Perspective.* New York: Cambridge University Press, 2004.

Finocchiaro, Maurice A., ed. *The Galileo Affair: A Documentary History.* Berkeley: University of California Press, 1989.

Fischer, Daniel. *Mission Jupiter: The Spectacular Journey of the Galileo Spacecraft.* New York: Copernicus Books, 2001.

Fischer, Daniel, and Hilmar W. Duerbeck. *Hubble Revisited: New Images from the Discovery Machine.* New York: Copernicus Books, 1998.

Fisher, Richard B. *Edward Jenner, 1741-1823.* London: Andre Deutsch, 1991.

Flowers, Lawrence O., ed. *Science Careers: Personal Accounts from the Experts.* Lanham, Md.: Scarecrow Press, 2003.

Ford, Brian J. *The Leeuwenhoek Legacy.* London: Farrand, 1991.

_____. *Single Lens: The Story of the Simple Microscope.* New York: Harper & Row, 1985.

Fournier, Marian. *The Fabric of Life: Microscopy in the Seventeenth Century.* Baltimore: The Johns Hopkins University Press, 1996.

Fowler, A. C. *Mathematical Models in the Applied Sciences.* New York: Cambridge University Press, 1997.

Foyer, Christine H. *Photosynthesis.* New York: Wiley-Interscience, 1984.

Frängsmyr, Tore, ed. *Linnaeus: The Man and His Work.* Canton, Mass.: Science History Publications, 1994.

Franklin, Benjamin. *Autobiography of Benjamin Franklin.* New York: Buccaneer Books, 1984.

French, A. P., and P. J. Kennedy, eds. *Niels Bohr: A Centenary Volume.* Cambridge, Mass.: Harvard University Press, 1985.

French, Roger. *William Harvey's Natural Philosophy.* New York: Cambridge University Press, 1994.

Fridell, Ron. *DNA Fingerprinting: The Ultimate Identity.* New York: Scholastic, 2001.

Friedlander, Michael W. *Cosmic Rays.* Cambridge, Mass.: Harvard University Press, 1989.

Friedman, Meyer, and Gerald W. Friedland. *Medicine's Ten Greatest Discoveries.* New Haven, Conn.: Yale University Press, 2000.

Friedman, Robert Marc. *Appropriating the Weather: Vilhelm Bjerknes and the Construction of a Modern Meteorology.* Ithaca, N.Y.: Cornell University Press, 1989.

Friedrich, Wilhelm. *Vitamins.* New York: Walter de Gruyter, 1988.

Friedrichs, Günter, and Adam Schaff. *Microelectronics and Society: For Better or for Worse, a Report to the Club of Rome.* New York: Pergamon Press, 1982.

Frist, William. *Transplant.* New York: Atlantic Monthly Press, 1989.

Fuchs, Thomas. *The Mechanization of the Heart: Harvey and Descartes.* Rochester, N.Y.: University of Rochester Press, 2001.

Gallo, Robert C. *Virus Hunting: AIDS, Cancer, and the Human Retrovirus: A Story of Scientific Discovery.* New York: Basic Books, 1993.

Galston, Arthur W. *Life Processes of Plants.* New York: Scientific American Library, 1994.

Gamow, George. *The New World of Mr. Tompkins.* Cambridge, England: Cambridge University Press, 1999.

Gani, Joseph M., ed. *The Craft of Probabilistic Modeling.* New York: Springer-Verlag, 1986.

García-Ballester, Luis. *Galen and Galenism: Theory and Medical Practice from Antiquity to the European Renaissance.* Burlington, Vt.: Ashgate, 2002.

Gardner, Eldon J., and D. Peter Snustad. *Principles of Genetics.* 7th ed. New York: John Wiley & Sons, 1984.

Gardner, Robert, and Eric Kemer. *Science Projects About Temperature and Heat.* Berkeley Heights, N.J.: Enslow Publishers, 1994.

Garner, Geraldine. *Careers in Engineering.* 3d ed. New York: McGraw-Hill, 2009.

Gartner, Carol B. *Rachel Carson.* New York: Frederick Ungar, 1983.

Gasser, James, ed. *A Boole Anthology: Recent and Classical Studies in the Logic of George Boole.* Dordrecht, the Netherlands: Kluwer, 2000.

Gay, Peter. *The Enlightenment: The Science of Freedom.* New York: W. W. Norton, 1996.

_____. *Freud: A Life for Our Time.* New York: W. W. Norton, 1988.

Gazzaniga, Michael S. *The Social Brain: Discovering the Networks of the Mind.* New York: Basic Books, 1985.

Geison, Gerald. *The Private Science of Louis Pasteur.* Princeton, N.J.: Princeton University Press, 1995.

Gell-Mann, Murray. *The Quark and the Jaguar: Adventures in the Simple and the Complex.* New York: W. H. Freeman, 1994.

Georgotas, Anastasios, and Robert Cancro, eds. *Depression and Mania.* New York: Elsevier, 1988.

Gerock, Robert. *Mathematical Physics.* Chicago: University of Chicago Press, 1985.

Gest, Howard. *Microbes: An Invisible Universe.* Washington, D.C.: ASM Press, 2003.

Gesteland, Raymond F., Thomas R. Cech, and John F. Atkins, eds. *The RNA World: The Nature of Modern RNA Suggests a Prebiotic RNA.* 2d ed. Cold Spring Harbor, N.Y.: Cold Spring Harbor Laboratory Press, 1999.

Gigerenzer, Gerd, et al. *The Empire of Chance: How Probability Theory Changed Science and Everyday Life.* New York: Cambridge University Press, 1989.

Gilder, Joshua, and Anne-Lee Gilder. *Heavenly Intrigue: Johannes Kepler, Tycho Brahe, and the Murder Behind One of History's Greatest Scientific Discoveries.* New York: Doubleday, 2004.

Gillispie, Charles Coulston, Robert Fox, and Ivor Grattan-Guinness. *Pierre-Simon Laplace, 1749-1827: A Life in Exact Science.* Princeton, N.J.: Princeton University Press, 2000.

Gingerich, Owen. *The Book Nobody Read: Chasing the Revolutions of Nicolaus Copernicus.* New York: Walker, 2004.

_____. *The Eye of Heaven: Ptolemy, Copernicus, Kepler.* New York: Springer-Verlag, 1993.

Glashow, Sheldon, with Ben Bova. *Interactions: A Journey Through the Mind of a Particle Physicist and the Matter of This World.* New York: Warner Books, 1988.

Glass, Billy. *Introduction to Planetary Geology.* New York: Cambridge University Press, 1982.

Gleick, James. *Chaos: Making a New Science.* New York: Penguin Books, 1987.

_____. *Isaac Newton.* New York: Pantheon Books, 2003.

Glen, William. *The Road to Jaramillo: Critical Years of the Revolution in Earth Science.* Stanford, Calif.: Stanford University Press, 1982.

Glickman, Todd S., ed. *Glossary of Meteorology.* 2d ed. Boston: American Meteorological Society, 2000.

Goddard, Jolyon, ed. *National Geographic Concise History of Science and Invention: An Illustrated Time Line.* Washington, D.C.: National Geographic, 2010.

Goding, James W. *Monoclonal Antibodies: Principles and Practice.* New York: Academic Press, 1986.

Godwin, Robert, ed. *Mars: The NASA Mission Reports.* Burlington, Ont.: Apogee Books, 2000.

_____. *Mars: The NASA Mission Reports.* Vol. 2. Burlington, Ont.: Apogee Books, 2004.

_____. *Space Shuttle STS Flights 1-5: The NASA Mission Reports.* Burlington, Ont.: Apogee Books, 2001.

Goetsch, David L. *Building a Winning Career in Engineering: 20 Strategies for Success After College.* Upper Saddle River, N.J.: Pearson/Prentice Hall, 2007.

Gohlke, Mary, with Max Jennings. *I'll Take Tomorrow.* New York: M. Evans, 1985.

Gold, Rebecca. *Steve Wozniak: A Wizard Called Woz.* Minneapolis: Lerner, 1994.

Goldsmith, Donald. *Nemesis: The Death Star and Other Theories of Mass Extinction.* New York: Berkley Publishing Group, 1985.

Goldsmith, Maurice, Alan Mackay, and James Woudhuysen, eds. *Einstein: The First Hundred Years.* Elmsford, N.Y.: Pergamon Press, 1980.

Golinski, Jan. *Science as Public Culture: Chemistry and Enlightenment in Britain, 1760-1820.* Cambridge, England: Cambridge University Press, 1992.

Golthelf, Allan, and James G. Lennox, eds. *Philosophical Issues in Aristotle's Biology.* Cambridge, England: Cambridge University Press, 1987.

Gooding, David, and Frank A. J. L. James, eds. *Faraday Rediscovered: Essays on the Life and Work of Michael Faraday, 1791-1867.* New York: Macmillan, 1985.

Gordin, Michael D. *A Well-Ordered Thing: Dmitrii Mendeleev and the Shadow of the Periodic Table.* New York: Basic Books, 2004.

Gornick, Vivian. *Women in Science: Then and Now.* New York: Feminist Press at the City University of New York, 2009.

Gould, James L., and Carol Grant Gould. *The Honey Bee.* New York: Scientific American Library, 1988.

Gould, Stephen Jay. *Time's Arrow, Time's Cycle: Myth and Metaphor in the Discovery of Geological Time.* Cambridge, Mass.: Harvard University Press, 1987.

_____. *Wonderful Life: The Burgess Shale and the Nature of History.* New York: W. W. Norton, 1989.

Govindjee, J. T. Beatty, H. Gest, and J.F. Allen, eds. *Discoveries in Photosynthesis.* Berlin: Springer, 2005.

Gow, Mary. *Tycho Brahe: Astronomer.* Berkeley Heights, N.J.: Enslow, 2002.

Graham, Loren R. *Science, Philosophy, and Human Behavior in the Soviet Union.* New York: Columbia University Press, 1987.

Grattan-Guinness, Ivor. *The Norton History of the Mathematical Sciences.* New York: W. W. Norton, 1999.

Gray, Robert M., and Lee D. Davisson. *Random Processes: A Mathematical Approach for Engineers.* Englewood Cliffs, N.J.: Prentice-Hall, 1986.

Greene, Mott T. *Geology in the Nineteenth Century: Changing Views of a Changing World.* Ithaca, N.Y.: Cornell University Press, 1982.

Gregory, Andrew. *Harvey's Heart: The Discovery of Blood Circulation.* London: Totem Books, 2001.

Gribbin, John. *Deep Simplicity: Bringing Order to Chaos and Complexity.* New York: Random House, 2005.

_____. *Future Weather and the Greenhouse Effect.* New York: Delacorte Press/Eleanor Friede, 1982.

_____. *The Hole in the Sky: Man's Threat to the Ozone Layer.* New York: Bantam Books, 1988.

_____. *In Search of Schrödinger's Cat: Quantum Physics and Reality.* New York: Bantam Books, 1984.

_____. *In Search of the Big Bang.* New York: Bantam Books, 1986.

_____. *The Omega Point: The Search for the Missing Mass and the Ultimate Fate of the Universe.* New York: Bantam Books, 1988.

_____. *The Scientists: A History of Science Told Through the Lives of Its Greatest Inventors.* New York: Random House, 2002.

Gribbin, John, ed. *The Breathing Planet.* New York: Basil Blackwell, 1986.

Gutkind, Lee. *Many Sleepless Nights: The World of Organ Transplantation.* New York: W. W. Norton, 1988.

Hackett, Edward J., et al., eds. *The Handbook of Science and Technology Studies.* 3d ed. Cambridge, Mass.: MIT Press, 2008.

Hald, Anders. *A History of Mathematical Statistics from 1750 to 1950.* New York: John Wiley & Sons, 1998.

_____. *A History of Probability and Statistics and Their Applications Before 1750.* New York: John Wiley & Sons, 1990.

Hall, A. Rupert. *The Scientific Revolution, 1500-1750.* 3d ed. New York: Longman, 1983.

Halliday, David, and Robert Resnick. *Fundamentals of Physics: Extended Version.* New York: John Wiley & Sons, 1988.

Halliday, David, Robert Resnick, and Jearl Walker. *Fundamentals of Physics.* 7th ed. New York: John Wiley & Sons, 2004.

Hankins, Thomas L. *Science and the Enlightenment.* Reprint. New York: Cambridge University Press, 1991.

Hanlon, Michael, and Arthur C. Clarke. *The Worlds of Galileo: The Inside Story of NASA's Mission to Jupiter.* New York: St. Martin's Press, 2001.

Hanson, Earl D. *Understanding Evolution.* New York: Oxford University Press, 1981.

Hargittai, István. *Martians of Science: Five Physicists Who Changed the Twentieth Century.* New York: Oxford University Press, 2006.

Harland, David M. *Jupiter Odyssey: The Story of NASA's Galileo Mission.* London: Springer-Praxis, 2000.

_____. *Mission to Saturn: Cassini and the Huygens Probe.* London: Springer-Praxis, 2002.

_____. *The Space Shuttle: Roles, Missions, and Accomplishments.* New York: John Wiley & Sons, 1998.

Harland, David M., and John E. Catchpole. *Creating the International Space Station.* London: Springer-Verlag, 2002.

Harrington, J. W. *Dance of the Continents:* New York: V. P. Tarher, 1983.

Harrington, Philip S. *The Space Shuttle: A Photographic History.* San Francisco: Brown Trout, 2003.

Harris, Henry. *The Birth of the Cell.* New Haven, Conn.: Yale University Press, 1999.

Harrison, Edward R. *Cosmology: The Science of the Universe.* Cambridge England: Cambridge University Press, 1981.

Hart, Michael H. *The 100: A Ranking of the Most Influential Persons in History.* New York: Galahad Books, 1982.

Hart-Davis, Adam. *Chain Reactions: Pioneers of British Science and Technology and the Stories That Link Them.* London: National Portrait Gallery, 2000.

Hartmann, William K. *The Cosmic Voyage: Through Time and Space.* Belmont, Calif: Wadsworth, 1990.

_____. *Moons and Planets.* 5th ed. Belmont, Calif.: Brooks-Cole Publishing, 2005.

Hartwell, L. H., et al. *Genetics: From Genes to Genomes.* 2d ed. New York: McGraw-Hill, 2004.

Harvey, William. *The Circulation of the Blood and Other Writings.* New York: Everyman's Library, 1990.

Haskell, G., and Michael Rycroft. *International Space Station: The Next Space Marketplace.* Boston: Kluwer Academic, 2000.

Hathaway, N. *The Friendly Guide to the Universe.* New York: Penguin Books, 1994.

Havil, Julian. *Gamma: Exploring Euler's Constant.* Princeton, N.J.: Princeton University Press, 2003.

Hawking, Stephen W. *A Brief History of Time.* New York: Bantam Books, 1988.

Haycock, David. *William Stukeley: Science, Religion, and Archeology in Eighteenth-Century England.* Woodbridge, England: Boydell Press, 2002.

Hazen, Robert. *The Breakthrough: The Race for the Superconductor.* New York: Summit Books, 1988.

Headrick, Daniel R. *Technology: A World History.* New York: Oxford University Press, 2009.

Heath, Sir Thomas L. *A History of Greek Mathematics: From Thales to Euclid.* 1921. Reprint. New York: Dover Publications, 1981.

Heilbron, J. L. *The Dilemmas of an Upright Man: Max Planck As a Spokesman for German Science.* Berkeley: University of California Press, 1986.

_____. *Electricity in the Seventeenth and Eighteenth Centuries: A Study in Early Modern Physics.* Mineola, N.Y.: Dover Publications, 1999.

_____. *Elements of Early Modern Physics.* Berkeley: University of California Press, 1982.

_____. *Geometry Civilized: History, Culture, and Technique.* Oxford, England: Clarendon Press, 1998.

Heilbron, J. L., and Robert W. Seidel. *Lawrence and His Laboratory: A History of the Lawrence Berkeley Laboratory.* Berkeley: University of California Press, 1989.

Heisenberg, Elisabeth. *Inner Exile: Recollections of a Life with Werner Heisenberg.* Translated by S. Cappelari and C. Morris. Boston: Birkhäuser, 1984.

Hellegouarch, Yves. *Invitation to the Mathematics of Fermat-Wiles.* San Diego: Academic Press, 2001.

Henig, Robin Marantz. *The Monk in the Garden: The Lost and Found Genius of Gregor Mendel, the Father of Genetics.* New York: Mariner Books, 2001.

Henry, Helen L., and Anthony W. Norman, eds. *Encyclopedia of Hormones.* 3 vols. San Diego: Academic Press, 2003.

Henry, John. *Moving Heaven and Earth: Copernicus and the Solar System.* Cambridge, England: Icon, 2001.

Herrmann, Bernd, and Susanne Hummel, eds. *Ancient DNA: Recovery and Analysis of Genetic Material from Paleographic, Archaeological, Museum, Medical, and Forensic Specimens.* New York: Springer-Verlag, 1994.

Hershel, Sir John Frederic William, and Pierre-Simon Laplace. *Essays in Astronomy.* University Press of the Pacific, 2002.

Hillar, Marian, and Claire S. Allen. *Michael Servetus: Intellectual Giant, Humanist, and Martyr.* New York: University Press of America, 2002.

Hobson, J. Allan. *The Dreaming Brain.* New York: Basic Books, 1988.

Hodge, Paul. *Galaxies.* Cambridge, Mass.: Harvard University Press, 1986.

Hodges, Andrew. *Alan Turing: The Enigma.* 1983. Reprint. New York: Walker, 2000.

Hofmann, James R., David Knight, and Sally Gregory Kohlstedt, eds. *André-Marie Ampère: Enlightenment and Electrodynamics*. Cambridge, England: Cambridge University Press, 1996.

Holland, Suzanne, Karen Lebacqz, and Laurie Zoloth, eds. *The Human Embryonic Stem Cell Debate: Science, Ethics, and Public Policy*. Cambridge, Mass.: MIT Press, 2001.

Holmes, Frederic Lawrence. *Antoine Lavoisier, the Next Crucial Year: Or, the Sources of His Quantitative Method in Chemistry*. Princeton, N.J.: Princeton University Press, 1998.

Horne, James. *Why We Sleep*. New York: Oxford University Press, 1988.

Hoskin, Michael A. *The Herschel Partnership: As Viewed by Caroline*. Cambridge, England: Science History, 2003.

_____. *William Herschel and the Construction of the Heavens*. New York: Norton; 1964.

Howse, Derek. *Greenwich Time and the Discovery of Longitude*. New York: Oxford University Press, 1980.

Hoyt, William G. *Planet X and Pluto*. Tucson: University of Arizona Press, 1980.

Hsü, Kenneth J. *The Great Dying*. San Diego: Harcourt Brace Jovanovich, 1986.

Huerta, Robert D. *Giants of Delft, Johannes Vermeer and the Natural Philosophers: The Parallel Search for Knowledge During the Age of Discovery*. Lewisburg, Pa.: Bucknell University Press, 2003.

Hummel, Susanne. *Fingerprinting the Past: Research on Highly Degraded DNA and Its Applications*. New York: Springer-Verlag, 2002.

Hunter, Michael, ed. *Robert Boyle Reconsidered*. New York: Cambridge University Press, 1994.

Hynes, H. Patricia. *The Recurring Silent Spring*. New York: Pergamon Press, 1989.

Ihde, Aaron J. *The Development of Modern Chemistry*. New York: Dover, 1984.

Irwin, Patrick G. J. *Giant Planets of Our Solar System: Atmospheres, Composition, and Structure*. London: Springer-Praxis, 2003.

Isaacson, Walter. *Benjamin Franklin: An American Life*. New York: Simon & Schuster, 2003.

Jackson, Myles. *Spectrum of Belief: Joseph Fraunhofer and the Craft of Precision Optics*. Cambridge, Mass.: MIT Press, 2000.

Jacobsen, Theodor S. *Planetary Systems from the Ancient Greeks to Kepler*. Seattle: University of Washington Press, 1999.

Jacquette, Dale. *On Boole*. Belmont, Calif.: Wadsworth, 2002.

Jaffe, Bernard. *Crucibles: The Story of Chemistry*. New York: Dover, 1998.

James, Ioan. *Remarkable Mathematicians: From Euler to Von Neumann*. Cambridge, England: Cambridge University Press, 2002.

Janowsky, David S., Dominick Addario, and S. Craig Risch. *Psychopharmacology Case Studies*. 2d ed. New York: Guilford Press, 1987.

Jeffreys, Diarmuid. *Aspirin: The Remarkable Story of a Wonder Drug*. London: Bloomsbury Publishing, 2004.

Jenkins, Dennis R. *Space Shuttle: The History of the National Space Transportation System: The First 100 Missions*. Stillwater, Minn.: Voyageur Press, 2001.

Johanson, Donald, and B. Edgar. *From Lucy to Language*. New York: Simon and Schuster, 1996.

Johanson, Donald C., and Maitland A. Edey. *Lucy: The Beginnings of Humankind*. New York: Simon & Schuster, 1981.

Johanson, Donald C., and James Shreeve. *Lucy's Child: The Discovery of a Human Ancestor*. New York: William Morrow, 1989.

Johnson, George. *Strange Beauty: Murray Gell-Mann and the Revolution in Twentieth-Century Physics*. New York: Alfred A. Knopf, 1999.

Jones, Henry Bence. *Life and Letters of Faraday*. 2 vols. London: Longmans, Green and Co., 1870.

Jones, Meredith L., ed. *Hydrothermal Vents of the Eastern Pacific: An Overview*. Vienna, Va.: INFAX, 1985.

Jones, Sheila. *The Quantum Ten: A Story of Passion, Tragedy, and Science*. Toronto: Thomas Allen, 2008.

Jones, W. H. S., trans. *Hippocrates*. 4 vols. 1923-1931. Reprint. New York: Putnam, 1995.

Jordan, Paul. *Neanderthal: Neanderthal Man and the Story of Human Origins*. Gloucestershire, England: Sutton, 2001.

Joseph, George Gheverghese. *The Crest of the Peacock: The Non-European Roots of Mathematics*. London: Tauris, 1991.

Jungnickel, Christa, and Russell McCormmach. *Cavendish: The Experimental Life*. Lewisburg, Pa.: Bucknell University Press, 1999.

Kaplan, Robert. *The Nothing That Is: A Natural History of Zero.* New York: Oxford University Press, 2000.

Kargon, Robert H. *The Rise of Robert Millikan: Portrait of a Life in American Science.* Ithaca, N.Y.: Cornell University Press, 1982.

Katz, Jonathan. *The Biggest Bang: The Mystery of Gamma-Ray Bursts.* London: Oxford University Press, 2002.

Kellogg, William W., and Robert Schware. *Climate Change and Society: Consequences of Increasing Atmospheric Carbon Dioxide.* Boulder, Colo.: Westview Press, 1981.

Kelly, Thomas J. *Moon Lander: How We Developed the Apollo Lunar Module.* Washington, D.C.: Smithsonian Books, 2001.

Kemper, John D., and Billy R. Sanders. *Engineers and Their Profession.* 5th ed. New York: Oxford University Press, 2001.

Kepler, Johannes. *New Astronomy.* Translated by William H. Donahue. New York: Cambridge University Press, 1992.

Kermit, Hans. *Niels Stensen: The Scientist Who Was Beatified.* Translated by Michael Drake. Herefordshire, England: Gracewing 2003.

Kerns, Thomas A. *Jenner on Trial: An Ethical Examination of Vaccine Research in the Age of Smallpox and the Age of AIDS.* Lanham, Md.: University Press of America, 1997.

Kerrod, Robin. *Hubble: The Mirror on the Universe.* Richmond Hill, Ont.: Firefly Books, 2003.

_____. *Space Shuttle.* New York: Gallery Books, 1984.

Kevles, Bettyann. *Naked to the Bones: Medical Imaging in the Twentieth Century.* Reading, Mass.: Addison Wesley, 1998.

Kiessling, Ann, and Scott C. Anderson. *Human Embryonic Stem Cells: An Introduction to the Science and Therapeutic Potential.* Boston: Jones and Bartlett, 2003.

King, Helen. *Greek and Roman Medicine.* London: Bristol Classical, 2001.

_____. *Hippocrates' Woman: Reading the Female Body in Ancient Greece.* New York: Routledge, 1998.

Kirkham, M. B. *Principles of Soil and Plant Water Relations.* St. Louis: Elsevier, 2005.

Kline, Morris. *Mathematical Thought from Ancient to Modern Times.* New York: Oxford University Press, 1990.

Klotzko, Arlene Judith, ed. *The Cloning Sourcebook.* New York: Oxford University Press, 2001.

Knipe, David, Peter Howley, and Diane Griffin. *Field's Virology.* 2 vols. New York: Lippincott Williams and Wilkens, 2001.

Knowles, Richard V. *Genetics, Society, and Decisions.* Columbus, Ohio: Charles E. Merrill, 1985.

Koestler, Arthur. *The Sleepwalkers.* New York: Penguin Books, 1989.

Kolata, Gina Bari. *Clone: The Road to Dolly, and the Path Ahead.* New York: William Morrow, 1998.

Komszik, Louis. *Applied Calculus of Variations for Engineers.* Boca Raton, Fla.: CRC Press, 2009.

Kramer, Barbara. *Neil Armstrong: The First Man on the Moon.* Springfield, N.J.: Enslow, 1997.

Krane, Kenneth S. *Modern Physics.* New York: John Wiley & Sons, 1983.

Lagerkvist, Ulf. *Pioneers of Microbiology and the Nobel Prize.* River Edge, N.J.: World Scientific Publishing, 2003.

Lamarck, Jean-Baptiste. *Lamarck's Open Mind: The Lectures.* Gold Beach, Ore.: High Sierra Books, 2004.

_____. *Zoological Philosophy: An Exposition with Regard to the Natural History of Animals.* Translated by Hugh Elliot with introductory essay by David L. Hull and Richard W. Burckhardt, Jr. Chicago: University of Chicago Press, 1984.

Landes, Davis S. *Revolution in Time: Clocks and the Making of the Modern World.* Rev. ed. Cambridge, Mass.: Belknap Press, 2000.

Langone, John. *Superconductivity: The New Alchemy.* Chicago: Contemporary Books, 1989.

Lappé, Marc. *Broken Code: The Exploitation of DNA.* San Francisco: Sierra Club Books, 1984.

_____. *Germs That Won't Die.* Garden City, N.Y.: Doubleday, 1982.

La Thangue, Nicholas B., and Lasantha R. Bandara, eds. *Targets for Cancer Chemotherapy: Transcription Factors and Other Nuclear Proteins.* Totowa, N.J.: Humana Press, 2002.

Laudan, R. *From Mineralogy to Geology: The Foundations of a Science, 1650-1830.* Chicago: University of Chicago Press, 1987.

Lauritzen, Paul, ed. *Cloning and the Future of Human Embryo Research.* New York: Oxford University Press, 2001.

Leakey, Mary D. *Disclosing the Past.* New York: Doubleday, 1984.

Le Grand, Homer E. *Drifting Continents and Shifting Theories.* New York: Cambridge University Press, 1988.

Levy, David H. *Clyde Tombaugh: Discoverer of Planet Pluto.* Tucson: University of Arizona Press, 1991.

Lewin, Benjamin. *Genes III.* 3d ed. New York: John Wiley & Sons, 1987.

_____. *Genes IV.* New York: Oxford University Press, 1990.

Lewin, Roger. *Bones of Contention: Controversies in the Search for Human Origins.* New York: Simon & Schuster, 1987.

Lewis, Richard S. *The Voyages of Columbia: The First True Spaceship.* New York: Columbia University Press, 1984.

Lindberg, David C. *The Beginnings of Western Science.* Chicago: University of Chicago Press, 1992.

Lindley, David. *Degrees Kelvin: A Tale of Genius, Invention, and Tragedy.* Washington, D.C.: Joseph Henry Press, 2004.

Linzmayer, Owen W. *Apple Confidential: The Real Story of Apple Computer, Inc.* San Francisco: No Starch Press, 1999.

Lloyd, G. E. R., and Nathan Sivin. *The Way and the Word: Science and Medicine in Early China and Greece.* New Haven, Conn.: Yale University Press, 2002.

Logan, J. David. *Applied Mathematics.* 3d ed. Hoboken, N.J.: John Wiley & Sons, 2006.

Logsdon, John M. *Together in Orbit: The Origins of International Participation in the Space Station.* Washington, D.C.: National Aeronautics and Space Administration, 1998.

Longrigg, James. *Greek Medicine: From the Heroic to the Hellenistic Age.* New York: Routledge, 1998.

_____. *Greek Rational Medicine: Philosophy and Medicine from Alcmaeon to the Alexandrians.* New York: Routledge, 1993.

Lorenz, Edward. *The Essence of Chaos.* Reprint. St. Louis: University of Washington Press, 1996.

Lorenz, Ralph, and Jacqueline Mitton. *Lifting Titan's Veil: Exploring the Giant Moon of Saturn.* London: Cambridge University Press, 2002.

Loudon, Irvine. *The Tragedy of Childbed Fever.* New York: Oxford University Press, 2000.

Luck, Steve, ed. *International Encyclopedia of Science and Technology.* New York: Oxford University Press, 1999.

Lutgens, Frederick K., and Edward J. Tarbuck. *The Atmosphere: An Introduction to Meteorology.* 2d ed. Englewood Cliffs, N.J.: Prentice-Hall, 1982.

Lyell, Charles. *Elements of Geology.* London: John Murray, 1838.

_____. *The Geological Evidences of the Antiquity of Man with Remarks on Theories of the Origin of Species by Variation.* London: John Murray, 1863.

_____. *Principles of Geology, Being an Attempt to Explain the Former Changes of the Earth's Surface by Reference to Causes Now in Operation.* 3 vols. London: John Murray, 1830-1833.

Lynch, William T. *Solomon's Child: Method in the Early Royal Society of London.* Stanford, Calif.: Stanford University Press, 2000.

Ma, Pearl, and Donald Armstrong, eds. *AIDS and Infections of Homosexual Men.* Stoneham, Mass: Butterworths, 1989.

MacDonald, Allan H., ed. *Quantum Hall Effect: A Perspective.* Boston: Kluwer Academic Publishers, 1989.

MacHale, Desmond. *George Boole: His Life and Work.* Dublin: Boole Press, 1985.

Machamer, Peter, ed. *The Cambridge Companion to Galileo.* New York: Cambridge University Press, 1998.

Mactavish, Douglas. *Joseph Lister.* New York: Franklin Watts, 1992.

Mader, Sylvia S. *Biology.* 3d ed. Dubuque, Iowa: Win. C. Brown, 1990.

Magner, Lois. *A History of Medicine.* New York: Marcel Dekker, 1992.

Mahoney, Michael Sean. *The Mathematical Career of Pierre de Fermat, 1601-1665.* 2d rev. ed. Princeton, N.J.: Princeton University Press, 1994.

Mammana, Dennis L., and Donald W. McCarthy, Jr. *Other Suns, Other Worlds? The Search for Extrasolar Planetary Systems.* New York: St. Martin's Press, 1996.

Mandelbrot, B. B. *Fractals and Multifractals: Noise, Turbulence, and Galaxies.* New York: Springer-Verlag, 1990.

Mann, Charles, and Mark Plummer. *The Aspirin Wars: Money, Medicine and 100 Years of Rampant Competition.* New York: Knopf, 1991.

Marco, Gino J., Robert M. Hollingworth, and William Durham, eds. *Silent Spring Revisited.* Washington, D.C.: American Chemical Society, 1987.

Margolis, Howard. *It Started with Copernicus.* New York: McGraw-Hill, 2002.

Marshak, Daniel R., Richard L. Gardner, and David Gottlieb, eds. *Stem Cell Biology.* Woodbury, N.Y.: Cold Spring Harbor Laboratory Press, 2002.

Martzloff, Jean-Claude. *History of Chinese Mathematics.* Translated by Stephen S. Wilson. Berlin: Springer, 1987.

Massey, Harrie Stewart Wilson. *The Middle Atmosphere as Observed by Balloons, Rockets, and Satellites.* London: Royal Society, 1980.

Masson, Jeffrey M. *The Assault on Truth: Freud's Suppression of the Seduction Theory.* New York: Farrar, Straus and Giroux, 1984.

Mateles, Richard I. *Penicillin: A Paradigm for Biotechnology.* Chicago: Canadida Corporation, 1998.

Mayo, Jonathan L. *Superconductivity: The Threshold of a New Technology.* Blue Ridge Summit, Pa.: TAB Books, 1988.

McCarthy, Shawn P. *Engineer Your Way to Success: America's Top Engineers Share Their Personal Advice on What They Look for in Hiring and Promoting.* Alexandria, Va.: National Society of Professional Engineers, 2002.

McCay, Mary A. *Rachel Carson.* New York: Twayne, 1993.

McDonnell, John James. *The Concept of an Atom from Democritus to John Dalton.* Lewiston, N.Y.: Edwin Mellen Press, 1991.

McEliece, Robert. *Finite Fields for Computer Scientists and Engineers.* Boston: Kluwer Academic, 1987.

McGraw-Hill Concise Encyclopedia of Science and Technology. 6th ed. New York: McGraw-Hill, 2009.

McGraw-Hill Dictionary of Scientific and Technical Terms. 6th ed. New York: McGraw-Hill, 2002.

McGrayne, Sharon Bertsch. *Nobel Prize Women in Science: Their Lives, Struggles and Momentous Discoveries.* 2d ed. Washington, D.C.: Joseph Henry Press, 1998.

McIntyre, Donald B., and Alan McKirdy. *James Hutton: The Founder of Modern Geology.* Edinburgh: Stationery Office, 1997.

McLester, John, and Peter St. Pierre. *Applied Biomechanics: Concepts and Connections.* Belmont, Calif.: Thomson Wadsworth, 2008.

McMullen, Emerson Thomas. *William Harvey and the Use of Purpose in the Scientific Revolution: Cosmos by Chance or Universe by Design?* Lanhan, Md.: University Press of America, 1998.

McQuarrie, Donald A. *Quantum Chemistry.* Mill Valley, Calif: University Science Books, 1983.

Menard, H. W. *The Ocean of Truth: A Personal History of Global Tectonics.* Princeton, N.J.: Princeton University Press, 1986.

Menzel, Donald H., and Jay M. Pasachoff. *Stars and Planets.* Boston: Houghton Mifflin Company, 1983.

Merrell, David J. *Ecological Genetics.* Minneapolis: University of Minnesota Press, 1981.

Mettler, Lawrence E., Thomas G. Gregg, and Henry E. Schaffer. *Population Genetics and Evolution.* 2d ed. Englewood Cliffs, N.J.: Prentice-Hall, 1988.

Meyers, Robert A., ed. *Encyclopedia of Physical Science and Technology.* 18 vols. San Diego: Academic Press, 2005.

Meyerson, Daniel. *The Linguist and the Emperor: Napoleon and Champollion's Quest to Decipher the Rosetta Stone.* New York: Ballantine Books, 2004.

Middleton, W. E. Knowles. *A History of the Thermometer and Its Use in Meteorology.* Ann Arbor, Mich.: UMI Books on Demand, 1996.

Miller, Ron. *Extrasolar Planets.* Brookfield, Conn.: Twenty-First Century Books, 2002.

Miller, Stanley L. *From the Primitive Atmosphere to the Prebiotic Soup to the Pre-RNA World.* Washington, D.C.: National Aeronautics and Space Administration, 1996.

Mishkin, Andrew. *Sojourner: An Insider's View of the Mars Pathfinder Mission.* New York: Berkeley Books, 2003.

Mlodinow, Leonard. *Euclid's Window: A History of Geometry from Parallel Lines to Hyperspace.* New York: Touchstone, 2002.

Monmonier, Mark. *Air Apparent: How Meteorologists Learned to Map, Predict, and Dramatize Weather.* Chicago: University of Chicago Press, 2000.

Moore, Keith L. *The Developing Human.* Philadelphia: W. B. Saunders, 1988.

Moore, Patrick. *Eyes on the University: The Story of the Telescope.* New York: Springer-Verlag, 1997.

_____. *Patrick Moore's History of Astronomy.* 6th rev. ed. London: Macdonald, 1983.

Morell, V. *Ancestral Passions: The Leakey Family and the Quest for Humankind's Beginnings.* New York: Simon & Schuster, 1995.

Morgan, Kathryn A. *Myth and Philosophy from the Presocratics to Plato.* New York: Cambridge University Press, 2000.

Moritz, Michael. *The Little Kingdom: The Private Story of Apple Computer.* New York: Morrow, 1984.

Morris, Peter, ed. *Making the Modern World: Milestones of Science and Technology.* 2d ed. Chicago: KWS Publishers, 2011.

Morris, Richard. *The Last Sorcerers: The Path from Alchemy to the Periodic Table.* Washington, D.C.: Joseph Henry Press, 2003.

Morrison, David. *Voyages to Saturn.* NASA SP-451. Washington, D.C.: National Aeronautics and Space Administration, 1982.

Morrison, David, and Tobias Owen. *The Planetary System.* 3d ed. San Francisco: Addison Wesley, 2003.

Morrison, David, and Jane Samz. *Voyage to Jupiter*. NASA SP-439. Washington, D.C.: Government Printing Office, 1980.

Moss, Ralph W. *Free Radical: Albert Szent-Györgyi and the Battle over Vitamin C*. New York: Paragon House, 1988.

Muirden, James. *The Amateur Astronomer's Handbook*. 3d ed. New York: Harper & Row, 1987.

Mullis, Kary. *Dancing Naked in the Mind Field*. New York: Pantheon Books, 1998.

Mulvihill, John J. *Catalog of Human Cancer Genes: McKusick's Mendelian Inheritance in Man for Clinical and Research Oncologists*. Foreword by Victor A. McKusick. Baltimore: The Johns Hopkins University Press, 1999.

Nahin, Paul J. *Oliver Heaviside: Sage in Solitude*. New York: IEEE Press, 1987.

Ne'eman, Yuval, and Yoram Kirsh. *The Particle Hunters*. New York: Cambridge University Press, 1986.

Netz, Reviel. *The Shaping of Deduction in Greek Mathematics: A Study in Cognitive History*. New York: Cambridge University Press, 2003.

Neu, Jerome, ed. *The Cambridge Companion to Freud*. New York: Cambridge University Press, 1991.

North, John. *The Norton History of Astronomy and Cosmology*. New York: W. W. Norton, 1995.

Nutton, Vivian, ed. *The Unknown Galen*. London: Institute of Classical Studies, University of London, 2002.

Nye, Mary Jo. *Before Big Science: The Pursuit of Modern Chemistry and Physics, 1800-1940*. New York: Twayne, 1996.

Nye, Robert D. *Three Psychologies: Perspectives from Freud, Skinner, and Rogers*. Pacific Grove, Calif.: Brooks-Cole, 1992.

Oakes, Elizabeth H. *Encyclopedia of World Scientists*. Rev ed. New York: Facts on File, 2007.

Olson, James S., and Robert L. Shadle, ed. *Encyclopedia of the Industrial Revolution in America*. Westport, Conn.: Greenwood Press, 2002.

Olson, Steve. *Mapping Human History: Genes, Races and Our Common Origins*. New York: Houghton Mifflin, 2002.

Ozima, Minoru. *The Earth: Its Birth and Growth*. Translated by Judy Wakabayashi. Cambridge, England: Cambridge University Press, 1981.

Pagels, Heinz R. *The Cosmic Code*. New York: Simon & Schuster, 1982.

_____. *The Cosmic Code: Quantum Physics As the Law of Nature*. New York: Bantam Books, 1984.

_____. *Perfect Symmetry: The Search for the Beginning of Time*. New York: Simon & Schuster, 1985.

Pai, Anna C. *Foundations for Genetics: A Science for Society*. 2d ed. New York: McGraw-Hill, 1984.

Pais, Abraham. *The Genius of Science: A Portrait Gallery of Twentieth-Century Physicists*. New York: Oxford University Press, 2000.

Palmer, Douglas. *Neanderthal*. London: Channel 4 Books, 2000.

Parker, Barry R. *The Vindication of the Big Bang: Breakthroughs and Barriers*. New York: Plenum Press, 1993.

Parslow, Christopher Charles. *Rediscovering Antiquity: Karl Weber and the Excavation of Herculaneum, Pompeii, and Stabiae*. New York: Cambridge University Press, 1998.

Parson, Ann B. *The Proteus Effect: Stem Cells and Their Promise*. Washington, D.C.: National Academies Press, 2004.

Pedrotti, L., and F. Pedrotti. *Optics and Vision*. Upper Saddle River, N.J.: Prentice Hall, 1998.

Peitgen, Heinz-Otto, and Dietmar Saupe, eds. *The Science of Fractal Images*. New York: Springer-Verlag, 1988.

Peltonen, Markku, ed. *The Cambridge Companion to Bacon*. New York: Cambridge University Press, 1996.

Penrose, Roger. *The Emperor's New Mind: Concerning Computers, Minds, and the Laws of Physics*. New York: Oxford University Press, 1989.

Persaud, T. V. N. *A History of Anatomy: The Post-Vesalian Era*. Springfield, Ill.: Charles C. Thomas, 1997.

Peterson, Carolyn C., and John C. Brant. *Hubble Vision: Astronomy with the Hubble Space Telescope*. London: Cambridge University Press, 1995.

_____. *Hubble Vision: Further Adventures with the Hubble Space Telescope*. 2d ed. New York: Cambridge University Press, 1998.

Pfeiffer, John E. *The Emergence of Humankind*. 4th ed. New York: Harper & Row, 1985.

Piggott, Stuart. *William Stukeley: An Eighteenth-Century Antiquary*. New York: Thames and Hudson, 1985.

Pike, J. Wesley, Francis H. Glorieux, David Feldman. *Vitamin D*. 2d ed. Academic Press, 2004.

Plionis, Manolis, ed. *Multiwavelength Cosmology*. New York: Springer, 2004.

Plotkin, Stanley A., and Edward A. Mortimer. *Vaccines*. 2d ed. Philadelphia: W. B. Saunders, 1994.

Polter, Paul. *Hippocrates*. Cambridge, Mass.: Harvard University Press, 1995.

Popper, Karl R. *The World of Parmenides: Essays on the Presocratic Enlightenment*. Edited by Arne F. Petersen and Jørgen Mejer. New York: Routledge, 1998.

Porter, Roy. *The Greatest Benefit to Mankind: A Medical History of Humanity, from Antiquity to the Present*. New York: W. W. Norton, 1997.

Porter, Roy, ed. *Eighteenth Century Science*. Vol. 4 in *The Cambridge History of Science*. New York: Cambridge University Press, 2003.

Poundstone, William. *Prisoner's Dilemma*. New York: Doubleday, 1992.

Poynter, Margaret, and Arthur L. Lane. *Voyager: The Story of a Space Mission*. New York: Macmillan, 1981.

Principe, Lawrence. *The Aspiring Adept: Robert Boyle and His Alchemical Quest*. Princeton, N.J.: Princeton University Press, 1998.

Prochnow, Dave. *Superconductivity: Experimenting in a New Technology*. Blue Ridge Summit, Pa.: TAB Books, 1989.

Pullman, Bernard. *The Atom in the History of Human Thought*. New York: Oxford University Press, 1998.

Pycior, Helena M. *Symbols, Impossible Numbers, and Geometric Entanglements: British Algebra Through the Commentaries on Newton's Universal Arithmetick*. New York: Cambridge University Press, 1997.

The Rand McNally New Concise Atlas of the Universe. New York: Rand McNally, 1989.

Rao, Mahendra S., ed. *Stem Cells and CNS Development*. Totowa, N.J.: Humana Press, 2001.

Raup, David M. *The Nemesis Affair: A Story of the Death of Dinosaurs and the Ways of Science*. New York: W. W. Norton, 1986.

Raven, Peter H., and George B. Johnson. *Biology*. 2d ed. St. Louis: Times-Mirror/Mosby, 1989.

Raven, Peter H., Ray F. Evert, and Susan E. Eichhorn. *Biology of Plants*. 6th ed. New York: W. H. Freeman, 1999.

Reader, John. *Missing Links: The Hunt for Earliest Man*. Boston: Little, Brown, 1981.

Reichhardt, Tony. *Proving the Space Transportation System: The Orbital Flight Test Program*. NASA NF-137-83. Washington, D.C.: Government Printing Office, 1983.

Remick, Pat, and Frank Cook. *21 Things Every Future Engineer Should Know: A Practical Guide for Students and Parents*. Chicago: Kaplan AEC Education, 2007.

Repcheck, Jack. *The Man Who Found Time: James Hutton and the Discovery of the Earth's Antiquity*. Reading, Mass.: Perseus Books, 2003.

Rescher, Nicholas. *On Leibniz*. Pittsburgh, Pa.: University of Pittsburgh Press, 2003.

Reston, James. *Galileo: A Life*. New York: HarperCollins, 1994.

Rhodes, Richard. *The Making of the Atomic Bomb*. New York: Simon & Schuster, 1986.

Rigutti, Mario. *A Hundred Billion Stars*. Translated by Mirella Giacconi. Cambridge, Mass.: MIT Press, 1984.

Ring, Merrill. *Beginning with the Presocratics*. 2d ed. New York: McGraw-Hill, 1999.

Riordan, Michael. *The Hunting of the Quark*. New York: Simon & Schuster, 1987.

Roan, Sharon. *Ozone Crisis: The Fifteen-Year Evolution of a Sudden Global Emergency*. New York: John Wiley & Sons, 1989.

Robinson, Daniel N. *An Intellectual History of Psychology*. 3d ed. Madison: University of Wisconsin Press, 1995.

Rogers, J. H. *The Giant Planet Jupiter*. New York: Cambridge University Press, 1995.

Rose, Frank. *West of Eden: The End of Innocence at Apple Computer*. New York: Viking, 1989.

Rosenthal-Schneider, Ilse. *Reality and Scientific Truth: Discussions with Einstein, von Laue, and Planck*. Detroit: Wayne State University Press, 1980.

Rossi, Paoli. *The Birth of Modern Science*. Translated by Cynthia De Nardi Ipsen. Oxford, England: Blackwell, 2001.

Rowan-Robinson, Michael. *Cosmology*. London: Oxford University Press, 2003.

Rowland, Wade. *Galileo's Mistake: A New Look at the Epic Confrontation Between Galileo and the Church*. New York: Arcade, 2003.

Rudin, Norah, and Keith Inman. *An Introduction to Forensic DNA Analysis*. Boca Raton, Fla.: CRC Press, 2002.

Rudwick, Martin, J. S. *The Great Devonian Controversy: The Shaping of Scientific Knowledge Among Gentlemanly Specialists*. Chicago: University of Chicago Press, 1985.

Rudwick, M. J. S. *The Meaning of Fossils: Episodes in the History of Paleontology*. Chicago: University of Chicago Press, 1985.

Ruestow, Edward Grant. *The Microscope in the Dutch Republic: The Shaping of Discovery.* New York: Cambridge University Press, 1996.

Ruse, Michael. *The Darwinian Revolution: Science Red in Tooth and Claw.* 2d ed. Chicago: University of Chicago Press, 1999.

Ruspoli, Mario. *Cave of Lascaux.* New York: Harry N. Abrams, 1987.

Sagan, Carl. *Cosmos.* New York: Random House, 1980.

Sagan, Carl, and Ann Druyan. *Comet.* New York: Random House, 1985.

Sandler, Stanley I. *Chemical and Engineering Thermodynamics.* New York: John Wiley & Sons, 1998.

Sang, James H. *Genetics and Development.* London: Longman, 1984.

Sargent, Frederick. *Hippocratic Heritage: A History of Ideas About Weather and Human Health.* New York: Pergamon Press, 1982.

Sargent, Rose-Mary. *The Diffident Naturalist: Robert Boyle and the Philosophy of Experiment.* Chicago: University of Chicago Press, 1995.

Sauer, Mark V. *Principles of Oocyte and Embryo Donation.* New York: Springer, 1998.

Schaaf, Fred. *Comet of the Century: From Halley to Hale-Bopp.* New York: Springer-Verlag, 1997.

Schatzkin, Paul. *The Boy Who Invented Television: A Story of Inspiration, Persistence, and Quiet Passion.* Silver Spring, Md.: TeamCon Books, 2002.

Schiffer, Michael Brian. *Draw the Lightning Down: Benjamin Franklin and Electrical Technology in the Age of the Enlightenment.* Berkeley: University of California Press, 2003.

Schlagel, Richard H. *From Myth to Modern Mind: A Study of the Origins and Growth of Scientific Thought.* New York: Peter Lang Publishing, 1996.

Schlegel, Eric M. *The Restless Universe: Understanding X-Ray Astronomy in the Age of Chandra and Newton.* London: Oxford University Press, 2002.

Schliemann, Heinrich. *Troy and Its Remains: A Narrative of Researches and Discoveries Made on the Site of Ilium and in the Trojan Plain.* London: J. Murray, 1875.

Schneider, Stephen H. *Global Warming: Are We Entering the Greenhouse Century?* San Francisco: Sierra Club Books, 1989.

Schofield, Robert E. *The Enlightened Joseph Priestley: A Study of His Life and Work from 1773 to 1804.* University Park: Pennsylvania State University Pres, 2004.

Scholz, Christopher, and Benoit B. Mandelbrot. *Fractals in Geophysics.* Boston: Kirkäuser, 1989.

Schonfelder, V. *The Universe in Gamma Rays.* New York: Springer-Verlag, 2001.

Schopf, J. William, ed. *The Earth's Earliest Biosphere.* Princeton, N.J.: Princeton University Press, 1983.

Schorn, Ronald A. *Planetary Astronomy: From Ancient Times to the Third Millennium.* College Station: Texas A&M University Press, 1999.

Schwinger, Julian. *Einstein's Legacy.* New York: W. H. Freeman, 1986.

Sears, M., and D. Merriman, eds. *Oceanography: The Past.* New York: Springer-Verlag, 1980.

Seavey, Nina Gilden, Jane S. Smith, and Paul Wagner. *A Paralyzing Fear: The Triumph over Polio in America.* New York: TV Books, 1998.

Segalowitz, Sid J. *Two Sides of the Brain: Brain Lateralization Explored.* Englewood Cliffs, N.J.: Prentice-Hall, 1983.

Segrè, Emilio. *From X-Rays to Quarks.* San Francisco: W. H. Freeman, 1980.

Sekido, Yataro, and Harry Elliot. *Early History of Cosmic Ray Studies: Personal Reminiscences with Old Photographs.* Boston: D. Reidel, 1985.

Sfendoni-Mentzou, Demetra, et al., eds. *Aristotle and Contemporary Science.* 2 vols. New York: P. Lang, 2000-2001.

Shank, Michael H. *The Scientific Enterprise in Antiquity and the Middle Ages.* Chicago: University of Chicago Press, 2000.

Sharratt, Michael. *Galileo: Decisive Innovator.* Cambridge, Mass.: Blackwell, 1994.

Shectman, Jonathan. *Groundbreaking Scientific Experiments, Investigations, and Discoveries of the Eighteenth Century.* Westport, Conn.: Greenwood Press, 2003.

Sheehan, John. *The Enchanted Ring: The Untold Story of Penicillin.* Cambridge, Mass.: MIT Press, 1982.

Shilts, Randy. *And the Band Played On: Politics, People, and the AIDS Epidemic.* New York: St. Martin's Press, 1987.

Silk, Joseph. *The Big Bang.* Rev. ed. New York: W. H. Freeman, 1989.

Silverstein, Arthur M. *A History of Immunology.* San Diego: Academic Press, 1989.

Simmons, John. *The Scientific Hundred: A Ranking of the Most Influential Scientists, Past and Present.* Secaucus, N.J.: Carol, 1996.

Simon, Randy, and Andrew Smith. *Superconductors: Conquering Technology's New Frontier.* New York: Plenum Press, 1988.

Simpson, A. D. C., ed. *Joseph Black, 1728-1799: A Commemorative Symposium.* Edinburgh: Royal Scottish Museum, 1982.

Singh, Simon. *Fermat's Enigma: The Epic Quest to Solve the World's Greatest Mathematical Problem.* New York: Anchor, 1998.

Slayton, Donald K., with Michael Cassutt. *Deke! U.S. Manned Space: From Mercury to the Shuttle.* New York: Forge, 1995.

Smith, A. Mark. *Ptolemy and the Foundations of Ancient Mathematical Optics.* Philadelphia: American Philosophical Society, 1999.

Smith, G. C. *The Boole-De Morgan Correspondence, 1842-1864.* New York: Oxford University Press, 1982.

Smith, Jane S. *Patenting the Sun: Polio and the Salk Vaccine.* New York: Anchor/Doubleday, 1991.

Smith, Robert W. *The Space Telescope: A Study of NASA, Science, Technology and Politics.* New York: Cambridge University Press, 1989.

Smyth, Albert Leslie. *John Dalton, 1766-1844.* Aldershot, England: Ashgate, 1998.

Snider, Alvin. *Origin and Authority in Seventeenth-Century England: Bacon, Milton, Butler.* Toronto: University of Toronto Press, 1994.

Sobel, Dava. *Galileo's Daughter: A Historical Memoir of Science, Faith, and Love.* New York: Penguin Books, 2000.

_____. *Longitude: The True Story of a Lone Genius Who Solved the Greatest Scientific Problem of His Time.* New York: Penguin Books, 1995.

Spangenburg, Ray, and Diane Kit Moser. *Modern Science: 1896-1945.* Rev ed. New York: Facts on File, 2004.

Spilker, Linda J., ed. *Passage to a Ringed World: The Cassini-Huygens Mission to Saturn and Titan.* Washington, D.C.: National Aeronautics and Space Administration, 1997.

Stanley, H. Eugue, and Nicole Ostrowsky, eds. *On Growth and Form: Fractal and Non-Fractal Patterns in Physics.* Dordrecht, the Netherlands: Martinus Nijhoff, 1986.

Starr, Cecie, and Ralph Taggart. *Biology.* 5th ed. Belmont, Calif.: Wadsworth, 1989.

Stefik, Mark J., and Vinton Cerf. *Internet Dreams: Archetypes, Myths, and Metaphors.* Cambridge, Mass.: MIT Press, 1997.

Stein, Sherman. *Archimedes: What Did He Do Besides Cry Eureka?* Washington, D.C.: Mathematical Association of America, 1999.

Steiner, Robert F., and Seymour Pomerantz. *The Chemistry of Living Systems.* New York: D. Van Nostrand, 1981.

Stewart, Ian, and David Tall. *Algebraic Number Theory and Fermat's Last Theorem.* 3d ed. Natick, Mass.: AK Peters, 2002.

Stigler, Stephen M. *The History of Statistics.* Cambridge, Mass.: Harvard University Press, 1986.

Stine, Gerald. *AIDS 2005 Update.* New York: Benjamin Cummings, 2005.

Strathern, Paul. *Mendeleyev's Dream: The Quest for the Elements.* New York: Berkeley Books, 2000.

Streissguth, Thomas. *John Glenn.* Minneapolis, Minn.: Lerner, 1999.

Strick, James. *Sparks of Life: Darwinism and the Victorian Debates over Spontaneous Generation.* Cambridge, Mass.: Harvard University Press, 2000.

Strogatz, Steven H. *Nonlinear Dynamics and Chaos: With Applications to Physics, Biology, Chemistry and Engineering.* Reading, Mass.: Perseus, 2001.

Struik, Dirk J. *The Land of Stevin and Huygens: A Sketch of Science and Technology in the Dutch Republic During the Golden Century.* Boston: Kluwer, 1981.

Stryer, Lubert. *Biochemistry.* 2d ed. San Francisco: W. H. Freeman, 1981.

Stukeley, William. *The Commentarys, Diary, & Common-Place Book & Selected Letters of William Stukeley.* London: Doppler Press, 1980.

Sturtevant, A. H. *A History of Genetics.* 1965. Reprint. Woodbury, N.Y.: Cold Spring Harbor Laboratory Press, 2001.

Sullivan, Woodruff T., ed. *Classics in Radio Astronomy.* Boston: D. Reidel, 1982.

_____. *The Early Years of Radio Astronomy. Reflections Fifty Years After Jansky's Discovery.* New York: Cambridge University Press, 1984.

Sulston, John, and Georgina Ferry. *The Common Thread: A Story of Science, Politics, Ethics, and the Human Genome.* Washington, D.C.: Joseph Henry Press, 2002.

Sutton, Christine. *The Particle Connection.* New York: Simon & Schuster, 1984.

Suzuki, David T., and Peter Knudtson. *Genethics.* Cambridge, Mass.: Harvard University Press, 1989.

Swanson, Carl P., Timothy Merz, and William J. Young. *Cytogenetics: The Chromosome in Division, Inheritance, and Evolution.* 2d ed. Englewood Cliffs, N.J.: Prentice-Hall, 1980.

Swetz, Frank, et al., eds. *Learn from the Masters.* Washington, D.C.: Mathematical Association of America, 1995.

Tanford, Charles. *Franklin Stilled the Waves.* Durham, N.C.: Duke University Press, 1989.

Tarbuck, Edward J., and Frederick K. Lutgens. *The Earth: An Introduction to Physical Geology.* Columbus, Ohio: Charles E. Merrill, 1984.

Tattersall, Ian. *The Last Neanderthal: The Rise, Success, and Mysterious Extinction of Our Closest Human Relatives.* New York: Macmillan, 1995.

Taub, Liba Chaia. *Ptolemy's Universe: The Natural Philosophical and Ethical Foundations of Ptolemy's Astronomy.* Chicago: Open Court, 1993.

Tauber, Alfred I. *Metchnikoff and the Origins of Immunology: From Metaphor to Theory.* New York: Oxford University Press, 1991.

Taubes, Gary. *Nobel Dreams: Power, Deceit and the Ultimate Experiment.* New York: Random House, 1986.

Taylor, Michael E. *Partial Differential Equations I: Basic Theory.* 2d ed. New York: Springer, 2011.

Taylor, Peter Lane. *Science at the Extreme: Scientists on the Cutting Edge of Discovery.* New York: McGraw-Hill, 2001.

Thomas, John M. *Michael Faraday and the Royal Institution: The Genius of Man and Place.* New York: A. Hilger, 1991.

Thompson, A. R., James M. Moran, and George W. Swenson, Jr. *Interferometry and Synthesis in Radio Astronomy.* New York: John Wiley & Sons, 1986.

Thompson, D'Arcy Wentworth. *On Growth and Form.* Mineola, N.Y.: Dover, 1992.

Thoren, Victor E., with John R. Christianson. *The Lord of Uraniborg: A Biography of Tycho Brahe.* New York: Cambridge University Press, 1990.

Thrower, Norman J. W., ed. *Standing on the Shoulders of Giants: A Longer View of Newton and Halley.* Berkeley: University of California Press, 1990.

Thurman, Harold V. *Introductory Oceanography.* 4th ed. Westerville, Ohio: Charles E. Merrill, 1985.

Tietjen, Jill S., et al. *Keys to Engineering Success.* Upper Saddle River, N.J.: Prentice-Hall, 2001.

Tillery, Bill W., Eldon D. Enger, and Frederick C. Ross. *Integrated Science.* New York: McGraw-Hill, 2001.

Tiner, John Hudson. *Louis Pasteur: Founder of Modern Medicine.* Milford, Mich.: Mott Media, 1990.

Todhunter, Isaac. *A History of the Mathematical Theory of Probability: From the Time of Pascal to that of Laplace.* Sterling, Va.: Thoemmes Press, 2001.

Tombaugh, Clyde W., and Patrick Moore. *Out of Darkness: The Planet Pluto.* Harrisburg, Pa.: Stackpole Books, 1980.

Toulmin, Stephen, and June Goodfield. *The Fabric of the Heavens: The Development of Astronomy and Dynamics.* Chicago: University of Chicago Press, 1999.

Townes, Charles H. *How the Laser Happened: Adventures of a Scientist.* New York: Oxford University Press, 1999.

Traill, David A. *Schliemann of Troy: Treasure and Deceit.* London: J. Murray, 1995.

Trefil, James S. *The Dark Side of the Universe. Searching for the Outer Limits of the Cosmos.* New York: Charles Scribner's Sons, 1988.

_____. *From Atoms to Quarks: An Introduction to the Strange World of Particle Physics.* New York: Charles Scribner's Sons, 1980.

_____. *Space, Time, Infinity: The Smithsonian Views the Universe.* New York: Pantheon Books, 1985.

_____. *The Unexpected Vista.* New York: Charles Scribner's Sons, 1983.

Trefil, James, and Robert M. Hazen. *The Sciences: An Integrated Approach.* New York: John Wiley & Sons, 2003.

Trefil, James, ed. *The Encyclopedia of Science and Technology.* New York: Routledge, 2001.

Trento, Joseph J. *Prescription for Disaster: From the Glory of Apollo to the Betrayal of the Shuttle.* New York: Crown, 1987.

Trinkhaus, Eric, ed. *The Emergence of Modern Humans: Biocultural Adaptations in the Later Pleistocene.* Cambridge, England: Cambridge University Press, 1989.

Trounson, Alan O., and David K. Gardner, eds. *Handbook of In Vitro Fertilization.* 2d ed. Boca Raton, Fla.: CRC Press, 1999.

Tucker, Tom. *Bolt of Fire: Benjamin Franklin and His Electrical Kite Hoax.* New York: Public Affairs Press, 2003.

Tucker, Wallace H., and Karen Tucker. *Revealing the Universe: The Making of the Chandra X-Ray Observatory.* Cambridge, Mass.: Harvard University Press, 2001.

Tunbridge, Paul. *Lord Kelvin: His Influence on Electrical Measurements and Units.* London, U.K.: P. Peregrinus, 1992.

Tuplin, C. J., and T. E. Rihll, eds. *Science and Mathematics in Ancient Greek Culture.* New York: Oxford University Press, 2002.

Turnill, Reginald. *The Moonlandings: An Eyewitness Account.* New York: Cambridge University Press, 2003.

United States Office of the Assistant Secretary for Nuclear Energy. *The First Reactor.* Springfield, Va.: National Technical Information Service, 1982.

University of Chicago Press. *Science and Technology Encyclopedia.* Chicago: Author, 2000.

Van Allen, James A. *Origins of Magnetospheric Physics.* Expanded ed. 1983. Reprint. Washington, D.C.: Smithsonian Institution Press, 2004.

Van Dulken, Stephen. *Inventing the Nineteenth Century: One Hundred Inventions That Shaped the Victorian Age.* New York: New York University Press, 2001.

Van Heijenoort, Jean. *From Frege to Gödel: A Source Book in Mathematical Logic, 1879-1931.* Cambridge, Mass.: Harvard University Press, 2002.

Verschuur, Gerrit L. *Hidden Attraction: The History and Mystery of Magnetism.* New York: Oxford University Press, 1993.

_____. *The Invisible Universe Revealed: The Story of Radio Astronomy.* New York: Springer-Verlag, 1987.

Villard, Ray, and Lynette R. Cook. *Infinite Worlds: An Illustrated Voyage to Planets Beyond Our Sun.* Foreword by Geoffrey W. Marcy and afterword by Frank Drake. Berkeley: University of California Press, 2005.

Viney, Wayne. *A History of Psychology: Ideas and Context.* Boston: Allyn & Bacon, 1993.

Vogt, Gregory L. *John Glenn's Return to Space.* Brookfield, Conn.: Millbrook Press, 2000.

Von Bencke, Matthew J. *The Politics of Space: A History of U.S.-Soviet/Russian Competition and Cooperation in Space.* Boulder, Colo.: Westview Press, 1996.

Wagener, Leon. *One Giant Leap: Neil Armstrong's Stellar American Journey.* New York: Forge Books, 2004.

Wakefield, Robin, ed. *The First Philosophers: The Presocratics and the Sophists.* New York: Oxford University Press, 2000.

Waldman, G. *Introduction to Light.* Englewood Cliffs, N.J.: Prentice Hall, 1983.

Walker, James S. *Physics.* 2d ed. Upper Saddle River, N.J.: Pearson Prentice Hall, 2004.

Wallace, Robert A., Jack L. King, and Gerald P. Sanders. *Biosphere: The Realm of Life.* 2d ed. Glenview, Ill.: Scott, Foresman, 1988.

Waller, John. *Einstein's Luck: The Truth Behind Some of the Greatest Scientific Discoveries.* New York: Oxford University Press, 2002.

_____. *Fabulous Science: Fact and Fiction in the History of Science Discovery.* Oxford, England: Oxford University Press, 2004.

Walt, Martin. *Introduction to Geomagnetically Trapped Radiation.* New York: Cambridge University Press, 1994.

Wambaugh, Joseph. *The Blooding.* New York: Bantam Books, 1989.

Wang, Hao. *Reflections on Kurt Gödel.* Cambridge, Mass.: MIT Press, 1985.

Watson, James D. *The Double Helix: A Personal Account of the Discovery of the Structure of DNA.* Reprint. New York: W. W. Horton, 1980.

Watson, James D., and John Tooze. *The DNA Story.* San Francisco: W. H. Freeman, 1981.

Watson, James D., et al. *Molecular Biology of the Gene.* 4th ed. Menlo Park, Calif.: Benjamin/Cummings, 1987.

Weber, Robert L. *Pioneers of Science: Nobel Prize Winners in Physics.* 2d ed. Philadelphia: A. Hilger, 1988.

Weedman, Daniel W. *Quasar Astrophysics.* Cambridge, England: Cambridge University Press, 1986.

Wells, Spencer. *The Journey of Man: A Genetic Odyssey.* Princeton, N.J.: Princeton University Press, 2002.

Westfall, Richard S. *Never at Rest: A Biography of Isaac Newton.* New York: Cambridge University Press, 1980.

Wheeler, J. Craig. *Cosmic Catastrophe: Supernovae and Gamma-Ray Bursts.* London: Cambridge University Press, 2000.

Whiting, Jim, and Marylou Morano Kjelle. *John Dalton and the Atomic Theory.* Hockessin, Del.: Mitchell Lane, 2004.

Whitney, Charles. *Francis Bacon and Modernity.* New Haven, Conn.: Yale University Press, 1986.

Whyte, A. J. *The Planet Pluto.* New York: Pergamon Press, 1980.

Wilford, John Noble. *The Mapmakers.* New York: Alfred A. Knopf, 1981.

_____. *The Riddle of the Dinosaur.* New York: Alfred A. Knopf, 1986.

Wilkie, Tom, and Mark Rosselli. *Visions of Heaven: The Mysteries of the Universe Revealed by the Hubble Space Telescope.* London: Hodder & Stoughton, 1999.

Will, Clifford M. *Was Einstein Right?* New York: Basic Books, 1986.

Williams, F. Mary, and Carolyn J. Emerson. *Becoming Leaders: A Practical Handbook for Women in Engineering, Science, and Technology.* Reston, Va.: American Society of Civil Engineers, 2008.

Williams, Garnett P. *Chaos Theory Tamed*. Washington, D.C.: National Academies Press, 1997.

Williams, James Thaxter. *The History of Weather*. Commack, N.Y.: Nova Science, 1999.

Williams, Trevor I. *Howard Florey: Penicillin and After*. London: Oxford University Press, 1984.

Wilmut, Ian, Keith Campbell, and Colin Tudge. *The Second Creation: The Age of Biological Control by the Scientists That Cloned Dolly*. London: Headline, 2000.

Wilson, Andrew. *Space Shuttle Story*. New York: Crescent Books, 1986.

Wilson, Colin. *Starseekers*. Garden City, N.Y.: Doubleday, 1980.

Wilson, David B. *Kelvin and Stokes: A Comparative Study in Victorian Physics*. Bristol, England: Adam Hilger, 1987.

Wilson, Jean D. *Wilson's Textbook of Endocrinology*. 10th ed. New York: Elsevier, 2003.

Windley, Brian F. *The Evolving Continents*. 2d ed. New York: John Wiley & Sons.

Wojcik, Jan W. *Robert Boyle and the Limits of Reason*. New York: Cambridge University Press, 1997.

Wolf, Fred Alan. *Taking the Quantum Leap*. San Francisco: Harper & Row, 1981.

Wollinsky, Art. *The History of the Internet and the World Wide Web*. Berkeley Heights, N.J.: Enslow, 1999.

Wolpoff, M. *Paleoanthropology*. 2d ed. Boston: McGraw-Hill, 1999.

Wood, Michael. *In Search of the Trojan War*. Berkeley: University of California Press, 1988.

Wormald, B. H. G. *Francis Bacon: History, Politics, and Science, 1561-1626*. New York: Cambridge University Press, 1993.

Yen, W. M., Marc D. Levenson, and Arthur L. Schawlow. *Lasers, Spectroscopy, and New Ideas: A Tribute to Arthur L. Schawlow*. New York: Springer-Verlag, 1987.

Yoder, Joella G. *Unrolling Time: Huygens and the Mathematization of Nature*. New York: Cambridge University Press, 2004.

Yolton, John W. ed. *Philosophy, Religion, and Science in the Seventeenth and Eighteenth Centuries*. Rochester, N.Y.: University of Rochester Press, 1990.

Zeilik, Michael. *Astronomy: The Evolving Universe*. 4th ed. New York: Harper & Row, 1985.

Index

SUBJECT INDEX

Note: Page numbers in **bold** indicate main discussion

3G technology 223
21-Centimeter Radiation 597
1000 Genomes Project 87

A

AARP 310, 311
Abbott Laboratories 111
Abel, John Jacob 437, 554
aberrant protein 105
AbioCor artificial heart 119
abortion 473, 709
abrasion 512, 701, 702, 703, 704,
    705
absolute dating 38
absolute pressure sensors 69
absolute zero 392, 670, 671
absorption 102, 104, 170, 215, 230,
    272, 273, 277, 320, 401, 553,
    561, 562, 628, 629, 651, 652,
    653, 699
absorption spectroscopy 651
academic research 43, 65, 87, 219,
    302, 323, 348, 446, 519, 655,
    720
accelerometers 228, 308
accident litigation 99
accuracy 7, 32, 33, 70, 99, 195, 308,
    317, 358, 359, 382, 390, 391,
    394, 402, 417, 422, 444, 567,
    576, 660
acne 170, 171, 172, 173, 174
Acoustics **1-9**
acoustic neuroma 62
acoustic pressure 69
acoustic window 706
acquired immunodeficiency syn-
    drome (AIDS) 188, 295, 356,
    728
actinic keratosis 172
active galaxies 599
active noise control 7, 9
activities of daily living (ADLs) 305

actuarial science 159
adaptive optics 688
additives 76, 128
adeno-associated viruses 727
adhesion 322, 341, 679, 703
ADLs (Activites of Daily Living)
    306, 366, 578, 611, 612
adoption study 152
adrenal glands 238
adrenaline 109, 238
adsorption 628, 703, 724
Advanced Cell Technologies 142
advanced ceramics 127, 128, 129,
    130, 131
Advanced Energy Initiative of 2006
    208
advertising 87, 162, 263, 346, 348,
    434, 467, 588
aeration 628
aerial archaeology 39
aerial photography 370, 380
aerodynamics 572, 576
aeronautical engineering 416
aerosols 53, 56, 58, 59, 246, 247,
    406
aerospace engineering 69, 72, 416,
    430, 433, 638, 688
Affymetrix 302
Afghanistan and Iraq war 421, 508,
    509
age-related macular degeneration
    444, 488, 491, 502
aggregation 453, 632
Agilent 654
agribusiness 10
Agricola, Georgius 198, 425, 476
Agricultural Science **10-15**
agricultural parasitology 531, 534
Agricultural Research Service 35,
    36, 37, 631
agricultural science 10, 11, 12, 13,
    14, 24, 285
Agricultural Science 111

Agroforestry **16-20**
agronomy 10, 11, 21, 24, 25, 26,
    291, 369, 375, 627, 631
Agronomy **21-26**
Air Force Research Laboratory 7,
    423
airline industry 269, 404
air masses 54, 401, 402, 403
airports 5, 58, 404, 576
air-quality monitoring 258
air route traffic control centers 404
air temperature 630
air travel 643
air warfare 419
Alaska 135, 404, 428
Alcoa 130, 208
alerting devices 65
alfuzosin hydrochloride 714
algae 80, 92, 397, 398, 484, 629
algebra 81, 88, 94, 234, 235, 302,
    718
algorithms 83, 84, 85, 86, 90, 91, 93,
    145, 146, 147, 148, 176, 576,
    604, 605, 608
alkaloids 397, 398
allergies 62, 171, 172, 275, 314, 516,
    519, 537
allotransplantation 730
alloys 5, 127, 180, 181, 199, 200,
    207, 226, 229, 426, 576, 670,
    671, 675
ALS (amyotrophic lateral sclerosis)
    449, 662
altered state theories 348
alternating current (AC) 232, 233,
    334, 335, 671, 712
alternative fuels 396
aluminum 5, 197, 199, 200, 202,
    203, 206, 207, 208, 209, 229,
    230, 280, 281, 282, 426, 427,
    429, 670, 690, 702, 703

Alzheimer's disease 301, 305, 308, 309, 310, 449, 450, 452, 585, 588, 606, 665

amateur radio 598, 601, 602

Amazon 221, 222, 223, 236

America COMPETES Act 124

American Academy of Cosmetic Surgery (AACS) 174

American Academy of Dermatology 174, 175, 176

American Academy of Orthotics and Prosthetics 582

American Academy of Physical Education 364

American Academy of Sleep Medicine 634, 635, 637

American Board of Internal Medicine 277, 441, 595, 596

American Chemical Society 113, 250, 251, 655

American Civil War 458, 459, 685

American College of Cardiology 118, 119

American College of Clinical Engineering 134, 135, 136, 137

American Community Survey 160

American Dental Association 166, 168, 169

American Dietetic Association 263, 463, 467, 468

American Heart Association 118, 119, 263

American Institute of Hydrology 343

American Medical Association 715, 716

American Meteorological Society 58, 60, 73, 407

American Nurses Association 458, 462

American Optometric Association 500, 503, 504

American Planning Association 377, 380, 381

American Psychological Association 311, 348, 349, 350

American Red Cross 458

American Society for Microbiology 256

American Society of Clinical Hypnosis 347, 348, 349, 350

American Society of Mechanical Engineers 73, 614, 615

American Superconductor Corporation 675

American Vacuum Society 678

amino acids 108, 109, 316, 327, 328, 398, 412, 464, 656, 726

ammonia 58, 109, 256, 490, 630

ammonification 630

ammonites 527

amniocentesis 184, 471

Amontons, Guillaume 701, 704

ampere 357, 393

amplification 61, 63, 65, 185, 186, 187, 232, 493, 496, 560, 561

amplifiers 129, 561

amputation 579, 581, 582, 685

anabolic steroids 241

analogue modulation 561

analogue signal 61

analogue television 601

analyte 651

Anastas, Paul 246

Andrews, Julie 661

anemia 204, 274, 295, 313, 314, 315, 317, 318, 436, 438, 439, 470, 544

aneroid barometer 68

anesthesia 27, 28, 29, 30, 31, 165, 166, 383, 470, 517, 542, 556, 682, 683, 722

anesthesia technicians 30

anesthesia workstations 29

anesthesiologist assistants 30

anesthesiologists 30

anesthesiology 27, 28, 29, 30, 31

Anesthesiology **27-31**

anesthetists 30, 459, 460, 461

angiography 117, 382, 384, 593, 606, 607

angioplasty 114, 117, 119

angles 99, 104, 147, 178, 180, 212, 235, 357, 358, 388, 389, 410, 425, 494, 500, 605, 690, 714

angular resolution 597, 690, 691

Animal Breeding and Husbandry **32-37**

anodization 678

anorexia 718

anthelmintic 718

anthropogenic changes 59

anthropology 38, 43, 44, 83, 151, 158, 162, 191, 195, 299, 305, 349, 523, 633

anthrosphere 245, 248

antibiotics 62, 66, 76, 108, 111, 112, 113, 140, 164, 194, 263, 275, 288, 294, 518, 519, 624, 720, 721

antibodies 78, 108, 109, 110, 141, 175, 295, 314, 319, 321, 322, 329, 351, 352, 353, 354, 543, 618, 622, 623, 724, 725, 731

anticytokines 731

antidepressant medications 588

antigenicity 109

antigen preparation 109

antiparasite drugs 531

antipsychotic medications 587

antithyroid drugs 240

anxiolytics (antianxiety medications) 587

aperture 408, 430, 597, 599, 688

aperture synthesis 597, 599

apheresis 316

Apollo program 566, 646

Apollo-Soyuz Test Project (ASTP) 647

Apple 222, 223

applied kinesiology 363

applied mathematics 93, 95, 394

applied physics 49, 107, 183, 219, 231, 581, 705

applied research 8, 11, 14, 310, 418, 558

applied science 38, 75, 78, 97, 114, 363, 372, 408, 482, 499, 703

aquaculture 138

Archaeology **38-44**

Archimedes 69

architectural acoustics 2, 9

Areology **45-51**

Argonne National Laboratory 414

Aristotle 2, 97, 158, 160, 320, 364, 375, 401, 633
arm prostheses 579, 580
Army Corps of Engineers 43, 340, 343, 345, 417, 423
ARPANET (Advanced Research Projects Agency Network) 220, 221, 419
Arrhenius, Svante August 205, 206, 246
Arrowsmith (Lewis) 725
Artemisinin 397
arthritis 174, 505, 508, 624, 626
arthroscopy 505, 507, 684
artifacts 38, 39, 40, 41, 42, 43, 44, 384, 584
artificial cloning 138
artificial gemstones 281
artificial hearts 114, 116, 117, 118
artificial insemination 34, 35, 140, 616, 617, 619, 620
artificial intelligence 224, 235, 416, 565, 581, 714
artificial neural networks 442, 714
artificial organs 82, 119, 126, 441, 583, 596, 615, 735
artificial satellites 673
artillery 417, 420
arts-based rehabilitation programs 550
aseptic technique 682
Asia 10, 20, 81, 85, 87, 160, 300, 317, 397, 446, 557, 563, 661, 721
assisted reproductive technology (ART) 82, 239, 242, 546, 547
assistive device manufacturers 65
assistive technology 662
Association of Medical School Pharmacology Chairs (AMSPC) 557
Association of Sleep Disorders Centers 634
asteroids 45, 565, 600, 639, 642
asthma 591, 593, 595
astrobiology 45, 49, 565, 566, 570, 638, 639
astrodynamics 638
astronautics 576, 577, 638, 640, 650

astronomical spectroscopy 652
astrophysics 565, 566, 638, 639, 643, 644, 652, 688, 693, 694
atherosclerosis 145, 605
athletes 99, 100, 240, 241, 365, 366, 367, 368, 507, 509.
Atmospheric Sciences 52-60
atomic absorption spectroscopy (AAS) 652
atomic emission spectroscopy (AES) 652
atomic force microscope 408, 411, 680, 702
atomic scale 701, 702
atomic theory 211, 392, 393, 702
atomic weapons 418
atonia 634
attenuated total reflectance (ATR) 652, 680
audio engineering 2, 3, 232
audiologists 65, 518, 612
Audiology and Hearing Aids 61-67
Auger electron 215, 216, 219
Auger electron spectroscopy (AES) 216, 219
auscultation 591, 592, 622
Australia 8, 49, 63, 111, 117, 135, 141, 142, 199, 256, 282, 288, 296, 331, 379, 428, 546, 567, 631, 661, 733
autoimmune disease 625
autoimmunity 319, 625
automated cell-counting instruments 315
avatar 48
Aventis CropScience 636
aviation 148, 404, 563
Avodart (dutasteride) 714
Avogadro, Amedeo 358, 392
Avogadro constant 358, 392

B

baby boomers 174, 305, 306, 307, 309, 310, 311, 490
Bacillus thuringiensis 193, 287
bacterial spore-detection system 48
Bakewell, Robert 33
balances 391

ballistics 416
balloon angioplasty 115, 117
Bardeen, John 670, 674
bariatric surgery 240, 272, 276
Barnard, Christiaan 115
barograph 69
barometric pressure 73
Barometry 68-74
basal cell carcinoma 172
base units 357, 358, 388
basic research 10, 11, 13, 37, 106, 142, 143, 186, 188, 208, 249, 276, 354, 416, 421, 422, 527, 643, 668, 728
batching and mixing 128
bauxite 128, 199, 207, 426, 427, 429
BCS Theory 670
beaches 48
bearings 127, 218, 421, 576, 701, 703, 704
beauty and cosmetic industry 490
Beccaria, Cesare 153, 156
Beckman Coulter 654
Becquerel, Antoine-Henri 604
Bednorz, J. Georg 671, 674, 675
beer 25, 292
Beer's Law 651
behavioral optometry 499, 502
behavioral pharmacology 555
Behring, Emil von 352
benign prostatic hyperplasia 711
Benninghoven, Alfred 678
Bentham, Jeremy 153, 156
Berlin school of physiologists 102
Berners-Lee, Tim 221
Bernoulli, Daniel 2
Bertillon, Louis-Adolphe 159
beryl 201, 280
Berzelius, Jöns Jacob 204, 425
beverages 12, 25, 48, 76, 263, 466, 546, 547
Bezos, Jeff 223
Bhopal disaster, India 250
bicycles 99
Bigelow Aerospace 642, 648
Bina, Eric 221
bioartificial organs 123
bioassay 240
bioassessment 252, 256

biochemical engineering 76, 82, 399, 400
biochemical pathways 395, 396, 398
biodiesel 23, 398
biodiversity 86, 375
bioenergy sources 23
bioenergy technologies 304
Bioengineering **75-82**
bioethics 88
BioExpress System 110
biofilms 92, 254
bioindicators 256
bioinformaticians 83, 85, 86, 88
Bioinformatics **83-89**
bioinstrumentation 75, 78, 85, 88, 96
biological applications 653
biological databases 86
biological diversity 288
biological gradient 260
biological modeling 94
biological positivism 153
biological processes 75, 82, 84, 86, 90, 123, 628
biological psychiatry 585
biologics 329, 623, 624, 653
biology-based theories 152
biomagnification 245
biomarkers 108, 110, 113, 264, 439
biomass 21
biomaterials 117, 118, 119, 173, 442, 680
Biomathematics **90-95**
biomechanical engineering 82, 126, 271, 448, 481
Biomechanics **96-101**
biomedical equipment technicians (BMETs) 133, 136
biomedical imaging 81, 82, 124, 446, 448
biomedical ocular research 499, 501
biomedical optics 563
biomedical research 84, 104, 124, 187, 254, 259, 300, 320, 323, 412, 621, 723
biomolecular structures 104
bionanotechnology 78

bionics and biomedical engineering 448, 510, 596
Biopharmaceutics Classification System (BCS) 554
biophotonics 562, 563
Biophysical Society 106, 107
Biophysics **102-107**
bioprocess engineering 113, 399
bioproduct companies 88
biopsies 173, 308, 319, 321, 535, 607
Bio-Rad Laboratories 654
bioreactors 124, 126, 254, 396, 397
biosensors 78, 79, 208, 228, 257, 445
biostatistics 83, 181, 260, 264, 475
Biosynthetics **108-113**
biotech corporations 124
Biotechnology Industry Organization 194, 400
bipolar disorder 584
birth control 471, 472, 473, 617
bisphenol A 626
black holes 599
Black Sea 257
bladder, urinary 436, 445, 706, 711
Blaschko, Hermann Karl Felix "Hugh" 108
BLAST (Basic Local Alignment Search Tool) 91
blast furnace 426
blastocyst 618, 664, 665, 666
blending 252
Bloch, Felix 383
blood banking 535
blood clotting 110, 317, 464
blood pressure 5, 70, 71, 115, 117, 134, 238, 242, 262, 434, 437, 438, 468, 470, 500, 518, 624
blood tests 171, 623, 624
B lymphocytes 351, 352
BMI (body mass index) 275, 466
body armor 79, 130, 420
boiling point 213, 392
bombs 418, 419, 642
bone marrow transplant 316
Boolean algebra 234
Boole, George 235
Borelli, Giovanni-Alfonso 97

Borgia, Lucrezia 699
Botox 170, 174, 490, 661, 716
botulinum toxin (Botox) 171, 173, 174, 175, 658, 661, 716
botulism 174
Bourdon tubes 71
Bourgelat, Claude 719
Boyer, Herbert 292
Bozzini, Philip 712
braces 78, 165, 166, 578
Bragg equation 178
Bragg, William Henry 178
Bragg, William Lawrence 178
brain-controlled interfaces (BCI) 444
Brayton thermodynamic cycle 572
Brazil 208, 223, 336, 631, 661, 692
bread 25, 292, 396
breast cancer 98, 193, 260, 261, 321, 398, 535, 537, 539, 603, 606, 607
breast milk 543
breasts 580
Breathalyzer test 308
bridge design and barodynamics 423
bridges 129, 166, 199, 416, 419, 420, 630
bright-field microscopy 409
bright light therapy 635
bronchodilators 593
brown tide 92, 93
Bt-corn 140, 287
buoys 57
Bureau of Labor Statistics 112, 118, 162, 270, 461, 480, 481, 614, 715
Bureau of Land Management 43, 200, 379, 380, 381
Bureau of Reclamation, Department of the Interior 336, 343, 345
Burnet, Frank MacFarlane 733
burning 54, 56, 58, 59, 198, 248, 274, 275, 406, 428, 526, 574
burn Treatment 733
Bush, George W. 208, 568, 667
butanol 397

C

cabazitaxel (Jevtana) 714
cadmium 200, 205, 208, 245, 248, 427
calcification 580
calcium 178, 197, 201, 203, 214, 238, 241, 280, 281, 282, 424, 428, 464, 627, 629, 722
calculi 711
calculus 2, 24, 81, 88, 90, 94, 106, 130, 131, 155, 164, 209, 235, 438, 694, 718
California Institute of Technology 49, 219
caloric-restrictive diets 311
Canadian Light Source 656
Canadian Sleep Society 636
Canadian Space Agency 647, 648, 693
canals 45, 46
candela 358, 393, 495
capillary electrophoresis 655
carbohydrates 463, 467
carbon-14 (radiocarbon) dating 41
carbon dioxide 41, 46, 52, 53, 56, 80, 102, 105, 194, 212, 245, 247, 248, 257, 273, 403, 406, 433, 484, 526, 591, 593, 630, 684, 691
carbon fiber 130, 579, 580, 613
carbon monoxide 65, 226, 246, 680
carbon nanotubes 48, 130, 208, 323
cardiac arrest 116, 308, 544
cardiac arrhythmias 673
cardiac researchers 119
cardiac technicians 119
Cardiology 114-119
cardiorespiratory monitor 544
cardiovascular disease 239, 242, 262, 438, 466, 732
cardiovascular pharmacology 556
cardiovascular system 114, 315
card-not-present (CNP) transactions 222
cargo 576
carrier protein 109
Carson, Rachel 25, 246, 482

cartilage 79, 80, 120, 122, 123, 322, 323, 506, 507, 661
case-control studies 260, 264
cassava plant 142
Cassini, Giovanni Domenico 46
casting 148, 335, 579, 580
catadioptric telescopes 690
catalysts 77, 104, 129, 200, 248, 395, 427, 679
catalytic converters 200
cataracts 488, 489, 496, 499, 502, 562, 624
catheters 117, 673, 685, 711
cathode 130, 145, 146, 204, 205, 206, 207, 211, 229, 232, 233, 564, 604
cathode ray tube (CRT) 229
cation exchange capacity 627
CC (CopyCat) 141
CDC (Centers for Disease Control and Prevention) 167, 263, 265, 354, 356, 478, 480, 531, 534, 612, 615, 621, 687, 714, 729
CDs (compact discs) 348, 431
CD spectroscopy 103
Celera Genomics 300, 302
Cell and Tissue Engineering 120-126
cell biology 83, 102, 106, 125, 138, 211, 291, 296, 299, 317, 319, 351, 408, 414, 488, 489, 556, 651, 668, 724, 727
cell-counting instruments 315
cell matrices 123
cellular biology 90, 92, 125, 505, 511, 610, 614, 734
cellular movements 92
cellular pharmacology 556
cellular scaffolds 123
cellulose 396, 398
Celsius scale 392
cement 127, 129, 201, 428, 507, 630
census 158, 159, 160, 161, 162, 163, 612
Centers for Disease Control and Prevention (CDC) 167, 263, 265, 354, 356, 478, 480, 531, 534, 612, 615, 621, 687, 714, 729

Ceramics 127-132
cerebrospinal fluid 449, 451, 511
cesarean sections 470, 471, 542, 683, 686
chain of custody 696
chain reactions 193
Challenger Deep 483
channel rays 232
Charcot, Jean-Martin 347, 450
charge-coupled devices 415, 689, 691
cheese 12
chemical analysis 216, 678, 679, 697
chemical assays 554
chemical etching stage 227
chemical industries 250
chemical pathology (biochemistry) 535, 536, 537
chemical peeling 174
chemical pollution 246
chemical properties 22, 412, 414, 553, 627, 628
chemical vapor deposition 678
chemosynthetic, described 108
Chicago Area Project 155
chickens 12, 32, 33, 34, 35, 294, 728, 731
childbirth 28, 29, 260, 348, 470, 472, 474, 547, 616
Chile 428, 694
chlorine 53, 205, 245, 428, 629
cholera 141, 253, 259, 260
cholesterol 87, 115, 242, 262, 397, 398, 464, 466, 467, 605
chorionic villus sampling 471
chromatography 246, 554, 555
chromium 200, 426, 464
chronic diseases 263, 293, 307, 367, 467, 733
chronic obstructive pulmonary disease (COPD) 591, 593, 595
chronological age 307
chronology 41, 42, 487
Cialis (tadalafil) 714
cinematography 364
ciprofloxacin 714
circadian rhythm sleep disorders 635
circle 328

circuit boards 233

circuitry, electronic 7, 597

circular dichroism 103, 653

Cisco Systems 221

city planning boards 377

civil engineering 334, 336, 343, 357, 361, 388, 394, 416, 627

Civil War, American 458, 459, 685

cladistics 523

Claudius (Emperor of Rome) 494, 699

clays 627, 628, 629, 630, 631, 632

Clean Air Act of 1970 248

clean energy 302

Clean Energy Research Centers 130

clean room technology 229

cleft palate 519, 520, 661

C-Leg prosthesis 77

Cleopatra (Queen of Egypt) 170, 699

climate control 580

climate engineering 60, 82

climate modeling 374, 529, 571

climate variability 59, 405, 486

climatologists 568

Clinical Engineering **133-137**

clinical genetics 325, 330

clinical history 450

clinical pathologists 537

clinical pharmacology 555, 558, 559

clinical sleeping aids 635

clinical trials 109, 113, 118, 135, 261, 264, 308, 310, 452, 453, 454, 555, 557, 607, 666, 686, 715, 726, 727, 728, 729

Cloning **138-144**

cloning vectors 139

clotting factors 140, 141, 315, 317

clouds 52, 54, 56, 80, 247, 484, 566, 642, 689

coaching 100, 363, 367

coagulation 247, 313, 315, 464, 682

coal 23, 54, 127, 129, 130, 197, 200, 201, 202, 248, 252, 424, 428, 429, 526, 527, 528

coastal engineering 487

Coast Guard, U.S. 478

coastlines 568

cobalt 128, 413, 429, 464, 485

cocaine 28

cochlea 3, 4, 61, 62, 63, 65, 77, 78, 444, 447, 580

Code Division Multiple Access (CDMA) technology 223

Code of Ethics for Nurses 458

Code of Hammurabi 476

CODIS (Combined DNA Index System) 184

cognition 310, 589, 610

cognitive ergonomics 266, 267

Cohen, Stanley 292, 552

coherence 493, 494

cohort studies 260, 261, 262

cold gas thruster 574

collaborative interventions, nursing 459

Colladon, Jean-Daniel 707

Collins, Francis S. 300, 303

colonoscopy 272, 273, 274

Columbia University 332, 374

combat engineering 82, 416, 419, 420, 695

combinatorial chemistry 109, 112

combustion 72, 217, 218, 235, 338, 406, 572, 573, 574, 575, 576, 577, 703

combustors 576

comets 565, 568, 569, 691

commercials 588

commercial uses of algae 484

companion animals 718, 719, 723

complementary medicine 346

Composite Mirror Applications 433

computational fluid dynamics 48, 70

Computed Tomography **145-150**

computer models 4

computer networks 224

computer programming 94, 106, 220, 232, 416, 565, 601

computer software 182, 187, 300, 502, 554, 598, 599, 604

concave mirrors 430

concert halls 2, 8, 9

concrete 5, 129, 148, 201, 265, 335, 336, 419, 428, 594, 672

condensed matter 672

condensed milk 12

condensers 207

conductive hearing loss 62

conductivity 79, 198, 207, 411, 672

confinement operations 33

confocal Raman microscopy 679

congenital malformations 543, 547

conjoined twins 543

consciousness 25, 27, 28, 104, 346, 347, 437, 449, 450, 589

conservationists 484

construction equipment 5

construction firms 43

construction materials 5, 199, 428

Consultative Group on International Agricultural Research 13, 15

consumer electronics 234

consumer marketing 195, 588

consumer products 162, 655

consumer-to-business (c2B) commerce 221

consumer-to-consumer (C2C) commerce 221

contact lenses 80, 111, 413, 433, 488, 495, 499, 501, 502, 503

containers, glass 653

continuous improvement 538

continuous positive airway pressure (CPAP) 593, 635

contraception 471, 472, 616, 619

converters 71, 200

convex lenses 431

convulsive therapy 586

Cooperative Extension Service 11, 36

Cooper, Leon N. 670

Cooper pairs 672

COPD (chronic obstructive pulmonary disease) 591, 592, 593, 594, 595, 596

copper 197, 198, 199, 200, 202, 204, 205, 229, 234, 281, 334, 335, 426, 427, 428, 429, 464, 485, 629, 670, 671, 675

copper porphyry deposits 197, 198

copper wire 334, 335

coral 484

core sampling 41

Coriolis Effect 52

cork 320

corn 12, 13, 21, 22, 23, 25, 140, 193, 287, 288, 292, 294, 404, 464

Cornell University Ergonomic Web 270

coronary angiography 117

coronary artery bypass surgery 117

coronary vessels 114

Corrections Corporation of America 551

corrosion 207, 677

corticosteroids 453, 623, 624, 731

cosmetic dermatology 170, 176

cosmic rays 639, 640

cosmology 639

cost-benefit analysis 153

cotton 11, 13, 21, 22, 23, 25, 62, 193, 287, 288, 294

coulomb 357

Coulomb, Charles-Augustin de 702

Coulson, Alan R. 192

counseling 191, 277, 310, 325, 330, 331, 332, 346, 347, 349, 469, 476, 549, 550, 584, 595, 635

coupling gel 706

course work. See "Careers and Course Work" section within each article

Cousteau, Jacques 483

Covenant Heart and Vascular Institute 118

CPAP (continuous positive airway pressure) 593, 635

crackles 591, 592

cranial osteopathy 512, 515

craniosacral therapy 512

craniotomy 682

creatine 109, 385

credit cards 222

Crick, Francis 103, 180, 184, 191, 193, 286, 292, 300, 326

crime 99, 151, 152, 153, 154, 155, 156, 181, 188, 189, 282, 309, 331, 412, 425, 527, 537, 548, 549, 550, 551, 653, 656, 696, 697, 698, 699

Crimean War 458

crime investigation 43

crime laboratories 188

criminal rehabilitation 548

criminologists 151, 152, 155, 412

Criminology **151-157**

critical care anesthesiology 29

critical care medicine 591, 594, 596

Critical Path Initiative 557

Crohn's disease 110, 273, 276, 329, 353

crop rotation 16, 22

cross-sectional studies 260, 261

crude oil 80, 200, 252, 254

Cruise, Tom 709

crust (of Earth) 45, 197, 252, 352, 424, 427, 523, 525

cryogenics 670, 671, 675

cryopreservation 616

cryosurgery 171

cryptorchidism 714

crystallization 227

crystallography 103, 104, 177, 178, 179, 180, 182, 226, 279, 424, 670

crystals 3, 103, 104, 177, 178, 179, 180, 182, 197, 226, 227, 229, 230, 282, 408, 412, 413, 424, 425, 562, 564, 624, 672, 679, 706, 707

CT microscopy 147

Culp, Connie 685

cultural anthropology 43

cultural resource management 42

Curie, Pierre 707

currents, ocean 483

cyanobacteria 249, 629

cyclones 402, 403

cystic fibrosis 243, 299, 330, 331, 539, 541, 544, 591, 618, 726, 732

cystitis 714

cystoscopy 684, 686

cytogenic analysis 315

cytotechnologists 317

D

dairy products 12, 464

Damadian, Raymond 383, 384

dams 79, 129, 335, 336, 337, 342, 343, 344

Daniell cell 206

Darwin, Charles 40, 286, 301, 524

DASH (dietary approach to stop hypertension) 467

data analysts 88

databases 83, 84, 85, 86, 87, 89, 109, 111, 160, 182, 192, 194, 195, 235, 302, 332, 378, 380, 527

data collection 47, 56, 158, 380, 381, 566

data processing 75, 386

data storage 6, 85, 219, 562, 607

dating (radiocarbon) 38

Daviel, Jacques 489

da Vinci, Leonardo 97, 99, 339, 364, 701, 704

Davisson, Clinton 678

Davis, Theodore 712

Davy, Sir Humphry 204

Dayhoff, Margaret 84

D&C (dilatation and curettage) 472

DDT (dichloro-diphenyl-trichloro-ethane) 245

death sentence 549, 551

De Beers Group 282

Debye, Peter 178, 205

decibels (dB) 1, 4, 383

decision making 267, 305, 381, 460, 539

de Coulomb, Charles-Augustin 702

deductive method of theory development 152

deep brain stimulation 443, 444

deep space 47, 574, 692, 703

Defense Advanced Research Projects Agency. See DARPA

defibrillators 81, 114, 117, 544

degenerative diseases 142, 734

dehydration 212

Dell 224

delusions 584, 586

dementia 305, 309, 311, 450, 452, 453, 455, 584, 588, 660

Dement, William C. 634, 637

Demography and Demographics **158-163**

dendochronology 41

Denmark 111, 649, 719

dental floss 165

Dentistry **158-163**
dentures 165
deoxyribonucleic acid (DNA) 177, 207, 306. *See* DNA
depth profiling 678
derivative 173
Dermatology and Dermatopathology **170-176**
dermoscopy 171, 174
desalination 344
Descartes, René 434, 494
designer baby concept 621
destructive distillation 127
destructive interference 1
detergents 201, 248
deterrence 151, 548, 549
detonation 574
Detrol LA (tolterodine tartrate) 714
developed countries 13, 35, 37, 182, 193, 249, 253, 366, 398, 503, 624, 630, 721
developing countries 13, 23, 35, 112, 249, 270, 273, 276, 277, 287, 465, 503, 532, 721
developmental disorders 120
DeVries, William C. 115
Diagnostic and Statistical Manual of Mental Disorders 588
dialysis 123, 436, 437, 438, 439, 440
diamagnetism 670, 671
diamonds 129, 201, 279, 280, 281, 282, 283, 427, 485, 654
dideoxynucleoside triphosphates 192
dielectric materials 207
dietary reference intake (DRI) 463
dietitians 459, 467, 468
differential equations 90, 91, 92, 94, 106, 177
diffraction 177, 178, 179, 180, 181, 182, 280, 425, 493, 494, 568, 656, 678, 688
Diffraction Analysis **177-183**
digestion 272, 351, 449
digestive system 467
digital artists 527
digital cameras 223, 415, 496, 564
digital communication 221
digital devices 228, 234

digital electronics 237
digital hearing aids 65
digital imaging 409
digital logic 228
digital modulation 561
digital signal processing 61
digital signatures 220, 224
Dionne quintuplets 472, 546
dioxin 247, 699
diphtheria 261, 352, 353, 354, 541
disabilities 166, 263, 502, 503, 505, 507, 610, 611, 612, 613, 615, 660
Disability and Rehabilitation Research Coalition 613, 614
disease detection 438
disease diagnosis 147, 607
disease-modifying antirheumatic drugs (DMARDs) 623
disease treatment 303, 452
disinfection 253, 256
display screens 564
dissociation theories 347
Ditropan XL (oxybutynin chloride) 714
diving saucer 483
DNA Analysis **184-190**
DNA Sequencing **191-196**
dNTPs (deoxynucleoside triphosphates) 192
documentary evidence 41
Dokuchayev, Vasily Vasilyevich 628
Dolly the sheep 328, 329
Donald, Ian 709
donor organs 592
doping 227
Doppler effect 1
Doppler radar 56
Doppler technique 706
doripenem (Doribax) 714
Dornier, Claude 715
dose-response curve 259
drug testing 321, 698, 699
dual energy X-ray absortiometry 241
Duchenne muscular dystrophy 647
ducks 12
DuPont 81, 130, 250
Dussik, Karl 707

dust storms 567, 631
dutasteride (Avodart) 714
dwarf planets 565
dynamic light scattering 653
dynamic-pricing models 222
dynamics 48, 52, 53, 54, 70, 73, 76, 96, 107, 158, 181, 349, 373, 402, 404, 477, 572, 576, 599, 632, 704

E

ear canal 6, 61, 62, 63, 518
ear disorders 61, 516, 520, 521
ear infections 62
ear, nose and throat specialty (otorhinolaryngology) 67, 462, 516, 520, 521, 547, 596, 658, 662, 663, 687
Earth-Moon differences 54
earth sciences 45, 339, 645, 649
Earthwatch Institute 379
earthworms 629
ear trumpets 61
earwax 62, 518
Eastman Chemical 250
Eastman Kodak Company 271
eBay 221, 222, 223, 236
ebola hemorrhagic fever 728
e-books 220
echocardiogram 114, 706
echogenicity 706
E. coli 263, 396, 397, 398
ecological model 154
ecological parasitology 531, 533
ecology 21, 38, 43, 75, 154, 245, 249, 252, 256, 264, 291, 339, 344, 369, 370, 371, 372, 373, 374, 530, 533, 627
e-commerce 220, 221, 222, 223, 224, 225
Economic Geology **197-202**
economics 289, 720
ecosystems 22, 90, 91, 251, 254, 257, 289, 297, 369, 370, 373, 379, 486, 698, 699
eczema 171, 173, 175
edaphologists 628
Edison, Thomas 431

egg production 12

Egypt 10, 43, 48, 168, 342, 389, 661, 683, 701, 719

Ehrlich, Paul 314, 352, 556, 711

Einstein, Hans Albert 340

Eisenhower, Dwight 642

electrical energy 203, 205, 673

electrical engineers 234, 694

electrical gauges 71

electrical irregularities 116

electrical resistance 391, 670

electrical sensors 69

electric circuits 4

electric field 3, 146, 191, 203, 205, 494, 706

electricity 264

electric power 282, 393, 485, 670, 675

electric power generation 670

electrocardiogram (EKG) 115, 116, 708

electrochemical cell 203, 204, 393

Electrochemistry 203-210

electroconvulsive therapy 586, 587

electrodeposition 207, 208

electrodes 4, 79, 119, 205, 206, 210, 231, 443, 444, 445, 446, 447, 518, 614

electrodynamic tether 575

electroencephalograms (EEGs) 451

electroforming 207

electrolysis 203, 204, 205, 207

electrolyte processes 205

electrolytic cell 205

electromagnetic radiation 145, 560, 597, 599, 600, 603, 651, 652, 688, 689, 690, 692

electromagnetic spectrum 54, 56, 215, 216, 383, 392, 409, 493, 555, 560, 563, 597, 603, 604, 605, 652, 655, 688, 689

electromagnetic waves 104, 493, 672, 688

electrometallurgy 207

electromotive force 205

electromyograms 451

Electron Microscopy 211-214

Electron Spectroscopy Analysis 215-219

electronic command and control 422

Electronic Commerce 220-225

Electronic Materials Production 226-231

electronic retailing (e-tailing) 221

Electronics and Electronic Engineering Endocrinology 232-237

electronics technology 218, 236

electron-phonon interaction 670

electrons 103, 145, 146, 177, 178, 179, 205, 207, 211, 212, 213, 215, 216, 217, 218, 232, 233, 321, 357, 358, 392, 393, 408, 410, 411, 425, 493, 494, 495, 560, 561, 603, 605, 651, 670, 671, 672, 673, 678

electron spin resonance 216, 219

electrophoresis 105, 184, 185, 186, 187, 192, 203, 293, 655

electrophysiological Studies 451

electroplating 6, 203, 207

electropolishing 678

electrostatic tether 575

electrosurgery 171

electrowinning 207

Elmiron (pentosan polysulfate) 714

e-mail 195, 220, 222, 236

embryos 34, 36, 77, 138, 139, 140, 141, 142, 241, 243, 331, 473, 546, 547, 616, 617, 618, 621, 665, 667, 668

emergence 27, 28, 30, 257, 368, 370, 416

emergency medicine 457, 506, 594

emergency situations 117, 505

emigration 159, 161

emission spectroscopy 651

Emory University 557

encapsulation 154, 308, 731

encoding 397, 621, 726

endangered species 139, 140, 143, 288, 620, 721, 734

Endocrinology 238-244

endodontics 166

end-of-life care 275

energy balance 52, 54

energy efficiency 208, 246, 563

England 20, 50, 57, 58, 107, 149, 156, 186, 210, 219, 250, 258, 259, 265, 289, 352, 379, 406, 480, 504, 513, 527, 528, 547, 559, 573, 617, 681, 695, 700, 709, 719

entertainment 28, 225, 606

Environmental Chemistry 245-251

environmental geology 424, 428, 528

Environmental Microbiology 252-258

Enzyme-linked immunosorbent assay (ELISA) 109

Epidemiology 259-265

epigenetics 264

epilepsy 446, 449, 453, 454, 455, 541

epitaxial enhancement 227

equilibrium 71, 97, 377, 482

erectile dysfunction 711, 712, 713, 714, 715

Ergonomics 266-271

Erickson, Milton H. 347, 349, 350

erosion 11, 16, 18, 79, 274, 339, 375, 377, 630, 631, 632, 703, 704

error 160, 488, 489, 490, 499, 500, 502, 503, 561, 661, 725

erythrocytes 313

Esaki, Leo 672

Escherichia coli 109, 141, 193, 263, 396, 397

esophagus 272, 273, 274, 277, 543, 544, 661

estrogen 238, 240, 241, 260, 546, 617, 619

ethanol 21, 22, 23, 25, 105, 302, 397, 398, 399

ether 652

ethical issues 87, 303, 734

ET: The Extra-Terrestrial (film) 709

Eulexin (flutamide) 714

European Free Trade Area 118

European Space Agency 49, 569, 645, 647, 648, 649, 693

European Synchrotron Radiation Facility 656
European Union 118, 161, 293, 297
euthanasia 718, 720
eutrophication 248
evoked potentials 451, 453
evolution 1, 2, 38, 40, 50, 91, 191, 194, 195, 249, 291, 300, 311, 397, 523, 524, 525, 527, 529, 530, 531, 533, 565, 566, 594, 609
Ewald, Paul Peter 180
e-wallets 220, 222
Excellence in Clinical Engineering Leadership Award 134
exobiology 638, 639, 643
exploration 45, 48, 49, 50, 72, 198, 201, 229, 282, 380, 427, 428, 482, 526, 528, 567, 568, 569, 570, 586, 639, 642, 646, 647
Explorer I 639
ex situ remediation 254
external forces 475
extinction 35, 50, 143, 249, 301, 378, 523, 524, 526, 527, 529
extracorporeal shock wave lithotripsy 715
extraterrestrial Intelligence 599
extravehicular activity 645
extravehicular mobility unit 645
ex vivo xenotransplantation 731
Exxon Valdez oil spill 253
eyeglasses 430, 431, 433, 444, 496, 501
eye, human 104, 361, 391, 408, 409, 491, 497, 498, 599, 691
Eysenck, Hans 153

F

fabrics 580
Fabry-Perot interferometry 72
Factor VIII 726, 727
factory farming methods 37
Fahrenheit scale 392
family planning 160
Faraday, Michael 204, 206, 233
Faraday's Law 203

farm animals 10, 12, 32, 34, 36, 37, 56, 141, 142, 287
farming 10, 14, 17, 18, 21, 22, 24, 25, 26, 37, 79, 80, 296, 329, 372, 485, 486, 632, 647, 719
fats, dietary 464
fat-soluble vitamins 464
fatty acid 464
Fauchard, Pierre 164
faults (geologic) 199
FDG-PET 606
Federation of Analytical Chemistry and Spectroscopy Societies 655, 657
Federation of Landscape Architects 380, 381
FedEx 224
feed grains 23
feedstocks 398
Fermilab 673
Ferri, Enrico 153
ferries 403, 420
fertility 12, 16, 34, 36, 140, 158, 160, 161, 244, 469, 472, 616, 617, 619, 620, 621, 631
fertilizers 10, 12, 13, 21, 22, 25, 56, 79, 93, 378, 429, 465, 484, 627, 628, 629, 630, 632
fesoterodine (Toviaz) 714
fetal monitoring 471
fiber crops 23
fibers 10, 12, 25, 127, 129, 148, 296, 322, 365, 412, 449, 562, 564, 578, 726
fillings, dental 200
filtration 253, 436, 437, 438, 630
finance 13, 24, 135, 461
fingerprints 329, 412
fish 7, 18, 139, 168, 194, 287, 295, 337, 341, 404, 443, 464, 467, 483, 484, 485, 486, 528, 533, 567
fitness 37, 100, 310, 365, 366, 367, 368, 467, 472, 480
flash 42, 56, 630
flash floods 42, 56, 630
flavonoids 397
flax 12, 23
flight vehicle 572

flock 33, 140, 720, 722
Flomax (tamsulosin) 712
flooding 82, 335, 337, 340, 378, 404, 693
floodplains 375
Florida Heart Research Institute 118
flow cytometry 170, 315
flow-duration curves 340
fluid dynamics 48, 70, 73, 704
fluorescence spectroscopy 653
fluorescent staining 321
fluoridation 167
flutamide (Eulexin) 714
flu vaccine 624
flux 72, 402, 597, 598, 601, 604, 670, 675
Flvr Savr tomato 294
flyby mission 47
FM (frequency modulation) radio 65
Focault, Jean Bernard Leon 652
follicle-stimulating hormone 616
food-borne illness 263
Food production management 567
Food Safety and Inspection Service 534
forced expiratory volume (FEV) 593
forced vital capacity (FVC) 593
Ford Motor Company 673
forensic engineering 211
forensic genetics 190, 325, 331
forensic investigation 219
forensic mineralogists 425, 428
forensic pathology 535, 536, 537
forestry 11, 13, 16, 17, 18, 19, 20, 339, 369, 372, 375, 627
fossils 41, 168, 299, 384, 523, 524, 525, 526, 529, 568
foundations of buildings 627
Fourier transform infrared spectroscopy (FTIR) 652
fractal geometry 107
fracture care 506
fragile X syndrome 187
fragrances 398
Framingham Heart Study 261

France 8, 44, 49, 53, 81, 105, 111, 117, 160, 174, 206, 207, 213, 242, 293, 309, 340, 347, 360, 388, 390, 425, 458, 546, 573, 576, 636, 649, 654, 719, 727, 734

Franklin, Benjamin 233, 443, 715

Franklin, Rosalind 103, 180

Fraunhofer Institute for Solar Energy Systems 125

Fraunhofer lines 652

French Academy of Sciences 358

freon 245

Freud, Sigmund 347, 585, 587

functional age 307

functional magnetic resonance imaging (fMRI) 606

fungus 16

G

Galen 450, 489, 683, 685

galvanization 207

gamete intrafallopian transfer 618

garbage archaeology 42

Garcia, Manuel 517, 658

gardens 375

gases 28, 52, 53, 54, 56, 59, 68, 76, 79, 177, 178, 179, 245, 248, 373, 392, 403, 405, 406, 417, 482, 485, 497, 567, 572, 573, 574, 575, 640, 697

gasoline processing and production 529

gastric ulcers 272

Gastroenterology **272-278**

gauge pressure sensors 69

gearbox 574

gears 228, 572, 703, 705

gel electrophoresis 105, 184, 185, 186, 187, 192, 293

gemcitabine hydrochloride (Gemzar) 714

Gemology and Chrysology **278-284**

GenBank 84, 85, 86, 329

Genentech 292, 295, 331, 399

general anesthesia 28, 30, 166, 542

General Electric (GE) 607

generators 9, 233, 282, 335, 337, 380, 420, 518, 673

gene therapy 121, 140, 287, 289, 295, 297, 325, 330, 332, 447, 452, 725, 733

genetically modified (GM) food production 144, 290, 333, 468

genetic change 32, 33, 34

genetic counseling 191, 325, 330, 331, 332, 469

genetic databases 332

genetic diagnostics 193

genetic disorders 104, 184, 187, 289, 295, 325, 330, 331, 544, 733

Genetic Engineering **291-298**

genetic epidemiology 264

genetic manipulation 126, 286, 287, 292, 328, 396, 397

Genetically Modified Organisms **285-290**

Genographic Project 194, 196

genome-wide association study 617

Genomics **299-304**

genotype 32

geochemistry 45, 197, 251, 424, 428, 565

geodesy 45, 49

geographers 343

geological dating 525

geological time scale 523, 524

geologists 50, 179, 197, 198, 199, 201, 337, 424, 425, 428, 528, 566, 567, 570, 638, 654, 656, 673

geomagnetic storms 638

geometric optics 493, 495

geometry 24, 81, 90, 107, 364, 394, 430, 433, 574, 575

geophagy 631

geophysical methods 198

geophysicists 425, 428

geosynchronous orbit 572

GERD (gastroesophageal reflux disease) 273, 274

Geriatrics and Gerontology **305-312**

Germer, Lester 678

germ-line genetic engineering 77

germ-line therapy 330

gerontechnology 269

Gerontological Society of America 305, 310, 312

Giaever, Ivar 672

glaciers 341

glass electrodes 206

glaucoma 445, 488, 490, 491, 499, 501, 562

Global Age-Friendly Cities (WHO) 307

Global Observing System 57

Global Ocean Sampling expedition 87

global warming 48, 52, 56, 57, 82, 247, 248, 249, 289, 433, 523, 643

glycerin 398

glycerol 398

Golden Rice 140, 141, 193

Google 200, 236, 377

Google Earth 200, 377

Gorer, Peter 731

goserelin (Zoladex) 714

Götz of the Iron Hand 578

graphene 230, 231, 236

graphic arts 375, 380

gravel, for construction 428

gravity 45, 335, 482, 574

gravity assist 574

green chemistry 245, 246, 250

green energy 338

greenhouse effect 52, 54, 245, 565

green machining 128

green photonics 563

green revolution 11

green technology 433

ground warfare 420

Gruelle, John 728

Grzybowski, Józef 526

guano 427

guided biopsy 147

gunpowder 417, 423

guns 212, 419, 420, 703

Gutzmann, Hermann 659

gynecologists 238, 240, 469, 470, 471, 473, 682, 683, 684, 707, 708, 711 gynecology 238, 242, 469, 470, 471, 472, 474, 513, 616, 682, 706, 708, 709, 710

H

habitat modeling 379
habitats in space 647
Hadley cell 53
Haiti 631
halite 197, 201
hallucinations 584, 586
Hammurabi 476
handheld devices 210, 422, 564
hantavirus 728
hardware 85, 148, 175, 220, 221, 224, 386, 570, 608, 614, 636, 648
Harris, Chapin 165
harvesting 16, 17, 23, 331, 372, 379, 484, 709, 734
Hatch Act of 1887 11, 630
Haüy, René Just 180
hay 23
Hayden, Horace 165
head 61, 164, 166, 167, 181, 188, 209, 269, 302, 382, 385, 443, 451, 471, 477, 501, 503, 516, 517, 518, 519, 521, 545, 546, 547, 694, 703, 707, 708, 733
headache 446, 455
health issues 160, 161, 167, 259, 296, 310, 311, 474, 475, 476, 539, 614, 623
health physics 604
Healthy People 2010 program 167
hearing disorders 612
hearing loss (HL) 4, 5, 6, 9, 61, 62, 63, 65, 66, 447, 477, 516, 517, 518, 612
hearing protection 65, 66
heart 115, 117, 118, 119, 261, 263, 468, 667
Heart Disease Research Institute 118, 119
Heart Institute 118
heart-lung machine 117, 684
heart valves 81, 92, 117, 118, 129, 578, 580, 687, 730

heat engine 572, 574
heavy metals 247, 248, 630, 632
Helicobacter pylori 274
helicopters 39, 127
heliocentric model 566
Hematology **313-318**
hematopoiesis 313, 314
Hemholtz, Hermann von 2, 102
hemophilia 295, 315, 726, 727
Hennig, Willi 525
Henry, Joseph 643
herbicides 22, 294
heritability 32
Herodotus 506
Herschel, William 46, 603
hertz (Hz) 1, 70, 597, 604, 706
heterosphere 53
high-temperature superconductivity 672
high-throughput bioinformatics 85
high-throughput screening 109
highways 5, 42, 311
Hill's postulates 260
HIPAA (Health Insurance Portability and Accountability Act) 167
hip joint 505, 580
Hippocrates 259, 450, 476, 506, 517, 585, 591
histocompatibility 730
Histology **319-324**
historic preservation 380
Hitachi 148, 213, 414, 654, 680
HIV (human immunodeficiency virus) 93, 111, 186, 194, 295, 356, 532, 714, 728
holding tanks 6
holgraphic technology 564
Holmes, Katie 709
holography 560
home dialysis machines 438
homosphere 53
Hooke, Robert 253, 320, 524
Hopkins, Harold 712
Hopkins rod lens system 712
Hopps, John A. 116
HORIBA Group 654
Howard Hughes Medical Institute 88

HPV (human papillomaviruses) 172
HTML (hypertext markup language) 224
Hua Tuo 683
Hubble Space Telescope 433, 434, 435, 496, 497, 639, 691, 693
Hückel, Erich 205
human chorionic gonadotrophin 616
human factors engineering 75, 135, 266, 268, 270
Human Genetic Engineering **324-333**
human genome 83, 84, 85, 87, 92, 111, 121, 142, 165, 187, 188, 189, 192, 193, 194, 196, 264, 295, 296, 300, 301, 303, 304, 313, 314, 325, 326, 329, 331
Human Microbiome Project 87
Human Proteome Project 194, 195
humidity 58, 73, 401, 402, 403, 526, 541, 542, 544, 635
hunger 13, 465
hurricanes 56, 130, 377, 402, 405
husbandry 10
Huygens, Christiaan 494
hybrid rockets 574
hydration and irrigation 22
hydroacoustics 4
hydrocarbonoclastic bacteria 254
hydrodynamics 2
Hydroelectric Power Plants **334-338**
hydrogenation 464
Hydrogen fuel cells 208
hydrogen gas 205, 392, 599
hydrologic cycle 339, 340
Hydrology and Hydrogeology **339-345**
hydrophilicity 109
hydrophones 4
hydrosphere 248
hydrotherapy 586
hypertension 1, 238, 436, 438, 452, 467
hyperthyroidism 238, 239, 240
Hypnosis **346-350**
hypnotherapy 346, 347, 348
hypnotic analgesia 348

hypothalamus 238
hypothyroidism 238, 241, 242, 586
hysterectomy 469
hysteroscopy 469, 684, 687

I

illumination 397, 563, 564, 712
imaging, medical 39, 78, 107, 147,
148, 149, 242, 382, 544, 598,
603, 604, 606, 607, 608, 609
immigration 159, 161
immune cells 351, 354
Immunology and Vaccination
351-356
immunosuppression 730, 731
incapacitation 452, 549
incarceration 151, 154, 156, 548,
549, 550, 551, 552, 589
incineration 247
incubators 221, 241, 544, 546
Indian Ocean 402
induced pluripotent stem cell 664
inductively coupled plasma mass
spectrometer (ICP-MS) 426
industrial archaeologists 39
industrial ecology 245
industrial engineering 266
industrial processes 209, 218, 344
infant formulas 543, 544
infants 29, 63, 65, 118, 149, 274,
275, 470, 472, 541, 542, 543,
544, 545, 546, 547, 634
infection 62, 110, 115, 117, 194,
239, 263, 274, 306, 319, 352,
438, 450, 460, 472, 489, 518,
530, 531, 532, 533, 535, 537,
543, 544, 624, 682, 712, 714,
722, 724, 726, 728, 731
infertility 238, 239, 240, 469, 473,
616, 617, 618, 619, 684, 709,
710, 711, 714, 716, 721
inflammatory diseases 172, 173
informatics 84, 460, 461
informed consent 667
infrared radiation 229, 406, 433,
604, 689
infrared spectroscopy 229, 230, 652
infrared telescopes 689

ingot, electronic metals production
227
injury prevention 99, 100, 363
inner cell mass (ICM) 138, 140,
664, 665, 666
Innes, P.D. 677
inorganic chemistry 203, 218, 250,
651, 677
insects 22, 24, 62, 80, 191, 193, 245,
247, 248, 287, 288, 341, 412
insemination 34, 35, 140, 616, 617,
618, 619, 620
Institute for Applied Ecology 380,
381
insulation 5, 23, 48, 80, 129, 712
insulin 110, 140, 238, 239, 240, 243,
262, 286, 287, 292, 295, 325,
326, 328, 331, 396, 398, 586,
679, 727, 731, 733
insulin-coma therapy 586
insurance 22, 159, 196, 242, 303,
306, 368, 386, 460, 476, 479,
509, 607, 614, 623, 685, 686, 727
IntegraMed 620
Intel Corporation 270
intelligence, artificial 224, 235, 416,
565, 581, 714
intelligent agents 222
intelligibility, defined 658
interface science 231, 677, 678,
679, 680, 705
interference 7, 104, 105, 177, 178,
286, 289, 391, 493, 494, 496,
560, 598, 601, 602, 640, 670,
678, 690, 691
interferometry 72, 408, 599, 678,
688, 690, 691
interferon 396
Intergovernmental Panel on
Climate Change 59, 60
internal medicine residency 595
International Association for
Landscape Ecology 369, 370,
372
International Association of
Hydrological Sciences 343
International Association of
Logopedics and Phoniatrics
662, 663

International Certification
Commission for Clinical
Engineering and Biomedical
Technology 134
International Council of Nurses
457, 458, 462
International Ergonomics
Association 266, 270, 271
International Fund for Agricultural
Development 13, 15
International HapMap Project 194,
195, 196
International Labour Organization
475, 479, 480
International Organization for
Standardization 271
international organizations 12, 13,
379
International Society for Microbial
Ecology 256, 258
International System of Units
357-362
International Union of Pure and
Applied Biophysics 105, 107
Interstate Technology Regulatory
Council 380
interstellar medium 599
intrafallopian transfer 618
intraocular lenses 491
intubation 593
invasive radiology 685
in vitro, defined 120
in vitro engineering 121, 123
in vitro fertilization (IVF) 469, 618
in vivo, defined 120
in vivo engineering 121, 123
iodine 239, 240
ionic theory 205, 206
ionizing radiation 103, 145, 146,
215, 216, 382, 603, 604, 605,
606, 608, 609, 706, 707
ion propulsion 574
ion scattering spectroscopy 680
Iowa State University 33, 131, 719
iPad 222, 564
iPhone 223
IP (Internet Protocol) 221
Iran 142
iridium 411

irrigation 21, 22
ischemia 452
isoprenoids 397, 398
Israel 42, 43, 93, 142, 282, 344, 387, 661, 667

J

Jalyn (dutasteride and tamsulosin) 714
James Webb Space Telescope 433, 435, 496, 497, 693
Jansky, Karl G. 598, 601
Janssen, Hans 409, 431
Janssen, Zacharias 409, 431
Japan Aerospace Exploration Agency 56, 647, 648
Jarvik, Robert K. 115
JavaScript 224
Java virtual machine 224
J. Craig Venter Institute 88, 112, 301, 302, 303, 304
Jefferson, Thomas 40, 376
Jeffreys, Alec 185
Jenner, Edward 352, 355, 725, 728
JEOL 213, 214, 414, 680
jet propulsion 48, 50, 570, 573, 577, 602
Jet Propulsion Laboratory 48, 50, 570, 577, 602
Jevtana (cabazitaxel) 714
Josephson, Brian 672
Josephson effect 672
Josephson junctions 673
journalism 195
journals 134, 152, 264, 348, 542, 636, 667, 672, 715
judgmental sampling 38
Jupiter 565, 566, 567, 569, 575, 601, 674, 688, 691
Jupiter II (ship) 674
juvenile justice 155, 548, 549, 550

K

Kamerlingh Onnes, Heike 671, 675
Kamrava, Michael 546
Kelvin temperature scale 68, 357, 388, 392, 670

Kepler, Johannes 46, 494, 500
kidneys 80, 93, 109, 321, 331, 385, 436, 437, 439, 683, 687, 707, 711, 714, 715, 731
kinematics 96
Kinesiology **363-368**
kinetic energy 572
kinetics 96, 107, 203
Kirchoff's current law 233
Knoll, Max 103, 211
Koch, Robert 352, 536, 724
Koop, C. Everett 542
Korea 117, 142, 667
Kraepelin, Emil 585
Kyoto Protocol 248

L

labeling 185, 557, 654
laboratory equipment 107, 112, 697
labor unions 476
Laliberté, Guy 649
land cover evaluation 378
landfills 42, 56, 380
Landis, Floyd 242
Landscape Ecology **369-374**
land surface sensing 378
Land Use Control Information System 377
Land-Use Management **375-381**
land-use regulations 340
Lane-Claypon, Janet 260
Langevin, Paul 709
language disorder, defined 658
laparoscopy 469, 472, 543, 682, 684, 687
laparotomy 682
lapping process 227
Large Sky Area Multi-Object Fiber Spectroscopy Telescope (China) 693
Larson, Swede Arne 613
laryngectomy 661
laryngology 516
LASEK (laser assisted sub-epithelium keratomileusis) 489
laser ablation 519
laser, defined 493

LASIK (laser-assisted in situ keratomileusis) 562
lassa fever 728
latitude 56, 402
Lauterbur, Paul 383
law enforcement 151, 156, 188, 537, 548, 549, 551
law of reflection 431
laws and regulations 377
LCDs (liquid crystal displays) 564
lead-acid battery 206
Ledley, Robert 84
legal issues 501
legs 28, 174, 228, 268, 365, 385, 411, 506, 507, 512, 578, 579, 580, 581, 613
Leiter, Joseph 712
lentiviruses 726
leprosy 173
lesions 170, 171, 172, 173, 174, 274, 443, 450, 451, 452, 453, 531, 625, 706, 725
leukemia 313, 314, 315, 316, 317, 318, 727
leukocytes 313, 351
Levaquin (levofloxacin) 714
Levitra (vardenafil) 714
Lewis, Sinclair 725
licensed practical nurses (LPNs) 461
Liebig, Justus von 628
lifelong learning 309
life-span development (geriatrics and gerontology) 305
life-support systems 375, 646
lightning 403, 630
limbs, prosthetic 578, 579, 610
linear algebra 94, 302
lipids 173
liposuction surgery 174
Lippershey, Hans 431, 494, 689, 693
liquid chromatography 554
liquid-fueled rockets 574
liter unit of measurement 358, 359, 360, 397, 398
lithium 281, 288
live (attenuated) vaccines 353

livestock 10, 12, 16, 17, 18, 23, 36, 37, 286, 288, 294, 484, 530, 531

loam soil 627

lobotomy 586

local anesthesia 27, 28

Lombroso, Cesare 153

longitude 390

long-term care 273, 275, 683

Lorentz, Hendrik A. 672

Louis, Pierre 260

low Earth orbit 572, 638, 645

Lowell, Percival 45

low-throughput bioinformatics 85

lumbar puncture 451

luminous intensity 357, 358, 359, 388, 389, 391, 393

lung cancer 263, 330, 591, 594, 595, 596

lung sounds 592

lung transplants 591, 594

lung volume reduction surgery 592

Lupski, James R. 301

lupus 110, 302, 321, 353, 537, 623, 625

Lush, Jay 33

lymphocytes 313, 315, 351, 352, 724, 728

M

MacFarlane, Sir Frank 733

macronutrients 629

macular degeneration 444, 488, 490, 491, 502, 693

magnesium 197, 203, 209, 281, 424, 426, 464, 627, 629

magnetic resonance angiography 384

Magnetic Resonance Imaging **382-387**

magnetic tape 598

magnetometers 639, 670, 673, 674

magnification 103, 212, 213, 269, 280, 320, 321, 408, 409, 412, 494, 495, 500, 689, 692, 712

Maiman, Theodore 496

malaria 112, 412, 530, 531, 532, 533

malnutrition 275, 277, 465

Malthus, Thomas 159, 163

mammography 603, 606

manganese 202, 429, 464, 629

manic-depressive disorder 585

manometers 69

Marconi, Guglielmo 233

Mariana Trench 70, 483, 486, 702, 704

marine aerosols 247

marine biology 482

Mariner 4 (spacecraft) 46, 47

marketplace, global 480

Marshall, John 715

Mars Pathfinder lander/rover mission 47

Massachusetts Institute of Technology 49, 50, 63, 81, 235, 563, 655, 673, 675

mass production 234

mass ratio 572

material evidence 40, 41

material objects 44, 702

materials engineering 127, 130, 131

Materials Research Society 655, 681

materials science 69, 73, 75, 76, 177, 209, 211, 213, 215, 219, 229, 230, 408, 416, 442, 576, 651, 653, 654, 670, 675, 680, 681, 704

Mathematical Biosciences Institute 93, 94, 95

mathematical models 56, 92, 94, 97, 442, 443, 495

mathematical principles 237

mathematical tools 90, 91, 93, 94

matter-antimatter propulsion systems 574

Maxam and Gilbert Method 192

Max Planck Institute for Evolutionary Anthropology 300

Maxwell, James Clerk 604

Mayo Clinic 65, 119, 733, 734

m-business (mobile business) 223

McCarthy, John 235

measles 62, 260, 261, 353, 355, 725, 727

Measurement and Units **388-394**

Medawar, Peter 733

medical applications 93, 287, 687, 708

medical clinics 65

medical director 439, 538

medical field 135, 136, 238, 421, 433, 450, 507, 685

medical imaging 39, 78, 107, 147, 148, 149, 242, 382, 544, 598, 603, 604, 606, 607, 608, 609

medical laboratories 240, 296, 322, 323

medical laboratory scientists 316

medical laboratory technicians 538

medical office equipment 543

medical parasitology 530, 533

medical physics 102, 145, 148, 386, 608

medical records 267

medical research 214, 394, 421, 422, 466, 473, 495, 613, 624, 668, 686

medical technologists 317

medical technology 9, 126, 133, 134, 136, 416, 421, 508, 538, 613

medical toxicology 698, 700

Medicare 440, 622, 624, 625

MedImmune 331

Meissner effect 672

Meissner, Walther 672

melanoma 171, 172, 173, 174, 175

melatonin 635

membrane structure and transport 104

Ménière, Prosper 517

Ménière's disease 518

menopause 238, 240, 241, 244, 469, 471, 472, 474

mental function 366

mental illness 449, 584, 585, 586, 587, 588, 589, 590

mental status examinations 584

mercantilism 159

Merck Serono 331, 620

mercury 54, 58, 68, 69, 71, 200, 206, 245, 247, 248, 427, 476, 501, 671, 712

Mesmer, Franz Anton 347, 349

mesopause 53

mesosphere 53
Mesozoic (middle life) era 525
messenger RNA (mRNA) molecule 299
meta-analysis 551
Metabolic Engineering **395-400**
metabolism 311, 363, 395, 463, 553, 556
metabolites 108, 385, 396, 398, 554, 555, 556
metagenome technology 252
metagenomics 87, 88, 252, 254, 257
meteorites 54, 565, 567, 568, 569, 639, 654
meteorological optics 495
meteorologists 402, 403, 404, 567
Meteorology **401-407**
meter 68, 72, 99, 177, 357, 358, 359, 388, 390, 391, 392, 403, 483, 573, 597, 604, 613
methane 56, 197, 217, 245, 248, 254, 257, 569, 691
methimazole 240
metric system 357, 359, 360, 361, 388, 390, 392, 394
metrology 232, 322, 361, 391, 393, 394, 704
Mexico City 40
microbes 86, 186, 396
Microbial Genome Project 194
microchips 413
microCT 147
microelectromechanical system 69
microelectronics 228
microfossils 525, 526, 528, 529
micrometeoroid 638
microns 217, 218, 568, 656, 678, 679
micronutrients 629
micro-optics 562
micropaleontology 523, 525, 528
microphones 635
Microphotonics Center 563
microprocessors 72, 508
micropumps for drug delivery 110
microscopic anatomy 319
Microscopy **408-415**
Microsoft 224
Microsoft Access 224

microsurgery 412, 519, 617, 682
microsystems 322, 414, 444
microtomy 322
microwave sounding unit 406
midwives 458, 460, 461, 470, 472, 546
migraines 173, 513
migration 158, 159, 161
Military Audiology Association 64
Military Sciences and Combat Engineering **416-423**
milk 12, 32, 33, 34, 35, 93, 112, 141, 263, 292, 293, 294, 295, 464, 467, 543, 635
miltefosine 532
Mineralogy **424-429**
miniaturization 207, 444, 713
mini magnetospheric plasma propulsion 575
mining industry 372
Minsky, Marvin 235
Mirrors and Lenses **430-435**
missiles 199, 416, 417, 418, 419, 420, 422, 432, 572, 574, 575, 639, 642, 674, 692
Mississippi River 341, 390
mixed hearing loss 62
mobile business (m-business) 223
modularized space station 645
mole 172, 215, 357, 358, 392, 393
molecular cloning 138, 143, 728
molecular epidemiology 264
molecular pharmacology 556
molecular visualization and modeling 86
molecules and cells 91
molybdenum 464, 629
momentum thrust 573
monitoring technologies 29
monoclonal antibodies 108, 314, 354
monocytes 313
Monsanto 13, 24, 81, 142, 288
Montreal protocol 53
mood stabilizers 587
Moore's law 182, 208, 209, 230
morphine 29, 398, 623
mortality 159, 169, 260, 262, 438, 439, 472, 635

Moseley, Henry 180
motion pictures 3, 38, 601
motorcycles 679
motors, electric 200, 233, 617
mouth 28, 87, 164, 165, 166, 167, 168, 169, 172, 173, 257, 341, 516, 520, 530, 533, 593, 623, 624, 659, 720, 722
mouthwash 165
Mouton, Gabriel 358
movement analysis 364
movement disorders 79, 449, 455
MPEG-1 audio layer 3 (MP3) open encoding standard 4
mRNA (messenger RNA) 299, 327, 328, 330
Müller, Karl 671, 674, 675
Mullis, Kary 185, 188, 193
multiple births 473, 542, 546, 547
multiple sclerosis 78, 110, 353, 384, 446, 449, 450, 455, 537, 605, 611
multipotent stem cell 664
mumps 62, 261, 353, 355, 725
Murrell, Hywel 266, 267
muscle energy technique 513
musculoskeletal disorders 267, 268, 269, 605
musical instruments 2, 4, 7, 8, 9
mutations 83, 86, 87, 93, 184, 185, 187, 191, 193, 245, 295, 299, 313, 315, 330, 331, 353, 356, 535, 537, 539, 617, 618, 625, 725
myoelectric limbs 579
My Pyramid 467

N

NAND (not-AND) gates 234
nanobiology 104, 105
nanobots 204
nanoIR 656
nanomaterials 111, 181, 203, 209, 679, 681
nanomedicine 105
nanometers (nm) 78, 145, 177, 212, 411, 413, 493, 603, 654, 655, 677

nanoparticles 105, 111, 129, 208, 230, 246, 413, 680

nanoplasmonics 563, 564

nanotribology 701, 705

NanoZoomer Digital Pathology 322

nasal cavity 516, 518, 659, 661

National Association of University Forest Resources Programs 20

National Biomedical Research Foundation 84

National Cancer Institute 174, 263, 265, 274

National Center for Health Statistics 521

National Climatic Data Center 57, 60

National Cooperative Soil Survey 628

National Geographic Society 194, 486, 487

National Institute for Occupational Safety 267, 268, 271, 428, 429, 478, 480

National Institute of Biomedical Imaging and Bioengineering 81, 82, 124, 446, 448

National Institute of Corrections 155, 551

National Institute of Dental and Craniofacial Research 167

National Institute of Diabetes and Digestive and Kidney Diseases 242, 278, 317, 441, 545, 714

National Institute of Environmental Health Sciences 263, 265

National Institute of Justice 155

National Institute of Standards and Technology 106, 124, 361, 394, 655

National Institute on Aging 306, 310

National Institute on Deafness and Other Communication Disorders 63, 67, 519, 522

National Park Service 43

National Radio Astronomy Observatory 598, 599, 602, 691, 694, 695

national security 47

National Society of Genetic Counselors 331, 333

Native Americans 200

natural disasters 41, 168

natural history 50, 261, 454

Natural Resources Conservation Service 631, 632

Nature Conservancy 380

Naval Research Laboratory 7, 423

naval warfare 417

Neanderthals 300, 329

near-Earth object 565, 638

Near-field scanning optical microscopes 411

neck 61, 164, 166, 167, 238, 268, 269, 382, 385, 442, 507, 513, 516, 517, 519, 521, 578, 708

needle-free drug delivery systems 110

Needleman-Wunsch 86

negatively charged ions (anions) 203, 629

negative-pressure ventilation 593

Negev Desert 42

Neolithic era 585, 683

neonatal intensive care unit 472, 542, 543, 544

neonatologists 542

Nephrology **436-441**

nerve block 27

nerve conduction studies 451

nervous system 124, 320, 333, 353, 365, 385, 442, 443, 444, 447, 449, 450, 451, 452, 453, 454, 455, 511, 518, 555, 556, 726

Neural Engineering **442-448**

neural prosthetics 444

neuroaugmentation 443

neurochemicals 584, 587

neurological examination 450, 519

Neurology **449-456**

neuromodulation 442, 443, 446

neuromuscular stimulation 444

neuropharmacology 555, 556

neuroprosthetics 578, 580

neurotransmitters 322, 555, 584, 587

neurourology 713

neutron-diffraction analysis 178

neutron stars 599

nevus (warts) 172

Newton's Laws of Motion 97

New York State Foundation for Science, Technology and Innovation 208

New Zealand 8, 288, 531, 661, 734

NextGen Sciences 110

nickel 3, 172, 200, 202, 205, 485, 629, 678

NICU (neonatal intensive care unit) 541, 544

Nielsen (data collection) 160, 162, 399, 400, 582

Nightingale, Florence 458, 459

Nintendo Wii 309

nitrification 630

nitrous oxide 28, 30, 56, 109, 165

Nitze, Maximilian 712

Nixon, Richard M. 568

nondestructive evaluation 145, 147

nongovernmental organizations (NGOs) 17, 18, 19, 20, 246, 247, 289, 373, 474, 648

nonionizing radiation 603

nonlinear equations 91

nonprofit organizations 118, 715

nonrapid eye movement (non-REM) 633

nonrenewable resources 197, 218, 395, 429

nonsmall cell cancers 594

NSAIDs (nonsteroidal anti-inflammatory drugs) 274, 506

nuclear imaging 555, 605, 606, 609

nuclear technology 130

nucleic acid amplification testing 186

nucleotides 35, 84, 104, 184, 185, 186, 187, 191, 192, 299, 326, 327, 395

Nurse Practice Acts 459

Nursing **457-462**

Nutrition and Dietetics **463-468**

Nutrition Labeling and Education Act 467
nylon 165, 705

O

Obama, Barack 569, 665, 667
obstetricians 238, 470, 707, 708
Obstetrics and Gynecology **469-474**
obstructive airway diseases 591, 593
Occupational Health **475-481**
occupational therapy 305, 309, 506, 508, 611, 623
Ocean Optics 654, 655
Oceanography **482-487**
octuplets 546, 619
ocular motility examination 500
Oersted, Hans Christian 233
Office of Biological and Environmental Research 88
Office of Naval Research 208, 487
Office of Science and Technology 106, 571
office workers 268, 480
oganizational ergonomics 266, 267
Ohm, Simon 3
Ohm's resistance law 233
oil whirl 701
omega-3 fatty acids 294
oncologists 317
oncology 90, 191, 193, 272, 316, 317, 319, 323, 455, 457, 460, 469, 509, 516, 541, 545, 683, 715
oneirology 633
online businesses 221, 224
Onnes, Heike Kamerlingh 671, 675
Onsager, Lars 205
OP-AMP (operational amplifiers) 232, 234
open land management 379
operating systems 224
operational meteorologists 58
Opthalmology **488-492**
optical disks 431
optical microscopes 410, 411, 414, 563
optical networks 562, 564

optical telescopes 598, 639, 689, 690, 691, 692, 693, 695
Optics **493-498**
Optometry **499-504**
Oracle software company 224
oral and ethnographic evidence 41
oral and maxillofacial surgery 166
oral health 164, 166, 167, 168, 169
oral health care products 167
ore 128, 197, 198, 199, 202, 426, 428, 673
organ failure 731
organic chemistry 24, 94, 108, 113, 120, 177, 203, 209, 218, 245, 250, 355, 395, 503, 514, 531, 591, 651, 653, 718
Organization of Petroleum Exporting Countries (OPEC) 201
Origin of Species (Darwin) 286
orthodontics 129, 166, 658
Orthopaedic Research Society 508, 613, 615
Orthopedics **505-510**
orthosis 578, 613
orthotics 578, 579, 582, 583
osseointegration 578
Osteopathy **511-515**
osteoporosis 147, 238, 240, 241, 623, 624
otitis treatment 517
Otorhinolaryngology **516-522**
otoscopy 516, 518
outbreeding 32
outcrossing 34, 285
outpatient surgery 273, 684, 686
output devices 220
ovaries 238, 241, 469, 471, 617, 618, 707
overactive bladder 714, 715
oxidation 205, 207, 254, 390, 464, 678
oxybutynin chloride (Ditropan XL) 714
ozone 52, 53, 56, 207, 218, 246, 247, 249, 403, 405, 406, 643

P

Pacific Institute 344
packet switching 221
pain management 27, 29, 31, 452, 513
Paleontology **523-529**
Paleozoic (ancient life) era 525
pancreas 108, 124, 125, 238, 243, 272, 708
Panum, Peter 260
papyrus 41, 280, 556, 683
Paracelsus (Swiss physician) 554, 697
parallel computing 86
Parasitology **530-534**
parasomnias 633
parenteral nutrition 275
Park-Burgess model 154
Parkinson's disease 443, 444, 446, 449, 450, 452, 453, 455, 665, 666, 730, 733
parks 5, 154, 311, 376
parole 155, 548, 549, 550, 551
paroxysmal Disorders 453
parthenogenesis 138
partial differential equations 177
particulate matter 58
pascal (unit) 68
passivation 218
pastel colors 550
Pasteur, Louis 352, 355, 536, 724
patch cords 65
pathogens 48, 79, 104, 167, 170, 172, 253, 254, 256, 257, 263, 287, 289, 412, 414, 518, 731, 734
pathologists 167, 175, 213, 519, 535, 536, 537, 538, 539
Pathology **535-540**
Pathway Genomics 195
patriarchal structures 350
Pauli exclusion principle 217
Pauling, Linus 178, 204
payload 48, 573
PCBs (polychlorinated biphenyls) 247, 254
pediatric dentistry 166

pediatric endocrinology 238, 242, 243, 541
pediatric gastroenterology 274
Pediatric Medicine and Surgery **541-548**
pedodontics 166
pedologists 628
penicillin 76, 111, 246
penitentiaries 549
Penology **548-552**
pentosan polysulfate (Elmiron) 714
peptic ulcers 273, 276
peptide synthesis 109
percussion 591, 592, 622
perennial plants 12
performance-enhancing drugs 698
perinatologist 542
periodontics 166
peripheral vascular disease 579, 581
perlite 629, 631
permeability 76, 342, 554
perovskite 670
personalized medicine 88, 89, 140, 191, 193, 195, 557
pest control 22
pesticides 10, 13, 22, 25, 79, 80, 81, 93, 194, 246, 247, 248, 249, 251, 287, 294, 297, 378, 465
Pfizer 106, 111, 125, 354, 636, 712
phantom pain 581
pharmacogenetics 86, 140
pharmacological toxicology 698
Pharmacology **553-559**
pharmacy 305, 311, 553, 558
phenotypes 33, 86, 264, 300, 325, 332
Philips 148, 608
phlebotomists 538
Phoenix (Martian lander) 570
phoniatrics 612, 658, 661, 662, 663
phonons 670
photodetection 561
photodiodes 561
photodynamic therapy 174, 490
photoelectrochemistry 207
photoelectron spectrometer 216
photoelectron spectroscopy 216, 678

photographers 155, 496
Photonics **560-565**
photons 72, 145, 146, 211, 493, 494, 560, 561, 562, 564, 605, 606, 651, 653
photothermolysis 174
phylogenetic trees 91
phylogeny 524
physical activity 266, 267, 276, 311, 363, 365
physical chemistry 177, 218, 245, 250, 651, 656, 670
physical ergonomics 266, 267
physical optics 493
physical properties 1, 22, 24, 58, 218, 280, 283, 357, 390, 413, 429, 494, 526, 553, 554, 556, 628, 630, 645, 666, 671, 677
physical rehabilitation 505, 578, 610
Physick, Philip Syng 715
physiological acoustics 2, 3
phytoplankton 484
phytoremediation 249
picture archiving and communication systems 607
piezoelectric Effect 706
piezoelectric transducers 3
pig iron 127
pipelines 42, 207, 334, 640
pitch 3, 6, 66, 592, 662
Pitocin 471
placebo effect 346
placers 427
plague 262, 335, 352, 527, 725
Planck, Max 105, 257, 300, 494, 560
Planck's constant 493, 494
plane 102, 145, 279, 357, 358, 363, 382, 388, 419, 424, 430, 495, 575, 603
Planetology and Astrogeology **565-571**
plankton 197, 484
plaque assay 724
plasma 104, 110, 280, 313, 339, 351, 352, 354, 426, 436, 575, 576, 640, 652, 673, 678, 679, 680, 681, 686
plasma membrane 104, 313

plasma sails 575
plasmids 110, 285
plasmodium 531, 532
plastics 148, 149, 169, 204, 207, 213, 226, 229, 257, 296, 412, 579
plastic surgery 170, 661
platelets 76, 313, 314, 315
Plato 158, 160
Pluto 565, 566, 569, 570
Pneumocystis carinii pneumonia (PCP) 624
pneumothorax 591, 592
p-n junction 561
poisons 259, 520, 537, 696, 697, 699
polio vaccine 353, 727
politics 156, 265, 676
pollination 294, 297
polyacrylamide 105, 192
Polyakov, Valeri 649
polycystic ovary syndrome 617
polyelectrolytes 206, 208
polyethylene glycol 722
polymorphism 184, 185
PolyPlus 130
polysilicon 226
polysomnography 633, 634
Population Association of America 161, 163
population dynamics 158
population trends 616
porous silicon 229, 230
porphyry copper deposits 197
positively charged ions (cations) 203, 627, 629
positive-pressure ventilation 593
Positivist School of thought 155
positron emission tomography (PET) 167, 605
potable 339, 341, 343
Potrykus, Ingo 141
potting mix 629
Pott, Percival 476
poultry products safety 721
poultry science 11
powder metallurgy 576
powder X-ray diffraction 178
power grid 334, 335, 336, 419
precious metals 200, 279, 280, 281, 282, 283, 427

prediction, in probability 40
prefabricated space station 645
preimplantation genetic diagnosis 616
prenatal ultrasound 542, 543
presbycusis 62
pressure-sensitive paints 69, 71
pressure sensors 69, 71, 72, 73
pressure switches 71
pressure thrust 573
preventive health care 479
Priceline 221, 222, 223, 224
primary cell 205
primary metabolites 108
primers 139, 185, 186, 187
prison management 548
probabilistic sampling 38
probation 155, 548, 549, 550, 551
problem solving 84, 566
programming 65, 83, 85, 86, 88, 89, 94, 106, 155, 195, 220, 221, 232, 302, 390, 391, 416, 434, 565, 589, 601
Project Apollo 568
Project SCORE 641, 642
propellers 76, 572
Propulsion Technologies **572-577**
propylthiouracil 240
prostate 323, 539, 684, 708, 711, 712, 713, 714, 715, 716, 727
prostate cancer 539, 712, 714, 715, 716, 727
prostate-specific antigen (PSA) test 712
Prosthetics **578-583**
prosthodontics 166
Protein Information Resource 85
proton exchange membrane fuel cells 208
protozoa 253, 530, 531, 629
PSA (prostate-specific antigen test) 712
psoriasis 110, 170, 171, 173, 174, 175, 302
psoriatic arthritis 329
Psychiatry **584-590**
psychoanalysis 346, 585, 587
psychodynamic approach 585
psychological acoustics 2, 3, 8

psychological positivism 153
psychology-based theories 152
psychotherapy 346, 350, 584, 587, 635
Ptolemy 494, 500
public key 220
public relations 264
publishing 11
pulsars 599
Pulsed detonation engines 574
pulsed light 174
pulse oximeter 544
pulse sequences 604
pumice 631
punch biopsies 173
Purcell, Edward Mills 383
pyramids 431

Q

quality assurance 31, 112, 135, 136, 322, 538, 556
quality control 112, 133, 135, 136, 195, 227, 229
quality management 250, 461, 704
quantum computing 561, 670
quantum dots 208, 236
quantum mechanics 106, 177, 216, 410, 495, 670, 672, 702
quantum optics 493, 498, 564
quantum theory 226, 233, 564
quartz 3, 128, 197, 199, 279, 280, 281, 425, 427, 707
quasars 599, 690, 692
quinhydrone electrode 206
Quintuplets 472, 546

R

rabies 353, 355, 724, 725
radian 358
radiant warmers 544
radiation budget 54
radiation pasteurization. *See* irradiation
Radio Astronomy **597-602**
radioactive iodine 239, 240
radiocarbon dating 38
radiologists 147, 167, 605, 607, 608

Radiology and Medical Imaging **603-609**
radiometers 406
radionuclides 605, 609
radio telescopes 597, 598, 599, 600, 601, 602, 639, 688, 690
radiotherapy 102, 242, 594, 608
radio waves 53, 78, 419, 451, 597, 599, 603, 688, 689, 690, 706
rain gauges 56, 403
rales 591, 592
Raman, Sir Chandrasekhara Venkata 652, 655
Raman spectroscopy 653, 654, 680
Ramazzini, Bernardino 267, 476
ramjets 574
Ramón y Cajal, Santiago 442, 450
randomized controlled trials 261
Rankine scale 392
rapid eye movement (REM) 633
rare earth elements 200
Rashevsky, Nicolas 90
Rathje, William 42
rational choice theory 549
Raytheon 8, 432
reading disabilities 660
Reagan, Ronald 542, 641
reagent 535, 696
real-time polymerase chain reaction 654
Reber, Grote 598, 601
rebound sleep 634
Rechtschaffen and Kales system 634
recidivism 548
recommended dietary allowance 467
reconstructive surgery 166, 516, 517, 521, 682, 713
red blood cells 188, 313, 314, 315, 322, 339, 708
redox reactions 205
reduction 8, 78, 79, 145, 146, 174, 205, 248, 249, 288, 346, 347, 348, 394, 437, 452, 494, 562, 589, 592, 593, 635, 660, 678, 708
refining 23, 130, 229, 281, 282

reflection 5, 78, 321, 322, 406, 431, 434, 493, 495, 496, 498, 500, 562, 690, 691, 706
reflection telescopes 690
reflectivity studies 526
refracting telescopes 690, 693
refraction 280, 430, 493, 494, 495, 498, 500, 504, 560, 562, 690
refractive index 279
refractometer 280
refrigeration 53, 670, 673
regenerative medicine 121, 141, 323, 664, 730
regional anesthesia 28, 29, 683
registered nurses (RNs) 457
regulations 29, 99, 124, 149, 153, 167, 246, 248, 250, 340, 377, 378, 385, 391, 459, 460, 476, 478, 479, 480, 557, 721
regurgitation 115, 274
rehabilitation, criminal 548
Rehabilitation Engineering 610-615
reinforced plastics 149
relative dating, archaeological 41
relays 63, 77
remote sensing 56, 566, 640
renewable energy 194, 249, 302, 337, 379
reproductive cloning 138, 140, 621
reproductive endocrinology 238, 242, 469, 706, 709
Reproductive Science and Engineering 616-621
reproductive technology, assisted 82, 239, 242, 546, 547
reservoirs 50, 72, 79, 263, 532
resonant frequency 604
Resource Recovery Act of 1970 247
respiratory rate 544, 719
restriction enzymes 724
restriction fragment length polymorphism 184
resuscitation 27, 29, 542, 544
retainers (dental) 165
retinal bioengineering 444
retinal cell biology 488
retrovirus 730
reverse-auction model 221
reverse-price model 222

revetments 419
Rheumatology 622-626
rhonchi 592
rhythm method 619
rice 12, 23, 56, 141, 193, 294
riparian forest systems 18
risk assessment 475
roads 40, 129, 130, 161, 199, 201, 340, 416, 419, 428, 608, 630
Robb, Hunter 472, 685
Robinson, Kim Stanley 45, 46
Rochester Regional Photonics 563
rocketry 638, 639
Rohrer, Heinrich 411
Röntgen, Wilhelm 216
root canal therapy 166
Rous, Peyton 728
rovers (Mars missions) 48
Royal Society of London 322
rubies 201, 279, 280, 281, 282, 283
Ruska, Ernst 103, 211, 213, 411
Russian Federal Space Agency 647, 648, 650
Russian soils 628

S

Sabin, Albert 727
Sabin vaccine 725
Saccharomyces cerevisiae 396, 399
safety engineering 475
salamanders 139
Salk, Jonas 353, 727
Salk vaccine 725, 727
salmon 294, 337
salt 28, 66, 80, 126, 187, 197, 198, 202, 392, 425, 428, 429, 490
salvage archaeology 40, 42
salvarsan 711
Salyut program 647
Sambrook, Joseph 184
Sanctura (trospium) 714
sand 128, 197, 201, 202, 246, 340, 424, 532, 627, 628, 630, 638
Sandia National Laboratories 181, 399, 680, 681
sandpaper 170
Sanger, Frederick 192, 193
Sanger sequencing method 192

sanitation 33
Saturn V rocket 573, 575, 642
Sauveur, Joseph 2
scaffolds, cell and tissue engineering 123
scalpels 685
scanning Auger microscopy 219
scanning polarization force microscopy 247
scanning probe microscopy 230, 408, 701, 702
scanning tunneling microscopes 410, 411
scattering 177, 178, 179, 315, 494, 560, 651, 652, 653, 655, 680
schizophrenia 584
Schrieffer, John R. 670
Schrödinger, Erwin 103
scintigraphy 239, 544
Scott, F. Bradley 712
scramjet 574
Seattle 73, 115, 223, 437, 515
seawater 80, 207
secondary cell 205
secondary metabolites 108
second life 46
second (time) 217
sedimentation 340, 524, 623
seeds 10, 11, 22, 79, 81, 193, 287, 288, 464, 476, 629
seedstock 32, 34
seismology 334, 336
seizures 445, 450, 453, 454, 586, 587, 673
selective serotonin reuptake inhibitors 588
Senior Corps 307
sensorineural hearing loss 62
seriation 41
SETI Institute 600, 602
severe acute respiratory syndrome 263
shelterbelt systems 18
shingles (herpes zoster) 175, 624
shopping bot 222
short takeoff and landing aircraft 577
short tandem repeats 186, 187, 189
shovelbums 39

Siemens 148, 212, 213, 607
sigmoidoscopies 273, 274
signal processing 23, 61, 561
silage 23
sildenafil citrate (Viagra) 712, 714
SI (Le Système International d'Unités) 357, 390
silica 280, 281, 406, 630
silicon 72, 128, 129, 200, 204, 226, 227, 228, 229, 230, 231, 232, 233, 279, 408, 411, 413, 424, 426, 445, 673
silicon chips 226, 227, 228, 229, 230, 231
silicon wafers 228, 673
silviculture 16, 17, 19, 20, 339
silvopasture 18
simple random sampling 40
Sims, J. Marion 470, 617, 685
single crystal silicon 228
single-lens reflex camera 431
single nucleotide polymorphism 184
single photon emission computed tomography 605, 661
Sinsheimer, Robert 85
sintering 128, 129
sinus 517, 519, 520
skin cancer 170, 173, 406
skin substitute 110
Skylab 642, 646
slag 247
sleep apnea 166, 520, 591, 592, 593, 635
sleep deprivation 634
Sleep Research Center 634
sleep (somnology) 5, 7, 27, 28, 63, 166, 242, 346, 349, 446, 449, 455, 520, 521, 545, 584, 591, 592, 593, 594, 595, 633, 634, 635, 636, 637
slit lamp examination 501
sludge 629
slurry 6, 227
small cell lung cancers 594
smallpox vaccine 111, 352, 725
smart cards 220, 222, 224
smelting 127
Smith, Fred W. 224

Smith, Stephen 707
Smith-Waterman algorithm 91
Snellen chart 500
Snow, John 260
soaps 173
social-disorganization theory 155
social-ecology approach 154
social issues 14, 194, 297, 303, 619
social role theories 347
Social Security Act 305
social theories of aging 306
Society for Analytical Chemists of Pittsburgh 655
Society for Investigative Dermatology 174, 175, 176
Society of American Foresters 20
sociological positivism 153
sociology-based theories 152
Socrates 699
sodium chloride 177, 178, 179, 205
sodium metal 205
software engineering 85, 86, 220
soil 79
soil biota 628
soil conservation 631
soil fungi 629
soil health 22
soil mechanics 627
Soil Science 627-632
Sojourner rover 47
solar backscatter ultraviolet radiometers 406
solenoids 410
solid-fueled rockets 573, 574
solifenacin (Vesicare) 714
somatic cell gene therapy 330
somatic stem cells 664, 665, 666, 667, 668
somatotyping criteria 154
Somnology 633-637
sonic booms 5
South Africa 282, 631
southern blotting 184
Soviet Union 97, 418, 573, 639, 640, 641, 642, 647, 648
soybeans 21, 23, 287, 288, 294, 303
Soylent (software) 235
Soyuz spacecraft 642
spacecraft engineering 638

space engineering, defined 645
space exploration 45, 49, 50, 229, 568, 570, 642, 647
space radiation 647
Space Science 638-644
Space Stations 645-650
space tourism 638, 642, 643, 648
space weather 405, 638, 640, 641, 643
Space Weather Prediction Center 405, 641
SPAM 411
spasmodic dysphonia 661
spatial analysis 374, 377
specific impulse 572, 573, 574, 576
spectrographs 691
spectrometers 426, 639, 653, 655, 656
spectroscopic ellipsometry 653
Spectroscopy 651-657
Spectroscopy Society of Pittsburgh 655
speech acoustics 2, 4, 6
Speech Research Laboratory 8
Speech Therapy and Phoniatrics 658-663
Spemann, Hans 139
spheres 161
sphygmomanometers 71
spinal cord function 666
spinal stabilization 507
spirometry 593
sports engineering 266
Sputnik II 639
squamous cell carcinoma 172
SQUID (superconducting quantum interference device) 670
stainless steel 200
Stanford prison experiment 550
Stanford University 49, 156, 286, 348, 383, 607, 634, 637, 675
stapedectomy 65
Staphylococcus aureus 172, 261
Stargardt's macular dystrophy 666
stars 392, 565, 566, 599, 640, 643, 672, 688, 689, 690, 691, 693, 694
state agricultural experiment stations 36

statics 96

statistical analysis 40, 159, 162, 261, 393, 548, 551

statistical modeling 554

statistical process control 390, 393, 704

Stem Cell Research and Technology **664-670**

stenosis 115, 543, 708

stents 114, 118, 315, 454

stereo microscopy 321

sterilization 473, 616, 682, 711

Stern, Maximilian 712

steroids 241, 242, 720

sterols 398

stethoscopes 543, 719

Still, Andrew Taylor 511

stomata 526

Stoney, George Johnstone 206

strain theory 152

stratified random sampling 40

stratigraphy 41, 197, 523

stratopause 53

stratosphere 53, 59, 207, 643

streets 154

strength training equipment 507

Strutt, John William 2, 652

substrates 227, 396

sugars 23, 102, 302, 396, 464

Suleman, Nadya 546, 619

sulfur 80, 127, 246, 247, 257, 403, 406, 427, 629

sulfur compounds 201

Superconductiviy and Superconducting Devices **670-676**

Surface and Interface Science **677-681**

surface enhanced Raman spectroscopy 680

surface ships 418, 674

surface tension 212, 341

surface water 343, 568, 632

Surgery **682-687**

surrogacy 619

surveillance 155, 418, 419, 420, 421, 422, 460, 476, 557, 639, 692, 694, 695

surveying 357, 388, 394

Sutherland, Edwin H. 151

suture, defined 682

synchrotron 651

Syncom 3 (satellite) 641

synthetic biology 104, 105, 108, 299, 300, 301, 302, 332, 399

synthetic fiber 682

synthetic polymers 248

synthetic soils 629

syphilis 172, 711

systematic sampling 40

systemic lupus erythematosis 623

T

tadalafil (Cialis) 714

Taiwan 81

talk therapy 585

tamsulosin (Flomax) 712, 714

taphonomy 523, 528

targeted muscle reinnervation 580

taungya 16

TCP/IP (Transmission Control Protocol/Internet Protocol) 220

Technology Innovation Program 124

teeth 28, 43, 53, 129, 164, 165, 166, 167, 168, 169, 434, 464, 524, 543, 633, 658, 659, 665, 683

Teflon 437, 702, 703, 705

telegraph 206, 233, 642

telemedicine 647

Telescopy **688-695**

teratomas 543

terrorism 168, 187

Tesla, Nikola 233, 382

testicles 711, 714

tethers 575

textiles 41, 48, 177, 414, 653

theme parks 311

therapeutic cloning 138, 141, 142, 331

therapeutic proteins 110

thermal ionization mass spectrometers 426

thermal maturation 526

Thermo Fisher Scientific 654, 680

thermoluminescent dating 41

thermosphere 54

The Sea Around Us (Carson) 482, 486

The Silent World (Cousteau) 483

The Story of Utopias (Mumford) 376

thoracic surgeons 683

thoracotomy 682

throat 65, 273, 411, 516, 517, 518, 519, 520, 521, 635, 658

thrust (propulsion) 572

thyroid disease 240

thyroid gland 238, 239, 240

thyroid hormone 240, 241

tides 482, 483, 485

timber 12

timbre (audio) 3, 4

time measurement 359, 392

time-of-flight secondary ion mass spectrometry 678

tinnitus treatment 518

TIROS 1 402, 640

tissue cultures 321

tissue engineering 322, 323, 332

tissues, substitute 120

titanium 199

toilets 129

tolterodine tartrate (Detrol LA) 714

tomatoes 12

tonometry 500, 501

tonsillectomy 517, 519

toothbrushes 165

toothpaste 167

topography 50, 103, 377, 378, 379, 410, 413, 501, 600, 677, 678

torque 96

Torricelli barometer 68

Torricelli, Evangelista 52, 69, 70

Toshiba 148, 607

total internal reflection fluorescence microscopy 321

total lung capacity 593

tourism 638, 643, 648

tourmaline 281

Toviaz (fesoterodine) 714

toxicant, defined 696

toxicodynamics 696

toxicogenomics 325

toxicokinetics 696

Toxicology **696-700**
toxin, defined 696
tracheosophageal puncture 661
trade fairs 563
trade winds 483
traditional ceramics 128
traffic noise 5
trains 127, 309, 403, 428, 576, 670, 703
Train the Trainer 309
transcutaneous electrical nerve stimulation 446
transducers 3, 7, 8, 71, 72, 444, 604, 706, 707, 708
transfection 724
transfemoral prosthesis 579
transformers 128, 337
transgenesis 731
transhumanism 143
transition temperature 670
transmission control protocol and internet protocol (TCP/IP) 220
transmission electron microscopy 211, 212, 415
transportation industry 235
transtibial prosthesis 579
trauma 29, 61, 62, 65, 166, 173, 268, 444, 454, 460, 488, 503, 506, 507, 509, 512, 517, 521, 578, 581, 586, 712, 713, 716, 730
trepanning 585
tribology 218, 576, 701, 702, 703, 704, 705
trigonometry 81, 394
tRNA (transfer RNA) 192, 327, 328
Troll, Carl 370
Tropical Rainfall Monitoring Mission 56
tropopause 53
troposphere 53, 54, 401
tropospheric aerosols 56
tropospheric ozone 56
trospium (Sanctura) 714
trucks 127
Tsiolkovsky, Konstantin 639
tungsten 128, 146, 200, 202, 211, 411, 429, 605, 712
tunnels 69, 71, 114, 174, 629, 630

turbochargers 576
turbojets 574
turbomachines 574
Turing, Alan 235
TurtleSkin 48
twin studies 152
two-dimensional images 98, 145, 146, 213, 384, 712
typologies 155

U

ulcers 272, 273, 274, 275, 276, 277, 624
ultra-high vacuum techniques 678
ultrasonic cleaners 5
Ultrasonic Imaging **706-710**
ultraviolet photoelectron spectroscopy 216
ultraviolet visible spectroscopy 652
unconscious state 28
underwater archaeology 39
underwater navigation 7
underwater sound 2, 5
uniformitarianism 40
Union Carbide 250
United States Pharmacopeia 557, 559
United States Renal Data System 439
University of Western Australia 63
unmanned spacecraft missions 566
unmanned vehicles 422, 646, 647
unsaturated fats 464
urban archaeologists 38, 39
urbanization 340
Urban Land Institute 380
urban sociologists 154
ureter 436
urethra 711
urinalysis 438
urinary incontinence 713, 714, 715, 716
urinary tract infection 537, 714
urine 110, 193, 239, 240, 241, 242, 308, 436, 437, 438, 470, 537, 624, 654, 711, 715, 719
URL (uniform resource locator) 221

Urology **711-717**
Uroxatral (alfuzosin hydrochloride) 714
U.S. Air Force 419, 423, 674
U.S. Bureau of Labor Statistics 112, 118, 162, 270, 461, 480, 481, 614, 715
U.S. Bureau of Land Management 200
U.S. Bureau of Reclamation, Department of the Interior 343
U.S. Department of Health and Human Services 263
U.S. Department of Justice 157, 188, 552
U.S. Department of Labor 24, 428, 429, 478, 479, 480
U.S. Department of the Interior 336, 338, 379, 381
U.S. Forest Service 20, 379, 380, 381
U.S. Navy 377
U-tube manometer 68

V

vaccinia 726, 727
vacuum tubes 61, 116, 233, 234
valves 81, 92, 114, 115, 116, 117, 118, 129, 578, 580, 687, 730
vanadium 199, 200, 202
Van Allen, James 640
Van Allen radiation belts 639, 640
van Helmont, Jan 628
van Leeuwenhoek, Antoni 252, 314, 320, 322, 409
variable number tandem repeats 185, 186, 189
Varian 608
vascular diseases 452
vas deferens 619, 711
vasectomy 619, 711
Vaux, Calvert 376
velocity 2, 69, 97, 567, 575, 576, 599, 702
Venter, J. Craig 88, 112, 188, 300, 301, 302, 303, 304, 326, 397
Venus 46, 565, 566, 567, 575, 600

verbal communication 658
Verdezyne 302
vermiculite 629, 631
vertigo treatment 518
Very Large Array 599, 602, 691
very long baseline interferometry 688
Vesicare (solifenacin) 714
Veterinary Science **718-723**
Viagra (sildenafil citrate) 712
vibrational spectroscopy 652
vibrations 1, 2, 3, 5, 9, 61, 670
Viking 1 and Viking 2 craft 47
viral infections 142, 330, 728
Virchow, Rudolf 314, 320, 536
Virgin Galactic 642
Virology **724-729**
visceral osteopathy 513
viscosity 117, 341, 413
visual field examination 500
vitamins 12, 24, 108, 294, 311, 398, 437, 463, 464, 466, 467, 543
vitiligo 173
vitrinite 526
voice treatments 519
volatile organic compounds 79
Volta, Alessandro 204, 206, 233
voltaic cell 205
voltaic pile 233
von Braun, Wernher 46, 573
von Fraunhofer, Joseph 652
von Laue, Max 178, 180
von Liebig, Justus 628
von Ramm, Olaf 707
Voyager 1 646

W

Wackenhut Corporation 551
wafers, silicon 228, 673
warts 170, 173, 175, 353
waste biotreatment 254
waste heat 673
waste management 42, 82, 250, 253, 568
waste removal 296, 711
waterborne diseases 253
Water Quality Association 343
water-soluble vitamins 464

water vapor 52, 53, 56, 57, 245, 340, 401, 631, 691
Watson, James D. 103, 184, 191, 193, 286, 292, 300, 326
watt 494
Wells, Horace 28, 165
wheezes 592
white blood cells 142, 188, 313, 314, 315, 329, 351, 352
white light interferometry 678
white Mars 45, 46
whiteware 129
wildlife management 721
windbreaks 11, 16, 18, 632
wind erosion 631, 632
wine 28, 39, 93, 174, 282, 685
Winogradsky, Sergei 253
wireless capsule endoscopy 712
WITec 654
WMDs (weapons of mass destruction) 416
Wollaston, William Hyde 652
Woods Hole Oceanographic Institution 486, 487
word processing 477
work-related musculoskeletal disorders 266
World Data Center 403
World Gastroenterology Organization 276
World Meteorological Organization 57, 60, 407
World Sleep Federation 635, 637
worms 220, 530, 531, 718
Wright brothers 573
writing
  documentary evidence 41
Wyeth 111, 727

X

xenobiotic 696
Xenotransplantation **730-735**
xenozoonosis 730
XML (extensible markup language) 222, 224
X-ray fluorescence 426, 568, 653
X-ray photoelectron spectroscopy 216, 678

X-ray photoelectron spectrum 677

Y

YBCO (yttrium-barium-copper oxide, or 1-2-3) 670, 671
yeast 139, 192, 193, 292, 296, 302, 396, 397
yogurt 467
Young, Hugh Hampton 712, 715
Young, Thomas 178, 652, 655
Yushchenko, Viktor 699

Z

Zeiss, Carl 321, 322, 409, 414
zeolites 180
zero-gravity environments 645
zinc 66, 198, 199, 200, 202, 204, 205, 207, 427, 428, 464, 485, 490, 629
zirconia 128
Zoladex (goserelin) 714
zone refining 229
zoning 342, 377
zoo animals 720
zwitterions 206
zygote intrafallopian transfer 618